Air Pollution and Health

Air Pollution and Health

Edited by

STEPHEN T. HOLGATE
Southamptom General Hospital, Southampton, UK

JONATHAN M. SAMET
Johns Hopkins University, Baltimore, Maryland, USA

HILLEL S. KOREN
US Environmental Protection Agency, Research Triangle Park, North Carolina, USA

ROBERT L. MAYNARD
Department of Health, London, UK

ACADEMIC PRESS
San Diego   London   Boston
New York   Sydney   Tokyo   Toronto

Copyright © 1999, by
ACADEMIC PRESS

ISBN 0–12–352335–4

ACADEMIC PRESS
24–28 Oval Road
LONDON NW1 7DX
http://www.hbuk.co.uk/ap/

ACADEMIC PRESS
525 B Street, Suite 1900, San Diego,
California 92101–4495, USA
http://www.apnet.com

A catalogue record for this book is available from the British Library

Typeset by M Rules, London
Printed in Great Britain by The University Printing House, Cambridge
99 00 01 02 03 04 CU 9 8 7 6 5 4 3 2 1

Preface

The Industrial Revolution brought to Western Civilization prosperity and social changes that altered the direction of nations. The downside of such industrial proliferation was the extensive use of coal as an energy source and the severe air pollution that accompanied this. The impact that this pollution had on health was profound and led to the introduction of stringent emission control measures in many countries with attendant improvement in these health indices.

As we approach the second millennium, air pollution of a different kind is with us again – that derived from the combustion of oil productions and, in particular, vehicle fuels. The proliferation of both personal and commercial vehicles in all countries across the world and the impact that this is having on our towns and cities is taxing all governments, at both national and local levels. The answer will lie in the introduction of new transport policies, the generation of cleaner combustion engines and the introduction of alternative clean energy sources.

The nature of the air pollution problem relating to vehicle use varies widely from country to country and from one town or city to another, dictated not just by the volume of traffic but also by the prevailing weather conditions. The range of pollutants produced is also wide although oxides of nitrogen, ozone, carbon monoxide, polyaromatic hydrocarbons and suspended particulates appear to be the most important. The introduction of new methods of detecting and quantifying these pollutants and the establishment of rural and urban air pollution networks to monitor regional and local patterns has provided invaluable information that has been used to build up a profile of air pollutant problems. While many of the vehicle-related pollutants will have adverse effects on many biological systems including trees, grassland, arable crops and farm animals, it is their impact on human health, both alone and in combination, that has caused greatest concern.

The 44 chapters contained in this book bring together expertise from across the world to review current knowledge about the impact of air pollution on human health. A better understanding of atmospheric chemistry, new methods for pollutant monitoring, detailed epidemiology and the use of controlled exposure studies to examine single and mixtures of pollutants, provide a firm framework against which to devise new legislation to control this intractable problem. This book also contains sections that deal mechanistically with such issues as cancer, pulmonary and cardiovascular diseases. Special emphasis is placed on new data obtained from many sources on the impact of inhaled small particles on cardiovascular and respiratory morbidity and mortality, as well as the interaction of air pollutants with other factors such as the effects of allergen exposure in asthma and rhinitis.

In addition to outdoor air pollution, this volume emphasizes the importance of the indoor environment as a source of pollutants. Since humans spend much of their life indoors, whether in the home or the workplace, air pollutant exposure in these settings is clearly important when considering total exposure.

The principal purpose of this book is to bring together issues concerning air pollution and health in an easily accessible manner and in a single volume. The multidisciplinary nature of the book and the wide range of issues covered should provide the reader with access to up-to-date information on all issues relating to ambient air pollution and should, therefore, be of use to those interested in the monitoring of air pollution as well as those wishing to know more about its impact on indices of human health.

In the preparation of this book I have been greatly guided by my co-editors, Jonathan Samet, Hillel Koren and Robert Maynard, and the enormous expertise they brought to the selection of topics covered. However, we are all especially grateful to the many contributors who have expended so much time in preparing their chapters to create a unique book that we hope will be of value to those wishing to find out more about this important problem.

The knowledge gained from a greater focus on the impact of human activities on the environment will lead to a better understanding of the issues involved in assessing the health impact of air pollution and its complexity. Just as legislation and voluntary steps were introduced to reduce pollution from the industrial and domestic use of coal and its products, so new ways will be found to deal with vehicle-related pollution. These will not only require government intervention, but also the involvement and co-operation of the motor and petroleum manufacturing industries and the public itself.

Stephen T. Holgate

Contributors

Ursula Ackermann-Liebrich, Institute for Social and Preventive Medicine, University of Basel, Steinengraben 49, CH-4051, Basel, Switzerland

H. Ross Anderson, Department of Public Health Sciences, St George's Hospital Medical School, Cranmer Terrace, London SW17 0RE, UK

David V. Bates, University of British Columbia, 4891 College Highroad, Vancouver, BC V6T 1G6, Canada

Peter Brimblecombe, School of Environmental Sciences, University of East Anglia, Norwich NR4 7TJ, UK

Philip A. Bromberg, Center for Environmental Medicine and Lung Biology, University of North Carolina, Chapel Hill, NC 27599-7310, USA

Lilian Calderón-Garcidueñas, Curriculum in Toxicology, University of North Carolina, Chapel Hill, NC, USA

Kathleen Cameron, Chemicals and Biotechnology Division, Department of the Environment, Transport and the Regions, Room 3/F8, Ashdown House, 123 Victoria Street, London SW1E 6DE, UK

Lauraine G. Chestnut, Stratus Consulting, Inc., 1881 Ninth Street, Suite 201, Boulder, CO 80302, USA

Aaron J. Cohen, Health Effects Institute, Suite 700, 955 Massachusetts Avenue, Cambridge, MA 02139, USA

Daniel L. Costa, Pulmonary Toxicology Branch (MD-82), Experimental Toxicology Division, National Health and Environmental Effects Research Laboratory, US Environmental Protection Agency, Research Triangle Park, NC 27711, USA

Stephanie M. Coster, Department of the Environment, Transport and the Regions, Room 4/H16, Ashdown House, 123 Victoria Street, London SW1E 6DE, UK

J. Michael Davis, National Center for Environmental Assessment, RTP Division (MD-52), US Environmental Protection Agency, Research Triangle Park, NC 27711, USA

Richard G. Derwent, Atmospheric Process Research, Room 156, Meteorological Office, London Road, Bracknell GR12 2FZ, UK

W. Fred Dimmick, Office of Air Quality Planning and Standards (MD-13), US Environmental Protection Agency, Research Triangle Park, NC 27711, USA

Douglas W. Dockery, Associate Professor of Environmental Health, Harvard School of Public Health, and Associate Professor of Medicine (Epidemiology), Harvard Medical School, 665 Huntington Avenue, Boston MA 02115, USA

Kenneth Donaldson, Department of Biological Sciences, Napier University, 10 Colinton Road, Edinburgh, UK

W. Michael Foster, Department of Environmental Health Sciences, School of Hygiene and Health, Johns Hopkins University, 615 North Wolfe Street/Suite 7006, Baltimore, MD 21205-2179, USA

Jane E. Gallagher, Human Studies Division (MD-58c), National Health and Environmental Effects Research Laboratory, US Environmental Protection Agency, Research Triangle Park, NC 27711, USA

Andrew J. Ghio, National Health and Environmental Effects Research Laboratory (MD-58d), US Environmental Protection Agency, Research Triangle Park, NC 27711, USA

Lester D. Grant, National Center for Environmental Assessment, RTP Division (MD-52), US Environmental Protection Agency, Research Triangle Park, NC 27711, USA

Roy M. Harrison, Environmental Health, Institute of Public and Environmental Health, The University of Birmingham, Edgbaston, Birmingham B15 2TT, UK

Milan J. Hazucha, Department of Medicine, Division of Pulmonary Diseases and Center for Environmental Medicine and Lung Biology, University of North Carolina, Chapel Hill, 104 Mason Farm Road, Chapel Hill, NC 27599-7310, USA

Claire Holman, SENCO, Brook Cottage, Elberton, Olveston, Bristol BS35 4AQ, UK

Kazuhiko Ito, Department of Environmental Medicine, New York University School of Medicine, Nelson Institute of Environmental Medicine, Tuxedo, NY 10987, USA

Jouni J.K. Jaakkola, Department of Epidemiology, School of Hygiene and Public Health, Johns Hopkins University, 615 North Wolfe Street/Suite W6041, Baltimore, MD 21205-2179, USA

T. Elise Jackson, Chemical Industry Institute of Toxicology, PO Box 12137, 6 Davis Drive, Research Triangle Park, NC 22709-2137, USA

Peter K. Jeffery, Lung Pathology Unit, Department of Histopathology, National Heart and Lung Institute, Imperial College, Sydney Street, Royal Brompton Hospital, London SW3 6NP, UK

Frank J. Kelly, The Rayne Institute, St Thomas' Campus, Kings College London, Lambeth Palace Road, London SE1 7EH, UK

Urmila P. Kodavanti, Pulmonary Toxicology Branch (MD-82), Experimental Toxicology Division, National Health and Environmental Effects Research Laboratory, US Environmental Protection Agency, Research Triangle Park, NC 27711, USA

Hillel S. Koren, Human Studies Division (MD-58A), National Health and Environmental Effects Research Laboratory, US Environmental Protection Agency, Research Triangle Park, NC 27711, USA

Michal Krzyzanowski, WHO European Centre for Environmental Health, Bilthoven Division, PO Box 10, 3730 AA De Bilt, The Netherlands

Morton Lippmann, Nelson Institute of Environmental Medicine, New York University School of Medicin

William MacNee, ELEGI, Colt Research Labs., Wilkie Building, Medical School, Teviot Place, Edinburgh EH8 9AG, UK

Michael C. Madden, Human Studies Division (MD-58C), National Health and Environmental Effects Research Laboratory, US Environmental Protection Agency, Research Triangle Park, NC 27711, USA

David Maddison, Centre for Social and Economic Research Global Environment, University College London, Gower Street, London WC1E 6BT, UK

Robert L. Maynard, Department of Health, Room 658C, Skipton House, 80 London Road, London SE1 6LH, UK

Roger O. McClellan, Chemical Industry Institute of Toxicology, PO Box 12137, 6 Davis Drive, Research Triangle Park, NC 27709-2137, USA

Glenn R. McGregor, School of Geography and Environmental Sciences, The University of Birmingham, Edgbaston Park Road, Birmingham B15 2TT, UK

David M. Mills, Stratus Consulting Inc., 1881 Ninth Street, Suite 201, Boulder, CO 80302, USA

Milagros Milne, School of Engineering, Materials and Minerals Engineering Division, University of Wales Cardiff, Queen's Buildings, The Parade, PO Box 685, Cardiff CF2 3TA, UK

Kristen Nikula, Biopersistent Particle Center, Lovelace Respiratory Research Institute, PO Box 5890, Albuquerque, NM 87185, USA

Terry L. Noah, Department of Pediatrics and the Center for Environmental Medicine and Lung Biology, University of North Carolina, Chapel Hill, NC 27599-7310, USA

Bart D. Ostro, Office of Environmental Health Hazard Assessment, California Environmental Protection Agency, Berkeley, CA, 94611 USA

Halûk Özkaynak, Department of Environmental Health, Harvard School of Public Health, 655 Huntington Avenue, Boston, MA 02115, USA

Renee C. Paige, Department of Anatomy, Physiology and Cell Biology, University of California Davis, Davis, CA 95616, USA

David Pearce, Centre for Social and Economic Research Global Environment, University College London, Gower Street, London WC1E 6BT, UK

David B. Peden, Department of Pediatrics and Center for Environmental Medicine and Lung Biology, University of North Carolina School of Medicine, 104 Mason Farm Road, Chapel Hill, NC 27599-7310, USA

Charles G. Plopper, Department of Anatomy, Physiology and Cell Biology, University of California Davis, Davis, CA 95616, USA

Frederick D. Pooley, School of Engineering, Materials and Minerals Engineering Division, University of Wales Cardiff, Queen's Buildings, The Parade, PO Box 685, Cardiff CF2 3TA, UK

C. Arden Pope III, Department of Economics, Brigham Young University, 142 Faculty Office Building, PO Box 22363, Provo, UT 84602-2363, USA

Regula Rapp, Institute for Social and Preventive Medicine, University of Basel, Steinengraben 49, CH-4051 Basel, Switzerland

Roy Richards, School of Biosciences, Cardiff University, Museum Avenue, PO Box 911, Cardiff CF1 3US, UK

Lesley Rushton, MRC Institute for Environment and Health, University of Leicester, 94 Regent Road, Leicester, LE1 7DD, UK

James M. Samet, National Health and Environmental Effects Research Laboratory, US Environmental Protection Agency, Research Triangle Park, NC 27711, USA

Jonathan M. Samet, Department of Epidemiology, School of Hygiene and Public Health, Johns Hopkins University, 615 North Wolfe Street/Suite W6041, Baltimore, MD 21205-2179, USA

Richard B. Schlesinger, Department of Environmental Medicine, New York University School of Medicine, 57 Old Forge Road, Tuxedo, NY 10987, USA

Dieter Schwela, World Health Organization Division of Operational Support in Environmental Health, Urban Environmental Health, 20 Avenue Appia, CH 1211 Geneva 27, Switzerland

Chon R. Shoaf, National Center for Environmental Assessment (MD-52), US Environmental Protection Agency, Research Triangle Park, NC 27711, USA

Frank E. Speizer, Channing Laboratory, Department of Medicine, Brigham and Women's Hospital, Harvard Medical School, 181 Longwood Avenue, Boston, MA 02115-5804, USA

Peter T. Thomas, Director of Toxicology, Covance Laboratories, Madison, WI USA

George D. Thurston, Department of Environmental Medicine, New York University School of Medicine, Nelson Institute of Environmental Medicine, Tuxedo, NY 10987, USA

Andrew Wadge, Department of Health, Room 642B, Skipton House, 80 London Road, London SE1 6LH, UK

Robert Waller, 72 King William Drive, Charlton Park, Cheltenham GL53 7RP, UK

Ann M. Watkins, US Environmental Protection Agency, 401M St., SW, Washington, DC 20460, USA

Albert H. Wehe, Office of Air Quality Planning and Standards (MD-13), US Environmental Protection Agency, Research Triangle Park, NC 27711, USA

John Widdicombe, Sherrington School of Physiology, St Thomas' Hospital (UMDS), Lambeth Palace Road, London SE1 7EH, UK

Martin L. Williams, AEQ, Department of the Environment, Transport and the Regions, 4/E14 Ashdown House, 123 Victoria Street, London SW1E 6DE, UK

Judith T. Zelikoff, Department of Environmental Medicine, New York University School of Medicine, New York, NY, USA

Contents

RESPIRATORY TRACT DETERMINANTS OF AIR POLLUTION EFFECTS

GENERAL METHODOLOGICAL AGENTS OF AIR POLLUTANT HEALTH EFFECTS

CARBON MONOXIDE, LEAD AND AIR TOXICS

ESTIMATING HEALTH AND COST IMPACTS

AIR QUALITY STANDARDS AND INFORMATION NETWORKS

The colour plate section appears between pages 178 and 179.

1

Introduction

DAVID V. BATES

University of British Columbia, Vancouver, BC, Canada

It is not surprising that the study of the effects of air pollution on human populations began with major episodes of increased mortality, in which the cause and effect relationship between the dramatic episode and its consequences could not be doubted. The episodes in the Meuse Valley in 1930, in the small town of Donora in Pennsylvania in 1948, and the London episode of December 1952 provided unequivocal evidence of that kind. The city Ordinance against air pollution passed in Pittsburgh in 1946 was carried through by concerned citizens, but no scientific evidence of the impact of current air pollution levels on the population had been secured (Bates, 1994).

The development of epidemiological studies can be properly dated from the London episode; it was natural to ask the question of what effects, other than acute mortality, the air pollution might be causing. A linkage between air pollution levels in different parts of the London Metropolitan region, and the occurrence of bronchitis quickly emerged, but it was not until 1965 that Holland and Reid established the model for many future studies. This was a cross-sectional comparison of lung function of postal workers in London compared with that of others in different country towns in which the pollution level was known to be much lower. The socioeconomic level of the workers was the same in both locations; non-smokers, ex-smokers, and smokers at three different levels of intensity were characterized; and there were no climatic differences between the different regions. The results showed a clear decrement of function in all of these categories between the London and provincial workers.

We now know that the forced expiratory volume in 1 s (FEV_1), which with the peak expiratory flow rate (PEFR), was the test used in that study to characterize function, is closely related to survival; hence we can now infer that residence in more polluted regions of Britain involved a lower survival expectancy. Recent studies of non-smoking women in Beijing, and early cross-sectional studies in France, both showed decrements of FEV_1 in

AIR POLLUTION AND HEALTH
ISBN 0-12-352335-4

those living in more polluted regions, but we still do not have a precise interpretation of what this finding means in structural terms. Is there a higher degree of airway responsiveness? Is the induced bronchitis responsible for small airway disease? Is the degree of emphysema more severe? It is also quite possible that the lower FEV_1 in adults in more polluted regions is due to the fact that growing up in such locations meant that lung growth was altered, so that the initial FEV_1 (say at age 18) was never as high as it was in those growing up in cleaner locations.

It was the concomitant increase in cigarette smoking in most countries that confused the understanding of air pollution effects. Indeed, the relevant literature of the period might be interpreted as showing that all chronic lung disease was attributable to cigarette smoking, and that air pollution only exerted an effect by increasing the mortality in acute episodes. Although this is biologically inherently unlikely to be true, it was the study of Holland and Reid that first indicated that air pollution might be causing long-term chronic effects of some significance. We can now summarize the effects of air pollution due to uncontrolled coal burning as being responsible for enhancing the risk of chronic obstructive pulmonary disease in smokers and its severity, increasing the prevalence of chronic bronchitis and sputum production, and possibly, as discussed below, increasing the risk of lung cancer. Surprisingly, the prevalence of asthma does not seem to be related to this type of air pollution. Acute episodes in which particulates and acid aerosols are at high levels increase mortality from respiratory disease in all age groups.

It was in 1952, the year of the major London episode, that Hagen-Smidt in Los Angeles showed that tropospheric ozone was formed when oxides of nitrogen and hydrocarbons were both in the air and subjected to sunlight. He was investigating the adverse effects of photochemical air pollution on citrus fruit. The study of the acute effects of ozone on lung function did not start until 10 years later, but by 1970 it had become clear that ozone was an intensely irritant gas, and that normal subjects showed a wide variation in sensitivity to it. Ten years after that, a series of studies of children at summer camps documented the fall in lung function that commonly occurred in summer outdoor conditions. Two other developments were crucial in establishing the importance of ozone as an air pollutant. One was the demonstration that an induced fall in forced vital capacity (FVC) after ozone breathing was accompanied by evidence from bronchial lavage that inflammation had occurred in the lung; and the other was the study of large banks of hospital admission data that showed a significant association between summer ozone levels and hospital admissions for acute respiratory disease. In the northeast of North America, where these studies were conducted, ozone was closely correlated with aerosol sulfates in the summer.

The difficulty with ozone is that any biological effect or mechanism is theoretically possible. Increased airway reactivity, small airway inflammation, pneumonia, damage from oxygen radicals, and even induced neoplasia can all be postulated. Interference with normal lung growth is also a possibility. The main acute effects that have been demonstrated are acute reductions in lung function, aggravation of asthma (an effect still ignored in much contemporary literature), an increased risk of pneumonia in the elderly, and hospital admissions for acute respiratory disease in all age groups, including those in infants under the age of 1 year. As far as chronic effects are concerned, we have evidence from one study of incoming Berkeley students that lifetime ozone exposure might be associated with a significant reduction in terminal airflow velocity (Kunzli et al., 1997). More work on possible chronic effects of ozone exposure must be done before we can be confident that our knowledge is complete.

Finally, the past 10 years have seen a remarkable 'avalanche' of studies incriminating urban particles in the respirable range (less than 10 μm in diameter, or PM_{10}). The first data showed that in time-series analysis, there was an association between daily mortality, excluding accidents and suicides, and the level of PM_{10} prior to the relevant date. This association has now been shown to be robust to different methods of accounting for weather variations; in many different populations (over 30 at last count); and when other pollutants such as SO_2 or ozone or acid aerosols are virtually absent. There is also striking coherence, in that PM_{10} levels have been shown to be associated with function test decline in children, hospital admissions for respiratory disease, aggravation of asthma, increased school absences, and lower lung function in children. There is general evidence – not complete because of the relative scarcity of monitoring data – that all of these associations are stronger if $PM_{2.5}$ instead of PM_{10} is considered.

In contrast to the situation with regard to ozone, the mechanism of these effects is not precisely understood. Although the associations have been shown to occur when pollution levels never exceed a PM_{10} value of 150 μg/m³ for any hour in the monitoring period, these are still very low levels of exposure compared with those to which workers in many occupations are exposed. Indoor PM_{10} is increased considerably when there is a cigarette smoker in the house, and there is a possibility that it is the particulate component which is responsible for these effects.

The observation that exposure to passive cigarette smoke increases the risk of lung cancer in non-smokers is one of the pieces of information that suggests that outdoor exposure to particles derived from combustion products emitted from vehicles might also increase the risk of lung cancer. So powerful is the effect of cigarette smoking in increasing the risk of lung cancer that the unequivocal demonstration of an enhancement of risk by air pollutants is difficult. The evidence on which such an opinion must be based is drawn together in this book.

The study of the adverse health effects of air pollution has come a long way since 1952. Much more powerful tools are available for data collection and analysis; far more data are available; there is a much greater understanding of the power and limitations of statistical methods; and the economic costs of increasingly strict regulation are such that there is a willingness (albeit reluctant) to invest in research programs designed to answer some of the outstanding important questions. Although it can point to an honorable past, environmental epidemiological studies of air pollution and its effects can be expected to have a distinguished future. One of the reasons why research in this field is such an interesting challenge is that the field has such breadth that many disciplines are involved in its full understanding. This is well exemplified by the multi-disciplinary focus of the different sections in this book. In some scientific fields, the contemporary focus seems to get narrower and narrower; but an understanding of the effects of air pollutants on the human respiratory, and possibly also cardiovascular systems, necessitates a broadening of the approach as the complexity of the questions becomes apparent. It is this aspect of the field that ensures its continuing interest.

REFERENCES

Bates DV (1994) *Environmental Health Risks and Public Policy; Decision-making in Free Societies.* Seattle: Washington University Press, p. 117.

Holland WW and Reid DD (1965) The urban factor in chronic bronchitis. *Lancet* **1**: 445–448.

Kunzli N, Lurmann F, Segal M *et al.* (1997) Association between lifetime ambient ozone exposure and pulmonary function in college freshmen – results of a pilot study. *Environ Res* **72**: 8–23.

2

Air Pollution and Health History

PETER BRIMBLECOMBE

School of Environmental Sciences, University of East Anglia, Norwich, UK

INTRODUCTION

Recent times have seen startling changes in the way we view our environment. This has made it easy to forget that concern over air pollution and health is not restricted to the late twentieth century. A study of the environmental problems of the past is useful because it brings new perspectives to the issue. The unfamiliar historical context can often help to throw the causal features into sharper relief.

This chapter focuses on the historical developments in the UK where there is, unfortunately or not, a long history of environmental contamination. The account will stop in the 1950s with the development of modern research, much of which was initiated in response to the deadly London smog of 1952 and the growing problems in Los Angeles.

EVIDENCE OF INDOOR AIR POLLUTION IN ANTIQUITY

It is likely that indoor air pollution has a history much longer than documentary records. Archaeological evidence suggests that it was widely experienced in the distant past. Mummified lung tissue provides the most useful source of information on prehistoric exposure to particulate materials, but this can be found only where it has been preserved by tanning, freezing or desiccation. Such mummification may be deliberate or accidental, but the geographical extent is broad, covering many drier and colder regions.

AIR POLLUTION AND HEALTH
ISBN 0-12-352335-4

Palaeopathological samples of lung tissue can be examined after rehydration (Reyman and Dowd, 1980), which allows subsequent treatment to be similar to that for fresh tissue. Thus, microscopic examination enables solid deposits in the lung to be readily identified.

The dry climate of Egypt with its associated high concentrations of wind-blown sand may have been responsible for the occurrence of pneumoconiosis in the mummies examined by Cockburn *et al.* (1975), although these authors are careful to point out that these individuals are unlikely to have suffered from any resultant disability. Early examples of industrial lung diseases caused by exposure to mineral dusts (pneumoconiosis and silicosis) are found in the lung tissue of a sixteenth century Peruvian miner and among East Anglian flint-knappers (Shaw, 1981). The more general occurrence of anthracotic deposits in ancient lung tissue is assumed to be the result of lifelong exposure to smoke indoors.

Less direct evidence of polluted interiors may be found in skeletal materials. Wells (1977) examined many skulls from early burial grounds in the British Isles. Although diseases of the maxillary antrum or sinus had long been recognized by palaeopathologists, technical problems meant that the incidence and range had remained uncertain. The development of the antroscope allowed sinusitis to be detected in complete skulls by looking for osteitic changes in the floor of the antrum. These are seen as a roughening of the bone, which in severe cases becomes pitted with holes some 1–3 mm in diameter. Maxillary sinusitis is a common disease today and was routinely present in ancient populations. In ancient Egypt and Nubia it may have been aggravated by the inhalation of wind-blown sand, but we more often associate the disease with cool damp climates where crowded rooms with poor ventilation enhanced exposure to droplet infection and indoor smoke particles.

The incidence of sinusitis determined from 387 skulls recovered from British burial grounds spanning the Bronze Age to the Middle Ages are presented in Fig. 2.1. An estimate from the remains of 500 Indians massacred in South Dakota in the early fifteenth century is given for comparison. The low incidence of sinusitis for all but the Anglo-Saxon

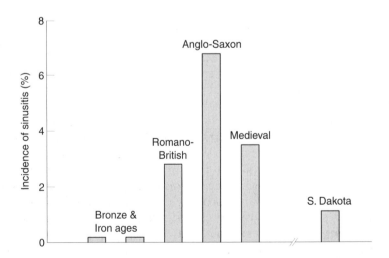

Fig. 2.1   Sinusitis frequency in skeletal material from various English sites and one in South Dakota, USA.

population means that the number of cases in each sample is statistically small. However, Wells examined many more skulls before his death and said these supported the temporal distribution noticed in his earlier work. No cases of sinusitis were found in Bronze and Iron Age materials, but the incidence increased in the Romano-British Period and reached a peak in Anglo-Saxon times. The high incidence may have been related to facial structure, as the typical Anglo-Saxon face was characterized by a proportionally longer and narrower nose than that of Bronze Age populations. However, the aggravating effects of poorly ventilated huts cannot be ignored. Chimneys were rare, so smoke from the central hearth rose and escaped from the simplest holes in the roof. High levels of indoor pollution in reconstructed Dark Age huts has been described from studies in Sweden (Edgren and Herschend, 1982). By contrast, cooking in Romano-British times was likely to be outdoors or away from the living areas.

More dramatic osteological changes can be found as tumours involving the maxillary sinus, although invasions by the primary growth may have occurred from extensions nasopharyngeal, palatal or other nearby lesions. The destruction of part of the palate and wall of the maxillary sinus is found in an Egyptian skeleton from 5000 BC. Wells (1977) suggested these could develop from prolonged inhalation of carcinogenic smoke particles in unventilated interiors, but this is unlikely to be the sole explanation.

Further evidence of the indoor pollution experienced by people in the past comes from contemporary studies of polluted interiors of huts in less developed parts of the world. In some locations, the conditions can parallel those of prehistoric times. Studies have been made in many countries that hint at the relationship between indoor pollution generated by cooking fires and health (Smith, 1987, 1993). Fuel types vary widely from yak dung or charcoal, through to biogas, so variation in emission patterns is to be expected, although ventilation rate and combustion temperature/efficiency are also important issues.

URBAN POLLUTION IN ANTIQUITY

Outdoor pollution becomes significant with the development of populous cities or large industrial activities. In the past cities were often small in terms of population, but the inhabitants lived at a high density, which could easily lead to pollutants becoming concentrated. Additionally, while some industrial processes took place in forests where fuel was abundant or near mining operations, small-scale air pollution sources were also found scattered within cities.

From the earliest times, urban air pollution was seen as a potential source of health problems. In ancient Greece, sources of odour, from rubbish, etc., were meant to be kept at a distance from the town and the astynomoi (controllers of the town) had responsibility for control of these kinds of nuisance. Civil suits over smoke pollution were brought before Roman courts and there were some attempts to keep polluting industries outside the wealthy suburbs (Brimblecombe, 1987a). Roman civil servants, such as Frontinus (De Aquis Urbis Romae II.88), saw the duty of supplying water to the city as implying broader sanitary concerns, which included pure air.

The best-known descriptions of the medical links between health and air quality are found in The Hippocratic Corpus, which contains a well-known book on *Air, Water and*

Places. It tells of the importance of climate, the properties of air that come with various winds and the quality of water, especially its metalliferous content. These environmental factors were taken as important in the treatment of disease. Other writings within the corpus emphasize the role of weather in epidemics.

Individual physicians appreciated the effects that air might have on sensitive individuals. Emperor Nero's tutor and adviser, Lucius Annaeus Seneca, was often in poor health and suffered from asthma, and in a letter to Lucilius (Epistulae Morales CIV) he wrote that his doctor had ordered him to leave Rome. No sooner had he escaped its oppressive atmosphere and awful culinary stenches than he found his health improving. Galen (c.AD 130–200), the court physician to Marcus Aurelius, offered a rational view of the aetiology of pestilence that survived through to the early modern period. Pestilence was described as a disease which attacked a great number and arose from the corruption of the air. Such ideas are found much later, for instance, among the writings of the Anglo-Saxon historian Bede. He was familiar with both the Latin writers and the Hippocratic Corpus, though in Latin translations. It is likely the concepts were better known at this time in the Arab world, where the prevalence of miasmatic theories of disease made it easy for air pollution and health to be linked at an early date (Gari, 1987).

THE GROWTH OF COAL AS A FUEL

The cities of antiquity used wood as a fuel, and it was not until the thirteenth century, when the depletion of forests caused a wood shortage, that London became the first city to use large quantities of coal. Coal found some minor uses throughout thirteenth century Britain at particular locations, so we find that Eleanor of Provence, the wife of Henry III, complained of coal smoke during work on Nottingham Castle in 1257. As the fuel became increasingly used in London for the production of mortar and some metal working, the smoke problem became severe enough, by the 1280s, to need regulation. A typical document of the period asks a group of officials:

> to enquire touching certain lime kilns constructed in the city and suburbs of London and Southwerk, of which it is complained that where as formerly the lime used to be burnt with wood it is now burnt with seacoal, whereby the air is infected and corrupted to the peril of those frequenting and dwelling in those parts.
>
> (Cal.Pat.Rolls, 20th May 1285)

Accounts like this show the readiness of medieval administrators to link health with air pollution. Smoke itself, may not have been regarded as a serious threat to health because some writers in the following centuries argued that wood smoke indoors contributed to a healthy family. It is more likely that the unfamiliar smell of the sulfurous gases from coal provoked concern. A strong negative public reaction slowed the adoption of coal as a domestic fuel for many centuries. Furthermore, wider domestic use required an effective chimney, but these were common only in better houses of the early medieval period.

Medieval authorities had few options when trying to control air pollution from coal. They could exert fines for using coal or simply ban it as a fuel. There were also attempts to place smoky industries in the lee of cities beyond their walls. Economic issues were also relevant and although coal imports were taxed, the revenues gained were probably not used to repair the damage caused by air pollution until the 1600s (Brimblecombe, 1992).

None of these regulatory measures did much to alter the long-term growth in coal use. Growth may have been slow in the fourteenth century, as the Black Death allowed some reforestation, but Elizabethan times saw an increase in the domestic use of coal. This started among poor people (Stowe, 1598), but with King James I, who had used coal in Scotland, on the throne even the nobility began to use the fossil fuel.

The climate of the seventeenth century was very different from that of the Middle Ages when coal was first introduced to London. The weather had become much colder and the period has been called The Little Ice Age. Low temperatures led to increased demand for domestic fuel at a time when imports from woodlands were difficult as wet weather made the roads impassable. Seaborne coal rapidly became the fuel of choice, but this meant the effects of its smoke were soon evident. In the early 1600s these relate to the deposits of soot around the city and on clothes and gardens (Platt, 1603). Even though King James I used coal he was concerned about the effects the smoke had on the buildings of London, and Archbishop Laud fined brewers for the damage their use of coal caused to St Paul's cathedral.

Laud's student, Sir Kenelme Digby, wrote a small book entitled *A Discourse on Sympathetic Powder*, which is mostly remembered for an unusual approach to the treatment of wounds (or perhaps non-treatment). It also relates an atomic theory of coal burning adapted from Margaret Cavendish's *Poems and Fancies* (1653). Digby was convinced that the sharp atoms present in coal smoke contributed to half of the deaths in London which derive from 'pstisicall' and pulmonary distempers. Digby's claim seems much exaggerated, as an examination of contemporary London Bills of Mortality, would have shown him. Fortunately his fellow member of the Royal Society, John Graunt, was more thorough and used these records in one of the founding works of demography, *Natural and Political Observations* (1662). He concluded that the high death rate in London was as much a result of burning coal as other factors relating to urban life. While we cannot be so sure his conclusions were fully justified, clearly links between air pollution and health were regularly made by scientists of the seventeenth century.

The diarist, John Evelyn, was also a fellow of the Royal Society and the tract *Fumifugium, or the Smoake of London Dissipated* is found amongst his many environmental writings. This booklet has been reprinted often and it illustrates the degree to which Evelyn was concerned with environmental issues. He redirected attention to the industrial sources of pollution, arguing that domestic fires were not significant contributors. He fought for cleaner air with a casual disregard for economic reality that doomed his well-meaning plans to failure. However, his unwillingness to adopt the passive stance towards smoke evident among others in later centuries shows that he brought a particularly moral stance to the issue. He recognized that certain individuals would be confronted by higher exposure to pollutants by virtue of their trade, but argued that we should not commend such high exposures. Furthermore, he pointed out, some individuals could be particularly sensitive to contaminants in the air and might never feel well in cities, although the majority showed little effect.

Medical description improved significantly in the seventeenth century. With a spirit of questioning awareness fostered by the age of rationality, new diseases were described. Rickets was of particular interest to physicians, such as Glisson and Whistler, who puzzled about whether the disease was novel or its increasing incidence arose because it had not been previously properly characterized. The links between vitamin D deficiency and the effects on calcium metabolism and bone softening were not understood by scientists at this time. However, it is hard not to wonder whether air pollution contributed to its occurrence. Besides dietary deficiency among the seventeenth century population of London, the lack of ultraviolet light in the cold smoky winters would have reduced the synthesis of vitamin D. Rickets was to reach epidemic proportions in Victorian England, where one can be more certain of the roles of dietary deficiency and reduced sunlight (Howe, 1972).

The medical scientists of the seventeenth century could not arrive at the link between sunlight and rickets, but they tried hard to identify harmful components in the air. Lacking a germ theory they considered that toxic elements such as airborne lead, arsenic and antimony played an important role in health. The harmful effects of these elements had long been known from the airborne dispersal about mining operations. Evelyn, Hooke and others thought that low concentrations of these elements could have subtle effects on some of the urban population (Brimblecombe, 1987b). In Italy the pioneer, Bernardino Ramazzini, while writing on the health of workers, *de Morbis Artificum* (1713), was also concerned with general issues of environmental health.

The medical writers who discussed air pollution in the eighteenth century do not seem to have been as imaginative as those of the preceding century. John Arbuthnot (of Alexander Pope's *Epistle to Dr Arbuthnot*) catalogued much earlier learning in *Concerning the Effects of Air on Human Bodies* (1733). Noting the corrosive nature of urban air, he attributed the damage to combustion-derived sulfurous steams. He observed a high death rate among infants and suggested that asthmatics be removed to the country, but this has little conviction and is hardly novel. Huxham (1772) and Walker (1777) wrote on the effect of weather and air on health, the latter believing that lungs became accustomed to the polluted air of the city. A visiting clergyman, William Jones, described how little sympathy his sufferings evoked:

> Twas a misery to hear the wretching Londoners wheezing and coughing and gasping for breath as they walked the streets, and I myself was a fellow-sufferer for the fogs and damps and night air by no means suit my lungs; medical jockeys would pronounce me thick-winded if not touched or broken winded.

Perhaps the most significant advance of this period was the discovery of chemically induced cancers. Evelyn had argued a century earlier that the effects of air pollution might not be immediate and even wondered if some pollutants might not penetrate through the skin. John Hall (1761), who was also interested in diseases around smelters (Hall, 1750), wrote *Cautions Against the Immoderate Use of Snuff*. This work was followed by Percival Pott's *Chirurgical Observations* . . . (1775), which is usually regarded as the earliest work on occupational carcinogenesis.

Writers before the nineteenth century, and here Evelyn is an exception, were generally passive in their response to air pollution. Few seemed to think that active solutions to the problems were very important. Doctors noted the debilitating effect of urban air, but merely advised their sensitive patients to leave for the country, as seen in the popular medical poem of the eighteenth century (Armstrong, 1744): 'Fly the city, shun its turbid air; Breathe not the chaos of eternal smoke . . .'. Such advice is little improvement on what Seneca's physician had given more than 1600 years earlier. The importance of fresh country air became deeply ingrained by the nineteenth century mind with the development of parks as the 'lungs of London', and an increasing emphasis on the desirability of trips to the country to escape the city smoke.

TECHNOLOGICAL CHANGE

Two important technological changes over the early modern period are especially relevant to the impact of air pollution on health.

The first of these was the chimney. Although it is not too difficult to bear the limited smoke from a fuel like charcoal in poorly ventilated rooms, bituminous coal smoke would hardly be tolerable. By Elizabethan times effective chimneys were far more common, which meant that coal could be adopted as a domestic fuel by the poor. Harrison (1577) wrote of the increase in chimneys since his youth, with guarded enthusiasm, because indoor smoke was considered important in hardening the timbers of the house and warding off disease among its residents. The problems of indoor pollution might have been solved with stoves, which grew popular on the continent. Although there were periods where this method of reducing exposure to smoke was encouraged (e.g. Justel, 1686–87; Arnott, 1855), it never took hold in England.

Steam engines were the second technological development. They had far-reaching impacts on the structure of our society and cities. In the early days there were many objections to the noise and smoke, but in reality the air pollution problems were localized while there were limited numbers. The smoke initially soiled houses and damaged plants nearby, but as the number grew there were serious concerns at regulation. In some rapidly industrializing cities administrative reaction began before 1800 (Bowler and Brimblecombe, 1999). In Manchester local enthusiasts set up a voluntary Board of Health under Thomas Percival which regarded smoke abatement as an important issue. The developments here anticipated the sanitary reforms that became more general in Britain and Europe and North America by the mid nineteenth century.

The rapid urbanization meant that cities in England had high concentrations of coal smoke. This had broad effects on fashion and business. Patterns were required to keep the hems of dresses off the soot-covered ground and black umbrellas allowed protection from ink-coloured rain. Clothing became soiled as it hung to dry after being washed and thriving businesses were set up to refurbish clothes that had been smoked (Grosley, 1772). English women did not favour white clothes (Allen, 1971) and cream was much in vogue by the end of the nineteenth century. Buildings suffered too: inside hangings were not popular and wallpapers were often dark-coloured in Victorian times. Outdoor damage was rapid also. The soot begrimed buildings and sulfur dioxide weathered carved stone and corroded iron rapidly. Leases of some buildings in the eighteenth century required them to be repainted every 3 years to hide the effects of smoke (Brimblecombe, 1987b).

TRENDS IN AIR POLLUTION

The early nineteenth century saw a growing interest in the health of towns because they had become the locus of population as people shifted into cities, stimulated by rapidly increasing employment opportunities. Urban medical topographies became common and often drew attention to the increased mortality in urban areas. The interest in health was broad and included a considerable range of urban reform to improve the sanitary conditions of urban life. These changes brought legislation such as the Health of Towns Act (1853).

This sanitary legislation was responsible for most legislative pressures on air pollution. There were some attempts at direct control of industrial emissions via the Alkali Act (1863), but attempts at air pollution control in nineteenth century Britain are most frequently found as smoke abatement clauses within various Acts concerned with health and town improvement. This is perhaps most clearly seen within the important Public Health Act (1875). Here new health laws, with their smoke abatement clauses, differed from previous legislation by trying to operate on a regional or national level. They were intended to go beyond being the simple local rules that had typified earlier attempts to improve the urban environment (Diedericks and Juergens, 1990).

Thus the role of local government was formalized under this type of legislation and its administrative structure began to escape from the clumsy courts and assizes it had inherited from the medieval period (Bowler and Brimblecombe, 1999). Overlapping responsibilities, bias and lack of expertise rendered these ancient elements of local government ineffective against the growing problem of industrial pollution. The statutory recognition of the importance of the local Medical Officer of Health and sanitary inspectors were important changes, as they allowed professionalism to grow. These changes were also paralleled in central government, which began to recognize the role scientists could play in policy development and administration, e.g. Angus Smith (Alkali Inspector), Sir Lyon Playfair (Member of the Royal Commission on the State of Large Towns) and Sir Napier Shaw (Director of the Meteorological Office).

The passage of numerous laws concerning smoke abatement did not necessarily abate smoke. In the nineteenth century so much emphasis was placed on economic progress that industry occupied a privileged position (Melosi, 1980). It is claimed that environmental laws rarely developed where they impeded industrial progress. However, on the individual level Medical Officers and Sanitary Inspectors were often very active (Brimblecomber and Bowler, 1992). Rapidly industrializing cities, such as Manchester, for example, were often the most active in initiating their own smoke abatement policies (Bowler and Brimblecombe, 1999).

However, the emerging regulation lacked substantive powers. Even where there was enthusiasm, there were administrative and technical barriers to the abolition of smoke. Some British laws of the late nineteenth century tried to ensure that effective administrative mechanisms allowed the developing legislation to work, e.g. the Public Health Acts of 1875 required local governments to appoint an Inspector of Nuisances and not to rely on police officers to police nuisance. Although the administrative procedures were well defined and often diligently followed, the lack of appropriate smoke control technology seemed to prevent both the administrators and industrialists from achieving a real reduction in emissions (Brimblecombe and Bowler, 1990).

Throughout the eighteenth and nineteenth centuries, it is likely that the products of coal burning – smoke and sulfur dioxide, increased in concentration in the air of English cities. Control measures were ineffective, so any improvement may have derived from geographical changes. The development of public transport systems, for example, meant that cities could grow rapidly in size. Thus population density could decrease and lower the areal emission strength. Industry sometimes relocated to the outskirts of cities and lowered its impact on the urban centre.

One marker of changes in urban air quality over time is fog. The particles from coal

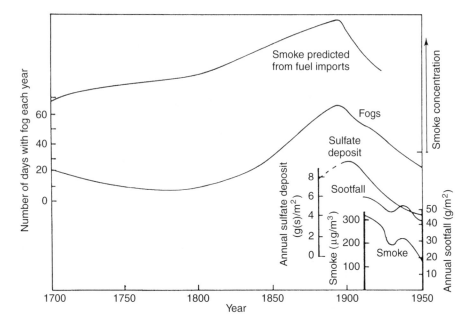

Fig. 2.2 Air pollution in London since the seventeenth century comparing predicted values with fogs and later measurements. From Brimblecombe (1987b), with permission.

burning, particularly when associated with sulfuric acid, an oxidation product of sulfur dioxide, provide excellent nuclei for fog droplets. Thus the atmospheres of coal burning towns can be very foggy. Fogs in London increased throughout the nineteenth century to the point where the Victorian London fog became a literary icon for the metropolis. However, in Edwardian times they were much reduced and people wondered at the mysterious disappearance of the fog (Bernstein, 1975). These changes and estimates of air pollution concentration modelled from fuel use hint at the diluting effect urban expansion had on air pollution (Fig. 2.2).

FOGS AND HEALTH

Fogs have long been related to poor health. The Arab al-Razi (AD 850–925) said 'when fog dominates in a town . . . smallpox, measles and epidemics will occur' (Gari, 1987). There is some evidence that increases in mortality in seventeenth century London were associated with weeks of intense fog. Figure 2.3 shows a winter period in 1679 when we know from meteorological diaries that there were some remarkably dense fogs. Weeks with intense fog also show substantial increases in deaths among the elderly, with remarkable increases ascribed to phthisic. Records such as this may be indicators of early impact of air pollution on urban mortality.

The London fogs through the nineteenth century were the most visible evidence of worsening air pollution. Advocates of cleaner air drew attention to them in the many

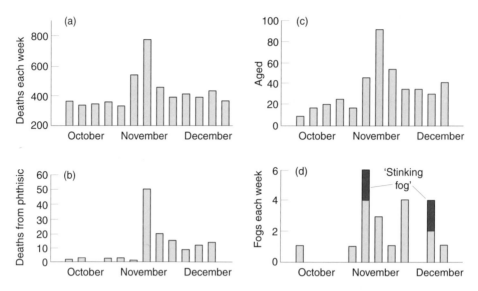

Fig. 2.3 Death rates each week during severe fogs in London, 1679: (a) total deaths; (b) deaths from phthisic; (c) aged deaths. These are compared with (d) the number of fogs and 'great' or 'stinking' fogs (shaded) each week. Modified from Brimblecombe (1987b), with permission.

pamphlets that focused on fog, e.g. *Smoke in Relation to Fogs in London* (Russell, 1887) and *Preventible Causes of Impurity in London's Air* (Galton, 1880). These writers were convinced that there was a relation between air pollution and health, but showed little sophistication. Russell (1887), for example, merely guessed that one in a hundred working months were lost in Victorian London from illnesses related to air pollution. Some perceptive individuals noted that death rate increased after fogs. The detective writer Robert Barr wrote a story that has London's inhabitants suffocated by a huge persistent fog and talks of the well-known increase in death rates, typical of foggy weather (Brimblecombe, 1987b).

Buchan and Mitchell (1876) published an interesting account of weather-related health in Victorian London, but took a very general view of the relation between the air quality and mortality. Russell (1895) was convinced that the smoky fogs of London caused heavy mortality. He was concerned about the many deaths from bronchitis, which he claimed resulted from imprudence in clothing and diet, particularly alcoholic excess, and was especially the product of dusty and smoky air. However, it was really only in the present century that Hill, Russell and Smith (Russell, 1924) began carefully to relate records of fog frequency with the incidence of death. *The Lancet* took an interest in pollution and took a lead in the development of the earliest air pollution monitoring network at the opening of the twentieth century (Brimblecombe, 1987b).

TWENTIETH CENTURY

The legislative advances of the nineteenth century did not bring the sharp improvements in air quality the advocates of sanitary reform had hoped for. Population growth and increasing wealth brought with them an increasing need for energy. Cities burnt more fuel than ever, though this might, to some extent, have been ameliorated by the dilution that occurred as cities expanded.

The lack of improvement in air quality can be explained partly as a lack of resolve in the actions of administrators and the relatively primitive smoke abatement equipment. Administrators had to balance improving the air and retaining the prosperity that industrial growth brought. Local government was often dominated by businessmen whose sympathy with industry is understandable. Nevertheless, the Victorian period saw notable advances in the way smoke abatement was approached and many cities benefited from the activities of diligent nuisance inspectors (Brimblecombe and Bowler, 1992, Bowler and Brimblecombe, 1999). Sometimes the authorities believed there were improvements, but in reality these came only slowly. In the end continual pressure pushed manufacturers to upgrade furnaces and move to automatic stoking. It was probably this that contributed to the gradual decline in smoke emission from furnaces in UK cities during the first half of the twentieth century.

The twentieth century had begun with a broad acceptance of the notion that air pollution was bad for health. Statistical evidence for the effects of polluted air on health had improved, yet causal mechanisms were poorly understood. Unlike water pollution, it was hard to assign definite diseases to air pollution, rather the effects of air pollution showed themselves in general increases in mortality and morbidity. Some understanding could be reached from studies of occupational exposure, but clearly these did not provide ideal guidance to population exposure; scientists realized that occupational exposure is intermittent, controllable and to a narrowly defined population. Historically, a distinction between occupational exposure and general urban exposure had been acknowledged by John Evelyn (1661) in his tract three centuries earlier.

The main source of information, in the early twentieth century, came from extreme episodes. As we have seen, the Victorian medical experts were interested in the statistics of death rates during periods of intense fogs. The twentieth century allowed detailed studies to be made on some particularly severe episodes: the Meuse Valley, Belgium (1930), Donora, US (1948), Pozo Rico, Mexico (1950), and London (1952). This last event probably represented the best model of an urban exposure, yet each of these episodes gave information about the effects of short duration exposure to high concentrations of air pollutants. In London it was generally agreed that about 4000 excess deaths resulted from the 1952 fog. Those who were severely ill had histories of chest complaints, and in some cases hospital admissions for respiratory complaints quadrupled within a single institution (for some key references, see Heimann, 1961).

Some believed that the high death rates associated with these episodes might obscure a more subtle general increase in morbidity that occurred more regularly. However, this was never easy to establish (e.g. Bouhuys et al., 1978; Lave and Seskin, 1977), so the evidence associating more typical air pollution levels experienced in urban areas with health failed to be convincing until well into the second half of the twentieth century.

Better progress was made with studies of individual pollutants at high concentrations. Research on carbon monoxide, much of which started in the 1920s and 1930s, was particularly successful. The studies came partly as a recognition of important occupational exposure, but many scientists were aware that more general exposures to carbon monoxide arose from automobiles (for some key references, see Heimann, 1961). This work led to an early understanding that carbon monoxide affected the oxygen-carrying capacity of the blood and that its effects were acute rather than chronic.

The 1930s saw an odd preoccupation with the effect of air pollution on light. This may have been influenced by the Victorian experience of a high incidence of rickets among children of urban Britain. The early documentary film *The Smoke Menace* (1937) shows J.B.S. Haldane much concerned with health problems derived from sunlight reduction in smoky atmospheres, and Marsh's classic book, *Smoke* (1947), readdresses the same concern.

Post-war years saw work on more direct health effects of sulfur dioxide and smoke. These were driven by the need to understand health problems in the atmospheres of coal burning cities in the years after the London smog. The two pollutants became coupled, with an implied synergism in early regulations.

PHOTOCHEMICAL OXIDANTS

Perhaps the most important change in twentieth century air pollution has resulted from the shift away from coal as a source of energy. This occurred early in the USA where oil was adopted more rapidly and the automobile emerged as an important source of pollution. This change has now become widespread, so the mobile combustion of fossil fuels is the dominant source of urban air pollution.

This has moved attention away from primary pollutants such as sulfur dioxide towards nitric oxide and carbon monoxide. However, more significantly, it has also made us aware of the importance of reactions in the atmosphere that generate secondary pollutants. Such photochemical smog was first observed in Los Angeles in the 1940s. The novel origin of this form of air pollution was not recognized until the 1950s when Haagen-Smit realized that it was essentially oxidative and involved the formation of ozone from the action of sunlight on volatile organic materials.

This new type of pollution was accompanied by novel health effects. The most immediately noticeable were irritation of the eyes, nose and throat. The oxidized form of the irritant pollutants was known from the work of Littman *et al.* (1956). Although some suspected ozone was not the principle irritant, it took a little time before the role of specific oxidation products, such as peroxyacetylnitrate, were established.

CONCLUSION

Work since the 1950s has become increasingly sophisticated. It has had to recognize the need to integrate data from the widest possible range of sources: laboratory studies on animal and human exposure, epidemiological studies and biochemical work. The sensitivity of experiments is such that it has been increasingly evident that tolerable levels of pollutants in air are often lower than the cruder, earlier experiments would have suggested were acceptable.

We have moved into a new regime. The pollution in cities is very different from that in times past. The smogs of London 50 years ago have long gone from most cities. The synergism between smoke and sulfur dioxide that seemed so evident then is no longer clear. It has gradually been superseded by a simple association between mortality and PM_{10} (particles of diameter up to 10 μm) in atmospheres now less polluted by sulfur dioxide

REFERENCES

Allen W (1971) *Translantic Crossing*. London: Heinemann.

Armstrong J (1774) *The Art of Preserving Health*. London.

Arnott N (1855) *Smokeless Fireplaces, Chimney Valves and Other Means, Old and New, of Obtaining Healthful Warmth and Ventilation*. London: Longmans.

Bernstein HT (1975) The mysterious disappearance of Edwardian London fog. *The London Journal* **1**: 189–206.

Bouhuys A, Beck GJ and Schoenberg JB (1978) Do present levels of air pollution outdoors affect respiratory health? *Nature* **276**: 466–471.

Bowler C and Brimblecombe P (1999) Air pollution control in Manchester up to the passage of the 1875 Public Health Act. *Environ History* (submitted).

Brimblecombe P (1987a) The antiquity of smokeless zones. *Atmos Environ* **21**: 2485.

Brimblecombe P (1987b) *The Big Smoke*. London: Methuen.

Brimblecombe P (1992) A brief history of grime. In: Webster RGM (ed.) *Stone Cleaning*. London: Donhead, pp. 53–62.

Brimblecombe P and Bowler C (1990) Air pollution history, York 1850–1900. In: Brimblecombe P and Pfister C (eds) *The Silent Countdown*. Berlin: Springer-Verlag. pp. 182–195.

Brimblecombe P and Bowler C (1992) The history of air pollution in York, England. *J Air Waste Manag Assoc* **42**: 1562–1566.

Buchan A and Mitchell A (1876) The influence of weather in mortality from different diseases and at different ages. *J Scot Met Soc* **4**: 187–265.

Cockburn A, Barraco RA, Reyman TA and Peck WH (1975) Autopsy of an Egyptian mummy. *Science* **187**: 1555–1160.

Diedericks H and Juergens C (1990) The environment in the Netherlands in the 19th century. In: Brimblecombe P and Pfister C (eds) *The Silent Countdown*. Berlin: Springer-Verlag, pp. 167–181.

Edgren B and Herschend F (1982) Ektorp för fjärde gaongeng. *Forskning och Framsteg* **5**: 13–19.

Evelyn J (1661) *Fumifugium, or The Inconvenience of the Aer and Smoak of London Dissipated . . . .* London: printed by W. Godbid for Gabriel Bedel and Thomas Collins.

Galton D (1880) *Preventible Causes of Impurity in London Air*. London: Sanitary Institute of Great Britain.

Gari L (1987) Notes on air pollution in Islamic heritage. *Hamdard* **30**: 40–48.

Grosley PJ (1772) *A Tour of London*.

Hall H (1750) On noxious and salutiferous fumes. *Gentleman's Magazine*, p. 20.

Hall J (1761) *Cautions Against the Immoderate Use of Snuff*. London.

Harrison J (1577) *Holinshed's Chonicles* **III**: 16.

Heimann H (1961) Effects of air pollution on human health. In: *Air Pollution*. Geneva: WHO, pp. 159–220.

Howe GM (1972) *Man, Environment and Disease in Britain*. Newton Abbot: David and Charles.

Huxham J (1772) *Observations de Aere*. London.

Justel H (1686–87) An account of an engine . . . . *Phil Trans* **16**: 78.

Lave LB and Seskin EP (1977) *Air Pollution and Human Health*. Baltimore, MD: Johns Hopkins University Press.

Littman FE, Ford HW and Endow N (1956) Formation of ozone in the Los Angeles atmosphere. *Ind. Engng Chem.* **48**: 1492–1497.

Melosi MV (ed.) (1980) Environmental crisis in the city. In: *Pollution and Reform in American Cities 1870–1930*. Austen: University of Texas Press.

Platt H (1603) *A New Cheape and Delicate Fire of Cole-Balles . . . .* London.

Reyman TA and Dowd AM (1980) Processing mummified tissue for histological examination. In: Cockburn A and Cockburn E (eds) *Mummies, Disease and Ancient Cultures*. Cambridge: Cambridge University Press, pp. 258–273.

Russell R (1887) *Smoke in Relation to Fogs in London*. London: National Smoke Abatement Institute.

Russell R (1895) The atmosphere in relation to health and human life and health. *Smithsonian Institute, Annual Report*, pp. 203–348.

Russell WT (1924) The influence of fog on mortality from respiratory diseases. *Lancet* 335–339.

Shaw AB (1981) Knapper's rot – silicosis in East-Anglian flint-knappers. *Med History* **25**: 151.

Smith KR (1987) *Biofuels, Air Pollution and Health*. New York: Plenum Press.

Smith, KR (1993) Fuel combustion, air pollution exposure and health. *Ann Rev Energy Environ* **18**: 529–566.

Stowe J (1598) *A Survey of London*.

Walker A (1777) *A Philosophical Estimate of the Cause, Effects and Cure of Unwholesome Air in Large Cities*. London: Robson.

Wells C (1977) Disease of the maxillary sinus in antiquity. *Med Biol Illust* **27**: 173–178.

GEOGRAPHICAL, ATMOSPHERIC AND GROUND DETERMINANTS OF AIR POLLUTION

3

Basic Meteorology

GLENN R. MCGREGOR

School of Geography and Environmental Sciences, The University of Birmingham, Birmingham, UK

INTRODUCTION

Meteorology, or the science of weather, is in many ways at the heart of the relationship between air pollution and health (Fig. 3.1). This is because variations in the physical and dynamic properties of the atmosphere, on time scales from hours to days, can play a major role in influencing air quality. The three-dimensional wind field and its related turbulence is important for the dispersion and diffusion of pollutants both horizontally and vertically. Vertical temperature gradients and how they affect the stability of the atmosphere also determine the extent to which pollutants are diffused through the atmospheric column and the rates of dry and wet deposition. Meteorological conditions, by controlling reaction rates, also influence the chemical and physical processes involved in the formation of a variety of secondary pollutants. Indirectly, changes in atmospheric circulation and therefore the weather can affect emissions as the onset of cold or warm spells may increase heating and cooling needs and therefore the requirement for electricity generation. Certain weather spells also encourage the use of motor vehicles and therefore indirectly affect emissions. The average weather or climate of a location is also important. Some locations, because of their general climate and topographical setting, are predisposed to poor air quality. This is because the general climate is conducive to chemical reactions that lead to the transformation of emissions, while the topography and associated local circulation system restricts the dispersion of pollutants.

Weather and its long-term counterpart, climate, are also important for health. This is

AIR POLLUTION AND HEALTH
ISBN 0-12-352335-4

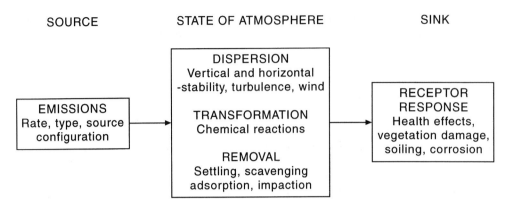

Fig. 3.1 The role of the atmosphere in the air pollution source sink relationship. Source: Oke, T.R. 1987 *Boundary Layer Climates*, Methuen, London.

because many people are weather sensitive. Very cold or hot spells markedly increase mortality and morbidity, especially amongst the elderly. Such spells of weather, in combination with elevated pollution levels, may also act in a synergistic way to exacerbate either the weather or air quality sensitivity of individuals. Weather, air pollution and health are therefore undoubtedly linked in a complex tripartite relationship.

The purpose of this chapter is to provide an introduction to the basic weather and climate processes that play a role in the relationship between air quality and health. First the relationship between air quality and basic meteorology will be assessed; more specifically the meteorological processes associated with the dispersion, diffusion and removal of pollutants from the atmosphere will be described. This will be followed by an outline of ways in which health may be affected by weather. The complex issue of the three-way relationship between weather, air pollution and health will be addressed briefly in the final parts of the chapter.

AIR POLLUTION METEOROLOGY

The field of air pollution meteorology is concerned with understanding the meteorological processes associated with air pollution. In discussions of air pollution meteorology the words dispersion and diffusion are often used interchangeably. They in fact mean different things. Dispersion refers to the movement or transport of pollutants horizontally or vertically by the wind field, while diffusion refers to the dilution of pollutants. Dispersion in the vertical direction is largely controlled by the stability of the atmosphere, whereas horizontal dispersion is determined by wind direction. In contrast diffusion is largely a result of turbulence in the atmosphere and is dependent on the variability characteristics of the wind regime. This section will therefore focus mainly on a consideration of the roles played by atmospheric stability and wind in controlling the dispersion and diffusion of pollutants. The meteorological processes associated with removal of pollutants will also be introduced briefly. Because the atmospheric stability state and the wind regime at a location at any given time is often a response to the large-scale meteorological situation, the relationship between weather systems and air pollution will also be considered. This section will close with a brief assessment of the role of modelling for understanding the meteorology of air pollution.

Atmospheric Stability

It is useful to begin with the concepts of stability, adiabatic processes and lapse rates as they relate to 'parcels of air'. An air parcel, defined as a sandwich-like slice of the atmosphere, may have one of four possible stability states. If forces acting on an air parcel are balanced then the air parcel will remain at rest. This is referred to as a state of equilibrium. If an air parcel at equilibrium is subject to some force, it can respond in a number of ways. If it returns to its original position after being displaced it is said to be in a state of stable equilibrium. If, however, it keeps moving away from its original position, then it is in a state of unstable equilibrium. Finally, the state of neutral equilibrium refers to an air parcel that moves to a new position and remains there.

Adiabatic processes are processes that take place without the exchange of heat with the environment. For example, the change of volume or pressure of air and/or its temperature may occur without the flow of heat into or out of it. Many pressure and temperature changes in the atmosphere are adiabatic, for the following reasons: air is a poor conductor of heat; the mixing of air with its surroundings is very slow; and radiative processes produce only small changes during short periods.

When an unsaturated (dry) parcel of air rises or sinks, it expands or contracts adiabatically. It does this when it enters a region of lower (or higher) pressure. As the air parcel expands or contracts, it cools or warms at a rate of $9.8°C/km$ (often rounded to $10°C/km$). This rate of temperature change is referred to as the dry adiabatic lapse rate (DALR) because the rate of temperature change is equal to that of dry air. In the case of moist air, the air must remain unsaturated if it is to cool at the DALR. Adiabatic cooling of saturated air, or air with a humidity of 100%, results in condensation of the moisture contained in it, leading eventually to cloud formation. As the process of condensation is a heat-releasing process (latent heat is released), this partly counteracts the adiabatic cooling. Consequently the rate of cooling of saturated air is less than that of unsaturated air. For this reason the saturated adiabatic lapse rate (SALR) is less than the DALR. The magnitude of the SALR depends on the temperature of the air as warm air is able to hold more moisture than cool air. The higher the temperature of the saturated air, the less the SALR, because of the greater heat released during the condensation process. For example, the SALR is approximately $4°C/km$ in the equatorial regions where the temperatures are around $30°C$, whereas in the high latitudes, with air temperatures around $0°C$, the SALR is about $7°C/km$.

Because the atmosphere is not always dry or saturated, the actual measured or environmental lapse rate (ELR) is often less than the DALR and SALR; the global average is $6.5°C/km$. The ELR may, however, be greater or less than the DALR or SALR for a parcel of dry air or saturated air, respectively. Differences between the ELR and the two adiabatic lapse rates determine whether the atmosphere will be in a state of stability or instability. These relationships are summarized in Table 3.1.

The relationships displayed in Table 3.1 apply to both ascent and descent of air. Perhaps these are best understood by considering some graphical illustrations (Fig. 3.2).

Table 3.1 Consequences of differences between ELR and adiabatic lapse rates (ALR) for unsaturated and saturated air

Lapse rate relationship	Unsaturated air	Saturated air
ELR>ALR	Unstable	Unstable
ELR<ALR	Stable	Stable
ELR=ALR	Neutral	Neutral

For an unsaturated air parcel with a temperature of 20°C at ground level rising to a height of 1 km, the air parcel temperature will fall at the DALR to 10°C (Fig. 3.2a). If the ELR is 8°C/km, then the surrounding air at 1 km above ground level will be 12°C. The rising parcel of air will therefore have a temperature less than that of the surrounding air; it will be denser. Consequently it will sink back down to the ground. The same holds true for an unsaturated air parcel with a temperature of –4°C descending from 3 km to 2 km and warming at the DALR. Its temperature on arriving at the 2 km level will be 6°C and therefore it will be warmer and lighter than the surrounding air at 4°C. The parcel will rise back to its original position at 3 km. In both cases the atmosphere is stable as there has been no net movement of the air parcel (Fig. 3.2a). Clearly this situation is not optimal for the vertical dispersion of pollutants.

The result is different in Fig. 3.2b, where the ELR is now 11°C/km. When the rising air parcel, which is cooling at 10°C/km, arrives at 1 km, its temperature (10°C) will be higher than that of the surrounding air (9°C). As a result its density will be less than that of the surrounding air and it will continue to rise. In the case where an air parcel is forced to descend from 3 km, where its temperature is –13°C, to 2 km, it will be colder than the surrounding air. It will therefore continue to descend. In both cases shown in Fig. 3.2b – ascending and descending – the air is unstable and continues to move away from its original position. Conditions of instability are therefore conducive to the vertical dispersion of pollutants.

The examples given in Fig. 3.2 also apply to the case of a saturated parcel of air rising or sinking. In the case of ascent, as long as the air parcel's temperature as a consequence of cooling at the SALR is less (or greater) than the surrounding air, then the air parcel will be stable (or unstable). In the atmosphere the ELR is not always greater or less than both

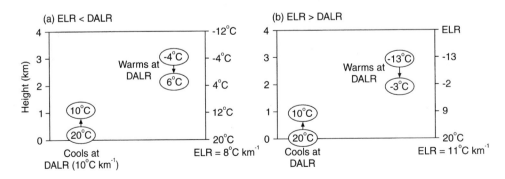

Fig. 3.2 Relationship between environmental and adiabatic lapse rates and atmospheric stability

the SALR and the DALR. Consider the situation where the ELR is 8°C/km. For an unsaturated air parcel this would mean stability (ELR < DALR), but for a saturated air parcel this would mean instability (ELR > SALR). Such a situation (SALR < ELR < DALR) is called the conditional state, or conditional stability or instability, since the final outcome depends on the moisture content of the rising air parcel.

Normally atmospheric temperature profiles are measured by radiosondes. These are launched attached to a balloon on at least a daily basis from selected meteorological stations. The radiosondes contain small instruments for recording temperature, humidity, pressure, wind speed and wind direction. This information is transmitted to a ground receiving station as the instrument ascends, thus giving a complete record of the thermal, dynamic and moisture structure of the atmosphere. This information is then used to determine the state of atmospheric stability using the simple relationships outlined above (Fig. 3.3). Not all locations have this information available to them, so proxy measures of atmospheric stability, based on standard ground level meteorological observations, have been developed. Perhaps the most widely used of these is Pasquill's stability classes. The original scheme (Pasquill, 1961), classifies stability according to wind speed and intensity of insolation for the daytime and wind speed and cloud cover for night-time. Altogether there are six stability classes: A, extremely unstable; B, moderately stable; C, slightly unstable; D, neutral; E, slightly stable; F, extremely stable (Table 3.2). Class A is applicable to strong surface heating, as experienced in the summer, and low ventilation

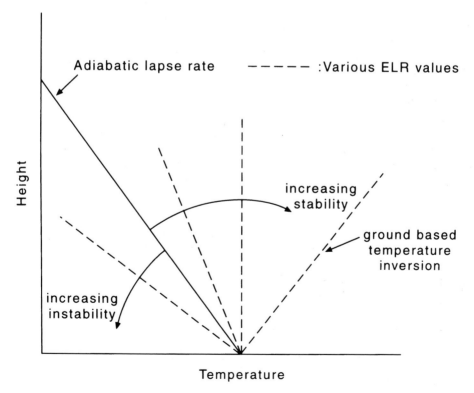

Fig. 3.3 Stability as related to the relationship between the adiabatic lapse rate and various environmental lapse rates.

Table 3.2 Pasquill's stability classes (A to F) and defining meteorological conditions. Note cloudiness is expressed in or as eighths.

Surface wind speed (m/s)	Daytime insolation			Night-time cloudiness	
	Strong	Moderate	Slight	Thin, overcast or > 4/8 low cloud	Cloudiness < 3/8
<2	A	A–B	B	–	–
2–3	A–B	B	C	E	F
3–4	B	B–C	C	D	E
4–6	C	C–D	D	D	D
>6	C	D	D	D	D

conditions. Class D represents heavy overcast day or night conditions, while F only applies at night-time for conditions typified by patchy cloudiness to clear skies with a slight breeze. Joint classes such as A–B represent average values for A and B, and so on. Various attempts have been made to relate the Pasquill stability classes to quantities that characterize turbulence and thus stability in the boundary layer (Atwater and Londergan, 1985; Golder, 1972; Luna and Church, 1971).

So far only the situation of falling temperatures with height has been considered. This is the case for a positive ELR. However, temperature may sometimes be found to increase with height; the ELR is negative. Such an atmospheric thermal structure is referred to as an inversion. An inversion is therefore defined as the situation where warm air overlies cool air; the normal vertical temperature profile has been inverted. In such a situation atmospheric conditions are extremely stable. Inversions perhaps represent the worst meteorological scenario for the vertical dispersion of pollutants because turbulence is suppressed and vertical motion is all but eliminated. There are various types of inversions. These reflect the processes involved in their formation, the outcome of which is warm air overlying cool air.

Before discussing the various inversion types and their processes of formation it is useful to define the terms used for describing inversion characteristics. The inversion base is that altitude where the temperature profile reverses, i.e. cold temperatures give way to warmer temperatures with height; this is the inflection point in an atmospheric temperature profile. The inversion base may be located at the ground, in which case temperatures increase away from the ground (a ground-based inversion). Alternatively the base may be above the ground. In this case a positive lapse rate near the surface gives way to a negative lapse rate at some height above the surface, producing an elevated or capping inversion. These are most frequently formed above the atmospheric boundary layer where turbulent diffusion of pollutants normally takes place. Capping inversions act as a lid to the atmospheric boundary layer and are very effective at preventing the vertical diffusion of pollutants. The inversion top is defined as the altitude where the positive lapse rate becomes negative and temperature falls with height. Inversion strength is the difference in temperature between the top and the base of the inversion, while inversion depth is the height difference between the inversion base and top (Fig. 3.4).

Subsidence inversions are formed as a result of subsiding or sinking air. If the air is unsaturated it will warm at the DALR, a result of increasing pressure and compression.

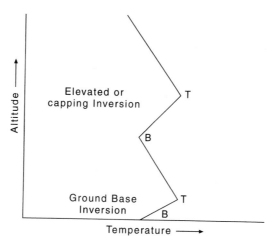

Fig. 3.4 Inversion terminology. T is inversion top, B is inversion base. When B is at ground level this is a ground-based inversion. When B is above the ground the inversion is an elevated one.

Subsidence warming usually occurs over a wide area within anticyclonic or high-pressure systems. Subsidence inversions are a notable feature of the climate of locations lying under the influence of the semipermanant subtropical anticyclones. Many of the pollution problems of subtropical locations such as Los Angeles, Mexico City, Shanghai and Johannesburg relate to such a climate setting. In the mid-latitudes where the weather systems are more migratory than the subtropics, travelling anticyclones can stall and persist for periods for up to 2–3 weeks owing to changes in the upper air circulation. Locations under the influence of such blocking anticyclones therefore experience prolonged periods of subsidence. This leads to a general stagnation of the air and the development of an intense subsidence inversion which in effect caps the atmospheric column. Below the inversion pollutant concentrations build, as in the case of the subtropical anticyclones, and both air quality and visibility deteriorate rapidly. Such situations are often referred to as periods of anticyclonic gloom; at the surface conditions are very murky while at height clear blue skies may exist. Subsidence inversions may also develop on the lee side of mountains, especially in clear cool winter conditions when there can be great heat loss over several days. This leads to a ponding of cool air against the foot of the mountain range. If atmospheric circulation conditions are such that air is travelling from upstream of the mountain range, then this air will descend in the lee of the mountain range and warm adiabatically. The outcome is a layer of warm subsident air overlying the cool ponded air at the foot of the mountain range. Good examples of locations where lee-side subsidence inversions may bring about offensive air quality conditions include areas east of the North American Rockies, the Canterbury Plains in the lee of New Zealand's Southern Alps, and the Fohn wind belt areas of the European Alps.

Radiation inversions are the result of excessive cooling of a surface. This type of inversion is usually ground-based but may also occur as an elevated inversion if the cooling surface is a cloud layer or even a pollution layer. Ground-based radiation inversions find their best expression in the hours surrounding sunrise during clear calm winter conditions. Under such conditions there is a large loss of heat from the surface. This is because, in the presence of clouds, outgoing heat in the form of long-wave radiation is absorbed by clouds and re-emitted back to the surface, which results in surface warming. Windy conditions are effective at mixing warm air with cool air, equilibrating vertical temperatures and thus eroding inversions. A noteworthy phenomenon associated with nocturnal ground-based radiation versions is the nocturnal jet. This is a region of high-velocity wind found just above the nocturnal surface inversion (Kraus *et al.*, 1985). It is therefore a region which offers great potentail for the vertical and horizontal diffusion of pollutants. However, as the height of this is usually well above effective emission height for most point sources, this potential may only be realized for point sources with emissions that possess high initial values of momentum and buoyancy.

Advection inversions are caused by the horizontal flow (advection) of air. Such advected air either displaces warm air vertically (cool air advection) or results in warm air moving over cooler air or a cool surface (warm air advection). Cool air advection often happens in weather situations described by meteorologists as cold fronts. This occurs when cool air is replacing warm air. The front is the boundary between the cold and warm air. In such situations warm air is wedged up by the cold air because of density differences with the result that the cold air slides in beneath the warm air creating an elevated inversion. As the frontal surface, or the slope between the cold and warm air, is rather steep at about 1 : 50 to 1 : 100 (ratio of vertical rise to horizontal distance), the further away from the front the greater the distance the warm air is from the ground. Therefore in the case of a cold front the fairly high inversion base rises and the mixing depth increases as a cold front passes over a location. As this process can happen within a day, such cold front inversions often pose little problem in terms of air quality. This situation contrasts with that of a warm front.

By definition a warm front is the situation when warm air moves into an area and displaces cooler more dense air by over-running it so that warm air sits above cold air. In this case the frontal surface has a much shallower slope compared with that of the cold front – around 1 : 200. The inversion base is therefore much closer to the ground in the case of a warm front. Moreover, because the frontal surface slopes upward and ahead of a warm front, with an approaching warm front the inversion base comes increasingly closer to the ground, thus reducing mixing depths. As warm fronts are relatively sluggish, compared with their cold counterparts, this set of meteorological circumstances can mean that air quality conditions can deteriorate over a day or two as a warm front moves into an area. Only with the passage of the warm front will mixing depths increase and dispersion conditions improve.

Advection inversions can also occur where warm air moves over a cool surface. This often happens when air flowing from the sea or a lake crosses a cold wintertime land surface. The advected warm air is cooled from below with the result that warm air lies above cool surface air. If the moist lower layers of the advected warm air are cooled sufficiently and condensation nuclei are not limited, condensation is likely to occur producing fog which, in combination with ground level pollution, especially particulate matter, can produce acidic smog.

Inversions have been studied extensively because of their importance for air pollution with emphasis on the relationship between ground and low-level elevated inversions and air pollution (Holzworth, 1972; Remsburg *et al.*, 1979). The characteristics of inversions have been established using a variety of data sources such as radiosondes (Preston-Whyte *et al.*, 1977), acoustic sounders (Prater and Collis, 1981) and meteorological towers (Baker *et al.*, 1969). Using a 5-year radiosonde data set, Milionis and Davies (1992) have identified the majority of the inversion types discussed above for the Hemsby area in the UK. For example subsidence capping inversions, formed as a result of anticyclonic subsidence, were found to have their greatest frequency at heights of around 500–2000 m. Subsidence capping inversions were also noted to strengthen and deepen in the direction of the ground, occasionally merging with ground-based radiation inversions during calm night conditions. Turbulence capping inversions, formed in the upper parts of the atmospheric boundary layer due to adiabatic expansion, were found to be most frequent from just above the surface to around the 500 m level. Cool air moving off the sea in daytime sea breezes was also found to enhance the development of inversions. Frontal inversions were noted as occurring well above the surface between the 2000 m and 5500 m levels.

Clearly inversions are important meteorological phenomena in terms of air pollution because they limit the vertical dispersion of pollutants. They also have implications for the horizontal transport of pollutants as they act as effective conduits for long-range pollution transport, especially if a stable atmosphere with a well-developed inversion covers a large geographical area. Such long-range transport may only be achieved if the wind field conditions are suitable.

Wind

The wind field is often perceived as two-dimensional, but it is really three-dimensional and serves to transport and diffuse pollutants both horizontally and vertically. Within the atmospheric boundary layer, where the majority of pollution dispersion and diffusion takes place, the interplay of three forces determines horizontal wind speed and direction. These forces are the pressure gradient force, the Coriolis force and surface friction. A pressure gradient force will exist wherever there is a difference in pressure between two locations. If such a force exists, then air will flow from the place of high pressure to low pressure. The greater the pressure gradient, the greater the wind speed. Once air starts to flow in response to a pressure gradient force it comes under the influence of the Coriolis force. This is a deflective force due to the earth's rotation. In the northern hemisphere it tends to deflect airflow to the right, while in the southern hemisphere airflow is to the left. As the Coriolis force is greatest at the poles and least at the equator, winds of equal strength will be deflected at a greater angle away from the pressure gradient for a high compared to a low latitude situation. Furthermore, as the strength of the Coriolis force is also related to wind speed as well as latitude, high-velocity winds will be deflected more than low velocity winds. Surface friction is a result of the retarding influence of the earth's irregular surface. Wind flowing over a rough surface is slowed down by frictional drag; this causes it to alter its direction. In aerodynamic terms the ocean surface is smooth compared with the land so that, on average, the deflection caused by surface friction effects is only 5–10° over the oceans compared with 15–35° over the land.

Wind direction as determined by the three forces described above controls the direction of pollutant transport. A persistent wind direction will result in pollutants being transported in a well-defined direction, whereas variable wind directions, as often experienced in near-calm conditions, will result in pollutants being broadcast over a wide area. The interaction of wind direction with pollutant source is also important. Where several sources are aligned downwind cumulative loading may occur, with the result that concentrations increase downwind. With regard to this an understanding of the local wind climatology is important when planning the siting of polluting activities. Often it is assumed in planning that the best location for atmospheric polluters is on the lee side of towns or built-up areas as defined by the prevailing wind. However, this is often not the best approach. This is because the prevailing wind is usually associated with turbulent air-flows and thus low average surface concentrations. The most severe pollution episodes, however, usually occur during times when the prevailing weather situations are replaced by ones characterized by atmospheric stability conditions quite different from normal. Such conditions may generate different airflow patterns from that of the prevailing wind. It is therefore wiser to locate atmospheric polluters in downstream areas, as defined by wind direction patterns during stable atmospheric conditions.

Surface friction, as well as influencing wind direction, plays an important role as far as the vertical wind structure is concerned. When the wind, like any other fluid, flows over a surface, the air parcels close to the surface are slowed down relative to parcels well above the surface. Consequently wind speed at the surface is less than that above the surface; a wind speed gradient exists such that up to a height where the surface no longer exerts a frictional drag on the wind flow over it, there is a rapid increase in wind speed. Furthermore, the rougher the surface, the deeper is the layer influenced by frictional drag. As a result of frictional drag air parcels flowing above those retarded by surface friction are also slowed down, giving rise to shearing forces. Two types of shear forces occur, namely viscous shear and turbulent shear. The former occurs in the case of very slow laminar flow. In faster, more turbulent flow where rotating three-dimensional parcels, often referred to as turbulent eddies, are interacting with each other, turbulent shear exists. This is the situation that prevails for the majority of the time in the boundary layer.

Turbulence is the atmospheric characteristic with most influence on the diffusion of pollutants. Turbulence is caused by fluctuations in wind speed and direction. These fluc-tuations are imposed on the mean flow of the wind, so in discussing wind flow characteristics we talk about both mean and fluctuating motions. This is perhaps best understood by looking at a wind trace (Fig. 3.5). The trace shows two types of fluctuation. These can be broken down into a low-frequency wave-like fluctuation, the mean flow, on which is superimposed high-frequency rapid oscillations, or turbulent motions. With regards to a puff of pollutant the longer-period fluctuations tend to transport it while the short-period rapid fluctuations tear it apart.

Turbulence is dependent upon three factors: mechanical effects of objects protruding into an air stream; the vertical rate of increase of wind speed; and the vertical temperature profile (Lyons and Scott, 1990). Mechanical turbulence has its origins at the surface, is related to wind speed and decreases with height. If the vertical rate of increase of wind speed is great, strong vertical shear exists. In such a case air parcels arriving at a different level will impart some of their momentum properties to that layer, causing a momentary change in the flow properties of that layer; turbulence will be increased. Therefore if strong vertical shear is present a chaotic field of vertical and horizontal turbulence will be created (Lyons and Scott, 1990) and any pollutants will be transported and diffused in the direction of the turbulence and pollutant gradient.

Atmospheric stability, or vertical temperature profile characteristics as discussed in the previous section, also has important effects on turbulence. This relates to the rela-tive roles of free and forced convection. The former is a product of surface heating while the latter results from air being mechanically forced over an object. Generally

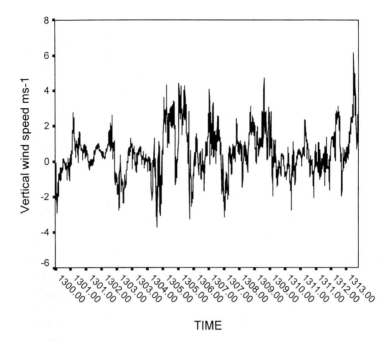

Fig. 3.5 Trace of vertical wind velocity. Note velocity fluctuations at both high and low frequencies. Positive speeds indicate ascending air and negative speeds indicate descending air.

mechanically generated eddies are smaller than thermally forced ones and serve to dilute a pollution plume rapidly as clean air is mixed with polluted air. Eddies arising from free convection tend to transport pollutants up and down with little dilution. Dilution in this case is only achieved when mechanically generated eddies eat away at the edges of the plume.

During cloudy and windy periods strong thermal stratification of the atmosphere is absent. This is because clouds reduce surface heating/cooling and strong winds mix cool upper air with warm surface air, thus eroding any temperature gradients. Under such neutral conditions forced convection dominates as a result of surface friction effects. The turbulent eddies in this case are circular and increase in diameter with height (Oke, 1987). In contrast during unstable conditions, when strong thermal gradients exist, the vertical movement of eddies is enhanced. Although at the surface mechanical effects continue to dominate, away from the surface thermal effects play a greater role, the outcome of which is a vertical stretching of the eddies and a reduction of the wind speed gradient. This is because vertical exchange of momentum is promoted over a deep layer; faster flowing upper air is mixed downwards with slower moving lower air and vice versa. The situation for stable conditions is opposite to that for unstable conditions. Vertical movements are suppressed, the eddies are compressed and the vertical wind speed gradient is steepened (Fig. 3.6).

The interaction between atmospheric stability and turbulence results in a number of plume dispersion patterns as seen downwind from point sources (Fig. 3.7). These patterns are looping, coning, fanning, lofting and fumigation (Oke, 1987; Scorer, 1990). Looping is typical of unstable conditions in which the associated eddies are large compared with

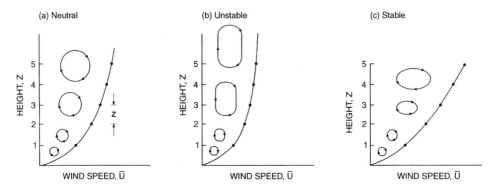

Fig. 3.6 Turbulence and its relationship to atmospheric stability. Source: Oke, T.R. 1987 *Boundary Layer Climates*, Methuen, London

the width of the instantaneous plume (Scorer, 1990). As the plume is transported up and down by the turbulent eddies, the plume takes on a sinuous shape when viewed in profile (Fig. 3.7a). Little dilution of the plume occurs as the turbulent eddies are transported intact. The sinuosities travel with the wind and grow with distance from the source. As looping can become very accentuated downstream from the source, the relatively undiluted plume may come into temporary contact with the ground, resulting in instantaneous increases in pollutant concentrations. Mechanically generated eddies are usually responsible for diluting the plume by lateral erosion. The greater the wind speed and the more convoluted the surface, the greater the rate of plume diffusion. Lateral erosion eventually leads to plume disintegration.

Coning is typical of conditions bordering on neutral stability. Although this may occur at any time during the diurnal cycle and season, it is most common in the evening when the lowest layers of the atmosphere become stable due to rapid cooling. The spread of eddies to greater heights is thus restricted (Scorer, 1990). The cone shape of the plume is a result of the absence of either stability or instability (Fig. 3.7b). In such neutral conditions the vertical and horizontal spreading of the plume, mainly by mechanically generated eddies, is almost even so that a cone shape is formed (Oke, 1987). Downwind the plume diameter grows by diffusion but vertical transport remains limited.

In a strongly stable atmosphere, such as that might occur under anticyclonic conditions with a capping inversion, fanning is typical (Fig. 3.7c). Such daytime conditions are enhanced at night. In a strongly stable atmosphere vertical diffusion both upwards and downwards is restricted as turbulence is almost non-existent, so ground level concentrations may be close to zero if the stack is not very short or the plume does not intersect with rising ground in the distance. Horizontal spreading of the plume may be caused by light but variable winds to produce a fan shape when viewed from above. In the absence of variable wind directions the plume may remain as a single or meandering ribbon with concentrations remaining unchanged for up to 100 km (Oke, 1987).

Lofting occurs if a rising plume manages to punch through a weak inversion or the plume is emitted at a height greater than the inversion top (Fig. 3.7d). Under such conditions the plume will continue to rise through an unstable atmosphere. Downward transport of the plume is restricted because of the presence of the inversion. This favourable set of circumstances is often found as the nocturnal ground-based radiation

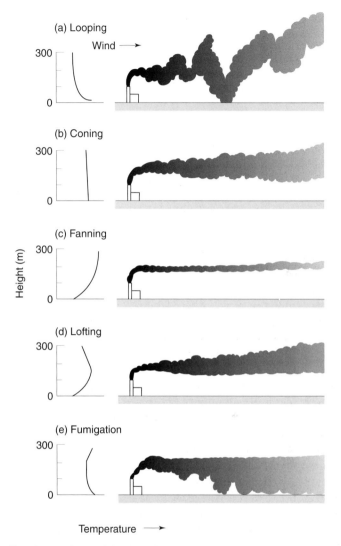

Fig. 3.7 The effect of atmospheric stability on pollution dispersion patterns from a single stack. Source: Oke, T.R. 1987 *Boundary Layer Climates*, Methuen, London.

inversion builds. If, however, the inversion builds to a height greater than the height of the plume, the plume changes to the fanning type (Oke, 1987).

Whereas a plume emitted through or above an inversion is the most favourable scenario for dispersion, the most dangerous situation is realized with diametrically opposed conditions. When a plume becomes trapped in a well-mixed layer below an intense capping inversion, fumigation occurs (Fig. 3.7e). Fanning is often the precursor to fumigation with the transition occurring soon after sunrise. This is because with sunrise surface heating is initiated and along with it mixing of the surface layer. As the mixing height grows the thermally induced eddies reach the bottom of the fanning plume. Pollutants are then mixed downwards with the result that ground level pollution levels will rise. If the fanning cone has travelled a considerable distance overnight then fumigation conditions may be experienced over a wide area (Oke, 1987).

Removal of Pollutants

Gravitational settling, dry deposition and precipitation scavenging are the main processes by which pollutants are removed from the atmosphere. Gravitational settling occurs when atmospheric turbulence is unable to keep particulate matter suspended; individual particles become too heavy and fall out. The rate of removal of particulate matter by gravitational settling is related to the particulate size, density and wind speed. Aggregation of small particles into larger ones can increase the chances of settling. Gases may also become absorbed onto particles and be removed along with the particulate matter. Atmospheric aerosols formed during photochemical processes may also experience gravitational settling (Oke, 1987).

Precipitation scavenging is often misinterpreted as referring to the process by which gaseous or particulate matter is swept up by falling precipitation. This is only one of the precipitation scavenging processes and is referred to as washout. The other process, called rainout or snowout, is an in-cloud scavenging process (Oke, 1987) related to precipitation formation processes. Particles carried up into the atmosphere often become condensation nuclei and grow by the Bergeron–Findeissen process into raindrops (Wallace and Hobbs, 1987). As the raindrop grows by condensation it often absorbs gases or captures other suspended particles. Raindrop growth continues until the raindrop is no longer able to remain suspended in the cloud mass. The raindrop or snowflake then falls, carrying with it the original condensation nucleus, absorbed gases and any captured particulate matter.

Dry deposition, sometimes called surface absorption, occurs as a result of pollutants being transferred to the surface from the atmosphere by turbulence. The surface therefore becomes a pollutant sink. The rate of material deposition is related to the deposition velocity, which is dependent on the intensity of the turbulence or wind speed (Oke, 1987).

Weather Systems and Air Pollution

Implicit in the above discussions is the assumption that the state of the atmosphere at both the synoptic (large) and meso scales determines the transportation and diffusion of pollutants. At the synoptic scale, daily variations in the atmospheric circulation, as manifest by the passage of fronts and cyclonic and anticyclonic systems, modulates the meso-scale meteorological processes that influence air quality (Crespi *et al.*, 1995). At the meso scale the wind field is important. Certain weather situations therefore provide the requisite meteorological conditions for pollution episodes.

The field of study that attempts to find and explain linkages between air quality and the nature of the synoptic-scale atmopsheric circulation or weather systems is synoptic climatology. Synoptic climatologists in assessing atmosphere air pollution relations may take two broad approaches. These are the circulation to environment and environment to circulation approaches (Yarnal, 1993). In the circulation to environment approach a classification of the main atmospheric circulation types is made. This classification is independent of air quality and no air quality criteria are used in developing the classification. The second step is the evaluation of the relationship between the spectrum of circulation types and air quality, the purpose of which is to identify those circulation types that produce poor air quality. In the environment to circulation approach some air quality criteria are used to select days on which pollution levels are high. For example in the case of ozone, a critical level of a mean daily concentration of 120 ppb could be chosen. All days possessing concentrations above this limit then are analysed in terms of their main atmospheric circulation features. The outcome of this approach is hopefully the identification of a set of atmospheric characteristics that are common to most, if not all, the high ozone concentration days. Of these two approaches it is the circulation to environment approach that has been utilized most frequently when analysing relationships between weather and air pollution.

Classifications of atmospheric circulations may be developed using a variety of methods. These include manual, correlation, eigenvector, compositing, indexing and specification techniques (Yarnal, 1993). Of these the manual and eigenvector-based techniques have been applied most widely in synoptic climatological studies of air pollution. The traditional approach taken by most British climatologists when analysing relationships between weather and air quality has been the application of the manually derived Lamb's weather types (LWT) (McGregor *et al.*, 1996). Briefly the Lamb classification is made up of seven major circulation types and 27 subtypes, including an unclassifiable type (Lamb, 1972). Perhaps one of the best applications of LWT to air pollution including acid rain has been that of Davies *et al.* (1991), who for three sites in the UK evaluated the proportion of the annual ion content of precipitation attributable to the various LWT. Davies *et al.* (1991) found that Lamb's cyclonic, westerly and anticyclonic types play the dominant roles. These types are important as they offer ideal trajectories from the major clean or offensive pollution source areas to the sites studied. Although central European synoptic climatologists have developed a variety of manual circulation classification schemes, of which the Grossweterlaggen of the German Weather Service is best known, there appears to have been little application of such classifications to the analysis of air pollution.

Perhaps the most widely applied manual atmospheric circulation classification scheme in the USA is the Muller classification. This classification describes eight typical weather patterns over the contiguous USA (Muller, 1977). For Shreveport, Louisiana, Muller and Jackson (1985) applied the Muller scheme and identified the Gulf High, Pacific High, Continental High and Coastal Return types, which occurred approximately one-third of the time, as producing the greatest pollution potential. This is because these were associated with low-level night-time inversions, high radiation inputs conducive to photochemical activity and airflows from photochemical precursor regions to the north. Conversely, Gulf Return and Frontal Gulf Return types, which also occurred roughly one-third of the time, produced better air quality due to their greater overnight mixing heights, strong horizontal dispersion from the southwest to southeast and low radiation receipts due to daytime cloud cover. Based on their findings Muller and Jackson (1985) recommended a southwest to southeast location for industries producing noxious atmospheric effluent.

Although manual classifications such as the LWT (Lamb, 1972) and Muller (1977) schemes demonstrate a reasonable degree of utility in assessing relationships between circulation and air quality, there has been a move away from such schemes to semi-automated statistically derived classification schemes. The impetus for this approach came with the early work of Christensen and Bryson (1966), who developed objective weather typing methods based on the use of principal components analysis. This methodology was later tested and improved on by Ladd and Driscoll (1980). Based on the methodological philosophy of these papers Kalkstein and Corrigan (1986), using a combination of principal components and cluster analyses, demonstrated clearly the application of semiautomated synoptic climatological classification schemes in air pollution climatology studies. This approach has subsequently been developed further and applied mainly in the USA (Davis and Gay, 1993; Davis and Kalkstein, 1990; Eder *et al.*, 1994; Yarnal, 1993). Although these studies have shown clearly that such an automated synoptic typing approach works well and has the ability to replicate results consistently and can be applied quickly and easily, there has been little application of this approach in the UK and Europe. However, by applying the synoptic typing methodology to the Birmingham area in the UK, McGregor and Bamzelis (1995) identified and described six major air mass types in terms of their climatological, meteorological and pollution characteristics. For extreme pollution events a subpolar North Sea mixed maritime continental anticyclonic type (Type 5) with warm, subhumid and calm conditions stood out as the most important for NO_2, O_3, NO, CO and PM_{10}. A related summer binary northern continental anticyclonic southern cyclonic type (Type 4) characterized by moderate warmth and moisture levels and low cloud cover, as well as high radiation inputs and calm to weak northerly and easterly flow, was also found to be of importance for O_3 and SO_2. McGregor and Bamzelis (1995) explained the high incidence of extreme pollution events for these two anticyclonic types with reference to their inherent air mass characteristics: weak winds which limit transport and diffusion, high solar radiation receipts which enhance photochemical activity, and clear skies conducive to the development of nocturnal inversions and the suppression of mixing layer heights. For SO_2 events a blocking mixed maritime continental anticyclonic type characterized by temperate, moist cloudy to foggy conditions with light northeasterly veering to southwesterly flows was found to be the most important. Meteorologically this resembles closely an anticyclonic gloom type. McGregor and Bamzelis (1995) consider this type to be important for extreme SO_2 events as northeasterly flows may transport SO_2 into the Birmingham area from the north and east where distant pollution sources such as coal burning power generation plants exist. That the synoptic-scale situation can provide the requisite conditions for long-range transport of pollutants and thus pollution events at locations a considerable distance from the source(s) is clear when air mass back-trajectories are calculated (Fig. 3.8).

In additon to large-scale circulation systems, meso-scale systems as typified by sea and land breezes and valley and mountain winds may also provide the requisite conditions for high pollution levels. This is because local wind systems are poor ventilators owing to their weak flows, demonstrate a diurnal reversal of wind direction and are closed circulation systems.

Local wind systems are of importance in terms of a location's pollution climatology where they occur frequently and regularly. This is the case in many areas where strong diurnal temperature variations are conspicuous, as found especially in subtropical latitudes and Mediterranean-type climates. The thermal changes between day and night are the main driving force of the diurnal wind systems. Other favourable circumstances are the typically small pressure gradients and low wind velocities of the general circulation, which reduce large-scale turbulence and allow the rapid formation of local pressure differences.

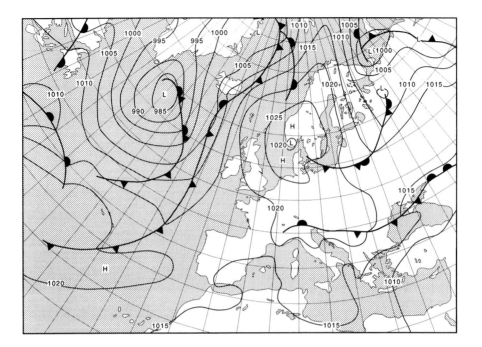

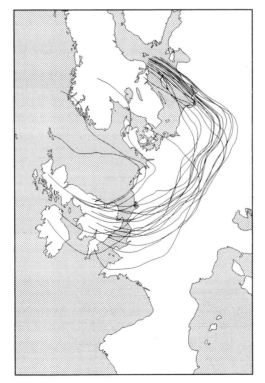

Fig. 3.8   Back trajectories of air for 11 August 1995, when high ozone levels were recorded across the British Isles. Note how the trajectories pass over industrialized areas in western Europe where ozone precursors are abundant.

The occurrence of fronts and strong depressions hinders the development of diurnal wind systems. In the mid-latitudes they are therefore best developed during periods dominated by anticyclonic weather systems or slack pressure gradients at the synoptic scale. Because of their limited duration, diurnal wind systems usually are effective only over relatively small areas. They can, but rarely do, extend far from their regions of origin and by their very nature show many local variations. Generalizing broadly, there are two main types of location of diurnal winds: coastal regions, both along the sea and near large lakes, where systems of land and sea (or lake) breezes occur frequently; and areas of variable relief, where different types of valley and mountain winds can develop.

Thermal differences between land and water surfaces are the main cause of coastal wind systems as typified by sea and land breezes. During the day, the land heats up rather quickly under the influence of solar radiation, while water surfaces remain cooler, because the heat is dissipated over thicker layers of water by turbulence and waves, and by direct penetration and absorption. As a result, a small convection cell develops, with winds near the earth's surface blowing towards the land – the sea breeze. At night, the land cools off rapidly due to long-wave radiation (heat) loss. The water, because of its thermal inertia, remains at about the same temperature as during the day. Consequently the daily pressure pattern is reversed and a land breeze is formed as relatively cool land air moves down the small local pressure gradient to the area of lower pressure over the sea. As a consequence there is a reversal of wind direction.

The sea breeze is usually the stronger of the two winds. It can, under favourable conditions, reach speeds of 4–8 m/s, and the thickness of the air layer involved can be as much as 1000 m. The sea breeze in the tropics can reach inland as far as 100 km, up to 80 km further than its mid-latitude counterpart. In some locations, the sea breeze may be so well developed it may push over coastal topographic barriers and penetrate inland. At some distance inland air rises in the ascending arm of the sea breeze convection cell and returns towards the sea at about 1500–3000 m, thus forming a closed circulation system.

The sea breeze usually starts, near the coast, a few hours after sunrise, typically mid to late morning, but expands landward and seaward during the day. It attains its maximum development when sea–land temperature contrasts are at about their maximum. In some places this can be in the morning, as opposed to the afternoon, as land temperatures in the afternoon may be depressed owing to cloud development related to the sea breeze. The sea breeze usually continues until shortly after sunset, but its circulation at higher levels may persist a few hours longer. Seasonally, the sea breeze and its lake breeze equivalent are strongest when insolation is intense. It is therefore best developed during the dry season in the tropics and in the summer in extra-tropical locations.

The land breeze is weaker than the sea breeze in most climates. This is because the land–sea temperature difference due to daytime heating is much greater than that due to night-time cooling. Its main cause is the rapid cooling of the land surface during the night. This cooling influence is limited to a thin surface layer of air. Moreover, this layer is strongly affected by friction. Therefore the land breeze rarely exceeds 3 m/s; however, it may be enhanced by katabatic flows (see below). The thickness of the moving air layer in a land breeze is usually only a few hundred metres. The land breeze does not normally reach more than 15–20 km seaward. It generally starts about three hours after sunset, increasing in strength until sunrise, and at times continuing beyond sunrise.

Land breezes are best developed in areas where the water surface is relatively warm: in equatorial regions, near warm ocean currents and relatively shallow lakes. Long and clear nights, occurring during dry seasons and, in the outer tropics, during the winter also favour land breezes.

Although local coastal circulations may provide welcome relief from daytime summer heat, they do have their disadvantages as far as pollution dispersion is concerned because they are essentially closed circulation cells. This is because pollutants emitted during the day, although diffused vertically in the rising landward limb of the sea breeze cell, will be returned to land in the descending seaward and landward lower branches. Furthermore, because alternating sea and land breezes result in wind direction reversals, pollutants carried landward and vertically during the day may be returned at night due to subsidence of air over the land. This in effect produces a slopping back and forth of pollutants.

Over areas with large differences in relief, diurnal wind systems often develop. Their basic origins are heating and cooling of air on slopes. During a sunny day mountain slopes heat up rapidly owing to large radiation receipts. The free atmosphere over the lowlands remains less affected by these large insolation inputs and is slightly cooler than air over the mountain slopes. Mountain slope air therefore becomes unstable and tends to rise up the slope. This type of upslope flow is called 'valley' wind or anabatic flow. It can be easily recognized as it is often accompanied by the formation of cumulus clouds near mountain tops or over escarpments and slopes. At night, a reverse temperature difference develops, as the highlands cool off rapidly because of long-wave radiation loss. This cooler dense air then moves downslope under the influence of gravity and is called a 'mountain' wind or katabatic flow.

Anabatic flows are usually stronger and more persistent than katabatic ones. They frequently continue well after sunset and this tendency is particularly strong in the outer tropics during the summer, when insolation is very intense and nights rather short. Under these circumstances the anabatic winds, if developed on a large scale, can continue throughout the night. Katabatic winds are normally weaker than daytime anabatic winds because thermal differences are usually smaller and friction reduces wind speeds near the earth's surface. Katabatic winds, however, can be just as strong if not stronger than anabatic winds in some cases. This is especially true for high elevation mountain environments where, because of elevation effects, night-time cooling can be extremely rapid under clear sky situations. Given this set of conditions, downslope topographically channelled katabatic flows can be very strong, thus producing good dispersion conditions. In lower elevation situations cool descending air may result in the formation of valley and basin fogs as katabatic flows cool valley air to its dewpoint. This set of circumstances also gives rise to inversions owing to valley bottom ponding of cool air. This is potentially a hazardous situation; fumigation can result, as often experienced in many mountain valley and basin locations in the winter when smoke from wood and coal fires is vertically confined and remains trapped beneath the nighttime inversion near the valley or basin bottom (Kaiser, 1996).

Modelling Air Pollution

Air pollution is modelled using air quality models (AQM). These can be considered as a numerical laboratory that simulates the real world. For air quality managers AQMs can provide guidance for regulatory decision making by simulating future emission scenarios that may be realized at some point in the future. For large military, nuclear power and chemical facilities AQM play a central role in emergency response dose assessment systems (Lyons and Tremback, 1993; Williams and Yamada, 1990). For the scientist AQMs are essential for developing a scientific understanding of transportation and dispersion processes in the atmosphere (Lyons and Tremback, 1993; Lyons et al., 1995). AQMs, however, vary in their level of sophistication. This is related to the way in which the complex processes of transport, dispersion, diffusion, dry and wet deposition and chemical transformations are represented or parameterized in AQMs. Furthermore, atmospheric pollution involves different scales and the meteorological processes that control pollution vary with these scales. At the near-source scale the nature of the thermal plume and local turbulence dominate the dispersion processes. At the urban scale turbulent mixing and chemical reactions are important processes. At the meso scale, chemical reactions and wind transportation are the controlling factors, while at the continental scale wind transportation and deposition are the major processes.

Of all the AQMs, the Gaussian straight-line plume dispersion models are the simplest ones in terms of the parameterization of meteorology. Although these models have been used for approximately three decades by air quality managers and emergency response authorities, the inherent assumptions of Gaussian models compromise their utility for providing realistic air pollution predictions for all but the simplest meteorological and terrain conditions. In addition to the assumption of steady-state (constant) meteorological conditions, other Gaussian model assumptions include: a Gaussian (normal) distribution of pollutant concentration in both the vertical and cross-wind directions; instantaneous transport of the plume through the atmosphere; purely horizontal transport and a horizontally homogeneous surface environment usually defined by a single meteorological measurement site (Lyons et al., 1995).

Because of their inherent problems, over the last decade or so there has been a move away from the simple Gaussian models to integrated modelling systems. These comprise three-dimensional meteorological and air quality models; the meteorological model provides the meteorological input into the air quality model (Lyons et al, 1995; Pielke et al, 1991, 1992; Pilinis et al, 1993). The success of such modelling systems has been largely due to the ability of the meteorological models to model the full three-dimensional nature of meteorological fields, especially vertical meso-scale motion. This meteorological parameter is not only one of the most difficult parameters to measure but plays a critical role in the vertical transport and diffusion of the pollutant plume. For example Lyons et al (1995) used a prognostic meso-scale meteorological model called RAMS (regional atmospheric modelling system), combined with a Lagrangian particle dispersion model, to investigate the nature of meso-scale dispersion in the coastal zone where the meteorology is characterized by strong vertical ascent, subsidence and wind shear, spatially variable mixing heights and re-circulating wind fields – conditions not resolved by Gaussian models. They found that because of unexpectedly strong vertical motion, plumes released along a shoreline may be vertically translocated out of the sea breeze flow in the region of the sea breeze front. Furthermore, their research demonstrated that within sea breeze return flows re-circulating plumes may bifurcate into distinct branches with varying amounts of pollutants entrained back into the inflow layer of the sea breeze, giving rise to complex surface concentration and dispersion patterns as a result of multiple fumigation (Fig. 3.9). For the greater Athens area similar approaches have also been applied to the investigation of this location's air pollution meteorology. Using the RAMS prognostic model in combination with a three-dimensional air quality model, Pilinis et al

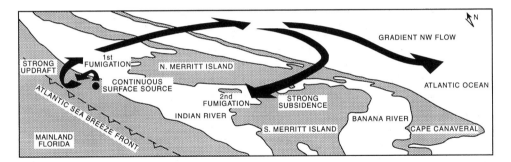

Fig. 3.9   The trajectory of pollutants as modelled using a meso-scale meteorological model in conjunction with a Lagrangian particle dispersion model. Note primary and secondary fumigation (see Fig. 3.7). Source: Lyons, W.A., Pielke, R.A., Tremback, C.J., Walko, R.L., Moon, D.A. and Keen, C.S. 1995 *Atmospheric Environment*, 29, 283–301.

(1993) demonstrated how the daily evolution of the sea breeze could account for spatial variations in photochemical pollution. Under sea breeze conditions it was shown that the main Athens urban area exhibited high nitrogen oxide concentrations related to automobile emission whereas the suburbs to the northeast of the urban area experienced high ozone concentrations, a result of ozone precursors being transported to this area by the sea breeze.

As noted by Lyons *et al.* (1995), the use of prognostic meteorological models in conjunction with air quality and dispersion models now makes it possible to understand the complex meteorology of air pollution episodes. More importantly, such integrated modelling strategies may be applied not only in research but also to real-time air quality forecasting situations.

WEATHER AND HEALTH

There is considerable evidence that human physiological and biochemical processes are partially related to changes in atmospheric conditions such as pressure, temperature and humidity (Curson, 1996). For example, a range of medical responses have been reported in the case of the passage of major cold fronts (Fig. 3.10), amongst which are increases in the number of fatal heart attacks (Anto and Sunyer, 1986; Bucher and Hasse, 1993; Sanchez *et al.*, 1990). Extremes of heat and cold, as experienced during heat and cold waves, are also known, through their effects on the human heat balance, to induce thermal stress and important physiological changes (Figs. 3.11 and 3.12). However, while the independent effects on health of individual weather elements such as temperature at a variety of temporal scales (daily, monthly, annual) has become reasonably well established (Bull and Morton, 1975; Frost, 1993; Gyllerup *et al.*, 1991; Jendritzky, 1996; Katsouyanni *et al.*, 1993; Kunst *et al.*, 1993; Langford and Bentham, 1995; Larsen, 1990; Matzarakis and Mayer, 1991; Rose, 1966; Saez *et al.*, 1995), little is understood about the interactive effects of weather elements (temperature, atmospheric moisture, wind speed and direction, cloud cover, and atmospheric pressure) on health. This is important as the human body

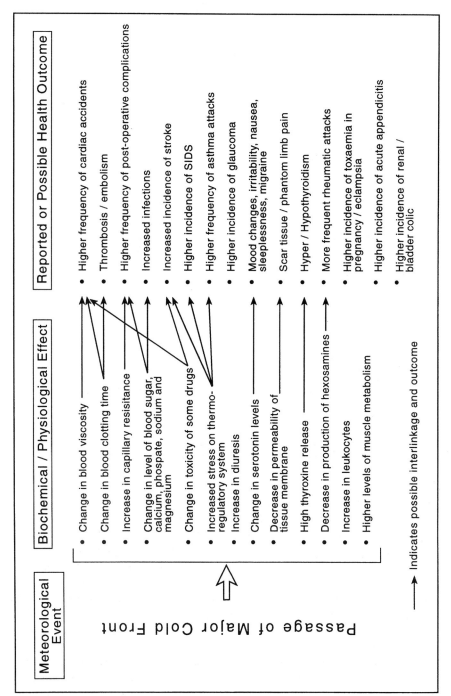

Fig. 3.10 Health responses due to frontal passage. Source: Curson, P. 1996 In Henderson-Sellers, A. and Giambelluca, T. (eds). *Climate Change: Developing Southern Hemisphere Perspectives.* John Wiley and Sons, Chichester.

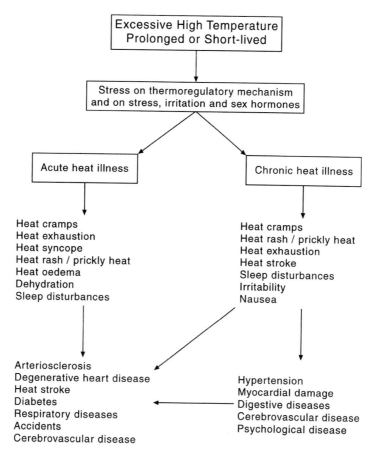

Fig. 3.11 Physiological responses to extreme heat. Source: In Henderson-Sellers, A. and Giambelluca, T. (eds). *Climate Change: Developing Southern Hemisphere Perspectives.* John Wiley and Sons, Chichester.

responds, at both short and long time-scales, to the synergistic or interactive behaviour of the weather elements that make up the physical atmospheric environment.

The synergistic behaviour of atmospheric elements is embodied in air masses. These are large volumes of air with homogeneous physical properties. The long-term temporal and spatial behaviour of the air masses that accompany weather systems produces the climate of an area, while their short-term variability brings, on a daily basis, the weather and air quality changes experienced at a location. Given the fact that air masses encapsulate the holistic behaviour of the physical atmospheric environment, recent research by bio-meteorologists on weather and health relationships has focused on identifying those air mass types that engender health outcomes.

In an analysis of 10 US cities, Kalkstein (1991) showed that elevated summertime mortality events for the majority of cities were associated with one particular air mass type; a very warm moist tropical type. Interestingly, this air mass type was inoffensive in terms of pollution levels, indicating that daily mortality was more sensitive to air mass type than to pollution levels. More recently, Greene and Kalkstein (1996) have applied a

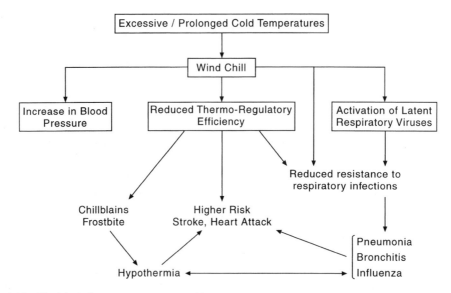

Fig. 3.12 Physiological responses to extreme cold. Source: In Henderson-Sellers, A. and Giambelluca, T. (eds). *Climate Change: Developing Southern Hemisphere Perspectives.* John Wiley and Sons, Chichester.

continental-scale air-mass-based classification to the analysis of daily mortality in the US and have found that the sensitivity of mortality is dependent on the background climatic or air mass conditions. Mortality in locations with considerable variations in summer air mass type was found to be more sensitive to climatic variations than mortality in locations where summer air mass types vary little. This finding is of considerable importance and provides some corroborative evidence for the assertion that the impact of weather on mortality is climate dependent.

For the UK, McGregor (1999) has analysed the relationships between daily winter ischaemic heart disease (IHD) mortality and winter air mass types. He found that IHD mortality is sensitive to two air mass types: a cold polar continental type which brings cold damp conditions to the UK from the east, and a moderately warm blustery humid maritime type associated with storms moving over the UK off the Atlantic from the west.

Whether studied in terms of single weather elements or air mass types it is clear that there are, at times, close associations between weather and health. However, health is also closely linked to air quality. For example, respiratory response has been shown to be related to air quality (Pope and Kalkstein, 1996; Schwartz, 1994) but, as outlined in *Air Pollution Meteorology* above, air quality is closely linked to the daily variations in meteorological conditions. Unravelling this complex association between weather, air pollution and health is one of the most difficult research issues in contemporary biometeorology and epidemiology.

AIR POLLUTION, WEATHER AND HEALTH

In addition to mortality and morbidity being analysed in terms of its weather sensitivity, the role of pollution in determining mortality rates has been widely considered. However, most studies on climate air pollution and health have focused mainly on the independent effects of either climate or pollution on health. The separation and quantification of climate-induced and pollution-induced health effects has been relatively neglected (McMichael *et al.*, 1996). Studies that have attempted to differentiate between the impacts of air pollution and weather upon mortality (Ito *et al.*, 1993; Katsouyanni *et al.*, 1993; Pope and Kalkstein, 1996; Schwartz and Dockery, 1992; Shumway *et al.*, 1988; Thurston, 1989) have revealed that weather plays a secondary role to that of fine particulate matter and acid aerosols. In contrast, for some locations pollution has been found to have little impact on daily mortality in comparison to the effect of short-term weather events (Biersteker and Evendijk, 1976; Kalkstein, 1991; Kunst *et al.*, 1993; Mackenbach *et al.*, 1993).

That both weather and air pollution should not be treated separately when assessing links between weather, air pollution and health is clear from the results of a very limited number of studies. These show clearly that short-term elevated mortality may be related to the interactive effects of extreme weather events with high pollution levels (Katsouyanni *et al.*, 1993; Shumway and Azari, 1992; Shumway *et al.*, 1988). Regarding respiratory tract related responses, McGregor *et al.* (1999) have shown for winter in the Birmingham area, UK, that two general weather modes are associated with high respiratory hospital admission rates. One is a cool moist mode associated with either northwesterly flows off the North Atlantic or anticyclonic gloom conditions with little airflow. Unlike the former synoptic-scale situation, the latter is associated with high particulate matter concentrations, suggesting that this particular weather and air quality situation may precipitate a distinct health response due to weather–air pollution interactions. In contrast, elevated admissions on cool moist northwesterly days appear to be related purely to meteorological effects similar to that found for the USA by Goldstein (1980). The second weather mode associated with high daily respiratory admissions is cold dry weather. Like the moist anticyclonic gloom situation, this weather situation also provides the requisite meteorological conditions (little ventilation, combined subsidence and radiation inversions and restricted boundary layer mixing heights) for high concentrations of particulate matter. Cold dry weather may therefore also act in a synergistic way with poor air quality to engender elevated hospital respiratory admissions.

Although an understanding of the short-term interactive effects of weather and poor air quality on health is being developed through biometeorological and epidemiological research, little is known about the health impacts of long-term exposure to poor air quality climates. This is an area in which research is required (McMichael *et al.*, 1996).

CONCLUSION

Weather and its longer-term counterpart climate play a key role in the relationship between air pollution and health. The dispersion and diffusion of air pollution is heavily dependent on the three-dimensional physical and dynamical properties of the atmosphere. Atmospheric stability, which refers to the thermal structure of the atmosphere, controls the extent to which pollutants may be dispersed and diffused vertically. Of the various stability states, absolute stability poses the greatest potential for the accumulation of pollutants in the boundary layer owing to the presence of inversions. Although there are a variety of inversion types, inversions in general act as a physical barrier to the vertical mixing of pollutants. Wind generally serves to transport and disperse pollutants. However, high-frequency variations of wind speed and direction generate atmospheric turbulence in the form of turbulent eddies. Diffusion of pollutants occurs when the size of the turbulent eddies is smaller than the pollution plume, whereas transport of pollution occurs if the eddies are larger than the plume.

Boundary layer atmospheric stability and wind regime are often closely linked to the large or synoptic-scale weather situation. This, at times, can directly provide the requisite meteorological conditions for the development of air pollution episodes, for example strongly subsident conditions in the case of a persistent anticyclonic system. At other times the synoptic-scale situation may indirectly enhance the development of local circulation systems such as sea (lake) and land breezes and mountain and valley winds. In air pollution meteorology local circulation systems are of special interest as they are poor pollution dispersers. It is to such circulation systems that prognostic meso-scale meteorological and air quality models may be applied in order to understand the complex meteorology of air pollution episodes associated with these systems. Meteorological modelling may also provide the basis for real-time air quality forecasting and thus weather, air quality and health watch warning systems.

Because people can be sensitive to both weather and air pollution, certain weather and air quality configurations may engender distinct health outcomes. Unravelling these complex interactive effects is a challenge. Air mass types may provide a particularly useful basis for analysing variations in morbidity and mortality as a result of the interactive effects of weather and air quality. This is because air masses represent the holistic nature of the physical atmospheric environment and their characteristics are quantifiable. Climate thresholds beyond which human health problems increase markedly can therefore be determined. Such determinations are of interest to environmental epidemiologists. Furthermore, traditional epidemiological techniques can be applied to the investigation of variations of mortality, morbidity and air quality within an air mass type. This approach facilitates the determination of the relative sensitivity of health outcomes to intrinsic air mass physical and chemical properties and sheds light on the possible meteorological and air quality factors that may account for variations in mortality and illness within and between air mass types.

Although not addressed explicitly in this chapter, the issue of climate change has been implicit in the above discussions. Such change holds a number of implications for the issues addressed in this book. Increases in temperatures related to climate change may well enhance the rate of formation of secondary pollutants such as ozone, especially during summer in the mega-cities of the future. Climate change could therefore bring about an increased frequency of extreme weather events and high pollution concentrations (McMichael et al., 1996). Related to such a scenario is the possibility that the combined exposure to weather and air pollution extremes may have significant health impacts especially for the young and elderly as well as those who are sensitive to either weather and/or air pollution. Increases in acute or short-term mortality and morbidity may therefore result. However, statements such as this will remain only as scenarios until the exact nature of the synergistic impacts of weather extremes and offensive air quality conditions on health are firmly established. This represents one of the most challenging research issues in biometeorology and environmental epidemiology.

REFERENCES

Anto JM and Sunyer J (1986) Environmental health: a point source asthma outbreak. *Lancet* **1**: 900–903.

Atwater MA and Londergan RJ (1985) Differences caused by stability class on dispersion in tracer experiments. *Atmosp Environ* **19**: 1045–1051.

Baker DG, Enz JW and Paulus HJ (1969) Frequency, duration, commencement time and intensity of temperature inversions at St Paul Mineapolis. *J Appl Meteorol* **8**: 747–753.

Biersteker K and Evendijk JE (1976) Ozone temperature and mortality in Rotterdam in the summers of 1974 and 1975. *Environ Res* **12**: 214–217.

Bucher K and Hasse C (1993) Meteorotropy and medical-meteorological forecasts. *Experientia* **49**: 759–768.

Bull GM and Morton J (1975) Environment temperature and death rates. *Age and Aging* **7**: 210–224.

Christiansen AM and Bryson RA (1966) An investigation of the potential of component analysis for weather classification. *Monthly Weather Review* **94**: 697–709.

Curson P (1996) Human health climate and climate change. In: Henderson-Sellers A and Giambelluca T (eds) *Climate Change: Developing Southern Hemisphere Perspectives*. Chichester: John Wiley & Sons.

Crespi SN, Artinano B and Cabal H (1995) Synoptic classification of mixed-layer height evolution. *J Appl Meteorol* **34**: 1666–1677.

Davies TD, Dorling S, Pierce C *et al.* (1991) The meteorological control on the anthropogenic ion content of precipitation at three sites in the UK: utility of Lamb Weather Types. *Int J Climatol* **11**: 797–807.

Davis RE and Gay DA (1993) An assessment of air quality variations in the south western United States using an upper air synoptic climatology. *Int J Climatol* **13**: 755–781.

Davis RE and Kalkstein L (1990) Development of an automated spatial synoptic climatological classification. *Int J Climatol* **10**: 769–794.

Eder BK, Davis JM and Bloomfield P (1994) An automated classification scheme to better elucidate the dependence of ozone on meteorology. *J Appl Meteorol* **33**: 1182–1199.

Frost DB (1993) Myocardial infarct death: the population at risk and temperature habituation. *Int J Biometeorol* **37**: 46–51.

Golder R (1972) Relations among stability parameters in the surface layer. *Boundary Layer Meteorol* **3**: 47–58.

Goldstein IF (1980) Weather patterns and asthma epidemics in New York and New Orleans, USA. *Int J Biometeorol* **24**: 329–339.

Greene JS and Kalkstein LS (1996) Quantitative analysis of summer air masses in the eastern United States and an application to human mortality. *Climate Res* **7**: 43–53.

Gyllerup S, Lanke J, Lindholm LH and Schersten B (1991) High coronary mortality in cold regions of Sweden. *J Internal Medicine* **230**: 470–485.

Holzworth GL (1972) Vertical temperature structure during the Thanksgiving week air pollution episode in New York City. *Monthly Weather Rev* **100**: 445–450.

Ito K, Kowaoi T, Kindo T and Hiruma F (1993) Associations of London England daily mortality with particulate matter, sulphur dioxide and acidic aerosol pollution. *Arch Environ Health* **48**: 213–220.

Jendritzky G (1996) Komplexe unwelteinwirkungen: Klima (Complex environmental effects: climate). In: Beyer A and Eis D (eds) *Praktische Umweltmedizin, 4 Nachleferung* (Practical Environmental Health, 4th Supplement). Heidelburg: Springer-Verlag, pp. 1–30.

Kalkstein LS and Corrigan P (1986) A synoptic climatological approach for geographical analysis: Assessment of sulfur dioxide concentrations. *Annals of the Association of American Geographers* **76**: 381–395.

Kaiser A (1996) Analysis of the vertical temperature and wind structure in alpine valleys and their effects on air pollutants. *Proceedings of the 24th International Conference on Alpine Meteorology*, Bled, Slovenia, 9–13 September 1996, pp. 312–319.

Kalkstein LS (1991) A new approach to evaluate the impact of climate upon human mortality. *Environ Health Perspect* **96**: 145–150.

Kalkstein LS and Davis RE (1989) Climate effects on human mortality: an evaluation of demographic and interregional responses in the United States. *Ann Assoc Am Geog* **79**: 44–64.

Katsouyanni K, Pantazopoulou A, Touloumi G *et al.* (1993) Evidence for the interaction between air pollution and high temperature in the causation of excess mortality. *Arch Environ Health* **48**: 235–242.

Kraus H, Malcher J and Schaller E (1985) Nocturnal low-level jet during PUKK. *Boundary Layer Meteorol* **31**: 187–195.

Kunst AE, Looman CWN and Mackenbach BP (1993) Outdoor air temperature and mortality in the Netherlands: a time series approach. *Am J Epidemiol* **137**: 331–341.

Ladd JW and Driscoll DM (1980) A comparison of objective and subjective means of weather typing – an example from West Texas. *Journal of Applied Meteorology* **19**, 691–704

Lamb HH (1972) British Isles weather types and a register of daily sequence of circulation patterns. *Geophys Mem* **116** (London).

Langford IH and Bentham G (1995) The potential effects of climate change on winter mortality in England and Wales. *Int J Biometeorol* **38**: 141–148.

Larsen U (1990) The effects of monthly temperature fluctuations on mortality in the United States from 1921 to 1985. *Social Biol* **37**: 172–187.

Luna E and Church HW (1971) A comparison of turbulence intensity and stability ratio measurements to Pasquill stability types. *Conference on Air Pollution Meteorology*, American Meteorological Society, p. 90.

Lyons T and Scott B (1990) *Principles of Air Pollution Meteorology.* London: Belhaven Press.

Lyons WA, Pielke RA, Tremback CJ *et al.* (1995) Modelling impacts of mesoscale vertical motion upon coastal zone air pollution dispersion. *Atmos Environ* **29**: 283–301.

Lyons WA and Tremback CJ (1993) A prototype operational mesoscale air dispersion forecasting system using RAMS and HYPACT. Paper 93-TP-26B01. *86th Annual Meeting and Exhibition, Denver, Air and Waste Management Association.* Pittsburgh, 16 pp.

Mackenbach JP, Looman CWN and Kunst AE (1993) Air pollution, lagged effects of temperature and mortality: The Netherlands 1979–1987. *J Epidemiol Community Health* **47**: 121–126.

Matzarakis A and Mayer H (1991) The extreme heat wave in Athens in July 1987 from the point of view of human biometeorology. *Atmos Environ* **25B**: 203–211.

McGregor GR (1999) Winter ischaemic heart disease mortality in Birmingham UK: a synoptic climatological analysis. *Climate Res* (in press).

McGregor GR and Bamzelis D (1995) Synoptic typing and its application to the investigation of weather air pollution relationships, Birmingham, UK. *Theor Appl Climatol* **51**: 223–256.

McGregor GR, Cai X, Thorne JE *et al.* (1996). In: Genard AJ and Slater T (eds) *Managing a Conurbation.* Warwickshire: Brewin Books, pp. 115–128.

McGregor GR, Walters S and Wordley J (1999) Investigating the relationship between daily hospital respiratory admissions and weather using winter airmass types. *Int J Biometeorol* (in press).

McMichael AJ, Haines A, Sloof R and Kovats RS (eds) (1996) *Climate Change and Human Health*, WHO/EHG/967. Geneva: WHO.

Milionis AE and Davies TD (1992) A five year climatology of elevated inversions at Hemsby (UK). *Int J Climatol* **12**: 205–216.

Muller RA (1977) A synoptic climatology for environmental base line analysis: New Orleans. *J Appl Meteorol* **16**: 20–33.

Muller RA and Jackson AL (1985) Estimates of climatic air quality potential at Shreveport, Lousiana., *J Climate Appl Meteorol* **24**: 293–301.

Oke TR (1987) *Boundary Layer Climates.* London: Methuen.

Pasquill F (1961) The estimation of the dispersion of windborne material *Meteorol Mag* **90**: 33–49.

Pielke RA, Lyons WA, McNider *et al.* (1991) Regional and mesoscale meteorological modelling as applied to air quality studies. Van Dop, H and Styn, D.G. (eds.) *Air Pollution Modelling and its Applications VIII.* New York: Plenum Press.

Pielke RA, Cotton WR, Walko RL *et al.* (1992) A comprehensive meteorological modelling system – RAMS. *Meteorology and Atmospheric Physics* **49**: 69–91.

Pilinis C, Kassomneos P and Kallos G (1993) Modelling of photochemical pollution in Athens, Greece: application of the RAMS-CALGRID modelling system. *Atmos Environ* **27B**: 353–370.

Pope CA and Kalkstein LS (1996) Synoptic weather modelling and estimates of the exposure–response relationship between daily mortality and particulate air pollution. *Environ Health Persp* **104**:

414–420.

Prater BE and Collis JJ (1981) Correlations between acoustic sounder dispersion estimates, meteorological parameters and pollution concentrations. *Atmos Environ* **15**: 793–798.

Preston-Whyte RA, Diab RD and Tyson PD (1977) Towards an inversion climatology of Southern Africa: Part 2. Non-surface inversions in the lower atmosphere. *South African Geographer* **59**: 45–59.

Remsburg EE, Buglia JJ and Woodbury GE (1979) The nocturnal inversion and its effect on the dispersion of carbon dioxide at ground level in Hampton, VA. *Atmos Environ* **13**: 443–447.

Rose G (1966) Cold weather and ischaemic heart disease. *Br J Prevent Social Med* **20**: 97–100.

Saez M, Sunyer J, Castellsague J *et al.* (1995) Weather temperature and mortality: A time series analysis approach in Barcelona. *Int J of Epidemiol* **24**: 576–582.

Sanchez JL *et al.* (1990) Forecasting particulate pollutant concentrations in a city from meteorological variables and regional weather patterns. *Atmospheric Environment* **24A**: 1509–1519.

Schwartz J (1994) Air pollution and daily mortality: a review and meta-analysis. *Environmental Research* **64**: 36–52.

Schwartz J and Dockery DW (1992) Increased mortality in Philadelphia associated with daily air pollution concentrations. *Am Rev Resp Dis* **145**: 600–604.

Scorer RS (1990) *Meteorology of Air Pollution: Implications for the Environment and its Future.* Chichester: Ellis Horwood.

Shumway RH and Azari AS (1992) Structural modelling of epidemiological time series. *Final Report to Research Division California Air Resources Board, Sacramento CA 95812* (Contract ARB A833–136 available on request).

Shumway RH, Azari AS and Pawitan Y (1988) Modelling mortality fluctuations in Los Angeles as functions of pollution and weather effects. *Environmental Research* **45**: 224–241.

Thurston GD (1989 Reexamination of London, England, mortality in relation to exposure to acidic aerosols during 1963–1972 winters. *Environ Health Perspect* **79**: 73–82.

Wallace JM and Hobbs PV (1987) *Atmospheric Science: An Introductory Survey.* London: Academic Press.

Williams M and Yamada T (1990) A microcomputer based forecasting model: potential applications for emergency response plans and air quality studies. *J Air Waste Manag Assoc* **40**: 1266–1270.

Yarnal B (1993) *Synoptic Climatology in Environmental Analysis.* London: Belhaven Press.

4

Atmospheric Chemistry

R.G. DERWENT

Atmospheric Processes Research, Meteorological Office, Bracknell, UK

INTRODUCTION

A pollutant is usually understood to be a substance which, between the point of its emission into the atmosphere and its ultimate removal, causes harm to a target, whether this is an ecosystem or human health. Air pollution arises from the competition between emission processes which increase pollutant concentrations and dispersion and advection processes in the atmosphere which reduce them. However, not all air pollution phenomena conform to this simple characterization. An important category of air pollutants consists of substances that are not emitted into the atmosphere themselves in any significant quantity, but are formed there by chemical reactions. These latter pollutants are termed secondary pollutants to distinguish them from the primary or emitted pollutants (Seinfeld, 1986; Brimblecombe, 1996). This chapter is about the atmospheric transformation processes which convert particular primary pollutants, or precursors, into secondary pollutants.

Despite the significant improvements made in air quality over the last three decades, there are three major secondary pollution problems which potentially may exert some public health impact in our urban and industrialized population centres worldwide (WHO, 1987). These problem pollutants include: nitrogen dioxide (NO_2), ozone (O_3) and suspended particulate matter. The transformations which produce them form the subject matter of this chapter.

During wintertime pollution episodes, NO_2 concentrations in urban areas may exceed internationally accepted air quality criteria set for the protection of human health. This somewhat unexpected phenomenon has been reported in some large industrial and urban population centres in northwest Europe, particularly in the UK (Bower *et al.*,

AIR POLLUTION AND HEALTH
ISBN 0-12-352335-4

1994) during stagnant wintertime weather conditions. The sole precursor of the elevated NO_2 levels is the nitric oxide (NO) emitted by motor traffic and by stationary combustion sources, such as industrial, commercial and domestic boilers fuelled by coal, gas or oil.

During summertime pollution episodes, photochemical reactions driven by sunlight lead to the conversion of organic compounds and oxides of nitrogen into photochemical oxidants, in particular ozone (Leighton, 1961). This phenomenon was first reported in Los Angeles in the 1940s and has subsequently been observed in almost all urban and industrial population centres worldwide. These photochemical reactions also lead to the oxidation of sulfur dioxide (SO_2) into a fine haze of sulfuric acid (H_2SO_4) aerosol. This photochemically generated aerosol contributes to suspended particulate matter and gives rise to visibility reduction and the loss of distant horizons (Hidy, 1986). Elevated urban concentrations of ozone and suspended particulate matter may exceed internationally accepted environmental criteria levels set to protect human health (WHO, 1996), particularly in northwest Europe and North America.

WINTERTIME URBAN POLLUTION EPISODES AND ELEVATED NITROGEN DIOXIDE CONCENTRATIONS

Sources of Nitrogen Dioxide in Urban Areas

The oxides of nitrogen are ubiquitous urban air pollutants whose main sources include exhausts of both petrol- and diesel-engined motor vehicles, stationary combustion in boilers fired by coal, oil or gas, and industrial processes involving the use or production of nitric acid. Nitric oxide (NO) is by far the most important nitrogen-containing species emitted into the atmosphere, on a mass basis, from human activities involving motor traffic and combustion in power stations, in the home or in industrial processes. Few adverse environmental impacts are associated directly with NO and most concerns have been associated with the atmospheric transformation products that are formed following its release into the ambient atmosphere.

Transformations Involving NO and NO_2

The main fate of emitted NO is to react with the ozone (O_3), ubiquitously present in the lower atmosphere, as in reaction (1). This is the case for the NO generated by motor traffic, power stations, microbial processes in soils, lightning, biomass burning and high-flying aircraft. Reaction (1) has been exceedingly well studied under laboratory conditions and is one of the best understood of all the atmospheric chemical processes:

$$NO + O_3 = NO_2 + O_2 \tag{1}$$

Under typical atmospheric conditions the reaction takes place within a few seconds, leading either to the complete conversion of all of the O_3 to nitrogen dioxide (NO_2) with an excess of unreacted NO, or to the conversion of all of the NO to NO_2 with an excess of unreacted O_3. In highly polluted atmospheres, or close to individual pollution sources, the former behaviour is typically observed. This is because although ozone is widely distributed in the lower atmosphere, its concentration is not usually high compared with that of NO, and hence ozone concentrations become rapidly depleted.

In general NO is initially present in a concentrated plume, whether of motor vehicle exhaust or in a power station plume, with NO concentrations perhaps reaching 1–1000 ppm (1 ppm represents a mixing ratio of 1 volume in a million volumes of air). At the edges of the plume the reaction between NO and O_3 occurs, producing NO_2 at a rate limited by the concentration of O_3 in the ambient atmosphere, generally 10–50 ppb (1 ppb represents a mixing ratio of 1 volume in a billion (10^9) volumes of air), and by the rate of entrainment of background air into the plume. As the plume of material moves downwind, it expands in size and plume centre-line concentrations of NO steadily decline. NO_2 concentrations towards the edges of the plume steadily rise to approach those of the ambient O_3 background and begin to encroach into the body of the plume from the edges. Eventually, the processes of atmospheric diffusion and dispersion prevail and the plume begins to merge into the ambient background. Under these conditions, comparable concentrations of NO and NO_2 coexist at levels comparable to or below those of the O_3 background concentrations. For a typical power station, these processes may take several hours and hundreds of kilometres travel, whereas for a motor vehicle exhaust, several minutes and hundreds of metres may suffice.

In stagnant weather conditions, particularly during cold wintertime conditions, a further oxidation route converting NO into NO_2 may come into play through reaction (2):

$$NO + NO + O_2 = NO_2 + NO_2 \tag{2}$$

Because of its second-order dependence on NO concentrations, this process only becomes of any significance when NO concentrations are exceedingly high. This process explains the occurrence of the unprecedented concentrations of NO_2 observed in some urban areas during wintertime. In one such episode in central London during December 1991, hourly peak concentrations of NO_2 as high as 423 ppb were reported in association with NO concentrations in excess of 1 ppm (Bower *et al.*, 1994).

Air Quality Data for NO_2 and NO_x in Urban Areas

To illustrate the complexity of the interplay between the oxides of nitrogen in the urban atmosphere, a complete year of hourly air quality data for NO_2 and NO_x is presented in Fig. 4.1. The air quality data are taken from a continuous chemiluminescence NO_x monitor sited in a mobile laboratory adjacent to a roadside location in London. The monitor sampled air at a height of 5 m above ground, 5 m from the edge of Exhibition Road, which carries a modest amount of traffic (1500–1750 vehicles per hour during the daytime). The hourly NO_x concentrations have been segregated into 10 ppb bins and the mean of the simultaneous hourly NO_2 concentrations for each NO_x bin has been plotted in Fig. 4.1. Over the annual period, which included the most severe wintertime pollution episode for decades, a systematic relationship is revealed between the simultaneous NO_2 and NO_x concentrations which is driven by the rapid transformation processes involving NO and NO_2 (Derwent *et al.*, 1995).

At low NO_x concentrations with excess ozone the photostationary state is established quickly during daytime, and for NO_x concentrations less than 20 ppb NO_2 concentrations are typically twice those of NO. As NO_x concentrations rise, the fraction of the NO_x present as NO increases and the fraction present as NO_2 decreases. The concentrations of NO and NO_2 are roughly equal for NO_x concentrations between 70 and 80 ppb. Figure 4.1 shows that at NO_x concentrations between 100–200 ppb the NO_2 concentration curve rises less steeply than at low NO_x concentrations. For NO_x concentrations of 200 ppb, NO_2 concentrations are typically less than one half of the NO concentrations. In the region with NO_x concentrations in the range 0–200 ppb, the behaviour of NO_2 is largely accounted for by the photostationary state during daylight and by the availability of ozone during night-time.

Figure 4.1 shows that for NO_x concentrations above 200 ppb other processes come into play. Nitrogen dioxide concentrations do not remain at the 40 ppb levels seen at NO_x concentrations of 200 ppb but begin to increase more rapidly, particularly at the highest NO_x concentrations. Since there is never enough ozone available, particularly during wintertime, to produce such elevated NO_2 concentrations, other oxidation routes must come into play. At the high NO_x levels seen in December 1991 in Central London (see Fig. 4.1), the main oxidation route producing NO_2 is reaction (2). The second-order nature of this reaction causes the upwards swing in the NO_2 vs NO_x concentration plot in Fig. 4.1.

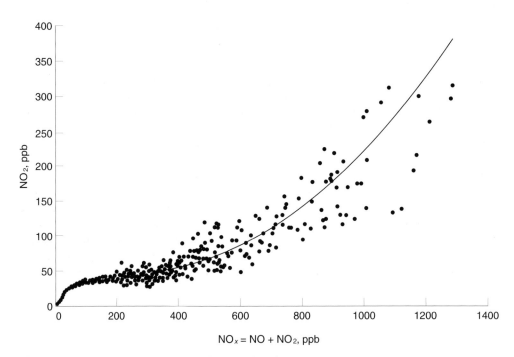

Fig. 4.1   Average hourly nitrogen dioxide concentrations for each increasing hourly mean $NO_x = NO + NO_2$ concentrations during the period July, 1991 to July, 1992 at a roadside location in central London. ◆, observation; —, fit.

SUMMERTIME REGIONAL POLLUTION EPISODES AND ELEVATED OZONE CONCENTRATIONS

Photostationary State and Ozone Production

In addition to its reactions with O_3 and oxygen (O_2), the other important atmospheric reactions of nitric oxide are with the hydroperoxy (HO_2) in reaction (3) and the organic peroxy (RO_2) radicals in reaction (4):

$$HO_2 + NO = NO_2 + OH \tag{3}$$

$$RO_2 + NO = NO_2 + RO \tag{4}$$

All of the peroxy radical reactions exemplified by reactions (3) and (4) above result in the conversion of NO to NO_2. This conversion process is an essential element of the photochemical generation of ozone. It is of crucial importance in photochemical smog formation in the polluted urban boundary layer, in the formation of greenhouse gases in the lower atmosphere, and in the chemistry of the stratospheric ozone layer (Crutzen, 1974).

Nitrogen dioxide is the first and most immediate reaction product of the atmospheric oxidation of the NO emitted by human activities. During daylight, the main fate of the NO_2 so formed is to absorb solar ultraviolet (UV) radiation and to undergo photolysis, reforming the NO and O_3 from which it was made, through reactions (5) and (6):

$$NO_2 + \text{radiation (wavelengths 200–420 nm)} = NO + O \tag{5}$$

$$O + O_2 + M = O_3 + M \tag{6}$$

where M represents an oxygen O_2 or nitrogen N_2 molecule which acts as a so-called 'third body'. The rate of photolysis of NO_2 depends on the solar actinic irradiance; this in turn depends on the height of the sun in the sky and hence time of day and season, as well as on the amount and height of any cloud and haze which may obscure the sun. For much of the daylight portion of the year the lifetime of NO_2 is only a matter of minutes before it is photolysed back to NO.

The reaction of NO with O_3 and the photolysis of NO_2 form a cycle of reactions which occurs rapidly over a timescale of minutes in the sunlit atmosphere. Together, this cycle of reactions generates a mixture of NO and NO_2 in the sunlit atmosphere from initially either pollutant present on its own. This ensures that under most atmospheric conditions NO and NO_2 coexist as a mixture. The mixture is conveniently referred to as $NO_x = NO + NO_2$. This reaction system is the so-called 'photostationary state' system which, when established, leads to a rather constant relationship, equation (7), between the concentrations of NO, NO_2 and O_3 in the sunlit atmosphere:

$$\frac{[NO] \times [O_3]}{[NO_2]} = J / k_1 \tag{7}$$

where J is the photolysis rate coefficient for reaction (5) and k_1 is the first-order rate coefficient for reaction (1).

Under summertime conditions, in the mid-afternoon at mid-latitudes and clear sky conditions, J will typically have a value of about $7 \times 10^{-3} \, s^{-1}$, k_1 about $4 \times 10^{-4} \, (ppb^{-1} \, s^{-1})$. This leads to the relationship:

$$\frac{[NO] \times [O_3]}{[NO_2]} = 18 \tag{8}$$

which implies that $[NO_2] = 2 \times [NO]$ at about 30–40 ppb $[O_3]$
and: $[NO_2] = 5 \times [NO]$ at about 90 ppb $[O_3]$ in the sunlit atmosphere

In wintertime conditions and at night-time, the photolysis on NO_2 does not occur to anything like the same extent as during the summer. Thus NO_2 becomes a rather more stable product of the atmospheric oxidation of NO. Under these wintertime conditions almost any concentrations of NO and NO_2 can coexist and there is no simple relationship such as the 'photostationary state' relationship between them. In polluted atmospheres, then, there is a tendency for all of the O_3 to be depleted by reaction (1) since NO concentrations are generally greater than those of O_3. In rural atmospheres, because ozone concentrations are generally greater than concentrations of NO, the opposite tendency usually occurs and reaction (1) goes to completion, leaving the NO greatly depleted.

Mechanism of Photochemical Ozone Formation

During the warm, sunny summertime anticyclonic conditions associated with regional-scale ozone episodes, peroxy radical concentrations become elevated by photochemical activity and the local rate of ozone production increases dramatically through the reactions (4), (5), (6) and (9).

$$RO_2 + NO = NO_2 + RO \tag{4}$$

$$NO_2 + radiation = NO + O \tag{5}$$

$$O + O_2 + M = O_3 + M \tag{6}$$

$$RO_2 + O_2 = RO + O_3 \tag{9}$$

In these reactions the NO and NO_2 appear to be left unchanged, acting as catalysts. By shifting the balance in the NO–NO_2–O_3 photostationary state, the peroxy radicals become an important photochemical source of ozone. The peroxy radicals themselves are irreversibly degraded into alkoxy radicals (RO). Generally speaking, alkoxy radicals are highly reactive with oxygen, forming an HO_2 radical and a carbonyl compound:

$$RO + O_2 = HO_2 + R'COR'' \tag{10}$$

The HO_2 radical may go on to react with NO, producing another NO to NO_2 conversion step which results in the production of another ozone molecule:

$$HO_2 + NO = NO_2 + OH \tag{3}$$

The hydroperoxy radical is thus rapidly converted into a hydroxyl radical (OH).
    Photochemical ozone formation can therefore be described as the process by which organic peroxy radicals shift the balance in the NO–NO_2–O_3 photostationary state in

favour of ozone production. The peroxy radicals themselves are converted into carbonyl compounds such as aldehydes and ketones. To complete the picture, an explanation is required about the origins of the organic peroxy radicals which turn out to be essential in regional-scale ozone production.

Mechanism of Hydrocarbon Oxidation

Organic peroxy radicals are almost exclusively formed by the attack of the highly reactive hydroxyl radical on the organic compounds ubiquitously present in the polluted atmospheric boundary layer. These reactions may be represented as:

$$OH + RH = R + H_2O \tag{11}$$

$$R + O_2 + M = RO_2 + M \tag{12}$$

The detailed mechanisms of the reaction which convert organic compounds into their corresponding peroxy radicals depend on the structure of the individual organic compounds involved (Atkinson, 1994). Most organic compounds react with hydroxyl radicals either by H-abstraction or by addition, if they contain carbon–carbon multiple bonds. In the case of alkanes, cycloalkanes, carbonyls and oxygenated hydrocarbons, the hydroxyl radical removes a hydrogen atom originally connected to the carbon skeleton of the parent compound, forming a carbon radical and water vapour. These carbon radicals quickly react with oxygen to form the corresponding peroxy radical. For example, if the parent organic compound was methane, the organic radical formed would be the methyl radical. This radical rapidly combines with oxygen to form a methyl peroxy radical:

$$OH + CH_4 = CH_3 + H_2O \tag{13}$$

$$CH_3 + O_2 + M = CH_3O_2 + M \tag{14}$$

In the case of alkenes, alkynes and aromatics, the hydroxyl radical adds to the multiple bond, producing a carbon radical which again almost invariably reacts with oxygen in the analogous process to form a peroxy radical. For example, with ethylene (ethene) the reaction sequence would look like this:

$$OH + C_2H_4 + M = HOC_2H_4 + M \tag{15}$$

$$HOC_2H_4 + O_2 = HOC_2H_4O_2 \tag{16}$$

Coupling together the OH attack on the parent organic compound with the conversion of the peroxy radicals to alkoxy radicals, and of the alkoxy radicals to carbonyls, the sequence of ozone production begins to take shape. For methane this occurs as follows:

$$OH + CH_4 = CH_3 + H_2O \tag{13}$$

$$CH_3 + O_2 + M = CH_3O_2 + M \tag{14}$$

$$CH_3O_2 + NO = CH_3O + NO_2 \tag{17}$$

$$NO_2 + radiation = NO + O \tag{5}$$

$$O + O_2 + M = O_3 + M \tag{6}$$

$$CH_3O + O_2 = HO_2 + HCHO \tag{18}$$

$$HO_2 + NO = OH + NO_2 \tag{3}$$

$$NO_2 + radiation = NO + O \tag{5}$$

$$O + O_2 + M = O_3 + M \tag{6}$$

$$CH_4 + O_2 + O_2 = HCHO + O_3 + O_3$$

For ethylene (ethene) the process is:

$$OH + C_2H_4 + M = HOC_2H_4 + M \tag{15}$$

$$HOC_2H_4 + O_2 = HOC_2H_4O_2 \tag{16}$$

$$HOC_2H_4O_2 + NO = NO_2 + HOC_2H_4O \tag{19}$$

$$NO_2 + radiation = NO + O \tag{5}$$

$$O + O_2 + M = O_3 + M \tag{6}$$

$$HOC_2H_4O + O_2 = HO_2 + HCHO + HCHO \tag{20}$$

$$HO_2 + NO = OH + NO_2 \tag{3}$$

$$NO_2 + radiation = NO + O \tag{5}$$

$$O + O_2 + M = O_3 + M \tag{6}$$

$$C_2H_4 + O_2 + O_2 = HCHO + HCHO + O_3 + O_3$$

In this series of rapid consecutive reactions the OH radical is recycled, the nitric oxide and nitrogen dioxide are recycled, two molecules of ozone are produced and the parent organic compounds are converted into carbonyl compounds. In this way, a small steady-state concentration of the highly reactive hydroxyl radicals can degrade substantial concentrations of organic compounds, producing ozone as an important reaction product. To estimate the rate of ozone production, we need to understand the rate of degradation of the individual organic compounds, and hence we need to estimate the steady-state concentration of hydroxyl radicals.

The Fast Photochemical Balance in the Sunlit Atmosphere

The hydroxyl radical concentrations in the sunlit atmospheric boundary layer are established by the fast photochemical balance reactions which link together each of the sources and sinks of the major free radical species. There are two main pools of free radical species: the pool of hydroxyl radicals and the pool of peroxy radicals, both hydroperoxy and the organic peroxy radicals (Derwent, 1995). In addition, there are six major categories of free radical reactions which together make up the fast photochemical balance. These six categories comprise:

(1) sources of hydroxyl radicals,
(2) sources of HO_2 and RO_2 radicals,
(3) reactions which interconvert OH into HO_2 and RO_2,

(4) reactions which interconvert HO_2 and RO_2 into OH,
(5) sinks for hydroxyl radicals,
(6) sinks for HO_2 and RO_2 radicals.

The resultant of all these processes is the formation of a steady-state concentration of the free radical species in the sunlit atmospheric boundary layer. Typically, close to noon, under moderately polluted conditions in northwest Europe, these steady-state concentrations may approach the following values:

$$\begin{array}{lll} \text{OH} & 6.0 \times 10^6 & \text{molecule/cm}^3 \\ \text{HO}_2 & 1.2 \times 10^8 & \text{molecule/cm}^3 \\ \text{CH}_3\text{O}_2 & 3.0 \times 10^7 & \text{molecule/cm}^3 \end{array}$$

The rate of ozone production then can be estimated using:

$$d[O_3]/dt = k_{NO+HO_2}[NO][HO_2] + k_{NO+CH_3O_2}[NO][CH_3O_2]$$
$$+ \text{ sum of terms of the type } k_{NO+RO_2}[NO][RO_2]$$

and by substituting typical values of the terms:

$$d[O_3]/dt = 1 + 0.2 + 1.2 \text{ ppt s}^{-1}$$
$$= 8 \text{ ppb h}^{-1}$$

This analysis implies that to reach the ozone concentrations typically found in regional-scale ozone episodes of about 100 ppb, an elevation of about 50–70 ppb is required above the northern hemisphere background ozone level. On the basis of the above estimates, this would require 6–9 h of intense, sustained photochemical activity, which might imply 2 days total reaction time and of the order of 500 km of travel. Long-range transport is therefore anticipated to be an important dimension to regional-scale photochemical episodes.

SUMMERTIME REGIONAL POLLUTION EPISODES AND THE FORMATION OF SECONDARY PARTICULATE MATTER

Secondary Suspended Particulate Matter

The first clear and unambiguous evidence that the heat hazes frequently seen in fine, sunny weather in Europe were caused by human activities became available nearly three decades ago. Lovelock (1972) showed how regionally polluted and photochemically reacted air masses were advected from continental Europe to the remote Atlantic coast of Ireland. The occurrence of simultaneous elevations in turbidity and CFC-11, a unique manufactured halocarbon tracer, showed conclusively that the summertime heat haze, strongly reminiscent of Los Angeles smog, which is invariably associated with the large European anticyclonic weather systems, is a product of human activities. Subsequently, Cox et al. (1975) showed that these turbid photochemically reacted air masses also contained elevated ozone concentrations. Long-range transport can bring elevated ozone and suspended particulate matter concentrations to even the most remote regions of Europe (Guicherit and van Dop, 1977).

Airborne suspended particulate matter is responsible for the turbid nature of summertime anticyclonic air masses and hence for the visibility reduction associated with them. The suspended particulate matter in these turbid air masses consists of a wide range of different chemical substances with a wide range of physical properties. Suspended particulate matter can only be described relative to some measurement method and cannot be described in an absolute way. There are therefore many different ways of characterizing these regionally polluted turbid air masses.

Aerosol and cloud physicists define three particle size ranges or modes: nucleation, accumulation and coarse particle modes. Air quality scientists use terms such as coarse, fine, PM_{10} and $PM_{2.5}$. Generally speaking, the particles which are thought to be most injurious to human health are the fine particles of the air quality scientists (Pope *et al.*, 1995) or accumulation mode particles of the aerosol physicists (Hidy, 1986). Coarse particles are not thought to be so damaging to human health and act more as a nuisance through the soiling of surfaces from the accumulation of grit and dust. The terms PM_{10} and $PM_{2.5}$ refer to the size fractions of the suspended particulate matter with diameters less than 2.5 or 10 μm, respectively; both are usually reported in units of $\mu g/m^3$.

Fine particles, with a size range of less than 2.5 μm in diameter, $PM_{2.5}$, have a variable composition in time and space. During the summertime regional pollution episodes associated with heat haze and visibility reduction (Hidy, 1986), the main fine particle components appear to be:

- ammonium sulfate,
- sulfuric acid,
- ammonium nitrate,
- sodium nitrate,
- elemental carbon,
- ammonium chloride.

Ammonium sulfate appears to account for the largest fraction of the fine particulate mass. Elemental carbon is a primary pollutant species and so is given no further consideration in this chapter. The remainder of this section is given over to a description of the transformation processes by which SO_2, NO_x and ammonia (NH_3) emissions are converted into aerosol ammonium sulfate, nitrate and chloride and sulfuric acid.

Mechanisms of Secondary Aerosol Particle Formation

The only significant source of fine particulate ammonium sulfate is the chemical conversion process from SO_2 which occurs in the atmosphere, since there are few direct emissions of fine sulfate particles of any importance. The atmospheric formation process begins with the emission of sulfur dioxide from the burning of coal and oil in stationary and mobile sources. During regional pollution episodes in summertime, sunlight-driven photochemical reactions, fuelled by hydrocarbons and oxides of nitrogen emitted by human activities, lead to elevated concentrations of the extremely reactive hydroxyl OH radical. Hydroxyl radicals oxidize sulfur dioxide (SO_2) to sulfur trioxide (SO_3) in a two-step reaction (Pandis *et al.*, 1995) which also produces a hydroperoxy HO_2 radical:

$$OH + SO_2 + M = HOSO_2 + M \tag{21}$$

$$HOSO_2 + O_2 = HO_2 + SO_3. \tag{22}$$

Sulfur trioxide reacts very quickly with water vapour to form sulfuric acid (H_2SO_4) vapour:

$$SO_3 + H_2O = H_2SO_4 \tag{23}$$

The hydroperoxy radical reacts with nitric oxide NO to produce nitrogen dioxide NO_2 which is photolysed to produce ozone:

$$HO_2 + NO = NO_2 + OH \tag{3}$$

$$NO_2 + radiation = NO + O \tag{5}$$

$$O + O_2 + M = O_3 + M \tag{6}$$

There is therefore a close relationship between haze production, visibility reduction and photochemical ozone formation.

The hydroxyl radical is recycled in the photochemical oxidation of sulfur dioxide in the sunlit atmospheric boundary layer through the hydroperoxy radical and its subsequent reaction with NO. In this way, a small concentration of hydroxy radicals, of the order of one hundredth to one tenth of a ppt, can lead to a substantial SO_2 oxidation rate, approaching a few per cent per hour.

Sulfuric acid vapour readily nucleates on its own or with water molecules to form a fine aerosol of sulfuric acid droplets in the nanometre size range of the nucleation mode (Pandis *et al.*, 1995). These exceedingly small droplets will then grow by coagulation and coalescence with other sulfuric acid droplets or with pre-existing suspended particulates and droplets in the submetre particle size range. The end-product of these aerosol nucleation and growth processes is a dynamic distribution of sulfuric acid droplets and particles, with sizes varying from nanometres to micrometres. Freshly oxidized material produced by these gas-to-particle conversion processes is generally in the smallest size ranges and aged material in the larger submicrometre size range. By far the largest number of particles are in the nanometre range and by far the largest contribution to total particle mass is in the submicrometre range.

Secondary Ammonium Particle Formation

Ammonia is the only alkaline gas of any significance found in the atmosphere. It is emitted mainly from agriculture through the spreading and disposal of animal wastes and the use of nitrogeneous fertilizers. Atmospheric ammonia is taken up on the surface of sulfuric acid droplets and particles created by the gas-to-particle conversion processes described above. This take-up can be exceedingly rapid and may give ammonia an atmospheric lifetime of less than an hour. The end-point of the uptake of ammonia onto the surface of the sulfuric acid aerosol is a mixture of particles and droplets containing varying amounts of sulfuric acid and ammonium sulfate.

Ammonia also reacts with other acidic gases present in the atmosphere, such as nitric acid and hydrogen chloride, to produce neutral and low volatile ammonium compounds:

$$NH_3 + HNO_3 = NH_4NO_3 \tag{24}$$

$$NH_3 + HCl = NH_4Cl \tag{25}$$

Both ammonium nitrate and ammonium chloride may dissolve in pre-existing aerosol droplets or may be adsorbed onto the surface of any pre-existing aerosol particles. In this way, nitrate, chloride and ammonium species become incorporated as secondary pollutants into suspended particulate matter in the size range less than 2.5 or 10 μm.

Whereas hydrogen chloride (HCl) is a primary pollutant emitted by coal burning and incineration, nitric acid is the main secondary pollutant formed from the oxidation of NO_x emissions:

$$OH + NO_2 + M = HNO_3 + M \tag{26}$$

The major part of the nitrate present in suspended particulate matter is formed by the sea salt displacement reaction (Martens *et al.*, 1973), which occurs on the surface of sea salt particles:

$$NaCl_{(s)} + HNO_{3(g)} = NaNO_{3(s)} + HCl_{(g)} \tag{27}$$

REFERENCES

Atkinson R (1994) Gas-phase tropospheric chemistry of organic compounds. *J Phys Chem Ref Data Monogr* **2**: 1–216.
Bower JS, Broughton GF and Stedman JR (1994) A winter NO_2 smog episode in the UK. *Atmos Environ* **28**: 461–475.
Brimblecombe P (1996) *Air Composition and Chemistry*. Cambridge: Cambridge University Press.
Cox RA, Eggleton AEJ, Derwent RG *et al.* (1975) Long range transport of photochemical ozone in North Western Europe. *Nature* **255**: 118–121.
Crutzen PJ (1974) Photochemical reactions initiated by and influencing ozone in the unpolluted troposphere. *Tellus* **26**: 47–57.
Derwent RG (1995) Air chemistry and terrestrial gas emissions: a global perspective. *Phil Trans Roy Soc Lond,* **A351**: 205–217.
Derwent RG, Middleton DR, Field RA *et al.* (1995) Analysis and interpretation of air quality data from an urban roadside location in central London over the period from July 1991 to July 1992. *Atmos Environ* **29**: 923–946.
Guicherit R and van Dop H (1977) Photochemical production of ozone in Western Europe (1971–1975) and its relation to meteorology. *Atmos Environ* **11**: 145–155.
Hidy GM (1986) Definition and characterization of suspended particles in ambient air. In: Lee SD, Schneider LD, Grant LD and Verkerk PJ (eds) *Aerosols: Research, Risk Assessment and Control Strategies*. Michigan, USA: Lewis Publishers.
Leighton PA (1961) *Photochemistry of Air Pollution*. New York: Academic Press.
Lovelock JE (1972) Atmospheric turbidity and CCl_3F concentrations in rural southern England and southern Ireland. *Atmos Environ* **6**: 917–925.
Martens CS, Wesolowski JJ, Harriss RC and Kaifer RJ (1973) *J Geophys Res* **78**: 8778.
Pandis SN, Wexler AS and Seinfeld JH (1995) Dynamics of tropospheric aerosols. *J Phys Chem,* **99**: 9646–9659.
Pope CA, Thun MJ, Namboodiri MM *et al.* (1995) Particulate air pollution as a predictor of mortality in a prospective study of US adults. *Am J Respir Crit Care Med* **151**: 669–674
Seinfeld JH (1986) *Air Pollution*. New York: John Wiley & Sons.
WHO (1987) *Air Quality Guidelines for Europe*. Copenhagen, Denmark: World Health Organisation Regional Publications, European Series no. 23.
WHO (1996) Update and revision of the WHO *Air Quality Guidelines for Europe*. Bilthoven, Netherlands: European Centre for Environment and Health.

5

Measurements of Concentrations of Air Pollutants

ROY M. HARRISON

The University of Birmingham, Birmingham, UK

INTRODUCTION

Any estimation of the public health impact of air pollution exposure must depend heavily upon estimates of air pollutant concentrations. Such estimates may be derived either by measurement or by numerical modelling and are often a combination of the two, with model predictions being used to interpolate or provide data for areas where monitoring data are lacking.

Measurements of air pollutant concentrations are normally carried out at fixed site monitoring stations. An alternative for some pollutants is to attach personal monitors to volunteers to measure personal exposure directly. Such measurements can differ greatly from fixed site measurements of concentration since the populations spend the majority of their time indoors, where concentrations can differ appreciably from those out-of-doors. People also spend time in some highly polluted environments such as the interior of their car. This chapter will be restricted to the measurement of concentrations of air pollutants at fixed stations out-of-doors. Issues of personal exposure will be dealt with in Chapter 9 on exposure assessment. For non-carcinogenic air pollutants, much of our knowledge of public health impacts is derived from community-based epidemiological studies which relate health outcomes to concentrations of air pollution measured at fixed point outdoor monitoring stations. Exposure–response relationships derived from such studies can readily be combined with other outdoor fixed point concentration measurements and used in a predictive manner to estimate public health impacts of air pollution exposure. Routine outdoor measurements as opposed to personal exposure assessments therefore have a key role to play in the evaluation of the health effects of air pollution.

AIR POLLUTION AND HEALTH
ISBN 0-12-352335-4

THE AIR QUALITY MANAGEMENT PROCESS AND ROLE OF MONITORING

Measurements of air pollutant concentrations are best considered in the context of the entire air quality management process. In general the purpose of local air quality management is to safeguard human health, and hence an understanding of the overall process is valuable. Figure 5.1 sets out the key components of an air quality management strategy. The strategy uses monitoring to provide information on the present levels of air pollution and involves the use of air quality standards to provide a benchmark of acceptable air quality (Middleton, 1997). If measured concentrations are comfortably within health-based air quality standards, then little if any action may be required. However, in urban areas throughout the world, concentrations of some air pollutants are liable to exceed air quality standards and development of control strategies is essential. Such control strategies will entail the reduction of emissions from one or more categories of source, and before embarking on such a policy, the regulatory authority will wish to have quantitative knowledge of the likely benefits of the strategy and will probably have conducted a cost/benefit appraisal in which the cost-effectiveness of different control strategies can be compared. To do this requires development of an air quality management model, which is a numerical model of air pollution in which the primary inputs take the form of a source inventory and details of the local meteorology (Skouloudis, 1997). Provided the boundary conditions of the model are known, i.e. the concentrations of air pollution entering the model domain from outside, then a well-conceived model should be capable of predicting both annual average and short-term concentrations within the urban area. Monitoring still has an important role to play since monitoring data will be used to validate the predictions of the model for current air quality and will be used to evaluate trends in air pollutant concentrations, and hence to determine the effectiveness of the control strategy once it is implemented.

Another use of monitoring data is in informing the public of concentrations of air pollution, and where necessary, providing the basis for short-term remedial action. Thus, in some countries current and predicted air pollution concentration information is available to the general public through the media and through toll-free telephone lines with accompanying advice to potentially sensitive groups on how best to safeguard their health. Some countries have also adopted strategies whereby once air pollution concentrations exceed a specified threshold and are predicted to continue at high levels, action is taken to reduce emissions generally by restricting the use of private cars. This action was taken in Paris for a short period in 1997 and was regarded as generally quite effective.

In summary, monitoring of air pollutant concentrations has many purposes. These include the following:

- provision of information on current concentrations of air pollution;
- the means to evaluate whether air pollutant concentrations exceed health-based standards and guidelines, and to estimate public health impacts;
- the provision of information to the public;
- the basis on which to implement short-term remedial strategies;
- the basis from which to construct long-term strategies to improve air quality;
- the means of determining trends in air pollution and the effectiveness of long-term control strategies;

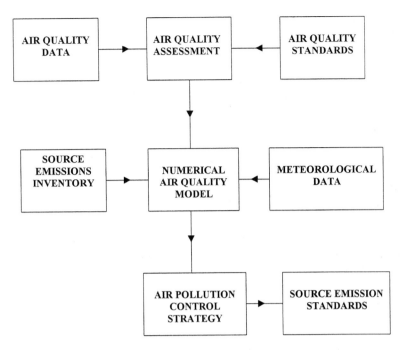

Fig. 5.1 Key components of an air quality management strategy.

- the ability to validate current predictions of numerical air quality management models.

THE POLLUTANTS WHICH ARE MEASURED

Most routine monitoring of air pollution throughout the world is conducted primarily for the protection of public health. As such, the pollutants which are monitored are those having the greatest potential health impact. A brief introduction to these pollutants and the properties which influence monitoring requirements are provided below. For more detail, the reader is referrred to Harrison (1996).

Sulfur Dioxide

This is a pollutant which arises predominantly from the combustion of sulfur-containing fuels. Its atmospheric distribution and temporal variation is strongly dependent on the nature of the fuel and the location of combustion. In less developed countries, much coal is still burnt in poorly controlled sources with emissions close to ground level. Consequently hour-to-hour variations in sulfur dioxide are generally not very great, but in periods of weather which are not conducive to pollutant dispersion high ground-level concentrations can build up. This was a major problem in Europe and

North America up to about 1960, after which controls on combustion sources led to the use of cleaner fuels for home heating in cities. In developed countries, emissions of sulfur dioxide tend to be predominantly from major point sources such as power stations burning coal or oil, industrial combustion plants and sulfuric acid works. These large sources are often fitted with flue gas desulfurization technology designed to limit sulfur dioxide releases. Emissions from such point sources are in the form of concentrated plumes of pollution which can lead to highly elevated ground-level concentrations over quite short periods (Seinfeld, 1986). In the kind of weather which leads to build-up of pollutants from ground-level sources such as motor traffic or coal burning in domestic premises, plumes from elevated point sources are unable to penetrate efficiently to ground level and their impact is likely to be very small. The implication of the potential for rapid fluctuations in sulfur dioxide due to point source emissions, combined with air quality standards and guidelines that can involve averaging periods of one hour (World Health Organization) or even 15 minutes (Expert Panel on Air Quality Standards), is that fast-response measurements are required to protect the public against short-term elevations in sulfur dioxide concentration. The role of such measurements is unlikely to be to provide immediate public information, but is rather to monitor the impact of control policies to ensure that health-based guidelines are not exceeded.

Suspended Particulate Matter

Figure 5.2 shows the size distribution of airborne particles (Whitby, 1978). These cover a very wide range of sizes. Most of the mass of particles is associated with the accumulation mode (approximately 0.1–2.5 μm diameter) and coarse particle mode (approximately 2.5–100 μm). Most currently operational monitoring networks determine suspended particulate matter either as black smoke or as PM_{10}. In the USA and UK some monitoring networks also determine a fraction of particles called $PM_{2.5}$. These measurements require some definition (see Mark 1998).

PM_{10}

This is the fraction of particles passing an inlet with a 50% cut-off efficiency at an aerodynamic diameter of 10 μm. The PM_{10} size is shown in Fig. 5.3 in relation to definitions for inhalable, thoracic and respirable particles. As may be seen from Fig. 5.2, the PM_{10} fraction includes the majority of atmospheric particles, excluding only the upper end of the coarse range of particles. In some environments, especially very dusty ones, these coarse particles which are outside the PM_{10} range may comprise an appreciable proportion of total particle mass. They are, however, believed to be of little health significance, and it was for this reason that sampling of PM_{10} was initiated. Previously, samplers collected what was known as total suspended particulates (TSP). This involved the use of a rather poorly designed sampler inlet (in comparison to the specially designed PM_{10} inlet) whose characteristics were dependent both on orientation and windspeed. TSP was not therefore a clearly defined particle fraction.

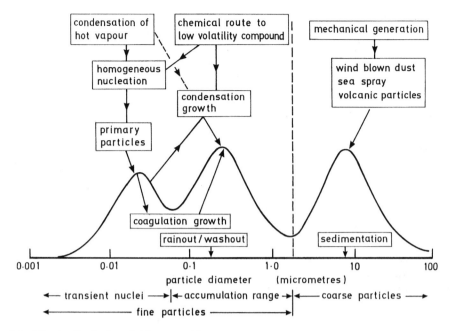

Fig. 5.2 The size distribution of airborne particles.

$PM_{2.5}$

Whilst PM_{10} corresponds approximately to a thoracic particle fraction, $PM_{2.5}$ is designed to sample fine particles, i.e. those from the transient nuclei and accumulation ranges. $PM_{2.5}$ corresponds approximately to a respirable particle fraction, i.e. particles capable of penetrating to the alveolar region of the lung (see Fig. 5.3). Sampling can be very similar to that for PM_{10}, but using a different size-selective inlet. Alternatively, both $PM_{2.5}$ and PM_{10} fractions can be sampled using a so-called dichotomous sampler (see below).

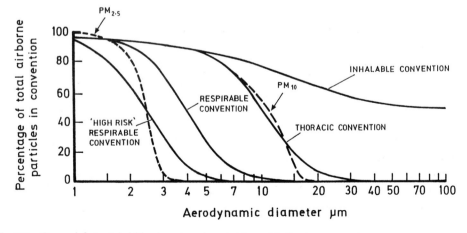

Fig. 5.3 Curves defining inhalable, thoracic and respirable particle fractions.

Black Smoke

This is a quite different kind of measurement in which particles are collected on a filter whose blackness is subsequently determined, generally by light reflection or sometimes by transmission. In the reflection method, light is shone on the particle-exposed area of a filter and the intensity of reflected light is measured. The reflectometer is previously set to 100% using an unexposed filter which allows the percentage of incident light absorbed by the exposed filter to be quantified. The filter reflectance is then related to a concentration of particles in the atmosphere by use of a standard calibration curve. These calibration curves were generally determined using samples of known mass of atmospheric particles collected many years ago when coal smoke was the predominant source of airborne particles. This has led to use of more than one calibration curve and, for example, British Standard Smoke is not identical to OECD Standard Smoke (as used by the European Union). Since the shape of the calibration curves is the same, however, the difference between the two measurements is represented by a constant numerical factor (0.85). The traditional black smoke calibration curves are no longer applicable and the relationship between airborne particles determined by the black smoke method and determined gravimetrically as PM_{10} is dependent on both site and season. Black smoke, although still a widely used measurement in many parts of the world, is therefore a rather crude measure of airborne particles and reflects primarily the elemental carbon content rather than the mass of the particles (QUARG, 1993).

Air quality standards for particulate matter are mostly set in terms of 24-h average or longer averaging periods (e.g. US EPA, 1997) . The exception is the UK EPAQS standard (EPAQS, 1995), which relates to rolling 24-h means that require hourly average measurements for their estimation. As will be described later, the most frequently used instrumentation provides 24-h average concentration data, but instruments are available for measuring hourly average or even more highly time-resolved data for particulate matter.

Carbon Monoxide

Carbon monoxide is a gas emitted from the incomplete combustion of carbon-containing fuels. Its predominant source in developed countries is from gasoline powered vehicles. Emissions from diesel engines and from stationary combustion plants are relatively small. It is present in the atmosphere as a gas, and air quality standards and guidelines typically refer to averaging periods of 1, 8 or 24 h, and therefore instrumentation capable of hourly measurements is required. There are no good chemical procedures for determination of carbon monoxide in air and physically based instrumental procedures are necessary.

Oxides of Nitrogen

The measurement of these species is made more complex because they are present in the atmosphere as both nitric oxide, NO, and nitrogen dioxide, NO_2. The two compounds may interconvert on a timescale of seconds in response to changes in sunshine or ozone concentration. Whilst emissions of NO_x are predominantly in the form of nitric oxide, nitrogen dioxide predominates in all but the most polluted environments, and this is the more toxic compound which is the subject of air quality standards. The main source of emissions is high-temperature combustion, with nitric oxide being formed predominantly from combination of atmospheric nitrogen and oxygen. With some fuels, oxidation of nitrogen in the fuel is also a source.

Volatile Organic Compounds

From the viewpoint of protection of public health, the compounds of concern are primarily benzene and 1,3-butadiene. These arise predominately from road traffic, although both also have industrial sources. They exist in the atmosphere as vapour. Since these compounds are genotoxic carcinogens, air quality standards and guidelines have been set in terms of long-term, generally annual average, concentrations (Harrison, 1998). A variety of techniques with widely varying averaging periods are available for their measurement and averaging periods of 1 h are quite feasible.

Polycyclic Aromatic Hydrocarbons

These compounds are classified as semivolatile, which means that they are partitioned in the atmosphere between material in the vapour phase and that adsorbed onto the surface of particles. For the two- and three-ringed polycyclic aromatic hydrocarbons (PAH), vapour phase material is generally strongly predominant, whereas for five- and six-membered rings there is typically at least 90% in the particle-associated form (Smith and Harrison, 1998). Comprehensive measurement of PAH therefore requires sampling of both particle and vapour phase, although the health consequences of the two phases will not be the same, and air quality standards may be set in terms only of the particle-associated phase. Health concern over PAH relates to the carcinogenicity of some members of the group, and air quality standards are likely to be set in terms of long-term averages. Typical measurement averaging periods are 24 h.

Ozone

Ozone differs from the other pollutants in that it is wholly secondary in origin, i.e. it is formed within the atmosphere rather than being emitted by sources. Because of its complex atmospheric chemistry, its concentrations tend to be suppressed close to sources of nitric oxide emissions such as road traffic (QUARG, 1993). In Europe where cities are relatively compact, rural concentrations of ozone therefore generally exceed urban concentrations by a considerable margin due to local destruction of ozone within the urban area. In the summer months ozone formation occurs in polluted air on the longer timescales associated with transport of air across rural regions, giving rise to pollution episodes. Ozone shows a strong diurnal variation in concentration in both urban and rural areas with concentration peaks in daytime. Health-based standards and guidelines are generally expressed in terms of 1-h or 8-h average concentrations, and hence continuous fast-response monitoring is both necessary and available.

METHODS OF AIR POLLUTANT ANALYSIS

A wide range of methods are available for the analysis of air pollutants. Broadly speaking these divide into chemical methods best suited to providing 24-h average concentrations and procedures based on physical properties of the pollutant which can frequently give continuous fast-response measurements on a timescale of minutes or less (Lodge, 1989). The latter are clearly advantageous, but not always available owing to considerably greater expense. Some of the chemical methods can be applied as low-cost passive sampling procedures, as with diffusion tubes yielding weekly average data for a tiny fraction of the cost of a continuous analyser. The reader requiring a detailed appraisal of this subject is referred to specialized texts. An introduction will be given here to the techniques most commonly used in national monitoring networks.

Sulfur Dioxide

By far the most widely used instrumental technique for analysis of sulfur dioxide is the gas phase fluorescence instrument (Fig. 5.4). Air is drawn continuously through a cell in which it is irradiated by light of wavelength in the region of 214 nm. This causes fluorescence, the intensity of which is related to the sulfur dioxide concentration. Potential interference from quenching of the SO_2 fluorescence by water vapour and from hydrocarbons capable of fluorescence of the same wavelength as SO_2 is overcome in commercial instruments by incorporation of diffusion dryers and hydrocarbon scrubbers. The instruments measure down to 1 ppb or less of sulfur dioxide with a response time of around one minute.

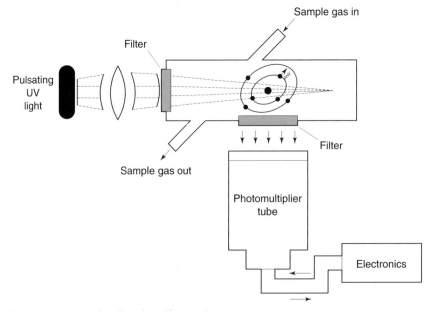

Fig. 5.4 An instrumental analyser for sulfur dioxide.

Passive diffusion tube samplers which are capable to determining sulfur dioxide over averaging periods of a week or more have been developed, but have yet to win widespread acceptance due to problems of poor precision and accuracy.

Suspended Particulate Matter

In North America the high volume sampler (Lodge, 1989) has been in use for many years, originally as a means of measuring total suspended particulates (see above). These instruments are now fitted with a size-selective inlet so as to measure PM_{10} (see Fig. 5.5). Particles passing the PM_{10} inlet are collected on a pre-weighed filter paper, typically of glass fibre or quartz, which is again weighed after collecting particles for 24 h at a rate of around 1 m^3 per minute. Because of the presence of hygroscopic salts in the airborne particles, there are protocols requiring equilibration of the filters at constant temperature and low relative humidity prior to weighings.

Also widely used in North America are the so-called dichotomous samplers (see Fig. 5.6). Air drawn into these instruments is divided into two streams at a virtual impactor. One stream of air continues in a straight line with the incoming air and carries coarse particles by virtue of their inertia onto the coarse filter. The second stream of air diverts at

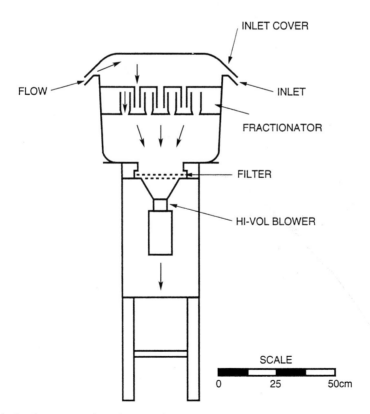

Fig. 5.5 A high volume air sampler with PM_{10} inlet.

right angles from the inlet stream and carries only fine particles, which are collected on the fine particle filter. The coarse particle filter collects also fine particles at the concentration present in the air, but these can be subtracted on the basis of measurements from the fine filter. If the instrument is equipped with a 10 μm size-selective inlet and the virtual impactor has a cut-off of 2.5 μm, the particles are collected as separate fine (less than 2.5 μm) and coarse (2.5–10 μm) fractions. The dichotomous sampler typically operates at an inlet flow rate of 16.7 litres per minute and 24 h are required to obtain samples suitable for weighing. Collection is typically on Teflon filters which are better suited to accurate weighings and to chemical analysis than the filter media used in the high volume sampler.

Continuous measurements of PM_{10} or $PM_{2.5}$ mass concentration with a response time of around 15 min can be achieved using the tapered element oscillating microbalance (TEOM) sampler (Fig. 5.7). In this instrument air is drawn in through a PM_{10} or $PM_{2.5}$ inlet and is pre-heated to 50°C to drive water out of the particles. The dried particles are collected on a small filter which is attached to the vibrating element of an oscillating microbalance. The vibrational frequency changes with the accumulation of particles and is determined continuously. Intercomparison studies have shown that the TEOM generally gives lower measurements than the high volume sampler due to the loss of semivolatile materials during the pre-heating stage. The great advantage of the instrument, however, is that its faster response time and real-time output give far greater insights into processes controlling particle concentrations and a far better source of public information than the long averaging periods and retrospective data reporting from the high volume method.

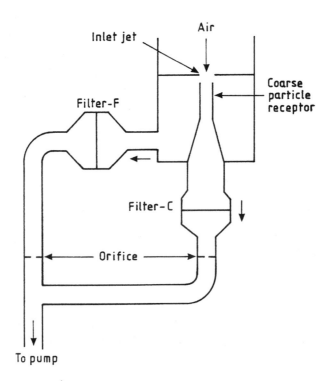

Fig. 5.6 A dichotomous sampler.

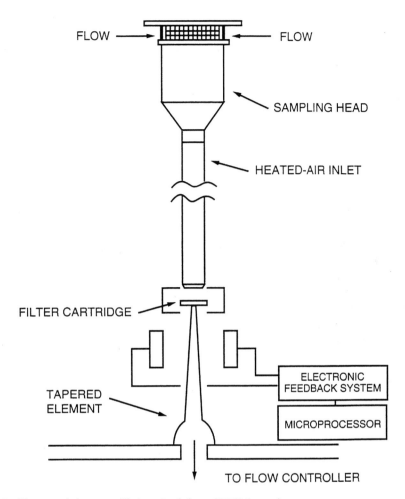

Fig. 5.7 The tapered element oscillating microbalance (TEOM) sampler.

Carbon Monoxide

As indicated above, chemical methods for carbon monoxide are generally unsatisfactory and instrumental techniques are the norm. A number of these are available, but most state-of-the-art instruments use either non-dispersive or gas filter correlation infrared methods. The gas filter correlation instrument (Fig. 5.8) involves broad band infrared radiation passing alternately through gas cells containing carbon monoxide and molecular nitrogen as they rotate beneath the source on a spinning wheel. These pulses of radiation then travel through a multipass optical cell through which ambient air is drawn. The sample and reference beam are separated in time but not space: the sample beam is produced when the infrared light passes through the nitrogen cell and the reference beam is produced when the light passes through the carbon monoxide cell. The reference beam is subject to absorption by all of the components in the sample cell other than carbon monoxide because of saturation absorption of the carbon monoxide frequencies by the carbon monoxide in the correlation wheel. The difference in signal between the two beams is therefore the result of absorption by carbon monoxide within the air sample. The instrument is capable of measuring down to about 0.1 ppm with a range up to 50 ppm and a response time of around 2 min.

Oxides of Nitrogen

The universally adopted instrumental method for measuring oxides of nitrogen makes use of the chemiluminescent analyser. The method is based on measurement of the light emission from the chemiluminescent reaction between nitric oxide in the air and ozone generated within the instrument. A typical instrument for ambient air measurements is capable of measuring down to 1 ppb with a response time of under a minute (Fig. 5.9). The instruments typically operate in two modes which may function alternately with one reaction chamber or simultaneously with two reaction chambers. In one mode ambient air is passed directly to the reaction chamber for mixture with ozone, and the nitric oxide (NO) content is measured directly. In the other mode, known as the NO_x mode, the air is passed through a heated stainless steel or molybdenum converter before reaching the reaction chamber. The converter is designed to convert nitrogen dioxide (NO_2) to NO, and hence the instrument measures NO_x. Nitrogen dioxide is determined by difference between the readings in the two modes. The measurement of nitric oxide is not subject to important interferences. However, the measurement of nitrogen dioxide is subject to interferences from other compounds which decompose to form nitric oxide under the same conditions as does nitrogen dioxide. There is generally some response from compounds like peroxyacetyl nitrate and nitric acid, although in polluted air these are unlikely to provide a major increment on the true concentration of nitrogen dioxide.

Many nitrogen dioxide measurements are nowadays made using an inexpensive technique based on the diffusion tube. The apparatus consists of a straight hollow tube of length about 7 cm and internal diameter 1 cm, closed at one end, which is placed vertically upwards with the open end at the bottom (see Fig. 5.10). Inside the closed end is a metal grid coated with triethanolamine, which acts as an adsorber and perfect sink for nitrogen dioxide. The rate at which nitrogen dioxide is collected by the triethanolamine is determined by the ambient air concentration of nitrogen dioxide, which moves through the tube by molecular diffusion. At the end of the sampling period, typically of 1 or 2 weeks, the collected nitrite is analysed to estimate the nitrogen dioxide concentration. For a number of reasons these tubes are rather imprecise and can be inaccurate. To overcome the first problem they are often deployed in multiples of around three to improve precision. Systematic errors arise from a reduced diffusion length in windy conditions and chemical reactions in the $NO-NO_2-O_3$ system. These cannot be readily addressed and in the urban environment have been shown to lead to an overestimation of the order of 30% (QUARG, 1993).

Volatile Organic Compounds

There are few networks that make routine measurements of volatile organic compounds. This is a difficult and potentially expensive analysis because of the wide range of

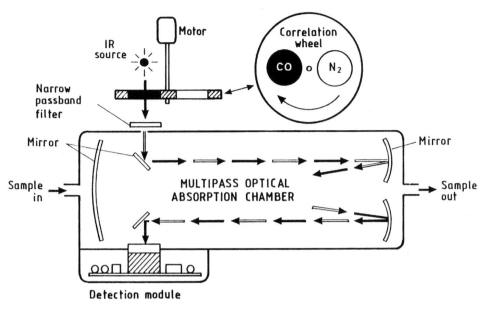

Fig. 5.8 A gas filter correlation analyser for carbon monoxide.

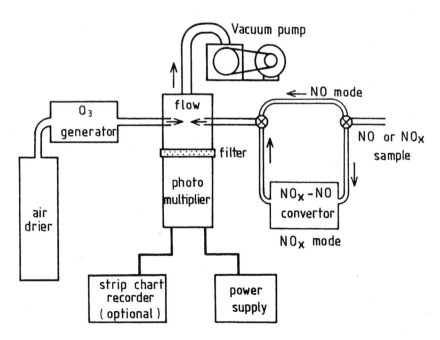

Fig. 5.9 A chemiluminescent analyser for oxides of nitrogen.

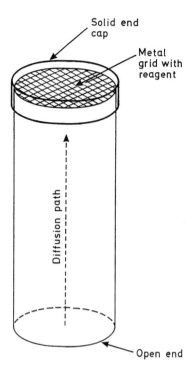

Solid end
cap

Metal
grid with
reagent

Diffusion path

Open end

Fig. 5.10   A diffusion tube sampler.

compounds present and the relatively low concentrations. Diffusion tubes are available and are reasonably well accepted for long-term sampling of hydrocarbons of C_6 and above. Thus, they can be used to quantify concentrations of benzene, but are unsuitable for measurement of 1,3-butadiene. Methods giving better time-resolved data generally require a pre-concentration of the air using an absorbent or freeze-out, or a combination of the two. Volatile organic compounds (VOCs) collected over a period of the order of an hour are then revolatilized into a gas chromatograph where they are separated and quantified individually with detection by a flame ionization detector or a mass spectrometer (see Fig. 5.11). The latter allows the identity of the compound to be confirmed, whereas the former can give problems of peak assignment, although in general, experienced operators are able to quantify individual compounds with confidence. With a suitable pre-concentration, the technique is capable of measuring concentrations down to 0.1 ppb or lower. Automated systems based on continuous cyclic gas chromatography have been developed and are used in the UK hydrocarbon network (Dollard *et al.*, 1995).

Polycyclic Aromatic Hydrocarbons

As mentioned above, a full determination of polycyclic aromatic hydrocarbon (PAH) requires sampling of both vapour phase and particle-associated forms. This can be

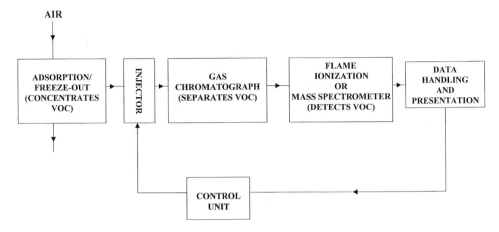

Fig. 5.11   An analytical system for VOCs in air.

achieved using a high volume sampler in which air is drawn first through a filter and subsequently through a vapour collector, which frequently takes the form of a plug of polyurethane foam. The latter gives a low flow resistance but provides a very effective collector of PAH vapour. After collection over a period of typically 24 h, both the filter and polyurethane foam plug are separately extracted into an organic solvent which is then analysed for its PAH content using one of a range of techniques, most commonly either high-performance liquid chromatography with detection by UV absorption and/or fluorescence, or gas chromatography with detection by a mass spectrometer. By this means concentrations of individual compounds can be determined.

Ozone

For many years chemical methods based on the oxidizing capacity of ozone were in widespread use. These were later shown to be unreliable due to a variable dependence of the oxidation reaction on the ozone concentration and an influence of the sampler itself upon the oxidation rate. Subsequently, chemiluminescent analysers based on the reaction of ozone with ethylene became popular, but nowadays UV photometric analysers are almost universally adopted for measurement of ozone. These utilize the strong UV absorption by ozone at 254 nm. This is measured in a long path absorption cell through which ambient air is drawn (see Fig. 5.12). The air entering the cell does so alternately through a direct inlet and an inlet which passes through an ozone scrubber, which is designed to decompose ozone but allow passage of other UV-absorbing materials such as mercury vapour and some hydrocarbons. The difference in UV absorption between the two inlet routes is due solely to ozone, and from knowledge of the absorption cross-section of ozone, the instrument computes the concentration. Response is relatively fast and readings can be taken at intervals of less than one minute.

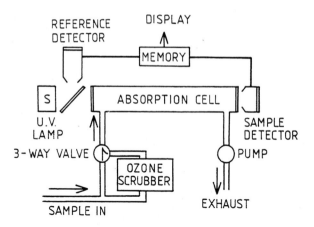

Fig. 5.12   A UV absorption analyser for ozone

INSTRUMENT DEPLOYMENT AND USE

Whilst the above instruments may appear to make air pollution measurements straight-forward and trouble-free, the reality is very different. Although instrument reliability is nowadays quite high, there are many operational details which require specialized knowledge before good data can be collected. Thus, for example, because many pollutants, including especially reactive substances like ozone, are adsorbed or decomposed in unsuitable sampling lines and inlets, the air pollution monitoring stations need to be very carefully designed and validated (Lodge, 1989). Suspended particulate matter is not readily transmitted through sample lines, and generally the inlet of the instrument is designed to be exposed directly to the outside atmosphere and sampling station design will need to accommodate this. The majority of instruments also require very careful calibration and zeroing, typically on a weekly basis, if good data are to be obtained. Even with all of the above requirements met, there is still a need for rigorous quality assurance programmes if high-quality data are to be guaranteed. This is a major challenge for monitoring network operators the world over.

Continuous fast-response monitoring instruments can generate massive volumes of data. Effective data logging facilities are therefore essential, and both commercially developed and home-built options are feasible. In general, data are acquired from the monitoring instruments at regular time intervals of say 1 or 2 min and stored temporarily for generation of 15-min or hourly average concentrations. These average values are then put into long-term data storage. Modem telemetric procedures make it easy to access instruments remotely and to transmit data in real time to a remote location. It is therefore possible both to view the data in near real time and to monitor the performance of the station through automated calibration routines without the need to visit. Weekly site visits are then typical and meet the need for instrument servicing and manual calibration.

Air pollution monitoring is potentially expensive, and to get best value for money it is essential to implement rigorous quality assurance regimes to ensure that data quality is always high (Sweeney *et al.*, 1997). This will involve the use of certified standards where available for instrument calibration and periodic independent checks of instrument function by comparison with other standards, or in the case of ozone, transfer instruments. In some countries responsibility for network quality assurance lies with contractors who are wholly independent of the network operators; this approach can have some benefits in ensuring high standards.

LOCATION OF MONITORING STATIONS

Even the best calibrated and prepared analysers will not generate good data if they are located in the wrong place. As a general principle, air pollution monitoring station inlets need to be well exposed to air from all directions. Unless there is a desire to sample within street canyons, samplers should not be located close to high buildings, but rather should be in open areas or on roofs so as to obtain good exposure to the atmosphere.

Broadly speaking there are two major types of sampling location, although within them there are inevitably subtypes. The major categories are as follows.

Background Stations

These are located so as not to be strongly influenced by any one source. They can be located in urban or rural areas, but in both cases they will be in an open space without immediate local sources. Ideally, they should be at least 100 m from any busy road, although in urban areas this is often not practicable, whilst in rural areas it is usually possible and always desirable for a background station to be further than this distance from a road. The advantage of background stations is that they do not represent extremes of concentration but represent the background contributed by the sum of many sources upon which local source contributions at other locations are superimposed. The majority of people live in background-type areas and hence these stations can represent the concentrations of air outside many people's homes. Additionally, because they are not subject to the strong effect of any individual local source, they are good for reflecting long-term trends in concentration.

Hotspot Locations

Most members of the population do spend some of their time in areas where air pollutant concentrations are especially high. Such environments are close to busy roads and at the point of maximum ground-level concentrations from a major point source. Most networks therefore include some hotspot sites, most typically alongside major roads or close to busy road intersections or within street canyons. Such samplers have the disadvantage of representing air quality only at their precise location. A roadside site cannot be taken as representative of a location 20 m away along the same road, or a position closer to or further from the road. In contrast, if a well-located background site were to be moved by 20 m in any direction, it should give essentially the same measurements. The value of the hotspot site is in reflecting the extremes of concentration, which is particularly useful in respect of the fast acting pollutants for which high, acute exposures are important.

Networks sometimes use rural sites as a means of determining air quality upwind or downwind of a city, or in the case of ozone, for measuring concentrations at locations where they are likely to be at a maximum. Rural sites are generally located well away from major sources of traffic, as under stable atmospheric conditions the pollution signature of a major road can be seen a kilometre or further downwind.

MONITORING NETWORKS

Air monitoring instruments are occasionally operated at single sites unconnected with any other, but most typically form part of a network. Such networks are sometimes given careful advanced planning, although many grow in a rather haphazard manner according to the availability of funds. Network design (Munn, 1982) in a city will depend strongly upon the purpose of the network. In the past the philosophy behind some networks was to establish a picture of the pollution climate of a city, and to do so, samplers were located at sites representative of particular kinds of land use. In particular, commercial, industrial and residential districts were used for locating samplers. In more recent years this approach has tended to be supplanted by designs of network aimed at providing public information and protection against adverse health effects. To do so requires strategic decisions on the kinds of location to be sampled. Most networks focus their activities predominantly upon urban background locations, since these represent the exposure of many people. Often roadside sites are included, but care must be exercised in utilizing the data from such sites as they represent extremes of concentration to which few people are exposed. Two basic approaches have been adopted in deciding the locations of samplers in centrally planned networks. The former involves using a grid of specified size and locating samplers at each grid intersection point. Such a network exists in the city of Berlin (CEC, 1992). Another approach is to model air pollution concentrations across the city and to place samplers at background sites within the areas predicted to receive the highest concentrations of air pollution. This is probably the most rational approach, although it requires considerable resources to generate a reliable numerical model.

Detailed consideration of sites for monitoring becomes less important as the number of sites increases. Therefore, large and dense networks of sites can be established when inexpensive techniques such as NO_2 diffusion tubes are used for sampling. This approach has been used extensively for mapping air pollution concentrations across cities (QUARG, 1993). Generally, a combination of relatively sparse high-technology monitoring stations with a denser network of low-technology stations is a rational solution. Mapping with inexpensive methods can also be useful in deciding locations for high-technology monitoring to meet predetermined criteria, such as points with high, or background concentrations.

REFERENCES

CEC (1992) *Handbook of Urban Air Improvement.* Brussels: Commission of the European Communities.

Dollard G, Davies TJ, Jones BMR *et al.* (1995) The UK hydrocarbon monitoring network. In: Hester RE and Harrison RM (eds) *Volatile Organic Compounds in the Atmosphere.* Cambridge: Royal Society of Chemistry, pp. 37–50.

EPAQS (1995) *Particles.* Expert Panel on Air Quality Standards. UK Department of the Environment. London: HMSO.

Harrison RM (ed.) (1996) Air pollution: sources, concentrations and measurements. In: *Pollution: Causes, Effects and Control.* Cambridge: Royal Society of Chemistry, pp. 144–168.

Harrison RM (1998) Setting air quality standards. In: Hester RE and Harrison RM (eds) *Health Effects of Air Pollutants.* Cambridge: Royal Society of Chemistry, pp. 57–73.

Lodge JP (1989) *Methods of Air Sampling and Analysis,* 3rd edn. Chelsea, MI: Lewis Publishers.

Mark D (1998) Atmospheric aerosol sampling. In: Harrison RM and Van Grieken R (eds) *Atmospheric Aerosols.* Chichester: Wiley, pp. 29–94.

Middleton DR (1997) Improving air quality in the United Kingdom. In: Hester RE and Harrison RM (eds) *Air Quality Management.* Cambridge: Royal Society of Chemistry, pp. 1–17.

Munn RE (1982) *The Design of Air Quality Monitoring Networks.* Basingstoke: Macmillan Press.

QUARG (1993) *Urban Air Pollution in the United Kingdom.* Report prepared at the request of the UK Department of the Environment. London: Quality of Urban Air Review Group.

Seinfeld JH (1986). *Atmospheric Chemistry and Physics of Air Pollution.* New York: Wiley.

Skouloudis AN (1997) The European auto-oil programme: scientific considerations. In: Hester RE and Harrison RM (eds) *Air Quality Management.* Cambridge: Royal Society of Chemistry, pp. 67–93.

Smith DJT and Harrison RM (1998) Polycyclic aromatic hydrocarbons in atmospheric particles. In: Harrison RM and Van Grieken RE (eds) *Atmospheric Aerosols.* Chichester: Wiley, pp. 253–294.

Sweeney B, Quincey PG, Milton MJT and Wood PT (1997) Quality assurance and quality control of ambient air quality measurements. In: Davison G and Hewitt CN (eds) *Air Pollution in the United Kingdom.* Cambridge: Royal Society of Chemistry, pp. 126–144.

US EPA (1997) National Ambient Air Quality Standards for Particulate Matter; Final Rule. *Federal Register* **62**(138): 40CFR Part 50.

Whitby KR (1978). The physical characteristics of sulphur aerosols. *Atmos Environ* **12**: 135–159.

6

Patterns of Air Pollution in Developed Countries

MARTIN L. WILLIAMS

Department of the Environment, Transport and the Regions, London, UK

INTRODUCTION

This chapter sets out to describe the patterns of air pollution in developed countries. There are several terms here which need to be defined, or at least explained in greater depth. First, there is the question of what patterns are to be described. Clearly to attempt to describe precise spatial patterns in one or more countries would be to run the risk of being overly specific by describing one country, or overly ambitious, and not especially enlightening, by describing such patterns in a large number of countries. Where it is appropriate to discuss spatial patterns of air pollution, the discussion will attempt to draw out general principles, illustrating these with specific examples where this is helpful.

There is then the question of temporal patterns, and insofar as this is synonymous with an analysis of trends in air pollution emission and concentrations, this is likely to be a more useful aspect of the discussion. Indeed, the concept of development implies a time dimension in itself, and it is in the time-series of air pollutants that we will illustrate the essential characteristics of air pollution in developed countries.

There is nonetheless an issue of what the term 'developed' means in this context. Although the conventional usage of the term relies broadly on some measure, or measures, of economic performance, a sociologist might define a country's state of development differently from an economist. Similarly, there may be reasons to consider a country as developed in the air pollution sense alone. In this chapter, therefore, we will not necessarily rely solely on economic definitions of the term 'developed', but as will become clear in the discussion below, we will define air pollution characteristics that will identify the state of development of a country.

AIR POLLUTION AND HEALTH
ISBN 0-12-352335-4

The discussion which follows has been written very much with the context of the health effects of air pollution in mind. However, there may in many cases be important implications for other sensitive environmental receptors – such as ecosystems – in measures which countries may have taken, or are taking, to abate health effects, and these will be discussed.

Patterns in emissions over time will be discussed first, followed by a consideration of what the implications of such emission patterns and trends have been for air quality, particularly in the context of potential effects on human health.

EMISSION PATTERNS

Trends in emission patterns have displayed similar characteristics in many developed countries in the last few decades. This has been most clearly demonstrated in the power generation and domestic sectors and in transport.

In the power generation sector there has, in most countries in Europe at least, been a shift from uncontrolled coal and oil-fired generation to a position where a much higher proportion of the output comes from cleaner fuels such as gas or nuclear power. Where coal and oil are still used, then they are often used in conjunction with some form of abatement technology such as flue gas desulphurization. The net result of these changing patterns is that for most industrial sectors, emissions of pollutants such as SO_2 have become decoupled from measures of economic output of the sector. This is illustrated for the UK in Fig. 6.1, which shows emissions of SO_2 as a function of electricity generated for the period 1970–92.

This separation of the trends in emissions from economic output or GDP could be used as an indicator of the degree of development of a country in an air pollution sense, as opposed to a purely economic sense. On this basis, countries could therefore in principle be considered to be developed in the air pollution sense even if they were not considered to be so in the conventional economic sense of the term.

The domestic sector has also seen major changes in fuel use and hence emissions over the past few decades. In most countries in north west Europe, and elsewhere, coal was at one time the predominant fuel. Following major smog incidents, such as those observed in the Meuse Valley in 1930, and in London in 1952 and 1962, coal use in the domestic sector declined substantially to the extent that, apart from some small remaining areas, in most EC countries it is now no longer used to any great extent in the domestic sector. Its use has been replaced by cleaner fuels such as electricity, gas and in some cases solid smokeless fuels. The profound consequences this has had for urban air quality will be discussed in later sections, both in terms of sulphur dioxide (SO_2) and, perhaps more importantly, in terms of smoke and particle concentrations.

Emissions of NO_x have been more difficult to reduce. This to a large extent is because, all other things being equal, NO_x emissions are proportional to the efficiency of combustion, and it is only in recent years that end-of-pipe abatement techniques have been widely used in the power generation sector. The switch to gas fuels does have some benefits in terms of NO_x emissions, even with no post-combustion clean-up, but additional measures, for example on conventional coal-fired power stations, such as SCR (selective catalytic reduction) can further reduce NO_x emissions. In the UK, for example, NO_x

emissions from power stations did not reduce by the same amount as those of SO_2 over the period between 1970 and 1980, while in the period after this, the use of low-NO_x burners and an increasing use of gas for power generation have reduced emissions significantly. Emissions of NO_x from power generation, and indeed from the rest of the industrial sector, should ideally not be discussed in isolation from NO_x emissions from transport, since both together make the dominant contribution to most developed countries' emissions of NO_x. This topic will be further explored below when transport emissions are discussed. The differences and similarities in behaviour of NO_x compared with SO_2 emissions from power generation can be illustrated with data from the UK in Fig. 6.1, which shows emissions from this sector from 1970 to 1996 compared with an output measure, in this case electricity generated.

Emissions of NO_x from transport are estimated to have increased significantly over the past three decades or so and have only more recently begun to become decoupled from activity or output measures (such as vehicle-kilometres driven), at least in northwest Europe. Up until now this has been largely due to the increasing use of catalytic converters on petrol cars, and the increased use of diesel cars, and is illustrated for the UK in Fig. 6.2. This trend is expected to continue as catalytic converters become more widely used, and as improved low-emission technologies become increasingly available for both diesel and petrol technologies. However historically, in most developed countries, NO_x (and other emissions) from motor vehicles have increased over the past two or three decades before this more recent decoupling of emissions and activity has occurred.

Although the implications of developments in fuel use and emission patterns for local air quality will be discussed in more detail below, it is worth making some relevant comments here. In many cases there were significant shifts in the patterns of electricity generation even though coal and oil remained the dominant fuels. In the UK for example, the smaller urban power stations which were predominantly located close to the centres of major urban areas, were gradually replace by much larger, and more efficient, plant located in rural areas large distances away from population centres with their emissions being released through much taller chimneys. Even though overall emission from the sector may not in the shorter term have changed very much, and the separation between emissions and economic output not necessarily realized, there were nonetheless very significant changes in local air quality as will be discussed below.

IMPLICATIONS FOR AIR QUALITY

The changing patterns of emissions have had significant implications for urban air quality. The movement away from coal in the domestic, commercial and smaller industrial sectors has in itself resulted in significant decreases in concentrations of pollutants such as SO_2 and particles. (The latter have traditionally been measured as so-called black smoke in most European countries.) This trend has been enhanced in many countries by the shifts in emission patterns from electricity generation, not only in moving to cleaner fuels and technologies, but also by the move to larger power plants (with taller stacks) in rural areas away from town centres.

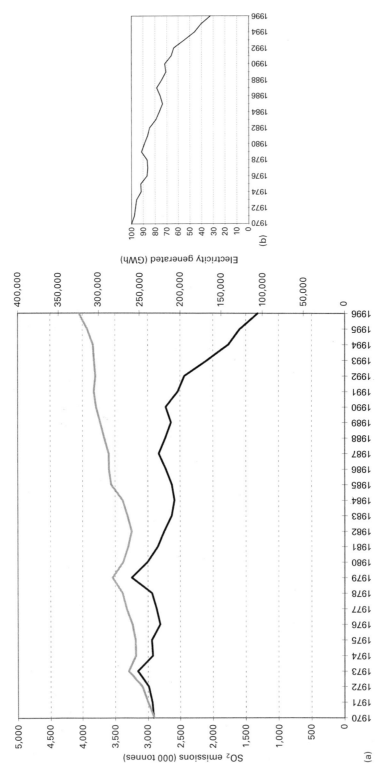

Fig. 6.1 (a) UK power station emissions of sulfur dioxide and electricity generated. (b) Index of UK power station SO_2 emissions per electricity generated (1970 = 100).

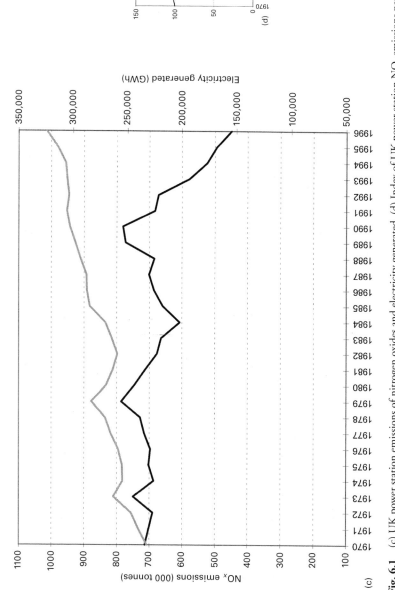

Fig. 6.1 (c) UK power station emissions of nitrogen oxides and electricity generated. (d) Index of UK power station NO_x emissions per electricity generated (1970 = 100).

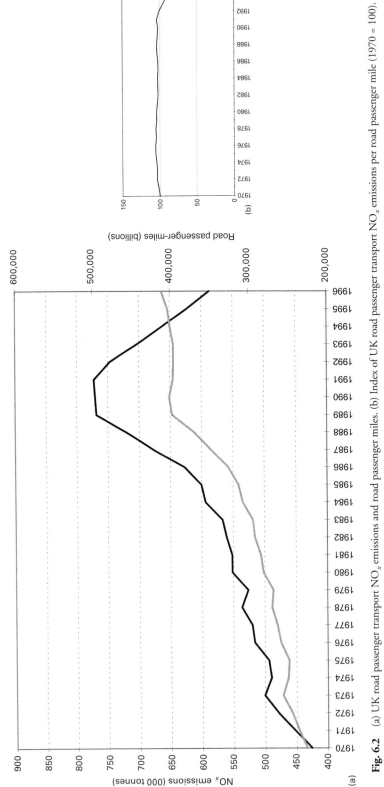

Fig. 6.2 (a) UK road passenger transport NO$_x$ emissions and road passenger miles. (b) Index of UK road passenger transport NO$_x$ emissions per road passenger mile (1970 = 100).

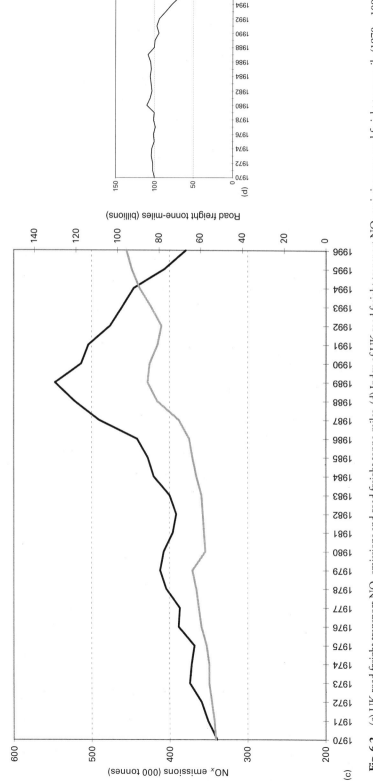

Fig. 6.2 (c) UK road freight transport NO_x emissions and road freight tonne-miles. (d) Index of UK road freight transport NO_x emissions per road freight tonne-mile (1970 = 100).

Sulphur Dioxide

These trends, or similar ones, have taken place in developed countries over the past three decades or so. Over this period at the same time, motor vehicle activity has increased considerably. Relative to stationary sources using solid and liquid fuels, motor vehicles are a relatively small source of SO_2 and the increase in vehicle activity and emissions of SO_2 has not counteracted the decrease in urban SO_2 concentrations from reduced coal and oil emissions. In terms of SO_2 concentrations in urban areas, however, it is interesting to note that there has been a significant change in the pattern of exposure. Average levels are no longer high and, as will be shown later, SO_2 levels do not tend to be elevated even during prolonged inversion conditions in winter episodes. High exposures to SO_2 currently tend to be from relatively short-lived peaks arising from individual combustion sources, rather than from extended exposures over days or months. This is illustrated in Fig. 6.3 which shows the hourly average SO_2 concentrations at a busy road kerbside site in Central London compared with an urban background location some 2 km distant. It is reasonable to assume that the major difference between the two sites in terms of local sources is the influence of traffic. The data for 1993–94 show no significant difference in terms of the timing or size of the hourly peaks, suggesting a common source, possibly large combustion plant in the East Thames corridor. However, it is interesting to note that the monthly mean values at the kerbside site are consistently higher by a few ppb.

Nitrogen Oxides

The changing patterns of NO_x concentrations are less clearly defined. This is mainly because of the absence of well-characterized measurements of oxides of nitrogen until the mid to late 1970s. Nonetheless, emission estimates are reasonably accurate, and these have suggested that, unlike SO_2, there has not been a significant decrease in NO_x emissions. Indeed, they suggest that there has been an increase in emissions of NO_x resulting from the increase in traffic activity over the past few decades. Although measurements of NO_x are not as plentiful as those of SO_2 over the past 30 years or so, there is some evidence of this trend in emissions in air quality measurements in urban areas. Fig. 6.4 shows a time series of NO_x measurements in Central London since 1977. The trend in mean levels is difficult to distinguish (in fact it is not statistically significant), but there is a clear upward trend in the 98th percentile of hourly average NO_x concentrations up to about 1991. This coincides quite closely with the emission estimates from road traffic, which were estimated to have peaked in 1989 in the UK as a whole. Since then, the increasing use of lower emission technologies such as diesel cars and petrol cars with three-way catalysts has been responsible for the decrease in emissions and reduction in ambient NO_x concentrations.

Changing patterns of NO_2 are more difficult to discern because of the secondary nature of the pollutant. Although some of the original emission of NO_x will be in the form of NO_2, most of the NO_2 measured in ambient air, away from the immediate source, arises from the oxidation of NO by ozone and the subsequent photostationary state in daylight hours. Changing patterns of NO_2 in urban background, suburban and rural locations will therefore be intimately linked with changes in ozone concentrations. At kerbsides and during periods of high pollution in winter, other mechanisms can give

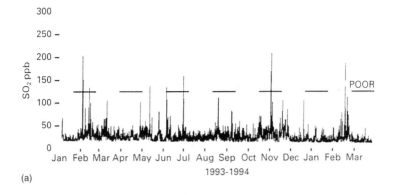

(a)

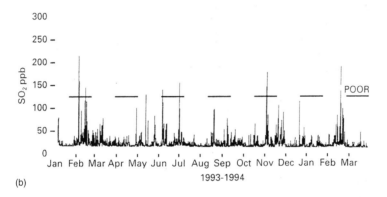

(b)

Fig. 6.3   Hourly average sulfur dioxide concentrations (ppb) at two London sites. (a) Cromwell Road (kerbside);
(b) Bridge Place (urban background).

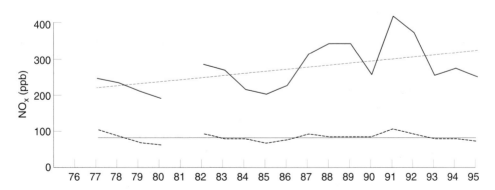

Fig. 6.4   Trends in Central London NO$_x$ concentrations (based on hourly means). ---, trend in 98th percentile;
—, trend in mean.

rise to elevated NO_2 concentrations. The different mechanisms can be illustrated in a plot of NO_2 concentrations against those of NO_x, as shown in Fig. 6.5, which displays data for a site in London. At very low NO_x levels, emissions of NO react with ambient ozone to form NO_2, and up to the ambient level of ozone (usually around 35 ppb or so in the UK) the increase in NO_2 is linear in NO_x. As the supply of ozone is exhausted, further increase in NO_x concentrations does not increase NO_2. However, at sufficiently high NO_x concentrations such as may occur at busy kerbsides (or even further afield in particularly severe winter episode meteorological conditions such as occurred in London in December 1991), a further mechanism can come into play in which two molecules of NO react with one of oxygen to produce two molecules of NO_2. The rate of this reaction is therefore proportional to the square of the NO concentration, and it is now thought that this reaction is the principal reason for the more rapid increase in NO_2 concentrations at very high NO_x concentrations, as shown in Fig. 6.5. Conversely, the squared-dependence on the NO concentration should mean that reductions in NO_x emissions (mainly NO) will have a more than proportional effect in reducing the very high NO_2 peaks occasionally measured in severe winter episodes.

Since most NO_2 concentrations are determined by the ambient ozone level, and since trends in average ozone levels have not been very marked, it has not been easy to demonstrate trends in NO_2. One successful example is a recent study in Sweden (Sjödin *et al.*, 1996) which demonstrated significant downward trends in NO_2 at a series of towns and cities in Sweden since 1986–87. In terms of identifying trends in NO_2 average concentrations at least, conditions were favourable for a successful outcome in that the average NO_2 concentrations at the beginning of the period were at most around 20 ppb, and were likely to be lower than the ambient ozone background in most cases, so that the NO_2–NO_x relationship would probably have been in the linear region of the curve in Fig. 6.5.

A fascinating insight into the way NO_x concentrations have developed since the days of the early London 'smogs' is afforded by some early measurements of NO and NO_2 reported by a Ministry of Technology report for the Warren Spring Laboratory site in Islington, London, for the period of the 1962 smog (Warren Spring Laboratory 1967). It should be mentioned that the measurement method is not described and the errors and

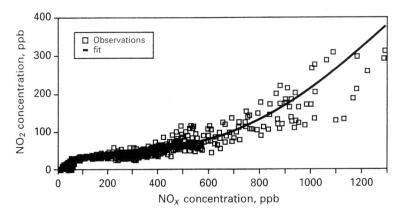

Fig. 6.5   The observed relationship between NO_2 and NO_x.

accuracy are therefore not well defined. The measurements were not made on a continuous basis but for several hours on each day, and show NO_2 concentrations ranging from 65 ppb (over 3.5 h on 8 December) to 197 ppb (over 2.3 h on 6 December). The highest total NO_x concentration measured was during the 2.3 h on 6 December and was 742 ppb. In contrast, during the December 1991 episode, hourly NO_2 levels reached 423 ppb, and total NO_x concentrations reached about 1500 ppb. Levels in the 1991 episode therefore appear to have been about double what they were in the 1962 smog in London. The earlier report also quotes daily average data for December 1961, a month unaffected by a major smog episode. The average values were NO 29 ppb and NO_2 20 ppb, so that total NO_x was 49 ppb. This can be compared with December averages for NO_x for the West London site in the national UK network for 1994 and 1995 of 113 ppb and 89 ppb, respectively. Again, average concentrations of NO_x appear to have doubled over this period. This is consistent with estimates of emissions from mobile sources which modelling studies show to be the main contributors to ambient urban NO_x concentrations.

There is very little data on other pollutants over such a long time period, but emission estimates suggest that other vehicle-related pollutants would also have increased substantially over the past three decades. As has already been noted, more recently – in the last five years or so – lower emission technologies have become more widely used and emissions from the vehicle sector are not increasing as rapidly as vehicle activity. There are some measurements over the past 10 years or so which illustrate the recent trends to lower emissions. In an analysis of carbon monoxide (CO) data from a remote site on the west coast of the Republic of Ireland, Derwent *et al.* (1998) have shown that, in polluted air masses originating on the continent of Europe before arriving at the measurement site, CO concentrations have shown a downward trend of around 13 ppb per year over the past 8 years or so. The conclusions drawn by the authors are that the use of lower emission technologies in the vehicle fleets of most European countries is responsible for this measurable decrease in CO. Such trends have also been observed at some individual urban locations. Figure 6.6 shows a plot of data from 1990 for two urban sites in the UK; the downward trend in the mean and peak values of CO is clear.

Ozone

A pollutant of particular importance in developed countries is ozone. Unlike most of the other pollutants considered here, ozone is not emitted directly in any significant amounts, but is formed in the atmosphere from the reactions in sunlight between NO_x and volatile organic compounds (VOCs). The processes which produce ozone in the atmosphere are complex and are discussed in an earlier chapter dealing with atmospheric chemistry (see Chapter 4). The important point is that high ozone levels measured at a given location may bear no relation to the emissions of precursors in the same area – a situation which tends to prevail in ozone episodes in northwest Europe where easterly wind flows in anticyclonic conditions can transport ozone-forming air masses over many hundreds, if not thousands, of kilometres. These were the conditions under which ozone was first identified in Europe, in measurements at a remote location on the west coast of Ireland, and it has long been recognized that control strategies for ozone in Europe would have to be international in character. This contrasts with the situation in the USA, where ozone problems were first identified in Los Angeles. In these situations the

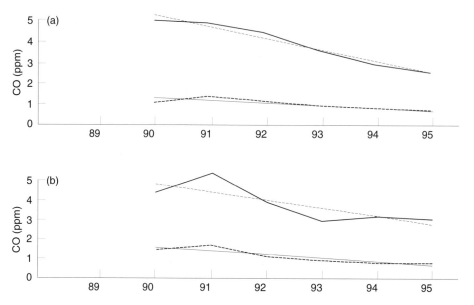

Fig. 6.6 Trends in carbon monoxide concentrations (based on hourly means) in 1995 in (a) Glasgow City Chambers and (b) West London. ---, trend in 98th percentile; —, trend in mean.

precursors to ozone formation can remain in the same geographical area for some days owing to the local characteristic flows associated with the juxtaposition of mountains and sea. It is only in recent years that attention has been focused on the longer-range transport aspects of abatement strategies for ozone in the USA. A situation similar to that in Los Angeles prevails in several large cities in southern Europe, and here abatement strategies appropriate to transboundary transport and formation of ozone may not necessarily be appropriate.

An important consideration is that ozone can have adverse effects not only on human health, but also on crops, vegetation and natural ecosystems, as well as on materials, particularly plastics and rubbers. Indeed, in the early years of ozone studies in the Los Angeles area, damage to rubber at ambient levels was regularly observed, and was used at one time as an approximate method for measuring ozone. In fact ozone is unique in the sense that the difference between the tropospheric baseline level and levels at which adverse effects can be demonstrated is very small. In the particular case of some crops, notably wheat, studies have shown the best correlation between loss in crop yields and accumulated ozone levels in excess of 40 ppb rather than other levels. At the latitudes of most of Europe, 40 ppb is virtually indistinguishable from the tropospheric baseline. Similarly in the context of health effects, the WHO have recently set their air quality guideline figure for ozone at 60 ppb over 8 h, and the UK Expert Panel on Air Quality Standards (EPAQS) recommended a standard (in a similar sense to the WHO) of 50 ppb over 8 h. In considering this close relationship between the tropospheric baseline and damage threshold, it should be noted that the tropospheric baseline is not completely natural in origin. There is now good evidence to suggest that this baseline may have increased by a factor of two or more since the turn of the twentieth century as a result of the increase in emissions on a global scale from human activities.

This means that there are two regimes of abatement strategies for the control of ozone levels. First, in order to reduce the peak ozone levels found in summertime photochemical episodes of the type prevalent in most urban/industrial European areas, emission controls in the region itself will be necessary. This region may be geographically large, and would probably cover Europe. Second, to have any significant influence on tropospheric baseline levels in the region of 40 ppb, where effects on vegetation have been demonstrated, considerations of global emission patterns would be necessary.

The identification of trends in ozone levels is even more difficult than for other, primary, pollutants since ozone levels and exposures can vary tremendously from one year to the next depending on the weather conditions, even if emissions change very little. Nevertheless, there is some evidence that reductions in precursor emissions in Europe have begun to decrease peak ozone levels. Anecdotally this is suggested by the fact that in the very hot summer of 1976, hourly values of 250 ppb and more were observed in the southern UK, while in recent years, concentrations over 150 ppb have been rare. A more systematic analysis was undertaken by the UK Photochemical Oxidants Review Group in its Fourth Report, shown in Fig. 6.7 (Department of the Environment, Transport and the Regions, 1997). However on a global scale the picture appears to be different; this is shown in Fig. 6.8 where an analysis of data from Mace Head on the west coast of Ireland is shown. The ozone data have been separated into those observed on days when the air mass originated in Europe ('polluted') and those originating from remote regions of the northern hemisphere ('unpolluted'). The downward trend in ozone formed from European air confirms the previous analysis, while the northern hemisphere ozone shows a small but positive increase over the period. As strategies are evolved aiming at levels in the region of 50–60 ppb, it is clear that increasing account will have to be taken of global influences on ozone concentrations. Looking at Europe as a whole, the pattern of concentrations is such that exceedences of human health-related thresholds increase along a gradient from the UK towards Central Europe. Similarly, exceedences of these thresholds in Scandinavia are among the lowest in Europe. An analysis in terms of exceedences of the standard recommended by the UK Expert Panel on Air Quality Standards (50 ppb running 8-h mean) is shown in Fig. 6.9.

Another important feature of ozone, particularly in considering its effects on human health, is the fact that because it is a secondary pollutant it will always occur alongside other pollutants. Perhaps the most important in this regard, at least on the basis of health effect evidence to date, are fine particles. During photochemical episodes, the same processes which lead to the formation of ozone also lead to the oxidation of SO_2 and NO_x emissions through to sulphuric and nitric acids, respectively, and their subsequent aerosol formation. (This is the so-called 'heat-haze' visible on hot sunny days during photochemical episodes, even in rural areas. It is predominantly sulphate and nitrate aerosol formed from emissions of SO_2 and NO_x produced by human activities possibly many hundreds or thousands of kilometres away.) These aerosol particles are around 1 μm or less in size and thus not only can they live for a long time in the atmosphere and be transported great distances, but they can also penetrate the respiratory tract and potentially cause adverse effects on health. Similarly, in photochemical episodes the atmospheric processes can also give rise to elevated levels of NO_2, typically of a similar order to the ozone concentrations, potentially reaching levels of the order of 100 ppb.

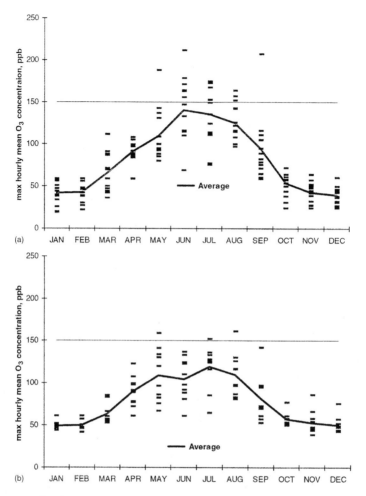

Fig. 6.7 Maximum hourly ozone concentration observed at any site during each month of years (a) 1972–1985 and (b) 1986–1995.

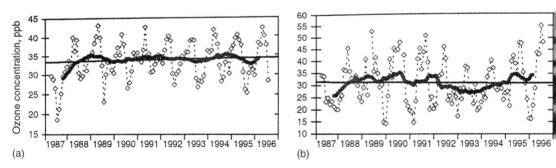

Fig. 6.8 Monthly mean and 12 month running mean ozone concentrations at Mace Head (western Ireland) with a linear regression on the 12-month running means, in (a) unpolluted air masses and (b) polluted air masses. In (a) the trend is 0.0924 ppb/year; in (b) it is –0.0372 ppb/year. --◊--, monthly mean; —, 12 month running mean; — linear regression.

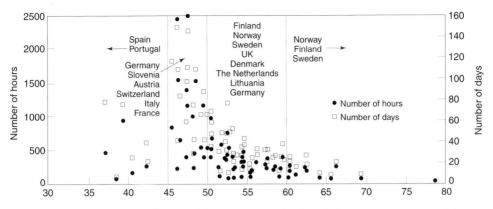

Fig. 6.9 Average of the number of hours and days per year on which the eight hour running mean exceeds 50 ppb, for 1988 to 1994 at EMEP ozone monitoring sites, plotted against site latitude (only years with at least 75% data capture are included).

Particulate Matter

Perhaps the most important pollutant in terms of adverse effects on human health is particulate matter. This is discussed in more detail elsewhere in this book and it is certainly not the intention to discuss the evidence for the effects here. Suffice it to say at this stage that associations have been widely demonstrated between various measures of particles and adverse effects. There is still much work under way attempting to elucidate the mechanisms of action of particles, and at present therefore it is not possible to be clear as to the harmful agent or agents in the particle mix. This means that little is known of the precise size range of importance, or the chemical components of the aerosol which might be important.

One thing that is very clear, in the historical context with which this chapter began, is that the overall mass loading of particles in the atmosphere of most large urban areas in Europe (and elsewhere) has reduced enormously. Equally, it is clear that the extensive adverse effects on human health no longer occur on the scale that they once did in the pollution episodes of the 1950s and 1960s and earlier. This is not surprising in the light of the reductions in emissions of particles and other pollutants associated with the widespread coal burning that was prevalent during that period. Whether or not the nature of the effects is different in the present situation is an unresolved issue and one which is outside the scope of this chapter. The discussion here will focus on the current nature of the urban aerosol, insofar as it is known, distinguishing where possible the changes which have occurred over the past few decades and the reasons for them.

The changes result primarily from a situation in 1962 when, for example, almost 9 million tonnes of coal and over 2 million tonnes of coke were burnt in London, to a situation today where the consumption of these fuels is almost absent. The most important source of particles emitted directly into the atmosphere of large urban areas is currently the motor vehicle. In London a recent study by the London Research Centre has estimated that of the PM_{10} emitted in London, 12% arises from petrol vehicles and 67% from diesels (London Research Centre, 1997). The nature of this emission is currently the subject of much study. In terms of particle size, the emission from motor vehicles is generally in the submicrometre range, with peaks in the mass distribution generally around 0.1 μm or so. There is still much research needed to define the size distribution of the emissions from motor vehicles, and to understand the dynamics of the process as the freshly emitted very small particles grow into the so-called accumulation mode.

The crucial question here is, given the evidence for adverse effects of particles on human health, what will the effects of control measures have on these effects in future? Here of course it is important to have some idea about the nature of the damaging component of the particle distribution. One line of thought has suggested that, since they are capable of penetrating to the full extent of the respiratory system, smaller particles may be more important than larger ones. Indeed, the USEPA has recently promulgated national ambient air quality standards for $PM_{2.5}$ as well as for PM_{10}. Additionally, there have been studies, notably from Oberdörster's group, which have shown changes in toxicity of particles of titanium dioxide on reducing the particle size to 20 nm (Oberdörster *et al.*, 1994). These considerations led Seaton and colleagues to present a hypothesis that the important quantity was the number of ultrafine particles, rather than their mass, or even possibly their composition (Seaton *et al.*, 1995).

It is becoming clear that the finer particles (less than 2.5 μm, for example) are associated with emissions from human activities, predominantly from combustion sources, while the coarser fractions have a much bigger component from natural dusts such as wind-blown soils and other mechanically generated particles. This is shown in Fig. 6.10, which shows the good correlation between hourly concentrations of $PM_{2.5}$ and NO_x at a site in Birmingham, UK, for the winter period October 1994 to March 1995, while the relationship between PM_{10} and NO_x is very much weaker. It is interesting to note that the

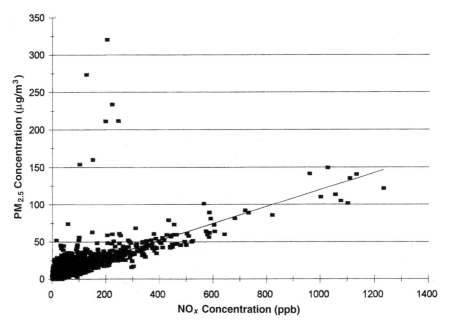

Fig. 6.10   Relationship of hourly $PM_{2.5}$ and NO_x concentrations, Birmingham Hodge Hill, October 1994–March 1995. The line drawn gives a linear fit to the data. From Harrison *et al.* (1997) *Atmospheric Environment* **31**: 4103–4117, with permission.

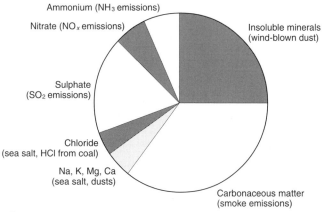

(a) Total

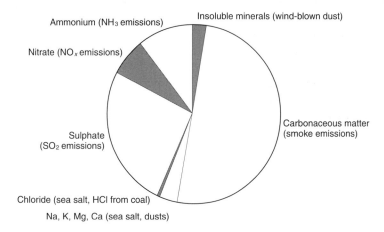

(b) Fine fraction

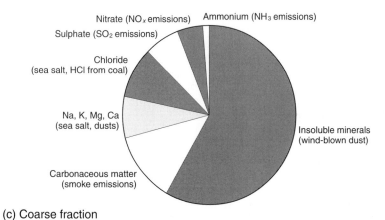

(c) Coarse fraction

Fig. 6.11 Approximate composition of urban particles in the UK.

corresponding plot of $PM_{2.5}$ in summer at the same site shows a much poorer correlation, suggesting that there is another source of fine particles in the summer. This turns out to be the so-called secondary aerosol, composed mainly of sulphates and nitrates, referred to earlier in the discussion on ozone. The reason that there is a much weaker correlation on an hourly basis is that the secondary aerosol arises from a wide spectrum of sources, many of which will be many hundreds of kilometres away, whereas the NO_x concentrations arise principally from motor vehicle emissions in the nearby urban area.

The question of chemical composition of the aerosol is also important and some patterns are beginning to emerge, albeit at a fairly coarse level of resolution at this stage. The composition of aerosols in the UK has been discussed in the Third Report of the UK Quality of Urban Air Review Group, and the results are shown in Fig. 6.11 (Department of the Environment, 1996). The fine fraction (below about 2.5 μm) is composed mainly of organic carbonaceous material, presumably arising largely from vehicle combustion, although other local and more distant sources may also contribute, and secondary aerosol, which appears to be mainly ammonium sulphate with some nitrate also present. The coarse fraction also has these components in smaller percentage terms, and is predominantly composed of inorganic mineral material arising from wind-blown dust and other similar sources. It is interesting to note that these broad features are also found in a recently reported analysis of fine particles during inversion conditions in a rural valley in Utah, with a small urban centre and no heavy industry (Mangelson et al., 1997), as shown in Fig. 6.12. The similarities are quite striking, with the main difference being in the composition of the secondary aerosol. In the Utah analysis the secondary aerosol shows a much bigger fraction of nitrate relative to sulphate compared with the UK data, where the sulphate predominates. This is likely to be a pointer to an emerging pattern in Europe where reductions in sulphur emissions have been, and are likely to continue to be, larger than those of NO_x, so that the relative proportion of nitrate in European aerosols is likely to increase relative to sulphate.

There tends to be a reasonable correlation between PM_{10} and $PM_{2.5}$. This is reassuring since there is a large body of evidence demonstrating associations between the former and adverse effects on health, while the latter is the metric of particles which is the more likely of the two to respond to control technologies reducing particle emissions from human

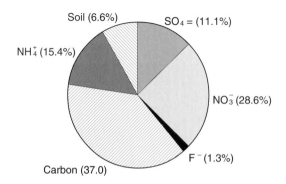

Fig. 6.12   Average fine particle (<25 μm) composition for two inversion episodes in February 1993 in Cache Valley, Utah, USA. From Mangelson et al. (1997) *Journal of The Air Waste Management Association* **47**: 167–175, with permission.

activities. Indeed, in the absence, as yet, of clear evidence on the most harmful aspects of the aerosol, one might envisage some form of no regrets control measures which might even invoke the precautionary principle. This approach might be viable if one could be sure that control measures would drive particle emissions in the right direction, but some recent work (Silva and Prather, 1997) gives pause for thought. A study of motor vehicle emissions on some catalyst and non-catalyst equipped cars showed that the emissions from the catalyst equipped vehicles were shifted significantly to the smaller size ranges, showing number density peaks in the range 0.1–0.2 µm compared with 0.5–2.0 µm in the non-catalyst cars, as shown in Fig. 6.13.

There is clearly a need for more information on the size distributions and composition of the atmospheric aerosol. It could be argued that further analysis should await more information from the health effects community, but the development of some exciting new techniques for analysing particles suggests that advances in atmospheric science and health effects would be more effectively pursued as complementary activities. Recent work of Richards and co-workers at Cardiff University in the UK (BéruBé *et al.*, 1997) uses electron microscopy techniques not only to view the morphology of fine particles, but also to analyse chemically the bulk profile of the particles. This sophisticated and powerful analysis technique in combination with studies of mechanisms of biological activity

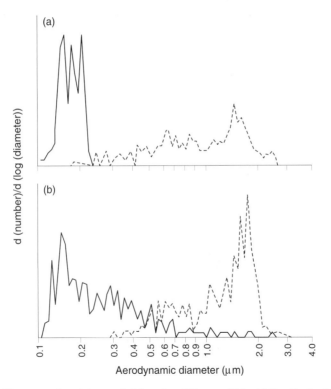

Fig. 6.13 Size histograms of particles sampled from four different vehicles (a) Size distributions of particles from 1973 Chevrolet Blazer (---) and 1991 Nissan Sentra (—). (b) Size distributions of particles for 1994 Acura Integra (---) and 1996 Mazda Millenia (—). Silva and Prather (1997) *Environmental Science and Technology* **31**: 3074–3080, with permission.

at the cellular level shows promise for elucidating the mechanisms of biological action and the important components of the particle distribution.

Another technique that has recently begun to provide a wealth of detail on atmospheric particles is aerosol time-of-flight mass spectrometry (ATOFMS). This was used by Prather and co-workers in the study of vehicle emissions referred to above and provides size distribution data and associated chemical analyses (Silva and Prather, 1997). This technique was used in the study of vehicle emissions to show that the chemical composition of the particle fractions above and below about 1 μm (see Fig. 6.13) were quite different, with predominately organic species in the submicrometre mode and inorganic compounds above 1 μm, with the technique showing the potential to identify large numbers of individual compound in the different size ranges.

Patterns of atmospheric aerosols have clearly changed enormously since the days when coal burning was widespread and the particles were mainly coal-derived carbonaceous particles associated with sulphuric acid and other irritant compounds. It is interesting to note however that such estimates as exist for number concentrations from that era suggest that in terms of this metric the changes have not been as large as they have been in mass terms. Measurements made by Waller (1967) using electron microscopy revealed concentrations of the order of 10^4/cm^3 at background sites, 3–5×10^4/cm^3 in street samples, and up to 16×10^4/cm^3 in samples taken in road tunnels and in urban fogs. Measurements of number concentrations taken recently near a busy road and in central Birmingham, UK, are shown in Fig. 6.14. These suggest that number concentrations may not have changed much since the 1960s, even though mass concentrations have decreased considerably.

As a final thought on the current patterns of airborne particles and the mechanisms by which they might affect human health, it is interesting to compare the numbers of particles in urban aerosols with the numbers of molecules of other pollutants which might also display adverse effects. Theoretically, to sustain a mass concentration of 10 μg/m^3 with 0.02 μm particles would require about 2.4×10^6 particles/cm^3, and with 0.1 μm particles it would require about 19 000 particles/cm^3 (assuming a density of 1 g/cm^3). We have already seen that measurements suggest that number concentrations of the order of 10^5/cm^3 can be measured near roads, and unpublished work of D. Booker and J. McAughey (personal communication, 1997) suggests that in heavily trafficked regions of a large city, short-term peak levels of the order of 10^6/cm^3 could be observed. To put this in context, during the high pollution episode in London in December 1991, hourly averages of NO$_2$ and CO reached about 400 ppb and about 15 ppm respectively. These levels are equivalent to 1×10^{13} molecules/cm^3 and 3.8×10^{14} molecules/cm^3. An ozone episode at the upper end of levels currently observed in northwest Europe of 100 ppb would have a concentration of 2.5×10^{12} molecules/cm^3. In terms of the presentation in the air of potentially polluting entities to the human body, the numbers of particles are therefore several orders of magnitude smaller than the numbers of molecules of other pollutants which are present in the air at the same time. The structure and composition of the individual particles is therefore crucial in understanding the mechanism(s) by which particles produce adverse health effects. The significance of evolving patterns of both particulate and gaseous pollution, particularly for human health, is a major challenge for both atmospheric and medical science.

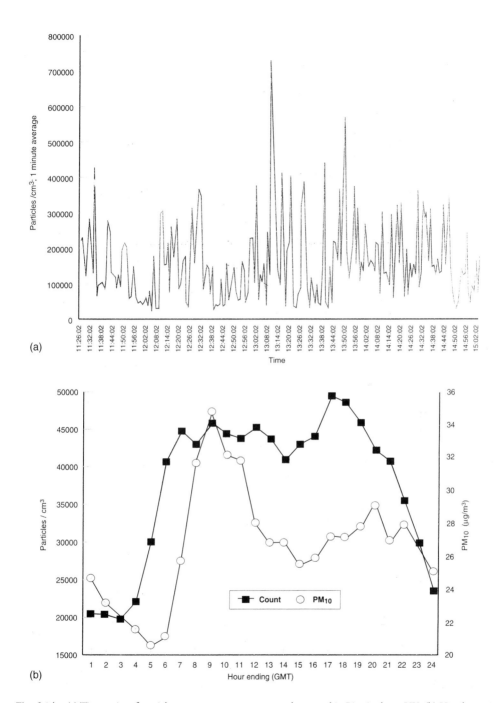

Fig. 6.14 (a) Time series of particle count measurements near a busy road in Birmingham, UK. (b) Hourly mean diurnal particle count and PM$_{10}$ mass, Central Birmingham, UK. From Harrison *et al*, 1997.

REFERENCES

BéruBé KA, Jones TP and Williamson BJ (1997) Electron microscopy of urban airborne particulate matter. *Microsc Anal* September: 11–13.

Department of the Environment (1996) *Airborne Particulate Matter in the United Kingdom*, Third Report of the Quality of Urban Air Review Group. London: DoE.

Department of Environment, Transport and the Regions (1997) *Ozone in the United Kingdom*, Fourth Report of the Photochemical Oxidants Review Group. London: DETR.

Derwent RG, Simmonds PG, O'Doherty S *et al.* (1998) *Atmos Environ* **32** (in press).

Harrison RM, Deacon AR, Jones MR and Appleby RS (1997) Sources and processes affecting concentrations of PM_{10} and $PM_{2.5}$ particulate matter in Birmingham (UK), *Atmos Environ* **31**: 4103–4117.

London Research Centre (1997) *London Atmospheric Emissions Inventory*. London: LRC.

Mangelson NF, Lewis L, Joseph JM *et al.* (1997) The contribution of sulphate and nitrate to atmospheric fine particles during winter inversion fogs in Cache Valley, Utah. *J Air Waste Manag Assoc* **47**: 167–175.

Oberdörster G, Ferin J and Lehnert BE (1994) Correlation between particle size, *in vivo* particle persistence and lung injury. *Environ Health Perspect* **102**: 173–177.

Seaton A, MacNee W, Donaldson K and Godden D (1995) Particulate air pollution and acute health effects. *Lancet* **345**: 176–178.

Silva, PJ and Prather, KA (1997) On-line characterization of individual particles from automobile emissions. *Environ Sci Technol* **31**: 3074–3080.

Sjödin Å, Sjöberg K, Svanberg PA and Backström H (1996) Verification of expected trends in urban traffic NO_x emissions from long-term measurements of ambient NO_2 concentrations in urban air. *Sci Total Env* **189/190**: 213–220.

Waller RE (1967) Studies on the nature of urban air pollution. International Institute for the Conservation of Works of Art, London Conference on Museum Climatology, pp. 65–69.

Warren Spring Laborarory (1967) Ministry of Technology, *The Investigation of Atmospheric Pollution 1958–1966*, Thirty-Second Report. London: HMSO.

7

Patterns of Air Pollution in Developing Countries

MICHAL KRZYZANOWSK

WHO, European Centre for Environmental Health, Bilthoven, The Netherlands

DIETER SCHWELA

WHO, Geneva, Switzerland

INTRODUCTION

Rapid growth of the population in cities, development of industry and intensification of road traffic pose significant challenges to natural resources, in particular to ambient air quality. In many developing countries basic needs, such as cooking, household heating or moving around, have to be fulfilled despite limited availability of clean fuels and clean technologies, and effective environment protection programmes are often missing. In consequence, air quality is poor in many cities of developing countries. This results in the ambient concentrations of air pollutants tending to be highest in countries in the early stages of development and their being likely to decline only when higher levels of development are reached (WHO, 1997a).

The assessment of levels and trends of air pollutants in the world's biggest cities in the 1980s was the subject of a comprehensive analysis published in 1992 (WHO/UNEP, 1992). The analysis was produced by Global Environmental Monitoring System (GEMS) – a joint programme of WHO and UNEP established in 1973 and closed in 1995. Of the 20 cities included in the assessment, most of those from developing countries had 'serious problems' with ambient air quality: concentrations of one or more of the evaluated pollutants exceeded the WHO Air Quality Guideline levels by more than a factor of 2. Such serious pollution of ambient air creates serious risks for population health. These

AIR POLLUTION AND HEALTH
ISBN 0-12-352335-4

risks are often combined with the burden of other hazards characteristic of poor living conditions found in developing and overcrowded conurbations. National reviews of air quality and its health effects, based on studies conducted in the 1980s and early 1990s, are available from some countries; they confirm that high levels of air pollution exist not only in the biggest cities (WHO/EHG, 1995), but also in other parts of developing countries.

This chapter provides an overview of air quality in cities of developing countries from Africa, Asia, and America in the 1990s. In order to reflect the risk to public health associated with poor air quality, the analysis focuses on information about ambient air quality which can, potentially, indicate the levels of population exposure to air pollutants.

AIR QUALITY DATA

The main source of information used in the analysis is the Air Management Information System (AMIS) established and maintained by WHO as a continuation of GEMS/Air (WHO, 1997b). The present system is based on voluntary reporting of data by Member States. Information on basic statistics relating to ambient concentrations of selected pollutants are collected on a database. Annual (arithmetic) mean and 95th percentiles of daily mean concentrations of sulfur dioxide, nitrogen dioxide, ozone, carbon monoxide, suspended particulate matter and lead are the basic parameters. In principle, data from three types of monitoring stations are stored: 'industrial', reflecting levels in areas affected by emissions from industry; 'city centre/commercial', which will be mostly affected by traffic; and 'residential', which should best reflect the basic level of population exposure. Until now, the coverage of the system has been limited, but the intention is to collect data from at least a few major cities in each country. New data are systematically added to the system with the aim of acquiring current information from some 300 cities by the end of 2000. Owing to the open and voluntary nature of the reporting, the more strict and uniform criteria of data acceptance applied by GEMS/Air are not imposed by AMIS. It is assumed that national quality assurance systems are sufficiently effective to provide valid data and also to allow international data exchange. However, it is likely that the monitoring methods, siting of the monitors and completeness of the data relating to the reported period may vary between locations, potentially affecting the comparability of the information between cities. Though methods used in individual cities are likely to have changed significantly over the reporting period, it is assumed that data from consecutive years do reflect the long-term patterns of air pollution.

1996 is the most recent year for which data from most of the chosen cities were available for this analysis. For trend analysis, the annual averages of pollution concentrations between 1990 and the most recent year with data were used. The concentration of pollutants in 'residential' areas of the cities was chosen as the basis of the assessment. Information on concentrations of pollutants close to industry was not used as pollution levels are usually higher in these locations than in the rest of the city but affect the average exposure of the city population only to a limited extent. For a few cities, data from monitors located in city centres were used if no separate information on air quality in 'residential' areas was available.

This analysis is based on the data from 32 cities in 14 developing countries of Asia, Africa or America for which data on annual mean levels of pollution in residential areas of cities (or in city centres) and relating to the 1990s was available at the time of data retrieval. Annual mean concentrations of sulfur dioxide, total suspended particulates (TSP), or particles with an aerodynamic diameter of less than 10 μm (PM_{10}), as well as nitrogen dioxide, were analysed. For most of the cities data from more than one year in the 1990s were available, allowing evaluation of pollution trends.

An additional source of information used in this analysis are publications in international scientific journals. These papers allow deeper insight into the air quality and determinants of population exposure to air pollution in developing countries. However, the relevant reports are scarce and there is certainly a need for more comprehensive assessment of air quality, using standardized methods and focusing on population exposure to the main air pollutants.

AIR POLLUTION LEVELS AND TRENDS

Sulfur dioxide

In most cities for which data were analysed, the annual mean concentration of sulfur dioxide in residential areas did not exceed 50 μg/m^3, i.e. the level of the WHO Air Quality Guideline (Fig. 7.1). Notable exceptions were several cities in China, where the sulfur dioxide concentration reached 330 μg/m^3 in Chongqing and 100 μg/m^3 in Beijing in 1994. Interestingly, in the Chinese cities the levels reported from 'residential' locations exceeded those from 'commercial' regions of the city and are comparable with the levels in industrial zones. This may reflect the impact of combustion of sulfur-containing coal for domestic heating and cooking. High levels of sulfur dioxide may also be seen in other developing countries, especially in those with cold winters, as illustrated by a report from Nepal (Sharma, 1997). Daily mean concentrations of sulfur dioxide were in the range 273–350 μg/m^3 in residential areas of Katmandu in September–December 1993. At monitoring sites close to main roads the reported range was 310–875 μg/m^3, indicating the influence of emissions from traffic. More than half the vehicles registered in the city are equipped with two-stroke engines and many are old and ill maintained.

In most of the cities with data allowing trend assessment, a decline in annual mean sulfur dioxide concentrations was seen during the 1990s. The most dramatic reduction in concentrations of sulfur dioxide was reported from Mexico City, where the concentration in various residential areas fell from 100–140 μg/m^3 in 1990–91 to 32–37 μg/m^3 in 1995–96. In the most polluted Chinese cities the rate of decline was between 1 and 10% per year.

Suspended particulate matter

The most commonly monitored and reported indicator of this type of air pollution is the mass concentration of total suspended particulates (TSP). In most of the cities, the TSP annual mean concentration exceeded 100 μg/m^3, with levels exceeding 300 μg/m^3 in several cities of China and India (Fig. 7.2). There is no evidence of any overall consistent,

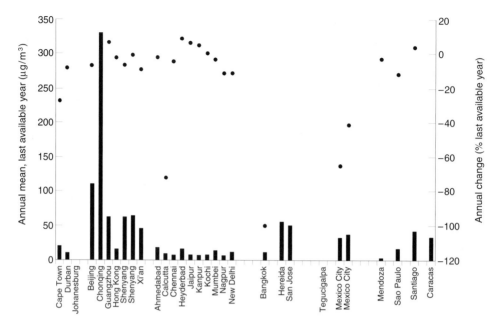

Fig. 7.1 Annual mean in last available year (bars) and annual change of sulfur dioxide concentrations (•) in residential areas of cities in developing countries. Annual change is given as a percentage of the last available year mean.

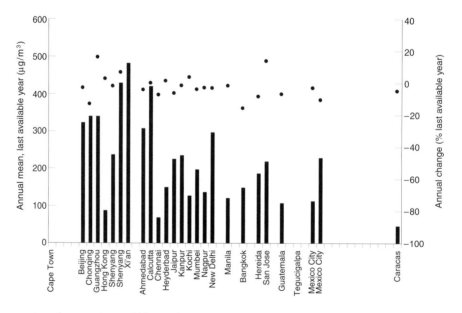

Fig. 7.2 Annual mean in last available year (bars) and annual change of total suspended particulate matter concentrations (•) in residential areas of cities in developing countries. Annual change is given as a percentage of the last available year mean.

systematic and significant change in TSP levels: the data from the 1990s show increasing and decreasing trends in a similar number of cities. Relatively the most apparent decrease in TSP concentrations is shown by the data from Bangkok but, there too, progress has not been steady. More consistent, though with a smaller relative rate, is the decrement in TSP concentrations in Mexico City. The opposite tendency can be seen in some Chinese cities, with the most rapid increase of TSP concentration being found in Guangzhou: from less than 150 $\mu g/m^3$ in 1990–92 to more than 300 $\mu g/m^3$ in more recent years.

In a limited number of cities reporting data to AMIS, the mass concentration of particles with aerodynamic diameters of less than 10 μm (PM_{10}) is also measured. The most commonly registered annual average PM_{10} level ranged from 50 to 100 $\mu g/m^3$ in the years 1995 to 1996 (Fig. 7.3). The highest concentrations, exceeding 200 $\mu g/m^3$, were observed in Calcutta and New Delhi. In most towns with high PM_{10} annual average concentrations for the last year an increase in concentration was seen during the 1990s. In most cases this increase occurred even when a decrement in TSP was reported. An opposite trend, and a decrement in PM_{10} level, was seen in the central and Southern American cities. In Mexico City, the relative decrement in PM_{10} was greater than that of TSP.

Though limited, this information on size-specific particulate pollution allows comparison of mass concentrations of TSP and PM_{10}. For most sites in years with data on both indicators, the ratio of PM_{10} to TSP was between 0.4 and 0.8. However, in a few cases the ratio exceeded one, indicating that the measurements reported to AMIS might have been done at different locations or during different periods. In a southeastern part of Mexico City, the ratio remained between 0.25 and 0.32 during 1991–96, whilst in the south-western part of the city it was consistently between 0.44 and 0.55. More specific

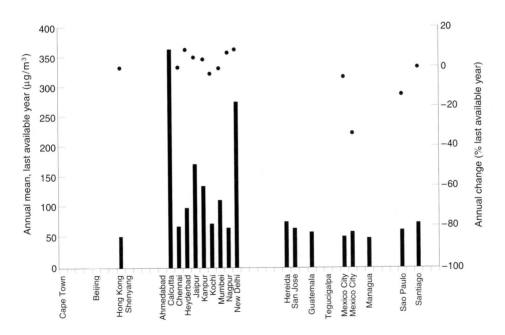

Fig. 7.3 Annual mean in last available year (bars) and annual change of respirable particulate matter (PM_{10}) concentration (•) in residential areas of cities in developing countries. Annual change is given as a percentage of the last available year mean.

studies of size distribution of airborne particles conducted in the northern cities of China in the mid-1980s indicate that some 70% of mass concentration of TSP was contributed by PM_{10} (Ning *et al.*, 1996). During the heating season, particles with diameters of less than 2 µm were found to make up some 30–50% of TSP. Elemental analysis of the particles confirmed that human activities were the main source of the fine fraction of the particulate matter. Similar results were reported from Jakarta, where the particles of diameter less than 7.2 µm contributed more than 80% of TSP (Zou and Hooper, 1997). Traffic-related material contributed significantly to the overall particulate mass, and especially to the fine fractions.

Nitrogen dioxide

Annual mean concentrations of nitrogen dioxide remained moderate or low, not exceeding 40 µg/m³ in most of the cities reporting to AMIS (Fig. 7.4). However, in Mexico City and Cape Town, an annual average of 70 µg/m³ has been exceeded regularly in the 1990s. A paper based on data from centrally located monitors in São Paulo indicates an annual mean of 240 µg/m³ in 1990–91 (Saldiva *et al.*, 1995). Trends vary between cities but a 5–10% annual increase was more common than a decrease in concentrations of this pollutant.

The observed pattern is consistent with the volume of motor vehicle traffic in each city. The highest pollution levels, and the increasing trends, are observed in cities with high and increasing motor vehicle traffic density. In Southern Asia and in Latin America,

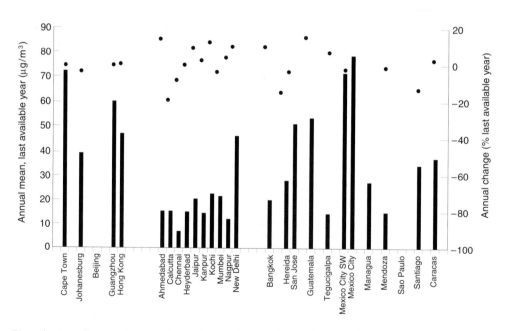

Fig. 7.4 Annual mean in last available year (bars) and annual change of nitrogen dioxide concentrations (•) in residential areas of cities in developing countries. Annual change is given as a percentage of the last available year mean.

high nitrogen dioxide concentrations combined with intensive ultraviolet (UV) radiation results in photochemical smog with high ozone levels. This is illustrated by the analysis of temporal and spatial patterns of tropospheric ozone in New Delhi (Singh *et al.*, 1997), where the build-up of ozone over the day is faster than scavenging of ozone by nitrogen dioxide. The mixture of high nitrogen dioxide emission from gasoline combustion and strong sunlight in Mexico City is the cause of the notorious photochemical smog found there. According to the data reported to AMIS, ozone concentrations exceed the WHO Air Quality Guideline (120 μg/m^3 8-h average) on more than 300 days each year in 1994–96 and the 95th percentile of maximum daily 1-h average ozone concentrations was around 500 μg/m^3. Some decrement was seen, however, in respect of the annual mean ozone concentrations, indicating slow improvement of air quality on non-extreme days.

RELEVANCE OF THE OBSERVED POLLUTION PATTERNS FOR HEALTH IN DEVELOPING COUNTRIES

The reported levels of air pollution cause concern for population health in many cities of developing countries. Suspended particulate matter is the most widespread air pollutant which is likely to cause adverse health effects, increasing morbidity and mortality in a large part of the population of these countries. The predominant sources of information relating to the effects of this type of air pollution are studies conducted in the USA and Europe (WHO/ECEH, 1999) but several investigations conducted in Mexico, Chile, Brazil and China confirm that the elevated levels of particulate matter also have adverse effects on health in developing countries (Borja-Aburto *et al.*, 1997; Ostro *et al.*, 1996, Saldiva *et al.*, 1995; Zhaoyi *et al.*, 1995). Although most of the monitoring networks measure and report TSP as the particle pollution indicator, the limited data on PM$_{10}$ and the special studies of dust composition indicate high levels of respirable dust, confirming the health relevance of the pollution measured as TSP.

In the developing countries, important sources of suspended particulate matter are also frequently found indoors. The use of open fires or inadequately ventilated stoves leads to indoor concentrations of pollution which are often higher than those found outdoors (WHO, 1997a). Since people spend up to 90% of their time indoors, most of the exposure occurs in these heavily polluted environments. One of the important effects of exposure is the increased risk of acute respiratory infections (ARI), the principal cause of infant and child mortality in developing countries (Bendahmane, 1997). It has been estimated that a 20% reduction in indoor air pollution from biomass combustion could reduce mortality from ARI by 4–8%: this is comparable with that expected from other interventions such as reduction of low birth weight and improvement of vaccination programmes. In urban areas and in countries where indoor sources are less common, indoor levels of the main air pollutants, especially PM$_{2.5}$, have been found to be comparable with concentrations outdoors (Lee *et al.*, 1997). Therefore, ambient air quality remains an important risk factor.

In addition to measures taken to reduce emissions from industry and traffic, interventions to reduce levels of particles should include banning or open burning of solid wastes and agricultural refuse. Forest fires also increase air pollution and affect exposure to particulate matter of populations who may live far from the fires. During the notorious uncontrolled forest fires in Indonesia in August/September 1997 the concentration of PM_{10} in Malaysia reached 931 $\mu g/m^3$ in Kuching and 431 $\mu g/m^3$ in Kuala Lumpur (Brauer, 1997). These pollution episodes may promote adverse effects on health and need a careful evaluation with respect to their impacts on mortality and on the incidence of acute respiratory symptoms. However, it might be expected that, based on the data on pollution in many cities, the main burden on health is caused by regular emissions and pollution levels and not by extreme events such as major fires.

Health concerns related to elevated sulfur dioxide levels are limited to communities using large quantities of fuels contaminated by sulfur. In Chinese cities with annual levels of sulfur dioxide exceeding 300 $\mu g/m^3$, one can expect serious health effects, although the associated high level of particulate pollution makes it difficult to identify the relative roles of these pollutants. Therefore the programme to improve air quality must focus on both types of pollution.

Concerns with nitrogen dioxide levels in the developing countries are focused on cities with very intensive and growing motor vehicle traffic, and especially on those experiencing many sunny days, with intensive UV radiation promoting ozone formation. Several studies have been conducted in Mexico City, where UV radiation and thus ozone levels are enhanced by the high altitude. These studies demonstrate effects of high ozone levels on pulmonary function of exercising children (Casteillejos *et al.*, 1995) and on the frequency of lower respiratory symptoms in children with mild asthma. The increased frequency of symptoms has also been confirmed by the increase in visits to clinics indicating a need for medical consultation and treatment (Romieu *et al.*, 1995, 1996).

CONCLUSIONS

Levels of air pollutants, and especially of particulate matter, are a serious problem in many cities of developing countries and should be a matter of serious concern to public health authorities. Comprehensive, long-term air quality management programmes are needed in these cities to ensure that tolerable levels of pollution are not exceeded and that the population exposure is reduced to acceptable levels. Such a programme involves a wide range of local authorities, industry, transport and local communities. Efforts must not be limited to combatting only outdoor pollution. Improvement of cooking and heating facilities will greatly reduce exposures occurring indoors and, through improvement of fuel use efficiency, will also benefit ambient levels of pollution.

Growing numbers of motor vehicles, often old and ill-maintained and using poor quality fuel, will pose an increasing risk to health through emissions of large quantities of fine particles. These emissions will increase the population exposure to lead, affecting neurological development of children (see Chapter 34). This emerging risk, possibly replacing or adding to the present impacts, should be avoided through programmes promoting the use of cleaner vehicles and the identification of other approaches that merge economic development and protection of both health and the environment.

REFERENCES

Bendahmane DB (1997) *Air Pollution and Child Health: Priorities for Action.* Environmental Health Project, Activity Report no. 38. Washington, DC: US AID.

Borja-Aburto VH, Loomis DP, Bangdivala SI *et al.* (1997) Ozone, suspended particulates and daily mortality in Mexico City. *Am J Epidemiol* **145**: 258–268.

Brauer M (1997) *Assessment of Health Implications of Haze in Malaysia.* Mission Report. Manila: WHO Regional Office for the Western Pacific, RS/97/0441.

Castillejos M, Gold DR, Damokosh AI *et al.* (1995) Acute effects of ozone on the pulmonary function of exercising schoolchildren from Mexico City. *Am J Respir Crit Care Med* **152**: 1501–1507.

Lee HS, Kang BW, Cheong JP and Lee SK (1997) Relationships between indoor and outdoor air quality during the summer season in Korea. *Atmos Environ* **31**: 1689–1693.

Ning DT, Zhong LX and Chung YS (1996) Aerosol size distribution and elemental composition in urban areas of Northern China. *Atmos Environ* **30**: 2355–2362.

Ostro BD, Sanchez CA and Eskeland G (1996) Air pollution and mortality: results from a study in Santiago, Chile. *J Exp Anal Environ Epidemiol* **6**: 97–114.

Romieu I, Meneses F, Sienra-Monge JJL *et al.* (1995) Effects of urban air pollutants on emergency visits for childhood asthma in Mexico City. *Am J Epidemiol* **141**: 546–553.

Romieu I, Meneses F, Ruiz S *et al.* (1996). Effects of air pollution on the respiratory health of asthmatic children living in Mexico City. *Am J Crit Care Med* **154**: 300–307.

Saldiva PHN, Pope CA III, Schwartz J *et al.* (1995). Air pollution and mortality in elderly people: a time-series study in São Paulo, Brasil. *Arch Environ Health* **50**: 159–163.

Sharma CK (1997) Urban air quality of Katmandu Valley 'Kingdom of Nepal'. *Atmos Environ* **31**: 2877–2883.

Singh A, Sarin SM, Shanmugam P *et al.* (1997) Ozone distribution in the urban environment of Delhi during winter months. *Atmos Environ* **31**: 3421–3427.

WHO (1997a) *Health and Environment in Sustainable Development. Five Years after the Earth Summit.* Geneva: World Health Organization.

WHO (1997b) *Healthy Cities Air Management Information System* (AMIS CD ROM). Geneva: World Health Organization.

WHO/ECEH (1999) *Air Quality Guidelines for Europe,* second edn. Copenhagen: WHO Regional Office for Europe (in press).

WHO/EHG (1995) *Air Pollution and Its Health Effects in China: A Monograph.* International Programme on Chemical Safety. Geneva: WHO/EHG/PCS/95.21.

WHO/UNEP (1992) *Urban Air Pollution in Megacities of the World.* Oxford: Blackwell Publishers, published on behalf of WHO and UNEP.

Zhaoyi X, Chen B-H, Kjellström T *et al.* (1995). Study of severe air pollution and mortality in Shenyang, China. In: *Air Pollution and Its Health Effects in China: A Monograph.* International Programme on Chemical Safety. Geneva: WHO/EHG/PCS/95.21, pp. 47–88.

Zou LY and Hooper MA (1997) Size-resolved airborne particles and their morphology in central Jakarta. *Atmos Environ* **31**: 1167–1172.

8

Sources of Air Pollution

CLAIRE HOLMAN

SENCO, Bristol, UK

INTRODUCTION

Air pollution comes from a wide variety of sources. For the classical pollutants – sulfur dioxide (SO_2), nitrogen oxides (NO_x), carbon monoxide (CO), volatile organic compounds (VOCs), and particulate matter (PM) – the single most important source is generally the combustion of fossil fuels. Of particular importance is the burning of fuels for road transport and electricity generation.

Some pollutants are emitted directly into the atmosphere and are known as primary pollutants. Others are formed in the air as a result of chemical reactions with other pollutants and atmospheric gases; these are known as secondary pollutants. Carbon monoxide and sulfur dioxide are examples of primary pollutants, while ozone is an example of a secondary pollutant. Some pollutants, such as nitrogen dioxide (NO_2) and some particulate matter, are both primary and secondary pollutants. They are both emitted directly into the atmosphere, and formed from other pollutants. In the case of nitrogen dioxide some is emitted directly from power stations and vehicle exhaust, while most is formed by the oxidation of nitric oxide (NO) in the air. Fine particulate matter is emitted directly from a number of natural and anthropogenic sources as well as being formed within the atmosphere. Secondary PM_{10} is formed mainly from the oxidation of sulfur dioxide and nitrogen dioxide. A less important source is the reaction of ammonia with hydrochloric acid gas (from the combustion of coal and municipal incineration). In some areas the reactions of hydrocarbons in the air to produce organic particles is an important source of secondary PM_{10}.

The distinction between primary and secondary pollutants is important for understanding air pollution and devising pollution abatement strategies. For primary pollutants

AIR POLLUTION AND HEALTH
ISBN 0-12-352335-4

there is a proportional relationship between emissions and ambient concentrations. However, for a secondary pollutant reducing emissions of the precursor does not lead to a proportional reduction in its ambient concentration. For some pollutants in some circumstances a reduction of precursor emissions may lead to an increase in concentrations. This is most notable in the case of urban ozone, where reducing local emissions of nitrogen oxides can lead to an increase in local ozone concentrations.

Quantitative information on the rate of emission of air pollutants from different sources is vital for the development of appropriate abatement strategies. By identifying the major sources of both primary pollutants and the precursors of secondary pollutants, the best means of improving air quality can be identified. Where there are several measures of varying cost and ease of implementation, pollution control authorities need to know which measure, or package of measures, will be most cost-effective in improving air quality to the desired level.

Annual emission inventories, that is a database of the emissions of primary pollutants from all known sources for a defined geographical area and year, are routinely compiled at national level, and increasingly at the urban level. National emission inventories are used both within a country for prioritizing pollution control measures and internationally in the negotiations for pollution control treaties, such as the protocols under the United Nations Economic Commission for Europe's Convention on Long Range Transboundary Air Pollution (CLRTAP). Urban inventories are an essential input for air quality modelling studies designed to predict the concentration of various air pollutants under different emission reduction scenarios. However, for this purpose the annual inventories need to be converted into hourly estimates of emissions and require a fine degree of spatial detail.

These inventories provide the information necessary to identify the main sources of primary pollutants and the precursors of secondary pollutants. Typically they contain information on the amount of pollution released from major industrial plants, road transport, and other sources, and average figures for the emissions from smaller sources throughout the area. Their accuracy depends on the data used to compile them. Emissions of air pollutants are estimated from knowledge of the processes that form them. For some pollutants this is relatively straightforward, as emissions are dependent largely on the composition of the fuel. Information on fuel consumption and composition and knowledge of the presence and type of pollution abatement technology is sufficient to produce a good estimate of emissions from many sources. For other pollutants, emissions depend on combustion conditions, such as temperature and pressure, as well as on pollution abatement technology, and are thus more difficult to estimate. Emission inventories for such pollutants are therefore liable to be less accurate.

Emissions from power stations are known fairly accurately as there are a relatively small number of large sources. However, estimating emissions from a large number of small sources is much more difficult and these estimates are less accurate. For example, in the UK there are about 20 million households and 25 million vehicles. The latter pose a particular problem in the estimation of emissions because there are a large number of factors that influence their emissions, including how the vehicles are driven, traffic conditions, types of vehicle and emission control systems fitted.

Inventories can only give an indication of the relative contribution of different sources to ambient concentrations of air pollution. Large point sources tend to emit at a much greater height above the ground than area sources such as road traffic and domestic heating. These lower sources are likely to play a more important role, per tonne emitted, in determining local air quality than those from large point sources. Sources of pollution outside an inventory area may also contribute to pollution within the area.

This chapter discusses the major sources of the main primary air pollutants including the contribution of the different sources to total emissions of a range of pollutants in the European Union (EU), the USA and selected urban areas in the UK. This is followed by a description of how emission inventories are constructed using the UK's National Atmospheric Emissions Inventory as an example. Finally, as road transport is the single most important source of pollution in many urban areas in Europe, North America and elsewhere, the factors influencing their emissions are discussed.

SOURCES OF AIR POLLUTION

Introduction

This section briefly describes the main sources of the classical primary air pollutants – SO_2, NO_x (oxides of nitrogen: generally NO_2 + NO), CO, VOCs and PM. In the section on VOCs the sources of two compounds – benzene and 1,3-butadiene – are included. The main sources of ammonia are also described because of its important role in the formation of secondary PM. This is followed by a brief description of the main sources of selected persistent organic compounds and metals. Other air pollutants including methane (CH_4), nitrous oxide (N_2O), carbon dioxide (CO_2), chlorofluorocarbons (CFCs), hydrochlorofluorocarbons (HCFCs), hydrofluorocarbons (HFCs) and per-fluorocarbons (PFCs) are not included in this chapter because they are not directly toxic to human health, being principally greenhouse gases. The halocarbons, mainly CFCs, are also responsible for the depletion of stratospheric ozone.

The two main sources of the classical air pollutants are power stations and road transport. The contribution of different sources, however, varies between countries and within countries. In the EU in 1994, combustion in energy and transformation industries were responsible for 61% of total SO_2, and 19% of NO_x emissions. Road transport contributed 49% of NO_x and 4% of SO_2 (European Environment Agency, 1997a). In the UK the contributions from power stations and road transport were very similar to the EU averages (Department of the Environment, 1996). However, in several Eastern European countries road transport contributes only about 10–20% of the total NO_x emissions (European Environment Agency, 1996), reflecting the lower traffic levels in these countries.

Comparisons between the EU and the USA are not always straightforward as different source categories and methodologies for estimating emissions are used. However, it appears that power stations are a more important source of SO_2 and NO_x in the USA than in the EU. In 1994 70% and 33% of the total SO_2 and NO_x emissions came from this source, whilst 1% and 32%, respectively, came from road transport (US Environmental Protection Agency, 1995a).

Typically urban areas have a higher contribution from road traffic and a lower contribution from power stations than nationally. This is because most power stations and other large stationary sources are no longer located in cities in developed countries; instead they have been moved to rural locations. For example, in the British conurbations of Greater London, West Midlands and Greater Manchester, road transport was responsible for 63–85% of the total NO_x emissions in 1995 compared with 46% nationally (National Environmental Technology Centre, 1997).

For the classical pollutants, data are provided in this section on the contribution of different sources in USA, EU, UK and the British conurbations of Greater London, West Midlands and Greater Manchester. The main anthropogenic source categories used are those from the CORINAIR* 1994 inventory for Europe with waste treatment and disposal; agriculture, forestry and land use changes combined as 'other' sources. Data for the USA, UK and the British conurbations has been roughly fitted into these source categories, and are not strictly comparable. The data are for 1994 except for the British cities, for which the data come from 1995. Despite the different years and the source categories, the tables give a good indication of the relative contribution of the difference sources in different areas.

Sulfur Dioxide (SO₂)

The major source of sulfur dioxide is the combustion of fossil fuels containing sulfur, in particular from power stations burning coal and heavy fuel oil. Natural gas, petrol and diesel fuels have a relatively low sulfur content.

In the past the main source of sulfur dioxide was the burning of coal in homes, factories, offices, schools and other buildings. Today in most Western towns and cities buildings are heated by natural gas and electricity. The latter is generated in large power stations typically situated in rural areas rather than close to towns. There remain, however, a few cities in Western Europe where there are no supplies of natural gas, and coal and heating oil remain important sources of space and water heating.

Table 8.1 shows the relative contributions from these sources to total SO_2 emissions in the USA, EU, UK and selected British cities. Industrial combustion, especially for the generation of electricity, is the dominant source in all areas.

Although road transport is a minor source of sulfur dioxide at the national level, in some urban areas it can be important. Raised concentrations of sulfur dioxide have been detected alongside busy roads. However, as a result of controls on the permitted sulfur levels in automotive fuels introduced in the mid-1990s both in the USA and EU, it is likely that these elevated concentrations at the roadside will disappear. In the EU, further limits on the sulfur content of diesel and petrol have been agreed for the year 2000. These restrictions were not primarily introduced to reduce emissions of SO_2 from road transport, but instead to facilitate the reduction in emissions of other pollutants. Sulfur in diesel contributes to the exhaust PM, whilst in petrol it affects the performance of catalytic converters and other components of the pollution control system.

---

* This is the atmospheric emission inventory of the CORINE (Coordination d'Information Environmentale) programme, set up for the collection of information on the state of the environment and natural resources in the EU. Inventories for 1985, 1990 and 1994 have been completed. The 1995 inventory is currently being compiled by the European Environment Agency.

Table 8.1 Estimated contributions of the main sources to total SO_2 emissions

Source[a]	Contribution (%)					
	EU	USA	UK	Greater London	West Midlands	Greater Manchester
Combustion in energy and transformation industries	61	70	70	27	{61}	{95}
Non-industrial combustion	8	1	6	10		
Combustion in manufacturing industry	19	{23}	19	18		
Production processes	5		0	14	15	2
Extraction and distribution of fossil fuels	0		0	0	0	0
Solvent and other product use	0	0	0	0	0	0
Road transport	4	1	2	23	16	3
Other mobile sources and machinery	2	1	2	6	3	<1
Other	1	<1	0	2	5	0

[a] These are the main anthropogenic source categories used in the CORINAIR 1994 inventory for Europe (waste treatment and disposal; agriculture, forestry and land use changes have been combined as other). Data for USA, UK and the British conurbations has been fitted to these source categories, and are not strictly comparable. The data is for 1994 except for the British cities which is for 1995. However, the table gives a good indication of the relative contribution of the difference sources in different areas.

Sources: European Environment Agency (1996), US Environmental Protection Agency (1995a), National Environmental Technology Centre (1997), London Research Centre (1996, 1997, 1998).

Nitrogen Oxides (NO_x)

Nitrogen oxides are a key pollutant because their emission leads to elevated concentrations of nitrogen dioxide, and is a precursor of ozone formed in the troposphere. They also contribute to the atmospheric fine particulate matter burden as a result of oxidation to form nitrate aerosols.

Nitrogen oxides (NO_x) are formed in the atmosphere by lightning, forest fires and bacterial activity in soils. These sources are important on the global scale but in Europe and North America anthropogenic sources dominate.

NO_x is formed during high-temperature combustion, largely from the nitrogen and oxygen present in air, but also from the oxidation of nitrogen contained in fuels. The main sources are internal combustion engines, fossil fuel-fired power stations and industrial combustion. Table 8.2 shows that road transport contributes about one half of all emissions in the EU, but considerably more in British urban areas. In the USA industrial combustion is responsible for about half of all emissions.

Almost all NO_x is emitted as nitric oxide (NO) which is then rapidly oxidized to the more toxic nitrogen dioxide (NO_2).

Carbon Monoxide (CO)

Most anthropogenic carbon monoxide is generated in combustion processes. Internal combustion engines, both in on-road vehicles and in diverse off-road uses, comprise the

Table 8.2   Estimated contributions of the main sources to total NO_x emissions

Source[a]	Contribution (%)					
	EU	USA	UK	Greater London	West Midlands	Greater Manchester
Combustion in energy and transformation industries	19	33	24	2	⎱	⎱
Non-industrial combustion	4	2	5	11	10	33
Combustion in manufacturing industry	8	⎱	10	3	⎰	⎰
Production processes	2	16	0	1	1	1
Extraction and distribution of fossil fuels	1	⎰	4	0	0	0
Solvent and other product use	0		0	0	0	0
Road transport	49	32	49	75	85	63
Other mobile sources and machinery	15	13	7	7	2	3
Other	1	3	0	1	3	1

[a] These are the main anthropogenic source categories used in the CORINAIR 1994 inventory for Europe (waste treatment and disposal; agriculture, forestry and land use changes have been combined as other). Data for USA, UK and the British conurbations has been fitted to these source categories, and are not strictly comparable. The data is for 1994 except for the British cities which is for 1995. However, the table gives a good indication of the relative contribution of the difference sources in different areas.

Sources: European Environment Agency (1996), US Environmental Protection Agency (1995a), National Environmental Technology Centre (1997), London Research Centre (1996, 1997, 1998).

Table 8.3   Estimated contributions of the main sources to total CO emissions

Source[a]	Contribution (%)					
	EU	USA	UK	Greater London	West Midlands	Greater Manchester
Combustion in energy and transformation industries	1	<1	0	<1	⎱	⎱
Non-industrial combustion	12	4	7	1	1	4
Combustion in manufacturing industry	6	⎱	2	1	⎰	⎰
Production processes	5	5	0	<1	<1	<1
Extraction and distribution of fossil fuels	0	⎰	1	0	0	0
Solvent and other product use	0		0	0	0	0
Road transport	62	65	88	97	98	95
Other mobile sources and machinery	7	16	0	1	<1	1
Other	6	11	1	0	<1	0

[a] These are the main anthropogenic source categories used in the CORINAIR 1994 inventory for Europe (waste treatment and disposal; agriculture, forestry and land use changes have been combined as other). Data for USA, UK and the British conurbations has been fitted to these source categories, and are not strictly comparable. The data is for 1994 except for the British cities which is for 1995. However, the table gives a good indication of the relative contribution of the difference sources in different areas.

Sources: European Environment Agency (1996), US Environmental Protection Agency (1995a), National Environmental Technology Centre (1997), London Research Centre (1996, 1997, 1998).

principal sources. The majority of the carbon in automotive fuels is oxidized to carbon dioxide, while a small fraction is incompletely oxidized to carbon monoxide.

Table 8.3 shows that the contribution of road transport ranges from 62% in the EU and USA to nearly 100% in some urban areas.

Volatile Organic Compounds (VOCs)

VOCs comprise a wide range of chemical compounds including hydrocarbons (alkanes, alkenes, aromatics), oxygenates (alcohols, aldehydes, ketones, and ethers) and halogen-containing species. Methane is an important component of VOCs but its environmental impact derives mainly from its contribution to global warming, and it is increasingly being considered separately from the other VOCs. The major environmental impact of non-methane VOCs (NMVOCs) is their role in the formation of tropospheric ozone. The impact of individual VOCs varies considerably. All are precursors of ozone but their ability to form ozone varies markedly. In addition, a few VOCs are considered toxic to human health, such as benzene and 1,3-butadiene. The main sources of these two pollutants are described below.

The sources of anthropogenic NMVOCs are more diverse than those of many other pollutants, but predominantly come from solvent use, road transport and industrial processes. This is shown in Table 8.4. Important sources of VOCs from solvent use include surface coatings, glues and adhesives, printing and degreasing. These sources are often difficult to control as there are a large number of mainly small sources. In road transport, petrol vehicles without catalysts emit considerably more VOCs than comparable

Table 8.4 Estimated contributions of the main sources to total NMVOC emissions

Source[a]	Contribution (%)					
	EU	USA	UK	Greater London	West Midlands	Greater Manchester
Combustion in energy and transformation industries	1	<1	0	<1	0	2
Non-industrial combustion	4	3	2	2	<1	
Combustion in manufacturing industry	0		0	<1	0	
Production processes	6	48	19	<1	<1	59
Extraction and distribution of fossil fuels	5		11	11	3	1
Solvent and other product use	25		31	24	49	14
Road transport	32	27	29	60	46	20
Other mobile sources and machinery	5	10	2	2	<1	<1
Other	20	13	5	1	1	4

[a] These are the main anthropogenic source categories used in the CORINAIR 1994 inventory for Europe (waste treatment and disposal; agriculture, forestry and land use changes have been combined as other). Data for USA, UK and the British conurbations has been fitted to these source categories, and are not strictly comparable. The data is for 1994 except for the British cities which is for 1995. However, the table gives a good indication of the relative contribution of the difference sources in different areas.

Sources: European Environment Agency (1996), US Environmental Protection Agency (1995a), National Environmental Technology Centre (1997), London Research Centre (1996, 1997, 1998).

diesel vehicles (Quality of Urban Air Review Group, 1993). Emissions from petrol cars with catalysts are similar to those from diesel cars. Important industrial sources for VOC emissions include the chemical industry, food and drink industry, and the storage, handling and processing of products (Warren Spring Laboratory, 1993).

Natural sources of NMVOCs can also be important, particularly in summer when ozone pollution episodes occur. Vegetation, primarily forests, emit large amounts of VOCs when temperatures are high. These biogenic emissions consist of a wide variety of species including isoprene, terpenes and oxygenated compounds. Isoprene is the most reactive and has been found to play a significant role in ozone formation in the USA.

Whilst the EU emission inventory includes natural sources, there is considerable uncertainty over the results obtained in predicting ozone formation. To undertake air quality modelling of the formation of ozone in Europe using the European Monitoring and Evaluation Program/Meteorological Synthesizing Centre-West (EMEP MSC-W) ozone model, the emissions of natural NMVOCs have been re-estimated using the 'Biogenic Emissions Inventory System' developed in the US (EMEP, 1994). This takes account of vegetation cover and the effects of temperature and solar radiation on emissions. Estimates on a yearly basis suggest that anthropogenic sources of VOC may be greater than those of isoprene for virtually all European countries. However, this is not always the case on a monthly basis. Isoprene emissions in the hotter countries, e.g. Spain, may exceed anthropogenic VOC emissions during the summer months. There remains considerable uncertainty over these estimates of isoprene emissions. It has been suggested that the uncertainties may be as high as a factor of 5–10 during particular ozone episodes. In addition, preliminary estimates suggest that emissions of the so-called 'unidentified' biogenic VOCs may be of the same order as those of isoprene.

Benzene

The main sources of benzene in the atmosphere are the production, distribution and use of automotive fuels. In the UK in 1994 road transport contributed 72% of all benzene emissions. In the USA, a greater number of sources contribute to benzene emissions. In 1990, road transport has been estimated to contribute 43%, with off-road vehicles, oil and gas production, wood burning, forest fires and prescribed burning, e.g. of waste, contributing another 50% (US Environmental Protection Agency, 1996a). In the British conurbations of Greater London, West Midlands and Greater Manchester, road transport was responsible for 82, 99.5 and 93% respectively in 1995 (London Research Centre, 1996, 1997, 1998).

Petrol vehicles emit more benzene in exhaust than diesel vehicles, even when catalytic converters are used to control emissions from petrol cars. Benzene is present in petrol and can evaporate into the air from fuel storage containers and during refuelling. Currently the maximum benzene content of petrol permitted in the EU is 5% by volume, although there is considerable variation across Europe. In the UK it is typically around 2%. In the US the maximum benzene level in reformulated gasoline is 1%. Levels in standard gasoline are not permitted to be higher than those in 1990 when the average level was 1.6%. Benzene is also formed in the combustion process from other aromatics. The aromatic content is not explicitly controlled in either USA or EU automotive fuel specifications, although the EU Fuel Quality Directive will introduce a maximum total aromatic content of between 42% from 2000, and 35% from 2005. It will also introduce a 19% maximum for benzene.

1,3-Butadiene

There are no known natural sources of 1,3-butadiene. It is used as an industrial chemical particularly in the manufacture of synthetic rubber for tyres, and some liquefied petroleum gas (LPG) contains up to 8% by volume (Expert Panel on Air Quality Standards, 1994). The main source of 1,3-butadiene in the atmosphere is the combustion of petrol and diesel fuel, but some also comes from the burning of other fossil fuels. In the British conurbations it has been estimated that over 95% comes from road transport (London Research Centre, 1996, 1997, 1998). In the USA in 1990, mobile sources (on and off road) were estimated to contribute over 75% of total 1,3-butadiene emissions, the remainder largely coming from forest fires and prescribed burning (US Environmental Protection Agency, 1996a).

Petrol and diesel contains little 1,3-butadiene, but it is formed during combustion in the engine. The exhaust from a petrol car (fitted with a three-way catalyst) contains less 1,3-butadiene than that from a diesel car. This is because catalytic converters are efficient at removing 1,3-butadiene (Quality of Urban Air Review Group, 1993).

In some areas forest fires can be a significant source of 1,3-butadiene in the atmosphere (Pechin and Associates, 1994).

Particulate Matter

Particulate matter is not a single pollutant, but is made up of particles of many different sizes and chemical composition, from a wide range of natural and anthropogenic sources.

The concentration of airborne particles may be measured in several different ways, which is reflected in the way that emissions of particles are estimated. In the past estimates have tended to be of total suspended particles (TSP) and/or black smoke (BS)*, but increasingly estimates of PM_{10} (particles generally less than 10 μm aerodynamic diameter) emissions are being made. In the UK, for example, annual estimates of emissions of both black smoke and PM_{10} are made, while in the USA, annual emissions of both TSP and PM_{10} are estimated. For many sources, however, quantifying emission rates is extremely difficult. Examples include wind-blown soil, street dust re-suspended by traffic, and sea spray. It is also difficult to define the original source of secondary particles. These are mainly ammonium sulfate and nitrate, arising from the oxidation of sulfur and nitrogen oxides. The acids formed are neutralized in the atmosphere by ammonia. Certain organic compounds can also be precursors to secondary PM.

Table 8.5 shows the contribution of the difference sources to total PM_{10} emissions in different areas. The US inventory for PM_{10} shows that the vast majority of emissions do not come from combustion processes, but come from 'other' sources. These are mainly fugitive dust including those formed from paved and unpaved roads, and agriculture and forestry. The European inventories suggest that the dominant source of PM_{10} is fossil fuel combustion in industry and road transport. This discrepancy arises because

---

* Black smoke emissions are calculated by multiplying the particle mass by a factor representing the relative blackness of the particles (from 1.0 for coal to 3.0 for diesel particles).

Table 8.5   Estimated contributions of the main sources to total PM_{10} emissions

Source[a]	Contribution (%)					
	EU	USA	UK	Greater London	West Midlands	Greater Manchester
Power generation	29	1	15	4	0	⎫
Industrial combustion	14	⎱2⎰	18	2	27	⎬ 45
Small combustion sources	13		13	2	9	⎭
Production processes	14		26	4	<1	23
Road transport	17	1	25	77	56	31
Other mobile sources	1	1	3	5	1	<1
Other	11	95	1	6	7	<1

[a] These are the main anthropogenic source categories used in the TNO PM inventory for Europe (25 European countries, excludes the former Soviet Union). Data for USA, UK and the British conurbations has been fitted to these source categories, and are not strictly comparable. The data is for 1993 for Europe, 1994 for USA, and 1995 for the UK and British cities. The table gives a good indication of the relative contribution of the difference sources in different areas.

Sources: TNO (1997), US Environmental Protection Agency (1995a), National Environmental Technology Centre (1997), London Research Centre (1996, 1997, 1998).

the US inventory includes a number of sources that are difficult to quantify, such as natural wind erosion, which are excluded or treated differently in the European inventories.

Fugitive dust emissions and wind erosion together contributed about 75% of the total US 1994 emissions. None of the European inventories includes natural sources, such as sea spray and wind blown dust and re-suspension of dust from roads because of the great uncertainty of the emission rates for these sources. The data used to derive emissions estimates for these sources come largely from the USA where the climate is often dryer and natural sources are more important. Direct extrapolation of USA emission estimates to Europe is difficult. Even in the USA, the emission estimates for wind-blown dust and re-suspended road dust have not been regarded as reliable and the methodology for dealing with these sources is currently being revised.

Estimates of emissions of particulate matter due to wind erosion are very sensitive to regional soil conditions and weather. For example, the US EPA have estimated that in 1988 emissions from wind erosion were 16 million tonnes, while in 1994 it was about 2 million tonnes. The lack of precipitation prior to the planting of spring crops, especially in central and western USA, contributed to greater wind erosion (US Environmental Protection Agency, 1995a)

The differences in these PM_{10} inventories illustrate the need to use consistent methodologies if comparisons between different areas are to be made.

Airborne PM_{10} is often divided into fine and coarse PM_{10}. Fine particles are often regarded to be the $PM_{2.5}$ fraction (i.e. those particles with an aerodynamic diameter generally less than 2.5 μm) and the coarse particles to be $PM_{10-2.5}$. The major sources of the fine particles are fossil fuel combustion, vegetation burning, and the smelting or other processing of metals. Secondary PM are also fine particles. The coarse particles typically contain aluminosilicate and other oxides of crustal elements, and is mainly fugitive dust from roads, industry, agriculture, construction and demolition, and fly ash from fossil fuel combustion.

Table 8.6 shows the contribution of different source to the total emissions of PM of three different sizes (PM_{10}, $PM_{2.5}$ and $PM_{0.1}$) in Europe for 1993 (TNO, 1997). The total mass of $PM_{0.1}$ is estimated to be an order of magnitude less than that of PM_{10} and the contribution from road transport and industrial processes is much higher. The contribution from stationary combustion sources is much lower. Road transport is the single most important source of the very fine primary particles.

Table 8.6 Estimated contributions of the main sources of PM_{10}, $PM_{2.5}$, and $PM_{0.1}$ emissions in Europe (excluding the former Soviet Union) in 1993

Source	Contribution (%)		
	PM_{10}	$PM_{2.5}$	$PM_{0.1}$
Power generation	29	32	17
Industrial combustion	14	11	5
Small combustion sources	13	11	7
Production processes	14	15	24
Road transport	17	20	41
Other mobile sources	1	1	3
Agriculture	9	8	0
Waste processing plants	2	3	4
Total (kt)	4 800	2 700	390

Source: TNO (1997).

Emission inventories can provide reasonably good estimates of particle emissions from fuel combustion and industry, but a different approach is needed to quantify the role of secondary PM and fugitive sources such as re-suspended road dust, wind-blown dust and sea spray. For this, receptor modelling techniques have been used. These are based on investigating the chemical composition of the PM, and the relationships between the different PM components and gaseous pollutants, using meteorological data to infer sources.

In the UK, various approaches to source apportionment have led to the view that there are three principal sources of PM_{10} in British cities: road traffic, secondary PM and re-suspended surface dusts and soils. Construction and demolition activities can be important sources locally (Deacon *et al.*, 1997). In the winter, episodes of elevated PM_{10} concentrations are associated mainly with vehicle exhaust emissions, while in the summer there is a greater contribution from secondary PM (Quality of Urban Air Review Group, 1996). At a site in Birmingham where both PM_{10} and $PM_{2.5}$ are measured, $PM_{2.5}$ contributes about 80% of the PM_{10} in the winter months. In the summer there is a far larger contribution from larger particles, with $PM_{10-2.5}$ contributing on average almost half the PM_{10}.

Most of the PM_{10} from road transport comes from diesel vehicles. Cars running on leaded petrol can also be an important source, particularly in Europe where a significant number of cars still use this fuel.

Similar source apportionment techniques used in the USA suggest that in western states fugitive dust, motor vehicles and wood smoke are the major sources of airborne PM, while in the eastern states stationary combustion and fugitive dust are the main sources (US Environmental Protection Agency, 1995c).

Attempts have been made in the USA to produce an emission inventory including both primary emissions and secondary formation of particles. Pechin and Associates (1994) have estimated that secondary $PM_{2.5}$ exceeds primary $PM_{2.5}$ emissions by a factor of four.

Ammonia

Little is known about the sources of ammonia and their emission rates compared with our knowledge of some of the other gaseous pollutants. The main source is believed to be agriculture, mainly livestock wastes, with a small contribution from fertilizer application. Industrial processes are also thought to be a significant source. Other sources such as uncultivated soils, human respiration and traffic are generally considered to be small. It has been suggested that low-temperature combustion, for example domestic bonfires and decaying vegetation, may make a significant contribution.

The spatial distribution of ammonia emissions is quite different from those of SO_2 and NO_x, reflecting the distribution of livestock, particularly cattle. Ammonia emissions have increased in recent decades with more intensive husbandry.

In the EU in 1994 it is estimated that 95% of the ammonia emissions were from agriculture, with road transport, waste treatment and production processes contributing to the other 5% (European Environment Agency, 1997a). In the UK, non-agricultural sources are estimated to contribute about 12.5 per cent of the total emissions. Within the agriculture sector cattle are estimated to contribute 50% of the total emission, pigs and poultry 10% each, and sheep 5%. The remainder comes from the use of fertilizers. The total emission in the UK in 1995 was estimated to be 320 kt. This estimate is very uncertain and research to improve it suggests that the figure may need to be revised downwards (National Environmental Technology Centre, 1997).

Persistent Organic Compounds

Polycyclic Aromatic Hydrocarbons (PAH)

Polycyclic aromatic hydrocarbons are a group of chemical compounds with two or more fused aromatic rings. Their major source is the incomplete combustion of organic material. They are emitted in both the vapour phase and absorbed onto particles. There are few data available on the emissions from different sources and, according to the UK National Atmospheric Emission Inventory, estimates are only accurate to within an order of magnitude (National Environmental Technology Centre, 1997).

PAHs and their derivatives come mainly from stationary sources, particularly wood burning (residential heating and wild fires), open tyre burning, coal combustion and primary aluminium production. In the UK in 1994 nearly half was estimated to come from domestic combustion of coal and wood, a quarter from non-combustion industrial processes and a quarter from road transport. In a US Environmental Protection Agency (1996b) estimate of emissions of seven PAHs the major sources identified were residential wood burning, forest fires and prescribed burning, together contributing 75% of the total emissions. In an estimate of emissions of 16 PAHs residential wood burning and gasoline distribution contributed 75% of the total emissions.

Dioxins

The main source of dioxins (polychlorinated dibenzo-*p*-dioxins, PCDD) and furans (polychlorinated dibenzofurans, PCDF) is the combustion of chlorine-containing compounds. In particular, the incineration of municipal solid waste at insufficient temperatures is thought to be the major source of these compounds. Other potential sources include vehicles, domestic and industrial coal combustion.

In the UK in 1993 the largest source of dioxins and furans was waste incineration, which accounted for 75% of the total emissions (National Environmental Technology Centre, 1997). A 1990 US inventory of dioxins (adjusted to the toxic equivalent of 2,3,7,8-TCDD) showed that waste incineration (municipal, medical and other biological incineration) contributed a similar percentage of total emissions (US Environmental Protection Agency, 1996b).

Polycholorinated biphenyls (PCBs)

PCBs were made industrially from 1929 until the late 1970s for use in electrical transformers and capacitors. They were also used in carbon-less papers and inks. Although they are no longer manufactured, PCBs continue to be emitted into the atmosphere from PCB-filled transformers and capacitors still in use, the scrapping of redundant PCB-contaminated electrical goods, and during the recycling of contaminated scrap metal.

Metals

Lead

There are several natural sources of lead in the atmosphere including weathering of rocks, volcanic activity and the uptake and subsequent release by plants. Other natural sources include sea spray, wind-blown dust and forest fires.

The major source of lead in the urban atmosphere is road traffic. Since the 1920s lead compounds have been added to petrol to improve its performance. This source is declining as a result of the increased use of unleaded petrol (which is not completely lead-free) and a reduction in the maximum permitted level in leaded petrol. Although emissions have declined by 80% since 1970, road transport is still responsible for about 75% of all emissions in the UK (National Environment Technology Centre, 1997). This will decline as the new EU Fuel Quality Directive essentially bans its use in petrol from the year 2000. Lead is also released into the atmosphere during the mining and smelting of ores, the production, use, recycling and disposal of lead-containing products and the burning of fossil fuels and wood.

Other metals

There are trace amounts of many metals in the atmosphere from natural processes such as the weathering of rocks, wind erosion and suspension of soils, and volcanic activity.

There are three main anthropogenic sources of metals in the atmosphere: burning of fossil fuels, waste incineration and the metal production and processing industries. The relative importance of each varies with the different metals emitted.

The burning of fossil fuels is an anthropogenic source of antimony, arsenic, beryllium, cadmium, cobalt, chromium, copper, manganese, mercury, molybdenum, nickel, vanadium and zinc. Waste incineration is important for cadmium and mercury, while the smelting of metals is an important source for arsenic, cadmium, copper and zinc. Chromium and manganese come from processes involved in the production of iron and steel.

In urban areas, close to busy roads, vehicle tyre and component wear can be a significant source of a range of metals including platinum released from catalysts.

ESTIMATING EMISSIONS

Introduction

It is not possible to measure emissions from all sources, or even from all the different source types. In practice, estimates of atmospheric emissions are based on the results of measurements made at selected or representative samples of the main source categories. The methodologies used to estimate emissions are constantly being refined and improved.

In an inventory area there will often be millions of individual sources of pollution. Emissions are calculated, generally from a measure of activity of sources related to the emissions. These are typically fuel consumption for stationary sources and distance travelled for mobile sources. Emission factors, derived from measurements of individual or representative sources, are then multiplied by the appropriate statistic, to give the emission rate:

$$\text{Emission rate} = \text{Chosen statistic} \times \text{Emission factor}$$

For many pollutants the major source of emission is the combustion of fossil fuels. Consequently fuel consumption statistics are widely used in the estimation of emissions. When using fuel consumption statistics it is important that consumption data rather than delivery data are used, as fuels are often stockpiled or used outside the inventory area. The latter is particularly important when considering aviation emissions. Most emission inventories only consider aircraft activity during landing and take-off cycles and exclude cruising emissions. The rate of emission is greatest during the landing and take-off cycles and thus of most relevance to air quality.

Emissions from sources such as industrial processes, gas pipelines, and solvent and petrol evaporation require different types of activity statistics. These might be process output, or sales of a product. For some industrial processes emission estimates are based on an emission measurement over a period of time and the number of such periods that occur in the required estimation period.

In practice, the calculation of emissions tends to be more complicated than outlined above but the principles remain the same. Of all the sectors, estimating road transport emissions is the most complex as emissions come from such a large number of different sources, each with different emission characteristics depending on vehicle and fuel technology, and on how the vehicle is maintained and driven. The approach used is described later in this chapter.

Emission inventories may contain data for three types of sources:

- line sources including roads and railways;
- area sources including space and water heating for buildings;
- point sources, typically large power stations or industrial plant.

In some inventories, emissions from line sources are included separately along the line of the road, railway track, etc. In other inventories, line sources are included in area sources. Area sources are generally defined as those sources that individually emit relatively little but collectively result in significant emissions. In an emissions inventory these small or diffuse sources are averaged over a defined area, which may be a local administrative area (e.g. District Council area in the UK) or a regular grid (for example the EMEP 50 × 50 km grid).

For point sources emission estimates are provided for each individual plant or emission source within a plant, usually in conjunction with data on location, capacity or throughput, operating conditions, etc. The tendency is for more sources to be included as point sources as legislative requirements extend to more source types and pollutants and openness provides more such relevant data. Many inventories have suffered from authorities refusing to divulge confidential industrial information.

The detailed methodology used in constructing inventories depends on the availability of suitable information. In some countries data are readily available; in others, there are few reliable data available and many assumptions have to be made. Often, emission factors developed in other countries are used; this may or may not be appropriate.

As more information becomes available inventories have become more reliable. Figure 8.1 shows how estimates of NO_x emissions for the UK for the year 1980 have changed as a result of refinements to the inventory methodology. The x-axis gives the year in which revisions to the methodology where introduced, and the y-axis shows the resulting change in estimated emissions.

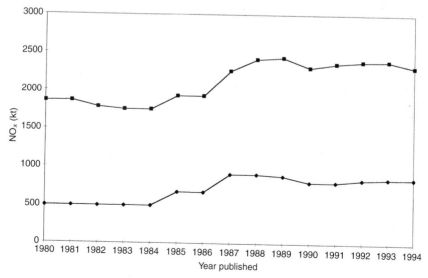

Fig. 8.1 NO_x estimates for 1980. ◆, road transport; ■, total.
Source: National Environmental Technology Centre (personal communication, 1997).

UK's National Atmospheric Emission Inventory (NAEI)

The UK's National Atmospheric Emission Inventory (NAEI) is briefly described below
to illustrate some of the factors taken into account when constructing emission inventories. Details of the NAEI inventory is given in *UK Emissions of Air Pollutants
1970–1995* (National Environmental Technology Centre, 1997). It currently covers
more than 30 pollutants including all the classical air pollutants, PAHs, PCBs, and a
selection of halocarbons, pesticides and heavy metals. Where possible, estimates are
made for the period 1970 to 1995, but for some pollutants there is insufficient information to produce an annual time-series. In this summary of the NAEI methodology
only the main classical pollutants are discussed. The report gives details of the emission
factors used. A database of emission factors for the main industrial and transport
sources used in compiling emission inventories for British cities is also available from
the London Research Centre's web page (http://www.london-research.gov.uk).
Information for compiling other inventories is available from the US Environmental
Protection Agency (1995b), IPCC/UNEP/ OECD/IEA (1995) and EMEP/CORI-
NAIR (1996).

In the NAEI, emissions for most sectors/pollutants are estimated from fuel consumption figures for each fuel and economic sector. For non-combustion sources estimates of
emissions are based on knowledge of individual sources, obtained largely from UK
research. The NAEI is reported in the EMEP/CORINAIR format, which is less detailed
than the original NAEI categories used to calculate the emissions.

The inventory is divided into a number of modules for different types of source. In this
chapter the description of the NAEI focuses on those sources that contribute significantly
to emissions of the classical pollutants, and ignores those sources which are major sources
of the greenhouse gases such as methane (e.g. gas leakage and landfill).

Stationary combustion

In NAEI the base combustion module is used to estimate emissions from the majority of
stationary sources including:

- domestic
- public service
- refineries
- iron and steel
- other industry
- gas production
- agriculture.

The base combustion module uses the following equation to calculate emissions:

$$E(p, s, f) = A(s, f) \times e(p, s, f)$$

where $E(p, s, f)$ is the emission of pollutant p from source s from the combustion of fuel
f (kg); $A(s, f)$ is the combustion of fuel f in source s (kg or therms); and $e(p, s, f)$ is the
emission factor of pollutant p from source s from combustion of fuel f.

The emissions factors are expressed in terms of kg pollutant/t for solid and liquid fuels
and g/therm for gases. For gases, NAEI factors are based on the gross calorific value of

the fuel, not the net calorific value as in some other inventories (e.g. CORINAIR and IPCC).

For power stations the NO_x emissions from coal and oil combustion are based on estimates from individual power stations, and are not based on estimates from fuel consumption. Emission estimates for coal-fired plant include those from the fuel oil used to light up the boilers. For PM_{10}, the emission estimates of the operators are used where available. These take account of the operation of electrostatic precipitators. Where these data are incomplete, emission rates for the appropriate fuel are used. Included in the power station category are emissions from generators using landfill and sewage gas.

The emission from the production of fuels requires careful allocation between the process that produces the fuel and the final burning of the fuel. In the NAEI the carbon content of the coal, or coke, consumed in the conversion processes, and the carbon content of the product (e.g. coke, blast furnace gas, town gas) is calculated to give an estimate of the carbon emissions during the process. Emissions of other pollutants are estimated from a factor reflecting the ratio of the total carbon input to the process to the amount of carbon emitted.

Industrial Processes

For industrial processes emission factors are typically based on the mass of emission per mass of product. For example, estimates of emissions of NO_x from the manufacture of nitric acid use an emission factor of 3.98 t per kt of 100% acid produced.

There are a very large number of processes which emit NMVOCs. Even within a single industry the variation in the processes used and the extent and type of abatement techniques used make it difficult to apply a single emission factor. Often emissions from sources are small individually, but important collectively. In addition, many different compounds are released into the atmosphere, and measurement techniques may not respond equally well to all compounds.

In the NAIE NMVOC emissions are estimated for the following industries:

- oil refineries
- iron and steel
- road construction
- chemical industry
- bread baking
- alcoholic beverages
- other food

The NAEI estimates of NMVOC emissions have poor precision because data for some individual processes and solvent use are incomplete. The UK total NMVOC emissions are currently estimated to be precise to ±50%.

Offshore Oil and Gas

The emission factors used in calculating emissions from offshore oil and gas exploitation are based on emissions per platform or per terminal. These are based on work commissioned by the UK Offshore Operators Association.

Construction

The NAIE estimates the order of magnitude of annual emission of PM_{10} from construction sites. In 1995 this was 0.5–8.3 kt. Difficulties in estimating these emissions do not allow a greater degree of precision. The emissions depend on the size of site, the amount of construction and traffic activity on the site, and on the weather conditions. Material carried outside the site and deposited on nearby roads may be entrained either by passing traffic or the wind and may become a significant local source. The NAEI uses data on the number of houses and the amount of commercial and industrial floor-space constructed each year, as well as construction time, to make an estimate of the total area under construction.

An emission factor for total particulate emissions of 2.69 t/ha has been suggested from an American study (US Environmental Protection Agency, 1995b). Assuming that only 20% of TSP is PM_{10}, this gives an emission factor of 0.538 t/(ha month). The number of days in the UK with rain is about 150 to 200 per year, suggesting that the emission factor should be no more than half that derived from US measurements, where there is a considerably drier climate. These factors together produce an estimate of the annual PM_{10} emissions in the range 0.5–8.3 kt.

Solvent use

NMVOC emissions for solvent use are estimated for the following sectors:

- decorative paint
- industrial coating
- film coating
- surface cleaning
- dry cleaning
- agrochemicals
- printing
- industrial adhesives
- aerosols
- non-aerosol consumer products
- seed oil extraction
- leather coating
- leather degreasing
- textile coating manufacture
- rubber goods manufacture
- wood impregnation
- other solvent use.

The emission estimates are largely based on data on the use of solvents supplied by the relevant industry associations and manufacturers. As noted above, there is great uncertainty about the NMVOC emission estimates because of missing data from solvent use and some industrial processes.

Road Transport

For road transport the emissions of CO_2, SO_2 and black smoke are estimated from petrol and diesel consumption. The sulfur dioxide emissions factors vary annually and are based on the sulfur content of the fuels. Estimates of black smoke emissions are derived using soiling factors for different fuels. For example, emissions from diesel combustion are considered three times more, and from petrol combustion 0.43 less dark per unit mass than coal smoke. Emissions of these pollutants can be broken down by vehicle type based on their estimated fuel consumption.

The emissions of NO_x, CO and NMVOC are calculated from emission factors based on performance rather than fuel consumption. Emissions from cold starts, hot engines and evaporative losses are each calculated separately.

Emissions from motor vehicles depend on the emission regulations in force when they were built, the type of vehicle and the fuel used. For estimating the hot emissions, that is those from a fully warmed up engine, an estimate of the number of vehicle kilometres driven by each vehicle type on each road type are needed. This is then multiplied by the appropriate emission factor to give the hot emission estimate. This requires the following information:

- Vehicle kilometres driven, split into four road types:
  - urban
  - rural single carriageway
  - rural dual carriageway
  - motorway

- Vehicle categories:
  - cars (split into petrol and diesel)
  - light goods vehicles (split into petrol and diesel)
  - heavy duty vehicles (spit into small and large according to number of axles)
  - buses and coaches
  - motorcycles

- Vehicle type split, that is the number in each vehicle class built to each regulation:
  - petrol cars   pre ECE
    ECE 15.00
    ECE 15.01
    ECE 15.02
    ECE 15.03
    ECE 15.04
    EC Stage 1 (current three-way catalyst)
  - diesel cars   Pre EC stage 1
    EC Stage 1
  - petrol LGV   Pre EC stage 1
    EC Stage 1
  - diesel LGV   Pre EC stage 1
    EC Stage 1
  - small HGV   Old
    Pre EC Stage 1

- large HGV
- buses & coaches
- motorcycles

                                    EC stage 1
                                    Old
                                    Pre EC Stage 1
                                    EC stage 1
                                    Old
                                    Pre EC Stage 1
                                    EC stage 1
                                    Current <50 cc
                                    Current > 50 cc 2-stroke
                                    Current > 50 cc 4-stroke

- Emission factors

The vehicle type split is produced from data on vehicle sales and the age distribution of vehicles on the road using a fleet model. This gives, for each year, the fraction of each type of vehicle in the fleet that were made to each regulation.

For petrol cars, account is taken of the decrease in annual mileage with increasing vehicle age.

The emission factors are based on measurements of emissions from vehicles, mainly using on-road measurements. The units are g/km driven for each type of vehicle. There are large variations from vehicle to vehicle (of the same order as the measurements themselves), and the emissions factors used in calculations are means of the measured emissions. For some groups of vehicles, e.g. buses and motorcycles, there are few data and the uncertainties are large.

The emission factors include a number of implicit assumptions. For example, the impact of load and road gradient on emissions from goods vehicles are not explicitly included. It is also assumed that the vehicles are not grossly abused.

The methodology for estimating road transport emissions is most advanced for cars. The results of emission measurements from a large number of cars have been used to produce speed–emission curves for different technology cars (see next section). Data on the vehicle–speed distribution for each of the four road types is combined with the appropriate point on the speed emission curve, to give an estimate of emissions from all cars driven on each type of road. The emissions from all vehicles and road types are then summed to give total emissions. The emissions from other forms of road transport are not so well known and single emission factors for each road type are used.

When an engine is cold its emissions are greater than when it has warmed up. Cold start emission factors are derived from the difference in emissions measured from cars with cold and warmed up engines. These tests are usually undertaken at a temperature of about 25°C and so do not take account of the effect of ambient temperature on cold start emissions, which is known to be significant. The number of cold starts is estimated from information on the average trip length and the total number of vehicle kilometres. This gives the number of trips, and it is assumed that two-thirds of these are from cold. The number of cold starts is estimated from this and multiplied by the extra cold start emission. Cold start emission factors are included for diesel cars and light goods vehicles but are unavailable for heavy goods vehicles and buses.

Evaporative emissions of NMVOC are based on an emission factor of 5.9 g/kg fuel. This factor is applied to all petrol used in the UK and is considered to be appropriate for the temperatures and fuel volatility found in the UK. This does not take account of seasonal changes in fuel volatility, ambient temperature nor of the use of abatement technology such as carbon canisters.

Estimating PM_{10} from road transport is done slightly differently than for the other pollutants. For petrol cars and light goods vehicles the emission factors depend on whether the vehicle is run on leaded or unleaded petrol and whether it is fitted with a three-way catalyst. The same value is used irrespective of road type. For diesel vehicles emission factors are based on measurement data as for the other pollutants.

Aircraft

Aircraft emissions are estimated for landing and take-off and flight up to a height of about 1000 m. This roughly represents the boundary layer and the emissions into it. The emissions are estimated from the number of aircraft movements at British airports and emission factors derived from studies at Heathrow and Gatwick airports.

Other Mobile Sources

Emissions from the other mobile sources: railways, coastal shipping and fishing are estimate using the base combustion module as described in the section above on stationary combustion.

Off-road sources were explicitly included in the NAEI in 1995 for the first time. Prior to that these emissions were included under other sectors. As emissions from other sources are declining these uncontrolled sources are beginning to make a significant contribution to total emissions of some pollutants. These include a range of mobile equipment powered by diesel or petrol fuelled engines including agricultural equipment such as tractors and combine harvesters, construction equipment such as bulldozers and excavators, lawn mowers, aircraft support equipment, and industrial machines such as portable generators and compressors. The 1995 estimates are based on a modification of the methodology given in EMEP/CORINAIR (1996).

Precision of Emission Estimates

The current state-of-the-art inventory methodologies still require many assumptions and extrapolation of data, and thus there remains considerable uncertainty over the estimated total emissions. For some pollutants the uncertainty is much greater than for others.

Table 8.7 shows the estimated precision of the UK NAIE. It is necessary to make assumptions about emissions due to the insufficient measurement of emissions from industrial plants and motor vehicles. For example, the results of short measuring programmes are often used to infer annual average emissions. Fuel consumption statistics are often inaccurate because of the common practice of reporting deliveries rather than consumption. In addition, there can be considerable uncertainty in the secondary data used. For example, vehicle kilometres are derived from sample surveys of road traffic which are themselves subject to uncertainties.

Table 8.7 Precision of the UK National Atmospheric Emission Inventory

Pollutant	Estimated precision (%)
Carbon dioxide	±5
Methane	±30–40
Sulfur dioxide	±10
Nitrogen oxides	±30
NMVOC	±50
Carbon monoxide	±40
Black smoke	±20–25

Source: National Environmental Technology Centre (1997).

FORECASTING EMISSIONS

It is often useful to be able to forecast how emissions may change in coming years, both for 'business as usual' scenarios and for various pollution abatement scenarios. In particular, this allows the impact of potential new abatement measures to be assessed and the likelihood of meeting pre-defined targets (such as targets in the Protocols to CLRTAP) to be determined.

Levels of economic activity, demographic influences and the use of abatement measures all affect the trends in anthropogenic emissions of air pollution. Up until the 1950s emissions tended to grow as the economy (and population) increased and to decline in periods of recession. Since then the relationship between economic growth and emissions has become more complex as many processes and products have become more energy efficient, and a wide range of abatement techniques have been utilized. In addition, consumer choice of products has also influenced emissions. For example, the change from coal to gas and electricity for space and water heating played an important role in reducing emissions of sulfur dioxide and smoke in London from the 1960s onwards. In countries such as France, with high usage of diesel cars*, the emissions of PM are higher than in countries with similar traffic levels, but where diesel cars are less popular.

The prediction of future emissions requires forecasts of how the relevant activity statistic and emission factors will change over time: official government forecasts of energy use and traffic levels are often used. For example, in the UK the national road transport forecasts are used to forecast future transport emissions. Some emission forecasting models incorporate modules which can define future demand for travel, goods, etc. based on factors such as gross domestic product (GDP), population and household size. An example of the latter is the FOREMOVE model developed by the Aristotle University of Thessaloniki for the European Commission for forecasting future road transport emissions.

Emission forecasting models must take account of both mandatory and voluntary changes in processes and product specifications as these may influence emissions. The impact of such measures can be difficult to estimate. For example, technology used to meet new emissions standards for cars which have been agreed by the legislators but not yet introduced into the marketplace may affect the cold start emissions, hot emissions, or both. The allocation of the emission benefits between these types of emission is often fairly

* About half of new cars in France run on diesel.

arbitrary. Typically, emissions estimates for future technology vehicles are reduced in proportion to the emission limits for current and future vehicles.

HARMONIZATION OF INVENTORY METHODOLOGIES

The construction of emission inventories requires detailed information about each source, including its level of activity, the composition of the fuel used, and the impact of any pollution abatement technology, as outlined above in the description of the NAEI. When there are literally millions of sources this is not a trivial exercise, and different countries have over a number of years developed their own estimation methodologies. The adoption of international agreements requiring the reporting of national emissions, such as CLRTAP and the UN Framework Convention on Climate Change, has led to the need to develop consistent methodologies to ensure that realistic comparisons of emissions from different countries can be made.

In Europe the European Environment Agency (EEA) oversees the collation of data for the CORINAIR inventory. This is the atmospheric emissions inventory of the CORINE programme, set up to gather, coordinate and ensure consistency of information on the state of the environment and natural resources in the European Union. CORINAIR currently covers 27 countries both within and outside the European Union and eight pollutants (SO_2, NO_x, NMVOCs, NH_3, CO, CH_4, N_2O and CO_2). The inventory includes 260 source categories at different levels of detail. The establishment of a common European methodology has been evolving over the past 15 years, beginning in 1983 with an Organization for Economic Cooperation and Development (OCED) project to produce an inventory for 17 member countries. In 1985 the European Commission's Environment Directive (DGXI) funded the compilation of an emission inventory for the 12 Member States for 1980 and 1983. The inventory covered SO_2, NO_x, VOC and particulate matter. The first CORINAIR inventory was for 1985 covering SO_2, NO_x and VOCs. Subsequently, emission inventories for 1990 and 1994 have been compiled with the intention of producing an annual inventory for Europe. The 1995 CORINAIR inventory is being compiled as this chapter is being written.

The Cooperative Programme for Monitoring and Evaluation of the Long-Range Transmission of Air Pollutants in Europe (EMEP), formed by a Protocol under CLRTAP, has, with the EEA, produced guidelines for estimation and reporting of emission data in 1996 (EMEP/CORINAIR). Reporting of emission data to the Executive Body of the Convention is required to fulfil obligations regarding strategies and policies in compliance with the implementation of Convention's Protocols. Parties are required to submit annual national emissions of SO_2, NO_x, NMVOC, CH_4, CO and NH_3 using the 11 main source categories agreed with CORINAIR by 31 December following each year. Parties are also required to provide EMEP periodically with emission data within a defined 50×50 km grid. This is used as input to integrated assessment models such as the Regional Air Pollution Information and Simulation (RAINS) model developed by the International Institute for Applied Systems Analysis (IIASA, Austria) which is used to predict the impact of emission reduction strategies on acidification, eutrophication (excessive deposition of nitrates) and tropospheric ozone concentrations across Europe.

The IPCC Guidelines for National Greenhouse Gas Inventories was published in

1995 (IPCC/UNEP/OECD/IEA, 1995). This lays down a common framework for reporting emissions. It recognizes that IPCC needs to accommodate other existing inventory programmes and gives information on how to convert from a CORINAIR inventory to an IPCC one. The IPCC Guidelines contain default methodologies and assumptions for the estimation of greenhouse gas emissions and removals. Users are encouraged to use these only when national data are not available. Methodologies for estimating all anthropogenic emissions of CO_2 and CH_4 are described in detail, and background information is provided on the estimation of emissions of N_2O and tropospheric ozone precursors, i.e. CO, NO_x and NMVOC.

FACTORS INFLUENCING ROAD TRANSPORT EMISSIONS

The major source of air pollution in most urban areas in developed countries, and increasingly in developing countries, is road transport. Due to the importance of this source in terms of urban air quality, factors which influence emissions from this source are briefly described in this section.

As discussed above, in urban areas the contribution to total emissions from road transport is generally greater than indicated by national emission data. This is because there is typically more traffic and less industry in urban areas. To develop effective control strategies for reducing urban emissions it is important to know which type of vehicles contribute most to the total road traffic emissions. Table 8.8 shows the relative contribution of seven categories of vehicle in London. Petrol cars are the single most important source of CO, NO_x and NMVOC, while lorries are an important source of SO_2 and PM_{10}. These vehicles are also an important source of NO_x, contributing about a quarter of all emissions in London.

Table 8.8   Percentage contribution of different types of vehicles to total emissions in London in 1995

Source	Contribution (%)				
	CO	NO_x	SO_2	PM_{10}	NMVOC
Cars (petrol)	88	42	4	11	53
Cars (diesel)	<1	2	2	5	<1
Vans (petrol)	4	3	<1	1	2
Vans (diesel)	<1	2	2	5	1
Lorries	1	23	12	50	2
Buses	3	4	2	6	<1
Motorcycles	1	<1	<1	<1	2
Total road transport	97	75	23	78	60

Source: London Research Centre (1998).

In London the majority of cars are petrol fuelled, whilst virtually all lorries are diesel fuelled. High emissions of CO and VOC are characteristic of petrol cars without three-way catalytic converters (these were not mandatory in the UK until January 1993, so the majority on the road in 1995 were non-catalyst cars). PM_{10} production is characteristic of diesel fuelled vehicles. The higher emissions of sulfur dioxide from lorries compared with cars is due to diesel containing more sulfur than petrol.

For many vehicles little is known about their on-road emissions under real-world conditions. For heavy duty vehicles much of the emission testing has been carried out using engines on bench dynamometers rather than in vehicles. This is because heavy duty engines can be used for a variety of applications, and the end use of an engine is often not known when it is manufactured. As a consequence, the legislation controlling emissions requires tests to be carried out using bench dynamometers. However, a number of emissions testing laboratories have recently acquired chassis dynamometers for use with large vehicles such as lorries and buses, and better data are becoming available. In addition there has been some on-road testing of emissions from large vehicles.

Technology

Probably the greatest influence on emissions from road transport is the technology used in the construction of the vehicle. Over the past 30 years emissions from road vehicles have been controlled by increasingly stringent legislation. Today, emissions of the main gaseous pollutants from petrol cars are less than 1% of the levels emitted before emissions legislation was introduced around 1970. Even now, with the use of controlled three-way catalysts on all new cars in the US and EU, further reductions in emissions are possible through improvements to the combustion process, and increasing the efficiency and durability of the catalyst. Most of the emissions during the regulatory driving cycles occur during the first few minutes when the catalyst is warming up. A number of techniques including close coupled catalysts (i.e. situated close to the engine) and electrically heated catalysts have been developed to reduce these emissions.

The improvements in emissions from diesel vehicles have been less dramatic. There is a trade-off between measures to improve combustion efficiency, which reduce PM emission, and those that reduce NO_x emissions. Three-way catalysts cannot be used on diesel vehicles because there is too much oxygen in the exhaust for the reduction of NO_x to nitrogen to occur. It has been suggested that PM emissions should be reduced using post-combustion after-treatment, such as particle traps, and that engine technology improvements be devoted to reducing the NO_x emissions. Although there have been many successful field trials with particle traps, they have not generally been widely adopted. They have been used mostly in countries where diesel oil with very low levels of sulfur is widely available (e.g. Sweden).

There are some significant differences in the emissions from the two types of diesel technology currently used. Indirect injection (IDI) diesel engines are used mainly for light duty vehicles, whilst direct injection technology (DI) is used mainly for heavy duty vehicles. The fuel consumption of DI engines is less than IDI, but the NO_x and PM emissions and noise are higher. Much effort is being devoted by motor manufacturers to reduce the emissions of small DI engines. Currently in Europe these vehicles are allowed to emit more than their IDI counterparts, but this derogation is due to be removed by the turn of the century.

The method of aspiration also has a significant impact on emissions. Turbo-charging reduces emissions of CO, NO_x and PM compared with naturally aspirated diesel engines.

Large reductions in NO_x emissions from diesel vehicles are likely to depend on the development of a DeNO$_x$ catalyst capable of efficiently removing NO_x in an oxygen-rich environment. DeNO$_x$ devices are being developed, and are currently produced by some Japanese car makers, but have yet to prove to be both durable and to have high emission removal efficiency for long periods in service in other countries. In Japan they have been fitted to vehicles with direct injection gasoline engines (a new form of lean-burn petrol engine). This technique has the advantage of improved fuel consumption, but as the combustion runs lean, three-way catalysts cannot be used to reduce emissions. Current DeNO$_x$ catalysts do not have a very high efficiency compared with three-way catalysts. Prototype DeNO$_x$ catalysts that are more efficient tend to be more sensitive to sulfur in the fuel. Japanese petrol has very low sulfur compared with petrol in the USA and EU, and with diesel. To obtain large reductions in NO$_x$, DeNO$_x$ catalyst technology requires very low sulfur fuel to be widely available.

Increasing fuel efficiency of cars is very important, especially in Europe, where legislators are putting increasing pressure on motor manufacturers to improve the fuel consumption of their products to reduce CO$_2$ emissions. This may encourage a move towards DI diesel engines for cars. Similar pressures are driving petrol technology in the same direction, such that in the next decade direct injection petrol cars may become relatively commonplace. At least one Japanese manufacturer has a direct injection petrol car on the European market.

Fuel

Two types of internal combustion engine are used in road vehicles: spark ignition (petrol) and compression ignition (diesel). Petrol engines are generally used in motorcycles, passenger cars and vans, while diesel engines are used in all types of vehicle except motorcycles. In the USA and some European countries, such as Sweden, there are few diesel cars in use. In other European countries, diesel cars have become very popular in recent years. In France, half of all new cars sold use diesel, and in the UK sales have increased from under 6% in 1990 to around 20% by the mid-1990s. It is not thought that the popularity of diesel cars in the UK will reach that in France and there is some evidence that the proportion of diesel powered new cars is declining.

Comparison between Diesel and Petrol

Diesel cars have lower emissions of the gaseous regulated pollutants (CO, NO$_x$ and VOC) than petrol cars without catalysts, and evaporative emissions are low. However, they have significantly higher emissions of PM. Emissions from petrol cars are dramatically reduced using three-way catalytic converters (TWC). In comparison with TWC cars, diesel cars emit more NO$_x$, but less CO and VOC. They also emit less CO$_2$.

PM emissions from petrol cars are not routinely measured, and are not controlled in either EU or US emissions legislation. The mass of particles emitted from diesel cars may be more than an order of magnitude higher than those from petrol cars with three-way catalysts. PM emissions from cars running on leaded petrol are greater than from those using unleaded petrol (without a catalyst), as the lead contributes to the PM burden of the exhaust.

A recent UK study investigating emissions of particles from vehicles of different technologies suggests that for light duty vehicles the different technologies can be ranked, based on PM emissions, as follows (ETSU, 1997):

old technology IDI diesel> DI with catalyst = old technology IDI with particulate trap
   >current IDI diesel>>non-catalyst petrol with leaded petrol>catalyst petrol

Many different VOC species are emitted from vehicle exhausts and each exhibits a different toxicity and photochemical ozone creation potential. Petrol cars tend to emit more of the lighter VOCs (C_1–C_6) while diesel cars emit more heavier VOCs (C_{13+}) (Quality of Urban Air Review Group, 1993). Benzene emissions are about three times higher from petrol cars with catalysts than from diesel, while toluene emissions are an order of magnitude higher. Aldehyde emissions are higher from diesel cars. The carcinogens 1,3-butadiene, and polycyclic aromatic hydrocarbons (PAH) are, however, present in higher concentrations in diesel exhaust than in that from a petrol car with a catalyst. A comparison of emissions from petrol and diesel cars, with and without catalysts, is given in Table 8.9.

Table 8.9   Comparison of emissions from petrol and diesel cars

Pollutant	Petrol without catalyst	Petrol with three-way catalyst	Diesel without catalyst	Diesel with oxidation catalyst
NO_x	****	*	**	**
VOC	****	**	***	*
CO	****	***	**	*
PM	**	*	****	***
Aldehydes	****	**	***	*
Benzene	****	***	**	*
1,3-Butadiene	****	**	***	*
PAH	***	*	****	**
SO_2	*	*	****	****
CO_2	***	***	*	**

Key: Asterisks indicate which type of car has typically the highest emissions. * Lowest emissions, **/*** intermediate *** highest emissions. This table only indicates the relative order of emissions between the different types of cars. No attempt has been made to quantify the emissions. The difference in emissions between, say **** and *** may be an order of magnitude, or much less.

Source: Quality of Urban Air Review Group, 1993

Reformulated Fuels

The relationships between fuel properties, engine performance and exhaust emissions are often complex. There may be trade-offs in that changes in the fuel composition to benefit one pollutant may have an adverse effect on others. It can also be difficult to isolate the effect on emissions of one fuel parameter from others. For example, reducing PAHs in diesel reduces emissions of these compounds. However, these compounds increase the fuel density and the effect may be due to this physical change.

   In recent years there have been two major collaborative programmes between the oil and motor industries to investigate these effects. The first, the US Auto Oil Air Quality Improvement Research Program (AQIRP), concentrated on petrol, while the European Programme on Emissions, Fuels and Engine Technologies (EPEFE) investigated both petrol and diesel properties.

   EPEFE (ACEA/EUROPIA) showed that the key parameters affecting emissions from diesel vehicles are sulfur, cetane number, density and the amount of polycyclic aromatic compounds in the fuel. There is general agreement that there is a linear relationship between fuel sulfur content and PM emissions from diesel engines. The size of the effect may be small for indirect injection diesel light duty vehicles. Increasing the cetane number reduces CO and VOC emissions, but increases PM emissions from light duty engines. For heavy duty vehicles increasing cetane number also reduced the NO_x emissions. Reducing the diesel density lowers NO_x emissions in heavy duty engines but increases the emissions in light duty vehicles. PM emissions from light duty vehicles are lowered. Reducing PAH in diesel reduces the NO_x and PM emissions from both types of vehicle.

One of the key pollutants emitted from diesel vehicles is NO_x. Since EPEFE, the European motor manufacturers association (ACEA and ESTU) have investigated the impact of total aromatic content and cetane index on NO_x emissions from diesel vehicles. ACEA believes that it has robust evidence to suggest that reducing total aromatics in diesel reduces NO_x emissions from both light and heavy duty vehicles (Signer, 1997; ACEA, 1997).

For petrol, the important parameters include aromatic content, Reid vapour pressure (RVP), distillation characteristics and oxygen content. The sulfur content of petrol is also important because it can act as a catalyst-poison, reducing catalyst efficiency, and interferes with the operation of oxygen sensors and on-board diagnostic (OBD) systems. Similar effects are likely to be observed with advanced light duty diesel vehicles fitted with oxidation catalysts and OBD systems. Other trace elements in gasoline, such as lead and manganese, exhibit similar effects.

Oxygenates (such as MTBE) in petrol reduce CO emissions and to a lesser extent HC, but may increase emissions of NO_x and aldehydes. RVP affects the amount of evaporative emissions. The effect of the fuel's distillation characteristics (E_{100} – the percentage that evaporates at 100°C) and aromatic content are interdependent. At a constant E_{100} reducing aromatics reduced VOC and CO emissions but increased NO_x emissions when measured over the urban and extra-urban driving cycles. This effect on NO_x is subject to some controversy as almost all other studies have shown a decrease in NO_x emissions with increasing aromatic content of petrol. Even in EPEFE, over the urban driving cycle, this effect was noted. The results have been explained by the motor manufacturers to be due to the use of three prototype vehicles poorly calibrated for the fuel.

The composition of the fuel can also affect the emissions of individual VOC compounds. Benzene emissions from petrol vehicles depend on the amount of benzene and aromatics in the fuel. About half the benzene emitted comes from benzene in the fuel; the remainder is formed during the combustion of other aromatics.

For cars without canisters (typically those without catalysts) there is an exponential increase in evaporative emissions with Reid vapour pressure and ambient temperature. For cars fitted with canisters evaporative emissions only occur when the canister becomes overloaded. The small canister currently fitted to petrol cars does not control refuelling losses, and increasingly 'stage 2' vapour control systems are being fitted to petrol pumps to reduce these emissions.

Fuel additives can also have an important impact on emissions. The classical example is emissions of lead compounds as a result of lead additives in petrol. Some fuels contain detergents to stop the build-up of sooty particles in engines.

Alternative Fuels

The role of alternative fuels to reduce emissions from traffic has received considerable attention in recent years. Many reports have been published comparing the relative merits of the different fuels (e.g. in the UK: Poulton, 1994; ETSU, 1996).

From all these studies a general consensus emerges. There may be a role for the gaseous fuels in the short term for some types of vehicle. Biofuels, hydrogen and electricity, however, are unlikely to be widely used in the next 10 years for a variety of reasons depending on the individual fuel. These include being economically uncompetitive, the unavailablity of technology suitable of meeting customer requirements in terms of performance, safety and durability, and uncertain security of fuel supply. Thus their use is likely to be restricted to small niche markets. The fuels considered most likely to make a major impact in the short term are compressed natural gas (CNG) and liquid petroleum gas (LPG).

There have been numerous studies measuring emissions from gas-fuelled vehicles and comparing the results with similar vehicles conventionally fuelled. It is difficult to make direct comparisons between different studies without investigating the many parameters that influence emissions from vehicles. In the past, gas-fuelled vehicles were often poorly converted from petrol vehicles, and emissions could be high compared with those from conventionally fuelled vehicles. However, the current state-of-the-art gas-fuelled vehicles can offer real emission benefits, especially when compared with diesel vehicles. Emissions of the two key diesel pollutants, PM and NO_x, are significantly lower.

Vehicle Maintenance

Poorly maintained vehicles consume more fuel and emit higher levels of CO and VOC than those that are regularly serviced. Factors such as running over-rich, poorly adjusted ignition timing and worn spark plugs can all affect emissions. The effect of correct tuning for cars without catalysts is particularly important, and a number of studies have shown that tuning, in general, results in a reduction in emissions of CO, VOC and PM. On the other hand, there is an increase in NO_x emissions.

For three-way catalyst cars the most frequent reasons for high emissions are first, the failure of the oxygen sensor, and second, failure of the catalyst itself. Often the driver is unaware that the pollution abatement system has failed.

Catalysts gradually degrade with use, making them less efficient. Old catalysts tend to take longer to reach the light-off temperature and deposits may build up on the surface of the catalyst. These effects can be accelerated by trace elements in petrol, such as lead and sulfur.

Diesel vehicles are more stable and require less servicing than petrol vehicles. However, it is important that vehicles are serviced according to the manufacturers' specifications: a recent survey of 63 in-use lorries in the Netherlands found that 70% were not (TNO, 1995). Emissions can increase as a result of deposits on the fuel injector or wear of cylinder or piston rings, leading to increased consumption of lubricants. Emissions of PM and VOC are likely to increase while those of NO_x are likely to remain constant or even decline as the efficiency of combustion falls. The emission of visible smoke indicates a poorly maintained vehicle.

There is some evidence that modern heavy duty diesel engines (built to Euro I emission standards or equivalent) are less inclined to deposits building up in the fuel injection system because the high operating pressure keeps them free of obstructions.

A number of steps have been taken to reduce emissions from malfunctioning vehicles. First, in many part of the USA and Europe inspection and maintenance schemes have been introduced as part of the periodical road-worthiness test. Second, in the USA on-board diagnostic systems are fitted to warn the driver of a malfunctioning pollution control system. Similar systems are due to be fitted to new EU petrol cars from the year 2000.

Operational Factors

The way a vehicle is driven has a large impact on its emissions. This is influenced by the type of road the vehicle is being driven on, the level of congestion, traffic management, and driver attitude and behaviour. In general, when an engine is operating under load, there is efficient combustion and emissions of VOCs are low and NO_x high, and when the vehicle is idling emissions of CO are high and NO_x low.

In diesel engines high emissions of VOC occur when an engine is started from cold, when idling or under part load, and may appear as 'white smoke' from the exhaust. At low loads emissions of NO_x are low.

Several processes contribute to the formation of particles in the engine. When the engine is operating at high loads, the carbon fraction is predominantly produced, while at low loads the VOC fraction, from both the fuel and lubricant, becomes more important.

Speed

Emissions from vehicles tend to be lowest when they are driven at moderate speeds and highest when driven at low or high average speeds, giving a characteristic U-shaped speed-emission curve. A speed-emission curve for NO_x is shown in Fig. 8.2 for a medium sized European petrol car with a three-way catalyst. It should be noted that the data are derived from average and not instantaneous speeds. Thus at the lowest speed this includes the stopping and starting characteristic of urban traffic.

Larger vehicles tend to emit more pollutants per kilometre driven than smaller vehicles. Thus generally cars emit less than vans which emit less than lorries. There can, however, be exceptions to this general rule. For example, emissions of NO_x from a small car driven at high speed are likely to be greater than those from a large car driving at the same road speed. This is because the large car will only be driving at part load, while the small car will be driving closer to full load.

Traffic Congestion

Congestion has a significant impact on petrol vehicle emissions. Measurements by the former Warren Spring Laboratory (Quality of Urban Air Review Group, 1993) suggest that where there is congestion this factor is more important than the cold start emission penalty. However, cold starts are more widespread – all cars are affected – whereas congestion is more localized and occurs mainly during the peak traffic hours. Emissions of

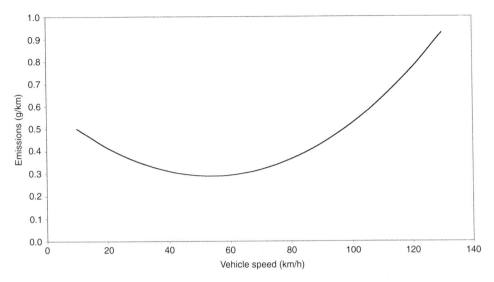

Fig. 8.2 NO$_x$ emissions from medium-sized petrol cars (1.4–2.0 litres) fitted with a three-way catalyst. Source: European Environment Agency (1997b).

CO, VOCs and NO$_x$ from petrol cars with catalysts increase with congestion.

Diesel cars appear to be less affected by congestion. CO and VOC emissions are lower under congested driving conditions. NO$_x$ emissions, in contrast, are considerably higher, similar to those of a petrol car without a catalyst, and much higher than a petrol car with a catalyst.

Aggressive Driving

A Dutch study has shown that when drivers are asked to drive 'aggressively' emissions increase significantly compared with those when they are asked to drive 'normally' (Heaton *et al.*, 1992). The average emissions of CO increased by a factor of about 3.5 and the NO$_x$ by a factor of about 2 from petrol cars with catalysts.

Vehicle Load

Emissions from goods vehicles depends on whether they are laden or not. On-road measurements in the UK suggest that emissions generally increase by a factor of between 1 and 2 when fully laden compared when empty.

Road Gradient

Road gradient affects the resistance of a vehicle to traction, which in turn affects vehicle emissions. In principle, the emissions of both light and heavy duty vehicles are affected, but the impact is small for light duty vehicles. However, because of their greater mass the influence of road gradient on emissions is much more significant for heavy duty vehicles (European Environment Agency, 1997b).

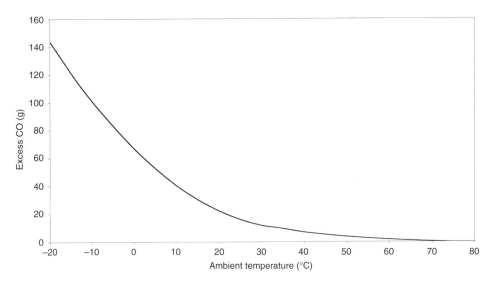

Fig. 8.3   Cold start CO emission penalty with ambient temperature.
Source: Personal communication from John Hickman of the Transport Research Laboratory (1997).

Cold Starts

Emissions, particularly of CO and VOCs, are greatest when the engine is cold. It has been estimated that in the UK in 1990 (when there were very few petrol cars with catalysts) approximately one-third of the total VOC and CO emissions from passenger cars was emitted from cold start (Holman *et al.*, 1993). Cold starts are relatively unimportant for NO_x emissions.

Cold start emissions are relatively more important for cars fitted with catalytic converters than non-catalyst petrol cars because it takes a few minutes for the catalyst to reach its operational temperature. Thus during those early minutes of a journey when emission from the engine are highest, the catalyst does not work efficiently.

Ambient temperature has an important affect on emissions, especially when the engine is started from cold. Data from the UK Transport Research Laboratory show that the 'cold start emission penalty' increases exponentially with decreasing ambient temperature. This is shown for CO in Fig. 8.3 (personal communication from John Hickman of the Transport Research Laboratory, 1997). For diesel cars the influence of ambient temperature on cold start emissions is less marked.

There is some evidence that ambient temperature also has a small impact on hot emissions.

CONCLUSIONS

Techniques to estimate sources of pollution have improved considerably over the years, but there remains considerable uncertainty, particularly for pollutants such as NMVOCs which come from a diverse range of mainly small sources, and PM which is both a primary and secondary pollutant. Emission estimates for anthropogenic sources of NMVOCs are probably within ±50%, and for PM ±25% of true values. To obtain the complete picture it is important to include natural as well as anthropogenic sources. For natural sources the estimates are much more uncertain.

National anthropogenic annual emission inventories can, on their own, give a misleading picture regarding the appropriate sources to control. What is important for most pollutants are those sources that make major contributions to elevated concentrations during pollution episodes. As most people live in cities, urban inventories (and air quality modelling) are important for identifying these sources.

The main source of air pollution is the burning of fossil fuels, particularly in power stations and motor vehicles. Emissions from both these sources have begun to decline in recent years owing to the introduction of efficient abatement measures. In most urban areas, road transport is the dominant source of pollution as measures to reduce wintertime smogs in earlier decades lead to the re-location of many large stationary sources to rural or semi-rural sites.

Road transport is the single most important source of most of the classical air pollutants. The exceptions are sulfur dioxide and volatile organic compounds. To develop effective control strategies for reducing urban emissions it is important to know which type of vehicles are contributing most to the total road traffic emissions.

In an inventory area there will typically be millions of individual sources of pollution. It is not possible to measure emissions from every single one. Instead, emissions are calculated, generally from a measure of activity relating to the emissions, such as fuel consumption for stationary sources and distance travelled for mobile sources. Emission factors, derived from measurements of individual or representative sources, are then multiplied by the appropriate statistic, to give the emission rate.

REFERENCES

ACEA (Association des Constructeurs Européens d'Automobiles) (1997) Influence of Diesel Fuel Quality on Heavy Duty Diesel Engine Emissions: Further Investigations of the EPEFE Results by the ACEA Heavy Duty Diesel Truck Manufacturers, Brussels.

ACEA/EUROPIA (undated) European Programme on Emissions, Fuels and Engine Technologies, Brussels.

Department of the Environment (1996) *Digest of Environmental Statistics*, no. 18. London: HMSO.

Deacon AR, Derwent RG, Harrison RM *et al.* (1997) Analysis and interpretation of measurements of suspended particulate matter at urban background sites in the United Kingdom. *Sci Total Environ* **203**: 17–36.

EMEP (1994) *Biogenic VOC Emissions in Europe*, Part 1: Emissions and Uncertainties. Oslo: MSC-W, Norwegian Meteorological Institute.

EMEP/ CORINAIR (1996) *Atmospheric Emission Inventory Guidebook*, vols 1–2. Copenhagen: EMEP, Co-operative Programme for Monitoring and Evaluation of the Long Range Transmission of Air Pollutants in Europe, and CORINAIR, The Atmospheric Emission Inventory for Europe, European Environment Agency.

ETSU (1996) *Alternative Road Transport Fuels – A Preliminary Life-cycle Study for the UK*, vols 1 and 2. Harwell: ETSU.

ETSU (1997) *UK Research Programme on the Characterisation of Vehicle Particulate Emissions*. A report produced for the Department of Environment, Transport and the Regions (DETR) and the Society of Motor Manufacturers and Traders (SMMT). London: HMSO.

European Environment Agency (1996) CORINAIR 1990 Summary Tables, Copenhagen.

European Environment Agency (1997a) CORINAIR 1994 Summary Tables, Copenhagen.

European Environment Agency (1997b) COPERT II: Computer Programme to Calculate Emissions from Road Transport – Methodology and Emission Factors, Final Draft Report, Copenhagen.

Expert Panel on Air Quality Standards (1994) 1,3-Butadiene, London: Department of the Environment, HMSO.

Heaton DM, Rijkeboer RC and van Sloten P (1992) Analysis of emission results from 1000 in-use passenger cars tested over regulation cycles and non-regulation cycles. Paper presented at *Traffic-Induced Air Pollution*, symposium held at Graz Technical University, 10 and 11 September 1992.

Holman C, Wade J and Fergusson M (1993) *Future Emissions from Cars 1990 to 2015: The Importance of the Cold Start Emissions Penalty*. Godalming, UK: World Wide Fund for Nature.

IPCC/UNEP/OECD/IEA (1995) *Guidelines for National Greenhouse Gas Inventories*, vols 1–3. Bracknell, UK IPCC/UNEP/OECD/IEA.

London Research Centre (1996) *West Midlands Atmospheric Emissions Inventory*. London: LRC.

London Research Centre (1997) *Greater Manchester Atmospheric Emissions Inventory*. London: LRC.

London Research Centre (1998) *Greater London Atmospheric Emissions Inventory*. London: LRC.

National Environmental Technology Centre (1995) *UK Emissions of Air Pollutants (1970–1993)*. A report of the National Atmospheric Emissions Inventory. Culham, UK: AEA Technology.

National Environmental Technology Centre (1997) *UK Emissions of Air Pollutants 1970–1995*. A Report of the National Atmospheric Emissions Inventory. Culham: AEA Technology.

Pechin PM and Associates (1994) National PM Study: OPPE Particulate Programs Evaluation System. A report to the US Environmental Protection Agency (with 1996 update) quoted in Wilson R and Spengler JD (eds) (1996) *Particles in Our Air: Concentrations and Health Effects*. Cambridge, MA: Harvard School of Public Health.

Poulton ML (1994) *Alternative Fuels for Road Vehicles*. Southampton, UK: Computational Mechanics Publications.

Quality of Urban Air Review Group (1993) *Diesel Vehicle Emissions and Urban Air Quality*, Second Report, prepared at the request of the Department of the Environment, Birmingham.

Quality of Urban Air Review Group (1996) *Airborne Particulate Matter in the United Kingdom*, Third Report, prepared at the request of the Department of the Environment, Birmingham.

Signer M (1997) *Influence of Fuel Quality on Diesel Engine Emissions*. World Fuel Conference, 21–23 May 1997, Brussels.

TNO (1995) *Final Report on Random Truck Sampling Programme, 1994–1995*, Delft.

TNO (1997) *Particulate Matter Emissions (PM_{10}–$PM_{2.5}$–$PM_{0.1}$) in Europe in 1990 and 1993*. Apeldoom: TNO Institute of Environmental Sciences, Energy Research and Process Innovation.

US Environmental Protection Agency (1995a) *National Air Pollutant Emission Trends*, 1990–1994. Research Triangle Park, NC: USEPA.

US Environmental Protection Agency (1995b) *Compilation of Air Pollutant Emission Factors*, AP-42. Research Triangle Park, NC: USEPA.

US Environmental Protection Agency (1995c) *Air Quality Criteria for Particulate Matter*. Research Triangle Park, NC: USEPA.

US Environmental Protection Agency (1996a) *Inventory of Sources of Emissions for Five Candidate Title III Section 112(k) Hazardous Air Pollutants: Benzene, 1,3–Butadiene, Formaldehyde, Hexavalent Chromium and Polycyclic Organic Matter*. Report prepared by Eastern Research Group, Research Triangle Park, NC.

US Environmental Protection Agency (1996b) *Emission Inventory for 112(c)(6) Pollutants: Polycyclic Organic Matter (POM), 2,3,7,8-Tetrachlorodibenzo-p-dioxin (TCDD)/2,3,7,8-Tetrachlorodibenzofuran (TCDF), Polychlorinated Biphenyl Compounds (PCBs), Hexachlorobenzene, Mercury and Alkylated Lead*. Research Triangle Park, NC: USEPA.

Warren Spring Laboratory (1993) *Emissions of Volatile Organic Compounds from Stationary Sources in the UK*. Stevenage: Warren Spring Laboratory.

9

Exposure Assessment

HALÛK ÖZKAYNAK*

Department of Environmental Health, Harvard School of Public Health, Boston, MA, USA

INTRODUCTION

Exposure assessment plays a key role in evaluating the risks of air pollution. Accurate exposure assessment is requisite for epidemiological studies and, as one of the critical elements of risk assessment, provides information on individual or population patterns of exposure to indoor and outdoor pollutants. Principles of exposure assessment have long been used in the workplace to evaluate occupational exposures; the extension to population exposures is more recent, arising with the recognition that exposures to the general population can have adverse health effects. Over recent decades, exposure assessment has emerged as a more formal discipline and researchers are now incorporating tools of exposure assessment explicitly into assessment of the impact of the environment on health. The history of exposure assessment and its evolution is set out in a 1991 National Research Council report entitled *Human Exposure Assessment For Airborne Pollutants*. The topic of exposure assessment is also discussed in the chapter by Samet and Jaakkola in this volume (see Chapter 20).

In evaluating risks of air pollution, both the shape of the distribution of exposures and the exposures at the highest end are of interest. The shape and measures of central tendency – i.e. mean and median – provide an indication of overall population risk. The upper end of the distribution comprises those individuals or groups of individuals in a population who are potentially at greatest risk. It is important to determine the source, transport, concentration and behavioral factors that may lead to substantially greater exposures for those

* Present affiliation: US Environmental Protection Agency, National Exposure Research Laboratory (MD-56), Research Triangle Park, NC 27711, USA

AIR POLLUTION AND HEALTH
ISBN 0-12-352335-4

at the higher end of the exposure distribution; the US Environmental Protection Agency (EPA) recognized this potential in its exposure assessment guidelines (USEPA, 1992). Consequently, exposure assessment for air pollutants characterizes the sources of variation in population exposures by such factors as location and time of day and from the differing pollutant concentrations in various indoor and outdoor environments.

Assessment of exposures to air pollutants refers to the analysis of various processes that lead to human contact with pollutants after release into the environment. The routes of exposure to air pollutants include not only inhalation, but also ingestion and dermal contact with pollutants that have deposited or settled on soil, surfaces, foods and other objects. In this chapter we focus on the air inhalation pathway only, although ingestion or dermal exposure routes may be quite relevant for some pollutants, such as lead, which contaminate either settled dust or residential indoor surfaces. Key definitions are provided in Table 9.1 (National Research Council, 1991).

Table 9.1   Definitions of terms

Term	Definition
Exposure	An event that occurs when there is contact at a boundary between a human and the environment with a contaminant of a specific concentration for an interval of time.
Total exposure	Accounts for all exposures a person has to a specific contaminant, regardless of environmental medium or route of entry (inhalation, ingestion and dermal absorption).
Dose	The amount of a contaminant that is absorbed or deposited in the body of an exposed organism for an increment of time, usually from a single medium.
Internal dose	Refers to the amount of the environmental contaminant absorbed in body tissue over a given time of interacting with an organ's membrane surface.
Biologically effective dose	The amount of a deposited or absorbed contaminant or its metabolites that has interacted with a target site over a given period so as to alter a physiological function.

National Research Council (NRC) & Committee on Advances in Assessing Human Exposure to Airborne Pollutants 1991.

As illustrated in Fig. 9.1, exposure assessment begins with identification of key sources of selected pollutants and their emission rates into the air. Outdoors, a variety of stationary, mobile (i.e. vehicles) and area sources contribute to emissions of gaseous and particulate pollutants. Similarly, in the indoor environment, household cooking and heating sources, building materials, consumer products and human activities result in intermittent or continuous emissions of many classes of pollutants, including carbon monoxide (CO), volatile organic compounds (VOCs), particulate matter (PM) and biological agents. All of these sources, either indoors or outdoors, contribute to air pollution concentrations in the various microenvironments where people spend time during the daily course of their activities. The term 'exposure' refers to length of contact with the pollutant (in the case of air pollutants, this is predominantly inhalation) during a specified time period (hours, day, months, etc.). Thus, exposure to ambient air pollution refers to contact with pollutant concentrations that an individual encounters during time spent outdoors as well as to contact with outdoor pollution that may have penetrated indoors. Units are expressed in $\mu g/m^3$ or ppm. Any assessment of exposure needs to take into account that concentrations may vary both spatially and temporally depending on the location and the characteristics of the ambient sources.

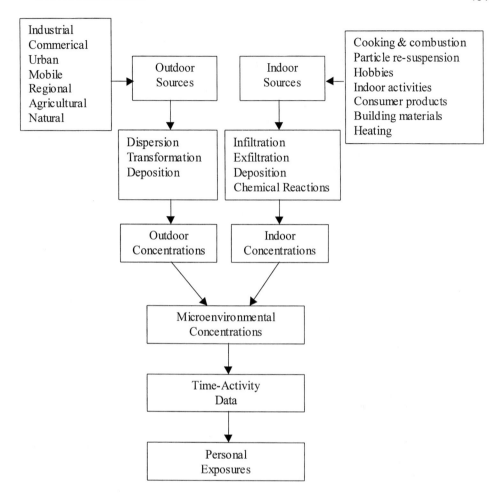

Fig. 9.1 Sources and factors influencing indoor, outdoor and personal exposures to air pollutants.

Pollutants from outdoor sources can also penetrate indoors, and indoor microenvironments may be a significant locus of exposure to outdoor pollutants. One key step in exposure assessment is, therefore, to predict (i.e. model) or directly measure the geographic and time profile of the pollution concentrations in the study region of interest. Various air pollution dispersion models – for example, the EPA's Industrial Source Complex-Short Term (ISCST) and mobile source model Mobile 5 – can be used to predict concentrations of air pollutants both in urban and suburban environments using information on characteristics of the sources in the area and the associated emission rates of the pollutants. These models allow determination of fate and transport of the pollutants in the environment and provide concentration values at different receptor points, i.e. locations, in the assessment area. Alternatively, indoor or outdoor fixed location monitoring can be conducted to obtain information on pollutant concentration profiles, and to derive empirical relationships between source emissions and microenvironmental concentrations.

Since the definition of exposure includes human presence and contact during the course of a selected time period, exposure assessment also involves the identification of the locations where individuals spend time and the amount of time spent in each of these locations. The term 'microenvironment' refers to a location (e.g. outdoors near home, commuting, walking, indoors at home, indoors at work, school, or shopping mall) in which the concentration is spatially uniform during the time spent in the microenvironment, even though the concentration may vary over time. In the microenvironmental model of exposure, personal exposures (E) are estimated by the combination of pollutant concentrations in selected microenvironments (C_i), with the fraction of time spent in each of these microenvironments (f_i):

$$E = \sum_{i=1}^{n} f_i C_i \tag{1}$$

Thus, knowledge of where people are and what they are doing during the course of a typical workday or weekend is essential for determining personal exposures. Moreover, time–activity profiles of different population subgroups (e.g. the elderly, individuals with asthma or chronic obstructive pulmonary disease, children) can also vary substantially, although limited information is available for these special groups. Time–activity diaries from a representative general population and certain subgroups have also been collected by various survey techniques, including direct diary entries and telephone-based 24-h recall diaries. The largest diary study of the US population, the National Human Activity Pattern Study carried out by the US EPA (Robinson and Nelson, 1995), has recently collected information on daily time–activity patterns from over 10 000 randomly selected persons across the USA.

Field studies have also been conducted involving direct monitoring using active or passive personal sampling devices worn by participants over the course of one day or more. Direct measurements of personal exposures are quite important, not only in developing information on actual distribution of personal exposures within a population group, but also for formulating and validating the microenvironmental exposure models defined by equation (1). A number of human exposure assessment field studies, particularly EPA's Total Exposure Assessment Methodology (TEAM) studies and the National Human Exposure Assessment Study (NHEXAS), have provided a foundation for building models of human exposures to CO, VOCs, pesticides, and particulate matter less than 10 μm in aerodynamic diameter (PM_{10}) (Özkaynak and Spengler, 1996; Wallace, 1987, 1996). These field studies have provided information on variation in indoor and outdoor concentrations and the correlation between indoor and outdoor concentrations, and the contributions of indoor and outdoor concentrations to personal exposure. The results of these studies have been generalized to other situations using the exposure models estimated with the field data.

All of the information on environmental concentrations (either directly measured or modeled from source emissions using meteorological inputs), demographic data (including geographic location and age distribution of the population), and contact times in different exposure microenvironments derived from time–activity diaries can be combined in an integrated exposure modeling framework. The basic models that are often used to predict personal exposures to commonly studied air pollutants, such as particulate matter and gaseous pollutants (e.g. CO, NO_2, VOCs), are described further below.

MEASUREMENT METHODS FOR AIR POLLUTION EXPOSURES

Measurements of personal exposures are best obtained directly with personal monitoring devices placed on people: this method is called the direct approach. The indirect approach, described below in *Personal Exposure Models*, is based on concentrations measured in microenvironments with stationary monitoring devices or on estimates of these concentrations. Concentrations in the principal microenvironments are then combined to derive personal exposure using applicable time–activity information according to equation (1).

Typical air monitoring instruments can be classified as either active, sampling air with a pump, or passive, sampling with sorption, entrapment, or some other principle. In time, active samplers can be continuous (automatic/direct reading), semicontinuous, or integrated (non-automatic). Passive samplers include diffusion tubes, badges and detector tubes.

For monitoring particle concentrations, both direct and indirect sampling methods are available. Direct reading instruments provide real-time information on the concentration and size distribution of aerosols. Instruments of this type are primarily based on either optical principles, electrical resonance oscillation, or β attenuation (Koutrakis and Sioutas, 1996). These devices, with the exception of some optical systems, are used primarily for stationary, rather than personal, monitoring. Optical instruments rely on scattering or absorption of light by particles. Calculations based on radioactive transfer physics are then used to convert the light signal data to size-specific number and mass concentration information. Electrical counters measure the ability of charged particles to cross an electrical field. Since the amount of charge received by particles depends on their size, particle size distribution can be obtained from particulate material collected on the walls of the device. A commonly used real-time particle measurement device, TEOM (tapered element oscillating microbalance), relies upon changes in the resonant frequency of an oscillating surface (tape) as a function of mass collected on the surface. Because this device requires heating of the inlet air stream, there is some loss of volatile species during measurement. β-attenuation instruments, used widely in Europe, draw air through an orifice, and the particles collect on a surface between a β source and a counter. The amount of β attenuation is related to particle mass concentration.

The direct reading instruments have a number of limitations, however. Chemical composition of the aerosols can significantly influence the reading. In addition, relative humidity, gas adsorption, and particle collection efficiency may also affect the results. Recently, Koutrakis and colleagues at Harvard (Koutrakis *et al.*, 1995) developed a continuous aerosol mass monitor (CAMM). The CAMM measures particulate mass concentrations based on continuous measurement of pressure drop across a Fluoropore filter. This device has now been miniaturized as a personal sampler for particles less than 2.5 μm in aerodynamic diameter ($PM_{2.5}$).

Conventional inertial impactors have been widely used to determine size-specific mass concentration of particles. In these devices, particles below a certain cut-off size (e.g. less than 10 μm in aerodynamic diameter, d_a) are separated from the air stream by their inertia and collected on a filter, such as a Teflon filter that has low background acid and sulfate concentrations. The performance of these inertial impactors has been well characterized. Depending on the location and purpose of the sampling program, either high or low volume impactors may be appropriate. High volume samplers are built for routine outdoor sampling purposes. Low volume impactors, such as the Harvard or the MSP impactors, have been designed for indoor and outdoor use. Likewise, a personal exposure monitor (PEM) was also designed to collect PM_{10} and $PM_{2.5}$ on a 37 mm Teflon filter at a 4 lpm flow rate (Marple et al., 1987; Thomas et al., 1993). The PEM system worn by the participants in the Particle Team (PTEAM) study (Özkaynak et al., 1996), consisted of a canvas bag containing the pump and the battery pack; it was worn on the hip, front, back, or over the shoulder.

Another particulate sampler similar to an impactor is a cyclone device. In a conical-shaped cyclone, air is drawn tangentially near the top. Larger particles collect on the cyclone walls as flow reverses and spirals upward within the device. Smaller size particles (e.g. d_a <2.5 µm, where d_a is aerodynamic diameter) are collected on a filter cassette placed at the top or exit of the flow. The European EXPOLIS study has recently used small cyclones manufactured by BGI, Inc., quite successfully for personal and micro-environmental sampling of $PM_{2.5}$ (Kenny and Gussman, 1997).

Another instrument designed to sample both aerosols and gases such as SO_2, NH_3 and HNO_3 is the impactor/denuder filter pack system for both personal and microenviron-mental sampling (Koutrakis et al., 1988). As described recently by Koutrakis and Sioutas (1996), the principle of operation for diffusion denuders is that gases diffuse to the walls much faster than particles. Consequently, these devices trap virtually all of a given type of pollutant gas while only a negligible fraction of particles is collected on the denuder walls. Different coating materials can be used to trap different gases. Sodium carbonate (Na_2CO_3), sodium hydroxide (NaOH), and potassium hydroxide (KOH) have been used extensively as coating substrates because they collect acidic gases such as HNO_3, HCl, HNO_2, SO_2, HCOOH and CO_3COOH very efficiently (Koutrakis et al., 1988, 1993). Citric acid has been used to collect basic gases such as NH_3 (Koutrakis et al., 1993).

Gaseous pollutant monitoring methods range from continuous electrochemical meth-ods for CO and NO_2 to diffusion badges for NO_2, O_3, SO_2, CO and HCHO sampling. For organic vapors, methods include activated charcoal badges for passive personal or microenvironmental monitoring and sorbent-packed tubes for active personal sampling. Portable or personal sampling instruments for active continuous CO monitoring have been used successfully in various field investigations (Ott, 1990). Real-time personal monitoring methods for other gaseous species, however, are more difficult to design and deploy. Available techniques for measuring human exposures to gaseous and particulate pollutants are discussed further in a number of references (Bower et al., 1997; Spengler and Samet, 1991; WHO, 1998).

PERSONAL EXPOSURE MODELS

Exposure models provide an analytic structure for combining different types of data col-lected from disparate studies that allows for more complete use than direct study methods of the existing information on a particular contaminant. Results of models can be used to evaluate the determinants of exposures and attendant doses at various points of the pop-ulation exposure distribution that cannot be measured directly because of limitations of monitoring methods or resources. Furthermore, models can be used to characterize sources of uncertainty in exposure assessments and these can then be incorporated formally into such models in order to estimate uncertainty about prediction endpoints (e.g. exposure, dose, or risk for a health outcome), and to identify the factors influencing prediction accuracy and precision. Validated models can then be used to investigate the efficacy of various strategies for managing the public health risks associated with exposures from environmental contaminants.

Modeling Personal Exposures

For air pollution, most exposure models utilizing the data gathered from these field investigations have focused on the inhalation pathway and two key microenvironments: indoors at home, and outdoors. A variety of physical and statistical modeling methods have been developed for predicting exposures to gases and particles (see, for example, Duan, 1982; Lioy, 1990; Ott, 1985; Ryan, 1991; Spengler and Soczek, 1984). As described by equation (1), in the physical models framework, total personal exposure is modeled as a sum of exposures encountered in various microenvironments. Numerically, daily personal exposures (E_i) of an individual 'i' are computed as the sum of time–activity weighted microenvironmental exposures:

$$E_i = \sum_{j=i}^{m} E_{ij} = \sum_{j=i}^{m} f_{ij} \times C_{ij} \qquad (2)$$

where E_{ij} is the exposure to individual 'i' in microenvironment j ($\mu g/m^3$), f_{ij} is the fraction of time spent by person 'i' in microenvironment 'j' during the 24-h prediction period (i.e. $t_{ij}/24$), C_{ij} is the average pollutant concentration ($\mu g/m^3$) in microenvironment 'j' when individual 'i' is present, and 'm' is the number of microenvironments considered in the model. As mentioned earlier, key assumptions in this microenvironmental modeling approach are that: (1) concentrations of pollutants are distributed uniformly in each of the microenvironments; (2) f_i and C_i are not correlated; and (3) a limited number of microenvironments are sufficient to characterize total personal exposures.

The validity and the precision of the microenvironmental exposure models depend on the number of different microenvironments that are needed to capture most of the variations in the concentrations affecting exposures. In addition, various exposure scenarios within these microenvironments or submicroenvironments also need to be considered. These may, for example, include smoking or tobacco smoke exposure indoors and in an automobile; cooking at home; heating with kerosene space heaters; and use of consumer products emitting VOCs. If activities or concentrations of studied pollutants do not vary greatly across different locations that individuals visit over the course of a day, a smaller number of microenvironments is sufficient to model personal exposures. For example, five principal microenvironments may be needed in exposure models for $PM_{2.5}$ and PM_{10} to represent exposure locations/activities associated with distinct PM exposures: outdoors, indoors at home during daytime, indoors at home during night-time, in-transit, and indoors not at home.

In practice, either measurements available from ambient air monitoring sites or modeled concentrations are used to estimate the outdoor pollution concentration profiles across a community. Concentrations in residences or workplaces are often estimated using semiempirical methods that account for penetration of outdoor pollutants indoors and contributions of indoor sources such as smoking, cooking, heating and vacuuming to indoor pollutant levels.

Since most people spend large portions of their time indoors, information is needed on indoor concentrations, either from measurements or models. A recent nationwide study of time budgets in the USA indicates that residents spend 87.2% of their time indoors, 7.2% in transit, and only 5.6% outdoors (Robinson and Nelson, 1995). Clearly, indoor environments have a key role in determining personal exposures since most of our time during the day is spent either indoors at home, at work, or at school. According to Klepeis et al. (1996), survey respondents reported that 69% of their time was spent in residential indoor environments, followed by 7% in school and public buildings, 6% in vehicles, 5% in offices/factories, and 2% in bars/restaurants. However, there was variation by age and gender in times spent in schools and public places, such as malls and in hospitals. For exposure assessment, these differences in time–activity profiles are important for determining the contributions of various microenvironments and for describing differences in exposures across groups.

In understanding the potential risks of outdoor pollutants that have penetrated indoors, information is needed on indoor/outdoor concentrations and on changes in chemical and toxicologic properties of particles after infiltration indoors. Both indoor and outdoor pollutant sources contribute to indoor concentrations. The level of protection offered by building characteristics can directly influence the resultant exposure to ambient pollutants, such as PM_{10} and $PM_{2.5}$. Building-specific parameters such as volume, air exchange rate, penetration efficiencies, types of buildings and furnishing materials, and room use patterns, as well as cleaning frequency, all affect indoor particle concentrations. The air exchange rate contributes to dilution of indoor PM source concentrations, such as from smoking or cooking, but it will not affect the physical penetration of ambient particles through the building envelope.

Indoor PM, NO_2, and VOC concentrations can be modeled following the methodology developed from the PTEAM study by Özkaynak et al. (1995a, 1996). A semiempirical physical model that assumes contributions to residential indoor pollution from outdoor, environmental tobacco smoke (ETS), cooking, and other unaccounted indoor sources is specified (Özkaynak and Spengler, 1996):

$$C_{in} = \frac{Pa}{a+k} C_{out} + \frac{N_{cig}S_{smk} + T_{cook}S_{cook}}{(a+k)Vt} + \frac{S_{other}}{(a+k)V} \tag{3}$$

where:

C_{in} = pollutant concentration indoors;
C_{out} = pollutant concentration outdoors;
P = penetration fraction (unitless);
a = air exchange rate (per h),
k = pollutant decay or deposition rate (per h);
N_{cig} = number of cigarettes smoked;
V = house volume (m^3);
T_{cook} = cooking time (h);
t = sampling period (h);
S_{smk} = estimated source strength for cigarette smoking (mg/cig);
S_{cook} = estimated source strength for cooking (mg/mm);
S_{other} = re-suspension or estimated source strength for other indoor sources (mg/h).

Indoor or personal pollutant concentrations (C_{per}) at which exposures occur can also be modeled using an empirical relationship, expressing indoor personal concentrations as a function of outdoor concentrations (C_{out}):

$$C_{in} \text{ or } C_{per} = \beta_o + \beta_1 C_{out} \tag{4}$$

This formulation facilitates statistical estimation of the influence of outdoor concentrations on either the indoor or personal pollutant concentrations. Using both types of exposure modeling formulations defined by equations (3) and (4), researchers have investigated indoor–outdoor relationships for selected key pollutants: PM_{10}, $PM_{2.5}$, O_3, NO_2, CO and SO_2.

For homes with no smokers, modeling of data from the PTEAM study shows that indoor/outdoor concentration ratios are about 60–70% for $PM_{2.5}$ and about 50% for PM_{10} (Özkaynak *et al.*, 1996; Wallace, 1996). The study also indicates that about 60% of outdoor PM_{10} is expected to contribute to personal PM_{10}. However, the concentrations and even the composition of the indoor particles are usually different from those of outdoor particles because contributions of indoor cooking, smoking, vacuuming, and other personal activities that either re-suspend or generate aerosols have physical and chemical characteristics unlike those of outdoor aerosols. Therefore, in modeling the indoor PM_{10} or $PM_{2.5}$ concentrations, estimates of source-specific PM emission rates are required. Özkaynak *et al.* (1996) provides information on estimated indoor PM source strengths and also on typical values for the deposition or decay rates for PM_{10} (0.65±0.28/h) and $PM_{2.5}$ (0.39±0.16/h).

Information on air exchange rates between indoor and outdoor environments are also needed in order to predict indoor pollutant concentrations using physical mass balance models. Air exchange rates vary by season and by the type of structure. In the USA the rates range from 0.1/h to 3/h with a median value of 1/h and a geometric standard deviation of 2/h (Pandian *et al.*, 1993). Using these values, researchers found that the average contribution of cooking and cigarette smoking to indoor PM levels ranged from 3% to 30%, depending on whether cooking or smoking was reported by the study participants. However, other or unidentified sources of indoor PM_{10} were also found to account for a large portion of indoor PM_{10}, or about 26% across all homes (Özkaynak *et al.*, 1996). An important finding from this study was that daytime personal exposures to PM_{10} were about 50% above that estimated from concurrent indoor and outdoor concentrations. It is assumed that personal activities such as cooking, living in a home with a smoker, and house-cleaning activities could contribute to elevated personal PM_{10} concentrations.

Indoor NO_2 levels are also influenced by outdoor concentrations, as well as by emissions from indoor gas combustion appliances. Typically, ambient NO_2 infiltrates indoors readily; about 60±10% of outdoor NO_2 penetrates indoors (Drye *et al.*, 1989; Özkaynak *et al.*, 1995b). NO_2 emissions from gas cooking appliances with no pilot lights contribute about 4–6 ppb to indoor NO_2 levels. Older types of gas stoves with continuously lit pilot lights add between 8–15 ppb to indoor NO_2 concentrations (Özkaynak *et al.*, 1995b). NO_2 is a moderately reactive gas and decays quickly indoors through chemical and surface-based reactions. The experimental decay rates are given in Özkaynak *et al.* (1982) and also in Nazaroff *et al.* (1993).

Indoor ozone concentrations are typically much lower than outdoor ozone levels because ozone is a highly reactive gas. A recent study in 43 southern California homes (Lee *et al.*, 1997) showed that mean indoor/outdoor ozone ratios in residences with open windows were 0.7±0.2. In homes with air conditioning, however, the mean indoor/outdoor ratio was quite small (0.1). Typically, indoor ozone levels in homes are 10% to 30% of outdoor values. The study by Lee *et al.* (1998) also estimated a mean indoor ozone decay rate of 2.8 ± 1.3/h.

Indoor VOC concentrations are typically several times higher than the outdoor levels (Wallace, 1987). According to Wallace (1987), indoor sources of VOCs include smoking (benzene, xylenes, ethylbenzene, styrene in breath), dry-cleaned clothes (tetrachloroethylene), air fresheners (limonene), house-cleaning materials, use of chlorinated water (chloroform), deodorizers (*p*-dichlorobenzene), and various occupational exposures and pumping gas (benzene). Most of the outdoor VOCs, however, penetrate indoors quite efficiently without much loss (Lewis, 1991). Therefore, for some VOCs, such as benzene, with fewer indoor sources (with cigarette smoking and attached garages as the primary indoor sources of benzene), indoor VOC levels are well correlated with outdoor levels and indoor/outdoor ratios are closer to 1. In contrast, new buildings have VOC levels about 100 times higher than outdoor levels and only decrease to 10 times the outdoor levels after 2–3 months of occupancy. Paints and adhesives are assumed to be the primary contributors to indoor VOC levels in new buildings.

Finally, pollutants such as CO and SO_2 also penetrate indoors efficiently and are not reactive. Aside from occasional emissions from kerosene space heaters, SO_2 typically is an outdoor pollutant, as is CO. Motor vehicle and gasoline combustion engines are the dominant sources of ambient CO. Gas cooking or heating appliances and smoking, however, also contribute to indoor CO concentrations. In general, indoor CO levels are slightly higher than the outdoor values for homes with gas cooking appliances (Wilson *et al.*, 1995).

Averaging Times for Exposure Assessment

Appropriate averaging periods for exposure monitoring and modeling and the minimum number of microenvironments should be determined on a pollutant-by-pollutant basis and the biological averaging time of interest. For some pollutants (e.g. CO), real-time monitoring is possible, by measuring concentrations at fixed locations or in proximity to people to estimate personal exposure. However, for either exposure modeling purposes or assessment of health effects, this level of temporal resolution is rarely needed. For pollutants that produce effects acutely, such as CO, substantial temporal resolution may be needed to capture biologically relevant profiles of exposure. Since human activities and microenvironments may vary in time intervals greater than a minute, requirements for modeling dictate modeling time steps of minutes to an hour for CO.

However, for VOCs, particles and metals that are typically linked with subacute and chronic health effects, the required time interval for estimating personal exposures may range from an hour to 24 h. Because of limitations of instrumentation, exposure models developed for these pollutants have assumed either 24-h or 12-h average concentrations as inputs. In contrast, exposure models, such as SHAPE (for CO), THEM (for particles), NEM (for CO and ozone) and BEADS (for benzene), have selected alternative time intervals in estimating personal or population exposures. In general, the averaging period of interest for health effects dictates the optimum resolution time required for modeling. Models such as SHAPE and NEM are time-dependent, requiring real-time (i.e. minute-by-minute) knowledge of time activity and microenvironmental concentration data. The BEADS model for benzene (MacIntosh *et al.*, 1995) is time-independent in that daily or 12-h average activity and concentration profiles are used in the simulation of population exposures. The differences between time-dependent and time-independent methods have important implications for input requirements, model run-time, general applicability, and desired accuracy and precision in the predictions.

Alternative Exposure Models and Their Use in Health Effects Investigations

A number of different exposure assessment approaches have typically been used in epidemiologic investigations of air pollution. In increasing order of sophistication, these include: (1) classification of individual exposures (high vs low); (2) measured or modeled outdoor concentrations; (3) measurement of indoor and outdoor concentrations; (4) estimation of personal exposures using indoor, outdoor and other microenvironmental concentrations along with time–activity diaries; (5) direct measurement of personal exposures; and (6) measurement of breath and other biomarkers of exposures. The least sophisticated approach with regard to classification of exposure groups, using a categorical variable (e.g. homes with gas vs electric cooking stoves for NO_2 impact assessment), could lead to bias from exposure misclassification (Özkaynak *et al.*, 1986).

Many environmental health studies, however, are based on ambient or community surveillance monitoring data. Aside from the usual spatial variations in outdoor pollutant concentrations, human exposures to many pollutants also involve pollutant exposure sources and locations other than outdoor pollutants and monitored ambient environments (e.g. for PM, NO_2 and VOCs). For reactive pollutants, such as ozone, indoor pollution levels are significantly lower than the outdoor concentrations. Consequently, owing to the greater amount of time spent indoors, personal ozone exposures are more closely related to indoor ozone concentrations than to outdoor ozone levels. In general, therefore, exposure models based on ambient data are only less accurate than microenvironmental models that combine indoor and outdoor concentration measurements (or predictions) with time–activity data, and source and household characteristics information.

Several studies support these general assertions. For example, Xue *et al.* (1993) showed that the estimated R^2 (a measure of model fit) increased from 0.28 to 0.74 when different exposure models were applied to 2-day average ambient bedroom and personal NO_2 data. These results showed that the predictive power of the NO_2 personal exposure models was quite poor ($R^2 = 0.28$) if no ambient or indoor NO_2 measurements were used aside from the home characterization questionnaire (HCQ) variables, such as use of gas cooking appliances and air conditioners. The NO_2 exposure models with either outdoor or indoor NO_2 measurements in addition to the HCQ variables had greater predictive power ($R^2 = 0.6$). The full, time–activity weighted microenvironmental model, however, had a much higher R^2 value ($R^2 = 0.74$). Clearly, ideal exposure models should combine outdoor measurements with indoor concentration measurements or predictions proportional to the fraction of time spent in each of these two key microenvironments. These findings have implications for the design of monitoring studies in support of health effects investigations.

SUMMARY AND CONCLUSIONS

Application of exposure models requires collection and use of different types of data as inputs to models (such as multimedia source strengths, time/activities and dispersion and removal rates, etc.) and sophisticated statistical techniques for implementing the relationships between various model parameters, time/activities, and concentration distributions. Model input requirements have implications for developing and implementing well-designed monitoring strategies and programs. Often, a detailed temporal and spatial profile of ambient and personal concentrations is necessary for complete characterization of human exposures to air pollutants. Increasing the resolution of ambient monitors and applying atmospheric dispersion models enhance the validity of personal exposure models. However, assessing personal exposures to outdoor pollutants during commuting, for example, requires more curbside, in-vehicle, and personal exposure monitoring. Studying exposures of a greater number of subjects, collecting time–activity diaries, and measuring indoor pollution emission or source strengths in different types of microenvironments are areas where more scientific data are needed. The real success of any present or future exposure models will, of course, strongly depend on results from carefully constructed field validation studies. Without proper validation, exposure models will either remain as interpretive tools or as complementary information to measurement data. The challenge for the future is to increase the accuracy, precision, and use of exposure models in developing effective source and exposure control strategies for the pollutants that pose the greatest risks to human health.

REFERENCES

Bower J, Plews JV, McGinlay J and Charlton A (1997) *A Practical Guide to Air Quality Monitoring.* UK: National Environmental Technology Centre.

Drye E, Özkaynak H, Burbank B *et al.* (1989) Development of models for predicting the distribution of indoor nitrogen dioxide concentrations. *J Air Pollut Control Assoc* **39**: 1169–1177.

Duan N (1982) Models for human exposure to air pollution. *Environ Int* **8**: 305–309.

Kenny LC and Gussman RA (1997) Characterization and modeling of a family of cyclone aerosol pre-separators. *J Aerosol Sci* **28**: 677–688.

Klepeis NE, Tsang AM and Behar JV (1996) Analysis of the National Human Activity Pattern Survey (NHAPS) respondents from a standpoint of exposure assessment. ORD Report EPA/600/R-96/074. Washington, DC: US Environmental Protection Agency.

Koutrakis P and Sioutas C (1996) Physico-chemical properties and measurement of ambient particles. In: Spengler J and Wilson R (eds) *Particles in Our Air, Concentrations and Health Effects.* Cambridge, MA: Harvard University Press.

Koutrakis P, Wolfson JM and Spengler JD (1988) An improved method for measuring aerosol strong acidity: results from a nine-month study in St Louis, Missouri, and Kingston, Tennessee. *Atmos Environ* **22**: 157–162.

Koutrakis P, Sioutas C, Ferguson S *et al.* (1993) Development and evaluation of a glass honeycomb denuder/filter pack system to collect atmospheric particles and gases. *Environ Sci Technol* **27**: 2497–2501.

Koutrakis P, Wolfson M and Sioutas C (1995) Continuous ambient particle monitor. US patent no. 5 571 945.

Lee KY, Spengler JD and Özkaynak H (1998) Ozone decay rates in residences. *J Air Waste Manag Assoc* (submitted).

Lewis CW (1991) Sources of air pollutants indoors: VOC and fine particulate species. *J Expo Anal Environ Epidemiol* **1**: 31–44.

Lioy PJ (1990) Assessing total human exposure to contaminants, a multidisciplinary approach. *Environ Sci Technol* **24**: 938–945.

MacIntosh DL, Xue J, Özkaynak H *et al.* (1995) A population-based exposure model for benzene. *J Expo Anal Environ Epidemiol* **5**(3): 375–403.

Marple V, Rubow KL, Turner W and Spengler JD (1987) Low flow rate and impactor for indoor air sampling design and calibration. *J Air Pollut Control Assoc* **37**: 1303–1307.

National Research Council (1991) Committee on Advances in Assessing Human Exposure to Airborne Pollutants. *Human Exposure Assessment for Airborne Pollutants: Advances and Opportunities.* Washington, DC: National Academy Press.

Nazaroff WW, Gadgil AJ and Weschler CJ (1993) Critique of the use of deposition velocity in modeling indoor air quality. 04-012050-17.

Ott WR (1985) Total human exposure. *Environ Sci Technol* **19**: 880–886.

Ott WR (1990) Total human exposure: basic concepts, EPA field studies, and future research needs. *J Air Waste Manag Assoc* **40**(7): 966–975.

Özkaynak H and Spengler JD (1996) The role of outdoor particulate matter in assessing total human exposure. In: Wilson R and Spengler JD (eds) *Particles in Our Air, Concentrations and Health Effects.* Cambridge, MA: Harvard University Press, pp. 63–84.

Özkaynak H, Ryan PB, Allen GA and Turner WA (1982) Indoor air quality modeling: compartmental approach with reactive chemistry. *Environ Int* **8**: 461–471.

Özkaynak H, Ryan PB, Spengler JD and Laird NM (1986) Bias due to misclassification of personal exposures in epidemiologic studies of indoor and outdoor air pollution. *Environ Int* **12**: 389–393.

Özkaynak H, Xue J, Spengler J and Billick IH (1995b) Errors in estimating children's exposures to NO_2 based on week-long average indoor NO_2 measurements. In: Morawska L, Bofinger ND, Maroni M (eds) *Indoor Air – An Integrated Approach.* Oxford: Elsevier Science Ltd, pp. 43–46.

Özkaynak H, Xue J, Weker R *et al.* (1995a) *The Particle TEAM (PTEAM) Study: Analysis of Data*, vol. III. Final report of the Environmental Protection Agency. Research Triangle Park, NC: US Environmental Protection Agency.

Özkaynak H, Xue J, Spengler J, *et al.* (1996) Personal exposure to airborne particles and metals: results from the Particle Team Study in Riverside, California. *J Expo Anal Environ Epidemiol* **6**: 57–78.

Pandian MD, Ott WR and Behar JV (1993) Residential air exchange rates for use in indoor air and exposure modeling studies. *J Expo Anal Environ Epidemiol* **3**: 407–416.

Robinson C and Nelson WC (1995) *National Human Activity Pattern Survey Data Base.* Research Triangle Park, NC: US Environmental Protection Agency.

Ryan PB (1991) An overview of human exposure modeling. *J Expo Anal Environ Epidemiol* **1**: 453–473.

Spengler JD and Samet JM (1991) A perspective on indoor and outdoor air pollution. In: Samet JM and Spengler JD (eds) *Indoor Air Pollution. A Health Perspective.* Baltimore, MD: Johns Hopkins University Press, pp. 1–29.

Spengler JD and Soczek MC (1984) Evidence for improved ambient air quality and the need for personal exposure research. *Environ Sci Technol* **8**: 268A–280A.

Thomas KW, Pellizzari ED, Clayton CA *et al.* (1993) Particle Total Exposure Assessment Methodology (PTEAM) 1990 study: method performance and data quality for personal, indoor, and outdoor monitoring. *J Expo Anal Environ Epidemiol* **3**: 203–226.

US Environmental Protection Agency (USEPA) (1992) Guidelines for exposure assessment. Fed Reg **57**: 22888–22938.

Wallace LR (1987) The total exposure assessment methodology (TEAM) study: summary and analysis. Washington, DC: Office of Research and Development, US Environmental Protection Agency.

Wallace L (1996) Indoor particles: a review. *J Air Waste Manage Assoc* **46**: 98–126.

Wilson AL, Colome SD and Tian Y (1995) *California Residential Indoor Air Quality Study*, vol. 3: Ancillary and exploratory analysis. Irvine, CA: Integrated Environmental Services.

World Health Organization (WHO) (1998) Environmental health criteria document on human exposure assessment. International Program on Chemical Safety. Geneva: World Health Organization.

Xue J, Özkaynak H, Ware JH and Spengler JD (1993) Alternative estimates of exposure to nitrogen dioxide and their implications to epidemiologic study design. *Proc Indoor Air* **3**: 343–348.

RESPIRATORY TRACT DETERMINANTS OF AIR POLLUTION EFFECTS

Animal Models to Study for Pollutant Effects

URMILA P. KODAVANTI and DANIEL L. COSTA

Pulmonary Toxicology Branch, Experimental Toxicology Division, National Health and Environmental Effects Research Laboratory, US Environmental Protection Agency, Research Triangle Park, NC, USA

This article has been reviewed by the National Health and Environmental Effects Research Laboratory, US Environmental Protection Agency and approved for publication. Approval does not signify that the contents necessarily reflect the views and the policies of the Agency nor mention of trade names or commercial products constitute endorsement or recommendation for use.

INTRODUCTION

Understanding of human pathobiology can be gained using laboratory animal models that have been developed to reflect human conditions. Studies involving animal models provide important information on biological mechanisms of initiation, progression, and resolution of toxicant-induced tissue injury. In the context of air pollution health effects studies, models are used to understand pollutant deposition and clearance as well as the mechanisms of biological action (Brain *et al.*, 1988a; Reid, 1980; Stuart, 1976). Depending on the human condition being modeled, appropriate healthy or susceptible laboratory animals can be selected to estimate human health risks from inhaled pollutants (Slauson and Hahn, 1980). Much of our understanding of the toxicity of major air pollutants, such as tropospheric ozone (O_3), sulfur dioxide (SO_2), nitrogen dioxide (NO_2), particulate matter (PM), carbon monoxide (CO) and other 'air toxic' pollutants (e.g. phosgene and metal/acid aerosols) has derived from studies using laboratory animal

AIR POLLUTION AND HEALTH
ISBN 0-12-352335-4

models (Committee of the Environmental and Occupational Health Assembly of the American Thoracic Society, 1996a, b).

Although a variety of inbred or outbred strains of mice, rats, hamsters, guinea pigs, rabbits, dogs and primates have been used in inhalation toxicology studies (Stuart, 1976), most studies have employed healthy animals to examine the basic mechanisms underlying pollutant-induced injury. Recently it has become recognized that human susceptibilities to air pollution effects vary dramatically depending on the pre-existent conditions of the host, such as diseases, nutritional deficiencies (Barnes, 1995; Dockery *et al.*, 1993; Hatch, 1995) or genetic differences (Kleeberger, 1995; Kodavanti *et al.*, 1997a). In some instances, susceptibilities to pollutant effects are increased beyond the range covered by the typical uncertainty factors that are considered in health risk estimates to protect susceptible subgroups (Dockery *et al.*, 1993; Pope *et al.*,1995; Schwartz, 1994). The experimental studies performed in the past to understand altered responses to air pollutants due to these pre-existing conditions of the host have been spotty and the underlying mechanisms are not clear. Thus, as risk-based toxicology progresses, the inclusion of animal models of varying susceptibilities (Fig. 10.1) will provide sound experimental evidence for estimating the risks and for better understanding the toxic mechanisms.

Pulmonary structure and pattern of air flow vary widely between animals and human, and therefore, it is critical that the animal model is chosen wisely. Selection of the appropriate model depends upon the condition being modeled and how closely the model mimics the human in terms of its biological handling of the pollutant (Warheit, 1989). In general, a range of human conditions – lung disease, genetic alteration/segregation, age,

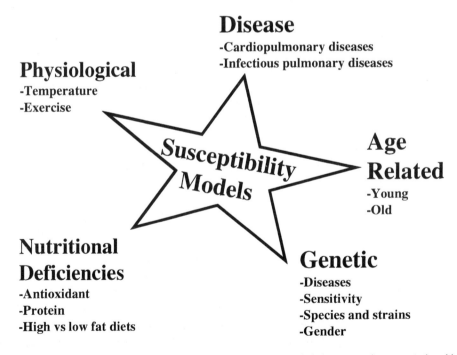

Fig. 10.1 Varying host conditions that influence the susceptibility to inhaled or ingested toxicant-induced lung injury.

malnutrition, or altered physiological profile due to systemic disease – encompasses most potential susceptible subgroups (Brain *et al.*, 1988a) and these can be modeled in selected laboratory animal species. In this chapter, we consider the application of both healthy and susceptible animal models for the evaluation of air pollution health effects. Major focus is given, however, to animal models of varying susceptibilities because the literature of the healthy animal response is large and has been detailed elsewhere (Gardner *et al.*, 1988; Lee and Schneider, 1995).

LABORATORY ANIMALS IN HEALTH EFFECTS STUDIES OF AIR POLLUTANTS

The choice of animal species (large or small) in inhalation toxicology studies of air pollutants is frequently driven by factors ranging from their relevance to humans, their availability and expense, to ethical concerns about animal use. Recently, greater interest has been shown in the potential to extrapolate from animal to human based on the genetic technology to identify specific host variables. At present, rodents are most widely used in inhalation toxicological studies while dogs are largely used for cardiovascular studies, monkeys are preferred for selected chronic inhalation studies of pollutants, e.g. the pulmonary effects of ozone (Harkema *et al.*, 1993) and the nervous system effects of manganese (Bird *et al.*, 1984). The advantages and disadvantages associated with use of laboratory animals for air pollution studies are highlighted below.

The advantage of using mice is the availability of a wide variety of immunological reagents and detailed information on their genetic backgrounds. However, the more than 200 strains of laboratory mice vary dramatically in their sensitivity to chemicals. Thus, the selection of given strain may *determine* the 'toxicity' of an inhaled substance or the interpretation thereof (Stuart, 1976). Mice have been used in earlier studies involving pulmonary retention of inhaled radionuclides to gain information on comparative toxicities. However, because the particle deposition pattern in mice can be dramatically different from that in humans, the target organ dose can vary between the mouse and the human (Snipes *et al.*, 1989; Warheit, 1989). Because of their low spontaneous occurrence of tumors, some mouse strains have been shown to be useful in carcinogenicity studies of radiation, the influenza virus and ozonized gasoline, to which they are more susceptible (Dagle and Sanders, 1984; Pott and Stober, 1983). Mice have also been used widely in allergy and asthma research, not so much because of their similarity to humans in terms of the pathobiology of allergy and asthma, but because of their ease of use, the ready access to large numbers of animals (except for special transgenics), and the information available on their genetics and immunology.

Because of their larger size, rats have the advantage of being amenable to more physiological measures and to adequate blood and tissue sampling. The rat's longer life-span also is an advantage in chronic studies (Stuart, 1976); however, as in the case of the mouse, strain-related differences in sensitivity may need to be considered in model selection. Unlike the mouse, immunological reagents and detailed genetic information are less available for rats. It is hoped that comparable information on rat molecular genetics and immunology will soon become more widely available, because the rat has long been the preferred test species in most inhalation and toxicology studies.

The extensive use of the rat in inhalation studies has provided considerable basic information about its morphologic and physiologic responses to injury, thereby supporting its continued use. The rat model has been adopted as the standard for most bioassessments (e.g. National Toxicology Program) and remains the species of choice in studies of inhalant-induced fibrosis and carcinoma. In the 1970s, a number of studies of inhaled radionuclides yielded important information on lung deposition and dose variability (Bianco *et al.*, 1980) in the rat. Acute and chronic inhalation studies of occupationally hazardous substances have been conducted using rats (Leach *et al.*, 1973). Recent studies addressing the potential long-term consequence of particles such as diesel, soot, carbon black, toner, etc. have demonstrated that the rat may be uniquely sensitive to clearance overload and progressive accumulation of particles in the lung over time (Brockmann *et al.*, 1998; Mauderly *et al.*, 1994; Oberdorster, 1995; Valberg and Watson, 1996). This accumulation occurs at relatively high particle concentrations and appears to predispose the rat to tumorigenesis and fibrosis. The mouse and hamster do not appear to respond to the same extent, although they also accumulate the particles. Coal miners who have high lung burdens of dust also do not appear to develop tumors. The reasons for these discrepancies are unclear and may relate to concomitant epithelial turnover and inflammation in the rat. Whatever the explanation, these differences in pathogenesis emphasize the need for cautious interpretation of any results and care in the selection of animal species in inhalation studies. Nevertheless, most biological responses appear reasonably comparable or can be interpreted in light of the extensive database and basic understanding of the species differences.

Because hamsters develop relatively few spontaneous lung tumors and have a high resistance to infection, they were frequently chosen for carcinogenicity studies in the 1960s and 1970s (Heinrich *et al.*, 1986). Hamsters also have been used in chronic studies of emission by-products such as diesel exhaust and most notably, cigarette smoke (Stuart, 1976). As the rat grew in favor as the standard animal for inhalation exposure studies (in large part due to the availability of standardized inbred strains), the hamster was used primarily for the development of animal models of clinical disease (e.g. emphysema) and studies of drug-induced lung pathobiology (e.g. bleomycin-induced fibrosis) (Gurujeyalakshmi *et al.*, 1998; Qian and Mitzner, 1989). However, the use of hamsters in studying these disorders with air pollution has been limited, possibly due to the small database on the basic biology and health effects of air pollution in this species.

Guinea pigs also have a long history of use in inhalation toxicology. Their sensitive bronchoconstrictive response to irritant inhalants and antigens has sustained their use in acute and subacute studies of episodic air pollution exposures (Hatch *et al.*, 1986a,b; Hegele *et al.*, 1993). However, because they grow quickly to adult size, their use has been limited in chronic inhalation studies. Like humans, they have eosinophils in broncho-alveolar lavage fluid, and thus their inflammatory responses to inhaled pollutants are comparable to those of humans, although the eosinophilia is somewhat overexpressed (Hernandez *et al.*, 1994). Also, the guinea pig model of ascorbate deficiency closely resembles the human situation because, unlike other rodents, it requires dietary ascorbate supplementation (Hatch *et al.*, 1986b). The unique placement of the guinea pig on the evolutionary tree has raised questions about its appropriateness. However, some of its features make it an ideal species for acute and allergic pulmonary reactions. The greatest limitation to the use of the guinea pig is the virtual lack of immunologic reagents if one wishes to conduct state-of-the-art molecular biology studies.

Larger laboratory animal species such as the dog have been used in cardiopulmonary studies of air pollution because their lung structure and size resemble that of humans (Plate 1), and they rarely develop spontaneous lung tumors (Heyder and Takenaka, 1996; Reif *et al.*, 1970). The dog's basic lung physiology also is known to be similar to that of the human. The mechanism by which pulmonary interstitial edema occurs in dogs following myocardial infarction also appears to be much like that in the human (Slutsky *et al.*, 1983). The hematopoietic system of the beagle dog and its development of humoral and cell-mediated immunity also parallels that of humans (Bloom *et al.*, 1987; Hahn *et al.*, 1991). Both the size and the anatomy of the dog lung provide advantages for studies of particle deposition, especially when attempting to address questions of size-dependent distal lung injury (Cohen, 1996; Fang *et al.*, 1993). The dog also has been used to model chronic human bronchitis induced by SO_2 inhalation; the model has been shown to closely resemble the human disease (Drazen *et al.*, 1982; Greene *et al.*, 1984).

For many years, the dog has been used to assess acute and longer-term effects, including radionuclides and particle-induced tumor formation (Heyder and Takenaka, 1996). Recently, the dog also has been employed to study the cardiological effects of environmental particulate exposure to support epidemiological findings in humans (Godleski *et al.*, 1997). However, despite these advantages, the use of the dog in toxicological research has declined over the years because of stricter husbandry regulations, difficulties in procurement, greater expense, handling difficulties and the arousal of public sentiment against its use in experiments.

Inert dust aerosol deposition and translocation studies in miniature swine have yielded data similar to that for humans. This species is similar to humans in many regards, including their size, diet, gastrointestinal tract, skin characteristics, and their long life span (Stuart, 1976). Swine and bovine tissues also can be available for development of *in vitro* models, because large amounts of tissue can be obtained from the meat industry. Bovine pulmonary parenchyma and artery endothelial cells have been used extensively for studies not only of pollutants, but also of pharmaceuticals (Fukui *et al.*, 1996; Madden *et al.*, 1987; Ochoa *et al.*, 1997). The equine lung also has been shown to be very similar to the human lung in terms of gross anatomy and intermediate alveolarization of the distal airways. Studies have been conducted in horses to evaluate deposition and clearance of radiolabelled particles, and in donkeys to study the effects of inhaled cigarette smoke and SO_2 on particle transport (Stuart, 1976).

Monkeys are frequently preferred for research on heart diseases, for vaccine development, and for AIDS and anesthesia research (Ghoniem *et al.*, 1996; Petry and Luke, 1997). Monkeys also have been used to study the pulmonary effects of chronic SO_2, acid mists and O_3 (Alarie *et al.*, 1975). The monkey has been shown to be particularly useful for studies of the end airway morphometric changes associated with long-term O_3 exposure (Harkema *et al.*, 1993). Squirrel monkeys exposed to NO_2 have exhibited responses ranging from slight pathology to mortality after challenges with influenza viruses (Henry *et al.*, 1970). Of all the laboratory animals, the monkey has the lung structure most similar to that of humans. For this reason, studies of disease pathogenesis associated with air pollution exposure in this species provide the most convincing data regarding potential human health effects. Although the monkey can be a very useful animal model for many pulmonary studies, their current use in research is generally limited to very specific applications such as infectious diseases and drug testing, because of difficulties associated with their availability and expense, husbandry demands, ethics, and the need for large numbers for statistical power.

AIR POLLUTION AND ANIMAL MODELS FOR THE STUDY OF VARYING SUSCEPTIBILITIES

The Clean Air Act of 1970 mandates that regulations for specific air pollutants be sufficiently stringent to protect susceptible subpopulations (US Environmental Protection
Agency, 1987). Epidemiological studies over the last decade have suggested that children,
asthmatics and elderly people with pre-existing cardiopulmonary diseases may be more
susceptible to air pollution-induced injury (Dockery *et al.*, 1993; Pope *et al.*, 1995;
Schwartz, 1994). This susceptibility concern has triggered interest among toxicologists to
seek a better understanding of biologically plausible mechanisms of cardiopulmonary
impairments using laboratory animal models of varying susceptibilities. Below are descriptions of a number of animal models that reflect susceptible human subgroups and selected
studies involving air pollutants.

Age

While it might be expected that young and old humans would react differently to air pollutants, most animal studies have not included these potentially susceptible subgroups.
Inhalation studies typically focus on the effects in young adult rodents as a standardized
model. There have been only a few studies where oxidant gases such as O_3, NO_2 and
oxygen (O_2) have been investigated using animal models from different age groups
(Montgomery *et al.*, 1987; Weinstock and Beck, 1988). It is apparent from these studies
that very young rats are more tolerant to damage induced by oxidant gases (Mauderly *et
al.*, 1987; Mustafa *et al.*, 1985), whereas very young mice appear to be more susceptible
(Sherwin and Richters, 1985). Older rats, on the other hand, have been shown to be
more susceptible to O_3-induced lung injury (Stiles and Tyler, 1988; Vincent and
Adamson, 1995). Some of these studies have yielded equivocal results, thus, there
remains controversy in the assessment of age sensitivity, especially with regards to longer-
term outcomes. It is presumed but not confirmed that structural attributes of the lung
and the inductiveness of antioxidative mechanisms play critical roles in determining the
oxidant resistance of neonatal and young rodent models (Mustafa *et al.*, 1985; Tyson *et
al.*, 1982).

A study reported by Mauderly *et al.* (1987) in which rats were exposed to diesel particles during maturation (from birth through weaning) suggested that adults were more
susceptible than the young in terms of degree of pulmonary injury, the efficiency of lung
clearance of radiolabeled particles, and collagen accumulation. However, it is not known
whether young mice respond differently to these pollutants than they do to O_3. The particle deposition patterns can vary with the stage of lung development, since alveolarization
and dimensional changes occur in the respiratory tree after birth in both laboratory animals and humans (Weinstock and Beck, 1988). Age-related susceptibility also may
depend upon the type of pollutant and the affected target, since site-specific cellular proteins are modified during development and the spectrum of gene expression evolves.
Since the incidences of spontaneous cancer are increased in most aging animals and
humans, it is likely that older animals may show different responses in terms of the
carcinogenic effects of inhaled pollutants. The knowledge available from existing studies

on young adult animals provides an important reference point when deciding what dose levels should be used for comparable studies in different age subgroups.

Gender

Because of basic physiological differences between men and women, pollutant health effects may differ (Beck and Weinstock, 1988). Comparative studies involving animal models representing both genders can help us understand whether differences in responsiveness are due to these innate biological dissimilarities. For this reason standardized bioassays such as those regulating test requirements incorporate both genders. However, past studies of air pollutants have only occasionally included males and females. Since human gender differences are known to exist in the case of cigarette smoke and O_3 effects, the limited prospective provided by animal studies is unfortunate (Beck and Weinstock, 1988; Bush et al., 1996; Seal et al., 1993). It appears from the epidemiology of cigarette smoke health effects that differences in relative lung size and density of tracheobronchial mucus secretory cells between men and women play a critical role in their susceptibility to smoke-induced bronchitis (reviewed in Beck and Weinstock, 1988). Differences in the structure of the lung and its subcomponents could influence deposition and clearance of inhaled substances in a gender-related manner. In general, epidemiological findings suggest the fact that females are more resistant to the harmful effects of chronic cigarette smoke inhalation than males (Enjeti et al., 1978; Tager and Speizer, 1976). However, acute O_3 exposure studies designed to evaluate gender differences with regard to effects on forced vital capacity and forced expiratory volume in 1 second, tidal volume and breathing frequency have failed to reveal any differences between men and women (Messineo and Adams, 1990). Analogously, the chronic effects of O_3 in Fischer 344 male and females rats appear to be similar with regard to resultant pathology and functional changes (Stockstill et al., 1995). The differences in chronic responsiveness of men and women to cigarette smoke and lack thereof in those humans and animals exposed to O_3 may reflect the complexity of cigarette smoke and its component effects on various cell types of the lung.

When tracheal epithelial cells from male and female rats are exposed to cigarette smoke, it has been shown that cells from females secrete more mucus than those from males. Mucus production may relate to more efficient removal of harmful smoke particles from conducting airways and therefore the reduced vulnerability to bronchitis seen in females (Hayashi et al., 1978). This hypothesis has not been tested experimentally using in vivo animal models. The production of mucus has been shown to be influenced by hormonal changes in females, e.g. postmenopausal women do not exhibit estrous cycle-related changes in the mucus-secreting cells of the airways (Chalon et al., 1971). It is possible that with some pollutants or animals species, the gender-related differences may not be significant enough to make adjustments in regulatory decisions, however, understanding these differences is critical in making meaningful evaluations of health risk.

Pregnancy, which can be considered a temporary physiological condition in females, may result in increased susceptibility to pollutant-induced pulmonary injury. Associated with pregnancy are fetal growth and development, which may be directly influenced by the pollutant or indirectly affected through decrements in the pulmonary health of the mother. Developmental effects of inhaled pollutants, especially O_3, have been investigated in mice (Bignami *et al.*, 1994). Moderate effects on selected neurobehavioral tests were noted in newborn mice when dams were exposed to O_3 during pregnancy and the neonatal period. O_3 also has been shown to be more toxic to pregnant and lactating rats when compared with age-matched non-pregnant controls (Gunnison *et al.*, 1992; Gunnison and Finkelstein, 1997). CO is also well studied in terms of the mechanism by which early fetal mortality and low birthweight occurs in exposed individuals, especially smokers (Acevedo and Ahmed, 1998; Seker-Walker *et al.*, 1997). Studies have shown that placental blood flow can be compromised by cigarette smoke. It is not yet known if some of these effects are nicotinic or secondary irritant responses (Economides and Braithwaite, 1994). The limited data on air pollution effects during pregnancy warrants further investigations to evaluate possible health effects of pollutants in pregnant animal models and developing fetuses.

Species/Strain

A variety of laboratory animal species and strains are used in the assessment of pollutant-induced pulmonary health effects. Selection of an appropriate animal species may depend largely upon how relevant it is to the human, because ultimately extrapolation is required to make a fair evaluation of the human health risks of air pollutants (Brain *et al.*, 1988b; Warheit, 1989). It is recommended that the response of the selected animal model species to the test material is similar to that of the human (Weil, 1972). In the case of inhaled pollutants, the deposition, clearance, metabolism, absorption, storage, and other potentially species-based physiological aspects of the animal model should be appreciated. However, in most instances our knowledge on all these aspects is not available *a priori*. Ideally, it is recommended that more than one species be used for the initial characterization of any toxic response (Brain *et al.*, 1988b). The obvious disadvantage to using multiple species is the greater time, effort and expense that is required. However, this disadvantage should be outweighed by the information that can be provided for improving the risk assessment process, and better understanding of the pathobiological mechanisms of injury and disease (Brain *et al.*, 1988b).

There are marked morphological and morphometrical structural differences between the human and laboratory animal lung (reviewed in Warheit, 1989). The branching patterns of the conducting airways differ: human lungs are dichotomous and essentially symmetrical, while in non-primate animals the lungs are highly asymmetrical and monopodal. As a result, particle deposition in humans is less uniform, while in animals a more uniform distribution is achieved (Lippmann and Schlesinger, 1984; Phalen and Oldham, 1983). There are also marked differences in airway structures among animal species that may influence the impact of inhaled materials. For example, the distal airway structures of dogs, cats and macaque monkeys are somewhat similar to those in humans in having respiratory bronchioles; however, in most small rodents the terminal bronchioles terminate directly into alveolar units (Tyler, 1983). Marked species differences are also apparent in the cell populations distributed throughout the lung (Phalen *et al.*, 1989). In the sheep, mucus goblet cells are the predominant secretory cells, but in the mouse, Clara cells are the primary secretory cells (Plopper, 1983).

Descriptions of each animal species and strains that have been used in studies of air pollution are beyond the scope of this chapter. The salient issues related to species differences in the health effects of air pollutants have been presented by Brain *et al.* (1988b). An excellent review comparing the lung responses to inhaled particles and gases across many commonly used species has been provided by Warheit (1989). Some examples are presented below in the context of varied species and strain-related responses.

Rodent models of genetic or strain-related susceptibility to O_3 have provided an important tool for understanding the biological basis of variable individual responses to environmental pollutants (Kleeberger, 1995). The mouse strain C57B6 is susceptible to O_3-induced neutrophilic inflammation, whereas the C3H/HeJ is not. It has been predicted that single gene inheritance at chromosomal locus *Inf* is responsible for the differing susceptibility between these strains (Kleeberger, 1995). Recently we have shown that combustion particles are fibrogenic and cause fibronectin gene expression in Sprague-Dawley rat, while the Fischer 344 rat is relatively less sensitive (Kodavanti *et al.*, 1997a). Species differences have also been noted in a number of studies in terms of their pulmonary antioxidant pools and their responsiveness to air pollutants (Hatch *et al.*, 1986a; Hatch, 1992). It is likely that the observed differences may reside in genetic strain-related susceptibility or resistance in these rats. The more understanding we have about the species-associated genetic susceptibilities of commonly used laboratory animal models and humans, the better our extrapolations will be.

Nutrition

Since most air pollutants induce injury through oxidative mechanisms, nutritional deficiencies of antioxidant vitamins constitute a major concern regarding susceptibility (Colditz *et al.*, 1988; Hatch, 1995; Menzel, 1992; Pryor, 1991; Shakman, 1974). Table 10.1 summarizes notable studies of air pollution health effects in nutritionally compromised animal models. Animal models of altered nutritional status can be produced in most instances by dietary manipulation. In the guinea pig, a deficiency in vitamin C – a critical pulmonary antioxidant – can be achieved by feeding a deficient diet for 2–3 weeks, since guinea pigs, like humans, cannot synthesize their own vitamin C (Hatch *et al.*, 1986b). However, this is not the case for vitamin C deficiency in the rat, since rats produce endogenous vitamin C. A rat model of vitamin E deficiency can be developed by dietary restriction (Chow *et al.*, 1979; Goldstein *et al.*, 1970). It has been shown that O_3 and NO_2-induced pulmonary injuries are exacerbated in vitamin C deficient guinea pigs, especially at relatively low concentrations of O_3 (Kodavanti *et al.*, 1995a,b, 1996a; Slade *et al.*, 1989). Similarly, vitamin E deficiency in the rat has been associated with greater pulmonary injury from O_3 (Goldstein *et al.*, 1970; Sato *et al.*, 1976) and from NO_2 (Ayaz and Csallany, 1978; Elsayed and Mustafa, 1982; Menzel, 1979) at relatively lower concentrations. This lipid-soluble membrane-bound antioxidant is thought to be critical in scavenging lipid peroxides produced by free radicals at the lung's surface (Hatch, 1995; Pryor, 1991).

Vitamin A is important in the maintenance, differentiation and proliferation of epithelial cells, activities common to both normal lungs and during injury (Takahashi *et al.*, 1993). Severe vitamin A deficiency alone has been shown to cause bronchiolitis and pneumonia in diet-restricted animal models (Bauernfeind, 1986). Decreased labeling of alveolar and bronchiolar epithelial cells have been reported following O_3 exposure in rats deficient in vitamin A (Takahashi *et al.*, 1993). Similarly, a rat model of dietary

Table 10.1 Air pollution studies using rodent models of nutritional manipulation[a]

Nutritional manipulation	Species	Pollutant	Outcome	References
Selenium-deficient	Rats	O_3	Increased lung injury and lipid peroxidation	Eskew *et al.*, 1986
	Mice	O_3	No stimulation of glutathione shunt enzymes	Elsayed *et al.*, 1983
Vitamin E-deficient	Rats	O_3	Pulmonary injury from 0.1 ppm O_3. No effect from longer-term O_3.	Goldstein *et al.* (1970), Sato *et al.* (1976)
	Rats	NO_2	Increased lipid peroxidation	Menzel (1979), Elsayed and Mustafa (1982)
	Mice	NO_2	Suppressed blood and lung glutathione peroxidase	Ayaz and Csallany (1978)
Vitamin C deficiency	Guinea pigs	O_3	Marked increase in pulmonary injury with short term and moderate increase with longer-term O_3	Hatch *et al.* (1986b), Kodavanti *et al.* (1995a,b, 1996a)
Glutathione deficiency	Rats	O_3	Increased pulmonary fibrosis	Sun *et al.* (1988)
Vitamin B-$_6$ deficiency	Rats	O_3	Mortality due to O_3 and no increase in lysyl oxidase and collagen synthesis in vitamin B$_6$-deficient rats	Myers *et al.* (1986)
Vitamin A deficiency	Rats	O_3	Increased epithelial damage	Takahashi *et al.* (1993)
Food restriction	Rats	O_3	Only a modest increase or a decrease in O_3 toxicity	Dubick *et al.* (1985), Kari *et al.* (1997)
Protein deficiency	Rats	O_3	No effects on O_3 toxicity	Dubick *et al.* (1985)

[a] This table is not meant to include all available nutritional deficiency/excess models; the models used for air pollution studies in the past are listed. Also the table does not provide an exhaustive list of those studies; rather, major studies are listed.

vitamin B$_6$ deficiency has shown impaired collagen cross-linking following O_3 exposure. The mechanism appears to involve the role of vitamin B$_6$ in the action of lysyl oxidase, a rate-limiting enzyme in collagen cross-linking (Myers *et al.*, 1986). Perinatal vitamin B$_6$ deficiency also causes increased mortality in O_3-exposed rat pups (Myers *et al.*, 1986).

Glutathione (GSH) deficiency has been achieved by treating rats with buthionine sulfoxamine, which inhibits its synthesis (Sun *et al.*, 1988). The pulmonary tissue levels of GSH can also be depleted by treating animals with diethyl maleate without affecting its synthesis pathway. Isolated perfused lungs from an animal injected with diethyl maleate have been used as an *ex vivo* animal model to study the mechanism of oxidant-induced lung injury in GSH deficiency (Joshi *et al.*, 1986, 1988). Glutathione (GSH)-deficient rats have been shown to be more susceptible to O_3-induced fibrosis, suggesting that GSH may function as an antioxidant in oxidant-induced lung injury (Sun *et al.*, 1988).

Selenium is another essential nutrient which has been shown to affect the toxicity of oxidant air pollutants (Elsayed *et al.*, 1983). GSH peroxidase, involved in GSH utilization and the removal of free radicals, has four selenium residues that are critical to enzyme activity. It has been shown that selenium-deficient rats are more susceptible to O_3-induced lung injury (Eskew *et al.*, 1986). Likewise, the impact of O_3 on the activities of GSH recycling enzymes in the lungs of mice can be directly influenced by dietary selenium deficiency (Elsayed *et al.*, 1983). It is reasonable to assume that many other pollutants may act in an analogous manner; however, more studies are needed to understand the role of selenium in oxidative lung damage.

Nutritional deficiencies in the human population are common and alone may constitute a major modifier of air pollutant responsiveness (reviewed in Hatch, 1995). Inner city populations under socioeconomic stress are often deficient in many essential antioxidants, while also enduring the highest potential for exposure to urban pollution. However, little attention has been paid to the potential health risks of air pollutants and the influence of nutritional status. How nutritionally deficient animal models can provide data for such consideration remains to be determined.

Physiological Status

Exercise can directly increase three major cardiopulmonary parameters: pulmonary ventilation (tidal volume and respiratory rate), cardiac output and blood flow (Brain *et al.*, 1988c). It can also profoundly affect both total pollutant deposition and pollutant distribution at sites in the lung. Most humans, and especially children, are likely to be more heavily exposed to toxic materials than has been appreciated because of varied activity profiles or work-related exercise. Thus, exercise is typically used in human exposure studies (e.g. acute O_3) to attain maximal sensitivity for detecting effects (Devlin *et al.*, 1991; Koren *et al.*, 1989). Exercise imposition has been occasionally applied to rodent studies of air pollutants, with the bulk of the data coming from studies evaluating changes in the deposition pattern of particles and gases associated with altered breathing patterns (Brain *et al.*, 1988c; Hatch *et al.*, 1994; Mautz *et al.*, 1988; Tepper *et al.*, 1990, 1991).

It has been shown that the deposition of inhaled $^{35}SO_2$ in the nose and along the airways of dogs can be profoundly affected when flow rate is increased (reviewed in Brain *et al.*, 1988c). For example, a 10-fold increase in a flow rate produces a 320-fold increase in the amount of $^{35}SO_2$ presented to the trachea. When the increased flow rate was directed through the mouth, as would be expected in exercising humans, the penetration increases to 660-fold (Frank *et al.*, 1969; Brain 1970). Syrian hamsters similarly have been shown to deposit 2–3 fold more ^{99m}Tc aerosols when minute ventilation (i.e. oxygen consumption) is doubled on an exercise wheel (Harbison and Brain, 1983). Other experimental manipulations have been attempted in animal studies to mimic the increased ventilation of exercising humans (e.g. increased CO_2 concentrations in the breathing air) (Hatch *et al.*, 1994; Tepper *et al.*, 1990). These and other studies support the contention that exercise or elevated ventilation can be a significant contributor to susceptibility to air pollutants. This effect needs to be evaluated further to ascertain the potential risk that physical activity may pose for humans. It should be understood, however, that putative susceptible human subgroups (e.g. those with chronic cardiopulmonary ailments) are less likely to have high exercise profiles.

Respiratory disease models

Epidemiological evidence linking air pollution to increased mortality and morbidity in chronic obstructive pulmonary disease (COPD) patients and increased hospitalization among asthmatics supports the belief that pre-existent disease imposes an increased risk of air pollutant health effects (Dockery *et al.*, 1993; Pope *et al.*, 1995; Schwartz, 1994). The collective epidemiological evidence for effects in both healthy and susceptible groups has led the Environmental Protection Agency to revise the current National Ambient Air Quality Standard (NAAQS) for PM air pollution, since effects have been observed at or below the levels previously thought to be safe (McClellan and Miller, 1997). Critics of this decision suggest that the observations lack biological plausibility and that the animal toxicity database is neither sufficient nor confirmatory enough to support the revised standards (McClellan and Miller, 1997; Vendal, 1997). While the health effects of air pollutants are noted specifically in susceptible individuals (such as those with pre-existent cardiopulmonary diseases), a question has been raised about the validity of using toxicity data derived from healthy animals to make risk estimations. Although the susceptibility of individuals with pre-existent disease to the adverse effects of pollutants is generally acknowledged, less attention has been paid to the use of animal disease models in addressing this issue because of the complexity associated with variability in responses and extrapolation of data to humans (Kodavanti *et al.*, 1998). In this section, we provide an overview/description of those rodent respiratory disease models that have been employed or that have the potential for use in studies of air pollutant susceptibility. We then look at their actual application in air pollution toxicity studies. Finally, we address several issues associated with disease models use, and consider the future of research in this field.

Disease models can be developed by chemical, surgical or genetic manipulation of the animal species. Many animal models of respiratory disease are well established and relevant reviews and papers have been published dealing with criteria, pathogenesis and their limited use in air pollution studies (Gilmour and Koren, 1999; Kodavanti *et al.*, 1998; Kumar, 1995; Reid, 1980; Slauson and Hahn, 1980; Sweeney *et al.*, 1988). A recently published review paper focuses on a broad spectrum of rodent models of cardiopulmonary diseases (Kodavanti *et al.*, 1998). Below is a brief description of several well-established and characterized respiratory disease models that may be useful for the study of air pollution susceptibility. Selected examples of air pollution health effect studies using animal models of cardiopulmonary disease are given in Table 10.2.

Asthma/Allergy

Asthma is a complex obstructive lung disease characterized in humans by reversible bronchoconstriction, mucus production, airways inflammation, and hyperreactivity to pharmacologic bronchoconstrictor agonists (McFadden and Hejal, 1995). Frequently there is an allergic component to the disease (atopy), which may be reflected in increased serum IgE antibody titer (Arm and Lee, 1992; Frew, 1996). The incidence of asthma in humans has been rising in developed countries and is a major health concern worldwide (Cookson and Moffatt, 1997). Asthma typically is episodic but can lead to chronic lung disease and, when triggered acutely, it may even be life-threatening. It is believed that environmental and genetic interactions are involved in expression (initiation and elicitation)

Table 10.2 Animal models of pulmonary disease employed in selected air pollution studies to determine susceptibility[a]

Human disease type	Rodent species used for model	Experimental manipulations	Air pollutants studied	Reference (selected air pollution studies)
Bronchitis	Rat, hamster	SO_2, 4–8 wks, 200–500 ppm	Aerosol, PM	Sweeney et al. (1995), Godleski et al. (1996)
	Dog	SO_2, 4–8 wks, 250–500 ppm	–	–
	Guinea pig	Polymyxin B	–	–
Emphysema	Rat, hamster, mouse	Pancreatic elastase	Aerosol, PM, NO_2, SO_2, O_3, cigarette smoke, O_2, ammonium sulfate	Raub et al. (1983), Costa and Lehmann (1984), Loscutoff et al. (1985), Sweeney et al. (1987), Yokoyama et al. (1987), Smith et al. (1989), Mauderly et al. (1989, 1990)
	Mouse	Genetic	–	–
Asthma/allergy	Mouse, guinea pig, Brown Norway rat	Ovalbumin	O_3, NO_2, SO_2, PM	Matsumura et al. (1972), Reidel et al. (1988), Selgrade and Gilmour (1994), Gilmour (1995), van Loveren (1996), Gilmour and Koren (1999), Takano et al. (1997)
	Brown Norway rat	Dust-mite	NO_2	Gilmour et al. (1996)
COPD	Hamster	Elastase + SO_2	–	–
Pulmonary fibrosis	Rats, hamster, mouse	Bleomycin	PM	Sweeney et al. (1983), Costa and Lehmann (1984), Adamson and Hedgecock (1995), Adamson and Prieditis (1995), Kodavanti et al. (1996b)
Pulmonary infections viral	Mouse, rat	Influenza, respiratory syncytial virus (RSV)	O_3, PM, phosgene	Gardner (1982), Selgrade et al. (1988, 1989), Jakab and Himieleski (1988), Jakab and Bassett (1990), Ehrlich and Burleson (1991), Lebrec and Burleson (1994), Selgrade and Gilmour (1994), Gilmour and Koren (1999)
Pulmonary infections bacterial	Mouse, rat	S. zooepidemicus, H. influenzae, S. aureus, P. aeruginosa	PM, O_3, NO_2, phosgene, SO_2, metals	Goldstein et al. (1971), Ehrlich et al. (1979), Hatch et al. (1985), Gilmour et al. (1993 a,b), Jakab (1993), Lebrec and Burleson (1994), Selgrade and Gilmour (1994), Gilmour and Koren (1999)
Pulmonary vasculitis/ hypertension	Rat	Monocrotaline, 40–60 mg/kg, i.p., or s.c., 10–15 days	PM	Costa et al. (1994), Killingsworth et al. (1997), Gardner et al. (1997), Kodavanti et al. (1997b), Watkinson et al. (1998)

[a]This table is not meant to include all available rodent pulmonary disease models; the models that are more popular are included. Also the table does not provide an exhaustive list of all air pollutant studies. A list of major studies, and in some cases pertinent review papers are cited. –, not known to be used in air pollution studies.

of the disease. Several genes have been identified in humans that are linked to asthma susceptibility (Kim *et al.*, 1998; Ober, 1998).

Animal models of asthma have been widely characterized and used in research (Cassee and van Bree, 1996; Gilmour, 1995; Karol *et al.*, 1985; Selgrade *et al.*, 1997; Selgrade and Gilmour, 1994; van Loveren *et al.*, 1996). The most popular models of laboratory rodents include ovalbumin sensitized and subsequently challenged mice, Brown Norway rats and guinea pigs (Elwood *et al.*, 1991; Gilmour, 1995; Selgrade *et al.*, 1997; Selgrade and Gilmour, 1994). The model development protocols in general include initial intraperitoneal or subcutaneous injection of an allergen (ovalbumin in most cases) in the host (sensitization) followed by (2–3 weeks later) large inhalation challenge (elicitation) with the allergen (Gilmour, 1995; Selgrade and Gilmour, 1994). A house dust mite allergen has also been used successfully to develop allergic rat and mouse models (Cheng *et al.*, 1998; Gilmour and Selgrade, 1996). One or several challenge doses of allergen increases the expression of pulmonary T helper cell 2 (TH$_2$) cytokines and eosinophilic as well as neutrophilic airways inflammation, injury and airway hyperresponsiveness (Selgrade *et al.*, 1997). Serum and bronchoalveolar lavage fluid levels of IgE are also increased (Selgrade and Gilmour, 1994). These acute changes are analogous to human asthma and may persist for one to several weeks, depending on the sensitization and challenge protocols used. Large animal models have also been developed (such as the dog and the sheep) using biological antigens, e.g. Ascaris protein (Matsumoto and Ashida, 1993; Reynolds *et al.*, 1997). However, these large animal models are generally limited to mechanistic studies because of the size of the animals and hence the limitation in sample numbers.

Each asthma model in air pollution studies has its own advantages and disadvantages. The mouse models are amenable for immunological and molecular investigations because of the variety of probes that are available, but unlike humans, their bronchoconstriction response is weak (Corry *et al.*, 1996). Guinea pigs, in contrast, have a strong bronchoconstrictive response to pharmacologic agonists, but the limited availability of guinea pig-specific immunological and molecular probes is problematic (Savoie *et al.*, 1995). Use of the Brown Norway rat is growing since it develops eosinophilia and airway hyperresponsiveness after sensitization and challenge; both are analogous to the reactions in atopic human asthmatics (Bellofiore and Martin, 1988). Access to probes is not as common as for the mouse, however. The use of the mouse and other asthma models is limited to acute studies, because none has yet developed the 'chronicity' of airways inflammation, fibrosis, and recurrent congestion that exist in humans. A number of studies have been done of the interactions of environmental pollutants with allergen sensitization and the relative susceptibilities of these rodent asthma models (Gilmour, 1995; Selgrade and Gilmour, 1994). However, much remains to be investigated in terms of biological mechanisms.

Bronchitis

Human bronchitis is associated with chronic cigarette smoking and is characterized by increased mucus secretion and plugging, airways inflammation, airways fibrosis, damage to the ciliary epithelium, airway obstruction and hyperresponsiveness (Wilson and Reyner, 1995). Bronchitis-like disease can be induced in animals by extended exposure to high levels of SO$_2$. The rat and dog are the preferred species (Drazen *et al.*, 1982; Shore *et al.*, 1987, 1995), although guinea pigs and hamsters have also seen limited use

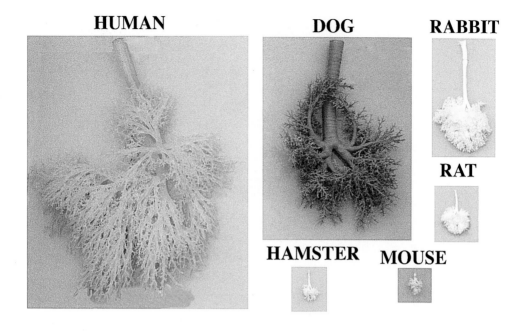

Plate 1. Species comparison of latex lung casts showing size and structural variabilities.

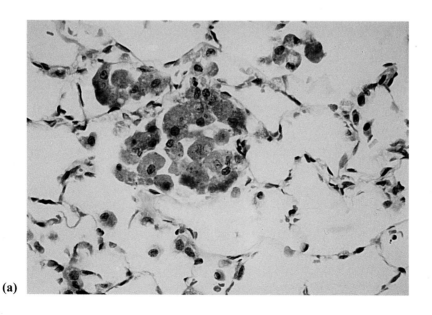

(a)

Plate 2 (a). See legend overleaf.

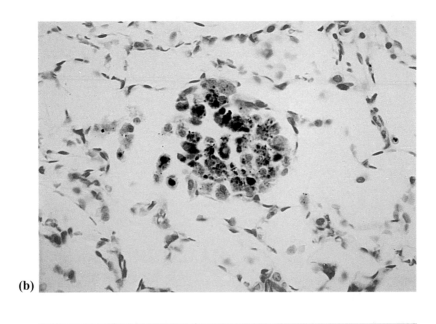

(b)

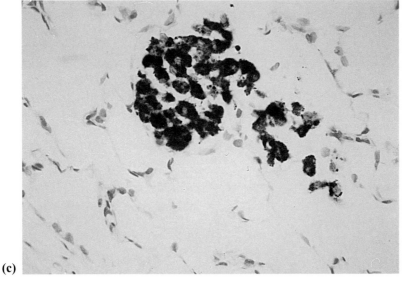

(c)

Plate 2. Iron stain of rat lung tissue after exposure to a carbonaceous component of an ambient air pollution particle (magnification approximately 200x). Two filters from Salt Lake City were agitated in aqueous solution at pH 8.5. After centrifugation, the supernatant was acidified to a pH of 2.0. The pellet was lyophilized and instilled into rats under halothane anesthesia (2-5%). The rats were euthanized immediately (a), at 2 weeks (b) and 4 weeks (c). The metal (blue stain) is in greater concentration in the macrophages that have phagocytosed the pigmented carbonaceous material isolated from the air pollution particles. This uptake of iron is not immediate and the metal sequestered in the rat lung appears to be endogenous.

(Goldring *et al.*, 1970; Ito *et al.*, 1995). Exposure of rats and dogs to SO_2 at concentrations of 200–600 ppm, usually 4–6 h/day, five days a week for a total of 4–8 weeks, has been shown to induce bronchitic lesions that resemble the human pathophysiology (Drazen *et al.*, 1982; Shore *et al.*, 1995). Although the dog respiratory physiology and structure perhaps more closely resembles that of humans, SO_2 causes a decrease in airway responsiveness to bronchoconstrictor agonists, while in humans and in rats airways responsiveness is increased with the disease (Shore *et al.*, 1995).

Tracheobronchial submucosal glands and epithelial goblet cells contribute to the mucus layer in humans, in contrast to the rat in which Clara cells and those limited goblet cells that exist produce the mucus (Goco *et al.*, 1963). With initial challenge, the goblet cells proliferate and contribute the bulk of the excess (Basbaum *et al.*, 1990). Unlike that in the human, the lesion in the rat readily reverses when SO_2 is removed (Snider, 1992a; US Department of Health, Education and Welfare, 1969). Although some improvement in bronchitic patients can be achieved by removal of the causative agents (e.g. smoking), the time required for reversal can extend to several months or years, and in humans can be confounded by infection as well as by frequent use of medication (Wilson and Reyner, 1995). Since the rat model of SO_2-induced bronchitis differs from human disease in this regard, the reversibility aspect should be carefully considered when selecting these models to study air pollution susceptibility, especially with longer-term pollution exposures.

Repeated exposure to lipopolysaccharide (endotoxin) in rats results in bronchitis analogous to human disease in terms of mucus hypersecretion, airways inflammation, fibrosis and epithelial cell hyperplasia (Harkema and Hotchkiss, 1992). The essential difference between SO_2 and endotoxin models is the degree of airway inflammation (greater in endotoxin model) and mucus hypersecretion (greater in sulfur dioxide model). The guinea pig model of polymyxin B-induced bronchitis has been characterized as primarily eosinophilic inflammation but the goblet cell pathology is not well defined (Ogawa *et al.*, 1994). Thus far only the SO_2 model of bronchitis has been used in air pollution studies.

Emphysema

In humans, emphysema is most often associated with cigarette smoking and the genetic deficiency of α-1 antiprotease (Surgeon General, 1984). Exposures to cadmium, NO_2 and hyperoxia as well as starvation and copper deficiency, can also induce emphysema-like disease, but there are pathological differences based on the type of injury (Snider, 1992b). The most common form of emphysema in humans is centriaciner, with the disease being centered in bronchio-alveolar duct junctions consistent with a pattern of inhaled toxicant deposition (Snider, 1992b). In contrast, emphysema associated with α-1 antiprotease deficiency and elastase overload appears as permanent enlargement of airspaces distal to the terminal bronchioles and is described as panlobular emphysema (Snider *et al.*, 1986). Destruction of alveolar walls and airspace enlargement constitute the morphological hallmarks of emphysema regardless of its distribution. In laboratory animals, emphysema can be readily produced by a single intratracheal exposure to elastase. It most closely resembles the panlobular disease (Snider *et al.*, 1986).

Purified preparations of bovine pancreatic elastase have been commonly used to produce emphysemic animals, although reports utilizing human neutrophil or pancreatic elastase, or even papain can be found (Snider *et al.*, 1986). Hamsters have been used widely to establish the elastase-induced emphysema model because of their sensitivity to elastolytic enzymes. However, rats, mice and rabbits have also been frequently employed in this regard (Deschamps *et al.*, 1995; Karlinsky and Snider, 1978; Snider *et al.*, 1986). Intratracheal instillation of elastase results in acute alveolitis characterized by degradation of elastin, rupture of alveolar epithelium, pulmonary edema, hemorrhage, and inflammation followed by airspace enlargement. The initial inflammation resolves within a few weeks, followed by synthesis of new elastin and collagen; however, the distortion and derangement of alveolar structures are permanent (Snider, 1992b).

Spontaneously occurring genetic models of emphysema are also available. The emphysema of the tight-skin and pallid mice results from a genetic deficiency in α-1 antiprotease (de Santi *et al.*, 1995; Szapiel *et al.*, 1981). Progressive overload of elastolytic activity can be observed from the first months of life. Unlike the exogenous elastase model, total lung elastin appears to decrease over time and inflammation is not evident (as determined by lavage fluid analysis). This model may be a choice in determining emphysema-associated susceptibility without coexistent inflammation. Another strain, the blotchy mouse, also exhibits panlobular emphysema but its genetic impairment is in copper absorption which is critical for lysyl oxidase, an essential connective tissue cross-linking enzyme (Ranga *et al.*, 1993). This model of emphysema may be unrelated to that in humans, since it does not involve α-1 antiprotease dysfunction which is thought to be critical to pathogenesis in humans. As with the bronchitis model, a few air pollution studies have been conducted using emphysema animal models. Those studies involving emphysema models generally show little difference from healthy controls. The lack of underlying inflammation in those models may partly explain their lack of sensitivity.

Chronic Obstructive Pulmonary Disease (COPD)

Chronic obstructive pulmonary disease represents a group of related respiratory diseases presenting with chronic airways obstruction, mucus hypersecretion, chronic inflammation and injury to the pulmonary airways and parenchyma (O'Connor *et al.*, 1989; Repine *et al.*, 1997). Right ventricular hypertrophy and *cor pulmonale* often develop as result of abnormalities in the blood flow through damaged or remodeled and diseased pulmonary vasculature (MacNee, 1994). Typically the term COPD is used to refer collectively to chronic bronchitis and emphysema (Higgins and Thom, 1989). These patients suffer frequent and persistent infections because of their inability to clear infectious agents from the respiratory tract (Jansen *et al.*, 1995; White, 1995). In humans, COPD has been associated with chronic cigarette smoking, chronic asthma or even recurrent infection (Cugell, 1988). Although animal models of bronchitis, emphysema, asthma and airways infections are well characterized as separate entities, there is no one model available that mimics this pathophysiology. Occasionally, rodent models of pulmonary injury/vasculitis and bronchitis are presented as COPD models; however, neither adequately represents the typical COPD characteristics in terms of both pathology and physiology.

Pulmonary Fibrosis

Idiopathic pulmonary fibrosis is relatively rare in humans; however, occupational exposure to various metals, mineral and organic dusts as well as certain drugs have been associated with the disease (Crouch, 1990). The pathogenesis of the disease is characterized by initial pulmonary injury and/or inflammation, and by concomitant repair and subsequently fibrogenesis. The pattern of fibrotic lesions depends upon the causative agent, the type of insult, and the route of exposure (Hepleston, 1991). A number of animal models involving drug-, chemical-, or mineral dust-induced fibrosis have also been described, however, usually at high doses (Crouch, 1990; Hay et al., 1991; Hepleston, 1991). Chronic endotoxin exposure has also been reported to result in airways fibrosis in animals (Harkema and Hotchkiss, 1992). Depending upon the study objectives, models can be selected and produced simply by inhalation, intratracheal or even oral exposure of animals to drugs or mineral dusts.

The most extensively used pulmonary fibrosis model is that induced by bleomycin. Bleomycin, an antineoplastic agent that can be administered systemically or by intratracheal instillation, results in development of alveolar fibrosis in 2–3 weeks in most laboratory animals; the rat and hamster are the most widely used models (Crouch, 1990; Hay et al., 1991; Raghow et al., 1985). The pathogenesis in these models includes initial lung injury, inflammation, upregulation of number of cytokines, fibroblast proliferation and hypertrophy (2–10 days) (Raghow et al., 1985). The initial inflammatory response begins to regress after this period, with a resolution of edema and increased collagen deposition in the lung parenchyma (Hernnas et al., 1992). We and others have used this model in determining particulate-induced lung injury (Adamson and Hedgecock, 1995; Adamson and Prieditis, 1995; Kodavanti et al., 1996b).

Respiratory Infection

Respiratory infection is a common human ailment and is frequently a complicating, if not a direct contributor to most respiratory-associated deaths (Babiuk et al., 1988; Murray and Lopez, 1997; Wilson, 1997). Most infections are self-limiting and are cleared up quickly in healthy individuals (Wells et al., 1981). The outcome or severity of the infection depends on the virulence of the infectious microorganism and the susceptibility of a host. For example, in patients with AIDS, cystic fibrosis or COPD, respiratory infections clear only slowly and are recurrent (Jansen et al., 1995; Waxman et al., 1997; White, 1995). Although animal models of bacterial and viral infection have been used in a variety of studies, their major limitation is that the organisms that infect humans have variable species-dependent virulence in the animal models. And since these infections are often cleared very quickly in animals, their relevance to compromised humans where infections are long lasting is questioned (O'Reilly, 1995).

Acute airways viral infections cause epithelial necrosis, increased bronchial epithelial and endothelial permeability, and inflammatory cell influx. T helper cell 1 (TH_1), and in some infections TH_2, are thought to be involved, as the production of relevant cytokines is increased (Holtzman et al., 1996; Cookson and Moffatt, 1997). Mice and rats are the rodents most frequently used to develop viral infection models through inoculation of infectious viruses in the airway (Green, 1984; Lebrec and Burleson, 1994). A highly virulent and lethal influenza strain (influenza A/Hong Kong/8/68, H3N2 virus) adapted to B6C3F1 and CD mice, a less virulent strain (A/Port Chalmers/1/73, H3N2) adapted to CD mice, and a similar non-lethal strain adapted to rats have all been used in the development of influenza infection models in mice and rats (Lebrec and Burleson, 1994). The lethal mouse model terminates in extensive pneumonia and lung consolidation, thereby limiting its relevance to detailed study of infectious disease; however, the non-lethal rat and mouse models exhibit airways epithelial damage, inflammation and antibody formation, perhaps through an interferon-γ mediated TH_1 response analogous to that in humans (Lebrec and Burleson, 1994). Virus is cleared within a week and recovery is apparent within 2 weeks (Jakab and Hmieleski, 1988; Lebrec and Burleson, 1994). The BALB/c mouse model of human respiratory syncytial virus yields similar pathology to the human, but relatively milder lesions recover over 2–3 weeks (Taylor et al., 1984). A neonatal rat model of parainfluenza (Sendai virus) infection, on the other hand, results in airway pathology and hyperresponsiveness which lasts for several weeks following infection (Sorkness et al., 1991).

Bacterial infections, although not as common in humans as viral infections, can be more pathogenic and may occur more frequently in patients with COPD, AIDS and other respiratory diseases (O'Reilly, 1995). In healthy individuals, pathogenesis is associated with initial proliferation followed by phagocytic cell-mediated clearance, acute lung injury and inflammation. In general, rat and mouse models develop signs similar to those in humans, but the pathology in rodents tends to be relatively mild (Gilmour and Koren, 1999). Rat models include those of *Streptococcus zooepidemicus, Haemophilus influenzae, Pseudomonas aeruginosa* and others (Beaulac et al., 1996; Gilmour and Selgrade, 1993; Maciver et al., 1991). To represent the pathology associated with chronic bacterial infection in humans, attempts have been made to injure the airways by chemical means or by encapsulation of the bacteria in agar beads prior to infection to prolong the presence of bacteria in the airways (reviewed in O'Reilly, 1995). Interaction of air pollutants in modulating the immune system has been relatively well studied using infection models. Some of these will be described in the following applications section.

Pulmonary Vasculitis/Hypertension

Idiopathic pulmonary hypertension in humans is rare, however chronic pulmonary diseases or congenital heart diseases also result in enlargement of the right ventricle and pulmonary hypertension (Robinovitch, 1989). Highlanders also develop a mild form of pulmonary hypertension, but it is different in its pathobiology than the idiopathic form (Hultgren and Grover, 1968). Two rat models of pulmonary vasculitis/hypertension have been extensively characterized. They include monocrotaline-induced pulmonary vasculitis/hypertension (Meyrick, 1991; Robinovitch, 1991; Roth and Ganey, 1988; Wilson et al., 1992) and hypoxia-induced pulmonary hypertension (King et al., 1994; Maruyama et al., 1991). A single intraperitoneal or subcutaneous injection of monocrotaline (40–120 mg/kg body weight for rats) results in progressive development of pulmonary vascular endothelial injury and hypertension, which is associated with pulmonary alveolar injury and inflammation within the first 2 weeks of injection. At approximately 2 weeks post injection, pulmonary hypertension develops and ventricular weight begins to increase; the lung lesions subside but arteriolar remodeling persists and progresses over time following monocrotaline treatment (Huxtable, 1993; Meyrick, 1991). This model differs from the idiopathic human disease by virtue of its significant lung inflammation, which is lacking in the human. The model is reminiscent of adult respiratory distress syndrome.

Exposure of rats to hypoxia (nitrogen dilution) or a simulated high altitude of 16 000 feet (4800 m) (hypobaric) for ~3 weeks results in progressive lung vascular muscularization, pulmonary hypertension and right cardiomegaly (Maruyama et al., 1991; McKenzie et al., 1994). Unlike the monocrotaline model, little if any pulmonary edema and inflammation are present in the model. Acute hypoxia-induced pulmonary arteriole constriction increases resistance to blood flow though the lung which eventually is established via vessel remodeling (Vyas-Somani *et al.*, 1996). Polycythemia is associated with this model, but not with monocrotaline model (Resta *et al.*, 1994; Zhao *et al.*, 1991). Structural changes associated with muscularization of pulmonary arterioles differ in these models; mono-crotaline causes selective endothelial damage in contrast to that of hypoxia-induced endothelial stimulation (Heath, 1992; King *et al.*, 1994). However, in both cases pulmonary hypertension is apparent. Recently we and others have employed this model for air pollution susceptibility studies (Costa *et al.*, 1994; Killingsworth *et al.*, 1997; Kodavanti *et al.*, 1997b; Watkinson *et al.*, 1998).

Transgenic Animals

The availability of transgenic and knockout mice to study the tissue-specific effects of unique genes has yielded significant information regarding the evaluation of lung-specific roles of transgenes in pathology (Glasser *et al.*, 1994; Ho, 1994). Transgenic mice, over-expressing a variety of markers in a lung-specific manner, use lung-specific promotor enhancer elements expressed in a highly selective manner in certain types of mammalian lungs. Five prime flanking regions of the surfactant proteins A, B, and C as well as Clara cell secretory protein genes have been isolated and used to direct specific genes in respiratory epithelium in mice (Mossman *et al.*, 1995; Stripp *et al.*, 1990). Transgenic mice (developed using various promoters in a selective and temporal manner for certain lung cell types) have allowed the expression of a variety of regulatory genes that influence the pulmonary system. Applications of a transgenic mouse with disrupted cystic fibrosis transmembrane conductance regulator (CFTR) for cystic fibrosis research and a knockout for high-affinity IgE receptors in asthma research are examples of how transgenic mice can serve as valuable tools in understanding the mechanisms of disease pathogenesis and in drug development strategies (Chow *et al.*, 1997; Dombrowicz *et al.*, 1993; Grubb and Gabriel, 1997).

Whether specific transgenic or knockout mice develop pathophysiological changes that resemble known human respiratory disease remains to be established. Mouse models of diseases associated with known genetic deficiencies are likely to produce similar symptomology (e.g., the CFTR knockout mouse as an analogue of human cystic fibrosis) (Chow *et al.*, 1997). However, many knockouts or overexpressers of cytokines are represented with a pathobiology that may not relate to any specific human respiratory disease. For example, surfactant protein A knockout mice are very susceptible to infection because clearance of the microorganisms mediated by macrophages is impaired (LeVine *et al.*, 1997). However, this is not likely to represent a pathobiology of COPD, AIDS or cystic fibrosis, in which susceptibility to infection is similarly increased but induced by different mechanisms. At this time, transgenics offer great opportunity to understand basic toxicological mechanisms and in some instances to understand the roles of specific transgenes in defining susceptibility to air pollutants. However, their usefulness as respiratory disease models remains to be determined.

Applications of Respiratory Disease Models in Air Pollution Studies

The potential susceptibility of humans with respiratory diseases to air pollution effects is widely recognized. However, only sporadic attempts have been made to use animal models of diseases to assess enhanced pulmonary injury by inhaled pollutants. One of the first studies was reported in 1976 in which healthy and a variety of diseased rodent models were exposed to simulated urban air pollution at different levels for durations up to 18 months (Hartroft *et al.*, 1976). Although there are limitations to interpreting these data, this study clearly demonstrated pollution level-dependent mortality in rats maintained on a thrombogenic diet, as well as in those rendered hypertensive by unilateral constriction of a left renal artery. Surprisingly, emphysemic hamsters were not very sensitive. This study has not been followed by more detailed or replicative experiments, perhaps because of the time and the effort involved, and/or complications associated with interpretation of the results.

Particle deposition (radiolabeled) has been found to be increased in bronchitic rats and fibrotic hamsters exposed to ^{99m}Tc-labeled insoluble aerosols where focal central areas in the lung appeared most affected (Sweeney *et al.*, 1983, 1995). In contrast, initial deposition of inhaled ^{239}Pu in rats with pulmonary fibrosis has been shown to be similar to controls, however, particle retention time was increased due to pre-existent disease (Lundgren *et al.*, 1991). In a mouse model of bleomycin-induced pulmonary fibrosis, carbon uptake was increased in interstitial macrophages only during active epithelial injury (3 days post bleomycin instillation), unlike the less marked deposition in mice with resolved bleomycin-induced injury/inflammation but with well-formed fibrotic lesions (4 weeks post bleomycin) (Adamson and Hedgecock, 1995). Subsequently it was shown that increased deposition of carbon particles, 3 days post bleomycin, was not associated with more severe fibrotic lesions (Adamson and Prieditis, 1995). Our use of a rat model 7 days post bleomycin to represent the inflammatory phase and 4 weeks post bleomycin to represent the stabilizing fibrosis phase suggests that the impact of particulate exposure is highly dependent on the pathological stage (especially inflammatory) of a disease (Kodavanti *et al.*, 1996b). The few abovementioned studies of particle deposition and injury with the bleomycin model of fibrosis suggests that altered deposition may also be a factor in their susceptibility to air pollutants.

Several air pollution studies have been conducted with emphysema models established in the rat, hamster and guinea pig (Costa and Lehmann, 1984; Lafuma *et al.*, 1987; Loscutoff *et al.*, 1985; Mauderly *et al.*, 1989, 1990; Raub *et al.*, 1983; Smith *et al.*, 1989; Sweeney *et al.*, 1987; Yokoyama *et al.*, 1987). Although particle deposition in emphysemic animals appears to be quantitatively similar or even less than in controls, the studies have emphasized the heterogeneity in the deposition pattern and its possible impact on the toxic outcome (Sweeney *et al.*, 1987). Emphysemic rats have been reported to be no more susceptible to diesel particles and NO_2 than healthy rats (Mauderly *et al.*, 1989, 1990). On the other hand a mixture of O_3 and SO_2 has been shown to cause a less dramatic decrease in diffusion capacity in emphysematic hamsters when compared with normals (Raub *et al.*, 1983), suggesting that there may be pollutant-specific responsiveness in this model. No studies to date have demonstrated overt susceptibility of emphysemic animals models to any air pollutants.

Recent studies conducted at the EPA and the Harvard School of Public Health have used rodent models of monocrotaline-induced pulmonary vasculitis/hypertension and SO_2-induced bronchitis to examine the impact of emission particulate material (Costa *et al.*, 1994; Gardner *et al.*, 1997; Godleski *et al.*, 1996; Killingsworth *et al.*, 1997; Kodavanti *et al.*, 1997b; 1998; Watkinson *et al.*, 1998). Although significant effects (e.g. mortality and edema) have been observed in both models following emission particulate exposures, the results were not consistent between laboratories and exposure protocols.

The reasons for this discrepancy are unclear, and perhaps underlie differences in model development, particle type or exposure scenarios. Responses of animal disease models can be quite variable despite their common genetic backgrounds. Clearly, more relevant disease models need to be studied with regards to major air pollutants. Pollutants are likely to interact with tissue components in pollutant type, disease type and severity, and exposure protocol-specific ways.

Pollutant interactions with models of allergic sensitization are somewhat better studied and understood than those of chronic respiratory disease (reviewed in Gilmour, 1995; Gilmour and Koren, 1999; Karol *et al.*, 1985; Selgrade and Gilmour, 1994; van Loveren *et al.*, 1996). However, how the incidence of asthma and the exacerbation of its symptoms are related to ambient air pollution is unclear (Cookson and Moffatt, 1997). In general, pollutants exposures have been shown to exacerbate allergic responses in mouse, rat and guinea pig models regardless of the specific allergic sensitization and pollutant exposure protocols. Guinea pigs exposed to high levels of O_3, NO_2 and SO_2 prior to sensitization with ovalbumin exhibit elevated specific antibody titer and anaphylactic responses following challenge with the same antigen (Matsumura *et al.*, 1972; Reidel *et al.*, 1988). Likewise, O_3 pre-exposure has also been shown to enhance the allergic response to inhaled platinum in monkeys (Biagini *et al.*, 1986). Similarly, NO_2 has been shown to augment the response to house dust mite antigens in a rat model (Gilmour *et al.*, 1996). For a number of years, diesel particles have been thought to increase allergic diseases and rhinitis in individuals exposed to automobile air pollution by acting as an adjuvant in allergic sensitization (reviewed in Selgrade *et al.*, 1997; Takano *et al.*, 1997). This enhancement has been modeled in the mouse, where it is shown that antigen-induced airway inflammation and local cytokine production can be enhanced by diesel particles (Takano *et al.*, 1997). Since diesel particles are composed of a number of organic as well as inorganic constituents, it remains to be shown what components are responsible for this response. Clearly, pollutants can interact with the immune system in such a way that allergic or immediate hypersensitivity responses to a known antigen are increased (likely through a shift of T lymphocytes to the TH_2 pathway) (Selgrade *et al.*, 1997). However, there remains great uncertainty regarding the enhancement of actual sensitization by pollutants – at least in humans.

Interactions between environmental pollutants and infection have been appreciated for many years and constitute the best studied of all types of respiratory disease models. These studies have recently been reviewed (Gilmour and Koren, 1999; Lebrec and Burleson, 1994; Selgrade and Gilmour, 1994). A number of experiments involving infection models have shown that pollutant gases and PM can interfere with the normal host defense mechanisms of the lung and can enhance susceptibility to infections. Using a mouse model of *Streptococcus zooepidemicus*, Coffin and Blommer (1967) showed that prior exposure to irradiated automobile exhaust increased susceptibility to infections. Using the same mouse model, Hatch *et al.* (1985) have shown that a variety of particulates from different sources enhanced mortality from infection by more than 50%. Interestingly, Jakab (1993) has reported that a mixture of carbon black and vapors of acrolein or formaldehyde results in reduced intrapulmonary inactivation of *Staphylococcus aureus* and increased the pathology associated with influenza virus infection. Acute O_3 exposure prior to infection with influenza has been shown to enhance mortality in mice, although the response is highly dependent on experimental protocol (Selgrade *et al.*, 1988).

Exposure to O_3 *after* infection has been shown to reduce virus pathogenesis in a mouse model (Jakab and Hmieleski, 1988). Similarly a number of studies with O_3 and other pollutants have shown differential effects on viral host defense (Ehrlich and Burleson, 1991; Gardner, 1982; Jakab and Bassett, 1990; Selgrade *et al.*, 1989). The bacterial infection models have been shown to be exquisitely sensitive to O_3 (Ehrlich *et al.*, 1979; Goldstein *et al.*, 1971). Perturbation of macrophage ability or capacity to clear infectious agents by pollutants has been postulated to contribute to this interaction (Gilmour *et al.*, 1993a,b). In animal models of compromised host defense, the interactions between pollutants and infectious organisms may be even more likely (e.g. in bronchitis), though also more complex. Nevertheless, this area of research is particularly important and has a history of congruence between animals and humans.

Issues Related to the Use of Respiratory Disease Models in Air Pollution Studies

The use of respiratory disease animal models in air pollution health effects and the interpretation of the resultant data for human health risk estimation can be exceedingly complex (Brain *et al.*, 1988a; Sweeney *et al.*, 1988). There are a variety of diseases of varying levels of severity, not to mention innate individual human differences. On the other side is the complexity of air pollution, comprising particles as mixtures of many components, photochemical oxidants, sulfates and other airborne materials, each of which may itself influence susceptibility by multiple mechanisms. Epidemiology is clearly limited when addressing these issues. Studies involving animal disease models can be conducted and can provide an additional tool with which to address some of these variables in a controlled manner, but extrapolation of the results to the human case can be very complicated because of differences in disease causation and pathogenesis in humans and animal models. Acute or subacute onset of the disease in the model versus a relatively chronic onset in humans, the time course of reversibility in the induced animal disease versus the human disease also are important considerations. As with all toxicological studies, there still remain hurdles of extrapolation related to species differences in host responsiveness, pollutant dosimetry, and the impact of age and gender. When using animal models of disease, individual variability is encountered at two levels: first during the development of a disease and second during exposure to the pollutant, where individual responses to each factor will vary. However, because the difficulties in conducting experiments with human patients can be formidable, animal models of disease may offer the only reasonable approach to studying enhanced risk and the mechanisms of that susceptibility. The best possible understanding of the pathogenesis in the animal models and the human in terms of the differences in causation, initiation, progression and regression of the disease are critical in using disease models for determination of air pollution health effects.

FUTURE PERSPECTIVES

The US Environmental Protection Agency's Clean Air Act mandates that most susceptible subpopulations be protected from the harmful health effects of air pollutants (US Environmental Protection Agency, 1987). Epidemiology, however, associates air pollution with increased morbidity and mortality of susceptible subpopulations, and the support from animal experimentation is poor (McClellan and Miller, 1997; Pope *et al.*, 1995).

This chapter emphasizes the use of animal models that can be rendered more suscepti- ble to air pollution health effects. While cardiopulmonary disease and nutritional ailments appear to be the most important targets for pollutant impacts, use of laboratory animal models created to reflect these susceptible human conditions need to be expanded with better understanding of the associated mechanisms. This under- standing needs to be more than hazard identification, whereby the biological mechanism(s) underlying the increased air pollution susceptibilities are identified. First, models of cardiopulmonary diseases need to be better defined in terms of molecular pathogenesis and their chronicity as it relates to the human condition. It is critical that animal models be studied at environmentally relevant doses of air pollutants, singly or as a mixture, since health effects in humans are observed at low levels of air pollution. At present, it is not known whether there is a threshold of air pollutant effects in extremely susceptible humans (McClellan and Miller, 1997). It is also not known whether there is a linear relationship between concentration of PM in the air and health effects seen in humans, especially at low ambient levels (McClellan and Miller, 1997). Air pollution susceptibility as it relates to pre-existent cardio- pulmonary disease, nutritional deficiencies, age or other host attributes is likely to receive increased emphasis in the future. The coordinated application of toxicological approaches combined with epidemiology and clinical studies is likely to be the most fruitful way to address these complex questions of susceptibility.

ACKNOWLEDGEMENTS

We thank Drs Linda Birnbaum, Ian Gilmour and Kevin L. Dreher of the US EPA for critical review of the manuscript.

REFERENCES

Acevedo CH and Ahmed A (1998) Hemeoxygenase-1 inhibits human myometrial contractility via carbon monoxide and is upregulated by progesterone during pregnancy. *J Clin Invest* **101**: 949–955.

Adamson IYR and Hedgecock C (1995) Patterns of particle deposition and retention after instillation to mouse lung during acute injury and repair. *Environ Lung Res* **21**: 695–709.

Adamson IYR and Prieditis HL (1995) Response of mouse lung to carbon deposition during injury and repair. *Environ Health Perspect* **103**: 72–76.

Alarie YC, Krumm AA, Busey WM *et al.* (1975) Long-term exposure to sulfur dioxide, sulfuric acid mist, fly ash, and their mixtures. Results of studies in monkeys and guinea pigs. *Arch Environ Health* **30**: 254–262.

Arm JP and Lee TH (1992) Pathobiology of bronchial asthma. *Adv Immunol* **51**: 323–382.

Ayaz KL and Csallany AS (1978) Long-term NO_2 exposure of mice in the presence and absence of vitamin E. II. Effect of glutathione peroxidase. *Arch Environ Health* **33**: 292–296.

Babiuk LA, Lawman MJ and Ohmann HB (1988) Viral-bacterial synergistic interaction in respiratory disease. *Adv Virus Res* **35**: 219–249.

Barnes PJ (1995) Air pollution and asthma: molecular mechanisms. *Mol Med Today* **1**: 149–155.

Basbaum C, Gallup M, Gum J *et al.* (1990) Modification of mucin gene expression in the airways of rats exposed to sulfur dioxide. *Biorheology* **27**: 485–489.

Bauernfeind JC (ed.) (1986) *Vitamin A Deficiency and Its Control.* Orlando, FL: Academic Press.

Beaulac C, Clement MS, Hawari J and Lagace J (1996) Eradication of mucoid *Pseudomonas aeruginosa* with fluid liposome-encapsulated tobramycin in an animal model of chronic infection. *Antimicrob Agents Chemother* **40**: 665–669.

Beck BD and Weinstock S (1988) Gender. In: Brain JD *et al.* (eds) *Variations in Susceptibility to Inhaled Pollutants. Identification, Mechanisms, and Policy Implications.* Baltimore, MD: The Johns Hopkins University Press, pp. 127–141.

Bellofiore S and Martin JG (1988) Antigen challenge of sensitized rats increases airway responsiveness to methacholine. *J Appl Physiol* **65**: 1642–1646.

Biagini RE, Moorman WJ, Lewis TR and Bernstein IL (1986) Ozone enhancement of platinum asthma in a primate model. *Am Rev Respir Dis* **134**: 719–725.

Bianco A, Bassi P, Belvisi M and Tarroni G (1980) Inhalation of a radioactively labeled monodisperse aerosol in rats for the assessment of regional deposition and clearance. *Am Ind Hyg Assoc J* **41**: 563–567.

Bignami G, Musi B, Dell'Omo G *et al.* (1994) Limited effects of ozone exposure duringpregnancy on physical and neurobehavioral development of CD-1 mice. *Toxicol Appl Pharmacol* **129**: 264–271.

Bird ED, Anton AH and Bullock B (1984) The effect of manganese inhalation on basal ganglia dopamine concentrations in rhesus monkey. *Neurotoxicology* **5**: 59–65.

Bloom JC, Lewis HB, Sellers TS and Deldar A (1987) The hematological effects of cefonicid and cefazedone in the dog: a potential model of cephalosporine hematotoxicity in man. *Toxicol Appl Pharmacol* **90**: 135–142.

Brain JD (1970) The uptake of inhaled gases by the nose. *Ann Otol Rhinol Laryngol* **79**: 529–539.

Brain JD, Beck BD, Warren AJ, and Shaikh RA (eds) (1988a) *Variations in Susceptibility to Inhaled Pollutants. Identification, Mechanisms, and Policy Implications.* Baltimore, MD: The Johns Hopkins University Press.

Brain JD, Valberg PA, and Mensah GA (1988b) Species differences. In: Brain JD et al. (eds) *Variations in Susceptibility to Inhaled Pollutants. Identification, Mechanisms, and Policy Implications.* Baltimore, MD: The Johns Hopkins University Press, pp. 89–103.

Brain JD, Skornik WA, Spaulding GL and Harbison M (1988c) The effects of exercise on inhalation of particles and gases. In: Brain JD *et al.* (eds) *Variations in Susceptibility to Inhaled Pollutants. Identification, Mechanisms, and Policy Implications.* Baltimore, MD: The Johns Hopkins University Press, pp. 204–220.

Brockmann M, Fisher M and Muller KM (1998) Exposure to carbon black: a cancer risk? *Int Arch Occup Environ Health* **71**: 85–99.

Bush ML, Asplund PT, Miles KA *et al.* (1996) Longitudinal distribution of O_3 absorption in the lung: gender differences and intersubject variability. *J Appl Physiol* **81**: 1651–1657.

Cassee F and van Bree L (1996) Host responsiveness in health and disease: a brief overview of animal models. In: Lee J and Phalen R (eds) *Proceedings of the Second Colloquium on Particulate Air Pollution and Human Health.* Salt Lake City, UT: University of Utah, pp. 4.63–4.69.

Chalon J, Lowe DAY and Orkin LR (1971) Tracheobronchial cytologic changes during the menstrual cycle. *J Am Med Assoc* **218**: 1928–1931.

Cheng KC, Lee KM, Krug MS *et al.* (1998) House dust mite-induced sensitivity in mice. *J Allergy Clin Immunol* **101**: 51–59.

Chow CK, Plopper CG and Dungworth DL (1979) Influence of dietary vitamin E on the lungs of ozone-exposed rats. *Environ Res* **20**: 309–317.

Chow YH, O'Brodovich H, Plumb J *et al.* (1997) Development of an epithelial-specific expression cassette with human DNA regulatory elements for transgene expression in lung airways. *Proc Natl Acad Sci USA* **94**: 14695-14700.

Coffin DL and Blommer EJ (1967) Acute toxicity of irradiated auto exhaust. Its indication by enhancement of mortality from streptococcal pneumonia. *Arch Environ Health* **15**: 36–38.

Cohen BS (1996) Particle deposition in human and canine tracheobronchial casts: a determinant of radon dose to the critical cells of the respiratory tract. *Health Phys* **70**: 695–705.

Colditz GA, Stampfer MJ and Green LC (1988) Diet. In: Brain JD *et al. Variations in Susceptibility to Inhaled Pollutants. Identification, Mechanisms, and Policy Implications.* Baltimore, MD: The Johns Hopkins University Press, pp. 314–331.

Committee of the Environmental and Occupational Health Assembly of the American Thoracic Society (1996a) Health effects of outdoor air pollution. Part 1. *Am J Respir Crit Care Med* **153**: 3–50.

Committee of the Environmental and Occupational Health Assembly of the American Thoracic Society (1996b) Health effects of outdoor air pollution. Part 2. *Am J Respir Crit Care Med* **153**: 477–498.

Cookson WOCM and Moffatt MF (1997) Asthma: An epidemic in the absence of infection? *Science* **275**: 41–42.

Corry DB, Folkesson HG, Warnock ML *et al.* (1996) Interleukin 4, but not interleukin 5 or eosinophils, is required in a murine model of acute airway hyperreactivity. *J Exp Med* **183**: 109–117.

Costa DL and Lehmann JR (1984) Emphysema and fibrosis: risk factors in responsiveness to air pollution. In: *Proceedings of 15th Annual Conference on Environmental Toxicology.* University of California, Irvine, CA. Wright-Patterson Air Force Base, Ohio: AFAMRL-TR-84-002, pp. 23–35.

Costa DL, Lehmann JR, Frazier LT *et al.* (1994) Pulmonary hypertension: a possible risk factor in particulate toxicity. *Am J Respir Crit Care Med* **149**: A480.

Crouch E (1990) Pathobiology of pulmonary fibrosis. *Am J Physiol* **259** (*Lung Cell Mol Physiol* **3**), L159–L184.

Cugell DW (1988) A brief introduction for behavioral scientists. In: McSweeney AJ and Grant I (eds) *Chronic Obstructive Pulmonary Disease. A Behavioral Perspective.* New York: Marcel Dekker, pp. 1–16.

Dagle GE and Sanders CL (1984) Radionuclide injury to the lung. *Environ Health Perspect* **55**: 129–137.

de Santi MM, Martorana PA, Cavarra E and Lungarella G (1995) Pallid mice with genetic emphysema. Neutrophil elastase burden and elastin loss occur without alteration in the bronchoalveolar lavage cell population. *Lab Invest* **73**: 40–47.

Deschamps C, Farkas GA, Beck KC *et al.* (1995) Experimental emphysema. *Chest Surg Clin N Am* **5**: 691–699.

Devlin RB, McDonnell WF, Mann R *et al.* (1991) Exposure of humans to ambient levels of ozone for 6.6 hours causes cellular and biochemical changes in the lung. *Am J Respir Cell Mol Biol* **4**: 72–81.

Dockery DW, Pope CA III, Xu X *et al.* (1993) An association between air pollution and mortality in six US cities. *N Engl J Med* **329**: 1753–1759.

Dombrowicz D, Flamand V, Brigman KK *et al.* (1993) Abolition of anaphylaxis by targeted disruption of the high affinity immunoglobulin E receptor alpha chain gene. *Cell* **75**: 969–976.

Drazen JM, O'Cain CF and Ingram Jr RH (1982) Experimental induction of chronic bronchitis in dogs. Effects on airway obstruction and responsiveness. *Am Rev Respir Dis* **126**: 75–79.

Dubick MA, Heng H and Rucker RB (1985) Effects of protein deficiency and food restriction on lung ascorbic acid and glutathione in rats exposed to ozone. *J Nutr* **115**: 1050–1056.

Economides D and Braithwaite J (1994) Smoking, pregnancy and the fetus. *J R Soc Health* **114**: 198–201.

Ehrlich JP and Burleson GR (1991) Enhanced and prolonged pulmonary influenza virus infection following phosgene inhalation. *J Toxicol Environ Health* **34**: 259–273.

Ehrlich R, Findlay JC and Gardner DE (1979) Effects of repeated exposures to peak concentrations of nitrogen dioxide and ozone on resistance to Streptococcal pneumonia. *J Toxicol Environ Health* **5**: 631–642.

Elsayed NM and Mustafa MG (1982) Dietary antioxidants and the biochemical response to oxidant inhalation. I. Influence of dietary vitamin E on the biochemical effects of nitrogen dioxide exposure in rat lung. *Toxicol Appl Pharmacol* **66**: 319–328.

Elsayed NM, Hacker AD, Kuehn K *et al.* (1983) Dietary antioxidants and the biochemical response to oxidant inhalation. II. Influence of dietary selenium on the biochemical effects of ozone exposure in mouse lung. *Toxicol Appl Pharmacol* **71**: 398–406.

Elwood W, Lotvall JO, Barnes PJ and Chung KF (1991) Characterization of allergen-induced bronchial hyperresponsiveness and airway inflammation in actively sensitized Brown-Norway rats. *J Allergy Clin Immunol* **88**: 951–960.

Enjeti S, Hazelwood B, Permutt S *et al.* (1978) Pulmonary function in young smokers: male–female differences. *Am Rev Respir Dis* **118**: 667–676.

Eskew ML, Scheuchenzuber WJ, Scholz RW *et al.* (1986) The effects of ozone inhalation on the immunological response of selenium- and vitamin E-deprived rats. *Environ Res* **40**: 274–284.

Fang CP, Wilson JE, Spektor DM and Lippmann M (1993) Effect of lung airway branching pattern and gas composition on particle deposition in bronchial airways: III. Experimental studies with radioactively tagged aerosol in human and canine lungs. *Exp Lung Res* **19**: 377–396.

Frank NR, Yoder RE, Brain JD and Yokoyama E (1969) SO$_2$ (^{35}S-labeled) absorption by the nose and mouth under conditions of varying concentration and flow. *Arch Environ Health* **18**: 315–322.

Frew AJ (1996) The immunology of respiratory allergies. *Toxicol Lett* **86**: 65–72.

Fukui M, Yasui H, Watnabe K *et al.* (1996) Hypoxic contraction of contractile interstitial cells isolated from bovine lung. *Am J Physiol* **270**: L962–L972.

Gardner DE (1982) Effects of gases and airborne particles on lung infections. In: McGrath JJ and Barnes EB (eds) *Air Pollution: Physiological Effects*. New York: Academic Press, pp. 47–79.

Gardner DE, Crapo JD and Massaro EJ (eds) (1988) *Toxicology of the Lung*, Target Organ Toxicology Series, Dixon RL (ed). New York: Raven Press.

Gardner SY, Terrell D, McGee JK *et al.* (1997) Particle-induced pulmonary ventilatory abnormalities in a rat model of pulmonary hypertension. *Am Rev Respir Crit Care Med* **155**: A247.

Ghoniem GM, Shoukry MS and Monga M (1996) Effects of anesthesia on urodynamic studies in the primate model. *J Urol* **156**: 233–236.

Gilmour MI (1995) Interaction of air pollutants and pulmonary allergic responses in experimental animals. *Toxicology* **105**: 335–342.

Gilmour MI and Koren HS (1999) Interaction of inhaled particles with the immune system. In: Gehr P and Heyder J (eds) *Particle-Lung Interactions*. New York: Marcel and Dekker. In press.

Gilmour MI and Selgrade MJK (1993) A comparison of the pulmonary defenses against Streptococcal infection in rats and mice following O$_3$ exposure: differences in disease susceptibility and neutrophil recruitment. *Toxicol Appl Pharmacol* **123**: 211–218.

Gilmour MI and Selgrade MJK (1996) A model of immune-mediated lung disease in rats sensitized to house dust mite and upregulation of immunity following exposure to nitrogen dioxide. *Chest* **109** (Suppl. 3): 69S.

Gilmour MI, Park P, Doerfler D and Selgrade MJK (1993a) Factors that influence the suppression of pulmonary antibacterial defenses in mice exposed to ozone. *Exp Lung Res* **19**: 299–314.

Gilmour MI, Park P and Selgrade MK (1993b) Ozone-enhanced pulmonary infection with *Streptococcus zooepidemicus* in mice. The role of alveolar macrophage function and capsular virulence factors. *Am Rev Respir Dis* **147**: 753–760.

Gilmour MI, Park P and Selgrade MJK (1996) Increased immune and inflammatory responses to dust mite antigen in rats exposed to 5 ppm NO$_2$. *Fundam Appl Toxicol* **31**: 65–70.

Glasser SW, Korfhagen TR, Wert SE and Whitsett JA (1994) Transgenic models for study of pulmonary development and disease. *Am J Physiol* **267** (*Lung Cell Mol Physiol* **11**): L489–L497.

Goco RV, Kress MB and Brantigan OC (1963) Comparison of mucus glands in the tracheal bronchial tree of man and animals. *Ann NY Acad Sci* **106**: 555–571.

Godleski JJ, Sloutas C, Katler M *et al.* (1996) Death from inhalation of concentrated ambient air particles in animal models of pulmonary disease. In: Lee J and Phalen R (eds) *Proceedings of the Second Colloquium on Particulate Air Pollution and Human Health*. Salt Lake City, UT: University of Utah, pp. 4136–4143.

Godleski JJ, Sloutas C, Verrier RL *et al.* (1997) Inhalation exposure of canines to concentrated ambient air particles. *Am J Respir Crit Care Med* **155**: A246.

Goldring IP, Greenburg L, Park S-S and Ratner IM (1970) Pulmonary effects of sulfur dioxide exposure in the Syrian hamster. II. Combined with emphysema. *Arch Environ Health* **21**: 32–37.

Goldstein BD, Buckley RD, Cardenas R and Balchum OJ (1970) Ozone and vitamin E. *Science* **169**: 605–606.

Goldstein E, Tyler WS, Hoeprich PD and Eagle C (1971) Adverse influence of ozone on pulmonary bactericidal activity of murine lung. *Nature* **229**: 262–263.

Green GM (1984) Similarities of host defense mechanisms against pulmonary infectious diseases in animals and man. *J Toxicol Environ Health* **13**: 471–478.

Greene SA, Wolff RK, Hahn FF *et al.* (1984) Sulfur dioxide-induced chronic bronchitis in beagle dogs. *J Toxicol Environ Health* **13**: 945–958.

Grubb BR and Gabriel SE (1997) Intestinal physiology and pathology in gene-targeted mouse models of cystic fibrosis. *Am J Physiol* **273**: G258–G266.

Gunnison AF and Finkelstein I (1997) Rat lung phospholipid fatty acid composition in prepregnant, pregnant, and lactating rats: relationship to ozone-induced pulmonary toxicity. *Lung* **175**: 127–137.

Gunnison AF, Weideman PA and Sobo M (1992) Enhanced inflammatory response to acute ozone exposure in rats during pregnancy and lactation. *Fundam Appl Toxicol* **19**: 607–612.

Gurujeyalakshmi G, Hollinger MA and Giri SN (1998) Regulation of transforming growth factor-beta1 mRNA expression by taurine and niacin in the bleomycin hamster model of lung fibrosis. *Am J Respir Cell Mol Biol* **18**: 334–342.

Hahn J, Kolb HJ, Schumm M *et al.* (1991) Immunological characterization of canine hematopoietic progenitor cells. Ann. Hematol. 63, 223-226.

Harbison ML and Brain JD (1983) Effects of exercise on particle deposition in Syrian golden hamsters. *Am Rev Respir Dis* **128**: 904–908.

Harkema JR and Hotchkiss JA (1992) *In vivo* effects of endotoxin on intraepithelial mucosubstances in rat pulmonary airways. Quantitative histochemistry. *Am J Pathol* **141**: 307–317.

Harkema JR, Plopper CG, Hyde DM *et al.* (1993) Response of macaque bronchiolar epithelium to ambient concentrations of ozone. *Am J Pathol* **143**: 857–866.

Hartroft PM, Kuhn III CC, Freeman SV *et al.* (1976) Effects of chronic, continuous exposure to simulated urban air pollution on laboratory animals with cardiovascular and respiratory diseases. The Institute of Electrical and Electronics Engineers, Inc., USA Annals, No. 75CH1004-I34-5, pp. 1–5.

Hatch GE (1992) Comparative biochemistry of airway lining fluid. In: Parent RA (ed.) *Comparative Biology of the Normal Lung*. Boca Raton, FL: CRC Press, pp. 617–632.

Hatch GE (1995) Asthma, inhaled oxidants, and dietary antioxidants. *Am J Clin Nutr* **61**: 625S–630S.

Hatch GE, Boykin E, Graham JA *et al.* (1985) Inhalable particles and pulmonary host defense: *In vivo* and *in vitro* effects of ambient air and combustion particles. *Environ Res* **36**: 67–80.

Hatch GE, Slade R, Stead A and Graham JA (1986a) Species comparison of acute inhalation toxicity of ozone and phosgene. *J Toxicol Environ Health* **19**: 43–53.

Hatch GE, Slade R, Selgrade MJK and Stead A (1986b) Nitrogen dioxide exposure and lung antioxidants in ascorbic acid-deficient guinea pigs. *Toxicol Appl Pharmacol* **82**: 351–359.

Hatch GE, Slade R, Harris LP *et al.* (1994) Ozone dose and effect in humans and rats. A comparison using oxygen-18 labeling and bronchoalveolar lavage. *Am J Respir Crit Care Med* **150**: 676–683.

Hay J, Shahzeidi S and Laurent G (1991) Mechanisms of bleomycin-induced lung damage. *Arch Toxicol* **65**: 81–94.

Hayashi M, Sornberger GC and Huber GL (1978) Differential response in the male and female tracheal epithelium following exposure to tobacco smoke. *Chest* **73**: 515–518.

Heath D (1992) The rat is a poor model for the study of human pulmonary hypertension. *Cardiosci* **3**: 1–6.

Hegele RG, Robinson PJ, Gonzalez S and Hogg JC (1993) Production of acute bronchiolitis in guinea-pigs by human respiratory syncytial virus. *Eur Respir J* **6**: 1324–1331.

Heinrich U, Muhle H, Takenaka S *et al.* (1986) Chronic effects on the respiratory tract of hamsters, mice, and rats after long-term inhalation of high concentrations of filtered and unfiltered diesel engine emissions. *J Appl Toxicol* **6**: 383–395.

Henry MC, Findlay J, Spangler J and Ehrlich R (1970) Chronic toxicity of NO_2 in squirrel monkeys. 3. Effect on resistance to bacterial and viral infection. *Arch Environ Health* **20**: 566–570.

Hepleston AG (1991) Minerals, fibrosis and the lung. *Environ Health Perspect* **94**: 149–168.

Hernandez A, Daffonchio L, Brandolini L and Zuccari G (1994) Effect of a mucoactive compound (CO 1408) on airway hyperreactivity and inflammation induced by passive cigarette smoke exposure in guinea pigs. *Eur Respir J* **7**: 693–697.

Hernnas J, Nettelbladt O, Bjermer L *et al.* (1992) Alveolar accumulation of fibronectin and hyaluronan precedes bleomycin-induced pulmonary fibrosis in rats. *Eur Respir J* **5**: 404–410.

Heyder J and Takenaka S (1996) Long-term canine exposure studies with ambient air pollutants. *Eur Respir J* **9**: 571–584.

Higgins MW and Thom T (1989) Incidence, prevalence, and mortality: intra- and intercountry differences. In: Hensley MJ and Saunders NA (eds) *Clinical Epidemiology of Chronic Obstructive Pulmonary Disease*. New York: Marcel Dekker, pp. 23–43.

Ho YS (1994) Transgenic models for the study of lung biology and disease. *Am J Physiol* **266**: L319–L353.

Holtzman MJ, Sampath D, Castro M *et al.* (1996) The one-two of T helper cells: does interferon-γ knock out the Th2 hypothesis for asthma? *Am J Respir Cell Mol Biol* **14**: 316–318.

Hultgren HN and Grover RF (1968) Circulating adaptation to high altitude. *Ann Rev Med* **19**: 119–152.

Huxtable RJ (1993) Hepatic nonaltruism and pulmonary toxicity of pyrrolizidine alkaloids. In: Gram TE (ed.) *Metabolic Activation and Toxicity of Chemical Agents to Lung Tissue and Cells.* Oxford: Pergamon Press, pp. 213–237.

Ito M, Kaniwa T, Horiuchi H *et al.* (1995) Effect of clenbuterol on sulfur dioxide-induced acute bronchitis in guinea pigs. *Res Commun Mol Pathol Pharmacol* **87**: 199–209.

Jakab GJ (1993) The toxicologic interactions resulting from inhalation of carbon black and acrolein on pulmonary antibacterial and antiviral defenses. *Toxicol Appl Pharmacol* **121**: 167–175.

Jakab GJ and Bassett JP (1990) Influenza virus infection, ozone exposure, and fibrogenesis. *Am Rev Respir Dis* **141**: 1307–1315.

Jakab GJ and Hmieleski RR (1988) Reduction of influenza virus pathogenesis by exposure to 0.5 ppm ozone. *J Toxicol Environ Health* **23**: 455–472.

Jansen HM, Sachs PE and Alphen LVP (1995) Predisposing conditions of bacterial infections in chronic obstructive pulmonary disease. *Am J Respir Crit Care Med* **151**: 2073–2080.

Joshi UM, Dumas M and Mehendale HM (1986) Glutathione turnover in perfused rabbit lung. Effect of external glutathione. *Biochem Pharmacol* **35**: 3409–3412.

Joshi UM, Kodavanti PRS and Mehendale HM (1988) Glutathione metabolism and utilization of external thiols by cigarette smoke-challenged, isolated rat and rabbit lung. *Toxicol Appl Pharmacol* **96**: 324–335.

Kari F, Hatch G, Slade R *et al.* (1997) Dietary restriction mitigates ozone-induced lung inflammation in rats: a role for endogenous antioxidants. *Am J Respir Cell Molec Biol* **17**: 740–747.

Karlinsky JB and Snider GL (1978) Animal model of emphysema. *Am Rev Respir Dis* **117**: 1109–1133.

Karol MH, Stadler J and Magreni C (1985) Immunotoxicologic evaluation of the respiratory system: animal models for immediate- and delayed-onset pulmonary hypersensitivity. *Fundam Appl Toxicol* **5**: 459–472.

Killingsworth CR, Alessandrini F, Krishna Murthy GG *et al.* (1997) Inflammation, chemokine expression, and death in monocrotaline-treated rats following fuel oil fly ash inhalation. *Inh Tox* **9**: 541–565.

Kim HS, Tsai PB and Oh CK (1998) The genetics of asthma. *Curr Opin Pulm Med* **4**: 46–48.

King A, Smith P and Heath D (1994) Ultrastructural difference between pulmonary arteriolar muscularization induced by hypoxia and monocrotaline. *Exp Mol Pathol* **61**: 24–35.

Kleeberger SR (1995) Genetic susceptibility to ozone exposure. *Toxicol Lett* **82/83**: 295–300.

Kodavanti UP, Hatch GE, Starcher B *et al.* (1995a) Ozone-induced pulmonary functional, pathological and biochemical changes in normal and vitamin C deficient guinea pigs. *Fundam Appl Toxicol* **24**: 154–164.

Kodavanti UP, Costa DL, Dreher KL *et al.* (1995b) Ozone-induced tissue injury and changes in antioxidant homeostasis in normal and ascorbate deficient guinea pigs. *Biochem Pharmacol* **50**: 243–251.

Kodavanti UP, Costa DL, Richards J *et al.* (1996a) Antioxidants in bronchoalveolar lavage fluid cells isolated from ozone-exposed normal and ascorbate-deficient guinea pigs. *Exp Lung Res* **22**: 435–448.

Kodavanti UP, Costa DL, Jaskot R *et al.* (1996b) Influence of preexisting pulmonary disease on residual oil fly ash particle-induced toxicity in the rat. *Am J Respir Crit Care Med* **153**: A542.

Kodavanti UP, Jaskot RH, Su W-Y *et al.* (1997a) Genetic variability in combustion particle-induced chronic lung injury. *Am J Physiol* **272** (*Lung Cell Mol Physiol* **16**): L521–L532.

Kodavanti UP, Jackson M, Gardner SY *et al.* (1997b) Particle-induced lung injury in hypertensive rats. *Am Rev Respir Crit Care Med* **155**: A247.

Kodavanti UP, Costa DL and Bromberg P (1998) Rodent models of cardiopulmonary disease: their potential applicability in studies of air pollutant susceptibility. *Environ Health Perspect* **106** (Suppl. 1): 111–130.

Koren HS, Devlin RB, Graham DE *et al.* (1989) Ozone-induced inflammation in the lower airways of human subjects. *Am Rev Respir Dis* **139**: 407–415.

Kumar RK (1995) Experimental models in pulmonary pathology. *Pathology* **27**: 130–132.

Lafuma C, Harf A, Lange F et al. (1987) Effect of low-level NO_2 chronic exposure on elastase-induced emphysema. *Environ Res* **43**: 75–84.

Leach LJ, Yuile CL, Hodge HC et al. (1973) A five-year inhalation study with natural uranium dioxide (UO_2) dust. II. Postexposure retention and biologic effects in the monkey, dog and rat. *Health Phys* **25**: 239–258.

Lebrec H and Burleson GR (1994) Influenza virus host resistance models in mice and rats: utilization for immune function assessment and immunotoxicology. *Toxicol* **91**: 179–188.

Lee SD and Schneider T (eds) (1995) Fourth US–Dutch International Symposium. *Comparative Risk Analysis and Priority Setting for Air Pollution Issues*. Pittsburgh, PA: Air and Waste Management Association.

LeVine AM, Bruno MD, Huelsman KM et al. (1997) Surfactant protein A-deficient mice are susceptible to group B streptococcal infection. *J Immunol* **158**: 4336–4340.

Lippmann M and Schlesinger RB (1984) Interspecies comparisons of particle deposition and mucociliary clearance in tracheobronchial airways. *J Toxicol Environ Health* **13**: 441–469.

Loscutoff SM, Cannon WC, Buschbom RL et al. (1985) Pulmonary function in elastase-treated guinea pigs and rats exposed to ammonium sulfate or ammonium nitrate aerosols. *Environ Res* **36**: 170–180.

Lundgren DL, Mauderly JL, Rebar AH et al. (1991) Modifying effects of preexisting pulmonary fibrosis on biological responses of rats to inhaled $^{239}PuO_2$. *Health Physics* **60**: 353–363.

Maciver I, Silverman SH, Brown MRW and O'Reilly T (1991) Rat model of chronic lung infections caused by non-typable *Haemophilus influenzae*. *J Med Microbiol* **35**: 139–147.

MacNee W (1994) Pathophysiology of *cor pulmonale* in chronic obstructive pulmonary disease. *Am J Respir Crit Care Med* **150**: 833–852.

Madden MC, Eling TE and Friedman M (1987) Ozone inhibits endothelial cell cyclooxygenase activity through formation of hydrogen peroxide. *Prostaglandins* **34**: 445–463.

Maruyama K, Ye C, Woo M et al. (1991) Chronic hypoxic pulmonary hypertension in rats and increased elastolytic activity. *Am J Physiol* **261** (*Heart Circ Physiol* **30**): H1716–H1726.

Matsumoto T and Ashida Y (1993) Inhibition of antigen-induced airway hyperresponsiveness by a thromboxane A2 receptor antagonist (AA-2414) in *Ascaris suum*-allergic dogs. *Prostaglandins* **46**: 301–318.

Matsumura Y, Mizuno K, Miyamoto T et al. (1972) The effects of ozone, nitrogen dioxide, and sulfur dioxide on experimentally induced allergic respiratory disorder in guinea pigs. IV. Effects on respiratory sensitivity to inhaled acetylcholine. *Am Rev Respir Dis* **105**: 262–267.

Mauderly JL, Bice DE, Carpenter RL et al. (1987) Effects of inhaled nitrogen dioxide and diesel exhaust on developing lung. *Res Rep Health Eff Inst* **8**: 3–37.

Mauderly JL, Bice DE, Cheng YS et al. (1989) Influence of experimental pulmonary emphysema on the toxicological effects from inhaled nitrogen dioxide and diesel exhaust. *Res Rep Health Eff Inst* **30**: 1–47.

Mauderly JL, Bice DE, Cheng YS et al. (1990) Influence of preexisting pulmonary emphysema on susceptibility of rats to inhaled diesel exhaust. *Am Rev Respir Dis* **141**: 1333–1341.

Mauderly JL, Snipes MB, Barr EB et al. (1994) Pulmonary toxicity of inhaled diesel exhaust and carbon black in chronically exposed rats. Part I: Neoplastic and non-neoplastic lung lesions. *Res Rep Health Eff Inst* **68**: 1–75.

Mautz WJ, Kleinman MT, Phalen RF and Crocker TT (1988) Effects of exercise exposure on toxic interactions between inhaled oxidant and aldehyde air pollutants. *J Toxicol Environ Health* **25**: 165–177.

McClellan RO and Millar FJ (1997) An overview of EPA's proposed revision of the particulate matter standard. *CIIT Activities* **17**: 1–22.

McFadden ER and Hejal R (1995) Asthma. *Lancet* **345**: 1215–1220.

McKenzie JC, Kelly KB, Merisko-Liversidge EM et al. (1994) Developmental pattern of ventricular atrial natriuretic peptide (ANP) expression in chronically hypoxic rats as an indicator of the hypertrophic process. *J Mol Cell Cardiol* **26**: 753–767.

Menzel DB (1979) Nutritional needs in environmental intoxication: vitamin E and air pollution, an example. *Environ Health Perspect* **29**: 105–114.

Menzel DB (1992) Antioxidant vitamins and prevention of lung disease. *Ann NY Acad Sci* **669**: 141–155.

Messineo TD and Adams WC (1990) Ozone inhalation effects in females varying widely in lung size: comparison with males. *J Appl Physiol* **69**: 96–103.

Meyrick B (1991) Structure function correlates in the pulmonary vasculature during acute lung injury and chronic pulmonary hypertension. *Toxicol Pathol* **19**: 447–457.

Montgomery MR, Raska-Emery P and Balis JU (1987) Age-related difference in pulmonary response to ozone. *Biochim Biophys Acta* **890**: 271–274.

Mossman BT, Mason R, McDonald JA and Gail DB (1995) Advances in molecular genetics, transgenic models, and gene therapy for the study of pulmonary diseases. *Am J Respir Crit Care Med* **151**: 2065–2069.

Murray CJL and Lopez AD (1997) Mortality by cause for eight regions of the world: Global burden of disease study. *Lancet* **349**: 1269–1276.

Mustafa MG, Elsayed NM, Ospital JJ and Hacker AD (1985) Influence of age on the biochemical response of rat lung to ozone exposure. *Toxicol Ind Health* **1**: 29–41.

Myers BA, Dubick MA, Gerriets JE *et al.* (1986) Lung collagen and elastin after ozone exposure in vitamin B-6-deficient rats. *Toxicol Lett* **30**: 55–61.

Ober C (1998) Do genetics play a role in pathogenesis of asthma? *J Allergy Clin Immunol* **101**: S417–S420.

Oberdorster G (1995) Lung particle overload: implications for occupational exposures to particles. *Regul Toxicol Pharmacol* **21**: 123–135.

Ochoa L, Waypa G, Mahoney Jr JR *et al.* (1997) Contrasting effects of hypochlorous acid and hydrogen peroxide on endothelial permeability: prevention with cAMP drugs. *Am J Respir Crit Care Med* **156**: 1247–1255.

Ogawa H, Fugimura M, Saito M *et al.* (1994) The effect of the neurokinin antagonist FK-224 on the cough response to inhaled capsaicin in a new model of guinea-pig eosinophilic bronchitis induced by intranasal polymyxin B. *Clin Auton Res* **4**: 19–27.

O'Connor GT, Sparrow D, Weiss ST (1989) The role of allergy and nonspecific airway hyperresponsiveness in the pathogenesis of chronic obstructive pulmonary disease. *Am Rev Respir Dis* **140**: 225–252.

O'Reilly T (1995) Relevance of animal models for chronic bacterial airway infections in humans. *Am J Respir Crit Care Med* **151**: 2101–2108.

Petry H and Luke W (1997) Infection of macaque monkeys with simian immunodeficiency virus: an animal model for neuro-AIDS. *Inter Virology* **40**: 112–121.

Phalen RF and Oldham MJ (1983) Tracheobronchial airway structure as revealed by casting techniques. *Am Rev Respir Dis* **128**: S1–S4.

Phalen RF, Yeh H-C and Prasad SB (1995) Morphology of the respiration tract. In: McClellan RO and Henderson RF (eds) *Concepts in Inhalation Toxicology*. Washington, DC: Taylor and Francis, pp. 129–149.

Plopper CG (1983) Comparative morphologic features of bronchiolar epithelial cells: the clara cell. *Am Rev Respir Dis* **128**: S37–S41.

Pope III CA, Dockery DW and Schwartz J (1995) Review of epidemiological evidence of health effects of particulate air pollution. *Inhal Toxicol* **7**: 1–18.

Pott F and Stober W (1983) Carcinogenicity of airborne combustion products observed in subcutaneous tissue and lungs of laboratory rodents. *Environ Health Perspect* **47**: 293–303.

Pryor WA (1991) Can vitamin E protect humans against the pathological effects of ozone in smog? *Am J Clin Nutr* **53**: 702–722.

Qian SY and Mitzner W (1989) *In vivo* and *in vitro* lung reactivity in elastase-induced emphysema in hamsters. *Am Rev Respir Dis* **140**: 1549–1555.

Raghow R, Lurie S, Seyer JM and Kang AH (1985) Profile of steady state levels of messenger RNAs coding for type I procollagen, elastin, and fibronectin in hamster lungs undergoing bleomycin-induced interstitial pulmonary fibrosis. *J Clin Invest* **76**: 1733–1739.

Ranga V, Grahn, D and Journey TM (1993) Morphologic and phenotypic analysis of an outcross line of blotchy mouse. *Exp Lung Res* **4**: 269–279.

Raub JA, Miller FJ, Graham JA *et al.* (1983) Pulmonary function in normal and elastase treated hamsters exposed to a complex mixture of olefin-ozone-sulfur dioxide reaction products. *Environ Res* **31**: 302–310.

Reid LM (1980) Needs for animal models of human diseases of the respiratory system. *Am J Pathol* **101**: S89–S101.

Reidel F, Kramer M, Scheibenbogen C and Reiger CHL (1988) Effects of SO_2 exposure on allergic sensitization in the guinea pig. *J Allergy Clin Immunol* **82**: 527–534.

Reif JS, Rhodes WH and Cohen D (1970) Canine pulmonary disease and the urban environment. I. The validity of radiographic examination for estimating the prevalence of pulmonary disease. *Arch Environ Health* **20**: 676–683.

Repine JE, Bast A and Lankhorst I (1997) Oxidative stress in chronic obstructive pulmonary disease. *Am J Respir Crit Care Med* **156**: 341–357.

Resta TC, Russ RD, Doyle M *et al.* (1994) Cardiovascular responses to hemorrhage during acute and chronic hypoxia. *Am J Physiol* **267** (*Regulatory Interactive Comp Physiol* **36**): R619–R627.

Reynolds PN, Rice AJ, Reynolds AM *et al.* (1997) Tachykinins contribute to the acute airways response to allergen in sheep actively sensitized to *Ascaris suum*. *Respirology* **2**: 193–200.

Robinovitch M (1989) Pulmonary hypertension. In: Moss AJ, Adams FM and Emmanoluilides GC (eds) *Heart Disease in Infants, Children, and Adolescents*. Baltimore: Williams & Wilkins, pp. 856–885.

Robinovitch M (1991) Investigational approaches to pulmonary hypertension. *Toxicol Pathol* **19**: 458–469.

Roth RA and Ganey PE (1988) Platelets and puzzles of pulmonary pyrrolizidine poisoning. *Toxicol Appl Pharmacol* **93**: 463–471.

Sato S, Kawakami M, Maeda S and Takishima T (1976) Scanning electron microscopy of the lungs of vitamin E-deficient rats exposed to a low concentration of ozone. *Am Rev Respir Dis* **113**: 809–821.

Savoie C, Plant M, Zwikker M et al. (1995) Effect of dexamethasone on allergen-induced high molecular weight glycoconjugate secretion in allergic guinea pigs. *Am J Respir Cell Mol Biol* **13**: 133–143.

Schwartz J (1994) What are people dying of on high air pollution days? *Environ Res* **64**: 26–35.

Seal Jr E, McDonnell WF, House DE *et al.* (1993) The pulmonary response of white and black adults to six concentrations of ozone. *Am Rev Respir Dis* **147**: 804–810.

Secker-Walker RH, Vacek PM, Flynn BS and Mead PB (1997) Smoking in pregnancy, exhaled carbon monoxide, and birth weight. *Obstet Gynecol* **89**: 648–653.

Selgrade MJK and Gilmour MI (1994) Effects of gaseous air pollutants on immune responses and susceptibility to infectious and allergic disease. In: Dean JH et al. (eds) *Immunotoxicology Immunopharmacology*. New York: Raven Press, pp. 395–441.

Selgrade MJK, Illing JW, Starnes DM *et al.* (1988) Evaluation of effects of ozone exposure on influenza infection in mice using indicators of susceptibility. *Fundam Appl Toxicol* **11**: 169–180.

Selgrade MJK, Starnes DM, Illing JW *et al.* (1989) Effects of phosgene exposure on bacterial, viral, and neoplastic lung disease susceptibility in mice. *Inhal Toxicol* **1**: 243–259.

Selgrade MK, Lawrence DA, Ullrich SE *et al.* (1997) Modulation of T-helper cell populations: potential mechanisms of respiratory hypersensitivity and immune suppression. *Toxicol Appl Pharmacol* **145**: 218–229.

Shakman RA (1974) Nutritional influences on the toxicity of environmental pollutants. *Arch Environ Health* **28**: 105–113.

Sherwin RP and Richters V (1985) Effect of 0.3 ppm ozone exposure on type II cells and alveolar walls of newborn mice: an image-analysis quantification. *J Toxicol Environ Health* **16**: 535–546.

Shore S, Kariya ST, Anderson K *et al.* (1987) Sulfur-dioxide-induced bronchitis in dogs. Effects on airway responsiveness to inhaled and intravenously administered methacholine. *Am Rev Respir Dis* **135**: 840–847.

Shore S, Kobzik L, Long NC *et al.* (1995) Increased airway responsiveness to inhaled methacholine in a rat model of chronic bronchitis. *Am J Respir Crit Care Med* **151**: 1931–1938.

Slade R, Highfill JW and Hatch GE (1989) Effects of depletion of ascorbic acid or nonprotein suflhydryls on the acute inhalation toxicity of nitrogen dioxide, ozone and phosgene. *Inhal Toxicol* **1**: 261–271.

Slauson DO and Hahn FF (1980) Criteria for development of animal models of diseases of the respiratory system. *Am J Pathol* **101**: 103–122.

Slutsky RA, Peck WW and Higgins CB (1983) Pulmonary edema formation with myocardial infarction and left atrial hypertension: intravascular and extravascular pulmonary fluid volumes. *Circulation* **68**: 164–169.

Smith LG, Busch RH, Buschbom RL *et al.* (1989) Effect of sulfur dioxide or ammonium sulfate exposure alone or combined, for 4 or 8 months on normal and elastase-impaired rats. *Environ Res* **49**: 60–78.

Snider GL (1992a) Animal models of chronic airways injury. *Chest* **101** (Suppl. 3): 74S–79S.

Snider GL (1992b) Emphysema: The first two centuries and beyond. A historical overview, with suggestions for future research: part 2. *Am Rev Respir Dis* **146**: 1615–1622.

Snider GL, Lucey EC and Stone PJ (1986) Animal models of emphysema. *Am Rev Respir Dis* **133**: 149–169.

Snipes MB, McClellan RO, Mauderly JL and Wolff RK (1989) Retention patterns for inhaled particles in the lung: comparisons between laboratory animals and humans for chronic exposure. *Health Phys* **57**: 69–77.

Sorkness R, Lemanske Jr RF and Castleman WL (1991) Persistent airway hyperresponsiveness after neonatal viral bronchiolitis in rats. *J Appl Physiol* **70**: 375–383.

Stiles J and Tyler WS (1988) Age-related morphometric differences in responses of rat lungs to ozone. *Toxicol Appl Pharmacol* **92**: 274–285.

Stockstill BL, Chang LY, Manache MG *et al.* (1995) Bronchiolarized metaplasia and interstitial fibrosis in rat lungs chronically exposed to high ambient levels of ozone. *Toxicol Appl Pharmacol* **134**: 251–263.

Stripp BR, Whitsett JA and Lattier DL (1990) Strategies for analysis of gene expression: pulmonary surfactant proteins. *Am J Physiol* **259**: L185–L197.

Stuart BO (1976) Selection of animal models for evaluation of inhalation hazards in man. In: Aharonson EF, Ben-David A and Klingberg MA (eds) *Air Pollution and the Lung*. New York: John Wiley & Sons, pp. 268–288.

Sun JD, Pickrell JA, Harkema JR *et al.* (1988) Effects of buthionine sulfoximine on the development of ozone-induced pulmonary fibrosis. *Exp Mol Pathol* **49**: 254–266.

Surgeon General (1984) In: *The Health Consequences of Smoking: Chronic Obstructive Lung Disease*, US Department of Health and Human Services. Washington, DC: US Government Printing Office.

Sweeney TD, Brain JD, Tryka AF and Godleski JJ (1983) Retention of inhaled particles in hamsters with pulmonary fibrosis. *Am Rev Respir Dis* **128**: 138–143.

Sweeney TD, Brain JD, Leavitt SA and Godleski JJ (1987) Emphysema alters the deposition pattern of inhaled particles in hamsters. *Am J Pathol* **128**: 19–28.

Sweeney TD, Brain JD and Godleski JJ (1988) Preexisting disease. In: Brain JD, Beck BD, Warren AJ and Shaikh RA (eds) *Variations in Susceptibility to Inhaled Pollutants. Identification, Mechanisms, and Policy Implications*. Baltimore, MD: The Johns Hopkins University Press, pp. 142–158.

Sweeney TD, Skornik WA, Brain JD *et al.* (1995) Chronic bronchitis alters the pattern of aerosol deposition in the lung. *Am J Respir Crit Care Med* **151**: 482–488.

Szapiel SV, Fulmer JD, Hunninghake GW *et al.* (1981) Hereditary emphysema in the tight-skin (Tsk/+) mouse. *Am Rev Respir Dis* **123:** 680–685.

Tager IB and Speizer FE (1976) Risk estimates for chronic bronchitis in smokers: a study of male–female differences. *Am Rev Respir Dis* **113**: 619–625.

Takahashi Y, Miura T and Takahashi K (1993) Vitamin A is involved in maintenance of epithelial cells on the bronchioles and cells in the alveoli of rats. *J Nutr* **123**: 634–641.

Takano H, Yoshikawa T, Ichinose T *et al.* (1997) Diesel exhaust particles enhance antigen-induced airway inflammation and local cytokine expression in mice. *Am J Respir Crit Care Med* **156**: 36–42.

Taylor G, Stott EJ, Hughes M and Collins AP (1984) Respiratory syncytial virus infection in mice. *Int Immun* **43**: 649–655.

Tepper JS, Wiester MJ, Weber MF and Menache MG (1990) Measurements of cardiopulmonary response in awake rats during acute exposure to near-ambient concentrations of ozone. *J Appl Toxicol* **10**: 7–15.

Tepper JS, Wiester MJ, Weber MF *et al.* (1991) Chronic exposure to a stimulated urban profile of ozone alters ventilatory responses to carbon dioxide challenge in rats. *Fundam Appl Toxicol* **17**: 52–60.

Tyler WS (1983) Comparative subgross anatomy of lungs. Pleuras, interlobular septa, and distal airways. *Am Rev Respir Dis* **128**: S32–S38.

Tyson CA, Lunan KD and Stephens RJ (1982) Age-related differences in GSH-shuttle enzymes in NO_2- or O_3-exposed rat lungs. *Arch Environ Health* **37**: 167–176.

US Department of Health, Education, and Welfare (1969) *Toxicological Effects of Sulfur Oxides on Animals*, Publication no. AP-50. Washington, DC: Public Health Service, National Air Pollution Control Administration.

US Environmental Protection Agency (1987) National Ambient Air Quality Standards for Particulate Matter: Final Rule. Washington, DC: Fed. Reg. 52, 24638–24669.

Valberg PA and Watson AY (1996) Lung cancer rates in carbon-black workers are discordant with prediction from rat bioassay data. *Regul Toxicol Pharmacol* **24**: 155–170.

van Loveren H, Steerenberg PA, Garssen J and van Bree L (1996) Interaction of environmental chemicals with respiratory sensitization. *Toxicol Lett* **86**: 163–167.

Vedal S (1997) Ambient particles and health: lines that divide. *J Air Waste Manag Assoc* **47**: 551–581.

Vincent R and Adamson IY (1995) Cellular kinetics in the lungs of aging Fischer 344 rats after acute exposure to ozone. *Am J Pathol* **146**: 1008–1016.

Vyas-Somani AC, Aziz SM, Arcot SA *et al.* (1996) Temporal alterations in basement membrane components in the pulmonary vasculature of chronically hypoxic rat: Impact of hypoxia and recovery. *Am J Med Sci* **312**: 54–67.

Warheit DB (1989) Interspecies comparisons of lung responses to inhaled particles and gases. *Crit Rev Toxicol* **20**: 1–29.

Watkinson WP, Campen MJ and Costa DL (1998) Cardiac arrhythmia induction after exposure to residual oil fly ash particles in a rat model of pulmonary hypertension. *Toxicol Sci* **41**: 209–216.

Waxman AB, Goldie SJ, Brett-Smith H and Matthay RA (1997) Cytomegalovirus as a primary pulmonary pathogen in AIDS. *Chest* **111**: 128–134.

Weil CS (1972) Guidelines for experiments to predict the degree of safety of a material for man. *Toxicol Appl Pharmacol* **21**: 194–199.

Weinstock S and Beck B (1988) Age and nutrition. In: Brain JD, Beck BD, Warren AJ and Shaikh RA (eds) *Variations in Susceptibility to Inhaled Pollutants. Identification, Mechanisms, and Policy Implications.* Baltimore, MD: The Johns Hopkins University Press, pp. 204–220.

Wells MA, Albrecht P and Ennis F (1981) Recovery from a viral respiratory infection. I. Influenza pneumonia in normal and T-deficient mice. *J Immunol* **126**: 1036–1041.

White DA (1995) Pulmonary infection in the immunocompromised patient. *Semin Thorac Cardiovasc Surg* **7**: 78–87.

Wilson R (1997) Bacterial infections of the bronchial tree. *Curr Opin Pulm Med* **3**: 105–110.

Wilson R and Reyner CF (1995) Bronchitis. *Curr Opin Pulm Med* **1**: 177–182.

Wilson DW, Segall HJ, Pan LC *et al.* (1992) Mechanisms and pathology of monocrotaline toxicity. *Crit Rev Toxicol* **22**: 307–325.

Yokoyama E, Nambu Z, Ichikawa I *et al.* (1987) Pulmonary response to exposure to ozone of emphysematous rats. *Environ Res* **42**: 114–120.

Zhao L, Winter RJD, Krausz T and Hughes JMB (1991) Effects of continuous infusion of atrial natriuretic peptide on the pulmonary hypertension induced by chronic hypoxia in rats. *Clin Sci* **81**: 397–385.

11

Novel Approaches to Study Nasal Responses to Air Pollution

LILIAN CALDERÓN-GARCIDUEÑAS

Curriculum in Toxicology, University of North Carolina, Chapel Hill, NC, USA

TERRY L. NOAH

Department of Pediatrics and the Center for Environmental Medicine and Lung Biology, University of North Carolina, Chapel Hill, NC, USA

HILLEL S. KOREN

US Environmental Protection Agency, National Health and Environmental Effects Research Laboratory, Human Studies Division, Research Triangle Park, NC, USA

This document has been reviewed in accordance with US Enviromental Protection Agency policy and approved for publication. Mention of trade names or commercial products does not constitute endorsement or recommendation for use.

INTRODUCTION

The human nasal epithelium is a valuable sentinel of exposure to a myriad of toxic and carcinogenic substances, allergens, particulate matter and infectious agents. The nasal passages can be seen as windows to the respiratory system, windows that are readily accessible and which, once altered, may be compromised in their ability to protect the lower respiratory tract from exposure to harmful substances.

The nose is a complex structure that provides access of ambient air to the lower respiratory tract. While the nose performs its vital physiological functions it can be a target for environmental pollutants and may undergo alteration after acute or chronic exposure to noxious agents. In the last few decades, investigators have concentrated on understanding the effects that air pollutants and xenobiotics exert upon the nasal tissues, the role the nose plays as a component of the respiratory system, and the consequences when its defense mechanisms are altered.

In this chapter, we summarize the methods currently used to study the nasal passages in relation to air pollutants, and we review some of the control and field studies that apply these methods and currently available study techniques. We provide the basic background needed to study the nasal passages, and an overview of the range of studies that are available to assess human populations.

ROLE AND STRUCTURE OF THE NOSE

The nasal cavity and mucosa both warm, moisten and filter incoming air and also contain receptors for olfaction. Together with the lower airways, they maintain an alveolar environment in which temperature and humidity are optimal for the main alveolar function: oxygen and carbon dioxide exchange (Proctor and Andersen, 1982).

The nose conditions between 10 000 and 20 000 liters/day of inspired air and is the most important portal of entry for the gases, particulates, allergens, and microorganisms present in the environment (Cole, 1993; Henderson *et al.*, 1993). The inspiratory passage of ambient air through a healthy nose is capable of eliminating a large portion of those health hazards.

The depth of penetration of particles depends on their size and velocity, the turbulence of the airstream, and the anatomical configuration of the air passages. Early data on nasal deposition has been reviewed by Lippmann (1970). Fry and Black (1973) investigated the sites of particle deposition and the clearance rates in the human nose using monodispersed particles of polystyrene, labelled with ^{99m}Tc, and reported that at least 45% of the retained material was deposited in the anterior region of the nasal passages. The site of maximal deposition was 2–3 cm behind the tip of the nose. Particles deposited on the ciliated epithelium were rapidly removed by the mucociliary clearance mechanisms, while particles deposited in the unciliated region were removed relatively slowly (Hilding, 1963).

The predominant deposition mechanism of particles in the upper respiratory tract is inertial impaction; changes in the inhaled airstream direction or in the magnitude of air velocity streamlines or eddy components are not followed by airborne particulates because of their inertia. The airways of the human head are major deposition sites for the largest inhalable particles (>10 μm aerodynamic diameter) as well as for the smallest particles (<0.1 μm). Soluble materials deposited on the nasal mucosa will be accessible to underlying cells if they can diffuse to them through the mucus prior to removal via mucociliary transport. Dissolved substances may be subsequently translocated into the bloodstream following movement within intercellular pathways between epithelial cell tight junctions or by active or passive transcellular transport mechanisms.

As mentioned previously, the nasal passages have an extensive vasculature. Uptake into the blood from this region thus can occur rapidly (Proctor, 1977, Proctor *et al.*, 1982, Rasmussen, *et al.*, 1990; Swift, 1976, Swift and Proctor, 1988; Swift *et al.*, 1992). Major factors that affect the disposition (deposition, uptake, distribution, metabolism and elimination) of inhaled particles in the respiratory tract include the physicochemical characteristics of the inhaled aerosol (particle size, distribution, solubility, hygroscopicity), as well as anatomic (architecture and size of airways passages) and physiological (ventilation rates, clearance mechanisms) characteristics. The anterior nares, in which 45% of inspired particles are deposited, is non-ciliated, and thus clearance is achieved by wiping or blowing the nose (Koening and Pierson, 1984).

The nasal cavity is lined by surface epithelial cell populations with specific roles in conducting and maintaining the normal functions of the nose. These cell populations include squamous, transitional, respiratory and olfactory epithelium. The respiratory epithelium covers over 80% of the nasal airway and is pseudostratified and ciliated with mucus secreting cells. It is the primary cellular component in the dynamic mucociliary apparatus (Fig. 11.1). This apparatus is the first line of defense for both the upper and lower respiratory tracts, eliminating potentially injurious, inorganic and organic particles and gases. Inhaled agents are deposited on the constantly renewed mucous layer, and are propelled by the synchronized cilia beating from the airway to the digestive tract, where the agents are eliminated from the body. For interactions to occur between inhaled materials and the epithelium, the materials must penetrate the airway lining fluid (ALF) (Leopold, 1995). ALF is a thin layer of a complex viscoelastic fluid composed of a mucus 'gel' layer on the surface of a 'sol' layer bathing the cilia; high molecular weight glycoconjugates are the major contributors to the high viscocity and the gel-like properties of the mucus. The main mucus glycoproteins are mucins, polyanions associated with cations and positively charged proteins. Their acidic character is due to the presence of sulfate groups, a feature observed in epithelial goblet cells which secrete more acidic mucin than their counterparts in submucosal glands (Stahl and Ellis, 1973). Four proteins compose 40–60% of the total protein in human nasal secretions: serum albumin, lysozyme, immunoglobulin A and lactoferrin. The total protein content of human nasal mucus ranges from 1.6 to 2.5 mg/ml (Brofeldt *et al.*, 1986). Lipids (predominantly cholesterol and lesser amounts of phospholipids), ions (sodium, chloride and potassium), small molecular weight organics, antioxidants, and inflammatory mediators also are present in the nasal fluids.

Antioxidants play a critical role in protecting the epithelium against air pollutants and xenobiotics. Humans have high levels of uric acid and low levels of ascorbic acid in nasal lavage (NAL) fluids, and levels of these antioxidants could affect the relative susceptibility of the nasal epithelium to oxidant pollutants (Hatch, 1992; Housley *et al.*, 1995a,b; Peden *et al.*, 1990, 1993; Raphael *et al.*, 1989). Maintenance of the mucus layer through various active transport processes, regulation of the secretory rate of mucin, and epithelial integrity are important factors contributing to the normal function of the mucociliary apparatus.

Secretory immunity is also a defense system of the airway mucosa. Specific antibodies mainly belong to secretory immunoglubulin A (SIgA). Secretory IgM (SIgM) and serum-derived and locally produced IgG also contribute to epithelial surface protection, but to a lesser extent (Brandtzaeg, 1995). Winther *et al.* (1987) estimate the lymphocyte to monocyte/macrophage ratio in nasal mucosa from healthy adults to be approximately 10:1, the T cell to B cell ratio to be 3:1, and the T helper/inducer cell to T

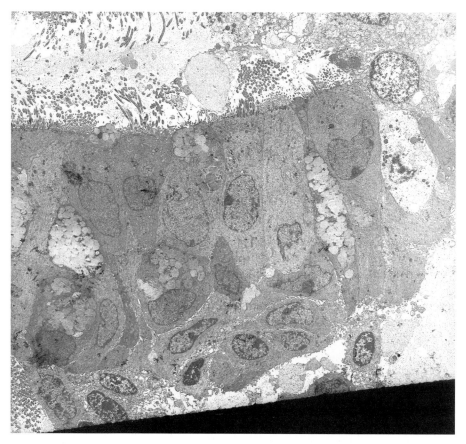

Fig. 11.1   Electron micrograph of human normal nasal epithelium. Ciliated cells, goblet cells containing mucus, and basal undifferentiated cells are seen. × 660. Courtesy of Dr Johnny Carson.

suppressor/cytotoxic cell ratio is 2.5:1. In addition, the authors observed regional differences with a relatively increased number of T suppressor/cytotoxic cells around submucosal glands, and a relatively large number of B cells in lymphocyte aggregates in the lamina propria. The HLA-DR antigen was expressed in epithelial cells, suggesting involvement of the surface epithelium in local immune responses.

The human olfactory system is organized into epithelial and lamina propria components and is limited to a very small area, approximately 8% of the total nasal epithelium (DeSesso, 1993). The olfactory cells are bipolar neurons that exhibit stereocilia; these lie on the epithelial surface and house the receptors for olfaction. Humans can recognize approximately 100 000 scents and some neurons in the nose are spatially segregated according to the scents they detect. The olfactory mucosa is exposed to a wide spectrum of nasal toxicants, including metals, solvents and pesticides. Understanding the spectrum of lesions induced by these toxicants facilitates our assessment of the health risks posed by these materials to humans (DeSesso, 1993; Morgan, 1991). There is, however, little information available in humans regarding dosimetry of inhaled chemicals and the causes of olfactory toxicity.

NASAL SAMPLING TECHNIQUES

A wide range of techniques is available to researchers to assess and quantify nasal mucosal damage in human subjects. There are several methods of obtaining nasal cells and different techniques sample different compartments of the nasal mucosa. A summary of these methods follows.

Blown Secretions

This consists of blowing nasal secretions onto wax-paper or directly onto a glass slide, followed by air drying and staining of the cells. A modification of this technique involves instillation of 1 ml of normal saline solution into each nostril, in sequence, followed by blowing of the nostril being studied, while the other is gently blocked with a finger. The technique is specially suitable for a quick view of the cell population in nasal secretions in such clinical situations as allergic rhinitis. The disadvantages include obtaining only those cells present in nasal secretions, dependence on the presence of secretions at the time of specimen acquisition, and restriction of sampling to the anterior regions of the nose.

Nasal Smears

These are obtained with cotton-wool swabs gently moved along the anterior to the posterior part of the inferior and/or the middle turbinates. The swab is then either smeared over a glass slide and the specimen is fixed and stained or it is immersed in a transport fluid for microbiology culture techniques. This method yields cells from nasal secretions and the superficial portion of the nasal mucosa. The results are not uniform and are difficult to evaluate from different subject populations. One use of this technique is to determine the presence or absence of a specific cell type (e.g. eosinophils) as a relative proportion of total cells. This technique has also proven useful for detection of upper respiratory tract carriage of *Streptococcus pneumoniae* and *Haemophilus influenzae* in healthy children (Capeding, *et al.*, 1995).

Imprints

In this method, plastic strips covered with 1% albumin are introduced into the nasal cavity under direct vision and are gently pressed over a nasal mucosa surface. The strips are then fixed, stained and examined under the microscope. The material obtained contains nasal secretions and superficial epithelium. The disadvantages of this technique include the presence of mucus on the smears that obliterates the cell morphology owing to heavy mucopolysaccharide staining, and the small numbers of epithelial cells obtained (Pipkorn and Enerback, 1984).

Brush

In this technique, epithelium is harvested with a small brush made of plastic-coated steel wire with nylon bristles. The brush is introduced into the nasal cavity under direct vision and is rotated while being introduced and removed. The brush is then submerged into a saline-buffered solution or transport media if cell culture is the final choice (Black *et al.*, 1998). The brush is shaken so the harvested cells go into suspension. The advantages of the procedure are the quantification of the total number of cells per volume (since the initial fluid volume is known) by hemocytometer, staining of multiple slides for quantification of different cell types, and the use of cell pellets for biochemical analysis. The disadvantages include a mild and transient nasal discomfort and possible limitation of the number of cells obtained, depending on the dexterity of the operator and the brush surface.

Nasal Scrapings

A specimen of nasal secretions and epithelium is obtained by using a plastic curette (Rhino-Probe, ASI, Arlington, TX) to scrape the nasal mucosa under direct vision. The inferior and middle turbinates are easily reached with the curette and an expert operator can obtain a quarter of a million cells with good tolerance from the subject and minimal discomfort (sneezing and tearing in the ipsilateral eye). The scraping can then be smeared on a slide, submerged in neutral formaldehyde, or used for tissue culture in an appropriate culture media. The advantages of this technique are the specificity of sampling site, ease of repetition, good numbers of cells and minimal discomfort. The disadvantage is that the samples are fragmented and small, thus if the tissue is abnormal to begin with, it is difficult to keep its architecture intact when the histotechnician embeds the tissue in paraffin and cuts the material. Therefore, this technique requires a skilled person to handle the tiny specimens and orient the histology cuts.

Nasal Biopsies

Both the inferior and the middle turbinates are accessible for biopsy procedures and control of possible complications. The procedure requires local anesthesia by topical application of a decongestant such as oxymetazoline or phenylephrine, followed by a local anesthetic (lidocaine or tetracaine). Some subjects require injecting the biopsy site with a solution of 1% lidocaine and 1:100 000 epinephrine. The biopsy specimen can be obtained with a punch, Geristma or Takahashi forceps, and post biopsy bleeding is controlled by a vasoconstrictor. The tissue can be frozen or fixed for molecular, immunohistochemical or microscopic studies. The major advantages of biopsy specimens is that all layers of the nasal mucosa are sampled from the epithelium to the basal membrane to the submucosa, and one is able to sample macroscopically abnormal areas and include adjacent unremarkable areas for comparison. Any number of special techniques can be used with these tissues and it is possible to obtain several dozen slides from each biopsy. The disadvantages include discomfort for the patient, the need for anesthesia, limitation of the procedure to adults, and the limited number of specimens that can be obtained from the same person. Possible complications include bleeding, pain and synechiae formation. A good clinical history should be taken first by a physician with special attention to previous bleeding disorders, consumption of drugs that alter bleeding time, previous family and personal history of local anesthetic reactions, and use of prescribed or non-prescribed medications or drugs that may alter the nasal mucosa. This procedure should be done by a physician with previous experience in this technique.

Nasal Lavage

Nasal lavage is used extensively for assessment of nasal inflammatory cell influx, antioxidants, eicosanoid mediators, neuropeptide release, nasal glandular products, increased vascular permeability products, cytokines, and other products from cells such as eosinophils, neutrophils and mast cells (Graham *et al.*, 1988; Graham and Koren, 1990; Houser *et al.*, 1995; Koren *et al.*, 1990; Noah *et al.*, 1995a,b; Peden, 1996; Ramis *et al.*, 1988; Tonnesen and Hindeberg, 1988). A common lavage technique requires introduction of 5 cm^3 of buffer solution (Ringer's, phosphate-buffered saline) in each nostril while asking the subject to tilt the head back and close the soft palate against the posterior pharyngeal wall. The subject keeps the solution in the nose for 10 seconds, after which the head is bent forward and the lavage solution expelled into a plastic receptacle. The recovered fluid is usually 80% of the volume originally introduced. The volume of the lavage is recorded, and the total number of cells is counted with a hemocytometer, thus obtaining the total number of cells collected. The lavage is placed on ice until centrifugation at 500 rpm for 10–15 min, at 4°C. The supernatant is separated and saved for measurement of soluble mediators, and the cell pellet can be resuspended in buffer or culture media and used for cytocentrifuge slide preparations. The slides are then fixed and stained to allow enumeration of different cell types. This technique samples cells in nasal secretions and has the advantage of being easy to perform, reproducible, painless and repeatable. Measurements can be undertaken at frequent intervals to follow the kinetics of cellular and biochemical changes. Harvested cells can be used for *ex vivo* studies and there is usually a high cellular yield (McCaffrey, 1997).

The nasal lavage method is limited by the fact that the lavage fluid does not remain in contact with the mucosa for long periods. Most variations of the technique require the subject to tilt the head back, which contaminates the nasal lavage with nasopharyngeal secretions. Housley *et al.* (1995a,b), developed a method that allows for prolonged nasal lavage and avoids contamination from the nasopharynx. Subjects are asked to gently blow their noses and then report which nasal cavity had the greatest resistance to airflow. This cavity is recorded as open and the opposite one as closed. The balloon of a modified Foley catheter is then inflated inside one of the nostrils to occlude it. Subjects are asked to lean forward at 45° and 5 ml of 0.9% sterile saline at room temperature are passed into the cavity and left for periods of up to 10 min. The procedure is then repeated for the other cavity and samples are collected.

A modification of the NAL technique involves the insufflation of a small volume of saline solution by nasal sprayer. The subject sprays the solution into his nostril five consecutive times and then expels the NAL fluid into a cup. The subject then repeats the entire cycle six times in each nostril. One advantage is that the study subject is in control.

The disadvantages are that the final volume depends on the force used in the insufflation, and the sampling is limited to the anterior part of the nose (Peden, 1996).

IN VITRO STUDIES

Tissue Culture

Human airway epithelial cell cultures offer a convenient *in vitro* model system for the study of the epithelium in the etiology of airway diseases and the underlying mechanisms of damage. Several authors have demonstrated that nasal epithelial cells can be cultured *in vitro* from surgical specimens and biopsies (Chevillard *et al.*, 1993; Devalia, *et al.*, 1990; Steele and Arnold, 1985; Werner and Kissel, 1995; Wu, 1985; Wu *et al.*, 1985). Over the past few years, the substratum on which cells are seeded and the constitution of the culture medium have been manipulated to influence cellular differentiation. Epithelial culture medium previously required serum, but more recently this has been replaced by serum-free medium supplemented with growth factors and hormones. Cells from nasal polyps or turbinates were grown in Ham's F_{12} medium supplemented with insulin, transferrin, epidermal growth factor, hydrocortisone, T_3, cholera toxin and bovine hypothalamus extract (Wu, 1985; Wu *et al.*, 1985). Organ culture of human nasal tissue with an air–liquid interface reproduces physiological conditions *in vitro* and allows for studies of effects of environmental factors, pharmacological agents and such microbiological agents as bacteria and viral infections (Jackson *et al.*, 1996). When human nasal epithelial cells are grown on porous membranes at the air–liquid interface, there is good development of ciliated cells. Submerged cultures grown on collagen matrices give a cuboidal epithelial-like morphology (Schierhorn *et al.*, 1995; Schmidt *et al.*, 1996). Suspension culture systems have been used for the study of ciliogenesis *in vitro* (Jorissen *et al.*, 1990).

Exposure of Human Ciliated Nasal Cells to Air Pollutants

Mucociliary transport has been studied in vitro using different air pollutants. Kienast *et al.* (1993) quantified ciliary beat frequency (CBF) with video-interference microscopy and used nasal ciliated cells from 12 healthy volunteers. Cells were placed on a polycarbonate membrane in contact with the surface of a reservoir filled with RPMI 1640 or Ringer's solution, allowing the cells to be supplied by capillarity. In an exposure chamber the cells were exposed for 30 min to SO_2 2.5 or 12.5 ppm at a temperature of 37°C and a 100% air humidity. SO_2 induced a dose-dependent decrease in CBF of the cells cultured in Ringer's solution. This change correlated with a pH decrease in the solution, while cells cultured in RPMI reduced their CBF only moderately after exposure to SO_2 at 12.5 ppm. These results suggested that the highly water-soluble SO_2 reversibly eliminates CBF in association with a decrease in pH.

Nitrogen dioxide, a common pollutant in photochemical smog, has a documented toxicity to the pulmonary airways (Heller and Gordon, 1986, Ranga and Kleinerman, 1981). Acute, low-level NO_2 exposure was investigated by Carson *et al.* (1993), who obtained nasal respiratory epithelium from young males exposed to clean air or to 2 ppm of NO_2 for 4 h. Minor alterations in the luminal border membranes of ciliated cells and increases in a subpopulation of compound cilia were described and contrasted with animal chronic exposure findings, which include disintegration and destruction of the airway epithelium, dysmorphology and loss of cilia, fragmentation of tight junctional complexes, and changes in membrane-associated complexes (Heller and Gordon, 1986; Ranga and Kleinerman, 1981). Carson *et al.* (1993) have suggested that the adverse mucociliary function in normal humans exposed to low levels of NO_2 may be minimal. On the other hand, Wang *et al.* (1995) described the effects of a 6-h exposure to NO_2 on the early-phase nasal response to allergen challenge in patients with a history of seasonal allergic rhinitis. Subjects were exposed in a randomized single blind study to either air or 400 ppb NO_2 for 6 h and then either challenged with allergen or without it. Allergen challenge after exposure to both air and NO_2 significantly increased levels of mast cell tryptase, while allergen challenge after exposure to NO_2 alone significantly increased the levels of eosinophil cationic protein. The authors concluded that acute exposure to NO_2 at concentrations found at the curbside in heavy traffic during episodes of pollution may 'prime' eosinophils for subsequent activation by allergen in those individuals with a history of seasonal allergic rhinitis.

The effects of near-ambient ozone concentrations on the expression of intercellular adhesion molecule-1 (ICAM-1) and the synthesis of cytokines by human nasal epithelial have been investigated by Beck *et al.* (1994). Increased expression of ICAM-1 and constitutive expression of interleukin-8 (IL-8) implicated the nasal epithelium as a source of cytokines that mediate local inflammatory responses. Adhesion molecules promote cell–cell interactions and antigen-specific recognition by T lymphocytes. They also stimulate the effector mechanisms of activated lymphocytes. The authors suggest that ozone may initiate airway inflammation as a consequence of early interactions with respiratory epithelial cells, by stimulating the release of multiple cytokines.

CONTROLLED AIR POLLUTANT EXPOSURE STUDIES

Human exposure to pollutants under controlled chamber conditions results in important information regarding the health effects associated with individual pollutants. The results are valid for the investigated group and the data can be extrapolated to conditions in the real environment. Ozone is the most widely studied pollutant to date in relation to nasal effects. Dosimetry models provide estimates of the amount of inhaled O_3 that is delivered to the target tissues. The distribution of site-specific responses in the nasal passages is probably affected by a series of factors, including local clearance mechanisms (blood, mucus), cell-specific responses and regional dose or uptake patterns (Kimbell, 1995; Morgan and Monticello, 1990a,b). In addition, regional uptake patterns of an inhaled gas can be influenced by nasal airflow patterns, interactions that occur between the gas and components of the airway lining at the air/ALF interface, and the disposition of the inhaled substance within ALF and tissues (Kimbell *et al.*, 1993).

In studies in which volunteers are exposed to ozone and in epidemiological studies at elevated concentrations of photochemical air pollution, the number of polymorphonuclear leukocytes (PMN) is increased in NAL fluid. Graham *et al.* (1988) reported increased numbers of PMN in the NAL fluid of 21 volunteers exposed to 0.5 ppm O_3 at rest for 4 h on two consecutive days, compared to air exposure. NAL was performed immediately before and after each exposure, as well as 22 h after the last exposure. Nasal lavage fluid contained significant increases of PMN at all post-exposure times tested, with peak values occurring immediately prior to the second day of exposure. Graham and Koren (1990) compared inflammatory mediators present in the NAL and BAL fluids of humans exposed to ozone. The authors found significant increases in PMN in NAL taken immediately after exposure and on the next day. Increases in NAL and BAL fluids were similar, demonstrating a qualitative similarity between changes in the lower airways and the upper respiratory tract. NAL may thus be useful to study the acute inflammatory effects of ozone, and may reflect ozone's effects in the lower airways. Devlin (1993) therefore proposed using NAL as a surrogate for the more costly and invasive BAL in most epidemiological studies.

Epidemiological studies have attempted to assess the effect of recurrent ozone exposures in humans and to identify sensitive subpopulations. Asthmatics may have relatively increased sensitivity to ozone. Asthmatic adults are more susceptible to upper airway inflammation at O_3 concentrations that do not affect lung function; they also appear to be susceptible to disease exacerbation during summertime 'smog' episodes (Koren and Bromberg, 1995). Using a bolus lavage technique, Bascom *et al.* (1990) demonstrated that ozone induces both neutrophil and eosinophil influx into the nasal airways of subjects with allergic rhinitis. However, ozone did not appear to enhance allergen-induced eosinophil influx in these subjects. MacBride *et al.* (1994) compared nasal lavage and lung and nasal function between ten asthmatic and eight healthy subjects. Subjects were exposed in a head dome to clean air, 0.12 or 0.24 ppm O_3, for 90 min during intermittent moderate exercise. Leukocyte counts and chemotactic factors, leukotriene B_4, platelet-activating factor and IL-8 were analyzed in NAL. A significant increase in NAL PMN was detected in the asthmatic subjects immediately upon exposure and 24 h after exposure to 0.24 ppm O_3; there was a positive correlation between IL-8 and PMN after exposure to ozone.

Variability in intra-individual nasal PMN counts requires careful consideration in both the design and the interpretation of studies using NAL (Hauser *et al.*, 1994). Steerenberg *et al.* (1996) performed repeated NAL in 12 healthy volunteers not exposed to pollutants, over the course of two months, and recorded relatively high intra-individual variance in PMN. To account for within-subject variability, Hauser *et al.* (1994) suggested expressing the change in an individual's PMN count as a percentage of the individual's mean PMN count for the trial in the case of chamber-controlled studies.

Nasal lavage has also been useful in sulfur dioxide (SO_2) exposure studies. Levels of S-sulfonate measured in NAL from subjects exposed to SO_2 have been found to be an effective short-term marker of exposure to this pollutant. Bechtold *et al.* (1993) determined the levels of S-sulfonate by treating NAL fluid protein with cyanide to cleave the S–S linkage and release the sulfite. In two experiments, humans were exposed to air or 1 ppm SO_2 for 10 min, and to air or 7 ppm SO_2 for 20 min, with lavage immediately after exposure. S-Sulfonate levels were significantly higher in the exposed groups compared with the control groups. It is important to note that the levels of S-sulfonates observed in NAL fluid were almost three orders of magnitude higher that those measured in plasma following similar SO_2 exposures.

Other end-points have been explored in nasal lavage samples in relation to air pollutants. The antioxidant content of ALF is a potentially important parameter because antioxidants may provide an initial defense against environmental toxins. The major antioxidants in ALF include uric acid, mucin, albumin, ascorbic acid and reduced glutathione (Cross *et al.*, 1994). Housley *et al.* (1995b) examined the urate/oxidant interactions in NAL following *in vitro* exposure to ozone concentrations of 50, 100 and 200 ppb. They collected NAL from eight healthy volunteers using a modified Foley catheter technique which permits prolonged contact of the isotonic saline with the anterior nasal cavity. Nasal lavage samples in multiwell plates were exposed to ozone and samples were removed at intervals from 15 to 240 min following exposure and then assayed for uric acid depletion. Uric acid concentrations in nasal lavage fell from 8.52 μm at time zero to 3.99, 0.05 and 0.07 μm after 240 min at the three different O_3 concentrations. The authors concluded that uric acid in NAL samples is scavenged by ozone in a dose- and time-dependent manner. Secretion of uric acid is increased by cholinergic stimulation and correlates well with the secretion of lactoferrin.

Peden *et al.* (1990, 1993) suggest that the principal source of uric acid in nasal secretions is plasma and that uric acid is taken up, concentrated, and secreted by nasal glands. Gender differences in the concentrations of uric acid in NAL have been reported by Housley *et al.* (1996); they compared NAL from 15 men, aged 20–68 years, and 11 women, aged 20–59 years. Women were found to have lower levels of uric acid both in NAL and plasma, and the authors proposed that factors such as gender could be important for the nasal responses to inhaled oxidants such as ozone.

The concentrations of outdoor air pollutants belonging to the so-called 'criteria pollutants' (air pollutants regulated by the US Environmental Protection Agency) are subjected to stringent legal limits based on scientific evidence for their health effects. Indoor air pollutants, on the other hand, are not subjected to health or safety laws and, although societies such as The American Society of Heating, Refrigeration and Air Conditioning Engineers have issued new standards for indoor ventilation since 1989, there is still a long way to go in the regulation of indoor pollutants. Air samplings of new buildings (built with an emphasis in energy conservation and marked reduction in ventilation rates) show detectable levels of volatile organic compounds (VOC) derived from the off-gassing of structural materials and furnishings. Formaldehyde is also frequently found in indoor environments from sources such as fabrics and new carpeting. Secondary contamination can arise from various human activities, from outdoor pollutant infiltration, indoor combustion products and accumulation of bioaerosols. Tobacco smoke is a dominant indoor air pollutant, in spite of increasing restrictions in public and work places. The World Health Organization (WHO) definition of the 'sick building syndrome' includes irritation of the eyes, nose and throat, as well as dry mucous membranes and skin. Such symptoms increased in buildings with higher occupation rates and high fleece factors, and while women have higher symptom rates than men, managers are affected less than clerical-secretarial workers. Bascom (1995) summarizes the problem of indoor air pollution. According to Bascom (1991), the baseline prevalence of nasal symptoms among building occupants is 20% on average, but in some studies it has been as high as 50–60%. Koren *et al.* (1992) suggested that VOC present in the synthetic materials used in homes and office buildings contribute to some of the 'sick building syndrome' components. In a study by the same authors, 14 subjects were exposed to a mixture of VOC (25 mg/m^3; Molhave and Moller, 1979) and NAL was used to monitor PMN influx into the nasal passages. A significant increase in PMN in NAL was found immediately after a 4-h exposure to VOC and 18 h post exposure, confirming that the nasal epithelium is also a target of VOC.

Numerous xenobiotics have been shown to damage the nasal mucosa of laboratory animals (Barrow, 1986; Harkema and Hotchkiss, 1995; Harkema *et al.*, 1987; Henderson *et al.*, 1993; Monticello *et al.*, 1989). The alterations to the nasal mucosa can include acute inflammation with a PMN's influx into the lamina propria and surface epithelium; degeneration and necrosis of the surface epithelium; epithelial ulceration; and transformation of the normal epithelium to a morphologically different, adaptive epithelium (Harkema and Hotchkiss, 1995). Ozone has been used in monkeys to examine its effects on nasal tissues. In a study by Harkema *et al.* (1987), monkeys were exposed to 0.15 or 0.30 ppm O_3, 8 h/day, for 6 or 90 days. Lesions of the nasal mucosa were restricted to the very anterior aspect of the nasal passages and involved the transitional epithelium and the adjacent respiratory epithelium. Ozone-induced lesions included an intraepithelial influx of PMN (in the 6-days exposed animals), attenuation of cilia, degeneration of ciliated cells and necrosis, an increase in the numbers of epithelial cells and mucous cell hyperplasia. Interestingly, the magnitude of the pathological changes was not dependent on the ozone concentration, but rather on the duration of the exposure. Hatch *et al.* (1989) evaluated the dose of oxygen-18-labeled O_3 in the noses of F344 rats and monkeys and found that excess ^{18}O in the monkey was highest in the lateral wall transitional epithelium and ethmoid turbinates; the former anatomical region has a counterpart in the human nose and therefore it also is likely to be an area of ozone deposition in humans.

In ozone-exposed Fisher (F-344/N) rats, the lesions are restricted to the transitional epithelium, where there is a true metaplastic response; mucous metaplasia taken place in a squamous epithelium devoid of goblet cells, along with attenuation of cilia in the lateral walls of the nasopharynx. In rats, the newly formed intraepithelial mucopolysaccharides (induced after O_3 exposure) act as a protective shield for the nasal epithelium and underlying lamina propria, preventing further injury by the continued ozone exposure. Hotchkiss *et al.* (1989) compared the acute ozone effects in nasal and pulmonary inflammatory responses in rats. They examined the cellular inflammatory responses in the nasal cavity and lower respiratory tract by means of NAL and BAL and morphometric quantitation of PMN within the nasal mucosa and pulmonary terminal bronchoalveolar duct regions. Rats were exposed to 0.0, 0.12, 0.8 or 1.5 ppm O_3 for 6 h and were sacrificed immediately or 3, 18, 42 or 66 h following exposure. The main finding was that the number of PMN recovered by NAL and BAL accurately reflected the tissue neutrophilic response at sites within the nasal cavity and lung that were injured by acute ozone exposure. Further, the results suggested that at high ozone concentrations (0.8 and 1.5 ppm), the acute nasal inflammatory response was attenuated by a simultaneous, competing inflammatory response within the centriacinar region of the lung. This attenuation in NAL neutrophilia also has been described in children exposed to high O_3 concentrations in Mexico City (Calderón-Garciadueñas *et al.*, 1995). Hotchkiss *et al.* (1989) concluded that analysis of NAL for changes in cellular composition may be a useful indicator of acute exposure to ambient levels of ozone, but at higher ozone levels the nasal cellular inflammatory response may underestimate the effects of ozone both on nasal and pulmonary epithelia. Increased DNA synthesis is seen in the nasal squamous and transitional epithelia of O_3-exposed rats and coincides with the onset of metaplastic changes (Johnson *et al.*, 1990). Harkema and Hotchkiss (1995) suggest that while dosimetry may play a role in the observed pattern of ozone-induced DNA synthesis, most of the observed response is likely due to inherent differences in epithelial susceptibility to O_3-induced injury.

Mucociliary function has also been studied in O_3-exposed rats (Harkema *et al.*, 1994) and there is a clear-cut positive correlation between decreased mucociliary activity (as determined by measurement of flow rates of mucus *in vitro* through video-motion analysis), marked mucus cell metaplasia, and exposure to high O_3 concentrations (0.5 and 1.0 ppm for 20 months, 6 h/day, 5 days/week). Altered mucociliary clearance leaves the more distal pulmonary airways vulnerable to potentially injurious concentrations of inhaled xenobiotics or infectious agents, a point of relevance for human populations.

Formaldehyde is an important chemical widely used in the building, insulation and textile industries. Inhalation exposures (6 h/day, 5 days/week) for 2 years to 14.3 ppm induced a 50% incidence of squamous cell carcinoma in the nasal cavities of F-344 rats (Monticello *et al.*, 1996). Formaldehyde-induced nasal lesions in rats and monkeys suggest that the distribution of the lesions is related to regional uptake patterns and not to tissue susceptibility (Morgan, 1997). In monkeys, the distribution of nasal lesions attributed to nasal airflow patterns is further supported by molecular dosimetry studies in which formaldehyde-induced DNA–protein cross-links correlates with sites of lesions (Casanova *et al.*, 1991). Klein-Szanto *et al.*, (1992) explored the effects of silastic devices containing various concentrations of formaldehyde upon xenotransplanted human adult nasal respiratory epithelium using irradiated athymic nude mice. They found areas of epithelial erosion, inflammation and focal hyperplastic-metaplastic areas with a high proliferative index. In the urban environment, combined exposure to ozone and formaldehyde can occur, Reuzel *et al.* (1990) have studied the interactive effects of ozone and formaldehyde on the nasal respiratory lining epithelium in rats. Ozone at 0.4 or 0.8 ppm or formaldehyde at 3 ppm enhanced nasal cell proliferation and produced histopathological nasal changes. Since increased cell proliferation is known to play an important role in chemical carcinogenesis, and particularly in formaldehyde carcinogenesis, the observed synergism may indicate an increased cancer risk in humans simultaneously exposed to formaldehyde and ozone.

Gasoline includes a variety of products that differ according to origin and type of oil, refinery technologies, and additives. The combustion of gasoline in motor engines produces exhausts composed partly of gasoline vapors and combustion products, whose composition varies with the gasoline composition and the engine type (Maltoni and Soffritti, 1995). Inhalation exposure of rats to vapors of high isoparaffin-content gasoline produces both benign and malignant renal tumors. Toxicokinetics of oxygenated additives such as methyl-tertiary-butyl ether (MTBE) and tertiary butyl alcohol (TBA) have been performed in 10 healthy male volunteers by Johanson *et al.* (1995). Uptake and disposition were studied by measuring MTBE and TBA in inhaled and exhaled air, blood and urine. Low uptake, high post-exposure exhalation, and low blood clearance indicated slow metabolism of MTBE relative to many other solvents. Nasal parameters such as peak expiratory flow, acoustic rhinometry and inflammatory markers in NAL indicated minimal effects of MTBE. Gasoline service workers in the Nordic countries have shown an increased 3.5-fold risk of nasal cancer and a 30% risk for kidney cancer. The cohort studied by Lynge *et al.* (1997) was exposed to gasoline vapors with benzene levels estimated to be 0.5–1 mg/m^3, suggesting that exposure to benzene and polyaromatic hydrocarbons in the work environment increases the risk for nasal neoplasms, a situation that could be applicable to highly polluted urban environments.

AIR POLLUTANT FIELD STUDIES

Field and epidemiological studies addressing the effects of predominant air pollutants and complex mixtures on nasal and lung parameters in exposed humans populations is of utmost relevance. Researchers should be aware, however, of the difficulty of estimating or controlling several factors, including actual pollutant exposures, levels of temperature, relative humidity, allergens, and activity patterns in volunteers.

Studies with Nasal Inflammation End-points

Field studies have demonstrated an association between ozone exposure at ambient levels and alterations in lung function in both children and adults (Spektor *et al.*, 1988; Castillejos *et al.*, 1992; Koren *et al.*, 1991). Ozone exposure results in nasal congestion, increased levels of histamine, PMNs, and mononuclear cells in nasal lavage fluid (Koltai, 1994). In Western Europe where ozone constitutes a major air pollutant, summer O_3 concentrations frequently exceed the NAAQS and there is concern about the health effects for exposed populations, specifically school children. Frischer *et al.* (1993) studied upper airway inflammation in 44 children by repeated NAL from May to October 1991. Days with high (>180 ppb) and low (<140 ppb) ozone exposures were compared. There was a significant increase in intra-individual mean PMN counts on high ozone days and linear regression analysis of log-PMN counts yielded a significant effect for ozone. The same group (Frischer *et al.*, 1997) explored the *in vivo* hydroxyl radical attack in nasal lavage samples from the 44 children studied in the previous paper. The focus of the paper was based on historical data showing that ozone reacts with water and gives rise to reactive hydroxyl radicals capable of oxidizing a wide range of biomolecules. *Ortho*-tyrosine was significantly higher following days with high ozone exposure. Based on their findings, Frischer *et al.* (1997) suggested that hydroxyl radical attack subsequent to ambient ozone takes place in the upper airways of healthy children, and relates to lung function decrements.

Residents of Mexico City are chronically exposed to high ozone concentrations (4 h/day with >0.08 ppm O_3). Calderón-Garciadueñas *et al.* (1992) characterized nasal mucosal changes in short-term (<30 days) and long-term (>60 days) residents, and in residents of a low-pollution area on the Gulf of Mexico as controls. Compared with controls, residents of high-ozone areas had loss of cilia, basal cell hyperplasia, squamous metaplasia and submucosal vascular proliferation. Similar changes, as well as infiltration of epithelium by neutrophils, were seen in a subsequent study of preadolescent children from the same high-pollution area (Calderón-Garciadueñas *et al.*, 1995). These studies demonstrate the relevance of nasal tissues as sentinels of exposure to polluted urban environments, as well as attesting to the feasibility of field studies in pediatric populations.

Fuel oil ash, a known respiratory irritant, contains vanadium pentoxide. Boilermakers and utility workers are exposed to fuel oil ash in their working environments. Hauser *et al.* (1994) studied nasal inflammatory responses in 37 workers with a baseline NAL after an average period of 114 days away from work and 3 days after the NAL basal study. They estimated the PM_{10} and respirable vanadium dust with a personal sampling device for respirable particles. Their results showed a significant increase in NAL PMN in non-smokers, but not in smokers, suggesting that either smokers have a diminished inflammatory response or that the smoking masks the effect of exposure to fuel oil.

Diaz-Sanchez *et al.* (1994, 1996) studied the effects of diesel exhaust particles (DEP) on the levels of messenger RNA for cytokines in nasal lavage cells and reported an increase in nasal cytokine expression with concomitantly enhanced local IgE production. They suggested that these two factors may play a role in the increased incidence of asthma and allergic rhinitis. However, Morgan *et al.* (1997) reviewed the literature relating to the health effects of diesel emissions in relation to nasal irritation and concluded there was not sufficient evidence of a permanent anatomical effect.

Certain common respiratory viruses, including respiratory syncytial virus (RSV), rhinovirus and parainfluenza virus, can directly stimulate cultured human airway epithelial cells to produce cytokines and chemokines (Becker *et al.*, 1991; Choi and Jacoby, 1992; Noah and Becker, 1993). To determine whether similar cytokine induction might occur *in vivo* in childhood infections, Noah *et al.* (1995a,b) performed serial nasal lavages and superficial subturbinate epithelial biopsies in infants and young children in a daycare center, around the time of naturally acquired viral upper respiratory illnesses (URI). Concentrations of cytokines in nasal lavage fluids (NAL) were measured using specific enzyme-linked immunosorbent assay (ELISA). Significant increases in NAL levels of cytokines IL-1β, IL-6, IL-8, and TNF-α were noted during acute URI compared with pre-illness baselines in the same children. At a follow-up time point 2–4 weeks after the onset of symptoms, levels of most cytokines had fallen back to near baseline. Paired superficial nasal epithelial biopsies were obtained for some children during acute URI and at follow-up. RNA from these cells was reverse transcribed and the resulting cDNA amplified by PCR using primers for the same cytokines. Most subjects had increases in epithelial mRNA abundance for IL-1B and IL-8 during acute URI, suggesting that the epithelium was at least one cellular source of the increases in these cytokines noted in NAL fluid. Noah *et al.* (1995a,b, 1997) have also used NAL to characterize baseline levels of inflammation and cytokines in children with asthma and cystic fibrosis (CF). In a study of school-age children, subjects were characterized on the basis of history, immediate hypersensitivity allergy skin tests, airway reactivity, and lung function as being either non-allergic/non-asthmatic, allergic but non-asthmatic, or allergic and asthmatic (Noah *et al.*, 1995a,b). These children underwent a single nasal lavage during an asymptomatic period in the summer. NAL levels of IL-8 and ECP were greater in asthmatics compared with levels in the other two groups. This finding is consistent with 'activation' of epithelial IL-8 as either a contributing cause or a consequence of the eosinophil-dominated inflammation characteristic of asthmatic airways.

These studies illustrate the safety and potential usefulness of nasal lavage in the study of respiratory inflammation. The results may be important for prediction of pediatric populations at greater risk from exposure to inhaled pollutants which trigger inflammation, as well as for the design of pollutant studies in these populations.

Studies using Nasal DNA as an End-point

A new field of interest to molecular epidemiologists is the exploration of DNA damage-inducible genes as biomarkers for exposures to environmental agents. The field is extremely relevant to respiratory epithelia and air pollutants and there is a particular interest in the use of the nasal epithelium as an accessible tissue in direct contact with air pollutants (Flato *et al.*, 1996; Johnson *et al.*, 1997).

The genetic damage found in cancer cells is of two sorts: dominant, with proto-oncogenes as targets, and recessive, where the targets are tumor suppressor genes. p53 belongs to the second category and the 20-year history of p53-related investigations is a paradigm for cancer research. It illustrates the convergence of previously parallel lines of basic, clinical and epidemiological investigation and the rapid transfer of research findings from the laboratory to clinical and to field studies (e.g. air pollutant epidemiological research). The fact that loss of p53 function contributes to more than half of all human cancers illustrates its importance (Hansen and Oren, 1997). p53 is induced in response to DNA damage to prevent the continued proliferation of genetically impaired cells. DNA damage primarily causes the activation and stabilization of latent p53, present at low levels in normal cells. The carboxy terminal domain of p53 recognizes certain types of DNA damage, including short single strands which also activate the sequence-specific DNA binding function of p53. By binding to specific sites within their promoters, p53 activates the transcription of various genes, including p21, Gadd 45, mdm2, bax, thrombospondin 1 and cyclin G (Ko and Prives, 1996). p53 also facilitates DNA repair, and indeed, loss of p53 can reduce the rate and efficiency of nucleotide excision. Harris (1993) suggested that the spectrum of mutations in p53 induced in human cancer can help identify particular carcinogens and define the biochemical mechanisms responsible for the genetic lesions in DNA that cause human cancer. Moreover, the frequency and type of p53 mutations can also act as a molecular dosimeter of carcinogen exposure and thereby provide information about the molecular epidemiology of human cancer risk. Progress in the field of molecular carcinogenesis and molecular epidemiology increases our ability to accurately assess cancer risk and will allow researchers in the pollutant health effects field to use these new tools to define more clearly the health consequences of pollutant exposures.

The expression of DNA damage-inducible genes is being investigated as a biomarker of radon progeny in epithelial cells from the anterior nasal cavity (Johnson *et al.*, 1997). Exposures resulted in a significant increase in the number of cells in the G_1 phase and a decrease in S phase cells, along with an increase in the number of cells containing DNA strand breaks. Studies by Johnson *et al.* (1997), suggest that chemical and physical agents induce different expression markers of p53, p21 and Gadd 153 proteins; these could be used to discriminate between toxic and non-toxic materials such as asbestos and glass microfiber.

REFERENCES

Barrow CS (ed.) (1986) *Toxicology of the Nasal Passages.* New York: Hemisphere.

Bascom R (1991) The upper respiratory tract: mucous membrane irritation. *Environ Health Perspect* **95**: 39–44.

Bascom R (1995) Human susceptibility to indoor contaminants. *Occup Med* **10**: 119–132.

Bascom R, Pipkorn U, Lichtenstein LM and Naclerio RM (1988) The influx of inflammatory cells into nasal washings during the late response to antigen challenge. *Am Rev Respir Dis* **138**: 406–412.

Bascom R, Naclerio RM, Fitzgerald TK *et al.* (1990) Effect of ozone inhalation on the response to nasal challenge with antigen of allergic subjects. *Am Rev Respir Dis* **142**: 594–601.

Bechtold WE, Waide JJ, Sandstrom T *et al.* (1993) Biological markers of exposure to SO_2: S-sulfonates in nasal lavage. *J Exp Anal Environ Epidemiol* **3**: 371–382.

Beck NB, Koenig JQ, Luchtel DL *et al.* (1994) Ozone can increase the expression of intercellular adhesion molecule-1 and synthesis of cytokines by human nasal epithelial cells. *Inhal Toxicol* **6**: 345–357.

Becker S, Quay J and Soukup J (1991) Cytokine (tumor necrosis factor, IL-6, IL-8) production by respiratory syncytial virus-infected human alveolar macrophages. *J Immunol* **147**: 4307–4312.

Black H, Yankaskas JR, Johnson L and Noah TL (1998) Interleukin-8 production by cystic fibrosis nasal epithelial cells after tumor necrosis factor-alpha and respiratory syncytial virus stimulation. *Am J Respir Cell Mol Biol* **19**: 210–215.

Brandtzaeg P (1995) Immunocompetent cells of the upper airway: functions in normal and diseased mucosa. *Eur Arch Otorhinolaryngol* **252** (Suppl. 1): S8–S21.

Brofeldt S, Mygind N, Sorensen CH *et al.* (1986) Biochemical analysis of nasal secretions induced by methacholine, histamine, and allergen provocations. *Am Rev Respir Dis* **133**: 1138–1142.

Calderón-Garcidueñas L, Osorno-Velazquez A, Bravo-Alvarez H *et al.* (1992) Histopathological changes of the nasal mucosa in Southwest Metropolitan Mexico City inhabitants. *Am J Pathol* **140**: 225–232.

Calderón-Garcidueñas L, Rodriguez-Alcaraz A, Garcia R *et al.* (1995) Nasal inflammatory responses in children exposed to a polluted urban atmosphere. *J Toxicol Environ Health* **145**: 427–437.

Capeding MR, Nohynek H, Sombrero LT *et al.* (1995) Evaluation of sampling sites for detection of upper respiratory tract carriage of *Streptococcus pneumoniae* and *Haemophilus influenzae* among healthy Filipino infants. *J Clin Microbiol* **33**: 3077–3079.

Carson JL, Collier AM, Hu SS and Devlin RB (1993) Effect of nitrogen dioxide on human nasal epithelium. *Am J Respir Cell Mol Biol* **9**: 264–270.

Casanova M, Morgan KT, Steinhagen WH *et al.* (1991) Covalent binding of inhaled formaldehyde to DNA in the respiratory tract of Rhesus monkeys: Pharmacokinetics, rat-to-monkey interspecies scaling, and extrapolation to man. *Fundam Appl Toxicol* **17**: 409–428.

Castillejos M, Gold DR, Dockery D *et al.* (1992) Effects of ambient ozone on respiratory function and symptoms in Mexico City school children. *Am Rev Respir Dis* **145**: 276–282.

Chevillard M, Hinnrasky J, Pierrot D *et al.* (1993) Differentiation of human surface upper airway epithelial cells in primary culture on a floating collagen gel. *Epith Cell Biol* **2**: 17–25.

Choi AMK and Jacoby DB (1992) Influenza virus A infection induces interleukin-8 gene expression in human airway epithelial cells. *FEBS Lett* **309**: 327–329.

Cole P (1993) *The Respiratory Role of the Upper Airways.* St Louis, MO: Mosby Year Book.

Cross CE, van der Vliet A, O'Neill CA *et al.* (1994) Oxidants, antioxidants and respiratory tract lining fluids. *Environ Health Perspect* **102** (Suppl. 10): 185–191.

DeSesso JM (1993) The relevance to humans of animal models for inhalation studies of cancer in the nose and upper airways. *Qual Assur Good Prac Reg Law* **2**: 213–231.

Devalia JL, Sapsford RJ, Wells CW *et al.* (1990) Culture and comparison of human bronchial and nasal epithelial cells *in vitro. Respir Med* **84**: 303–312.

Devlin RB (1993) Identification of subpopulations that are sensitive to ozone exposure: use of end points currently available and potential use of laboratory-based end points under development. *Environ Health Perspect* **101** (Suppl. 4): 225–230.

Diaz-Sanchez D, Dotson AR, Takenaka H and Saxon A (1994) Diesel exhaust particles induce local IgE production in vivo and alter the pattern of IgE messenger RNA isoforms. *J Clin Invest* **94**: 1417–1425.

Diaz-Sanchez D, Tsien A, Casillas A *et al.* (1996) Enhanced nasal cytokine production in human beings after in vivo challenge with diesel exhaust particles. *J Allergy Clin Immunol* **98**: 114–123.

Flato S, Hemminki K, Thunberg E and Georgellis A (1996) DNA adduct formation in the human nasal mucosa as a biomarker of exposure to environmental mutagens and carcinogens. *Environ Health Perspect* **104** (Suppl. 3): 471–473.

Frischer TM, Kuehr J, Pullwitt A *et al.* (1993) Ambient ozone causes upper airways inflammation in children. *Am Rev Respir Dis* **148**: 961–964.

Frischer TM, Pulwitt A, Kuhr J *et al.* (1997) Aromatic hydroxylation in nasal lavage fluid following ambient ozone exposure. *Free Radic Biol Med* **22**: 201–207.

Fry FA and Black A (1973) Regional deposition and clearance of particles in the human nose. *J Aerosol Sci* **4**: 113–124.

Graham D and Koren HS (1990) Biomarkers of inflammation in ozone-exposed humans: comparison of the nasal and bronchoalveolar lavage. *Am Rev Respir Dis* **142**: 152–156.

Graham D, Henderson F and House D (1988) Neutrophil influx measured in nasal lavages of humans exposed to ozone. *Arch Environ Health* **43**: 228–233.

Hansen R and Oren M (1997) p53; from inductive signal to cellular effect. *Curr Opin Geneti Dev* 7: 46–51.

Harkema JR and Hotchkiss JA (1995) Ozone-induced proliferative and metaplastic lesions in nasal transitional and respiratory epithelium: comparative pathology. In: Miller FJ (ed.) *Nasal Toxicity and Dosimetry of Inhaled Xenobiotics.* London: Taylor & Francis, p. 187.

Harkema JR, Plopper CG, Hyde DM *et al.* (1987) Response of the macaque nasal epithelium to ambient levels of ozone: A morphologic and morphometric study of the transitional and respiratory epithelium. *Am J Pathol* **128**: 29–44.

Harkema JR, Morgan KT, Gross EA *et al.* (1994) Consequences of prolonged inhalation of ozone on F344/N rats: collaborative studies. Part VII: Effects on the nasal mucociliary apparatus. Res *Rep Health Eff Inst* **65**(Pt 7): 3–26.

Harris CC (1993) p53: At the crossroads of molecular carcinogenesis and risk assessment. *Science* **262**: 1980–1981.

Hatch GE (1992) Comparative biology of the normal lung. In: Parent RA (ed.) *Treatise on Pulmonary Toxicology*, vol. 1. Boca Raton, FL: CRC Press, pp. 617–632.

Hatch GE, Wiester MJ, Overton JH and Aissa M (1989) Respiratory tract dosimetry of ^{18}O-labeled ozone in rats: implications for a rat–human extrapolation of ozone dose. In: Schneider T, Lee SD, Wolters GTR and Grant LD (eds) *Atmospheric Ozone Research and its Policy Implications.* New York: Elsiever, pp. 553–560.

Hatch GE, Harkema JR, Plopper CG and Harris L (1995) Ozone dosimetry studies in the nose and eye using oxygen-18. In: Miller FJ (ed.) *Nasal Toxicity and Dosimetry of Inhaled Xenobiotics.* London: Taylor & Francis, pp. 125–134.

Hauser R, Garcia-Closas M, Kelsey KT and Christiani DC (1994) Variability of nasal lavage polymorphonuclear leukocyte counts in unexposed subjects: its potential utility for epidemiology. *Arch Env Health* **49**: 267–272.

Heller RF and Gordon RE (1986) Chronic effects of nitrogen dioxide on cilia on hamster bronquioles. *Exp Lung Res* **10**: 137–152.

Henderson RF, Hotchkiss JA, Chang IY *et al.* (1993) Effect of cumulative exposure on nasal response to ozone. *Toxicol Appl Pharmacol* **119**: 59–65.

Hilding AC (1963) Phagocytosis, mucous flow and ciliary action. *Arch Environ Health* **6**: 67–79.

Hotchkiss JA, Harkema JR, Sun JD and Henderson RF (1989) Comparison of acute ozone-induced nasal and pulmonary inflammatory responses in rats. *Toxicol Appl Pharmacol* **98**: 289–302.

Houser R, Elreedy S, Hoppin JA and Christiani DC (1995) Upper airway response in workers exposed to fuel oil ash: nasal lavage analysis. *Occup Environ Med* **52**(5): 353–358.

Housley DG, Eccles R and Richards RJ (1995a) Antioxidants and biotransformation potential in human nasal lavage. In: Miller FD (ed.) *Nasal Toxicity and Dosimetry of Inhaled Xenobiotics.* London: Taylor & Francis.

Housley DG, Mudway I, Kelly FJ *et al.* (1995b) Depletion of urate in human nasal lavage following in vitro ozone exposure. *Int J Biochem Cell Biol* **27**: 1153–1159.

Housley DG, Eccles R and Richards RJ (1996) Gender difference in the concentration of the antioxidant uric acid in human nasal lavage. *Acta Otolaryngol (Stockh)* **116**: 751–754.

Jackson AD, Rayner CF, Dewar A *et al.* (1996) A human respiratory-tissue organ culture incorporating an air interface. *Am J Respir Crit Care Med* **153**: 1130–1135.

Johanson G, Nihlen A and Lof A (1995) Toxicokinetics and acute effects of MTBE and EBTE in male volunteers. *Toxicol Lett* **82–83**: 713–718.

Johnson NF, Hotchkiss JA, Harkema JR and Henderson RF (1990) Proliferative responses of rat epithelia to ozone. *Toxicol Appl Pharmacol* **103**: 143–155.

Johnson NF, Carpenter TR, Jaramillo RJ and Liberati TA (1997) DNA damage-inducible genes as biomarkers for exposures to environmental agents. *Environ Health Perspect* **105** (Suppl. 4): 913–918.

Jorissen M, Van der Scheren B, Van den Berghe H and Cassiman JJ (1990) Ciliogenesis in cultured human nasal epithelium. *ORL J Otorhinolaryngol Relat Spec* **52**: 368–374.

Kienast K, Knorst M, Riechelmann H *et al.* (1993) *In vitro* studies of the beat frequency of ciliary cell cultures after short-term exposure to SO_2 and NO_2. *Med Klin* **88**(9): 520–524.

Kimbell JS (1995) Issues in modeling dosimetry in rats and primates. In: Miller FJ (ed.) *Nasal Toxicity and Dosimetry of Inhaled Xenobiotics.* London: Taylor & Francis, pp. 73–83.

Kimbell JS, Gross EA, Joyner DJ *et al.* (1993) Application of computational fluid dynamics to regional dosimetry of inhaled chemicals in the upper respiratory tract in the rat. *Toxicol Appl Pharmacol* **121**: 253–263.

Klein-Szanto AJ, Ura H, Momiki S *et al.* (1992) Effects of formaldehyde on xenotransplanted human respiratory epithelium *Res Rep Health Eff Inst* **51**: 1–17.

Ko LJ and Prives C (1996) p53: puzzle and paradigm. *Genes Dev* **10**: 1054–1072.

Koening JQ and Pierson WE (1984) Nasal responses to air pollutants. *Clin Rev Allergy* **2**: 255–261.

Koltai PJ (1994) Effects of air pollution on the upper respiratory tract of children. *Otolaryngol Head Neck Surg* **111**: 9–11.

Koren HS and Bromberg PA (1995) Respiratory responses of asthmatics to ozone. *Int Arch Allergy Immunol* **107**: 236–238.

Koren HS, Hatch GE and Graham DE (1990) Nasal lavage as a tool is assessing acute inflammation in response to inhaled pollutants. *Toxicology* **60**: 15–25.

Koren HS, Devlin RB, Becker S *et al.* (1991) Time-dependent changes of markers associated with inflammation in the lungs of humans exposed to ambient levels of ozone. *Toxicol Pathology* **19**: 406–411.

Koren HS, Graham DE and Devlin RB (1992) Exposure of humans to a volatile organic mixture. III. Inflammatory response. *Arch Environ Health* **47**: 39–44.

Leopold DA (1995) Nasal toxicity: end points of concern in humans. In: Miller FJ (ed.) *Nasal Toxicity and Dosimetry of Inhaled Xenobiotics.* London: Taylor & Francis, p. 23.

Lippman M (1970) Deposition and clearance of inhaled particles in the human nose. *Ann Otol Rhinol Laryngol* **79**: 519–528.

Lynge E, Andersen A, Nilsson R *et al.* (1997) Risk of cancer and exposure to gasoline vapors. *Am J Epidemiol* **145**: 449–458.

MacBride DE, Koenig JQ, Luchtel DL *et al.* (1994) Inflammatory effects of ozone in the upper airways of subjects with asthma. *Am J Respir Crit Care Med* **149**: 1192–1197.

McCaffrey TV (1997) *Rhinology Diagnosis and Treatment.* New York: Thieme Medical Publishers.

Maltoni C and Soffritti M (1995) Gasoline as an oncological problem. *Toxicol Ind Health* **11**: 115–117.

Molhave L and Moller J (1979) The atmospheric environment in modern Danish dwellings. In: Faenger PO and Valbjorn O (eds) *Indoor Climate.* Horsholm, Denmark, pp. 171–186.

Monticello TM, Morgan KT, Everitt JI and Popp JA (1989) Effects of formaldehyde gas on the respiratory tract of Rhesus monkeys: Pathology and cell proliferation. *Am J Pathol* **134**: 515–527.

Monticello TM, Swenberg JA, Gross EA *et al.* (1996) Correlation of regional and nolinear formaldehyde-induced nasal cancer with proliferating populations of cells. *Cancer Res* **56**: 1012–1022.

Morgan KT (1991) Approaches to identification and recording of nasal lesions in toxicology studies. *Toxicol Pathol* **19**: 337–351.

Morgan KT (1997) A brief review of formaldehyde carcinogenesis in relation to nasal pathology and human health risk assessment. *Toxicol Pathol* **25**: 291–307.

Morgan KT and Monticello TM (1990a) Formaldehyde toxicity: respiratory epithelial injury and repair. In: Thomassen DG and Nettsheim P (eds) *Biology, Toxicology and Carcinogenesis of Respiratory Epithelium.* New York: Hemisphere, pp. 155–171.

Morgan KT and Monticello TM (1990b) Airflow, gas deposition and lesion distribution in the nasal passages. *Environ Health Perspect* **85**: 209–218.

Morgan WK, Reger RB and Tucker DM (1997) Health effects of diesel emissions. *Ann Ocup Hyg* **41**: 643–658.

Noah TL and Becker S (1993) Respiratory syncytial virus-induced cytokine production by a human bronchial epithelial cell line. *Am J Physiol* **265**: L472–L478.

Noah TL, Henderson FW, Wortman IA *et al.* (1995a) Nasal cytokine production in viral acute upper respiratory infection of childhood. *J Infect Dis* **171**: 584–592.

Noah TL, Henderson FW, Henry MM *et al.* (1995b) Nasal lavage cytokines in normal, allergic and asthmatic school-age children. *Am Respir Crit Care Med* **152**: 1290–1296.

Noah TL, Black HR, Cheng P-W *et al.* (1997) Nasal and bronchoalveolar lavage fluid cytokines in early cystic fibrosis. *J Infect Dis* **175**: 638–647.

Peden DB (1996) The use of nasal lavage for objective measurement of irritant-induced nasal inflammation. *Regul Toxicol Pharmacol* **24**: S76–S78.

Peden DB, Hohman R, Brown ME *et al.* (1990) Uric acid is a major antioxidant in human nasal airway secretions. *Proc Natl Acad Sci USA* **87**: 7638–7642.

Peden DB, Swiersz M, Ohkubo K *et al.* (1993) Nasal secretion of the ozone scavenger uric acid. *Am Rev Respir Dis* **148**: 455–461.

Pipkorn U and Enerback L (1984) A method for the preparation of imprints from the nasal mucosa. *J Immunol Methods* **73**: 133–137.

Proctor DF (1977) The upper airways: I. Nasal physiology and defense of the lungs. *Am Rev Respir Dis* **115**: 97–129.

Proctor DF and Andersen IB (1982) *The Nose: Upper Airway Physiology and the Atmospheric Environment*. Amsterdam: Elsevier Science.

Ramis I, Serra J, Rosello J *et al.* (1988) PGE_2 and PGF_2 in the nasal lavage fluid from healthy subjects: methodological aspects. *Prost Leuko Essent Fatty Acids* **34**: 109–112.

Ranga V and Kleinerman J (1981) A quantitative study of ciliary injury in the small airways of mice. *Exp Lung Res* **2**: 49–55.

Raphael GD, Jeney EN, Baranjuk KI *et al.* (1989) Pathophysiology of rhinitis-lactoferrin and lysozyme in nasal secretions. *J Clin Invest* **84**: 1528–1535.

Rasmussen TR, Swift DL, Hilberg O and Pedersen OF (1990) Influence of nasal passage geometry on aerosol particle deposition in the nose. *J Aerosol Med* **3**: 15–25.

Reuzel PGJ, Wilmer JWGM, Woutersen RA *et al.* (1990) Interactive effects of ozone and formaldehyde on the nasal respiratory lining epithelium in rats. *J Toxicol Environ Health* **29**: 279–292.

Schierhorn K, Brunnee T, Paus R *et al.* (1995) Gelatin sponge-supported histoculture of human nasal mucosa. *In Vitro Cell Develop Biol Anim* **31**: 215–220.

Schmidt D, Hubsch U, Wurzer H *et al.* (1996) Development of an in vitro human nasal epithelial (HNE) cell model. *Toxicol Lett* **88**: 75–79.

Spektor DM, Lippman M, Lioy PJ *et al.* (1988) Effects of ambient ozone on respiratory function in active normal children. *Am Rev Respir Dis* **137**: 313–320.

Stahl GH and Ellis DB (1973) Biosynthesis of respiratory tract mucins. A comparison of canine epithelial goblet-cell and submucosal-gland secretions. *Biochem J* **136**: 845–850.

Steele VE and Arnold JT (1985) Isolation and long term culture of rat, rabbit and human nasal turbinate epithelial cells. *In Vitro Cell Dev Biol* **21**: 681–687.

Steerenberg PA, Fisher PH, Gmelig MF *et al.* (1996) Nasal lavage as a tool for health effect assessment of photochemical air pollution. *Hum Exp Toxicol* **15**: 111–119.

Swift DL (1976) Design of the human respiratory tract to facilitate removal of particulates and gases. *AIChE Symp Ser* **72**(156): 137–144.

Swift DL and Proctor DF (1988) A dosimetric model for particles in the respiratory tract above the trachea. *Ann Occup Hyg* **32** (Suppl. 1): 1035–1044.

Swift DL, Montassier N, Hopke PK *et al.* (1992) Inspiratory deposition of ultrafine particles in human nasal replicate cast. *J Aerosol Sci* **23**: 65–72.

Tonnesen P and Hindeberg I (1988) Serotonin in nasal secretion. *Allergy* **43**: 303–309.

Wang JH, Devalia JL, Dudlle JM *et al.* (1995) Effect of six-hour exposure to nitrogen dioxide on early-phase nasal response to allergen challenge in patients with a history of seasonal allergic rhinitis. *J Allergy Clin Immunol* **96**: 669–676.

Werner U and Kissel T (1995) Development of a human nasal epithelial cell culture model and its suitability for transport and metabolism studies under in vitro conditions. *Pharm Res* **12**: 565–571.

Winther B, Innes DJ, Mills SE *et al.* (1987) Lymphocyte subsets in normal airway mucosa of the human nose. *Arch Otolaryngol Head Neck Surg* **113**: 59–62.

Wu R (1985) In vitro differentiation of airway epithelial cells. In: Schiff LJ (ed.) *In Vitro Models of Airway Epithelial Cells*. Boca Raton, FL: CRC Press, pp. 1–26.

Wu R, Yankaskas J, Cheng E *et al.* (1985) Growth and Differentiation of human nasal epithelial cells in culture. *Am Rev Respir Dis* **132**: 311–320.

Effects of Cigarette Smoke and Air Pollutants on the Lower Respiratory Tract

PETER K. JEFFERY

Lung Pathology Unit, Histopathology, National Heart and Lung Institute, Imperial College, School of Medicine, London, UK

INTRODUCTION

Our atmosphere normally sustains life by providing oxygen and carbon dioxide and protects us from cosmic radiation and meteorites. Human activities in the developing world have introduced noxious materials into an atmosphere which now has the capacity to injure the respiratory tract: this poses an especial and increasing threat to worldwide human health.

The environmental causes of airway and alveolar injury are multiple (Table 12.1). Historically, domestic (i.e. from coal and wood fires in the home) and industrial pollutants were major causes of lung injury in the UK, but due to more strict regulation of home and industrial emissions their impact has been much reduced. Now, by default, self-pollution by cigarette smoke and vehicle exhaust emission have become the leading causes of pollutant-induced injury. For some individuals exposure to a variety of agents encountered in their workplace is also an important cause of lung damage.

The volume of air entering the airways each day at rest is of the order of 10 000–15 000 litres. The pathway and pattern of airflow through the nose or mouth (referred to as the upper respiratory tract) and the tracheobronchial tree (referred to as the lower respiratory tract) affects the pattern of particle deposition in the lung. The larynx is conventionally considered to be the boundary between the upper and lower respiratory tracts: the upper extends from the external nares to the larynx and the lower from the larynx to the visceral pleura as a system of airways that branch dichotomously and asymmetrically (Jeffery,

AIR POLLUTION AND HEALTH
ISBN 0-12-352335-4

Table 12.1 Pollutants

Gases	inorganic (e.g. SO_2, NO_x, O_3)
	organic (e.g. tobacco smoke)
Particulates	inert (e.g. carbon, diesel exhaust)
	allergens (e.g. pollens, house-dust mite)
	living (e.g. bacteria, viruses)

1995). The airways of the respiratory tract may be conveniently considered as either conductive or respiratory, with the respiratory bronchioles forming a transitional zone. Airways are usually designated by structure and order of division: those distal to the trachea with cartilage in their walls are by definition *bronchi*. In the trachea, supportive cartilage is present in the form of irregular, sometimes branching, crescentric rings (16–20 in humans), all of which are incomplete dorsally where they are bridged by connective tissue and bands of smooth muscle. In large bronchi the cartilages are irregular in shape but frequent enough to be found in any plane. In small bronchi they are less frequent and may be missed in transverse section. Airways distal to the last cartilage plate are termed *bronchioli*. The last bronchiolar divisions have their ciliated lining epithelium interrupted by alveoli and are referred to as *respiratory bronchioli*: the generation proximal to the first-order respiratory bronchioles are referred to as terminal bronchioli. *Terminal bronchioli* form the last purely conductive airways and the respiratory bronchiolus the site where gaseous exchange begins. There are generally three orders of respiratory bronchioli. A single terminal bronchiolus with its succeeding respiratory bronchioli, two to nine orders of alveolar ducts and alveolar sacs together form the *respiratory acinus*, which is about 1 cm in diameter and forms the basic respiratory unit of the lung.

The extent and nature of the lungs response to pollutants will vary upon whether the nasal passages are by-passed, such as in cigarette smoking, and the nature (physical/chemical), magnitude and duration of exposure to the pollutant. In regard to inhaled atmospheric gaseous pollutants, their solubility in water (i.e. as in highly soluble gases) is important in determining whether the nasal passages, large (i.e. proximal bronchi) conducting airways or small (distal non-cartilagenous bronchioli) conducting and respiratory airways or alveoli (air sacs) are the primary sites of injury: the more water-soluble the gas, the more proximal the airway affected. For particulate pollutants size is a critical determinant: as particles of the order of 10 micrometres (μm) or above are usually deposited in the nasal passages, whereas those less than 10 μm (referred to as PM_{10}) deposit in the lower airways. Particulate/gas and gas/water interactions also occur which alter markedly their size and deposition pattern based on their sedimentation, impaction and diffusion as they pass from turbulent through linear to diffusive flow more peripherally. Respiratory rate and reflex airway constriction is also important in determining the site of deposition: the more constricted the airway, the more proximal (i.e. central) is the deposition. The site of deposition, in turn, determines in large part the nature of the lungs response and the long-term structural consequences (pathology) of chronic exposure.

The way ambient air moves frequently and freely over the lining of the airways and alveoli make it particularly susceptible to environmental pollutants. The first site of host contact with the pollutant is the delicate, moist mucosa which lines the entire respiratory tract from nose to alveolus. It is the mucosa which initiates immune responses and reactions and the host's response to acute or chronic irritation. Figures 12.1a–c show that

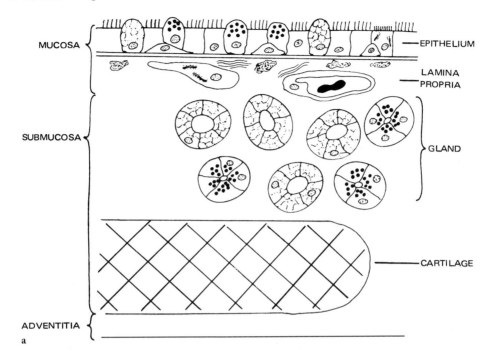

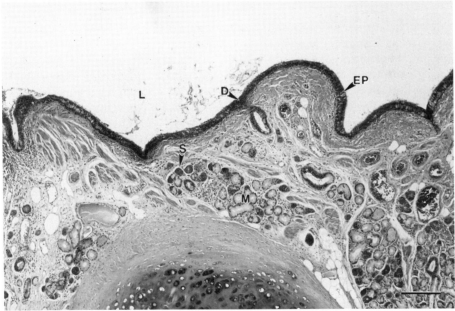

Fig. 12.1 The airway wall: (a) a diagrammatic representation of its overall structure; (b) its appearance by light microscopy of a haematoxylin and eosin (H&E)-stained section showing the airway lumen (L), surface lining epithelium (EP), a gland duct (D) opening into the airway lumen with its underlying mucous (M) and serous (S) secretory acini. Blood vessels of the systemic system are present (arrow) below the epithelium (scale bar = 500 μm); (c) scanning electron micrograph of surface epithelium. Fracture plane shows the lateral surfaces of the cells and pseudostratified structure of the ciliated, columnar epithelium. The apical surface shows a complete covering of cilia: all cells are attached to a roughly contoured basement membrane (arrows). Supporting collagen (CO) lies beneath. (Scale bar = 50 μm).

c

the airway wall comprises epithelial, lymphoid, muscular, vascular and nervous elements interspersed in a pliable connective tissue support arranged as (1) a lining mucosa of surface epithelium, basement membrane and supporting elastic lamina propria, (2) a submucosa in which lie mucus-secreting glands, muscle and cartilage plates, and (3) a relatively thin adventitia coat. The conductive airways perform many functions beyond mere conduction of inspired and expired gases: they warm, humidify and cleanse inhaled air of potentially harmful dust particles, gases, bacteria and other living organisms. The more distal respiratory zone is therefore kept free of pollution and infection by airway defence mechanisms which include:

- nervous reflexes leading to bronchoconstriction and/or cough
- ciliary activity
- secretion of mucus, lysozyme, lactoferrin and secretory IgA
- cellular immune response and reactions
- detoxification by specialized epithelial cells (e.g. Clara cells)

The bulk of the respiratory region of adult human lungs contain about 300 million alveoli: each alveolus measures about 250 µm in diameter when expanded (Fig. 12.2a). Small holes, pores of Kohn, are found in the alveolar walls of many species including humans (Fig. 12.2b). There are from one to seven pores in each alveolus and their diameter ranges from 2 to 13 µm. The pores are not present at birth but develop after the first year of life when they form an alternative route for gas entry to adjacent alveoli by collateral ventilation. Expansion of the lung is facilitated by surface tension reducing lipids, collectively known as pulmonary surfactant, secreted by alveolar (type II) lining cells.

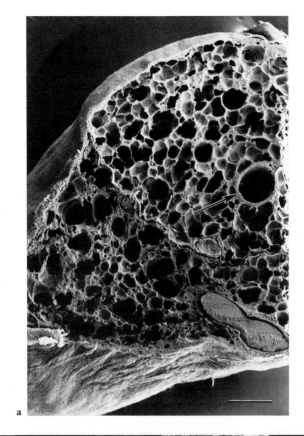

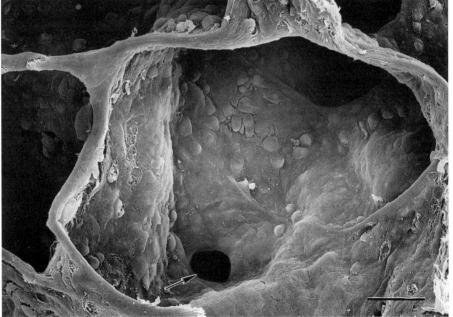

Fig. 12.2 Air sacs (alveoli) in human lung as seen by SEM: (a) a small non-cartilagenous terminal bronchiole (arrow) and all its subsequent divisions together with approximately 2000 alveoli form a unit referred to as the respiratory acinus. There are estimated to be 3×10^6 alveoli in the adult lung (scale bar = 200 μm); (b) each alveolus, approximately 250 microns (μm) in diameter, normally connects with its neighbouring alveoli via small pores (pore of Kohn: arrow) which facilitate collateral ventilation (Scale bar = 50 μm).

The nasal passages normally condition the air we breathe efficiently by virtue of their shape, character of the airflow and nature of the mucosa and its vasculature. However, allergic rhinitis may occur in atopic individuals exposed to common aeroallergens either in the atmosphere or workplace. A heightened sensitivity of the nasal mucosa may also be induced by a range of non-specific irritants including cigarette smoke, traffic fumes, domestic sprays, perfumes and bleach. In particular, traffic exhaust has been considered to be a major factor in the increasing prevalence of hayfever but this is much debated (Rusznak et al., 1994; Von Mutius et al., 1992). Whilst the upper respiratory tract may be much affected by inhaled pollutants, this chapter will focus on the effects seen in the lower respiratory tract.

Chronic exposure of the lower respiratory tract to pollutants induces an inflammatory reaction and may alter the proportions of the various epithelial and airway wall cell types and their functions. More seriously, pollutants may cause ulceration of the mucosa, inappropriate repair or uncontrolled proliferative responses. In these conditions there may be destruction of tissue or inappropriate re-modelling and growth leading to distinct pathologies. The inflammatory reactions and structural changes are now considered in more detail. There are, of course, many pollutants which could be considered but only those commonly found and of particular current interest are considered herein. These include: self-pollution by cigarette smoke and major constituents of vehicle exhaust including nitrogen dioxide, ozone and diesel exhaust.

CIGARETTE SMOKE

Much is now known concerning the effects of tobacco products on health. Tobacco has been smoked for centuries by the native populations of America, although the Chinese claim that they grew and used it long before the discovery of America. Subsequent to its introduction into Europe in the middle of the sixteenth century, tobacco smoking spread throughout the world. The World Health Organization estimates there are about 1.1 thousand million smokers in the world. Whilst first used for medicinal purposes, the harmful effects of smoking were soon recognized. In the twentieth century, strong associations were found between cigarette smoking and chronic obstructive lung disease and cancer of the lung. Significant but weaker associations were also found with cancer of the alimentary tract, bladder and cervix, myocardial infarction and peripheral vascular disease (Davis and Kaliner, 1983; IARC, 1986; Royal College of Physicians, 1977). Throughout middle age death rates of current cigarette smokers are more than twice the rates of non-smokers. For six disorders the evidence suggests that the difference in mortality between smokers and life-long non-smokers is due largely to tobacco use: (1) chronic obstructive pulmonary disease (COPD), (2) cancer of the lung, (3) aortic aneurysm, (4) ischaemic heart disease, (5) respiratory heart disease, and (6) peripheral vascular disease. COPD, lung cancer and aortic aneurysm account for 85% of smoking-related deaths. Smoking may also contribute to hypertension, cerebrovascular disease, arteriosclerosis and peptic ulceration. At present it is estimated that tobacco kills over 3 million people per year. In developed countries alone, annual deaths from smoking number about 0.9 million in 1965, 1.3 million in 1975, 1.7 million in 1985 and 2.1 million in 1995. If the current increasing rate continues, the death toll from smoking will rise to 10 million per year by the year 2020 (Peto *et al.*, 1992). More than half these deaths will be in middle age (i.e. aged 35–69 years) and constitute an average shortening of an individual's life by approximately 23 years. This will greatly exceed the expected deaths from AIDS, tuberculosis, maternal mortality, car accidents, suicide and homicide combined.

Over 4000 different chemicals have been identified in cigarette smoke (CS) (Dube and Green, 1982; Stedman, 1968) and this has made it difficult to identify any individual agent as the prime cause of a particular lung condition. The many biologically active agents in CS include: pyridine alkaloids (e.g. nicotine), carbon monoxide, acetaldehyde, oxides of nitrogen, hydrogen cyanide, ammonia, acrolein, phenols, *N*-nitrosamines, benzo compounds (such as the polycyclic aromatic hydrocarbon benzopyrene, radioactive components and trace metals. The composition of the smoke depends very much on the way the tobacco is smoked, the chemical properties of the leaf, the cigarette blend, the wrapper and the filter. The majority of the components of CS are produced in an oxygen-deficient, hydrogen-rich environment by a process of pyrolysis and distillation in a zone immediately behind the tip which burns at temperatures up to 950°C (Dube and Green, 1982). Mainstream, sidestream smoke (the latter given off by the tip of the cigarette) and exhaled smoke give rise to smoke of differing composition. Whilst the effects of direct inhalation of mainstream smoke are of major importance, sidestream smoke is usually inhaled through the nose by the smoker and passively by the non-smoker and its effects have been insufficiently considered.

Mainstream undiluted smoke leaving the mouthpiece contains up to 1.3×10^{10} mixed particles per cm^3, ranging from 0.2 to 1.0 μm in diameter (Dube and Green, 1982). The particles bear a slight charge and the mainstream smoke has a short-lived reducing capacity with two types of free radical: one in the particulate phase (probably a quinone–hydroquinone complex) and one in the gaseous phase which is highly reactive, e.g. the oxidation of nitrogen oxide to the dioxide (NO_2) (Church and Pryor, 1986). The method by which the tobacco is cured affects its sugar content, which in turn contributes largely to the acidity of the smoke. The pH influences the proportion of nicotine and other basic components in the vapour phase. At pH 5.4 all the nicotine is in the particulate phase: hence, in the more acidic smoke of cigarettes (as compared with cigars and pipe tobacco), less nicotine is absorbed through the buccal mucosa and there is a greater need to inhale deeply to maximize nicotine absorption (Armitage and Turner, 1970). In contrast, the more alkaline smoke of cigars and cigarettes made of Burley or black (air-cured) tobacco allows nicotine to be easily absorbed through the buccal mucosa. The majority of the mitogenic and carcinogenic agents reside in the particulate phase. Depending on the design of the cigarette, the mainstream smoke from a single cigarette may contain up to 30 mg 'tar' and 3.0 mg nicotine or, if efficiently filtered, as little as 0.5 mg 'tar' and 0.05 mg nicotine. Radioactive elements in CS include polonium, radium and thorium (Bodgen *et al.*, 1981; Harley *et al.*, 1980); these α-emitters may interact with the more moderate concentrations of radon daughters present in indoor air under poor conditions of ventilation to more than double the concentration of radioactive elements inhaled (Bergman *et al.*, 1984). The chemical composition and gas-to-particulate ratio of sidestream CS may be different to mainstream due to its prolonged time in air and its cooling. Sidestream smoke contains carbon monoxide, ammonia, formaldehyde, benzene, nicotine, acrolein and an assortment of potentially genotoxic and carcinogenic organic compounds (Dye and Adler, 1994; Lofroth, 1989; Proctor and Smith, 1989).

The varying effects of CS on the lung depend not only on its composition and method of smoking, but also on a variety of complex host-defence mechanisms.

ACUTE AND SUBACUTE EFFECTS OF CIGARETTE SMOKE

Reflexes

The fastest responses to inhaled CS and other irritants involve neural reflexes that probably originate in receptors in the surface epithelium. Sensory receptors (enteroceptors) in the nose, larynx, tracheobronchial tree and respiratory portion of the lung all participate (Richardson, 1984) and the overall effect depends on which reflexes predominate and the site in the respiratory tract at which the reflex is initiated (Jeffery, 1994). In humans there is a slight transient bronchoconstriction and deepening (and often slowing) of breathing. The fall in airway conductance is maximal at 20 s after inhalation and lasts little more than 40 s, and the effect is dose-related (Rees *et al.*, 1982). The slowing and deepening appear to arise from reflexes initiated in upper airways and is triggered by both the particulate and vapour phases of CS. Thus, by-passing the upper airways results in rapid shallow breathing, coughing and a fall in airways conductance.

Metabolism of arachidonic acid

In common with many other organs, the lung is capable of both synthesizing and inactivating the prostaglandin (PG) and thromboxane metabolites of arachidonic acid. Experiments with rats show that CS promotes marked lipolytic activity. Prostacyclin (PGI_2) production is decreased (increasing the likelihood of platelet aggregation and affecting vascular tone and haemostasis), whilst PGF_2 and PGE_2 production is markedly increased (Bakhle, 1984): high tar cigarettes have the greatest effect. Increased PGE_2 production is thought to be due to nicotine rather than carbon monoxide (Berry *et al.*, 1979; Wennmalm, 1977). However, nicotine does not appear to be implicated in the CS-induced decrease in PGI_2 (Sant'Ambrogio, 1982). CS has also been shown to decrease the *inactivation* of PGE_2 (Mannisto *et al.*, 1981). These changes are probably important to both the pulmonary and cardiovascular effects of CS.

Induction of lung antioxidants

Cigarette smoke contains oxidants that severely deplete intracellular antioxidants in lung cells by increasing oxidant stress (Bridges *et al.*, 1993; Chow, 1993) both *in vitro* (Bridgeman *et al.*, 1991; Rahman and MacNee, 1994) and *in vivo* (Cotgreave *et al.*, 1987; Rostu *et al.*, 1988): the effect can be detected in the human plasma of smokers (Rahman *et al.*, 1996a,b). Acute exposure *in vivo* to CS also decreases the glutathione (GSH) concentration in bronchoalveolar lavage fluid (Cotgreave *et al.*, 1987) in both rat (Moldeus *et al.*, 1985) and rabbit lungs (Rostu *et al.*, 1988). However, the decrease is not associated with oxidation of GSH to GSSG or protein thiolation but is associated with the reduction in the activities of γ-glutamylcysteine synthetase (a precursor of GSH), glutathione peroxidase and glucose-6-phosphate dehydrogenase in alveolar epithelial cells both *in vitro* (Rahman and MacNee, 1994) and in whole lungs (Moldeus *et al.*, 1985; Rostu *et al.*, 1988). The role of GSH in protection against CS-induced detachment of type II alveolar cells *in vitro* (a test used to mark cell damage) is unclear. Depletion of GSH enhances CS-induced epithelial leakage *in vivo* (Li *et al.*, 1994; MacNee *et al.*, 1991). Glutathione is elevated in the bronchoalveolar fluid of chronic cigarette smokers (Cantin *et al.*, 1987) whereas vitamin E is reduced (Pacht *et al.*, 1986). The elevation of GSH in bronchoalveolar lavage fluid of chronic smokers seems to be an adaptative response, since chronic exposure of alveolar epithelial cells to CS condensate induces γ-glutamylcysteine synthetase gene expression *in vitro* (Rahman *et al.*, 1996). However, in spite of the increased levels of GSH found in bronchoalveolar lavage fluid, it appears to be insufficient to protect against CS-induced epithelial cell damage.

Secretion of mucus

Animal studies show that diluted CS elicits the release of sugar-rich mucins (glycoproteins and glycosaminoglycans) into conducting airways: the effect is probably due to the stimulation by nictone of ganglia located within the airway wall that innervate airway mucus-secreting tissue (Peatfield *et al.*, 1986; Rogers *et al.*, 1987). The earliest subacute effects in humans are known to include early morning coughing and production of sputum. Sputum volume increases with the number of cigarettes smoked (Higginbottam and Borland, 1984).

Mucosal permeability

The permeability of the airway epithelial lining to macromolecules is determined by pinocytotic activity and the 'leakiness' of selectively permeable (so-called 'tight') junctions, which form sealing belts extending around the apico-lateral surfaces of adjacent epithelial cells (Godfrey, 1997; Godfrey *et al.*, 1992; Jeffery, 1990) (Fig. 12.3). In humans the passage through the mucosa of a small (mol wt 492 Da) inhaled, radiolabelled molecule, DTPA (diethylene triamine pentacetic acid), is increased by both the gas and particulate phases of CS (Higginbottam and Borland, 1984). Similar results have been found experimentally in guinea pig where the mucosa may become permeable to a 40 000 mol wt marker (Hulbert *et al.*, 1981). There is also experimental evidence that in response to CS, albumin and other serum proteins rapidly leak across the bronchial mucosa (Boucher *et al.*, 1980). These effects seem to be mediated via production of oxidants rather than nicotine (Minty *et al.*, 1984), although this is controversial (Higginbottam and Borland, 1984). There is controversy as to whether the increased permeability correlates with structural rearrangements of the tight junctional complexes evident by electron microscopy (Hulbert *et al.*, 1981; Walker *et al.*, 1984).

Fig. 12.3 Freeze-fracture replica of the apicolateral tight junction of a ciliated cell in a human bronchus, showing the sealing strands and their cross-bridges (arrows), which act as a barrier to the bulk flow of fluid, ions and macromolecules. Airway lumen (L). (Reproduced by permission of RWA Godfrey). (Scale bar = 1.0 μm.)

Regulatory peptides

The classical view that airways are regulated by a balance of cholinergic (excitatory) and adrenergic (inhibitory) nerves has been challenged by the functional and immunohistochemical demonstration of nerves which are neither adrenergic nor cholinergic in type (Jeffery, 1994). The so-called 'NANC' nerves comprise fibres with neurotransmitters which include a variety of regulatory peptides such as substance P located in sensory nerves. Irritation of these fibres by a variety of substances, including cigarette smoke, causes the release of substance P, producing nasal vasodilatation, local oedema due to increased vascular permeability and secretion of mucus (Lundbland and Lundberg, 1984: Lundberg and Saria, 1983; Lundberg *et al.*, 1984; Rogers *et al.*, 1989). Many of the immediate local effects of CS are mediated, therefore, by its effect on the release of several regulatory peptides located in sensory nerves.

Mucociliary function

It is uncertain whether cigarette smoking depresses mucociliary function in the short term: mucociliary clearance in humans immediately after smoking one or more cigarettes is variously reported to be increased (Albert *et al.*, 1975; Camner and Philipson, 1974), unchanged (Goodman *et al.*, 1978; Pavia *et al.*, 1971; Yeates *et al.*, 1975) or reduced (Nakhosteen *et al.*, 1982). It has been shown that in healthy non-smoking volunteers CS taken in via the mouth and exhaled via the nose does not inhibit either mucociliary transport in the nose (as assessed by transport of saccharine to the throat) or ciliary beat frequency of nasal mucosal brushings (Stanley *et al.*, 1986). However, habitual smokers who have abstained from smoking 2 h prior to the saccharine test show reduced nasal mucociliary transport. Discrepancies between research groups are probably the result of differences in methodology and dose of exposure. Short-term exposure of animals to CS has shown impairment of mucociliary function in some (Albert *et al.*, 1969; Dalham, 1966; Iravani, 1972; Stupfel, 1974) but not all studies (Bair and Dilley, 1967; Labelle *et al.*, 1966).

An early and sustained inhibition of adenylate kinase activity in ciliated tracheal cells has been shown in hamsters exposed to CS (Mattenheimer and Mohr, 1975). Inhibition of the enzyme results in a decrease of intracellular adenosine triphosphate, which is required for ciliary movement. CS has also been shown *in vivo* to decrease, reversibly, the electrical potential difference across tracheal epithelium; *in vitro*, the particulate phase decreases the short-circuit current by inhibition of chloride secretion with minimal effects on sodium absorption (Welsh, 1983). As active ion transport across airway mucosa is thought to play an important part in regulating the low viscosity (periciliary) fluid in which cilia beat, this may, in part, explain the abnormalities of mucociliary clearance often observed. Experimental exposure of rats to CS for 2 weeks does not appear to alter pharmacological receptor number, part of the mechanism by which ciliary beat frequency may be regulated (Sharma and Jeffery, 1987).

In contrast, there is evidence indicating that long-term exposure to CS (one year or more) reduces mucociliary function in humans (Camner *et al.*, 1973). The separate effects of gas and particulate phases have been reviewed (Dalham, 1966; Pettersson *et al.*, 1982): both would appear to be active. Nicotine has a biphasic effect (stimulatory at low concentrations and blocking at high concentrations), which is probably related to its effect on ganglionic nicotine receptors (Hybbinette, 1982).

EFFECTS OF CIGARETTE SMOKE ON INFLAMMATORY CELLS

Immune function

There is increasing evidence, first recognized in the early 1960s, that cigarette smoking may have widespread detrimental effects on the capacity of the lung to generate a cellular immune response (Holt, 1987; Holt and Keast, 1977). Early animal studies demonstrated that CS-exposed alveolar macrophages have impaired capacity to phagocytose and kill bacteria (Holt and Keast, 1973; Rylander, 1971, 1973): the effect is short-lived and returns to normal in spite of prolonged exposure (Rylander, 1973, 1974).

It has been demonstrated that functional complement receptors (i.e. C3b) are decreased on human smokers' macrophages, but their phagocytosis and bacterial killing has been shown to be normal (Cohen and Cline, 1971; Warr and Martin, 1977). However, their interaction with lymphocytes appears to be abnormal (Laughter *et al.*, 1977) and they respond poorly to lymphocyte migration inhibition factor (Warr, 1979). Smokers' lymphocytes also respond poorly to mitogens (Daniele *et al.*, 1977; Silverman *et al.*, 1975). Much of these findings may relate to the decrease in cell viability associated with CS exposure (Blue and Janoff, 1978; Holt *et al.*, 1974) or to the depression of enzyme function (Green *et al.*, 1977) and protein synthesis (Low, 1974). Prolonged CS exposure at low concentration may, however, stimulate these processes (Holt and Keast, 1973; Lentz and Diluzio, 1974).

Smokers' leukocytosis is characterized by a 30% increase in peripheral blood leukocyte count (Miller *et al.*, 1982; Yeung and De Bunico, 1984). Neutrophils (increased by 44%) appear to be normal in respect of chemotactic, microbicidal and secretory activity. Monocytes are largely functionally normal but may be defective in their capacity to kill intracellular *Candida*. Basophils may decrease in number and degranulate. Natural killer cell activity is depressed (Hughes *et al.*, 1985). The total number of T lymphocytes is increased in smokers: and analysis of T cell subsets by flow cytometry shows that T helper cells are increased in light smokers but decreased in heavy smokers with a complementary increase in the CD8 T suppressor/cytotoxic subset in both blood and bronchoalveolar lavage fluid (BAL). Associated functional changes appear to be very much age-related (see Holt, 1987). In man, the effect of CS on humoral immunity has been studied less intensively. Most studies show a decrease in serum immunoglobulins (Ig), (Anderson *et al.*, 1982) decreased agglutinin levels to sheep erythrocytes (Fletcher *et al.*, 1969) and altered antibody response to influenza virus after natural infection and vaccination (Finklea *et al.*, 1971). Acutely, CS appears to exert its effect close to the point of entry but with prolonged exposure depresses not only local but also systemic antibody responses. In a study of 1600 white Americans there were indications that smoking led, directly or indirectly, to an increase in serum IgE, which qualitatively may be different to that seen in non-smokers (Bahna *et al.*, 1983a; Burrows *et al.*, 1981). Similar findings have been observed for concentrations of serum IgD (Bahna *et al.*, 1983b). Maternal smoking increases also the levels of fetal IgE (as determined by measurement of cord serum) and increases the risk of future allergic disease (Magnusson, 1986). A relationship between a history of smoking and development of allergy in relation to specific occupational exposure has also been suggested (Finnegan *et al.*, 1991). *In vitro* studies indicate that CS extracts may directly activate the complement pathway and increase the levels of acute-phase reactants (Kew *et al.*, 1985). Anti-nuclear and rheumatoid factors are also increased more frequently in smokers than non-smokers (Mathews *et al.*, 1973). In contrast there are also studies which demonstrate that CS exposure may suppress radiation-induced inflammation in the lung (Bjermer *et al.*, 1993). There appear therefore to be smoking-related alterations of both cellular and humoral immune function but the relevance of these findings to the development of chronic disease is as yet unclear.

Inflammatory changes in bronchitis and chronic obstructive pulmonary disease (COPD)

Inflammatory cells infiltrate the bronchial mucosa of smokers and these relatively large airways have been the focus of recent biopsy studies conducted in volunteers using flexible fibreoptic bronchoscopy. The biopsy studies of smokers' airways are of particular interest as they allow comparison with the biopsy changes reported in asthma (Jeffery, 1991, 1998). It is already well known that in atopic and non-atopic asthma there is an inflammatory infiltrate comprised of activated (CD25+), T helper (CD4+) lymphocytes and activated (EG2+) eosinophils associated with gene expression and secretion of interleukins (IL) 4 and IL-5, IL-10, and the pro-inflammatory cytokines GM-CSF and TNFα (Azzawi *et al.*, 1990; Bradley *et al.*, 1991; Djukanovic *et al.*, 1990; Humbert *et al.*, 1996; Jeffery *et al.*, 1989; Robinson *et al.*, 1992, 1996). The production of IL-4 and IL-5 but not IL-2 and interferon γ is referred to as the T helper type 2 (TH$_2$) phenotype.

There is also evidence of inflammation in bronchial biopsies of smokers with stable COPD and those obtained following exacerbations of bronchitis (Di Stefano *et al.*, 1994; Lacoste *et al.*, 1993; Ollerenshaw and Woolcock, 1992; O'Shaughnessy *et al.*, 1997; Saetta *et al.*, 1993; Vignola *et al.*, 1993) (Fig. 12. 4). Lymphocytes, macrophages and plasma cells form the predominant cell types with scanty neutrophils, and in contrast to asthma there are relatively few eosinophils (in the absence of an exacerbation

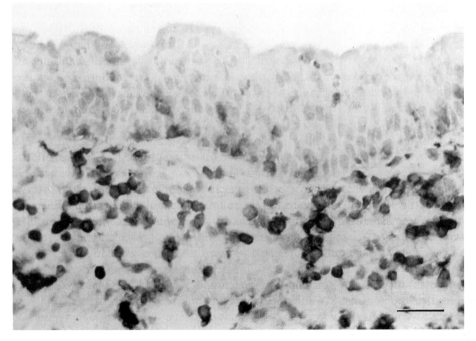

Fig. 12.4 Histological section of a bronchial mucosal biopsy taken by flexible fibreoptic bronchoscopy from a patient with an exacerbation of bronchitis. The section has been immunostained for CD45+ leucocytes shown here as black. There are large numbers of CD45+ leucocytes infiltrating the subepithelial zone, and fewer within the squamoid surface epithelium of this smoker (specimen obtained by Dr M Saetta and stained by Drs Li and Wang). (Scale bar = 30 μm.)

of infection). There are significant increases in the numbers of CD45 (total leukocytes), CD3 (T lymphocytes), CD25 activated and VLA-1 (late activation) positive cells and of macrophages (Saetta *et al.*, 1993; O'Shaughnessy *et al.*, 1997). The numbers of tissue eosinophils are increased when there is an exacerbation of bronchitis (Saetta *et al.*, 1993). However, it has been suggested that, in contrast to asthma, the tissue eosinophils found in COPD do not degranulate (Lacoste *et al.*, 1993). Increases in the cell surface adhesion molecules associated with such inflammation are described (Di Stefano *et al.*, 1994; Vignola *et al.*, 1993). O'Shaughnessy and colleagues (1997) have demonstrated few neutrophils and eosinophils in bronchial biopsies of stable bronchitic smokers with or without COPD: as airflow limitation progressively worsens T lymphocytes and neutrophils increase in the surface epithelium, as do T lymphocytes and macrophages in the subepithelium. Interestingly, it is the CD8+ lymphocyte subset which increases in number and proportion in COPD and the increase of CD8+ cells shows a significant association with measurements of decline in lung function (i.e. forced expiratory volume in 1 s, or FEV_1) (O'Shaughnessy *et al.*, 1997). This contrasts with the predominance and activation of the CD4+ T cell subset which is the characteristic change of mild atopic asthma. The ratio of peripheral blood CD4 and CD8 cells appears to be under genetic control (Amadori *et al.*, 1995) and, together with the observations in bronchial biopsies, has led to the hypothesis that individuals with a low CD4 : CD8 ratio may be particularly susceptible to the effects of CS (O'Shaughnessy *et al.*, 1997). The high numbers of neutrophils found in BAL from subjects with COPD (Thompson *et al.*, 1989) is not reflected in their numbers in the bronchial mucosa, at least in the subepithelial zone (often referred to as the lamina propria), which is the zone usually quantified in bronchial biopsies (Lacoste *et al.*, 1993; O'Shaughnessy *et al.*, 1997). However, O'Shaughnessy and colleagues (1996) and Saetta *et al.* (1997) have demonstrated the preferential accumulation of neutrophils in both the surface epithelium and submucosal mucus-secreting glands, respectively. In the last mentioned study there is also a significant decrease in the CD4/CD8 ratio, confirming the earlier report of O'Shaughnessy and colleagues.

The inflammation found in mucus-secreting glands is associated with the volume of sputum produced: the relationship is a direct one and more convincing than that of gland size (hypertrophy) with sputum volume (Mullen *et al.*, 1985). In COPD, inflammation appears to be a feature throughout the bronchial tree, particularly in the respiratory regions (i.e. respiratory bronchioli and alveolar walls) of the lung. Niewoehner and co-workers (1974) and Cosio and colleagues (1980) were among the first to describe the inflammation of the respiratory region in smokers dying suddenly: inflammation in bronchioles and a respiratory bronchiolitis and alveolitis consisting of pigmented macrophages (Wright *et al.*, 1988). These inflammatory changes to small airways appear to be related most closely to clinical airflow obstruction in COPD (Hogg *et al.*, 1968; Snider, 1986; Thurlbeck, 1985). As the inflammation of small airways worsens, there is decline in FEV_1 and there are also increases in the numbers of CD8+ T lymphocytes accompanied by increases of smooth muscle mass (Saetta, 1998). The destruction of the alveolar wall and respiratory bronchioles in smoker's emphysema is also considered to be the result of an inflammatory reaction involving both lymphocytes and neutrophils, and likely initiated by inhalation of cigarette smoke (Finkelstein *et al.*, 1995; Jeffery, 1990; MacNee and Selby, 1993).

In respect of the circulating neutrophil, their average diameter is greater than 7.0 μm which necessitates their deformation as they squeeze through alveolar capillary segments of 5 μm diameter. Neutrophil traffic through the capillaries of the lung is normally slower (i.e. there is a higher transit time) than that of red blood cells (RBCs) as they are 700 times less deformable than RBCs (MacNee and Selby, 1993). Recent studies with radioactively labelled neutrophils have demonstrated that the normal delay in neutrophil transit is further exaggerated during smoking, although transiently, even in healthy subjects (MacNee *et al.*, 1989). Exposure of neutrophils to cigarette smoke acutely *in vitro* and *in vivo* results in decreased deformability associated with polymerization of actin microfilaments (Drost *et al.*, 1992; MacNee and Selby, 1993). This is one likely mechanism of the observed cigarette smoke-induced increase in neutrophil transit time in the periphery of the lung which may be associated with the increased release of enzymes, elastolysis and destruction of alveolar walls. Lymphocytes have also been demonstrated to form a significant component of the alveolar wall inflammatory infiltrate in COPD: the greater their number, the greater is the loss of alveolar tissue (Finkelstein *et al.*, 1995). Thus, whilst neutrophils and macrophages clearly play a role in tobacco smoke-induced lung disease, there is now much interest in the CD8+ T lymphocyte in initiating and controlling this process, which may involve both inappropriate tissue re-modelling and alveolar destruction (see *Structural Changes and Pathology of Common Lung Conditions Associated with Smoking and Air Pollution* below).

AIR POLLUTION

One of the most dramatic examples of the effects of air pollution on health was seen in an episode that started on 4 December 1952 in London and (in that month) resulted in 4000 excess deaths (HMSO, 1954; Logan, 1953). It was public concern about this, rather than pressure from the medical profession, that led to the first clean air legislation in Britain (Bates, 1994). Since then drastic efforts have been made to reduce air pollution: these have been effective and greatly beneficial to the health of the nation.

The main air pollutants measured in the earlier years were sulfur dioxide (SO_2) and particulate matter (PM in the form of black smoke) from the burning of domestic coal. Early animal-based studies investigated the effects of relatively high doses (e.g. 50–400 ppm) of inhaled SO_2 on airway epithelium and its proliferative response, and demonstrated ulceration of large airway epithelium, increases of epithelial cell mitoses, and a goblet cell hyperplasia and increase in gland size which mimicked the changes observed in human bronchitis (Lamb and Reid, 1968). However as a result of national and international measures directed to control the most important sources of emission (i.e. power stations and refineries), sulfur dioxide levels have decreased greatly in the UK and several other countries in Europe. Instead, in developed countries a new form of pollution has emerged resulting from increased use of liquid petroleum gas or kerosine in industry and increased use of motor vehicles. Now ozone and respirable particular matter (e.g. PM_{10}) often exceed recommended standards and pose new threats to human health. These pollutants include oxides of nitrogen (NO_x), ozone and diesel exhaust.

Worldwide, about 480 million people are being exposed to increased levels of ozone (Schwela, 1996), of which at least 150 million are in Europe (Sivertsen and Clench-Aas,

1996). Air pollution in Western Europe is mainly traffic-related, and therefore ozone and NO_x are particularly important. In Eastern Europe pollution is mainly related to the combustion of fossil fuels, and so particulate matter and sulfur dioxide are the key constituents. It is important to distinguish between those air pollutants which are causative and increase the prevalence of disease and those which predispose or *exacerbate* and increase the severity of existing disease (such as allergic conditions). Clinical and experimental studies allow us to obtain basic information about mechanisms, the inflammatory, cellular and molecular mechanisms of the toxicity of pollutants, such as nitrogen dioxide, ozone and particulates. The following three sections focus on the inflammatory changes which occur as the host initially mounts what is designed to be a protective response to the effects of these pollutants on the lung.

EFFECTS OF NITROGEN DIOXIDE (NO₂)

Nitrogen dioxide is a highly reactive, nitrogen-centre free radical and the most toxic of the nitrogen compounds (Morrow, 1975). Two parts per million (2 ppm) is the exposure limit for an 8 h work shift in most European countries and may be encountered in shipyards, mines, garages and tunnels. It is a poorly water-soluble gas and is therefore deposited far more peripherally in the airspaces than the highly water-soluble sulfur dioxide (SO_2) (Miller *et al.*, 1992; Postlethwait and Bidani, 1994). Previous human and animal studies have suggested that the major target sites for the action of NO_2 are the terminal bronchioli (Overton, 1984). The effects of NO_2 include impairment of epithelial permeability and mucociliary function (acutely) or goblet cell hyperplasia and alveolar/emphysematous change following chronic exposure (Evans *et al.*, 1973, 1975; Giordano *et al.*, 1972; Stephens *et al.*, 1972). Since NO_2 is a potent oxidant, its main mechanism of pulmonary toxicity has been suggested to involve lipid peroxidation of cell membranes (Patel and Block, 1986) and various actions of free radicals (Proctor and Reynolds, 1984). The airway and alveolar epithelium is covered by a thin layer of epithelial lining fluid (referred to as ELF) rich in antioxidant defences such as glutathione, uric acid, ascorbic acid and α-tocopherol (Kelly *et al.*, 1995; Slade *et al.*, 1993). Thus NO_2 is unlikely to diffuse in unreacted form through ELF (Postlethwait and Bidani, 1994) and at low levels of exposure the potential for oxidative injury to the respiratory epithelium is thereby reduced (Kelly *et al.*, 1995). Antioxidants such as ascorbic acid and α-tocopherol also appear to play major roles in protecting the airway and alveolar tissues from the effects of NO_2 (Kelly *et al.*, 1996; Mohsenin, 1994).

Airway inflammation induced by NO_2 has been studied in animal models by histology and BAL. In animals, freeze-fracture and electron microscopic studies of hamsters exposed to relatively high concentrations (30 ppm) for 5–9 months demonstrated an NO_2-associated disruption of bronchiolar and alveolar epithelial tight junctions (Gordon *et al.*, 1986). In other studies the inflammatory reaction involves mainly an increase in the number of neutrophils, macrophages, lymphocytes and mast cells associated with the development of emphysematous change (Glasgow *et al.*, 1987; Stephens *et al.*, 1972; Thomas *et al.*, 1967). Bronchoscopy with BAL following carefully controlled experimental exposures has provided the opportunity to assess the effects of NO_2 on human airways. Several such investigations have failed to confirm the increases in the number of neutrophils in 'pooled' large (150–240 ml) volume BAL (Frampton *et al.*, 1989a,b; Watt *et al.*, 1995). However, a small yet significant increase in neutrophils has been reported after exposure of 2 ppm for 6 h (Frampton *et al.*, 1991). At concentrations below 2 ppm, no changes of inflammatory cell numbers have been found (Frampton *et al.*, 1989a,b). However, dividing the lavage samples into two fractions is useful in distinguishing between the airway (i.e. proximal) and alveolar contribution of inflammatory cells to BAL inflammation (Rennard *et al.*, 1990). On this basis, the use of separate analyses of instilled aliquots has demonstrated an increase in the number of neutrophils which is restricted to the most proximal airways (Becker *et al.*, 1993; Helleday *et al.*, 1994). For higher levels of exposure there are dose-dependent increases in BAL mast cells and lymphocytes found after a single exposure to 2.25–5.5 ppm of NO_2 which resolve within 72 h of cessation of exposure (Sandstrom *et al.*, 1991; Watt *et al.*, 1995).

The effects of repeated exposure to NO_2 have also been investigated. Rubenstein and co-workers exposed healthy volunteers to 0.60 ppm of NO_2 for 2 h with intermittent mild to moderate exercise on four separate days within a 6-day period. Apart from a slight increase in the percentage of natural killer (NK) cells, no other changes in BAL lymphocyte subtypes were seen (Rubenstein *et al.*, 1991). In contrast, repeated exposure to 1.5 and 4 ppm of NO_2 for 20 min every second day produced a BAL cell response after six exposures when the numbers of CD19+ (B cells) and CD16+ +CD56+ (NK cells) decreased and the CD4+/CD8+ cell ratio was altered (Sandstrom *et al.*, 1992a,b).

Markers of inflammation such as albumin, fibronectin, hyaluronan, angiotensin converting enzyme (ACE), β_2-microglobulin, leukocyte elastase, lactate dehydrogenase, total protein, leukotriene B_4 (LTB_4), prostaglandin E_2 (PGE_2), thromboxane A_2 (TxA_2) and tumour necrosis factor α (TNF-α) have each been assayed and reported to be unaffected by NO_2 exposure (Frampton *et al.*, 1989a; Sandstrom *et al.*, 1990). However, Kelly and colleagues (1996) have compared normal subjects exposed (in a sealed chamber) to either air or 2 ppm NO_2 for 4 h, an exposure equivalent to that likely to be experienced in some occupational environs. Those exposed to NO_2 demonstrated time-dependent, transient decreases of BAL and bronchial wash (BW) ascorbic acid and uric acid, which likely act as sacrificial antioxidants to remove NO_2 from the inspired air. In addition, Blomberg and co-workers (1997) found an increase in the numbers of neutrophils in the BW following this single exposure, which was accompanied by an increase in IL-8. In bronchial mucosal biopsies, a study by Jörres and co-workers (1992) did not find significant changes of inflammatory cell numbers in the biopsies (or BAL) following a single 3 h exposure to 1 ppm of NO_2. Also, biopsy samples of large (central) airways demonstrate no signs of inflammatory cell recruitment after a single exposure to 2 ppm NO_2 for 4 h, nor is there upregulation of adhesion molecules of relevance to the inflammatory response (Blomberg *et al.*, 1997). Repeated exposure of normal subjects over 4 days to 2 ppm NO_2 significantly decreases intra-epithelial neutrophils (Blomberg, 1998).

Many of the airway wall inflammatory effects are likely initiated by the epithelium that lines the airways. Experimental studies of cultured cells have been informative in this regard. *In vitro* studies of human bronchial epithelial cells indicate that exposure of epithelial cells to 0.4–0.8 ppm of NO_2 for 6 h can induce the synthesis of pro-inflammatory cytokines such as granulocyte/macrophage colony-stimulating factor (GM-CSF), interleukin (IL)-8 and TNF-α (Devalia *et al.*, 1993). Non-stimulated alveolar macrophages maintained *in vitro* constituitively release pro-inflammatory cytokines which appear not to be modulated by exposure to NO_2. However, there is a dose-dependent

NO$_2$-induced reduction of the secretion of IL-1α, IL-6, IL-8 and TNF-α produced by macrophages previously stimulated by LPS.

Ozone

Ozone (O$_3$) is mainly produced in the troposphere by a series of sunlight-driven reactions involving nitric oxides and volatile organic compounds arising largely from human activities of combustion (Bascom *et al.*, 1996; Sandstrom, 1995). During hot summers, ground-level concentrations of ozone may exceed 0.2 ppm in Central Europe and other areas of the world, such as California (Sandstrom, 1995). Ozone may also be produced in high concentrations in workplaces such as welding plants and paper mills. Ozone is virtually insoluble in water and, unlike NO$_2$, is deposited along the entire airway. In resting subjects, approximately 90% of inspired O$_3$ is absorbed with substantial uptake and effects in both the upper and lower airways (Gerrity *et al.*, 1988; Hiltermann *et al*, 1998; Hu *et al.*, 1992). Mathematical models have suggested that the tissue dose of inhaled O$_3$ is greatest at the position of the bronchoalveolar junction (Miller *et al.*, 1985; Overton *et al.*, 1987), and this region has been shown to be very sensitive to O$_3$-induced damage, as shown by histopathological studies in animals (Barry *et al.*, 1988; Carlsson *et al.*, 1996). However, as O$_3$ is highly reactive it is likely that little, if any, reaches and reacts directly with alveolar epithelium O$_3$ (Langford *et al.*, 1995; Pryor, 1992). The majority of the effects of O$_3$ are probably mediated by a cascade of secondary free-radical-derived products and cellular damage occurs when antioxidant defences are overwhelmed (Kelly *et al.*, 1995; Mustafa, 1990). Its toxic effects include stimulation of airway irritant receptors, alterations of epithelial permeability, ciliary damage, alterations to Clara cells (responsible for the detoxification of many pollutants) and epithelial denudation each of these very much dependent on dose (Bhalla and Crocker, 1986; Bhalla *et al.*, 1990; Schwartz *et al.*, 1976). In some cases, in humans, there may however be *increases* of mucociliary clearance (Foster *et al.*, 1987). In rats and monkeys experimentally exposed to O$_3$, there appears to be re-modelling of the respiratory units with transformation of proximal alveolar ducts into respiratory bronchioles (Barr *et al.*, 1988; Eustis *et al.*, 1981). There may be important synergistic effects as O$_3$ toxicity has been shown to decrease clearance of a number of other toxins, in particular inhaled fibres (Pinkerton *et al.*, 1989).

In 1986, Seltzer *et al.* carried out the first BAL study in normal human subjects exposed to ozone. Exposure to 0.4 and 0.6 ppm of O$_3$ for 2 h induced an increase in the number of neutrophils recovered and there were increased concentrations of PGE$_2$, PGF$_2$ and TxB$_2$. Koren and collaborators (1989) also exposed healthy non-smoking male subjects to 0.4 ppm of O$_3$ for 2 h with intermittent heavy exercise and conducted BAL 18 h later. Following ozone exposure, there were major changes in inflammatory markers together with an eight-fold increase in the percentage of neutrophils and a small but significant increase in neutrophil elastase. The concentrations of protein, albumin and immunoglobulin G (IgG) were also elevated, suggesting increased vascular permeability. Furthermore, elevated levels of PGE$_2$ and fibronectin were found following O$_3$ exposure, whereas the level of leukotriene B$_4$ was unaffected. In another study, exposure to 0.22 ppm of O$_3$ for 4 h not only induced a BAL neutrophilia, but also increased the numbers of lymphocytes, eosinophils and mast cells in healthy subjects (Frampton *et al.*, 1997). The vascular endothelial cell adhesion molecule ICAM-1 was also also found to be significantly upregulated 18 h after exposure to 0.2 ppm of O_3 for 4 h, together with trends for increased numbers of neutrophils and eosinophils (Balmes *et al.*, 1997). Exposure to ozone also increases the levels of the neutrophil chemoattractant IL-8 in the proximal lavage and GM-CSF and α_1-antitrypsin in the BAL (Aris *et al.*, 1993). Blomberg and colleagues (personal communication) have confirmed the BW neutrophilia and the increase of BAL neutrophils and lymphocytes, the last represented by trends to increased CD3+, CD8+ and CD45Ro T-cell subsets, the last a member of antigen-primed or 'memory' T-cells.

The first human study to use bronchial biopsies to study O_3-induced inflammation demonstrated increased neutrophil numbers in the bronchial mucosa, described 18 h after exposure to 0.4 ppm of O_3 for 4 h (Aris *et al.*, 1993). As shown previously, the numbers of neutrophils and concentrations of LDH, total protein, albumin and fibronectin increased in BAL fluid after ozone exposure. Blomberg and co-workers have recently shown in bronchial biopsies that, following exposure of normal subjects to 0.12 ppm O_3 for 2 h, there was only an upregulation of P selectin expression on vascular endothelium with no significant change to the numbers of inflammatory cells or lung function. However, after a 2-h exposure to 0.20 ppm (in a separate study) these researchers demonstrated upregulated expression of the vascular adhesion molecules LP selectin and ICAM-1, together with increased numbers of neutrophils (in both the surface epithelium and sub-epithelial zone) (Fig. 12.5a,b) and increased numbers of mast cells in the subepithelium.

Both ambient ozone and allergen exposure can be associated with airways inflammation, albeit the response may be qualitatively different (Hiltermann *et al.*, 1997). The synergistic effects of exposure to O_3 have also been evaluated in asthmatics and smokers. Basha and co-workers (1994) compared the exposure of healthy subjects and patients with asthma to 0.20 ppm of O_3 for 6 h during moderate exercise and bronchoscopy performed 18 h after exposure. The results showed that the asthmatic subjects had increased levels of neutrophils, IL-6 and IL-8 compared with the normal subjects. These initial results have been confirmed recently when asthmatic subjects were exposed to 0.2 ppm of O_3 for 4 h and compared with a group of healthy subjects. The results showed that asthmatic subjects had significantly greater O_3-induced increases in neutrophils and total protein concentration than O_3-exposed healthy subjects (Scannell *et al.*, 1996). Asthmatics may thus be at risk of developing more severe O_3-induced airway inflammation than normal subjects. These findings may be of relevance to the increase in asthma morbidity associated with episodes of ozone pollution reported in epidemiological studies. These findings *in vivo* are supported by *in vitro* and animal studies in which exposure to air pollutants induces increased responses to allergen.

In vitro studies of human alveolar macrophage function have been investigated *in vitro* after exposure to 0.1–1.0 ppm of O_3 for 2–4 h. There was an O_3-induced concentration-independent increase in the release of PGE_2 (Becker *et al.*, 1991). In another study the exposure of alveolar macrophages to similar O_3 concentrations induced elevated gene expression and secretion of TNFa, IL-1β, IL-6 and IL-8 (Arsalane *et al.*, 1995). Human epithelial cells have also been shown to release the arachidonic acid metabolites TxB_2, PGE_2, LTC_4, LTD_4 and LTE_4 in response to exposure to 0.1–1.0 ppm of O_3, suggesting that airway epithelial cells may also be an important source of eicosanoids following ozone exposure (McKinnon *et al.*, 1993). Airway epithelial cells have been exposed to 0.1 ppm of O_3 and the supernatants analysed for cytokine and fibronectin content. The

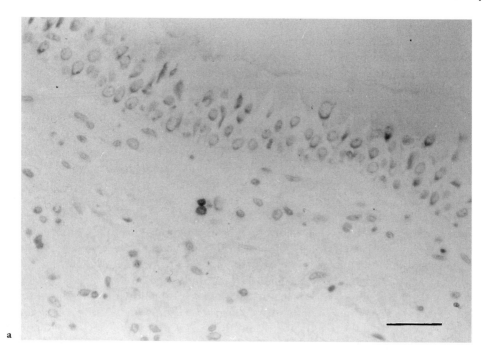

a

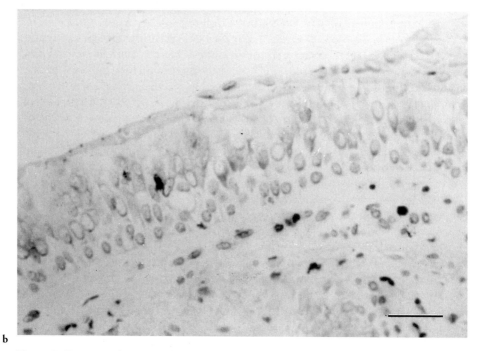

b

Fig. 12.5 Experimental exposure of healthy human volunteers to low levels of air pollutants such as ozone and diesel exhaust induces an inflammatory response in the proximal airways. This shows bronchial biopsies immunostained for neutrophils: (a) after sham (air) exposure with very few neutrophils present in the airway mucosa and (b) after exposure to 0.20 ppm ozone for 2 h which induces neutrophilic infiltration of the mucosa (Scale bar for both = 50 μm). (Reproduced by kind permission of Dr A. Blomberg)

epithelial cells were found to produce substantial amounts of all these proteins and increased activity was detected as early as 1 h following exposure (Devlin *et al.*, 1994). Oxidative stress can be an important regulator of IL-8 gene expression (DeForge *et al.*, 1993). Exposure of epithelial cells to 0.1 ppm of O_3 induced an increase in the levels of IL-8 mRNA and IL-8 protein. The IL-8 gene is regulated by the transcription factors nuclear factor (NF)-κB, NF-IL-6 and possibly activator protein 1 (AP-1) and each of these transcription factors has been shown to be activated after O_3 exposure (Jaspers *et al.*, 1997). In mice, it has also been suggested that the chemotactic protein macrophage inflammatory protein 2 (MIP-2), a member of the same family of cytokines as IL-8, may be of importance in the recruitment of neutrophils into the murine lung following exposure to O_3 (Driscoll *et al.*, 1993). Interestingly a recent study in humans suggests that neutrophil elastase may not be required to initiate the O_3 induced response (Hiltermann *et al.*, 1998).

Diesel exhaust

Diesel engines are widely used because of their efficiency, robustness and low running costs. However, for equivalent applications, diesel engines emit approximately 10 times more particles than petrol engines without catalytic converters and up to 100 times more than petrol engines with catalytic converters (Nauss *et al.*, 1995; Zweidinger, 1982). They are, therefore, a major contributor to atmospheric pollution by particulate matter. The main gases which result from combustion of diesel fuel are carbon monoxide (CO), oxides of nitrogen (NO, NO_2) and sulfur dioxide (SO_2), while a small but significant percentage of the fuel is polymerized, pyrolysed, cracked, oxidized, sulfonated and nitrated into several hundreds of compounds (Scheepers and Bos, 1992). The physical and chemical nature of the particles, their distribution in the respiratory tract and the biological events occurring in response to the particles determine their biological effects. Particle deposition in the respiratory tract is dependent on both particle size and the pattern of breathing. Most inhaled particles with an aerodynamic diameter of more than 5 μm are deposited in the upper or larger airways, whereas smaller particles are more prone to be deposited in the small airways (bronchioli) and alveoli. In addition, mouth breathing increases the upper cut-off size for particles reaching the lower airways from about 10 to 15 μm and results in greater particle deposition in the lower airways (Brain and Valberg, 1979; Chow, 1995). Lately, attention has focused on PM_{10} and concern has also been raised regarding the health effects of environmental exposure to ultrafine particles which have a diameter less than 0.05–0.10 μm. Ultrafine particles are always present in urban atmosphere in very large numbers (>0.5 × 10^5 particles/cm^3), can be highly reactive (Oberdorster *et al.*, 1996) and may penetrate through airway epithelium and vascular walls. Hospital admissions for bronchitis or asthma have been shown to be associated with atmospheric levels of PM_{10} (Dockery and Pope, 1994; Dockery *et al.*, 1993; Koren, 1995). Most studies have suggested that a 10 μg/m^3 increase in PM_{10} on the day of the hospital visit or one to two days before the visit was typically associated with a 1–4% increase in the number of hospital visits and with an increase in daily mortality equal to 0.5–1.5% (Pope *et al.*, 1995). Increased levels of PM_{10} have been linked to worsened peak flow, increased inhaler usage and respiratory symptoms in asthmatic children, appearing with a time-lag of 24 h (Pope *et al.*, 1991). In adults with severe asthma, the same association has been found, but with a lag period of 4 days (Walters *et al.*, 1994). Analysis of the cause of mortality has shown the strongest association with respiratory and, secondarily, with cardiovascular mortality (Schwartz and Dockery, 1992).

In humans diesel exhaust particles (DEPs) are able to induce a heightened IgE response in the upper airways after intranasal challenge. Moreover, the *in vivo* increase in IgE production is associated with an increase in the number of IgE-secreting cells (Diaz Sanchez *et al.*, 1994). Diaz-Sanchez and collaborators have performed a study in which subjects were challenged intranasally with saline or DEPs (Hazucha *et al.*, 1983). Following saline, nasal lavage cells had barely detectable levels of interferon (IFN)-γ, IL-2 and IL-13 mRNA. After challenge with DEPs, detectable mRNA levels were found for IL-2, IL-4, IL-5, IL-6, IL-10, IL-13 and IFNγ, and those present at baseline were increased. There was also enhanced production of IL-4 protein. The recent studies of Salvi and colleagues cited by Blomberg (1998) have demonstrated that exposure of normal subjects to 300 μg/m^3 diesel exhaust in the presence of 1.6 ppm NO_2 for 1 h produces a pronounced inflammatory response in the airways. In bronchial biopsies taken 6 h after cessation of exposure there was increased expression of vascular ICAM-1 and VCAM-1: in complementary fashion there was a significant increase in the number of LFA-1+ cells (e.g. neutrophils) and a trend towards an increase of VLA4+ cells (e.g. eosinophils). There was also a T cell response with CD4+ (T helper) cells predominating and increased expression of mRNA for IL-8 and a similar trend for IL-5. Mast cells showed a a four-fold increase in biopsies with elevated concentrations of methylhistamine in BAL fluid.

In animal studies DEPs aggravate ovalbumin-induced airway inflammation in sensitized mice characterized by infiltration of eosinophils and lymphocytes and increases in the number of goblet cells in the bronchial epithelium (i.e. goblet cell hyperplasia) (Takano *et al.*, 1997). There is also enhanced local expression of IL-5, IL-4, GM-CSF and IL-2 in both lung tissue and BAL. These experimental findings support the hypothesis that DEPs enhance the manifestations of allergic asthma.

Taken together, these experimental and human studies suggest that DEPs have the capacity to encourage allergic inflammation and may also play a part in contributing to the increasing incidence of allergy and asthma seen in the developed Western world (Diaz Sanchez, 1997; Diaz Sanchez *et al.*, 1996).

STRUCTURAL CHANGES AND PATHOLOGY OF COMMON LUNG CONDITIONS ASSOCIATED WITH SMOKING AND AIR POLLUTION

The pathological consequences of chronic exposure to cigarette smoking and low-level pollutants include four major diseases of the lungs: chronic bronchitis (syn. 'mucous hypersecretion'), chronic adult bronchiolitis (also referred to as 'small airways disease' or chronic obstructive bronchiolitis), emphysema and cancer. The first three are often considered clinically under one heading, variously termed chronic airflow limitation (CAL) or obstruction (CAO), chronic obstructive pulmonary (COPD) or lung (COLD) disease, or obstructive airways (OAD) disease (Fletcher and Pride, 1984; Jeffery, 1990; Snider, 1986). Asthma is often a complicating feature, which although not caused by atmospheric pollutants or cigarette smoke may be aggravated by them (Fig. 12.6). Several of the components of atmospheric pollution, in particular SO_2, NO_2 and O_3, may induce bronchitic and/or emphysematous changes similar to those induced by chronic cigarette smoke. The following considers the effects of cigarette smoke by way of example.

The relationship between cigarette smoking and COPD was one of the first to be recognized (Greaves and Colebatch, 1986). Early effects of CS may begin in the womb as the fetus is exposed to bloodborne metabolites from the mother (Helms, 1994). The neonate and infant may be exposed passively to sidestream and exhaled mainstream smoke in the home. There is an increased prevalence of respiratory infection in children less than 2 years old who are passively exposed to CS (Colley *et al.*, 1974) and infections at this time are associated with predisposition to the development of chronic bronchitis in later life (Burrows *et al.*, 1977). Cigarette smoking often begins in the early teenage years (often underestimated) and continues into adulthood when, during the third decade, there develop signs of early morning 'smokers' cough' and throat clearing. In the fourth decade recurrent attacks of respiratory infection begins which in the fifth decade cause concern and first visits to the general practitioner. In the sixth there is breathlessness and referral to a specialist hospital. The disease often progresses, leading to death due from respiratory, cardiac failure or pneumonia. When compared with non-smokers, there is, in smokers, a decline in lung function and performance (Fletcher, 1984; Fletcher and Peto, 1977) leading to disability and early death. Epidemiological evidence leaves no reasonable doubt that CS is the major causative agent of COPD, with atmospheric and occupational pollution as additional contributory factors. Not all who smoke, however, show deterioration in lung function. It seems that a subgroup of about 15–20% of smokers are particularly susceptible: the reasons for this are not clear but may involve genetic predisposition including perhaps the genetic control of CD4/CD8 cell ratio (see above), differences in depth or pattern of inhalation, variations in cellular and biochemical response, differences in immune or regenerative capacity of lung cells, childhood infections and early passive exposure (Barter and Campbell, 1975; Bates, 1973; Fletcher and Peto, 1977). Three conditions are recognized as contributing to COPD: chronic bronchitis, chronic bronchiolitis and emphysema. The last two contribute in varying degree to the rapid decline in lung function (FEV_1) characteristic of COPD.

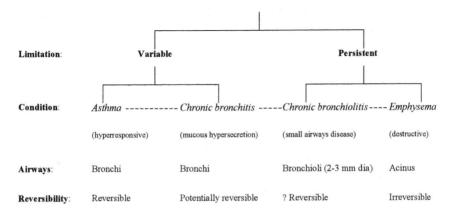

Fig. 12.6 Interrelationships between inflammatory conditions of the lung associated with inhaled pollutants.

Chronic bronchitis (CB)

Chronic bronchitis is diagnosed clinically as persistent cough with the production of sputum (by convention persisting for more than three months of two consecutive years) (Fletcher and Pride, 1984; Medical Research Council, 1965; Thurlbeck, 1977). Cough and sputum production are the symptoms most frequently experienced by smokers: both mechanisms are effective in clearing large proximal airways (down to about the sixth generation of branching), acting to protect the more distal respiratory portion of the lung from damage. Sputum and respiratory tract secretions are a mixture of epithelial and serum-derived constituents. Normally, respiratory tract secretions probably amount to less than 100 ml/day (Toremalm, 1960) and consist primarily of glycosaminoglycans (Coles *et al.*, 1984; Lopez-Vidriero and Reid, 1985). As we have seen, repeated irritation by CS causes inflammation, and in addition there are increases in the number and activity of secretory cells in the mucosa (Fig. 12.7a,b) and an enlargement of submucosal glands (by an increase in both the number and size of their cells). Mucous gland enlargement and hyperplasia of secretory cells have been considered to be the histological hallmarks of CB (Reid, 1954) but the increase in sputum production may be mediated by the CS-related inflammatory process (Mullen *et al.*, 1987).

The enlargement of the submucosal glands is associated with more rapid synthesis and discharge of intracellular mucins (Sturgess and Reid, 1972). Parallel animal studies show that following 2–6 weeks exposure to CS, laryngo-tracheal submucosal glands and surface

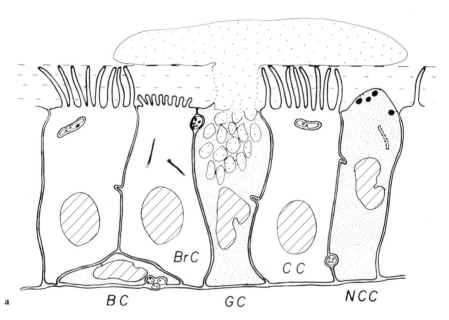

Fig. 12.7 Airway surface epithelium: (a) diagrammatic representation showing ciliated (CC) non-ciliated (NCC), goblet (GC), basal (BC) and brush cells (BrC) of the epithelium with mucus overlying the cilia; (b) Transmission electron micrograph (TEM) of mucous cells (MC) each filled with relatively large and confluent electron-lucent secretory granules. L = airway lumen (scale bar = 10 μm); (c) SEM of the lumenal surface of goblet cells showing intracellular secretory granules pressing against the internal aspect of the apical plasma membrane. The borders of each cell are outlined by increased numbers of apical microvilli which appear bright (arrows). (Scale bar = 3.0 μm).

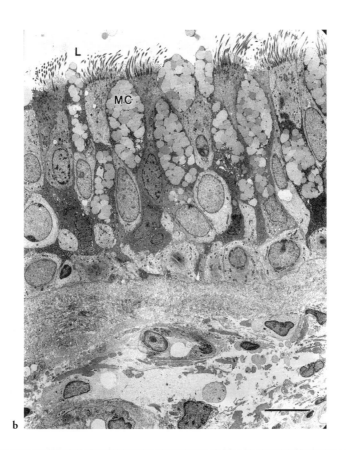

b

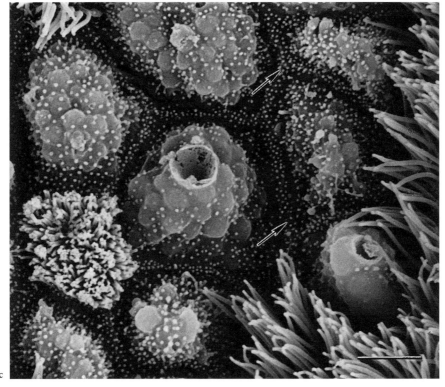

c

secretory cells synthesize and discharge epithelial mucins at a faster rate than non-exposed controls (Coles *et al.*, 1979; Jeffery *et al.*, 1984; Rogers *et al.*, 1987). Experimental exposure of the specific pathogen-free rat to CS increases the number of epithelial secretory cells at all airway levels of the bronchial tree, with the greatest effect in the proximal airways (Fig. 12.8) (Jeffery and Reid, 1981; Jones *et al.*, 1972; Rogers and Jeffery, 1986). These changes are not specific to CS as similar changes occur with a range of irritants including SO_2 and NO_2. As well as the overall number of secretory cells being increased, those containing acidic mucin increase at the expense of the neutral mucin-containing cells (Jones *et al.*, 1973; Rogers and Jeffery, 1986). Sulfate esters and carboxyl groups confer acidic properties on the large molecular weight (MW = $7-20 \times 10^6$ Da) glycoproteins by their addition at terminal or near terminal positions on the sugar side chains of the protein core. Increases in acidic mucins are also characteristic of smokers and bronchitic patients (Kollestrom *et al.*, 1977). It is thought that these chemical changes may affect the rheological properties of mucus and hence the ease with which it is cleared by mucociliary transport. These changes do not appear to be due to the nicotine in CS as blood levels far in excess of those achieved by inhalation of CS do not result in mucous cell hyperplasia *per se* (Rogers *et al.*, 1986).

Experimental studies indicate that at least two mechanisms are responsible for the increase in mucous cells: mucous transformation of serous cells and proliferation of existent or newly formed mucous cells (Ayers and Jeffery, 1982, 1988; Bolduc *et al.*, 1981; Boren and Paradise, 1978; Jeffery and Reid, 1981). The stimulus to cell division is rapid, with a peak of proliferation seen between 1 and 3 days, which decreases rapidly in spite

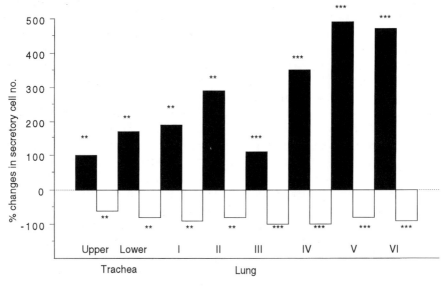

Fig. 12.8 Experimental exposure of specific pathogen-free rats to an atmosphere of cigarette smoke, generated automatically from cigarettes and given for approximately 4 h daily for two weeks. Subacute inhalation of cigarette smoke increases the airway number of secretory cells containing mucin. Counting the numbers containing acidic (■) or neutral (□) glycoprotein shows that at each successive airway level the smoke induces an increase in the percentage of the acidic and reduction in the neutral mucin-containing cells (expressed here as mean percentage change of control values: from Rogers and Jeffery 1986). I–VI are generations of airway progressively more distal.** and *** = $P < 0.01$ and < 0.001 respectively.

of continued exposure to levels found in unexposed animals (i.e. tolerance develops). If the smoke exposure is interrupted, tolerance is lost, and on re-exposure there is a further proliferative burst (see Ayers and Jeffery, 1988).

Measurements show that epithelial thickening in the experimental animal is an early response to inhaled CS and that the thickening is not due to stratification but rather to cell hypertrophy (Jeffery and Reid, 1981). There is an increase in the length of mitochondria of ciliated cells and also of their apical microvilli. Cilia remain normal in both structure and density whilst the proportion of all epithelial cells which are ciliated increases. The increase in surface coverage by cilia is seen clearly when the epithelium is viewed by scanning electron microscopy (Jeffery et al., 1988). These results are in contrast to the late stages of the disease in humans which are characterized by atrophy of epithelium (Wright and Stuart, 1965), focal squamous metaplasia (Kleinerman and Boren, 1974) and decreases of both ciliated cell number and mean ciliary length (Chang, 1957; Misokovitch et al., 1974; Wanner, 1977). Ultrastructural changes in cilia such as the development of compound cilia have been attributed directly to the effects of cigarette smoke (Ailsby and Ghadially, 1973), but in the author's opinion these changes are non-specific or more likely consequences of complicating exacerbations of infection and due to bacterial products known to be ciliotoxic (Wilson et al., 1987). The need for adequate controls and caution in the interpretation of ciliary abnormalities seen by electron microscopy has been stressed (Fox et al., 1981). In vitro brushings of ciliated cells obtained from bronchitic patients or from the nose of healthy volunteers following exhalation of smoke through the nose show no abnormality of ciliary beat frequency (Carstairs et al., 1984; Stanley et al., 1986).

Impairment of mucociliary transport may, however, be due to alterations in the volume and viscoelastic profile of the secretions, which are known to interact in a very specific way with the tips of the cilia (Figs. 12.9a,b). The physical characteristics of the bronchial sections in part depend upon their chemical composition (Lopata et al., 1974; Lopez-Vidriero and Reid, 1978), the degree of purulence (Lethem et al., 1987), and the complexing of mucin with other molecules such as secretory IgA (Harbitz et al., 1980), cationic proteins (e.g. lactoferrin) and lipid and DNA (Jeffery, 1987). There is evidence that once a smoker has developed chronic bronchitis, the impairment of mucociliary function is greater than that seen in smokers without bronchitis and that, in the former, cessation of smoking does not entirely reverse the impairment of mucociliary transport. (Agnew et al., 1982; Goodman et al., 1978). These CS-induced effects on the mucociliary system may also contribute to accumulation and pooling of secretions in the tracheobronchial tree, placing a greater reliance on cough as a clearance mechanism.

Whilst the volume of sputum produced shows a strong correlation with smoking, infective episodes and individual forced expiratory volume in 1 s (FEV_1), it does not, however, correlate with the progressive and accelerated decline in FEV_1 seen with age (Fletcher, 1984; Peto et al., 1983). Thus, whilst excessive secretions in airways may obstruct airflow, this does not of itself result in chronic progressive deterioration of lung function.

Chronic obstructive bronchiolitis

Airflow limitation, as determined by FEV_1, usually occurs late in the course of CS-related events, whereas inflammation in small airways (i.e. bronchioli less than 2–3 mm

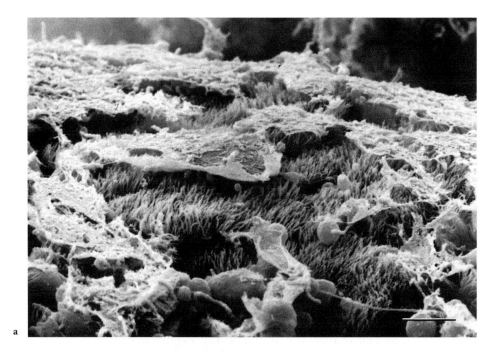

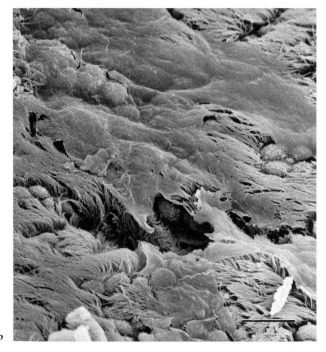

Fig. 12.9 (a) SEM of the surface of human bronchial mucosa showing a field of cilia with flakes of mucus at their tips (scale bar = 10.0 μm). (b) SEM of bronchial surface showing the experimental effects of subacute exposure to cigarette smoke: there is now almost a near complete sheet of mucus covering the cilia. (Scale bar = 10.0 μm).

in diameter) occurs relatively early and may be detected physiologically well before the age of 30 years (Buist *et al.*, 1979; Nemery *et al.*, 1981). The small airway defect is characterized by persistent airflow limitation which may show progressive deterioration in the absence of emphysema. Whilst the site of the lesion and diagnosis is, as yet, difficult to pinpoint by lung function, experimental physiologists have indicated that the dominant site lies in bronchioli of less than 3 mm diameter (Allansmith *et al.*, 1978; Hogg *et al.*, 1968). Early smoking-related changes have been described in studies comparing lungs of young smokers and controls of similar age from a group who had experienced sudden non-hospital deaths (Cosio *et al.*, 1980; Mitchell *et al.*, 1976; Niewoehner *et al.*, 1974) (Table 12.2). It is suggested that the early lesion in cigarette smokers is progressive airway, a macrophage bronchiolitis and alveolitis. There is then inflammation leading to peribronchiolar fibrosis and loss of alveolar attachments (Figs. 12.10a,b) (Saetta *et al.*, 1994). The resultant narrowing of small bronchioli has been well demonstrated by Bignon and colleagues (Fig. 12.11) (Bignon *et al.*, 1969). The peribronchiolar inflammation may predispose to the development of centrilobular emphysema and may be responsible for subtle abnormalities detected by lung function.

In bronchioli, secretory and ciliated cells are the main cell types (Jeffery and Corrin, 1984; Jeffery and Reid, 1975) and of them the Clara cell is the major secretory and progenitor cell. It has been suggested that the Clara cell normally produces a hypophase component of bronchiolar surfactant (Gil and Weibel, 1971), and a low molecular weight protease inhibitor (syn. antileukoprotease or bronchial mucosal protease inhibitor: Mooren et al., 1983), as well as a polychlorinated biphenyl (PCB)-binding protein which might be of toxicological significance (Anderson et al., 1994; Stripp et al., 1996). The last is the main anti-elastase of sputum and prevents autolytsis of airway tissues (Kramps et al., 1984). Clara cells are replaced by mucous cells in smokers (Ebert and Terracio, 1975); mucus appears in peripheral airways and its secretion is abnormally increased therein (Ebert and Hanks, 1981). Recent BAL data in humans support the suggestion of loss of Clara cells and show that alterations to CC10 and CC16, markers for this cell type, may be reduced in smokers and individuals exposed to pollutants (Bernard et al., 1994a,b). Peripheral extension of mucous cells is also a histological feature of CS-induced 'bronchitis' in the rat (Jones et al., 1973; Rogers and Jeffery, 1986) and the mechanism by which this occurs is likely to be the transformation of Clara cells (Jeffery and Reid, 1977).

Table 12.2 Small airway changes & consequences

Inflammation (i.e. respiratory bronchiolitis alveolitis)
Mucous metaplasia & plugging
Fibrosis
Bronchial smooth muscle enlargement
Loss of alveolar attachments
Stenoses & airway collapse
Collateral ventilation
Gas mixing abnormalities
Hypoxic vasoconstriction
Pulmonary hypertension
Cor pulmonale

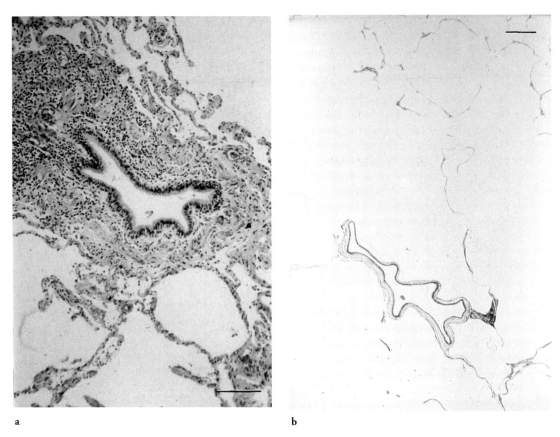

a　　　　　　　　　　　　　　　　　　　　　　　　b

Fig. 12.10　(a) Histological section of alveolar region in a case of COPD in which there are enlarged alveolar spaces surrounding a small airway with marked peribronchiolitis consisting primarily of lymphocytes (appearing as black dots within the airway wall). Stain: haematoxylin and eosin (scale bar = 200 μm). (b) Section of emphysematous lung in which there is destruction of alveolar attachments to the bronchiolar wall, resulting in its tortuous appearance and early collapse during expiration. Stain: haemotoxylin and eosin. (Scale bar = 0.5 cm).

What are the consequences of mucous metaplasia? One mechanism by which CS and pollutants may act is by reducing bronchiolar anti-proteases, leading to proteolytic digestion and tissue damage. Such changes in respiratory bronchioli underlie the development of centrilobular emphysema. Mucus cannot easily be cleared from peripheral bronchioli by coughing and replacement of the surfactant lining by mucus leads to instability of these small airways and hence to early airway closure during expiration (Macklem *et al.*, 1970). As the cross-sectional area of the bronchiolar zone of the lung is normally large in relation to the bronchial divisions (Horsfield, 1981), breathlessness and airflow limitation due to small airway disease are detectable only late in the course of the condition. This means that, once detected, relatively severe, progressive changes are already well established. Non-ciliated bronchiolar or Clara cells are also the major site of cytochrome P-450-dependent enzymes which act to detoxify many of the reactive xenobiotics present in cigarette smoke and atmospheric pollution. Their replacement and loss would likely render the lung more susceptible to pollutant-induced injury.

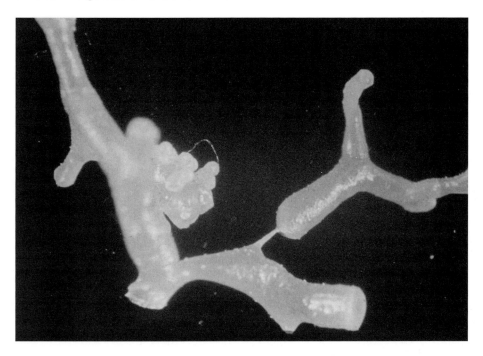

Fig. 12.11   Bronchial cast of human small airways illustrating the marked focal constriction of one airway branch in COPD (courtesy of Professor J Bignon).

Emphysema

Emphysema is defined as 'a condition of the lung characterized by abnormal, permanent enlargement of the airspaces distal to the terminal bronchiole (i.e. the acinus), accompanied by destruction of their walls, and without obvious fibrosis' (Snider *et al.*, 1985). Early changes have been thought to include subtle disruption to elastic fibres, with accompanying loss of elastic recoil, bronchiolar and alveolar distortion, and the appearance of fenestrae which enlarge (Gillooly and Lamb, 1993) (compare Figs. 12.2a and 12.12b), eventually leading to loss of interalveolar septa. In smokers with emphysema there is loss of alveolar wall tissue even in regions remote from obvious macroscopic lesions; recent data have shown that this is accompanied by a net *increase* in the mass of collagen. This suggests that, contrary to the current internationally accepted definition (see above), there is active alveolar wall fibrosis in emphysema (Lang *et al.*, 1994).

Two main forms of emphysema are described. They are distinguished by the part of the acinus affected. *Centriacinar* emphysema is characterized by focal destruction restricted to respiratory bronchioli and the central portions of the acinus, each focus surrounded by areas of grossly normal lung parenchyma. This form of emphysema is usually more severe in the upper lobes of the lung (Fig. 12.13). *Panacinar* emphysema involves some degree of destruction of the walls in a fairly uniform manner of *all* the air spaces beyond the terminal bronchiolus. This form of emphysema is characteristic of patients who develop smoking-related emphysema relatively early in life and, in contrast to the centriacinar form, has a tendency to involve the lower lobes more than the upper. In the familial form

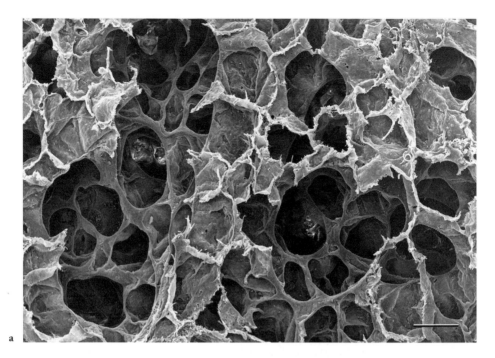

a

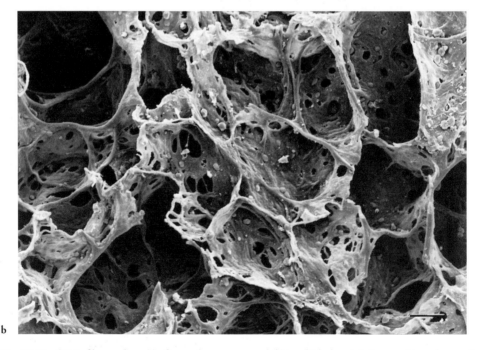

b

Fig. 12.12   SEM of human lung (a) of normal appearance and (b) in which there is 'microscopic' emphysema in which the alveolar walls are peppered with fenestrae. Such early lesions probably result in loss of lung elastic recoil. (Scale bars = 150 μm and 50 μm respectively).

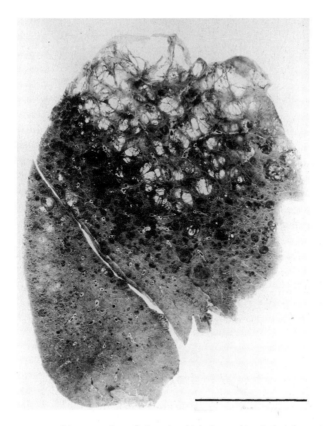

Fig. 12.13 Gross appearance of the cur surface of a lung in which the smoking-induced centri-acinar emphysema is restricted to the upper aspect of each lobe (courtesy of Professor B Heard). (Scale bar = 10 cm).

of panacinar emphysema it is usually associated with α_1-antitrypsin deficiency (Eriksson, 1964). The loss of alveolar wall attachments to peripheral airways may result in loss of 'radial traction' and consequent increased tortuosity and narrowing of small airways (Saetta *et al.*, 1985). This is thought to be responsible for the decrease in expiratory flow seen in patients with emphysema (Anderson and Foraker, 1962; Linhartova *et al.*, 1977; Saetta *et al.*, 1994).

Epidemiological studies have demonstrated a significant relationship between cigarette smoking and severity of emphysema (Auerbach *et al.*, 1972) but the mechanism(s) by which CS causes such damage is still the subject of much research and debate. The current working hypothesis is that emphysema is the result of an imbalance between proteolytic enzymes and protease inhibitors in the lung, favouring an excess of enzyme and in particular elastases. In addition, the imbalance between oxidants and antioxidants contribute by allowing an excessive oxidant burden to degrade the normal protease inhibitor screen (Cantin and Crystal, 1985; Gadek *et al.*, 1979). The proposed mechanism involves interactions between CS, T (CD8+) cells, alveolar macrophages, chemoattractants, neutrophils, elastases, endogenous and exogenous oxidants, protease inhibitors, antioxidants and lung connective tissue, primarily elastin, which undergoes repeated destruction, synthesis and degradation (Kimbel, 1985). The *in vitro* effects of CS on pulmonary connective tissue are consistent with the protease/antiprotease and oxi-dant/antioxidant hypotheses. In spite of this, experimental animal models of CS-induced emphysema have proved difficult to develop.

Lung cancer

Of the total of 2.1 million tobacco-attributed deaths in 1995, 590 000 involved lung cancer (Peto *et al.*, 1992). The evidence for smoking as the main cause of upper respira-tory tract and lung cancer meets the generally accepted criteria for causal association and derives from both prospective and retrospective studies in humans, supported by experi-mental studies in animals (Mohr and Reznik, 1978). Other affected organs include bladder, kidney, digestive tract and pancreas. The magnitude of the problem, its poor prognosis (the 5-year survival rate of lung cancer is less than 10%) have caused researchers to focus particular attention on the lung where at least 85–90% of cancers are attributa-ble to smoking. Most lung cancer arises in the lining epithelium of the airways (Fig. 12.13) (Jeffery, 1987). There are four major types of carcinoma with many subdivisions: (1) squamous cell, (2) adeno, (3) small cell, and (4) large cell undifferentiated carci-noma. The detailed histological classification of lung cancer is not within the remit of this chapter and is dealt with elsewhere (McDowell, 1987; McDowell *et al.*, 1978).

Pathogenesis

Squamous cell (or 'epidermoid') carcinoma is the most common form and the one most often associated with smoking in humans (Askin and Kaufman, 1985; Spencer, 1985) (Figure 12.14a). This form of tumour tends to be located near the hilum of the lung, has squamous differentiation with cytologic and nuclear features of atypia, and char-acteristically infiltrates the lung. The cells contain keratin and when well differentiated form keratin pearls (Fig. 12.14b). As the lining epithelium of airways is not normally composed of squamous cells, squamous carcinoma is presumed to arise at sites of squa-mous metaplasia, with dysplasia and carcinoma *in situ* representing intermediate stages. Such premalignant changes have been linked to smoking history (Auerbach *et al.*, 1961, 1962a,b; Gouveia *et al.*, 1982; Spencer, 1985). A proposed sequence of events involves: squamous metaplasia, stratification, atypia, 'carcinoma *in-situ*' and ultimately invasive cancer. However, the cell of origin is still unproven and it need not be the basal cell, which morphologically most closely resembles this form of tumour cell. The pro-liferative and metaplastic potential of the goblet (mucus-secreting) cell is now well recognized and it may well prove to be the stem cell which undergoes squamous meta-plasia and malignant change (Jeffery, 1987; Jeffery and Reid, 1981; McDowell and Trump, 1983).

The key to the ease with which tumours can be induced seems to lie with the prolonged irritative, proliferative and metaplastic effects of CS, alone or in combination with other pollutants, and the interaction of its many constituents. For example, whilst it may be dif-ficult to induce tumours by experimental inhalation of benzo[a]pyrene alone, its combination with the inhaled pollutant sulfur dioxide, which induces proliferation and squamous metaplasia, allows tumours to be induced relatively easily (Kuschner, 1985).

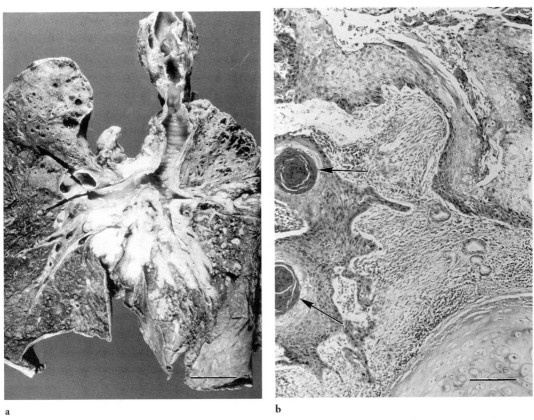

a   b

Fig. 12.14 (a) Gross appearance of left upper lobe resected from a 63 year old smoker. The lobe is almost completely replaced by a large, poorly differentiated squamous carcinoma which appears pale (scale bar = 4 cm). (b) Section through bronchial wall which is infiltrated by a well differentiated squamous carcinoma. Islands of tumour show pronounced kerati whorls (arrows) (Scale bar = 150 μm). (Courtesy of Professor B. Corrin).

Cell proliferation is an essential phenomenon for the fixation and expression of neoplastic change and the induction of cell proliferation may be a rate limiting step in the process (Saffiotti *et al.*, 1985). Proliferation is required to replace turnover of the cells comprising the lung: turnover is normally relatively slow and the factors which control cell proliferation are little understood (Ayers and Jeffery, 1988; Boren and Paradise, 1978). Exposure of experimental animals to CS shows quite clearly that this is a powerful stimulus to the rapid induction of epithelial cell proliferation. Cell sloughing or necrosis induced by CS may be important proliferative triggers, as might be the potent stimulation of quanylate cyclase activity with concomitant rises in quanosine $3',5'$-monophosphate (cGMP): the effect is reasonably specific as the cellular levels of adenosine $3',5'$-monophosphate (cAMP) do not appear to change (Arnold *et al.*, 1977; Klas, 1980). In regard to proliferative potential of airway lining cells, the mucous, serous and basal cells of the trachea and bronchus, the Clara cell of the bronchiolus and the type II pneumonocyte of the alveolus are the stem cells which respond to the irritation and damage due to a variety of irritants, including CS (Jeffery, 1987).

Synergism

In this way many of the constituent molecules in CS and other inhaled pollutants are likely to interact synergistically to induce proliferation and the multifocal squamous metaplasia on which malignant change is then superimposed. In the case of squamous metaplasia, nutritional factors may also be of particular importance. Squamous metaplasia may be induced by CS alone (Davis et al., 1975; Smith et al., 1978), vitamin A deficiency alone (Anzano et al., 1980) or by a combination of the two (Keenan, 1987; Meade et al., 1979). Experimental studies show that vitamin A deficiency markedly modifies the 'bronchitic' response to subacute inhalation of CS by augmenting its metaplastic effects and inducing a thickened surface epithelium which is squamous, stratified and atypical. Several epidemiological studies show that the incidence of epithelial tumours is increased in subjects with either low plasma retinol (Kark et al., 1981; Wald et al., 1980) or low dietary intake of the vitamin (Bjelke, 1975; Peto et al., 1981; Shekelle et al., 1981). One mechanism which may in part explain the synergism is the observed increase in binding of benzo[a]pyrene to hamster tracheal DNA when there is a deficiency of vitamin A (Genta et al., 1974; Kaufman et al., 1974). Other interactions between CS and environmental factors are well recognized. For example, epidemiological and experimental studies demonstrate synergistic effects between asbestos and smoking. Compared with the non-smoking general population, the increased risk of bronchogenic carcinoma in non-smoking asbestos workers is about four-fold, whereas in those who also smoke it is 80- to 90-fold (Mossman et al., 1985). In regard to radiation exposure in miners, it has been calculated that the cancer risk is 10 times greater for smokers than for non-smokers, with shorter induction latent periods and earlier age of onset in the smokers (Archer, 1985; Kennedy and Little, 1978; Masse et al., 1984). Experimentally, substances which are thought to stimulate microsomal enzyme systems may similarly shorten the latent period for induction of lung tumours following exposure to radioactive radon daughters (Queval et al., 1979).

The events described are all relatively early events in a multistage process of long latency which when associated with a history of chronic smoking culminates in the emergence of a malignant cell line.

SUMMARY AND CONCLUSION

The associations between smoking, atmospheric and occupational pollution, and lung inflammation and disease are strong and are made even more convincing when one examines the improvement of life expectancy and its quality on giving up the smoking habit and improving the air we breathe. Considering lung function and cellular responses, there are acute effects on airway dynamics, mucociliary clearance, epithelial and vascular permeability and immune function. Cells known to be affected include especially airway lining epithelial cells, while there are differential affects on lymphocyte subsets, alveolar macrophages, neutrophils, mast cells, fibroblasts and platelets. Molecular and biochemical effects include: (1) metabolism of a variety of exogenous chemicals by microsomal mixed function oxidases and altered binding of metabolized products to DNA, (2) alterations of DNA (synthesis, damage and repair), (3) stimulation of lipolysis, arachidonic acid metabolism (and altered degradation of its products), increased production of intra-cellular messengers, transcription factors and increased release of proinflammatory cytokines, (4) increased release and synthesis of antioxidants by cells, (5) release of factors chemo-attractant for inflammatory cells, (6) release of proteases (elastases) and inhibition of antiproteases in lung epithelial lining fluid and blood, and (7) effects on circulating levels of immunoglobulins and complement. Self- and passive exposure to tobacco smoke and atmospheric and occupational pollution all interact and may, in a susceptible group, overcome our normal defences to initiate and perpetuate chronic inflammation and the development of several lung conditions, all of which have a major impact on world health.

ACKNOWLEDGEMENTS

I am extremely grateful to Dr Anders Blomberg for allowing me to preview his early results and those of his colleagues in which they have investigated the experimental effects of atmospheric pollutants in humans (personal communication). I would also like to give a special thank you to Miss Leone Oscar for her patience and care in the preparation of this manuscript, and also to Mr Andy Rogers and Dr T Brain for their willing assistance with the illustrations.

REFERENCES

Agnew JE, Little F, Pavia D and Clarke SW (1982) Mucus clearance from the airways in chronic bron-chitis: smokers and ex-smokers. *Bull Eur Physiopathol Respir* **18**: 473–484.
Ailsby RL and Ghadially FN (1973) Atypical cilia in human bronchial mucosa. *J Pathol* **109**: 75–77.
Albert RE, Spiegelman JR, Schatsky S and Lippmann M (1969) Effect of acute exposure to cigarette smoke on bronchial clearance in the miniature donkey. *Arch Environ Health* **8**: 30–41.
Albert RE, Peterson HT Jr, Bohning DE and Lippmann M (1975) Short term effects of cigarette smok-ing on bronchial clearance in humans. *Arch Environ Health* **30**: 361–367.
Allansmith MR, Grener JV and Baird RS (1978) Number of inflammatory cells in the normal con-junctiva. *Am J Ophthalmol* **86**: 250–259.
Amadori A, Zamarchi R, De Silvestro G *et al.* (1995) Genetic control of the CD4/CD8 T-cell ratio in humans. *Nat Med* **1**: 1279–1283.
Anderson AE and Foraker AG (1962) Relative dimensions of bronchioles and parenchymal spaces in lungs from normal subjects and emphysematous patients. *Am J Med* **32**: 218–226.
Anderson O, Noack G, Robertson B *et al.* (1994) Ontogeny of a human polychlorinated biphenyl-bind-ing protein. *Chest* **105**: 17–22.
Anderson P, Pederson DF, Bach D and Bonde GJ (1982) Serum antibodies and immunoglobulins in smokers and non-smokers. *Clin Exp Immunol* **47**: 467–473.
Anzano MA, Olsen JA and Lamb AJ (1980) Morphologic alterations in the trachea and the salivary gland following the induction of rapid synchronous vitamin A deficiency in rats. *Am J Pathol* **98**: 717–732.
Archer VE (1985) Enhancement of lung cancer by cigarette smoking in uranium and other miners. In: Mass NJ, Kaufman DJ, Siegfried JN *et al.* (eds.) *Cancer of the Respiratory Tract: Predisposing Factors*, vol. 8. New York: Raven Press, pp. 23–77.
Aris RM, Christian D, Hearne PQ *et al.* (1993) Ozone-induced airway inflammation in human sub-jects as determined by airway lavage and biopsy. *Am Rev Respir Dis* **148**: 1363–1372.

Armitage AK and Turner DM (1970) Absorption of nicotine in cigarette and cigar smoke through the oral mucosa. *Nature* **226**: 1231–1232.

Arnold WP, Aldred R and Murad F (1977) Cigarette smoke activates guanylate cyclase and increases guanosine 3′,5′-monophosphates in tissues. *Science* **198**: 934–936.

Arsalane K, Gosset P, Vanhee D *et al.* (1995) Ozone stimulates synthesis of inflammatory cytokines by alveolar macrophages *in vitro. Am J Respir Cell Mol Biol* **13**: 60–68.

Askin FB and Kaufman DJ (1985) Histomorphology of human lung cancer. In: Mass MJ, Kaufman DJ, Siegfried JM *et al.* (eds) *Carcinogenesis*, vol. 8. New York: Raven Press, pp. 1–15.

Auerbach O, Stout AP, Hammond EC and Garfinkel L (1961) Changes in bronchial epithelium in relation to cigarette smoking and in relation to lung cancer. *N Engl J Med* **265**: 253–267.

Auerbach O, Stout AP, Hammond EC and Garfinkel L (1962a) Bronchial epithelium in former smokers. *N Engl J Med* **267**: 119–125.

Auerbach O, Stout AP, Hammond EC and Garfinkel L (1962b) Changes in bronchial epithelium in relation to sex, age, residence, smoking and pneumonia. *N Engl J Med* **267**: 11–119.

Auerbach O, Hammond EC, Garfinkel L and Benante C (1972) Relation of smoking and age to emphysema. *N Engl J Med* **286**: 853–858.

Ayers M and Jeffery PK (1982) Pathways of goblet cell hyperplasia – the response of bronchial epithelium to tobacco smoke. *J Pathol* 90–91.

Ayers M and Jeffery PK (1988) Proliferation and differentiation in adult mammalian airway epithelium: a review. *Eur Respir J* **1**: 58–80.

Azzawi M, Bradley B, Jeffery PK *et al.* (1990) Identification of activated T lymphocytes and eosinophils in bronchial biopsies in stable atopic asthma. *Am Rev Respir Dis* **142**: 1407–1413.

Bahna SL, Heiner DC and Myhre DA (1983a) Immunoglobulin E pattern in cigarette smokers. *Allergy* **38**: 57–64.

Bahna SL, Heiner DC and Myhre BA (1983b) Changes in serum IgD and cigarette smokers. *Clin Exp Immunol* **51**: 624–630.

Bair JW and Dilley JV (1967) Pulmonary clearance of $^{59}Fe_2O_3$ and $^{51}Cr_2O_3$ in rats and dogs exposed to cigarette smoke. In: Davies CH (ed.) *Inhaled Particles and Vapours*, vol. 2. New York: Pergamon Press, pp. 251–268.

Bakhle YS (1984) Effects of cigarette smoke on the metabolism of arachidonic acid and prostaglandins in the lung. In: Cumming G and Bonsignore G (eds) *Smoking and the Lung*, vol. 17. New York: Plenum, pp. 305–318.

Balmes JR, Aris RM, Chen LL *et al.* (1997) Effects of ozone on normal and potentially sensitive human subjects. Part I: Airway inflammation and responsiveness to ozone in normal and asthmatic subjects. Research Report – Health Effects Institute; **78**: 1–37.

Barr BC, Hyde DM, Plopper CG and Dungworth DL (1988) Distal airway remodeling in rats chronically exposed to ozone. *Am Rev Respir Dis* **137**: 924–938.

Barry BE, Mercer RR, Miller FJ and Crapo JD (1988) Effects of inhalation of 0.25 ppm ozone on the terminal bronchioles of juvenile and adult rats. *Exp Lung Res* **14**: 225–245.

Barter CE and Campbell AH (1975) Relationship of constitutional factors in cigarette smoking to decrease in 1-second forced expiratory volume. *Am Rev Respir Dis* **113**: 305–314.

Bascom R, Bromberg PA, Costa DA *et al.* (1996) Health effects of outdoor air pollution. *Am J Respir Crit Care Med* **153**: 3–50.

Basha MA, Gross KB, Gwizdala CJ *et al.* (1994) Bronchoalveolar lavage neutrophilia in asthmatic and healthy volunteers after controlled exposure to ozone and filtered purified air. *Chest* **106**: 1757–1765.

Bates DV (1973) The fate of the chronic bronchitic: report of the 10 year follow-up in the Canadian Department of Veterans Affairs coordinated study of chronic bronchitis. *Am Rev Respir Dis* **108**: 1043–1065.

Bates DV (1994) Setting the stage: critical risks. In: *Environmental Health Risks and Public Policy. Decision Making in Free Societies.* Seattle: University of Washington Press, pp. 6–56.

Becker S, Madden MC, Newman SL *et al.* (1991) Modulation of human alveolar macrophage properties by ozone exposure *in vitro. Toxicol Appl Pharmacol* **110**: 403–415.

Becker S, Devlin R, Horstman D *et al.* (1993) Evidence for mild inflammation and change in alveolar macrophage function in humans exposed to 2 ppm NO_2. *Indoor Air: Health Effects (Helsinki)* **1**: 471–476.

Bergman H, Edling C and Axelson O (1984) Indoor radon daughter concentrations and passive smoking. In: Berglund B, Lindvall T and Sundell J (eds) *Proceedings of the 3rd International Conference on Indoor Air Quality and Climate. Radon Passive Smoking Particulates and Housing Epidemiology.* Stockholm: Stockholm Swedish Council for Building Research, pp. 79–84.

Bernard AM, Roels HA, Buchet JP and Lauwerys RR (1994a) Serum Clara cell protein: an indicator of bronchial cell dysfunction caused by tobacco smoking. *Environ Res* **66**: 96–104.

Bernard AM, Gonzalez-Lorenzo JM, Siles E *et al.* (1994b) Early decrease of serum Clara cell protein in silica-exposed workers. *Eur Respir J* **7**: 1932–1937.

Berry CN, Hoult JRS, Littleton JM *et al.* (1979) Nicotine causes prostaglandin efflux from isolated perfused rat lung. *Br J Pharmacol* **66**: 101P.

Bhalla DK and Crocker TT (1986) Tracheal permeability in rats exposed to ozone. *Am Rev Respir Dis* **134**: 572–579.

Bhalla DK, Rasmussen RE and Tjen S (1990) Interactive effects of ozone, cytochalasin D, and vinblastine on transepithelial transport and cytoskeleton in rat airways. *Am J Respir Cell Mol Biol* **3**: 119–129.

Bignon J, Khoury F, Evan P *et al.* (1969) Morphometric study in chronic obstructive bronchopulmonary disease. *Am Rev Respir Dis* **99**: 669–695.

Bjelke E (1975) Dietary vitamin A and human lung cancer. *Int J Cancer* **15**: 561–565.

Bjermer L, Cai Y, Nilsson K *et al.* (1993) Tobacco smoke exposure suppresses radiation-induced inflammation in the lung: a study of bronchoalveolar lavage and ultrastructural morphology in the rat. *Eur Respir J* **6**: 1173–1180.

Blomberg A (1998) Medical dissertation: Inflammatory and antioxidant responses in the airways to oxidative and particulate air pollution. Umea: Arbetslivsinstitutets tryckeri.

Blomberg A, Krishna MT, Bocchino V *et al.* (1997) The inflammatory effects of 2 ppm NO_2 on the airways of healthy subjects. *Am J Respir Crit Care Med* **156**: 418–424.

Blue ML and Janoff A (1978) Possible mechanisms of emphysema in cigarette smoking: release of elastase from human polymorphonuclear leucocytes by cigarette smoke condensate in vitro. *Am Rev Respir Dis* **117**: 317–325.

Bodgen JD, Kemp SW, Busemthind IF *et al.* (1981) Composition of tobaccos in countries with high and low incidences of lung cancer. I selenium polonium-210, alternaria, tar and nicotine. *J Natl Cancer Inst* **66**: 27–31.

Bolduc P, Jones R and Reid L (1981) Mitotic activity of airway epithelium after short exposure to tobacco smoke and the effect of the anti-inflammatory agent phenylmethyloxadiazole. *Br J Exp Pathol* **62**: 461–468.

Boren HG and Paradise LJ (1978) Cytokinetics of lung. In: Harris CC (ed) *Pathogenesis and Therapy of Lung Cancer*, vol. 10. New York: Marcel Dekker, pp. 369–418.

Boucher RC, Johnson J, Inoue S *et al.* (1980) The effect of cigarette smoking on the permeability of guinea pig airways. *Lab Invest* **43**: 94–100.

Bradley BL, Azzawi M, Jacobson M *et al.* (1991) Eosinophils, T-lymphocytes, mast cells, neutrophils and macrophages in bronchial biopsies from atopic asthmatics: comparison with atopic non-asthma and relationship to bronchial hyperresponsiveness. *J Allergy Clin Immunol* **88**: 661–674.

Brain JD and Valberg PA (1979) Deposition of aerosol in the respiratory tract. *Am Rev Respir Dis* **120**: 1325–1373.

Bridgeman MME, Marsden M, Drost E, *et al.* (1991) The effect of cigarette smoke on lung cells. *Am Rev Respir Dis* **143**: A737.

Bridges AB, Scott NA, Parry GJ and Belch JJ (1993) Age, sex, cigarette smoking and indices of free radical activity in healthy humans. *Eur J Med* **2**: 205–208.

Buist S, Ghezzo H, Anthonisen NR *et al.* (1979) Relationship between the single breath N_2 test and age, sex and smoking habits in three North American cities. *Am Rev Respir Dis* **120**: 305–318.

Burrows B, Knudson RJ and Lebowitz MD (1977) The relationship of childhood respiratory illness to adult obstructive airway disease. *Am Rev Respir Dis* **115**: 751–760.

Burrows B, Halonen M, Barbee RA and Lebowitz MD (1981) The relationship of serum immunoglobulin E to cigarette smoking. *Am Rev Respir Dis* **124**: 523–525.

Camner P and Philipson K (1974) Mucociliary clearance. *Scand J Environ Health* **90**: 45–48.

Camner P, Mossberg B and Philipson K (1973) Tracheobronchial clearance and chronic obstructive lung disease. *Scand J Respir Dis* **54**: 272–281.

Cantin A and Crystal RG (1985) Oxidants, antioxidants and the pathogenesis of emphysema. *Eur J Respir Dis* **66** (Suppl. 139): 7–17.

Cantin AM, North SL, Hubbard RC and Crystal RG (1987) Normal alveolar epithelial lung fluid contains high levels of glutathione. *J Appl Physiol* **63**: 152–157.

Carlsson LM, Marklund SL, Edlund T and Sandstrom T (1996) Increased neutrophilic inflammation of the airways after ozone exposure in mice lacking extracellular-superoxide dismutase (EC-SOD). *Eur Respir J* **9**(772): 114s (abstract).

Carstairs JR, Nimmo AJ and Barnes PJ (1984) Autoradiographic localisation of Beta adrenoreceptor in human lung. *Eur J Pharmacol* **105**: 189–190.

Chang SC (1957) Microscopic properties of whole mounts and sections of human bronchial epithelium of smokers and non-smokers. *Cancer* (Phila) **10**: 1246–1262.

Chow CK (1993) Cigarette smoking and oxidative damage in the lungs. *Ann NY Acad Sci* **686**: 289–299.

Chow JC (1995) Measurement methods to determine compliance with ambient air quality standards for suspended particles. *J Air Waste Manag Assoc* **45**: 320–382.

Church DF and Pryor WA (1986) The free radical chemistry of cigarette smoke and its toxicological implications. *Environ Health Perspect* **64**: 111–126.

Cohen AB and Cline MJ (1971) The human alveolar macrophage: isolation, cultivation in vitro and studies of morphologic and functional characteristics. *J Clin Invest* **50**: 1390–1398.

Coles SJ, Levine LR and Reid L (1979) Hypersecretion of mucus glycoproteins in rat airways induced by tobacco smoke. *Am J Pathol* **94**: 459–472.

Coles SJ, Bhaskar KR, O'Sullivan BD *et al.* (1984) Airway mucus: composition and regulation of its secretion by neuropeptides in vitro. In: *Mucus and Mucosa*, Ciba Foundation Symposium 109. London: Pitman Medical, pp. 40–60.

Colley JRT, Holland WW and Corkhill RT (1974) Influence of passive smoking and parental phlegm on pneumonia and bronchitis in early childhood. *Lancet* **2**(7888): 1031–1034.

Cosio MG, Hale KA and Niewoehner DE (1980) Morphologic and morphometric effects of prolonged cigarette smoking on the small airways. *Am Rev Respir Dis* **122**: 265–271.

Costabel U, Bross KJ, Reuter C *et al.* (1986) Alterations in immunoregulatory T-cell subsets in cigarette smokers. A phenotypic analysis of bronchoalveolar and blood lymphocytes. *Chest* **89**: 39–44.

Cotgreave IA, Johansson U, Moldeus P and Brattsand R (1987) The effect of acute cigarette smoke inhalation on pulmonary and systemic cysteine and glutathione redox states in the rat. *Toxicology* **45**: 203–212.

Dalham T (1966) Effect of cigarette smoke on ciliary activity. *Am Rev Respir Dis* **93**: 108–114.

Daniele RP, Dauber JH, Altose MD *et al.* (1977) Lymphocyte studies in asymptomatic cigarette smokers. A comparison between lung and peripheral blood. *Am Rev Respir Dis* **116**: 997–1005.

Davis BR, Whitehead JK, Gill ME *et al.* (1975) Response of rat lung to inhaled tobacco smoke with or without prior exposure to 3,4-benzpyrene (BP) intra-tracheal installation. *Br J Can* **31**: 469–484.

Davis PB and Kaliner M (1983) Autonomic nervous system abnormalities in cystic fibrosis. *J Chron Dis* **36**: 269–278.

DeForge LE, Preston AM, Takeuchi E *et al.* (1993) Regulation of interleukin 8 gene expression by oxidant stress. *J Biol Chem* **268**: 25568–25576.

Devalia JL, Campbell AM, Sapsford RJ *et al.* (1993) Effect of nitrogen dioxide on synthesis of inflammatory cytokines expressed by human bronchial epithelial cells in vitro. *Am J Respir Cell Mol Biol* **9**: 271–278.

Devlin RB, McKinnon KP, Noah T *et al.* (1994) Ozone-induced release of cytokines and fibronectin by alveolar macrophages and airway epithelial cells. *Am J Physiol* **266**: L612–L619.

Di Stefano A, Maestrelli P, Roggeri A *et al.* (1994) Upregulation of adhesion molecules in the bronchial mucosa of subjects with chronic obstructive bronchitis. *Am J Respir Crit Care Med* **149**: 803–810.

Diaz Sanchez D (1997) The role of diesel exhaust particles and their associated polyaromatic hydrocarbons in the induction of allergic airway disease. *Allergy* **52**: 52–56.

Diaz Sanchez D, Dotson AR, Takenaka H and Saxon A (1994) Diesel exhaust particles induce local IgE production in vivo and alter the pattern of IgE messenger RNA isoforms. *J Clin Invest* **94**: 1417–1425.

Diaz Sanchez D, Tsien A, Casillas A *et al.* (1996) Enhanced nasal cytokine production in human beings after in vivo challenge with diesel exhaust particles. *J Allergy Clin Immunol* **98**: 114–123.

Djukanovic R, Roche WR, Wilson JW *et al.* (1990) Mucosal inflammation in asthma. *Am Rev Respir Dis* **142**: 434–457.

Dockery DW and Pope AC (1994) Acute respiratory effects of particulate air pollution. *Ann Rev Public Health* **15**: 107–132.

Dockery DW, Pope AC, Xu X *et al.* (1993) An association between air pollution and mortality in six US cities. *N Engl J Med* **329**: 1753–1759.

Driscoll KE, Simpson L, Carter J *et al.* (1993) Ozone inhalation stimulates expression of a neutrophil chemotactic protein, macrophage inflammatory protein 2. *Toxicol Appl Pharmacol* **119**: 306–309.

Drost EM, Selby C, Lannan S *et al.* (1992) Changes in neutrophil deformability following in vitro smoke exposure: mechanisms and protection. *Am J Respir Cell Mol Biol* **6**: 287–295.

Dube MF and Green CR (1982) Methods of collection of smoke for analytical purposes. *Recent Adv Tobacco Sci* **8**: 42–102.

Dye JA and Adler KA (1994) Effects of cigarette smoke on epithelial cells of the respiratory tract. *Thorax* **49**: 825–834.

Ebert RV and Hanks PB (1981) Mucus secretion by the epithelium of the bronchioles of cigarette smokers. *Br J Dis Chest* **75**: 277–282.

Ebert RV and Terracio MJ (1975) The bronchiolar epithelium in cigarette smokers: observations with the scanning electron microscope. *Am Rev Respir Dis* **111**: 4–11.

Eriksson S (1964) Pulmonary emphysema and alpha 1-antitrypsin deficiency. *Acta Med Scand* **175**: 197–205.

Eustis SL, Schwartz LW, Kosch PC and Dungworth DL (1981) Chronic bronchiolitis in non-human primates after prolonged ozone exposure. *Am J Pathol* **105**: 121–137.

Evans MJ, Cabral-Anderson LJ and Stephens RJ *et al.* (1973) Renewal of alveolar epithelium in the rat following exposure to NO_2. *Am J Pathol* **70**: 175–198.

Evans MJ, Cabral LJ, Stephens RJ *et al.* (1975) Transformation of alveolar type 2 cells to type 1 cells following exposure to NO_2. *Exp Mol Pathol* **22**: 142–150.

Finkelstein R, Fraser RS, Ghezzo H and Cosio MG (1995) Alveolar inflammation and its relation to emphysema in smokers. *Am J Respir Crit Care Med* **152**: 1666–1672.

Finklea JF, Hasselblood V, Riggan WB *et al.* (1971) Cigarette smoking and haemagglutination inhibition response to influenza after natural disease and immunisation. *Am Rev Respir Dis* **104**: 368–376.

Finnegan MJ, Little S, Gordon DJ *et al.* (1991) The effect of smoking on the development of allergic disease and specific immunological responses in a factory workforce exposed to humidifier contaminants. *Br J Ind Med* **48**: 30–33.

Fletcher CM (1984) Chronic bronchitis and decline in pulmonary function with some suggestions on terminology. In: Cumming G and Bonsignore G (eds) *Smoking and the Lung*, vol. 17. New York: Plenum Press, pp. 397–420.

Fletcher CM and Peto R (1977) The natural history of chronic airflow obstruction. *Br Med J* **1**: 1645–1649.

Fletcher CM and Pride NB (1984) Definition of emphysema, chronic bronchitis, asthma and airflow obstruction: twenty-five years on from the CIBA symposium. *Thorax* **39**: 81–85.

Fletcher RD, Sumney DL, Langkamp HH and Platt D (1969) The ability of human serum to agglutinate sheep erythrocytes and the effect of tobacco mosaic virus. *Am Rev Respir Dis* **100**: 92–94.

Foster WM, Costa DL and Langenback EG (1987) Ozone exposure alters tracheobronchial mucociliary function in humans. *J Appl Physiol* **63**: 996–1002.

Fox B, Bull TB, Makey AR and Rawbone R (1981) The significance of ultrastructural abnormalities of human cilia. *Chest* **80**: 796–799.

Frampton MW, Finkelstein JN, Roberts NJ Jr *et al.* (1989a) Effects of nitrogen dioxide exposure on bronchoalveolar lavage proteins in humans. *Am J Respir Cell Mol Biol* **1**: 499–505.

Frampton MW, Smeglin AM, Roberts NJ Jr *et al.* (1989b) Nitrogen dioxide exposure in vivo and human alveolar macrophage inactivation of influenza virus in vitro. *Environ Res* **48**: 179–192.

Frampton MW, Smeglin AM, Roberth NJ *et al.* (1991) Characterization of the inflammatory response to NO_2 exposure using bronchoalveolar lavage in humans. *Am Rev Respir Dis* **4**: A89 (abstract).

Frampton MW, Morrow PE, Torres A *et al.* (1997) Effects of ozone on normal and potentially sensitive human subjects. Part II: Airway inflammation and responsiveness to ozone in non-smokers and smokers. Research Report – Health Effects Institute (June); **78**: 39–72. (See discussion pp. 81–99.)

Gadek JE, Fells GA and Crystal RG (1979) Cigarette smoking induces functional antiprotease deficiency in the lower respiratory tract of humans. *Science* **206**: 1315–1316.

Genta VM, Kaufman BG, Harris C *et al.* (1974) Vitamin A deficiency enhances binding of benzo(a)pyrene to tracheal epithelial DNA. *Nature* **247**: 48–49.

Gerrity TR, Weaver RA, Berntsen J *et al.* (1988) Extrathoracic and intrathoracic removal of O_3 in tidal-breathing humans. *J Appl Physiol* **65**: 393–400.

Gil J and Weibel E (1971) Extracellular lining of bronchioles after perfusion-fixation of rat lungs for electron microscopy. *Anat Rec* **169**: 185–200.

Gillooly M and Lamb D (1993) Microscopic emphysema in relation to age and smoking habit. *Thorax* **48**: 491–495.

Giordano Jr AM, Morrow MS and Morrow PE (1972) Chronic low-level nitrogen dioxide exposure and mocociliary clearance. *Arch Environ Health* **25**: 443–449.

Glasgow JE, Pietra GG, Abrams WR *et al.* (1987) Neutrophil recruitment and degranulation during induction of emphysema in the rat by nitrogen dioxide. *Am Rev Respir Dis* **135**: 1129–1136.

Godfrey RWA (1997) Human airway epithelial tight junctions (Review). *Microsc Res Tech* **38**: 488–499.

Godfrey RWA, Severs NJ and Jeffery PK (1992) Freeze-fracture morphology and quantification of human bronchial epithelial tight junctions. *Am J Respir Cell Mol Biol* **6**: 453–458.

Goodman RM, Bergen BM, Lander JF *et al.* (1978) Relationship of smoking history and pulmonary function tests to tracheal mucus velocity in non-smokers, young smokers, ex-smokers and patients with chronic bronchitis. *Am Rev Respir Dis* **117**: 205–214.

Gordon RE, Solamo D and Kleinerman J (1986) Tight junction alterations of respiratory epithelium following long-term nitrogen dioxide exposure and recovery. *Exp Lung Res* **11**: 179–193.

Gouveia J *et al.* (1982) Degree of bronchial metaplasia in heavy smokers and its regression after treatment with a retinoid. *Lancet* **i**: 710–712.

Greaves IA and Colebatch HJH (1986) Observations on the pathogenesis of chronic airflow obstruction in smokers: implications for the detection of 'early' lung disease. *Thorax* **41**: 81–87.

Green GM, Jakab GJ, Low RB and Davis GS (1977) Defence mechanisms of the respiratory membrane. *Am Rev Respir Dis* **115**: 479–514.

Harbitz O, Jensson AO and Smidsrod O (1980) Quantitation of proteins in sputum from patients with chronic obstructive lung disease: 1. Determination of immunoglobulin A. *Eur J Respir Dis* **61**: 84–94.

Harley NH, Cohen BS and Tso TC (1980) Polonium-210: a questionable risk factor in smoking-related carcinogenesis. In: Gori GB and Bock FG (eds) *A Safe Cigarette?* New York: Coldspring Harbor, pp. 93–104.

Hazucha MJ, Ginsberg JF, McDonnell WF *et al.* (1983) Effects of 0.1 ppm nitrogen dioxide on airways of normal and asthmatic subjects. *J Appl Physiol* **54**: 730–739.

Helleday R, Sandstrom T and Stjernberg N (1994) Differences in bronchoalveolar cell response to nitrogen dioxide exposure between smokers and non-smokers. *Eur Respir J* **7**: 1213–1220.

Helms PJ (1994) Lung growth: implications for the development of disease. *Thorax* **49**: 440–441.

Higgenbottam T and Borland C (1984) The gas phase of tobacco smoke and the development of lung disease. In: Cumming G and Bonsignore G (eds) *Smoking and the Lung*, vol. 17. New York: Plenum Press, pp. 353–380.

Hiltermann TJN, de Bruijne CR, Stolk J *et al.* (1997) Effects of photochemical air pollution and allergen exposure on upper respiratory tract inflammation in asthmatics. *Am J Resp Crit Care Med* **156**: 1765–1773.

Hiltermann TJN, Lapperre TS, van Bree L *et al.* (1998) Non-invasive techniques to study airway inflammation in ozone-exposed subjects with asthma. See medical dissertation: Air pollution and asthma: epidemiological and clinical experimental studies with ozone, Leider (NL). Cip-Gegevens Koninklijke Bibliotheek, Den Haag.

Hiltermann TJN, Peters EA, Alberts B *et al.* (1998) Ozone-induced airway hyperresponsiveness in patients with asthma: role of neutrophil-derived serine proteinases. In press.

HMSO (1954) Mortality and morbidity during the fog of December 1952. London: HMSO.

Hogg JC, Macklem PT and Thurlbeck WM (1968) Site and nature of airway obstruction in chronic obstructive lung disease. *N Engl J Med* **278**: 1355–1360.

Holt PG (1987) Immune and inflammatory function in cigarette smokers. *Thorax* **42**: 241–249.

Holt PG and Keast D (1973) Acute effects of cigarette smoke on murine macrophages. *Arch Environ Health* **26**: 300–304.

Holt PG and Keast D (1977) Environmentally induced changes in immunological function: acute and chronic effects of inhalation of tobacco smoke and other atmospheric contaminants in man and experimental animal. *Bacteriol Rev* **41**: 205–214.

Holt PG, Bartholomaus WM and Keast D (1974) Differential toxicity of tobacco to various cell types including those of the immune system. *Aust J Exp Biol Med Sci* **52**: 211–214.

Horsfield K (1981) The structure of the tracheobronchial tree. In: Scadding JG, Cumming G and Thurlbeck WM (eds) *The Structure of the Tracheobronchial Tree.* London: William Heinemann, pp. 54–70.

Hu SC, Ben Jebria A and Ultman JS (1992) Longitudinal distribution of ozone absorption in the lung: quiet respiration in healthy subjects. *J Appl Physiol* **73**: 1655–1661.

Hughes DA, Haslam PL, Townsend PJ and Turner-Warwick M (1985) Numerical and functional alterations in circulatory lymphocytes in cigarette smokers. *Clin Exp Immunol* **61**: 459–466.

Hulbert WC, Walker DC, Jackson A and Hogg JC (1981) Airway permeability to horseradish peroxidase in guinea pigs: the repair phase after injury by cigarette smoke. *Am Rev Respir Dis* **123**: 320–326.

Humbert M, Durham SR, Ying S *et al.* (1996) IL-4 and IL-5 mRNA and protein in bronchial biopsies from atopic and non-atopic asthma. Evidence against 'intrinsic' asthma being a distinct immunopathological entity. *Am J Respir Crit Care Med* **154**: 1497–1504.

Hybbinette J-C (1982) A pharmacolorgical evaluation of the short-term effect of cigarette smoke on muco-ciliary activity. *Acta Otolaryngol* **94**: 351–359.

International Agency for Research on Cancer (IARC) (1986) Tobacco smoking. In: *Monograph on the Evaluation of the Carcinogenic Risk of Chemicals to Humans.* Lyon, France.

Iravani J (1972) Effects of cigarette smoke on the respiratory epithelium of rats. *Respiration* **29**: 480–487.

Jaspers I, Flescher E and Chen LC (1997) Ozone-induced IL-8 expression and transcription factor binding in respiratory epithelial cells. *Am J Physiol* **272**: L504–L511.

Jeffery PK (1987) Structure and function of adult tracheobronchial epithelium. In: McDowell EM (ed.) *Lung Carcinomas.* London: Churchill Livingstone, pp. 42–73.

Jeffery PK (1987) The origins of secretions in the lower respiratory tract. *Eur J Respir Dis* **71**: 34–42.

Jeffery PK (1990) Microscopic structure of normal lung. In: Brewis RAL, Gibson GJ and Geddes DM (eds) *Textbook of Respiratory Medicine.* London: Baillière Tindall, pp. 57–78.

Jeffery PK (1990) Tobacco smoke-induced lung disease. In: Cohen RD, Lewis B, Alberti KGMM and Denman AM (eds) *The Metabolic and Molecular Basis of Acquired Disease.* London: Baillière Tindall, pp. 466–495.

Jeffery PK (1991) Morphology of the airway wall in asthma and chronic obstructive pulmonary disease. *Am Rev Respir Dis* **143**: 1152–1158.

Jeffery PK (1994) Innervation of the airway mucosa: Structure, function and changes in airway disease. In: Goldie R (ed.) *Immunopharmacology of Epithelial Barriers*, vol. 8 of the Handbook of Immunopharmacology (series ed. C Page). London: Academic Press, pp. 85–118.

Jeffery PK (1995) Microscopic structure of normal lung. In: Brewis RAL, Corrin B, Geddes DM and Gibson GJ (eds) *Respiratory Medicine*, 2nd edn. London: Baillière Tindall, pp. 54–72.

Jeffery PK (1998) Structural and inflammatory changes in COPD: a comparison with asthma. *Thorax* **53**: 129–136.

Jeffery PK and Corrin B (1984) Structural analysis of the respiratory tract. In: Bienenstock J (ed.) *Immunology of the Lung.* New York: McGraw-Hill, pp. 1–27.

Jeffery PK and Reid L (1975) New observations of rat airway epithelium: a quantitative electron microscopic study. *J Anat* **120**: 295–320.

Jeffery PK and Reid L (1977) The respiratory mucous membrane. In: Brain JD, Proctor DF and Reid L (eds) *Respiratory Defence Mechanisms*, vol. 3 of Lung Biology in Health and Disease. New York: Marcel Dekker, pp. 193–246.

Jeffery PK and Reid L (1981) The effect of tobacco smoke with or without phenylmethyloxadiazole (PMO) on rat bronchial epithelium: a light and electron microscopic study. *J Pathol* **133**: 341–359.

Jeffery PK, Rogers DF, Ayers M and Shields PA (1984) Structural aspects of cigarette smoke-induced pulmonary disease. In: Cumming G and Bonsignore G (eds) *Cigarette Smoking and the Lung*, vol. 17, Ettore Majorana Life Science Series. New York: Plenum Press, pp. 1–32.

Jeffery PK, Brain APR, Shields PA *et al.* (1988) Response of laryngeal and tracheo-bronchial surface lining to inhaled cigarette smoke in normal and vitamin A-deficient rats: a scanning electron microscopic study. *Scanning Electron Microsc* **2**: 545–552.

Jeffery PK, Wardlaw A, Nelson FC *et al.* (1989) Bronchial biopsies in asthma: an ultrastructural quantification study and correlation with hyperreactivity. *Am Rev Respir Dis* **140**: 1745–1753.

Jones R, Bolduc P and Reid L (1972) Protection of rat bronchial epithelium against tobacco smoke. *Br Med J* **2**: 142–144.

Jones R, Bolduc P and Reid L (1973) Goblet cell glycoprotein and tracheal gland hypertrophy in rat airways: the effect of tobacco smoke with or without the anti-inflammatory agent phenylmethyloxydiazole. *Br J Exp Pathol* **54**: 229–239.

Jörres R, Nowak D, Grimminger F *et al.* (1992) The effects of 1 ppm nitrogen dioxide on broncho-alveolar lavage cells and bronchial biopsy specimens in normal and asthmatic subjects. *Am Rev Respir Dis* **145**: A456 (abstract).

Kark JD, Smith AH, Switzer BR and Hames CG (1981) Serum, vitamin A (retinol) and cancer incidence in Evans county, Georgia. *J Natl Cancer Inst* **66**: 7–16.

Kaufman DJ, Genta VM and Harris CC (1974) Studies on carcinogen binding in vitro in isolated hamster tracheas. In: Karbe E and Park JS (eds) *Lung Cancer: Carcinogenesis and Bioassays.* Springer-Verlag, pp. 564–574.

Keenan KP (1987) Cell injury and repair of the tracheobronchial epithelium. In: McDowell EM (ed.) *Lung Carcinomas.* Edinburgh: Churchill Livingstone, pp. 74–93.

Kelly FJ, Mudway I, Krishna MT and Holgate ST (1995) The free radical basis of air pollution: focus on ozone. *Respir Med* **89**: 647–656.

Kelly FJ, Blomberg A, Frew A *et al.* (1996) Antioxidant kinetics in lung lavage fluid following exposure of humans to nitrogen dioxide. *Am J Respir Crit Care Med* **154**: 1700–1705.

Kennedy AR and Little JB (1978) Radiation carcinogenesis in the respiratory tract. In: Harris CC (ed.) *Pathogenesis and Therapy of Lung Cancer*, vol. 10. New York: Marcel Dekker, pp. 189–261.

Kew RR, Ghebrehiwet B and Janoff A (1985) Cigarette smoke can activate the alternative pathway of complement in vitro by modifying the third component of complement. *Clin Invest* **75**: 1000–1007.

Kienast K, Knorst M, Muller Quernheim J and Ferlinz R (1996) Modulation of IL-1 beta, IL-6, IL-8, TNF-alpha and TGF-beta secretions by alveolar macrophages under NO_2 exposure. *Lung* **174**: 57–67.

Kimbel P (1985) Proteolytic damage and emphysema pathogenesis. In: Petty TL (ed.) *Chronic Obstructive Pulmonary Disease*, vol. 28. New York: Marcel Dekker, pp. 105–128.

Klass DJ (1980) Cigarette smoke exposure in vivo increases cyclic GMP in rat lung. *Arch Environ Health* **35**: 347–350.

Klienerman J and Boren HG (1974) Morphologic basis of chronic obstructive lung disease. In: Baum GL (ed.) *Textbook of Pulmonary Disease.* Boston, MA: Little, Brown & Company, p. 571.

Kollestrom N, Lord PW and Whimster WF (1977) A difference in the composition of bronchial mucus between smokers and non-smokers. *Thorax* **32**: 155–159.

Koren HS (1995) Associations between criteria air pollutants and asthma. *Environ Health Perspect* **103**: 235–242.

Koren HS, Devlin RB, Graham DE *et al.* (1989) Ozone-induced inflammation in the lower airways of human subjects. *Am Rev Respir Dis* **139**: 407–415.

Kramps JA, Franken C and Dijkman JH (1984) ELISA for quantitative measurement of low-molecular-weight bronchial protease inhibitor in human sputum. *Am Rev Respir Dis* **129**: 959–963.

Krishna MT, Blomberg A, Biscione GU *et al.* (1987) Short-term ozone exposure upregulates P-selectin in normal human airways. *Am J Respir Crit Care Med* **155**: 1798–1803.

Kuschner M (1985) Perspective on pathologic predisposition to lung cancer in humans. In: Masse MJ, Kaufman DJ, Siegel JM *et al.* (eds) *Cancer of the Respiratory Tract: Predisposing Factors*, vol. 8. New York: Raven Press, pp. 17–21.

Labelle CW, Bevilicqua DM and Brieger H (1966) The influence of cigarette smoke on lung clearance. *Arch Environ Health* **12**: 588–596.

Lacoste J-Y, Bousquet J, Chanez P *et al.* (1993) Eosinophilic and neutrophilic inflammation in asthma, chronic bronchitis, and chronic obstructive pulmonary disease. *J Allergy Clin Immunol* **92**: 537–548.

Lamb D and Reid L (1968) Mitotic rates, goblet cell increase and histochemical changes in mucus in rat bronchial epithelium during exposure to SO_2. *J Pathol Bacteriol* **96**: 97–111.

Lang MR, Fiaux GW, Gilooly M *et al.* (1994) Collagen content of alveolar wall tissue in emphysematous and non-emphysematous lungs. *Thorax* **49**: 319–326.

Langford SD, Bidani A and Postlethwait EM (1995) Ozone-reactive absorption by pulmonary epithelial lining fluid constituents. *Toxicol Appl Pharmacol* **132**: 122–130.

Laughter AH, Martin RR and Twomey JJ (1977) Lymphoproliferative responses to antigens mediated by human pulmonary alveolar macrophages. *J Lab Clin Med* **89**: 1322–1326.

Lentz PE and Diluzio NR (1974) Transport of alpha-aminoisobutyric acid by alveolar macrophages incubated with cigarette smoke and nicotine. *Arch Environ Health* **28**: 333–335.

Lethem MI, James SL and Marriott C (1990) The role of mucous glycoproteins in the rheological properties of cystic fibrosis sputum. *Am Rev Respir Dis* **142**: 1053–1058.

Li XY, Donaldson K, Rahman I and MacNee W (1994) An investigation of the role of glutathione in increased epithelial permeability induced by cigarette smoke in vivo and in vitro. *Am J Respir Crit Care Med* **149**: 1518–1525.

Linhartova A, Anderson AE Jr and Foraker AG (1977) Further observations on lumenal deformity and stenosis of non respiratory bronchioles in pulmonary emphysema. *Thorax* **32**: 50–53.

Lofroth G (1989) Environmental tobacco smoke: overview of chemical composition and genotoxic components. *Mutation Res* **222**: 73–80.

Logan WPD (1953) Mortality in the London fog incident. *Lancet* **1**: 336–339.

Lopata M, Barton AD and Lourenco RV (1974) Biochemical characteristics of bronchial secretions in chronic obstructive pulmonary disease. *Am Rev Respir Dis* **110**: 730–739.

Lopez-Vidriero MT and Reid L (1978) Chemical markers of mucus and glycosaminoglycans and their relation to viscosity in mucoid and purulent sputum from various hypersecretory diseases. *Am Rev Respir Dis* **117**: 465–477.

Lopez-Vidriero MT and Reid L (1985) Bronchial mucus in asthma. In: Weiss EB, Segal MS and Stein M (eds) *Bronchial Asthma: Mechanisms and Therapeutics.* Boston, MA: Little, Brown & Company, pp. 218–235.

Low RB (1974) Protein biosynthesis by the pulmonary alveolar macrophage: conditions of assay and the effect of cigarette smoke extracts. *Am Rev Respir Dis* **110**: 466–477.

Lundberg JM and Saria A (1983) Capsaicin-induced desensitization of the airway mucosa to cigarette smoke, mechanical and clinical irritants. *Nature* **302**: 251–253.

Lundberg JM, Lundbland L, Saria A and Anggard A (1984) Inhibition of cigarette smoke-induced oedema in the nasal mucosa by capsaicin pretreament and a substance P antagonist. *Naunym Schmiedebergs Arch Pharmacol* **326**: 181–185.

Lundbland L and Lundberg JM (1984) Capsaicin sensitive sensory neurons mediate the response to nasal irritation induced by the vapour phase of cigarette smoke. *Toxicology* **33**: 1–7.

Macklem PT, Proctor DF and Hogg JC (1970) The stability of peripheral airways. *Resp Physiol* **8**: 191–203.

MacNee W and Selby C (1993) New perspectives on basic mechanisms in lung disease: 2. Neutrophil traffic in the lungs: role of haemodynamics, cell adhesion and deformability. *Thorax* **48**: 79–88.

MacNee W, Wiggs B, Belzberg AS and Hogg JC (1989) The effect of cigarette smoking on neutrophil kinetics in human lungs. *N Engl J Med* **321**: 924–928.

MacNee W, Bridgeman MME, Marsden M *et al.* (1991) The effects of N-acetylcysteine and glutathione on smoke-induced changes in lung phagocytes and epithelial cells. *Am J Med* **91**: 60S–66S.

Magnusson CGM (1986) Maternal smoking influences cord serum IgE and IgD levels and increases the risk for infant allergy. *J Allergy Clin Immunol* **78**: 898–904.

Mannisto J, Kuusisto T, Matintalo M and Uotila P (1981) The activation of prostaglandin E_2 is decreased in rat isolated lungs by cigarette smoke ventilation. *Acta Physiol Scand* **112**(2): 32A.

Masse R, Chameaud J and Lafuma J (1984) Co-carcinogenic effect of tobacco smoke in rats. In: Cumming G and Bonsignore G (eds) *Smoking and the Lung,* vol. 17. New York: Plenum Press, pp. 61–76.

Mathews JD, Hooper BM, Wittingham S *et al.* (1973) Association of autoantibodies with smoking, cardiomuscular morbidity and death in the Busselton population. *Lancet* **ii**: 754–758.

Mattenheimer H and Mohr U (1975) Ciliotoxicity of cigarette smoke and adenylate kinase. *Z Klin Chem Biochem* **13**(7): 325–326.

McDowell EM (1987) Bronchogenic carcinomas of the lung. In: McDowell EM (ed.) *Lung Carcinomas*. London: Churchill Livingstone, pp. 255–285.

McDowell EM and Trump BF (1983) Conceptual review: histogenesis of preneoplastic and neoplastic lesions in tracheobronchial epithelium. *Surv Synth Pathol Res* **2**: 235–279.

McDowell EM, Becci PJ, Barrett LA and Trump CF (1978) Morphogenesis and classification of lung cancer. In: Harris CC (ed.) *Pathogenesis of Therapy of Lung Cancer*, vol. 10. New York: Marcel Dekker, pp. 445–519.

McKinnon KP, Madden MC, Noah TL and Devlin RB (1993) *In vitro* ozone exposure increases release of arachidonic acid products from a human bronchial epithelial cell line. *Toxicol Appl Pharmacol* **118**: 215–223.

Meade PD, Yamashiro S, Harada T and Basrur PK (1979) Influence of vitamin A on the laryngeal response of hamsters exposed to cigarette smoke. *Prog Exp Tumour Res* **24**: 320–329.

Medical Research Council (1965) Definition and classification of chronic bronchitis for clinical and epidemiological purposes. A report to the Medical Research Council by their Committee on the etiology of chronic bronchitis. *Lancet* **i**: 775–780.

Miller FJ, Overton Jr JH, Jaskot RH and Menzel DB (1985) A model of the regional uptake of gaseous pollutants in the lung. I. The sensitivity of the uptake of ozone in the human lung to lower respiratory tract secretions and exercise. *Toxicol Appl Pharmacol* **79**: 11–27.

Miller FJ, Overton JH, Kimbell JS and Russell ML (1992) Regional respiratory tract absorption of inhaled reactive gases. In: Gardner DE, Crapo JD and McClellan RO (eds) *Toxicology of the Lung*. New York: Raven Press, p. 485.

Miller LG, Goldstein G, Murphy M and Ginns LC (1982) Reversible alterations in immunoregulatory T cells in smoking. Analysis by monoclonal antibodies and flow cytometry. *Chest* **82**: 526–529.

Minty BD, Royston D, Jones JG and Hulands GH (1984) The effect of nicotine on pulmonary epithelial permeability. *Chest* **86**: 72–74.

Misokovitch G, Appel J and Szule J (1974) Ultrastructural changes of ciliated columnar epithelium and goblet cells in chronic bronchitis biopsy material. *Acta Morphol Acad Sci Hung* **22**: 91–103.

Mitchell RS, Stanford RE, Johnson JM *et al.* (1976) The morphologic features of the bronchi, bronchioles and alveoli in chronic airway obstruction: a clinicopathologic study. *Am Rev Respir Dis* **114**: 137–145.

Mohr U and Reznik G (1978) Tobacco carcinogenesis. In: Harris CC (ed.) *Pathogenesis and Therapy of Lung Cancer*, vol. 10. New York: Marcel Dekker, pp. 263–367.

Mohsenin V (1994) Human exposure to oxides of nitrogen at ambient and supra-ambient concentrations. *Toxicology* **89**: 301–312.

Moldeus P, Berggren M and Grafstrom R (1985) N-acetylcysteine protects against the toxicity of cigarette smoke and cigarette smoke condensates in venous tissues and cells in vivo. *Eur J Respir Dis* **66**: 123–129.

Mooren HWD, Kramps JA, Franken C *et al.* (1983) Localisation of a low-molecular weight bronchial protease inhibitor in the peripheral human lung. *Thorax* **38**: 180–183.

Morrow PE (1975) An evaluation of recent NO_x toxicity data and an attempt to derive an ambient air standard for NO_x by established toxicological procedures. *Environ Res* **10**: 92–112.

Mossman BT, Cameron GS and Yotti LP (1985) Co-carcinogenic and tumour promoting properties of asbestos and other minerals in tracheo-bronchial epithelium. In: Mass NJ, Kaufman DJ, Siegfried JN *et al.* (eds.) *Cancer of the Respiratory Tract: Predisposing Factors*, vol. 8. New York: Raven Press, pp. 217–238.

Mullen JBM, Wright JL, Wiggs BR *et al.* (1985) Reassessment of inflammation of airways in chronic bronchitis. *Br Med J* **291**: 1235–1239.

Mullen JBM, Wright JL, Wiggs BR *et al.* (1987) Structure of central airways in current smokers and ex-smokers with and without mucus hypersecretion. *Thorax* **42**: 843–846.

Mustafa MG (1990) Biochemical basis of ozone toxicity. *Free Radic Biol Med* **9**: 245–265.

Nakhosteen JA, Lindemann L and Viera J (1982) Mucociliary clearance. Data for smokers, nonsmokers and patients with respiratory disease. *Deutsche Medizinische Wochenschrift* **107**: 1713–1716.

Nauss KM, Busby Jr WJ, Cohen AJ *et al.* (1995) Critical issues in assessing the cardiogenicity of diesel exhaust: a synthesis of current knowledge. In: Health Effects Institute (ed.) *Diesel Exhaust – A Critical Analysis of Emission, Exposure and Health Effects* (a special report on the Institute's Diesel Working Group). Cambridge, MA: Health Effects Institute, pp. 13–18.

Nemery B, Moavero NE, Brasseur L and Stanescu DC (1981) Significance of small airways test in middle-aged smokers. *Am Rev Respir Dis* **124**: 232–238.

Niewoehner DE, Klienerman J and Rice D (1974) Pathologic changes in the peripheral airways of young cigarette smokers. *N Engl J Med* **291**: 755–758.

O'Shaughnessy TC, Ansari TW, Barnes NC and Jeffery PK (1996) Inflammatory cells in the airway surface epithelium of smokers with and without bronchitic airflow obstruction. *Eur Respir J* **9** (Suppl. 23): 14s.

O'Shaughnessy TC, Ansari TW, Barnes NC and Jeffery PK (1997) Inflammation in bronchial biopsies of subjects with chronic bronchitis: inverse relationship of CD8+ T lymphocytes with FEV₁. *Am J Respir Crit Care Med* **155**: 852–857.

Oberdorster G, Finkelstein J, Ferin J *et al.* (1996) Ultrafine particles as a potential environmental health hazard. Studies with model particles. *Chest* **109**: 68.

Ollerenshaw SL and Woolcock AJ (1992) Characteristics of the inflammation in biopsies from large airways of subjects with asthma and subjects with chronic airflow limitation. *Am Rev Respir Dis* **145**: 922–927.

Overton JH (1984) Physiochemical processes and the formulation of dosimetry models: fundamentals of extrapolation modelling of inhaled toxicant, ozone and nitrogen dioxide. In: Miller FJ and Menzel DB (eds) *Fundamentals of Extrapolation Modelling of Inhaled Toxicants, Ozone and Nitrogen Dioxide*. Washington: Hemisphere.

Overton JH, Graham RC and Miller FJ (1987) A model of the regional uptake of gaseous pollutants in the lung. II. The sensitivity of ozone uptake in laboratory animal lungs to anatomical and ventilatory parameters. *Toxicol Appl Pharmacol* **88**: 418–432.

Pacht ER, Kaseki H, Mohammed JR *et al.* (1986) Deficiency of vitamin E in the alveolar fluid of cigarette smokers. *J Clin Invest* **77**: 789–796.

Patel JM and Block ER (1986) Nitrogen dioxide-induced changes in cell membrane fluidity and function. *Am Rev Respir Dis* **134**: 1196–1202.

Pavia D, Thomson ML and Pocock SJ (1971) Evidence for temporary slowing of mucociliary clearance in the lung caused by tobacco smoking. *Nature* **231**: 325–326.

Peatfield AC, Davies JR and Richardson PS (1986) The effect of tobacco smoke upon airway secretion in the cat. *Clin Sci* **21**: 179–187.

Peto R, Doll R, Buckley JD and Sporn MB (1981) Can dietary B-keratin materially reduce human cancer rates? *Nature* **290**: 201–208.

Peto R, Speitzer FE, Cochrane AL *et al.* (1983) The relevance in adults of airflow obstruction, but not of mucus hypersecretion, to mortality from chronic lung disease. *Am Rev Respir Dis* **128**: 491–500.

Peto R, Lopez AD, Boreham J *et al.* (1992) Mortality from tobacco in developed countries: indirect estimation from national vital statistics. *Lancet* **339**: 1268–1279.

Pettersson B, Curvall M and Enzell CR (1982) Effects of tobacco smoke compounds on the ciliary activity of the embryo chicken trachea in vitro. *Toxicology* **23**: 41–55.

Pinkerton KE, Brody AR, Miller FJ and Crapo JD (1989) Exposure to low levels of ozone results in enhanced pulmonary retention of inhaled asbestos fibers. *Am Rev Respir Dis* **140**: 1075–1081.

Pope CA, Dockery DW, Spengler JD and Raizenne ME (1991) Respiratory health and PM₁₀ pollution. A daily time series analysis. *Am Rev Respir Dis* **144**: 668–674.

Pope CA, Bates DV and Raizenne ME (1995) Health effects of particulate air pollution: time for reassessment? *Environ Health Perspect* **103**: 472–480.

Postlethwait EM and Bidani A (1994) Mechanisms of pulmonary NO₂ absorption. *Toxicology* **89**: 217–237.

Proctor CJ and Smith G (1989) Considerations of the chemical complexity of ETS with regard to inhalation studies. *Exp Pathol* **37**: 164–169.

Proctor PH and Reynolds ES (1984) Free radicals and disease in man. *Physiol Chem Phys Med NMR* **16**: 175–195.

Pryor WA (1992) How far does ozone penetrate into the pulmonary air/tissue boundary before it reacts? *Free Radic Biol Med* **12**: 83–88.

Queval P, Beaumatin J, Morin M *et al.* (1979) Inducibility of microsomal enzymes in normal and precancerous lung tissue. Synergistic action of 5-6 benzoflavone or methyl-cholanthrene in radiation induced carcinogenesis. *Biomedicine* **31**: 182–186.

Rahman I (1996a) Role of oxidants and antioxidants in smoking-induced lung diseases. *Free Radic Biol Med* **21**: 669–681.

Rahman I (1996b) Systemic oxidative stress in asthma, COPD and smokers. *Am J Respir Crit Care Med* **154**: 1055–1060.

Rahman I and MacNee W (1994) Glutathione and its redox system in alveolar epithelial cells in response to oxidative stress. *Am J Respir Crit Care Med* **149**: A457.

Rahman I, Smith CA, Lawson M *et al.* (1996) Induction of gamma-glutamylcysteine synthetase by cigarette smoke is associated with AP-1 in human alveolar epithelial cells. *Febs Lett* **396**: 21–25.

Rees PJ, Chowienczyk PJ and Clark TJH (1982) Immediate response to cigarette smoke. *Thorax* **37**: 417–422.

Reid L (1954) Pathology of chronic bronchitis. *Lancet* **1**: 275–279.

Rennard SI, Ghafouri M, Thompson AB *et al.* (1990) Fractional processing of sequential bronchoalveolar lavage to separate bronchial and alveolar samples. *Am Rev Respir Dis* **141**: 208–217.

Richardson PS (1984) The reflex effects of cigarette smoking. In: Cumming G and Bonsignore G (eds) *Smoking and the Lung*, vol. 17. New York: Plenum, Ettore Majorana, International Science Series, pp. 33–46.

Robinson DS, Hamid Q, Ying S *et al.* (1992) Predominant TH2-like bronchoalveolar T-lymphocyte population in atopic asthma. *N Engl J Med* **326**: 298–304.

Robinson DS, Tsicopoulos A, Meng Q *et al.* (1996) Increased interleukin-10 messenger RNA expression in atopic allergy and asthma. *Am J Respir Cell Mol Biol* **14**: 113–117.

Rogers DF and Jeffery PK (1986) Inhibition by oral *N*-acetylcysteine of cigarette smoke-induced 'bronchitis' in the rat. *Exp Lung Res* **10**: 267–283.

Rogers DF, Williams DA and Jeffery PK (1986) Nicotine does not cause 'bronchitis' in the rat. *Clin Sci* **70**: 427–433.

Rogers DF, Turner NC, Marriott C and Jeffery PK (1987) Cigarette smoke-induced 'chronic bronchitis': a study in situ of laryngo-tracheal hypersecretion in the rat. *Clin Sci* **72**: 629–637.

Rogers DF, Aursudkij B and Barnes PJ (1989) Effect of tachykinins on mucus secretion in human bronchi in vitro. *Eur J Pharmacol* **174**: 283–286.

Rostu UM, Kodavanti PRS and Mehendale HM (1988) Glutathione metabolism and utilisation of external thiols by cigarette smoke-challenged, isolated rat and rabbit lungs. *Toxicol Appl Pharmacol* **96**: 324–335.

Royal College of Physicians Report (1977) *Smoking or Health?* The Third Report. London: Pitman Medical Publishing.

Rubenstein I, Reiss TF, Bigby BG *et al.* (1991) Effects of 0.60 ppm nitrogen dioxide on circulating and bronchoalveolar lavage lymphocyte phenotypes in healthy subjects. *Environ Res* **55**: 18–30.

Rusznak C, Devalia JL, Herdman MJ and Davies RJ (1994) Effect of six hours exposure to 400 ppb nitrogen dioxide (NO_2) and/or 200 ppb sulphur dioxide (SO_2) on inhaled allergen response in mild asthmatic subjects. *Clin Exp Allergy* **4**: 186 (abstract).

Rylander R (1971) Lung clearance of particles and bacteria. *Arch Environ Health* **23**: 321–326.

Rylander R (1973) Toxicity of cigarette smoke components: free lung cell response in acute exposures. *Am Rev Respir Dis* **108**: 1279–1282.

Rylander R (1974) Pulmonary cell responses to inhaled cigarette smoke. *Arch Environ Health* **29**: 329–333.

Saetta M (1998) CD8+ T-lymphocytes in peripheral airways of smokers with chronic obstructive pulmonary disease. *Am J Respir Crit Care Med* **157**: 822–826.

Saetta M, Ghezzo H, Wong Dong Kim *et al.* (1985) Loss of alveolar attachments in smokers. A morphometric correlate of lung function impairment. *Am Rev Respir Dis* **132**: 894–900.

Saetta M, Di Stefano A, Maestrelli P *et al.* (1993) Activated T-lymphocytes and macrophages in bronchial mucosa of subjects with chronic bronchitis. *Am Rev Respir Dis* **147**: 301–306.

Saetta M, Di Stefano A, Maestrelli P *et al.* (1994) Airway eosinophilia in chronic bronchitis during exacerbations. *Am J Respir Crit Care Med* **150**: 1646–1652.

Saetta M, Finkelstein R and Cosio MG (1994) Morphological and cellular basis for airflow limitation in smokers. *Eur Respir J* **7**: 1505–1515.

Saetta M, Turato G, Facchini FM *et al.* (1997) Inflammatory cells in the bronchial glands of smokers with chronic bronchitis. *Am J Respir Crit Care Med* **156**: 1633–1639.

Saffiotti U, Stinson SF, Keenan KP and McDowell EM (1985) Tumour enhancement factors and mechanisms in the hamster respiratory tract carcinogenesis model. In: Masse MJ, Kaufman DJ, Siegel JM *et al.* (eds) *Cancer of the Respiratory Tract: Predisposing Factors*, vol. 8. New York: Raven Press, pp. 63–91.

Sandstrom T (1995) Respiratory effects of air pollutants: experimental studies in humans. *Eur Respir J* **8**: 976–995.

Sandstrom T, Andersson MC, Kolmodin-Hedman B *et al.* (1990) Bronchoalveolar mastocytosis and lymphocytes after nitrogen dioxide exposure in man: a time-kinetic study. *Eur Respir J* **3**: 138–143.

Sandstrom T, Stjernberg N, Eklund A *et al.* (1991) Inflammatory cell response in bronchoalveolar lavage fluid after nitrogen dioxide exposure of healthy subjects: a dose–response study. *Eur Respir J* **4**: 332–339.

Sandstrom T, Helleday R, Bjermer L and Stjernberg N (1992a) Effects of repeated exposure to 4 ppm nitrogen dioxide on bronchoalveolar lymphocyte subsets and macrophages in healthy men. *Eur Respir J* **5**: 1092–1096.

Sandstrom T, Ledin MC, Thomasson L *et al.* (1992b) Reductions in lymphocyte subpopulations after repeated exposure to 1.5 ppm nitrogen dioxide. *Br J Ind Med* **49**: 850–854.

Sant'Ambrogio G (1982) Information arising from the tracheobronchial tree of mammals. *Physiol Rev* **62**: 531–569.

Scannell C, Chen L, Aris RM *et al.* (1996) Greater ozone-induced inflammatory responses in subjects with asthma. *Am J Respir Crit Care Med* **154**: 24–29.

Scheepers PT and Bos RP (1992) Combustion of diesel fuel from a toxicological perspective. I. Origin of incomplete combustion products. *Int Arch Occup Environ Health* **64**: 149–161.

Schwartz J and Dockery DW (1992) Increased mortality in Philadelphia associated with daily air pollution concentrations. *Am Rev Respir Dis* **145**: 600–604.

Schwartz W, Dungworth DL, Mustafa MG *et al.* (1976) Pulmonary responses of rats to ambient levels of ozone: effects of 7-day intermittent or continuous exposure. *Lab Invest* **34**: 565–578.

Schwela D (1996) Exposure to environmental chemicals relevant for respiratory hypersensitivity: global aspects. *Toxicol Lett* **86**: 131–142.

Seltzer J, Bigby BG, Stulbarg M *et al.* (1986) O_3-induced change in bronchial reactivity to methacholine and airway inflammation in humans. *J Appl Physiol* **60**: 1321–1326.

Sharma RK and Jeffery PK (1987) The effects of cigarette smoke on autonomic receptor density in rat lung. *Clin Sci* **73**: 36–37P.

Shekelle RB, Lepper M, Lui S *et al.* (1981) Dietary vitamin A and the risk of cancer in the western electric study. *Lancet* **ii**: 1185–1189.

Silverman NA, Potvin C, Alexander JC and Chretien PB (1975) In vitro lymphocyte reactivity and T cell levels in chronic cigarette smokers. *Clin Exp Immunol* **22**: 285–292.

Sivertsen B and Clench-Aas J (1996) Exposure to environmental chemicals relevant for respiratory hypersensitivity: European aspects. *Toxicol Lett* **86**: 143–153.

Slade R, Crissman K, Norwood J and Hatch G (1993) Comparison of antioxidant substances in bronchoalveolar lavage cells and fluid from humans, guinea pigs and rats. *Exp Lung Res* **19**: 469–484.

Smith G, Wilton LV and Binns R (1978) Sequential changes in the structure of the rat respiratory system during and after exposure to tobacco smoke. *Toxicol Appl Pharmacol* **46**: 579–591.

Snider GL (1986) Chronic obstructive pulmonary disease – a continuing challenge. *Am Rev Respir Dis* **133**: 942–944.

Snider GL, Kleinerman J and Thurlbeck WM (1985) The definition of emphysema. Report of a National Heart and Blood Institute, division of lung diseases, Workshop. *Am Rev Respir Dis* **132**: 182–185.

Spencer H (1985) Carcinoma of the lung. In: *Pathology of the Lung*, 4th edn. Oxford: Pergamon Press, pp. 872–884.

Stanley PJ, Wilson R, Greenstone MA *et al.* (1986) Effect of cigarette smoking on nasal mucociliary clearance and ciliary beat frequency. *Thorax* **41**: 519–523.

Stedman RL (1968) The chemical composition of tobacco and tobacco smoke. *Chem Reviews* **68**: 153–207.

Stephens RJ, Freeman G and Evans MJ (1972) Early response of lungs to low levels of nitrogen dioxide. Light and electron microscopy. *Arch Environ Health* **24**: 160–179.

Stripp BR, Lund J, Mango GW *et al.* (1996) Clara cell secretory protein: a determinant of PCB bioaccumulation in mammals. *Am J Physiol* **271**: L656–L664.

Stupfel M (1974) Penetration of pollutants in the airways. *Bull Eur Physiopathol Respir* **10**: 481–509.

Sturgess J and Reid L (1972) The organ culture study of the effects of drugs on the secretory activity of the human bronchial submucosal gland. *Clin Sci* **43**: 533–543.

Takano H, Yoshikawa T, Ichinose T *et al.* (1997) Diesel exhaust particles enhance antigen-induced airway inflammation and local cytokine expression in mice. *Am J Respir Crit Care Med* **156**: 36–42.

Thomas HV, Mueller PK and Wright R (1967) Response of rat lung mast cells to nitrogen dioxide inhalation. *J Air Pollut Control Assoc* **17**: 33–35.

Thompson AB, Daughton D, Robbins RA *et al.* (1989) Intraluminal airway inflammation in chronic bronchitis. Characterization and correlation with clinical parameters. *Am Rev Respir Dis* **140**: 1527–1537.

Thurlbeck WM (1977) Aspects of chronic airflow obstruction. *Chest* **72**: 341–349.

Thurlbeck WM (1985) Chronic airflow obstruction. Correlation of structure and function. In: Petty TL (ed.) *Chronic Obstructive Pulmonary Disease*, 2nd edn. New York: Marcel Dekker, pp. 129–203.

Toremalm NH (1960) The daily amount of tracheobronchial secretions in man: a method for continuous tracheal aspiration in largyngectomized and tracheostimized patients. *Acta Otolaryngol* **158**: 43–53.

Vignola AM, Campbell AM, Chanez P *et al.* (1993) HLA-DR and ICAM-1 expression on bronchial epithelial cells in asthma and chronic bronchitis. *Am Rev Respir Dis* **148**: 689–694.

Von Mutius E, Fritzsch C, Weiland SK *et al.* (1992) Prevalence of asthma and allergic disorders among children in united Germany: a descriptive comparison. *Br Med J* **305**: 1395–1399.

Wald N, Idle N, Borum J and Bailey A (1980) Low serum vitamin A and subsequent risk of cancer. *Lancet* **ii**: 813–815.

Walker DC, MacKenzie A, Wiggs BR *et al.* (1984) The structure of tight junctions may not correlate with permeability. *Cell Tissue Res* **235**: 607–613.

Walters S, Griffiths RK and Ayres JG (1994) Temporal association between hospital admissions for asthma in Birmingham and ambient levels of sulphur dioxide and smoke. *Thorax* **49**: 133–140.

Wanner A (1977) Clinical aspects of muco-ciliary transport. *Am Rev Respir Dis* **116**: 73–125.

Warr GA (1979) The biology of normal human bronchoalveolar cells. *Bull Eur Physiopathol Respir* **15**: 23–34.

Warr GA and Martin RR (1977) Immune receptors of human alveolar macrophages: comparison between cigarette smokers and non-smokers. *J Reticulo Endothelial Soc* **22**: 181–187.

Watt M, Godden D, Cherrie J and Seaton A (1995) Individual exposure to particulate air pollution and its relevance to thresholds for health effects: a study of traffic wardens. *Occup Environ Med* **52**: 790–792.

Welsh MJ (1983) Cigarette smoke inhibition of ion transport in canine tracheal epithelium. *J Clin Invest* **71**: 1614–1623.

Wennmalm A (1977) Nicotine stimulates prostaglandin formation in the rabbit heart. *Br J Pharmacol* **59**: 91–100.

Wilson R, Pitt T, Taylor G *et al.* (1987) Pyocyanin and 1-hydroxyphenazine produced by *Pseudomonas aeruginosa* inhibit the beating of human respiratory cilia *in vitro*. *J Clin Invest* **79**: 221–229.

Wright JL, Hobson JE, Wiggs B *et al.* (1988) Airway inflammation and peribronchiolar attachments in the lungs of nonsmokers, current and ex-smokers. *Lung* **166**: 277–286.

Wright RR and Stuart CM (1965) Chronic bronchitis with emphysema: a pathological study of the bronchi. *Medicina Thoracalis* **22**: 210.

Yeates DB, Aspin H, Levison H *et al.* (1975) Mucociliary tracheal transport rates in man. *Appl Physiol* **39**: 487–495.

Yeung MC and De Buncio A (1984) Leukocyte count, smoking and lung function. *Am J Med* **76**: 31–37.

Zweidinger RB (1982) Emission factors from diesel and gasoline powered vehicles: correlation with Ames test. In: Lewtas J (ed.) *Toxicological Effects of Emissions from Diesel Engines*. Amsterdam: Elsevier, pp. 83–96.

13

Structure–Function Relationships

PHILIP A. BROMBERG

Center for Environmental Medicine and Lung Biology, University of North Carolina, Chapel Hill, NC, USA

INTRODUCTION

This chapter is based on the notion that most inhaled gaseous and particulate toxicants are absorbed in or deposit on the surface of the airways and alveoli. The structures comprising this surface are therefore the site of initial host–pollutant interactions and are accessible to sampling in human subjects. The anatomy and physiology of respiratory surfaces which constitute the interface between inhaled air and the organism is the principal topic of this chapter.

The dimensions and branching pattern of the airways and larynx are of course determinants of the anatomic dead space and the fluid mechanics of inspiratory and expiratory air flow. These geometrical features also determine the axial distribution of air flow impedance, regional surface area and surface/volume ratios, regional transit times, and deposition of inhaled particles by impaction, sedimentation and Brownian diffusion during inspiration, breathholding and expiration. A number of models of normal human airways and lung parenchyma have been proposed based on measurements of casts and of fixed, inflated lungs (which therefore neglect the effects of lung disease and of dynamic fluctuations in airways caliber). These have been widely used to predict inhaled particle deposition and the axial dosimetry (e.g., per unit surface area) of inhaled soluble or reactive gases. A number of reviews are available, providing functional as well as geometrical information deduced from such models (Chang, 1989; Darquenne and Paiva, 1998; International Commission of Radiological Protection, 1994; Robertson, 1998; Ultman, 1988; Weibel, 1963; Weibel 1989) and comparing experimental animals with humans (McBride, 1992; Mercer and Crapo, 1992). These topics will not be specifically addressed in this chapter.

AIR POLLUTION AND HEALTH
ISBN 0-12-352335-4

EPITHELIAL FEATURES

Cartilaginous Airways (Trachea and Bronchi)

The epithelium of cartilaginous airways (and also the nasal respiratory epithelium) is of the pseudostratified columnar type with a preponderance of ciliated cells (also bearing microvilli) and about 20% goblet (mucin-producing) cells. Small numbers of other cell types are also present but serous cells are rare in adult humans (Basbaum *et al.*, 1990). The pseudostratification of the nuclear profiles is due to the presence of flattened, pyramidal, undifferentiated, cytokeratin-containing basal cells attached to the basement membrane, but not reaching the airway lumen. These cells are progenitor cells for regrowth of injured epithelium. However, secretory cells also appear to be able to give rise to ciliated cells either by dedifferentiation, division and redifferentiation of daughter cells, or perhaps by non-mitotic transdifferentiation (Basbaum *et al.*, 1990; Robbins and Rennard, 1997). Paracrine communication between surface epithelial cells and the basal epithelial cells is probably required for coordinated mucosal responses to injury of surface cells by inhaled reactive toxicants. Chronic injury of the epithelium (e.g. cigarette smoking) can produce various types of morphologic alterations including increased goblet cell numbers, squamoid metaplasia and dysplasia, and malignant transformation.

The epithelium is firmly attached to underlying basement membrane by specific ligands and receptors and the cells are attached to one another near their apical margins by a complex junctional apparatus and to basal cells by desmosomes. Nevertheless, in the presence of mucosal edema aqueous spaces appear between the basolateral surfaces of adjacent cells. The basement membrane appears to be porous to proteins as well as to small molecules and ions. Sensory nerve fibers penetrate the basement membrane and terminate between the basolateral aspects of adjacent epithelial cells. Epithelial cells come into close contact with surface macrophages at their apical surface and are very close to normal mucosal residents such as mast cells, neuroendocrine cells, dendritic cells, lymphocytes, macrophages and fibroblasts on their basal aspect. Some of these cells actually appear to be intraepithelial. In species with well-developed foci of mucosal follicular mononuclear and lymphoid aggregates (BALT) the overlying epithelium is flattened and nonciliated (Bienenstock, 1984). The importance of BALT in human airways mucosa is not clear, however.

Bronchioles

In the more distal airways, the epithelium is thinner and more cuboidal, while basal cells disappear. Mucus-secreting cells are replaced by another specialized secretory type, the Clara cell (see reviews by Plopper, 1993; Plopper *et al.*, 1997). This cell is particularly vulnerable to certain toxicologic insults (4-ipomeanol, naphthalene), perhaps because of its high level of cytochrome P-450 activity (Boyd, 1982). Clara cells also produce and secrete specific small proteins (e.g. CC10) whose functions are still not well understood but may be associated with binding of certain xenobiotics (Stripp *et al.*, 1996), as well as certain surfactant-associated proteins and secretory leukoprotease inhibitor. Since mucins appear to be absent in the normal bronchiolar epithelium and submucosal glands are absent in these airways, there is no mucus gel layer floating on the periciliary liquid. The Clara cells also appear to serve as progenitor cells for repair of damaged small airways epithelium.

Alveolar Regions

The cuboidal, ciliated and Clara cell-dominated epithelium of the terminal bronchioles transitions to flat, thin, alveolar type I cells in the respiratory bronchioles which become increasingly alveolated as they approach the alveolar ducts. A rapid expansion of both volume and surface area occurs in this region with surface/volume ratio increasing as the diameter of successive generations decreases. Most (~90%) of the gas-exchange surface (~75 m^2) is covered by very thin type I alveolar cells (average surface area per cell ~5000 μm^2) (Schneeberger, 1997). In addition, there are even more numerous cuboidal type II alveolar cells which cover only 10% of the surface but are specialized for production of the lipid and protein components of surfactant, and also serve as progenitor cells. Variable numbers of alveolar macrophages (one or more per alveolus) are present on the surface and are obtainable in large numbers by bronchoalveolar lavage. This region of the lung subserves the all-important function of diffusive gas exchange. The surface liquid layer is extremely thin and covered with a layer of phospholipid surfactant. The total path length for diffusion of gas from alveolar air to capillary blood is only 0.1–1 μm.

In the fetus the epithelium secretes chloride, accompanied passively by sodium and water, and maintains fluid-filled airway lumens that are continuous with the amniotic fluid. Peri- and post-natally, however, ion transport is dominated by active sodium resorption from the alveolar surface (with passive chloride and water flux following sodium) (Boucher, 1994). The alveolar surfaces therefore have very little aqueous surface liquid. Physiologic saline instilled into air spaces (e.g. during bronchoalveolar lavage) is rapidly cleared by active absorption.

Epithelial Cell Polarity

Throughout the airspaces, the respiratory epithelium is characterized by so-called tight junctions near the apical margin of adjoining cells. These structures, whose molecular anatomy is beginning to yield its secrets (Mitic and Anderson, 1998), impose a significant barrier to solute movement between the surface liquid and the intercellular space which communicates with the submucosa across a relatively porous basement membrane.[*] In addition, the tight junctional apparatus effectively separates the apical and basolateral membranes of each epithelial cell, thus allowing for selective insertion of various receptors and channels that confer the polarity underlying vectorial transcellular transport and selective secretion of epithelial cell products to one or the other face of the epithelium.

---

[*] Nevertheless, there appear to be some communication pathways that allow submucosal exudative fluid including molecules up to about 200 kDa to non-selectively reach the airway surface (Persson, 1994). The normal concentration of these proteins in airway surface liquid has been estimated to be about 10% of their serum levels (Davis and Pacht, 1997). Submucosal edema associated with increased permeability of the post-capillary venules in the airways lamina propria can rapidly add proteins like complement, α-1-antiprotease, albumin and kininogen to the airway surface liquid (Persson, 1994). Furthermore, inflammatory cells can traverse the intercellular junctions thus reaching the lumenal surface.

SURFACE LIQUID

Airways

The apical surfaces of the epithelial cells which line the lumen of the airways and air spaces are covered by a continuous layer of extracellular airways surface liquid (ASL) which contains about 93% water (Boucher, 1994). In the cartilaginous airways the ASL consists of a periciliary 'sol' layer and a more lumenally located mucus 'gel' layer. The thickness of the sol layer is appropriate to the height of the cilia (~6 µm in large airways; ~4 µm in small airways) so that during their power stroke the cilia contact the undersurface of the gel layer. Mucin is produced and stored in condensed form as membrane-enclosed granules in the apical portion of specialized secretory cells in bronchial epithelia and in the mucous glands (Jeffery, 1994). Mucin is a linear protein with a high serine and threonine content that is heavily glycosylated (Boat *et al.*, 1994). Of the multiple mucin protein genes that have been cloned, the MUC 5 gene is prominent in respiratory epithelia, but other MUC genes are also expressed. The oligosaccharide side chains are covalently attached in the Golgi apparatus to the many serine and threonine hydroxyl residues of the protein backbone by *O*-glycosidic linkage with UDP-activated *N*-acetylgalactosamine, followed by sequential addition of other carbohydrates (*l*-fucose, *d*-galactose and *N*-acetylglucosamine) and often terminate in sulfate or sialic acid residues which confer a high density of fixed negative charge and are responsible for Alcian blue staining. Unlike mannose-rich serum glycoproteins and uronic acid-rich proteoglycans, mucins contain no mannose or uronic acid. Proteoglycans are secreted by serous cells, especially in inflammatory states, and also are found in the glycocalyx that coats the surface of epithelial cells. Certain mucin carbohydrate residues may bind bacteria (e.g. *S. aureus*: Trivier *et al.*, 1997; *Pseudomonas aeruginosa*: Baltimore *et al.*, 1989; Carnoy *et al.*, 1993; Ramphal *et al.*, 1991; Sajjan *et al.*, 1992), or virus (e.g. influenza: Higa *et al.*, 1985), thus immobilizing the organisms and ensuring their clearance via the mucociliary escalator (Leigh, 1998). Disruption of mucociliary clearance, however, might allow such bacteria to proliferate, adhere to receptors on epithelial cell surfaces, and finally cause infection (Widdicombe, 1995a).

A thin osmiophilic, surfactant-like layer has also been demonstrated by electron microscopy in appropriately fixed specimens (Gehr *et al.*, 1996). This appears to lie between the gel and sol layers in large airways and on the surface of the periciliary fluid in smaller airways. This layer may be important in determining the fate of particles deposited on small airways and also in facilitating the movement of the mucus gel in larger airways, especially during cough.

Alveoli

In the alveoli and small bronchioles surfactant coats the air–liquid interface. Surfactant lowers air/liquid surface tension (especially as surface area shrinks) which is essential for maintenance of normal lung compliance and avoidance of alveolar and small airways collapse. The kinetic surface-active properties (hysteresis) of surfactant appear to vary along the airways axis so that the inflation–deflation cycle generates surface pressure gradients which could move material from alveoli onto bronchiolar surfaces (Gehr *et al.*, 1996). Surfactant is synthesized by alveolar type II cells and stored as intracellular lamellar bodies prior to secretion. It consists of dipalmitoyl phosphatidyl choline and other phospholipids in association with apoproteins (SP-A, SP-B, SP-C) that have been characterized and whose genes have been cloned (Mason and Shannon, 1997). SP-B (Whitsett *et al.*, 1995) and SP-C are hydrophobic proteins that are essential for normal surfactant spreading and surface-tension lowering function (Johansson and Curstedt, 1997). SP-A, however, may not be vital to surfactant function (Ikegami *et al.*, 1997; Korfhagen *et al.*, 1996; McCormack, 1997) and is also synthesized by submucous glands in airways. SP-A receptors on macrophages support surfactant uptake into membrane-enclosed vacuoles in which enzymatic destruction occurs. SP-A can also serve as an opsonizing agent for particulate matter and microbes (Kabha *et al.*, 1997; McCormack *et al.*, 1997; Pasula *et al.*, 1997) and SP-A-deficient mice are susceptible to experimental infection with various organisms (LeVine *et al.*, 1997).

The location of surfactant makes it a target for inhaled reactive gases. Fortunately, the lipid components are rich in saturated fatty acid residues which are much less sensitive to oxidative attack than unsaturated fatty acids (UFA) and polyunsaturated fatty acids (PUFA). However, the protein components, and particularly SP-A (or SP-A receptors on cells) may be vulnerable, and this could lead to impaired macrophage defense against microbes without major effects on lung compliance.

Glandular Secretion

In human cartilaginous airways, there is a substantial submucosal component of multi-acinar secretory glands (which is even more prominent in patients with chronic airways inflammation). These are of mixed cell composition (serous cells in the most distal acinar region, mucous cells more proximally, columnar cells in the collecting ducts, ciliated cells in the terminal ducts leading to the airway surface). The gland lumens can be emptied by contraction of peri-acinar myoepithelial cells under neural control. Gland secretion is under neurochemical control and includes a variety of proteins such as lysozyme, lactoferrin, SP-A, sIgA, secretory leukoprotease inhibitor (SLPI) and proteoglycans (all from the serous cells) (Basbaum *et al.*, 1990), and mucins from the mucous cells, as well as small molecules like urate and lactate. The collecting duct cells have the capacity to reabsorb electrolytes and (as in sweat gland ducts) may be relatively water impermeable, leading to hypotonicity of the secretion as it emerges onto the airway surface (Knowles *et al.,* 1997). These glandular secretions contribute significantly to the surface liquid in larger airways, but glands are absent in non-cartilaginous airways (bronchioles).

Antioxidants

Since a number of inhaled toxicants are thought to function as oxidants, the antioxidant resources of the ASL are of interest. The analytical values cited by Hatch (1992) are not corrected for dilution and the bronchoalveolar lavage (BAL) procedure samples alveoli as well as medium size bronchi and small airways. Assuming a 1:100 dilution for BAL, his values for protein, ascorbate, urate and glutathione (GSH) were 810, 1.5, 1.5 and 21 (mg/100 ml), respectively. (Nasal lavage was much richer in urate and much poorer in GSH.) For GSH, this BAL value is equal to about 0.7 mmol/l, which is not too different from the range of 0.3–0.45 mmol/l cited by Davis and Pacht (1997) in what they term 'alveolar epithelial lining fluid' or 'ELF'. For protein the value of 8 mg/ml derived from Hatch is similar to the 7 mg/ml for ELF cited by Davis and Pacht (of which 50% is albumin). The concentration of ascorbate (about 80 μmol/l) is similar to that of urate in human ASL, but is considerably higher in guinea pig and (especially) rat lung lavage. Since both urate and ascorbate react rapidly and preferentially with ozone (O_3) *in vitro* (Cross *et al.*, 1992), and since urate can be replaced by gland secretion while dehydro-ascorbate (by analogy with other cells) (Rumsey and Levine, 1998) probably can be taken up and reduced to ascorbate intracellularly and then resecreted by epithelial cells, these reactive and renewable small molecules may contribute significantly to the uptake and buffering of inhaled ozone. Indeed, using a bolus technique Ultman's group showed that inhalation of O_3 through the (urate-rich) nose (Peden *et al.*, 1991) produced much more upper airway uptake than did similar inhalation through the mouth (Kabel *et al.*, 1994).

A number of serum-derived (albumin, ceruloplasmin, transferrin) and gland secretion-derived (lactoferrin) proteins in the ASL may have antioxidant or anti-Fenton chemistry functions. In addition, extracellular forms of Cu–Zn superoxide dismutase (SOD) and of catalase have been described (Davis and Pacht, 1997).

Sampling

The technique of BAL and its variants allows one to obtain variably diluted samples of airways surface liquid for analysis. Other techniques (e.g. absorption of liquid by filter paper wicks positioned bronchoscopically: Knowles *et al.*, 1997) provide direct samples but this is limited to large airways in which reflex stimulation of bronchial gland secretion by the procedure may dilute the fluid and alter its composition. Bronchoscopy also provides access to the trachea and bronchi where one may safely obtain epithelium by brushing and mucosal biopsies with forceps (Ghio *et al.*, 1998).

CILIARY–SURFACE LIQUID INTERACTION AND PARTICLE CLEARANCE

Clearance Kinetics

Insoluble particles deposited on ciliated airway surfaces are cleared relatively rapidly by the mucociliary escalator with a half-time less than 6 h. Remaining particles after 24 h have classically been considered to reside in alveoli where they have been rapidly phagocytosed by resident and newly recruited alveolar macrophages which clear very slowly, with a half-life of about 100–300 days (see reviews by Lippmann and Schlesinger, 1984; Raabe, 1982; Schlesinger, 1995; Snipes, 1995). In addition, deposited particles can enter epithelial cells and even penetrate to the submucosa, particularly under exposure conditions leading to particle 'overload' (Ferin *et al.*, 1994; Mermelstein *et al.*, 1994;

Oberdörster *et al.*, 1994). If particle number (rather than mass) is a critical feature in determining 'overload', then for a given inhaled mass, smaller particles would be more likely to escape normal clearance mechanisms and penetrate the epithelium (Churg, 1996; Oberdörster *et al.*, 1994).

Newer work suggests that certain insoluble particles deposited on small ciliated airways (bronchioles) exhibit an intermediate clearance rate with a half-time of the order of 10–20 days (Anderson *et al.*, 1995; Bennett *et al.*, 1998; Scheuch *et al.*, 1996). Since the bronchioles lack a mucin gel layer, it is possible that for given aerodynamic properties (which depend on density as well as diameter) small, dense particles settle through the surfactant-coated surface liquid layer to the epithelial surface where they are engulfed by airway macrophages and are not cleared by the cilia, whereas larger, less dense particles remain on the surface where they can be moved relatively rapidly in a cephalad direction and cleared by ciliary beating. Particles in the former class, if they caused airway surface macrophages to become activated, could provoke an inflammatory reaction by secreted macrophage cytokines such as TNF-α.

Cilia

The coordinated beating of the 200 or so cilia on the surface of each epithelial cell, at a frequency of about 10 Hz, within the ASL results in orad movement of the surface liquid at linear velocities of up to 10 mm/min in the trachea, but lower velocities in more distal airways. The cephalad 'power stroke' of each extended cilium in a plane perpendicular to the surface (the recovery stroke involves bending of the cilium so that it moves through the sol layer parallel to the surface) interacts with the viscoelastic gel layer which is normally dominated by tangled anisotropic mucin molecules that are further lengthened by S–S bonds.* The larger-scale multicellular coordination of epithelial ciliary activity must require cell–cell communication mechanisms such as movement of calcium ions through gap junctions, and others. The molecular architecture of respiratory tract cilia, and their biochemistry and biomechanics are outside the scope of this chapter (See reviews by Satir and Sleigh, 1990; Wanner *et al.*, 1996). The presence of cilia enormously increases apical surface area per cell and may play a role in explaining the apparent vulnerability of ciliated cells to inhaled toxicants.

Sol–Gel Interactions and Clearance of Soluble Materials

It is often assumed that only the most superficial ('gel') layer of the surface liquid moves in this manner, being contacted by the coordinated power stroke of the tips of the cilia, while the deeper ('sol') fluid is stationary. Radiolabeled aerosolized DTPA and albumin would be expected to distribute throughout the airways surface liquid and thus be in contact with the epithelial surface. As anticipated, the relatively absorbable DTPA was cleared more rapidly (gamma camera) than albumin (Bennett and Ilowite, 1989). When clearance of ^{99m}Tc-labeled albumin and DTPA from a localized bronchial deposition site (injection

---

* Inflammation and cell disruption can add significant amount of viscous DNA and actin to the surface liquid, with deleterious effects on mucus clearance.

of 6 μl through a bronchoscopically positioned catheter fitted with a microspray nozzle) was compared in dogs with that of insoluble submicrometer-sized sulfur colloid particles, the soluble compounds cleared from the deposition site more slowly than the insoluble particles (Lay *et al.*, 1997; John Lay, personal communication). Furthermore, globs of radioactivity that appeared to be moving cephalad by mucociliary clearance showed trailing radioactivity, suggesting that solute was redistributing from a patch of mucus gel to the underlying sol layer. Blood radioactivity levels indicated some absorption of ^{99m}Tc-DTPA (but not of labeled albumin). With ^{99m}TcO$_4^-$, which is rapidly taken up across epithelium (Man *et al.*, 1985), disappearance from the deposition site was even more rapid than for the insoluble particles undoubtedly due to transepithelial absorption from the sol layer (Lay *et al.*, 1995). Thus, one can hypothesize that a non-absorbed solute would ultimately be cleared by the mucociliary mechanism, albeit more slowly than an insoluble particle, as solute dissolved in an immobile periciliary sol layer equilibrates with fresh mucus moving cephalad in the gel layer.

However, using well-ciliated primary cultures of human airways epithelium, Matsui *et al.* (1998) have recently shown (with confocal and conventional microscopy of fluorescent microspheres and photoactivated fluorescent dyes) that the sol layer ('periciliary' liquid) was transported just as rapidly as surface mucus. This periciliary liquid movement was dependent on the presence of mucus and indicates a coupling between the surface mucus layer and the remaining periciliary fluid. This coupling was more effective than would be predicted by shear-driven flow alone (i.e. a slab of superficial mucus moving along its interface with an immobile plane of periciliary sol), and may depend on ciliary mixing of the periciliary liquid and exchange of solutes between the sol and gel layers. These observations suggest that even solutes (aerosolized solutions or leached from solid particulates) that are initially deposited on the surface mucus will be swept in an orad direction in the presence of effective mucociliary clearance, and will not simply remain at the site of penetration into the periciliary liquid (unless the solutes are bound to underlying cells or are absorbed across the epithelium).

Regulation of Surface Liquid Volume

The ASL fluid and electrolyte balances along the airways axis are complex. The fact that the aggregate surface area of each 'generation' of airways decreases in the orad direction implies that fluid might 'puddle' to increasing depths as surface fluid is swept along by ciliary action. Indeed, there is some increase in depth of the surface liquid layer, which parallels increased ciliary length as one moves from small to progressively larger airways. In addition, the velocity of surface liquid flow as judged by mucus transport velocity is greater in larger airways, so that the effect of decreasing surface area is compensated for to some extent. Finally, the airways epithelial cells are equipped with the appropriate apical and basolateral ion channels and pumps to permit secretion or reabsorption of sodium and chloride, thus generating an osmotic gradient to support water secretion or reabsorption. Under basal conditions, Na$^+$ absorption dominates while Cl$^-$ and water follow passively. However, Cl$^-$ secretion can be stimulated if there is an appropriate electrochemical gradient for movement of cellular Cl$^-$ across the apical membrane and if Cl$^-$ channels are open. There are several apical membrane Cl$^-$ channels, one being the CFTR protein which is deficient in the disease cystic fibrosis (Gabriel and Boucher, 1997). In addition, the tracheobronchial glands secrete a somewhat hypotonic fluid and are under autonomic control. The precise integration and control of this complex homeostatic mechanism is not fully understood, although a good deal is known about individual airways epithelial ion channels. It is tempting to speculate that mechanical stresses generated by cilia beating against a viscoelastic load are transmitted to their basal bodies at the apical margin and to the underlying cytoskeleton, and transduced into appropriate responses – e.g. purine secretion, increased ionized calcium, cAMP and NO synthesis. These responses then serve to regulate the depth of the periciliary surface liquid so as to match the length of the extended cilia during the active force-generating portion of their cycle, and also to alter ciliary beat frequency.

CONDITIONING OF INSPIRED AIR TEMPERATURE AND HUMIDITY

Water and Heat Fluxes

The trachea and bronchi (in addition to the upper airway) also serve a 'conditioning' function by warming inspired air to core temperature and humidifying it to a relative humidity of 100%. This requires substantial volumes of water which must be supplied by the mucosa (and ultimately the submucosal microvasculature) in these airways. Small osmotic gradients presumably drive the water (through aqueous pores and paracellularly) from the submucosal interstitial fluid through the epithelium and into the airway surface liquid where vaporization can occur. The large heat of vaporization for water causes some cooling of the mucosa but the regional circulation also provides a source of warming. Bronchial blood flow is known to increase during dry air inhalation (Wagner, 1997). Nevertheless, the tracheal and upper airway mucosal surfaces are at a lower temperature than core temperature and this allows expired alveolar air ($T=37°C$, relative humidity 100%) passing over these surfaces to transfer some of its heat and water before exiting from the oronasal orifices. Since it does not seem that airway surface liquid osmolality deviates very much from isotonicity (Knowles *et al.*, 1997), the water conductances across the epithelium must be relatively high.

Exercise-induced Bronchospasm

Inhalation of large volumes of dry air causes bronchospasm in asthmatic individuals ('exercise-induced bronchospasm')(McFadden, 1997; McFadden and Gilbert, 1994), although the precise roles of mucosal cooling and post-hyperventilation rewarming, bronchial mucosal hyperemia, and possible changes in surface liquid and mucosal tonicity and ion composition remain unclear. Leukotrienes play a significant role in mediating the broncho-constriction while prostaglandins may be responsible for the refractory period that follows resolution of bronchoconstriction (O'Byrne, 1997). The phenomenon of SO_2-induced bronchospasm in asthmatics bears a number of resemblances to exercise-induced bronchospasm and, indeed, some degree of increased ventilation is required to elicit this phenomenon (Sheppard, 1986).

Humidification and Hygroscopic Inhaled Particles

The humidification of inspired air containing hygroscopic particles leads to rapid particle growth and can therefore potentially alter particle deposition profiles (Martonen *et al.*, 1989). Since the aerodynamic behavior of globular particles (quasi-spherical) is dependent on their aerodynamic diameter, which for particles > 1 μm geometrical diameter (d) is directly proportional to d and to the square root of the density (ρ), hygroscopic particle swelling tends to reduce the effect of swelling on aerodynamic diameter if initial particle density is greater than that of water. For example, hygroscopic doubling of the geometrical diameter of a spherical particle with an initial density 3 times that of water simultaneously reduces mean particle density to only 1.25 units since volume has increased 8-fold and 7/8 of this volume is water with unit density. The ratio of the final (f) to initial (i) aerodynamic diameters, $d_f \sqrt{\rho_f} \big/ d_i \sqrt{\rho_i}$, is therefore only 1.3 rather than 2.0.

SUBSURFACE ELEMENTS

Vascular

Immediately below airways basement membrane and extending to the smooth muscle layer is an abundant microvasculature. This is a key structural element not only for tissue nutrition and water and metabolite exchanges, but for the development of inflammatory responses directed by mediators released by activated epithelial cells, resident subepithelial cells and neural elements. Vasodilation, plasma exudation and the endothelial adherence, arrest and extramicrovascular migration of various types of circulating blood cells are the elements in such responses.

The microvasculature in the conducting airways is supplied by branches of the (systemic) bronchial arteries except in the trachea, where branches of other systemic arteries perform this function. The post-capillary venules from trachea ultimately empty into the right heart, whereas those from the intrapulmonary bronchi anastomose with pulmonary venules. The bronchial vasculature is innervated by adrenergic, cholinergic and non-adrenergic non-cholinergic (NANC) fibers and responds like other systemic vascular beds to neurotransmitters and to autacoids and various mediators with vasoconstriction or vasodilation. The bronchial vasculature also supports extensive neovascularization and anastomosis with pre-capillary pulmonary arterioles in chronic suppurative airways disease (e.g. cystic fibrosis), in chronic pulmonary artery occlusion or atresia, and in bronchogenic neoplasms.

The afferent blood supply to gas exchange regions normally is from the pulmonary arteries and arterioles and drainage is via the pulmonary venules and veins. The normal adult (low pressure, low resistance) pulmonary circulation does not respond to all mediators or stimuli in the same manner as systemic vessels and also seems to have little angiogenic potential, although distal extension of arteriolar smooth muscle can occur.

Smooth Muscle

Airways smooth muscle is generally arranged so that contraction causes increased rigidity and elasticity of airway walls as well as decrease of airway and lumenal caliber. Lumenal caliber is also determined by passive forces such as lung elastic recoil, the degree of inter-dependance between lung parenchyma and the intrapulmonary airways, and transmural pressure differences, as well as by the thickness of the submucosa and the presence of intralumenal secretions. Decreased lumenal caliber will of course have significant effects on the fluid dynamics of air flow, on regional ventilation and gas exchange, and will alter standard lung function parameters if the changes are sufficiently widespread. Chemicals delivered to the airway surfaces (e.g. inhaled aerosol of methacholine or albuterol) are capable of causing contraction or relaxation by interaction with M3 or β-adrenergic muscle receptors, respectively. Thus, by direct as well as indirect pathways, airway surface events can affect smooth muscle tone. By contrast, parenchymal (or pleural) smooth muscle is sparse and does not appear to be functionally important. Thus, lung compliance is largely determined by passive pressure–volume characteristics of parenchymal archi-tectural elements and the alveolar air–liquid surface tension.

HANDLING OF GASES

Unlike the 'physiologic' gases, O_2 and CO_2, and the air pollutant, CO, which are taken up or excreted from the capillary blood exclusively at the alveolar level by diffusion down a gradient of partial pressure between alveolar air and capillary blood, the other 'criteria' air pollutant gases, O_3, NO_x and SO_2, are taken up along the airways surface by chemical reactions that are essentially irreversible. The direct actions of the latter pollutant gases are therefore confined to the respiratory system.

O_2 and CO are both very poorly soluble in aqueous media. They both (reversibly) ligate Fe(II) heme (as in hemoglobin and myoglobin) although O_2 is potentially capable of accepting an electron from the Fe atom, with formation of superoxide anion ($O_2^{·-}$) and methemoglobin (Fe(III)). NO, which like CO is an endogenous metabolic product as well as sometimes being present in ambient air, exhibits a very high affinity for heme groups and can also react with –SH groups.

Carbon monoxide

At any given (low) $P_I CO$, the CO uptake rate ($V CO$) by the capillary blood is limited by the lung diffusing capacity for CO ($D_L CO$) which reflects the processes of CO diffusion across the alveolar–capillary membranes in series with the chemical reaction of CO with hemoglobin in pulmonary capillary red cells. $V CO = P_A CO \times D_L CO$ when initial blood pCO is negligible. This makes CO at low partial pressures an excellent test gas for the measurement of $D_L CO$, which is itself determined by the area and thickness of the alveo-lar capillary membrane, the amount of hemoglobin (Hb) in the pulmonary capillaries, and the alveolar pO_2 (since O_2 and CO compete for ligation to Hb). At a given $F_I CO$ and $F_I O_2$, the uptake (or excretion) kinetics (but not the ultimate steady-state COHb level) are affected by alveolar ventilation, since increased V_A maintains $P_A\text{CO}$ closer to $P_I\text{CO}$, and by capillary blood volume (which affects $D_L\text{CO}$), but not by pulmonary capillary blood flow. The steady-state level of COHb in the blood depends principally on $F_I\text{CO}$ and endogenous CO production (Coburn et al., 1965) (oxidation of a methene bridge in the protoporphyrin ring of the heme group by constitutive or inducible heme oxygenases).* The equilibrium saturation ($S\text{CO}$) of Hb with CO depends on the $p\text{O}_2$ as well as $p\text{CO}$ since these ligands compete for the same binding sites.

$$\frac{S_{\text{CO}}}{S_{\text{O}_2}} = M\,\frac{p\text{CO}}{p\text{O}_2},$$

where M is a quasi-constant whose value is about 220 for Hb, indicating that CO has a much higher affinity for Hb than does O_2.

In the absence of CO in the ambient air, steady-state blood COHb levels reflect endogenous CO production which normally equals CO excretion. However, during airways inflammation local CO production (via inducible heme oxygenase) may elevate expired CO levels. In normal individuals the COHb level is 0.5–1.0%, but in patients with hemolytic anemia or other states associated with rapid heme turnover, the COHb level may be considerably higher. When CO is present in the inspired air and $F_I\text{O}_2$ is 0.21, one can approximate the steady-state blood COHb level by adding 1% COHb for each increment of 7 ppm in $F_I\text{CO}$. Thus, at 100 ppm $F_I\text{CO}$, for example, one would expect a normal individual to ultimately attain a COHb level of about 14%.

Under low $F_I\text{CO}$ conditions, the rate of mixing of the blood compartment is rapid compared with the gain (or loss) of CO across the alveolar capillary barrier. Thus, arterial and peripheral venous blood have essentially the same S_{CO}. During rapid CO uptake with high $F_I\text{CO}$, however, arterial $S\text{CO}$ will be significantly higher than simultaneous peripheral venous $S\text{CO}$ until a steady state is attained (Benignus et al., 1994).

Oxygen

Oxygen is breathed at much higher partial pressures than CO. Thus O_2 uptake is ordinarily not diffusion-limited, and is perfusion (pulmonary capillary blood flow) dependent. Unlike CO, whose uptake at low levels of $F_I\text{CO}$ is diffusion-limited, $p\text{O}_2$ normally reaches equilibrium between alveolar air and capillary blood well within the transit time for blood flowing through the capillary (i.e. $P_A\text{O}_2 = pc'\text{O}_2$ where c' represents 'end-capillary' blood). Thus, more O_2 can be taken up if flow increases or if mixed venous $S\text{O}_2$ falls while maintaining equilibration of partial pressure across the alveolar–capillary membrane. Under steady-state conditions of gas exchange the $p\text{O}_2$ of arterial blood normally is 7–15 mmHg lower than that of the 'ideal' alveolar air ($P_A\text{O}_2$) ($P_A\text{O}_2 \cong P_I\text{O}_2 - (p_a\text{CO}_2/\text{respiratory exchange ratio})$) because of heterogeneity of ventilation/perfusion ratios in the normal lung, and especially the presence of regions with low ventilation/perfusion ratios, but not because of diffusion limitation in individual gas exchange units. (Nevertheless, limitation of O_2 uptake by diffusion can be demonstrated in normal individuals with

---

* The dihalomethanes (organic solvents) are xenobiotics that are metabolized to CO by cytochrome P-450 (2E1). Individuals inhaling dihalomethane vapors can develop significant elevations of blood COHb levels, especially if CYP2E1 activity has been induced by other factors (Pankow, 1996).

increased O_2 uptake (exercise) in the presence of reduced P_IO_2 (e.g. high altitude), and in individuals with diffuse interstitial lung disease, especially during exercise, under normobaric conditions.)

Carbon dioxide

Carbon dioxide, the major gaseous product of oxidative metabolism, is about 20 times as soluble in aqueous media as O_2 at 37°C. Its flux at the alveolar–capillary interface therefore is extremely unlikely to be diffusion-limited. CO_2 absorption or excretion across airway walls is, however, thought to be negligible. Thus, all expired CO_2 is deemed to originate in alveoli that are perfused as well as ventilated. Under steady-state conditions, alveolar (and arterial) pCO_2 reflects the ratio of metabolic CO_2 production ($\dot{V}CO_2$) to alveolar ventilation ($\dot{V}_A$). Furthermore the 'physiologic deadspace' ventilation (V_D) – i.e. the portion of ventilation devoted to conducting airways ('anatomic' dead space) plus poorly-perfused or non-perfused alveoli – can be estimated as a fraction of total expired ventilation ($\dot{V}_E$) from the expression:

$$\dot{V}_D/\dot{V}_E = 1 - (P_ECO_2/p_aCO_2)$$

In normal individuals the 'physiologic' and 'anatomic' dead spaces are essentially identical and (including the upper airway) measure about 150 ml in resting adults, which constitutes 25–30% of resting tidal volume. During exercise, tidal volume increases markedly, so that despite some dilatation of the intrathoracic airways at higher lung volumes which increases anatomic dead space, the dead space-to-tidal volume ratio falls to as low as 10%. In individuals with severe lung disease, especially when many ventilated areas are non-perfused, resting $\dot{V}_D/\dot{V}_E$ can be as high as 60–80%, due to increase in 'physiologic' (not anatomic) dead space.

Nitrogen monoxide (Nitric oxide)

The gas NO represents a special case since it is produced by airways mucosa and by NANC motor nerves, and probably serves as a regulatory mediator of airways function. NO has even higher affinity for Hb heme iron than does CO. This makes Hb an excellent scavenger for NO. Although NO is an important physiological mediator in many tissues, is produced by endothelial cells lining blood vessels, and is also involved in pathologic processes (e.g. endotoxic shock), the very high affinity of circulating Hb for NO makes it seem unlikely that much NO enters the alveolar air from pulmonary capillary blood (Byrnes et al., 1997; Gustafsson 1997; but see Hyde et al., 1997). Thus, exhaled NO probably originates principally in the airways, especially in the upper airways (nose and paranasal sinuses). The lower airways also produce NO that reaches the airway lumen, and airways epithelium expresses inducible (type II) nitric oxide synthase (iNOS) (Gaston et al., 1994). Asthmatic airways inflammation is associated with increased NO excretion from lower airways which is probably due to increased levels of iNOS in epithelial cells and can be downregulated by inhaled glucocorticoids (Gaston et al., 1997). It is not clear, however, that neutrophilic, non-asthmatic inflammation is associated with increased NO excretion. Indeed, patients with suppurative airways and sinus disease associated with cystic fibrosis and primary ciliary dyskinesia typically have reduced levels of nasal and lower airway NO, although levels of dissolved nitrate and nitrite appear to be elevated in cystic fibrosis sputum (Linnane *et al.*, 1998).

The NO excretion from normal nasal airways is approximately 500 nl/min, but is increased in the presence of allergic rhinitis. The NO excretion from normal lower airways is about 100 nl/min, but is several-fold higher in asthmatics. Despite its reputation as a toxic gas, F_INO levels of up to 20–30 ppm are used therapeutically for selective vasodilation of the pulmonary vasculature, and for certain other clinical problems, apparently without serious toxic reactions even after many hours of inhalation, although there is a potential for generation of met-Hb and for 'rebound' vasoconstriction after cessation of treatment.

Reactive and Inert Gases

Chemically reactive gases of relatively high aqueous solubility (e.g. NH_3, SO_2) are largely absorbed in the upper airways and exert their toxic effects at that level. Chemically reactive gases of lower solubility (e.g. O_3, NO_2) will largely escape absorption in the upper airway and thus reach the lower airways and even parenchyma where they may cause toxicity (Hanna *et al.*, 1989). None of these gases, however, manage to survive respiratory transit and be absorbed into capillary blood. The kinetics of the processes that underlie reactive gas absorption into respiratory tissue are often considered to be first-order (or pseudo first-order), thus allowing dosimetric values to be normalized for inspired concentration. However, inspiratory flow rate (V_I/t_I) is an important determinant of axial dosimetry. Increasing V_I/t_I displaces reactive gas uptake distally. Thus, exercise-induced increases in ventilation not only increase the total inhaled dose rate, but also displace the fractional uptakes along the airways axis toward the periphery. Because of their chemical reactivity these gases are absorbed largely at the respiratory surfaces – i.e. the surface liquid, the subadjacent epithelium, and surface macrophages (Ultman *et al.*, 1988; Bidani and Postlethwait, 1998).

Chemically unreactive ('inert') gases do not cause respiratory system toxicity. Their handling by the respiratory system depends upon their aqueous solubility. Very insoluble gases (e.g. He, SF_6) can be, and are, used to estimate alveolar gas volume (by dilution) since the dissolution and blood flow losses are negligible. Moderately soluble but unreactive inhaled gases (e.g. C_2H_2 and anesthetics) are lost from the alveolar air space by dissolution in pulmonary tissue, and especially by dissolution in capillary blood and transport out of the lung. Their uptake into capillary blood is never diffusion-limited. Such gases can be used to estimate pulmonary capillary blood flow (slope of the disappearance curve) and also pulmonary tissue volume (the latter by estimating loss from the gas phase at 'zero time' in a rebreathing system where back-extrapolation of simultaneous CO disappearance is used to define 'zero time') (Sackner *et al.*, 1980). Highly soluble, but non-reactive gases (e.g. dimethyl ether) are also taken up by airways mucosa. West and Wagner (1997) have established procedures for estimating the quantitative distribution of $\dot{V}_A/\dot{Q}$ ratios in human lung by intravenously infusing a saline solution of six inert gases selected to cover a broad range of solubility (blood/gas partition coefficients, λ), whose fractional ventilatory excretions (and circulatory retentions) under steady-state conditions are therefore sensitive to the entire range of possible $\dot{V}_A/\dot{Q}$ ratios. For any individual gas exchange unit characterized by a certain $\dot{V}_A/\dot{Q}$ ratio,

$$\frac{P_A}{P_{\bar{v}}} = \frac{P_{\acute{c}}}{P_{\bar{v}}} = \frac{\lambda}{\lambda + \dot{V}_A/\dot{Q}}$$

This method has been applied widely in studies of normal physiology and disease.

Ozone

Ozone is a poorly soluble but highly reactive gas. It is partially depleted in the upper airways when inhaled but a major fraction does reach the lower airways. O_3 can react with uric acid, which is secreted by human submucosal airway glands and is present in near-mmol/l concentrations in nasal surface liquid. O_3 also can react with surface liquid components like ascorbate, proteins and unsaturated lipids. Pryor and his colleagues have proposed that some of the toxic products of the latter reaction (e.g. hydroxyhydroperoxides, hydroxyaldehydes) are important intermediate mediators of ozone effects on underlying epithelium (Pryor, 1994; Pryor and Church, 1991; Pryor et al., 1995, 1996), and Pryor (1992) has calculated that inhaled O_3 per se does not even reach the epithelial cell apical membrane in conducting airways. Leikauf et al. (1995) estimated that for F_1O_3 of 120 ppb, hydroxyhydroperoxide formation rates resulting from ozone attack on ASL unsaturated fatty acid residues, if confined to a thickness of 10 μm, would result in a concentration of this product adequate to activate cellular (epithelial) phospholipases, thus causing rapid synthesis of multiple bioactive lipid mediators (Kafoury et al., 1998; Leikauf et al., 1993; Wright et al., 1994).

Models of ozone uptake (F.J. Miller et al., 1985) based on airways geometry, fluid mechanics and assumptions concerning chemical reactivity with airway surface liquid components predict a fairly stable total dose per cm^2 area for combined surface liquid and epithelium from trachea to respiratory bronchiole. The proportion of uptake attributed to surface liquid (nearly 100% in large airways which agrees with Pryor's calculations) decreases progressively as surface liquid thins and its reactivity with ozone diminishes, so that the highest epithelial *tissue* dose is predicted for the terminal bronchiole–respiratory bronchiole region. This is indeed a site of damage in O_3-exposed animals. Using an orally inhaled small volume bolus of ozone, Ultman's group (Hu et al., 1994) finds very little O_3 in the expirate even at bolus penetration volumes less than the anatomic dead space, suggesting that O_3 uptake is already substantial within the conducting airways, at least during relatively quiet breathing. Bronchoscopic sampling along the large airways using a rapid O_3 analyzer of the type developed by Ben-Jebria and Ultman (Gerrity et al., 1995) also indicates that a substantial fraction (35%) of orally inspired O_3 is taken up in the upper airway and trachea of resting individuals, and that O_3 in exhaled air is limited to the initially expired volume representing airways dead space. That O_3 inhalation produces toxicity in large airways is supported by evidence of ciliated cell loss and increased epithelial mitotic index in small animals, neutrophilic inflammation in humans (Aris et al., 1993), increased bronchial artery blood flow in sheep (Schelegle et al., 1990) and by the symptoms of cough and of substernal pain exacerbated by deep inspiration in humans (Passannante et al., 1998).

The Miller–Overton model (1985) predicts a 3 log unit drop in dose per cm^2 surface (of which almost all is assigned to tissue rather than surface liquid) as the inspired air moves from generation 17 (first-order respiratory bronchioles) to generation 23 (alveoli). This is partly accounted for by increasing surface/volume ratios and by dilution of incoming O$_3$-containing air in the FRC, but even more importantly by the progressively decreasing residual O$_3$ concentration in the air entering successive generations. Nevertheless, even at inspired concentrations <1 ppm, O$_3$ must penetrate to the alveolar area since alveolar type I cells are damaged and alveolar macrophage function is disturbed (Devlin and Koren, 1990). Furthermore, BAL shows enrichment with ^{18}O atoms after exposure to ^{18}O$_3$ (Hatch *et al.*, 1994). Increased inspiratory flow rates undoubtedly displace the distribution of fractional uptakes of inhaled O$_3$ along the airways toward the periphery because of decreased transit times in the airways in spite of some decrease in resistance to lateral gas diffusion to the airways surface liquid layer attributable theoretically to thinning of the gas boundary layer at the airway wall.

Sulfur dioxide

Sulfur dioxide is moderately soluble in aqueous media and is then rapidly hydrated to form H$_2$SO$_3$, which in turn dissociates to H$^+$ and HSO$_3^-$.* The latter (unlike its CO$_2$ homologue, HCO$_3^-$) reacts further with R—S—S—R′ groups to form R—S—SO$_3^-$ and R′—SH derivatives. SO$_2$ is considerably more soluble than CO$_2$ and is clearly taken up in large airways (including nasopharynx) where it has irritant effects. Increased inspiratory flow allows deeper penetration of inhaled SO$_2$ to the lower airways and increased minute ventilation increases total exposure dose. With development of oral breathing as ventilation increases, by-passing the nose, and with rapid inspiratory flow rates, enough SO$_2$ reaches lower airways to cause reflex bronchoconstriction in asthmatic subjects, even at a F$_I$SO$_2$ of as little as 0.25 ppm (Sheppard, 1986). High levels of SO$_2$ (e.g. 200–300 ppm) can cause extensive injury to airways with near-total deciliation and replacement by a secretory cell-dominated epithelium. This provides a useful animal model of airways injury (M.L. Miller *et al.*, 1985).

More quantitatively, if we assume that 50% of inspired SO$_2$ is absorbed in upper airways and that uptake of the remaining 50% is limited to the first three generations of lower airways, with a combined surface area of ~125 cm^2 (Weibel model), and further that the depth of penetration of absorbed SO$_2$ is initially limited to the airway surface liquid (10 μm) and the subadjacent epithelial cell layer (10 μm), then for $\dot{V}_I$ = 45 l/m (STPD) of 1 ppm SO$_2$ for 1 minute we would have 22.5 μl (1 μmol) SO$_2$ distributed in a volume of 250 μl. Expressed as a concentration, 1 μmol SO$_2$/250 μl ≅ an average of 4 mmol/l which is a significant load of H$^+$ and HSO$_3^-$. Within 2½ minutes (long enough to develop bronchoconstriction in asthmatic subjects) the cumulated concentration of HSO$_3^-$ would be 10 mmol/l. Thus, even for low inspired concentrations of reactive/soluble gases a surprisingly large airway surface concentration of products can be achieved in principle if the absorption surface and depth (i.e. volume) are limited.

---

* The fixed negative charge on mucin in the ASL of cartilaginous airways implies that counterions will distribute preferentially in mucin-rich layers. Thus, protons should be overrepresented in the mucus gel and therefore produce a low pH whereas the HSO$_3^-$ anions would be rejected into the periciliary liquid and toward the epithelial surface.

COMMUNICATION OF SURFACE EVENTS TO DEEPER AIRWAY STRUCTURES

How do inhaled, reactive pollutants cause alteration in airways function if their direct actions are limited to the airways surface? Multiple mechanisms can be suggested for which there is more or less evidence. Some of these are briefly discussed below.

Epithelial Secretion of Paracrine Mediators

Injured or environmentally stressed airways epithelial cells are capable of enzymatically liberating large amounts of arachidonate from the sn-2 position of phospholipids. The principal products from free arachidonate and from the residual lysophospholipids are PGE_2 and $PGF_{2\alpha}$ (cyclooxygenase pathway) and 5-, 12- and 15-hydroxyeicosatetraenoates (HETE) (lipoxygenase pathways), and platelet activating factor (PAF), respectively (Churchill et al., 1989; Henke et al., 1988; Holtzman, 1992; Hunter et al., 1985; Madden et al., 1994; McKinnon et al., 1993; Samet et al., 1992). However, they do not produce leukotrienes (LTB_4 and LTC_4, D_4, E_4), apparently because of lack of LTA_4 synthase and of glutathione-S-transferase activity (Bigby et al., 1989). Epithelial cells also can be activated to produce inducible nitric oxide synthase and generate large amounts of NO (Gaston et al., 1994).

Furthermore, recognition of the ability of epithelial cells to synthesize and secrete a variety of small proteins (cytokines) has broadened appreciation of epithelial function far beyond its classical roles as the purveyor of mucociliary transport in the conducting airways and of surfactant secretion in the alveolar region, and as a passive barrier between inspired air components and the host. Levine (1995) has categorized airway epithelial cytokine products into several groups:

(1) chemotactic factors for mobile inflammatory cells – which mostly belong to the α(CXC) and β(CC) families of chemokines (Rollins, 1997);

(2) colony-stimulating factors (e.g. GMCSF) which promote survival, activation and differentiation of mobile inflammatory cells (e.g. neutrophils, monocytes, eosinophils) and of certain resident cells (e.g. dendritic cells) (Lim et al., 1996; McWilliams et al., 1996);

(3) pleiotropic cytokines like IL-6, IL-11 and TNF-α. IL-6 is an important factor for B and T lymphocyte maturation and proliferation (DiCosmo et al., 1994) in addition to its role in stimulating hepatic synthesis of acute phase proteins;

(4) growth factors like TGF-β (Kelley, 1993; Ryan and Finkelstein, 1993; Sacco et al., 1992) and fibronectin (Adachi et al., 1997; Aoshiba et al., 1997; Devlin et al., 1994).

These various lipid mediators and cytokines bind to specific receptors on the membranes of target cells within airways walls and lung parenchyma and activate signal transduction mechanisms in these cells. The activated cells express their own repertoire of surface adhesion molecules, mediators and cytokines which creates a complex, interacting network of fixed and mobile cells in the airway and lung tissue. The epithelial cells themselves display surface receptors (e.g. for TNF-α, IFN-γ, PGE_2, PAF, kinins, ATP) which modulate their own function, and adhesion molecules such as ICAM-1 which are important for interactions with mobile inflammatory cells. In addition, it is possible that normal (but not injured) epithelial cells constitutively produce mediators that down-regulate the function of nearby resident cells such as mast cells (Peden *et al.*, 1997) or dendritic cells. The critical role of the subepithelial microvascular network in these events has been alluded to previously.

Thus, primary injury of epithelial cells is capable of leading to an orchestrated process of inflammation, repair, apoptosis and possible remodelling, involving mobile as well as resident cells (see reviews by Adler *et al.*, 1994; Finkelstein *et al.*, 1997; Leikauf *et al.*, 1995; Shelhamer *et al.*, 1995; as well as papers by Adachi *et al.*, 1997; DiCosmo *et al.*, 1994; Kawamoto *et al.*, 1995; Mio *et al.*, 1998; Zhao *et al.*, 1998). The precise composition and temporal pattern of the mixture of mediators released by injured or stressed epithelial cells probably depends on the specific insult which activates appropriate signal transduction pathways – e.g. phosphorylation cascades, NFκB, apoptotic events, etc.

Epithelial Barrier Function

The barrier properties of the epithelium, if breached by pollutant or other injury may allow penetration of inhaled, deposited antigens or other chemicals to deeper sites in the airway wall. Increased epithelial permeability to DTPA following ozone exposure, for example (Kehrl *et al.*, 1987), could offer some basis for post-ozone enhancement of acute reactivity to specific antigen inhalational challenge (Jörres *et al.*, 1996) or to methacholine, although there are other possible explanations.

Neural Communication

Sensory, unmyelinated neurons are found within or immediately subadjacent to the airways epithelium. They are variably abundant depending on species and are probably relatively less abundant in humans than in certain animals. Many of these fibers are C fibers which are non-myelinated and arborize widely within the airway wall, thus supplying branches to glands, smooth muscle, vasculature, intrinsic nerve ganglia, and certain mucosal cells, as well as projecting as a vagal axon whose cell body is in the nodose ganglion and which continues on to synapse in the brainstem. Other superficial unmyelinated sensory nerve endings are particularly abundant at carinas. Their fibers acquire a thin coat of myelin (Aδ fibers) and also ascend in the vagus (see Coleridge and Coleridge, 1994).

The bronchial C fibers subserve pain sensation and are sensitized or stimulated by mediators such as PGE$_2$ and bradykinin and by substances such as capsaicin acting through specific receptors that modulate membrane ion channel function. When activated they release tachykinins (substance P, neurokinins) and other peptides from their terminals, thus providing these nociceptive afferent fibers with motor functions. The distal arborization of the axon of these fibers permits so-called 'axon reflexes' which are non-synaptic and cause activation of the airway wall structures thus innervated via NK-1 (substance P-specific) and NK-2 (neurokinin A-specific) receptors. This unusual sensori-motor mechanism is important in the genesis of neurogenic inflammation and may contribute to the induction of a state of hyperreactivity to bronchoconstrictive agents in airways smooth muscle (Reynolds *et al.*, 1997; Spina *et al.*, 1998) which is a feature of acute ozone exposure.

Tissues in the airway expressing tachykinin receptors, including epithelial cells, exposed to tachykinin-secreting C fibers also express neutral endopeptidase on their surface. This enzyme degrades tachykinins and other small peptides, thus downregulating the effects of these agonists. Its activity is decreased by various inhaled toxicants (e.g. O_3 and toluene di-isocyanate), and may be upregulated by inhaled glucocorticoids (Baraniuk *et al.*, 1995; Leikauf *et al.*, 1995; Nadel, 1991; Sont *et al.*, 1997; van der Velden *et al.*, 1998).

The nociceptive bronchial C fibers are decorated with opioid receptors and their function can be suppressed by opioid administration (e.g. Passannante *et al.*, 1998). Pain sensation is probably also subserved by some of the Aδ fibers, but these are not thought to be opioid suppressible. Aδ fibers subserve the cough reflex (Widdicombe, 1995b) and are sensitive to inflation and deflation of the lung, especially when lung compliance is reduced (Coleridge and Coleridge, 1994). They are known as 'rapidly adapting' receptors. Through their central connections in the medulla these sensory fibers also cause rapid, shallow breathing, which is a characteristic finding in humans and animals exposed to inhaled irritants (e.g. ozone) that penetrate beyond the larynx. The altered breathing pattern can modify the distribution of deposition of inhaled reactive gases and particles. (Laryngeal sensory stimulation, however, is likely to cause apnea and laryngospasm.) Capsaicin-sensitive C fibers are also present in the lung parenchyma (J receptors). These receptors generally respond to different autacoids than do bronchial C fibers. They are stimulated by parenchymal congestion and serve as afferent neurons for synaptic reflexes which cause bradycardia, hypotension, inhibition of somatic motor activity and rapid, shallow breathing. Whether they also have a local motor role, analogous to bronchial C fibers, is not clear.

The third type of airways sensory nerve is heavily myelinated and originates in stretch receptors within airway smooth muscle. These are 'slowly adapting' inflation receptors. Their fibers ascend in the vagus nerve as well and are responsible for the Hering–Breuer reflex. Their location deep within the airway wall makes it unlikely that they would be directly affected by inhaled toxicants.

The preganglionic, cholinergic vagal motor fibers coming from cell bodies in the medulla synapse in intrinsic ganglia within the subadventitial regions of the airway walls. The ganglia are complex structures with a variety of neurotransmitters. The postganglionic fibers innervate multiple structures in the airway wall (blood vessels, submucosal glands, smooth muscle) and exhibit a variety of transmitters in addition to acetylcholine (NO, VIP, etc.) (Fischer and Hoffman, 1996; Fischer *et al.*, 1996). The regulation of vagal motor effects on these airway wall structures is therefore very complex but bronchoconstriction, bronchial gland secretion and increased bronchial mucosal blood flow are common reflex outcomes of stimulation of airway C fibers and irritant receptors (Aδ fibers). The postganglionic fibers are normally downregulated by acetylcholine via an inhibitory M-2 receptor whose function is sensitive to a variety of inflammatory processes and whose inactivation can therefore contribute to the genesis of bronchial hyperreactivity (Costello *et al.*, 1997; Schultheis *et al.*, 1994; Shelhamer *et al.*, 1995).

Immune Responses

The subepithelial pulmonary dendritic cell network has recently emerged as a key element in the immunologic response to inhaled antigens (Holt *et al.*, 1997; Lambrecht *et al.*, 1996; McWilliams *et al.*, 1996). Cytokines and other mediators (eicosanoids, NO) produced by epithelial cells, surface macrophages, and C-fiber tachykinins, all can affect the turnover and function of dendritic cells. A review of these emerging mechanisms is beyond the scope of this chapter but it seems likely that inhalational immunotoxicology will be shown to involve these key cells, both in the determination of the primary immunologic responses to inhaled potential antigens or haptens, and in the airways response to inhalational challenge with antigens after immune sensitization has been established.

Thus, there is no paucity of mechanisms and pathways that allow surface events to exert widespread effects on underlying structures, and even to cause systemic effects by absorption of locally produced cytokines (e.g. IL-6 effects on liver cells) as well as through the nervous system and the immune system.

REFERENCES

Adachi Y, Mio T, Takigawa K *et al.* (1997) Mutual inhibition by TGF-β and IL-4 in cultured human bronchial epithelial cells. *Am J Physiol* (*Lung Cell Mol Physiol*) **273**: L701–L708.

Adler KB, Fischer BM, Wright DT *et al.* (1994) Interactions between respiratory epithelial cells and cytokines: relationships to lung inflammation. In: Chignard M, Pretolani M, Renesto P and Vargaftig BB (eds) *Cells and Cytokines in Lung Inflammation*, vol. 725 of Annals of the New York Academy of Sciences. New York: New York Academy of Sciences, pp. 128–145.

Anderson M, Philipson K, Svartengren M *et al.* (1995) Human deposition and clearance of 6 μm particles inhaled with an extremely low flow rate. *Exp Lung Res* **21**: 187–195.

Aoshiba K, Rennard SI and Spurzem JR (1997) Fibronectin supports bronchial epithelial cell adhesion and survival in the absence of growth factors. *Am J Physiol* (*Lung Cell Mol Physiol*) **273**: L684–L693.

Aris RM, Christian D, Hearne PQ *et al.* (1993) Ozone-induced airway inflammation in human subjects as determined by airway lavage and biopsy. *Am Rev Respir Dis* **148**: 1363–1372.

Baltimore RS, Christie CDC and Smith GJW (1989) Immunohistopathologic localization of *Pseudomonas aeruginosa* in lung from patients with cystic fibrosis. *Am Rev Respir Dis* **140**: 1650–1661.

Baraniuk JN, Ohkubo K, Kwon OJ *et al.* (1995) Localization of neutral endopeptidase (NEP) mRNA in human bronchi. *Eur Respir J* **8**: 1458–1464.

Basbaum CB, Berthold J and Walter EF (1990) The serous cell. *Annu Rev Physiol* **52**: 97–113.

Benignus VA, Hazucha MJ, Smith MV and Bromberg PA (1994) Prediction of carboxyhemoglobin formation due to transient exposure to carbon monoxide. *J Appl Physiol* **76**: 1739–1745.

Bennett WD and Ilowite JS (1989) Dual pathway clearance of 99m-DTPA from the bronchial mucosa. *Am Rev Respir Dis* **139**: 1132–1138.

Bennett WD, Scheuch G, Zeman KL *et al.* (1998). Bronchial airway deposition and retention of particles in inhaled boli: effect of anatomic dead space. *J Appl Physiol* **85**: 685–694.

Bidani A and Postlethwait EM (1998) Kinetic determinants of inhaled reactive gas absorption. In: Hlastala MP and Robertson HT (eds) *Complexity in Structure and Function of the Lung*, vol. 121 of *Lung Biology in Health and Disease*. New York: Marcel Dekker, chap. 9, pp. 243–296.

Bienenstock J (1984) Bronchus-associated lymphoid tissue. In: Bienenstock J (ed.) *Immunology of the Lung and Upper Respiratory Tract*. New York: McGraw-Hill, pp. 96–118.

Bigby TD, Lee DM, Meslier N and Gruenert DC (1989) Leukotriene A$_4$ hydrolase activity of human airway epithelial cells. *Biochem Biophy Res Commun* **164**: 1–7.

Boat TF, Cheng PW and Leigh MW (1994) Biochemistry of mucus. In: Takishima T and Shimura S (eds) *Airway Secretion: Physiological Bases for the Control of Mucous Hypersecretion*, vol. 72 of *Lung Biology in Health and Disease*. New York: Marcel Dekker, pp. 217–282.

Boucher RC (1994) Human airway ion transport. *Am J Respir Crit Care Med* **150**: 271–281 (Part 1), 581–593 (Part 2).

Boyd MR (1982) Metabolic activation of pulmonary toxins. In: Witschi H and Nettesheim P (eds) *Mechanisms in Respiratory Toxicology*, vol. II. Boca Raton, FL: CRC Press, pp. 85–112.

Byrnes CA, Dinarevic S, Busst C *et al.* (1997) Is nitric oxide in exhaled air produced at airway or alveolar level? *Eur Respir J* **10**: 1021–1025.

Carnoy C, Ramphal R, Scharfman A *et al.* (1993) Altered carbohydrate composition of salivary mucins from patients with cystic fibrosis and the adhesion of *Pseudomonas aeruginosa. Am J Respir Cell Mol Biol* **9**: 323–334.

Chang HK (1989) Flow dynamics in the respiratory tract. In: Chang HK and Paiva M (eds) *Respiratory Physiology: An Analytic Approach*, vol. 40 of *Lung Biology in Health and Disease*. New York: Marcel Dekker, chap. 2, pp. 57–138.

Churchill L, Chilton FH, Resau JH *et al.* (1989) Cyclooxygenase metabolism of endogenous arachidonic acid by cultured human tracheal epithelial cells. *Am Rev Respir Dis* **140**: 449–459.

Churg A (1996) The uptake of mineral particles by pulmonary epithelial cells. *Am J Respir Crit Care Med* **154**: 1124–1140.

Coburn RF, Forster RE and Kane PB (1965) Considerations of the physiological variables that determine the blood carboxyhemoglobin concentration in man. *Man J Clin Invest* **44**: 1899–1910.

Coleridge HM and Coleridge JCG (1994) Pulmonary reflexes: neural mechanisms of pulmonary defense. *Annu Rev Physiol* **56**: 69–91.

Costello RW, Schofield BH, Kephart GM *et al.* (1997) Localization of eosinophils to airway nerves and effect on neuronal M2 muscarinic receptor function. *Am J Physiol* **273**: L93–103.

Cross CE, Motchnik PA, Bruener BA *et al.* (1992) Oxidative damage to plasma constituents by ozone. *FEBS Letters* **298**: 269–272.

Darquenne C and Paiva M (1998) Gas and particle transport in the lung. In: Hlastala MP and Robertson HT (eds) *Complexity in Structure and Function of the Lung*, vol. 121 of *Lung Biology in Health and Disease*. New York: Marcel Dekker, chap. 10, pp. 297–323.

Davis WB and Pacht ER (1997) Extracellular antioxidant defenses. In: Crystal RG, West JB *et al.* (eds) *The Lung: Scientific Foundations*, 2nd edn. Philadelphia: Lippincott-Raven Publishers, pp. 2271–2278.

Devlin RB and Koren HS (1990) The use of quantitative two-dimensional gel electrophoresis to analyze changes in alveolar macrophage proteins in humans exposed to ozone. *Am J Respir Cell Mol Biol* **2**: 281–288.

Devlin RB, McKinnon KP, Noah T *et al.* (1994) Ozone-induced release of cytokines and fibronectin production by alveolar macrophages and airway epithelial cells. *Am J Physiol* **266**: L612–L619.

DiCosmo BF, Geba GP, Picarella D *et al.* (1994) Airway epithelial cell expression of interleukin-6 in transgenic mice. Uncoupling of airway inflammation and bronchial hyperreactivity. *J Clin Invest* **94**: 2028–2035.

Ferin J, Oberdörster G, Soderholm SC and Gelein R (1994) The rate of dose delivery affects pulmonary interstitialization of particles in rats. *Ann Occup Hyg* **38**: 289–293.

Finkelstein JN, Johnston C, Barrett T and Oberdörster G (1997) Particulate-cell interactions and pulmonary cytokine expression. *Environ Health Perspect* **5**: 1179–1182.

Fischer A and Hoffman B (1996) Nitric oxide synthase in neurons and nerve fibers of lower airways and in vagal sensory ganglia of man. Correlation with neuropeptides. *Am J Respir Crit Care Med* **154**: 209–216.

Fischer A, Canning BJ and Kummer W (1996) Correlation of vasoactive intestinal peptide and nitric oxide synthase with choline acetyltransferase in the airway innervation. *Annu N Y Acad Sci* **805**: 717–722.

Gabriel SE and Boucher RC (1997) Ion channels. In: Crystal RG, West JB *et al.* (eds) *The Lung: Scientific Foundations*, 2nd edn. Philadelphia: Lippincott-Raven Publishers, pp. 305–318.

Gaston B, Drazen JM, Loscalzo J *et al.* (1994) The biology of nitrogen oxides in the airways. *Am J Respir Crit Care Med* **149**: 538–551.

Gaston B, Kobzik L and Stamler JS (1997) Distribution of nitric oxide synthase in the lung. In: Zapol WM and Bloch KD (eds) *Nitric Oxide and the Lung*, vol. 98 of *Lung Biology in Health and Disease*. New York: Marcel Dekker, pp. 75–86.

Gehr P, Green FHY, Geiser M *et al.* (1996) Airway surfactant, a primary defense barrier: mechanical and immunological aspects. *J Aerosol Med* **9**: 163–181.

Gerrity TR, Biscardi F, Strong A *et al.* (1995) Bronchoscopic determination of ozone uptake in humans. *J Appl Physiol* **79**: 852–860.

Ghio AJ, Bassett M, Chall AN *et al.* (1998) Bronchoscopy in healthy volunteers. *J Bronchology* **5**: 185–194.

Gustafsson LE (1997) Exhaled nitric oxide production by the lung. In: Zapol WM and Bloch KD (eds) *Nitric Oxide and the Lung*, vol. 98 of *Lung Biology in Health and Disease*. New York: Marcel Dekker, pp. 185–201.

Hanna LM, Frank R and Scherer PW (1989) Absorption of soluble gases and vapors in the respiratory system. In: Chang HK (ed.) *Respiratory Physiology*, vol. 40 of *Lung Biology in Health and Disease*. New York: Marcel Dekker, pp. 277–315.

Hatch GE (1992) Comparative biochemistry of airway lining fluid. In: Parent RA (ed.) *Comparative Biology of the Normal Lung*. Boca Raton, FL: CRC Press, pp. 617–634.

Hatch GE, Slade R, Harris LP *et al.* (1994) Ozone dose and effect in humans and rats: a comparison using O-18 labeling and bronchoalveolar lavage. *Am J Respir Crit Care Med* **150**: 676–683.

Henke D, Danilowicz RM, Curtis JF *et al.* (1988) Metabolism of arachidonic acid by human nasal and bronchial epithelial cells. *Arch Biochem Biophys* **267**: 426–436.

Higa HH, Rogers GN and Paulson JC (1985) Influenza virus hemagglutinins differentiate between receptor determinants bearing *n*-acetyl, *n*-glycolyl, and *n,o*-diacetylneuraminic acid. *Virology* **144**: 279–282.

Holt PG, Macaubas C, Cooper D *et al.* (1997) Th-1/Th-2 switch regulation in immune responses to inhaled antigens: role of dendritic cells in the aetiology of allergic respiratory disease. *Adv Exp Med Biol* **417**: 301–306.

Holtzman MJ (1992) Arachidonic acid metabolism in airway epithelial cells. *Annu Rev Physiol* **54**: 303–329.

Hu S-C, Ben-Jebria A and Ultman JS (1994) Longitudinal distribution of ozone absorption in the lung: effects of respiratory flow. *J Appl Physiol* **77**: 574–583.

Hunter JA, Finkbeiner WE, Nadel JA *et al.* (1985) Predominant generation of 15-lipoxygenase metabolites of arachidonic acid by epithelial cells from human trachea. *Proc Natl Acad Sci USA* **82**: 4633–4637.

Hyde RW, Geigel EJ, Olszowka AJ *et al.* (1997) Determination of production of nitric oxide by lower airways of humans – theory. *J Appl Physiol* **82**: 1290–1296.

Ikegami M, Korfhagen TR, Bruno MD *et al.* (1997) Surfactant metabolism in surfactant protein A-deficient mice. *Am J Physiol* **272**: L479–L485.

International Commission of Radiological Protection. (1994) *Human Respiratory Tract Model for Radiological Protection: A Report of a Task Group of the ICRP*, ICRP publication 66. Oxford: Elsevier Science Ltd.

Jeffery PK (1994) Microscopic structure of airway secretory cells: variation in hypersecretory disease and effects of drugs. In: Takishima T and Shimura S (eds) *Airway Secretion: Physiological Bases for the Control of Mucous Hypersecretion*, vol. 72 of *Lung Biology in Health and Disease*. New York: Marcel Dekker, pp. 149–215.

Johansson J and Curstedt T (1997) Molecular structures and interactions of pulmonary surfactant components. *Eur J Biochem* **244**: 675–693.

Jörres R, Nowak D, Magnussen H *et al.* (1996) The effect of ozone exposure on allergen responsiveness in subjects with asthma or rhinitis. *Am J Respir Crit Care Med* **153**: 56–64.

Kabel JR, Ben-Jebria A and Ultman JS (1994) Longitudinal distribution of ozone absorption in the lung: comparison of nasal and oral quiet breathing. *J Appl Physiol* **77**: 2584–2592.

Kabha K, Schmegner J, Keisari Y and Parolis HI (1997) SP-A enhances phagocytosis of klebsiella by interaction with capsular polysaccharides and alveolar macrophages. *Am J Physiol* **272**: L344–L352.

Kafoury RM, Pryor WA, Squadrito G *et al.* (1998) Lipid ozonation products (LOP) initiate signal transduction by activating phospholipases A$_2$, C and D. *Toxicol Appl Pharmacol* **150**: 338–349.

Kawamoto M, Romberger DJ, Nakamura Y *et al.* (1995) Modulation of fibroblast type I collagen and fibronectin production by bovine bronchial epithelial cells. *Am J Respir Cell Mol Biol* **12**: 425–433.

Kehrl HR, Vincent LM, Kowalsky RJ *et al.* (1987) Ozone exposure increases respiratory epithelial permeability in humans. *Am Rev Respir Dis* **135**: 1124–1128.

Kelley J (1993) Transforming growth factor-β. In: Kelley J (ed.) *Cytokines of the Lung*, vol. 61 of *Lung Biology in Health and Disease.* New York: Marcel Dekker, pp. 101–137.

Knowles MR, Robinson JM, Wood RE *et al.* (1997) Ion composition of airway surface liquid of patients with cystic fibrosis as compared with normal and disease-control subjects. *J Clin Invest* **100**: 2588–2595.

Korfhagen TR, Bruno MD and Ross GF (1996) Altered surfactant function and structure in SP-A gene targeted mice. *Proc Natl Acad Sci USA* **93**: 9594–9599.

Lambrecht BN, Pauwels RA and Bullock GR (1996) The dendritic cell: its potent role in the respiratory immune response. *Cell Biology Int* **20**: 111–120.

Lay JC, Berry CR, Kim CS and Bennett WD (1995) Retention of insoluble particles following local intrabronchial deposition in dogs. *J Appl Physiol* **79**: 1921–1929.

Lay JC, Stang MR, Fisher PE *et al.* (1997) Retention of soluble vs. insoluble materials in the conducting airways. *Am J Respir Crit Care Med* **155**: A954.

Leigh MW (1999) Airway secretions. In: Yankaskas JR and Knowles MR (eds) *Cystic Fibrosis in Adults.* Philadelphia: Lippincott-Raven Publishers, pp. 69–92.

Leikauf GD, Zhao Q, Zhou S and Santrock J (1993) Ozonolysis products of membrane fatty acids activate eicosanoid metabolism in human airway epithelial cells. *Am J Respir Cell Mol Biol* **9**: 594–602.

Leikauf GD, Simpson LG, Santrock J *et al.* (1995) Airway epithelial cell responses to ozone injury. *Environ Health Perspect* **103**: 91–95.

LeVine AM, Bruno MD, Huelsman KM *et al.* (1997) Surfactant protein A-deficient mice are susceptible to group B streptococcal infection. *J Immunol* **158**: 4336–4340.

Levine SJ (1995) Bronchial epithelial cell–cytokine interactions in airway inflammation. *J Invest Med* **43**: 241–249.

Lim TK, Chen GH, McDonald RA and Toews GB (1996) Granulocyte-macrophage colony-stimulating factor overrides the immunosuppressive function of corticosteroids on rat pulmonary dendritic cells. *Stem Cells* **14**: 292–299.

Linnane SJ, Keatings VM, Costello CM *et al.* (1998) Sputum nitrate/nitrite but not exhaled nitric oxide is elevated in adult patients with pulmonary exacerbations of cystic fibrosis. *Am J Respir Crit Care Med.* **158**: 207–212.

Lippmann M and Schlesinger RB (1984) Interspecies comparisons of particle deposition and mucociliary clearance in tracheobronchial airways. In: Miller FJ and Menzel DB (eds) *Fundamentals of Extrapolation Modeling of Inhaled Toxicants.* Washington, DC: Hemisphere, pp. 441–469.

Madden MC, Smith JP, Dailey LA and Friedman M (1994) Polarized release of lipid mediators derived from phospholipase A2 activity in a human bronchial cell line. *Prostaglandins* **48**: 197–215.

Man SFP, Ahmed IH, Man GCW and Nguyen A (1985) Characteristics of pertechnetate movement across the canine tracheal epithelium. *Am Rev Respir Dis* **131**: 90–93.

Martonen TB, Hofmann W, Eisner AD and Ménache MG (1989) The role of particle hygroscopicity in aerosol therapy and inhalation toxicology. In: Crapo JD, Miller FJ, Smolko ED *et al.* (eds) *Extrapolation of Dosimetric Relationships for Inhaled Particles and Gases.* San Diego, CA: Academic Press, pp. 303–316.

Mason RJ and Shannon JM (1997) Alveolar type II cells. In: Crystal RG, West JB *et al.* (eds) *The Lung: Scientific Foundations,* 2nd edn. Philadelphia: Lippincott-Raven Publishers, pp. 543–555.

Matsui H, Randell SH, Peretti SW *et al.* (1998) Coordinated clearance of periciliary liquid and mucus from airway surfaces. *J Clin Invest* **102**: 1125–31.

McBride JT (1992) Architecture of the tracheobronchial tree. In: Parent RA (ed.) *Treatise on Pulmonary Toxicology: Volume I. Comparative Biology of the Normal Lung.* Boca Raton, FL: CRC Press, chap. 5, pp. 49–61.

McCormack F (1997) The structure and function of surfactant protein-A. *Chest* **111**: 114S–119S.

McCormack FX, Festa AL, Andrews RP *et al.* (1997) The carbohydrate recognition domain of surfactant protein A mediates binding to the major surface glycoprotein of *Pneumocystis carinii.* *Biochemistry* **36**: 8092–8099.

McFadden Jr ER (1997) Airway function: regulation by physical factors. In: Crystal RG, West JB *et al.* (eds) *The Lung: Scientific Foundations*, 2nd edn. Philadelphia: Lippincott-Raven Publishers, pp. 1333–1344.

McFadden Jr ER and Gilbert IA (1994) Exercise-induced asthma. *N Engl J Med* **330**: 1362–1367.

McKinnon KP, Madden MC, Noah TL and Devlin RB (1993) In vitro ozone exposure increases release of arachidonic acid products from a human bronchial epithelial cell line. *Toxicol Appl Pharmacol* **118**: 215–223.

McWilliams AS, Napoli S, Marsh AM *et al.* (1996) Dendritic cells are recruited into the airway epithelium during the inflammatory response to a broad spectrum of stimuli. *J Exper Med* **184**: 2429–2432.

Mercer RR and Crapo JD (1992) Architecture of the acinus. In: Parent RA (ed.) *Treatise on Pulmonary Toxicology: Volume I. Comparative Biology of the Normal Lung*. Boca Raton, FL: CRC Press, chap. 10, 109–119.

Mermelstein R, Kilpper RW, Morrow PE and Muhle H (1994) Lung overload, dosimetry of lung fibrosis and their implications to the respiratory dust standard. *Ann Occup Hyg* **38**: 313–322.

Miller FJ, Overton Jr JH, Jaskot RH and Menzel DB (1985) A model of the regional uptake of gaseous pollutants in the lung. I. The sensitivity of the uptake of ozone in the human lung to lower respiratory tract secretions and exercise. *Toxicol Appl Pharmacol* **79**: 11–27.

Miller ML, Andriga A, Rafaelef L *et al.* (1985) Effect of exposure to 500 ppm SO_2 on the lungs of the ferret. *Respiration* **48**: 346–354.

Mio T, Liu X-D, Adachi Y *et al.* (1998) Human bronchial epithelial cells modulate collagen gel contraction by fibroblasts. *Am J Physiol* **274**: L119–L126.

Mitic LL and Anderson JM (1998) Molecular architecture of tight junctions. *Annu Rev Physiol* **60**: 121–142.

Nadel JA (1991) Neutral endopeptidase modulates neurogenic inflammation. *Eur Respir J* **4**: 745–754.

Oberdörster G, Ferin J, Soderholm S *et al.* (1994) Increased pulmonary toxicity of inhaled ultrafine particles: due to lung overload alone? *Ann Occup Hyg* **38**: 295–302.

O'Byrne PM (1997) Exercise-induced bronchoconstriction: elucidating the roles of leukotrienes and prostaglandins. *Pharmacotherapy* **17**: (1 Pt 2): 31S–38S.

Pankow D (1996) Carbon monoxide formation due to metabolism of xenobiotics. In: Penney DG (ed.) *Carbon Monoxide*. Boca Raton, FL: CRC Press, pp. 25–43.

Passannante AN, Hazucha MJ, Bromberg PA *et al.* (1998) Nociceptive mechanisms modulate ozone-induced human lung function decrements. *J Appl Physiol* **85**: 1863–1870.

Pasula R, Downing JF, Wright JR *et al.* (1997) Surfactant protein A (SP-A) mediates attachment of *Mycobacterium tuberculosis* to murine alveolar macrophages. *Am J Respir Cell Mol Biol* **17**: 209–217.

Peden DB, Hohman R, Brown ME *et al.* (1991) Uric acid is a major antioxidant in human nasal airway secretions. *Proc Natl Acad Sci USA* **87**: 7638–7642.

Peden DB, Dailey L, Wortman I *et al.* (1997) Epithelial cell-conditioned media inhibits degranulation of the RBL-2H3 rat mast cell line. *Am J Physiol* (*Lung Cell Mol Physiol*) **272**: L1181–L1188.

Persson CGA (1994) Airway mucosal exudation of plasma. In: Takishima T and Shimura S (eds) *Airway Secretion: Physiological Bases for the Control of Mucous Hypersecretion*, vol. 72 of *Lung Biology in Health and Disease*. New York: Marcel Dekker, pp. 415–467.

Plopper CG (1993) Pulmonary bronchiolar epithelial cytotoxicity: microanatomical considerations. In: Gram TE (ed.) *Metabolic Activation and Toxicity of Chemical Agents to Lung Tissue and Cells*. Oxford: Pergamon Press, pp. 1–24.

Plopper CG, Hyde DM and Buckpitt AR (1997) Clara cells. In: Crystal RG, West JB *et al.* (eds) *The Lung: Scientific Foundations*, 2nd edn. Philadelphia: Lippincott-Raven Publishers, chap. 35, pp. 517–533.

Pryor WA (1992) How far does ozone penetrate into the pulmonary air/tissue boundary before it reacts? *Free Radicals Biol Med* **12**: 83–88.

Pryor WA (1994) Mechanisms of radical formation from reactions of ozone with target molecules in the lung. *Free Radicals Biol Med* **17**: 451–465.

Pryor WA and Church DF (1991) Aldehydes, hydrogen peroxide, and organic radicals as mediators of ozone toxicity. *Free Radicals Biol Med* **11**: 41–46.

Pryor WA, Squadrito GL and Friedman M (1995) The cascade mechanism to explain ozone toxicity: the role of lipid ozonation products. *Free Radicals Biol Med* **18**: 935–941.

Pryor WA, Bermudez E, Cueto R and Squadrito GL (1996) Detection of aldehydes in bronchoalveolar lavage of rats exposed to ozone. *Fundam Appl Pharmacol* **34**: 148–156.

Raabe OG (1982) Deposition and clearance of inhaled aerosols. In: Witschi H and Nettesheim P (eds) *Mechanisms in Respiratory Toxicology*, vol. I. Boca Raton, FL: CRC Press, pp. 27–76.

Ramphal R, Carnoy C, Fievre S *et al.* (1991) *Pseudomonas aeruginosa* recognizes carbohydrate chain containing type 1 (Galβ1-3GlcNAc) or type 2 (Galβ1-4GlcNAc) disaccharide unit. *Infect Immun* **59**: 700–704.

Reynolds PN, Holmes MD and Scicchitano R (1997) Role of tachykinins in bronchial hyper-responsiveness. *Clin Exp Pharmacol Physiol* **24**(3–4): 273–280.

Robbins RA and Rennard SI (1997) Biology of airway epithelial cells. In: Crystal RG, West JB *et al.* (eds) *The Lung: Scientific Foundations*, 2nd edn. Philadelphia: Lippincott-Raven Publishers, pp. 445–457.

Robertson HT (1998) Measurement of regional ventilation by aerosol deposition. In: Hlastala MP and Robertson HT (eds) *Complexity in Structure and Function of the Lung*, vol. 121 of *Lung Biology in Health and Disease*. New York: Marcel Dekker, chap. 12, pp. 379–399.

Rollins BJ (1997) Chemokines. *Blood* **90**: 909–928.

Rumsey SC and Levine M (1998) Absorption, transport, and disposition of ascorbic acid in humans. *J Nutr Biochem* **9**.

Ryan RM and Finkelstein JN (1993) Growth factors and pulmonary epithelia. In: Brody JS, Center DM and Tkachuk VA (eds) *Signal Transduction in Lung Cells*, vol. 65 of *Lung Biology in Health and Disease*. New York: Marcel Dekker, pp. 263–291.

Sacco O, Romberger D, Rizzino A *et al.* (1992) Spontaneous production of transforming growth factor-β2 by primary cultures of bronchial epithelial cells. *J Clin Invest* **90**: 1379–1385.

Sackner MA, Markwell G, Atkins N *et al.* (1980) Rebreathing techniques for pulmonary capillary blood flow and tissue volume. *J Appl Physiol* **49**: 910–915.

Sajjan U, Corey M, Karmali M and Forstner J (1992) Binding of *Pseudomonas cepacia* to normal human intestinal mucin and respiratory mucin from patients with cystic fibrosis. *J Clin Invest* **89**: 648–656.

Samet JM, Noah TL, Devlin RB *et al.* (1992) Effect of ozone on platelet-activating factor production in phorbol-differentiated HL60 cells, a human bronchial epithelial cell line (BEAS S6), and primary human bronchial epithelial cells. *Am J Respir Cell Mol Biol* **7**: 514–522.

Satir P and Sleigh MA (1990) The physiology of cilia and mucociliary interactions. *Annu Rev Physiol* **52**: 137–155.

Schelegle ES, Gunther RA, Parsons GH *et al.* (1990) Acute ozone exposure increases bronchial blood flow in conscious sheep. *Respir Physiol* **82**: 325–336.

Scheuch G, Stahlhofen W and Heyder J (1996) An approach to deposition and clearance measurements in human airways. *J Aerosol Med* **9**: 35–41.

Schlesinger RB (1995) Deposition and clearance of inhaled particles. In: McClellan RO and Henderson RF (eds) *Concepts in Inhalation Toxicology*, 2nd edn. Washington, DC: Taylor & Francis, pp. 191–224.

Schneeberger EE (1997) Alveolar type I cells. In: Crystal R, West JB *et al.* (eds) *The Lung: Scientific Foundations*, 2nd edn. Philadelphia: Lippincott-Raven Publishers, pp. 535–542.

Schultheis AH, Bassett DJ and Fryer AD (1994) Ozone-induced airway hyperresponsiveness and loss of neuronal M2 muscarinic receptor function. *J Appl Physiol* **76**: 1088–1097.

Shelhamer JH, Levine SJ, Wu T *et al.* (1995) Airway inflammation. *Ann Intern Med* **123**: 288–304.

Sheppard D (1986) Mechanisms of airway responses to inhaled sulfur dioxide. In: Loke J (ed.) *Pathophysiology and Treatment of Inhalation Injuries*, vol. 34 of *Lung Biology in Health and Disease*. New York: Marcel Dekker, pp. 49–65.

Snipes MB (1995) Pulmonary retention of particles and fibers: biokinetics and effects of exposure concentrations. In: McClellan RO and Henderson RF (eds) *Concepts in Inhalation Toxicology*, 2nd edn. Washington, DC: Taylor & Francis, pp. 225–255

Sont JK, van Krieken JH, van Klink HC *et al.* (1997) Enhanced expression of neutral endopeptidase (NEP) in airway epithelium in biopsies from steroid- versus nonsteroid-treated patients. *Am J Respir Cell Mol Biol* **16**: 549–556.

Spina D, Page CP and Morley J (1998) Sensory neuropeptides and bronchial hyperresponsiveness. In: Said SI (ed.) *Proinflammatory and antiinflammatory peptides*, vol. 112 of *Lung Biology in Health and Disease*. New York: Marcel Dekker, pp. 89–146.

Stripp BR, Lund J, Mango GW *et al.* (1996) Clara cell secretory protein: a determinant of PCB-bioaccumulation in mammals. *Am J Physiol* (*Lung Cell Mol Physiol*) **271**: L656–L664.

Trivier D, Houdret N, Courcol RJ *et al.* (1997) The binding of surface proteins from *Staphylococcus aureus* to human bronchial mucins. *Eur Respir J* **10**: 804–810.

Ultman JS (1988) Transport and uptake of inhaled gases. In: Health Effects Institute (ed.) *Air Pollution, the Automobile and Public Health*. Washington, DC: National Academy Press, pp. 323–366.

van der Velden VH, Naber BA, van der Spoel P *et al.* (1998) Cytokines and glucocorticoids modulate human bronchial epithelial cell peptidases. *Cytokine* **10**: 55–65.

Wagner EM (1997) Bronchial circulation. In: Crystal RG, West JB *et al.* (eds) *The Lung: Scientific Foundations*, 2nd edn. Philadelphia: Lippincott-Raven Publishers, pp. 1093–1105.

Wanner A, Salathe M and O'Riordan TG (1996) Mucociliary clearance in the airways. *Am J Respir Crit Care Med* **154**: 1868–1902.

Weibel ER (1963) *Morphometry of the Lung*. New York: Springer Verlag.

Weibel ER (1989) Lung morphometry and models in respiratory physiology. In: Chang HK and Paiva M (eds) *Respiratory Physiology: An Analytic Approach*, vol. 40 of *Lung Biology in Health and Disease*. New York: Marcel Dekker, chap. 1, pp. 1–56.

West JB and Wagner PD (1997) Ventilation-perfusion relationships. In: Crystal RG, West JB *et al.* (eds) *The Lung: Scientific Foundations*, 2nd edn. Philadelphia: Lippincott-Raven Publishers, pp. 1693–1710.

Whitsett JA, Nogee LM, Weaver TE and Horowitz AD (1995) Human surfactant protein B: structure, function, regulation, and genetic damage. *Physiol Rev* **75**: 749–757.

Widdicombe J (1995a) Relationships among the composition of mucus, epithelial lining liquid, and adhesion of microorganisms. *Am J Respir Crit Care Med* **151**: 2088–2092.

Widdicombe J (1995b) Neurophysiology of the cough reflex. *Eur Respir J* **8**: 1193–1202.

Wright DT, Adler KB, Akley NJ *et al.* (1994) Ozone stimulates release of platelet activating factor and activates phospholipase in guinea pig tracheal epithelial cells in primary culture. *Toxicol Appl Pharmacol* **127**: 27–36.

Zhao Q, Simpson LG, Driscoll KE *et al.* (1998) Chemokine regulation of ozone-induced neutrophil and monocyte inflammation. *Am J Physiol* **274** (*Lung Cell Mol Physiol* **18**): L39–L46.

14

Deposition and Clearance of Inhaled Particles

W. MICHAEL FOSTER

Department of Environmental Health Sciences, School of Hygiene and Public Health, The Johns Hopkins University, Baltimore, MD, USA

INTRODUCTION

There has been much recent interest in particulate air pollution and its impact upon public health. The goal of this chapter is to review information, experimental and theoretical, relevant to particulate exposure. The chapter is divided into two sections and only reviews information based upon human investigation. The first section is concerned with the mechanisms and factors that influence particle deposition in the respiratory system, while the second section summarizes current information on mechanisms, pathways and kinetics of particle clearance. The focus is primarily upon particles that are relatively insoluble to the respiratory epithelia and are thus removed by mucociliary function and cellular mechanisms. Only moderate emphasis is placed on clearance mechanisms relevant to soluble particles that are removed by gaining entry into the bronchial, pulmonary and systemic circulations.

PARTICLES INTO THE RESPIRATORY TRACT

General Concepts

The membrane surfaces of the respiratory tract represent the largest interface between humans and the environment. These surfaces are constantly exposed to a host of particles suspended in the inspired air. The fraction of respired particulate that deposits within

AIR POLLUTION AND HEALTH
ISBN 0-12-352335-4

oronasal, pharyngeal, and lower respiratory tract regions has been investigated and modeled for a broad range of particle sizes and ventilatory patterns (Chan and Lippmann, 1980; Gerrity *et al.*, 1979; Martonen, 1983; Task Group on Lung Dynamics, 1966). The regional distribution of particles which have deposited within the upper and lower respiratory tracts and their potential to interact with, and/or overburden, local clearance and defense mechanisms is also of importance (Churg, 1996).

Reports in the epidemiological and clinical literature since 1970 have frequently linked chronic exposure to airborne particulate to pulmonary injury. However, in general these associations are described for subject populations that have been drawn from occupational settings, primarily the mining industry with generally high dust levels (Attfield and Hodous, 1992; Kennedy *et al.*, 1985). Smoking also has been identified as a major risk factor for the development of lung disease and cigarette smoke is a major contributor to air toxics found in the indoor environment (Huchon *et al.*, 1984). To extrapolate risk to the general population from airborne particulate based upon these reports or epidemiologic analyses and associations (Dockery and Pope, 1994) requires some caution. Additional factors need to be considered and include: (a) inhalable particulate present in urban air frequently coexists with other respiratory irritants such as oxidant gases like ozone, or acidic aerosols that in themselves convey risk and susceptibility to airway disease; and (b) pre-existent airway disease in an individual enhances the normal deposition mechanisms of particulate in the lung and may increase susceptibility as well.

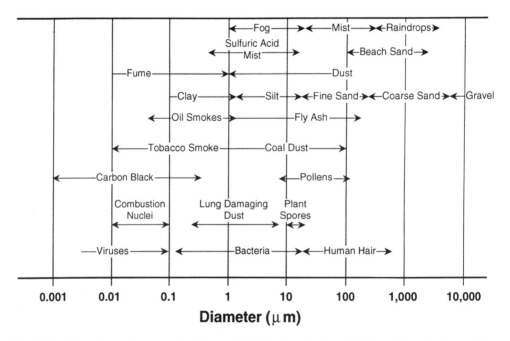

Fig. 14.1   Approximate size ranges for airborne inhalation hazards (from the Mine Safety Appliances Co., Pittsburgh, PA).

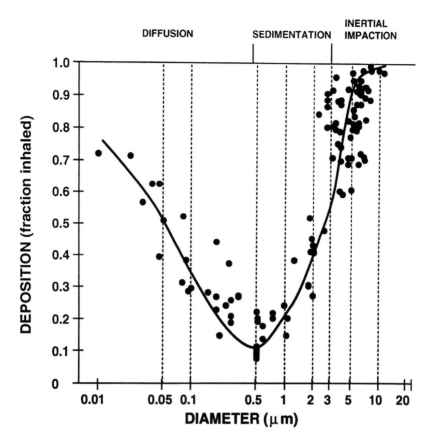

Fig. 14.2 Compact particle size and lower respiratory tract deposition in humans. The potential mechanisms and deposition probability for particles of different diameter are indicated. Solid circles represent individual or mean data points of deposition; smooth curve is predicted deposition and curvilinear fit to theoretical and experimental data (Chan and Lippmann, 1980; Task Group on Lung Dynamics, 1966). Deposition is expressed as the fraction of the inhaled particles (mouth breathing at rest) that deposit within the lower respiratory tract.

In interpreting these findings, consideration needs to be given to the characteristics of particles and to patterns of deposition and clearance in the lung. Compact particles are usually characterized by size and aerodynamic properties and particles with diameters less than 10 µm are considered to be respirable by humans; ultrafine particles with diameters between 0.001 and 0.1 µm have also become of recent concern (Peters *et al.*, 1997). Examples of the ambient particles found within these size ranges are included in Fig. 14.1. For compact particles, the diameter of the particle inhaled is the dominant variable determining deposition in the respiratory tract. This is demonstrated in Fig. 14.2 where a generalized curve is presented on the relationship between particle size and deposition in the human respiratory tract. The curve, a summary of theoretical and empirical data (Chan and Lippmann, 1980; Task Group on Lung Dynamics, 1966), indicates for compact particles that the particle size leading to minimal deposition in the respiratory tract has a diameter of about 0.5 µm.

Mechanisms of Particle Deposition

The primary mechanisms responsible for particle deposition in the respiratory tract are: (a) settling, whereby particles with densities greater than that of air experience a downward force due to gravity; (b) inertial impaction, whereby particles moving down an airstream that suddenly changes direction will continue to travel in the initial direction of the airstream for a short distance, and may encounter and impact on a surface; and (c) diffusion, whereby random collisions between gas molecules and submicrometer-sized particles push the particles about in an irregular manner that is called Brownian movement. Secondary processes of deposition include interception, which usually applies to fibers, and the electrostatic attraction between charged particles and the image charge on the airway wall (Melandri *et al.*, 1977). Each of these mechanisms is influenced by the pattern of respiratory breathing. For example, an increase in the depth of the tidal volume tends to bring a larger fraction of the particles inhaled into contact with more surface area of the respiratory epithelium, and thus increases the potential for deposition by gravitational settling and diffusion. Respiratory flow rates during inspiratory and expiratory phases of the breath cycle affect particle deposition by velocity-dependent mechanisms, i.e. inertial impaction, which usually occurs for particles with diameters larger than 3 μm. For particles with diameters < 3 μm, time-dependent mechanisms such as sedimentation and diffusion are the most significant mechanisms for deposition of particles. Thus the respiratory pattern (depth of the tidal volume and the breathing rate) has a strong

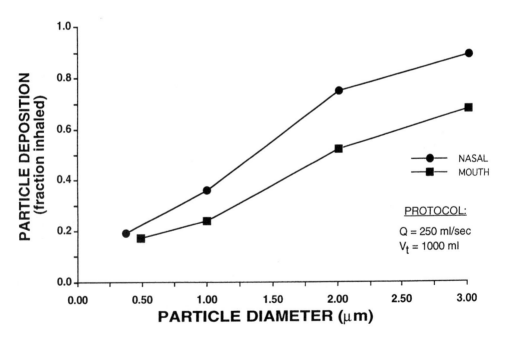

Fig. 14.3   Deposition of compact particles in the human respiratory tract and mode of breathing. Deposition probabilities are compared for nasal and mouth breathing modes of particle inhalation. Four particle sizes are evaluated for each mode of breathing using the tidal volume (V_t) and mean inspiratory flow rate (Q) indicated. Deposition is expressed as the fraction of the inhaled particles that deposit within the upper and lower respiratory tract. Adapted from Heyder *et al.* (1975).

influence over the size of the particle that deposits, the regional location of deposition, and mechanisms that favor deposition in the respiratory tract. Physical activity with attendant changes in the breathing pattern affects particle deposition. Even light physical activity or exercise and the resulting increase in respiratory minute ventilation affects deposition. At rest, breathing is predominantly through the nose, but oral breathing increases with exertion. The mode of breathing (switch-over from nasal to oronasal breathing strategies) is frequently switched when minute ventilation increases; for adults the switch-over point generally occurs at about a mean ventilation rate of 35 l/min (Niinimaa et al., 1980). Particles larger in diameter than 1 μm deposit in the nose and upper respiratory tract by impaction; therefore, at constant flow rate an increase in particle diameter causes deposition to increase and if particle size is fixed, a rise in the mean flow rate leads to an increase in deposition (Swift, 1991). The influence of breathing mode upon respiratory tract particle deposition is demonstrated in Fig. 14.3 (Heyder et al., 1975); these data clearly indicate that the nose (nasal breathing mode) is an effective filter when particle diameter exceeds 1 μm.

Several investigators have assessed breathing pattern and particle deposition. By varying the mean respiratory flow rate Heyder and colleagues investigated the influence of time-dependent factors upon particle deposition within the respiratory tract. The data presented in Fig. 14.4 show deposition following inhalation of non-hygroscopic particles,

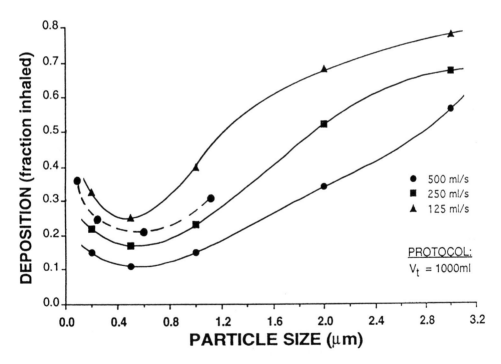

Fig. 14.4 Effect of time-dependent factors on deposition of compact particles. Deposition probability is compared for particles inhaled at three different mean inspiratory flow rates. Particles inhaled by mouth with tidal volume (V_t) and mean inspiratory flow rates indicated. Smooth lines fit by eye to data points of respective inspiratory flow rate: dashed line is fit to data points for tidal volume of 1000 ml and mean inspiratory flow rate of 500 ml/s (see text). Deposition expressed as the fraction of the inhaled particles that deposit within the lower respiratory tract. Adapted from Blanchard and Willeke (1994) and Heyder et al. (1975).

using a tidal volume of 1 liter and three different mean respiratory flow rates (125, 250 and 500 ml/s). For particles that would be primarily influenced by sedimentation and diffusion, for example over the size range of 0.2 to 3.0 µm diameter, a low mean respiratory flow rate, i.e. 125 ml/s, provided a longer respiratory period and particle deposition was enhanced over that found with higher mean flow rates, i.e. 250 and 500 ml/sec. A trough or a limiting particle diameter at which deposition was minimal occurred between 0.4 and 0.5 µm diameter. For compact particles sedimentation and diffusional deposition mechanisms become minimal at this size diameter. Using hygroscopic-type particles Blanchard and Willeke found a similar pattern of limited deposition and their confirmatory results have been added to Fig. 14.4 after correction of particle diameters due to the effects of relative humidity (Blanchard and Willeke, 1984).

Muir and Davies (1967) investigated the influence of varying the depth of the tidal volume, i.e. penetration volume, on particle deposition. These investigators utilized a 0.5 µm diameter particle and, as shown in Fig. 14.5, particle deposition was enhanced by increasing the tidal volume (depth) at which particles were inhaled and penetrated the respiratory tract. An increase in the tidal volume would enlarge the surface area available for particle contact and prolong the time of the respiratory period (and time for deposition) as well. Heyder and colleagues demonstrated this same effect (depth of tidal volume) for a range of particle sizes both below and above the 0.5 µm diameter trough in particle size (Heyder *et al.*, 1975). The corollary to the depth of breathing is the frequency of the

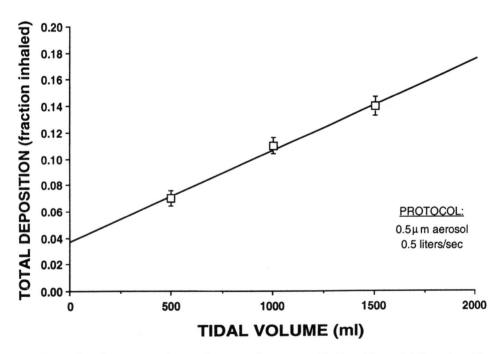

Fig. 14.5 Effect of penetration volume on deposition of compact particle. Deposition probability evaluated for 0.5 µm diameter particle inhaled with three different tidal volumes. Particle inhaled by mouth with the mean inspiratory flow rate indicated; mean deposition (±SE) expressed as the fraction of inhaled particles that deposit within the lower respiratory tract. Adapted from Mair and Davies (1967).

breathing; thus if the mean ventilatory flow rate is increased by raising the frequency of breathing, deposition of fine-sized particles in the respiratory tract is decreased since there is less time available for settling and diffusion of particles to take place (Muir and Davies, 1967). This is demonstrated in Fig. 14.6 for inhalation at two different tidal volumes (penetration depths). Therefore changes in the respiratory breathing pattern as may occur during physical activity or exercise can impact upon the primary mechanisms responsible for particle deposition – inertial impaction, settling, and diffusion – and can thereby modify total or regional particulate burden to the respiratory tract.

Less obvious factors that influence particle deposition are ambient conditions such as increases in temperature or the presence of co-pollutants that can also modify the respiratory pattern and mode of breathing. For example, during exposure to ambient concentrations of ozone, neural afferents in the lung and bronchial musculature are stimulated by degranulation products of reactive oxygen species (generated by ozonolysis of unsaturated fatty acids), such as hydrogen peroxide. This leads to airway responses, e.g. shallow breathing and bronchoconstriction. The effects of co-pollutants on particle deposition are considered in a subsequent section of the review.

The volume of the lung at rest position, i.e. the functional residual capacity, is an additional factor affecting particle deposition mechanisms. Davies and colleagues (1972) evaluated this effect by having healthy subjects inhale test particles (0.5 μm diameter) at specified volumes above and below their normal functional residual capacity. Particle

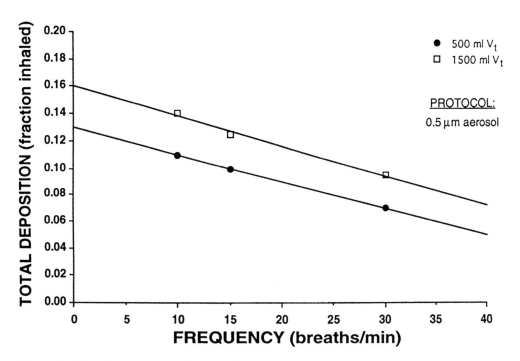

Fig. 14.6   Effect of mean inspiratory flow rate on deposition of compact particle. Deposition probability of 0.5 μm diameter particle inhaled at three different breathing frequencies (breaths/min) are compared for two penetration volumes. Particles inhaled by mouth with tidal volumes (V_t) indicated; deposition expressed as the fraction of inhaled particles that deposit within the lower respiratory tract. Adapted from Mair and Davies (1967).

deposition results are presented in Fig. 14.7. As the lung volume at which particles were inhaled was maintained below or above the normal functional residual capacity of the lung, the potential for deposition increased or decreased, respectively. A decrease in deposition as the functional residual capacity is increased suggested that there was less penetration of the aerosol into a larger resting lung volume. Thus the potential for fine particles, near to 0.5 μm in diameter, to deposit by sedimentation and diffusional mechanisms decreases when the functional residual capacity is increased and the mean airspace size enlarges.

The deposition fraction of ultrafine particles within the human respiratory tract has also been investigated. Heyder and colleagues (1986) utilizing non-hygroscopic-type particles 0.007 to 0.4 μm in diameter found that deposition increased with decreasing diameter. Anderson, Hiller and Mazumder (1988) characterized deposition of ultrafine sized particles in healthy subjects for a variety of respiratory patterns; fractional deposition data are presented in Fig. 14.8, including comparisons for five particle sizes inhaled with three tidal volumes. This investigation also found that the probability for deposition increased as the diameter of ultrafine sized particles decreased. By experimental design the mean flow rate and minute ventilation were held constant and thus when tidal volume was varied, i.e. enlarged, the total residence time and penetration of ultrafine particles into the respiratory tract increased and led to an enhancement of particle deposition.

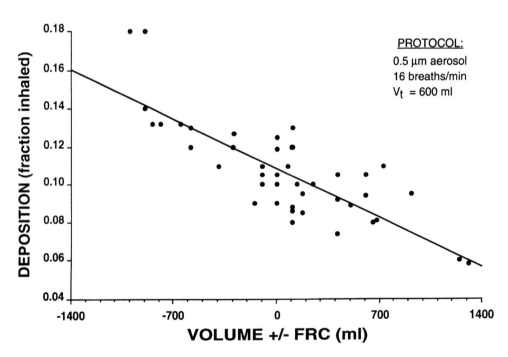

Fig. 14.7 Effect of functional residual volume on deposition of compact particle. Deposition probability evaluated for a 0.5 μm diameter particle inhaled with the lung volume held at volumes above (+) and below (−) the normal volume of the functional residual capacity (0). Particle inhaled by mouth at indicated breathing frequency and tidal volume (V_t). Deposition expressed as the fraction of inhaled particles that deposit within the lower respiratory tract. Solid line represents fit of data points to a linear regression. Adapted from Davies *et al.* (1972).

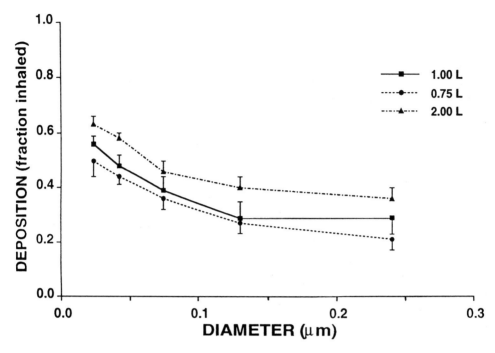

Fig. 14.8 Effect of penetration volume on deposition of ultrafine particles. Deposition probability is compared for ultrafine particles (range of 0.02 to 0.24 μm diameter) inhaled by mouth with three different tidal volumes. Particles inhaled with a mean inspiratory flow rate of 1000 ml/s and minute ventilation of 12 l/min. Solid and broken lines fit to deposition data points for respective penetration volumes; deposition expressed as the fraction of inhaled particles that deposit within the lower respiratory tract. Adapted from Anderson *et al.* (1988).

Effective filtration of ultrafine particles (diameters between 0.007 and 0.2 μm) by the upper respiratory tract has not been uniformly observed (Heyder *et al.*, 1986; Swift *et al.*, 1994); in part inconsistencies may be related to the small number of subjects that were evaluated in these investigations. However, Swift and colleagues, using physical models that were replicated from infant and adult nasal passages, reported that particles with size characteristic of unattached radon progeny aerosol (0.001 μm diameter) had nasal deposition efficiencies ranging from 60 to 95% (Cheng *et al.*, 1988). For larger sized ultrafine particles and a nasal flow of 20 l/min (comparable to that for normal breathing at rest), inspiratory nasal deposition efficiencies were 16 and 40% for 0.01 and 0.005 μm particles, respectively (Tu and Knutson, 1984).

Regional Deposition of Particles in the Respiratory Tract

The information thus far has focused on deposition mechanisms of compact particles in the human respiratory tract and has summarized empirical data obtained while varying the respiratory breathing pattern and resting lung volume. Many inhaled substances, including both pollutant and therapeutic aerosols, change size in a humid environment such as the respiratory tract because of their hygroscopic properties. Descriptions of these

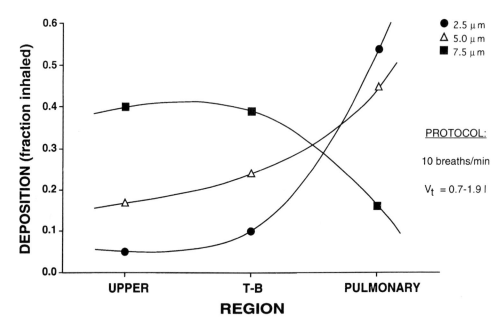

Fig. 14.9 Regional deposition of compact particles in the human respiratory tract. Deposition probability within three regions of the respiratory tract: upper, tracheobronchial (T-B) and pulmonary are compared for three particle sizes: 2.5, 5.0 and 7.5 μm diameter. Particles inhaled by mouth with breathing frequency and tidal volume (V_t) indicated; solid lines are curvilinear fit to regional deposition data. Deposition expressed as the fraction of inhaled particles that deposit within the lower respiratory tract. Adapted from Foord *et al.* (1972).

factors are beyond the scope of this review but predictions for hygroscopic growth of ambient aerosols and the influence on particle deposition in the human respiratory tract can be found in two recent reviews by Feron and colleagues (1989) and by Hiller (1991).

For assessing injury from inhaled particles and potential health consequences, the region or location where deposition has occurred is relevant. The sites of particle deposition and injury in the respiratory tract ultimately define the barriers to be engaged, rates of clearance, and potential for retention or dissolution of particulate into respiratory tissues. This type of information has been gained for the human respiratory tract using radiolabeled aerosols as surrogates for respirable particulate. Isolated or spatial radiation detectors over the thorax and upper respiratory tract are employed to gauge regional particle deposition and retention, i.e. nasopharyngeal, tracheobronchial and pulmonary regions. Early work by Foord and colleagues (1972) demonstrated that for particles within the respirable range, for example 2.5 to 7.5 μm diameter, deposition varied within these 3 principal respiratory regions. Regional particle deposition data from their investigation are presented in Fig. 14.9. Over the size range of particle diameters evaluated, i.e. 2.5 to 7.5 μm, particle burdens within respective respiratory regions reflect the deposition mechanisms with highest potential for each region. Hence larger diameter particles influenced primarily by inertial impaction and velocity-dependent mechanisms deposit in the upper (nasopharyngeal) and tracheobronchial regions and smaller diameter, fine particles are influenced more by sedimentation mechanisms and are more likely to deposit within the pulmonary regions.

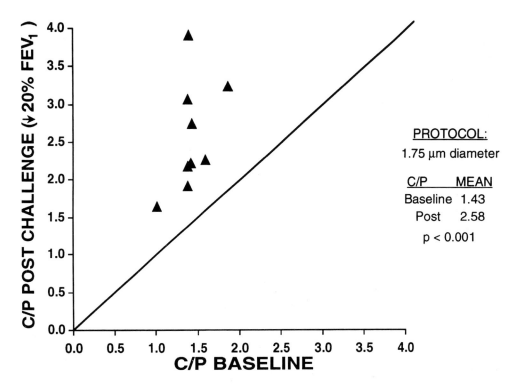

Fig. 14.10 Bronchospasm and regionalization of deposition in asthmatic humans. Compact particles inhaled by mouth at baseline and post-induction of bronchospasm. Regional deposition is characterized by the ratio of particle deposition in central to peripheral airways (C/P index); each data point represents the relationship of C/P index at baseline to C/P index post-bronchospasm in a given subject. All data points positioned above the line of identity (solid line) and thus C/P index increased in value following challenge due to centralization of particle deposition. Bronchospasm of asthmatics sufficient to reduce the forced expiratory volume in 1 s (FEV_1) by 20% from baseline. Mean values of C/P index of subjects ($n=9$) at baseline and post-bronchospasm are indicated with statistical P value. Adapted from O'Riordon *et al.* (1993).

Chamberlain and colleagues demonstrated for fine particle sizes, approximately 1 µm in diameter, that fewer particles tended to deposit in the lung apices and that this vertical gradient in deposition closely paralleled estimates of regional ventilation (Chamberlain *et al.*, 1983). Subsequently others have shown in healthy subjects that for larger sized particles, 2.4 µm and 6.0 µm in diameter, a similar gradient in deposition exists and correlates to the regional gradient in lung volume that is present at functional residual capacity (Groth *et al.*, 1989). Although investigations with particles in the ultrafine range (~0.1 µm) have provided information on uniformity of particle deposition in the respiratory tract, the studies to date have not considered deposition on a regional basis (Lloyd *et al.*, 1994).

For fine sized particles, ranging in diameter between 0.7 and 2.5 µm, it was found that during exercise when minute ventilation increased (breathing frequency and tidal volume both increase) the total fraction of particles that deposit per breath did not change from the fraction that deposited during breathing at rest (Bennett *et al.*, 1985; Morgan *et al.*, 1984). This suggested that over this range of particle sizes, particulate burden to the respiratory tract during exercise is not increased over the burden delivered at rest. However, it is important to note that the regional deposition was altered during exercise such that fewer particles penetrated and deposited peripherally simultaneous with an increase in particle deposition centrally in bronchial airways (Bennett *et al.*, 1985).

Influence of Lung Disease upon Particle Deposition

Epidemiologic reports have associated particulate air pollution with increased mortality among older subjects having cardiopulmonary and respiratory diseases (Dockery and Pope, 1994). Although Bennett and colleagues demonstrated that particle deposition during restful, spontaneous breathing was independent of age in healthy subjects (Bennett *et al.*, 1996), differences are found when deposition is evaluated in representative older subjects with respiratory disease. For example, lung deposition of 1.0 µm diameter particles was evaluated in groups of subjects with varying levels of airway obstruction: asymptomatic smokers, smokers with small airway disease, asthmatics, and patients with chronic obstructive airway disease (Kim and Kang, 1997). During rebreathing of particles all subject groups utilized similar breathing patterns. For the smokers with small airway disease and patients with obstructive lung disease the fractional deposition of particulate, i.e. the fraction retained within the lower respiratory tract, exceeded the fractional deposition found for normal subjects by 49% and 103%, respectively. Particle deposition appeared to increase proportionately with the severity of airway obstruction. Love and Muir evaluated intrapulmonary deposition of 1 µm particles in workers occupationally exposed to particulate, and also found deposition to increase with severity of airway obstruction (Love and Muir, 1976).

With respect to particles in the ultrafine size range (0.02 to 0.24 µm diameter), the potential for deposition increases as the diameter decreases in size. Thus fractional deposition in healthy subjects ranges between 17% for the larger 0.24 µm particles, to 47% for the smaller 0.02 µm diameter particles. In obstructive lung disease patients, particle deposition for ultrafines (diameters of 0.02, 0.04, 0.08, 0.13 and 0.24 µm) is enhanced in comparison to the fractional deposition observed in the normal lung. The average increase in deposition across all particle sizes was 54% but the range of the increases was wide, e.g. 15% for the larger 0.24 µm diameter particle versus 129% for the smaller 0.02 µm diameter particle (Anderson *et al.*, 1990). Potential mechanisms that would enhance deposition in this form of lung disease and suggested by the authors include increased transit time of particles, abnormal expiratory collapse of airways due to flow limitation, and turbulent diffusion as a result of decreased airway caliber.

Following inhalation of radiolabeled test particles, chronic obstructive lung disease patients have demonstrable lung scans with focal, hyperdeposition of particles (Foster *et al.*, 1982; Santolicandro and Giuntini, 1979; Smaldone *et al.*, 1993). Frequently the hyperdeposition occurs in lung regions containing large segmental bronchi that have been identified as sites of expiratory airflow limitation (Smaldone and Smith, 1985). In progressive airway disease, for fine sized particles this effect appears to be independent of particle diameter. Preferential deposition of fine particles within segmental bronchi of obstructed subjects suggests that centralization of deposition may result from some aspect of airway disease, e.g. dynamic collapse and airway flow limitation by these bronchi on expiration (Foster *et al.*, 1988; Smaldone and Messina, 1985). Whether central airway localization of deposition also occurs with ultrafine particles in obstructed patients (as noted above, the deposition fraction of these particles is increased in subjects with chronic airway obstruction) has not been evaluated by regional analysis.

Influence of Bronchospasm and Co-pollutants upon Particle Deposition

The presence of airway hyperreactivity has also been associated with regionalization of particle deposition. As with more severe forms of obstructive airway disease, particle deposition is enhanced within central airways (Backer and Mortensen, 1992; Chung *et al.*, 1988; Clague *et al.*, 1983; O'Riordan *et al.*, 1993; Svartengren *et al.*, 1986). Following an acute non-specific aerosol challenge, asthmatic subjects demonstrate an acute centralization of particle deposition; this is demonstrated in Fig. 14.10. Under ambient conditions, subjects with hyperreactive airways and increased airway tone could be expected to have a similar response and concentrate deposition of ambient, inhalable particulate onto epithelial surfaces of central airways.

Co-pollutants in ambient air and their direct effects on airway and respiratory tissues can also impact upon particulate deposition. For example, during the acute period that follows exposure to an ambient pollutant gas such as ozone, if test particles are inhaled in the 3 μm size range, the fractional deposition of inhaled aerosol that is retained becomes increased as a result of the pre-exposure to ozone. Enhanced deposition within the respiratory tract resulted during normal breathing patterns and was present before ozone-induced inflammatory effects to pulmonary tissues could completely develop. Therefore factors such as non-homogeneous ventilation and mucus hypersecretion, responses that evolve soon after inhalation of ozone, were suggested as more likely factors leading to enhancement in particle deposition (Foster *et al.*, 1993).

Since epidemiologic investigations have implicated acid sulfate aerosols in the exacerbation of respiratory disease, there has been interest in determining if the toxicity of respirable carbon or fly ash particles is enhanced in the presence of acidic air pollutants. However, in a single study to date, healthy and asthmatic subjects exposed to a 1 h period of 0.5 μm carbon aerosol at a concentration of ~250 μg/m^3, plus ~100 μg/m^3 of ultrafine H$_2$SO$_4$ aerosol that was generated from fuming sulfuric acid, did not exhibit an increase in their respiratory irritancy to H$_2$SO$_4$ aerosol. Functional end-points and airway resistance changes were used as outcome variables to evaluate irritancy to sulfate aerosols (Anderson *et al.*, 1992).

CLEARANCE OF PARTICLES OUT OF THE RESPIRATORY TRACT

General Concepts

For particles that deposit onto the epithelial surfaces of the respiratory tract, a complex defense system exists which is capable of solubilization and clearance of particles. The defense system is also stimulated by the ability of particles to induce cellular responses. Clearance is an essential component of the defense system and involves a series of interrelated mechanisms that includes: solubility, macrophage function, mucociliary transport, cellular endocytosis and intercellular sieving, and lymph and capillary blood flow. The fate of particles entering into this system depends upon the deposition site, solubility and the infectious, chemically reactive, or bland nature of the material, as well as physical characteristics of size and shape. Complete reviews are available on nasal physiology and the clearance of deposited particulate from the upper respiratory tract (Fry and Black, 1973; Proctor, 1977; Rusznak et al., 1994); thus the information contained in the remainder of this review will pertain to the lower respiratory tract.

It has been generally accepted that insoluble particles which deposit onto epithelial surfaces of the lower respiratory tract are cleared during two phases (Langenback et al., 1990). The initial phase thought to have clearance half-times between 3 and 12 h and identified as tracheobronchial, concludes by convention after 24 h. The second phase, slow in comparison to the tracheobronchial phase, requires several months to complete and has been identified as alveolar. The early phase is believed to represent clearance of particles which have deposited onto the ciliated surfaces of the conducting airways and which, if insoluble, are removed by mucociliary activity. Particle clearance during the delayed phase is thought to be related to phagocytic activity, lymphatic drainage, and dissolution of particles that have deposited onto respiratory surfaces distal to the respiratory bronchioles. For alveolar clearance, individual mechanisms have not been precisely described and thus physical and cellular routes are frequently considered in aggregate. It is clear that neither mucociliary clearance nor alveolar macrophage forms a perfect defense, and thus some fraction of deposited particles are taken up and sequestered by the airway and alveolar epithelium (Churg, 1996). With regards to the temporal aspects of particulate clearance, the following concepts evolved from *in vivo* studies of particle deposition and clearance. Human investigations between 1955 and 1970 (Albert and Arnett, 1955; Booker et al., 1955; Lippmann and Albert, 1969; Morrow et al., 1968) established the following: (a) after deposition, insoluble particles are rapidly removed from the lung via the tracheal airway during an initial 24 h clearance period; the fraction of particles cleared during this period, expressed as a percentage of the amount of particulate initially deposited, increases with particle diameter, i.e. larger sized particles penetrate less and clear more completely; and (b) post-24 h, the clearance of particulate from the lung is minimal, i.e. the daily rate of particle removal during the first week of clearance commencing on the 2nd day after deposition, averaged 5% or less of the fraction of particles cleared during the first 24 h.

More recent investigations of insoluble particle clearance have tended to support the earlier concepts. Studies designed to evaluate the initial fast phase of clearance (Lourenco et al., 1971; Poe et al., 1977; Sanchis et al., 1972) have demonstrated in healthy subjects that particle clearance concludes after 24 h with little measurable removal of retained particulate during 2–3 days of additional measurement. In a study by Stahlhofen et al. (1980), in which subjects were evaluated with a range of particle sizes (2–9 μm diameters) and two breathing patterns, the fast phase of clearance was found to conclude within 24–34 h of particle deposition. When particles were inhaled at low respiratory flow rates, which would favor sedimentation over impaction mechanisms, the tendency was for the fast phase of particle removal to be extended 6–10 h beyond the conventional 24 h endpoint. After the conclusion of the fast phase (24–34 h) lung retention levels for all particle sizes and breathing patterns remained constant during consecutive measurements for an additional 2- to 5-day period. Additional evidence to support the view that the fast phase of particle clearance immediately apparent after aerosol inhalation concludes after 24 h was provided in replicate studies of healthy subjects which demonstrated repeatability of particle retention found after the initial 24 h period, and uniformity of retention during consecutive measurement periods 2–4 days after deposition (Camner and Philipson, 1978).

Considered together, these investigations affirm that in healthy subjects the fast phase of particle clearance from the lower respiratory tract appears to be complete within 24–34 h of deposition for a variety of insoluble particles above 2 μm in diameter. This clearance phase is presumed to represent clearance of particles from ciliated airways by mucociliary function and expectoration, when present. However, in only one of the investigations was an attempt made to describe the regional distribution (inner, middle and outer areas of two-dimensional lung image) of the 24 h retention (although the effects of regional lung volume on particle retention were not considered)(Sanchis et al., 1972).

The percentage of particles retained within the lower respiratory tract at 24 h post-deposition, i.e. the practical end-point of the fast phase of clearance, also has utility as an index of particle penetration and peripheral airway deposition. For instance, low values of the 24-h retention index are interpreted to indicate that a smaller percentage of the particles inhaled were able to penetrate the lung and deposit in peripheral airways. Thus it is not surprising that functional predictors of airway obstruction, a lung disease which tends to centralize deposition and reduce peripheral penetration of inhaled particles (Dolovich et al., 1976), are associated with lower values of the 24-h lung retention index (Greening et al., 1980; Pavia et al., 1977; Poe et al., 1977).

A number of human investigations have utilized long-lived radiolabeled particles to evaluate the kinetics of the slow phase of lung clearance. The results generally support the concept that in healthy subjects after an initial fast clearance phase the retention of particles in the lung can be described over the next 15–20 days by a single time constant (Stahlhofen et al., 1980). Characteristic and replicable half-times of clearance are reported for a variety of particle species (Camner and Philipson, 1978; Stahlhofen et al., 1981) and for some studies, retention was followed for over 300 days (Bailey et al., 1982; Bohning et al., 1982). This slow phase of lung clearance is proposed to represent particles that had initially deposited onto epithelial surfaces of the respiratory tract distal to terminal bronchioles and their removal by endocytosis, lymphatic drainage and particle dissolution. However, regional distribution of particle retention was not analyzed and thus it is possible that when the slow phase commenced (approximately 1 day after particle inhalation) a component of the clearance may reflect delayed airway clearance or transport of particles from slowly clearing, distal ciliated airways into central bronchi (Stahlhofen et al., 1994).

Overall the investigations of short-term mucociliary clearance and long-term alveolar clearance support the general concept that clearance of insoluble particles from the respiratory tract of healthy subjects occurs in two phases: a fast, or 'tracheobronchial', phase that requires about 24 h to complete; followed by a slow, or 'alveolar', phase that can require several months to complete. Since there was no direct sampling of retained particles in respiratory tract tissues, nor regional analysis of particle distributions at the conclusion of the fast phase or during the slow phase, it is still unclear whether the kinetics of the fast phase of clearance can be used to predict and model the retention time of particles that initially deposit onto the ciliated tracheobronchial airways.

Factors Influencing Particle Clearance

The primary mechanism for removal of insoluble particulate from the lower respiratory tract makes use of the liquid covering of the tracheobronchial airways, which is dynamically transported to the larynx. This process is referred to as mucociliary clearance; in addition to the removal of particulate, sloughed cells and mucus secretions are transported. The airways are largely covered by a thin liquid lining of mucus. Mucus is a viscoelastic secretion that protects the underlying mucosa from dehydration, and contains inhaled particulate that comes into contact with it. Current understanding is that the liquid lining of the airway surface is a two-fluid model in which the upper layer is a viscoelastic gel (mucus, cross-linked glycoproteins) that overlays a sol layer (serous). The serous layer bathes the cilia that protrude from the epithelial surface; thus mucus is thought to be propelled by ciliary beating and flows above the sol layer. Mucus glycoproteins (mucins) are the principal components that confer viscoelasticity. These are produced and released by specialized cells in the epithelium, including the serous cell, the goblet cell, possibly the ciliated cells, and by seromucus glands in the submucosa (Rogers, 1994). Direct stimulation of the airway surface by exogenous agonists and antagonists of the autonomic nervous system has dramatic effects upon airway secretion and the clearance of mucus (Foster et al., 1976; Groth et al., 1991; Marom et al., 1981).

As target-type cells, epithelia can respond to irritant stimuli with alterations in defense or barrier function, e.g. secretion of mucus and alteration of permeability. The prostanoids, for example prostaglandin $PGF_{2\alpha}$, which are known to increase mucus secretion of human respiratory tissues (Shelhamer et al., 1980), are found to be increased in sera of subjects after laboratory exposures to the pollutant gas, ozone (Schelegle et al., 1989) and in lung lavage fluids 3–18 h post exposure to ozone (Devlin et al., 1991; Seltzer et al., 1986). As effector-type cells, epithelia can respond to irritant stimuli by synthesis or release of secondary mediators of inflammation, i.e. prostanoids, oxygen free radicals, and cytokines. Cytokines are a group of soluble polypeptides and glycoproteins that are similar to hormones but are not necessarily endocrine-derived. Cytokines act primarily in the local milieu, usually in autocrine and paracrine fashion, although several have been identified at static levels in the systemic circulation. A number of cell types, including endothelial and epithelial cells, macrophage, and monocytes, are capable of releasing cytokines. Proinflammatory cytokines, e.g. tumor necrosis factor-α (TNF-α) interleukin-6 (IL-6) and IL-8 and other mediators such as $PGF_{2\alpha}$ (as mentioned above) and reactive oxygen intermediates are known to be increased *in vivo* when assayed in lavage fluids from subjects previously exposed to oxidant irritants (Devlin et al., 1991); although the source of $PGF_{2\alpha}$ and the cytokines *in vivo* remains uncertain. Several of these cytokines are reported to regulate immune responses and have been associated with the recruitment, activation, and persistence of polymorphonuclear leukocytes and other cells at airway and respiratory sites of inflammation (Levine, 1995; Torres et al., 1997). The importance or relationship of these cytokines, present in liquids that bathe epithelial surfaces, to injury or impairment of integrative barrier functions (epithelial permeability and particle clearance) of the respiratory epithelium have not been investigated in the *in vivo* setting of the human lung. However, preliminary clearance data collected by our laboratory *in vivo* are presented in Fig. 14.11 and support an association between systemic levels of proinflammatory cytokines, e.g. IL-6, during exposure to an ambient oxidant pollutant such as ozone, and rates of mucociliary clearance assessed

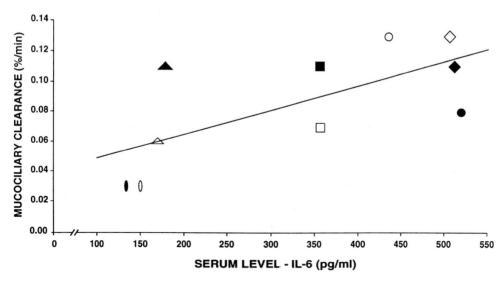

Fig. 14.11 Relationship of serum cytokine to airway clearance of insoluble particles. Mucociliary clearance of particles from peripheral airways was assessed over a 4-h period (initial 2 h included chamber exposure to 300 ppb ozone). Each symbol (solid and open) represents duplicate clearance studies acquired on two separate study days in a given subject (*n*=5). Serums collected at end of the ozone exposure period and analyzed for IL-6. Line of regression (solid line) is indicated (*r*=0.68).

during and after the exposure period to ozone. Regionally this effect predominated in the smaller peripheral airways of the lung.

The interplay between fluid transfer across the airway epithelium (permeability) and the efficiency of particle clearance by the mucociliary apparatus along with respiratory secretions and cellular debris atop the ciliated epithelium have often been postulated as essential components of epithelial barrier function (Nathanson and Nadel, 1984). An association between epithelial permeability and the kinetics of particle clearance from the epithelial surfaces has not been firmly established *in vivo*; however in this connection it is known that smokers have increased parenchymal permeability (O'Brodovich and Coates, 1987) and abnormalities in mucociliary clearance of particulate from the peripheral airways (Foster *et al.*, 1985).

Reorganization of the normal epithelial topography of the airways, as is known to occur in chronic bronchitis, may lead to stasis of mucus flow and prolong retention of deposited particulate within small bronchial airways. The delayed clearance of particles from peripheral bronchi in chronic bronchitis is also consistent with alterations in mucosal permeability (O'Brodovich and Coates, 1987) and inflammatory cellular infiltrates found in respiratory bronchioles of early smokers. Cigarette smoking is known to lead to increased oxidant release by lung phagocytes (Hoidal and Niewoehner, 1983). Inflammatory exudates as a result of reactive oxygen species released onto epithelial surfaces of peripheral airways may alter surfactant and surface tension, and lead to airway narrowing (Kulle *et al.*, 1984). Thus the stability of the airway wall and lumen of peripheral airways (Macklem *et al.*, 1970) required for continuous clearance of particles by mucociliary clearance may falter and lead to increased retention of particulate at epithelial surfaces.

Clearance of Soluble Particles

Up to this point in the review the clearance of particulate from epithelial surfaces of the respiratory tract has focused upon insoluble particles and their clearance by mucociliary function and phagocytotic mechanisms. However, for clearance of soluble particulate, the transfer of materials through epithelial membranes and passage into the bronchial and pulmonary circulations are the primary exit routes from respiratory tissues. An excellent review by Widdicombe on the permeability characteristics of the respiratory epithelium is available (Widdicombe, 1997). Many of the studies to characterize epithelial integrity have utilized small, soluble, radiolabeled hydrophilic solutes, e.g. diethylenetriamine pentaacetate aerosol (DTPA, 392 Da). Clearance of solutes is believed to follow para-cellular pathways, although a fraction of the clearance will be dependent upon solubility within lipid cellular membranes and clearance intracellularly from the apical epithelial surface. Utilizing small markers such as DTPA to gauge permeability of epithelium at distal bronchiolar and alveolar surfaces of humans, soluble particle clearance has been shown to be altered by chronic cigarette smoking. However, following a period of smoking abstention abnormalities in epithelial barrier function are reversible and clearance rates for soluble particles return towards normal (Minty et al., 1981). Acute exercise in healthy subjects also alters epithelial permeability and leads to an increase in clearance of soluble particles from peripheral airway surfaces (Lorino et al., 1989). For cigarette smokers peripheral inflammation is believed to be an important factor in the abnormal clearance kinetics of soluble particles. In the case of exercise, less is known about the mechanism responsible for the increase in particle clearance. Initially increased minute ventilation and prooxidant effects of exercise on the respiratory epithelium were hypothesized as factors causal to the enhancement of soluble particle clearance, but subsequently this was disproved in evaluations of clearance with and without antioxidant intervention (Lorino et al., 1994). Clearance of hydophilic soluble particles has also been found to be extremely rapid in acute lung injury syndromes, e.g. adult respiratory distress syndrome. Although the cause of these changes is unknown, deficiencies in surfactant, injury to alveolar ducts and bronchioles, and the presence of oxidizing agents all may exert an influence (Groth, 1991; O'Brodovich and Coates, 1987).

From an occupational injury standpoint, soluble particle clearance through the respiratory epithelium is abnormally increased in both smoking and non-smoking retired coal miners as compared with subjects unexposed to coal dust (Susskind et al., 1988). The mechanism for persistence of this injury without continued exposure to coal dust (mean time since last exposure was 2 years) is unknown, although the investigation also identified separately the presence of elevated levels in pulmonary uptake of gallium citrate, indicative of an active inflammatory process.

Particle Clearance in the Abnormal Lung

The major risk factor for chronic obstructive lung disease is cigarette smoking; the prevalence of chronic bronchitis (CB) in the US is ~12.5 million people. Smoking is associated with a marked oxidant/antioxidant imbalance in blood and increased levels of lipid peroxidation products in plasma (Rahman et al., 1996). The postmortem study of young smokers by Niewoehner and associates (1974) was the first to demonstrate that smoking produces an inflammatory response that leads to structural changes in the peripheral airways. Studies of bronchial biopsy specimens in chronic bronchitis (CB) suggest that when airflow obstruction is present, a greater number of mucosal T cells and macrophage are present (Distefano *et al.*, 1996). A significant relationship has been found in young smokers between increased generation of oxidants by activated phagocytes selectively removed from peripheral blood and degree of impairment in spirometric function *in vivo* (Richards *et al.*, 1989). Bronchial hypersecretion is an important early feature of CB. We have found that particle transport is usually impaired within peripheral airways in CB as compared to peripheral airway particle clearance of age-matched normal subjects. This is demonstrated in Fig. 14.12 for subjects with normal lung function and limited smoking histories (Foster *et al.*, 1985). In asymptomatic smokers the transport velocity of particulate emanating from stem bronchi is equal to rates of particle transport observed in non-smokers; however, for the peripheral airways distal to stem bronchi, particle clearance half-times are >60% slower in asymptomatic smokers. Overall the peripheral airway effect predominates and therefore, as shown in the figure, the airway clearance of particles in smokers was delayed. Morphologic studies support the concept that the initial lesion of smokers is in the peripheral bronchi, and precedes loss of elastic recoil of central bronchial or parenchymal tissue (Niewoehner *et al.*, 1974).

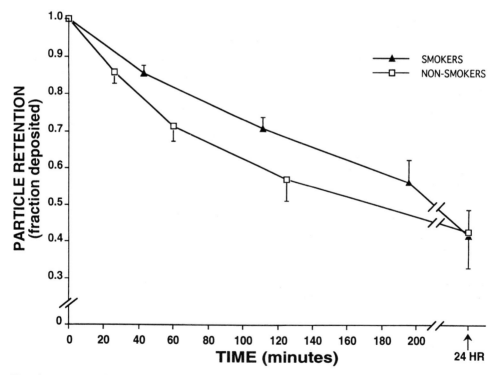

Fig. 14.12 Airway clearance of insoluble particles in subjects with bronchitis. Mucociliary clearance of particles from peripheral airways assessed over 3-h period and compared between age-matched healthy subjects and asymptomatic cigarette smokers. Mean (±SE) data points at indicated times represent clearance of test marker particles from peripheral airways; particle retentions at 24 h end-point represent particles initially deposited onto epithelial surfaces (bronchiolar and alveolar) distal to ciliated airways and clear with a different time course. Adapted from Foster *et al.* (1985).

The major adjunct to the mucociliary function system of particulate clearance from the airways is coughing. Although the physiopathology of chronic cough is not certain, in bronchial biopsies from subjects (non-asthmatic) with chronic cough a number of consistent observations have been found, including increased epithelial desquamation and the presence of inflammatory cells, submucosal fibrosis, squamous cell metaplasia and loss of cilia (Boulet et al., 1994). The effectiveness of cough in the removal of insoluble particulate along with respiratory secretions from airway surfaces is not clearly defined. On the one hand, voluntary coughing is found to assist clearance of mucus from central airways of some patients with obstructive lung disease (Camner et al., 1979; Puchelle et al., 1980; Yeates et al., 1975), but not all (Goodman et al., 1978). On the other hand, frequent episodes of voluntary cough in healthy subjects without abnormal tracheobronchial physiology significantly enhances particle clearance. This response was found to be related to airflow, and independent of the direction of flow; for example maneuvers that raised inspiratory airflows were equally as effective in stimulating clearance of particles from the airways as increases in expiratory airflow (Bennett et al., 1990). More recently, Foster and colleagues reported that elderly subjects with obstructive airway disease and spontaneous cough have significant improvement in particle clearance in the short term (several hours). However, all subjects, including those with impairment of clearance initially (and did not cough), removed similar amounts of particles from the lower respiratory tract by the traditional 24-h end-point of the fast clearance phase (Groth et al., 1997).

For elderly subjects with severe chronic airflow obstruction (COPD) a situation opposite to that found in early bronchitis is present and particle clearance from the epithelial surfaces of the airway is reduced within central, dependent bronchial airways. Previously it had been demonstrated that particle clearance was deficient in COPD (Wanner et al., 1977); and using regional analysis techniques, particle clearance has now been found to be disproportionately affected by the disease process in central airways (Smaldone et al., 1993). In COPD, which diffusely affects lung parenchyma and airways, there was no reason to suspect a priori that the major defect to particle clearance would have been in central airways. This recent evidence of a clearance defect in central bronchi in COPD (Smaldone et al., 1993) and the preponderance of lung injury, such as bronchogenic carcinoma of the central airways, have led to speculation that in COPD clearance of carcinogenic materials may be deficient from tracheobronchial regions.

It is not certain whether the fast or tracheobronchial phase of clearance can be considered complete in persons with lung disease after the initial 24-h period of clearance (Agnew et al., 1981). Sequestration of particles by airway epithelial cells as suggested by animal models of particle clearance (Gore and Patrick, 1982) would also serve to delay tracheobronchial clearance of particles; however, this process is not a uniform occurrence in all animal models (Lay et al., 1995; Velasquez and Morrow, 1984), nor is extrapolation to the human airway certain. The presence of a non-continuous mucous layer or of surfaces of the airway mucoid film in which cephalad movement is continuously absent (Van As, 1977) would also delay clearance of particles from the tracheobronchial airways. However, a number of particle clearance studies designed to evaluate particle retention in the airways of COPD patients have some bearing on this controversy. These investigations, in which the retention measurements extended beyond the conventional 24-h post-deposition end-point, found the tracheobronchial phase to be complete after 1 day (Lin and Goodwin, 1976; Poe et al., 1977; Sanchis et al., 1972; Thomson and Short, 1969). We have described lung clearance of insoluble particles for patients with obstructive lung disease in whom a central pattern of lung deposition was observed for particles inhaled at resting tidal volumes (Foster et al., 1982). Only a small percentage of the subjects had normal rates of particle clearance during an initial measurement period 3 h post-inhalation of the test particles; but 80% of the subjects had zero retention of particles at 24 h post-inhalation. Lung disease in these patients centralized aerosol deposition, but zero retention at 24 h post-inhalation demonstrated that airway clearance of particulate, although impaired and delayed perhaps initially, had been effective. We have also found clearance to occur over a 24 h period in a more recent investigation of subjects with obstructive airway disease who likewise had preferential deposition of inhaled test particles within lobar and segmental bronchi. As a percentage of the particles initially deposited, a mean retention at 24 h post-deposition of 9.6% was observed and ranged between virtually 0% in several subjects to a high of 40% (Groth et al., 1997). A recent evaluation of clearance designed to investigate airway clearance activity beyond the initial 24-h period further established that airway clearance of particles is complete in older subjects with CB or COPD within 72 h of inhalation; in fact only about an additional 3% of the retained particles cleared between the 48 h and 72 h measurement periods (Ericsson et al., 1995).

Markedly abnormal small airways are present in some workers with mineral dust exposure, such that pathologic observation of this lesion is an indicator of dust exposure; the presence of this lesion is associated with abnormalities of airflow (exceeding those induced by smoking alone) (Churg et al., 1985). A recent investigation of particle deposition and clearance from the lower respiratory tract airways of asbestos cement workers suggests that these workers have abnormalities of particle deposition and clearance consistent with obstructive airway disease. For example, workers with effects related to asbestos demonstrated significantly lesser penetration of test particles into bronchiolar and alveolar regions, followed by significantly slower clearance of the particles from ciliated airways in comparison to workers never exposed to respiratory irritants (Lorenzo et al., 1996).

Effects of Ambient Pollutants on Clearance

Major air pollutants that are frequently present in ambient air, along with inhalable particles, include ozone gas and acid sulfur-based aerosol. Ozone's influence upon the mucociliary system of particle clearance has been evaluated in healthy humans. Dose–response characteristics of this system to ozone suggested that regionally the smaller peripheral bronchioles were more sensitive than stem bronchi to ambient concentrations of ozone, i.e. 200 ppb (Foster et al., 1987). Ozone's effect on particle clearance from the lung airways in aggregate are presented in Fig. 14.13. A more recent investigation in humans has suggested that the sensitivity of the mucociliary system during exposure to ozone is short-lived and effects on particle clearance dissipate within a 4 h period post-exposure; however, for this evaluation heavy exercise was a co-treatment with ozone exposure and the experimental design did not investigate the effect of ozone exposure alone (Gerrity et al., 1993). Effects of exercise also merit consideration, as studies in humans have shown increased clearance of both soluble and insoluble particles from epithelial surfaces of the respiratory tract (Bennett et al., 1985; Lorino et al., 1989, 1994; Wolff et al., 1997). A potential stimulus for the acute increase in insoluble particle clearance induced by exposure to ozone is the release by epithelial cells of prostanoid mediators

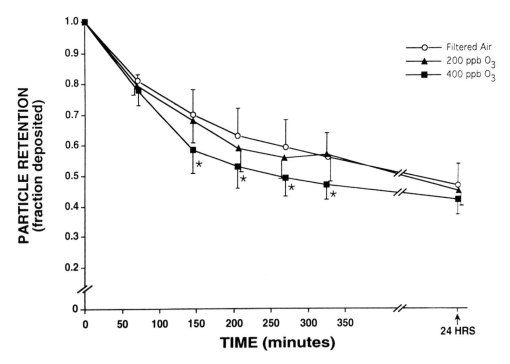

Fig. 14.13 Effect of chamber exposure to ozone on insoluble particle clearance. Mucociliary clearance of particles from lung airways assessed over 5.5-h periods and compared on three separate study days during and after exposures to filtered air or ozone at two concentrations: 200 and 400 ppb. Mean (±SE) data points at indicated times represent clearance of test marker particles from lung airways; particle retentions at 24-h end-point represent particles initially deposited onto epithelial surfaces (bronchiolar and alveolar) distal to ciliated airways and clear with a different time course. Exposure periods to filtered air and ozone were of 2 h duration; * indicates a mean retention level significantly different from the corresponding retention level of filtered air exposure. Adapted from Foster *et al.* (1987).

during and after exposure. For example, prostaglandin $PGF_{2\alpha}$, which is known to enhance release of respiratory secretions by human tissue *in vitro* (Shelhamer *et al.*, 1980), is also found to be increased in the sera of subjects exposed to ozone (Schelegle *et al.*, 1989) and present in fluids lavaged from the lower respiratory tract of humans 3–18 h post-exposure (Devlin *et al.*, 1991; Seltzer *et al.*, 1986).

Investigations *in vivo* of particle clearance from the lower respiratory tract have clearly established sensitivity of healthy subjects to acid sulfate aerosol (Leikauf *et al.*, 1984). The effects of exposure to H_2SO_4 on the clearance of insoluble particles from the airways of healthy subjects are demonstrated for sulfate aerosol at two dose levels in Fig. 14.14. Particle clearance became delayed from the airways after a 1 h nasal inhalation of air containing 100 $\mu g/m^3$ of H_2SO_4 aerosol (0.5 μm diameter). In comparison to an earlier investigation (Leikauf *et al.*, 1981), the distal ciliated airways appeared to be more sensitive to the 100 $\mu g/m^3$ dose than larger, more central bronchi. Causative factors in this response have been investigated *in vitro*. Ciliary activity of epithelial cells in primary culture and sampled from the airway of normal healthy subjects is significantly impaired by exposure to H_2SO_4 pH 5 conditions (to mimic potential effects of acid aerosol pollution)

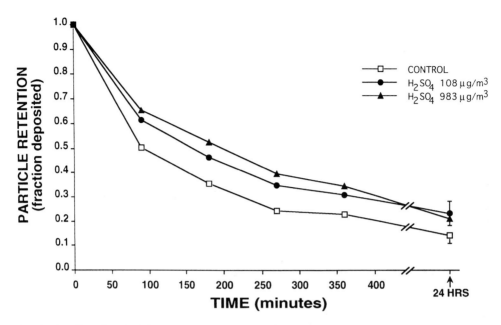

Fig. 14.14 Effect of nasal inhalation of H_sSO_4 aerosol on soluble particle clearance. Mucociliary clearance of particles from lung airways assessed over 6-h periods and compared on three separate study days: control or following 1-h exposures to 108 or 983 µg/m³ H_sSO_4 aerosol. Mean data points at indicated times represent clearance of test marker particles from lung airways; particle retentions at 24-h end-point represent particles initially deposited onto epithelial surfaces (bronchiolar and alveolar) distal to ciliated airways which clear with a different time course. Adapted from Leikauf *et al.* (1981).

(Hastie *et al.*, 1997). Impaired ciliary activity is reversible to normal after a 1 h recovery period at pH 7.4. Significant impairment *in vitro* was also observed for airway epithelium from allergic and asthmatic subjects but recovery did not occur and suggested enhanced susceptibility to H_2SO_4. Exposure to H_2SO_4 pH 6 produced non-significant changes in ciliary function for all subject groups and supported the hypothesis that the H^+ may be the effective component rather than SO_4 in airway injury (Balmes *et al.*, 1989; Schlessinger, 1989). As reviewed above in the first section, the addition of carbon aerosol to sulfuric acid fume did not increase the irritancy of the fume. However, with respect to particle clearance mechanisms, a high effective concentration of carbon dust, when inhaled for a short period as a challenge aerosol, acts as a respiratory irritant and acutely enhances mucociliary clearance of particles from large bronchial airways (Camner *et al.*, 1973). Although the mechanism(s) responsible for increasing mucus clearance are uncertain the inhalation of chemically inert dust particles (coal dust, or charcoal powder) can lead acutely to increases in airway and pulmonary resistance and a slight decrement of lung compliance (Dubois and Dautrebande (1958; Widdicombe *et al.*, 1962). These mechanical lung responses are in part reflexly mediated and dependent upon parasympathetic innervation of the airway (Widdicombe *et al.*, 1962), and the duration of these responses during repetitive exposures to dust particles requires additional investigation (Andersen *et al.*, 1979).

With respect to soluble particle clearance and the influence of co-pollutants commonly found in ambient air, only ozone's effects have been described to date. In laboratory studies, exposure to ozone increases soluble particle clearance from the lower respiratory tract (Kehrl et al., 1987). This effect appears to be dose-dependent, at least during the immediate post-exposure phase when a high, effective dose of ozone, e.g. 400 ppb, and vigorous exercise are used during the exposure. Exposures using mild exercise and more relevant concentrations of ozone, 200–250 ppb, do not influence clearance in the acute period after exposure of healthy or even of asthmatic, subjects (Laube et al., 1996). However if epithelial integrity of peripheral airways is assessed, for example at a later time point post-exposure, i.e. ~24 h later, clearance of soluble hydrophilic particles is significantly altered by exposure to an ambient level of ozone. Regionally, the lung apices demonstrate the largest response: particle clearance increased by 16% above baseline values, followed by peripheral lung regions (encompassing both apical and dependent lung regions but excluding the lung hilum) with an 8% increase (Foster and Stelkiewicz, 1996). These changes in soluble particle clearance through the respiratory epithelium coincide with the inflammatory changes (cellular and biochemical) that are observed by others (Devlin et al., 1991) in lavage fluids collected from the lower respiratory tract and collected at similar times points (~24 h post-exposure) and document the passage of low molecular weight particles from the airway lumen into blood. Exercise, which is frequently used during laboratory exposure to pollutants to raise minute ventilation and hence the dose of the respirable pollutant being evaluated, in itself causes an increase in the clearance of soluble particles from the parenchymal surfaces of the lung (Lorino et al., 1989, 1994). The mechanisms responsible for this exercise effect are unknown.

SUMMARY

Inertial impaction, sedimentation and diffusion are the primary mechanisms which govern particle deposition in the human lung. During particle exposures, ventilatory parameters known to affect regional lung deposition include: the depth of the tidal volume, which determines particle penetration into small airways; mean inspiratory airflow, which is an index of forces governing aerosol impaction onto airway walls; and the duration of the breath cycle, which determines respiratory residence time and influences sedimentation and diffusion of particles. In healthy subjects, exposure to ozone acutely enhances the lung deposition of fine particles. The presence of obstructive lung disease centralizes deposition of inhaled particulate and enhances the deposition of fine and ultrafine particles. Clearance of particles from the lower respiratory tract is dependent upon location and solubility of the particle. For insoluble particles removal is a dynamic process involving two phases: a short-term airway phase and a long-term alveolar component. Neither phase is a complete or perfect defense against deposited particulate. In bronchitis and chronic obstructive lung disease, abnormalities in the short-term phase of particle clearance are present in peripheral and central airways, respectively. Air pollutants, such as ozone and H_2SO_4 aerosol, which alter secretion of fluids onto airway surfaces by mediator release and H^+ effects, influence clearance of insoluble particles acutely. Retention time of soluble particles is dependent upon the integrity of the epithelial membranes and clearance is rapid in comparison to that of insoluble particles. Passage of soluble particles into bronchial and pulmonary circulations is enhanced by exercise and exposure to ozone in healthy subjects and by lung disease associated with inflammation and oxidant stress to epithelial tissues. Additional investigations that validate nasal and regional deposition of ultrafine sized particles in the lower respiratory tract (Anderson *et al.*, 1990; Cheng *et al.*, 1988), and clearance kinetics from the respiratory epithelium (Roth *et al.*, 1994) are warranted. Abnormalities and delays in the removal of particles from respiratory epithelia may predispose at-risk members of the population, i.e. those with inflammatory or obstructive airway disease, to pollution-induced increases in morbidity. As laboratory investigations have become increasingly sophisticated this should lead to a better understanding of particle dosimetry and cellular and molecular phenomena associated with inhalation of ambient particulate.

ACKNOWLEDGEMENTS

Dr W. Michael Foster is supported by awards from the Center for Indoor Air Research (#97–11, Linthicum, MD) and the National Institutes of Environmental Health Sciences and Heart, Lung and Blood (#ES–03819, National Institutes of Health, Washington, DC). The assistance of Ms Kristen Macri in the preparation of the artwork for this chapter and for technical support is gratefully acknowledged.

REFERENCES

Agnew JE, Pavia D and Clarke SW (1981) Airways penetration of inhaled radioaerosol: an index to small airways function. *Eur J Respir Dis* **62**: 239–255.
Albert RE and Arnett LC (1955) Clearance of radioactivity dust from the lung. *Arch Environ Health* **12**: 99–106.
Andersen T, Lunqvist GR, Proctor DF and Swift DL (1979) Human response to controlled levels of inert dust. *Am Rev Respir Dis* **119**: 616–627.
Anderson KR, Avol EL, Edwards SA *et al.* (1992) Controlled exposures of volunteers to respirable carbon and sulfuric acid aerosols. *J Air Waste Manag Assoc* **42**: 770–776.
Anderson PJ, Hiller FC and Mazumder MK (1988) Deposition of 0.02–0.2 μm particles in the human respiratory tract: effects of variations in respiratory pattern. *Ann Occup Hyg* **32** S1: 91–100.
Anderson PJ, Wilson JD and Hiller FC (1990) Respiratory tract deposition of ultrafine particles in subjects with obstructive or restrictive lung disease. *Chest* **97**: 1115–1120.
Attfield MD and Hodous TK (1992) Pulmonary function of U.S. coal miners related to dust exposure estimates. *Am Rev Respir Dis* **145**: 605–609.
Backer V, and Mortensen J (1992) Distribution of radioactive aerosol in the airways of children and adolescents with bronchial hyper-responsiveness. *Clin Physiol* **12**: 575–585.
Bailey MR, Fry FA and James AC (1982) The long term clearance kinetics of insoluble particles from the human lung. *Ann Occup Hyg* **26**: 273–290.
Balmes JR, Fine JM, Gordon T and Sheppard D (1989) Potential bronchoconstrictor stimuli in acid fog. *Environ Health Perspect* **79**: 163–166.
Bennett WD, Messina M and Smaldone GC (1985) Effect of exercise on deposition and subsequent retention of inhaled particles. *J Appl Physiol* **59**: 1046–1054.
Bennett WD, Foster WM and Chapman WF (1990) Cough-enhanced mucus clearance in the normal lung. *J Appl Physiol* **69**: 1670–1675.
Bennett WD, Zeman KL and Kim C (1996) Variability of fine particle deposition in healthy adults: Effect of age and gender. *Am J Respir Crit Care Med* **153**: 1641–1607.
Blanchard JD and Willeke K (1984) Total deposition of ultrafine sodium chloride particles in human lungs. *J Appl Physiol* **57**: 1850–1856.

Bohning DE, Atkins HL and Cohen SH (1982) Long-term particle clearance in man: normal and impaired. *Ann Occup Hyg* **26**: 259–271.

Booker DV, Chamberlain AC, Rundo J *et al.* (1955) Elimination of 5 micron particles from the human lung. *Nature* **215**: 30–33.

Boulet L-P, Milot J, Boutet M *et al.* (1994) Airway inflammation in nonasthmatic subjects with chronic cough. *Am J Respir Crit Care Med* **149**: 482–489.

Camner P and Philipson MS (1978) Human alveolar deposition of 4 micron teflon particles. *Arch Environ Health* **33**: 181–185.

Camner P, Helstrom P-A and Philipson K (1973) Carbon dust and mucociliary transport. *Arch Environ Health* **26**: 294–296.

Camner P, Mossberg B, Philipson K and Strandberg K (1979) Elimination of test particles from the human tracheobronchial tract by voluntary coughing. *Scand J Respir Dis* **60**: 56–62.

Chamberlain MJ, Morgan WKC and Vinitski S (1983) Factors influencing the regional deposition of inhaled particles in man. *Clin Sci* **64**: 69–78.

Chan TL and Lippmann M (1980) Experimental measurements and empirical modelling of the regional deposition of inhaled particles in humans. *Am Ind Hyg Assoc J* **41**: 399–409.

Cheng Y-S, Yamada U, Yeh H-C and Swift DL (1988) Diffusional deposition of ultrafine aerosols in a human nasal cast. *J Aer Sci* **19**: 741–751.

Chung KF, Jeyasingh K and Snashall PD (1988) Influence of airway calibre on the intrapulmonary dose and distribution of inhaled aerosol in normal and asthmatic subjects. *Eur Respir J* **1**: 890–895.

Churg A (1996) The uptake of mineral particles by pulmonary epithelial cells. *Am J Respir Crit Care Med* **154**: 1124–1140.

Churg A, Wright JL, Wiggs B *et al.* (1985) Small airways disease and mineral dust exposure. *Am Rev Respir Dis* **131**: 139–143.

Clague H, Ahmad D, Chamberlain MJ *et al.* (1983) Histamine bronchial challenge: effect on regional ventilation and aerosol deposition. *Thorax* **38**: 668–675.

Davies CN, Heyder J and Subba Ramu MC (1972) Breathing of half-micron aerosols I. Experimental. *J Appl Physiol* **5**: 591–600.

Devlin RB, McDonnell WF, Mann R *et al.* (1991) Exposure of humans to ambient levels of ozone for 6.6 hr causes cellular and biochemical changes in the lung. *Am J Respir Cell Mol Biol* **4**: 72–81.

DiStefano A, Turato G, Maestrelli P *et al.* (1996) Airflow limitation in chronic bronchitis is associated with T-cell and macrophage infiltration of the bronchial mucosa. *Am J Res Crit Care Med* **153**: 629–632.

Dockery DW and Pope CA (1994) Acute respiratory effects of particulate air pollution. *Ann Rev Pub Health* **15**: 107–132.

Dolovich MB, Sanchis J, Rossman C and Newhouse MT (1976) Aerosol penetrance: a sensitive index of peripheral airways obstruction. *J Appl Physiol* **40**: 468–471.

Dubois AR and Dautrebande L (1958) Acute effects of breathing inert dust particles and of carbachol aerosol on the mechanical characteristics of the lungs in man. Changes in response after inhaling sympathomimetic aerosols. *J Clin Invest* **37**: 1746–1755.

Ericsson CH, Svartengren K, Svartengren M *et al.* (1995) Repeatability of airway deposition and tracheobronchial clearance rate over three days in chronic bronchitis. *Eur Respir J* **8**: 1886–1893.

Feron GA, Oberdorster G and Henneberg R (1989) Estimation of the deposition of aerosolized drugs in the human respiratory tract due to hygroscopic growth. *J Aer Med* **3**: 271–284.

Foord N, Black A and Walsh M (1972) Pulmonary deposition of inhaled particles with diameters in the range of 2.5 to 7.5 μm. In: Walton WH (ed.) *Inhaled Particles IV*. Oxford: Pergamon Press, pp. 137–149.

Foster WM and Stetkiewicz (1996) Regional clearance of solute from the respiratory epithelia: 18–20 hours postexposure to ozone. *J Appl Physiol* **81**: 1143–1149.

Foster WM, Bergofsky EH, Bohning DE *et al.* (1976) Effect of adrenergic agents and their mode of action on mucociliary clearance in man. *J Appl Physiol* **41**: 146–152.

Foster WM, Langenback EG and Bergofsky EH (1982) Lung mucociliary function in man: interdependence of bronchial and tracheal mucus transport velocities with lung clearance in bronchial asthma and healthy subjects. *Ann Occup Hyg* **26**: 227–244.

Foster WM, Langenback EG and Bergofsky EG (1985) Disassociation in the mucociliary function of central and peripheral airways of asymptomatic smokers. *Am Rev Respir Dis* **132**: 633–639.

Foster WM, Costa DL and Langenback EG (1987) Ozone exposure alters tracheobronchial mucociliary function in humans. *J Appl Physiol* **63**: 996–1002.

Foster WM, Langenback EG, Smaldone GC and Bergofsky EH (1988) Flow limitation on expiration induces central particle deposition and disrupts effective flow of airway mucus. *Ann Occup Hyg* **32** S1: 101–111.

Foster WM, Silver JA and Groth ML (1993) Exposure to ozone alters regional function and particle dosimetry in the human lung. *J Appl Physiol* **75**: 1938–1945.

Fry FA and Black A (1973) Regional deposition and clearance of particles in the human nose. *J Aer Sci* **4**: 113–124.

Gerrity TR, Lee PS, Hass FJ *et al.* (1979) Calculate deposition of inhaled particles in the airway generations of normal subjects. *J Appl Physiol* **46**: 867–873.

Gerrity TR, Bennett WD, Kehrl H and DeWitt P (1993) Mucociliary clearance of inhaled particles measured 2 h after ozone exposure in humans. *J Appl Physiol* **74**: 2984–2989.

Goodman RM, Yertin BM, Landa JF *et al.* (1978) Relationship of smoking history and pulmonary function tests to tracheal mucus velocity in nonsmokers, young smokers, ex-smokers, and patients with chronic bronchitis. *Am Rev Respir Dis* **11**: 205–214.

Gore DJ and Patrick G (1982) A quantitative study of the penetration of insoluble particles into the tissue of the conducting airways. *Ann Occup Hyg* **26**: 149–162.

Greening AP, Miniati M and Fazio F (1980) Regional deposition of aerosols in health and in airways obstruction: a comparison with krypton-81m ventilation scanning. *Bull Eur Physiopath Respir* **16**: 287–298.

Groth ML, Bennett WD, Cartagena M *et al.* (1989) Regional ventilation and breathing pattern (not particle size): key factors for apex/base gradient in lung deposition of mono- and poly-disperse aerosols. *Am Rev Respir Dis* **139**: A245.

Groth ML, Langenback EG and Foster WM (1991) Influence of inhaled atropine on lung mucociliary function in humans. *Am Rev Respir Dis* **144**: 1042–1047.

Groth ML, Macri K and Foster WM (1997) Cough and mucociliary transport of airway particulate in chronic obstructive lung disease. *Ann Occup Hyg* **41**(S1): 515–521.

Groth S (1991) Pulmonary clearance of ^{99m}Tc-DTPA. *Dan Med J* **38**: 101–113.

Hastie AT, Everts KB, Zangrilli J *et al.* (1997) HSP27 elevated in mild allergic inflammation protects airway epithelium from H_2SO_4 effects. *Lung Cell Mol Physiol* **17**: L401–L409.

Heyder J, Armbuster L, Gebhart J *et al.* (1975) Total deposition of aerosol particles in the human respiratory tract for nose and mouth breathing. *J Aer Sci* **6**: 311–328.

Heyder J, Gebhart J, Rudolf G *et al.* (1986) Deposition of particles in the human respiratory tract in the size range 0.005–15 μm. *J Aer Sci* **5**: 811–825.

Hiller FC (1991) Health implications of hygroscopic particle growth in the human respiratory tract. *J Aer Med* **4**: 1–23.

Hoidal JR and Niewoehner DE (1983) Lung phagocyte recruitment and metabolic deterioriation induced by cigarette smoke in humans and hamsters. *Am Rev Respir Dis* **126**: 548–552.

Huchon GJ, Russell JA, Barritault LG *et al.* (1984) Chronic air-flow limitation does not increase respiratory epithelial permeability assessed by aerosolized solute, but smoking does. *Am Rev Respir Dis* **130**: 457–460.

Kehrl HR, Vincent LM, Kowalsky RJ *et al.* (1987) Ozone exposure increases respiratory epithelial permeability in humans. *Am Rev Respir Dis* **135**: 1124–1128.

Kennedy SM, Wright JL, Mullen JB *et al.* Pulmonary function and peripheral airway disease in patients with mineral dust or fume exposure. *Am Rev Respir Dis* **132**: 1294–1299.

Kim CS and Kang TC (1997) Comparative measurement of lung deposition of inhaled fine particles in normal subjects and patients with obstructive airway disease. *Am J Respir Crit Care Med* **155**: 899–905.

Kulle TJ, Milman JH, Sauder LR *et al.* (1984) Pulmonary function adaptation to ozone in subjects with chronic bronchitis. *Environ Res* **34**: 55–63.

Langenback EG, Bergofsky EH, Halpern JG and Foster WM (1990) Supramicron-sized particle clearance from alveoli: route and kinetics. *J Appl Physiol* **69**: 1302–1308.

Laube B, Kurian V and Foster WM (1996) Fine particle deposition and epithelial clearance after ozone exposure in normal and asthma subjects. *Am J Respir Crit Care Med* **153**: A700.

Lay JC, Berry CR, Kim CS and Bennett WD (1995) Retention of insoluble particles after local intrabronchial deposition in dogs. *J Appl Physiol* **79**: 1921–1929.

Leikauf G, Yeates DB, Wales KA *et al.* (1981) Effects of sulfuric acid aerosol on respiratory mechanics and mucociliary particle clearance in healthy nonsmoking adults. *Am Ind Hyg Assoc J* **42**: 273–282.

Leikauf G, Spektor DM, Albert RE and Lippmann M (1984) Dose-dependent effects of submicrometer sulfuric acid aerosol on particle clearance from ciliated human lung airways. *Am Ind Hyg Assoc J* **45**: 285–292.

Levine S (1995) Bronchial epithelial cell–cytokine interactions in airway inflammation. *J Invest Med* **43**: 241–249.

Lin MS and Goodwin DA (1976) Pulmonary distribution of an inhaled radioaerosol in obstructive pulmonary disease. *Radiology* **118**: 645–651.

Lippmann M and Albert RE (1969) The effect of particle size on the regional deposition of inhaled aerosols in the human respiratory tract. *Am Ind Hyg Assoc J* **30**: 257–275.

Lloyd JJ, James JM, Shields RA and Testa HJ (1994) The influence of inhalation technique on technegas particle deposition and image appearance in normal volunteers. *Eur J Nucl Med* **21**: 394–398.

Lorenzo LD, Mele M, Pegorari MM *et al.* (1996) Lung cinescintigraphy in the dynamic assessment of ventilation and mucociliary clearance of asbestos cement workers. *Occup Environ Med* **53**: 628–635.

Lorino AM, Meignan M, Bouissou P and Atlan G (1989) Effects of sustained exercise on pulmonary clearance of aerosolized ^{99m}Tc-DTPA. *J Appl Physiol* **67**: 2055–2059.

Lorino AM, Paul M, Cocea L *et al.* (1994) Vitamin E does not prevent exercise-induced increase in pulmonary clearance. *J Appl Physiol* **77**: 2219–2223.

Lourenco RV, Klimek MF and Borowski CJ (1971) Deposition and clearance of 2 micron particles in the tracheo-bronchial tree of normal subjects – smokers and nonsmokers. *J Clin Invest* **50**: 1411–1420.

Love RG and Muir DCF (1976) Aerosol deposition and airway obstruction. *Am Rev Respir Dis* **114**: 891–897.

Macklem PT, Proctor DF and Hogg JC (1970) The stability of peripheral airways. *Respir Physiol* **8**: 191–203.

Marom Z, Shelhamer JH and Kaliner M (1981) Effects of arachidonic acid and prostaglandins on the release of mucous glycoproteins from human airways *in vitro*. *J Clin Invest* **67**: 1695–1702.

Martonen T (1983) Deposition of inhaled particulate matter in the upper respiratory tract, larynx, and bronchial airways: mathematical description. *J Toxicol Environ Health* **12**: 787–800.

Melandri C, Prodi V, Tarroni G *et al.* (1977) On the deposition of unipolarly charged particles in the human respiratory tract. In: Walton WH (ed.) *Inhaled Particles IV*. Oxford: Pergamon Press, pp. 193–201.

Minty BD, Jordan C and Jones JG (1981) Rapid improvement in abnormal pulmonary epithelial permeability after stopping cigarettes. *Br Med J* **282**: 1183–1186.

Morgan WKC, Ahmad D, Chamberlain MJ *et al.* (1984) The effect of exercise on the deposition of an inhaled aerosol. *Respir Physiol* **56**: 327–338.

Morrow PE, Gibb FR and Gazioglu KM (1968) A study of particulate clearance from the human lungs. *Am Rev Respir Dis* **96**: 1209–1221.

Muir DCF and Davies CN (1967) The deposition of 0.5 μ diameter aerosols in the lungs of man. *Ann Occup Hyg* **10**: 161–174.

Nathanson I and Nadel JA (1984) Movement of electrolytes and fluid across airways. *Lung* **162**: 125–137.

Niewoehner DE, Kleinerman J and Rice DB (1974) Pathologic changes in the peripheral airways of young cigarette smokers. *N Eng J Med* **291**: 755–758.

Niinimaa V, Cole P, Mintz S and Shephard RJ (1980) The switching point from nasal to oronasal breathing. *Respir Physiol* **42**: 61–71.

O'Brodovich H and Coates G (1987) Pulmonary clearance of Tc-DTPA: a non-invasive assessment of epithelial integrity. *Lung* **165**: 1–16.

O'Riordan TG, Walser L and Smaldone GC (1993) Changing patterns of aerosol deposition during methacholine bronchoprovocation. *Chest* **103**: 1385–1389.

Pavia D, Thomson MI, Clarke SW and Shannon HS (1977) Effect of lung function and mode of inhalation on penetration of aerosol into the human lung. *Thorax* **32**: 194–197.

Peters A, Wichmann HE, Tuch T *et al.* (1997) Comparison of the number of ultra-fine particles and the mass of fine particles with respiratory symptoms in asthmatics. *Ann Occup Hyg* **41** S1: 19–23.

Poe ND, Cohen MB and Yanda RL (1977) Application of delayed lung imaging following radioaerosol inhalation. *Radiology* **122**: 739–746.

Proctor DF (1977) The upper airways. I. Nasal physiology and defense of the lungs. *Am Rev Respir Dis* **115**: 97–129.

Puchelle E, Zahm JM, Girard F *et al.* (1980) Mucociliary transport *in vivo* and *in vitro*. *Eur J Respir Dis* **61**: 254–264.

Rahman I, Morrison D, Donaldson K and MacNee W (1996) Systemic oxidative stress in asthma, COPD, and smokers. *Am J Respir Crit Care Med* **154**: 1055–1060.

Richards GA, Theron AJ, Van der Merwe CA and Anderson R (1989) Spirometric abnormalities in young smokers correlate with increased chemiluminescence responses of activate blood phagocytes. *Am Rev Respir Dis* **139**: 181–187.

Rogers DF (1994) Airway goblet cell: responsive and adaptable front-line defenders. *Eur Respir J* **7**: 1690–1706.

Roth C, Scheuch G and Stahlhofen W (1994) Clearance measurements with radioactively labelled ultra-fine particles. *Ann Occup Hyg* **38**(S1): 101–106.

Rusznak C, Devalia JL, Lozewicz S and Davies RJ (1994) The assessment of nasal mucociliary clearance and the effect of drugs. *Respir Med* **88**: 89–101.

Sanchis J, Dolovich M, Chalmers R and Newhouse M (1972) Quantitation of regional aerosol clearance in the normal human lung. *J Appl Physiol* **33**: 757–762.

Santolicandro A and Giuntini C (1979) Patterns of deposition of labelled monodisperse aerosols in obstructive lung disease. *J Nucl Med All Sci* **23**: 115–127.

Schelegle ES, Adams WC, Giri SN and Siefkin AD (1989) Acute ozone exposure increases plasma PGF2α in ozone sensitive human subjects. *Am Rev Respir Dis* **140**: 211–216.

Schlessinger RB (1989) Factors affecting the response of lung clearance systems to acid aerosols: role of exposure concentration, exposure time, and relative acidity. *Environ Health Perspect* **79**: 121–126.

Seltzer J, Bigby BG, Stulbarg M *et al.* (1986) Ozone induced change in bronchial reactivity to methacholine and airway inflammation in humans. *J Appl Physiol* **60**: 1321–1326.

Shelhamer JH, Marom Z and Kaliner M (1980) Immunologic and neuropharmacologic stimulation of mucous glycoprotein release from human airways *in vitro*. *J Clin Invest* **66**: 1400–1408.

Smaldone GC and Messina MS (1985) Flow limitation, cough, and patterns of aerosol deposition in humans. *J Appl Physiol* **59**: 515–520.

Smaldone GC and Smith PL (1985) Location of flow-limiting segments via airway catheters near residual volume in humans. *J Appl Physiol* **59**: 502–508.

Smaldone GC, Foster WM, O'Riordan TG *et al.* (1993) Regional impairment of mucociliary clearance in chronic obstructive pulmonary disease. *Chest* **103**: 1390–1396.

Stahlhofen W, Gebhart J and Heyder J (1980) Experimental determination of the regional deposition of aerosol particles in the human respiratory tract. *Am Ind Hyg Assoc J* **41**: 385–398.

Stahlhofen W, Gebhart J, Heyder J *et al.* (1981) Intercomparison of regional deposition of aerosol particles in the human respiratory tract and their long-term elimination. *Exp Lung Res* **2**: 113–131.

Stahlhofen W, Scheuch G and Bailey MR (1994) Measurement of the tracheobronchial clearance of particles after bolus inhalation. *Ann Occup Hyg* **38** S1: 189–196.

Susskind H, Brill AB and Harold WH (1988) Clearance of Tc-99m aerosol from coal miners' lungs. *Ann Occup Hyg* **32** S1: 157–169.

Svartengren M, Philipson K, Linnman L and Camner P (1986) Regional deposition of particles in human lung after induced bronchoconstriction. *Exp Lung Res* **10**: 223–233.

Swift DL (1991) Inspiratory inertial deposition of aerosols in the human nasal airway replicate casts: implication for the proposed NCRP lung model. *Rad Protect Dosimetry* **38**: 29–34.

Swift DL, Cheng Y-S, Su Y-F and Yeh H-C (1994) Ultrafine aerosol deposition in the human nasal and oral passages. *Ann Occup Hyg* **38** S1: 77–81.

Task Group on Lung Dynamics (1966) Deposition and retention models for internal dosimetry of the human respiratory tract. *Health Phys* **12**: 173–207.

Thomson ML and Short MD (1969) Mucociliary function in health, chronic obstructive airway disease, and asbestosis. *J Appl Physiol* **26**: 535–539.

Torres A, Utel MJ, Morrow PE and Frampton MW (1997) Airway inflammation in smokers and non-smokers with varying responsiveness to ozone. *Am J Respir Crit Care Med* **156**: 728–736.

Tu KW, and Knutson EO (1984) Total deposition of ultrafine hydrophobic and hygroscopic aerosols in the human respiratory system. *Aer Sci Tech* **1**: 453–465.

Van As A (1977) Pulmonary airway clearance mechanisms: a reappraisal. *Am Rev Respir Dis* **115**: 721–726.

Velasquez DJ and Morrow PE (1984) Estimation of guinea pig tracheobronchial transport rates using a compartmental model. *Exp Lung Res* **7**: 163–176.

Wanner A, Salathe M and O'Riordan TG (1977) Mucociliary clearance in the airways. *Am Rev Respir Dis* **116**: 73–125.

Widdicombe J (1997) Airway and alveolar permeability and surface liquid thickness: theory. *J Appl Physiol* **82**: 3–12.

Widdicombe JG, Kent DC and Nadel JA (1962) Mechanism of bronchoconstriction during inhalation of dust. *J Appl Physiol* **17**: 613–616.

Wolff RK, Dolovich MB, Obminski G and Newhouse MT (1977) Effects of exercise and eucapnic hyperventilation on bronchial clearance in man. *J Appl Physiol* **43**: 46–50.

Yeates DB, Aspin N, Levison H *et al.* (1975) Mucociliary transport rates in man. *J Appl Physiol* **39**: 487–495.

15

Respiratory Reflexes

JOHN WIDDICOMBE

Sherrington School of Physiology, St Thomas' Hospital, London, UK

INTRODUCTION

At least 15 types of afferent ('sensory') nervous receptors have been identified in the respiratory tract and lungs (Table 15.1), and many of these are stimulated by inhaled pollutants. The central reflex responses to stimulation are similarly abundant, those on breathing (e.g. cough) being the most conspicuous, but there are many others involving motor systems in the autonomic nervous system. There are also reflex effects on skeletal muscles in the upper airways and larynx, and on somatic muscle tone via the spinal cord. When a pollutant in adequate concentration is inhaled, potentially very many of these reflexes will be activated, and the bodily response will be the intregral of reflexes. Experimentally the reflexes have usually been studied in isolation and in experimental animals, and the way they interact has received far less attention. Evidence of their role in humans is usually by analogy with animal results.

Inhaled pollutants may be irritant and cause damage to the tissues, almost by definition, although in low concentrations they could activate physiological mechanisms without tissue damage. This chapter will deal only with irritant pollutants and those, such as allergens, that set up tissue responses either in healthy or in allergic humans; non-irritant pollutants are unlikely to induce reflex responses. It is a useful convention to talk of two types of response: defensive, which tend to repel or extrude the invading irritant and thereby prevent it causing damage; and protective, when the reflexes will compensate for any damage done by any irritant which has not been kept out (Korpas and Tomori, 1979). In general terms there are two types of sensory receptors in the respiratory system that respond to tissue damage or potential damage and set up defensive or protective reflexes: C fibre receptors with non-myelinated afferent fibres, and rapidly adapting receptors (RARs) with thin myelinated Aδ afferent fibres. Both types of receptor are nociceptive and polymodal, responding to an enormous variety of irritants and inflammatory

AIR POLLUTION AND HEALTH
ISBN 0-12-352335-4

Table 15.1 Airway sensory receptors and responses to irritants

Site	Receptor	Response to irritants
Nose	Touch	?+
	Cold/flow	?0
	C fibre	?++
Pharynx	Touch	?
	C fibre	?++
Larynx	Pressure	0
	Cold/flow	0
	Drive	0
	Irritant	+++
	C fibre	+++
Trachea/bronchi	SAR	0
	RAR	+++
	C fibre	+++
	NEB	?0
Alveoli	C fibre	+++

+, positive response; 0, no response; SAR, slowly adapting stretch receptor; RAR, rapidly adapting stretch receptor; NEB, neuroepithelial body.

tissue mediators. In this respect the respiratory receptors resemble those of many other tissues: in skin, skeletal muscle and many viscera there are two groups of polymodal nociceptor, with C and Aδ afferent fibres, respectively. Although it may be an oversimplification, on the whole the Aδ receptors mediate the rapid defensive responses, whereas both groups of receptors initiate slower protective reflexes.

This chapter will first discuss briefly the local axon reflex responses to receptor stimulation by irritants; second, the central reflex responses to inhaled irritants for each part of the respiratory tract and lungs in turn; and finally, the possible reflex mechanisms that underly hyperreactivity to irritants. Although, as indicated above, this may not be the ideal approach, it is necessary because almost all the research on reflexes has been with isolation of the stimulus within the respiratory system and with experimental animals.

NEUROGENIC INFLAMMATION

Activation of sensory receptors in the airway mucosa by irritants will, in addition to setting up central reflexes, cause local axon reflex responses by the release of sensory neuropeptides (Fig. 15.1). The total local response is neurogenic inflammation (Barnes, 1986, 1995; McDonald, 1994). The main mediators are the tachykinins substance P (SP) and neurokinin A (NKA), and calcitonin gene-related peptide (CGRP). These are thought to come mainly from C fibre receptors, but the possible role of other sensory endings, such as RARs that have non-myelinated terminals in the epithelium, has not been ruled out. The neuropeptides diffuse through the interstitium to local target organs, in particular blood vessels, submucosal glands, epithelial cells and possibly airway smooth muscle. The different peptides will have different actions, at least quantitatively, on the

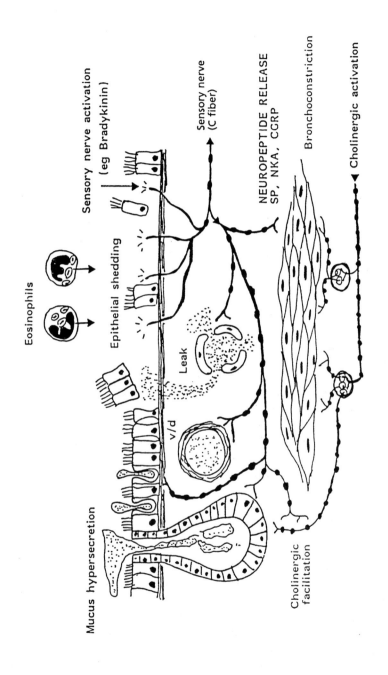

Fig. 15.1 Possible neurogenic inflammation in asthmatic airways via retrograde release of peptides from sensory nerves via an axon reflex. Substance P (SP) causes vasodilatation, plasma exudation and mucus secretion, whereas neurokinin-A (NKA) causes bronchoconstriction and enhanced cholinergic reflexes, and calcitonin gene-related peptide (CGRP) vasodilatation. Reproduced from Barnes (1995) with permission of Blackwell Science Ltd.

various targets, depending on whether the pharmacological receptors are NK_1 (SP-sensitive) or NK_2 (NKA-sensitive). Mucosal responses may be complex: for example, in the guinea pig, SP contracts bronchial smooth muscle (Karlsson, 1994) but relaxes that of the blood vessels (Widdicombe, 1993), causes secretion from submucosal glands (Fung and Rogers, 1997), and changes epithelial structure and function (McDonald, 1987).

Neurogenic inflammation can be prevented by local anaesthesia of the sensory nerves, or by depletion of their neuropeptides by large doses of the neuroexcitant and neurotoxic agent capsaicin (Karlsson, 1994; Maggi and Meli, 1988). In the airways the neuropeptides are broken down by mucosal peptidases, in particular neutral endopeptidase (NEP) (Nadel, 1991). This is found mainly at the base of the epithelium, but also at other sites in the mucosa. Its inhibition, for example by drugs such as phosphoramidon or thiorphan, potentiates neurogenic inflammation and some of the central reflexes associated with stimulation of the sensory nerves (this, for example, is true for cough: see below). In contrast, supplementation of the level of NEP inhibits the local responses to mediators such as SP.

There has been extensive research on the nervous mechanisms of neurogenic inflammation (see Karlsson, 1994), but convincing evidence for its existence in humans has not been presented (Joos *et al.*, 1995). This is more likely to be because convincing methods to demonstate its presence in humans have not been devised, than because it is absent. However, one result points to its absence for the human nose (Grieff *et al.*, 1995). Neurogenic inflammation is relevant to the reflex responses to irritants for three reasons.

(1) It is a reflex, not involving central pathways, set up by inhaled irritants.
(2) The sensory receptors that release the neuropeptides will themselves activate central reflexes. It is not known whether the axon reflex excitation can occur alone, without nerve terminal excitation reaching a threshold adequate to send impulses up to the central nervous system; such a mechanism can occur in the skin, in relation to the axon reflex 'triple response'. But potentially neurogenic inflammation will coexist with central reflex responses involving the same target organs and arising from the same receptors.
(3) The neuropeptides released by stimulation of the sensory nerves may in turn activate other sensory receptors which will themselves exert reflex actions, including on breathing and on the airways' effector tissues. This has been most studied in relation to cough (see below).

NASAL REFLEXES

Surprisingly, in view of the ease of access to the nose compared with the lungs, reflexes from the former have been studied less, and far less definitively, that those to the latter. This is partly because nasal pathology, although far more common than that of the lower airways and lungs, seldom kills and therefore attracts less research funding; and also because the surgical approach to the innervation of the nose, essential for the study of reflexes, is much more difficult than is that of the lower airways and lungs. It is also true that in many polluted environments breathing is through the mouth, either because the nose is blocked by vascular congestion and secretions, or because exercise is taking place.

This means that most or all of the irritants by-pass the nose and go straight to the larynx and lower respiratory tract.

Histologically the only nerve endings, presumed sensory, that have been described in the nasal mucosa are non-myelinated fibres in and under the epithelium (Cauna, 1982). Physiologically, recording from afferent nerve fibres shows groups responsive respectively to touch, to flow/cold and to irritants (Widdicombe *et al.*, 1988). There are no receptors to luminal air pressure, presumably because the nasal cavity is not distensible; however, respiratory reflex responses to nasal pressure changes, of unknown sensory origin but possibly from the nasopharynx, have been described (Widdicombe, 1988). It has been suggested that the touch receptors mediate pain, since touching the nasal mucosa is a painful stimulus. However, in the absense of definitive research, it is difficult to say which receptors mediate the various sensations, and which mediate various reflexes. In particular, we need to know the role of C and myelinated fibre afferents in the responses to irritants.

Irritants in the nose, including pollutants, set up variety of reflexes (Kratschmer, 1870; Lacroix and Lundberg, 1995; Widdicombe, 1988) (Table 15.2). The respiratory changes are sneeze and/or apnoea, but the neurology of the former has been little studied physiologically, partly because it is blocked by even light anaesthesia (Korpas and Tomori, 1979). Irritants in the nose may cause neurogenic inflammation, with vasodilatation, oedema and exudate, and glandular secretion (Baraniuk and Kaliner, 1995). The changes will tend to block the nose and, if not extreme, make it a better filter for the inhaled agents. In experimental studies, irritants in the nose usually cause apnoea together with laryngeal closure, and a variety of other changes: closure of the 'nasal valve', reflex nasal vasodilatation and mucus secretion (Widdicombe, 1988). Reflex cardiovascular changes

Table 15.2   Reflex respiratory responses to irritants

Site	Receptors	Response
Nose	Touch/C fibre	Sneeze, apnoea Mucus secretion Mucosal vasodilatation Laryngoconstriction Bronchoconstriction/dilatation
Larynx	Irritant/C fibre	Cough, expiration, apnoea Mucus secretion Mucosal vasodilatation Laryngoconstriction Bronchoconstriction
Trachea/bronchi	RAR/C fibre	Cough, apnoea, RSB Mucus secretion Mucosal vasodilatation Laryngoconstriction Bronchoconstriction
Alveoli	C fibre	Apnoea, RSB Mucus secretion Mucosal vasodilatation Laryngoconstriction Bronchoconstriction

RAR, rapidly adapting receptor; RSB, rapid shallow breathing.

are complex, but there is usually bradycardia and either hypotension or hypertension, the latter being due to vasoconstriction in skin, skeletal muscle, alimentary tract and kidneys. Nasal irritation causes reflex secretion of mucus and mucosal vasodilatation in the lower airways (Fung and Rogers, 1997; Widdicombe, 1993). Changes in smooth muscle tone are more uncertain, but there have been many studies that show either a reflex bronchoconstriction or a bronchodilatation (Coleridge *et al.*, 1989; Widdicombe and Wells, 1994). It seems reasonable to suppose that two different afferent pathways from the nose are responsible for the opposing bronchomotor reflex responses, but if so these pathways have not been identified. In humans, both healthy and with airways' diseases such as asthma and rhinitis, nasobronchial reflexes have almost always been described as bronchoconstrictor (Widdicombe and Wells, 1994).

While there is no doubt that irritant pollutants in the nose can set up a range of reflex respiratory responses, their importance in humans, apart from possibly blocking the nose and diverting airflow more directly to the lower airways, remains uncertain.

NASOPHARYNGEAL AND OROPHARYNGEAL REFLEXES

Mechanical stimulation of the nasopharynx causes reflex bronchodilatation, vasoconstriction with hypertension, tachycardia and airway mucus secretion, together with a respiratory response known as the aspiration reflex. This consists of a series of strong and brief inspiratory efforts (Tomori and Widdicombe, 1969; Widdicombe, 1988) (Table 15.2). There seems to be no reason to suppose that these reflexes are activated by inhaled irritants. However, one study has shown that the sensory receptors for the reflex assembly can be excited by allergens in sensitized animals (Nail, 1981), so the possibility of a response to mediators released by inflammatory changes remains. Similarly for the pharynx the responses to mechanical stimulation are swallowing, retching and a sense of nausea, but there is little evidence that these changes can be induced by pollutants, except perhaps for the sensation of nausea.

LARYNGEAL REFLEXES

The laryngeal mucosa has at least five different types of sensory receptor localized there (Sant' Ambrogio and Sant' Ambrogio, 1996; Sant' Ambrogio *et al.*, 1995; Widdicombe, 1986) (Table 15.1): three with myelinated afferent fibres that respond to pressure, cold/flow and touch respectively; their activity may be changed by inhaled irritants, but the latter do not seem to be primary stimuli. More important in this context are the normally silent 'irritant' receptors, also with myelinated afferent fibres, and possibly the C fibre receptors. The former resemble the tracheobronchial rapidly adapting (RAR) or irritant receptors, and are both exquisitely mechanosensitive and also respond to a variety of chemical irritants, such as ammonia, sulphur dioxide, various acids, non-isomolar liquids including distilled water, and cigarette smoke (Widdicombe, 1986). The C fibre receptors, although less numerous than for the lower respiratory tract, have been little studied, but they are stimulated by some chemical irritants including capsaicin, the

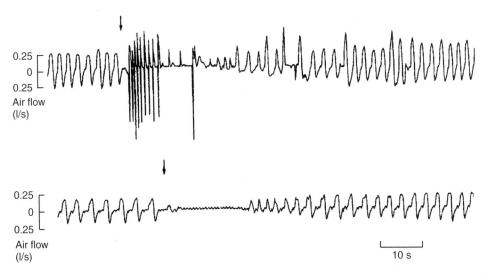

Fig. 15.2   Respiratory responses to laryngeal stimulation in the anaesthetized subject. The subject was anaesthetized with propofol. Laryngeal stimulation (water injection) was given at the arrow. In the upper record, light anaesthesia with coughing. In the lower record, with deeper anaesthesia the same stimulus caused apnoea. Modified from Nishino *et al.* (1996).

traditional test substance, albeit non-specific, for C fibre receptors. None of the types of receptor has been clearly defined histologically, although many nerves, presumed sensory afferents, have been described in the laryngeal mucosa. The reflex actions of the irritant and C fibre receptors have not been fully worked out, but in general it seems likely that the irritant receptors cause cough and the C fibre receptors apnoea, and both induce other reflex responses; more selective chemical and mechanical stimuli will be needed to test this concept. Possibly the irritant receptors respond more to mechanical and the C fibre receptors to chemical stimuli, but this distinction has not really been established.

Irritants in the larynx evoke many reflexes (Widdicombe, 1986, 1988) (Table 15.2) which were first studied comprehensively by Kratschmer in 1870. Much subsequent research confirms that they evoke coughing, expiratory efforts or apnoea, or all in sequence, with closure of the larynx, bronchoconstriction, airway mucus secretion, vasodilatation in the nose and the lower airways, and powerful cardiovascular responses (Sant' Ambrogio *et al.*, 1995). There is nearly always bradycardia, usually with hypertension and various changes in cardiac output. While it may be unwise to generalize about the reflex responses, *in toto* they seem to constrict the airways, which may lead to a more efficient filtering system and centralized deposition of irritants, and to concentrate the cardiac output to those regions, heart and brain, where vascular perfusion is especially crucial. An analogy with the diving reflex from the nose (Daly, 1986) may be instructive.

The reflexes from the larynx are profoundly influenced by sleep, which inhibits cough while retaining an apnoeic and bradycardic reflex (Sullivan *et al.*, 1979); and by even light general anaesthesia, which may convert a larygneal-originating cough to an apnoeic response (Nishino *et al.*, 1996) (Fig. 15.2).

TRACHEOBRONCHIAL REFLEXES

Four sensory pathways from the lower respiratory tract have been identified (Coleridge and Coleridge, 1986; Widdicombe, 1989) (Table 15.1).

(1) Slowly adapting pulmonary stretch receptors (SARs) modulate the pattern of breathing, and have important reflex effects in inhibiting bronchomotor tone and in increasing heart rate and cardiac output. However, they are insensitive to chemical stimuli (apart from a weak inhibition by acid and carbon dioxide), and there is no evidence that they play a role in responses to inhaled irritants.

(2) Sensory receptors associated with neuroepithelial bodies have been identified; their natural and exogenous stimuli and their reflex role have never been fully established (Lauweryns et al., 1985).

(3) Rapidly adapting receptors (RARs) are present, the terminals of which lie in and under the epithelium of the larger airways. Reflex respones to excitation of these receptors includes cough (from the trachea and larger bronchi) or deep inspirations (augmented breaths: from the intrapulmonary bronchi), broncho-constriction, laryngoconstriction, mucus secretion and airway vasodilatation (Coleridge and Coleridge, 1986, 1994) (Table 15.2).

(4) Bronchial C fibre receptors cause reflex apnoea and rapid shallow breathing, bronchoconstriction, laryngoconstriction, airway mucus secretion and vasodi-latation, hypotension and bradycardia and, at least in experimental animals, inhibition of spinal reflexes. They are probably the main neural mechanism that underlies neurogenic inflammation (see earlier).

RARs and C fibre receptors are excited by a variety of inhaled irritants, including ammonia, ozone, cigarette smoke, acid solutions and capsaicin (Coleridge and Coleridge, 1985; Karlsson et al., 1988). In this respect they resemble afferent systems in other organs such as skin, skeletal muscle and some viscera, which have two differ-ent pathways, Aδ and C fibre-mediated, that respond to inflammation and irritants. Qualitatively their reflex responses are strikingly similar (Table 15.2). Two exceptions are that RARs have not been shown to inhibit spinal reflexes (although this response does not seem to have been sought); and whereas RARs cause cough and augmented breaths, bronchial C fibre receptors cause apnoea and rapid shallow breathing (Widdicombe, 1996).

Receptor structure

The afferent terminals for the receptors that respond to irritants lie in, and probably under, the airway epithelium (Baluk et al., 1992, 1993; Widdicombe, 1994, 1996). They are concentrated at points of airway branching, the sites at which RARs can be localized by mechanical probing during single fibre recording. However the terminals presumably also include those of C fibre receptors, and the proportions of the two types of receptor have not been determined. Immunofluorescent microscopy shows that the nerves contain SP and CGRP (Baluk et al., 1992, 1993) (Fig. 15.3). Under the electron microscope some nerves lie close to the lumen just beneath the tight junctions between epithelial cells

(Laitinen, 1985) (Fig. 15.4), and immunofluorescent microscopy shows projections of the nerves from the basal region of the epithelium up to this site. In general the appearance of the nerves is consistent with a role in responding to inhaled irritants. Interestingly, mice and ferrets, animals which do not cough on tracheobronchial irritation, have few or no epithelial nerves (Karlsson *et al.*, 1988). Fibre recording studies show few RARs deep in the lungs or at the alveolar level (Widdicombe, 1954), and coughing is difficult or impossible to excite from these areas. C fibre receptors in the intrapulmonary bronchial wall have been studied (Coleridge and Coleridge, 1984) but, unlike those of the RARs, their populations have not been quantified.

The anatomical pattern of these receptors has not been mapped out, although one study showed that in dogs tracheal RARs have projections both in the epithelium and

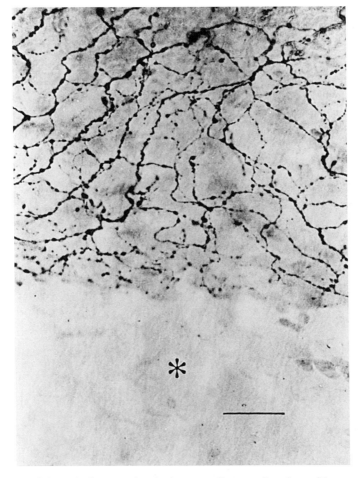

Fig. 15.3 A region of the tracheal mucosa directly above a cartilagenous ring of a rat. Nerves are exhibited by immunofluorescence for substance P. The density and orientation of the intraepithelial plexus of substance P-immunoreactive axons are similar to those found between the cartilagenous rings. The intraepithelial nerve plexus is absent in a region of the epithelium (*) which was accidentally removed during processing. Bar, 20 μm. Reprinted from Baluk *et al.* (1992), *Journal of Comparative Neurology*, by permission of Wiley-Liss, Inc., a subsidiary of John Wiley & Sons, Inc.

deeper in the mucosa (Mortola *et al.*, 1975). However, in another study stripping of the epithelium removed all histologically visible nerves, suggesting that few or none were deeper in the mucosa (Baluk *et al.*, 1992) (Fig. 15.3). Physiological studies with *in vitro* guinea pig trachea show that both RARs and C fibre receptors have a surface extent of only 0.5–1.0 mm (Fox *et al.*, 1993). The anatomical pattern of individual receptors is important in relation to neurogenic inflammation: the extent of local axon reflex responses will depend on the distribution of the nerve terminals and on the balance between diffusion of neuropeptides and their breakdown by peptidases.

Receptor stimuli

Both RARs and C fibre receptors are polymodal, and respond to a large number of mechanical and chemical irritants (Coleridge and Coleridge, 1986; Widdicombe, 1989, 1996). For the RARs, those in the trachea are more mechano-sensitive and less chemosensitive, and the opposite is true for those in the intrapulmonary bronchi (Widdicombe, 1954). There are other physiological distinctions between RARs at the two sites, for example those in the trachea adapt to a maintained stimulus more rapidly, and show an off-response. Similar differences have not been described for C fibre receptors. The RARs are far more sensitive to mechanical stimuli than are C fibre receptors. This applies to punctate stimuli and to natural stimuli such as inflation and deflation of the lungs; both

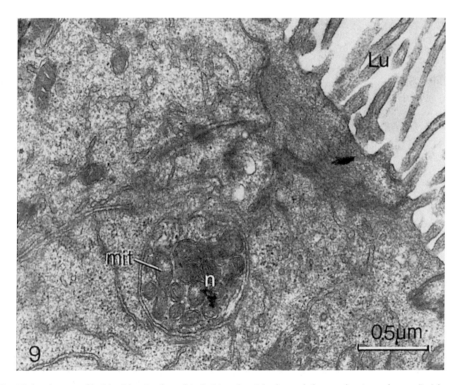

Fig. 15.4   Axon profile (n) with mitochrondria (mit) and vesicles located close to the airway lumen (Lu) between ciliated cells in human bronchus. From Laitinen (1985).

events powerfully activate RARs, but deflation has little effect on C fibre receptors, and inflation is a rather weak stimulus. 'Inert' carbon dust stimulates RARs (Sellick and Widdicombe, 1970), but this agent does not seem to have been tested on C fibre receptors.

There are also quantitative differences in their responses to chemical irritant stimuli, although in general any irritant in adequate concentration will activate receptors of both types. With the *in vitro* guinea pig tracheal preparation, C fibre receptors are sensitive to capsaicin, non-isosmolar solutions, acids and a variety of inflammatory mediators, for example (Fox, 1995, 1996). RARs in the same preparation do not respond to the mediators or to acid, and only a minority respond to capsaicin (Riccio *et al.*, 1996). However, *in vivo* the situation is quite different. Both C fibre receptors and RARs respond to an array of inflammatory mediators and to agents such as acids (including SO_2), non-isosmolar solutions, capsaicin and cigarette smoke (Widdicombe, 1996). The difference may be because *in vivo* RARs are stimulated by an increase in interstitial liquid volume, as occurs in response to inflammatory mediators and to sensory neuropeptides (Bonham *et al.*, 1996); this mechanism would be absent in the *in vitro* preparation. In terms of central reflex responses the mechanism of receptor activation is largely irrelevant, although the neural motor reactions will depend quantitatively on the pathophysiological conditions of the target organ.

Thus in general inhaled chemical irritants will excite both groups of receptor, but the balance in their activities will depend on the concentration of the irritant and probably on the site of its deposition.

Reflex responses

These have already been summarized and are shown in Table 15.2. Of the protective reflex responses, bronchoconstriction, mucosal vasodilatation (possibly together with increased interstitial liquid volume and luminal exudate) and mucus secretion will all tend to narrow the airways, and this may encourage central deposition of an inhaled irritant. Any beneficial role of laryngeal constriction, mainly in the expiratory phase, is more uncertain. Mucus secretion, as well as contributing to the narrowing of the airways, will have additional and possibly more important actions. The increased mucus volume will absorb any mechanical or chemical irritant, and help to remove it by mucociliary transport. In addition the increased thickness of mucus liquid overlying the epithelium will result in a lower concentration of the solution of any inhaled irritant; this in turn will cause a proportionally lower uptake of the agent, since the rate of diffusion into the tissues depends on the concentration difference across the epithelium (Widdicombe, 1997). Some experimental evidence supports this conclusion.

The most obvious reflex response to irritant pollutants in the lower airways is cough. While it is generally agreed that stimulation of RARs can cause cough, there is debate as to whether C fibre receptors also mediate cough (Fox, 1996; Karlsson 1996; Karlsson *et al.*, 1988; Widdicombe, 1996). The main evidence supporting this view is that in guinea pigs capsaicin-induced degeneration of airway sensory nerves inhibits cough due to acid or to capsaicin itself (Karlsson, 1993). However capsaicin is non-specific both in its degenerative action on sensory nerves (in large doses) and in its stimulant action on them (in small doses). Thus given either intravenously or as an aerosol it can stimulate both RARs and bronchial C fibre receptors, including in guinea pigs (Widdicombe, 1996). There is even some evidence that bronchial C fibre receptors can *inhibit* cough (Widdicombe, 1996), as do certainly the pulmonary ones (Tatar *et al.*, 1988; also see below). As already mentioned, RARs are stimulated by an increase in interstitial liquid volume, and this has been postulated to explain the action of SP in causing cough. Thus activation of C fibre receptors would not only cause release of SP and neurogenic inflammation and also central reflexes mediated by this pathway; but also the SP would induce plasma extravasation and excitation of RARs. This mechanism could explain the cough (Widdicombe, 1995, 1996) (Fig. 15.5).

Sensation

Inhalation of irritants can cause intense discomfort, even pain, localized to the larynx, trachea and lungs (Paintal, 1986). A good example is seen with riot-control (CS) 'gas'. This is usually a smoke of submicrometre-sized particles and when inhaled causes not only cough, but intense discomfort over the whole respiratory tract down to the lungs. With larger-particle aerosols of agents such as capsaicin, the discomfort is more restricted to the

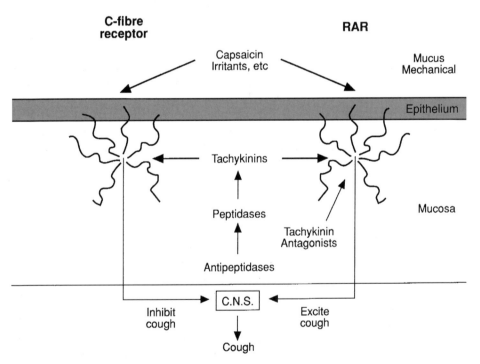

Fig. 15.5 Diagram for hypothetical role for tachykinins in cough. Tachykinins may be released from C fibre receptors and diffuse to RARs, which they stimulate to cause cough. They can be broken down by peptidases, which in turn can be inhibited by antipeptidases. Tachykinin antagonists can prevent tachykinins from acting on the RARs. If there is sufficient stimulation of C fibre receptors, these can cause a central inhibition of cough. C.N.S., central nervous system; RAR, rapidly adapting receptor. From Widdicombe (1995).

upper airways, larynx and trachea. Agents such as capsaicin or lobeline, given intravenously to humans and 'selective' for C fibre receptors, cause various unpleasant sensations in the laryngeal and tracheal regions and, if the dose is increased, cough as well (Raj et al., 1995). The cough could be a behavioural response to the laryngeal sensation. It is not seen in experimental animals, anaesthetized or unanaesthetized. In any event neither capsaicin nor lobeline are specific and both can excite RARs as well as C fibre receptors, at least in experimental animals (Raj et al., 1995).

ALVEOLAR REFLEXES

Histologically very few nerves can be seen in the alveolar walls, and pulmonary C fibre receptors have been studied mainly by fibre-recording experiments. It is thought that the terminals lie in the alveolar walls (they were originally called J receptors, for 'juxtapulmonary capillary') because they respond very rapidly to stimulant chemicals injected into the right heart (Paintal, 1983). However, this observation does not rule out a site in the bronchioles and other small airways.

Pulmonary C fibre receptors can be stimulated by a number of inhaled irritants, including cigarette smoke, chlorine, phosgene, volatile anaesthetics and a variety of inflammatory mediators (Coleridge and Coleridge, 1984); more common irritant pollutants do not seem to have been studied.

Activation of the receptors causes an array of reflexes similar to those from tracheobronchial receptors: laryngoconstriction, bronchoconstriction, mucus secretion, mucosal vasodilatation, hypotension and bradycardia. In addition there is inhibition of spinal reflexes. The respiratory response is apnoea and rapid shallow breathing. Cough has never been seen, and indeed it would not be effective against a stimulus at alveolar level. In fact activation of pulmonary C fibre receptors inhibits cough induced from the tracheobronchial tree or the larynx (Tatar et al., 1988). As indicated above, stimulation of pulmonary C fibre receptors may cause or contribute to the unpleasant sensation of inhaled irritants such as riot-control smoke; and in humans intravenous stimulants of pulmonary C fibre receptors cause discomfort in the laryngeal and tracheal regions (Raj et al., 1995).

REFLEX HYPERREACTIVITY

Inhaled irritants may damage or even destroy the epithelium. This, together with mucosal inflammation, could lead to hyperreactive airway nerves, or hyperalgesia. The only convincing evidence for this in humans is the hyperreactive cough reflex seen after inhalation of irritant pollutants, both spontaneous cough and that tested by cough challenge (Karlsson, 1993, 1996). Whether other central reflexes from airway and lung receptors are hyperreactive does not seem to have been studied. One study showed that RARs increased their activity over a period of hours while the animal preparation deteriorated (Mills et al., 1969), but nearly all experiments have been planned to avoid such deterioration. Asthma has been described as a form of hyperalgesia of the airways (Karlsson, 1996; Spina, 1996), but chiefly in relation to neurogenic inflammation. If the sensory receptors are hyper-sensitive, as seems likely, one would expect their central reflex actions to be enhanced; however, apart from cough, this possibility does not seem to have been explored.

CONCLUSIONS

There have been very extensive studies on the reflex responses to inhaled irritant pollu-tants. With the exception of work on cough and sensation, these studies have been almost entirely on experimental animals, because it is usually neither ethical nor practical to con-duct similar research on humans.

The animal experiments show that inhaled irritants can set up a range of reflexes from all parts of the respiratory tract and lungs, and the reflex responses involve not only breathing and many target organs in the respiratory tract, but also the cardiovascular system and spinal reflex activity. For humans, presumably the same mechanisms are pres-ent, but in general their importance remains to be assessed. Even the existence of neurogenic inflammation in the airways has not been convincingly demonstrated in humans, again possibly because the experimental methods required to test it have not been devised.

REFERENCES

Baluk P, Nadel JA and McDonald DM (1992) Substance P-immunoreactive sensory axons in the rat res-piratory tract: a quantitative study of their distribution and role in neurogenic inflammation. *J Comp Neurol* **319**: 586–598.

Baluk P, Nadel JA and McDonald DM (1993) Calcitonin gene-related peptide in secretory granules of serous cells in the rat tracheal epithelium. *Am J Cell Mol Biol* **8**: 446–453.

Baraniuk JN and Kaliner MA (1995) Functional activity of upper airway nerves. In: Busse WW and Holgate ST (eds) *Asthma and Rhinitis*. Oxford: Blackwell Scientific Publications, pp. 652–666.

Barnes PJ (1986) Asthma as an axon reflex. *Lancet* **2**: 234–255.

Barnes PJ (1995) Airway neuropeptides. In: Busse WW and Holgate ST (eds) *Asthma and Rhinitis*. Oxford: Blackwell Scientific Publications, pp. 667–685.

Bonham AC, Kott KS, Kappagoda and Joan JP (1996) Substance P contributes to rapidly adapting receptor responses to pulmonary venous congestion in rabbits. *J Physiol* **493**: 229–238.

Cauna N (1982) Blood and nerve supply of the nasal lining. In: Proctor DE and Andersen IB (eds) *The Nose: Upper Airway Physiology and the Atmospheric Environment*. Amsterdam: Elsevier Biomedical, pp. 45–69.

Coleridge HM and Coleridge JCC (1986) Reflexes evoked from the tracheobronchial tree and lungs. In: Cherniack NS and Widdicombe JG (eds) *Control of Breathing, Handbook of Physiology, Section 3, The Respiratory System*, vol. II. Bethesda, MD: American Physiological Society, pp. 395–429.

Coleridge HM and Coleridge JCC (1994) Pulmonary reflexes: neural mechanisms of pulmonary defense. *Ann Rev Med* **56**: 69–91.

Coleridge HM, Coleridge JCG and Schultz HD (1989) Afferent pathways involved in reflex regulation of airway smooth muscle. *Pharmacol Ther* **42**: 1–63.

Coleridge JCC and Coleridge HM (1984) Afferent vagal C-fibre innervation of the lungs and airways and its functional significance. *Rev Physiol Biochem Pharmacol* **99**: 1–110.

Coleridge JCC and Coleridge HM (1985) Lower respiratory tract afferents stimulated by inhaled irri-tants. *Am Rev Respir Dis* **131**: S51–S54.

Daly M de B (1986) Interactions between respiration and circulation. In: Cherniack NS and Widdicombe JG (eds) *Control of Breathing, Handbook of Physiology, Section 3, The Respiratory System*, vol. II. Bethesda, MD: American Physiological Society, pp. 529–594.

Fox AJ (1995) Mechanisms and modulation of capsaicin activity on airway sensory nerves. *Pulm Pharmacol* **8**: 207–215.

Fox AJ (1996) Modulation of cough and airway sensory fibres. *Pulm Pharmacol* 335–342.

Fox A, Barnes PJ, Urban L and Dray A (1993) An *in vitro* study of the properties of single vagal afferents innervating guinea pig airways. *J Physiol* **469**: 21–35.

Fung DCK and Rogers DF (1997) Airway submucosal glands: physiology and pharmacology. In: Rogers DF and Lethem MI (eds) *Airway Mucus: Basic Mechanisms and Clinical Perspectives*. Basel: Birkhauser-Verlag, pp. 179–210.

Grieff L, Svensson C, Andersson M and Persson CG (1995) Effects of topical capsaicin in seasonal allergic rhinitis. *Thorax* **42**: 779–783.

Joos GF, Germonpre PR and Pauwels RA (1995) Neurogenic inflammation in human airways: is it important? *Thorax* **50**: 217–219.

Karlsson J-A (1993) A role for capsaicin sensitive, tachykinin containing nerves in chronic coughing and sneezing but not in asthma: a hypothesis. *Thorax* **48**: 396–400.

Karlsson J-A (1994) Excitatory nonadrenergic, noncholinergic innervation of airways smooth muscle: role of peptides. In: Raeburn D and Giembycz MA (eds) *Airways Smooth Muscle: Structure, Innervation and Neurotransmission*. Basel: Birkhauser-Verlag, pp. 43–79.

Karlsson J-A (1996) The role of capsaicin-sensitive C-fibre nerves in the cough reflex. *Pulm Pharmacol* **9**: 315–322.

Karlsson J-A, Sant' Ambrogio G and Widdicombe JG (1988) Afferent neural pathways in cough and reflex bronchoconstriction. *J Appl Physiol* **65**: 1007–1023.

Korpas J and Tomori Z (1979) *Cough and Other Respiratory Reflexes*. Basel: Karger.

Kratschmer F (1870) Uber Reflexe von der Nassenschleimaut auf Athmung und Kreislauf. *Sitzungsker. Akad Wiss Wein Math Naturwiss Kl Abt* **62**: 147–170.

Lacroix JS and Lundberg JM (1995) Neural reflex pathways in rhinitis. In: Busse WW and Holgate ST (eds) *Asthma and Rhinitis*. Oxford: Blackwell Scientific Publications, pp. 686–690.

Laitinen A (1985) Autonomic innervation of the human respiratory tract as revealed by histochemical and ultrastructural methods. *Eur J Respir Dis* **140**: 1–42.

Lauweryns JM, Van Lommel AT and Dom R (1985) Innervation of rabbit intrapulmonary neuro-epithelial bodies. *J Neurol Sci* **67**: 81–92.

Maggi CA and Meli A (1988) The sensory-efferent function of capsaicin-sensitive sensory neurones. *Gen Pharmacol* **19**: 1–43.

McDonald DM (1987) Neurogenic inflammation in the respiratory tract: actions of sensory nerve mediators on blood vessels and epithelium of the airway mucosa. *Am Rev Respir Dis* **136**: S65–S71.

McDonald DM (1994) The concept of neurogenic inflammation in the respiratory tract. In: Kaliner MA, Barnes PJ, Kunkel GHH and Baraniuk JM (eds) *Neuropeptides in Respiratory Medicine*. New York: Marcel Dekker, pp. 321–350.

Mills JE, Sellick H and Widdicombe JG (1969) Activity of lung irritant receptors in pulmonary microembolism, anaphylaxis and drug-induced bronchoconstriction. *J Physiol* **203**: 337–357.

Mortola JP, Sant' Ambrogio G and Clement MG (1975) Localization of irritant receptors in the airways of the dog. *Respir Physiol* **24**: 107–114.

Nadel JA (1991) Neutral endopeptidase modulates neurogenic inflammation. *Eur Respir J* **4**: 745–754.

Nail BS (1981) Sensitization of polymodal airway receptors. In: Hutas J and Debreczeni LA (eds) *Advances in Physiological Sciences: Respiration*, vol. 10. Budapest: Akad Kido, pp. 479–484.

Nishino T, Togaito Y and Isono S (1996). Cough and other reflexes on irritation of airway mucosa in man. *Pulm Pharmacol* 285–292.

Paintal AS (1983) Lung and airway receptors. In: Pallot DJ (ed.) *Control of Respiration*. London: Croom Helm, pp. 78–107.

Paintal AS (1986) The visceral sensations – some basic mechanisms. In: Cervero F and Morrison JFB (eds) *Visceral Sensation*. Amsterdam: Elsevier, pp. 3–19.

Raj H, Singh VK and Paintal AS (1995) Sensory origin of lobeline-induced sensations: a correlative study in man and cat. *J Physiol* **482**: 235–246.

Riccio MN, Kummer W, Biglari B *et al.* (1996) Interganglionic segregation of distinct vagal afferent fibre phenotypes in guinea pig airways. *J Physiol* **496**: 521–530.

Sant' Ambrogio G and Sant' Ambrogio FB (1996) Sensory mechanisms in cough: role of laryngeal afferents in cough. *Pulm Pharmacol* **9**: 309–314.

Sant' Ambrogio G, Tsubone H and Sant' Ambrogio FB (1995) Sensory information from the upper airway: role in the control of breathing. *Respir Physiol* **102**: 1–16.

Sellick H and Widdicombe JG (1970) Stimulation of lung irritant receptors by cigarette smoke, carbon dust, and histamine aerosol. *J Appl Physiol* **31**: 15–19.

Spina D (1996) Airway sensory nerves: a burning issue in asthma. *Thorax* **51**: 335–337.

Sullivan CE, Kozar LF, Murphy E and Phillipson EA (1979) Arousal, ventilatory, and airway responses to bronchopulmonary stimulation in sleeping dogs. *J Appl Physiol* **47**: 17–25.

Tatar M, Webber SE and Widdicombe JG (1988) Lung C-fibre receptor activation and defensive reflexes in anaesthetized cats. *J Physiol* **402**: 411–420.

Tomori Z and Widdicombe JG (1969) Muscular, bronchomotor and cardiovascular reflexes elicited by mechanical stimulation of the respiratory tract. *J Physiol* **200**: 25–49.

Widdicombe JG (1954) Receptors in the trachea and bronchi of the cat. *J Physiol* **123**: 71–104.

Widdicombe JG (1986) Reflexes from the upper respiratory tract. In: Cherniack NS and Widdicombe JG (eds) *Control of Breathing, Handbook of Physiology, Section 3, The Respiratory System*, vol. II. Bethesda, MD: American Physiological Society, pp. 363–394.

Widdicombe JG (1988) Nasal and pharyngeal reflexes: protective and respiratory functions. In: Matthew OM and Sant' Ambrogio G (eds) *Respiratory Function of the Upper Airway*. New York: Marcel Dekker, pp. 233–258.

Widdicombe JG (1989) Nervous receptors in the tracheobronchial tree: airway smooth muscle reflexes. In: Coburn RF (ed.) *Airway Smooth Muscle in Health and Disease*. New York: Plenum Press, pp. 35–53.

Widdicombe JG (1993) Why are the airways so vascular? *Thorax* **48**: 290–295.

Widdicombe JG (1994) Vagal reflexes in the airways. In: Kaliner MA and Barnes PJ (eds) *The Airways: Neural Control in Health and Disease*. New York: Marcel Dekker, pp. 187–202.

Widdicombe JG (1995) Neurophysiology of the cough reflex. *Eur Respir J* **8**: 1193–1202.

Widdicombe JG (1996) Sensory mechanisms of cough. *Pulm Pharmacol* 383–387.

Widdicombe JG (1997) Airway surface liquid in man and other animals. *Atemw-Lungenkrkh Jahrgang* **23**: 425–427.

Widdicombe JG and Wells UM (1994) Vagal reflexes. In: Kaliner MA, Barnes PJ, Kunkel GHH and Baraniuk JM (eds) *Neuropeptides in Respiratory Medicine*. New York: Marcel Dekker, pp. 279–308.

Widdicombe JG, Sant' Ambrogio G and Mathew OP (1988) Nerve receptors of the upper airway. In: Mathew OP and Sant' Ambrogio G (eds) *Respiratory Function of the Upper Airway*. New York: Marcel Dekker, pp. 193–232.

16

Antioxidant Defences in the Extracellular Compartment of the Human Lung

FRANK J KELLY

The Rayne Institute, St Thomas' Campus, Kings College London, London, UK

ROY RICHARDS

School of Biosciences, Cardiff University of Wales, PO Box 911, Cardiff, UK

INTRODUCTION

As a major interface with the external environment, the lungs rank high in the level of importance to the body. This status however brings with it many problems, such as the need to deal with bacteria, xenobiotics, viruses, particles and toxic gases that enter the body via the respiratory tract (Fig. 16.1). Many of these give rise to oxidative stress (the excess generation of reactive oxygen species (ROS) in relation to available antioxidant defences). This can arise either through direct radical reactions (e.g. nitrogen dioxide) or via redox recycling reactions (e.g. paraquat). In addition, many entities that enter the lungs, even if not oxidants themselves, give rise to an inflammatory response. This results in the influx, and activation, of large numbers of inflammatory cells, particularly neutrophils. When activated, neutrophils give rise to, amongst other products, ROS (Fig. 16.1).

To combat ROS, lung tissue has an extensive range of antioxidant defences (Doelman and Bast, 1990; Menzel, 1992). For classification purposes these are divided into a range

AIR POLLUTION AND HEALTH
ISBN 0-12-352335-4

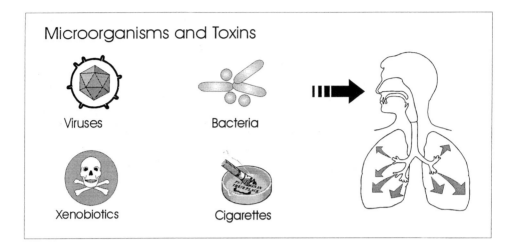

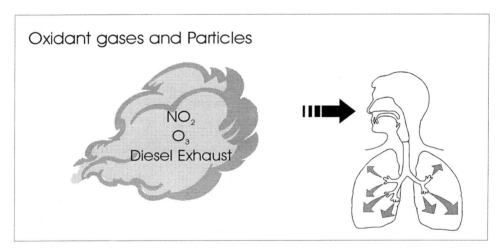

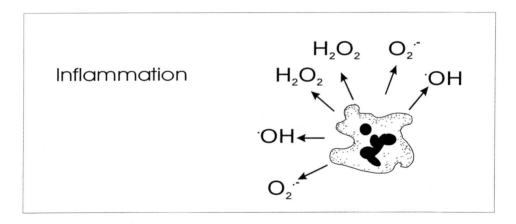

Fig. 16.1   Sources of oxidative stress in the lung.

of enzymatic and non-enzymatic defences. In addition to these intracellular antioxidant defences the lung also has a unique compartment of extracellular antioxidant defences (Davis and Pacht, 1991). These are found in the respiratory tract lining fluid (RTLF). RTLF is a thin amphipathic layer which overlays the respiratory epithelium of the lung, extending from the nasal cavity to the deepest alveoli. RTLF is therefore the first physical interface encountered by inspired gases, particles and xenobiotics entering the lung. As such, antioxidants present in RTLF play a pivotal role in screening and neutralizing potential harmful substances that enter the lung (Kelly et al., 1995). In this manner, the delicate surface of the lung is protected from oxidative injury. In this chapter we will review the information available regarding RTLF antioxidant defences in human subjects.

RTLF

RTLF is a two-layer structure comprising a lower, aqueous, sol phase and an upper, mucus, gel phase. The gel phase, derived mainly from submucosal glands (Quinton, 1979), is particularly good at trapping microorganisms and large particles from the airstream. These particles are then transported by mucociliary action to the posterior pharynx and swallowed. The gel phase of RTLF is therefore continually turning over. Although the gel phase probably has antioxidant properties, especially as it has many sulfydryl groups, little is presently known about this aspect of gel. The lower, sol, phase of RTLF bathes the lining epithelial cells and it is this aqueous layer which has been found to contain a wide range of antioxidant defences.

Not all aspects of RTLF are currently understood. For example, although it is appreciated that the depth of RTLF varies along the respiratory tree, the precise thickness of RTLF is still under debate (Duneclift et al., 1997; Widdicombe, 1997). In the upper airways RTLF may be 1–10 μm thick, whereas in the distal bronchoalveolar regions RTLF depth is only 0.2–0.5 μm and it contains higher concentrations of surfactant (Harwood and Richards, 1985; Hatch, 1992; Widdicombe, 1997). These differences may prove to be important factors in assessing the precise importance of RTLF antioxidants in combating inspired oxidants.

OBTAINING RTLF

RTLF samples can be obtained from human volunteers and patients by lavage. Nasal, proximal and distal bronchoalveolar lavage procedures are all now well established. For example, bronchoalveolar lavage (BAL) has been performed as a research and clinical tool for over 20 years (Reynolds and Newball, 1974). More recently, use of nasal lavage has also become popular (Harder et al., 1994; Hauser et al., 1995; Housley et al., 1995). As the nose is the primary portal of entry for air in humans, it is the first region of the respiratory tract that is in contact with inspired oxidants. Hence, if oxidants are going to have an impact on the respiratory tract, then effects should be detected first in the nasal passages. Nasal lavage (NL) is a simple and economical technique to perform. It is relatively non-invasive and, importantly, allows multiple sequential sampling of secretions from both nostrils of the same person.

Nasal lavage

There are several established procedures for carrying out NL. Brief descriptions of two approaches we have found suitable for antioxidant determinations are given below. The first involves inflation of the balloon of a modified Foley (FG) catheter inside one of the nasal vestibules to occlude it. Five ml of phosphate buffered saline (PBS) is then passed through the catheter and allowed to dwell in the cavity behind the balloon for periods up to 5 min. If required, the procedure can be repeated in the other nasal cavity. The recovered lavage fluid is then sieved through a 150 μm mesh and centrifuged at 300 *g* for 10 min. The resultant supernatant can then be utilized immediately for antioxidant analysis or stored for up to 6 months at –70°C. For a full description of this method, see Housley *et al.* (1995).

A second method that we have used successfully was originally described by Harder and colleagues (1994). Ten × 0.1 ml aliquots of PBS are sprayed into one nostril and recovered into a sterile plastic cup. This procedure is repeated five times. A similar procedure can then be carried out on the other nostril. The plastic cup is placed on ice and the recovered nasal lavage fluid is filtered and centrifuged prior to antioxidant analysis as described above. This method, in our hands, provides more concentrated samples than those obtained with the catheter approach. The advantage of the first method, however, is that it provides information on the secretory rate of various antioxidants.

Bronchial lavage

This procedure can be carried out alone or in combination with BAL (Helleday *et al.*, 1994). Typically, during fibreoptic bronchoscopy the tip of the bronchoscope is wedged in the apical section of the left lingual lobe or a middle lobe bronchus. BL is performed with 2 × 20 ml PBS. The fluid is infused and then gently suctioned back into a siliconized container placed in iced water. In general we tend to use the recovery from the first 20 ml of bronchial wash for analyses of cellular and soluble inflammatory components while the recovered fluid from the second 20 ml aliquot is used for antioxidant analysis. Recoveries are treated as described for NL fluid.

Bronchoalveolar lavage

We routinely perform BAL by instilling 3 × 60 ml PBS. This is usually carried out immediately following the bronchial wash procedure. As described above, the fluid is instilled with the tip of the bronchoscope carefully wedged into either a lingula lobe bronchus or a middle lobe bronchus and gently suctioned back into a siliconized container on ice. Recoveries are combined in this instance and then treated as described above.

Combination of all the above techniques, i.e. NL, BL and BAL, are possible within the same individual and thus can provide a window to RTLF composition in different regions of the respiratory tract in a single subject.

Problems/pitfalls to consider

Recovery of lavage fluid is never complete and becomes less effective the lower the region of the respiratory tract that is sampled. For example, whereas 80–90% of NL is recovered, BAL recovery is often as low as 50%. Furthermore, estimations of the extent of RTLF dilution is fraught with problems, especially with patients who may have increased microvascular permeability of their airways. As a result, it is presently extremely difficult to compare data arising from different laboratories – even if these focus on the same patient groups. A European Respiratory Society supported Task Group, reporting on the acellular components of BAL fluid, has recently addressed this problem and they have recommended that data be reported per millilitre of recovered BAL fluid (see (Baughman, 1997). With knowledge of the BAL volumes utilized (and recovered), adoption of this approach should help comparison of data from different laboratories and patient groups.

It is also important that lavage procedures are carried out carefully and the recovered fluid handled gently. This is because cells, resident in RTLF, easily become disrupted, and in doing so they release their intracellular antioxidants into the extracellular compartment. This will obviously modify the true levels of the compound(s) of interest in RTLF. This is a particularly important precaution when taking samples from patients in whom airway inflammation is present.

ANTIOXIDANTS PRESENT IN RTLF

Once lavage samples started becoming routinely available, the challenge to investigators moved to the laboratory bench where adaptations/improvements of existing techniques to quantify antioxidants took place. An important advance in recent years has been the development of sensitive techniques to determine antioxidants in dilute RTLF (BAL fluid) samples (see Kelly *et al.*, 1996). These techniques, which are mainly based on high-performance liquid chromatography (HPLC) provide both the specificity and sensitivity of antioxidant measurements. It should be noted, however, that certainly with respect to antioxidant determinations the current practice is to remove the gel phase of RTLF, hence measurements presently made in BAL fluid primarily represent sol concentrations only.

It has been appreciated for some time that antioxidants are present in BAL fluid obtained from a range of different animals (Jenkinson *et al.*, 1988; Skoza *et al.*, 1983; Slade *et al.*, 1993; Willis and Kratzing, 1974). Because of the limitations mentioned above, the extension of such studies to human subjects has only occurred in the last 10 years. Several groups have now shown that the sol phase of RTLF obtained from human subjects contains a wide spectrum of antioxidants (Cantin *et al.*, 1987; Pacht and Davis, 1988; Pacht *et al.*, 1986; Peden *et al.*, 1990a).

RTLF antioxidants of particular importance

Ascorbic acid

As a consequence of its high water solubility, L-ascorbate (vitamin C) is widely distributed throughout all aqueous compartments of the body, including RTLF. Ascorbate is an excellent reducing agent and scavenges a variety of free radicals and oxidants *in vitro*, including superoxide and peroxyl radicals, hydrogen peroxide, hypochlorous acid and singlet oxygen. During this scavenging activity, ascorbate loses one electron, resulting in the formation of the semi-dehydroascorbate radical (Buettner and Jurkiewicz, 1996). Importantly, this is a relatively unreactive free radical. Semi-dehydroascorbate can be reduced by glutathione (GSH) to dehyroascorbate with the contaminant production of GS·. As dehydroascorbate can subsequently be metabolized to the cytotoxic derivative oxalate, most cells contain the enzyme dehydroascorbic reductase which catalyses the generation of ascorbate from dehydroascorbate at the expense of the formation of oxidized glutathione (GSSG). In addition to the direct scavenging action of ascorbate, it also acts indirectly to prevent lipid peroxidation through its reaction with membrane tocopherol. In *in vitro* studies, it has been demonstrated that ascorbate is able to reduce the concentration of the tocopherol radical with the regeneration of its non-radical form, thereby restoring its scavenging activity (Frei *et al.*, 1989). This synergistic action of ascorbate, although clearly demonstrated *in vitro,* has, importantly, not yet been reported *in vivo.*

Uric acid

Uric acid is also an important antioxidant in RTLF. In the upper airways, partly as a result of its high concentration (see below), it is particularly important and is preferentially utilized compared with other antioxidants. Uric acid is an oxidized purine base which can directly scavenge hydroxyl radicals, oxyhaem oxidants formed between the reactions of haemoglobin and peroxy radicals, peroxyl radicals themselves and singlet oxygen (Becker, 1993). It acts in these reactions in a sacrificial mode, in that it is irreversibly damaged through the interaction to produce the oxidation product, allantoin (Ames *et al.*, 1981).

Glutathione

RTLF contains large concentrations of glutathione (L-γ-glutamyl-L-cysteinylglycine). Of particular note, RTLF glutathione is 100 times the concentration found in plasma, and in a predominately (>90% GSH) reduced form (Cantin *et al.*, 1987). Though the concentration of glutathione is high in RTLF, it is not known if the glutathione redox cycle operates in this environment. As such, glutathione probably functions in a sacrificial mode, reacting directly with ROS in the RTLF and generating GS·, which can subsequently be converted to GSSG through a radical transfer process. Glutathione is particularly good at defending against oxidants such as hypochlorous acid and hypobromous acid (Winterbourn, 1985), which are released from neutrophils and eosinophils, respectively. Therefore, RTLF glutathione may be particularly important in defending the extracellular surface of the lung against activated inflammatory cells.

α-Tocopherol

α-Tocopherol (vitamin E) is present within RTLF, albeit at relatively low concentrations. It is thought that this α-tocopherol is secreted by type II cells into RTLF along with surfactant (Rustow et al., 1993). α-Tocopherol is a powerful antioxidant, both in terms of its direct free radical scavenging activity and through its ability to terminate lipid peroxidation (Burton and Ingold, 1981; Nakamura et al., 1987; Niki et al., 1986; Witting, 1980). Vitamin E functions as a chain-breaking antioxidant in the lipid phase. Evidence of this function is indirect, focusing almost entirely on observations made using in vitro systems (Burton et al., 1983). It is thought that reactivity with organic peroxyl radicals accounts for the majority of the biological activity of α-tocopherol (Burton and Ingold, 1981; Witting, 1980). This reaction is of considerable importance because tocopherols reacting with lipid peroxyl radicals yield a relatively stable lipid hydroperoxide and a vitamin E radical, which effectively interrupts the lipid peroxidation chain reaction (McCay, 1985).

Other extracellular antioxidants

This chapter will concentrate on the four major extracellular antioxidants listed above because most information is available on these in human subjects. However, a variety of other compounds with antioxidant potential or with the ability to process antioxidants have been reported in the extracellular environment of the respiratory tract. These include aeruloplasm, lactoferrin, transferrin, superoxide dismutase, taurine, glucose and β-nicotamide adenine dinucleotide phosphate (NADPH) (Pacht and Davis, 1988).

NORMAL RTLF ANTIOXIDANT RANGES

There is still only limited information on RTLF antioxidant status in healthy individuals (Cantin et al., 1987; Housley et al., 1996; Kelly et al., 1996).

Distribution of antioxidants within the respiratory tract

Although it is clear that antioxidant concentrations vary between different regions of the respiratory tract, the extent of these differences is not yet clear. In the nasal cavity, uric acid (UA) is by far the most prevalent antioxidant (Housley et al., 1995, 1996; Peden et al. 1990a). When assessed by either the FG catheter approach or the multiple spray technique, values of 4–12 μmol/l are seen in normal individuals. It should be noted, however, that there are gender differences, with women having significantly lower levels of UA than men (Housley et al., 1996). A similar gender difference exists for plasma concentrations of UA, indicating that the major source of this antioxidant in NL is from plasma. However, Peden et al. (1990b) demonstrated that in the nasal cavity, RTLF sol phase UA concentration is increased following cholinergic stimulation of the airways. Furthermore, they found that the increase in UA correlated positively with lactoferrin concentration, but not with albumin. As lactoferrin is predominately derived from mucosal gland

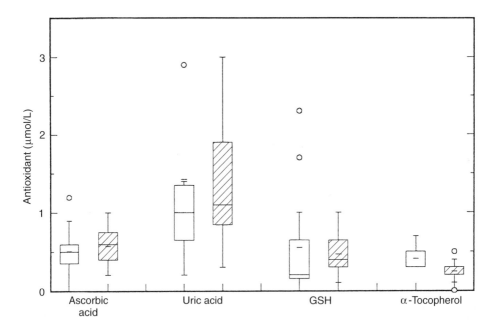

Fig. 16.2 Ascorbic acid, uric acid, glutathione and α-tocopherol antioxidant concentrations in bronchial wash (open boxes) and bronchoalveolar lavage (shaded boxes) from normal human volunteers. Data are presented as medians and interquartile ranges. Outliers, when present, are indicated as individual data points.

secretions (gel phase), this supports the contention that RTLF UA in the upper airways is derived, in part, from glandular secretions. Control of UA secretion in this compartment may therefore be closely related to mucus secretion. Glutathione and ascorbic acid are also present in nasal cavity RTLF, but at much lower concentrations (<1 μmol/l). Furthermore, although not always the case, low concentrations of α-tocopherol are sometimes observed.

In the peripheral airways (BL fluid) and bronchoalveolar region (BAL fluid) uric acid, ascorbic acid, GSH and α-tocopherol are usually present (Fig. 16.2). Although there is considerable intersubject variability in these antioxidants (Kelly *et al.*, 1996; Slade *et al.*, 1993), to date we have found reasonable associations between the levels of antioxidants in the different regions of the respiratory tract. It should be emphasized that whilst the BL fraction predominately reflects airway secretions within the terminal conducting airways, it will also contain some material from the more distal structures. Likewise, the BAL fluid sample, although predominately derived from the alveolar and proximal alveolar regions (Hatch, 1992), will contain components from the conducting airways.

SOURCE OF RTLF ANTIOXIDANTS

Little is known about the source of RTLF antioxidants, if and how their pool sizes are regulated, or importantly repleted following oxidative depletion. Information such as this is required to understand fully the role that RTLF antioxidants play in protecting the respiratory epithelium from oxidative injury, both in healthy individuals and in patients with chronic respiratory disease.

From our observations to date, we have noted that RTLF uric acid concentration seems to be associated closely with protein concentration. This appears to be the case at all levels of the respiratory tract. This finding suggests that RTLF uric acid comes from either a direct flux from the plasma pool (Housley *et al.*, 1996; Peden *et al.*, 1993) or, alternatively, some UA may be stored and then actively secreted from submucosal glands.

The source of RTLF glutathione is also not known but cellular secretion mechanisms are most likely involved as, unlike UA, plasma levels of glutathione are very low (2–3 µmol/l). A number of cell types such as alveolar macrophages, Clara cells and alveolar type II cells have been reported to contain high levels of glutathione (4.8, 3.2 and 0.54 mmol/l, respectively (Horton *et al.*, 1987). Poor reabsorption of glutathione from the respiratory tract is also important. The cell surface enzyme responsible for the uptake of glutathione from the extracellular space, γ-glutamyl transpeptidase, is present in much lower concentrations in the lung compared with other organs but may be located in specific epithelial cells such as Clara and type II cells (Dinsdale *et al.*, 1992). RTLF is also replenished relatively slowly, therefore any glutathione exported into this compartment is likely to remain there for prolonged periods.

Ascorbic acid as a low molecular weight, water-soluble compound is likely to be present in all such compartments in the body. RTLF ascorbate levels are such that they likely arise by free diffusion from the vasculature. In contrast α-tocopherol is present in relatively low concentrations in airways fluids and it has been suggested that it may be secreted from type II epithelial cells in association with surfactant (Rustow *et al.*, 1993).

RTLF ANTIOXIDANT REGULATION OF INHALED OXIDANTS

A major interest in RTLF antioxidants to a number of investigators (Housley *et al.*, 1995; Kelly *et al.*, 1995; Mustafa, 1990; Postlewait *et al.*, 1994; Pryor, 1991) is that they represent one of the first tiers of defence in the respiratory tract against inspired oxidants. The secondary pollutant gas, ozone, represents an excellent example in that through its powerful oxidizing ability it will react with numerous biomolecules. Despite the broad target specificity of ozone the rate at which reactions occur with different molecules can vary over several orders of magnitude (Pryor, 1992). Indeed, when ozone encounters an array of potential substrates it reacts initially with those with the highest intrinsic reactivities toward it. Uric acid, ascorbic acid and glutathione have all been shown (in defined settings) to display high intrinsic reactivities toward ozone (Giamalva *et al.*, 1985; Pryor, 1993; Pryor *et al.*, 1984). It has therefore been proposed that RTLF antioxidants function as sacrificial substrates for ozone (Kelly *et al.*, 1995; Pryor, 1993; Pryor *et al.*, 1995). By reacting with ozone they effectively remove it with only the production of ostensibly harmless products. Similarly, other gaseous pollutants such as nitrogen dioxide are probably dealt with in a similar manner (Postlethwait *et al.* 1991; Department of Health, 1993).

EVIDENCE THAT RTLF ANTIOXIDANTS SCAVENGE INSPIRED OXIDANTS

Although there is a sound theoretical basis, and good *in vitro* data (Housley *et al.*, 1995; Kelly and Tetley, 1997; Langford *et al.*, 1995; Mudway and Kelly, 1998; Postlethwait *et al.*, 1991) for assuming that RTLF antioxidants scavenge inspired oxidants, *in vivo* evidence to support this contention has only recently become available.

Nitrogen dioxide

In a series of studies focusing on nitrogen dioxide as an inspired oxidant, we examined the kinetics of nitrogen dioxide-induced antioxidant reactions (Kelly *et al.*, 1996). Healthy, non-smoking, asymptomatic subjects were exposed to filtered air and 2 ppm nitrogen dioxide, for 4 h on separate occasions at least 3 weeks apart. Bronchoscopy was performed 1.5 h, 6 h, or 24 h after each exposure. Exposure to nitrogen dioxide resulted in a rapid (1.5 h) loss of uric acid from the bronchial region, however by 6 h post-exposure it had increased significantly above control levels. This increase was only temporary, as bronchial wash uric acid concentration returned to control levels by 24 h post-exposure. A similar response of uric acid to nitrogen dioxide was seen in the bronchoalveolar region. Ascorbic acid concentration also fell in bronchial, and bronchoalveolar, lavage fluids 1.5 h following exposure to nitrogen dioxide, but these returned to control values by 6 h post-exposure.

In contrast, significant increases in reduced glutathione concentration were seen at 1.5 and 6 h in bronchial lavage fluid following exposure to nitrogen dioxide. Again, these responses were only temporary, with glutathione returning to control levels by 24 h post-exposure. No change in bronchoalveolar lavage fluid glutathione concentration, or malondialdehyde content, was seen following nitrogen dioxide exposure. Combined, these data support the view that antioxidants present in lung fluids react with nitrogen dioxide, and hence modulate its impact on the lung. In the upper respiratory tract (nasal cavity) uric acid seems to be the most important antioxidant, while in the upper airways (bronchial wash) increased production/release of glutathione seems to be the primary antioxidant response to nitrogen dioxide.

Ozone

Evidence is now also available in humans, supporting the extensive *in vitro* database (Housley *et al.*, 1995; Mudway and Kelly, 1998; Postlethwait *et al.*, 1991, 1994), which suggests that RTLF antioxidants scavenge inspired ozone. As with nitrogen dioxide, there is clear evidence of differential antioxidant responses to ozone at different levels of the respiratory tree. For example, when healthy volunteers are exposed to 200 ppb ozone, there is an immediate loss (during the course of exposure) of uric acid from the nasal cavity lining fluid, while glutathione and ascorbic acid levels remain unaltered (Mudway *et al.*, 1996a,b). Ninety minutes post-ozone exposure, while bronchial wash and bronchoalveolar lavage fluid ascorbate and uric acid concentrations are unchanged, glutathione concentrations are significantly increased. The increase in glutathione was found to coincide with a decrease in the number of alveolar macrophages present in the lavage fluid, suggesting that dying macrophages may release glutathione into the airways (Blomberg *et al.*, 1996). Six h post-ozone exposure, i.e. long past the period during which any direct interactions of ozone with airway antioxidants is feasible, bronchoalveolar lavage ascorbate concentration falls (Mudway *et al.*, 1997). This ascorbate response occurs at the same time as increased numbers of activated neutrophils are moving into the airways. As neutrophils are capable of releasing large quantities of superoxide radicals when activated, it is feasible that the loss of ascorbate is in response to the *secondary wave* of oxidative stress in the airways. Moreover, in 14/15 subjects in which this response was observed, there was also a decrease in blood ascorbate levels at this time suggesting that ascorbic acid was moving across from the blood compartment to the lung fluid compartment to make up for the neutrophil-induced loss of ascorbate. Such relationships are, of course, only speculative at present, and further studies are required to follow up these observations.

IMPACT OF OXIDANTS ON THE RESPIRATORY TRACT

Oxidants such as ozone elicit a broad spectrum of biological effects in both animal models and humans. These include airway inflammation (Kleeberger *et al.*, 1993; Schelegle *et al.*, 1991), decrements in lung function (Beckett, 1991; Folinsbee and McDonnell, 1988), induction of airway hyperresponsiveness (Brookes *et al.*, 1989; Li *et al.*, 1992) and tissue injury (Ibrahim *et al.*, 1980; Last *et al.*, 1984). For such pathological events to occur, based on the above *in vivo* studies demonstrating reactions between ozone and RTLF antioxidants, either: (1) not all oxidants are effectively scavenged by RTLF antioxidants; (2) in certain instances RTLF antioxidant defences become overwhelmed; or (3) oxidant interactions involve the generation of toxic species within RTLF (Kennedy *et al.*, 1992; Pryor, 1993; Pryor and Church, 1991; Pryor *et al.*, 1991, 1995). The importance of ozone reaction products as mediators of injury, as opposed to ozone itself, is supported both theoretically and experimentally (Leikauf *et al.*, 1988, 1989). Mathematical models of ozone transit time through RTLF, based on its aqueous diffusibility and reactivity toward resident substrates, suggest that it is unlikely that any inspired ozone will reach the airway epithelium (Pryor, 1992, 1994). The theory has thus developed that ozone (and other oxidizing pollutants) is scavenged in the respiratory tract by antioxidants present within RTLF. These reactions occur preferentially over those with macromolecular targets such as proteins and lipids. This hypothesis is supported by the finding that protein and lipid components of RTLF appear to be protected until most of the uric acid and ascorbic acid are consumed (Cross *et al.*, 1992).

Pioneering studies by the groups of Laukauff and Pryor have demonstrated a range of reactive products when ozone interacts with fluid and cell membranes. These include aldehydes, hydroxyhydroperoxides and hydrogen peroxide. These more stable secondary products of ozone are both pro-inflammatory and toxic to cells. Hence, it follows that the more inspired ozone is scavenged by RTLF antioxidants, the less will be available for the generation of toxic intermediates. In the *in vivo* studies reported above, no discernible markers of protein or lipid oxidation were observed in the lavage fluid obtained from any of the volunteers. This finding is not surprising given the levels of ozone and nitrogen dioxide utilized (0.2 and 2 ppm, respectively). These would not be expected to give rise to gross lung injury. There must however be some, perhaps more subtle, products produced which are responsible, in part, for the subsequent inflammatory response seen in most subjects. Moreover, it follows that low levels of RTLF antioxidants would leave individuals susceptible to, and perhaps more responsive to, oxidizing pollutants. In addition, these individuals may be less able to defend against subsequent oxidant stress arising through the influx of activated neutrophils to their airways.

SUMMARY

The external surface of the lung is covered with a thin layer of fluid, the RTLF. RTLF contains a wide range of compounds with distinct antioxidant properties. These antioxidants, especially the water-soluble compounds glutathione, ascorbic acid and uric acid, act as a first line of defence against inspired oxidants such as nitrogen dioxide and ozone. By reacting as sacrificial substrates with oxidizing gases, RTLF antioxidants protect the delicate outer surface of the lung. Studies in human volunteers have indicated that the distribution of antioxidants along the respiratory tract is not uniform. Uric acid is by far the most prevalent antioxidant in the nasal cavity while reduced glutathione is present in high concentrations in the lower airways. Marked differences in the amount of RTLF antioxidants between individuals has led to the proposal that those individuals with low RTLF antioxidant levels may be more susceptible to the impact of oxidizing pollutants. This hypothesis is currently under investigation in the author's laboratories.

ACKNOWLEDGEMENTS

The studies outlined in this manuscript were supported by the Medical Research Council (UK) and the Wellcome Trust.

REFERENCES

Ames BN, Cathcart R, Schwiers E and Hochstein P (1981) Uric acid provides an antioxidant defense in humans against oxidant and radical caused aging and cancer: a hypothesis. *Proc Natl Acad Sci USA* **78**: 6858–6862.
Baughman RP (1997) The uncertainties of bronchoalveolar lavage. *Eur Respir J* **10**: 1940–1942.
Becker RF (1993) Towards the physiological function of uric acid. *Free Radic Biol Med* **14**: 615–631.
Beckett WS (1991) Ozone, air pollution, and respiratory health. *Yale J Biol Med* **64**: 167–175.
Blomberg A, Mudway IS, Frew AJ *et al.* (1996) Dose-dependent antioxidant and cellular responses in the airways of human subjects exposed to ozone. *Eur Respir J* **9**: 147S.
Brookes KA, Adams WC and Schelegle ES (1989) 0.35 ppm O_3 exposure induces hyperresponsiveness on 24-h reexposure to 0.20 ppm O_3. *J Appl Physiol* **66**: 2756–2762.
Buettner GR and Jurkiewicz BA (1996) Chemistry and biochemistry of ascorbic acid. In: Cadenas E and Packer L (eds) *Handbook of Antioxidants*. New York: Marcel Dekker, pp. 91–115.
Burton GW and Ingold KU (1981) Autooxidation of biological molecules. 1. The antioxidant activity

of vitamin E and related chain breaking phenolic antioxidants *in vitro. J Am Chem Soc* **103**: 6472–6477.

Burton GW, Joyce A and Ingold KU (1983) Is vitamin E the only lipid-soluble, chain-breaking anti-oxidant in human plasma and erythrocyte membranes? *Arch Biochem Biophys* **221**: 281–290.

Cantin AM, North SL, Hubbard RC and Crystal RG (1987) Normal alveolar epithelial lining fluid contains high levels of glutathione. *J Appl Physiol* **63**: 152–157.

Cross CE, Motchnik PA, Bruener BA *et al.* (1992) Oxidative damage to plasma constituents by ozone. *FEBS Lett* **298**: 269–272.

Davis WB and Pacht ER (1991) Extracellular antioxidant defences. In: Crystal RG and West JB (eds) *The Lung: Scientific Foundations*. New York: Raven Press Ltd, pp. 1821–1827.

Department of Health (1993) Advisory Group on the Medical Aspects of Air Pollution Episodes. Third report, *Oxides of Nitrogen*. London: HMSO.

Dinsdale D, Green JA, Mason MM and Lee MJ (1992) The ultrastructural immunolocalization of γ glutamyltranspeptidase in rat lung: correlation with the histochemical demonstration of enzyme activity. *Histochem J* **24**: 14–152.

Doelman CJA and Bast A (1990) Oxygen radicals in lung pathology. *Free Radic Biol Med* **9**: 381–400.

Duneclift S, Wells U and Widdicombe J (1997) Estimation of thickness of airway surface liquid in ferret trachea *in vitro. Am J Physiol* **83**: 761–767.

Folinsbee LJ and McDonnell WF (1988) Pulmonary function and symptom responses after a 6.6 hour exposure to 0.12 ppm ozone with moderate excercise. *J Air Pollut Control Assoc* **38**: 28.

Frei B, Stocker R, England L and Ames BN (1989) Ascorbate, the most effective antioxidant in human blood plasma. In: Emerit *et al.* (eds) *Antioxidants in Therapy and Preventative Medicine*. New York: Plenum Press, pp. 155–163.

Giamalva DH, Church DF and Pryor WA (1985) A comparison of the rates of ozonation of biological antioxidants and oleate and linoleate esters. *Biochem Biophys Res Commun* **133**: 773–779.

Harder S, Peden DB, Koren HS and Devlin RB (1994) Comparison of nasal lavage techniques for analysis of nasal inflammation. *Am J Respir Crit Care Med* **151**: A565 (abstract).

Harwood JL and Richards R (1985) Lung surfactant. *Molec Aspects Med* **8**: 423–514.

Hatch GE (1992) Comparative biochemistry of airway lining fluid. In: Schlesinger RB (ed.) *Comparative Biology of the Lung*. Boca Raton, FL: CRC Press, pp. 617–632.

Hauser R, Elreedy S, Hoppin JA and Christiani DC (1995) Upper airway response in workers exposed to fuel oil ash: nasal lavage analysis. *Occup Environ Med* **52**: 353–358.

Helleday R, Sandstrom T and Stjernberg N (1994) Differences in bronchoalveolar cell response to nitrogen dioxide exposure between smokers and non smokers. *Eur Respir J* **7**: 1213–1220.

Housley DG, Mudway I, Kelly FJ *et al.* (1995) Depletion of urate in human nasal lavage following *in vitro* ozone exposure. *Int J Biochem Cell Biol* **27**: 1153–1159.

Housley DG, Eccles R and Richards RJ (1996) Gender difference in the concentration of the antioxidant uric acid in human nasal lavage. *Acta Otolaryngol (Stockh)* **116**: 751–754.

Horton JK, Meredith MJ and Bend JR (1987) Glutathione biosynthesis from sulphur-containing amino acids in enriched populations of Clara and Type II cells and macrophages freshly isolated from rabbit lung. *J Pharmacol Exp Ther* **240**: 376–380.

Ibrahim AL, Zee YC and Osebold JW (1980) The effects of ozone on the respiratory epithelium of mice II. Ultrastructural alterations. *J Environ Pathol Toxicol* **3**: 251–258.

Jenkinson SG, Black RD and Lawrence RA (1988) Glutathione concentrations in rat lung bronchoalveolar lavage fluid: effect of hyperoxia. *J Lab Clin Med* **112**: 345–351.

Kelly FJ and Tetley TD (1997) Nitrogen dioxide depletes uric acid and ascorbic acid but not glutathione from lung lining liquid. *Biochem J* **325**: 95–99.

Kelly FJ, Mudway I, Krishna TM and Holgate ST (1995) The free radical basis of air pollution: focus on ozone. *Respir Med* **89**: 647–656.

Kelly FJ, Blomberg A, Frew AJ *et al.* (1996) Antioxidant kinetics in lung lining fluid following exposure of human to nitrogen dioxide. *Am J Respir Crit Care Med* **154**: 1700–1705.

Kennedy CH, Hatch GE, Slade R and Mason RP (1992) Application of the EPR spin-trapping technique to the detection of radicals produced *in vivo* during inhalation exposure of rats to ozone. *Toxicol Appl Pharmacol* **114**: 41–46.

Kleeberger SR, Levitt RC and Zhang LY (1993) Susceptibility to ozone-induced inflammation I. Genetic control of the response to subacute exposure. *Am J Physiol* **264**: L15–L20

Langford SD, Bidani A and Postlethwait EM (1995) Ozone-reactive absorption by pulmonary epithelial lining fluid constituents. *Toxicol Appl Pharmacol* **132**: 122–130.

Last JA, Reiser KM, Tyler WS and Rucher RB (1984) Long term consequences of exposure to ozone.
Toxicol Appl Pharmacol **72**: 111–118.

Leikauf GD, Driscoll KE and Wey HE (1988) Ozone-induced augmentation of eicosanoid metabolism
in epithelial cells from bovine trachea. *Am Rev Respir Dis* **137**: 435–442.

Leikauf GD, Leming LM, O'Donnell JR and Doupnik CA (1989) Bronchial responsiveness of guinea
pigs exposed to acrolein. *J Appl Physiol* **66**: 171–178.

Li Z, Daniel EE, Lane C *et al.* (1992) Effect of an anti-Mo1 MAb on ozone indued airway inflammation and airway hyperresponsiveness in dogs. *Am J Physiol* **263**: L723–L726

McCay PB (1985) Vitamin E: interactions with free radicals and ascorbate. *Ann Rev Nutr* **5**: 323–340.

Menzel DB (1992) Antioxidant vitamins and prevention of lung disease. *Ann NY Acad Sci* **669**:
141–155.

Mudway IS and Kelly FJ (1998) Modelling the interactions of ozone with pulmonary epithelial lining
fluid antioxidants. *Toxicol Appl Pharmacol* **148**: 91–100.

Mudway IS, Housley D, Eccles R *et al.* (1996a) Differential depletion of human respiratory tract
antioxidants in response to ozone challenge. *Free Radic Res* **25**: 499–513.

Mudway IS, Kelly FJ, Krishna T *et al.* (1996b) Upper airways loss of uric acid during exposure to ozone
(O₃). *Eur Respir J* **9**: 415S.

Mudway IS, Krishna MT, Withers NJ *et al.* (1997) Consumption of lung lining fluid ascorbate by ozone
is associated with airway neutrophilia and pulmonary epithelium injury. *Am J Res Crit Care Med*
155: A835.

Mustafa MG (1990) Biochemical basis of ozone toxicity. *Free Radic Biol Med* **9**: 245–265.

Nakamura H, Takada SD, Smimabuka R *et al.* (1987) Effect of vitamin E on the response of lung
antioxidant enzymes in young rats exposed to hyperoxia. *Kobe J Med Sci* **33**: 53–63.

Niki E, Takahashi M and Komuro E (1986) Antioxidant activity of vitamin E in liposomal membranes.
Chem Lett **71**: 1573–1576.

Pacht ER and Davis WB (1988) Role of transferrin and ceruloplasmin in antioxidant activity of lung
epithelial lining fluid. *J Appl Physiol* **64** (5): 2092–2099.

Pacht R, Kaseki H, Mohammed JR *et al.* (1986) Deficiency of vitamin E in the alveolar fluid of cigarette smokers. *J Clin Invest* **77**: 789–796.

Peden DB, Brown ME, Wade Y *et al.* (1990a) Human nasal glandular secretion of a novel antioxidant
activity: cholinergic control. *Am Rev Respir Dis* **143**: 545–552.

Peden DB, Robert H and Brown ME (1990b) Uric acid is a major antioxidant in human nasal airway
secretions. *Proc Natl Acad Sci USA* **87**: 7638–7642.

Peden DB, Swiersz ZM, Ohkubo K *et al.* (1993) Nasal secretions of the ozone scavenger uric acid. *Am
Rev Respir Dis* **148**: 455–461.

Postlethwait EM, Langford SD and Bidani A (1991) Transfer of nitrogen dioxide through pulmonary
epithelial lining fluid. *Toxicol Appl Pharmacol* **109**: 464–471.

Postlethwait EM, Langford SD and Bidani A (1994) Determinants of inhaled ozone absorption in
isolated rat lungs. *Toxicol Appl Pharmacol* **125**: 77–89.

Pryor WA (1991) Can vitamin E protect humans against the pathological effects of ozone in smog? *Am
J Clin Nutr* **53**: 702–722.

Pryor WA (1992) How far does ozone penetrate into the pulmonary air/tissue boundary before it
reacts? *Free Radic Biol Med* **12**: 83–88.

Pryor WA (1993) Ozone in all its reactive splendor. *J Lab Clin Med* **122**: 483–486.

Pryor WA (1994) Mechanisms of radical formation from reactions of ozone with target molecules in the
lung. *Free Radic Biol Med* **17**: 451–465.

Pryor WA and Church DF (1991) Aldehydes, hydrogen peroxide, and organic radicals as mediators of
ozone toxicity. *Free Radic Res Commun* **11**: 41–46.

Pryor WA, Giamalva DH and Church DF (1984) Kinetics of ozonation. 2. Amino acids and model
compounds in water and comparisons to rates in non-polar solvents. *J Am Chem Soc* **106**:
7094–7100.

Pryor WA, Das B and Church DF (1991) The ozonation of unsaturated fatty acids: aldehydes and
hydrogen peroxide as products and possible mediators of ozone toxicity. *Chem Res Toxicol* **4**:
341–348.

Pryor WA, Squadrito GL and Friedman M (1995) The cascade mechanism to explain ozone toxicity: the role of lipid ozonation products. *Free Radic Biol Med* **19**: 935–941.

Quinton PM (1979) Composition and control of secretions from tracheal bronchial submucosal glands. *Nature* **279**: 551–552.

Reynolds HY and Newball HH (1974) Analysis of proteins and respiratory cells obtained from human lungs by bronchial lavage. *J Lab Clin Med* **84**: 559–573.

Rustow B, Haupt R, Stevens PA and Kinze D (1993) Type II pneumocytes secrete vitamin E together with surfactant lipids. *Am J Physiol* **265**: L133–L139.

Schelegle ES, Siefkin AD and McDonald RJ (1991) Time course of ozone-induced neutrophilia in normal humans. *Am Rev Respir Dis* **143**: 1353–1358.

Skoza L, Snyder A and Kikkawa Y (1983) Ascorbic acid in bronchoalveolar wash. *Lung* **161**: 99–109.

Slade R, Crissman K, Norwood J and Hatch GE (1993) Comparison of antioxidant substances in bronchoalveolar lavage cells and fluid from humans, guinea pigs, and rats. *Exp Lung Res* **19**: 469–484.

Widdicombe J (1997) Airway and alveolar permeability and surface liquid thickness: theory. *Am J Physiol* **82**: 3–12.

Willis RJ and Kratzing CC (1974) Ascorbic acid in rat lung. *Biochem Biophys Res Commun* **59**: 1250–1253.

Winterbourn, CC (1985) Comparative reactivities of various biological compounds with myeloperoxidase-hydrogen-peroxide-chloride, and similarity of the oxidant hypochlorate. *Biochem Biophys Acta* **840**: 204–210.

Witting LA (1980) Vitamin E and lipid antioxidants in free radical-initiated reactions. In: Pryor A (ed.) *Free Radicals in Biology*. London: Academic Press, pp. 295–319.

17

Air pollutants: modulators of pulmonary host resistance against infection

PETER T. THOMAS

Covance Laboratories, Madison, Wisconsin, USA

JUDITH T. ZELIKOFF

Department of Environmental Medicine, New York University School of Medicine, New York, New York, USA

INTRODUCTION

Clean air is essential to preserve the health of the respiratory tract. In order to ensure adequate host protection from infection, the lungs have numerous physical, innate, and adaptive immunologic defenses designed to deal effectively with the airborne assault of oxidant gases, vapors, and particulate aerosols that are part of modern life. Evidence is becoming increasingly clearer from experimental and epidemiological studies that exposure to these materials alters host defense and may result in increased susceptibility to infection. This chapter will briefly review the physical and immunological defenses that protect the respiratory tract from infectious insult, discuss the experimental laboratory evidence demonstrating reduced host defense resulting from exposure to gaseous and particulate air pollution, and summarize the epidemiological evidence that corroborates the laboratory results.

AIR POLLUTION AND HEALTH
ISBN 0-12-352335-4

PHYSICAL AND IMMUNOLOGICAL DEFENSES OF THE RESPIRATORY TRACT

The physical and immunologic mechanisms that confer protection to the lungs are diverse and closely interrelated. The physical protective mechanisms of the lung mainly consist of the mucociliary apparatus, pulmonary surfactant, and the non-specific activities of specialized phagocytic cells. These defenses play key roles in the three principal mechanisms thought to be operative for removal or absorption of foreign substances. The first is physical removal by the mucociliary escalator of the tracheobronchial tree; the second is phagocytosis, usually by pulmonary macrophages, dendritic cells, or other leukocytes; and the third is by absorption to the lymphatic system. For an in-depth review of this aspect of lung physiology and pathology, the reader is directed to several excellent reviews of the subject (Adler, 1994; Brian, 1985; Lopez-Vidriero, 1984; Schlesinger, 1990).

Aerosol/particle deposition in the lungs depends, in large part, on the size of the particle itself. In humans, particles ≥ 5 μm mass median aerodynamic diameter (MMAD) are usually deposited in the upper airways. Smaller particles (1–5 μm MMAD) are most likely deposited in the terminal airways or alveoli. Fine particles (< 1 μm MMAD) remain suspended in the inhaled air and reach the alveolar zone of the lung where they may be readily absorbed. The surface area of the alveolar zone is estimated to be between 50 and 100 m^2. Because of the large surface area, a high rate of blood flow, and the close proximity to the alveolar air, absorption of gases and liquid aerosols is easily facilitated.

The airways are lined by a ciliated mucus-secreting epithelium. In the trachea and bronchi, the epithelium is pseudostratified and columnar. As the airways continue to branch into bronchioles and become narrower, the epithelium decreases in height eventually consisting of just a single layer of cuboidal cells in the terminal bronchioli. A thin mucous film which protects the lungs against desiccation resulting from continuous air passage is the product of several sources including plasma transudation, serous and mucous secretions by submucosal glands in the bronchial wall, and secretions of cells on the bronchial surface epithelium (e.g. goblet, serous, and Clara cells). Apart from helping to maintain moisture content and fluidity in the lungs, the mucus also serves to trap inhaled foreign particles. In conjunction with ciliary beating, movement of mucus containing trapped particles occurs and the foreign matter is transferred up and out of the lungs, usually within a period of 24 h post-inspiration.

Pulmonary surfactant, a complex lipoprotein consisting of 90% lipid, 8% protein and 2% carbohydrate, plays an important role in maintaining the integrity of the respiratory tract. Phospholipids are one major component of surfactant, with dipalmitoyl phosphatidylcholine being the major form present. Generally, these phospholipid complexes, produced by Clara cells and macrophages, line the alveoli and small airways at thicknesses ranging from 0.1 to 0.2 μm. The functions of surfactant are primarily: to maintain proper humidity and stability of the alveolar wall (by reducing surface tension); to retard infiltration of capillary and interstitial fluids; and, to emulsify small inhaled particles, thus facilitating their phagocytic uptake and removal. Pulmonary surfactant proteins A and D (SP-A and SP-D), members of the collectin family of proteins, are important in the binding and opsonizing of pulmonary pathogens such as *Pneumocystis carinii*. In premature infants, inadequate levels of surfactant are often a major causative factor in poor respiratory health and contribute to an increased incidence of lung infections. Because surfactants are macrophage products, cytokines may influence their production; interferon-γ (IFN-γ) and tumor necrosis factor-α (TNF-α) have been shown to increase and decrease, respectively, SP-A mRNA levels in *in vitro* cell cultures (Ballard *et al.*, 1990; Wispe *et al.*, 1990).

Pulmonary macrophages (derived continuously from circulating monocytes) and polymorphonuclear leukocytes (PMN) play key roles in non-specific cellular defense mechanisms in the lung. The total number of pulmonary macrophages (PAM) in a normal healthy lung is 3–4 times that of lung PMN and PAM comprise > 80% of all the cells present in bronchoalveolar lavage fluid. There are at least three different types of macrophages in the lungs; the most prominent is the alveolar macrophage (AM), with the other types being airway and interstitial macrophages. AM are mobile, long-lived, actively adhere to respiratory structures, are relatively large (typically 20–40 μm in diameter), and are characterized by a ruffled surface membrane with numerous cytoplasmic processes and a large number of intracytoplasmic granules. AM express lineage-specific cytoplasmic membrane molecules (e.g. CD14) as well as surface receptors for a number of molecules, including the F_c region of IgG (CD16, CD32, and CD64), activated complement components (CD35), and various cytokines. Both AM and PMN are capable of engulfing opsonized particles coated with antibody and/or complement components. These same complement receptors can also recognize various microbial constituents; the leukocyte integrins CD11b/CD18 or CR3/Mac1 recognize and bind bacterial lipopolysaccharide. Binding of bacterial cells to these receptors facilitates their phagocytosis by leukocytes which, in turn, augments the lung protective response by triggering the release of pleiotropic cytokines by AM and PMN (reviewed by Ulevitch and Tobias, 1995).

Pulmonary macrophages produce a wide variety of cytokines and other effector molecules in response to pulmonary insult or disease. It is thought that the diversity of macrophage secretory products is the greatest of any cell in the immune system. These secretory products allow the macrophage to exert both pro- and anti-inflammatory effects and to regulate other cell types. Among secreted products are TNF-α, interleukin (IL)-1, -6, -12, and -15, and a number of chemoattractant cytokines (chemokines) involved in the orchestration of the immune response. In addition to macrophages, stromal cells and respiratory epithelial cells have the ability to produce and secrete pro-inflammatory cytokines, such as IL-1, -6 and -8, as well as growth factors that modulate the differentiation of other effector cells, such as granulocyte/monocyte-colony stimulating factor (GM-CSF), monocyte-colony stimulating factor (M-CSF), and granulocyte-colony stimulating factor (G-CSF) (Churchill *et al.*, 1992).

In addition to effector molecules, macrophages produce numerous biologically active substances responsible for bactericidal actions, including enzymes such as lysozyme catalase, and various proteases, and reactive oxygen and nitrogen species such as superoxide anion ($\cdot O_2^-$), hydrogen peroxide (H_2O_2) and nitric oxide (NO). Although large numbers of infectious agents are routinely deposited in the lungs, these tissues are usually sterile. It is due, in great part, to the phagocytic and lytic functions of macrophages and PMN that the lungs are able to remain sterile, even in the face of routine inspiration of large numbers of infectious agents, including bacteria, viruses, and other potential lung pathogens.

Other cell types contribute to the maintenance of pulmonary immunocompetence. It is becoming increasingly clear that respiratory epithelial cells play an important role as both 'target' and 'effector' cells during lung inflammation (reviewed in Adler, 1994). As target cells, epithelial cells respond to different exogenously and endogenously produced agents by altering their defense functions including mucus secretion, ciliary beating frequency, and ion transport. As effector cells, they can respond to these stimuli by synthesizing/releasing secondary inflammatory mediators, including eicosanoids, platelet activating factor, various free radicals, and cytokines.

In general, these coordinated pulmonary immune defenses are designed to be expeditious in protecting the respiratory system from environmental insult and infection. However, in some instances, immunologic processes activated following host exposure to an environmental pollutant or infectious agent can be damaging and result in a compromise of lung protective mechanisms. Normally, an acute phase/inflammatory response mediated by TNF-α, IL-1 and IL-6 released from macrophages, endothelial and epithelial cells allows the lung time to control the initial infection while an adaptive immune response is initiated. Under certain conditions, a prolonged inflammatory process occurring in response to an environmental agent or infectious challenge can alter non-specific pulmonary defenses.

In addition to numerous innate and physical defenses that protect the respiratory tract, the lung has a well-developed system for generating an adaptive immune response against infectious challenge. It is beyond the scope of this chapter to provide a detailed review of the structure and function of the entire mammalian immune system. Rather, the reader is directed to several excellent reviews and textbooks on the subject (Bellanti and Herscowitz, 1985; Fishman, 1988; Janeway and Travers, 1996, Paul, 1989; Roitt *et al.*, 1998; Thurlbeck and Churg, 1995).

There are numerous cell types that are actively involved in the cellular immunological responses of the respiratory tract. In addition to providing a first line of defense against respiratory infectious agents, lung macrophages act as 'professional' antigen-presenting cells (APC). Once bound to the APC/PAM, microorganisms are engulfed and degraded within cellular endosomes and lysosomes; peptide fragments which are generated as a result are then transferred to the cell surface in conjunction with class II molecules of the major histocompatibility complex (MHC) for presentation to circulating T lymphocytes. At the same time, MHC class II and B7 molecules are induced on the macrophage surface and result in a co-stimulatory signal required for T lymphocyte activation (Razi-Wolf *et al.*, 1992); as such, macrophage receptors that recognize microbial constituents are also likely to mediate the induction of co-stimulatory activities. Resembling PAM, but without non-specific esterase staining characteristics, dendritic cells in the trachea, bronchi and alveoli also serve as APC (Sertl *et al.*, 1986). In addition to these actively phagocytic cells, lymphocytes are present in the hilar and mediastinal lymph nodes and appear as single or cellular aggregates in the lung.

Antigen-specific T lymphocyte activation is the central event in the generation of adaptive immunity. The interaction of T lymphocytes with MHC class II molecule/antigen complexes is facilitated by the presence of co-receptors LFA-3, CD2, and ICAM-1. This interaction leads to a series of biochemical processes that result in T lymphocyte activation and the subsequently increased production of cytokines such as IL-1. Once activated, T lymphocytes produce IL-2, a pleiotropic growth-promoting cytokine with numerous autocrine and paracrine effects. After the cytokine cascade is initiated, the growth and differentiation of numerous cell types, including B lymphocytes and natural killer cells, ensues. The consequence of this is the generation of antigen-specific humoral and/or cell-mediated immune responses; this includes production of antibody that facilitates ingestion and destruction of the antigen as well as the specific and non-specific lysis of tumor cells. Furthermore, because specific immunologic memory is a result of the initial insult, upon re-exposure of the respiratory tract to the offending antigen, a greatly enhanced secondary host response will result.

An important antibody isotype present in the lungs is IgA. IgA, with a molecular weight of about 160 kDa, is found in respiratory secretions as a dimer. In addition, a separate 15 kDa polypeptide chain ('J chain') promotes polymerization by linking *C*-terminal cysteine residues which are found only on the secreted forms of the IgA 'A chains', a prerequisite for IgA transport though the pulmonary epithelia. Apart from its important role in preventing the adherence of bacteria, viruses, or other pathogens to epithelial cells (Fischetti and Bessen, 1989; Mostov, 1994), IgA is capable of activating complement via the alternative pathway. The importance of IgA to pulmonary host protection is best demonstrated by clinical studies wherein selective IgA deficiency, an immunoglobulin deficiency affecting approximately one in 800 people, has been shown to be correlated with an increased incidence of chronic lung infections in these patients.

EVIDENCE LINKING AIR POLLUTANTS TO REDUCED PULMONARY HOST DEFENSE: ANIMAL MODELS

Air pollution has long been considered a risk factor for respiratory infection (see Selgrade and Gardner, 1996; Selgrade and Gilmour, 1994). Epidemiologic studies (as described in the next section) suggesting a relationship between air pollution and increased incidence, severity, and duration of symptoms related to respiratory infections span four decades. Reactive gases have been most extensively studied and include those associated with fossil fuel emissions such as ozone (O_3), nitrogen dioxide (NO_2), and sulfur dioxide (SO_2). More recently, particulate pollutants, albeit not as well-studied, have also been shown to alter pulmonary host defense mechanisms and compromise local defenses against infectious agents.

While studying the effects of air pollutants directly upon human subjects offers a number of advantages, epidemiological studies and controlled clinical studies are limited by societal concerns, ethical and legal issues, as well as by cost. Because of these difficulties, animal models and *in vitro/ex vivo* systems have been utilized to aid in the prediction of human health risk to airborne pollutants.

Animal host-resistance models, many of which accurately reflect human disease, have been used to assess the effect of xenobiotics on altered host resistance and on the component immunological mechanism(s) important for defense against the selected infectious agent. While microorganisms such as respiratory syncitial virus (RSV), cytomegalovirus (CMV) and *Staphylococcus aureus* have been used for such studies, those infectious agents most commonly used as infectivity models for assessing pulmonary immunotoxicity include a variety of *Streptococcus* species, *Listeria monocytogenes*, *Klebsiella pneumoniae* and influenza virus (reviewed in Thomas and Sherwood, 1996).

The remainder of this section will review some of the experimental laboratory studies linking air pollutants (i.e. gases, metals and non-metal airborne particulates) to altered pulmonary host defense. Results from such studies performed *in vitro /ex vivo* will also be described. Since the purpose of this particular section is to provide an overview rather than an in-depth analysis of the effects of some of the more commonly occurring environmentally/occupationally inhaled pollutants on host resistance in animal models, the reader is referred to other book chapters and review articles for a more comprehensive review of specific air pollutants and their effects upon host immunocompetence (Albright and Goldstein, 1996; Chitano *et al.*, 1995; Selgrade and Gilmour, 1994; Selgrade and Gardner, 1996; Zelikoff and Thomas, 1998).

Metals

Heavy metals are ubiquitous in the biosphere, where they occur as part of the natural background of chemicals to which human beings are constantly exposed. Sources of environmental/occupational airborne metals includes those emitted from refineries, chemical plants, cement manufacturers, power plants, smelters, trash burning and tobacco smoke (reviewed in Zelikoff and Cohen, 1996).

Within the last few decades, experimental data have shown that high-dose as well as low-level exposure to certain metals induces subtle changes within a host, including altered immunological competence (reviewed in Zelikoff and Thomas, 1998). Environmental stressors like metals may act directly to kill the exposed organisms, or indirectly to exacerbate disease states by lowering resistance and allowing the invasion of infectious pathogens.

Although metal-induced effects are dependent upon such variables as host species, metal dose, and route and duration of exposure, the conclusion reached in most immunotoxicological studies is that heavy metals suppress immunocompetency. Moreover, the most consistent finding in experimental studies (as well as in epidemiological studies) evaluating the effects of metals on immune functions is a decreased host resistance to infectious agents (reviewed in Zelikoff and Cohen, 1996, 1997).

Arsenic (As)

Although the vast majority of human exposure comes from dietary intake, inhalation (as a result of industrial production) also represents a primary route of As exposure (Burns, 1998). Persons employed in copper, zinc and lead smelters, in the manufacture and spraying of pesticides, and in the production of wood treated with chromated copper arsenate are at particular risk of occupational exposure to airborne As (Nygren *et al.*, 1992). Moreover, the recent computer 'boom' has dramatically increased the numbers of people potentially exposed in the semiconductor industry to gallium arsenide (GaAs) and/or indium arsenide (InAs).

Most host resistance studies have used drinking water exposure to examine the immunotoxicological effects of As; however, inhalation of inorganic arsenicals has been shown in a few studies to alter host resistance and/or mortality in response to infectious bacterial agents. Aranyi *et al.* (1985) demonstrated that a single 3-h inhalation exposure of mice to insoluble arsenic trioxide (As_2O_3; at ≥ 270 µg As/m^3) produced a significant concentration-related decrease in resistance against subsequent infection with *Streptococcus zooepidemicus* and a decrease in pulmonary bactericidal activity; multiple exposures (i.e. 3 h/day, 5 day/week) to As_2O_3 at 500 µg/m^3 decreased pulmonary bactericidal activity and increased host mortality. In the same studies, *in vivo* bactericidal activity in the lungs of mice exposed to As_2O_3 and ^{35}S-labeled *Klebsiella pneumoniae* simultaneously was also suppressed. In contrast, intratracheal (IT) instillation of 200 mg GaAs/kg 24 h prior to bacterial challenge had no effect on *Listeria monocytogenes* infection and increased host resistance to infection with *Streptococcus pneumoniae* (Burns *et al.*, 1993). The investigators suggested that the observed effects were likely due to potential antibacterial properties of the As present in the serum (Burns *et al.*, 1996). The potent hemolytic poison arsine (AsH_3) gas selectively alters host resistance against infectious agents; studies by Rosenthal *et al.* (1989) demonstrated that inhalation of AsH_3 for 14 days (6 h/day, at 0.5, 2.5 and 5 ppm) increased host susceptibility to *L. monocytogenes* and *Plasmodium yoeli*, but had no effect upon host resistance to influenza virus.

Cadmium (Cd)

Although normally found in low concentrations in ambient air, its potent toxicity, means that Cd has significant effects upon the respiratory system and can increase an individual's risk for respiratory infection. Occupational exposure of workers to Cd, usually in the relatively water-insoluble form of cadmium oxide (CdO), generally occurs via inhalation of Cd fumes/dusts during smelting, battery manufacturing, soldering and pigment production (Koller, 1998). Moreover, because its biological half-life is estimated to be 10–25 years, Cd readily accumulates in the body.

The immunomodulating effects of Cd via non-respiratory routes of exposure have been well studied and the overall tendency is for Cd to induce various degrees of immunosuppression. Although inhalation of Cd represents a primary route of human exposure, relatively few studies have examined the effects of inhaled Cd upon immune function/host resistance against infectious agents. Moreover, only a handful of studies have employed insoluble CdO, the most common form of Cd found in outdoor and indoor air.

Gardner *et al.* (1977) demonstrated enhanced susceptibility to *S. zooepidemicus* in mice following a single 2-h exposure to $0.1–1.6$ mg Cd/m^3. An enhanced bacterially induced mortality following cadmium chloride ($CdCl_2$) exposure was accompanied by an impaired clearance of bacteria from the lungs and decreased numbers of AM recovered in bronchoalveolar lavage fluid immediately after exposure. In addition to the observed increase in pulmonary susceptibility to infection with Gram-positive bacteria, Cd also alters host susceptibility to Gram-negative organisms. A 15 min exposure to CdO (10 mg Cd/m^3) decreased pulmonary bacterial clearance and increased host mortality in rats infected by inhalation with *Salmonella enteritidis* and in mice infected with *Pasteurella multocida* (Bouley *et al.*, 1977).

While exposure to airborne Cd appears consistently to enhance pulmonary susceptibility to bacterial infection, protection against/enhancement of/no effect upon viral infection after Cd exposure have all been reported. Mice exposed for 15 min to CdO particles (10 mg Cd/m^3) and then infected 48 h later with influenza virus had significantly lower death rates than did the air-exposed controls (Chaumard *et al.*, 1983). Protection against viral infection was also observed in mice infected midway through a repeated exposure regimen (5 days/week for 4 weeks). This effect was unique to viral resistance as groups of similarly exposed mice challenged with *P. multocida* had a higher incidence of death. These protective effects of Cd against influenza were attributed to increased numbers of pulmonary inflammatory cells (Chaumard *et al.*, 1991). Alternatively, short-term inhalation of soluble $CdCl_2$ (2 h/day, for a total of 4 days) had no effect on mice infected with murine CMV on the first day of exposure (Daniels *et al.*, 1987); interestingly, when administered parenterally, CdCl$_2$ enhanced susceptibility to CMV infection and suppressed virus-augmented splenic natural killer cell activity. That mice might tolerate a higher dose of Cd administered intramuscularly as compared to via inhalation was thought to explain this discrepancy.

In conclusion, the evidence that aerosolized Cd exposure impairs host defenses to viral respiratory infections is equivocal. However, other interactions that might result from a combined viral infection and Cd exposure could also impact upon human health. For example, infection with Coxsackie virus B3 has been shown to alter Cd distribution in mice following IV infection (Ilback *et al.*, 1992). Thus, in a cyclical fashion, the effects that a combination of Cd exposure and viral infection may have on the availability/toxicity of Cd itself needs to be considered, i.e. as the availability of Cd is increased in the tissues, this may, in turn, result in further impairment of defenses against the virus, which would then be able to increase Cd availability and so on.

Nickel (Ni)

Human exposure to airborne Ni is ubiquitous from anthropogenic sources including combustion of fossil fuels and cigarette smoke (reviewed in Zelikoff and Cohen, 1996). Pulmonary absorption accounts for the major toxic route associated with Ni exposure, with gastrointestinal and dermal absorption being significantly lower. The toxicity of Ni is thought to arise because of its ability to interact with a variety of essential metal cations.

Exposure to Ni, like many other heavy metals, has been shown to alter host immunocompetence (reviewed in Zelikoff and Cohen, 1996, 1997). Since the primary route of Ni exposure in humans is via inhalation, a relatively large number of *in vivo* and *in vitro* studies have been performed examining the effects of inhaled or instilled soluble/insoluble Ni compounds upon immune cell function, particularly those of the AM (Smialowicz, 1998). In comparison, relatively little is known concerning the effects of Ni upon pulmonary host resistance against infection. Mice exposed for 2 h to soluble nickel chloride (NiCl$_2$; 500 μg Ni/m^3) or nickel sulfate (NiSO$_4$; 455 μg Ni/m^3) and infected 24 h later with *S. zooepidemicus* showed significantly enhanced mortality and decreased mean survival times compared with infected air controls (Adkins *et al.*, 1979). The pulmonary clearance of bacteria and the *ex vivo* capacity of AM obtained from these mice to ingest latex beads were also impaired. Interestingly, when mice were infected immediately (as opposed to 24 h after) following exposure, susceptibility to *S. zooepidemicus* was unaffected. In *ex vivo* studies, intracellular killing of *S. aureus* was depressed in AM recovered from rabbits exposed for 4 weeks (6 h/day, 5 days/week) to an aerosol of NiCl$_2$ (at 0.3 mg Ni/m^3) (Wiernik *et al.*, 1983). Other studies have suggested that Ni-induced alterations in extracellular bactericidal components (i.e. lysozyme) may contribute to an enhanced pulmonary susceptibility to bacteria (Lundborg and Camner, 1982).

Studies investigating the effects of inhaled Ni upon viral infection are limited. Daniels *et al.* (1987) demonstrated that exposure of mice to 100 μg Ni/m^3 (as NiCl$_2$) for 4 days (2 h/day) had no effect upon host challenge with CMV on the first day of exposure. Interestingly, exposure to NiCl$_2$ did enhance susceptibility to CMV and suppressed virus-augmented splenic NK cell activity when administered parenterally. As with Cd, the combination of viral infection and Ni exposure appears to bring about interactions which could easily impact upon human health. For example, *in vitro* exposure to NiSO$_4$ causes increased proliferation of Epstein–Barr virus (EBV)-positive lymphoblastoid cell lines and increased early antigen expression (Wu *et al.*, 1986). In certain 'high-risk' areas of China, EBV is associated with nasopharyngeal carcinoma; since high levels of Ni can be found in the environment of high-risk areas, it was proposed that Ni could contribute to the development of EBV-associated nasopharyngeal carcinoma. Nickel, like Cd, has also been shown to alter the distribution of Coxsackie virus B3 (Ilback *et al.*, 1992).

In summary, a number of studies have shown that inhalation exposure to airborne metals can increase host susceptibility to bacterial infection in the lung. Some of these same metals have also been shown to suppress AM function, ciliary beat frequency and splenic antibody-forming cell responses. All of these effects may contribute to enhanced mortality due to respiratory bacterial infections. However, while the evidence that metal exposure increases the risk of pulmonary bacterial infections is relatively consistent, the role that airborne metals may have in compromising host defenses against respiratory viral infection is unclear and requires further research.

Pollutant gases

Ozone

Ozone (O_3), an oxidizing air pollutant found as a result of photochemical reactions between hydrocarbon molecules, nitrogen oxides and the natural outdoor environment, has been shown in numerous studies to modulate pulmonary and systemic immune responses. Because interest in O_3-induced health effects began over four decades ago, a relatively large number of laboratory investigations have been performed to examine whether (and by what mechanisms) inhaled O_3 might affect pulmonary host susceptibility against infection with bacteria, viruses, protozoa and/or yeast.

Miller and Ehrlich (1958) demonstrated that mice exposed to 1.5 ppm O_3 either once for 4 h or repeatedly over 2 weeks (5 days/week) had a higher incidence of mortality and were more susceptible to infection with aerosolized *K. pneumoniae* than were control animals. In a later study, mice exposed chronically to 1.5 ppm O_3 (2 h/day, 5 days/week, for 3–12 months) and then challenged intravenously with *Mycobacterium tuberculosis* had a higher incidence of mortality than their unexposed infected counterparts; similar effects were not observed with shorter O_3 exposure durations (i.e. 2 months) (Thienes *et al.*, 1965).

Similar immunosuppressive effects in response to bacterial infection have also been demonstrated with lower O_3 concentrations. A single 3-h exposure to 0.1 ppm O_3 increased the incidence of mortality in mice challenged with *S. pyogenes* following exposure (Ehrlich, 1980); effects of co-exposure to O_3 and nitrogen dioxide (NO_2) on *S. pyogenes*-induced mortality were additive. In a series of studies by Gilmour, the investigators examined the effects of a single 3-h O_3 exposure upon host susceptibility to *S. zooepidemicus* infection in: mice and rats (Gilmour and Selgrade, 1993); two different mice strains (Gilmour *et al.*, 1993a); and in mice of different ages (Gilmour *et al.*, 1993b). In all cases, animals exposed to O_3 at 0.4 and/or 0.8 ppm O_3 were more susceptible to streptococcal infection than were infected filtered air-exposed controls. Mice were affected more severely than rats and 5 week-old mice proved more sensitive to O_3 than were their

9-week-old counterparts. Moreover, O_3-induced effects appeared to be strain-dependent in that mortality was greater in C_3H/HeJ than in C57Bl/6 mice. The authors attributed the observed effects to alterations in AM phagocytosis (Gilmour and Selgrade, 1993; Gilmour *et al.*, 1993a). Similar suppressive effects of O_3 have also been observed in rats challenged with Gram-positive *L. monocytogenes* (van Loveren *et al.*, 1988). In these studies, rats continuously exposed to O_3 (0.25–2.0 mg/m^3, for a period of 1 week) demonstrated suppressed phagocytosis and intracellular killing by AM resulting in reduced bacterial clearance from the lungs; acquired specific cellular immune responses needed for mediating respiratory listerial infection were also suppressed.

Unlike the consistent suppressive effects observed when O_3 exposure preceded bacterial infection, effects of O_3 upon ongoing pulmonary bacterial infections are equivocal. In studies by Goldstein *et al.* (1978), rats infected with aerosols of *S. aureus* prior to a 5-h exposure to 2.5 ppm O_3 had reduced levels of AM-associated lysosomal enzyme activity and a diminished rate of bacterial uptake and killing. In contrast, O_3 exposure (0.64 ppm for 14 or 28 days) of rats chronically infected with *Pseudomonas aeruginosa* had no effect upon the disease process (Sherwood *et al.*, 1984).

Effects of O_3 upon pulmonary viral infections have been poorly studied and the results appear contradictory. Mice exposed to 1 ppm O_3 for 5 consecutive days (3 h/day) and then infected with influenza virus following each of the individual exposures evidenced no effects upon mortality or mean survival time if infection took place after the first, third, fourth or fifth exposure; only mice infected after the second exposure had an increased incidence of mortality and a decrease in mean survival time. In no case were changes in pulmonary viral titres observed (Selgrade *et al.*, 1988); as such, it was concluded that exposure to O_3 prior to viral challenge had little effect upon antiviral defense mechanisms. Moreover, findings from an *in vitro* study in which human AM were exposed for 2 h to 1 ppm O_3 and then infected with RSV confirm these *in vivo* results in that no effect upon the percentage of infected AM or the amount of infectious RSV released by exposed cells (compared with control values) were observed (Soukup *et al.*, 1993).

Similar to the equivocal effects observed when exposure precedes viral infection, inhalation of O_3 following infection with influenza appears to either have no effect or to reduce viral-related pathologies. Exposure of influenza A-infected mice to 0.5 ppm O_3 reduced the involvement of respiratory epithelium in the infectious process and resulted in a less widespread infection of the alveolar parenchyma (Wolcott *et al.*, 1982); this O_3-induced alteration in viral antigen distribution was consistent with a reduced host mortality and prolonged survival time. In a later study, continuous exposure of influenza-infected mice to 0.5 ppm O_3 for 30 days was shown to have no effect upon viral proliferation *in situ* and appeared to mitigate virus-induced acute lung injury by ~50% (compared with that assessed in infected filtered-air controls); in contrast, chronic lung damage was exacerbated 30 days post-infection (Jakab and Bassett, 1990).

Acid Sulfur Oxides

There is increasing concern about the human health effects associated with the inhalation of ambient acid aerosols produced directly from fossil fuel combustion and indirectly from catalytic or photochemical oxidation of sulfur dioxide (SO_2). Despite the widespread occurrence of acidic sulfur oxides in the ambient environment and their potential risks to human health, the effects associated with pulmonary immune defenses, and in particular host resistance, have only been poorly studied and the results are inconsistent. Studies examining mice and rats exposed to atmospheres of SO_2 (MMAD = ≈1.3 μm) alone (100–170 μg SO_2/m^3) or in combination with monodispersed ferrous sulfate particles (MMAD = 2.4 μm) either 17 h prior to, or 4 h after infection with aerosols of *S. aureus* or *Streptococcus* (Group C), demonstrated no effects upon staphylococcal clearance or AM ingestion of bacteria (Goldstein and Lippert, 1979). Other studies have demonstrated that continuous exposure of mice to ≈10 ppm SO_2 for up to 3 weeks reduced resistance to infection with *K. pneumoniae*, increased host mortality, and shortened survival time (Azoulay-Dupuis *et al.*, 1982). Inasmuch as Ehrlich (1980) also observed no effects of SO_2 upon host resistance against another group C *Streptococcus* (i.e. *S. pyogenes*), it seems likely that SO_2 may selectively alter those pulmonary immune defense mechanisms specific for combating Gram-negative, rather than Gram-positive, infections.

Host resistance studies examining the effects of inhaled SO_2 upon viral infection have demonstrated mixed results. While studies by Fairchild (1977) proved that exposure of mice to SO_2 (as high as 6 ppm) produced no effect upon influenza virus replication, other studies (Fairchild *et al.*, 1972; Lebowitz and Fairchild, 1973; Ukai, 1977) have demonstrated either additive or synergistic effects of the two stressors on pulmonary inflammatory responses.

Nitrogen Oxides

Human activity, in particular fossil fuel combustion and vehicular traffic, is responsible for a large portion of the nitrogen oxides (NO_x) present in urban air. In the presence of sunlight, nitric oxide (NO) is converted to nitrogen dioxide (NO_2); of all the nitrogen oxides, NO_2 has been the most widely studied. While mostly examined in combination with other air oxidant gases (i.e. O_3), exposure to NO_2 appears consistently to suppress pulmonary host resistance against a variety of bacterial agents.

Ehrlich (1966) demonstrated that a single 2-h exposure of mice to 3.5 ppm NO_2 before or after respiratory challenge with aerosolized *K. pneumoniae* significantly increased host mortality; the same effect was observed in hamsters but at a 10-fold higher NO_2 concentration. Continuous exposure of mice to 0.5 ppm NO_2 for ≥ 3 months also decreased resistance against *K. pneumoniae* infection. These studies were extended by Henry *et al.* (1969) using squirrel monkeys and NO_2 concentrations between 10 and 35 ppm. Monkeys challenged with *K. pneumoniae* within 1 h following exposure had a higher incidence of death than infected unexposed controls. Moreover, at concentrations ≤ 35 ppm, bacterial clearance from the lungs of exposed monkeys was delayed or, in some cases, prevented; while *K. pneumoniae* was almost completely cleared from the lungs of infected unexposed monkeys, bacteria remained for as long as 19–51 days post-infection in the lungs of monkeys exposed to 10 ppm NO_2. The suppressive effects of NO_2 observed in these early studies were subsequently confirmed by a number of investigations. For example, Ehrlich (1980) demonstrated that a single 3-h exposure to NO_2 at ≥ 2.0 ppm resulted in excessive host mortalities in response to infection with either *K. pneumoniae* or *S. pyogenes*; continuous exposure to 5 ppm for 3 months (24 h/day, 7 days/week) or to 1.5 ppm for 8 h or longer also increased mortality in NO_2-exposed mice challenged with *K. pneumoniae*.

Similar suppressive effects of NO_2 have also been observed in response to infection with Gram-positive bacteria. In mice exposed to NO_2 peaks of 4.5 ppm for either 1, 3.5 or 7 h and then challenged with *S. zooepidemicus* immediately after exposure, the mortality rate was found to be directly related to the length of peak exposure and that all peak lengths increased mortality (Graham *et al.*, 1987). In the same study, it was shown that when bacterial challenge was delayed until 18 h after peak NO_2 exposures, the immunomodulating effects from the 3.5- and 7-h exposures still caused an increase in *Streptococcus*-induced mortality.

Since many air pollutants co-exist with NO_2 in urban ambient air, studies have examined the effects of either sequential or simultaneous exposure to combinations of NO_2 and other pollutants on pulmonary immunocompetence. Studies by Ehrlich *et al.* (1979) demonstrated that mice exposed (3 h/day, 5 days/week, for 2–6 months) to NO_2 (0.5 ppm) and O_3 (0.1 ppm) had reduced resistance (measured by mortality and mean survival time) to an aerosolized challenge with *S. pyogenes*. It was also noted that this decrease in resistance to infection occurred sooner when the mice continued to be exposed (for 14 days after infection) to both pollutants rather than to clean air. That the co-pollutant-exposed mice displayed decreased total lavageable cell counts and lower AM viabilities/phagocytic activities suggested that the observed reductions in host resistance may have been due to altered AM-mediated activities used in pulmonary *Streptococcus* infections. In other studies, mice were exposed to atmospheres containing 1.0 ppm NO_2 in combination with ~300 $\mu g/m^3$ respirable ferrous sulfate ($FeSO_4$) particles (MMAD = 0.4 μm) and then challenged with aerosols of *S. aureus* or Group C *Streptococci*. While exposure to the combined pollutants for 24 or 48 h had no effect upon pulmonary clearance of *S. aureus*, inhalation of either pollutant alone for 48 h, or to NO_2/$FeSO_4$ for 24 or 48 h, significantly decreased pulmonary clearance of inhaled Group C *Streptococci* (Sherwood *et al.*, 1981). These studies reveal demonstrate the importance of the challenge organism and the duration of exposure for demonstrating effects of relatively low levels of co-pollutants on host susceptibility.

Mycoplasma pulmonis has also been used as an infectivity model to assess the effects of inhaled NO_2 upon host resistance and pulmonary immune mechanisms. Mycoplasmal growth and pneumonic lesions were increased in the lungs of mice exposed to 10 ppm NO_2 at the time of, or just after, infection; serum antibody titres to the organism increased as a function of time post-infection regardless of the NO_2 concentration or the infection dose (Nisizawa *et al.*, 1988). Studies to examine the effects of exposure of mice to similar NO_2 levels for 4 h prior to exposure to *M. pulmonis* aerosols also indicated that NO_2 could potentiate murine respiratory mycoplasmosis (Parker *et al.*, 1989). The latter study concluded that NO_2 affected host lung defense mechanisms responsible for limiting the extent of *M. pulmonis* infection and that exposure levels required to produce potentiation varied with host genetic background, the number of organisms administered, and the end point measured. This viewpoint was supported by experiments which demonstrated that exposure of C57Bl mice to 5 or 10 ppm NO_2 for 4 h prior to infection with *M. pulmonis* reduced pulmonary clearance of these organisms and that this effect was due to impaired intrapulmonary killing of *M. pulmonis* rather than an NO_2-induced change in the rate of physical removal of the organisms from the lungs. The results of each study using *M. pulmonis* has been useful in establishing a direct link between decreased pulmonary clearance after NO_2 exposure and increased disease severity.

In contrast to the numerous studies demonstrating the suppressive effects of inhaled NO_2 upon host responses against bacterial infection, less is known concerning possible interactions between NO_2 and viral infections; among these few studies, the results are contradictory. Using influenza virus as the infectivity model, studies have demonstrated that inhalation exposure to NO_2 either has no effect (Lefkowitz *et al.*, 1986), increases (Henry *et al.*, 1970; Ito, 1971), or decreases (Buckley and Loosle, 1969) viral-induced host mortality. As with the other inhaled gaseous pollutants, differences in exposure regimens may account for these varying observations. Studies using viral agents other than influenza have fared no better. Exposure of mice to 5, 10 or 20 ppm NO_2 (4 h/day for 10 days) following infection with Sendai virus had no effect upon pulmonary viral titres (compared with unexposed infected mice), but did enhance lung pathology as evidenced by an increased lung wet/dry weight ratio, level of lavage fluid protein, and lavageable cell counts (Jakab, 1988). In another study, exposure of mice to 5 ppm NO_2 (6 h/day for 2 days) preceding and 4 days following IT-challenge with murine CMV resulted in a 10-fold increase in lung viral titres and more severe histopathologic changes compared with filtered-air-exposed infected controls (Rose *et al.*, 1988). Thus, it appears that experimental viral infection has not adequately addressed the issue raised by human epidemiologic studies: the possible role of NO_2 exposure in altered susceptibility to respiratory viral infections.

Particulate Pollutants (Other than Metals)

Although a considerable amount of information is currently available regarding the effects of gaseous pollutants upon host resistance to infection, there is a notable paucity of data pertaining to the effects of particulate pollutants (other than metals).

Single Pollutant Studies

Aranyi *et al.* (1983) investigated the effects of four coarse mode particles including quartz, ferric oxide (Fe_2O_3), calcium carbonate and sodium feldspar on host defenses against pulmonary bacterial infection. Mice IT-instilled with one of the aforementioned particulates (at 33 or 100 μg/instillate) and then exposed within 1h to aerosols of viable *Streptococcus* demonstrated increased pneumonia-related mortalities. Interestingly, no changes in host resistance were observed in response to inhaled *K. pneumoniae*, even at the higher particle instillate concentration.

The effects of sulfuric acid (H_2SO_4), an important atmospheric particulate formed from SO_2, upon pulmonary host defense mechanisms have also been examined. AM recovered from rabbits exposed for 3 h to H_2SO_4 at 1 mg/m³ had a reduced ability *ex vivo* to phagocytose and kill ingested *S. aureus* (Zelikoff *et al.*, 1997b). While the effects did not appear to be due to alterations in AM FcR expression, H_2SO_4-induced reductions in $^{\cdot}O_2^-$ production may have played a role, at least in part, in bringing about the observed effects upon intracellular killing given the importance of reactive oxygen intermediates for killing staphylococcal organisms. In conjunction with the results of earlier studies (Zelikoff and Schlesinger, 1992; Zelikoff *et al.*, 1994), it was concluded that H_2SO_4-induced changes in AM $^{\cdot}O_2^-$ and TNF-α production may be responsible for the changes observed in host resistance to infection.

Experiments designed to help explain the increased incidence of morbidity and/or mortality observed in elderly individuals exposed to particulate matter < 10 μm MMAD (PM_{10}) have examined New York City ambient PM_{10}-induced changes in host susceptibility to *S. pneumoniae* infection in the lungs of rats IT-infected at different time-points following concentrated PM_{10} exposure, as well as in rats infected with encapsulated *S. pneumoniae* 48 h prior to exposure (Zelikoff *et al.*, 1997a). Results indicated that a single exposure to concentrated ambient PM_{10} (100–300 $\mu g/m^3$) had little effect upon clearance of bacteria instilled immediately after exposure. In contrast, rats infected 48 h prior to a 5-h exposure to PM_{10} (65–150 $\mu g/m^3$) and then sacrificed at different time-points post-PM_{10} exposure, displayed significant effects compared with their infected air-exposed counterparts. These latter studies demonstrated that while numbers of pulmonary bacteria were approximately equal in the two exposure groups at the earliest post-exposure time-point (i.e. 4.5 h post-PM_{10} exposure), bacterial burdens in PM_{10}-exposed rats were higher (~10% above levels in controls) by 9 h post-exposure and by 18 h, bacterial burdens were elevated by >300%. (Fig. 17.1). At later post-PM_{10} exposure time-points (i.e. 24, 48, 72 and 120 h), the total number of bacteria (in terms of bacteria/g lung) in the lungs declined in the PM-exposed rats but was still above that measured in the lungs of control rats.

While no definitive conclusions can be reached at this time regarding the mechanism(s) by which concentrated ambient PM may be decreasing in host resistance, observed reductions in the levels of lavageable PMN (the primary cell type responsible for resolving pulmonary *S. pneumoniae* infections) and proinflammatory cytokines TNF-α and IL-1 (critical for the mobilization and activation of neutrophils in response to *S. pneumoniae* infection) suggest that compromise of that specific portion of the pulmonary immune defense system important for the removal/killing of *S. pneumoniae* may be responsible for the observed increases in pulmonary bacterial burdens in PM-exposed hosts.

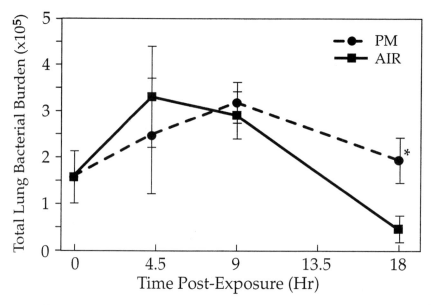

Fig. 17.1   Effects of inhaled particulate matter (PM_{10}) on pulmonary burdens of *Streptococcus pneumoniae*. Significantly different from filtered-air control ($p < 0.01$).

Particulate Mixtures

Co-exposure of mice to 10 mg/m^3 of carbon black (CB) and 2.5 ppm acrolein for 4 h/day for 4 days suppressed intrapulmonary killing of *S. aureus* and reduced the elimination of *L. monocytogenes* and influenza A virus from the lungs 24 h after exposure; however, intrapulmonary killing of *Proteus mirabilis* was enhanced and exposure to either pollutant alone had no effect upon the functional integrity of lung defenses against any of the four infectious agents (Jakab, 1993). It was suggested that CB particles may serve as acrolein carriers into the deep lung and, thus, bring about the enhanced biologic effect produced by this co-pollutant exposure. Additional studies demonstrated that co-exposure to 3.5 mg/m^3 CB and 2.5 ppm formaldehyde or 10 mg CB/m^3 and 5 ppm formaldehyde for 4 h after *S. aureus* infection had no effect upon intrapulmonary killing (Jakab, 1992).

Human exposure to airborne particulates most often occurs in combination with a variety of pollutant gases. One major contributor to particulate pollution around the world is the burning of wood for home heating. Based upon studies which indicated that children exposed to woodsmoke (WS) emissions have an increased incidence/severity of pulmonary infections, the effects of short- and long-term inhalation of WS on host resistance against respiratory challenge with an infectious bacterial agent have been examined (Zelikoff *et al.*, 1997c). Whole WS emissions (or particle-free effluents) from the burning of Red Oak were used to expose rats for 1 h/day for 4 consecutive days. During burning, the particulate concentration was maintained at 750 µg/m^3 (MMAD = 0.16 µm) and the carbon monoxide, NO$_x$, and polycyclic aromatic hydrocarbon levels at ~1.0, 0.004 and 1.03 ppt, respectively. Following inhalation, rats were: (a) IT-instilled with *S. aureus* to assess effects on pulmonary clearance; (b) sacrificed at various post-WS exposure time-points and bronchopulmonary lavage performed to assess lavage fluid parameters and AM functions; or (c) sacrificed and their lungs evaluated for any histopathological alterations. In rats IT-instilled with *S. aureus* at several different time-points following WS exposure, repeated inhalation of whole WS emissions gave rise to a progressive reduction in the *in vivo* killing of *S. aureus* (Fig. 17.2), even in the absence of any histopathological changes, pulmonary lung cell damage, or inflammation. Intrapulmonary killing/clearance of the bacteria was reduced to 60% of control values in hosts infected 3 h post-WS exposure, progressively declined to 2% (compared with control) after 5 days, and remained depressed for a period of up to 11 days post-exposure. Similar dramatic effects were not observed in rats exposed to particle-free effluents; this demonstrated the importance of particulates in bringing about the observed time-related effects of WS upon the host antibacterial response.

Results from *ex vivo* studies suggest that WS-induced reductions in production of $^{\cdot}O_2^{-}$, a mediator critical for the intracellular killing of ingested *S. aureus*, may at least partly be responsible for the observed decrease in host resistance (Zelikoff *et al.*, 1997c). Inasmuch as AM recovered from WS-exposed rats had a reduced ability *ex vivo* to kill the same *S. aureus* strain used *in vivo*, it appears that WS acts via direct effects upon AM function, rather than indirectly through neuroimmune mechanisms, to induce the changes in pulmonary antibacterial defense.

Studies have also investigated changes in host susceptibility in response to diesel engine emissions (DEE). In mice challenged with influenza virus after a 1, 3, or 6 month exposure to 2 mg/m^3 of DEE, coal dust (CD), or a combination of both, no effects upon viral titres, host mortality, IFN-γ levels, or hemagglutinin antibody response were observed

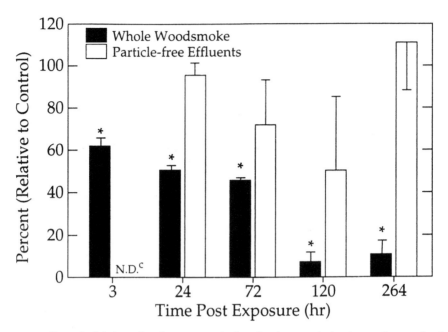

Fig. 17.2 Effects of inhaled woodsmoke emissions (with and without particulates) on pulmonary burdens of *Staphylococcus aureus*.
* Significantly different ($p < 0.05$). ND, studies with particle-free effluents not performed.

after 1 month of exposure to any of the particulates (alone or in combination) (Hahon *et al.*, 1985). However, the percentage of mice demonstrating lung consolidation was significantly higher in the 3-month groups exposed to DEE and CD/DEE than in the filtered air control groups; similarly, the percentages were twice that of the control for both DEE and CD/DEE-exposed animals in the 6-month exposure groups. These same animals also evidenced elevated pulmonary viral titres and a concomitant depression of IFN levels. Additional studies examining the relationship between cytochrome P-450 and resistance to influenza infection via IFN production demonstrated that exposure for 1 month to DEE particulates altered the normal relationship that exists between metabolic detoxification by P-450 enzymes and resistance to infection in normal mice (Rabovsky *et al.*, 1987). Findings from both of these studies suggest that DEE exposure can increase the severity of influenza virus infection and that effects may be mediated by DEE-induced changes in IFN.

EVIDENCE LINKING AIR POLLUTANTS TO REDUCED PULMONARY HOST DEFENSE: HUMAN DATA

The potential impact of air pollution upon human health, not only of just one individual but of entire communities, represents a major area of controversy and concern in the USA and worldwide. To put things in proper perspective, Hall *et al.* (1992) published a study assessing the health and economic impact of O_3 and PM_{10} pollution in the South Coast Air Basin of California. They concluded that the economic value of avoiding adverse effects associated with exposure to these and other pollutants would be ≈ $10 billion/year, with ≈1600 lives/year being saved. Numerous, excellent reviews of human experimental and epidemiological studies of the health effects of air pollution have been published; some of these include papers by Chitano *et al.* (1995), Choudhury *et al.* (1997), Dockery and Pope (1994), Dockery *et al.* (1992), Lebowitz (1996), Lippmann (1989), Keiding *et al.* (1995), Koenig and Luchtel (1997), and Oehme *et al.* (1996).

There have been numerous examples over the past 100 years that increased air pollution caused by industrial activity increases morbidity and mortality due, in part, to alterations in lung defense mechanisms. One notorious pollution-related event occurred in London in 1952. In this episode, the heavy use of coal for heating and cooking, coupled with a temperature inversion, gave rise to ambient concentrations of pollutants (primarily sulfur oxides (SO_x) and PM) which surpassed the already high levels normally found in this heavily polluted area. By its end, there had been a 2- to 3-fold increase in mortality (Ministry of Health, 1954), with most victims being the elderly or those who had pre-existing conditions (such as asthma or cardiac/respiratory diseases) which may have predisposed them to an increased risk. Similar episodes in the USA, although with a less dramatic morbidity, have occurred in Donora, PA, in 1948 and in New York City in the 1960s. While coal burning-derived pollutants have all but disappeared from the air of the majority of major cities worldwide, these pollutants have been replaced by agents which are largely generated by motor vehicles, including particulate matter (i.e. PM_{10}), O_3, NO_x, SO_x, and volatile organics. In recognition of the increasing threat to human health from exposure to these toxicants, in 1970 the US government established air quality standards for these and other hazardous air pollutants (HAPs) as a result of the passage of the Clean Air Act.

In recent years, studies have been conducted in several Eastern European countries investigating the association between air pollution and increased morbidity and mortality due to respiratory infection. In a study of asthmatic children from Sokolov, Czech Republic, exposure to a variety of air pollutants (including SO_2, PM_{10} and NO_2) during the winter of 1991–92 was found to be associated with decreased expiratory flow rates and an increased prevalence of fever (Peters *et al.*, 1997). It was concluded that exposure to these agents might have enhanced pulmonary symptomae that the children were experiencing during respiratory infections. Studies in Finland examining the comparative incidences of upper respiratory tract infections in children from polluted vs reference (relatively non-polluted) sites suggested children had an increased risk of infection at ambient concentrations of pollutants which were lower than those documented in Great Britain or central Europe (Jaakkola *et al.*, 1991).

Seemingly unrelated factors, such as ambient temperature, play an important but poorly understood role in pollutant-induced alterations in host susceptibility to respiratory infections. Choudhury *et al.* (1997) examined the relationship between the incidence of asthma and upper respiratory tract infection and PM_{10} exposure in the general population of Anchorage, Alaska. While a significant positive correlation was seen between morbidity and exposure, the relative risk was higher with respect to PM_{10} pollution during warmer days. Similarly, ambient levels of SO_2 and NO_2, as well as temperature, have displayed significant correlations with the number of upper respiratory infections and amount of absenteeism noted in schoolchildren in Helsinki (Ponka, 1990). When corrected for temperature, a weaker but statistically significant association remained between levels of SO_2 and the incidence of respiratory infections. Lebowitz (1996) performed a comprehensive review of many of these types of studies and noted that even with similar databases and locations, not all reports agreed with respect to what was the most significant pollutant and that many confounding factors (i.e. temperature, influenza outbreak, holidays, and seasons of the year) have strong effects on daily mortality and morbidity and may enhance or minimize the impact of any air pollution event.

In order to more completely understand the pulmonary host defense mechanisms responsible for an increased incidence of infection following exposure to oxidant and/or particulate air pollutants, long-term human exposure studies using healthy volunteers have been conducted with equivocal results. Human AM obtained following the bronchoalveolar lavage of NO_2-exposed individuals were only slightly less efficient inactivating influenza virus *in vitro* than were those from air-exposed controls (Frampton *et al.*, 1989). Other studies have noted that viral recovery rates and influenza-specific antibody titres in volunteers exposed to NO_2 for up to 3 years were not significantly different from those of controls (Goings *et al.*, 1989). In young volunteers exposed to moderate levels of O_3 (0.3 ppm, 6 h/day for 5 days) to determine whether or not susceptibility to experimental rhinovirus infection would be altered, no significant differences in rhinovirus nasal secretion titres, PMN recruitment, IFN production, or T lymphocyte antigenic responses were noted (Henderson *et al.*, 1988).

In addition to *in vivo* exposures, investigators have used *in vitro* exposure systems to evaluate the impact of oxidant pollutants (such as O_3) on human host defense mechanisms at the cellular and molecular level. In studies designed to determine whether or not O_3 altered pulmonary surfactant properties, human SP-A exposed *in vitro* to O_3 displayed an altered ability to inhibit alveolar type II cell phosphatidylcholine production (Oosting *et al.*, 1992). When incubated with this O_3-treated SP-A, human AM developed a reduced ability to produce $\cdot O_2^-$ and to engage in phagocytosis compared with control cells exposed to untreated SP-A. This impaired ability of SP-A to mediate alveolar effector cell functions may be responsible, in part, for reduced host defense following ozone exposure.

Becker *et al.* (1991) investigated the impact of O_3 exposure on human AM phagocytosis and production of several effector molecules including proinflammatory cytokines, prostaglandins, and $\cdot O_2^-$. The results of these studies suggested that the cell membrane was targeted by O_3 (in the absence of any effects upon protein synthesis) as receptor-mediated functions (e.g. phagocytosis, response to bacterial mitogens, and phorbol esters) became impaired in the exposed cells. These data may help explain the observations of a later study that noted that O_3 may have a greater adverse impact on individuals with a pre-existing respiratory infection than on those without infection (Ostro *et al.*, 1993).

Susceptibility of O_3-exposed human AM to infection with RSV was evaluated in an *in vitro* exposure system by Soukup *et al.* (1993). The data suggest that despite individual differences in cytokine responses resulting from exposure, *in vitro* infection of AM with different infection levels of RSV was not significantly impaired by exposure of up to 1.0 ppm O_3. In studies with respirable particulates, Steerenberg *et al.* (1998) reported alterations in proinflammatory cytokine production by cultured human bronchial epithelial cells exposed to diesel exhaust particles. They concluded that this may upset the immune homeostasis of the lung and compromise host defense.

CONCLUSIONS

It is clear that inhaled air pollutants modulate immune responses in the lungs. The results from more than 30 years of research under controlled laboratory conditions have categorically demonstrated that exposure of laboratory animals to the most common types of air pollutants (i.e. O_3, NO_x, SO_x, PM (organic and inorganic particles/complex mixtures)) results in an increased risk for pulmonary infections. The best understood consequence of this modulation is an increased susceptibility to infectious bacterial agents in laboratory rodents. The role that pollutant exposure may have in infectious viral disease is much less clear and the effects of air pollutants on viral replication and viral-associated immune responses (i.e. allergic or inflammatory responses associated with viral infection) have rarely been demonstrated. It appears that modulation of the functions of pulmonary immune cells (i.e. PAM and PMN), nascent factors (i.e. surfactant) and/or physical processes (i.e. ciliary beating, particle clearance) important for maintaining lung sterility play an important role in mediating these observed outcomes.

Differences in sensitivity between animal species and among exposure conditions make extrapolation from animal to human difficult. As a result, the overall risk factors for respiratory infections in humans still remain difficult to assess. For example, while associations between acute health effects and particulate air pollution have been commonly reported, it has also been shown that these effects could just as likely be due to meteorological conditions. Regardless of these potential confounding factors, the results from controlled human exposure studies and from retrospective epidemiology studies of air pollution events strongly suggest that the young and aged are most at risk and that asthma and other manifestations of hypersensitivity as well as an increased incidence of upper respiratory infections are the predominant symptoms.

Despite these uncertainties, the overall weight of human and animal evidence has prompted several government initiatives and controls that have resulted in an overall improvement in urban air quality. Much has been learned since the early studies by Miller and Ehrlich (1958) which demonstrated that NO_2-exposed mice exhibited increased susceptibility to aerosol infection. However, more research is still needed to define the molecular and cellular events/mechanisms underlying these alterations in pulmonary host defense in order for increasingly meaningful risk extrapolation to humans to be made.

REFERENCES

Adkins B Jr, Richards JH and Gardner DE (1979) Enhancement of experimental respiratory infection following nickel inhalation. *Environ Res* **20**: 33–42.

Adler KB (1994) Interactions between respiratory epithelial cells and cytokines: relationship to lung inflammation. In: Onlghard M, Prerclani M, Renesto P and Varghaig BB (eds) *Cells and Cytokines in Lung Inflammation. Ann NY Acad Sci* **725**: 128–145.

Albright JF and Goldstein RA (1996) Airborne pollutants and the immune system. *Otolaryngol Head Neck Surg* **114**: 232–238.

Aranyi C, Graf JL, O'Shea WJ *et al.* (1983) The effects of intratracheally-administered coarse mode particles on respiratory tract infection in mice. *Toxicol Lett* **19**: 63–72.

Aranyi C, Bradof JN, O'Shea WJ *et al.* (1985) Effects of arsenic trioxide inhalation exposure on pulmonary antibacterial defenses in mice. *J Toxicol Environ Health* **15**: 163–172.

Azoulay-Dupuis E, Bouley G and Blayo MC (1982) Effects of sulfur dioxide on resistance to bacterial infection in mice. *Environ Res* **29**: 312–319.

Ballard PL, Liley HG, Gonzales LW *et al.* (1990) Interferon-gamma and synthesis of surfactant components by cultured human fetal lung. *Am J Respir Cell Mol Biol* **2**: 137–143.

Becker S, Madden M, Newman S *et al.* (1991) Modulation of human alveolar macrophage properties by ozone. *Toxicol Appl Pharmacol* **110**: 402–415.

Bellanti JA and Herscowitz HB (1985) Immunology. In: Richard SM and Filkins JP (eds) *The Reticuloendothelial System: A Comprehensive Treatise*, vol. 6. New York: Plenum Press.

Bouley G, Dubreuil A, Despaux N *et al.* (1977) Toxic effects of cadmium microparticles on the respiratory system. *Scand J Work Environ Health* **3**: 116–121.

Brian JD (1985) Physiology and pathophysiology of pulmonary macrophages. In: Richard SM and Filkins JP (eds) *The Reticuloendothelial System: A Comprehensive Treatise*, vol. 7B. New York: Plenum Press, pp 315–337.

Buckley RD and Loosle CG (1969) Effects of nitrogen dioxide inhalation on germ-free mouse lungs. *Arch Environ Health* **18**: 588–595.

Burns LA (1998) Arsenic. In: Zelikoff JT and Thomas PT (eds) *Immunotoxicology of Environmental and Occupational Metals*. London: Taylor & Francis, pp. 1–26.

Burns LA, McCay JA and Munson AE (1993) Arsenic in the sera of gallium arsenide-exposed mice inhibits bacterial growth and increases host resistance. *J Pharmacol Exp Ther* **265**: 795–800.

Burns LA, Meade BJ and Munson AE (1996) Toxic effects of metals. In: Klaassen CD (ed.) *Casserett and Doull's Toxicology: The Basic Science of Poisons*. New York: McGraw-Hill, pp. 691–736.

Chaumard C, Quero AM, Bouley G *et al.* (1983) Influence of inhaled cadmium microparticles on mouse influenza pneumonia. *Environ Res* **31**: 428–439.

Chaumard C, Forestier F and Quero AM (1991) Influence of inhaled cadmium on the immune response to influenza virus. *Arch Environ Health* **46**: 50–55.

Chitano P, Hosselet JJ, Mapp CE and Fabbri LM (1995) Effect of oxidant air pollutants on the respiratory system: insights from experimental animal research. *Eur Respir J* **8**: 1357–1371.

Choudhury A, Gordian M and Morris S (1997) Associations between respiratory illness and PM_{10} air pollution. *Arch Environ Health* **52**: 113–117.

Churchill L, Friedman B, Schleimer P and Proud D (1992) Production of granulocyte macrophage colony stimulating factor by cultured human tracheal epithelial cells. *Immunology* **75**: 189–195.

Daniels MJ, Menache MG, Burleson GR *et al.* (1987) Effects of $NiCl_2$ and $CdCl_2$ on susceptibility to murine cytomegalovirus and virus-augmented natural killer cell and interferon responses. *Fundam Appl Toxicol* **8**: 443–453.

Dockery DW and Pope CA (1994) Acute respiratory effects of particulate air pollution. *Ann Rev Publ Health* **15**: 107–132.

Dockery DW, Schwartz J and Spengler JD (1992) Air pollution and daily mortality: associations with particulates and acid aerosols. *Environ Res* **59**: 362–373.

Ehrlich R (1966) Effect of nitrogen dioxide on resistance to respiratory infection. *Bacteriol Rev* **30**: 604–614.

Ehrlich R (1980) Interactions between environmental pollutants and respiratory infections. *Environ Health Perspect* **35**: 89–100.

Ehrlich R, Findlay JC and Gardner DE (1979) Effects of repeated exposures to peak concentrations of nitrogen dioxide and ozone on resistance to streptococcal pneumonia. *J Toxicol Environ Health* **5**: 631–642.

Fairchild GA (1977) Effects of ozone and sulfur dioxide on virus growth in mice. *Arch Environ Health* **32**: 28–33.

Fairchild GA, Roan J and McCarroll J (1972) Atmospheric pollutants and the pathogenesis of viral respiratory infection. *Arch Environ Health* **25**: 174–182.

Fischetti V and Bessen D (1989) Effect of mucosal antibodies to M protein in colonization by group A streptococci. In: Switalski L, Hook M and Beachery E (eds) *Molecular Mechanisms of Microbial Adhesion*. New York: Springer-Verlag, pp. 128–142.

Fishman AP (ed.) (1988) *Pulmonary Diseases and Disorders*, 2nd edn. New York: McGraw-Hill.

Frampton M, Smeglin A, Roberts N *et al.* (1989) Nitrogen dioxide exposure *in vivo* and human alveolar macrophage inactivation of influenza virus *in vitro*. *Environ Res* **48**: 179–192.

Gardner DE, Miller FJ, Illing JW and Kirtz JM (1977) Alterations in bacterial defense mechanisms of the lung induced by inhalation of cadmium. *Bull Eur Physiopathol Respir* **13**: 157–174.

Gilmour MI and Selgrade MJ (1993) A comparison of the pulmonary defenses against streptococcal infection in rats and mice following O_3 exposure: differences in disease susceptibility and neutrophil recruitment. *Toxicol Appl Pharmacol* **123**: 211–218.

Gilmour MI, Park P and Selgrade MJ (1993a) Ozone-enhanced pulmonary infection with *Streptococcus zooepidemicus* in mice. The role of alveolar macrophage function and capsular virulence factors. *Am Rev Respir Dis* **147**: 753–760.

Gilmour MI, Park P, Doerfler D and Selgrade MJ (1993b) Factors that influence the suppression of pulmonary antibacterial defenses in mice exposed to ozone. *Exp Lung Res* **19**: 299–314.

Goings S, Kulle T, Bascom R *et al.* (1989) Effect of nitrogen dioxide exposure on susceptibility to influenza A virus infection in healthy adults. *Am Rev Respir Dis* **139**: 1075–1081.

Goldstein E and Lippert W (1979) Effect of near-ambient exposures to sulfur dioxide and ferrous sulfate particles on murine pulmonary defense mechanisms. *Arch Environ Health* **34**: 424–432.

Goldstein E, Bartlema HC, van der Ploeg M *et al.* (1978) Effect of ozone on lysosomal enzymes of alveolar macrophages engaged in phagocytosis and killing of inhaled *Staphylococcus aureus. J Infect Dis* **138**: 299–311.

Graham JA, Gardner DE, Blommer EJ *et al.* (1987) Influence of exposure patterns of nitrogen dioxide and modifications by ozone on susceptibility to bacterial infectious disease in mice. *J Toxicol Environ Health* **21**: 113–125.

Hahon N, Booth JA, Green F and Lewis TR (1985) Influenza virus infection in mice after exposure to coal dust and diesel engine emissions. *Environ Res* **37**: 44–60.

Hall JV, Winer AM, Kleinman MT *et al.* (1992) Valuing the health benefits of clean air. *Science* **255**: 812–817.

Henderson F, Dubovi E, Harde S *et al.* (1988) Experimental rhinovirus infection in human volunteers exposed to ozone. *Am Rev Respir Dis* **137**: 1124–1128.

Henry MC, Ehrlich R and Blair WH (1969) Effect of nitrogen dioxide on resistance of squirrel monkeys to *Klebsiella pneumoniae* infection. *Arch Environ Health* **18**: 580–587.

Henry MC, Findlay J and Erlich R (1970) Chronic toxicity of NO_2 in squirrel monkeys. III. Effect on resistance to bacterial and viral infection. *Arch Environ Health* **20**: 566–570.

Ilback MG, Fohlman J, Friman G and Glynn AW (1992) Altered distribution of ^{109}cadmium in mice during viral infection. *Toxicology* **71**: 193–202.

Ito K (1971) Effect of nitrogen dioxide inhalation on influenza virus infection in mice. *Jpn J Hyg* **26**: 304–314.

Jaakkola J, Paunio M, Virtanen M and Heinonen O (1991) Low-level air pollution and upper respiratory infections in children. *Am J Publ Health* **81**: 1060–1062.

Jakab GJ (1988) Health Effects Institute Research Report No. 20: Modulation of Pulmonary Defense Mechanisms Against Viral and Bacterial Infections by Acute Exposure to Nitrogen Dioxide. Cambridge, MA: Health Effects Institute, pp. 1–34.

Jakab GJ (1992) Relationship between carbon black particulate-bound formaldehyde, pulmonary antibacterial defenses, and alveolar macrophage phagocytosis. *Inhal Toxicol* **4**: 325–342.

Jakab GJ (1993) The toxicologic interactions resulting from inhalation of carbon black and acrolein on pulmonary antibacterial and antiviral defenses. *Toxicol Appl Pharmacol* **121**: 167–175.

Jakab GJ and Bassett DJ (1990) Influenza virus infection, ozone exposure, and fibrogenesis. *Am Rev Respir Dis* **141**: 1307–1315.

Janeway CA and Travers P (eds) (1996) *Immunobiology: The Immune System in Health and Disease.* New York: Garland Publishing, Inc.

Keiding LM, Rindel AK and Kronborg D (1995) Respiratory illness in children and air pollution in Copenhagen. *Arch Environ Health* **50**: 200–206.

Koenig JQ and Luchtel D (1997) Respiratory responses to inhaled toxicants. In: Massaro EJ (ed.) *Handbook of Human Toxicology.* Boca Raton, FL: CRC Press, pp. 552–606.

Koller LD (1998) Cadmium. In: Zelikoff JT and Thomas PT (eds) *Immunotoxicology of Environmental and Occupational Metals.* London: Taylor & Francis, pp. 41–61.

Lebowitz MD (1996) Epidemiological studies of the respiratory effects of air pollution. *Eur Respir J* **9**: 1029–1054.

Lebowitz MD and Fairchild GA (1973) The effects of sulfur dioxide and A2 influenza virus on pneumonia and weight reduction in mice: an analysis of stimulus–response relationships. *Chem Biol Interact* **7**: 317–326.

Lefkowitz SS, McGrath JJ and Lefkowitz DL (1986) Effects of NO_2 on immune responses. *J Toxicol Environ Health* **17**: 241–248.

Lippmann M (1989) Health effects of ozone: a critical review. *J Air Pollut Control Assoc* **39**: 672–695.

Lopez-Vidriero M (1984) Lung secretions. In: Clark SW and Pavia D (eds) *Aerosols and the Lung: Clinical and Experimental Aspects*. London: Butterworths, pp. 19–48.

Lundborg M and Camner P (1982) Lysozyme levels in rabbit lung after inhalation of nickel, cadmium, cobalt, and copper chlorides. *Environ Res* **34**: 335–342.

Miller S and Ehrlich R (1958) Susceptibility to respiratory infections of animals exposed to ozone. I. Susceptibility to *Klebsiella pneumoniae*. *J Infect Dis* **103**: 145.

Ministry of Health (1954) Reports on Public Health and Medical Subjects no. 95: Mortality and Morbidity During the London Fog of December 1952. London: HMSO.

Mostov KE (1994) Transepithelial transport of immunoglobulins. *Ann Rev Immunol* **12**: 63–84.

Nisizawa T, Saito M, Nakayam K *et al.* (1988) Effects of nitrogen dioxide on *Mycoplasma pulmonis* infection and humoral immune responses in mice. *Jpn J Med Sci Biol* **41**: 175–187.

Nygren O, Nilsson CA and Lindahl R (1992) Occupational exposure to chromium, copper, and arsenic during work with impregnated wood in joinery shops. *Ann Occup Hyg* **36**: 509–517.

Oehme F, Coppock R, Mostrom M and Khan A (1996) A review of the toxicology of air pollutants: toxicology of chemical mixtures. *Vet Human Toxicol* **38**: 371–377.

Oosting R, van Iwaarden J, van Bree L *et al.* (1992) Exposure of surfactant protein A to ozone *in vitro* and *in vivo* impairs its interactions with alveolar cells. *Am J Physiol* **262**: 63–68.

Ostro B, Lipsett M, Mann J *et al.* (1993) Air pollution and respiratory morbidity among adults in southern California. *Am J Epidemiol* **137**: 691–700.

Parker RF, Davis JK, Cassell GH *et al.* (1989) Short-term exposure to nitrogen dioxide enhances susceptibility to murine respiratory mycoplasmosis and decreases intrapulmonary killing of *Mycoplasma pulmonis*. *Am Rev Respir Dis* **140**: 502–512.

Paul WE (ed.) (1989) *Fundamental Immunology*, 2nd edn. New York: Raven Press.

Peters A, Docker D, Heinrich J and Wichmann H (1997) Short-term effects of particulate air pollution on respiratory morbidity in asthmatic children. *Eur Respir J* **10**: 872–879.

Ponka A (1990) Absenteeism and respiratory disease among children and adults in Helsinki in relation to low-level air pollution and temperature. *Environ Res* **52**: 34–46.

Rabovsky J, Judy DJ, Rodak DJ and Petersen M (1987) Influenza virus-induced alteration of cytochrome P-450 enzyme activities following exposure of mice to coal and diesel particulates. *Environ Res* **40**: 136–144.

Razi-Wolf Z, Freeman G, Galvin F *et al.* (1992) Expression and function of the murine B7 antigen, the major co-stimulatory molecule expressed by peritoneal exudate cells. *Proc Natl Acad Sci USA* **89**: 4210–4214.

Roitt I, Brostoff J and Male D (eds) (1998) *Immunology*, 5th edn. London: C.V. Mosby.

Rose RM, Fuglestad JM, Skornik WA *et al.* (1988) The pathophysiology of enhanced susceptibility to murine cytomegalovirus respiratory infection during short-term exposure to 5 ppm nitrogen dioxide. *Am Rev Respir Dis* **137**: 912–917.

Rosenthal GJ, Fort MM, Germolec DR *et al.* (1989) Effect of subchronic arsine inhalation on immune function and host resistance. *Inhal Toxicol* **1**: 113–127.

Schlesinger RB (1990) The interaction of inhaled toxicants with respiratory tract clearance mechanisms. *CRC Crit Rev Toxicol* **20**: 257–286.

Selgrade MJ and Gilmour MI (1994) Effects of gaseous air pollutants on immune responses and susceptibility to infectious and allergic disease. In: Dean JH, Luster MI, Munson AE and Kimber I (eds) *Immunotoxicology and Immunopharmacology*. New York: Raven Press, pp. 395–411.

Selgrade MJ and Gardner DE (1996) Altered host defenses and resistance to respiratory infections following exposure to airborne metals. In: Chang LW (ed.) *Toxicology of Metals*. Boca Raton, FL: CRC Press, pp. 853–860.

Selgrade MJ, Illing JW, Starnes DM *et al.* (1988) Evaluation of effects of ozone exposure on influenza infection in mice using several indicators of susceptibility. *Fundam Appl Toxicol* **11**: 169–180.

Sertl K, Takemura T, Tschachler E *et al.* (1986) Dendritic cells with antigen-presenting capability reside in airway epithelium, lung parenchyma and visceral pleura. *J Exp Med* **163**: 436–451.

Sherwood RL, Lippert WE and Goldstein E (1981) Effect of ferrous sulfate aerosols and nitrogen dioxide on murine pulmonary defense. *Arch Environ Health* **36**: 130–135.

Sherwood RL, Kimur A, Donovan R and Goldstein E (1984) Effect of 0.64 ppm ozone on rats with chronic pulmonary bacterial infection. *J Toxicol Environ Health* **13**: 893–904.

Smialowicz RJ (1998) Nickel. In: Zelikoff JT and Thomas PT (eds) *Immunotoxicology of Environmental and Occupational Metals.* London: Taylor & Francis, pp. 163–194.

Soukup J, Koren HS and Becker S (1993) Ozone effect on respiratory syncytial virus infectivity and cytokine production by human alveolar macrophages. *Environ Res* **60**: 178–186.

Steerenberg P, Zonnenberg J, Dormans J *et al.* (1998) Diesel exhaust particles induced release of interleukin-6 and -8 by primed human bronchial epithelial cells *in vitro*. *Exp Lung Res* **24**: 85–100.

Thienes CH, Skillen RG, Hoyt A and Bogen E (1965) Effects of ozone on experimental tuberculosis and on natural pulmonary infections in mice. *Am J Ind Hyg Assoc J* **26**: 255–260.

Thomas PT and Sherwood RL (1996) Host resistance models in immunotoxicology. In: Smialowicz RJ and Holsapple MP (eds) *Experimental Immunotoxicology.* Boca Raton, FL: CRC Press, pp. 29–45.

Thurlbeck WM and Churg AM (eds) (1995) *Pathology of the Lung,* 2nd edn. New York: Thierne Medical Publishers, Inc.

Ukai K (1977) Effects of SO_2 on the pathogenesis of viral upper respiratory infection in mice. *Proc Soc Exp Biol Med* **154**: 591–596.

Ulevitch R and Tobias P (1995) Receptor-dependent mechanism of cell stimulation by bacterial endotoxin. *Ann Rev Immunol* **13**: 437–457.

van Loveren H, Rombout PJA, Wagenaar SC *et al.* (1988) Effects of ozone on the defense to a respiratory *Listeria monocytogenes* infection in the rat. *Toxicol Appl Pharmacol* **94**: 374–393.

Wiernik A, Hohansson A, Jarstrand C and Camner P (1983) Rabbit lung after inhalation of soluble nickel. I. Effects on alveolar macrophages. *Environ Res* **30**: 129–141.

Wispe JR, Clark JC, Warner BB *et al.* (1990) Tumor necrosis factor-alpha inhibits expression of pulmonary surfactant protein. *J Clin Invest* **86**: 1954–1960.

Wolcott JA, Zee YC and Osebold JW (1982) Exposure to ozone reduces influenza disease severity and alters distribution of influenza viral antigens in murine lungs. *Appl Environ Microbiol* **44**: 723–731.

Wu Y, Luo H and Johnson DR (1986) Effect of nickel sulfate on cellular proliferation and Epstein–Barr virus antigen expression in lymphoblastoid cell lines. *Cancer Lett* **32**: 171–179.

Zelikoff JT and Cohen MD (1996) Immunotoxicology of inorganic metal compounds. In: Smialowicz RJ and Holsapple MP (eds) *Experimental Immunotoxicology.* Boca Raton, FL: CRC Press, pp. 189–229.

Zelikoff JT and Cohen MD (1997) Metal immunotoxicology. In: Massaro EJ (ed.) *Handbook of Human Toxicology.* Boca Raton, FL: CRC Press, pp. 811–852.

Zelikoff JT and Schlesinger RB (1992) Modulation of pulmonary immune defense mechanisms by sulfuric acid: effects on macrophage-derived tumor necrosis factor and superoxide. *Toxicology* **76**: 271–281.

Zelikoff JT and Thomas PT (eds) (1998) *Immunotoxicology of Environmental and Occupational Metals.* London: Taylor & Francis.

Zelikoff JT, Sisco MP, Yang Z *et al.* (1994) Immunotoxicity of sulfuric acid aerosol: effects on pulmonary macrophage effector and functional activities critical for maintaining host resistance against infectious diseases. *Toxicology* **92**: 269–286.

Zelikoff JT, Fang K, Li Y *et al.* (1997a) Exposure to particulate matter (PM) suppresses pulmonary immune defense mechanisms important for host resistance against a pneumonia-producing bacteria. *Am J Respir Crit Care Med* **157**: A153.

Zelikoff JT, Frampton MW, Cohen MD *et al.* (1997b) Effects of inhaled sulfuric acid aerosols on pulmonary immunocompetence: a comparative study in humans and animals. *Inhal Toxicol* **9**: 731–752.

Zelikoff JT, Li Y, Nadziejko C *et al.* (1997c) PM_{10} generated from wood burning may be responsible for increased pulmonary infections: a toxicological model. *Fundam Appl Toxicol* **36**: 7.

Carcinogenic Responses to Air Pollutants

ROGER O. McCLELLAN

Chemical Industry Institute of Toxicology, Research Triangle Park, NC, USA

T. ELISE JACKSON

Chemical Industry Institute of Toxicology, Research Triangle Park and Curriculum in Toxicology, University of North Carolina, Chapel Hill, NC, USA

INTRODUCTION

Air pollutants are of major concern as a cause of carcinogenic responses in the respiratory tract and other organs. The ambient air contains gases, vapors, and particulate matter consisting of thousands of different chemical compounds with varied biological activity. Moreover, the concentration of some of the constituents may be elevated in the workplace, in homes, and in the ambient environment in urbanized or industrial locales, or as a result of natural phenomena such as fires and volcanoes. Some of the materials found in the air, e.g. asbestos, are known to be associated with an increased incidence of cancer; this has been shown by human experience with high-level exposures, most often in the workplace (IARC, 1987). For other materials, e.g. tetranitromethane, there is evidence from long-term *in vivo* studies in laboratory animals that the materials cause cancer, and they are presumed also to be human carcinogens (IARC, 1996). For yet other materials, e.g. analogs of benzidine ((1,1'-biphenyl)-4,4'-diamine), the evidence of carcinogenic hazard is more indirect and is based on (a) evidence of mutagenicity observed in bacteria or mammalian cells or (b) presumptions of mutagenicity and carcinogenicity based on relationships between chemical structure and biological activity (Ames *et al.*, 1975; Enslein *et al.*, 1994; Rosenkranz and Klopman, 1994).

AIR POLLUTION AND HEALTH
ISBN 0-12-352335-4

Large quantities of air must be respired to sustain life, and therefore every individual continually inhales many different chemicals. The quantities of gases and particles inhaled are a function of the quantity of air inhaled and the concentration of the specific gases and particulate matter in the air. The volumes inhaled are greater for those exercising vigorously while working or engaged in recreational activities. The concentration of specific materials in the gas varies markedly among the ambient, home, or workplace environments in different locales. The cancer risk to each individual is a function of the cancer-causing potency of the chemicals inhaled, the concentration of the chemical inhaled, the volume of air inhaled, the genetic susceptibility of the individual, and undoubtedly, complex interactions involving other agents and factors such as diet (Biesalski *et al.*, 1997).

The inhaled gases and particles interact with the respiratory tract. This very complex target has a huge surface area and contains many different types of cells. Some of these cells have the capacity to metabolize chemical compounds and thereby detoxify them or, in some cases, produce more chemically reactive products. Many of the cells of the respiratory tract, such as the bronchial epithelial cells, are reproductively viable and are continually dividing. Thus, the respiratory tract is a prime target for developing cancers from inhaled materials. Examples include nasal cavity cancers from hardwood dust, bronchogenic carcinomas from cigarette smoking, and mesothelioma from exposure to asbestos. In addition, many inhaled compounds are absorbed via the pulmonary circulation and transported to other tissues of the body where they or metabolites may produce cancer. For example, benzene may produce leukemia, and vinyl chloride may produce hemangiosarcomas in the liver of humans.

Cancer, which is actually a family of diseases characterized by uncontrolled cell growth, occurs with high frequency in long-lived populations, especially in developed countries (Landis *et al.*, 1998; WHO, 1996). In these countries, about one in three individuals will develop cancer, and one in four or five will die of cancer. Lung cancer is typically the most frequently observed cancer, primarily due to its association with cigarette smoking, as illustrated in Fig. 18.1. For example, in the USA in 1994, of 2 279 994 deaths, 534 000 were caused by cancer including 149,000 from lung cancer (Landis *et al.*, 1998; NCHS, 1997). The frequent occurrence of cancer, and especially lung cancer, raises justifiable concern for identifying the underlying etiological factors including the role of air pollutants.

The high incidence of lung cancer associated with smoking (on the order of 90% of the cases) complicates the identification of other air pollutants that may cause lung cancer by themselves or by interacting with cigarette smoking in some way. For example, radon exposure in combination with cigarette smoking results in increases in lung cancer greater than would occur with either radon or cigarette smoking alone. However, the cancer-causing potency of radon in non-smokers, especially at low levels of exposure, is a matter of intense debate (NRC, 1998a).

The earliest report of lung cancer is of special interest for a number of reasons, including the fact that it predates widespread smoking. In the sixteenth century, miners in Central Europe were dying of a wasting respiratory disease called Bergkrankheiten. We now know that the disease syndrome included lung cancer and was likely due to high-level exposure to metal ore dust and the decay products of uranium. Ironically, this experience was repeated more than two centuries later when miners in Colorado, New Mexico, and other regions were also exposed to high levels of radon and decay products. An increased incidence of lung cancer, especially in smokers, was observed before more rigorous controls, including installation of improved ventilation, were implemented (Lubin *et al.*,

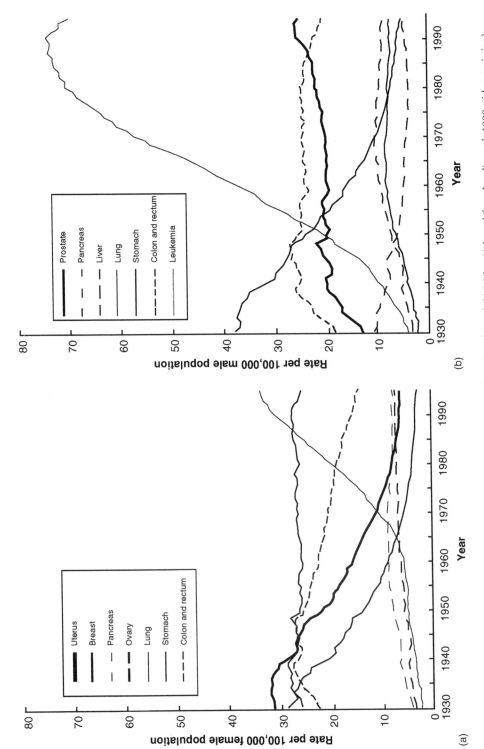

Fig. 18.1   Age-adjusted cancer death rates by site for the United States, 1930–1994. (a) Females and (b) Males. (Adapted from Landis *et al.*, 1998 with permission).

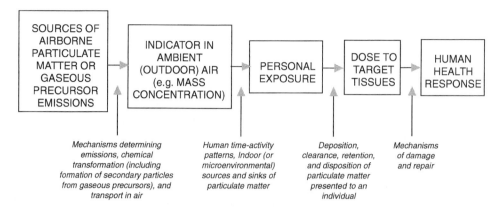

Fig. 18.2   Framework for integrating research and information on air pollutants extending from sources to health responses (NRC, 1998b with permission).

1994). As we approach the new millennium, concern for exposure to radon and its decay products extends to the typically lower concentrations found in homes (NRC, 1998a). Thus, the radon and lung cancer issue serves to illustrate the information linkage between occupational exposures with clear evidence of excess cancer risk with environmental exposures that are typically at much lower levels. The detection of an increased incidence of cancer in a human population represents an indictment of past societal practices involving failure to control the agent in question. Direct human evidence of carcinogenicity serves as a warning call for a better understanding of the causes of cancer and the need to reduce exposures to known or suspected carcinogenic agents.

A strategy for understanding the role of air pollutants in causing disease, including the induction of cancer, and the development of effective control strategies for reducing disease related to air pollution require information that extends on a continuum from sources of pollutants to health responses. This kind of integrated approach has been advocated most recently by the United States National Academy of Sciences Committee on Research Priorities for Airborne Particulate Matter, National Research Council (NRC) (NRC, 1998b) (Fig. 18.2).

This chapter focuses on potential carcinogenic responses to air pollution and emphasizes how this information is a part of the paradigm shown in Fig. 18.2. In the next section, we provide an overview of carcinogenesis and then move to a discussion of detecting carcinogens. Finally, we discuss regulatory and risk assessment frameworks for air pollutants and carcinogenic responses.

OVERVIEW OF CARCINOGENESIS

For more than 50 years, efforts to increase our understanding of the family of diseases classified as cancer have been at the forefront of global biomedical research. This emphasis on cancer research was stimulated by epidemiological detection of specific causative agents, an overall increase in cancer incidence largely attributable to lung cancer, and concern for the cancer-causing potential of increased industrialization, including the introduction of thousands of synthetic chemicals. In the USA, the level of attention given to cancer and potential cancer-causing agents was markedly increased in 1970 when President Richard Nixon declared a 'war on cancer'. This multipronged war involved increased funding for basic research; increased support for improved prevention, diagnosis, and treatment of cancer; and new legislation intended to have an impact on cancer. We will return to the legislative front later.

Knowledge of cancer etiology has been increasing at an accelerating rate, and there is great promise that further advances are close at hand. A review of those advances is outside the scope of this chapter. Suffice it to say that many of the advances stem from the study of lung cancer and the agents that cause it. At the same time, advances both in general knowledge of cancer and in cancers in other organs also affect our understanding of cancer caused by air pollution. The reader interested in more detailed coverage of carcinogenesis is referred to publications by Carbone (1997), Devereux *et al.* (1996), Pitot (1986, 1993), Pitot and Dragan (1996) and Vogelstein and Kinzler (1993).

Cancer is frequently used as a generic term for the family of neoplastic diseases. Neoplasia or the constituent lesion, a neoplasm, is defined as a heritably altered, relatively autonomous growth of tissue (Pitot, 1986; Pitot and Dragan, 1996). The critical elements of the definition are: (1) the heritable nature of neoplasia at the somatic, or germ cell, level; and (2) the relative autonomy of neoplastic cells, which reflects the abnormal regulation of gene expression inherent in the neoplastic cell or occurring in response to stimuli. Neoplasms may be classified as either benign or malignant based on their pattern of growth. A malignant neoplasia has the capacity for producing secondary growths that may be sustained at sites secondary to the primary neoplasm. Benign neoplasms are not capable of successful metastasis to other sites. Technically, cancers are malignant neoplasms. The term *tumor* is a general term applied to a space-occupying lesion that may or may not be neoplastic.

A carcinogen is an agent that causes or induces neoplasia. Pitot (1986) has noted that a carcinogen is defined more appropriately as an agent whose administration to previously untreated animals leads to a statistically significant increased incidence of neoplasms of one or more histogenetic types over the incidence in appropriate untreated animals. A key point to keep in mind in this definition is the emphasis given to an increase in incidence.

Cancer usually develops slowly over a long interval, called the latent period, between the initial exposure and the appearance of a malignant neoplasm. In the case of many cancer-causing air pollutants, prolonged exposure may be required to produce an increase in cancer above background. Cancer induced by cigarette smoking is an excellent example, with the lung cancer incidence increasing dramatically in individuals in their 60s who have smoked for decades. In males in the USA during 1992–94, invasive cancers of the lung and bronchus occurred with greater frequency in each age interval: birth to 3 years, 1 in 2500; 40 to 59 years, 1 in 72; and 60 to 79 years, 1 in 15 (Landis *et al.*, 1998; Ries *et al.*, 1997).

The development of cancer is a multistage process involving initiation, promotion, conversion, progression and, ultimately, clinical manifestation of disease (Boyd and Barrett, 1990; Harris, 1991; Pitot and Dragan, 1996; Vogelstein and Kinzler, 1993) (Fig. 18.3).

The key characteristics of the various stages are summarized in Table 1. From consideration of Fig. 18.3 and Table 18.1, it is apparent that various air pollutants may become part of the carcinogenic process at many different points in the multistage process. Moreover, a range of carcinogenic outcomes may occur involving different organs, tissues, and cells.

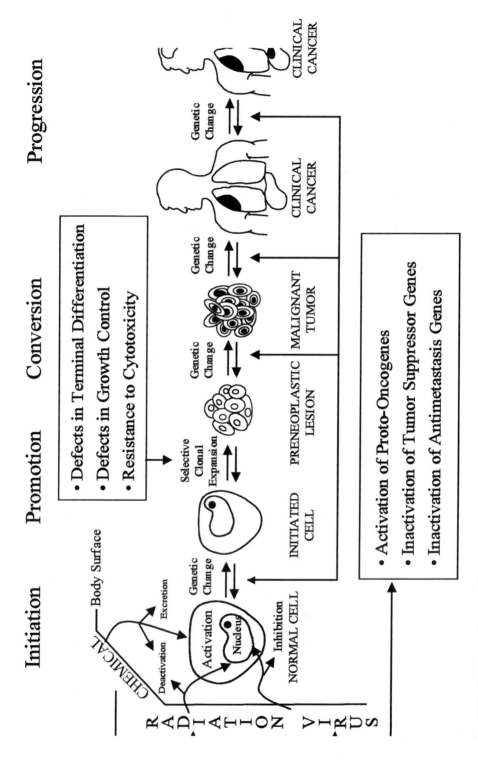

Fig. 18.3   Carcinogenesis as a multistage process (Harris, 1991, with permission).

Table 18.1 Morphological and biological characteristics of the stages of initiation, promotion and progression during carcinogenesis (from Pitot and Dragan, 1996)

Initiation	Promotion	Progression
Irreversible	Operationally reversible both at	Irreversible
Initiated 'stem cell' not morphologically identifiable	the level of gene expression and at the cellular level	Morphologically discernible alteration in cellular genomic structure resulting from karytopic instability
Efficiency sensitive to xenobiotic and other chemical factors	Promoted cell population existence dependent on continued administration of promoting agent	
Spontaneous (endogenous) occurrence of initiated cells	Efficiency sensitive to aging and dietary and hormonal factors	Growth of altered cells sensitive to environmental factors during early phase of this stage
	Endogenous promoting agents may effect 'spontaneous' promotion	
Requirement for cell division for 'fixation'		
Dose-response not exhibiting a readily measurable threshold	Dose-response exhibits measurable threshold and maximal effect	Benign or malignant neoplasms observed in this stage
Relative potency of initiators dependent on quantitation of preneoplastic lesions after defined period of promotion	Relative potency of promoters measured by their effectiveness in causing an expansion of the initiated cell population	'Progressor' agents advance promoted cells into this stage.

Care should be taken to avoid thinking of all pollutants as causing cancer in the same way. For example, some air pollutants, such as benzo[*a*]pyrene, may be initiators. Other air pollutants, such as irritant gases, may have no role in initiating cancer but may ultimately be shown to influence the carcinogenic process through an effect on promotion and progression. Other air pollutants might conceivably alter immune surveillance and thus the progression of cancer.

In recent years, increased insight has been gained into how different chemicals may cause cancer. Although we still do not know every step in the process, identifying the mode of action for many carcinogens is possible. The mode of action has been defined as the key obligatory process governing the action of the chemical in causing cancer (Butterworth *et al.*, 1995). For example, we have known for some time that the mode of action for some chemicals involves direct interaction between the chemical or its metabolites and DNA. Genetic damage is the result.

More recently, alternative modes of action have been identified, including cytotoxicity with compensatory cell regeneration, mitogenic stimulation, or receptor-mediated interaction with genetic control. Each of these different modes of action may have fundamental differences in the resulting relationship between exposure and cancer response. These differences need to be considered in developing quantitative estimates of cancer risk for different air pollutants.

Table 18.2   Short-term tests for carcinogen identification (from Pitot and Dragan, 1996)

Gene mutation assay
   Bacterial-Ames assay (histidine reversion assay in Salmonella)
     Mammalian mouse lymphoma thymidine kinase assay
       Chinese hamster ovary hypoxanthine-guanine
        phosphiphoribosyltransferase
Chromosome aberration
   In vitro assay in cell lines
   Mouse micronuclei
   Rat bone marrow cytogenetics
Primary DNA damage
   DNA adducts: ^{32}P postlabeling
   Strand breakage
   Induction of DNA repair
     Bacteria: SOS response
     Rat liver: unscheduled DNA synthesis (UDS) induction
   Sister chromatid exchange (SCE)
Morphological transformation
   Syrian hamster embryo (SHE)
   Balb/c 3T3

In the past, the most attention has been directed toward evaluating the initiating prop-erties of air pollutants by studying specific pollutants in short-term assays *in vitro* and *in vivo* that are designed to detect mutagenic changes rather than evaluate based on chronic whole animal bioassays. Perhaps the most widely used of these assays is the test developed by Ames *et al.* (1973) to detect the mutagenic capabilities of agents in a bacterial assay. Other assays involve the use of mammalian cells, including human cells. The US Environmental Protection Agency (US EPA, 1996a) has issued guidelines for detecting somatic cell mutations in culture. Some of the more frequently used short-term tests for identifying potential carcinogens are listed in Table 18.2. Some short-term assays are spe-cific for the respiratory tract and include the use of various respiratory tract cells in culture and lung tissue sections. Mutation assays have also been performed using intact animals exposed to the test material. Other knowledge of the carcinogenic properties of air pollutants has come from long-term studies of animals exposed to specific pollutants. The short-term studies provide insight into specific aspects of the carcinogenic process, while the long-term studies have the potential for actually demonstrating evidence of an increase in cancer incidence. Both kinds of studies contribute to an improved under-standing of the process of carcinogenesis and the role of specific air pollutants in that process. In the next section, we focus on detection of carcinogenic properties of air pollutants.

IDENTIFYING CARCINOGENIC AIR POLLUTANTS

A multipronged approach is useful for identifying carcinogenic air pollutants. This approach includes use of information from epidemiological studies, bioassays con-ducted in laboratory animal species, investigations with cells and tissues, and evaluations of chemical structure and biological activity. Ultimately, to move beyond detection of carcinogenic activity, an understanding of biological responses must be placed in a broader context that extends from sources of pollutants to biological responses (Fig. 18.2).

Epidemiological Approaches

Humans are the prime species of concern for assessing potential health risks from exposure to airborne materials. Therefore human data should obviously be used to the extent that it is available. Epidemiological approaches are discussed in Chapter 20 by Samet and Jaakkola and will only be discussed here within the context of assessing human risk. As noted earlier, studies on the carcinogenic risk of specific airborne materials are inevitably complicated by the substantial background of lung cancer risk attributed to cigarette smoking and also a lower level of risk from exposure to environmental tobacco smoke. The confounding factor of cigarette smoking places a premium on ascertaining the smoking history of the subjects in epidemiological studies. A related issue is the need for ascertaining the level of exposure to the air pollutant being investigated. Limitations in both areas frequently have an impact on the ability to acquire information on carcinogenic hazards and make it very difficult to quantify the cancer-causing potency of specific pollutants.

Air pollutants for which sufficient epidemiological data exist to characterize them as human carcinogens are listed in Table 18.3 which was compiled from multiple documents of the International Agency for Research on Cancer (IARC). The reader interested in the most recent classifications and updates is referred to that authoritative series, as exemplified by IARC (1997). Summary information is also available on the IARC website (www.iarc.fr).

A major issue for conducting and interpreting epidemiological studies, as well as for investigations with laboratory animals, is the relatively low sensitivity of our approaches owing to statistical limitations. Confidence in epidemiological findings is greatest when the observed ratio for the disease of interest in the exposed population relative to the control population is on the order of 2.0 or greater (a doubling of the disease incidence). When the odds ratio drops to 1.2 or even to 1.5, which represents a 20% or 50% increase in the disease, demonstrating a statistically significant increase in cancer becomes difficult because the necessary sample size becomes large. It also becomes difficult to be certain that the increase does not reflect uncontrolled effects of other factors. In laboratory animal studies with 100 animals per group, a doubling or more of cancer is required to have statistical confidence at the 0.05 level that the increase is exposure-related.

These detectable levels of increased risk contrast sharply with the need for policy decisions that protect against cancer risks on the order of 1 in 10 000 to 1 in 1 million persons. Thus, extrapolating downward is necessary, frequently over several orders of magnitude, from epidemiological and experimental observations to the levels of risk of concern for policy purposes. If laboratory animal data are used, a second extrapolation step is obviously involved in going from animals to humans. If cell or tissue data are used, a third extrapolation is required in extrapolating to the intact mammal.

Table 18.3 Agents causing cancer in the respiratory tract and/or entering via the respiratory tract and causing cancer and identified by epidemiological studies

Agent	Primary affected site(s)
4-Aminobiphenyl	bladder
Asbestos	bronchus, pleura, peritoneum
Arsenic	bronchus, skin, liver
2-Chloro-N-(2-chloroethyl)-N-methylethanimine hydrochloride (nitrogen mustard)	bronchus
Beryllium	lung
Bis[chloromethyl]ether	bronchus
Benzene	bone marrow
Benzidine	urinary bladder
(1,1'-Biphenyl)–4,4'-diamine (benzidine)	
Chromium VI	nasal sinus, bronchus
Coke over emissions	bronchus
Mustard gas	bronchus
Nickel compounds	nasal sinus, bronchus
Polynuclear aromatic hydrocarbons	bronchus (also skin and scrotum from direct contact)
Radon and decay products	bronchus
Silica crystalline silica (quartz or cristobalite)	lung
Tobacco smoking	bronchus, mouth, pharynx, larynx, bladder, esophagus
Vinyl chloride	liver
Wood dust	nasal sinus

Laboratory Animal Studies

Cancer Bioassays

Studies in laboratory animals are used to identify carcinogenic air pollutants in the absence of convincing human evidence. The approach for conducting carcinogenicity studies advocated by the US EPA has been coded in testing guidelines (US EPA, 1996b). Several review articles are available on methods for generating and characterizing test atmospheres (Cheng and Moss, 1995; Moss and Cheng, 1995; Wong, 1995) and exposing animals (Gardner and Kennedy, 1993; Phalen, 1997).

Ideally, bioassay studies should be conducted with two species (usually rats and mice) and both genders. Some of the issues to be considered in selecting species for study are listed in Table 18.4. In an attempt to reduce the duration and cost of cancer bioassays while also increasing their sensitivity, consideration is being given to using strains of mice with one or more genetic alterations that predispose them to development of cancer.

Many air pollutants are particulate in nature. The conduct of particulate matter studies

Table 18.4 Selecting species to use for chronic bioassay of suspected lung carcinogens (Hahn, 1993)

Requirements for ideal species	Species					
	Mice	Rats	Dogs	Syrian hamsters	Guinea pigs	Monkeys
Economical production and maintenance	++	++	−	++	−	−
Short normal life span	++	++	−	++	+	−
Can expose by inhalation	++	++	++	−	+	+
Respirable particles readily inhaled	+/−	+	++	+	+	++
Low incidence of spontaneous lung tumors	+	++	+	++	+	?
No endemic chronic respiratory disease	+	+	++	++	+	?
Tumors anatomically similar to humans	+	++	++	++	?	?
Defined genetics	++	+	−	−	−	−
Defined molecular biology	++	+	−	−	−	−

Scoring of meeting requirements: ++, well; +, adequate; −, does not; ?, unknown.

with animals raises concerns as to whether the particle size distribution is similar to that of concern for human exposure and whether particles of this size will be deposited in the respiratory tracts of the animal species. Rodent species are obligate nose breathers, and therefore larger particles are effectively deposited in the nares and do not penetrate to the lungs. In the mouse, a large portion of the particles greater than about 1 µm are deposited in the nares and will not reach the lower respiratory tract. In the rat, there is concern that particles greater than about 2 µm in aerodynamic diameter will not reach the pulmonary region. These concerns are based on recognition that larger particles may be inhaled, deposited, and cause adverse effects in people. Thus, since these large particles do not reach the pulmonary region of rodents to the extent they do in humans, results from rodent bioassays may underestimate potential toxic effects including the potential for cancer induction in humans. On the other hand, Mauderly (1997) and McClellan (1996) have both cautioned that rat models may not always be predictive of human responses for particulate matter such as diesel exhaust and carbon black that yields lung tumors in rats exposed at high concentrations. Studies in mice and Syrian hamsters have historically yielded a high percentage of false negatives.

As increased attention is given to using the tools of modern molecular biology in studies with laboratory animals, the level of knowledge of the animal species' genome and its ease of manipulation will be a factor in selecting specific species and strains. Special strains of mice have been developed with one or more genetic alterations that predispose them to cancer. These strains are currently being evaluated as more sensitive and inexpensive approaches to detecting carcinogens (Tennant *et al.*, 1996). Genetically altered animals have already been used to good advantage in expanding our knowledge of the effects of air pollutants. For example, Recio *et al.* (1996) have used mice with a recoverable gene marker to study somatic cell genetic alterations produced by inhaled 1,3-butadiene. Valentine *et al.* (1996) studied knockout mice that do not have P450 metabolizing enzymes to assess the role of these enzymes in metabolizing inhaled benzene.

Bioassay studies conducted to identify potential carcinogens typically involve the study of 50 animals of each gender at three exposure levels and control groups (air only) (US EPA, 1996b). Exposures usually begin early in life (6–8 weeks of age) and continue for 2 years or until only 10–20% of the animals are surviving. The highest exposure level typically corresponds to a maximum tolerated dose (MTD) with an intermediate level at one-half the MTD and the lowest level at one-fourth the MTD. This exposure regimen has been advocated for detecting carcinogens; it attempts to maximize the exposure, thereby maximizing the detection of carcinogens with a minimum number of animals. This approach has evoked much criticism since mechanisms of carcinogenesis may be dose-dependent (Mauderly, 1997; McClellan, 1996, 1997; McConnell, 1996; Morrow *et al.*, 1996; Oberdörster, 1996, 1997; Swenberg, 1995). Lewis *et al.* (1989) emphasized the importance of taking account of patterns of particle deposition and clearance when selecting and studying inhaled particulate matter. Dose selection in a bioassay can also influence the pattern of neoplasms induced, especially in lung carcinogenesis. Specifically, the incidence and multiplicity of neoplasms increases with increasing dose until a toxic dose is reached and then the incidence decreases. Also, the time for neoplasms to appear decreases with increasing dose. These considerations raise concern for the applicability of the findings from high-dose studies to assessing human risks at much lower doses. Concern has also been expressed regarding the limitations imposed on exposure–response modeling when data are obtained from only three exposure levels differing by a factor of four in air concentration.

Bioassay exposures are infrequently conducted for more than 2 years. It has been argued that the exposures should be stopped and the animals killed before substantial mortality has occurred and before spontaneous occurring neoplasms are observed in high frequency, thus increasing the likelihood of detecting cancers induced by the test agent. Others have argued that extending the exposures to enhance interactions that may occur between the pathogenesis of spontaneous neoplasia and toxicant-induced effects is appropriate. A study of rats exposed to diesel exhaust (Mauderly *et al.*, 1987) illustrated this point with more than 80% of the lung tumors detected after 24 months of exposure when exposures were continued to 30 months.

The bioassay studies typically involve whole-body exposure. In a few specialized cases, as when studying fibers or radioactive particles, the investigators have elected to conduct the studies using nose-only exposure techniques. This approach minimizes the potential for materials deposited on the animal's fur to be ingested during grooming. Obviously, the conduct of nose-only exposures is very labor-intensive since individual animals are placed in single tubes at the beginning of each daily exposure and removed at the end of each daily exposure. In addition, questions may be raised about the physiological well-being of restrained animals because of stress induced by confinement.

Many investigators have found it cost-effective to move beyond conducting only a strict bioassay by including animals for use in mechanistic studies of the pathogenesis of disease if toxicant-induced disease is observed. This is done by adding small groups of animals that can be exposed for a specified period of time such as 3 or 6 months. In some cases,

Table 18.5 Agents causing respiratory tract cancer in rodents, typically rats, exposed via inhalation

Antimony Trioxide	1,2-Epoxybutane
Asbestos	Ethylene Oxide
Benzo(a)pyrene	Formaldehyde
Beryllium Compounds	Hydrazine
Bis[chloromethyl]ether	Mustard Gas
1,3-Butadiene	Nickel Subsulfide
Cadmium Compounds	Nickel Carbonyl
Carbon Black	Nitrobenzene
Ceramic Fibers	N-Nitrosodimethylamine
Chromium VI	Radionuclides (alpha and beta-emitters)
Coal Tar	Radon and decay products
Crystalline Silica	Rockwool
Dichloromethane	Styrene Oxide
1,2-Dibromo-3-chloropropane	Tetranitromethane
1,2-Dibromoethane	Tobacco Smoke
Diesel Exhaust	Urethane
	Vinyl Chloride

animals are euthanized when the exposures are discontinued. In other cases, the animals may be held for some time without further exposure before being euthanized. Holding animals after exposure is terminated allows the potential for observing evidence of recovery. If animals are killed upon termination of exposure, specimens are available for detailed study, typically at times prior to the development of overt toxicity and the appearance of tumors. The conduct of mechanistic studies in parallel with the core bioassay is especially appropriate for studies involving inhalation exposure. The expense involved in preparing for a long-term inhalation study and the generation and characterization of the exposure atmosphere remain fixed regardless of the number of animals in the chamber. Being proactive in exposing additional animals with the first exposure series is better than having to go back and repeat the exposures to obtain valuable tissue specimens.

Agents identified in the IARC process as having sufficient or limited evidence of carcinogenicity in animal species are listed in Table 18.5. The IARC process will be discussed later in this chapter. The only agent identified as a human carcinogen (Table 18.3) and not identified as an animal carcinogen is the class of arsenic compounds. Arsenic compounds have been shown to be carcinogenic in laboratory animals when administered by intratracheal instillation and other routes. However, they have not been evaluated by the inhalation route. The reader interested in additional details on the most recent classifications and updates is referred to the IARC series of documents, as exemplified by IARC (1997). Mauderly (1997) used the IARC reviews as the basis for a critical assessment of the relevance of rat lung tumor data for assessing human lung cancer risk. While noting the general utility of the rat bioassay for assessing potential human cancer risks, he noted some of the issues involved with studies of particulate materials in rats.

Disposition and Kinetic Studies

An understanding of the disposition of the toxicant and of possible metabolites is crucial in understanding the mechanisms involved in the pathogenesis of any induced disease and the relevance of the findings to humans. Detailed information on the deposition, clearance, and retention of toxicants and metabolites in lung, lung-associated lymph nodes, and other tissues usually requires the design and conduct of studies that are complementary to the main bioassay studies. The studies should use an exposure atmosphere that is as similar as possible to that used for the bioassay and span concentrations that include the linear and saturable range for metabolism. The use of similar exposure concentrations is especially important when studying particulate materials for which small differences in particle size distribution can markedly influence the deposition and retention kinetics. Typically, provision is made for exposure of animals for various periods of time and then serial euthanasia of groups of animals to evaluate the disposition of the toxicant over time. If the material and potentially its metabolites have short half-lives for retention in the body, the exposure period and the time over which specimens will be obtained needs to be short. If, on the other hand, the retention half-lives in lung and other tissues are long, then lung exposure times extending over months may be necessary to understand the kinetics of lung burden build-up. In addition, studies conducted at graded levels of exposure may be necessary to evaluate whether the exposure concentration influences the kinetics of the toxicant or metabolite(s) (Medinsky and Klaassen, 1996).

Pathogenicity Studies

As noted earlier, traditional carcinogen detection bioassays do not always provide much insight into the mechanisms by which cancers are or are not produced by exposure to a specific toxicant. By increasing the number of animals available for study and with careful planning of ancillary studies, valuable information can be obtained to interpret the pathogenicity of toxicant-associated disease. This process can be illustrated with past research on two different air pollutants, ozone and diesel exhaust.

The Health Effects Institute (HEI), working in collaboration with the National Toxicology Program (NTP) of the National Institute of Environmental Health Sciences (NIEHS) in the USA, sponsored a multiple investigator study to evaluate the potential carcinogenicity of ozone in mice and rats (Catalano et al., 1995; NTP, 1995). There was no evidence of ozone-induced cancer in rats and limited evidence in mice, even though the highest exposure level was 1.0 ppm, just below levels causing acute toxicity in rodents and many times higher than typical ambient levels in the USA. The parallel and very detailed molecular, electron micrographic, biochemical, histological, and functional studies revealed minimal ozone-induced changes and did not reveal any changes that raised concern for potential carcinogenic effects.

The studies with diesel exhaust were conducted at the Lovelace organization in Albuquerque, New Mexico, and were part of an international effort to evaluate the carcinogenicity of diesel exhaust. The Lovelace studies involved both mice and rats exposed to the same test atmosphere (0, 0.35, 3.5, or 7.0 mg per m^3 of diesel exhaust particulate matter as an indicator of whole exhaust exposure). The studies in rats involved exposure and observation over a 30-month period. A significant increase in lung tumors was observed in the rats at the two highest exposure levels, while an effect related to diesel exhaust was not observed in mice (Mauderly *et al.*, 1987, 1996). Most importantly, detailed studies of the kinetics of diesel soot particles in the lungs (Wolff *et al.*, 1987) revealed that normal clearance processes were overloaded at the two highest exposure concentrations. In Chapter 32 of this book, Cohen and Nikula discuss the carcinogenicity of diesel exhaust.

At the two highest exposure levels, lung burdens of soot were much higher than predicted from kinetics at the lowest exposure level. Most significantly, the particle overload phenomenon had an associated inflammatory process, as evaluated by both cellular and biochemical indicators in lavage fluid (Henderson *et al.*, 1988) and by histopathological changes. Mild functional deficits were also associated with these levels (Mauderly *et al.*, 1983, 1988). There were striking differences in epithelial cell replication rates in mice and rats. The labeling index for rats was increased five-fold, while only a 20% increase was seen in mice (Mauderly, 1997). Nikula *et al.* (1997) also noted a striking difference in epithelial hyperplasia between rats and monkeys exposed to diesel soot or coal dust, with monkeys having a much more minimal response.

Subsequent studies have shown that exposure of rats to high levels of carbon black produces a lung tumor response similar to that seen with chronic high level exposure to diesel exhaust (Nikula *et al.*, 1995). Further mechanistic studies provide strong evidence that the carbon black effects occur via a non-genotoxic mechanism unique to the high levels of exposure (Driscoll, 1996; Driscoll *et al.*, 1996; Mauderly, 1997; McClellan, 1996; Oberdörster, 1997). McClellan and Mauderly have concluded that the lung tumors produced in rats by high-level exposure to diesel exhaust or carbon black are not relevant for assessing human risks from exposure to low ambient concentrations of diesel exhaust.

In some cases, detailed mechanistic studies are conducted subsequent to an initial bioassay that yields a carcinogenic response. For example, such studies were conducted at the Chemical Industry Institute of Toxicology (CIIT) to clarify the pathogenesis of nasal tumors observed in rats chronically exposed to high concentrations of formaldehyde (6 and 15 ppm) (Conolly *et al.*, 1995; Heck *et al.*, 1990). The CIIT mechanistic studies revealed a non-linear relationship between air concentration of formaldehyde and a measure of delivered dose, DNA–protein cross-links, cell-killing, and cell proliferation. The internal dosimetry data obtained in rats yield estimates of human cancer risk that are a factor of 10 lower than estimates based on traditional approaches. Using monkey data on the internal dosimetry of formaldehyde, the estimates of human risk are about 100 lower (McClellan *et al.*, 1993).

Mechanistic Human Studies

Valuable insights into the mechanisms by which air pollutants may cause cancer can come directly from the study of human populations. Obviously, human studies are much more opportunistic than controlled experiments with laboratory animals or *in vitro* studies. Human subjects cannot ethically be exposed to known human carcinogens except under the most sharply defined circumstances in which the subjects clearly will not be harmed. Therefore, studying humans under natural-exposure circumstances is usually necessary. If an exposure gradient is required to address a particular hypothesis, then studying subjects from different locales may be necessary.

Advances in the development and use of biomarkers related to cancer and, more broadly, to respiratory disease may allow hypotheses to be addressed that could not be addressed with old technologies (NRC, 1989). Future advances in molecular biology and informatics, and especially increased efficiency in analyzing for genomic differences and genetic alterations, hold promise for moving from studies with group sizes of a few dozen individuals to thousands. Such a shift in approach can be especially valuable for studying a family of diseases such as cancer, where individual cancer types (other than in smokers) may occur in low frequency and public policy concerns focus on risks of very low frequency.

RISK ASSESSMENT FRAMEWORK

Legislative Background

Clean Air Act

Public concerns have led to pressure for legislation to improve air quality. Public concerns have included potential impacts on health as well as the grossly evident effects of air pollution on visibility and soiling. The initial approaches to improving air quality were vested at local and state levels with a federal advisory role. In the USA, the enactment of the Clean Air Act (CAA) in 1970 shifted power for establishing standards to the federal level and delegated responsibility for achieving standards to the states (CAA, 1970). Subsequent amendments to the act have not markedly changed federal and state responsibilities. Two sections of the act focus on standards to protect human health, one dealing with criteria air pollutants and the other with hazardous air pollutants.

Criteria Air Pollutants

The criteria air pollutants (carbon monoxide, sulfur dioxide, nitrogen dioxide, lead, ozone and particulate matter) are widely distributed and are regulated because they may reasonably be anticipated to endanger public health or welfare. Primary National Ambient Air Quality Standards (NAAQS) are set for the criteria air pollutants to protect the public, including sensitive subpopulations, against adverse health effects with an adequate margin of safety. Although costs are not to be considered in setting the NAAQS, costs and technical feasibility can be considered in establishing the implementation schedule. Secondary standards are set to protect against welfare effects. The standards consist of four elements: an indicator (with the exception of particulate matter, the indicators are specific chemicals); a numerical concentration level; an averaging time for the level; and a statistical form. All the current primary NAAQS are based on health end-points other than cancer. There is no evidence that carbon monoxide, nitrogen oxide, sulfur dioxide or ozone initiates cancer. The irritant gases (nitrogen oxide, sulfur dioxide and ozone) may arguably serve as promoters for cancers of the conducting airways, their primary site of deposition and injury. However, the evidence is not compelling and has not been a consideration in setting NAAQS for these pollutants.

Lead has been identified as a possible human carcinogen by IARC (Category 2B)

(IARC, 1980). This evidence of carcinogenicity from high-dose animal studies has not been a factor in EPA's setting of the numerical level of the primary NAAQS for lead (US EPA, 1986a,b).

The indicator for particulate matter is currently a size-selective sample (PM_{10} and $PM_{2.5}$, particles with aerodynamic diameters not exceeding 10 μm and 2.5 μm, respectively) measured as mass without regard to chemical composition. This indicator is in contrast to the chemical-specific nature of all the other NAAQS. It is well recognized that many samples of PM include specific elements such as arsenic, cadmium, nickel and chromium compounds as well as polycyclic aromatic hydrocarbons that are known carcinogens or mutagens. The data linking PM air concentrations to increased cancer incidence are not compelling and have not been a factor to date in setting the numerical level of the primary NAAQS for PM (US EPA, 1996c, 1996d, 1997). To the extent that these specific constituents of PM pose a hazard, the general view has been that the numerical level of the PM standards has been set sufficiently low based on other health outcomes that it is also protective of cancer hazards.

Hazardous Air Pollutants

Hazardous air pollutants are defined in the Clean Air Act as amended in 1977 (CAA, 1977). An air pollutant is considered hazardous if it may reasonably be anticipated to result in an increase in mortality or an increase in serious irreversible or incapacitating reversible illness. This definition clearly includes cancer as a health end-point. Administration of this section of the Clean Air Act prior to the Amendments of 1990 involved setting a National Emission Standard for Hazardous Air Pollutants (NESHAP). The numerical level of each NESHAP was intended to protect against some unreasonable level of human health risk and take into account cost and technical feasibility. The courts ruled that the US EPA had to first set an acceptable risk level and then consider technical feasibility, cost, and other factors in a second step.

Efforts to establish NESHAPs were a major early stimulus to the risk assessment activities of the US EPA and especially the handling of cancer risks (Albert, 1994). The need to develop a framework for risk assessment led to the National Research Council preparation of the report *Risk Assessment in the Federal Government: Managing the Process* (NRC, 1983) and the issuance of the *Guidelines for Carcinogen Risk Assessment* (US EPA, 1996e). Inadequacies in data and controversy over the risk assessment process, especially in quantifying cancer risks, were major factors in slow progress being made in issuing of NESHAPs. Indeed, only seven NESHAPs were promulgated by 1990 (arsenic, asbestos, benzene, beryllium, mercury, vinyl chloride, and radionuclides – including radon in mines). Frustrated with the slow progress, the US Congress made marked changes in the approach to the control of hazardous air pollutants in 1990 Amendments to the Clean Air Act (CAA, 1990).

In the 1990 amendments Congress abandoned a health-based approach as a driver for regulating hazardous air pollutants and shifted to a two-phase approach. In the first phase, technology is the driver with maximum achievable control technology (MACT) imposed for various source categories to control emissions of 189 Hazardous Air Pollutants identified in the Act. In a second phase after installation of MACT, the sources are to be evaluated with regard to any residual risk.

The US Congress recognized the continuing controversy over the risk assessment process and specified in the 1990 Amendments to the CAA that the National Research Council of the National Academy of Sciences should evaluate the process. The result was the report *Science and Judgment in Risk Assessment* (McClellan, 1994; NRC, 1994). This report supported continued use of the basic risk assessment structure put forth in 1983. However, it emphasized the utility of using the risk assessment process to help identify research which, if performed, could reduce uncertainties in the risk assessment process (Fig. 18.4). The report also emphasized the value of developing specific science to replace default options which are generic approaches based on general scientific principles used in the absence of specific knowledge to the contrary. Two key default options are as follows.

(1) Laboratory animals are surrogates for humans in assessing cancer risks; humans are assumed to be as sensitive as the most sensitive species, strain, or gender evaluated in a bioassay.
(2) Chemicals act like radiation in causing cancer at low doses; intake of even one molecule has an associated probability for cancer induction that can be calculated.

The 1994 NRC recommendations stimulated the US EPA to propose revised carcinogen risk assessment guidelines (US EPA, 1996e). The guidelines have been reviewed but not formally issued. Key provisions involve increased use of specific scientific information, use of narrative characterizations of cancer hazard, and use of a two-step approach to characterize relationships between exposure and cancer response. The first step is to describe the relationship over the observable range and the second step to extrapolate downward. This extrapolation need not make use of the traditional linear, no-threshold model, but can propose alternative models. The alternatives can include the use of threshold models based on knowledge of the mode of action of specific chemicals. For example, the US EPA has proposed a maximum contaminant level goal for chloroform in drinking water of 300 ppm (US EPA, 1998). This is based on knowledge that chronic high-dose chloroform exposures causes cancer in rodents by a cell-killing mechanism with subsequent cell regeneration and that this mode of action would not be operative at low levels (ILSI, 1997). This contrasts with previous approaches that assume a calculable cancer risk down to the lowest levels of chloroform exposure. Presumably, the EPA will use the same basic approach for setting limits for airborne chloroform.

A key component of the overall risk assessment process is hazard identification (Fig. 18.4). Perhaps more so than for any other response, schemes have been developed for classifying agents as to their carcinogenicity.

CANCER CLASSIFICATION SCHEMES

IARC Carcinogenicity Classification

Overview of the Scheme

During the 1960s increased concern developed for cancer, especially for the potential role of occupational and environmental factors and personal habits in the causation of cancer. A strategy was envisioned that involved identification of human carcinogens and, once identified, their elimination. A major step in this direction was taken by the International

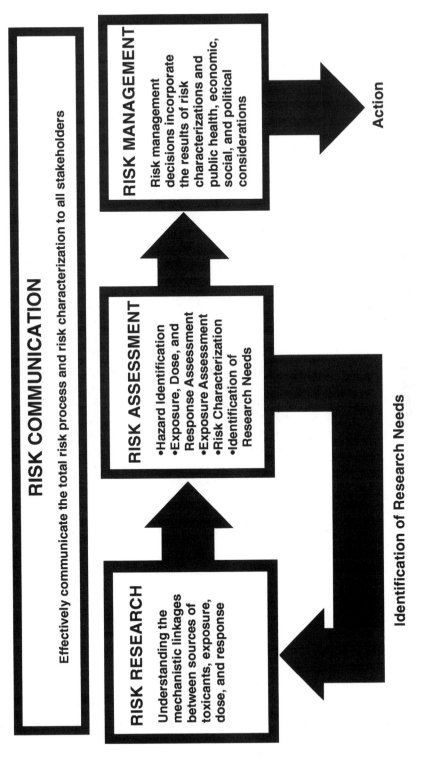

Fig. 18.4 Risk structure integrating research, assessment, management, and communication.

Agency for Research on Cancer (IARC) in 1969 when it created a program to evaluate the carcinogenic hazard of chemicals to humans and to prepare monographs on individual chemicals. Later, the program was expanded to include evaluation of industries and biological and physical agents.

The program has focused on evaluating the hazard potential of a material (that is, whether it can produce cancer in humans) and has not addressed risk, which would require quantitation of both the cancer-causing potency of an agent and exposure. The program has had wide impact because the monographs are viewed as authoritative and used as a basis for action by national and state regulatory bodies around the world.

The IARC approach in evaluating carcinogenic risks to humans is described in the preamble of each monograph (for example, see Volume 69, which addresses the carcinogenic risks of polychlorinated dibenzo-dioxins and dibenzofurans (IARC, 1997)). The IARC approach uses international working groups of experts with input from the IARC staff to carry out five tasks: (1) to ascertain that all appropriate references have been collected; (2) to select the data relevant for the evaluation on the basis of scientific merit; (3) to prepare accurate summaries of the data to enable the reader to follow the reasoning of the Working Group; (4) to evaluate the results of experimental and epidemiological studies; and (5) to make an overall evaluation of the carcinogenicity of the agent to humans.

In the monographs, the term *carcinogen* is used to denote an agent that is capable of increasing the incidence of malignant neoplasms. Traditionally, IARC has considered the evidence for carcinogenicity at all stages in the carcinogenic process independent of the underlying mechanisms involved. In 1991, IARC convened a group of experts to consider how mechanistic data could be used in the classification process. This group suggested a greater use of mechanistic data, including information relevant to extrapolation from laboratory animals to humans. The overall IARC evaluation process considers four types of data: (1) proposed data, (2) human carcinogenicity, (3) experimental carcinogenicity data, and (4) other data relevant to an evaluation of carcinogenicity and its mechanisms.

Human Data

Definitive evidence of human carcinogenicity can only be obtained from epidemiological studies, as described in Chapter 20 by Samet and Jaakkola. The epidemiological evidence is classified into four categories.

(1) *Sufficient evidence of carcinogenicity* is used when a causal relationship has been established between exposure to the agent and human cancer.

(2) *Limited evidence of carcinogenicity* is used when a positive association between exposure to an agent and human cancer is considered to be credible, but chance, bias, or confounding could not be ruled out with reasonable confidence.

(3) *Inadequate evidence of carcinogenicity* is used when the available studies are of insufficient quality, consistency, or statistical power to permit a conclusion regarding the presence or absence of a causal association or when no human data are available.

(4) *Evidence suggesting lack of carcinogenicity* is used when there are several adequate studies covering the full range of doses to which human beings are known to be exposed and these studies are mutually consistent in *not* showing a positive association between exposure and any studied cancer at any observed level of exposure.

Animal Data

The IARC evaluation process gives substantial weight to carcinogenicity data from laboratory animals. As has been noted (Wilbourn *et al.*, 1986), the first 41 IARC monograph volumes identified 44 agents were identified for which there is sufficient or limited evidence of carcinogenicity to humans; of these chemicals, all 37 agents that had been adequately tested experimentally produced cancer in at least one laboratory animal species. This observation led to the IARC conclusion that 'in the absence of adequate data in humans, it is biologically plausible and prudent to regard agents for which there is *sufficient evidence* of carcinogenicity in experimental animals as if they presented a carcinogenic risk to humans.' Mauderly (1997) reviewed the IARC evaluations through September 1996 and identified 56 agents with sufficient or limited evidence of human carcinogenicity. For 49 agents, sufficient or limited evidence of carcinogenicity was found in one or more animal species. Because of this strong concordance between human and laboratory animals, IARC classifies the strength of the evidence of carcinogenicity in experimental animals in a fashion analogous to that used for the human data.

The evidence of carcinogenicity in experimental animals is classified into four categories.

(1) *Sufficient evidence of carcinogenicity* is used when a working group considers that a causal relationship has been established between the agent and an increased incidence of malignant neoplasms or an appropriate combination of benign and malignant neoplasms in (a) two or more species of animals or (b) two or more independent studies in one species carried out at different times, in different laboratories, or under different protocols. A single study in one species might be considered under exceptional circumstances to provide evidence when malignant neoplasms occur to an unusual degree with regard to incidence, site, type of tumor, or age at onset.

(2) *Limited evidence of carcinogenicity* is used when the data suggest a carcinogenic effect but are limited for making a definitive evaluation.

(3) *Inadequate evidence of carcinogenicity* is used when studies cannot be interpreted as showing either the presence or absence of a carcinogenic effect because of major qualitative or quantitative limitations.

(4) *Evidence suggesting lack of carcinogenicity* is used when adequate studies involving at least two species are available that show that, within the limits of the tests used, the agent is not carcinogenic. Such a conclusion is inevitably limited to the species, tumors, and doses of exposure studied.

Other Relevant Data

Other evidence judged to be relevant to an evaluation of carcinogenicity and of sufficient importance to affect the overall evaluation is then described. This evidence may include data on preneoplastic lesions, tumor pathology, genetic and related effects, structure–activity relationships, metabolism and pharmacokinetics, physicochemical parameters, and analogous biological agents.

Data relevant to mechanisms of the carcinogenic action are also evaluated. The strength of the evidence that any observed carcinogenic effect is due to a particular mechanism is assessed, using terms such as weak, moderate, or strong. Then the Working Group assesses whether that particular mechanism is likely to be operative in humans. The strongest indications that a particular mechanism operates in humans come from data on humans or biological specimens obtained from exposed humans. The data may be considered to be especially relevant if they show that the agent in question has caused changes in exposed humans that are on the causal pathway to carcinogenesis. Such data may, however, never become available, because certain compounds may, at least conceivably, be kept from human use solely on the basis of evidence of their toxicity or carcinogenicity or both in experimental systems.

Overall Evaluation

Finally, the body of evidence is considered as a whole to reach an overall evaluation of the carcinogenicity to humans of an agent, mixture, or circumstance of exposure.

- Group 1: *The agent is carcinogenic to humans.* This category is used only when there is sufficient evidence of carcinogenicity in humans.
- Group 2: This category is used for a range of agents, from those for which the human evidence of carcinogenicity is almost sufficient to those for which no human data are available but for which there is experimental evidence of carcinogenicity.
- Group 2A: *The agent is probably carcinogenic to humans.* The category is typically used when there is limited evidence of carcinogenicity in humans and sufficient evidence of carcinogenicity in experimental animals.
- Group 2B: *The agent is possibly carcinogenic to humans.* This category is typically used when there is limited evidence in humans in the absence of sufficient evidence in experimental animals or when there is sufficient evidence of carcinogenicity in experimental animals in the face of inadequate evidence or no data in humans.
- Group 3: *The agent is not classifiable as to carcinogenicity in humans.* This category is used when agents do not fall into any other group.
- Group 4: *The agent is probably not carcinogenic to humans.* This category is typically used for agents for which there is evidence suggesting lack of carcinogenicity in humans, together with evidence suggesting lack of carcinogenicity in experimental animals.

As noted, the IARC categorization scheme does not address the potency of carcinogens. This poses serious constraints on the utility of IARC classification data for use beyond hazard identification. In short, a carcinogen is a carcinogen irrespective of potency. This lumping together of carcinogens irrespective of potency can be misleading to the non-specialist, including the lay public.

Other Classification Schemes

The US EPA (1986b) adopted a carcinogen classification scheme very similar to that of IARC. In the EPA scheme alpha designations are used rather than numbers, i.e. IARC Category 1 is an EPA Category A. Recently, the EPA proposed revision of the carcinogen risk assessment guidelines (US EPA, 1996e). A key provision of the proposed guidelines is to open the door to the use of narrative descriptions of the evidence of carcinogenicity. One of us has supported this proposed change and illustrated the impact of the change in describing the evidence on the carcinogenicity of carbon black and diesel exhaust (McClellan, 1996).

The EPA classifications, along with other information on the toxicity of specific chemicals can be accessed in the EPA's Integrated Risk Information Systems (IRIS) (Griffin, 1994). This database is available on line (www.epa.gov/iris). The IRIS database includes risk potency values when these have been developed. IRIS is especially valuable for comparing the potency of different agents and for estimating risk when the level of exposure is known or can be estimated.

The National Toxicology Program (NTP, 1998) is mandated by the US Congress to publish reports biennially that designate the carcinogenicity of chemicals. The designation scheme is even simpler than that used by IARC or EPA. It designated chemicals as (1) known to be a human carcinogen (29 entries in the Eighth Report), or (2) reasonably anticipated to be a human carcinogen (169 entries in the Eighth Report). The requirements for evidence for designation of a chemical in the second category are quite broad, especially based on recent changes that allow the use of mechanistic evidence of carcinogenicity in the absence of direct evidence of cancer in animals or humans. The designation of a large number of chemicals as reasonably anticipated to be human carcinogens based on variable evidence may be confusing to the public and regulators because the chemicals are likely to vary widely in their cancer-causing potency and, indeed, may ultimately prove to not even be carcinogens when subjected to more rigorous evaluation.

The issue of classifying chemicals as to carcinogenicity or other disease-producing potential such as neurotoxicity or reproductive effect deserves additional research and discussion as renewed attention is given to developing alternative toxicological methods that refine, reduce or replace the use of laboratory animals.

QUANTITATIVE ESTIMATES OF CARCINOGENIC POTENCY

The foregoing section on classification schemes is relevant to the hazard identification element of the risk assessment process (Fig. 18.4). To characterize risk, exposure to the agent and the carcinogenic potency of the agent both need to be quantified. The topic of exposure assessment is covered in an earlier chapter by Özkaynak (Chapter 9, this volume) and is the subject of EPA guidelines (US EPA, 1992). The development of quantitative estimates of carcinogenic potency is covered in the US EPA guidelines for carcinogen risk assessment (US EPA, 1986b, 1996e). The 1986 guidelines recommended the use of a linear, non-threshold exposure–response model. As discussed earlier, the new guidelines provide more latitude for using specific scientific knowledge to develop quantitative estimates of carcinogenic potency. The new guidelines emphasize the importance of using a two-step approach. In the first step, the relationship between exposure and cancer response is characterized in the range for which direct observations are available. In the second step, a downward extrapolation is made from the observations. This extrapolation step is especially contentious. Examples of inhalation unit risks estimates, i.e. lifetime cancer risk for continuous exposure to 1 $\mu g/m^3$, are given in the US EPA Integrated Risk Information System (IRIS) (Griffin, 1994).

Table 18.6 Data on the hazardous air pollutants (from NRC, 1994)

Chemical name	1991 TRI emissions[a] (tons/yr)	IUR[b] per μg/m³	OUR[c] per μg/l	EPA[d] WOE	RfC[e] (mg/m³)	IARC[f] WOE	In vivo S	In vivo G	In vitro M	In vitro C	Bacterial S	Bacterial E	R-D data[h]
Acetaldehyde	3540.5	2.2E6		B2	9.0E-3	2B	+		+		+	+	X
Acetamide	683.9					2B							X
Acetonitrile					UR								
Arylamide	32.1	1.3E-3	1.3E-4	B2	NV	2B	+				−		
Acrylamide	32.1	1.3E-3	1.3E-4	B2		2B							
Acrylonitrile	1094.4	6.8E-5	1.5E-5	B1	2.0E-3	2A	−	−	+	+−	+	+	X
4-aminobiphenyl						1							
Arsenic Compounds (inorganic including arsine)	95.2	1.3E-3		A	5.0E-5	1	+	−	−	+	−	−	
Asbestos	6.3			A		1	−		−+	+−	−	−	
Benzene	8737.2	8.3E-6	8.3E-7	A	UR	1	+−		−+	−	−	−	X*
Benzidine		6.7E-2	6.7E-3	A	NV	1	+−		+−	+	+	−	
Benzotrichloride	3.9		3.6E-4	B2	NV						+	+	
Benzyl chloride	13.4		4.9E-6	B2	NV		−		+−	+−	+	+	
Beryllium Compounds	0.1	2.4E-3	1.2E-4	B2	UR	2A	A		+	+	−	+	
beta-Propiolactone					NV	2B							
Bis(2-ethylhexyl) phthalate (DEHP)	521.7	4.0E-7		B2		2B	−	−	−	−	−	−	X
Bis(chloromethyl)ether	0.3	6.2E-2	6.2E-3	A	NV	1					+		X
Bromoform	0.1	1.1E-6	2.3E-7	B2						−	−		
1,3-Butadiene	1975.2	2.8E-4		B2		2A	+		+		+		X
Cadmium Compounds	34.7	1.8E-3		B1	UR	2A	−		+	+−	+	+	X

Table 18.6　*cont.*

Chemical name	1991 TRI emissions[a] (tons/yr)	IUR[b] per µg/m³	OUR[c] per µg/L	EPA[d] WOE	RfC[e] (mg/m³)	IARC[f] WOE	In vivo S	In vivo G	In vitro M	In vitro C	In vitro S	Bacterial E	R-D data[h]
							Genetic toxicity data[g]						
Carbon tetrachloride	773.4	1.5E-5	3.7E-6	B2		2B	−	−	−	−	−	+	X
Chloroform	9541.4	2.3E-5	1.7E-7	B2	UR	2B	−		−	−	−	−	X
Chloromethyl methyl ether	1.7			A	NV	1					+		
Chloroprene (2-chloro-1,3-butadiene)	735.3				7.0E-3	2B	+	+	−		+	+	X*
Chromium Compounds (+6 FOR IRIS)	278.2	1.2E-2		A	UR	1	+	+	+	+	+	+	
Coke Oven Emissions		6.2E-4		A		1							
1,2-Dibromo-3-chloropropane			4.0E-5	B2	2.0E-4	2B							
1,3-Dichloroproprene (Telone II)	10.2			B2	2.0E-2	2B							
1,4-Dichlorobenzene	168.1				8.0E-1	2B							
2,4-Dichlorophenoxy acetic acid (salts, esters)	8.1					2B							
DDE (p,p'-Dichlorodiphenyl dichloroethylene)		9.7E-6		B2		B2	+		+	+		−	
Diethyl sulfate	2.1				NV	2A	+		+	+	+	+	
3,3'-Dimethoxy benzidine		1.3E-5		B2	NV	2B			+	+	+		
3,3'-Dimethyl benzidine					NV	2B							
1,1-Dimethyl hydrazine	0.2			UR	NV	2B							

Table 18.6   *cont.*

Chemical name	1991 TRI emissions[a] (tons/yr)	IUR[b] per μg/m³	OUR[c] per μg/l	EPA[d] WOE	RfC[e] (mg/m³)	IARC[f] WOE	In vivo S	In vivo G	In vitro M	In vitro C	Bacterial S	Bacterial E	R-D data[h]
3,3′-Dimethoxy benzidine				B2		2B							
3,3-Dichlorobenzidine			1.3E-5	B2		2B							
Dimethyl sulfate	5.1			B2	NV	2A	+	+	+	+	+	+	
2,4-Dinitrotoluene	2.7		1.9E-5	B2	NV	2A	-		-+		+	-	X*
1,4-Dioxane (1,4-Diethyleneoxide)	359.3		3.1E-7	B2		2B			-	-	-		
1,2-Diphenylhydrazine		2.2E-4	2.2E-5	B2	NV						+	-	
Epichlorohydrin (1-chloro-2,3-epoxypropane)	229.6	1.2E-6	2.8E-7	B2	1.0E-3	2A	+-	-	+-	+	+	+	X
Ethyl acrylate	115.9			UR		2B	+		+		-		X
Ethyl carbamate (Urethane)	9.9				NV	2B							
Ethylene dibromide (Dibromoethane)	19.1	2.2E-4	2.5E-3	B2	2.0?E-4	2A	-		+	+	+	+	X*
Ethylene dichloride (1,2-dichloroethane)	1997.7	2.6E-5	2.6E-6	B2		2B	-	-	+	+	+	-	X
Ethylene oxide	896.5					1	+	+	+	+	+	+	X
Ethylene thiourea	0.3					2B	-	-	-	-	+	+	X
Formaldehyde	5109.2	1.3E-5		B1		2A	-/+	-	+	+	+	=	X*
Hexachlorobenzene	0.4	4.6E-4	4.6E-5	B2	NV	2B	-		-		-		
Hexamethylene-1,6-diisocyanate					1.0E-5	2B							
Hexamethylphosphoramide					7.0E-6	2B							X
Hydrazine	14.2	4.9E-3	8.5E-5	B2		2B	-		-		+	+	
Lead compounds	703.8			B2		2B	+/-		-/+		-	-	

Genetic toxicity data[9] — In vivo (S, G); In vitro (M, C); Bacterial (S, E)

Table 18.6 *cont.*

| Chemical name | 1991 TRI emissions[a] (tons/yr) | IUR[b] per µg/m³ | OUR[c] per µg/L | EPA[d] WOE | RfC[e] (mg/m³) | IARC[f] WOE | Genetic toxicity data[g] | | | | | | R-D data[h] |
| | | | | | | | In vivo | | In vitro | | Bacterial | | |
							S	G	M	C	S	E	
Lindane (gamma-hexachlorocyclo-hexane)	0.3				NV	2B							X*
Methylene chloride (Dichloromethane)	39669.2	4.7E-7	2.1E-7	B2	UR	2B	−		−	+		+	X
4,4'-Methylene bis(2-chloroaniline)	0.7				NV	2A							
Methylene chloride (dichloromethane)	39669.2	4.7E-7	2.1E-7	B2	UR	2B	−		−	+	+		X
4,4'-Methylenedianiline	6.6				UR	2B					+		
Nickel Compounds (subsulfide)	121.6	1.4E-2	1.4E-3	A	UR	1	−				−		X
Nitrobenzene	26.3			B2		2A			+	+	+	+	
2-Nitropropane	52.9				2.0E-2	2B							
N-Nitrosodimethylamine				B2		2A		−					
N-Nitrosomorpholine						2B							
N-Nitroso-N-methylurea					NV	2B							
Polychlorinated biphenyls (PCBs)				2B		2A							
1,3-Propane sultone						2B							
Propylene oxide	533.3	3.7E-6	6.8E-6	B2	3.0E-2	2A	+	−	+	+	+	+	
1,2-Propylenimine (2-Methyl aziridine)	0.2				NV	2B							
Radionuclides (including radon)				A		1							
Selenium Compounds	18.5			UR	1.0E+0	2B	+−		+		+		X*
Styrene	14238.2			UR	1.0E+0	2B	+−			+	+		X*

Table 18.6 *cont.*

Chemical name	1991 TRI emissions[a] (tons/yr)	IUR[b] per μg/m³	OUR[c] per μg/l	EPA[d] WOE	RfC[e] (mg/m³)	IARC[f] WOE	Genetic toxicity data[g]						R-D data[h]
							In vivo		In vitro		Bacterial		
							S	G	M	C	S	E	
Styrene oxide	0.8					2A	−		+		−	−	X
2,3,7,8-Tetrachlorodibenzo-p-dioxin						2B		+			−	−	X*
Tetrachloroethylene (perchloroethylene)	8343.7					2B		−	+		−	−	X
2,4-Toluene diamine	1.9				NV	2B							
2,4-Toluene	661.9				V	2B					−		
Toxaphene (chlorinated camphene)				B2	UR	2B	+	−	+−	+−	+	+	X
2,4,6-Trichlorophenol		3.1E-6	3.1E-7	B2	NV	2B			+	−			
Vinyl bromide (bromoethen)	1.8	8.4E-5	5.4E-5	A		2A		−			+		X*
Vinyl chloride	523.7	8.4E-5	5.4E-5	A		1	+		+		+	+	X

[a] TRI = 1991 Toxic Release Inventory data in tons/year. [b] IUR = Inhalation unit risk estimate per μg/m³. Source is the EPA Integrated Risk Information System (IRIS) database. [c] OUR = Oral unit risk per μg/L. Source is the EPA IRIS database. [d] EPA/WOE = EPA Weight-of-Evidence Cancer classification. Source is the EPA IRIS. [e] RfC Workgroup; V= verified, on IRIS = concentration given in mg/m³; NV = not verified; UR = under review. [f] ARC/WOE = International Agency for Research on Cancer Classification. [g] Genetic Toxicity Data; mammalian in vivo, S = somatic, G = germ cell; mammalian in vitro, M = mutation, c = chromosome aberration; bacterial, S = Salmonella type, E = *Escherichia coli* (EPA Genetic Activity Profile database provided by Dr. Michael Waters, EPA, current as of 1992). [h] Reproductive-Developmental Toxicity Data provided by Dr. John Vandenburg, EPA; X = data available, X* = some human data available. Adapted from National Research Council (NRC), 1994, Table 6, pp. 334–343 with permission.

Seventy-four of the chemicals out of a total of 189 listed as Hazardous Air Pollutants in the Clean Air Act Amendments of 1990 are shown in Table 18.6 (CAA, 1990). They are listed here because they have been classified by the International Agency for Research on Cancer or the US EPA as human carcinogens (13), probably human carcinogens (16) or possible human carcinogens (45). Inhalation unit risk estimates have been developed for many of these chemicals and are given in the table along with other key summary toxicity data. The toxic release emissions data shown are for 1991, as shown in the original table. For essentially all of these chemicals, emissions have continued to be reduced during the 1990s. In considering the emissions data, it should be recognized that this is only a crude index of potential exposure. Some of the chemicals are highly reactive and will quickly be degraded during atmospheric transport (Del Pup *et al.*, 1994). Other compounds are more stable and may survive during atmospheric transport and potentially serve as a source of human exposure.

The uncertainties associated with estimating both the exposure of human populations and the potency of potentially carcinogenic agents are substantial. Thus derived estimates of cancer risk for populations such as estimated cancers per year should be viewed with caution. Such estimates are perhaps of greatest value for comparing different risks and making decisions on prioritization of actions to reduce risks.

SUMMARY

The complexity of the respiratory tract makes the elucidation of the carcinogenic processes in this organ a great challenge. A major complication in identifying airborne agents that are carcinogenic is the high background level of lung cancer and other respiratory tract cancers caused by cigarette smoking. This is evident from the controversy surrounding estimates of lung cancers for residential exposure to the well-established carcinogens, radon and its progency (NRC, 1998b). The vast majority of the lung cancer cases estimated to be attributable to radon occur in smokers. There is a high degree of uncertainty associated with the lung cancer risk estimates for non-smokers. These same difficulties extend to assessing the cancer hazards of other air pollutants based on epidemiological observations.

Looking to the future, it is anticipated that identifying air pollutants that are carcinogenic, individually or as a class, will continue to be difficult. The likelihood of identifying additional carcinogens using epidemiological approaches will be reduced as increased attention is given to reducing human exposure to suspected carcinogens.

For new materials proposed for introduction to commence, increased attention must be given to proactive approaches to identifying potential carcinogens through laboratory animal and *in vitro* assays and structure–activity evaluations. The identification of such compounds early in development provides the opportunity to select alternative compounds that do not pose a carcinogenic hazard or to design engineering controls to limit human exposures.

REFERENCES

Aksoy M, Erdem S and Dincol G (1974) Leukemia in shoe-workers exposed chronically to benzene. *Blood* **44**: 837.

Albert RE (1994) Carcinogen risk assessment in the U.S. EPA. *Crit Rev Toxicol* **24**: 70–85.

Ames BM, Durston WE, Yamasaki E and Lee FD (1973) Carcinogens are mutagens: a simple test system combining liver homogenates for activation in bacteria for detection. *Proc Natl Acad Sci USA* **70**: 2281.

Ames BN, McCann J and Yamasaki E (1975) Methods for detecting carcinogens and mutagens with *Salmonella*/mammalian microsome mutagenicity test. *Mutat Res* **31**: 347–364.

Biesalski HK, Bueno de Mesquita B, Chesson A *et al.* (1997) European consensus statement on lung cancer: risk factors and prevention. *Eur J Cancer Prev* **6**: 316–322.

Boyd JA and Barrett JC (1990) Genetic and cellular basis of multistep carcinogenesis. *Pharmacol Ther* **46**: 469–486.

Butterworth BE, Conolly RB and Morgan KT (1995) A strategy for establishing mode of action of chemical carcinogens as a guide for approaches to risk assessments. *Cancer Lett* **93**: 129–146.

Carbone DP (1997) The biology of lung cancer. *Semin Oncol* **24**(4): 388–401.

Catalano PJ, Chang L-YL, Harkema JR *et al.* (1995) *Consequences of Prolonged Inhalation of Ozone on F344/N Rats: Collaborative Studies.* Part XI: Integrative Summary. Research Report Number 65. Cambridge, MA: Health Effects Institute.

Cheng Y-S and Moss OR (1995) Inhalation exposure systems. In: McClellan RO and Henderson RF (eds) *Concepts in Inhalation Toxicology*, 2nd edn. Washington, DC: Taylor & Francis, pp. 25–66.

CAA (1970) Clean Air Act Amendments of 1970, Public Law no. 91-604, 84 STAT, 1676.

CAA (1977) Clean Air Act Amendments of 1977, Public Law no. 95-95, 1977.

CAA (1990) Clean Air Amendments of 1990, Public Law no. 101-549, 104 STAT, 2399.

Conolly RB, Andjelkovich DA, Casanova M *et al.* (1995) Multidisciplinary, iterative examination of the mechanism of formaldehyde carcinogenicity: the basis for better risk assessment. *CIIT Activities* **15**(12): 1–11.

Del Pup J, Kmiecik J, Smith S and Reitman F (1994) Improvement in human health risk assessment utilizing site- and chemical-specific information: a case study. In: National Research Council, *Science and Judgment in Risk Assessment.* Washington, DC: National Academy Press, pp. 479–502.

Devereux TR, Taylor JA and Barrett JC (1996) Molecular mechanisms of lung cancer. Interaction of environmental and genetic factors. Giles F. Filley Lecture. *Chest* **109**(Suppl. 3): 14S–19S.

Driscoll KE (1996) Role of inflammation in the development of rat lung tumors in response to chronic particle exposure. *Inhal Toxicol* **8**: 139–153.

Driscoll KE, Carter JM, Howard BW *et al.* (1996) Pulmonary inflammatory, chemokine, and muta-genic responses in rats after subchronic inhalation of carbon black. *Toxicol Appl Pharmacol* **136**: 372–380.

Enslein K, Gombar VK and Blake BW (1994) Use of SAR in computer-assisted prediction of carcino-genicity and mutagenicity of chemicals by the TOPKAT program. *Mutat Res* **305**: 47–61.

Gardner DE and Kennedy GL (1993) Methodologies and technology for animal inhalation toxicology studies. In: Gardner DE, Crapo JD and McClellan RO (eds) *Toxicology of the Lung*, 2nd edn. New York: Raven Press, pp. 1–30.

Griffin WA (1994) Toward an improved information resource for risk assessment: EPA's Integrated Risk Information System (IRIS). *CIIT Activities* **14**(10): 1–7.

Hahn FF (1993) Chronic inhalation bioassays for respiratory tract carcinogenesis. In: Gardner DE, Crapo JD and McClellan RO (eds) *Toxicology of the Lung*, 2nd edn. New York: Raven Press, pp. 435–459.

Harris CC (1991) Chemical and physical carcinogenesis: advances and perspectives for the 1990s. *Cancer Res* **51**(Suppl.): 5023s–5044s.

Heck H'd'A, Casanova M and Starr TB (1990) Formaldehyde toxicity – a new understanding. *Crit Rev Toxicol* **20**: 397–426.

Henderson RF, Pickrell JA, Jones RK *et al.* (1988) Response of rodents to inhaled diluted diesel exhaust: biochemical and cytological changes in bronchoalveolar lavage fluid and in lung tissue. *Fundam Appl Toxicol* **11**: 546–567.

IARC (1980) IARC Monographs on the Evaluation of the Carcinogenic Risk of Chemicals to Humans, vol. 23: *Some Metals and Metallic Compounds.* Lyon, France: International Agency for Research on Cancer.

IARC (1987) IARC Monographs on the Evaluation of Carcinogenic Risks to Humans. Overall Evaluations of Carcinogenicity: An Updating of IARC Monographs vols 1 to 42, Suppl. 7. Lyon, France: International Agency for Research on Cancer.

IARC (1996) IARC Monographs on the Evaluation of Carcinogenic Risks to Humans, vol. 65: *Printing Processes and Printing Inks, Carbon Black and Some Nitro Compounds.* Lyon, France: International Agency for Research on Cancer.

IARC (1997) IARC Monographs on the Evaluation of Carcinogenic Risks to Humans, vol. 69: *Polychlorinated Dibenzo-*para*dioxins and Polychlorinated Dibenzofurans.* Lyon, France: International Agency for Research on Cancer.

ILSI (1997) *An Evaluation of EPA's Proposed Guidelines for Carcinogen Risk Assessment Using Chloroform and Dichloroacetate as Case Studies: Report of an Expert Panel.* Washington, DC: International Life Sciences Institute.

Landis SH, Murray T, Boldin S and Wingo PA (1998) Cancer statistics, 1998. *CA Cancer J Clin* **48**: 6–29.

Lewis TR, Morrow PE, McClellan RO *et al.* (1989) Establishing aerosol exposure concentrations for inhalation toxicity studies. *Toxicol Appl Pharmacol* **99**: 377–383.

Lubin JH, Boice JD Jr, Edling C *et al.* (1994) *Lung Cancer and Radon: A Joint Analysis of 11 Underground Miner Studies.* Report no. 94-3644. Bethesda, MD: US National Institutes of Health.

Mauderly JL (1997) Relevance of particle-induced rat lung tumors for assessing lung carcinogenic hazard and human lung cancer risk. *Environ Health Perspect* **105**: 1337–1346.

Mauderly JL, Benson JM, Bice DE *et al.* (1983) Life-span study of rodents inhaling diesel exhaust: results through 30 months. In: *Inhalation Toxicology Research Institute Annual Report,* LMF-107. Springfield, VA: National Technical Information Service, pp. 305–316.

Mauderly JL, Jones RK, Griffith WC *et al.* (1987) Diesel exhaust is a pulmonary carcinogen in rats exposed chronically by inhalation. *Fundam Appl Toxicol* **9**: 208–221.

Mauderly JL, Gillett NA and Henderson RF (1988) Relationship of lung structural and functional changes to accumulation of diesel exhaust particles. *Ann Occup Hyg* **32**: 659–668.

Mauderly JL, Banas DA, Griffith WC *et al.* (1996) Diesel exhaust is not a pulmonary carcinogen in CD-1 mice exposed under conditions carcinogenic to F344 rats. *Fundam Appl Toxicol* **30**: 233–242.

McClellan RO (1994) A Commentary on the NRC Report *Science and Judgment in Risk Assessment. Regul Toxicol Pharmacol* **20**: S142–S168.

McClellan RO (1996) Lung cancer in rats from prolonged exposure to high concentrations of carbonaceous particles: implications for human risk assessment. *Inhal Toxicol* **8**(Suppl.): 193–226.

McClellan RO (1997) Use of mechanistic data in assessing human risks from exposure to particles. *Environ Health Perspect* **105**(Suppl. 5): 1363–1372.

McClellan RO, Medinsky MA and Heck Hd'A (1993) Developing risk estimates for airborne materials. In: Garner DA, Crapo JD and McClellan RO (eds) *Toxicology of the Lung.* New York: Raven Press, pp. 603–651.

McConnell EE (1996) Maximum tolerated dose in particulate inhalation studies: a pathologist's point of view. *Inhal Toxicol* **8**(Suppl.): 111–123.

Medinsky MA and Klaassen CD (1996) Toxicokinetics. In: Klaassen CD (ed.) *Casarett and Doull's Toxicology: The Basic Science of Poisons,* 5th edn. New York, NY: McGraw-Hill, pp. 187–198.

Monticello TM, Morgan KT, Everitt JI and Popp JA (1989) Effects of formaldehyde gas on the respiratory tract of rhesus monkeys. Pathology and cell proliferation. *Am J Pathol* **134**: 515–527.

Morrow PE, Haseman JK, Hobbs CH *et al.* (1996) The maximum tolerated dose for inhalation bioassays: toxicity vs overload. *Fundam Appl Toxicol* **29**: 155–167.

Moss OR and Cheng Y-S (1995) Generation and characterization of test atmospheres: particles and droplets. In: McClellan RO and Henderson RF (eds) *Concepts in Inhalation Toxicology,* 2nd edn. Washington, DC: Taylor & Francis, pp. 91–126.

NCHS (National Center for Health Statistics) (1997) *Vital Statistics of the United States, 1994.* Washington, DC: Public Health Services.

NRC (1983) *Risk Assessment in the Federal Government: Managing the Process.* Washington, DC: National Academy Press.

NRC (1989) *Biologic Marker: Pulmonary Toxicology.* Washington, DC: National Academy Press.

NRC (1994) *Science and Judgment in Risk Assessment.* Washington, DC: National Academy Press.

NRC (1998a) *Research Priorities for Airborne Particulate Matter.* I. Immediate Priorities and a Long-Range Research Portfolio. Washington, DC: National Academy Press.

NRC (1998b) *Health Effects of Exposure to Radon,* BEIR VI. Washington, DC: National Academy Press.

National Toxicology Program (1995) *Technical Report on the Toxicology and Carcinogenesis Studies of Ozone and Ozone/NNK in Fischer-344/N Rats and B6C3F1 Mice.* NTP Technical Report no. 400. NIH Publication no. 93-3371. Research Triangle Park, NC: US Department of Health and Human Services, National Toxicology Program Information Office.

National Toxicology Program, US Department of Health and Human Services (1998) Eighth Edition of Biennial Report on Carcinogens. Rockville, MD: Technical Resources, Inc.

Nikula KJ, Snipes MB, Barr EB *et al.* (1995) Comparative pulmonary toxicities and carcinogenicities of chronically inhaled diesel exhaust and carbon black in F344 rats. *Fundam Appl Toxicol* **25**: 80–94.

Nikula KJ, Griffith WC, Avila KJ and Mauderly JL (1997) Lung tissue responses and sites of particle retention differ between rats and cynomolgus monkeys exposed chronically to diesel exhaust and coal dust. *Fundam Appl Toxicol* **37**: 37–53.

Oberdörster G (1996) Significance of particle parameters in the evaluation of exposure–dose–response relationships of inhaled particles. *Inhal Toxicol* **8**(Suppl.): 73–89.

Oberdörster G (1997) Pulmonary carcinogenicity of inhaled particles and the maximum tolerated dose. *Environ Health Perspect* **105**(5) (Suppl.): 1347–1355.

Phalen RF (ed.) (1997) *Methods in Toxicology.* Boca Raton, FL: CRC Press.

Pitot HC (1986) *Fundamentals of Oncology,* 3rd edn. New York: Marcel Dekker.

Pitot HC (1993) The molecular biology of carcinogenesis. *Cancer* **72**: 962–970.

Pitot HC and Dragan YP (1996) Chemical carcinogenesis. In: Klaassen CD (ed.) *Casarett and Doull's Toxicology: The Basic Science of Poisons.* New York: McGraw-Hill, pp. 201–269.

Recio L, Meyer KG, Pluta LJ *et al.* (1996) Assessment of 1,3-butadiene mutagenicity in the bone marrow of B6C3F1 lacI transgenic mice (Big Blue®): a review of mutational spectrum and lacI mutant frequency after a 5-day 625 ppm 1,3-butadiene exposure. *Environ Mol Mutagen* **28**: 424–429.

Ries LAG, Kosary CL, Hankey BF *et al.* (eds) (1997) *SEER Cancer Statistics Review, 1993–1994: Tables and Graphs.* NIH Publ. 97-2789. Bethesda, MD: National Cancer Institute.

Rosenkranz HS and Klopman G (1994) Structural implications of the ICPEMEC method for quantifying genotoxicity data. *Mutat Res* **305**: 99–116.

Stoker M (1996) Fundamentals of cancer cell biology. *Adv Cancer Res* **70**: 1–19.

Swenberg JA (1995) Bioassay design and MTD setting: old methods and new approaches. *Regul Toxicol Pharmacol* **21**(1): 44–51.

Tennant RW, Spalding J and French JE (1996) Evaluation of transgenic mouse bioassays for identifying carcinogens and noncarcinogens. *Mutat Res* **365**: 199–127.

Toxic Substances Control Act (1976) Public Law 94-469.

US EPA (1986a) *Air Quality Criteria Document for Lead,* vols III, IV. Prepared by the Office of Health and Environmental Assessment, Environmental Criteria and Assessment Office, Research Triangle Park, NC, for the Office of Air Quality Planning and Standards. EPA-600/8-83/028dF.

US EPA (1986b) Guidelines for Carcinogen Risk Assessment. *Federal Register* **51**: 33992–34003.

US EPA (1989) Evaluation of the Potential Carcinogenicity of Lead and Lead Compounds: In Support of Reportable Quantity Adjustments Pursuant to CERCLA Section 102. Prepared by the Office of Health and Environmental Assessment, Washington, DC. EPA/600/8-89-045A.

US EPA (1992) Guidelines for Exposure Assessment. *Federal Register* **57**: 22888–22938.

US EPA (1996a) *Health Effects Test Guidelines.* OPPTS870.5300, Detection of gene mutations in somatic cells in culture. EPA712-C-96-221.

US EPA (1996b) *Health Effects Test Guidelines.* OPPTS 870.4200, Carcinogenicity. EPA 712-C-96-211, June 1996, Public Draft.

US EPA (1996c) *Air Quality Criteria for Particulate Matter.* Washington, DC: US Environmental Protection Agency.

US EPA (1996d) *Review of the National Ambient Air Quality Standards for Particulate Matter: Policy Assessment of Scientific and Technical Information.* Office of Air Pollution Quality Standards. Report no. EPA/452/R-96/013. Washington, DC: US Environmental Protection Agency.

US EPA (1996e) Proposed Guidelines for Carcinogen Risk Assessment; Notice. *Federal Register* **61**(79): 17960–18011.

US EPA (1997) National Ambient Air Quality Standards for Particulate Matter; Final Rule. *Federal Register* **62**(138): 38652–38752.

US EPA (1998) National Primary Drinking Water Regulations: Disinfectants and Disinfection Byproducts Notice of Data Availability Proposed Rule. *Federal Register* **63**: 15673–15692.

Valentine JL, Lee SS-T, Seaton MJ *et al.* (1996) Reduction of benzene metabolism and toxicity in mice that lack CYP2E1 expression. *Toxicol Appl Pharmacol* **141**: 205–213.

Vogelstein B and Kinzler KW (1993) The multistep nature of cancer. *Trends Genet* **9**: 138–141.

Wilbourn J, Haroun L, Heseltine E *et al.* (1986) Response of experimental animals to human carcinogens: an analysis based upon the IARC Monograph Programme. *Carcinogenesis* **7**: 1853–1863.

Wolff RK, Henderson RF, Snipes MB *et al.* (1987) Alterations in particle accumulation and clearance in lungs of rats chronically exposed to diesel exhaust. *Fundam Appl Toxicol* **9**: 154–166.

Wong BA (1995) Generation and characterization of gases and vapors. In: McClellan RO and Henderson RF (eds) *Concepts in Inhalation Toxicology*, 2nd edn. Washington, DC: Taylor & Francis, pp. 67–90.

WHO (1996) *World Health Statistics Annual, 1995.* Geneva: World Health Organization.

GENERAL METHODOLOGICAL AGENTS OF AIR POLLUTANT HEALTH EFFECTS

19

Biomarkers of Exposure

MICHAEL C. MADDEN and JANE E. GALLAGHER

US Environmental Protection Agency, National Health and
Environmental Effects Research Laboratory, Human Studies
Division, Research Triangle Park, NC, USA

The research described in this paper has been reviewed by the National Health and Environmental
Effects Research Laboratory, US Environmental Protection Agency and approved for publication.
Approval does not signify that the contents necessarily reflect the views and policies of the Agency nor
does mention of trade names or commercial products constitute endorsement or recommendation for
use.

INTRODUCTION AND DEFINITIONS

A biological marker, commonly termed 'biomarker', is an indicator of a perturbation of
homeostasis within a whole organism, or a part of an organism (e.g. organ, tissue, cell,
organelle), that will ultimately lead to a disease in response to xenobiotic exposure
(adapted from the National Research Council (1989) and Ward and Henderson (1996)
definitions). A biomarker generally can be classified as a marker of exposure, effect, or susceptibility in humans (Fig. 19.1), although each class is not totally exclusive of each
other. A biomarker of exposure can be defined as the degree of individual exposure to
xenobiotics in a given environment by providing a measurement of internal dose. A relatively narrow definition of an exposure biomarker would be the level of the parent
compound or secondary products such as metabolites and/or reaction products in biological fluid, cells, or subcellular component, e.g. DNA and protein adducts (Fig. 19.2)
(Ward and Henderson, 1996). A biologically effective dose provides a link between the
exposure and biological responses (Kensler and Groopman, 1996). A broader definition
for a biomarker of exposure would include consideration of biological responses and the
magnitude of the responses induced by xenobiotics. This latter definition would include
a wide spectrum of biological responses ranging from somewhat limited whole organ

physiological responses (e.g. lung function decrements) to seemingly unlimited changes in the production of cellular mediators and gene activation. Some biological responses to pollutants are very common such as the lung neutrophilia induced by acute exposure to air pollutants (e.g. ozone, nitrogen dioxide, certain metal-containing particles, etc). In this chapter, we will use the stricter definition and not consider biological responses as exposure biomarkers, in order to maintain some pollutant-specificity in terms of studies evaluating just the parent compound or derivatives of the parent compound.

Ideally, the purpose of measuring biomarkers of xenobiotic exposure is ultimately to provide a better linkage between the exposure with biological effects and/or clinical disease. Thus biomarkers represent a spectrum between health and disease. Biomarkers of exposure may become critical in the prevention or treatment of a disease. Furthermore, exposure biomarkers may provide information regarding the mechanism(s) leading to a biological effect, and to determine if similar processes are occurring across species (Vine, 1996).

An important attribute of exposure biomarkers is that it takes into account individual determinants of the dosimetry and metabolism of the absorbed pollutant. The factors which affect these rest mainly with individual susceptibility and can include avoidance

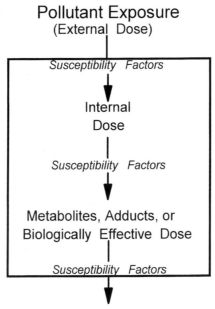

Pollutant Exposure
(External Dose)

Susceptibility Factors

Internal
Dose

Susceptibility Factors

Metabolites, Adducts, or
Biologically Effective Dose

Susceptibility Factors

Early Biological Effects; Altered Structure
and Function; Ultimately Clinical Disease

Fig. 19.1 Biomarkers of exposure in the continuum between pollutant exposure and biological effects. Exposure biomarkers (in box) are intermediates between the occurrence of the exposure and later biological effects produced as a consequence of the exposure. Under ideal circumstances the exposure biomarkers are a good dosimeter for the pollutant deposited in the lung, either by reflecting the parent compound or a derivative of the parent compound. At each step in the biomarker spectrum, susceptibility factors in individuals can affect the internal dose and biological responses.

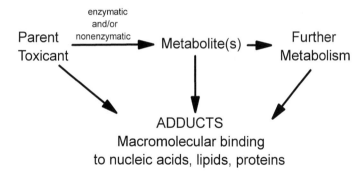

Fig. 19.2   Pathways of the transformation of a parent compound toxicant to other products that can be utilized as exposure biomarkers. Transformation of the parent pollutant can be through direct adduction to macromolecules or by enzymatic catabolism, or by non-enzymatic conversion to other products. Similar conversions can occur to the transformed products.

behavior, personal protective equipment, physiological and psychological adaptations, physical activity, metabolic factors and organ transport rates, among others (Vine, 1996). Thus measurement of the external exposure may not adequately correlate with the target tissue dose or internal dose due to these factors. Exposure biomarkers can be used to identify subjects who misclassify (accidentally or intentionally) their exposure history (Vine, 1996). Recently, much attention has been focused on the role of genetic polymorphisms as susceptibility factors in humans in modulating the formation of active and reactive metabolites of exposure and subsequent induction of biological effects (Perera, 1997). Genes that encode cytochrome P-450 1A1 (a phase I 'activating' enzyme) (Gonzalez and Gelboin, 1993) and glutathione S-transferase GSTM1 (a phase II 'detoxifying' enzyme) (Rebbeck, 1997) are examples of genetic polymorphisms that potentially influence the toxicity and/or carcinogenicity of particular polycyclic aromatic hydrocarbons (PAHs).

Exposure biomarkers have been recognized for centuries in forensic toxicology. Serullas in 1821 detected an arsenic derivative in the urine and stomach of poisoning victims (Polkis, 1996). Examination of mummified individuals by forensic specialists can shed light on the occupational history of an individual as well as air quality centuries or millennia ago. Antracose particles found in the lungs of Aleutian mummies (Zimmerman *et al.*, 1971) indicated particle exposures from cooking or pollution, and silicotic deposits in lungs from an Egyptian mummy suggested exposure to sandstorms (Cockburn *et al.*, 1975). Thus, particle concentration and composition can help develop an understanding of the activities of individuals within a society.

Epidemiologists continue to use biomarkers of exposure to classify the exposure levels of study populations. Industrial hygienists use exposure biomarkers to determine if safety levels for specific toxicants have exceeded the desired or permissible limits in the exposed workforce. Clinicians utilize biomarkers in order to support diagnoses and to recommend appropriate medical tests. More recently, biomarkers have been employed in identifying individuals at risk for developing a particular disease. At the time of writing, over 66% (3743 of 5640) of articles retrieved from a MEDLINE search of 'lung biomarkers' were linked with the phrase 'cancer', indicating a major focus on the respiratory tract as a route of exposure to xenobiotics and development of lung cancer.

Biomarkers of exposure need to be validated. That is, an association between the external dose and the internal dose should be established. Specificity and sensitivity are critical factors for the validation of biomarkers (National Research Council, 1989). Several different parent compounds that produce the same metabolite or biological reaction products would indicate low specificity. For example, to monitor exposure to environmental tobacco smoke (ETS), benzene and PAHs would not be very useful since there are many other sources for these compounds such as gasoline and dietary sources, among others (Scherer and Richter, 1997); thus the probability of obtaining a false positive is high. Conversely, if the specificity is high, such as may be seen by production of an ozonide from ozone exposure or cotinine from ETS, the probability of obtaining a false positive is relatively low. Sensitivity of the biomarker is the ability to detect a low level exposure to xenobiotics impacting the probability of obtaining a false negative. Sensitivity is enhanced by minimum levels of interfering substances (Ward and Henderson, 1996). Because most environmental exposures occur at low concentrations, high sensitivity for exposure biomarkers is generally required for biomarkers to be useful.

In order for a biomarker to be ideal for dosimetry studies it should be pollutant-specific, stable under biologic conditions, detectable at low concentrations, and provide a record of exposure over a significant length of time. The techniques used to analyze the biomarker should be inexpensive and provide a quick analysis, and, most importantly, the biomarker of interest should be retrievable by a non-invasive technique; this is especially important in human studies (Vine, 1996). Along with specificity and sensitivity, it is important that interindividual and experimental variability be established in order to predict the number of samples required to measure changes in the biomarker levels (statistically referred to as power). If the variability is low (relative to the mean) in the control population, then significant changes in the biomarker levels will be easier to detect (Ward and Henderson, 1996).

STRATEGIES FOR SAMPLING EXPOSURE BIOMARKERS

The timing for sampling exposure biomarkers in different organ systems is crucial to detecting the markers accurately. The human body contains compartments with differing rates of achieving maximum biomarker concentrations and differential rates of elimination (Henderson and Belinsky, 1993). Substances which originally enter the body via the respiratory system could reach extrapulmonary organ systems with different toxicokinetics. The ability to sample surrogate cells by non-invasive techniques is critical. For most human studies each toxicant reaches a different maximal level in each compartment. For instance, cotinine levels in urine are usually 10-fold higher than in blood (Vine *et al.*, 1993). Biomarkers of exposure can be classified into four groups in terms of persistence in the body (WHO Regional Office for Europe, 1995):

1 $t_{1/2}$ *of less than 12 h*: Generally these exposure biomarkers are found in the airway lumen, or re-equilibrate from blood fluid after the initial deposition. Many of these markers are usually retrieved from the lung and nose by lavage or breath collection.

2 $t_{1/2}$ *of 12–100 h*: Many markers in this category are ones that initially equilibrate into fat and then depurate, e.g. many solvents and also the PAH pyrene. The parent compound or metabolite will generally appear in the urine.

3 $t_{1/2}$ *of 100 h to 6 months*: Mainly DNA, hemoglobin, albumin and other protein adducts are found with these half-lives. The persistence of adducts in this category can vary widely; the $t_{1/2}$ of human serum albumin is less than one month while erythrocyte hemoglobin is approximately 4 months. Residues in the shaft of hair (e.g. nicotine from smoking) can remain indefinitely as a potential biomarker (Haley and Hoffman, 1985). However, most hair is either shed or cut, thus limiting the time frame for use of biomarkers in hair.

4 $t_{1/2}$ *over 6 months*: Following lung deposition, heavy metals such as uranium and mercury will be transported to other organ systems, with a portion of the metal dose eventually having a slow urinary excretion. The amount of the urinary metal can in some cases indicate the dose delivered to the lung such as for occu-pational uranium aerosol exposure (Schieferdecker *et al.*, 1985).

There is an interrelationship between the pollutant being examined, the timing of sampling for the biomarker and the technique that can be utilized for sampling (Fig. 19.3). The quantity of biological cells and fluid can be limited, which in turn would affect the number of times the fluids and/or cells can be retrieved. Additionally, sample storage and stability of the marker under investigation also must be taken into account.

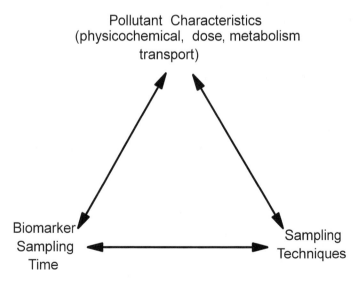

Fig. 19.3 Interrelationships among the characteristics of the pollutant of concern, the sampling technique that can be utilized, and the biomarker sampling time-points in measuring an exposure biomarker. Each of these factors will influence the other, dictated by the characteristics of the pollutant (including dose to the lung, volatility of the pol-lutant and/or metabolites, metabolism, reactivity, and transport within/outside the lung, etc.), the sampling techniques available to the investigator (budget, equipment, invasiveness allowable, etc.), and the sampling time (dependent on the subject compliance, cells and fluid to be obtained, etc.).

ILLUSTRATIVE EXAMPLES OF BIOMARKERS

Rather than attempt a comprehensive report on the array of exposure biomarkers used to measure lung exposure and/or toxicity, several illustrative examples will be presented below. These examples provide the reader with an overview of the classes of agents for which exposure biomarkers have been studied, including the types of markers examined and the timing of biological sampling. It should be stressed that generally (with the exception of high levels of occupational exposures) levels of the biomarker of interest are derived from many sources, e.g. via diet.

Benzene

The carcinogen benzene is an occupational exposure hazard, an atmospheric pollutant derived primarily from gasoline release, and also is a constituent of cigarette smoke constituent. Compared with most air pollutants, benzene biomarkers and related exposure markers have been relatively well studied (Bechtold and Henderson, 1993). The uptake and excretion of benzene itself has been examined using analyses of expired breath. These studies have shown that benzene has an initial rapid excretion followed by a slower breath excretion (Thomas *et al.*, 1991). In terms of monitoring blood and urine components, much effort has gone into determining the most specific metabolites of benzene to measure. Several hydroxy and polyhydroxy benzene metabolites can be derived from other sources (such as phenol) and lack good specificity for benzene (Ong and Lee, 1994). *Trans-trans* muconic acid (MA) is currently considered a good exposure biomarker for benzene in urine. More than 72% of children sampled in Baltimore, USA, had detectable levels of muconic acid (Weaver *et al.*, 1996), suggesting that benzene exposures are ubiquitous. In individuals occupationally exposed to benzene, urinary MA was shown to be elevated above levels of control subjects (Bechtold and Henderson, 1993) and correlated well with benzene exposure concentrations (Kivisto *et al.*, 1997) in occupational exposures. The monitoring of a benzene-specific metabolite, S-phenylcysteine-albumin adduct, was correlated with repeated exposures while the same metabolite attached to blood hemoglobin was not (Bechtold and Henderson, 1993). It is important to note that cigarette smoke contains benzene. MA levels in smokers were about three times higher than in non-smokers and a significant correlation between concentration of MA and levels of cotinine in smokers was observed. Interestingly, pregnant smoking women had higher levels than smoking non-pregnant women. These studies suggest that urinary MA can be utilized as a biochemical marker to quantitate benzene arising from exposure to cigarette smoking. Smoking could confound exposure analyses of individuals occupationally exposed to benzene (Melikan *et al.*, 1993, 1994.)

Ozone

Although ozone (O_3) has not been definitively linked to a clinical disease, there is sufficiently strong concern that chronic exposure may induce or contribute to increased lung diseases such as cancer, fibrosis, and increased rates of lung function decrements to warrant biomarker research. Additionally, O_3 exposure does induce a number of transient biological responses (e.g. lung neutrophilia, altered lung function capacity, host defense alterations) (US Environmental Protection Agency, 1996) that need to be linked to internal dose. A large database of biomarkers exists derived from many controlled chamber studies of human and rodent exposures as well as epidemiologic studies, but primarily related to biological effects; however, a number of exposure biomarker studies have been conducted. Owing to its highly oxidant nature, very little pure O_3 is believed to exist in nasal or lung lining fluid after deposition (Pryor, 1992). Ozone has been proposed to have a unique reaction scheme with unsaturated fatty acids and proteins (Santrock et al., 1992; Uppu and Pryor, 1994), leading in part to the formation of ozonides, carbonyls and hydrogen peroxide (H_2O_2) via breakage of carbon double bonds. Free radical processes also are associated with O_3 exposure (Byvoet et al., 1995). Using a strict exposure biomarker definition, some of the peroxidation-derived O_3 biomarkers may not be derived from the original O_3 oxygen atoms, but are formed through propagation of the peroxidation initiated by the O_3 oxygen atoms. Additionally, with in vivo exposures it is not clear which products are derived from the O_3 molecule as opposed to reactive oxygen species generated by neutrophils and/or by possibly activated resident phagocytes in response to the injury. With these caveats, some O_3 reaction products have been found in response to an acute in vivo exposure. For instance, an increase in H_2O_2 has been noted in exhaled breath of humans after 2 h exposure to O_3 with exercise (Madden et al., 1997). A transient increase in some aldehydes in the lavage fluid from rats that were exposed to O_3 has been reported (Pryor et al., 1996). An increase in a 4-hydroxynonenal-protein adduct was observed in murine cells recovered by lung lavage (Kirichenko et al., 1996). Frischer et al. (1997) have reported an increase in ortho-tyrosine, believed to be formed upon O_3-derived reactive oxygen species combining with tyrosine, in nasal lavage fluid of humans on days of 'high' O_3 concentrations in the ambient air. An increase in free radicals in lung lipid extracts of O_3-exposed rodents was detected by spin trapping techniques (Kennedy et al., 1992). An increase in 8-epi-prostaglandin $F_{2\alpha}$ (a lipid product formed through free radical processes) was reported in bronchoalveolar lavage fluid of human volunteers exposed to O_3 (Hazbun et al., 1993).

Polycyclic aromatic hydrocarbons

Proteins (hemoglobin and serum albumin) provide information with respect to distant versus recent exposures. However, the measurement of PAH-derived adducts is perhaps more compelling because they represent an early detectable and critical step in the chemical carcinogenesis process. The measurement of DNA adducts has important applications for assessment of internal dose, as these measurements reflect individual and species differences (including genetic predisposition, DNA repair, and mutations in oncogenes) which affect the relationship between exposure and dose. Thus, the measurement of DNA adducts has important applications for risk assessment (Fennell et al., 1996; La and Swenberg, 1996). More recently, studies have shown that diesel exhaust and associated PAH affect the human allergic response (Diaz-Sanchez et al., 1994, 1996). It is likely that some toxic and non-carcinogenic compounds can have non-genotoxic effects including oxidative damage and adverse health outcomes other than cancer which may be correlated with DNA adduct formation (Poirier and Weston, 1996).

Although it is preferable to measure adducts at steady state level when evaluating internal dose and estimating risk to the population at large, data from human populations are unlikely to represent uniform conditions of exposure even under chronic exposure circumstances. Thus a single DNA adduct measurement from only one time-point should be viewed with caution. Nevertheless, the presence and amount of specific DNA adducts that can be correlated with a chemical exposure are relevant for hazard identification and risk evaluation. DNA adduct analysis may also be one of the best tools available to characterize exposure to complex mixtures in epidemiological studies.

Benzo[a]pyrene (B[a]P) is one of the most widely studied carcinogenic PAH. B[a]P is often used as a surrogate for other PAH (Arlett et al., 1981). However, B[a]P only represents approximately 4% of the total polycyclic organic matter in most combustion-derived particle emissions (Möller et al., 1994). The concentration of B[a]P and other PAHs (as detected in coal tar residues and other PAH-containing mixtures) does not always correlate with the concentration of total or B[a]P-derived DNA adducts (Gallagher et al., 1990; Leadon et al., 1995) and underscores the importance of looking at the relative concentrations contribution of the myriad of individual DNA adducts induced by complex mixture exposures. Interestingly, studies by Li et al. (1996a) suggest that the in vitro induction of B[a]P-diol epoxide-DNA adducts in peripheral lymphocytes may be a susceptibility marker for human lung cancer. In two separate studies, DNA adduct levels were higher in breast cancer cases compared with non-cancer patients (Li et al., 1996b; Perera et al., 1995). Thus, recent interest has been generated in evaluating whether relative DNA adduct levels between similarly exposed individuals may provide important indicators of genetic susceptibility to cancer outcome.

Estimating the human cancer risk resulting from exposure to complex mixtures represents a formidable methodological problem. However, such exposures are thought to account for a large proportion of cancers, in particular because of widespread exposure to such mixtures within populations (Vainio et al., 1990). Biomarkers of exposures to diverse classes of chemicals and complex mixtures have only been recently developed at the DNA level and include diverse compounds such as alkylating agents, PAHs, heterocyclic PAH, nitro-PAH, aromatic amines, dyes, quinones, mycotoxins, chemotherapeutic agents, aminoimidazoazarens (food mutagens) and complex mixtures (cigarette smoke condensate, urban air, roofing, coke oven and foundry emissions) (Beach and Gupta, 1992).

Cigarette and Environmental Tobacco Smoke

Smokers have been extensively studied to validate biomarkers of xenobiotic exposure, especially using surrogate tissue markers (Hemminki, 1997). Urinary cotinine, a metabolite of nicotine, has been used as a classical marker of tobacco smoke exposure. A recent report utilized hair samples to determine cotinine levels in children (Knight et al., 1998). Hair analysis should allow a longer monitoring period than when using urinary products and thus reflect long-term smoking exposure. Cotinine levels in saliva also have been utilized as an exposure biomarker to ETS (Repace et al., 1998).

Cigarette smoke contains many reactive oxygen species (Church and Pryor, 1985), some of which can be measured directly in breath. Acetaldehyde in exhaled breath from smokers was shown to be increased immediately after smoking (McLaughlin et al., 1990).

An increase in breath acetaldehyde was not noted 15 min after smoking cessation, suggesting a very limited time period that this aldehyde can be used as a breath biomarker of cigarette smoke exposure. Shaskan and Dolinsky (1985) reported increased basal acetaldehyde breath concentrations in smokers compared with non-smokers that was enhanced in both groups by alcoholism. In smokers and non-smokers (who smoked one cigarette), reports have shown that breath alkanes (i.e. ethane, pentane, isoprene) derived via lipid peroxidation increased immediately after smoking, but returned to pre-exposure levels within hours (Euler *et al.*, 1996; Habib *et al.*, 1995), suggesting the transient nature of these biomarkers. These studies underscore the need to ascertain and account for what factors (e.g. metabolism, alternate source of toxicant) may affect the measurement of the biomarker of interest.

Total lung DNA 'bulky' adducts (i.e. derived from relatively large xenobiotics such as PAHs) are higher in smokers compared to non-smokers (Lewtas *et al.*, 1993; Phillips, 1996). Similar increases in total adducts also were observed in surrogate tissue of smokers – blood lymphocytes, placenta, and heart (Gallagher *et al.*, 1993; Lewtas *et al.*, 1993) – although the types and amounts of adducts in the lymphocytes can be highly variable (Jahnke *et al.*, 1990). Some studies have not shown increases in total DNA bulky adducts or specific smoke-related adducts in the placental tissue (e.g. Daube *et al.*, 1997). In terms of specific DNA adducts, 7-methylguanine (derived from ETS methylation) is higher in bronchial cells of smokers than in those of non-smokers (Mustonen *et al.*, 1993). These investigators also reported obtaining a good correlation between bronchial and peripheral blood lymphocyte levels suggesting that the extrapulmonary systems may be ideal as a surrogate tissue for estimating lung burdens from chronic smoking. Blood granulocytes from smokers had lower 7-methylguanine levels than lymphocytes, which was attributed to the more transient nature of the granulocytes (Mustonen and Hemminki, 1992). Non-smokers excrete urinary tobacco-specific nitrosamine metabolites from passive exposure to cigarette smoke (Brunnemann *et al.*, 1996). Data from a rodent study examining the kinetics of lung retention and local distribution of pyrene have shown metabolism within the airway epithelium due, in part, to slow clearance from the lung (Gerde *et al.*, 1998).

In addition to gas phase-derived exposure biomarkers, cigarette particle-derived markers also have been examined in a few studies. Churg and colleagues (1992a,b) have shown elevated levels of particles containing calcium, carbon and oxygen (probably calcium carbonate) in smokers' lungs (~23% of all particles) than in non-smokers' lungs (<1%).

Ambient Particulate Matter

Air pollution particles have a complex composition and the mechanism(s) by which these particles induce lung injury are not clear. As such, many hypotheses regarding the causative agent and approaches to monitor specific particulate matter constituents are under development and include biological material, organics, metals and acid content. With respect to studies of particulate-associated organics, most studies focus on the measurement of PAH-derived DNA and protein adducts. Instillation of an extract of air particles collected from an urban site in Italy (Izzotti *et al.*, 1996) increased levels of lung bulky DNA adducts, compared to adduct levels observed in unexposed rodents, or rodents instilled with an extract of rural air particles. Individuals in Italy inhaling urban air with much vehicular traffic have increased peripheral blood-DNA adducts and urinary excretion of a pyrene metabolite (Merlo *et al.*, 1998; Yang *et al.*, 1996).

Few studies have examined elevated human lung metal or total body burden of metals as it relates to linking exposure to air pollution levels. Lead body burden studies related to air pollution are more common. For example, blood lead levels in Uruguayan children were correlated with traffic levels near schools, but not with lead concentrations in drinking water or nearby soil (Schutz *et al.*, 1997). Maternal and cord blood lead levels were higher in Greek residents of urban areas with higher airborne lead levels than in residents from rural areas (Vasilios *et al.*, 1997). These latter two studies suggest that ambient air pollution may be an important source of the lead dose. The area of particulate matter-related biomarkers can be greatly assisted through development of analytical techniques such as scanning electron microscopy coupled with energy dispersive X-ray analysis (SEM/EDX). SEM/EDX can provide information regarding particle size, number, morphology and elemental composition in specific regions of the lung, thereby facilitating potential associations to be made between the particle characteristics collected from point, mobile, and urban air sources and those detected in the human lung.

FUTURE DIRECTIONS

Continued validation of biomarkers of exposure is warranted for studying pollutant-induced effects in the lung. Biomarkers for pollutants such as benzene and O_3 are still in the validation process and studies with these two pollutants that can evaluate with more rigor any association between exposure and observable biological responses are underway. The need to examine the US Environmental Protection Agency's 189 priority 'air toxics' pollutants underscores the need to validate a wide array of biomarkers (Olden and Guthrie, 1996). A major emphasis of the risk assessment/validation process will be pharmacokinetic modeling and extrapolation modeling between human and non-human species. Included in the extrapolation modeling between species would be empirical data derived from exposure studies that are relevant to environmental levels, mathematical modeling for non-human and human exposures, and their validation with human exposures studies from which only limited data can be obtained (Ward and Henderson, 1996). Because human data may be limited due to the inaccessibility of a tissue, studies in non-human models should include surrogate tissue that can also be utilized in human research. Exposure biomarkers also can be utilized to a limited extent in prospective follow-up and nested case–control epidemiological studies to better understand and characterize exposure levels and effect outcomes. Further assessment is needed on the extent to which genetic susceptibility and/or variability in personal exposures reflect interindividual differences in exposure biomarker levels.

In order to accomplish these end-points, validation of exposure biomarkers related to non-carcinogenic end-points needs further development, while those biomarkers related to carcinogenic end-points are being validated further. Human populations need to be identified in which the biomarker studies can be ultimately validated. Finally, ethical issues concerning safeguarding an individual's data related to that person's susceptibility to developing a disease must be resolved, especially when DNA specimens are being stored for current and future genetic testing. Biomarkers of exposure can play an important role in disease prevention, molecular epidemiological studies, clinical intervention trials, and studies aimed at identifying individuals at high risk for developing adverse health outcomes.

ACKNOWLEDGEMENTS

The authors thank Drs Pauline Mendola and Agasanur Prahalad for their careful reviews and helpful discussions of this manuscript.

REFERENCES

Arlett CF, Cole J, Broughton BC *et al.* (1981) Mutagenic effects in human and mouse cells by a nitropyrene. In: Tice RR, Costa DL and Schaich KM (eds) *Genotoxic Effects of Airborne Agents.* New York: Plenum, pp. 397–410.

Beach AC and Gupta RC (1992) Human biomonitoring and the ^{32}P-postlabeling assay. *Carcinogenesis* **13**: 1053–1074.

Bechtold WE and Henderson RF (1993) Biomarkers of human exposure to benzene. *J Toxicol Environ Health* **40**: 377–386.

Brunnemann KD, Prokopczyk B, Djordjevic MV and Hoffmann D (1996) Formation and analysis of tobacco-specific *N*-nitrosamines. *Crit Rev Toxicol* **26**: 121–137.

Byvoet P, Balis JU, Shelley SA *et al.* (1995) Detection of hydroxyl radicals upon interaction of ozone with aqueous media or extracellular surfactant: the role of trace iron. *Arch Biochem Biophys* **310**: 464–469.

Church DF and Pryor WA (1985) Free-radical chemistry of cigarette smoke and its toxicological implications. *Environ Health Perspect* **64**: 111–126.

Churg A and Stevens B (1992a) Calcium-containing particles as a marker of cigarette smoke exposure in autopsy lungs. *Exp Lung Res* **18**: 21–27.

Churg A, Wright JL, Stevens B and Wiggs B (1992b) Mineral particles in the human bronchial mucosa and lung parenchyma. II. Cigarette smokers without emphysema. *Exp Lung Res* **18**: 687–714.

Cockburn A, Barraco RA, Reyman TA and Peck WH (1975) Autopsy of an Egyptian mummy. *Science* **187**: 1155–1160.

Daube H, Scherer G, Riedel K *et al.* (1997) DNA adducts in human placenta in relation to tobacco smoke exposure and plasma antioxidant status. *J Cancer Res Clin Oncol* **123**: 141–151.

Diaz-Sanchez D, Dotson AR, Takenaka H and Saxon A (1994) Diesel exhaust particles induce local IgE production *in vivo* in humans and alter the pattern of IgE mRNA isoforms. *J Clin Invest* **94**: 1417–1425.

Diaz-Sanchez D, Tsien A, Casillas A *et al.* (1996) Enhanced nasal cytokine production in humans following *in vivo* challenge with diesel exhaust particles. *J Allergy Clin Immunol* **98**: 114–123.

Euler DE, Dave SJ and Guo H (1996) Effect of cigarette smoking on pentane excretion in alveolar breath. *Clin Chem* **42**: 303–308.

Fennell TT, Gallagher JE, Gorelick NJ *et al.* (1996) Toxicological significance of DNA adducts: summary of discussions with an expert panel. *Reg Toxicol Pharmacol* **24**: 9–18.

Frischer TH, Pullwitt A, Kühr J *et al.* (1997) Aromatic hydroxylation in nasal lavage fluid following ambient ozone exposure. *Free Radic Biol Med* **22**: 201–207.

Gallagher JE, Jackson MA, George MH and Lewtas J (1990) Dose-related differences in DNA adduct levels in rodent tissues following skin application of complex mixtures from air pollution sources. *Carcinogenesis* **11**: 63–68.

Gallagher J, Mumford J, Li X *et al.* (1993) DNA adduct profiles and levels in placenta, blood and lung in relation to cigarette smoking and smoky coal emissions. In: Phillips DH, Castegnaro M and

Bartsch H (eds) Postlabeling methods for the detection of DNA adducts, IARC Sci. Publ. 124, Lyon, France, pp. 283–292.

Gerde P, Muggenberg BA, Scott GG *et al.* (1998) Local metabolism in lung airways increases the uncertainty of pyrene as a biomarker of polycyclic aromatic hydrocarbon exposure. *Carcinogenesis* **19**: 493–500.

Gonzalez FL and Gelboin HV (1993) Role of human cytochrome P-450s in risk assessment and susceptibility to environmentally based disease. *J Toxicol Environ Health* **40**: 289–308.

Habib MJ, Clements NC and Garewal HS (1995) Cigarette smoking and ethane exhalation in humans. *Am J Respir Crit Care Med* **151**: 1368–1372.

Haley NJ and Hoffman D (1985) Analysis for nicotine and cotinine in hair to determine cigarette smoker status. *Clin Chem* **31**: 1598–1600.

Hazbun ME, Hamilton R, Holian A and Eschenbacher WL (1993) Ozone-induced increases in substance P and 8-*epi*-prostaglandin $F_{2\alpha}$ in the airways of human subjects. *Am J Respir Cell Mol Biol* **9**: 568–572.

Hemminki K (1997) DNA adducts and mutations in occupational and environmental biomonitoring. *Environ Health Perspect* **105**(Suppl. 4): 823–827.

Henderson R and Belinsky SA (1993) Biological markers of respiratory tract exposure. In: Gardner DE, Crapo JD and McClellan RO (eds) *Toxicology of the Lung*, 2nd edn. New York: Raven Press, pp. 253–282.

Izzotto A, Camoirano A, D'Agostini F *et al.* (1996) Biomarker alterations produced in rat lung by intratracheal instillations of air particulate extracts and chemoprevention with oral *N*-acetylcysteine. *Cancer Res* **56**: 1533–1538.

Jahnke GD, Thompson CL, Walker MP *et al.* (1990) Multiple DNA adducts in lymphocytes of smokers and nonsmokers determined by ^{32}P-postlabeling analysis. *Carcinogenesis* **11**: 205–211.

Kennedy CH, Hatch GE, Slade R and Mason RP (1992) Application of the EPR spin-trapping technique to the detection of radicals produced *in vivo* during inhalation exposure of rats to ozone. *Toxicol Appl Pharmacol* **114**: 41–46.

Kensler TW and Groopman JD (1996) Carcinogen-DNA and protein adducts: biomarkers for cohort selection and modifiable endpoints in chemoprevention trials. *J Cell Biochem Suppl* **25**: 85–91.

Kirichenko A, Morandi MT and Holian A (1996) 4-hydroxy-2-nonenal-protein adducts and apoptosis in murine lung cells after acute ozone exposure. *Toxicol Appl Pharmacol* **141**: 416–424.

Kivisto H, Pekari K, Peltonen K *et al.* (1997) Biological monitoring of exposure to benzene in the production of benzene and in a cokery. *Sci Total Environ* **199**: 49–63.

Knight JM, Eliopoulos C, Klein J *et al.* (1998) Pharmacokinetic predisposition to nicotine from environmental tobacco smoke: a risk factor for pediatric asthma. *J Asthma* **35**: 113–117.

La DK and Swenberg JA (1996) DNA adducts: biological markers of exposure and potential applications to risk assessment. *Mutat Res* **365**: 129–146.

Leadon SA, Sumerel J, Minton TA and Tischler A (1995) Coal tar residues produce both DNA adducts and oxidative DNA damage in human mammary epithelial cells. *Carcinogenesis* **16**: 3021–3026.

Lewtas J, Mumford J, Everson R *et al.* (1993) Comparison of DNA adducts from exposure to complex mixtures in various human tissues and experimental systems. *Environ Health Perspect* **99**: 89–97.

Li D, Wang M, Cheng L *et al.* (1996a) *In vitro* induction of benzo[*a*]pyrene diol epoxide-DNA adducts in peripheral lymphocytes as a susceptibility marker for human lung cancer. *Cancer Res* **56**: 3638–3641.

Li D, Wang M, Dhingra K and Hittelman WN (1996b) Aromatic DNA adducts in adjacent tissues of breast cancer patients: clues to breast cancer etiology. *Cancer Res* **56**: 287–293.

Madden MC, Hanley N, Harder S *et al.* (1997) Increased amounts of hydrogen peroxide in the exhaled breath of ozone-exposed human subjects. *Inhal Toxicol* **9**: 317–330.

Melikian AA, Prahalad AK and Hoffmann D (1993) Urinary *trans-trans*-muconic acid as an indicator of exposure to benzene in cigarette smokers. *Cancer Epidemiol Biomarkers Prev* **2**: 47–51.

Melikian AA, Prahahlad AK and Secker-Walker RH (1994) Comparison of the levels of the urinary benzene metabolite *trans-trans*-muconic acid in smokers and non-smokers, and the effects of pregnancy. *Cancer Epidemiol Biomarkers Prev* **3**: 239–244.

Merlo F, Andreassen A, Weston A *et al.* (1998) Urinary excretion of 1-hydroxypyrene as a marker for exposure to urban air levels of polycyclic aromatic hydrocarbons. *Cancer Epidemiol Biomarkers Prev* **7**: 147–155.

McLaughlin SD, Scott BK and Peterson CM (1990) The effect of cigarette smoking on breath and whole blood-associated acetaldehyde. *Alcohol* 7: 285–287.

Möller L, Schuetzle D and Autrup H (1994) Future research needs associated with the assessment of potential human health risks from exposure to toxic ambient air pollutants. *Environ Health Perspect* 102(Suppl. 4): 193–210.

Mustonen R and Hemminki K (1992) 7-methylguanine levels in DNA of smokers' and nonsmokers' total white blood cells, granulocytes and lymphocytes. *Carcinogenesis* 13: 1951–1955.

Mustonen R, Schoket B and Hemminki K (1993) Smoking-related DNA adducts: ^{32}P-postlabeling analysis of 7-methylguanine in human bronchial and lymphocyte DNA. *Carcinogenesis* 14: 151–154.

National Research Council (1989) *Biologic Markers in Pulmonary Toxicology*. Washington, DC: National Academy Press, pp. 1–42.

Olden K and Guthrie J (1996) Air toxics regulatory issues facing urban settings. *Environ Health Perspect* 104(Suppl. 5): 857–860.

Ong C-H and Lee B-L (1994) Determination of benzene and its metabolites: application in biological monitoring of environmental and occupational exposure to benzene. *J Chromatogr B* 660: 1–22.

Perera FP (1997) Environment and cancer: who are susceptible? *Science* 278: 1068–1073.

Perera FP, Estabrook A, Hewer A *et al.* (1995) Carcinogen adducts in human breast tissue. *Cancer Epidemiol Biomarkers Prev* 4: 233–238.

Phillips DH (1996) DNA adducts in human tissues: biomarkers of exposure to carcinogens in tobacco smoke. *Environ Health Perspect* 104(Suppl. 3): 453–458.

Poirier MC and Weston A (1996) Human DNA adduct measurements: state of the art. *Environ Health Perspect* 104(Suppl. 5): 883–893.

Polkis A (1996) Analytic/forensic toxicology. In: Klaassen CD (ed.) *Casarett and Doull's Toxicology: The Basic Science of Poisons*, 5th edn. New York: McGraw-Hill, pp. 951–967.

Pryor WA (1992) How far does ozone penetrate into the pulmonary air/tissue boundary before it reacts? *Free Radic Biol Med* 12: 83–88.

Pryor WA, Bermudez E, Cueto R and Squadrito GL (1996) Detection of aldehydes in bronchoalveolar lavage of rats exposed to ozone. *Fundam Appl Toxicol* 34: 148–156.

Rebbeck TR (1997) Molecular epidemiology of the human glutathione *S*-transferase genotypes *GSTM1* and *GSTT1* in cancer susceptibility. *Cancer Epidemiol Biomarkers Prev* 6: 733–743.

Repace JL, Jinot J, Bayard S *et al.* (1998) Air nicotine and saliva cotinine as indicators of workplace passive smoking exposure and risk. *Risk Anal* 18: 71–83.

Santrock J, Gorski RA and O'Gara JF (1992) Products and mechanism of the reaction of ozone with phospholipids in unilamellar phospholipid vesicles. *Chem Res Toxicol* 5: 134–141.

Scherer G and Richter E (1997) Biomonitoring exposure to environmental tobacco smoke (ETS): a critical reappraisal. *Hum Exp Toxicol* 16: 449–459.

Schieferdecker H, Dilger H, Doerfel H *et al.* (1985) Inhalation of U aerosols from UO_2 fuel element fabrication. *Health Phys* 48: 29–48.

Shaskan EG and Dolinsky ZS (1985) Elevated endogenous breath acetaldehyde levels among abusers of alcohol and cigarettes. *Prog Neuropsychopharmacol Biol Psychiat* 9: 267–272.

Schutz A, Barregard L, Sallsten G *et al.* (1997) Blood lead in Uruguayan children and possible sources of exposure. *Environ Res* 74: 17–23.

Thomas KW, Pellizzari ED and Cooper S (1991) A canister-based method for collection and GC/MS analysis of volatile organic compounds in human breath. *J Anal Toxicol* 15: 54–59.

Uppu RM and Pryor WA (1994) The reactions of ozone with proteins and unsaturated fatty acids in reverse micelles. *Chem Res Toxicol* 7: 47–55.

US Environmental Protection Agency (1996) Air Quality Criteria for Ozone and Related Photochemical Oxidants. EPA/600/P-93/004aF. Washington, DC: EPA, pp. 7:1–7:195.

Vainio H, Sorsa M and McMichael AJ (eds) (1990) *Complex Mixtures and Cancer Risk*. IARC *Sci Pub* 104, Lyon, France, pp. 441.

Vasilios D, Theodor S, Konstantinos S *et al.* (1997) Lead concentrations in maternal and umbilical cord blood in areas with high and low air pollution. *Clin Exp Obstet Gynecol* 24: 187–189.

Vine MF (1996) Biological markers of exposure: current status and future research needs. *Toxicol Ind Health* 12: 189–200.

Vine MF, Hulka BS, Margolin BH *et al.* (1993) Cotinine levels in semen, urine, and blood of smokers and nonsmokers. *Am J Public Health* **83**: 1335–1338.

Ward Jr JB and Henderson RE (1996) Identification of needs in biomarker research. *Environ Health Perspect* **104**(Suppl. 5): 895–900.

Weaver VM, Davioli CT, Heller PJ *et al.* (1996) Benzene exposure, assessed by urinary *trans,trans*-muconic acid, in urban children with elevated blood lead levels. *Environ Health Perspect* **104**: 318–323.

WHO Regional Office for Europe (1995) Guiding principles for the use of biological markers in the assessment of human exposure to environmental factors: an integrative approach of epidemiology and toxicology. *Toxicol* **101**: 1–10.

Yang K, Airoldi L, Pastorelli R *et al.* (1996) Aromatic DNA adducts in lymphocytes of humans working at high and low traffic density areas. *Chem Biol Interact* **101**: 127–136.

Zimmerman MR, Yeatman GW and Sprinz H (1971) Examination of an Aleutian mummy. *Bull NY Acad Med* **47**: 80–103.

The Epidemiologic Approach to Investigating Outdoor Air Pollution

JONATHAN M. SAMET and JOUNI J.K. JAAKKOLA

School of Hygiene and Public Health, The Johns Hopkins University, Baltimore, MD, USA

OVERVEIW

Introduction

The adverse effects of inhaled pollutants have been of substantial public health and regulatory concern. Over the last 50 years, substantial investigative effort has been directed not only at understanding mechanisms of disease pathogenesis, but also at providing the direct evidence on risks to human populations needed to develop risk management programs to protect the public's health. The resulting evidence of adverse health effects of air pollution, derived from population-based studies, has motivated broad regulatory programs and served to guide the development of standards for pollutant emissions and concentrations in outdoor air. This direct evidence has been gathered using epidemiologic research methods.

Epidemiology comprises the scientific methods used to study disease occurrence in human populations, including description of the occurrence of disease and identification of the causes of disease (Lilienfeld and Lilienfeld, 1980). This chapter reviews the epidemiologic approach for investigating the health effects of outdoor air pollution. Although the emphasis is on outdoor air pollution, the same concepts of exposure, study design and data interpretation have been extended to indoor air pollution as well. In fact, many studies address both indoor and outdoor air pollution, as exposures indoors and outdoors together determine respiratory health and substantial exposure to outdoor pollutants may occur indoors.

AIR POLLUTION AND HEALTH
ISBN 0-12-352335-4

To characterize the health effects of inhaled pollutants, complementary lines of investigation are employed, of which epidemiology is one: laboratory studies involving *in vivo* and *in vitro* systems, studies involving short-term pollutant exposures of human volunteers in the laboratory (frequently referred to as 'clinical studies'), and epidemiologic studies. Laboratory studies are primarily directed at understanding mechanisms of injury and describing exposure–response relationships; laboratory exposures may also be used to replicate biologically relevant patterns of human exposure while providing the capability of examining end-points that are not accessible in human populations, such as biomarker levels in bronchoalveolar lavage fluid or lung histopathology. In clinical studies, exposures and the circumstances of exposure can be carefully controlled, and sophisticated measures of outcome can be evaluated in a laboratory setting. However, pollutant exposures in clinical studies are generally brief and ethically limited to levels considered safe, and the laboratory exposures cannot fully replicate the complex mixtures found in outdoor and indoor air in the community.

Epidemiologic studies address the effects of inhaled pollutants as exposure occurs in the community. Their results can document the occurrence of adverse effects of air pollution and describe the relationship between exposure and response, and characterize effects on susceptible groups within the population, e.g. persons with asthma. In general, epidemiologic studies are carried out with the following objectives: (1) to determine if air pollution or a source of air pollution poses a hazard to human health; (2) to characterize the relationship between the level of exposure and the response; and (3) to examine responses of potentially susceptible populations to pollutant exposures. These objectives relate directly to the information needs of the policy-makers, providing answers to the complementary questions: (1) Does the pollutant pose a hazard to human health; (2) at what level of exposure are risks acceptable?; and (3) which groups need special consideration because of susceptibility? For example, in the USA, the Clean Air Act requires that the Administrator of the US Environmental Protection Agency (US EPA) set standards for selected pollutants that protect against adverse effects with an 'adequate margin of safety,' regardless of susceptibility. Epidemiologic evidence has figured centrally in setting standards for these pollutants.

Epidemiologic methodology has been used since the 1950s to investigate the health effects of air pollution. Although an obviously excessive number of deaths at times of high air pollution had been documented earlier in the century (Firket, 1936), the air pollution episodes in Donora, Pennsylvania, in 1948 and in London in 1952 dramatically called attention to the problem of outdoor air pollution (Shy *et al.*, 1978a). At the same time, rising mortality from lung cancer and chronic respiratory diseases raised concern that outdoor air pollution was responsible for the epidemic occurrence of these previously uncommon diseases.

British investigators conducted much of the initial work on the health effects of air pollution (Shy *et al.*, 1978a). Methods for conducting cross-sectional and longitudinal studies were developed that remain in use today (Samet, 1989). Standardized respiratory symptom and illness questionnaires were designed, and spirometers were used in the community setting to assess lung function. The early studies conducted in the UK showed that the prevalent widespread pollution, with particles and sulfur oxides, was associated not only with excess mortality but with respiratory symptoms and infections, reduced lung function, and exacerbation of the clinical status of persons with chronic respiratory diseases (Shy *et al.*, 1978a).

In the USA, early studies were conducted in several Pennsylvania communities and in other areas including Buffalo, New York, Nashville, Tennessee, Los Angeles, California, and Berlin, New Hampshire. In the early 1970s, the US EPA initiated a nationwide program of epidemiologic studies – the Community Health and Surveillance System – in an attempt to document comprehensively the health effects of air pollution (US EPA, 1974); this program, despite problems with data quality, represented one of the first large-scale efforts to address the health effects of air pollution using epidemiologic methods. In the mid-1970s, investigators at the Harvard School of Public Health initiated a landmark longitudinal investigation of the health effects of sulfur oxide and particulate pollution (Ferris *et al.*, 1979). This investigation, referred to as the Six Cities Study, involved about 20 000 children and adults recruited from six US cities that provided a gradient of pollution exposure at the time the investigation was planned (Ferris *et al.*, 1979). The investigators hypothesized that air pollution would adversely affect lung growth during childhood and accelerate lung function decline during adulthood. Participants have also been followed for mortality (Dockery *et al.*, 1993). During the 1980s, the same Harvard group implemented a second nationwide study, the 24 Cities Study, to investigate the health effects of acidic aerosols (Speizer, 1989). Such large-scale studies have been prompted by the need to address relatively smaller effects of air pollution than in past studies because ambient pollution levels have declined. Additionally, studies have become larger and more complex in response to the rising appreciation of the data requirements for estimating risks with reasonable precision and certainty.

While the initial focus of epidemiologic research was on outdoor air pollution, the potential public health significance of pollutant exposures in indoor environments was indicated by studies conducted during the late 1960s and early 1970s (Benson *et al.*, 1972; Samet *et al.*, 1987a, 1988; US EPA, 1993). These early studies showed that indoor sources could contaminate indoor air with the same pollutants found in outdoor air and suggested that indoor pollutants could produce adverse health effects. They also showed that outdoor air pollutants penetrated indoors and that indoor environments were consequently an important locus of exposure to outdoor pollutants. Early insights concerning the contribution of indoor pollution to total personal exposures to pollutants were gained from studies involving participants in the Six Cities Study (Spengler and Soczec, 1984). Assessments of the contributions of indoor and outdoor environments to total personal exposures showed that the predominant contributions of particles and nitrogen dioxide were from indoor pollution sources. The Total Exposure Assessment Methodology (TEAM) studies conducted by the US EPA yielded similar findings for the volatile organic compounds (Wallace, 1991). Surprisingly, even in communities with industries emitting volatile organic compounds, indoor sources accounted for most of the measured personal exposures in samples of the communities' residents. Based on these and other studies, the concept of total personal exposure has evolved and this concept remains fundamental in the design of studies on outdoor air pollution (NRC, 1991a).

During the 1990s, there has been rising use of time-series methods to evaluate the effect of air pollution on daily mortality and morbidity measures, such as numbers of hospitalizations or emergency room visits (American Thoracic Society, 1996a,b). These studies have been facilitated by new statistical methods for time-series analysis, as well as availability of suitable hardware and software. The databases are typically assembled from routinely collected information on mortality and morbidity, monitoring data collected for regulatory purposes, and publicly available information on confounding factors, such as weather. An extensive series of analyses, for example, has been conducted by Pope, Dockery and Schwartz on particulate air pollution (Schwartz *et al.*, 1995; Pope *et al.*, 1995). Analyses by this and other groups have indicated effects at air pollution levels found in many cities and below standards in a number of countries. These data have figured prominently in recent reviews of air pollution standards, as, for example, with the new standards of the US EPA for particulate matter.

Overview of the Epidemiologic Study Designs

The study designs used to investigate air pollution can be broadly grouped by the unit of observation: population groups or individual. Studies based on groups are referred to as ecological studies. Representative studies comprise comparisons of indicators of adverse health effects across geographic areas with different pollution levels and time-series studies of temporal associations between pollutant levels and outcome measures. The ecologic design has most often been applied to routinely collected morbidity and mortality data to assess the health effects of air pollution. Outcome measures have been compared across geographic regions with varying pollution or over time within a single geographic unit. For example, a number of analyses of mortality in London have addressed the effect of acidic aerosol concentration on mortality as outdoor levels have declined (US EPA, 1989). The ecologic design has well-characterized limitations: the use of data for a population group may misclassify the exposures of individuals in the group (the 'ecologic fallacy'), and it may be impossible to control for the effects of other potentially important exposures (Morgenstern, 1982).

The observational study designs having the individual as the unit of observation are the cross-sectional or descriptive study, the cohort study, and the case–control study; these are defined in Table 20.1. Each type of study has advantages and disadvantages for examining the effects of inhaled pollutants. For some types of sources, intervention studies may also be feasible.

Ecological studies, particularly if based on data already collected, are typically inexpensive and quickly conducted. Of the designs having the individual as the unit of observation and analysis, the cross-sectional study is generally the most economical and feasible approach, often used to compare health status of residents of more and less polluted areas. However, estimates of the effects of exposure may be biased by the tendency of more susceptible or more affected persons to reduce their level of exposure by leaving the polluted area. The temporal relationship between exposure and disease may thus be obscured or misrepresented in cross-sectional data. Cohort and case–control studies establish the proper sequence between exposure and disease.

In a cohort study, exposures of participants are assessed and they are followed for the development of the outcomes of interest. Cohort studies are labeled as prospective if the disease events will occur in the future, and retrospective if they have already taken place as the study is initiated. The Six Cities Study exemplifies the cohort approach: the subjects were enrolled from 1974 through 1976 and then followed with periodic measurements of lung function and respiratory symptoms and simultaneous monitoring of air pollution in the six communities (Ferris *et al.*, 1979). The cohort design has the advantage of permitting direct estimation of disease rates for exposed and non-exposed persons and the capability of prospectively accumulating comprehensive exposure information and accounting for changes in exposure over time (Table 20.2). The retrospective cohort design, often applied in the occupational setting but infrequently in studying air pollu-

Table 20.1   Glossary of epidemiologic terms

Association	Non-random occurrence of disease in relation to exposure
Bias	Error in the measurement of an exposure's effect
Case–control study	An analytical design involving selection of diseased cases and non-diseased controls followed by assessment of prior exposures
Clinical trial	An analytical design involving random assignment of exposure to two or more subject groups
Cohort study	An analytical design involving selection of exposed and non-exposed subjects with subsequent follow-up for development of disease
Confounding	Bias resulting from the contamination of an exposure's effect by that of another risk factor
Cross sectional study	Subjects are identified and exposure and disease status determined at one point in time
Incidence rate	Ratio of number of new cases to population at risk during a specified time period
Misclassification	Bias from error in determining exposure or disease status
Mortality rate	Ratio of number of deaths to population at risk during a specified time period
Prevalence	Proportion of population with disease at a particular time
Selection bias	Bias resulting from the technique used to select a study's subjects

Table 20.2   Data in a hypothetical cohort study with two exposure categories

	Exposed	Non-exposed
Involving population-time[a]		
New case during follow-up	a	b
Population time (person–years)	PY_1	PY_0
Involving cumulative risk[b]		
Disease during follow-up	a	b
No disease during follow-up	c	d

[a] Measure of association: ratio of incidence rates in exposed and non-exposed = $(a/PY_1)/(b/PY_0)$.
[b] Measure of association: risk ratio comparing exposed with non-exposed = $(a/a + c)/(b/b + d)$.

tion, can be used to evaluate rapidly the effects of a pollutant because exposure and disease have already taken place when the investigation is initiated. Disadvantages of the cohort design include potentially high costs and losses to follow-up.

The case–control study, which compares exposures of persons having the outcome of interest with those of controls, provides a measure of association between exposure and disease (Table 20.3). This design has been widely used for studying lung cancer, but infrequently for studying non-malignant respiratory diseases and air pollution. Thus, the case–control approach has been used to assess urban air pollution as a risk factor for lung cancer. For example, in a case–control study in New Mexico, the durations of urban residence by cases and controls were compared; length of residence in cities of various sizes was assumed to be a surrogate for exposure to pollutants in urban air (Samet *et al.*, 1987b). In comparison with the cohort study, the case–control study has the advantages of generally lower cost, greater feasibility, and a usually shorter time frame. The case–control study is

Table 20.3   Data in a hypothetical case–control study

	Exposed	Non-exposed
Case	a	b
Control	c	d

Measure of association: relative risk of disease in exposed compared to non-exposed = (a/b)(c/d) = ad/bc

the optimal approach for studying uncommon diseases. Bias from exposure assessment and, in some circumstances, from the selection of cases and controls, may limit this design.

The approach to studying effects of inhaled pollutants has varied with the health outcome of concern. The case–control and cohort designs have been used for respiratory and other cancers, whereas the cross-sectional and cohort designs have been used for effects other than malignancy in both the occupational and general environmental settings. For non-malignant outcome measures, both short-term and long-term cohort approaches may be appropriate. In 'panel studies,' a type of short-term cohort study, subjects – often persons like asthmatics with greater susceptibility – are enrolled and outcomes monitored intensively. Often, daily symptom diaries are recorded and peak flow measurements made at least daily to monitor respiratory status. Exposures may also be measured intensively using personal monitoring or on-site instrumentation. For example, a series of short-term cohort studies of summer camp participants (often referred to as 'camp studies') have been carried out to assess effects of oxidant and acidic pollutants on children who are spending most of their time outdoors and exercising vigorously (Lippmann, 1989). In some of these studies, continuously recording monitoring equipment has been sited at the camp. By contrast, long-term studies of many years, such as the Six Cities Study, are needed to assess the effect of air pollution on the growth and decline of lung function across the lifespan.

Although a distinction has been offered between ecologic studies and the other observational designs, the lines are blurred and some of the major air pollution studies have ecologic elements in their design. For example, in the Six Cities Study, exposure to air pollution is categorized for the participants at the most general level based on monitoring data from the particular city. Information on other exposures, e.g. indoor air pollution or occupation, is collected at the individual level. This type of design has been referred to as a multilevel design (Navidi *et al.*, 1994). It has application in investigating air pollution because of the difficulty of estimating exposures to outdoor air pollution for specific individuals.

Limitations of Epidemiological Studies of Air Pollution

The results of each type of observational epidemiologic study may be affected by biases, which can alter the relationship between exposure to a pollutant and the health outcome of concern. Selection bias refers to distortion of the exposure–outcome relationship by an effect of exposure or disease status on subject participation. For example, selection bias might arise if persons with heightened susceptibility to air pollution choose to leave locations associated with exposure.

Error in classifying either pollutant exposure or the health outcome is referred to as misclassification. If the error equally affects cases and controls in a case–control study or exposed and non-exposed subjects in a cohort study, the bias reduces associations toward the null value, i.e. no effect of air pollution. Such non-differential or random misclassification is of concern in most studies of inhaled pollutants and the lung; pollutant exposures are generally estimated using limited measurement data or surrogates for exposure, such as proximity to sources. Statistical power, the capability to detect exposure–outcome associations in a study, declines sharply as the degree of random misclassification increases (Gladen and Rogan, 1979; Shy *et al.*, 1978b). Strategies have been proposed for assessing random misclassification in studies of inhaled pollutants and adjusting for its effect (Armstrong and Oakes, 1982). If the extent of misclassification depends on subject status (case versus control) or exposure status (non-exposed versus exposed), then the resulting differential misclassification may increase or decrease associations. Differential misclassification or information bias is of particular concern in case–control studies that assess exposures with interviewer-administered questionnaires, as diseased and non-diseased subjects may not provide responses of comparable validity.

Bias from confounding occurs when the exposure of interest is associated with another risk factor. A confounding factor is itself a risk factor for the disease of interest and associated with the exposure under investigation in the study data. In studies of inhaled pollutants, particularly for those with weak effects, confounding by cigarette smoking and occupational exposures is always of concern. The effects of confounding can be controlled through exclusion of subjects with the potential confounding exposure, matching exposed and non-exposed subjects on potential confounding factors, or collection of data on potential confounding factors and adjustment in data analysis.

Interpretation of Epidemiologic Evidence

In the past, observational data have provided a clear indication of adverse effects of air pollution, sufficient to warrant action. The London Fog of 1952 was a dramatic call for action. However, at contemporary levels of air pollution in most developed countries, public health concern is directed at effects less dramatic than earlier air pollution episodes with evident excess mortality: exacerbation of the status of persons with chronic heart and lung diseases, effects on respiratory symptoms and lung function growth and decline, and contribution to the occurrence of chronic heart and lung diseases – asthma, chronic obstructive pulmonary disease (COPD), lung cancer and cardiovascular disease. Emphasis is placed on exposure–response relations as a basis for standard setting and on effects on susceptible groups within the population.

In this current context, epidemiologic data are interpreted within a holistic framework that draws in all relevant types of evidence. Guidelines for assessing causality of associations have been developed for this purpose in the pioneering work of Hill and others (Evans, 1993; Hill, 1965) and their application exemplified in the 1964 report of the Surgeon General's Committee (US Public Health Service, 1964). Subsequent guidelines for interpreting epidemiologic evidence reflect these pioneering efforts. The guidelines for causality applied to epidemiologic data (Table 20.4) offer points for evaluating the epidemiologic evidence and for assessing plausibility. Guidelines have also been proffered for evaluating epidemiologic studies which are providing data for regulatory and risk

Table 20.4 Criteria for assessing causality of associations

Strength of association	Strong associations considered to be more likely causal than weak associations
Consistency	Repeated observation of the association in different studies strengthens the likelihood of causality
Specificity	A cause is associated with a single effect
Temporality	Exposure precedes effect
Biologic gradient	An exposure–response relationship is present
Plausibility	The association should be consistent with relevant biologic data

Data from Hill (1965) and Rothman (1986).

assessment purposes (Federal Focus, 1996; US EPA, 1995). These guidelines emphasize principles of sound study design, control of bias, and adequate statistical power. All guidelines, however, are no more than points for evaluating evidence and they do not offer a rigid scorecard for evaluating data. Decision-making in the face of uncertainty seems to inevitably lead to questions concerning the soundness of available evidence as a basis for policy-making.

EXPOSURE ASSESSMENT

This section reviews methods for estimating pollutant exposures in epidemiologic investigations of air pollution.

Concepts of Exposure

Exposure can be defined as the contact of pollutant with a susceptible surface of the body, which for air pollutants includes the eyes, nose, mouth and throat, the airways and the skin (NRC, 1991b). The rate of exposure to an air pollutant is given as concentration, usually expressed as mass per unit volume or, for gaseous pollutants, also as a mixing ratio with air. It is useful to consider different parameters of exposure starting from instantaneous exposure, the exposure at a given point in time. The exposure of an individual over time can be characterized as a concentration–time function or, in other words, an exposure–time function. Depending on the type of health effects and the underlying mechanism, it may be appropriate to formulate exposure as average exposure over a specified time period, or as cumulative exposure over the lifetime or for a relevant period. The relevant exposure metric depends on the underlying biological mechanisms leading from exposure to health effect. For example, for ozone's short-term effects, both cumulative exposure and the exposure profile are relevant (Hazucha *et al.*, 1992; Lippmann, 1989).

The population exposure can be expressed as a distribution of individual exposures where the parameter of exposure can be either instantaneous exposure at a point in time, or average or cumulative exposure over time. The population exposure distribution can be used in assessing the public health impact of a given pollutant when an estimate of the exposure–response relation is available. Thus, information is needed on the exposure distribution for quantitative risk assessment.

Framework for Assessment of Exposure

In studies of the health effects of air pollution, the task is to capture an individual's exposure experience over a time period that is relevant for the development of the studied outcome. Ideally we would like to measure an individual's exposure continuously from birth to the end of life – perhaps also the mother's exposure during pregnancy or even the father's exposure during the maturation of semen – and reduce relevant exposure parameters from the exposure–time data. Although the methods of exposure assessment for air pollution have evolved substantially, moving from classification of exposure by such crude surrogates as residence location to the present techniques for monitoring exposures of individuals on a real-time basis, practically forces compromises.

Fig. 20.1 illustrates the determinants of exposure, dose and biologically effective dose that underlie the development of health effects. A studied chemical or microbiological pollutant may be produced both from *outdoor sources* and *indoor sources*. The strength of source can be expressed as the emission rate (mass per time). Dispersion of the pollutant is influenced by meteorological factors, such as wind speed and direction, air temperature and relative humidity, geographical factors such as topography, and chemical transformation in the air. Dispersion models have been developed to calculate concentration distributions around sources. The emissions will lead to concentrations in three-dimensional space; variation over time is the fourth dimension. The concept of the microenvironment has been developed for exposure assessment. A microenvironment is a space where the concentrations of the pollutants of interest are sufficiently homogeneous for the purposes of exposure assessment. Usually only indoor environments have been considered as microenvironments, but theoretically it is meaningful to include outdoor spaces with sufficiently homogeneous concentrations as microenvironments.

Exposure assessment approaches can be divided broadly into *direct* and *indirect* assessment methods. In direct assessment, each individual carries a personal monitor which registers the encountered concentrations continuously or integrated over a given time period. In indirect assessment, information is collected about (1) concentrations over time in different microenvironments and (2) time periods spent by individuals in each microenvironment (time–activity information). The exposure assessment is made by combining the available information either with mathematical formulas or by modeling. Fig. 20.1 illustrates this concept, showing the concentration–time matrix for different microenvironments: outdoors or indoors, at home or at work or, for example, in cars. The symbol m_i refers to a number (i) of different microenvironments, the symbol t_j to a number j of time periods spent in each microenvironment and the symbol c_{ij} refers to concentration of pollutant in each microenvironment at each of these time periods, based either on actual measurements or on estimates derived by modeling.

The total cumulative (integrated) exposure of an individual E_{tot} can be estimated by multiplying the pollution concentration in each microenvironment by the time spent in that environment and summing the time-weighted concentrations, as given by the formula:

$$E_{tot} = \Sigma \; c_{ij} t_{ij}$$

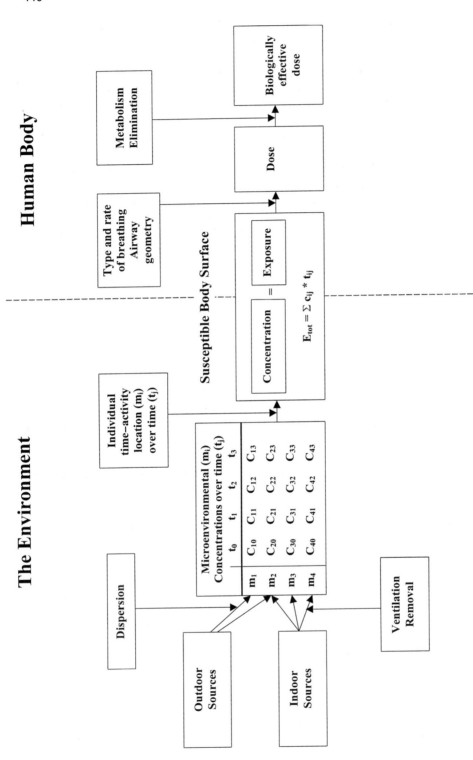

Fig. 20.1   The determinants of exposure, dose and biologically effective dose that underlie the development of health effects. Modified and completed from Jaakkola *et al.* (1994)

Time-specific exposures can be estimated similarly by focusing on relevant time periods. Theoretically, exposure estimates based on all microenvironments for infinitely short intervals will approach estimates based on personal monitoring.

The concentration at a susceptible surface is equivalent to exposure. However, the more mechanistically relevant dose and biologically effective dose depend on a number of factors beyond the exposure. For inhaled pollutants, dose is influenced by breathing rate (physical activity), type of breathing (mouth versus nose), airways geometry, and other factors. Metabolism and excretion after the uptake of a compound also affect the biologically effective dose.

In addition to direct and indirect assessment of exposure, biomarkers can be used to assess exposure to air pollution. A biomarker of exposure is an exogenous substance or its metabolite, the product of an interaction between xenobiotic agent and a target molecule in the body which can be measured in a compartment of the human body including tissues, cells, fluids, or expired air (NRC, 1989). There is a full continuum of biomarkers from indicators of exposure to early signs of health effects. Biomarkers are surrogates for exposure or dose; the relationship of the biomarker to exposure may be physiologically complex and variable among individuals. If the individual variation leading to variation in formation of biomarkers is also related to the susceptibility to adverse health effects, the exposure–response relations based on the biomarker will be biased in estimating risk.

Assessment of Exposure

Personal monitoring of concentrations at a susceptible surface represents the direct method for assessing exposure to air pollution; costs and acceptability limit the extent to which personal monitoring can be used in epidemiologic studies. Indirect methods can use one or several types of information on source strength, pollutant dispersion, concentrations from stationary monitoring, exposure, and time-activity from questionnaires, and biomarker concentrations in human samples (Table 20.5). An accurate, feasible, and

Table 20.5 Indirect methods of assessing exposure

Source of information	Type of information
Source strength	Emission rate (mass per time), traffic density
Geographical information	Distance of the place of residence from the source
Dispersion models	Spatio-temporal concentration distributions from modeling of emission rates, meteorology, air chemistry, geography
Outdoor–indoor penetration	Modeling from outdoor concentration, building and ventilation characteristics
Stationary monitoring	Concentration over time modeling from concentration of pollutants in microenvironments
Questionnaires and interviews	Source strength, distance from the source, time–activity
Personal monitoring	Continuous or cumulated concentrations over time
Human samples	Concentration of biomarkers of exposure in human tissues and hair
Toxicological models	Concentration and dose of pollutants in target organs modeling from concentration, breathing rate, metabolism

cost-effective exposure assessment is a major challenge in environmental epidemiology. In the best studies, a feasible indirect exposure assessment is combined with direct exposure assessment in a subgroup of the study population in order to estimate the validity of the indirect exposure estimates. For example, in the Six Cities Study, samples of the full complement have participated in studies of personal exposure to assess the contributions of indoor and outdoor concentrations to total personal exposure. In another population-based study in Tucson, Arizona, both indoor and outdoor sampling were performed for a number of pollutants (Lebowitz *et al*, 1985).

For outdoor air pollution, a simple exposure assessment can be based on the place of residence in a polluted or non-polluted community. The selection of the community could be based on the presence of a strong source, for example an industrial plant, data on emission rates, dispersion modeling or stationary monitoring of indicators of exposure or actual pollutants of interest. Often the unit of observation is the group rather than individual; i.e. individuals in the same residential area are included in the same exposure category. For example, Lunn and colleagues (1967) compared the respiratory health status of four groups of children living in Sheffield, England; the groups were selected from four areas within the city assumed to have contrasting exposures on the basis of the available monitoring data. Similarly, in a study of photochemical oxidant pollution, subjects were selected in Lancaster, California (low exposure), and in Glendora, California (high exposure) (Detels *et al.*, 1981, 1987). However, as shown by subsequent studies of personal exposure to air pollutants, data from a monitor located for regulatory purposes may not be highly correlated with personal exposures (Spengler and Soczec, 1984). In fact, even within a single urban area, the concentrations measured at one monitor may not be highly correlated with results from others (Goldstein and Landovitrz, 1977a,b). For pollutants with broad regional distributions, e.g. photochemical oxidants, centrally located monitors may suffice, but for pollutants found in multiple microenvironments, e.g. carbon monoxide or respirable particles, substantial misclassification may result from representing exposure solely by the results of central monitors.

Traffic density and proximity to roadways have also been used as exposure indicators. For example, Savitz and Feingold (1989) studied the relation between childhood cancer and residential traffic density. Traffic density was expected to be the best measure of exposure to benzene, lead and hydrocarbons, which were hypothesized to be causes of childhood cancer. On the basis of maps and lists of the streets and traffic volumes, the home was assigned a traffic density score of <500 vehicles per day or the recorded number of vehicles per day (ranging from 500 to over 100 000 per day).

Larssen and colleagues (1993) developed a mathematical model for car exhaust exposure in a study in downtown Oslo. In a similar approach, but on a shorter timescale, a total of 153 individuals participated in a 2-week study where they reported in a diary their location, activity, and other exposure data and symptoms hour-by-hour. The participants' home address and other places of visit indicated in the diary were positioned within a grid system. The road system was positioned within the same grid and the exhaust emissions were calculated from traffic data providing carbon monoxide and nitrogen dioxide concentrations in space and time. A standard dispersion model estimated hourly air pollution concentrations at receptor points of a given area. Ten subjects carried a portable monitor for continuous measurement of carbon monoxide and 15 subjects wore a passive NO_2 sampler for a week. These measurements were compared with the indirectly calculated exposure estimates.

Measurement instrumentation to support direct personal monitoring on continuous or short timescales is presently limited (Table 20.6) (NRC, 1991a). Two classes of personal monitoring instrumentation are available: (1) passive samplers that bind the pollutant to a reactive surface or matrix provide integrated measurements over relatively short periods of time (e.g. hours) and (2) continuous monitors that measure pollutant concentrations in real time and process and store the measurement in electronic memory.

While passive samplers are simple in design, construction and quantitative analysis they require strict quality control to maximize sensitivity for use in the community. Similarly, accurate measurements by continuous monitoring instrumentation necessitate careful and regular calibration against gas standards and, at times, extensive maintenance. The paucity of continuous data on personal exposures may be attributed to the high cost of electronic instrumentation and the labor required to maintain the monitors. For this reason, epidemiologic studies have tended to incorporate passive samplers to obtain exposure measurements.

Continuous monitoring has been limited to smaller numbers of subjects, often to validate integrated measurements from passive devices. The development of personal monitoring instrumentation has been constrained by the design requirements of portability, independent powering, sensitivity, accuracy in the low concentration range and expense (Wallace and Ott, 1982).

Although personal monitoring of some air pollutants is now technically possible, most epidemiologic studies have to apply indirect exposure assessment using different sources of data including biomarkers. The instruments used in personal monitoring can be used for assessing exposure in different microenvironments.

Biologic monitoring of chemical contaminants or metabolites provides another modality for estimating personal exposures (Table 20.7) (NRC, 1988, 1991a). For air pollutants

Table 20.6 Personal exposure monitoring instrumentation capable of quantitative measurements at ambient concentrations

Pollutant	Monitor type	Collection method	Analytical method
Particles			
Mass	Integrated, active	Pump/impactor	Gravimetric
Sulfates, nitrates, metals	Integrated, active	Pump/impactor	Gravimetric, chemical analysis, PIXE
Gases			
Carbon monoxide	Continuous active	Pump or diffusion	Electrochemical sensor
	Integrated, passive	Diffusion	Electrochemical
Formaldehyde	Integrated, active	Diffusion tube	Chromotropic acid
		Pump, treated filter	GC
Nicotine	Integrated, passive	Badge	GC
NO$_2$	Integrated, active	Pump/impingers	Colorimetric
	Integrated, passive	Diffusion tube	Colorimetric
	Continuous, active	Pump	Electrochemical sensor
O$_3$ (under development)	Integrated, passive	Badge	Electrochemical
Volatile organic compounds	Integrated, active	TEXAX cartridge	GC/MS

Adapted from Sexton K, Ryan PB (1998) Assessment of human exposure to air pollution: methods, measurements, and models. In Watson AY, Bates RR, Kennedy D (eds). Air Pollution, The Automobile and Public Health, pp. 207–238. Washington, DC: National Academy Press.

Table 20.7   Biologic markers of exposure to air pollutants

Pollutant	Marker
Allergens	Levels of specific antibodies
Carbon monoxide	Carboxyhemoglobin level; alveolar carbon monoxide level
Lead	Blood lead level
Polycyclic aromatic hydrocarbons	Levels of adducts in white blood cells
Volatile organic compounds	Concentration in exhaled air

of current concern, the available biomarkers reflect exposures over intervals of hours to days and not over longer averaging times that may be relevant for some health outcomes. Markers indicative of exposures over periods of several hours include carboxyhemoglobin or carbon monoxide in exhaled air, and levels of some volatile organic compounds in exhaled air.

HEALTH OUTCOME MEASURES

This section reviews the outcome measures that have been used in assessing the health effects of air pollution (Table 20.8). For the principal outcome measures, pathophysiologic mechanisms, accuracy and potential sources of bias are briefly considered.

Table 20.8   Health outcome measures in studies of air pollution

General
  Overall mortality
  Morbidity index

Respiratory
  Acute and chronic symptoms
  Acute infections
  Chronic respiratory diseases
  Degree of non-specific airways responsiveness
  Reduced level of lung function
  Increased rate of lung function decline
  Decreased rate of lung function growth
  Exacerbation of a chronic respiratory disease
  Hospitalization for a chronic respiratory disease
  Lung cancer
  Death secondary to a chronic respiratory disease

Neuropsychological
  Reduced performance on neurobehavioural testing
  Neuropsychological syndrome
  Neuropsychological disease

Overall and Cause-Specific Mortality

From the 1930s through the 1950s, episodes of excess mortality at times of extremely high outdoor air pollution provided dramatic evidence that air pollution can cause excess deaths (Shy *et al.*, 1978a). While overall mortality rates were increased during these episodes, the excessive number of deaths tended to be placed into cause-of-death categories for cardiovascular and respiratory diseases. Although such dramatic air pollution episodes are now infrequent in most developed countries, research continues into the effects of outdoor pollutants on overall and cause-specific mortality. In investigations of air pollution and mortality, routinely collected vital statistics data for specific geographic areas are used as the health outcome measure, while air pollution exposure of the areas' residents is estimated from outdoor monitoring sites assumed to be representative for the populations.

All-cause mortality is not subject to error from assignment of cause of death. However, pathophysiologic considerations typically lead to research hypotheses focused on cause-specific mortality, e.g. ischemic heart disease or COPD. Exposure to pollutants might cause death in persons with underlying COPD by exacerbating the clinical status of those with little functional reserve; for such patients, pollutant exposure, by diminishing the efficacy of host defenses, might also increase the incidence or severity of respiratory tract infections. Persons with ischemic heart disease are vulnerable to pollutants that impair oxygen delivery to tissues, e.g. carbon monoxide (Allred *et al.*, 1989).

Misclassification of the underlying cause of death by death certificate designation has been well documented (Kircher and Anderson, 1987); accuracy of cause-specific mortality data is potentially influenced by the extent of the population's contact with medical care, the diagnostic acumen of clinicians in the study areas, the accuracy of information on the death certificate, and the rate of error in coding the death certificate to a particular cause of death. The accuracy of death certificate information on the major respiratory diseases has not yet been systematically evaluated. Because of recent concern about increasing asthma mortality, the validity of death certificate designation of deaths as due to asthma has been examined in several countries (Sly, 1989). However, while the validity of death certificate data on respiratory cancer has been specifically evaluated (Percy *et al.*, 1981), comprehensive assessments of the quality of death certificate data for other major chronic respiratory diseases and for acute respiratory infections have not been performed. Misclassification of the underlying cause of death in vital statistics data would be expected to occur randomly, rather than differentially, in relation to the level of pollutant exposure.

All-cause and cause-specific mortality rates are also highly non-specific outcome measures. Mortality rates vary with the background distribution of risk factors determining the incidence of disease and with the survival rate of those who have developed disease. Thus, assessments of the effects of air pollutants on mortality can be sharpened if these other factors can be considered in data analysis.

Indexes of Morbidity

Epidemiologic studies of the health effects of air pollution have incorporated diverse indexes of general morbidity including absenteeism from school and work, days of restricted activity spent at home, and rates of utilization of outpatient medical facilities, visits to emergency rooms and hospitalization (NRC, 1985; Shy et al., 1978b). For example, in an investigation in Steubenville, Ohio, the relationship between the numbers of visits made to the principal hospital's emergency room and daily air pollution levels was assessed (Samet et al, 1981). Like mortality rates, the morbidity indexes are non-specific and subject to misclassification.

Respiratory Infections

Diverse microorganisms can cause respiratory tract infections including mycoplasma, viruses, bacteria and fungi (Graham, 1990). The spectrum of infecting organisms and the clinical manifestations of the infection vary from infants through the elderly (Monto and Ullman, 1974). Research on air pollution and respiratory infection has largely focused on infants and younger children, considered as susceptible to inhaled pollutants because their lungs are maturing and rates of respiratory infection are the highest of any age group (Mauderly, 1989; Monto and Ullman, 1974).

The occurrence of respiratory infections can be monitored using subject or parental reports of symptoms or illnesses or by using inpatient and outpatient records of clinical facilities. The usual clinical respiratory illness syndromes include upper respiratory tract infections ('colds'), otitis media and lower respiratory illnesses; the latter category includes croup, tracheobronchitis, bronchiolitis and pneumonia (Graham, 1990). Standardized and uniformly accepted clinical criteria have not been developed for these illnesses, and health-care practitioners typically develop their own operational criteria. In fact, there is no single unimpeachable 'gold standard' for establishing the presence of a respiratory infection; a clinical diagnosis and a positive culture for a pathogenetic organism represent the most valid basis for documenting infection.

In some studies of children and adults, illness histories have been obtained retrospectively by questionnaire. While such retrospective information can be readily collected, bias is likely with subjects who are symptomatic or ill at the time of data collection and more likely to report past illnesses (Samet et al., 1983). Prospective surveillance of illness avoids the potential problem of recall bias, but requires a more elaborate system for ascertaining the occurrence of illness. Surveillance approaches using calendar diaries for recording of symptoms have been successfully applied in community-based studies on respiratory illnesses (Dingle et al., 1964; Monto et al., 1971; Tager and Speizer, 1977), but have been used in only a few studies of inhaled pollutants. For example, in a cohort study on nitrogen dioxide and respiratory infections in Albuquerque, New Mexico, infants were enrolled shortly after birth and the occurrence of illness is ascertained by completion of a daily symptom diary and telephone contact every 2 weeks (Samet et al., 1992). To assess the validity of this system for illness ascertainment, a sample of ill children was evaluated by a nurse practitioner according to a standardized protocol (Lambert et al., 1994).

The occurrence of illness can also be documented by using diagnoses made by clinicians at the time of outpatient visits or hospital discharge diagnoses. However, illness rates based on contact with health care providers have potential determinants other than incidence, including patterns of access to health-care, the severity of the illnesses, and diagnostic practices of the clinicians. More severe illnesses are likely to prompt contact with a health-care provider, and thus illness rates based on clinical diagnoses are lower than those obtained by community-based surveillance. Thus, in the USA, community-based surveillance studies show that children have about two lower respiratory tract illnesses during the first year of life (Monto and Ullman, 1974); in contrast, 20–30% of children receive a physician's diagnosis as having a lower respiratory tract illness during this same age range (Denny and Clyde, 1986; Wright *et al.*, 1989). Nevertheless, studies of both indoor and outdoor air pollution have used indexes of respiratory infection derived from clinical encounters (Graham, 1990; US Department of Health and Human Services, 1986). However, confounding may be introduced into studies using such clinical indexes because both pollution exposure and patterns of health care utilization may be associated with demographic and socioeconomic factors that also determine illness rates (Graham, 1990).

Respiratory Symptoms

Standardized respiratory symptoms questionnaires, initially developed during the 1950s, are widely used in epidemiologic research for assessing the occurrence of the cardinal respiratory symptoms: cough, sputum production, wheezing and dyspnea (Samet, 1978). The presently used questionnaires have evolved from the questionnaire originally developed by the British Medical Research Council; like the first questionnaire, the currently available instruments emphasize chronic symptoms and are insensitive for detecting acute symptom responses. Limited data have been published on the validity and reliability of individual questions (Samet, 1978, 1989). In the USA, an American Thoracic Society committee initially adopted the Medical Research Council's questionnaire for adults in 1969. In 1978, the American Thoracic Society's Epidemiology Standardization Project published a revised questionnaire for adults and a new questionnaire for children (Ferris, 1978). More recently, asthma questionnaires have been developed by the International Union Against Tuberculosis and Lung Disease (IUATLD) (Burney *et al.*, 1989) and by the investigators conducting the International Study on Asthma and Allergy in Children (ISAAC) (Jenkins *et al.*, 1996). Proper use of these questionnaires reduces the potential for interviewer bias and assures comparability with data from other populations studied with the same techniques.

For pollutants with quickly changing concentrations and mechanisms of action producing acute symptom responses, short-term longitudinal studies ('panel studies') may be carried out to examine the relationship between pollutant levels and symptom occurrence on the time scale of a day or shorter periods. Typically, symptom status is tracked by asking subjects to complete a diary that covers such items as the occurrence of cough, sputum production, wheezing, sore throat, hoarseness and fever (Schwartz *et al.*, 1991). In studies involving controlled laboratory exposures, asthmatics are more susceptible to a number of inhaled pollutants than non-asthmatics (Bromberg, 1988). The diary approach has been applied to investigate the health effects of pollutant exposure on asthmatics and also on patients with COPD in the community setting (Lawther *et al.*, 1970; Whittemore and Korn, 1980). In studies of asthmatics, medication pattern and use of health-care services may be tracked in addition to symptom status. Standardized instruments for diary studies have not been published.

Pulmonary Function

Spirometry, involving the timed collection of exhaled air during the forced vital capacity maneuver, has been the most widely used technique for measuring lung function in epidemiologic studies of air pollution. Spirometers that are inexpensive, portable and durable are available for field use. Standardization of spirometry has long been advocated, and recent recommendations are available from the American Thoracic Society (American Thoracic Society, 1987; Ferris, 1978). These recommendations cover specifications for spirometers, testing protocols, and test interpretation. Data collected following these recommendations and using proper equipment have small within-subject variability (Bates, 1989; Samet, 1989). In a few studies, other types of measurements have been made, including the single-breath nitrogen test and lung volumes (Bates, 1989). However, these tests, as well as other types of testing used in clinical pulmonary function laboratories, have greater variability than spirometric measures of lung function and the equipment is more complex and expensive than a simple spirometer.

Spirometry provides measurements of the forced vital capacity, the total amount of exhaled air, and the volume of air exhaled in the first second or at other time points. A spirometer integrated with a microprocessor can measure flow rates at various lung volumes. These spirometric measures are sensitive to processes impairing ventilatory function of the lung, but injury cannot be inferred at specific anatomic loci because of particular patterns of abnormality of spirometric parameters (Mead et al., 1979). However, abnormalities of flow rates at lower lung volumes are associated with adverse effects on the small airways of the lung (Bates, 1989).

Although spirometry has proved effective for community-based studies, it cannot be readily used in large numbers of subjects to track function on a day-to-day basis. In many studies investigating the relationship between short-term variation in lung function and pollution exposure, peak expiratory flowrate (PEFR) has been measured using portable and inexpensive instruments that can be used by subjects themselves. PEFR measurement takes only a few minutes and can be performed several times throughout the day; measurements can be made before and after episodes of exposure. Accurate measurement of PEFR requires calibration of the peak flow meters and standardized protocols for subject training and data collection (Cross and Nelson, 1991; Quackenboss et al., 1991).

Non-specific Airways Responsiveness

Non-specific airways responsiveness refers to the extent of bronchoconstriction evoked by a non-antigenic stimulus (Sparrow and Weiss, 1989). The pharmacologic agents most widely applied to assess non-specific airways responsiveness are methacholine and histamine; exercise and hyperventilation with cold air may also be used. Asthmatics, by definition, have airways hyperresponsiveness. In populations, the distribution of non-specific airways responsiveness appears to be unimodal with skewing toward hyperresponsiveness (Sparrow and Weiss, 1989). In controlled exposure studies of asthmatics and healthy non-asthmatic subjects, non-specific airways responsiveness has often been one of the monitored outcome measures (Frank et al., 1985). In the community setting, assessment of non-specific airways responsiveness might provide a sensitive indicator of the effect of exposure to a complex mixture. The protocols for measuring non-specific airways responsiveness are time-consuming, however, and the possibility of adverse consequences of testing necessitates the presence of a physician. Thus, non-specific airways responsiveness has not yet been used in large-scale epidemiologic research on the health effects of air pollutants.

Neuropsychological Measure

Exposure to mixtures of volatile organic compounds in indoor air can be postulated to have neurobehavioral consequences (Molhave, 1990; Molhave *et al.*, 1990). In fact, volatile organic compounds have been postulated to be etiologic factors in the non-specific 'sick-building syndrome'. A variety of tests of neurobehavioral outcomes are available (Letz, 1991), and such tests have been applied in epidemiologic investigations (Kilburn *et al.*, 1985; Schenker *et al.*, 1982). However, standardized approaches for assessing neurobehavioral outcomes have not been developed (Letz, 1991).

ECOLOGICAL DESIGNS

Ecologic designs have long been used for assessing the health effects of air pollution. This design has had extensive application to air pollution because of the generally large degree of variation of air pollution levels over time and across geographic areas. The maintenance of monitoring networks for regulatory and surveillance objectives has facilitated the conduct of ecologic studies; use of routinely collected mortality and morbidity statistics as the health outcome measure further facilitates the conduct of ecologic studies. Two broad types of ecologic studies of air pollution can be distinguished: multiple group studies that have primarily been cross-sectional in design, and time-series studies which have generally been limited to single locations. Time-series studies have been conducted at the daily level and also at longer-term time frequencies. Newer studies, facilitated by advances in statistical methods and hardware and software, address time trends in multiple locations.

Methodologic limitations of the ecologic study design include the well-known ecologic fallacy, the potential bias in generalizing findings from groups to individuals, lack of information on potential confounding and modifying factors at the group level, and limitations of ecologic regression approaches for controlling confounding and estimating risks (Greenland and Morgenstern, 1989; Morgenstern, 1995). Nonetheless, ecologic designs have proved informative in indicating adverse health effects of air pollution. Time-series designs are less vulnerable to uncontrolled confounding and effect modification by other factors, particularly if directed at associations on a short-term basis, e.g. daily. At this level of observation many potential confounding factors are invariant; smoking, for example, does not change on a day-to-day basis.

Ecologic studies of straightforward design were extensively used in initial studies of the health effects of air pollution. Cross-sectional studies compared the health status of residents of communities having differing air quality (see Shy *et al.* (1978b) for a review of these early studies). Lave and Seskin (1973, 1977) reported a series of pioneering cross-sectional analyses of mortality in the USA. These investigators used regression methods to estimate the effects of air pollution on mortality while controlling for the effects of potentially modifying factors. Time-series designs were also used and applied to mortality and morbidity measures in such highly polluted cities as London and New York City during the 1950s and 1960s (see Shy *et al.* (1978b) for a review of these early studies). These studies used analytic methods that would be considered inadequate by today's standards; nonetheless, they tended to show effects which became undetectable as pollution levels declined (Holland *et al.*, 1979).

Much of the current concern with regard to the adverse health effects of particulate air pollution was ignited by findings of time-series studies of the association of daily mortality counts with air pollution levels. These new studies, largely published beginning in the early 1990s, assessed the effect of current or previous days' pollution on mortality, after taking account of temperature and other climatic variables (for recent reviews see American Thoracic Society (1996a,b) and Dockery and Pope (1994); the findings are also covered in the chapters on specific pollutants in this volume). These studies use new methods for time-series analysis that can more fully account for the correlated nature of time series data (Diggle *et al.*, 1994; Liang and Zeger, 1986). The statistical methods are based on regression techniques referred to as generalized additive models; these models allow for flexible specifications of relationships among variables in time. Newer algorithms for fitting variables based around the generalized estimating equation approach of Liang and Zeger (1986) accommodate the correlation structure of longitudinal data.

These time-series approaches are now being generalized to multiple locations. One model investigation, the APHEA (Air Pollution and Health: European Approach), is a parallel analysis of data from multiple European countries (Katsouyanni *et al.*, 1995). In this two-stage protocol, the investigators carry out data analysis of individual locations using common approaches so that meta-analysis of the findings is possible in a second stage. Dominici *et al.* (1999) have proposed a Bayesian approach that analyzes data within cities as the first stage and then combines the findings across cities using Bayesian regression methods. These multicity designs bring a uniform analytic approach to data from multiple locations; by combining data across cities, the precision of risk estimates can be enhanced and the sources of heterogeneity in effects across cities can be explored.

CROSS-SECTIONAL DESIGNS

While the time referent in a cross-sectional study is instantaneous by definition, most cross-sectional studies relate health outcome measures to usual or past exposures. Thus, in a typical cross-sectional study of air pollution, information is obtained on health status and on past exposures. Exposures may be estimated using monitoring data so as to estimate long-term average, cumulative or time-specific parameters of exposure. Residential history can also be used as an additional indicator of cumulative exposure in the past.

A cross-sectional study may also serve as a source of subjects for a case–control study and provide the initial participants for a cohort study. For example, the UCLA Population Studies of Chronic Obstructive Respiratory Disease began as cross-sectional surveys carried out in 1972–73 and 1977–78 (Detels *et al.*, 1981) and follow-up surveys were then carried out 5 years later (Detels *et al.*, 1987).

The type of air pollution, existing information on air pollution levels, and availability of population registries influence the optimal selection of the study population. The type of air pollution can broadly be categorized based on the primary source into point-source pollution, such as originates from industry or power plants, and widespread urban pollution from multiple point sources and mobile sources, and from pollution transported from outside the area of interest. The selection of a population can be population-based or based on prior knowledge of the levels of emissions or of air pollution concentrations.

In exposure-based or source-based selection of study population, prior knowledge of average air pollution levels in different geographical areas typically underlies selection. For example, the South Karelia Air Pollution Study assessed effects of malodorous sulfur compounds from pulp mills on respiratory and central nervous symptoms. In the first phase in 1987, three communities in South East Finland were chosen on the basis of location in relation to two pulp mills, calculated dispersion models, and some limited measurements of sulfur dioxide and needle sulfur (Jaakkola *et al.*, 1990). A self-administered questionnaire was distributed to all adults in defined residential areas and air pollution monitoring of hydrogen sulfide, sulfur dioxide and particulates was performed to validate the historical patterns of exposure. Such exposure-based selection of study communities can also be used in studies of urban air pollution from multiple sources. In The UCLA Population Studies of Chronic Obstructive Respiratory Disease, residential areas which had similar age, race and sex distributions, mean and median income, and home value according to the 1970 Census area were chosen on the basis of different levels of photochemical/oxidant type pollutants (Detels *et al.*, 1979).

In a population-based cross-sectional study, the study population consists of all the members or a random sample of the members of a given population usually living in a geographically defined area. Population-based samples can advantageously provide estimates of the population exposure distribution. In some countries, national population registries provide a good framework for population-based studies. The Finnish national population registry was used in the second phase of The South Karelia Air Pollution Study, which consisted of a cross-sectional study (Partti-Pellinin *et al.*, 1996) and a cohort study with a natural experiment of exposure redirection (Jaakkola *et al.*, 1998). Random samples of residents in selected communities were chosen for a postal survey and the registry also provided current addresses in electronic format. Recruitment of participants through schools has been widely used as this method can identify practically all children of a given age group in a geographically defined area. This approach has been used in several large-scale cross-sectional studies of the effects of air pollution on children's respiratory health such as the Kanawha Valley Health Study of the effects of volatile organic compounds (Ware *et al.*, 1993), the 24 Cities Study (Speizer, 1989), and a German study of road traffic pollution (Wjst *et al.*, 1993).

In a strictly cross-sectional study, the causal inference is based on a comparison of the prevalence of disease in subjects currently exposed and unexposed. The major weakness of this approach is the uncertainty in temporality, the most important criterion of causality: cause has to precede the effect. Obtaining past exposure data from other sources such as monitoring stations, emission records or traffic statistics may partly resolve this critical issue of time-sequence. For example, in a German cross-sectional study exposure assessment was solely based on data on car traffic in main streets and on permanent air monitoring stations, which were available for several years preceding the health data collection (Wjst *et al.*, 1993).

Selection bias is a concern of validity in particular in the cross-sectional study, because the level of exposure may have influenced people with the disease of interest or early symptoms or signs to leave the exposed residential area before the study. In the presence of serious exposure, the subjects most affected may have died. Both scenarios would lead to underestimation of the effect in a cross-sectional population. Any information on the past frequency of moving or in some situations on cause-specific mortality from the compared study population could help to assess the magnitude of potential selection bias.

Cross-sectional studies may be sensitive to information bias if outcomes, exposures or even key confounders are assessed subjectively. The presence of disease may influence the recall of exposure information or knowledge of past or current exposure may influence reporting of symptoms and conditions. Knowledge of exposure or outcome may also influence interviewers asking about or judging the outcome or exposure information. Knowledge or perception of air pollution exposure by the study subjects is sometimes difficult to avoid. Chemical air pollutants such as malodorous sulfur compounds and some volatile organic compounds have a distinct odor with a very low threshold of sensation, which means that odor is inherent with the exposure.

COHORT STUDIES

By contrast with the cross-sectional study design, cohort studies explicitly incorporate follow-up time; the cohort study permits the assessment of the association of incident events during follow-up with air pollution exposure. Two broad types of cohort studies of air pollution can be distinguished: the short-term studies, typically including small numbers of participants and intensive observation, referred to as 'panel' studies, and long-term studies, usually directed at chronic effects and mortality.

The term 'panel study' likely originated with early studies including follow-up of patient panels having chronic respiratory disease. For example, in an early study in London, Lawther and colleagues (1970) assessed daily health status in a group having COPD. A strong association observed initially weakened as air pollution levels declined over the course of the study.

Most contemporary panel studies include either groups considered to be susceptible to air pollution, e.g. persons with asthma, or groups with potentially informative patterns of exposure; typically, observations on exposures and outcomes are made on a daily basis. Exposure estimates may be derived from centrally sited monitors, from personal monitors, or based on the microenvironmental approach. The health outcome data need to vary on a day-to-day basis, as do air pollution concentrations. Panel studies often involve ascertainment of respiratory symptoms, medication use, and peak expiratory flow rate (PEFR).

The so-called 'camp' studies represent another application of the panel study design. These studies have involved groups of children attending summer camps. Typically, camp activities would be expected to increase lung pollutant dose because of heightened ventilation associated with exercise. By placing a monitor at the camp, researchers can estimate exposure and most time is spent outdoors or in very open indoor environments. This design has been extended to adults who exercise outdoors (Spektor *et al*, 1988). Findings of the camp studies have been particularly informative on the short-term effects of ozone (Lippmann, 1989), showing exposure-response relationships with PEFR quite comparable to the experimental findings.

Analysis of the panel study data has been facilitated by the new statistical methods for correlated time series. Schwartz *et al.* (1991) provide a review of approaches for diary data. The report by Neas *et al.* (1995) exemplifies the analysis of lung function data.

Few long-term cohort studies of air pollution and respiratory health have been conducted. The Berlin, New Hampshire study, initiated by Ferris in the 1960s (Ferris *et al.*, 1971) represents one of the pioneering designs. The Six Cities Study stands as the benchmark: a 17-year study including prospective data collection on exposures and outcomes (Dockery *et al.*, 1993; Ferris *et al.*, 1979).

The long-term cohort studies have been carried out to assess effects of air pollution on lung growth and aging, effects which can only be assessed longitudinally; on symptom and disease incidence; and on total and cause-specific mortality. Representative reports from the Six Cities investigators illustrate these applications of the cohort design (Dockery *et al.*, 1993; Duan, 1982; Neas *et al.*, 1995; Speizer, 1989; Speizer *et al.*, 1989). Few, major cohort studies of outdoor air pollution and respiratory health are now in progress.

OTHER STUDY DESIGNS

Ecological studies based on group and cross-sectional observations and cohort studies based on individuals as the units of observation are the most frequently used study designs in air pollution epidemiology. Case–control studies are experimental designs applying an intervention, and a change in exposure levels could offer some advantages, as a small number of previous studies demonstrate.

When studying the effects of air pollution on less frequent diseases such as lung cancer or other cancers, researchers find that a cohort study has several limitations. Ascertainment of a sufficient number of cases in a prospective cohort study would require hundreds of thousands of subjects (cancer incidence ~1 per 1000 person-years). The participants would have to be followed for several years to take into account the latency period from exposure to onset of disease or it would be necessary to collect retrospective exposure information for a well-defined cohort or dynamic population.

In a case–control study, the size of the study population can be optimized by choosing cases and an appropriate number of controls whose exposure to air pollutants need to be assessed. The cases can be selected from disease or mortality registries or from health-care facilities treating the patients with a disease of interest. The selection of appropriate controls is crucial for the validity of the study. Theoretically, the controls should be selected from the source population that produced the cases. In studies in which the cases come from population-based registries, selection of controls can be done in a straightforward fashion from the source population (a population-based or primary-based case–control study). In studies including cases from institutions (hospitals) without a clearly defined source population, the controls can be chosen from another patient group in the same institution. One of the few case–control studies of the health effects of air pollution was carried out by Jedrychowski and colleagues (1990) in the city of Cracow, Poland. They selected cases of lung cancer from a population-based mortality registry; controls were deceased subjects from other causes frequency-matched by age and sex. Exposure assessment was based on residential history in different parts of the city. Information on potential confounders was collected from next of kin.

A special type of case–control study is a prospective, cohort-based case–control study ('nested case–control study') where all new (incident) cases are selected from a cohort or from dynamic populations, and controls are drawn from the same source. This study combines strengths from both traditional cohort and case–control study designs. Information on some determinants of the disease of interest are collected prospectively, as in a cohort study, which guards against potential information bias and strengthens the causal inference as to temporality. Intensive assessment of disease and exposure can be directed at a limited number of cases and controls, rather than studying all the members of the cohort. This design was applied in a case–control study of bronchial obstruction in young children in Oslo, Norway, to study effects of indoor (Nafstad *et al.*, 1998) and outdoor air pollution (Magnus *et al.*, 1998). A total of 3754 children were followed from birth to 2 years and a total of 304 cases of bronchial obstruction were identified. The homes of the cases and age-matched controls underwent an inventory of potential sources of indoor pollution and measurements of indoor air quality.

A change in hypothesized cause followed by a change in the effect is strong evidence for causality. Therefore experimental and quasi-experimental study designs – studies where effects of a change in exposure take place without influence of the investigator (often called 'natural experiments) – could produce valuable information on the health effects of air pollution. The randomized, controlled, and blinded clinical trial has become a paradigm of a valid study of causality, because, in addition to the temporality, the design takes into account the comparability of populations (randomization) as to individual characteristics and extraneous effects, and comparability of information (blinding). A search of the literature to date does not yield any intervention studies in the context of outdoor air pollution studies. However, studies of the effect of indoor environment have applied controlled experiments where ventilation rate (Jaakkola *et al.*, 1991), proportion of air-recirculation (Jaakkola *et al.*, 1998) and humidification (Reinikainen *et al.*, 1992), have been altered in order to study their effects.

The investigators may not be able to influence the changes in exposure to air pollution, but there are examples of studies which have used a situation where changes have taken place for other reasons. Researchers in the South Karelia Air Pollution Study of the effects of emissions from pulp mills applied a cohort study with a 'natural experiment' (Jaakkola *et al.*, 1991). They were informed about future emission reductions from pulp mills due to installation of new technology. A cohort study was designed with a baseline data collection 4 months before the expected emission reduction in two communities in the vicinity of the pulp mills and in one reference community. A follow-up was carried out 14 months after the emission reduction. The study design allowed assessment of intraindividual change in the hypothesized outcomes including occurrence of respiratory infections and respiratory and neuro-psychological symptoms. Pope (1989) studied retrospectively the association of hospital admissions to particulate air pollution from steel mills, taking advantage of the closure and reopening of the local steel mill during the study period.

In summary, experimental designs provide a strong pool for studying the health effects of air pollution. Although ideal designs may not be possible, it is useful to imitate experimental settings to the maximum extent possible.

INTERPRETATION OF EPIDEMIOLOGIC DATA

In interpreting the results of an epidemiologic study, consideration needs to be given to the findings and limitations of the study and to the context set by other epidemiologic evidence and experimental data. An association between air pollution exposure and the occurrence of an adverse health effect may reflect chance, bias or cause. Statistical significance testing provides a measure of the likelihood that the association has occurred by chance.

Bias may exaggerate or weaken the degree of association. Misclassification and confounding are of particular concern in considering the findings of epidemiologic investigations of air pollution. Most often, random misclassification of exposure to air pollutants is anticipated, reflecting the use of exposure measures of limited validity to estimate personal exposures. Under most circumstances, random misclassification of exposure reduces the magnitude of associations; thus, the degree of association found in a study might have been greater if more valid measures of exposure were available. If the degree of misclassification can be estimated, then measures describing the effect of exposure can be adjusted (Kleinbaum *et al.*, 1982). In interpreting studies showing apparent lack of association, misclassification also merits consideration as an explanation for null findings.

Many factors may potentially confound the relation between air pollution exposure and health outcome (Table 20.8); confounding by these factors might increase or decrease the magnitude of association. In the USA and many other developed countries, the extent of air pollution exposure, both indoor and outdoor, is increasingly linked to occupational status and other socioeconomic factors that determine residence location and housing characteristics. Cigarette smoking is increasingly associated with education and income (US Department of Health and Human Services, 1989); thus, active smoking may confound associations with air pollution, and persons involuntarily exposed to environmental tobacco smoke may differ in potentially important characteristics from those not exposed.

The potential for confounding in epidemiologic studies of air pollution is well recognized. Investigators collect information on potentially confounding factors and then attempt to control for confounding as the data are analyzed. Alternatively, confounding may be controlled by excluding subjects exposed to the confounding factor of concern, e.g. excluding cigarette smokers from a study. Multivariate statistical methods are now widely employed to assess the effect of pollution exposure while controlling for the effects of confounding factors. Interpretation of the results of such models should consider the accuracy with which the confounding factors themselves can be accurately measured and the appropriateness of the model to represent the relationships of the confounding factors with the health outcome variable.

Criteria have been developed to assess the causality of associations found in epidemiologic investigations (Table 20.4) (Hill, 1965; Rothman, 1986). These criteria have been widely applied and have guided many important assessments of epidemiologic data; for example, the criteria were applied in the 1964 Surgeon General's Report that concluded that smoking caused lung cancer (US Department of Health and Human Services, 1964). On re-examination, however, it is apparent that the criteria cannot be applied in a rigid fashion, particularly in interpreting the often weak associations found in epidemiologic studies of air pollution. Clearly, exposure must precede outcome (Rothman, 1986), but meeting the other criteria cannot be regarded as necessary for judgment that an association is causal.

Interpretation of apparently negative studies also needs to consider bias, as both confounding and misclassification may reduce associations toward the null and reduce the statistical power afforded by the sample size. Statistical power refers to the probability that an investigation's results will be statistically significant for a particular level of effect with a particular sample size. Thus, in interpreting the results of a 'negative' study, statistical power should be calculated for effects of potential public health significance. Confidence limits, which describe the range of effect compatible with the data at a specified level of probability, should also be considered for studies showing no association.

CONCLUSIONS

Epidemiology will remain an important research method for addressing the health effects of air pollution. It provides evidence directly from human populations that is complementary to the findings of toxicologic studies. Although epidemiologic studies are subject to potentially limiting methodologic problems, new approaches for assessing exposure to pollutants and health outcomes should strengthen the epidemiologic approach to investigating the health effects of air pollution.

REFERENCES

Allred EN, Bleecker ER, Chaitman BR *et al.* (1989) Short-term effects of carbon monoxide exposure on the exercise performance of subjects with coronary artery disease. *N Engl J Med* **321**: 1426–1432.

American Thoracic Society (1987) Standardization of spirometry – 1987 update. *Am Rev Respir Dis* **136**: 1285–1298.

American Thoracic Society (1996a) Committee of the Environmental and Occupational Health Assembly: Bascom R, Bromberg PA, Costa DA *et al.* Health effects of outdoor air pollution. Part 1. *Am J Respir Crit Care Med* **153**: 3–50.

American Thoracic Society (1996b) Committee of the Environmental and Occupational Health Assembly: Bascom R, Bromberg PA, Costa DA *et al.* Health effects of outdoor air pollution. Part 2. *Am J Respir Crit Care Med* **153**: 477–498.

Armstrong BG and Oakes D (1982) Effects of approximation in exposure assessments on estimates of exposure–response relationships. *Scand J Work Environ Health* **8**(Suppl. 1): 20–23.

Bates DV (1989) 4, Altered physiologic states and associated syndromes. In: *Respiratory Function in Disease*. Philadelphia, Pennsylvania: W. B. Saunders Company.

Benson FB, Henderson JJ and Caldwell DE (1972) *Indoor–Outdoor Air Pollution Relationships: A Literature Review*. Pub. no. AP-112. Research Triangle Park, NC: US EPA.

Bromberg PA (1988) Asthma and automotive emissions. In: Watson AY and Bates RR (eds) *Air Pollution, the Automobile and Public Health*. Washington, DC: National Academy Press, pp. 465–498.

Burney PG, Laitinen LA, Perdrizet S *et al.* (1989) Validity and repeatability of the IUATLD (1984) Bronchial Symptoms Questionnaire: an international comparison. *Eur Respir J* **2**(10): 940–945.

Cross D and Nelson HS (1991) The role of the peak flow meter in the diagnosis and management of asthma. *J Allergy Clin Immunol* **87**: 120–128.

Denny FW and Clyde WAJ (1986) Acute lower respiratory tract infections in nonhospitalized children. *J Pediatr* **108**: 635–646.

Detels R, Rokaw SN, Coulson AH *et al.* (1979) The UCLA population studies of chronic obstructive respiratory disease. I. Methodology and comparison of lung function in areas of high and low pollution. *Am J Epidemiol* **109**(1): 33–58.

Detels R, Sayre JW, Coulson AH *et al.* (1981) The UCLA population studies of chronic obstructive respiratory disease. IV. Respiratory effect of long-term exposure to photochemical oxidants, nitrogen dioxide, and sulphates on current and never smokers. *Am Rev Respir Dis* **124**: 673–680.

Detels R, Tashkin DP, Sayre JW *et al.* (1987) The UCLA population studies of chronic obstructive pulmonary disease. *Chest* **92**: 594–603.

Diggle PJ, Liang KY and Zeger SL (1994) *Analysis of Longitudinal Data*. New York: Oxford University Press.

Dingle JH, Badger GF and Jordan WSJ (1964) *Illness in the Home. A Study of 25,000 Illnesses in a Group of Cleveland Families*. Cleveland, OH: Western Reserve University Press.

Dockery DW and Pope CA III (1994) Acute respiratory effects of particulate air pollution. *Ann Rev Public Health* **15**: 107–132.

Dockery DW, Pope CA III, Xu X *et al.* (1993) An association between air pollution and mortality in six U.S. cities. *N Engl J Med* **329**(24): 1753–1759.

Dominici F, Samet JM, Xu J and Zeger SL (1999) Air pollution and daily mortality. Combining results from multiple locations by hierarchical modeling. *J Am Stat Assoc* (submitted).

Duan N (1982) Models for human exposure to air pollution. *Environ Int* **8**: 305–309.

Evans AS (1993) *Causation and Disease: A Chronological Journey*. New York: Plenum Medical Books.

Federal Focus (1996) *Principles for Evaluating Epidemiologic Data in Regulatory Risk Assessment*. 0-9654148-0-9. Washington, DC: Federal Focus.

Ferris BG Jr (1978) Epidemiology standardization project. Part II. *Am Rev Respir Dis* **118**: 1–120.

Ferris BGJr, Higgins IT, Higgins MW *et al.* (1971) Chronic nonspecific respiratory disease, Berlin, New Hampshire, 1961–1967: a cross-sectional study. *Am Rev Respir Dis* **104**(2): 232–244.

Ferris BG Jr, Speizer FE, Spengler JD *et al.* (1979) Effects of sulfur oxides and respirable particles on human health: methodology and demography of populations in study. *Am Rev Respir Dis* **120**: 767–779.

Firket J (1936) Fog along the Meuse valley. *Trans Faraday Soc* **32**: 1192–1197.

Frank R, O'Neil JJ, Utell MJ *et al.* (1985) *Inhalation Toxicology of Air Pollution: Clinical Research Considerations*. Philadelphia, PA: American Society for Testing and Materials.

Gladen B and Rogan WJ (1979) Misclassification and the design of environmental studies. *Am J Epidemiol* **109**: 607–616.

Goldstein IF and Landovitz L (1977a) Analysis of air pollution patterns in New York City – I. Can one station represent the large metropolitan area? *Atmos Environ* **11**: 47–52.

Goldstein IF and Landovitz L (1977b) Analysis of air pollution patterns in New York City – II. Can one aerometric station represent the area surrounding it? *Atmos Environ* **11**: 53–57.

Graham NMH (1990) The epidemiology of acute respiratory infections in children and adults: a global perspective. *Epidemiol Rev* **12**: 149–178.

Greenland S and Morgenstern H (1989) Ecological bias, confounding, and effect modification. *Int J Epidemiol* **18**: 269–274.

Hazucha MJ, Folinsbee LJ and Seal EJ (1992) Effects of steady-state and variable ozone concentration profiles on pulmonary function. *Am Rev Respir Dis* **146**(6): 1487–1493.

Hill AB (1965) The environment and disease: association or causation? *Proc Roy Soc Med* **58**: 295–300.

Holland WW, Bennett AE, Cameron IR *et al.* (1979) Health effects of particulate pollution: reappraising the evidence. *Am J Epidemiol* **110**(5): 533–659.

Jaakkola JJ, Vilkka V, Marttila O *et al.* (1990) The South Karelia Air Pollution Study. The effects of malodorous sulfur compounds from pulp mills on respiratory and other symptoms. *Am Rev Respir Dis* **142**(6: Pt 1): t–50.

Jaakkola JJK, Heinonen OP and Seppänen O (1991) Mechanical ventilation in office buildings and the sick building syndrome. An experimental and epidemiological study. *Indoor Air* **1**: 111–122.

Jaakkola JJK, Partti-Pellinen K, Marttila O *et al.* (1998) The South Karelia Air Pollution Study: the effects of emission reduction of malodorous sulfur compounds from pulp mills on the occurrence of acute respiratory infection and symptoms. *Occup Environ Med* (accepted).

Jaakkola JJK, Tuomaala P and Seppänen O (1994) Air recirculation and sick building syndrome: a blinded crossover trial. *Am J Publ Health* **84**: 422–428.

Jedrychowski W, Becher H, Wahhrendorf J and Basa-Cierpialek Z (1990) A case–control study of lung cancer with special reference to the effect of air pollution in Poland. *J Epidemiol Community Health* **44**: 114–120.

Jenkins MA, Clarke JR, Carlin JB *et al.* (1996) Validation of questionnaire and bronchial hyperresponsiveness against respiratory physician assessment in the diagnosis of asthma. *Int J Epidemiol* **25**(3): 609–616.

Katsouyanni K, Schwartz J, Spix C *et al.* (1995) Short term effects of air pollution on health: a European approach using epidemiologic time series data: The APHEA protocol. *J Epidemiol Community Health* **50**: S12–S18.

Kilburn KH, Seidman BC and Warshaw R (1985) Neurobehavioral and respiratory symptoms of formaldehyde and xylene exposure in histology technicians. *Arch Environ Health* **40**: 229–233.

Kircher T and Anderson RE (1987) Cause of death. Proper completion of the death certificate. *J Am Med Assoc* **258**(3): 349–352.

Kleinbaum DG, Kupper LL and Morgenstern H (1982) *Epidemiologic Research. Principles and Quantitative Methods*. Belmont, CA: Lifetime Learning Publications.

Lambert WE, Samet JM, Skipper BJ *et al.* (1994) Nitrogen dioxide and respiratory illness in children. In: *Health Effects Institute Report Number 58. Part III: Quality assurance in an epidemiologic study*. Cambridge, MA: Health Effects Institute Report, pp. 1–33.

Larssen S, Tonnesen D, Clench-Aas J *et al.* (1993) A model for car exhaust exposure calculations to investigate health effects of air pollution. *Sci Total Environ* **134**(1–3): 51–60.

Lave LB and Seskin EP (1973) An analysis for the association between U.S. mortality and air pollution. *J Am Stat Assoc* **68**: 284–290.

Lave LB and Seskin EP (1977) *Air Pollution and Human Health*. Baltimore, MD: Johns Hopkins University Press.

Lawther PJ, Waller RE and Henderson M (1970) Air pollution and exacerbations of bronchitis. *Thorax* **25**: 525–539.

Lebowitz MD, Holberg CJ, Boyer B and Hayes C (1985) Respiratory symptoms and peak-flow associated with indoor and outdoor air pollutants in the Southwest. *J Air Pollut Control Assoc* **35**: 1154–1158.

Letz R (1991) Use of computerized test batteries for quantifying neurobehavioral outcomes. *Environ Health Perspect* **90**: 195–198.

Liang KY and Zeger SL (1986) Longitudinal data analysis using generalized linear models. *Biometrika* **73**(1): 13–22.

Lilienfeld AM and Lilienfeld DE (1980) *Foundations of Epidemiology*, 2nd edn. New York: Oxford University Press.

Lippmann M (1989) Health effects of ozone. A critical review. *J Air Pollut Control Assoc* **39**: 672–695.

Lunn JE, Knowelden J and Handyside AJ (1967) Patterns of respiratory illness in Sheffield infant school-children. *Br J Prev Soc Med* **21**: 7–16.

Magnus P, Nafstad P, Oie L *et al.* (1998) Exposure to nitrogen oxide and the occurrence of bronchial obstruction in children below 2 years. *J Epidemiol* (in press).

Mauderly JL (1989) Susceptibility of young and aging lungs to inhaled pollutants. In: Utell MJ and Frank R (eds) *Susceptibility to Inhaled Pollutants*. Philadelphia, PA: American Society for Testing and Materials, pp. 148–161.

Mead J (1979) Problems in interpreting common tests of pulmonary mechanical function. In: Macklem PT and Permutt S (eds) *The Lung in the Transition Between Health and Disease*. New York: Marcel Dekker, pp. 43–51.

Molhave L (1990) 15, Volatile organic compounds, indoor air quality and health. In: *Proceedings of the 5th International Conference on Indoor Air Quality and Climate*. Ottawa: Canada Mortgage and Housing Corp., pp. 33–76.

Molhave L, Each B and Pedersen OF (1990) Human reactions to low concentrations of volatile organic compounds. *Environ Int* **12**: 167–175.

Monto AS and Ullman BM (1974) Acute respiratory illness in an American community. *J Am Med Assoc* **227**: 164–169.

Monto AS, Napier JA, Metzner HL (1971) The Tecumseh study of respiratory illness. I. Plan of study and observations on syndromes of acute respiratory disease. *Am J Epidemiol* **94**: 269–279.

Morgenstern H (1982) Uses of ecologic analysis in epidemiologic research. *Am J Public Health* **72**: 11336–11344.

Morgenstern H (1995) Ecologic studies in epidemiology: concepts, principles, and methods. *Annu Rev Public Health* **16**: 61–81.

Nafstad P, Oie L, Mehl R *et al.* (1998) Residential dampness problems and development of bronchial obstruction in Norwegian children. *Am J Respir Crit Care Med* **157**: 410–414.

Navidi W, Thomas D, Stram D and Peters J (1994) Design and analysis of multilevel analytic studies with applications to a study of air pollution. *Environ Health Perspect* **102**(Suppl. 32).

Neas LM, Dockery DW, Koutrakis P *et al.* (1995) The association of ambient air pollution with twice daily peak expiratory flow rate measurements in children. *Am J Epidemiol* **141**(2): 111–122.

NRC (1985) National Research Council, Commission on Life Sciences, Board on Toxicology and Environmental Health Hazards *et al. Epidemiology and Air Pollution.* Washington, DC: National Academy Press, p. 1.

NRC (1988) National Research Council and Committee on the Biological Effects of Ionizing Radiation. *Health Risks of Radon and Other Internally Deposited Alpha-Emitters: BEIR IV.* Washington, DC: National Academy Press.

NRC (1989) *Biologic Markers in Pulmonary Toxicology.* Washington, DC: National Academy Press.

NRC (1991a) National Research Council (NRC) and Committee on Advances in Assessing Human Exposure to Airborne Pollutants. *Human Exposure Assessment for Airborne Pollutants: Advances and Opportunities.* Washington, DC: National Academy Press.

NRC (1991b) *Frontiers in Assessing Human Exposures to Environmental Toxicants.* Washington, DC: National Academy Press.

Partti-Pellinen K, Marttila O, Vilkka V *et al.* (1996) The South Karelia Air Pollution Study: effects of low-level exposure to malodorous sulfur compounds on symptoms. *Arch Environ Health* **51**(4): 315–320.

Percy C, Stanek E and Gloeckler L (1981) Accuracy of cancer death certificates and its effect on cancer mortality statistics. *Am J Public Health* **71**: 242–250.

Pope CA III (1989) Respiratory disease associated with community air pollution and a steel mill, Utah Valley. *Am J Public Health* **79**: 623–628.

Pope CA III, Dockery DW, Schwartz J (1995) Reveiw of epidemiologic evidence of health effects of particulate air pollution. *Imhal Toxicol* **7**: 1–18.

Quackenboss JJ, Lebowitz MD and Krzyzanoski M (1991) The normal range of diurnal changes in peak expiratory flow rates. Relationship to symptoms and respiratory disease. *Am Rev Respir Dis* **143**: 323–330.

Reinikainen LM, Jaakkola JJK and Seppanen O (1992) The effect of air humidification on symptoms and the perception of air quality in office workers. A six period cross-over trial. *Arch Environ Health* **47**: 8–15.

Rothman KJ (1986) *Modern Epidemiology,* 1st edn. Boston, MA: Little, Brown and Company.

Samet JM (1978) A historical and epidemiological perspective on respiratory symptoms questionnaires. *Am J Epidemiol* **108**: 435–446.

Samet JM (1989) Definitions and methodology in COPD research. In: Hensley M and Saunders N (eds) *Clinical Epidemiology of Chronic Obstructive Lung Disease.* New York: Marcel Dekker, pp. 1–22.

Samet JM, Speizer FE, Bishop Y *et al.* (1981) The relationship between air pollution and the emergency room visits in an industrial community. *J Air Pollut Control Assoc* **31**: 236–240.

Samet JM, Tager IB and Speizer FE (1983) The relationship between respiratory illness in childhood and chronic airflow obstruction in adulthood. *Am Rev Respir Dis* **127**: 508–523.

Samet JM, Marbury MC and Spengler JD (1987a) Health effects and sources of indoor air pollution. Part I. *Am Rev Respir Dis* **136**: 1486–1508.

Samet JM, Humble CG, Skipper BE and Pathak DR (1987b) History of residence and lung cancer risk in New Mexico. *Am J Epidemiol* **125**: 800–811.

Samet JM, Marbury MC and Spengler JD (1988) Health effects and sources of indoor air pollution. Part II. *Am Rev Respir Dis* **137**: 221–242.

Samet JM, Lambert WE, Skipper BJ *et al.* (1992) A study of respiratory illnesses in infants and NO_2 exposure. *Arch Environ Health* **47**: 57–63.

Savitz DA and Feingold L (1989) Association of childhood cancer with residential traffic density. *Scand J Work Environ Health* **15**(5): 360–363.

Schenker MB, Weiss ST and Murawski BW (1982) Health effects of residents in homes with urea formaldehyde foam insulation: a pilot study. *Environ Int* **8**: 359–363.

Schwartz J, Wypij D, Dockery DW *et al.* (1991) Daily diaries of respiratory symptoms and air pollution: methodological issues and results. *Environ Health Perspect* **90**: 181–187.

Shy CM, Goldsmith JR, Hackney JD *et al.* (1978a) Health effects of air pollution. *ATS News* **6**: 1–63.

Shy CM, Kleinbaum DG and Morgenstern H (1978b) The effect of misclassification of exposure status in epidemiological studies of air pollution health effects. *Bull N Y Acad Med* **54**: 1155–1165.

Sly RM (1989) Mortality from asthma. *J Allergy Clin Immunol* **84**(4: Pt 1): 421–434.

Sparrow D and Weiss ST (1989) Background. In: Weiss S and Sparrow D (eds) *Airway Responsiveness and Atopy in the Development of Chronic Lung Disease.* New York: Raven Press, pp. 1–19.

Speizer FE (1989) Studies of acid aerosols in six cities and in a new multi-city investigation: design issues. *Environ Health Perspect* **79**: 61–67.

Speizer FE, Fay ME, Dockery DW and Ferris BG Jr (1989) Chronic obstructive pulmonary disease mortality in six U.S. cities. *Am Rev Respir Dis* **140**: s49–55.

Spektor DM, Lippmann M, Thurston GD *et al.* (1988) Effects of ambient ozone on respiratory function in healthy adults exercising outdoors. *Am Rev Respir Dis* **138**(4): 821–828.

Spengler JD and Soczec MC (1984) Evidence for improved ambient air quality and the need for personal exposure research. *Environ Sci Technol* **8**: 268–280A.

Tager IB and Speizer FE (1977) Surveillance techniques for respiratory illness. In: Finkel AJ and Duel WC (eds) *Clinical Implications of Air Pollution Research*. Acton: Publishing Sciences Group, pp. 339–346.

US Department of Health and Human Services (1964) Report of the Advisory Committee to the Surgeon General: *Smoking and Health*. Washington, DC: US Government Printing Office. DHEW Publication no. [PHS] 1103.

US Department of Health and Human Services (1986) A Report of the Surgeon General: *The Health Consequences of Involuntary Smoking*. Washington, DC: US Government Printing Office.

US Department of Health and Human Services (1989) A Report of the Surgeon General: *Reducing the Health Consequences of Smoking. 25 Years of Progress*. Washington, DC: US Government Printing Office.

US EPA (1974) *Health Consequences of Sulphur Oxides: A Report from CHESS, 1970–1971*. EPA-650/1-74-004. Research Triangle Park, NC: US EPA.

US EPA (1989) US Environmental Protection Agency (EPA) and Office of Health and Environmental Assessment. An acid aerosols issue paper. Health effects and aerometrics. 600/8-88-005F. Washington, DC: US Government Printing Office.

US EPA (1993) *Indoor–Outdoor Air Pollution Relationships*, volume II: An Annotated Bibliography. Pub. no. AP-112b. Research Triangle Park, NC: US EPA.

US EPA (1995) Proposed guidelines for neurotoxicity risk assessment. Report no. 60(192). p. 52031.

US Public Health Service (1964) *Smoking and Health*. Report of the advisory committee to the surgeon general of the Public Health Service. Washington, DC: US Government Printing Office.

Wallace LA, Samet JM and Spengler JD (eds) (1991) Volatile organic compounds. In: *Indoor Air Pollution. A Health Perspective*. Baltimore, Maryland: Johns Hopkins University Press, pp. 252–272.

Wallace LA and Ott WR (1982) Personal monitors: a state-of-the-art survey. *J Air Pollut Control Assoc* **32**: 601–610.

Ware JH, Spengler JD, Neas LM *et al.* (1993) Respiratory and irritant health effects of ambient volatile organic compounds: the Kanawha County Health Study. *Am J Epidemiol* **137**(12): 1287–1301.

Whittemore AS and Korn EL (1980) Asthma and air pollution in the Los Angeles area. *Am J Public Health* **70**(7): 687–696.

Wjst M, Reitmeir P, Dold S *et al.* (1993) Road traffic and adverse effects on respiratory health in children. *Br Med J* **307**(6904): 596–600.

Wright AL, Taussig LM, Ray GC and Group Health Medical Associates (1989) The Tucson children's respiratory study. II. Lower respiratory tract illness in the first year of life. *Am J Epidemiol* **129**: 1232–1246.

21

Health Effects of Air Pollution Episodes

H. ROSS ANDERSON

St George's Hospital Medical School, London, UK

INTRODUCTION

The term 'air pollution episode' normally implies a short-term increase in ambient pollution which is greater than would be normally expected as part of day-to-day variation. In their most extreme form, air pollution episodes are accompanied by widespread public fear, disruption of day-to-day living, physical discomfort, illness and even death; they therefore belong to the family of environmental disasters which includes floods, earthquakes, volcanic eruptions, heatwaves and large-scale chemical or ionizing radiation escapes. At the other end of the scale, episodes may be no more than short periods during which pollution exceeds the usual range, unaccompanied by noticeable effects. The pollution is usually the direct or indirect consequence of burning fuel for transport, industry or domestic use, but may also come from other sources such as forest fires or volcanic eruptions. Episodes of pollution from the burning of fuel tend to occur not because of an increase in emissions, but because stagnant weather conditions impair their dispersal. Certain areas or towns may be prone to this because of their particular topography or climate. Early studies used the term 'fog' even when it was known to be fog associated with increased air pollution. The term 'smog' came into more general use in the latter half of the twentieth century, although according to Marsh (Wilkins, 1954) it was coined in 1905 by Dr H.A. Des Voex, a London physician and a founder of the Coal Abatement Society, to describe 'smoke-impregnated fog'. Because not all increases in air pollution are associated with fog, the term 'episode' is more appropriate if less emotive.

Over the last decade, most studies of short-term effects of air pollution have analysed time-series of daily data over long periods using increasingly sophisticated regression techniques. These techniques enable associations to be identified at lower levels than is possible with episode analysis techniques, while controlling for long- and short-term confounding factors such as season and temperature. As will become clear in this chapter,

AIR POLLUTION AND HEALTH
ISBN 0-12-352335-4

analysis of air pollution episodes is less sensitive to the detection of effects and more subject to uncontrolled confounding. In spite of this, the investigation of air pollution episodes remains important in air pollution epidemiology. The marked increases in mortality observed during episodes in Belgium in 1930 (Firket, 1936), Donora, PA, in 1948 (Shrenk *et al.*, 1949) and London in 1952 (Ministry of Health, 1954), were very influential in establishing that air pollution could have harmful short-term effects and in persuading governments to improve abatement strategies. The London fog of 1952 is a landmark in air pollution epidemiology because the effect was so large (an estimated 4000 deaths) and was easily identified without the necessity for subtle statistical methods. Without such evidence, the more recent results of time-series analyses of daily mortality would have less credibility. The investigation of episodes remains an important task for air pollution epidemiologists. Sometimes it is demanded by a concerned public or their politicians and the results may be still be influential in guiding policies and in educating the public to accept changes. They also retain scientific value by adding to case lore concerning the effects of mixtures of pollutants while at the same time indicating the possibility of effects at more usual levels of pollution.

The next section of this chapter will describe in narrative form the main published evidence on the health effects of air pollution episodes. It will cover winter and summer episodes as well as pollution from forest fires and volcanic eruptions. This will be followed by a discussion of methodological issues which are important for the carrying out and interpretation of studies.

WINTER EPISODES IN EUROPE (Table 21.1)

The majority of European evidence is from London, UK, though doubtless other cities were also affected by fogs. The size of London, together with the availability of relatively accessible data on daily deaths and weather factors (though routine air pollution data only became available in the 1940s), are among the reasons for it being a focus of interest.

London fogs

For a number of centuries, London was noted for its fogs (Brimblecombe, 1987). At the end of the nineteenth century Claude Monet, the French impressionist painter, found inspiration in the colours of the fogs and sunsets over the Thames and painted at least 100 works on these themes. What is now Greater London lies in the Thames Valley, which is bounded to the north and south by a range of low hills. These aid in the creation of temperature inversions and hamper the dispersal of fog. However, it also seems likely that the frequent fogs of the past were associated with the different climatic factors which prevailed during that time and that a change in climate is the main reason why fogs have declined in frequency during the twentieth century.

It was recognized in earlier centuries that air pollution might be harmful to health. In the seventeenth century John Evelyn, in attempting to excite interest in the effects of London smoke referred to it as the 'hellish and dismal cloud of sea coal which maketh the city of London resemble the suburbs of hell' (Evelyn, 1661). Although increased mortality during periods of fog had been noted in London and some other English cities since at least the eighteenth century it was not until the early twentieth century that serious attempts were made to quantify its effects. In an analysis of deaths occurring between October to March over 27 years Russell found a positive association between fogginess and respiratory mortality when associated with low temperatures and frosts (Russell, 1924, 1926). Although it was widely suspected that air pollution might be harmful, its contribution to mortality was not stressed in comparison with low temperature. For example, when Logan reported on an increase in mortality associated with the 1948 London fog he did not mention air pollution at all (though this was increased) (Logan, 1949).

Meuse Valley, Belgium 1930

This fog affected a narrow valley about 20 km long along the Meuse from 1 to 5 December 1930 (Firket, 1936). Ambient concentrations of air pollutants were not available at the time but probably included sulfur dioxide (SO_2), sulfuric acid, particles and various other pollutants including metals emitted from a number of factories. The fog was very unpleasant if not frightening for the population and a large proportion became unwell, mainly with retrosternal chest pain, coughing, shortness of breath and wheezing. Signs of cardiorespiratory failure occurred in several hundred patients with existing cardiorespiratory disease. There were 60 deaths, 10 times the number expected. The nature of the symptoms together with the results of 15 autopsies indicated toxic irritation of the entire respiratory tract. Symptoms did not become severe until the third day of the fog and resolved rapidly when it cleared. It was reported that many cattle had to be slaughtered, but the reasons for this are not recorded. In his paper, Firket noted prophetically that if such an event were to occur in London the number of deaths would be about 3200. Although affecting a small population, this episode is important as probably the first to show with some certainty the potential of air pollution to cause morbidity and death.

London 1952

This well-known episode occurred on 5 to 9 December 1952 and remains an important landmark in air pollution epidemiology because of the scale of effects – about 4000 extra deaths (Committee on Air Pollution, 1953; Logan, 1953; Ministry of Health, 1954; Wilkins, 1954). The episode occurred because a stationary high-pressure system situated over Western Europe led to a prolonged windless period in the Thames Valley. Aided by topographical factors, the calm conditions created a temperature inversion. Pollutants, mainly from the combustion of coal in domestic houses, could not disperse and dense fog covered the city to a height of 250–300 feet for a period of 4 days. Smoke levels were measured by the reflectance method (black smoke) in various parts of the city and reached a maximum concentration of 4460 $\mu g/m^3$ at County Hall and an average level over the four days of about 1600 $\mu g/m^3$ (about five times the normal level). SO_2 concentrations rose in parallel with those of smoke to a maximum of 1344 ppb at the County Hall with maximum values at different sites ranging from three to five times normal.

Table 21.1 Winter air pollution episodes in Europe[a]

Place and year	Dates	Reference	Maximum pollution levels[b]	Relative health effects[c]	Attributable health effects[d]
London 1873	9–12 Dec	(Ministry of Health, 1954)	n/a	Deaths all causes 1.4 / Deaths bronchitis 1.7	660 (LAC)
London 1880	26–29 Jan	(Ministry of Health, 1954)	n/a	Deaths all causes 1.5 / Deaths bronchitis 2.3	1180 (LAC)
London 1882	2–7 Feb	(Ministry of Health, 1954)	n/a	Deaths all causes 1.3 / Deaths bronchitis 1.6	610 (LAC)
London 1891	21–24 Dec	(Ministry of Health, 1954)	n/a	Deaths all causes 1.9 / Deaths bronchitis 2.6	1630 (LAC)
London 1892	28–30 Dec	(Ministry of Health, 1954)	n/a	Deaths all causes 1.4 / Deaths bronchitis 1.9	680 (LAC)
Glasgow 1925	17–23 Nov	(Ministry of Health, 1954)	n/a	Deaths all causes 1.6	
Meuse Valley 1930	1–5 Dec	(Firket, 1936)	n/a	Deaths all causes 6.0	50
London 1948	26 Nov–1 Dec	(Ministry of Health, 1954)	Smoke ($\mu g/m^3$): single max 2780, mean max 1086 SO_2 (ppb): single max 750, mean max 445	Deaths all causes 1.3 / Deaths bronchitis 2.0	800 (LAC)
London 1952	5–9 Dec	(Logan, 1953; Ministry of Health, 1954; Wilkins, 1954)	Smoke ($\mu g/m^3$): single max 4460, mean max 1600 SO_2 (ppb): single max 1340, mean max 700	Deaths all causes 2.6 / Deaths bronchitis 9.3 / Increase in GP visits / Increase in hospital admissions	1540 (LAC) 4000 (GL)
London 1956	3–6 Jan	(Logan, 1956)	Smoke ($\mu g/m^3$): single max 2830, mean max 1229 SO_2 ($\mu g/m^3$): max 1430	Deaths all causes 1.25 / Deaths bronchitis 2.23	1000 (GL)
London 1957	2–5 Dec	(Bradley et al., 1958; Martin, 1961)	Smoke ($\mu g/m^3$): max 2417 SO_2 ($\mu g/m^3$): max 3335	Deaths all causes 1.18 / Increase in applications for admission / Sickness benefit claims 1.5	676 (plus 87 in smog-related rail crash) (GL)
London 1959	26–31 Jan	(Martin 1961)	Smoke ($\mu g/m^3$): max 1800 SO_2 (ppb): max 600		225 (GL)
London 1962	1–7 Dec	(Marsh 1963; Scott, 1963; Waller and Commins, 1966)	Smoke ($\mu g/m^3$): max 2800 SO_2 ($\mu g/m^3$): max 4100		750 (GL)
Ruhr 1962	Dec 5 days	(Wichmann et al., 1989)	SP ($\mu g/m^3$) max 2400 SO_2 ($\mu g/m^3$) max 5000	Deaths all causes 1.15	

Table 21.1 *cont.*

London 1975	15–16 Dec	(Holland et al., 1979)	Smoke (μg/m³): max 546 SO₂ (μg/m³): max 994		150 (GL)
Ruhr 1979	Jan 1 day	(Wichmann et al., 1989)	SP (μg/m³) max 500 SO₂ (μg/m³) max 600	No increase in mortality	
Dublin 1982	11–14 Jan	(Kelly and Clancy, 1984)	Smoke (μg/m³): single max 1400, mean max 900 SO₂ (μg/m³): single max 350, mean max 250	Deaths in a single hospital: all causes 2.0	
Ruhr 1982	Jan 6 days	(Wichmann et al., 1989)	SP (μg/m³) max 500 SO₂ (μg/m³) max 600	No increase in mortality	
Ruhr 1985	17–21 Jan	(Wichmann et al., 1989)	SP (μg/m³) max 830 NO₂ (μg/m³) max 230	Deaths all causes 1.06 Hospital admissions 1.12 Outpatients 1.07 Ambulance transports 1.28 Physician contacts 0.98	
Netherlands 1985	16–21 Jan	(Dassen et al., 1986)	TSP (μg/m³): max 200–250 SO₂ (μg/m³): max 200–250	Lung function in primary school children reduced by 3–5%	
Netherlands 1987	14–22 Jan	(Brunekreef et al., 1989)	BS (μg/m³): max 100 SO₂ (μg/m³): max 300	Lung function in panel of primary school children reduced.	
Netherlands 1991	3–12 Feb	(Hoek and Brunekreef, 1993)	PM₁₀ (μg/m³): max 174 SO₂ (μg/m³): max 100	Lung function in 7 to 12-year-old children reduced. No increase in symptoms	
London 1991	12–15 Dec	(Anderson et al., 1995)	BS (μg/m³): single max 228, mean max 148 NO₂ (ppb): hourly max 423 SO₂ (ppb): hourly max 72	Deaths all causes 1.09 Deaths respiratory 1.18 Admissions respiratory 1.06	Deaths 127 Resp admissions 91

^{a} Main reference given in the second column. Some information was also obtained from (Holland et al., 1979).
^{b} Maximum at a single station during episode. Mean max refers to the maximum mean of several stations. BS = black smoke.
^{c} For the London studies this was usually based on the ratio of deaths in the episode week to those in the previous week. For the other studies, the method varies.
^{d} LAC = London Administrative County (Inner London) and GL = Greater London (Inner plus Outer London).

The effects on mortality were estimated by comparing deaths in the weeks ending 13 and 20 December with (1) corresponding weeks of 1947–51 (4075 deaths) and (2) the week ending 6 December 1952 (3412 deaths). It was concluded that a figure near to 4000 deaths was the best estimate. In proportional terms this was an increase in deaths in Central London (London Administrative County) and Outer London of 160% and 100%, respectively. The salient aspects of the mortality effects are shown in Fig. 21.1, which is reproduced from the original report (Ministry of Health, 1954). In the weeks before the episode, mortality was increasing gradually, as was usual at this time of the year. There was a sharp increase in mortality associated with the episode (lagged one day), followed by a fall which coincided with the end of the episode. This association is so strong that it cannot be confused with other variability in the time-series. One possible confounder was the effect of the low temperatures, but it is clear that the other 'Great Towns' in England showed only a small increase (4%) in comparison, making this explanation unlikely. An influenza epidemic followed the episode but exhaustive efforts failed to find evidence that there was any influenza at the time of the episode. It is interesting that no compensatory fall in deaths followed the episode, which would be consistent with short-term mortality displacement (harvesting). On the contrary, deaths remained higher than usual and while this could represent delayed effects of the fog, it is generally thought that the excess deaths in January and February were attributable to influenza. Between 80 and 90% of the deaths were from cardiorespiratory disease and the greatest relative increase was in bronchitis deaths, which rose by a factor of 9. While the majority of deaths were among those over 65, the proportional increase was similar in the 45–64, 65–74 and 75+ age groups (2.8, 2.8 and 2.7, respectively). Interestingly, the relative increase in deaths was considerable in younger age groups (neonatal deaths 1.8, post neonatal deaths 2.2, 1–14 years 1.3, 15–44 years 1.6. Autopsy findings were consistent with those from the 1930 Meuse episode in showing an increase in airways inflammation of toxic origin.

Attempts were made to examine effects on various indicators of morbidity. A questionnaire approach as used in the Donora episode (see later) was discarded on the grounds of impracticability and the possibility of respondent bias. Sickness Benefit Claims showed an increase following the episode, but not as great as the increase in mortality. Hospital admissions increased by 50% overall and respiratory disease admissions by 160%. The emergency bed service reported an unprecedented increase in applications for admission which peaked on 9 December, the day on which the episode abated (Abercrombie, 1953). Morbidity at a primary care level was available from one general practice in Outer London which reported a 70% increase in domiciliary visits but no change in attendances at the surgery overall (Fry, 1953). There was a noticeable increase in attendances for upper respiratory symptoms which seemed to be of an irritative nature, and of lower respiratory disorders suggestive of bronchial obstruction and irritation, infection and cardiac failure. It was noted that none of the many asthmatic children in the practice attended during this time. Another interesting feature was that the onset of severe symptoms tended to occur after several days of fog. This was also noted in the Meuse incident. Mention was made of the deaths of prize cattle at the Smithfield Show but it has been suggested that this could have been caused or at least aggravated by 'shipping fever', which tends to have greater effects on fat cattle (Lipfert, 1994).

The increase in deaths, especially from bronchitis, was considerably greater than experienced in previous London fogs (Table 21.1). Various reasons for this have been advanced.

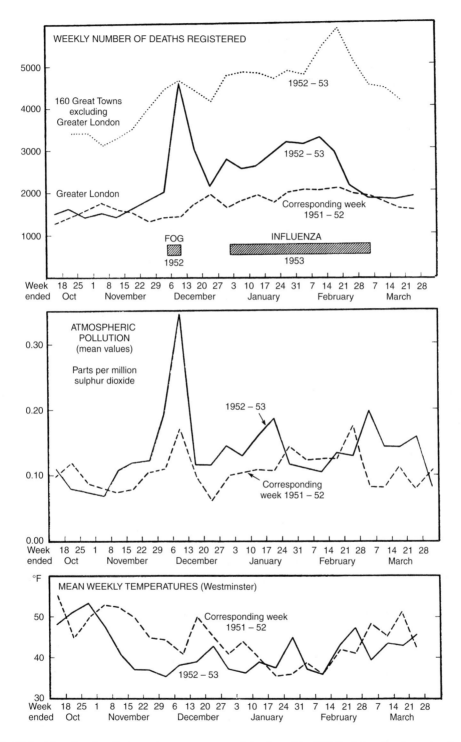

Fig. 21.1 Mortality, air pollution and temperature around the time of the 1952 fog in London. Reproduced with permission from the original report (Ministry of Health, 1954).

(1) The estimate of excess deaths may have been exaggerated owing to an underlying upward trend in deaths which was occurring anyway because of a respiratory disease outbreak (Lipfert, 1994).
(2) The intensity of the episode was underestimated. The amount of smoke was such that the filters became saturated and could not be assessed by the standard reflectance method. In fact the quoted maximum of 4460 $\mu g/m^3$ of smoke was estimated indirectly from SO_2 concentrations (Holland *et al.*, 1979).
(3) There may have been qualitative differences in the constituents of the pollution. There have been suggestions that it may have been accompanied by relatively greater concentrations of acid aerosols or carbon monoxide.
(4) Lastly, the population may have been more susceptible because it had not experienced the harvesting effect of influenza for some years (R. Waller, personal communication).

London 1956, 1957, 1959, 1962, 1975

The 1952 fog was important in promoting legislation, including the Clean Air Act 1956, to reduce emissions. Aided by these measures levels of smoke, and to a lesser extent of SO_2, fell markedly over the next decades. However, further episodes occurred, some of which have been documented (Table 21.1). In early January 1956, an episode occurred over 3 days in which smoke levels in Outer London (none are reported for inner London) reached 1229 $\mu g/m^3$. It was estimated that 1000 additional deaths occurred (Logan, 1956). Attribution of these effects to air pollution was simplified because the episode occurred in January, was well clear of the Christmas period and the early winter rise in respiratory mortality, was accompanied by normal rather than low temperatures and was clear of any incidence of influenza. There was no evidence for a harvesting effect. Notably, the effects were relatively greater in the young than the old. For all ages there was an increase of 25% in all cause mortality. For infants up to 4 weeks of age the increase was 67% compared with an increase of 30% in the 65+ age group.

During the 1950s research turned towards the possible effects of less severe episodes. Several studies reported associations between mortality and multiple episodes (Gore and Shaddick 1958; Martin, 1964; Martin and Bradley, 1960) both by visual inspection of time-series and by correlation analysis using deviations from 15-day moving averages to remove seasonal effects. They pointed towards the possibility that mortality might also be increased in non-episode conditions and laid the basis for modern studies of daily mortality which use computing and statistical techniques which were not available at that time. The last major episode accompanied by important effects on health occurred in 1962 (Marsh, 1963; Scott, 1963; Waller and Commins, 1966).

The association between episodes and morbidity was studied extensively using panels of chronic bronchitics (asthmatics were excluded) who recorded the status of their health condition daily over long periods of time (Lawther *et al.*, 1970). Visual inspection of the time-series was sufficient to show associations between morbidity and greater and lesser episodes over a number of years until 1969/70 when, at a time of much reduced pollution levels, any obvious association disappeared. It was concluded that the minimum pollution required to obtain a 'significant response' in this sensitive subgroup was 500 $\mu g/m^3$ SO_2 together with 250 $\mu g/m^3$ of smoke (as daily means).

Dublin 1982

During a 4-day period in January 1982, daily mean concentrations of smoke exceeded 1400 $\mu g/m^3$ and SO_2 exceeded 350 $\mu g/m^3$ in parts of Dublin. Case fatality in one large hospital was approximately doubled to 9.3% (120 deaths) compared with the January mean of 5.3% (64 deaths) (Kelly and Clancy, 1984). The increase in mortality coincided with the episode but low temperatures may also have contributed. Without comparison with surrounding non-polluted areas it is not possible to exclude an effect of cold weather.

The Ruhr, Germany, 1962, 1979, 1982, 1985

The effects of these episodes have been described by Wichmann and his colleagues (Wichmann *et al.*, 1989). The 5-day 1962 episode appears to have been severe with SO_2 reaching 5000 $\mu g/m^3$ and suspended particles 2400 $\mu g/m^3$. It was estimated that mortality increased by 15%. Lesser episodes in 1979 and 1982 were not associated with increased mortality.

In January 1985, in the Ruhr district of what was then West Germany, levels of SO_2 reached 830 $\mu g/m^3$, suspended particulates 600 $\mu g/m^3$, NO_2 230 $\mu g/m^3$ and carbon monoxide 8 $\mu g/m^3$. The episode lasted 5 days and about 6 million people were exposed. Mortality, hospital admissions, outpatient attendances, ambulance transports and GP consultations before and after the episode were compared with surrounding non-polluted areas of North Rhine–Westphalia. Compared with the non-polluted area, there was a relative increase of 6% in mortality, 12% in hospital admissions, 7% in outpatient consultations, and a decrease of 2% in physician contacts. There was an increase of 28% in ambulance transports but no data were available from the control area.

This episode was notable because the cloud of pollution moved westward over the Netherlands and into the UK, where it caused a modest rise in SO_2 to about 50% above usual levels. A study using a general practitioner recording system was unable to establish an association with increased respiratory visits in the UK Midlands (Ayres *et al.*, 1989).

The Netherlands 1985, 1987 and 1991

The Netherlands has experienced several moderate episodes of pollution as a result of transport from countries to the east. A series of panel studies in children gives insight into the acute effects of moderate episodes on the lung function of children.

The 1985 episode which affected the Ruhr moved west into the Netherlands where a panel of school children were already taking part in a longitudinal study of air pollution and lung function (Dassen *et al.*, 1986). Levels of particles and SO_2 were 2–3 times the baseline level, reaching about 250 $\mu g/m^3$. The episode lasted about 5 days during which the lung function of the 163 6- to 11-year-old children fell by 3–5% compared with baseline measurements. Lung function took about 3 weeks to return to baseline.

In 1987 another long-range transport of air pollution from the east resulted in an episode in which daily mean levels of SO_2 reached 300 $\mu g/m^3$, NO_2 100 ppb and black smoke 100 $\mu g/m^3$. In a panel of 6- to 12-year-old children, lung function showed a significant fall which persisted for several weeks (Brunekreef *et al.*, 1989).

Similar methods were used to investigate the effects of a small episode in 1991 in which SO_2 increased to 100 $\mu g/m^3$ and PM_{10} to 174 $\mu g/m^3$ (Hoek and Brunekreef, 1993). In a population sample of 7- to 12-year-old children, significant falls in lung function were observed unaccompanied by changes in symptoms.

London 1991

In December 1991, London experienced very still cold weather associated with a temperature inversion. One-hour concentrations of NO_2 rose to a record level of 423 ppb (5 × the seasonal average) and were above the WHO 1987 Guideline of 210 ppb for four successive days. Black smoke rose to 148 $\mu g/m^3$ (4 × seasonal average). SO_2 increased only slightly to 72 ppb (2 x seasonal average) (Bower et al., 1994). The predisposing weather conditions were typical of previous London episodes but in contrast the sources and nature of the pollution were very different. Whereas in the 1950s the main source of pollution was the burning of coal in domestic fires, in 1991 the main source of pollution was automobiles, with a lesser contribution from space heating (mainly using natural gas).

Inspection of time-series of the daily mortality and hospital admissions did not show clear indications of an association with the episode (Anderson et al., 1995). However, statistical analysis using control periods from the earlier week and four previous years revealed that all-cause mortality increased by 10%, cardiovascular by 14% and respiratory by 22%. In the elderly, hospital admissions for respiratory disease increased by 19% and obstructive lung diseases by 43%. In children there was little or no evidence for an increase in respiratory admissions or asthma. In order to control for the cold weather or a coincidental respiratory epidemic, London was compared with the rest of the South East of England. The relative excess risks for mortality and respiratory admissions remained almost unchanged (1.09 and 1.06, respectively, for all ages) but became non-significant. It was concluded that the increase was unlikely to be due to chance or to a coincidental factor such as a respiratory epidemic. It was not entirely explained by the associated cold weather. Air pollution was a plausible explanation, but whether NO_2, particles or the whole mixture was responsible could not be determined.

WINTER EPISODES IN THE US (Table 21.2)

Donora, PA, 1948

Donora was a small industrial town of 13 000 inhabitants situated in a narrow valley on the banks of the Mononahela River. During the last week of October 1948, anticyclonic conditions associated with little air movement lay over Pennsylvania and adjacent areas. The accompanying fog together with accumulated pollutants created a dense smog which in Donora lasted from 27 to 31 October. It was extremely unpleasant and after several days it was apparent that many people were adversely affected. There were even reports that seamen on passing river boats also became ill. The Federal Public Health Service was asked to investigate and this led to a comprehensive report (Shrenk et al., 1949). A large

Table 21.2 Air pollution episodes in the USA

Place and year	Dates	Reference	Maximum pollution levels	Relative health effects	Attributable health effects
Donora, PA, 1948	27–31 Oct	(Shrenk et al., 1949)	n/a	Deaths all causes 6.0	18
New York City 1953	12–21 Nov	(Greenburg et al., 1962a,b)	Smoke (CoH units): 8.0	Deaths all causes 1.09	200
New York City 1962	27 Nov–4 Dec (intermittent)	(Greenburg et al., 1963)	Smoke (CoH units): hourly max 8.6 SO_2 (ppm): hourly max 1400	Deaths all causes, no increase. Daily visits at emergency clinics: no increase. Visits to old peoples' homes: increase in visits for upper respiratory infections.	0
New York City 1966	23–25 Nov	(Becker et al., 1968; Glasser et al., 1967)	Smoke (CoH units): bihourly max 8.0, daily 6.1	Deaths all causes 1.1	168
Pittsburgh, PA, 1975	17–21 Nov	(Stebbings and Fogleman, 1979; Stebbings et al., 1976)	TSP (μg/m^3): 700 SO_2 (ppb): daily 130	No increase in lung function after the episode	
Steubenville, OH, 1978		(Dockery et al., 1982)	TSP (μg/m^3): 422 SO_2 (μg/m^3): 281	2% fall in FVC in panel of children	
Steubenville, OH, 1979		(Dockery et al., 1982)	TSP (μg/m^3): 271 SO_2 (μg/m^3): 455	1% drop in FVC	

community interview survey was conducted about a month after the smog cleared and found that over 40% of the population had experienced symptoms attributable to the smog; 10% were severely affected, more commonly among the elderly. The symptoms suggested irritation of exposed mucous membranes and the respiratory tract. Twenty deaths occurred during or shortly after the episode, 17 on day 4. This was about six times the number expected. Mortality was not increased in nearby areas not exposed to the episode. There was inconclusive evidence for an effect on domestic animals, especially dogs, and also reports of considerable deaths among farm animals.

It is likely that the pollution came from a build up of usual emissions rather than from a sudden abnormal discharge of toxic material. No measures of ambient pollution were available at the time but it was likely that the source pollution was mainly industrial, from a steel and wire plant and a zinc plant. These emitted sulfur oxides, nitrogen oxides, and various other potentially toxic substances, but no conclusion could be drawn about which were important. Particulate matter was not considered. This episode, though affecting a small population, served to increase awareness of the dangers of air pollution, soon to be confirmed on a large-scale in London and New York City.

New York City 1953, 1962, 1966

From 12 to 21 November 1953, an anticyclone associated with a temperature inversion dominated a large area of the eastern USA. The effects of this episode on mortality and morbidity in New York City was reported by Greenburg and colleagues (Greenburg *et al.*, 1962a,b). Stagnant air conditions led to an increase in air pollution over at least 5 days during which SO_2 (indicated by total acidity) increased to over 850 ppb on occasions, very much higher than usual levels. Smoke concentrations measured by the smoke shade method increased to over 8.0 CoH units, about four times higher than expected (1 CoH unit is approximately 100–150 $\mu g/m^3$ TSP). From November 15–24 in 1953 there were 2439 deaths from all causes; from the years 1950–56 (excluding 1953), 2235 deaths would have been predicted, which gives a 9% excess. All age groups were affected. Morbidity was investigated using data on visits to emergency clinics at four major New York hospitals during November 1950–56. Visits for upper respiratory infection, cardiac illnesses and asthma were analysed. Statistically significant increases in upper respiratory and cardiac visits were found in three and two hospitals, respectively. No evidence of an effect on asthma visits was found.

From 27 November to 4 December 1962 the city experienced intermittent episodes of air pollution (Greenburg *et al.*, 1963). Concentrations tended to increase during the night and dissipate during the day. Maximum smoke shade (8.6 CoH) and SO_2 (1400 ppb) were registered on the same day. Although deaths from 1–7 December showed no excess, there was an increase of 20% over predicted in deaths on the day after the peak concentrations. No increase in emergency room visits was recorded but there was a significant increase in visits to old people's homes for upper respiratory infections.

In November 1966 a stationary anticyclonic weather system lay over the eastern coast of the USA and led to an air pollution episode in New York City from 23 to 25 November. SO_2 rose to an hourly maximum of 1020 ppb and daily mean of 470 ppb (Glasser *et al.*, 1967). Smoke shade rose to a bi-hourly maximum of 8.0 and a daily mean of 6.1 CoH units. These were about two to three times usual levels. Unusually for such episodes, the weather was warmer than usual. In the analysis, control periods from the preceding days and previous 5 years were used. During the episode week the average daily deaths was 237; this compared with 261 in the episode week, an increase of 24 (10% increase). This amounts to an excess of 168 deaths during the whole episode. The increase affected all ages.

Morbidity was investigated by analysing data from emergency clinics at seven large hospitals; this revealed a rise in visits on the third day of the episode. However, interpretation is complicated by the Thanksgiving holiday on 24 November and because comparable data from other years are not available, nothing could be concluded. Another report on this episode used data obtained by interview from 2052 personnel employed by an insurance company (Becker *et al.*, 1968). The questionnaire was administered on 28 November, shortly after the episode. It concluded that symptoms indicating irritation of the eyes and respiratory system increased as air pollution increased. The effects were greater in subjects with existing respiratory disease.

Pittsburgh 1975

An air pollution episode occurred in Pittsburgh, 17–21 November 1975 (Stebbings *et al.*, 1976). The highest single daily mean value for TSP was 700 $\mu g/m^3$ and highest single daily mean for SO_2 was 130 ppb. Starting at the end of the episode, about 270 children in highly exposed and less exposed areas had daily measurements of lung function for 7 days, to test the hypothesis that any effect of the episode would be associated with an increase in lung function in the polluted area and no change in the less polluted area. No such upward trend or difference between the two areas was observed. In a subsequent reanalysis of these data it was found that there were significant effects on forced vital capacity (FVC) in a subgroup of children (Stebbings and Fogleman, 1979).

Steubenville, Ohio, 1978 and 1979

Steubenville was an industrial town of 31 000 people located in the valley of the Ohio River. It was known to have experienced frequent exceedances of the current air quality standard for TSP (260 $\mu g/m^3$) and occasional exceedances of the standard for SO_2 (365 $\mu g/m^3$). Pulmonary function was measured in nearly 200 children before and after two air pollution episodes in 1978 and 1979 (Dockery *et al.*, 1982). In the autumn of 1978, TSP reached 422 $\mu g/m^3$ and SO_2 281 $\mu g/m^3$. This was associated with a significant fall in forced vital capacity (FVC) (2% fall), but no reduction in forced expiratory volume in 0.75 s $(FEV_{0.75})$, which indicates a transient restrictive rather than obstructive defect. In 1979, TSP reached 271 $\mu g/m^3$ and SO_2 455 $\mu g/m^3$. This was associated with a significant drop in FVC (1%). As in the earlier episode, there was no effect on $FEV_{0.75}$. These workers went on to carry out regression analysis of air pollution on daily lung function and this indicated that lung function was reduced by about 1% over the range of pollution studied.

SUMMER EPISODES IN EUROPE

Nearly all episode studies have been concerned with winter episodes which were associated with stagnant weather conditions. Many of these were associated with abnormally cold weather which needed to be taken into account in the interpretation of the findings. Summer episodes are usually photochemical in origin with increased ozone levels and are often associated with unusually hot weather, which in itself can increase mortality and morbidity. These episodes tend to affect whole regions rather than localized areas such as cities (indeed, ozone levels are usually lower in cities), and it is perhaps for these reasons that summer air pollution episodes have rarely been investigated.

South East England 1976

In June 1976, London and the rest of the South East of England experienced an episode of ozone pollution in which 1-h concentrations reached 258 ppb in rural areas and 210 in Central London (United Kingdom Photochemical Oxidants Review Group, 1987). Between 22 June and 12 July, maximum hourly ozone levels were over 100 for 100 h in London and 160 h in rural areas. The weather was unusually hot but similar weather conditions are not usually associated with such high ozone levels. There is evidence to suggest that on this occasion photochemical reactions were enhanced by the transport of precursors (volatile organic compounds and oxides of nitrogen) from continental Europe. Mortality in London was increased by 40–60% (Macfarlane *et al.*, 1997). Hospital admissions for respiratory disease increased by 1.34 (H.R. Anderson, unpublished data) Because this was a regional episode, it was not possible to make comparisons with a geographical control area which was subject to the same high temperatures but without increased ozone levels; without such evidence the relative contributions of heat and air pollution cannot be distinguished.

Athens 1987

From 20 to 29 July 1987 Greece was affected by a heat wave during which temperatures were above 29°C each day (Katsouyanni *et al.*, 1993). It was estimated that 2000 excess deaths occurred in Athens. Daily mean concentrations of smoke ranged from 50 to 250 $\mu g/m^3$ and of ozone from 100 to 300 $\mu g/m^3$. Mortality increased in Athens (97%) more than in other urban areas (33%) and non-urban areas (27%). Episode analysis using previous years as controls found that mortality in Athens was higher than elsewhere after controlling for the effects of temperature. This interaction was confirmed by regression analysis, which indicated that the daily number of deaths increased when 24-h temperatures exceed 30°C. While the main effects of air pollution were not significant, there was evidence for interactions between temperature and SO_2, ozone and smoke. These results suggest that there is an interaction between high temperatures and air pollution in causing mortality.

Belgium 1994

This was also an investigation of mortality associated with a heat wave and increased pollution (Sartor *et al.*, 1995). The heat wave lasted from 27 June to 7 August and was associated with increased ozone (daily mean 35–112 $\mu g/m^3$), low humidity, a slight increase in smoke (32–164 $\mu g/m^3$) but with no increase in SO_2 and NO_2. Mortality was increased by 14% and this was predicted by the product of temperature with ozone (both lagged one day) which explained 41% of the variance in mortality.

FOREST FIRES

Forest fires may cause extensive air pollution episodes and when this covers populated areas there is often concern about health effects. Fires produce particles, carbon monoxide, nitrogen oxides, aldehydes and other hydrocarbons, and many other potentially toxic substances. The massive pollution which occurred in South East Asia in late 1997 was caused by annual burning off and forest clearance in Indonesia which became out of control because of the late arrival of the monsoon rains. This in turn was attributed to the El Niño effect. A similar but less extensive pollution episode occurred in the same region in 1994 (Nichol, 1997). If global warming increases the variability of world weather patterns, such incidents may become more common. In the previously mentioned paper on the 1953 New York City episode, it was mentioned in passing that on the 3rd November 1952 smoke from forest fires 'blotted out the sun' in New York City (Greenburg *et al.*, 1962b). On that day mortality was 20% above the daily average for the month.

The first major study of the health effects of forest fires appears to be by Duclos and colleagues, who investigated the 1987 forest fire disaster in California (Duclos *et al.*, 1990). There was high pollution from 8 August to 15 September with levels of TSP reaching over 1000 $\mu g/m^3$ and of PM_{10} 335 $\mu g/m^3$. Daily visits to emergency rooms at hospitals in the six most exposed counties were compared with those recorded during two reference periods. Significantly increased visits for asthma (+40%), COPD (+30%), laryngitis (+60%), sinusitis (+30%) and other respiratory infections (+50%) were observed during the fire period. It was concluded that these effects were moderate, considering the severity of the pollution, but that preventive action taken by susceptible persons (e.g. those with cardiorespiratory disease) may have ameliorated some potential effects.

In January 1994, the state of New South Wales, Australia, experienced one of the worst bushfires in its history (Smith *et al.*, 1996). Record levels of pollution were recorded in Sydney from 5 to 12 January. Nephelometry readings reached a maximum of over 10 beta $_{scat}$ units (a measure of the amount of light scattered by airborne particles <2.5 μm) and PM_{10} levels reached 250 $\mu g/m^3$, but SO_2 and ozone levels were within the usual range. Data were collected on presentations to emergency departments in the seven public hospitals serving Western Sydney for the period of pollution and control periods adjacent in time and in the previous year. There was no evidence of an absolute or relative increase in asthma presentations during the bushfire. No data on other respiratory presentations were presented. Judging by the PM_{10} levels, the intensities of the Californian and Australian pollution were rather similar, and it is not clear why there were effects on asthma in one fire but not the other.

VOLCANIC ERUPTIONS

Volcanic eruptions may cause air pollution from toxic gases such as carbon dioxide, SO_2, hydrochloric acid and hydrogen sulfide and from particles in the form of volcanic ash. These agents probably account for only a small proportion of deaths and morbidity from volcanic eruptions (Bernstein *et al.*, 1986). The explosive eruption of Mt St Helens from 18 May to 12 June 1980 was accompanied by a column of ash and gas that rose 20 km and deposited disruptive amounts of ash over hundreds of miles, affecting about 1.25 million people. In samples taken at different sites across Washington, peak levels of particles reached 30 000 $\mu g/m^3$. Ninety per cent of the particles were in the respirable range (<10 μm) and were composed largely of silica. Dangerous levels of toxic gases were not observed. Acute respiratory effects were investigated using hospital records and a case–control questionnaire survey in patients attending two large hospitals in the exposed area. There was a four-fold increase in ER visits for asthma and a two-fold increase for bronchitis, both of which coincided with the rapid increase in particle concentrations (Baxter *et al.*, 1981, 1983). The case–control study using neighbourhood controls established that the main risk factor was pre-existing asthma or bronchitis and that cases had not been unduly exposed to ash (in clean-up operations for example).

METHODOLOGICAL ASPECTS OF THE INVESTIGATION OF THE HEALTH EFFECTS OF AIR POLLUTION EPISODES

The preceding section presented a narrative account of the main published data on air pollution episodes with little critical comment on the methods used for determining an association with health outcomes or interpreting any association found. This section will deal with the steps involved in investigating and interpreting the health effects of episodes. These are: (1) defining the occurrence, timing, extent and intensity of the episode; (2) selecting appropriate health outcome data; (3) investigating whether an excess of mortality or morbidity occurred in association with the episode; and (4) evaluating the role of pollution and its constituents in causing any excess of adverse health effects.

Defining the episode

The occurrence of an episode is easy to identify when there is a noticeable increase in air pollution with sharp onset and offset. If this is not the case, it may be useful to use arbitrary criteria such as an exceedance of a guideline value or range based on statistical parameters. To avoid bias due to *post hoc* decisions, the timing of the episode should not be determined after looking at the time-series of health effects. The duration of an episode may sometimes be difficult to determine, especially if the onset and offset are not distinct, or if stations vary in their reported levels of pollutants – as is frequently the case. There is no standard way of describing the intensity of an episode; this leads to difficulty in estimating exposure response. Pollution levels may be reported as the maximum at a single station, or the mean of all the station maxima, or a mean over the duration of the episode, and in a variety of averaging times. Concentrations may be expressed in absolute terms or relative to an estimate of expected levels, usually the seasonal average. Generally, it is very difficult to define the levels of pollution so as to obtain a satisfactory measure of the exposure of the population in geographical terms. For these reasons it will normally be difficult to use data such as were presented in Tables 21.1 and 21.2 to draw conclusions about the effects of different types of episode or about exposure response. The final problem is to define the population exposed. This will usually be an administrative area often somewhat arbitrarily defined by the exposure profile or by the availability of routine data. In the case of summer ozone episodes, whole regions may be included.

Health outcome data

The three main categories of health data are mortality, direct measures of morbidity and indirect measures of morbidity from health care utilization. The most frequently used measure is all-cause mortality because it is nearly always available in some form, at least in developed countries. However, there may be difficulties in obtaining mortality data on a daily basis, by cause of death, by date of death (rather than registration), or by place of residence at death. Respiratory effects of air pollution might precipitate the death of individuals with other conditions, but because the International Classification of Diseases emphasizes the underlying cause of death these may be underestimated. Mortality studies have low statistical power to demonstrate small rises in mortality even in large cities such as London and New York, where daily deaths number only about 200 per day. Analysis of subgroups such as respiratory disease, which accounts for only about 15% of deaths, further reduces statistical power. If the hypothesis concerns the medical effects of air pollution, accidental deaths should be excluded. These may however be directly related to the incident, as illustrated by the 1956 London fog in which 87 people were killed in a rail accident attributable to the fog.

Data on the use of medical services are the most convenient source of information on morbidity. Sometimes hospital admissions are available from a routine system but there may be considerable delay before these are available. *Ad hoc* collection of utilization data is usually needed if a speedy investigation is required. An advantage of hospital or emergency room admissions is that the diagnosis is made at discharge by physicians with the aid of necessary investigations and is probably fairly accurate as medical diagnoses go. Some studies have identified hospital visits from log books and then examined a sample of notes to find out the timing of the onset of symptoms. Other outcomes in this category include primary care contacts, ambulance transports, sickness certification, and applications for emergency admissions. One difficulty with utilization data is that they are also subject to social and behavioural influences including the way in which health systems and professionals work. For example, there are well-known day of the week and holiday effects. Another is that the conditions predisposing to episodes and the smog itself may affect social organization and communications. Lastly, the publicizing of an episode might alter the way in which susceptible individuals take care of their medical condition, including the use of services. It is not clear whether the net effect of these factors would be to increase or decrease utilization in an episode. In the 1985 Ruhr episode, physician contacts did not increase, which led Wichmann to conclude that psychological factors were unlikely to be important (Wichmann *et al.*, 1989).

Direct measures of morbidity include symptoms of eye, upper or lower respiratory irritation or disease, use of medications and reduction in lung function. Because episodes are difficult to predict, it is rarely possible to obtain morbidity data prospectively. Notable exceptions are the panel studies in Steubenville and The Netherlands (described earlier in this chapter) in which children were having lung function and symptom assessments before the episode occurred. More commonly, data are collected soon after the event. Drawing conclusions from lung function series begun after the episode (by testing the hypothesis that if lung function were affected a rise would be detected) have not been convincing (Stebbings *et al.*, 1976), even where control areas are studied at the same time. The strong psychological impact of severe episodes is likely to bias response. Cohen and colleagues conducted telephone interviews with families during two moderate episodes which occurred in New York during the summer of 1970 (Cohen *et al.*, 1974). One of these episodes was publicized and the other was not. Irritative symptoms were reported to be increased in both episodes but not in a control area. There was no difference between the level of symptoms between the publicized and unpublicized episodes. While useful, this study probably does not predict the psychological effect of severe visible fog.

Estimating the health effects associated with the episode

It is necessary to define the period over which excess health effects might be expected. This will normally start on the first day of the air pollution episode, but its duration is problematic because little is known about the natural history of health problems aggravated by air pollution or the extent or timing of forward displacement of adverse outcomes such as mortality among those in which such an event was imminent (harvesting effect). The concept of harvesting applies mainly to mortality, but could also apply in part to hospital utilization. Too short or too long an episode period will underestimate the effects. It is likely that harvesting of vulnerable individuals will be occurring during the episode while at the same time individuals will be affected who will not experience the outcome until some time in the future. From inspection of the mortality curve, some episodes appear not to be associated with harvesting (e.g. London 1952 and 1956) whilst others do (Ruhr 1985). In practice, most investigators define the at-risk period as starting with the episode and ending several days after the end of the episode; this picks up effects delayed by a few days only but avoids dilution. In the case of utilization data such as hospital admissions, it is often advisable to include a whole week, to facilitate comparisons with control periods which cover the same days of the week. Other considerations include the need for statistical power: mortality and admissions are unlikely to provide sufficient numbers unless several days are grouped. As mentioned earlier, it is important to avoid a *post hoc* decision concerning the timing of the episode period, i.e. after inspection of the outcome data.

Having defined the episode period, the next task is to obtain the best estimate of the incidence of effects which would have occurred if the air pollution episode had not occurred. Various control strategies have been used.

(1) *Period immediately prior to the episode period.* The duration of this will depend on the situation. A long period will give stable estimates but may be biased by seasonal and other cyclical factors. Many episodes occur in the early winter against

a background of increasing respiratory disease. Coincidental factors such as extreme weather or respiratory epidemics may render the prior period unrepresentative.

(2) *Equivalent dates in adjacent years.* This is a strategy adopted by most major studies where data are available. Allowance must be made for annual trends. A convenient way to do this is by regression techniques (Anderson *et al.*, 1995; Katsouyanni *et al.*, 1993).

(3) *Post episode period.* This is sometimes used but there is a serious risk that the post episode period will be subject to a carry-over effect (which would bias the control estimate upwards) or to harvesting (which would bias the control estimate downwards).

(4) *Geographical control populations.* Including these in the analysis helps to separate the effects of pollution from other factors affecting the region generally, especially weather and respiratory epidemics. Ideally, these should be contemporaneous with data from the exposed population and have enough air pollution data to confirm a lower level of exposure.

Having assembled the health outcome data for the episode and control periods/areas, statistical analysis is directed at estimating the relative increase in health outcomes associated with the episode period. For counts, a suitable approach is log linear regression which gives a relative risk estimate (Anderson *et al.*, 1995).

Evaluating the role of air pollution

Once there is statistical evidence that the episode was associated with an increase in mortality or morbidity, the next question concerns causality. Episode studies are very vulnerable to confounding by weather, and because they comprise but one unit of analysis have no power to distinguish between different factors which might have the same effect – this is the task of daily time-series studies. However, since weather conditions such as temperature are usually regional, geographical controls with low exposure to pollution can help in separating the effects of pollution and temperature. In Fig. 21.1 taken from the report on the 1952 London fog, it is clear that the increase in mortality in 160 'great towns' of England was quite small (about 4%) compared with London. Since these were subjected to the same cold weather conditions as London (possibly more, since the heat island effect in London was relatively greater), it is likely that the increase in mortality during the episode in London was not attributable to the weather. A similar technique was used in the analyses of the Ruhr episode in 1985, Athens in 1987 and London in 1991. A potential difficulty with this approach lies in the possibility of an interaction between air pollution and low socioeconomic status in the large cities. Thus the vulnerability of Londoners to extreme weather might be greater than that of the surrounding South East of England.

Episode analyses are also prone to bias due to the coincidental occurrence of respiratory epidemics such as influenza. In England at least, it is known that these tend to have a similar time course across regions, which means that geographical control areas can be used to account for their effects.

Based on a single episode it is not possible to allocate effects to any particular pollutant, but attempts have been made to do this using the results of several episodes, especially the London fogs of 1949 to 1962. Following the 1952 fog smoke levels fell to a greater extent than those of SO_2. Subsequent fogs were associated with far fewer excess deaths and this was interpreted as indicating that smoke was more toxic than SO_2 (Holland *et al.*, 1979). This conclusion was probably simplistic in view of the imprecision in comparing exposure situations across episodes and the likelihood that the nature as well as the scale of pollution may also have changed. Little is known of trends in acid aerosols, for example. With the restriction of domestic consumption of coal, and growth of mobile sources, at least in relative terms, the nature of particulate pollution in terms of size and number distribution and in chemical composition may have changed also. Similarly, in 1991 there was great interest in the possible role of the high NO_2 levels in London, but it was possible to show, using meta analysis coefficients from daily mortality studies (Dockery and Pope, 1994), that the relatively modest increase in particles (80 $\mu g/m^3$ of black smoke) could alone have explained the 10% increase in mortality in the absence of NO_2. The occurrence of respiratory symptoms following the Mt St Helens eruption indicates that particles alone are capable of respiratory effects, since levels of pollutant gases from that eruption were low.

Similarly, there have been attempts to construct exposure–response curves using the results of episode analyses. In earlier years it was believed that effects were apparent only when levels increased by a factor of about four, or over 1000 $\mu g/m^3$ of particles or SO_2 (Martin, 1964). It can now be seen that this was due to the relative insensitivity of episode analyses. Daily time-series studies do not suggest a threshold within the ambient range and the upper end of the exposure–response curve tends to be less steep. Exposure–response relationships from episode studies, although likely to be very imprecise, tend to fit on the same curve. This indicates that there is probably no qualitative difference between the effects of high exposures during episodes and increases which occur day-to-day.

CONCLUSION

In extreme situations, air pollution episodes cause widespread public apprehension and are associated with measurable effects on health. In the developed world, episodes such as occurred in the post war decades are unlikely to be repeated, but there remains the risk of less severe episodes and of other disasters such as forest fires and volcanic eruption. This review of the main published evidence about the health effects of episodes provides clear evidence that air pollution episodes are harmful to people (and possibly animals as well). This serves to underpin the more subtle evidence from time-series studies of daily mortality and morbidity which, while addressing the question of confounding more effectively, are more open to dispute as to causality. As might be expected, given the variety of contexts in which episodes occur, there are inconsistencies, an important example being the variability in reported effects on asthma. Episodes give information about the effects of pollution mixtures in different contexts. Unfortunately, this heterogeneity has not been successfully exploited for identifying the most harmful constituents of pollution or for investigating exposure–response relationships. The epidemiological principles of investigating episodes are straightforward but in practice there are many difficulties and assumptions which make comparison of different studies or meta analysis very difficult.

This would be reduced if some of the methodological points outlined in this chapter were to be observed, but will never be eliminated. Nevertheless, the investigation of future episodes should be encouraged.

REFERENCES

Abercrombie GF (1953) December fog in London and the emergency bed service. *Lancet* **1**: 234–235.

Anderson HR, Limb ES, Bland JM *et al.* (1995) Health effects of an air pollution episode in London, December 1991. *Thorax* **50**: 1188–1193.

Ayres J, Fleming D, Williams M and McInnes G (1989) Measurement of respiratory morbidity in general practice in the United Kingdom during the acid transport event of January 1985. *Environ Health Perspect* **79**: 83–88.

Baxter PJ, Ing R, Falk H *et al.* (1981) Mount St Helens eruptions, May 18 to June 12, 1980. An overview of the acute health impact. *J Am Med Assoc* **246**: 2585–2589.

Baxter PJ, Ing R, Falk H and Plikaytis MS (1983) Mount St Helens eruptions: the acute respiratory effects of volcanic ash in a North American community. *Arch Environ Health* **38**: 138–143.

Becker WH, Schilling FJ and Verma MP (1968) The effect on health of the 1966 eastern seaboard air pollution episode. *Arch Environ Health* **16**: 414–419.

Bernstein RS, Baxter PJ and Buist AS (1986) Introduction to the epidemiological aspects of explosive volcanism. *Am J Public Health* **76**: 3–9.

Bower JS, Broughton GFJ, Stedman JR and Williams ML (1994) A winter NO_2 smog episode in the U.K. *Atmos Environ* **28**: 461–475.

Bradley EH, Logan WPD and Martin AE (1958) The London fog of December 2nd–5th, 1957. *Monthly Bull Ministry Health* **17**: 15–16.

Brimblecombe P (1987) *The Big Smoke: A History of Air Pollution in London Since Medieval Times.* London: Methuen.

Brunekreef B, Lumens M, Hoek G *et al.* (1989) Pulmonary function changes associated with an air pollution episode in January 1987. *J Air Pollut Control Assoc* **39**: 1444–1447.

Cohen AA, Nelson CJ, Bromberg SM *et al.* (1974) Symptom reporting during recent publicized and unpublicized air pollution episodes. *Am J Public Health* **64**: 442–449.

Committee on Air Pollution (1953) *Interim Report.* London: HMSO.

Dassen W, Brunekreef B and Hoek G (1986) Decline in children's lung function during an air pollution episode. *J Air Pollut Control Assoc* **36**: 1223–1227.

Dockery DW and Pope CA (1994) Acute respiratory effects of particulate air pollution. *Ann Rev Public Health* **15**: 107–132.

Dockery DW, Ware JH, Ferris BG Jr *et al.* (1982) Change in pulmonary function in children associated with air pollution episodes. *J Air Pollut Control Assoc* **32**: 937–942.

Duclos P, Sanderson LM and Lipsett M (1990) The 1987 forest fire disaster in California: assessment of emergency room visits. *Arch Environ Health* **45**: 53–58.

Evelyn J (1661) *Fumifugium: Or the Inconvenience of the Aer and Smoak of London Dissipated. Together With Some Remedies Humbly Proposed.* London: Gabriel Bedel.

Firket J (1936) Fog along the Meuse Valley. *Trans Faraday Soc* **32**: 1192–1197.

Fry J (1953) Effects of a severe fog on a general practice. *Lancet* **1**: 235–236.

Glasser M, Greenburg L and Field F (1967) Mortality and morbidity during a period of high levels of air pollution. *Arch Environ Health* **15**: 684–694.

Gore AT and Shaddick CW (1958) Atmospheric pollution and mortality in the county of London. *Br J Prev Soc Med* **12**: 104–113.

Greenburg L, Field F, Reed JI and Erhardt CL (1962a) Air pollution and morbidity in New York City. *J Am Med Assoc* **182**: 159–162.

Greenburg L, James MB, Droletti BM *et al.* (1962b) Report of air pollution incident in New York City, 1953. *Public Health Reports* **77**: 7–16.

Greenburg L, Erhardt C, Field F *et al.* (1963) Intermittent air pollution episode in New York City, 1962. *Public Health Reports* **78**: 1061–1064.

Hoek G and Brunekreef B (1993) Acute effects of a winter air pollution episode on pulmonary function and respiratory symptoms of children. *Arch Environ Health* **48**: 328–335.

Holland WW, Bennett AE, Cameron IR *et al.* (1979) Health effects of particulate pollution: reappraising the evidence. *Am J Epidemiol* **110**: 527–659.

Katsouyanni K, Pantazopoulou A, Touloumi G *et al.* (1993) Evidence for interaction between air pollution and high temperature in the causation of excess mortality. *Arch Environ Health* **48**: 235–242.

Kelly I and Clancy L (1984) Mortality in a general hospital and urban air pollution. *Irish Medical Journal* **77**: 322–324.

Lawther PJ, Waller RE and Henderson M (1970) Air pollution and exacerbations of bronchitis. *Thorax* **25**: 525–539.

Lipfert FW (1994) *Air Pollution and Community Health: A Critical Review and Data Source Book.* New York: Van Nostrand Rienhold.

Logan WPD (1949) Fog and mortality. *Lancet* **1**: 78.

Logan WPD (1953) Mortality in the London fog incident, 1952. *Lancet* **1**: 336.

Logan WPD (1956) Mortality from fog in London, January, 1956. *Br Med J* **1**: 722–725.

Macfarlane A, Haines A, Goubet S *et al.* (1997) Air pollution, climate and health: short-term effects and long term prospects. In: Charlton J and Murphy M (eds) *The Health of Adult Britain: 1841–1994*, vol. 1. London: The Stationery Office, pp. 187–204.

Marsh A (1963) The December smog, a first survey. *J Air Pollut Control Assoc* **13**: 384–387.

Martin AE (1961) Epidemiological studies of atmospheric pollution. *Monthly Bull Ministry Health* **20**: 42–49.

Martin AE (1964) Mortality and morbidity statistics and air pollution. *Proc Roy Soc Med* **57**: 969–975.

Martin AE and Bradley WH (1960) *Mortality, Fog and Atmospheric Pollution* (Section I – General). Bulletin issued from the Office of the Ministry of Health. London: Ministry of Health.

Ministry of Health (1954) *Mortality and Morbidity During the London Fog of December 1952.* Reports on Public Health and Medical Subjects no. 95. London: HMSO.

Nichol J (1997) Bioclimatic impacts of the 1994 smoke haze event in southeast Asia. *Atmos Environ* **31**: 1209–1219.

Russell WT (1924) The influence of fog on mortality from respiratory diseases. *Lancet* **2**: 335–339.

Russell WT (1926) The relative influence of fog and low temperature on the mortality from respiratory disease. *Lancet* **2**: 1128–1130.

Sartor F, Snacken R and Demuth CI (1995) Surmortalité associée à une vague de chaleur et à la pollution photochimique durant l'été 1994 en Belgique. Bruxelles: Institute d'hygiène et d'épidémiologie.

Scott JA (1963) The London fog of December, 1962. *The Medical Officer*: 250–253.

Shrenk HH, Heimann H, Clayton GD *et al.* (1949) *Air Pollution in Donora, PA.* Public Health Bulletin no. 306.

Smith MA, Jalaludin B, Byles JE *et al.* (1996) Asthma presentations to emergency departments in western Sydney during the January 1994 bushfires. *Int J Epidemiol* **25**: 1227–1236.

Stebbings JH, Fogleman DG, McClain KE and Townsend MC (1976) Effect of the Pittsburgh air pollution episode upon pulmonary function in schoolchildren. *J Air Pollut Control Assoc* **26**: 547–553.

Stebbings JH Jr and Fogleman DG (1979) Identifying a susceptible subgroup: effects of the Pittsburgh air pollution episode upon schoolchildren. *Am J Epidemiol* **110**: 27–40.

United Kingdom Photochemical Oxidants Review Group (1987) *Ozone in the United Kingdom.* London: Department of the Environment.

Waller RE and Commins BT (1966) Episodes of high pollution in London 1952–1966. In: Anonymous, *First International Clean Air Congress.* London: National Society for Clean Air, pp. 228–231.

Wichmann HE, Mueller W, Allhoff P *et al.* (1989) Health effects during a smog episode in West Germany in 1985. *Environ Health Perspect* **79**: 89–99.

Wilkins ET (1954) Air pollution and the London fog of December, 1952. *J Roy Sanitary Inst* **74**: 1–21.

OZONE

Epidemiological Studies of Ozone Exposure Effects

GEORGE D. THURSTON and KAZUHIKO ITO

Department of Environmental Medicine, New York University School of Medicine, Nelson Institute of Environmental Medicine, Tuxedo, NY, USA

Observational epidemiological studies provide especially relevant evidence as to whether ambient environmental factors, such as ozone (O_3) air pollution, can adversely affect the general public. This is because these studies consider the 'real-world' experiences of human populations as they are exposed to pollution in natural settings. Such observational epidemiological studies are termed 'ecological', if they consider large groups of people rather than individuals. They usually follow populations of people in a defined geographical area (e.g. a city) as they undergo varying everyday exposures to pollution over time, or from one place to another, and then statistically evaluate the variations in the total number of adverse health impacts that occur in these populations (e.g. city-wide respiratory hospital admissions counts) when higher (versus lower) concentrations of pollution are experienced.

In the case of O_3 air pollution, numerous recent observational epidemiological studies have yielded significant associations between variations in daily ambient concentrations of this pollutant and a wide range of adverse health outcomes. These include lung function decrements, the aggravation of pre-existing respiratory diseases (such as asthma), increases in daily hospital admissions and emergency department visits for respiratory causes, and excess mortality. These short-term epidemiological associations between O_3 and adverse health effects, combined with long-term O_3 exposure animal study results (e.g. see Chapter 24 of this book), raise concern that long-term exposure may also have a cumulative health impact. However, epidemiological studies investigating the potential chronic health effects in humans of long-term O_3 exposure are as yet very limited. As a result, this chapter will focus primarily on an evaluation of available epidemiological studies evaluating the acute effects of O_3 in the following categories: mortality studies; hospital admissions studies; emergency department (ED) visit studies; and individual

AIR POLLUTION AND HEALTH
ISBN 0-12-352335-4

studies (e.g. summer camp studies). A synthesis of these published studies is presented for each category in order to evaluate whether they collectively indicate statistically significant associations for that outcome. In addition, this synthesis will provide a multi-study combined estimate of the size of the 'best estimate' of the O_3 effect for that outcome. Thus, this chapter provides a concise review of the available observational epidemiological studies of air pollution and human health. We quantitatively evaluate whether the evidence indicates that exposures to O_3 at ambient levels can have adverse health effects in the general population.

TIME-SERIES MODELING AND METHODOLOGICAL ISSUES

Most of the studies in this review are longitudinal time-series statistical analyses, or those studies that follow a single population over time. Such time-series analyses present numerous statistical advantages over other types of studies. These include the fact that, by design, they obviate the need to control for individual-level confounding factors (e.g. education level, income, percentage of smokers, etc. in the population) that can confound other study designs (such as cross-sectional studies, which compare effects across different populations). This is the case because such intrinsic population characteristics are relatively constant from day-to-day. The population acts as its own 'control' in such a time-series model, since the incidence of health effects in a single population on higher pollution days is being compared with the incidence in that same population on lower pollution days.

Epidemiological studies also do not share the practical and ethical limitations of controlled human exposure studies, which are usually not able to include the most sensitive populations (e.g. those with severe pre-existing disease) and must consider pollution concentrations that are expected to result in only relatively mild responses. Instead, time-series epidemiological studies consider a community's entire population (including its most susceptible members), and the entire spectrum of ambient exposures experienced by that population. In addition to being more relevant to the general public, epidemiological studies also are extremely useful in capturing the most serious effects among the most sensitive members of the general population. They can have great statistical power, since they can consider extremely large populations (e.g. entire metropolitan areas) over many years.

Despite the above-noted strengths of time-series observational epidemiology, the application of such methods also presents important statistical challenges of their own, as discussed in detail by Thurston and Kinney (1995). In particular, shared long-term cycles in the health outcome (e.g. mortality) and the pollutants being analyzed can, if not adequately addressed, cause misleading associations (e.g. owing to shared winter-to-summer seasonal trends), and yield biased pollutant risk estimates. These cycles also can cause statistical autocorrelation and/or overdispersion in the model residuals that, if unaddressed, can bias pollutant significance tests. This problem is especially relevant to the time-series analysis of year-round and multi-year records of daily population counts of human morbidity and mortality, as these health outcome daily series usually exhibit strong seasonal variations over time. Moreover, it is difficult for such correlation-based models to separate the influences of other environmental factors that co-vary over time with the pollutant of interest in the study locale, potentially biasing the size estimate of the pollutant effect provided by such models. For example, O_3 is usually moderately to strongly associated with ambient temperature, representing a potential confounder in the elicitation of O_3 associations with morbidity and premature mortality. Thus, if not appropriately addressed, the influence of seasonal variations and temperature impacts on the incidence of health outcomes can confound the evaluation of the effect of ozone on human health provided by such time-series models.

Whether or not a published O_3 health effects study has addressed the potentially confounding influences of long-wave (e.g. seasonal) variations in the health outcome data has an especially large impact on the adverse health effects estimate for O_3, as it, too, exhibits large winter-to-summer variations in its concentration levels. The seasonality of morbidity and mortality is well documented, and was even explicitly mentioned in the writings of Hippocrates (400 BC). Furthermore, the need to address seasonal cycles in respiratory disease time-series (to avoid spurious long-wave dominated correlations) has been recognized for at least three decades (e.g. see Ipsen and Ingenito, 1969), but it often has been ignored or inadequately addressed in the literature.

Fortunately, a variety of time-series modeling options are now available to separate the short-wave 'signal' from the long-wave 'noise' superimposed on day-to-day variations. These include: Fourier techniques (i.e. fitting sine/cosine waves to the data); high-pass prefiltering; autoregression methods; the fitting of a smooth of mortality (e.g. via loess fits) over time; or the use of time (e.g. monthly) dummy variables in regressions. The results do not seem to depend on the choice of approach (Kinney *et al.*, 1995; Lipfert, 1994), but one of these methods must be applied to avoid confounding by seasonality. Thus, one important criterion for the evaluation of aggregate population time-series studies of the acute morbidity and premature mortality effects of O_3 is whether or not the authors have appropriately addressed long-wave periodicities in the data as part of their analysis.

Similarly, how the known acute effects of temperature extremes on human morbidity and mortality are handled can also affect estimates of the health effects association. This is especially true for O_3, which tends to experience peak concentrations on high temperature days, when many O_3 precursors are emitted at higher rates (e.g. via the greater vaporization of hydrocarbons) and their conversion to O_3 is most rapid. As noted in the US EPA Ozone Criteria Document: 'Ambient air temperature often exhibits a moderate to high correlation over time with O_3 in acute epidemiology studies due, in part, to the dependence of O_3 formation rate on light intensity. Among the studies reviewed . . . , correlations ranging from 0.06 to 0.90 (mean = 0.51) have been reported' (US EPA, 1996). Some older studies ignored temperature, which may have resulted in an overestimation of O_3 effects. Conversely, incorrectly modeling the associations between temperature and mortality as a linear relationship can cause the underestimation of the O_3 effect, owing to the serial intercorrelation of O_3 and temperature over time.

Although temperature and O_3 are moderately to highly correlated over time, differences in their respective relationships with health outcomes allows their effects to be discriminated via statistical methods. While the dose–response relationship between temperature and health effects is U-shaped, with increased adverse effects at either extremely high or extremely low temperatures, the relationship between O_3 and health effects is more linear, with adverse O_3 effects increasing as concentration increases. Thus, if these factors are properly specified in the model, their respective effect coefficients (i.e. that between the 'U-shaped' temperature specification and the linear O_3 term) will be less intercorrelated than their raw variables, allowing these model terms to be better discriminated (i.e. having lower intercorrelations between their coefficients), in spite of the moderate to high linear correlation between O_3 and temperature over time. Therefore we will, where possible, consider the sensitivity of the model estimates reported for ozone's effects on human health as a function of individual weather modeling approaches.

Another concern that has been raised regarding aggregate time-series analyses is that the collection of individual exposure data in such large populations is not practical. Instead central site monitor concentration data (either from a single site or from the average of multiple sites) are usually used to provide indices of day-to-day variations in population exposures. Environmental variables need to be representative of (i.e. correlated with) population exposures over time for time-series methods to be effective in detecting effects. Absolute bias in exposure measurements is not as problematic in time-series analyses as it would be in cross-sectional studies, since the comparisons are relative over time in a single community (i.e. relatively high days versus relatively low days), not between the average concentrations at different communities. Indeed, recent work by Mage and Buckley (1995) indicates that, while individual exposures are not always well character-ized by a central site monitor's data, such central site data are highly correlated with the average of individual exposures in a population. This is exactly what is required for the purposes of time-series analyses of aggregate population counts of health effects. This rela-tionship is especially true for O_3 pollution, which is almost entirely of outdoor origin and therefore not as confounded by indoor sources as some other pollutants are. Moreover, since the purpose of such time-series epidemiological analyses is usually to help set ambi-ent standards that will ultimately be monitored at a central site, the use of central site data in the original epidemiological studies simplifies the standard-setting process (i.e. thereby avoiding any extrapolation between individual exposures measured during research and central site monitor concentrations to be employed in standards-compliance monitoring). Any differences between personal and central site O_3 concentrations should not usually present a serious problem in the time-series studies of O_3 health effects that we consider here.

However, classical random errors which greatly reduce the correlation between measured and actual exposure (e.g. due to poor spatial correlations of a pollutant) will tend to reduce the significance of the pollutant effect, and bias pollutant effect regression estimates toward zero (e.g. Carroll et al., 1995). This may, in turn, reduce the ability of such time-series methods to detect the effects of some pollutants. The use of multiple-site averages in such aggregate epidemiological studies should reduce this problem. In the case of O_3 air pollution, which is highly correlated spatially across any metropolitan area, the chief concern is that the pollutant may not penetrate all buildings equally, e.g. in air-conditioned versus non-air-conditioned buildings. In warm climates (e.g. the US Southwest), the percentage of homes with air conditioning can exceed 90% (US Bureau of the Census, 1983), which may reduce the ability of reactive pollutants like O_3 to reach residents for much of the day. This may diminish the adverse health implications of outdoor O_3 air pollution in those cities, when compared with cities without extensive air conditioning. Thus, in hotter cities with widespread air conditioning (e.g. Houston, TX), the effect of O_3 may be reduced from that in areas having more limited use of air conditioning.

Other key factors that will be considered in this chapter are: (1) the extent to which other pollutants might affect the relationship between O_3 and health effects; (2) the influence of the choice of O_3 averaging time (e.g. daily 1-h or 8-h maximum, or 24-h average) on the size and significance of the O_3 effect estimate; and (3) whether there is qualitative and quantitative consistency in findings (i.e. coherence) across the various health outcomes with regard to the health implications of O_3 air pollution.

OZONE AND PREMATURE MORTALITY

While more recent studies of mortality and acute O_3 exposures have usually attempted to address potential confounders (such as influences of season, temperature and other pollutants) in their analyses, earlier studies of the possible association of O_3 with human mortality were usually flawed. Unlike studies of hospital admissions and emergency department visits, most studies of air pollution and mortality examine total daily counts, possibly because of a lack of by-cause data, and also because the small numbers of daily respiratory admissions limit the statistical power of such studies to detect by-cause effects, even if present. Most studies published in the 1950s and 1960s considered total daily mortality in Los Angeles, CA. Many of these older studies did not recognize and attempt to address seasonality in the data series. The California Department of Public Health study (1955), for example, was weakened by the qualitative treatment of the air pollution data. Newer studies (especially those conducted during the 1990s) have usually considered the various potential confounding factors to greater or lesser extents, but have still provided somewhat varying results. Most have addressed long-wave confounding satisfactorily, but not all have looked at the role of co-pollutants, considered the same O_3 averaging times (e.g. 1-h daily maximum value versus 8-h or 24-h average O_3 values), or modeled weather influences as carefully as possible. Thus, comparison of these studies is sometimes challenging.

Influence of Weather Specification on estimates of O_3 Mortality Effect

Unlike the modeling of associations between mortality and particulate matter (PM) air pollution, which appear to be relatively insensitive to the approach used to model the association between weather and mortality (e.g. see Samet *et al.*, 1997), the choice of time-series weather modeling approach may have a large effect on the estimate of O_3 mortality. This is because elevated O_3 concentrations are often highly correlated with high temperature, which has its own adverse effects on human health, especially when present in tandem with high humidity conditions (e.g. see Ellis, 1972). Thus, if both temperature and O_3 are simultaneously entered into a time-series as raw variables, they will compete with each other for the same mortality variations, resulting in unstable estimates of effect. However, as discussed above, the relationship between mortality and each of these two variables is quite different. Mortality has a non-linear U-shaped relationship with temperature (higher daily mortality at hot and cold extremes of temperature), while daily mortality rises with rising O_3, yielding a more linear relationship. This is shown in Fig. 22.1 for Detroit, MI, where temperature and relative humidity both have a moderately strong non-linear relationship with mortality. There is also an interaction between the effects of temperature and relative humidity. Figure 22.2 shows the relationship between O_3 concentration and relative risk of mortality in that same city, which appears to be more linear.

The non-linear nature of the dose–response relationship between temperature and

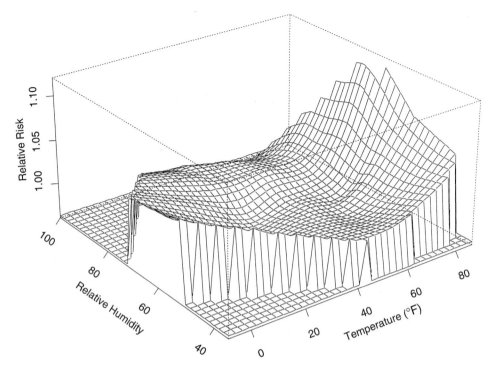

Fig. 22.1 Total mortality relative risk vs. daily temperature and relative humidity (adjusted for season and day-of-week) in Detroit, MI. From Thurston *et al.* (1999).

health effects therefore should be addressed when specifying a time-series model, except perhaps when the analysis is conducted in one season (either hot or cold only). A linear fit, although not ideal, may suffice in that case. Indeed, the way in which weather, and temperature in particular, is considered in a time-series model specification may be an important determinant of whether any distinct O_3 association has a chance of being discriminated by the analysis.

In order to evaluate the influence of a model's weather (i.e. temperature and relative humidity) specification on the resultant mortality effects estimate for O_3, results from the various recently published O_3 mortality studies are presented in Fig. 22.3, in terms of effects per 100 ppb of daily maximum 1-h average O_3 concentration, as a function of the weather modeling approach. All of these results are for a model specification where a PM concentration index also has been included, to reduce the chances that the reported O_3 relative risk (RR) incorporates the effects of another co-pollutant. A concentration of 100 ppb O_3 was chosen for these calculations to provide consistency across all study results, and because 100 ppb is on the order of the difference between the overall average and the maximum daily 1-h maximum concentration experienced in most urban areas. When the analysts used O_3 data averaged over another period (e.g. 8-h or 24-h averages), this is noted in the figure. Also, an adjustment was made to convert the reported RR in terms of a 1-h max. averaging period, based upon the ratio of the mean concentrations experienced for each averaging time in that study area. If this concentration information

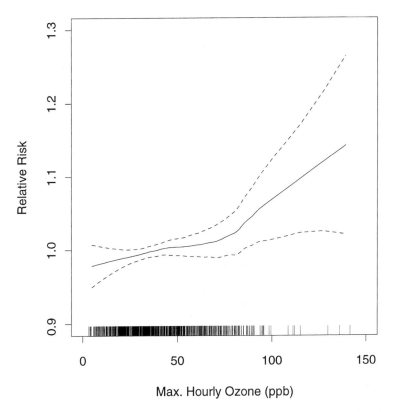

Fig. 22.2   Total mortality relative risk vs. ozone concentration (adjusted for weather, season, day-of-week and PM) in Detroit, MI. From Thurston *et al.* (1999).

was unavailable from a specific study, a conversion of 2.5 was used to convert 24-h average RRs, and a conversion of 1.33 was used to convert 8-h average RRs, based upon past experience (e.g. see Schwartz, 1997). Thus, Fig. 22.3 presents RRs for the various studies so that they are directly comparable, and in terms that should be more easily interpreted than other candidate pollution increments (e.g. the difference between the 25th and the 75th quartiles), since the 100 ppb RRs provide an index of the increased risk associated with a 'typical' high O_3 day versus an average day.

The results reported in Fig. 22.3 are consistent with the hypothesis that the choice of temperature specification employed in a total daily mortality time-series model can have an influence on the estimated O_3 effect. As would be expected, those studies that employed a linear temperature specification for year-round data (the left-most group of studies) had the lowest random effects model pooled O_3 RR estimate (pooled RR = 1.026; 95% CI, 1.016–1.036). To obtain this combined estimate, we used a two-stage random effects model approach, as suggested by DerSimonian and Laird (1986), to take into account the among-studies variance. However, as the model was improved to account for differing hot and cold extreme effects (center group of studies), the overall O_3 RR estimate increased somewhat (pooled RR = 1.033, 95% CI, 0.985–1.084), although it is statistically non-significant in these studies.

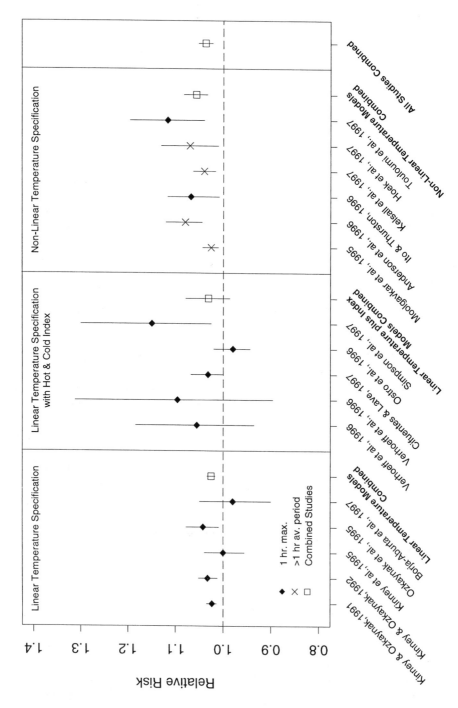

Fig. 22.3 Reported relative risks of mortality associated with a 100 ppb increase in 1-h maximum O_3 (with 95% CI). From Thurston *et al.* (1999).

In studies where the non-linear nature of temperature effects was most appropriately addressed, statistically significant O_3 RRs greater than 1.0 were consistently found, irrespective of the averaging time employed (pooled RR = 1.056; 95% CI, 1.032–1.081). Thus, the method of weather specification is an important factor in making an estimate of the O_3-mortality effect, with the better (i.e. more realistic) weather mortality models resulting in the largest O_3 RR estimate. Those studies that incorporate non-linear temperature specifications thus appear to have the greatest value in assessing the potential premature mortality effects of O_3.

Key Recent Mortality Studies

Moolgavkar and co-authors (1995) examined air pollution and daily mortality in Philadelphia for the period 1973 to 1988. Accidental deaths and suicides were excluded, but, in contrast to most other such studies, deaths in Philadelphia of non-residents or deaths of Philadelphia residents that occurred outside of Philadelphia were not excluded, which would cause some misclassification error over time. Twenty-four-hour means (spatially averaged over all monitors) of total suspended particulate matter (TSP), sulfur dioxide (SO_2) and O_3 were considered. Poisson regression modeling, including quintiles of temperature and indicators for years, was employed. In the full dataset analysis, in which all three pollutants were included in the model, daily mortality was significantly associated with the previous day's O_3 level (RR = 1.063; 95% CI, 1.018–1.108).

Anderson *et al.* (1996) examined air pollution and daily mortality in London during 1987–92. Eight-hour daytime average and daily 1-h maximum O_3 concentrations were employed, as measured at a single monitor in central London. The Poisson regression included temperature and humidity (but not their interaction), as well as adjustments for time and seasonal trends, day of the week, holidays, and an influenza epidemic that occurred during the study period. The authors addressed the U-shaped relation between daily temperature and daily mortality by fitting three separate linear terms for three different temperature ranges. When black smoke (BS) was included in the model with 8-h average O_3, the relative risk associated with an increase from the 10th to the 90th percentile concentration (24 ppb for the 8-h average, 31 ppb for the 1-h maximum) remained significant (RR = 1.027, 95% CI, 1.014–1.041).

Ito and Thurston (1996) analyzed the relationship between daily mortality and air pollution in Cook County, Illinois, from 1985 through 1990. Data on particulate matter mass less than 10 μm in diameter (PM_{10}), carbon monoxide (CO), SO_2 and O_3 were spatially averaged over multiple monitoring sites. Poisson regression models were applied that included weather and pollutant variables, sine/cosine series to adjust for long-term and seasonal trends, a linear time trend variable, and day-of-week dummy variables. A dual parabolic fit of hot and cold effects was employed. Among the pollution variables, PM_{10} and O_3 were most consistently associated with daily mortality. When PM_{10} and O_3 were in the model simultaneously, the relative risks associated with each were slightly smaller than for the individual pollutants, but still statistically significant for O_3. In this two-pollutant model, an increase of 100 ppb in the 2-day average of the 1-h daily maximum of O_3 was associated with an RR of 1.07 (95% CI, 1.01–1.12).

Kelsall *et al.* (1997) analyzed the effects of weather and multiple air pollutants on a time series of mortality from 1974 through 1988 in Philadelphia. Poisson regression models were employed that controlled for long-term time trends, season and weather. A non-linear adjustment for temperature was approximated by four linear terms corresponding to set temperature cut-points. When included individually in the model, the means of current and previous days' levels of TSP, SO_2 and O_3 had statistically significant effects on total daily mortality. A mortality relative risk of 1.019 (95% CI, 1.007–1.032) was estimated to be associated with an increase of one interquartile range in O_3 (20.2 ppb as a 24-h average) in the model that also included TSP.

Hoek and co-authors (1997) analyzed ambient air pollution and daily mortality in Rotterdam, the Netherlands, from 1983 to 1991. Total daily death counts were limited to deaths of residents of Rotterdam. Poisson regression models, including weather variables (temperature and relative humidity) and adjustments for long-term trends, seasonal trends, and influenza incidence, were employed. Non-parametric smoothers were used to adjust for temperature. Ozone was significantly associated with mortality in a model that also included TSP. In this model, the relative risk was 1.06 (95% CI, 1.01–1.11) for an increase from the 5th percentile to the 95th percentile O_3 concentration (34.2 ppb, as a 24-h average).

Touloumi and co-authors (1997) examined the short-term effects of ambient oxidants exposure on daily mortality as part of a collaborative research effort by several European countries (the Air Pollution and Health: A European Approach (APHEA) project). Six cities spanning Central and Western Europe were analyzed by each center separately following a standardized methodology to ensure comparability of results. Poisson autoregressive models allowing for overdispersion were fitted. Significant positive associations were found between daily deaths and both nitrogen dioxide (NO_2) and O_3. Increases of 50 $\mu g/m^3$ O_3 (1-h maximum) were associated with a 2.9% (95% CI, 1.0–4.9) increase in the daily number of deaths. The pooled estimate for the O_3 effect was only slightly reduced when two pollutant models (including black smoke) were applied.

Synthesis of O_3 Mortality Studies

As shown in Fig. 22.3, the estimated RR associated with a 100 ppb increase in 1-h max. O_3 (which is roughly equivalent to 75 ppb daily 8-h max., or 40 ppb 24-h average O_3) increases as the method of weather specification is improved. When all 15 studies are considered (after controlling for season, weather, day-of-week and PM co-pollutant effects), the combined analysis yielded an overall 100 ppb effect size of RR = 1.036 per 100 ppb increase in daily 1-h maximum O_3 (95% CI, 1.023–1.050).

A pooling of the seven studies with non-linear temperature terms (and simultaneously including a PM index term) reveals that the overall RR = 1.056 per 100 ppb increase in daily 1-h maximum O_3 (95% CI, 1.032–1.081), and that the various RRs are not statistically different from each other. Note, however, that none of these studies has considered the likely interaction of temperature and per cent relative humidity (%RH) in enhancing the model's fit of weather's influence on daily mortality (and making the weather specification less like O_3, further reducing intercorrelation between the pollutant and weather terms in the model). Thus, even those recent models that include non-linear temperature terms may still be underestimating the premature mortality effects of O_3.

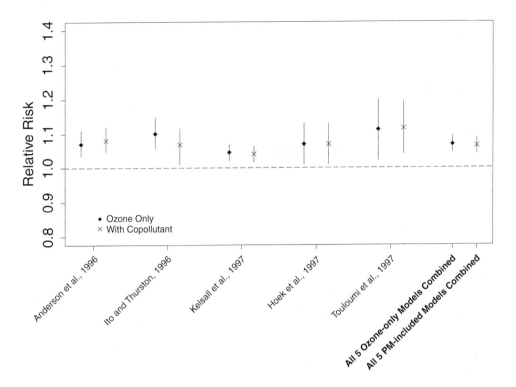

Fig. 22.4 Effect of including a PM index on estimates of the total mortality relative risk of a 100 ppb increase in daily 1-h maximum O_3 in studies using a non-linear temperature specification (with 95% CI). From Thurston *et al.* (1999).

In Fig. 22.4, the O_3 relative risk results from studies using non-linear fits of temperature are shown, both for the 'O_3 alone' model and for the same model with a PM air pollution index included. As indicated by these results, the inclusion of the PM index tends to change the combined estimate of the size of the O_3 RR only very slightly. Thus, the influence of a co-pollutant on the O_3 RR appears to be small.

The size of the premature mortality effects by O_3 indicated by this synthesis is greater than that assumed in most past assessments of O_3 mortality effects, especially in the case where the combined estimate from the non-linear temperature specification models is considered. For example, the recent US EPA Regulatory Impact Analysis used an O_3 mortality effect estimate that was equivalent to RR = 1.029 for a 100 ppb increase in 1-h max. O_3 (US EPA, 1997). This suggests that the US EPA estimates of the premature mortality benefits of implementing the new O_3 standard may have been too low by as much as a factor of two.

OZONE AND EXCESS MORBIDITY

Aggregate population time-series studies of associations between O_3 and morbidity have most often considered hospital admissions and ED visits. As Bates (1992) suggested in his discussion of the criteria to evaluate observational findings for consistency with causal inference, one would expect to see coherence among different health outcomes in their associations with an air pollutant. The types of hospital admissions that are analyzed in these time-series studies are 'emergency' or 'unscheduled' (as opposed to 'elective', or 'scheduled'). Therefore, the majority of the patients would likely have visited the emergency department first, and hospital admissions and ED visits should show coherence in their reported associations with O_3.

Since only a fraction of those who visit emergency departments actually are admitted, the number of emergency department visits is much larger than the number of hospital admissions. For example, at New York City hospitals, the ratio of emergency asthma ED visits to asthma hospital admissions among children is approximately 8 to 1 (Barton et al., 1993). Thus, ED visits provide a potentially more sensitive index of the effects of O_3 than admissions. However, data on ED visits are not as readily available in a computerized form, and there are far fewer published ED visit studies.

In the following sections, we briefly summarize recent hospital admission and ED visit studies that included O_3 as one of the explanatory variables. These reviews are followed by a statistical synthesis of the available studies, in order to yield combined quantitative estimates of the morbidity effects associated with present-day ambient O_3 exposures of the general population.

Ozone and Hospital Admissions

Most studies of hospital admissions and air pollution have focused specifically on the respiratory diagnosis categories, rather than on the total daily counts (as is usually the case with mortality studies). Various subcategories and age groups of respiratory admissions have been examined in these studies. Since the clinical conditions, risk factors and exposure patterns are probably different among different age groups, it is important to evaluate the study results for consistent groups of disease subcategories and age groups. The age distributions also are different among respiratory disease subcategories, and this can affect the statistical power to detect the size of a given effect. Seasonal temporal patterns associated with each age group also may differ. For example, seasonal cycles of emergency asthma admissions are generally more pronounced in younger age groups. The implication is that the model specification to control for seasonal cycles may vary depending on the age group. Younger age groups also are known to spend more time outdoors, resulting in higher O_3 exposure levels over longer periods. In the elderly age groups, there also may be a larger fraction of other chronic obstructive pulmonary diseases (COPD) misclassified as asthma. For these reasons, studies having similar disease subcategories and/or age groups provide the most useful comparisons.

A series of studies was conducted in Canada and USA (Bates and Sizto, 1983, 1987; Burnett et al., 1994, 1997a; Delfino et al., 1994; Thurston et al., 1992, 1994) to evaluate the effects of 'summer haze' air pollution. These all examined similar major respiratory categories: acute bronchitis and bronchiolitis (ICD 466); pneumonia (ICD 480–486); COPD excluding asthma (ICD 490–492, 494, 496); and asthma (ICD 493). The initial findings by Bates and Sizto suggested that 'summer haze' had a general effect on respiratory morbidity, as indicated by these admissions categories. The other studies therefore sought to identify those components of 'summer haze' (e.g. sulfates, acid and O_3) that may be responsible for the observed associations in the same outcomes. All of these studies focused on the summer months.

Among the Ontario studies, Bates and Sizto (1987) and Burnett *et al.* (1994) reported results by age groups. In Bates and Sizto's study, asthma admissions during summer for age group 0–14 years were not significantly associated with O_3 or other pollutants, while asthma and total respiratory admissions for the all-age group were associated with O_3 and sulfates. In Burnett *et al.*'s study (1994), asthma admissions were most significantly associated with O_3 and sulfate in age group 35–64 years, and admissions for infection (acute bronchitis/bronchiolitis and pneumonia) were associated with these pollutants only in the age group 0–1 year. Thus, there appears to be some heterogeneity in effects across age groups.

Thurston *et al.*'s studies in three New York State Metropolitan areas (1992) and Toronto, Ontario (1994) examined various gaseous and particulate matter air pollution indices. In both Buffalo and New York City, O_3 was a significant predictor of total respiratory and asthma admissions. O_3 was the most consistently significant predictor for both total respiratory and asthma admissions, even in models that also included temperature and a co-pollutant. In Delfino *et al.*'s (1994) study in Montreal, however, O_3 was not a significant predictor of respiratory and asthma admission. Burnett *et al.* (1997a) also examined multiple pollutants in Ontario during the years 1992–94, finding a monotonic relationship between O_3 and hospital admissions with no evidence of a threshold for effects, even at concentrations below an 80 ppb 1-h daily maximum. While other inter-correlated environmental variables (e.g. sulfates) may also have been cofactors in these associations, O_3 was the pollutant most consistently associated with both respiratory and asthma admissions in these studies.

Schwartz utilized Medicare data to study associations between air pollution and respiratory admissions among the elderly in various US cities, including: Birmingham, AL (1994a); Detroit, MI (1994b); Minneapolis-St. Paul, MN (1994c); Spokane, WA (1995a); New Haven CT (1995b); and Tacoma, WA (1995b). In all but Birmingham, O_3 was positively and significantly associated with pneumonia and COPD admissions. Asthma admissions were analyzed for Detroit, but no significant association with either PM_{10} or O_3 was found. This may have been due to the smaller daily counts for this category in the elderly (median = 1/day), which may not have been adequate to provide sufficient statistical power over the time period considered. Moolgavkar *et al.* (1997) reanalyzed the Medicare data in Minneapolis-St. Paul and Birmingham, using a data period (1986–91) that was two years longer than the period Schwartz analyzed (1986–89). They reported 'little evidence of association' between air pollution and respiratory hospital admissions (pneumonia and COPD combined) in Birmingham, but in Minneapolis-St. Paul, they reported that 'O_3 was most strongly associated with admissions, and this association was robust in the sense that it was little affected by the simultaneous consideration of other pollutants.' Thus, the strongest associations with O_3 air pollution among hospital admissions by the elderly are found for pneumonia and COPD admissions.

As a part of a more recent collaborative research effort by several European countries, the APHEA project, the short-term associations of multiple pollutants with adult (age 15–64 years) and elderly (age 65+) respiratory (ICD 460–519) hospital admissions were investigated in five cities: London, Amsterdam, Rotterdam, Paris and Milan (Spix *et al.*, 1998). A unique feature of this study was the use of relatively consistent methods of data selection and analysis for each city. Thus, the combined estimates would reduce the possibility of heterogeneity owing to difference in data selection and model specification.

Among the five pollutants examined (SO_2, BS, TSP, NO_2 and O_3), 'the most consistent and strong' associations with respiratory admissions were found for O_3. The finding was stronger for the elderly, and the estimated effect size was homogeneous over cities (RR = 1.038 per 50 $\mu g/m^3$ increase in max. 8-h average O_3 (95% CI, 018–1.058)). However, less consistent associations were found for other pollutants. Sunyer *et al.* (1997) reported APHEA analysis of asthma admissions in London, Paris, Barcelona (ED visits) and Helsinki. In the results combined for all these cities, O_3 was positively, but not significantly, associated with asthma admissions for the age group 15–64 years in winter and summer, while, in the age group <15 years, no association was observed for O_3. NO_2 and SO_2 were more significant predictors of asthma admissions. Anderson *et al.* (1997) also presented COPD admission analysis of the six APHEA cities, plus Barcelona, Spain. Among the pollutants examined (SO_2, NO_2, TSP, BS and O_3), they reported that 'the results for particles and O_3 are broadly consistent with those from North America'.

In some studies hospital admissions other than respiratory categories were analyzed in relation to multiple pollutants, including O_3. These studies may have been motivated by hypotheses different from those that investigated the effects of summertime air pollution on respiratory admissions, but may serve as evidence for specificity of O_3 effects. These include studies that investigated associations between air pollution and admissions for congestive heart failure (Burnett *et al.*, 1997b; Morris *et al.*, 1995) or for several cardiovascular categories (Schwartz and Morris, 1995). All these studies considered elderly subjects (age 65+). In the Morris *et al.* study, in which gaseous pollutants (CO, NO_2, SO_2 and O_3) were examined in seven large US cities, CO was most significantly and consistently associated with hospital admissions for congestive heart failure, while O_3 was the least significant predictor. In Burnett *et al.*'s analysis of the ten largest Canadian cities using gaseous pollutants (CO, NO_2, SO_2 and O_3) and the coefficient of haze (CoH) particle blackness index for the study period 1981–91, it was also CO that showed the 'strongest and most consistent' association with hospitalizations among the elderly for congestive heart failure. Other pollutants, including O_3, were more sensitive to simultaneous adjustment for multiple pollutants and weather model specifications. In Schwartz and Morris's study, in which PM_{10}, SO_2, CO and O_3 were analyzed for their associations with ischemic heart disease, dysrhythmias and congestive heart failure in Detroit, MI, both PM_{10} and CO showed 'independent associations' with congestive heart failure admissions, while 'O_3 was not associated with any of the outcomes'. PM_{10} was also associated with ischemic heart disease admissions. Thus, O_3 was not found to be associated with cardiovascular disease outcomes in the elderly.

Synthesis of O_3 hospital admissions studies

To combine the effect estimates reported in the hospital admission studies above, we grouped the estimates by outcome and age group. This resulted in three groups: (1) Canadian and New York State results for all age major respiratory categories; (2) Canadian and New York State results for all age asthma admissions; and (3) studies that used US Medicare (age 65+) data. These results, both individual and combined estimates, are shown in Fig. 22.5. The European multi-city study (APHEA) produced its own combined estimates by age group and outcomes; these are are also displayed in Fig. 22.5. Burnett *et al.*'s (1997a) analysis of 16 Canadian cities also provided a combined random effects model estimate.

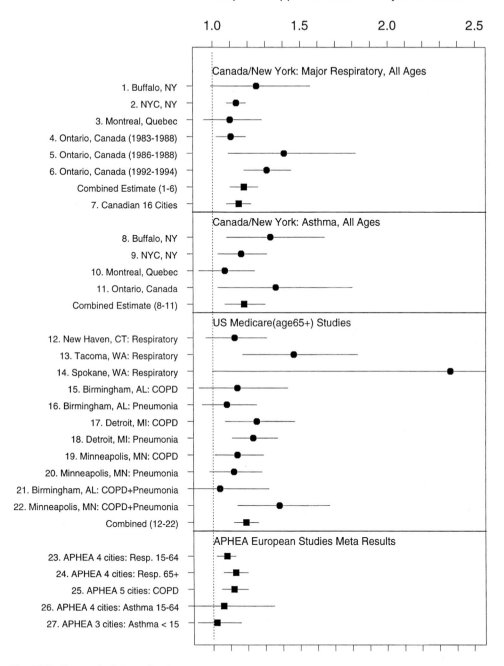

RR per 100 ppb increase in daily Max Ozone

Fig. 22.5 Reported relative risks of respiratory hospital admission associated with a 100 ppb increase in daily 1-h maximum O_3 (with 95% CI). From Ito and Thurston, (1999).

Study reference key: studies 1, 2, 8 and 9: Thurston *et al.* (1992); studies 3 and 10: Delfino *et al.* (1994); study 4: Burnett *et al.* (1994); studies 5 and 11: Thurston *et al.* (1994); study 6: Burnett *et al.* (1997c); study 7: Burnett *et al.* (1997a); studies 12 and 13: Schwartz (1995a); study 14: Schwartz (1995b); studies 15 and 16: Schwartz, (1994a); studies 17 and 18: Schwartz (1994b); studies 19 and 20: Schwartz (1994c); studies 21 and 22: Moolgavkar *et al.* (1997); studies 23 and 24: Spix *et al.* (1998); study 25: Anderson *et al.* (1997); studies 26 and 27: Sunyer *et al.* (1997).

Given the limited number of studies in each study-type group, it was not practical to examine the issue of temperature specification (i.e. linear vs. non-linear) influence on estimates of the effect on hospital admissions. In addition, many studies examine a single season (summer), during which a linear fit of the dose–response between temperature and morbidity would be expected to be better than it would be for year-round studies that include both cold and heat effects.

In order to show estimates of effect size in a consistent manner, we used the RR results as calculated for 100 ppb increase in daily 1-h maximum O_3. Most studies use this concentration averaging period to compute an estimated RR. Exceptions are: Delfino et $al.$'s (1994) model for major respiratory admissions; Schwartz's analyses of New Haven (1995b), Tacoma (1995a), Detroit (1994b) and Minneapolis (1994c); and Burnett et $al.$'s (1997a) analysis of Ontario data. In these cases, we used the ratio of the daily 1-h maximum to the other measures used in the original analyses to compute the increment comparable to a 100 ppb change in daily 1-h maximum (see legend to Fig. 22.5). The APHEA studies provided the results for both 8-h average O_3 and 1-h maximum O_3. Although 8-h average results were featured in those studies, because of their higher significance, we show the results for 1-h maximum O_3 for consistency with other studies. The increment used to compute RRs varied from study to study. Therefore, we used the coefficient and its standard error, if given in the original manuscripts, to recalculate the estimates and interval for a 100 ppb increase, or, when relative risks and confidence intervals (or t-ratios) were given, we back-calculated the coefficients using the increment, and recalculated the relative risks and confidence bands for a 100 ppb increase in O_3.

For the major respiratory hospital admission categories, there were six individual analyses. The three studies of Ontario data, Burnett et $al.$ (1994), Thurston et $al.$ (1994) and Burnett et $al.$ (1997a), cover different study periods (1983–88, 1986–88, and 1992–94, respectively). For this reason, we did not eliminate any of these studies. The original studies by Bates and Sizto (1983, 1987), as well as Lipfert and Hammerstrom's study (1992), were not included for the combined analysis only because they did not present relative risks for O_3. The combined random effects model estimate for major respiratory admissions for all ages was RR = 1.18 per 100 ppb increase in daily 1-h maximum O_3 (95% CI, 1.10–1.26). The combined estimate for asthma admission (all ages) for the subset of the same group of the studies was RR = 1.18 (95% CI, 1.07–1.30).

Most of the Medicare data analyses to date were conducted by Schwartz (1994a–c, 1995a,b). Moolgavkar et $al.$'s (1997) analysis of the extended data from the same location also employed similar approaches (i.e. generalized additive Poisson model with seasonal cycles and smoothed non-linear temperature). The outcomes analyzed in these studies were somewhat different (total respiratory, COPD and pneumonia separate, and COPD and pneumonia combined). However, since COPD and pneumonia account for a major fraction of the elderly hospital admissions, and since the COPD and pneumonia estimates were similar within each study, we calculated a combined estimate, RR = 1.19 (95% CI, 1.12–1.26), for all these categories together.

The APHEA studies reported combined estimates of multiple cities, but their diagnosis/age groups are different from those in the US/Canadian studies. The APHEA studies' estimates are somewhat smaller than those for the US/Canadian studies, especially the estimates for asthma. Finally, it should be noted that these estimates of effect size for respiratory hospital admissions are generally larger than those for total daily mortality studies. This is as might be expected, based upon the fact that total mortality includes categories of death less likely to be affected by air pollution than the specific respiratory categories considered by hospital admissions studies.

Ozone and Emergency Department (ED) Visits

Bates *et al.* (1990) examined asthma, pneumonia, other respiratory diseases (acute bronchitis, bronchiolitis, etc.) and total emergency visits for three age groups (0–14 years; 15–59; and 60+). Using a largely descriptive approach, they characterized the seasonal periodicities of respiratory emergency room visits in Vancouver, BC. They noted strong annual peaks in asthma visits in the autumn for age groups excluding 60+, but these long-wave peaks in respiratory visits were not found to be related to temperature, O_3 or NO_2. However, their sub-analysis of the warm season (May–October) included a dominant autumn asthma peak, which would obscure any summertime O_3 associations. Therefore, little can be inferred from this data analysis about the existence of an acute relationship between O_3 and Vancouver hospital visits for respiratory causes.

Cody *et al.* (1992) and Weisel *et al.* (1995) examined asthma emergency department visits (all ages) in nine central and northern New Jersey hospitals in relation to O_3 and temperature. In both studies, simultaneous inclusion of O_3 and temperature during the warm months (May–August) revealed a significant association between O_3 and asthma emergency department visits. The visits occurred 28% more frequently when the mean O_3 levels were above 60 ppb than when they were below 60 ppb.

White *et al.* (1994) abstracted clinical records for all children (age <16 years), predominantly black, with asthma or reactive airway disease in one large public hospital in Atlanta during the summer of 1990. After controlling for the effects of PM_{10}, day-of-week, and temperature influences in a Poisson regression model, the relative risk of visits for asthma or reactive airway disease was 1.50 (95% CI, 1.02–2.21) for high O_3 days (max. 1-h > 110 ppb) versus days below 80 ppb.

Romieu *et al.* (1996) also examined asthma ED visits for children (age < 16 years) in Mexico City between January and June, 1990. After adjustment for day-of-week, SO_2, temperature, and season in a Poisson model, the relative risk per 50 ppb increase in daily 1-h max. O_3 was 1.43 (95% CI, 1.24–1.66). In a separate analysis, with a dichotomous exposure variable, exposure to high O_3 levels (>110 ppb) for two consecutive days increased the number of asthma-related ED visits by 68% (95% CI, 1.30–2.17).

In Saint John (New Brunswick, Canada), Stieb and co-authors (1996) abstracted data on ED visits (all ages) presenting with a complaint of asthma for May through September, 1984–92. Controlling for seasonal cycles, day-of-week, temperature, SO_2 and NO_2, they found that O_3 (2-day lagged) was significantly associated with asthma ED visits. When the daily 1-h maximum O_3 level exceeded 75 ppb (the 95th percentile), the frequency of asthma ED visits was 33% higher (95% CI, 10–56%). Delfino *et al.* (1997) examined the relationship of daily ED visits for respiratory illness to air pollution in Montreal, Canada, from June through September, 1992 and 1993. Among the five age groups (<2 years; 2–18; 19–34; 35–64; and 65+) examined, the elderly group (65+) showed significant positive associations with 1-h max. O_3, PM_{10}, $PM_{2.5}$ and sulfates, but O_3 was most significantly associated with the ED visits. An increase of 36 ppb in 1-h max. O_3 was found to be associated with a 21% increase in ED visits (95% CI, 8–34%).

Thus, hospital ED visit studies show associations with O_3 that are qualitatively similar to those found for hospital admissions. However, a quantitative synthesis of ED visits was not attempted, owing to the limited number of ED studies available for evaluation, each looking at somewhat different population groups and outcomes, thus precluding any combined analysis by subgroup.

Individual-Level Studies

While the above-discussed daily mortality and morbidity time-series studies strongly indicate that O_3 air pollution is associated with significant adverse health effects in the general public, consistent adverse health effects observed at the individual level would rule out the possibility that the results from those various studies of aggregate population (e.g. city-wide) counts could be accounted for by some unaddressed ecological confounding. One complication is that the severe outcomes examined by time-series studies (e.g. hospital admissions and mortality) are probably exhibited primarily in a small segment of the population that is especially susceptible to those effects on those days. Thus, these especially susceptible people are not necessarily included in epidemiological studies of limited groups of individuals, even when groups likely to be affected are considered, such as children, the elderly and asthmatics. For this reason, these individual-level studies usually consider more common (and less severe) outcomes, such as reductions in lung function. Despite these limitations, such field epidemiologic studies of groups of individuals can still provide a useful coherence test as to whether evidence can be found of individual-level adverse health effects that are consistent with the associations seen by time-series studies for more severe outcomes in aggregate populations.

Investigators have studied the individual effects of O_3 air pollution by studying groups of individuals engaged in normal activities in their natural environments. The advantage of these field epidemiological studies over aggregate epidemiological studies is that they follow specific individuals over time, allowing for analyses of individual changes over time. In such studies, each person serves as his or her own control.

The most extensive individual-level epidemiologic database on pulmonary function responses to ambient O_3 comes from camp studies. Six recent key studies provide a combined database on individual exposure-response relationships for 616 children ranging in age from 7 to 17 years, each with at least six sequential measurements of forced expiratory volume in 1 s (FEV_1) and previous-hour O_3 exposures while attending summer camps (Avol et al., 1990; Higgins et al., 1990; Raizenne et al., 1989; Spektor et al., 1988a, 1991). Similar lung function effects have been noted in adults engaged in normal recreational activities (e.g. Korrick et al., 1998; Spektor et al., 1988b). Children at summer camps are especially suitable subjects for the study of health effects and air pollution, since their exposures and activity patterns can be well defined. As summarized in Table 22.1, these studies of children at summer camps in Canada and the USA have indicated that lung function declines are associated with summertime haze air pollution, especially O_3 (Kinney et al., 1996).

Studies of children with asthma at summer camps have shown that a worsening of asthma problems also are associated with exposure to summertime haze air pollution, especially O_3. Such 'asthma camps', in which children with asthma participate in outdoor activities under careful medical supervision in a campground environment, provide a suitable setting in which to search for firmer individual-level evidence of the associations

Table 22.1 Slopes from regressions of afternoon FEV$_1$ on ozone for six camp studies (Kinney *et al.*, 1996)[a]

Study	Slope (ml/ppb)	SE (slope)	*p*-value
Fairview Lake, 1984	−0.50	0.16	0.002
Fairview Lake, 1988	−1.29	0.27	0.0001
Lake Couchiching, 1983	−0.19	0.44	0.66
CARES, 1986	−0.29	0.10	0.003
San Bernardino, 1987	−0.84	0.20	0.0001
Pine Springs, 1988	−0.32	0.13	0.013
Overall	−0.50	0.07	0.0001

[a] Data for each study were fitted in a model with subject-specific intercepts and one pooled O$_3$ slope.

between air pollution and asthma. Thurston *et al.* (1997) followed a group of over 50 children aged 7–13 having moderate to severe asthma who attended a summer 'asthma camp' in Connecticut. They found that air pollution, and O$_3$ in particular, was significantly and consistently correlated with acute asthma exacerbations, chest symptoms, and lung function decrements. As shown in Fig. 22.6, the children's peak flow lung function declined on days of high O$_3$ concentration, while the incidence of asthma attacks (as indicated by the daily number of on-site doctor-diagnosed β-agonist medications prescribed) increased on high O$_3$ concentration days. Similarly, in a study by Ostro *et al.* (1995) of 83 African-American children, aged 7–12, the probability of experiencing shortness of breath was found to be associated with both daily PM and O$_3$ air pollution exposures.

Overall, results from individual-level field epidemiology studies are generally consistent with the experimental studies of laboratory animals and humans and with the aggregate epidemiological studies indicating that severe adverse health effects are associated with acute O$_3$ air pollution exposure.

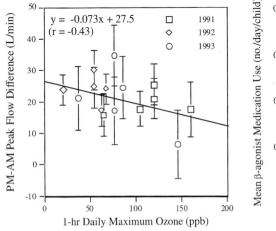
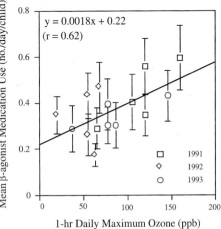

Fig. 22.6 Asthma camp associations between O$_3$ concentrations and both peak flow lung function change and asthma exacerbations (as indicated by β-agonist medication use). From Thurston *et al.* (1997).

Other Acute Morbidity

We have focused on the most severe health outcomes that may be associated with acute O_3 exposure in this chapter, but it is important to note that there are other morbidity effects experienced in the general public that are not routinely recorded, and therefore not usually as amenable to study and quantification. These include: minor restricted activity days (MRAD, a day on which normal activities are partially curtailed due to illness) (Ostro and Rothschild, 1989); acute respiratory symptom days (Krupnick *et al.*, 1990); and impaired athletic performance (Wayne *et al.*, 1967).

Although these other morbidity effects are less severe than ED visits, hospital admissions or premature mortality, they occur more frequently, and can involve a much larger fraction of the general population. Fig. 22.7 presents a recently made calculation of some of the present-day health impacts that could have been avoided by meeting the newly promulgated US O_3 standard in New York City (Thurston, 1997). Fig. 22.7 graphically displays the fact that the most severe O_3 effects are merely the 'tip of the iceberg' of the public health effects associated with O_3 air pollution.

Chronic Exposure Effects

Because of the above noted acute health effects associated with short-term O_3 exposures, combined with available long-term animal O_3 exposure study evidence (see Chapter 24

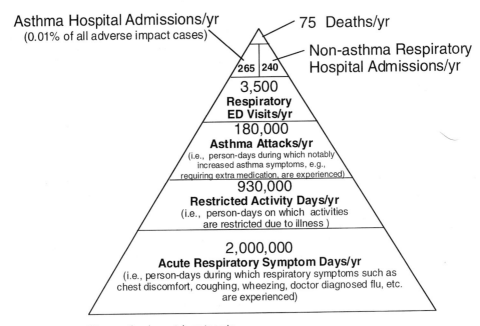

Fig. 22.7   Present-day adverse health effects estimated to be avoided by the meeting of the US EPA's 80 ppb 8-h average daily maximum standard in New York City, NY. From Thurston (1997).

of this book), there are concerns that long-term exposure also may have a cumulative human health impact. Evidence that long-term exposure could lead to permanent changes in health, or to the induction of new disease, would be of even greater public health concern than many of the short-term O_3 exposure morbidity effects noted above, which are uaually more transient in nature.

Chronic exposure epidemiological studies are of two types: cross-sectional studies (which compare health effects across different populations); and longitudinal studies (which follow changes in the prevalence of adverse health effects in a population over long periods of time). Longitudinal studies should have more power in detecting pollutant effects than cross-sectional studies, but long follow-ups of a single population are challenging, and losses in the surveyed population can be a serious problem.

Many of the older epidemiological studies of the health effects of chronic O_3 exposure suffered from exposure assessment problems. For example, the Six Cities Study (Dockery et al., 1993) was unable to detect chronic O_3 effects, but those cities all had fairly similar O_3 levels, weakening the ability of this study to discern any cross-sectional O_3 health effects differences across the particular cities considered.

Of the recent studies that have been conducted on the potential chronic exposure effects of O_3, the ongoing AHSMOG study of Seventh Day Adventists who have resided in California areas with differing oxidant levels provides some of the most useful information. For example, new results from this study indicate that long-term O_3 exposure can induce new cases of asthma. Nishino and co-authors (1996) report that there is a statistically significant ($p<0.05$) association between O_3 exposure and the development of asthma in men, but not in women, with O_3 being 'more important than any other pollutant'. In this study, the gender differences may be the result of males spending more time outside in summer than females (males, 18.4 h/wk; females, 10.9 h/wk, $p<0.001$). These results are consistent with the results previously published by this group that found that an increased risk of asthma development was significantly associated with increased ambient concentrations of O_3 exposure in men (RR = 3.12; CI, 1.61–5.85) (Greer et al., 1993).

In addition, a recent study by Kunzli and co-authors (1997) of freshmen at the University of California at Berkeley aged 17–21, who were lifelong residents of California, found that long-term exposure to O_3 was associated with declines in mid-expiratory lung function, even after addressing potentially confounding pollutants. For a 20 ppb increase in 8-h O_3, forced expiratory flowrate at 75% of expired lung volume ($FEF_{75\%}$) decreased 334 ml/s (95% CI, 11–657 ml/s), which corresponds to 14% of the population mean. The corresponding effect on $FEF_{25-75\%}$ was −420 ml/s (95% CI, +46 to −886, $p = 0.08$) or 7.2% of the mean. These O_3 exposure effects on baseline lung function were found to be independent of lifetime mean PM_{10}, NO_2, temperature or humidity.

Thus, while the small number of useful epidemiologic studies of chronic O_3 exposure on human health today provide only limited evidence, the most recent studies suggest that there may well be long-term health impacts of chronic O_3 exposure. These apparent effects include decrements in baseline lung function and the development of new cases of asthma. The latter of these two adverse outcomes (i.e. the induction of asthma) is of great public health policy and economic importance.

CONCLUSIONS

Recent aggregate population time-series epidemiology studies examining the acute effects of ambient O_3 have yielded significant associations with a wide range of adverse health outcomes, including lung function decrements, the aggravation of pre-existing respiratory disease, increases in daily hospital admissions and premature mortality. Individual-level camp studies clearly indicate that increases in the number of asthma exacerbations and their effects also are associated with increases in O_3 concentration, which corroborates the reported aggregate-level associations. In addition, less severe acute effects, such as increased numbers of restricted activity days, decreased lung function and diminished athletic performance, can occur even in healthy individuals. This indicates that the adverse effects of O_3 exposures are not limited to especially susceptible subpopulations (e.g. those with pre-existing respiratory disease), but that O_3 exposure can also adversely affect a large percentage of the general population. In addition, new long-term O_3 exposure studies suggest that both decreases in baseline lung function and the induction of new cases of asthma also may be associated with chronic O_3 exposure.

The emergency room visit and hospital admission studies collectively indicate that, when the major confounders are addressed (e.g. seasonality, temperature, day-of-week effects), consistent associations are seen between acute occurrences of respiratory morbidity and O_3 exposure. The evidence is especially strong for hospital admissions, since this association has been seen by numerous researchers at a variety of localities using a wide range of appropriate statistical approaches. Although the absolute size of the effect varies somewhat across localities and statistical approaches, these analyses collectively suggest that O_3 air pollution is associated with a substantial portion (on the order of 10–20%) of all respiratory hospital visits and admissions in the summer.

Our quantitative syntheses of the relevant epidemiologic studies provided combined estimates of the size of the O_3 effect for premature mortality and for various hospital admissions categories. The estimate provided by the combined random effects model for the effect of O_3 on major respiratory admissions for all ages was RR = 1.18 (95% CI, 1.10–1.26) per 100 ppb increase in daily 1-h maximum O_3 (or per 75 ppb as a maximum 8-h average, or per 40 ppb as a 24-h average). The combined estimate for asthma admission (all ages) for the subset of the same group of the studies was RR = 1.18 (95% CI, 1.07–1.30). Similarly, the most recent total mortality studies also collectively indicate significant increases in adverse health effects in association with ambient exposures to O_3, although the RRs are somewhat smaller (as would be expected, since the total mortality category includes causes of death less likely to be increased by acute air pollution exposures).

For all of the time-series studies of mortality and air pollution considered here, the combined analysis RR = 1.036 per 100 ppb increase in daily 1-h maximum O_3 (95% CI, 1.023–1.050), while the subset of studies that have most appropriately modeled the influences of weather yield a combined estimate of RR = 1.056 per 100 ppb increase in daily 1-h maximum O_3 (95% CI, 1.032–1.081). Overall, the observational epidemiologic evidence provides a coherent message: acute exposure to current ambient O_3 air pollution is associated with significant adverse health effects, including excess hospitalizations and premature deaths in the general population.

REFERENCES

Anderson HR, Ponce de Leon A, Bland JM et al. (1996) Air pollution and daily mortality in London: 1987–1992. Br Med J **312**: 665–669.

Anderson HR, Spix C, Medina S et al. (1997) Air pollution and daily admissions for chronic obstructive pulmonary disease in 6 European cities: results from the APHEA project. Eur Respir J **10**: 1064–1071.

Avol EL, Trim SC, Little DE et al. (1990) Ozone exposure and lung function in children attending a southern California summer camp. Paper 90-150.3. Proceedings of the 83rd Annual Meeting and Exhibition of the Air and Waste Management Association. Pittsburgh, PA: Air and Waste Management Association.

Barton DJ, Teets-Grimm K and Kattan M (1993) Emergency room visit patterns of inner city children for asthma. Am Rev Respir Dis **147**: A577.

Bates DV (1992) Health indices of the adverse effects of air pollution: the question of coherence. Environ Res **59**: 336–349.

Bates DV and Sizto R (1983) Relationship between air pollution levels and hospital admissions in Southern Ontario. Can J Public Health **74**: 117–133.

Bates DV and Sizto R (1987) Air pollution and hospital admissions in Southern Ontario: the acid summer haze effect. Environ Res **43**: 317–331.

Bates DV and Sizto R (1989) The Ontario air pollution study: identification of the causative agent. Environ Health Perspect **79**: 69–72.

Bates DV, Baker-Anderson M and Sizto R (1990) Asthma attack periodicity: a study of hospital emergency visits in Vancouver. Environ Res **51**: 51–70.

Borja-Aburto VH, Loomis DP, Bangdiwaia SI et al. (1997) Ozone, suspended particulates, and daily mortality in Mexico City. Am J Epidemiol **145**: 258–268.

Burnett RT, Dales RE, Raizenne ME et al. (1994) Effects of low ambient levels of ozone and sulfates on the frequency of respiratory admissions to Ontario hospitals. Environ Res **65**: 172–194.

Burnett RT, Brook JR, Yung WT et al. (1997a) Association between ozone and hospitalization for respiratory diseases in 16 Canadian cities. Environ Res **72**: 24–31.

Burnett RT, Dales RE, Brook JR et al. (1997b) Association between ambient carbon monoxide levels and hospitalization for congestive heart failure in the elderly in 10 Canadian cities. Epidemiology **8**: 162–167.

Burnett RT, Cakmak S, Brook JR and Krewski D (1997c) The role of particulate size and chemistry in the association between summertime ambient air pollution and hospitalization for cardiorespiratory diseases. Environ Health Perspect **105**: 614–620.

California Department of Public Health (1955) Clean Air for California: Initial report of the Air Pollution Study Project. San Francisco, CA: State of California, Department of Public Health.

Carroll RJ, Ruppert D and Stefanski LA (1995) Measurement Error in Nonlinear Models, Monographs on Statistics and Applied Probability 63. London: Chapman & Hall, p. 22.

Cifuentes LA and Lave L (1998) Association of daily mortality and air pollution in Philadelphia, 1983–1988. J Air Waste Manag Assoc (in press).

Cody RP, Weisel CP, Birnbaum G and Lioy PJ (1992) The effect of ozone associated with summertime photochemical smog on the frequency of asthma visits to hospital emergency departments. Environ Res **58**: 184–194.

Delfino RJ, Becklake MR and Hanley JA (1994) The relationship of urgent hospital admissions for respiratory illness to photochemical air pollution levels in Montreal. Environ Res **67**: 1–19.

Delfino RJ, Murphy-Moulton AM, Burnett RT et al. (1997) Effects of air pollution on emergency room visits for respiratory illnesses in Montreal, Quebec. Am J Respir Crit Care Med **155**: 568–576.

DerSimonian R and Laird N (1986) Meta-analysis in clinical trials. Control Clin Trials **7**: 177–188.

Dockery DW, Pope AC 3rd, Xu X et al. (1993) An association between air pollution and mortality in six U.S. cities. N Engl J Med **329**: 1753–1759.

Ellis FP (1972) Mortality from heat illness and heat-aggravated illness in the United States. Environ Res **5**: 1–58.

Greer JR, Abbey DE and Burchette RJ (1993) Asthma related to occupational and ambient air pollutants in nonsmokers. J Occup Med **35**(9): 909–915.

Higgins IT, D'Arcy JB, Gibbon DI *et al.* (1990) Effect of exposures to ambient ozone on ventilatory lung function in children. *Am Rev Respir Dis* **141**: 1136–1146.

Hippocrates (400 BC) *On Airs, Waters, and Places.* Translated and republished in *Medical Classics* **3**: 19–42 (1938).

Hoek G, Schwartz JD, Groot B and Eilers P (1997) Effects of ambient particulate matter and ozone on daily mortality in Rotterdam, The Netherlands. *Arch Environ Health* **52**: 455–463.

Ipsen J and Ingenito FE (1969) Episodic morbidity and mortality in relation to air pollution. *Arch Environ Health* **18**: 458–461.

Ito K and Thurston G (1996) Daily PM_{10}/mortality association: an investigation of at-risk populations. *J Expos Anal Environ Epidemiol* **6**: 79–96.

Ito K and Thurston G (1999) Epidemiological studies of acute ozone exposure and hospital visits/admissions. *J Expos Anal Environ Epidemiol* (submitted).

Kelsall JE, Samet JM, Zeger JE and Xu J (1997) Air pollution and mortality in Philadelphia 1974–1988. *Am J Epidemiol* **146**: 750–762.

Kinney PL and Özkaynak H (1991) Associations of daily mortality and air pollution in Los Angeles County. *Environ Res* **54**: 99–120.

Kinney PL and Özkaynak H (1992) Associations between ozone and daily mortality in Los Angeles and New York City. *Am Rev Respir Dis* **145**: A95.

Kinney PL, Ito K and Thurston GD (1995) A sensitivity analysis of mortality/PM-10 associations in Los Angeles. *Inhal Toxicol* **7**: 59–69.

Kinney PL, Thurston GD and Raizenne M (1996) The effects of ambient ozone on lung function in children: a reanalysis of six summer camp studies. *Environ Health Perspect* **104**: 170–174.

Korrick SA, Neas LM, Dockery DW *et al.* (1998) Effects of ozone and other pollutants on the pulmonary function of adult hikers. *Environ Health Perspect* **106**: 93–99.

Krupnick AJ, Harrington W and Ostro B (1990) Ambient ozone and acute health effects: evidence from daily data. *J Environ Econ Manag* **18**: 1–18.

Kunzli N, Lurmann F, Segal M *et al.* (1997) Association between lifetime ambient ozone exposure and pulmonary function in college freshmen – results of a pilot study. *Environ Res* **72**(1): 8–23.

Lipfert FW (1994) *Air Pollution and Community Health: A Critical Review and Data Sourcebook.* New York: Van Nostrand Reinhold.

Lipfert FW and Hammerstrom T (1992) Temporal patterns in air pollution and hospital admissions. *Environ Res* **59**: 374–399.

Mage DT and Buckley TJ (1995) The relationship between personal exposures and ambient concentrations of particulate matter. Paper no. 95-MP18.01. *Proceedings of the 88th Annual Meeting of the Air and Waste Management Association.* Pittsburgh, PA: Air and Waste Management Association.

Moolgavkar SH, Luebeck EG, Hall TA and Anderson EL (1995) Air pollution and daily mortality in Philadelphia. *Epidemiology* **6**(5): 476–484.

Moolgavkar SH, Luebeck EG and Anderson EL (1997) Air pollution and hospital admissions for respiratory causes in Minneapolis-St. Paul and Birmingham. *Epidemiology* **8**: 364–370.

Morris RD, Naumova EN and Munasnghe RL (1995) Ambient air pollution and hospitalization for congestive heart failure among elderly people in seven large US cities. *Am J Public Health* **85**: 1361–1365.

Nishino N, Abbey DE and McDonnell WF (1996) Long term ambient concentrations of ozone and development of asthma: the ASHMOG study. *Epidemiology* **7**: S31.

Ozkaynak H, Xue J, Severance P *et al.* (1995) Associations between daily mortality, ozone and particulate air pollution in Toronto, Canada. Paper presented at the *Colloquium on Particulate Air Pollution, Irvine, CA,* 24–25 January 1995. Irvine, CA: University of California.

Ostro BD and Rothschild S (1989) Air pollution and acute respiratory morbidity: an observational study of multiple pollutants. *Environ Res* **50**: 238–247.

Ostro BD, Lipsett MJ, Mann JK *et al.* (1995) Air pollution and asthma exacerbations among African-American children in Los Angeles. *Inhal Toxicol* **7**: 711–722.

Ostro BD, Sanchez JM, Aranda C and Eskeland GS (1996) Air pollution and mortality: results from a study of Santiago, Chile. *J Expos Anal Environ Epidemiol* **6**: 97–114.

Raizenne ME, Burnett RT, Stern B *et al.* (1989) Acute lung function responses to ambient acid aerosol exposures in children. *Environ Health Perspect* **79**: 179–185.

Romieu I, Meneses F, Ruiz S *et al.* (1996) Effects of air pollution on the respiratory health of asthmatic children living in Mexico City. *Am J Respir Crit Care Med* **154**: 300–307.

Samet JM, Zeger SL, Kelsall JE *et al.* (1997) *Particulate Air Pollution and Daily Mortality: Analysis of the Effects of Weather and Multiple Air Pollutants.* The Phase I.B Report of the Particle Epidemiology Evaluation Project. Cambridge, MA: Health Effects Institute.

Schwartz J (1994a) Air pollution and hospital admissions for the elderly in Birmingham, Alabama. *Am J Epidemiol* **139**: 589–598.

Schwartz J (1994b) Air pollution and hospital admissions for the elderly in Detroit, MI. *Am J Respir Crit Care Med* **150**: 648–655.

Schwartz J (1994c) PM10, ozone, and hospital admissions for the elderly in Minneapolis-St. Paul, Minnesota. *Arch Environ Health* **49**: 366–374.

Schwartz J (1995a) Short-term fluctuations in air pollution and hospital admissions of the elderly for respiratory disease. *Thorax* **50**: 531–538.

Schwartz J (1995b) Air pollution and hospital admissions for respiratory disease. *Epidemiology* **7**: 20–28.

Schwartz J (1997) Health effects of air pollution from traffic: ozone and particulate matter. In: Fletcher T and McMichael AJ (eds) *Health at the Crossroads: Transport Policy and Urban Health.* New York: John Wiley & Sons Ltd.

Schwartz J and Morris R (1995) Air pollution and hospital admissions for cardiovascular disease in Detroit, Michigan. *Am J Epidemiol* **142**: 23–35.

Simpson RW, Williams G, Petroeschevsky A *et al.* (1997) The association between outdoor air pollution and daily mortality in Brisbane, Australia. *Arch Environ Health* **52**: 442–454.

Spektor DM, Lippmann M, Lioy PJ *et al.* (1988a) Effects of ambient ozone on respiratory function in active normal children. *Am Rev Respir Dis* **137**: 313–320.

Spektor DM, Lippmann M, Thurston GD *et al.* (1988b) Effects of ambient ozone on respiratory function in healthy adults exercising outdoors. *Am Rev Respir Dis* **138**(4): 821–828.

Spektor DM, Thurston GD, Mao J *et al.* (1991) Effects of single- and multiday ozone exposures on respiratory function in active normal children. *Environ Res* **55**: 107–122.

Spix C, Anderson HR, Schwartz J *et al.* (1998) Short-term effects of air pollution on hospital admissions of respiratory diseases in Europe: a quantitative summary of APHEA study results. Air Pollution and Health: a European Approach. *Arch Environ Health* **53**: 54–64.

Stieb DM, Burnett RT, Beveridge RC and Brook JR (1996) Association between ozone and asthma emergency department visits in Saint John, New Brunswick, Canada. *Environ Health Perspect* **104**: 1354–1360.

Sunyer J, Spix C, Quenel P *et al.* (1997) Urban air pollution and emergency admissions for asthma in four European cities: the APHEA Project. *Thorax* **52**: 760–765.

Thurston GD (1997) Testimony at the hearings before the Committee on Environment and Public Works, United States Senate. Clean Air Act: Ozone and Particulate Matter Standards. Washington, DC: USGPO, pp. 118–127.

Thurston GD and Kinney PL (1995) Air pollution epidemiology: considerations in time-series modelling. *Inhal Toxicol* **7**: 71–83.

Thurston GD, Ito K, Kinney PL and Lippmann M (1992) A multi-year study of air pollution and respiratory hospital admissions in three New York State metropolitan areas: results for 1988 and 1989 summers. *J Exp Anal Environ Epidemiol* **2**: 429–450.

Thurston GD, Ito K, Hayes C *et al.* (1994) Respiratory hospital admissions and summertime haze air pollution in Toronto, Ontario: consideration of the role of acid aerosols. *Environ Res* **65**: 271–290.

Thurston GD, Lippmann M, Scott MB and Fine JM (1997) Summertime haze air pollution and children with asthma. *Am J Respir Crit Care Med* **155**(2): 654–660.

Thurston GD, Ito K and Gwynn RC (1999) Epidemiological studies of acute ozone exposure and mortality. *J Expos Anal Environ Epidemiol* (submitted).

Touloumi G, Katsouyanni K, Zmirou D *et al.* (1997) Short term effects of ambient oxidants exposure on mortality: a combined analysis within the APHEA Project. *Am J Epidemiol* **146**(2): 177–185.

US Bureau of the Census (1983) *County and City Data Book, 1983.* US Department of Commerce. Washington, DC: USGPO.

US EPA (US Environmental Protection Agency) (1996) *Air Quality Criteria for Ozone and Other Photochemical Oxidants.* Environmental Criteria and Assessment Office, EPA Report no. EPA-

600/P-93/004aF. Research Triangle Park, NC: US EPA.

US EPA (1997) *Regulatory Impact Analysis for the Particulate Matter and Ozone National Ambient Air Quality Standards and Proposed Regional Haze Rule.* Appendix J, Table 3. Office of Air Quality Planning and Standards. Research Triangle Park, NC: US EPA.

Verhoeff AP, Hoek G, Schwartz J and Wijnen JH (1996) Air pollution and daily mortality in Amsterdam. *Epidemiology* 7: 225–230.

Wayne WS, Wehrle PF and Carroll RE (1967) Oxidant air pollution and athletic performance. *J Am Med Assoc* **199**: 901–904.

Weisel CP, Cody RP and Lioy PJ (1995) Relationship between summertime ambient ozone levels and emergency department visits for asthma in central New Jersey. *Environ Health Perspect* **103** (Suppl. 2): 97–102.

White MC, Etzel RA, Wilcox WD and Lloyd C (1994) Exacerbations of childhood asthma and ozone pollution in Atlanta. *Environ Res* **65**: 56–68.

23

Controlled Exposure to Ozone, Nitrogen Oxides and Acids

MILAN J. HAZUCHA

The Center for Environmental Medicine and Lung Biology, The University of North Carolina at Chapel Hill, Chapel Hill, NC, USA

INTRODUCTION

NO_2 and O_3 frequently coexist in the ambient and indoor environment. Through complex chemistry, the two oxidants not only will react with each other, but also will react with other air contaminants as well. Some of the compounds generated may be even more toxic than the primary pollutants. Considering that the average American spends about 80% of his or her time indoors, some of the toxicants might be important not only in terms of indoor air quality, but also as potential health hazards. It is highly likely that elevated concentrations of these products in the indoor environment will result in the exposure of large and vulnerable subpopulations – including children, the elderly, asthmatics and individuals with chronic obstructive pulmonary disease (COPD) – any of whom may be especially susceptible to these pollutants. Surprisingly, few animal and even fewer human studies have investigated the potential adverse health effects of these gases in interaction. Considerable scientific information gaps thus exist in our understanding of a general lack of response reported in controlled laboratory studies with humans, versus the potentiating effects of these two gases (and related species) that are reported in epidemiological studies.

Since the literature is inconsistent about what constitutes an additive, cumulative, potentiating, synergistic or antagonistic effect, in the interpretation of the findings of the studies reviewed in this chapter these terms will be defined as follows. The *additive* (preferred usage) or *cumulative* response to a mixture is an algebraic sum of the effects induced by individual components of the mixture. Conceptually, the additive effect occurs only when the action of each pollutant is independent. When a pollutant that does not elicit

AIR POLLUTION AND HEALTH
ISBN 0-12-352335-4

a response when acting alone nevertheless increases the effect of another co-occurring pollutant, the effect is called *potentiation*. If the combined effect of the two pollutants is greater than the simple sum of effects elicited by each pollutant alone, the effect is interpreted as *synergistic*. The induction of effects that are smaller than the sum of individual changes is termed an *antagonistic* response. Most of the studies discussed in the subsequent sections employed these definitions.

To understand the nature and significance of the potential interaction effects of these pollutants better, the most important gas phase reactions and the interdependent aspects of dosimetry of O_3, NO_2, and nitric acid combinations will be outlined first. I then review animal studies of the pathophysiological effects of mixtures of these pollutants and the relevant dose–response studies. In the human health effects section, I discuss the findings of all published studies to date on the effects of O_3 in combination with other nitrogen oxides and derived pollutants. To aid in health risk assessment, I conclude with a discussion of possible pathophysiological mechanisms and the difficulties inherent in identifying adverse health effects.

GAS PHASE REACTIONS BETWEEN O_3 AND NO_2

Ambient and Indoor Air Chemistry

The primary photochemical reactions between O_3 and NO_2 produce many secondary pollutants, such as peroxyacetyl nitrate (PAN), nitric acid (HNO_3), nitrates (NO_3) and other nitrogen oxides (NO_y) (US EPA, 1993). Because of the photolytic instability of most of these species in sunlight, the concentrations of these pollutants are low. Although the seasonal, temporal and spatial patterns of O_3 and NO_2 in ambient air differ, under certain atmospheric conditions the two pollutants may undergo a chemical reaction outside the photochemical cycle. These reactions can produce a significant number of chemical species, some of which may be even more toxic than their precursors (Suzuki and Murashima, 1994; Suzuki *et al.*, 1995). Without irradiation, either during daytime or at night, substantial quantities of the nitrate radical (NO_3^-) may be formed. Under favorable conditions, the concentration of NO_3^- can be as high as 443 ppt (Weschler *et al.*, 1992). NO_3^-, in turn, will rapidly react with NO_2 to form dinitrogen pentoxide (N_2O_5). Potential formation of NO_3 and N_2O_5 in the exposure chamber atmosphere has been frequently suggested as the cause of synergistic responses observed in animal studies following exposure to combined O_3 and NO_2. Subsequent terminal reaction of N_2O_5 will produce HNO_3 (US EPA, 1996, 1993). The ambient concentration of HNO_3 in California has been reported to range from 0.5 to 56 $\mu g/m^3$ (Munger *et al.*, 1990). HNO_3 may be less toxic than O_3 alone, but its toxicity relative to NO_2 has not been studied sufficiently.

Although primarily an outdoor air pollutant, indoor O_3 concentrations may reach levels at which significant chemical reactions could occur between this oxidant and other constituents of indoor air, when O_3 is introduced either through infiltration of building structures or via a building ventilation air intake system. Without any internal sources of O_3, the indoor levels were typically about 20–30% of the outdoor concentration of O_3 in moderately ventilated buildings and 40–70% of the outdoor concentration of O_3 in highly ventilated structures (Thompson *et al.*, 1973; Weschler and Shields, 1997). The peak indoor concentration of O_3 could reach 0.17 ppm (Hales *et al.*, 1974; Weschler *et al.*, 1994). With electrostatic cleaners, poorly maintained corona discharge or UV light producing equipment operating in poorly ventilated rooms, the O_3 concentration can increase by as much as 0.2 ppm (Allen *et al.*, 1978).

In contrast to O_3, NO_x (NO, NO_2) are common in both outdoor and indoor environments. Because of their lower reactivity with substrates and their greater thermal stability, outdoor-to-indoor penetration of these species is higher than that found for O_3 at comparable ventilation rates. The indoor concentration of NO can range from 1 to 200 ppb, and that of NO_2 from 10 to 50 ppb (Weschler *et al.*, 1994).

Although the outdoor and indoor interactions between O_3 and NO_2 follow slightly different pathways, the intermediate nitrogen oxide species and the end-products formed without irradiation are the same. In daytime, either outdoors or indoors, O_3 will react rapidly with NO to form NO_2. Moreover, O_3 will react with NO_2 to generate the NO_3^- and subsequently N_2O_5, HNO_3 and free radicals (Weschler and Shields, 1997). Outdoors and at night, the major pathway of NHO_3 formation is through hydrolysis of N_2O_5. Indoors, NHO_3 is usually generated by NO_3 abstraction of H atoms from volatile organic compounds (VOC). Under favorable conditions, and particularly during summer months, the indoor concentration of NHO_3 can exceed the outdoor levels (Brauer *et al.*, 1991). Some reactions of NO_3 with VOC can lead to the formation of free radicals (Weschler and Shields, 1997; Weschler *et al.*, 1992). In addition to the formation of N_2O_5 and NHO_3, NO_2 can convert secondary amines to carcinogenic nitrosamines under certain conditions (Ichinose and Sagai, 1992). Some experimental evidence from animal studies implicates many of these secondary pollutants in the induction of pathophysiological changes. However, their relative toxicity is not yet well understood and some of these compounds have not yet been investigated.

Exposure Chamber Air Chemistry

The chemical interactions between O_3 and NO_2 in exposure chamber atmosphere was investigated in the very first studies of this mixture (Mustafa and Tierney, 1978). When NO_2 was introduced into an exposure chamber already containing O_3, the concentration of O_3 decreased substantially (Veninga and Lemstra, 1975). The chemistry of O_3 and NO_2 mixtures in the confined environment of a chamber recently has been investigated more quantitatively. Several studies have reported that to maintain the desired concentration of O_3 and NO_2 in a chamber atmosphere, the flow rate of both gases had to be increased above the level needed when only one pollutant was used for exposure (Gelzleichter *et al.*, 1992b; Mustafa *et al.*, 1984). Formation of nitrogen species is concentration-dependent when both O_3 and NO_2 are introduced into a dark exposure chamber. At low concentrations (0.2 ppm O_3 and 3.6 ppm NO_2), the O_3 loss was approximately 30%. At higher concentrations (0.8 ppm O_3 and 14.4 ppm NO_2), 62% of the initial O_3 was lost. The chemistry was also residence time-dependent; the higher the flow rate of gases into the chamber, the lesser the O_3 reaction with NO_2 (Gelzleichter *et al.*, 1992b). A second laboratory reported that a mixture of 0.35 ppm O_3, and 0.6 ppm NO_2 at 45% RH produced 49 $\mu g/m^3$ of HNO_3; 0.6 ppm O_3, and 2.5 ppm NO_2 at 85% RH produced 1900 $\mu g/m^3$ of HNO_3 in chamber atmosphere (Mautz *et al.*, 1988).

DOSIMETRY OF O_3, NO_x AND HNO_3

To understand the mechanisms that might be involved in eliciting pulmonary effects due to binary or sequential exposures to O_3, NO_x or other nitrogen species, we must first understand the dosimetry of these gases and acidic vapors (Hazucha, 1997). Currently, nothing is known about regional dosimetry of NO, an oxide of nitrogen usually present in the ambient atmosphere. The data available for NO_2 and O_3 suggest that these gases are not absorbed at the same rate and at the same airway sites. The dosimetry model of Miller and Overton (1982) for NO_2 predicts that terminal through respiratory bronchioles receive the highest dose of NO_2 per unit surface. Regional dosimetry models for O_3 predict a progressive decline in absorption from trachea to the periphery (Kabel *et al.*, 1994). Although several laboratory studies essentially validated these models for respective oxidants, it was not until Ben-Jebria *et al.* (1996) tested the predictive accuracy of the models when both gases were inhaled. They found that the pattern of absorption for these two gases in young healthy individuals is divergent, i.e. the O_3 fractional uptake decreased while NO_2 uptake increased as exposure progressed. More reactive O_3 was absorbed primarily by the mucosal layer of the conductive airways. Less reactive NO_2 penetrated deeper into the submucosa and epithelium of more peripheral airways. The study confirmed that the regional dose of NO_2 and O_3 does not appear to be influenced by presence of the other gas. Moreover, the authors have speculated that increased mucin secretion due to NO_2 might even ease absorption of O_3.

The chemistry of NO_2 and O_3 in the liquid phase is poorly understood. Inhaled NO_2 is quickly absorbed by the epithelial lining fluid of the airways (Postlethwait and Bidani, 1994; Postlethwait *et al.*, 1991). Once absorbed, in the presence of oxygen radicals various NO_x species enter a cycle of rapid chemical transformations leading to formation of NO_2^- and NO_3^- radicals. In the presence of superoxide (O_2^-), the preferential cycle in the airways appears to be formation of the toxic peroxinitrite radical ($OONO^-$) (Gaston *et al.*, 1994). Since O_3 is known to produce superoxide radicals, exposure to a mixture of NO_2 and O_3 may result in increased production of this species. Regardless of the reaction pathway, the final metabolic products are the same, i.e. nitrites, nitrates and nitric acid.

Dosimetry of HNO_3 has not been investigated. However, the model estimates of HNO_3 penetration and deposition into the airways show that because of neutralization of the acid in the airways by endogenously produced ammonium (NH_3), very little of the acid penetrates beyond the large conducting airways. Even under very unfavorable conditions for acid neutralization (very low production of NH_3, relatively high concentration of HNO_3 and high ventilation rates), the amount of acid reaching the trachea will be very low (Larson *et al.*, 1993). For this reason, HNO_3 is unlikely to be a major health hazard when inhaled either alone or in combination with other pollutants.

PATHOPHYSIOLOGICAL EFFECTS

Exposure to Mixtures of O_3 and NO_2

Organ Weight and Histopathology

Deviation from the normal weight is occasionally used as a gross index of toxic effect on the organs harvested from exposed animals. The time-dependent pattern of changes in the weight of the lung, thymus and spleen of mice did not substantially differ when combined and single gas exposures were compared. The average gain in lung weight and loss in thymus and spleen weights was intermediate between the large weight gains after O_3 exposure and the small weight losses found for the same organs during NO_2 exposure. The first 2 weeks of exposure usually had the greatest impact on the weight of the organs. Changes in weight were not correlated with changes in spleen function (Fujimaki, 1989; Mustafa *et al.*, 1984).

Pulmonary lesions during 9-week exposure of rats to a combined O_3 and NO_2 atmosphere seem to develop in three phases. The initial inflammation and epithelial and fibrotic changes stabilize after the first 3 weeks of exposure. The changes may even partially resolve during the subsequent 3-week period, before extensive remodeling, inflammation and progressive interstitial fibrosis takes place in the last 3 weeks of exposure. The histopathology and primary location of the lesions in the centriacinar region of the lung point to initial damage due to O_3 exposure. The subsequent spread of lesions to surrounding structures, with a progressive accumulation of collagen in the interstitium, is probably due to NO_2. Both the epithelial cell turnover rate and the rate of collagen synthesis follow the triphasic pattern of response. The causality of this relationship remains to be determined. Although termination of exposure will slow, stop or reverse some of the pathological processes during the subsequent recovery period, the nature and extent of the repair mechanisms have not been studied sufficiently. Additional experiments to define the progression of fibrosis during the recovery period have been inconclusive. The increase of collagen content appeared to reflect a repair process rather than the continued development of pulmonary fibrosis. Formation of reactive oxygen and nitrogen species, subsequent lipid peroxidation and a depletion of antioxidants have been suggested as the likely mechanisms for the observed pathophysiological changes. When coupled with the hypothesized formation of a highly toxic nitrogen pentoxide, this might account for some of the synergistic effects (Farman *et al.*, 1997).

Oxidative Stress and Antioxidants

In Vitro *Studies* The content of antioxidants in freshly obtained human plasma was rapidly depleted by exposure to a combination of high concentrations of O_3 (6 ppm) and NO_2 (10 ppm). When O_3 concentration was substantially decreased and NO_2 concentration increased under otherwise similar exposure conditions, the antioxidants were depleted even more. This suggests that the loss of antioxidants was primarily due to NO_2 exposure, although it is the weaker oxidant of the two. The overall oxidative damage in either mixture was less than additive. The authors suggested that the synergistic effects of the two gases found in other studies are not due to an increased oxidative damage in the extracellular fluids, but to a complex intracellular disruption of function. The oxidative damage to plasma constituents is an unlikely mechanism according to these investigators, since the ambient concentration of these oxidants must be in a toxic range for these gases to reach the concentrations used for plasma exposure (O'Neill *et al.*, 1995).

Ozone and NO_2 elicit different functional responses from directly exposed alveolar macrophages. The dose-dependent suppression by O_3 of the phagocytic activity of rat alveolar macrophages (AM) in a cell culture was much stronger than that caused by NO_2. The sensitivity of AM to oxidative damage by depletion of glutathione was greater for O_3 than for NO_2. Pre-incubation of AM with vitamin C provided more effective protection against NO_2 than against an equally toxic dose of O_3, while pre-incubation with α-tocopherol was about equally effective for both gases. These responses suggest that the reaction pathways and the reactive intermediate species leading to cellular damage might only partially coincide for these two oxidants. Moreover, these observations suggest that the sites of cellular damage for the respective oxidants may differ as well, since the protective action of glutathione and vitamin C take place in the cytoplasm, while α-tocopherol stabilizes the cellular membrane (Rietjens *et al.*, 1986, 1996).

In Vivo *Studies* In previous chapters, we have described the oxidative processes, the respective pathways and the antioxidative defense mechanisms for both gases. We know that formation of highly reactive hydroxyl radical (OH˙) leads to destruction of DNA, proteins, carbohydrates and membrane lipids. One typical mechanism of injury, membrane lipid peroxidation, leads to a self-promoting chain of events that produces peroxyl radicals and lipid peroxides. The conversion of membrane lipids to lipid peroxides (the product used as a marker of cellular damage) destabilizes the cell membrane and subsequently leads to cell dysfunction.

The extent of lipid peroxidation induced by the above radicals seems to be species-specific. Rats exposed to either O_3 or NO_2 or a combination of the two did not show any major changes in lipid peroxide levels (Ichinose and Sagai, 1989). A similar lack of lipid peroxidation was observed in hamsters (Ichinose *et al.*, 1988). Guinea pigs also did not show oxidative damage when exposed to single gases. Following exposure to a mixture of these oxidants, however, the lipid peroxides increased well above the control level. (Ichinose and Sagai, 1989). Mouse lung tissue appeared to be almost as susceptible to lipid peroxidation as guinea pig lung tissue (Ichinose *et al.*, 1988). Total phospholipid changes due to combined exposure were unremarkable for mouse, hamster and rat, but were clearly elevated in guinea pigs.

As in the case of lipid peroxidation, the post-exposure composition of fatty acid also revealed marked species differences. Prolonged exposure shifted the synthesis of polyunsaturated fatty acids (PUFA) from low polyunsaturated to high saturated fatty acids. Since these products are less sensitive to oxidation, this shift in synthesis could be seen as an adaptive mechanism to reduce peroxidation (Sagai *et al.*, 1987). Surprisingly, although DNA and protein are susceptible to oxidation, there were no significant differences in the total protein and DNA contents for exposure to only one or to a mixture of gases (O_3 and NO_2) of the species (Mustafa *et al.*, 1984; Ichinose and Sagai, 1989). During a lifetime exposure to near ambient concentrations of the combined gases, lipid peroxidation appeared to peak midway through both exposures in rats. The size of the synergistic effects induced by the mixtures was dose-dependent, driven primarily by NO_2 (Sagai and Ichinose, 1991).

Since oxygen is also a strong oxidizing agent and as such will generate reactive radicals and intermediaries, the lung is well prepared to minimize oxidative damage and to protect cell function. The primary defense mechanisms for countering oxidative stress are extra- and intra-cellular antioxidants. Both O_3 and NO_2 and a combination of the two have been shown to increase the antioxidative protective activity of a wide spectrum of enzymes and other antioxidants such as vitamins. Mixtures of these gases were much more effective in elevating antioxidant enzyme response than was each gas independently (Ichinose et al., 1988; Mustafa et al., 1984). The activity of these enzymes and the extent of subsequent protection was clearly species-dependent. The animals that were susceptible to lipid peroxidation also displayed lower activity for antioxidant enzymes than those animals that were not, e.g. rats or mice. The induction of antioxidant enzymes thus appears to be inversely related to cell damage, as assessed by lipid peroxidation. A similar inverse relationship with lipid peroxide levels was found for vitamin E (α-tocopherol) (Ichinose and Sagai, 1989; Ichinose et al., 1988; Sagai et al., 1987). Lifetime exposure of rats to near ambient concentration of $O_3 + NO_2$ transiently increased the content of non-protein sulfhydryl and vitamin E in lung homogenate. Antioxidants such as vitamin C and routinely measured antioxidant enzyme activities (SOD, G6PD, etc.) did not show any major changes from baseline. As observed in many previous studies, the occurrence of maximum response was assay-dependent. Additional measurement of ethane production (another indicator of peroxidation) in exhaled air showed a good association with the lipid peroxidation index. Thus, production of ethane potentially might serve as a non-invasive indicator of oxidative stress (Sagai and Ichinose, 1991).

Takahashi and colleagues have looked at the effects of prolonged exposure of rats to O_3 or NO_2 on the activity of pulmonary cytochrome P-450 microsomal monooxygenases, the system extensively involved in mixed monooxidation of lipophilic substrates. Ozone increased and NO_2 decreased both the activities and the content of cytochrome P-450 in microsomal fraction of lung tissue in a dose-dependent fashion (Takahashi and Miura, 1985). Addition of NO_2 to the O_3 exposure atmosphere substantially reduced the contents and the metabolizing activity of the cytochrome P-450 system. These investigators suggest that the attenuating effect of NO_2 on O_3 may be due not to oxidative processes in tissue *per se*, but instead to the formation of more toxic nitrates and nitrites (Takahashi and Miura, 1989). A strong dose-dependent negative association of the activity of microsomal NADPH-dependent cytochrome P-450 peroxidase and lipid peroxide production in rats exposed to a mixture of $O_3 + NO_2$ reflects the high sensitivity of this enzyme to membrane lipid peroxidation (Sagai and Ichinose, 1991).

Inflammation

Inflammation in rats chronically exposed to $O_3 + NO_2$ is exposure time-dependent. The initial rapid increase in number of inflammatory cells in the lung tissue leveled off midway through the exposure, but rose again near the end of exposure. Although the proliferation of cells was generalized, the macrophages, polymorphonuclear cells (PMN) and mast cells were present at greater numbers than other types of cells (Farman et al., 1997). The time-dependent disproportional rate of cell recruitment may reflect their differential activation by O_3 and NO_2 (Mochitate et al., 1992). The sites of accumulation for these cells differed as well, probably reflecting their specific roles in inflammation.

While macrophages and PMNs tended to aggregate within alveoli and the interstitium, mast cells clustered around bronchoalveolar duct junctions (Farman *et al.*, 1997).

Host Defense Mechanisms

Immunologically mediated responses to the inhalation of mixtures of oxidants have been studied only superficially. Fujimaki (1989) investigated the time-dependence of anti-body responses to T-cell-dependent and T-cell-independent antigens in the spleen of mice exposed to O_3 + NO_2. In the initial weeks of exposure, T-cell-dependent antibody responses were significantly suppressed, while T-cell-independent responses were enhanced. During the subsequent 6 weeks of exposure, the responses tended to return to baseline. The divergence of the antibody response is probably related to the activation of the helper thymus-derived lymphocytes (T cells), which is required for the cell-dependent antibody response, but not for the cell-independent response (Fujimaki, 1989).

Over the last decade, several animal studies have suggested that common air pollutants might influence carcinogenic processes such as dissemination of cancer cells and the pro-motion and progression of tumors. Mice that were intravenously injected with B16 melanoma cells showed significantly greater numbers of cancer cell colonies in the lung when exposed to ambient concentrations of a mixture of O_3 and NO_2 for 12 weeks than did animals exposed only to the individual gases. The additive effect was primarily due to NO_2, since O_3-exposed mice showed only slightly higher (not significant) colonization of cells than mice exposed to air. The natural cytostatic activity of spleen cells also was sig-nificantly reduced by exposure to O_3 + NO_2. The number of cancer colonies in mice that were injected with melanoma cells pre-treated *in vitro* with spleen cells from unexposed mice was significantly lower than that found in mice that received melanoma cells pre-treated by spleen cells obtained from O_3+NO_2 exposed animals (Richters, 1988).

The potential implications of these observations for humans are unclear. Although the proposed mechanism of a cancer metastasis is plausible, the applicability of this mode of dissemination to humans will require further study. The experimental evidence based on O_3, NO_2 and combined exposures of even the most susceptible species and strains of ani-mals (such as A/J mice) does not seem to implicate these two gases in carcinogenesis. Similarly, it is unclear whether these gases promote tumor development (Witschi, 1988). To date, no findings have been published on tumor promotion or development due to combined exposure to these gases.

Infectivity Models

Inhalation challenge with a bacterial aerosol (*Streptococcus pyogenes*) following repeated exposures of mice to either O_3, NO_2 or O_3+NO_2 over a range of concentrations, showed that the extent of respiratory infection was driven by the presence of O_3. Exposure to each gas alone decreased mouse survival time, increased mortality and slowed lung bacterial clearance only at higher concentrations of these pollutants. Inhalation of the mixture resulted in additive and synergistic effects at all combinations of the two gases except the lowest concentration (Ehrlich *et al.*, 1977, 1979). Much more complex exposure patterns were used to evaluate the severity of pulmonary bacterial infection in mice when chal-lenged with an aerosol of *Streptococcus zooepidemicus*. Because of the complexity of the design, interpretation of the findings is difficult. Excess mortality (over air control) did increase in a dose-dependent manner, reaching significance at concentrations above 0.1 ppm of either gas, including mixtures that showed synergism (Graham *et al.*, 1987).

Mortality

The mixture of O_3 + NO_2 was far more lethal than were the individual gases at comparable dose rates. Mortality of animals exposed to a mixture of the two gases at non-lethal concentrations was much higher than mortality after exposure of animals to either O_3 or NO_2 (Farman *et al.*, 1997). The authors of this and other studies have speculated that the formation of nitrogen pentoxide is the cause of the observed effects and higher death rates. However, no direct measurements of the toxic species that are potentially formed during exposure to binary mixtures have been reported in any animal studies.

Sequential Exposure to NO₂ and O₃

In a complex protocol alternating the sequence of exposures to 0.8 ppm O_3 and 14.4 ppm NO_2 as well as time of exposure (day vs. night) for each gas, the amount of total protein, PMNs and epithelial cells recovered from lavage fluid of rat lung was higher than that found after exposures to individual gases. The authors speculate that the enhanced response is due to other nitrogen species formed by dark reaction between O_2 and NO_2 (Gelzleichter *et al.*, 1992b).

Mice exposed for several days to O_3 (0.4 ppm) following an almost complete recovery from a 7-day exposure to NO_2 (14 ppm) had no change in body weight and only temporarily altered their eating and drinking behavior (Umezu *et al.*, 1993).

Exposure to Mixtures of O₃ and HNO₃

Short exposure of rats to high concentrations of O_3 and HNO_3 (Nadziejko *et al.*, 1992) and long-term exposure of rabbits to low concentrations produce a complex response (Schlesinger *et al.*, 1994). For most of the measured outcomes (LTB_4, LTC_4, LDH, TNFα, bronchopulmonary lavage protein, cell count and viability, elastase inhibitory capacity) the low concentration mixtures had an antagonistic effect. Any significant changes (as compared to single gas exposures) such as a decrease in alveolar macrophage and an increase in PMN count observed at high concentrations of the mixture were attributable to O_3. Superoxide generation by alveolar macrophages exposed to a mixture was antagonism in zymosan-stimulated and synergism in resting (spontaneously active) macrophages when compared to a production elicited by exposure to individual gases (Nadziejko *et al.*, 1992; Schlesinger *et al.*, 1994). The *in vitro* tracheal rings responsiveness to acetylcholine did not differ between exposure to each pollutant in isolation and to the mixture. The responsiveness of bronchial rings, however, was inconsistent (Schlesinger *et al.*, 1994). The relevance of these findings to human health remains to be determined. Many recent epidemiologic findings have shown an association between ambient acid levels and respiratory disease. However, the primary acidic species in those studies have been the products of sulfuric rather than nitric acid and the former is easily neutralized in the airways by HN_3.

DOSE–RESPONSE RELATIONSHIP

It is well known from studies on the effects of exposure to individual gases that apart from a species studied, the primary exposure determinants of a response are the gas concentration (C), exposure duration (T), and minute ventilation (V_E). The same variables also play a key role in eliciting the response to combined O_3 and NO_2. Exercizing rats had a significantly higher number of focal lesions in the terminal bronchioles and alveolar regions of the lung than did resting rats under the same exposure conditions. Exposure to mixed pollutants whether at rest or during exercise also had synergistic effects, as assessed by the number of lung parenchymal lesions (Mautz *et al.*, 1988).

Because of the increasing complexity of exposure regimens employed in various studies, it would be convenient to be able to express the overall exposure load in terms of a single exposure index that would incorporate the key determinants of exposure mentioned earlier. Attempts to develop and validate such an index have been only partially successful (Gardner, 1979). This concept was tested recently under very complex conditions of intermittent exposure of varied duration, either during the day or at night, to a mixture of O_3 and NO_2 at a wide range of concentrations (Gelzleichter *et al.*, 1992a,b; Last *et al.*, 1994). The combined cumulative dose was expressed as a sum of the respective products of concentration and exposure duration for O_3 and NO_2. The response was assessed by changes in bronchiolar lavage protein, epithelial cells and PMNs. Although the interpretation of these studies is difficult, several conclusions can be drawn about the relationship between $O_3 + NO_2$ mixture dose and the lung lavage constituents' response: (1) the response generally depends on concentration; (2) the same dose ($C \times T$= constant) will elicit approximately the same response only over a very narrow range of exposure conditions; (3) the outcome of a response (antagonistic, additive or synergistic) is exposure regimen and assay dependent; and (4) continuity of exposure may be an important determinant of response. In this animal model, the determining factor of the extent of lung injury was not the total dose of the mixture, but the repeated exposure–recovery sequence (Last *et al.*, 1994).

Rajini *et al.* (1993) exposed rats to four different mixtures of O_3 (0.2–0.8 ppm) and NO_2 (3.6–14.4 ppm) while keeping the dose constant. Of the three airway sites of interest (large intrapulmonary airways, peripheral airways and the alveolar regions), only the intrapulmonary airways showed a synergistic effect, as assessed by changes in the percentage of labeled epithelial cells (the labeling index). In the peripheral airways, the mixture with lowest concentration of gases had an antagonistic effect, while the highest concentrations had a synergistic effect, as measured by labeling index. In the alveolar region, the index value after exposure to the mixture was about the same as that found for single pollutants at all concentrations but the highest one, where the effect was additive. Under these experimental conditions, NO_2 appeared to be a more potent component of the mixture than O_3.

The rate of recovery and the extent of repair processes in the lung following exposure to mixtures appears to be dose-dependent. Lung collagen deposition paralleled the exposure dose rate of gases; the higher dose rate was followed by higher production of collagen (Farman *et al.*, 1997).

HUMAN HEALTH EFFECTS

Laboratory Exposure to Mixtures of O_3 and NO_2

The health effects of acute exposure (1–2 h) to a combined atmosphere of O_3 and NO_2 have been investigated in only eight studies; these will be discussed in the following sections. The O_3 concentration in these studies ranged from 0.12 ppm to 0.50 ppm and the NO_2 concentration from 0.15 ppm to 0.60 ppm. The study populations included young, adolescent, middle-aged and elderly males and females. Seven of the studies employed intermittent exercise during exposure; one study (Adams *et al.*, 1987) used a continuous exercise protocol. In addition to other physiologic tests, spirometry (FEV_1) was uniformly used to assess the pulmonary function changes.

Studies of Healthy Populations

Healthy adolescents exposed to air, O_3, NO_2 and $O_3 + NO_2$ showed small and inconsistent changes (6% of a baseline) in total airway resistance (R_T) and spirometric variables, none of which were statistically significant. The post-exposure follow-up for up to 6 hours did not show any time-related changes under any of the exposure conditions (Koenig *et al.*, 1988a,b).

Exposure of young males to the binary mixture under a range of temperatures (25–40°C) and relative humidity (45–85%) appeared to elicit about the same spirometric, plethysmographic and ventilatory pattern of changes (Folinsbee *et al.*, 1981) as the changes reported earlier from the same laboratory following exposures conducted under the same chamber atmosphere conditions with O_3 alone (Folinsbee *et al.*, 1977). Heavy continuous exercise of young trained males and females breathing O_3 and NO_2 did not induce greater effects as assessed by symptomatic response, spirometric lung function and ventilatory pattern than those found following O_3 exposure alone. However, there was a significant antagonism for specific airway resistance (SR_{aw}). The increases in resistance following exposure to the mixture were 50 and 80% smaller for males and females, respectively, than the increases induced by O_3 alone (Adams *et al.*, 1987).

Older men and women (51–76 years) responded less to O_3, NO_2 or $O_3 + NO_2$ than young subjects, with both age groups showing a considerable range of responsiveness. In both age groups, however, the mixture elicited slightly greater spirometric decrements on average (2–4% FEV_1) than O_3 alone, which dominated the response. The spirometric changes (FVC, FEV_1 and FEF_{25-75}) for both males and females were additive in young cohorts and synergistic in the old. Males were generally more responsive than females at all exposure conditions. The average symptomatic response of both young and old subjects was approximately the same after O_3 exposure. The mixture, however, induced more severe symptoms in the young and slightly weaker symptoms in the old age group, when compared with O_3 alone (Drechsler-Parks, 1987; Drechsler-Parks *et al.*, 1989). Combined exposure also significantly decreased the stroke volume, heart rate and cardiac output (15%) during exercise periods in older individuals when compared with either O_3 or NO_2 alone. This observation implies not only that exercise performance may be limited but, even more significant, that individuals with pre-existing cardiovascular disease could be at increased risk of aggravating the disease when exposed to the mixture

(Drechsler-Parks, 1995). Whether young individuals also respond with decreased cardiac output has not been studied.

Studies of Asthmatics

Spirometric lung function assessment has repeatedly demonstrated that adult asthmatics are more responsive to O_3 or NO_2 than healthy adults. No studies have been published to date on the health effects of combined O_3 and NO_2 exposure on adult asthmatics. In healthy and asthmatic adolescents exposed to O_3, NO_2 and $O_3 + NO_2$, the lung function changes in these exposure atmospheres were, on average, not significantly different from air exposure. A few spirometric and plethysmographic variables showed marginal but inconsistent changes. When compared with the baseline values, O_3 tended to decrease while NO_2 and $O_3 + NO_2$ tended to improve spirometric lung function when measured immediately post-exposure. The opposite was true for R_{aw} and R_T measures. These trends, however, disappeared within 25 min of the recovery period. As observed in healthy cohorts, several young asthmatics were hyperresponsive, but the variance estimates of the key spirometric variable (FEV_1) for the asthmatic and healthy cohorts have not differed. Thus, the proportion of susceptible individuals among healthy and asthmatic adolescents may be about the same (Koenig et al., 1985, 1987, 1988a,b; Linn et al., 1995). The results of these studies show that short-term exposures (<90 min) of moderately exercising adolescents with light asthma to mixtures of O_3 (≤ 0.2 ppm) and NO_2 (≤ 0.3 ppm) do not aggravate existing disease, and generally do not elicit any significant spirometric and plethysmographic lung function effects.

Sequential Exposure to NO_2 and O_3

In the most typical ambient pattern of oxidant pollutants in the urban areas, the morning peak of NO_2 is followed several hours later by formation of O_3, which peaks in the early afternoon. Surprisingly, only one study (Hazucha et al., 1994) sought to determine the health effects of such sequential exposure. A potential 'priming' effect of NO_2 on O_3-induced pulmonary function changes was studied in healthy young exercising females. A 2-h NO_2 (0.6 ppm) or air exposure was followed 3 h later by a 2-h exposure to O_3 (0.3 ppm). NO_2 exposure alone did not induce any significant symptomatic, spirometric, plethysmographic or ventilatory pattern changes. O_3 exposure alone elicited a typical symptomatic response to irritants, a decrease in spirometric and an increase in airway resistance variables, and an increase in airway responsiveness. Ventilatory parameters also showed a typical irritant response, a decrease in tidal volume and an increase in the frequency of breathing. Pre-exposure to NO_2 elicited significantly greater effects only in spirometric variables. The additive effects were attributed to O_3-induced inflammation of airways 'primed' by NO_2 pre-exposure.

Exposure to HNO_3 and O_3

Similarly unique laboratory studies were designed to evaluate the health effects of sequential (Aris et al., 1991) and combined HNO_3 (500 $\mu g/m^3$) and O_3 (0.2 ppm), at concentrations mimicking an urban air pollution episode (Aris *et al.*, 1993). The post-ozone spirometric lung function, SR_{aw}, symptoms and nonspecific bronchial responsiveness of healthy young but O_3-sensitive volunteers were all attenuated by previous exposure to fog containing HNO_3 (Aris *et al.*, 1991). Exposure of healthy young non-smokers to $HNO_3 + O_3$ induced a typical spirometric and plethysmographic response that was not significantly different from the effects caused by O_3 exposure alone. Analysis of the bronchoalveolar lavage (BAL) and lung biopsy samples obtained 18 h later showed a similar lack of interaction. Neither proximal nor bronchoalveolar lavage and biopsy findings showed any major differences in cellular and biochemical constituents between O_3 and the mixture. Although not statistically significant, inhalation of the mixture increased SR_{aw} much less than exposure to O_3 alone, suggesting a moderating effect of HNO_3 on bronchoconstriction when present in a mixture with O_3 (Aris *et al.*, 1993).

Exposures to PAN and O_3

Exposure to PAN (another photochemical oxidant) induced small spirometric decrements in healthy young subjects. When combined with O_3 (0.27 ppm PAN + 0.48 ppm O_3), the lung function effects were additive in both healthy young males and females. The reported subjective symptoms were more severe during combined exposure than with either gas exposure alone (Drechsler-Parks *et al.*, 1984; Horvath *et al.*, 1986). Lung function changes after exposure to PAN + NO_2 + O_3 were roughly the same as those after NO_2 + O_3 exposure. Older individuals were generally less responsive to these mixtures (Drechsler-Parks *et al.*, 1989). Repeated exposure to the mixture on five consecutive days resulted in adaptive pattern of response to O_3. The maximum spirometric decrements were observed on the second day of exposure. By the fifth exposure, the post-exposure lung function changes were close to baseline values. A single exposure to the mixture on the third day of the recovery period induced slightly greater spirometric effects than those measured after fifth exposure. Another single exposure on the seventh day of recovery produced spirometric decrements almost as large as those observed following the first day of exposure. These observations suggest a progressive loss of 'adaptation' over a period of seven days. The symptomatic response paralleled the pattern of lung function response (Drechsler-Parks, 1987).

Mechanisms of Physiological Response

O_3 and NO_2, both irritants and strong tissue oxidants, exert their toxic actions through many common mechanisms. As discussed earlier in the chapter, the regional doses and the primary sites of action of O_3 and NO_2 overlap but are not the same. Since these gases are relatively insoluble in water, they probably penetrate into the peripheral airways that are more sensitive to damage than better protected conducting airways. A recent study of the longitudinal distribution of absorption of these gases in the airways of healthy young individuals revealed that exposure to NO_2 increased the absorbing capacity of distal conducting airways for O_3 (Rigas *et al.*, 1997). Furthermore, with progression of an exposure (to each gas alone) the absorption fraction of O_3 tended to decrease, whereas the opposite was observed for NO_2. The authors suggest that NO_2-inflamed airways release additional substrates into the epithelial lining fluid that react with O_3, thus progressively removing O_3 from the airway's lumen (Rigas *et al.*, 1997). Since O_3 is a much stronger oxidant than NO_2, 'neutralization' of O_3 by this mechanism might explain many findings of antagonistic responses when the two gases are combined in an exposure atmosphere.

Accidental exposure to a high concentration of NO_2 typically induces a biphasic response: an early response that develops within 3–30 h after exposure and a relapse 2–3 weeks later. Since a similar pattern of response may be expected at lower NO_2 concentrations, it is not surprising that controlled laboratory studies reported no symptomatic effects immediately post-exposure and that lung function changes ranged from none to marginal. If toxic doses of NO_2 cause the first symptoms only several hours following the exposure, it is highly unlikely that ambient to low concentrations will elicit an immediate post-exposure response. Therefore, any symptoms reported immediately post-exposure to a mixture of NO_2 and O_3 are probably induced by O_3 alone. The same reasoning applies to the objective physiologic findings. Indeed, the in-depth review of the findings presented in the earlier sections shows that, except for a slightly more pronounced (+ 2–3%) though non-significant decline in the post-$(NO_2 + O_3)$ as compared with post-O_3 spirometric response of both young and old individuals, the observed decrements could be attributed solely to O_3.

Although spirometric changes were inconsistent and non-significant, two studies reported significant effects in seemingly unrelated cardiopulmonary variables such as airway resistance (Adams *et al.*, 1987) and cardiac output (Drechsler-Parks, 1995). These observations warrant further examination. The findings of Adams *et al.* (1987) (the only study to use continuous exercise) of 50% and 75% improvement in SR_{aw} of females and males, respectively, when NO_2 was present in the mixture, as compared to the respective changes post-O_3, suggest an antagonistic effect. The authors speculate that formation of nitric acid vapors may have induced these statistically significant changes by what could be interpreted as an attenuation of O_3 by NO_2.

A more recent study by Drechsler-Parks (1995) reported that in older individuals the exercise-induced increase in cardiac output (CO) during exposure to $NO_2 + O_3$ was significantly lower (by 15%) than CO measured under the same conditions during exposure to O_3, NO_2 or air. Although both the heart rate (HR) and the stroke volume (SV) decreased, the reduction in CO was primarily due to smaller SV. It is of interest that Folinsbee *et al.* (1981) also reported a decrease in HR during $NO_2 + O_3$ exposure as compared with exposure to air.

Recent advances in our understanding of the chemistry and reactivity of nitrogen oxides (NO_x) and the bioactive role played by some of these oxides (particularly NO) in regulation of multiple pathophysiological processes could explain some of these reported $NO_2 + O_3$ effects. In the liquid phase, 'exogenous' NO_x and that already present in the tissue undergoes (particularly in the presence of oxygen and O_3) a rapid series of reactions that generate multiple intermediate oxidative species of nitrogen, including NO (Gaston *et al.*, 1994). Production of NO through this mechanism may contribute to modulation of bronchial diameter and consequently to airway resistance. Studies on both animals (Brown *et al.*, 1994) and humans (Sanna *et al.*, 1994) reported a bronchodilator effect for inhaled NO. Brief inhalation of NO increased the baseline and prevented post-methacholine decrease in specific airway conductance (SG_{aw}) in healthy individuals. Similar modulation of bronchomotor tone by inhaled NO has been reported in asthmatics, but not in patients with COPD (Hogman *et al.*, 1993). The differential response of healthy individuals, asthmatics and COPD patients to NO might explain why COPD patients appear to have a more consistent and severe response to NO_2 than the other groups. It is tempting to speculate that if inhaled NO_2 stimulates the production of NO either directly or through incorporation into the NO_x metabolic cycle, then the interactive effect of NO_2 and the O_3 might be due to downregulation of O_3-activated cholinergic system and consequently to tempered increase in airway resistance (Hazucha, 1997).

Inflammation induced by O_3 might be another mechanism by which production of NO may be increased (Krishna *et al.*, 1996). It is well established that the inflammatory cells (mast cells, macrophages, PMN) and the airway epithelium are sources of NO (Barnes and Belvisi, 1993). The well-documented activation of PMN and release of cytokines due to O_3 may result in the switching on of inducible NO-synthase (iNOS) and the subsequent production of large amounts of NO. Several recent studies have reported increased production of NO following inhalation of pulmonary irritants, including O_3. Punjabi *et al.* (1994) showed that type II cells in O_3-exposed rats produced significantly more NO than did those cells in control animals. The activation of iNOS in another inflammatory model has been so consistent that Stewart *et al.* (1995) suggested using the concentration of NO in the exhaled air as an early marker of lung inflammation.

The typical inflammatory response due to O_3, observed consistently in both humans and animals, is the production of prostaglandins from arachidonic acid via activation of cyclooxygenase (COX) enzymes. Limited experimental evidence suggests that one of the inflammatory mechanisms might be coexpression of iNOS and COX (Sautebin *et al.*, 1995). Since prostaglandins also are potent bronchoactive agents, the balance between the production of NO and prostaglandins might be a mechanism that could explain the minimal or non-measurable change in airway resistance following $NO_2 + O_3$ exposure.

The third mechanism might be dependent on the metabolic end-products of NO, including nitrates and nitrites (Gaston *et al.*, 1994). However, neither sequential exposure to O_3 followed by HNO_3 (Aris *et al.*, 1991) nor exposure to a mixture of $O_3 + HNO_3$ (Aris *et al.*, 1993) produced any effects on the lung function response that were significantly different from the effect of exposure to O_3 alone. The same results were found for the cellular and mediator constituents of BAL, proximal airway lavage fluid, and the bronchial biopsy specimens, leading the authors to conclude that HNO_3 does not potentiate the inflammatory response induced by O_3 in healthy individuals.

Of the three potential mechanisms discussed above that might explain the observed amelioration of increased SR_{aw} by exposure to $NO_2 + O_3$, the upregulation of pulmonary iNOS by inflammatory processes and the release of inflammatory mediators seems the most plausible (Polzer *et al.*, 1994). It remains to be determined whether the smaller increase in R_{aw} attributed to the presence of NO_2 in a mixture with O_3 is beneficial to exposed individuals. Known pathophysiological mechanisms point in the opposite direction. Increased airway resistance is the body's defense response to irritants, presumably minimizing their penetration. Less constricted airways allow a greater penetration of pollutants into the periphery of the lung. Consequently, more sensitive areas of the lung might be open to injury from inhaled pollutants and co-pollutants, including particulate matter. Thus, NO_2 might 'prime' the respiratory system for subsequent challenge by other pollutants (Hazucha, 1997).

In its wide co-regulatory function, NO has also been implicated in a complex modulation of hemodynamics, including cardiac function and pulmonary and systemic vascular resistance. It is tempting to suggest that NO-mediated mechanisms are involved in both of the responses induced by $NO_2 + O_3$ exposure: decreased airway resistance and lower cardiac output during exercise. At present, however, no suitable mechanism(s) have been advanced by which NO_2 and O_3 could be involved in the generation of NO by vascular endothelial cells as well as in the consequent modulation of circulatory processes (Hazucha, 1997).

Drechsler-Parks (1995) has suggested that a reduction in the increase in cardiac output during exercise in an atmosphere of $NO_2 + O_3$ is due to the formation of nitrites and nitrates. This is consistent with the current assumption that the oxidation end-product of NO_2 in a liquid phase is nitrate NO_3^- (Gaston $et\ al.$, 1994). The hypothesis also is consistent with our understanding of the pathophysiological mechanisms of nitrate action on the cardiovascular system (Busse and Bassenge,1982). The hemodynamic effects of $NO_2 + O_3$ may be induced by this mechanism. The physiologic significance of Drechsler-Parks' finding, however, is unclear. Reduced capability to increase cardiac output during physical activity also could have negative consequences for some individuals under certain conditions.

REFERENCES

Adams WC, Brookes KA and Schelegle ES (1987) Effects of NO_2 alone and in combination with O_3 on young men and women. $J\ Appl\ Physiol$ **62**: 1698–1704.

Allen RJ, Wadden RA and Ross ED (1978) Characterization of potential indoor sources of ozone. $Am\ Ind\ Hyg\ Assoc\ J$ **39**(6): 466–471.

Aris R, Christian D, Sheppard D $et\ al.$ (1991) The effects of sequential exposure to acidic fog and ozone on pulmonary function in exercising subjects. $Am\ Rev\ Respir\ Dis$ **143**: 85–91.

Aris R, Christian D, Tager I $et\ al.$ (1993) Effects of nitric acid gas alone or in combination with ozone on healthy volunteers. $Am\ Rev\ Respir\ Dis$ **148**: 965–973.

Barnes PJ and Belvisi MG (1993) Nitric oxide and lung disease. $Thorax$ **48**: 1034–1043

Ben-Jebria A, Rigas ML and Ultman JS (1996) Effects of continuous exposure to O_3 and NO_2 and O_3 uptake in human lungs airways. $Am\ J\ Respir\ Crit\ Care\ Med$ **153**: A700.

Brauer M, Koutrakis P, Keeler GJ $et\ al.$ (1991) Indoor and outdoor concentrations of inorganic acidic aerosols and gases. $J\ Air\ Waste\ Manag\ Assoc$ **41**, 171–181.

Brown RH, Zerhouni EA and Hirshman CA (1994) Reversal of bronchoconstriction by inhaled nitric oxide – histamine versus methacholine. $Am\ J\ Respir\ Crit\ Care\ Med$ **150**: 233–237.

Busse R and Bassenge E (1982) Effect of nitrates, nitrate-like substances, calcium antagonists and beta-adrenergic receptor blockers on peripheral circulation. $Herz$ **7**: 388–405.

Drechsler-Parks DM (1987) Effect of nitrogen dioxide, ozone, and peroxyacetyl nitrate on metabolic and pulmonary function. $Res\ Rep\ Health\ Eff\ Inst$ **6**: 1–37.

Drechsler-Parks DM (1995) Cardiac output effects of O_3 and NO_2 exposure in healthy older adults. $Toxicol\ Ind\ Health$ **11**: 99–109.

Drechsler-Parks DM, Bedi JF and Horvath SM (1984) Interaction of peroxyacetyl nitrate and ozone on pulmonary functions. $Am\ Rev\ Respir\ Dis$ **130**: 1033–1037.

Drechsler-Parks DM, Bedi JF and Horvath SM (1989) Pulmonary function responses of young and older adults to mixtures of O_3, NO_2 and PAN. $Toxicol\ Ind\ Health$ **5**: 505–517.

Ehrlich R, Findlay JC, Fenters JD $et\ al.$ (1977) Health effects of short-term inhalation of nitrogen dioxide and ozone mixtures. $Environ\ Res$ **14**: 223–231.

Ehrlich R, Findlay JC and Gardner DE (1979) Effects of repeated exposures to peak concentrations of nitrogen dioxide and ozone on resistance to streptococcal pneumonia. $J\ Toxicol\ Environ\ Health$ **5**: 631–642.

Farman CA, Pinkerton KE, Rajini P $et\ al.$ (1997) Evolution of lung lesions in rats exposed to mixtures of ozone and nitrogen dioxide. $Inhal\ Toxicology$ **9**: 647–677.

Folinsbee LJ, Horvath SM, Raven PB *et al.* (1977) Influence of exercise and heat stress on pulmonary function during ozone exposure. *J Appl Physiol* **43**(3): 409–413.

Folinsbee LJ, Bedi JF and Horvath SM (1981) Combined effects of ozone and nitrogen dioxide on respiratory function in man. *Am Ind Hyg Assoc J* **42**: 534–541.

Fujimaki H (1989) Impairment of humoral immune responses in mice exposed to nitrogen dioxide and ozone mixtures. *Environ Res* **48**: 211–217.

Gardner DE (1979) Introductory remarks: session on genetic factors affecting pollutant toxicity. *Environ Health Perspect* **29**: 45–48.

Gaston B, Drazen JM, Loscalzo J *et al.* (1994) The biology of nitrogen oxides in the airways. *Am J Respir Crit Care Med* **149**: 538–551.

Gelzleichter TR, Witschi H and Last JA (1992a) Concentration-response relationships of rat lungs to exposure to oxidant air pollutants: a critical test of Haber's Law for ozone and nitrogen dioxide. *Toxicol Appl Pharmacol* **112**: 73–80.

Gelzleichter TR, Witschi H and Last JA (1992b) Synergistic interaction of nitrogen dioxide and ozone on rat lungs: acute responses. *Toxicol Appl Pharmacol* **116**: 1–9.

Graham JA, Gardner DE, Blommer EJ *et al.* (1987) Influence of exposure patterns of nitrogen dioxide and modifications by ozone on susceptibility to bacterial infectious disease in mice. *J Toxicol Environ Health* **21**: 113–125.

Hales CH, Rollinson AM and Shair FH (1974) Experimental verification of linear combination model for relating indoor–outdoor pollutant concentration. *Environ Sci Technol* **8**: 452–453.

Hazucha MJ (1997) Health effects due to interaction of nitrogen oxides and ozone. In: *Health Effects of Ozone and Nitrogen Oxides in an Integrated Assessment of Air Pollution*. Leicester: Institute for Environment and Health, pp 42–50.

Hazucha MJ, Folinsbee LJ, Seal E *et al.* (1994) Lung function response of healthy women after sequential exposures to NO_2 and O_3. *Am J Respir Crit Care Med* **150**: 642–647.

Hogman M, Frostell CG, Hedenstrom H *et al.* (1993) Inhalation of nitric oxide modulates adult human bronchial tone. *Am Rev Respir Dis* **148**: 1474–1478.

Horvath SM, Bedi JF and Drechsler-Parks DM (1986) Effects of peroxyacetyl nitrate alone and in combination with ozone in healthy young women. *J Air Pollut Control Assoc* **36**(3): 265–270.

Ichinose T and Sagai M (1989) Biochemical effects of combined gases of nitrogen dioxide and ozone. III. Synergistic effects on lipid peroxidation and antioxidative protective systems in the lungs of rats and guinea pigs. *Toxicology* **59**: 259–270.

Ichinose T and Sagai M (1992) Combined exposure to NO_2, O_3 and H_2SO_4-aerosol and lung tumor formation in rats. *Toxicology* **74**: 173–184.

Ichinose T, Arakawa K, Shimojo N *et al.* (1988) Biochemical effects of combined gases of nitrogen dioxide and ozone. II. Species differences in lipid peroxides and antioxidative protective enzymes in the lungs. *Toxicol Lett* **42**: 167–176.

Kabel JR, Ben-Jebria A and Ultman JS (1994) Longitudinal distribution of ozone absorption in the lung: comparison of nasal and oral quiet breathing. *J Appl Physiol* **77**: 2584–2592.

Koenig JQ, Covert DS, Morgan MS *et al.* (1985) Acute effects of 0.12 ppm ozone or 0.12 ppm nitrogen dioxide on pulmonary function in healthy and asthmatic adolescents. *Am Rev Respir Dis* **132**: 648–651.

Koenig, JQ, Covert DS, Marshall SG *et al.* (1987) The effects of ozone and nitrogen dioxide on pulmonary function in healthy and asthmatic adolescents. *Am Rev Respir Dis* 1152–1157.

Koenig JQ, Pierson WE, Covert DS *et al.* (1988a) The effects of ozone and nitrogen dioxide on lung function in healthy and asthmatic adolescents. *Res Rep Health Eff Inst* **6**: 5–24.

Koenig JQ, Covert DS, Smith MS *et al.* (1988b) The pulmonary effects of ozone and nitrogen dioxide alone and combined in healthy and asthmatic adolescent subjects. *Toxicol Ind Health* **4**: 521–532.

Krishna MT, Springall DR, Frew AJ *et al.* (1996) Mediators of inflammation in response to air pollution: a focus on ozone and nitrogen dioxide. *J R Coll Physicians Lond* **30**: 61–66.

Larson TV, Hanley QS, Koenig JQ *et al.* (1993) Calculation of acid aerosol dose. In: *Advances in Controlled Clinical Inhalation Studies*. Berlin: Springer-Verlag, pp 109–121.

Last JA, Sun WM and Witschi H (1994) Ozone, NO, and NO_2: oxidant air pollutants and more. *Environ Health Perspect* **102**(Suppl 10): 179–184.

Linn WS, Anderson KR, Shamoo D. *et al.* (1995) Controlled exposures of young asthmatics to mixed oxidant gases and acid aerosol. *Am J Respir Crit Care Med* **152**: 885–891.

Mautz WJ, Kleinman MT, Phalen RF *et al.* (1988) Effects of exercise exposure on toxic interactions between inhaled oxidant and aldehyde air pollutants. *J Toxicol Environ Health* **25**: 165–177.

Miller FJ and Overton JH (1982) Pulmonary dosimetry of nitrogen dioxide in animals and man. In: Schneider T and Grant L (eds) *Air pollution by nitrogen oxides.* Amsterdam: Elsevier, pp 377–387.

Mochitate K, Ishida K, Ohsumi T *et al.* (1992) Long-term effects of ozone and nitrogen dioxide on the metabolism and population of alveolar macrophages. *J Toxicol Environ Health* **35**: 247–260.

Munger JW, Collett J, Daube B *et al.* (1990) Fogwater chemistry at Riverside, California. *Atmos Environ* **24B**: 185–205.

Mustafa MG and Tierney DF (1978) Biochemical and metabolic changes in the lung with oxygen, ozone, and nitrogen dioxide toxicity. *Am Rev Respir Dis* **118**: 1061–1090.

Mustafa MG, Elsayed NM, von Dohlen FM *et al.* (1984) A comparison of biochemical effects of nitrogen dioxide, ozone, and their combination in mouse lung. *Toxicol Appl Pharmacol* **72**: 82–90.

Nadziejko CE, Nansen L, Mannix RC *et al.* (1992) Effect of nitric acid vapor on the response to inhaled ozone. *Inhal Toxicol* 343–358.

O'Neill CA, van der Vliet A, Eiserich JP *et al.* (1995) Oxidative damage by ozone and nitrogen dioxide: synergistic toxicity in vivo but no evidence of synergistic oxidative damage in an extracellular fluid. *Biochem Soc Symp* **61**: 139–152.

Polzer G, Lind I, Mosbach M *et al.* (1994) Combined influence of quartz dust, ozone and NO$_2$ on chemotactic mobility, release of chemotactic factors and other cytokines by macrophages in vitro. *Toxicol Lett* **72**: 307–315.

Postlethwait EM and Bidani A (1994) Mechanisms of pulmonary NO$_2$ absorption. *Toxicology* **89**: 217–237.

Postlethwait EM, Langford SD and Bidani A (1991) Interfacial transfer kinetics of NO$_2$ into pulmonary epithelial lining fluid. *J Appl Physiol* **71**: 1502–1510.

Punjabi CJ, Laskin JD, Pendino KJ *et al.* (1994) Production of nitric oxide by rat type II pneumocytes: increased expression of inducible nitric oxide synthase following inhalation of a pulmonary irritant. *Am J Respir Cell Mol Biol* **11**: 165–172.

Rajini P, Gelzleichter TR, Last JA *et al.* (1993). Alveolar and airway cell kinetics in the lungs of rats exposed to nitrogen dioxide, ozone, and a combination of the two gases. *Toxicol Appl Pharmacol* **121**: 186–192.

Richters A (1988) Effects of nitrogen dioxide and ozone on blood-borne cancer cell colonization of the lungs. *J Toxicol Environ Health* **25**: 383–390.

Rietjens IM, Poelen MC, Hempenius RA *et al.* (1986) Toxicity of ozone and nitrogen dioxide to alveolar macrophages: comparative study revealing differences in their mechanism of toxic action. *J Toxicol Environ Health* **19**: 555–568.

Rietjens IM, van Tilburg CA, Coenen TM *et al.* (1987) Influence of polyunsaturated fatty acid supplementation and membrane fluidity on ozone and nitrogen dioxide sensitivity of rat alveolar macrophages. *J Toxicol Environ Health* **21**: 45–56.

Rigas ML, Benjebria A and Ultman JS (1997) Longitudinal distribution of ozone absorption in the lung – effects of nitrogen dioxide, sulfur dioxide, and ozone exposures. *Arch Environ Health* **52**: 173–178.

Sagai M and Ichinose T (1991) Biochemical effects of combined gases of nitrogen dioxide and ozone. IV. Changes of lipid peroxidation and antioxidative protective systems in rat lungs upon life span exposure. *Toxicology* **66**: 121–132.

Sagai M, Arakawa K, Ichinose T *et al.* (1987) Biochemical effects on combined gases of nitrogen dioxide and ozone. I. Species differences of lipid peroxides and phospholipids in lungs. *Toxicology* **46**: 251–265.

Sanna A, Kurtansky A, Veriter C *et al.* (1994) Bronchodilator effect of inhaled nitric oxide in healthy men. *Amer J Respir Crit Care Med* **150**: 1702–1704.

Sautebin L, Ialenti A, Ianaro A *et al.* (1995) Modulation by nitric oxide of prostaglandin biosynthesis in the rat. *Br J Pharmacol* **114**: 323–328.

Schlesinger RB, Driscoll KE, Gunnison A. *et al.* (1990) Pulmonary arachidonic acid metabolism following acute exposures to ozone and nitrogen dioxide. *J Toxicol Environ Health* **31**: 275–290.

Schlesinger RB, Elfawal HAN, Zelikoff JT *et al.* (1994) Pulmonary effects of repeated episodic exposures to nitric acid vapor alone and in combination with ozone. *Inhal Toxicol* **6**: 21–4.

Stewart TE, Valenza F, Ribeiro SP *et al.* (1995) Increased nitric oxide in exhaled gas as an early marker of lung inflammation in a model of sepsis. *Am J Respir Crit Care Med* **151**: 713–718.

Suzuki H and Murashima T (1994) Ozone-mediated nitration of aromatic ketones and related compounds with nitrogen dioxide. *J Chem Soc Perkin Trans* **1/4**: 903–908.

Suzuki H, Tatsumi A, Ishibashi T *et al.* (1995) Ozone-mediated reaction of anilides and phenyl esters with nitrogen dioxide: Enhanced ortho-reactivity and mechanistic implications. *J Chem Soc Perkin Trans* **1/7**: 339–343.

Takahashi Y and Miura T (1985) In vivo effects of nitrogen dioxide and ozone on xenobiotic metabolizing systems of rat lungs. *Toxicol Lett* **26**: 145–152.

Takahashi Y and Miura T (1989) Effects of nitrogen dioxide and ozone in combination on xenobiotic metabolizing activities of rat lungs. *Toxicology* **56**: 253–262.

Thompson CR, Hensel EG and Kats G (1973) Outdoor–indoor levels of six air pollutants. *J Air Pollut Control Assoc* **23**: 881–886.

Umezu T, Suzuki AK, Miura T *et al.* (1993) Effects of ozone and nitrogen dioxide on drinking and eating behaviors in mice. *Environ Res* **61**: 51–67.

Umezu T, Suzuki AK, Miura T *et al.* (1994) Effects of ozone and nitrogen dioxide on drinking and eating behaviors in mice. In: *Neurobehavioral Methods and Effects in Occupational.* San Diego: Academic Press Inc, pp 949–965.

US EPA (1993) *Air Quality Criteria for Oxides of Nitrogen.* EPA/600/8-91/049cF, vol III. Washington DC: US EPA

US EPA (1996). *Air Quality Criteria for Ozone and Related Photochemical Oxidants.* EPA/600/AP-93/004cF, vol III. Washington DC: US EPA

Veninga T and Lemstra W (1975) Extrapulmonary effects of ozone whether in the presence of nitrogen dioxide or not. *Int Arch Arbeitsmed* **34**: 209–220.

Weschler CJ and Shields HC (1997) Potential reactions among indoor pollutants. *Atmos Environ* **21**: 3487–3495.

Weschler CJ, Brauer M and Koutrakis P (1992) Indoor ozone and nitrogen dioxide: a potential pathway to the generation of nitrate radicals, dinitrogen pentaoxide and nitric acid indoors. *Environ Sci Technol* **26**: 176–184.

Weschler CJ, Shields HC and Naik DV (1994) Indoor chemistry involving O_3, NO and NO_2 as evidenced by 14 months of measurements at a site in Southern California. *Environ Sci Technol* **28**: 2120–2132.

Witschi H (1988) Ozone, nitrogen dioxide and lung cancer: a review of some recent issues and problems. *Toxicology* **48**: 1–20.

Acute and Chronic Effects of Ozone in Animal Models

RENEE C. PAIGE and CHARLES G. PLOPPER

Department of Anatomy, Physiology and Cell Biology, School of Veterinary Medicine, University of California, Davis, CA, USA

INTRODUCTION

Experimental animal models have been used successfully to define the impact of exposure conditions and the pattern of the biological response to ozone in the respiratory system. Depending on the timing of exposure in relation to the stage of the response, exposure conditions may modify the biological response. The response pattern includes an acute injury phase for previously unexposed cells. In this phase, there is a significant and focal inflammatory response with temporal variations in the abundance and distribution of inflammatory cells. This is followed by an extensive repair process involving proliferation of affected epithelial populations and differentiation of the daughter cells participating in repair. Interstitial matrix components involved in repair also exhibit temporal variations. Experimental animal studies reveal substantial variation in the pattern of response depending on both the species and the strain within a given species, as well as the sites within the respiratory tract. The latter issue has stimulated interest in characterizing the variability in metabolic and cellular aspects of specific microenvironments within the respiratory system. The local dose of ozone also is influenced by the microenvironment, as defined by position within the branching airway tree.

This chapter addresses the major sites of injury (nasal cavity, trachea and proximal bronchi, and central acinar bronchioles and alveolar ducts) under three primary categories: cellular response, metabolic activity and physiological changes in respiratory function. This chapter also summarizes the impact of ozone exposure on developmental events occurring in postnatal animals. Because dose and exposure characteristics have a substantial impact on the nature of the response, details of these exposure conditions are listed for each study discussed. The numerous animal studies conducted with ozone have been summarized in a compendium by the U.S. Environmental Protection Agency (US EPA, 1996). In this chapter, we will highlight only those studies which either illustrate particular biological events related to specific exposure protocols or define specific aspects of the biological response. The reader is referred to the US EPA (1996) document for a complete listing of all studies in which animals have been employed to define the biological effects of ozone exposure.

CELLULAR RESPONSE

The majority of animal studies have relied on laboratory rodents, which despite being obligate nose breathers and otherwise physiologically distinct from humans and other primates, have provided considerable insight regarding the time course of injury and repair following ozone exposure. The primary sites of ozone injury are similar in primates and rodents: the epithelium of the nasal cavity, trachea and central acinar region. Ciliated cells in the nose and trachea and type I cells in alveolar ducts are the primary cell types injured. However, nasal cavity organization differs significantly in laboratory rodents and primates (Fig. 24.1) in terms of both abundance and distribution of turbinates and in types and distribution of lining epithelium and associated glands (Harkema, 1992). Tracheal epithelium also is extremely variable and species-specific (Mariassy, 1992), as is the architecture of the central acinar region (Plopper and Hyde, 1992).

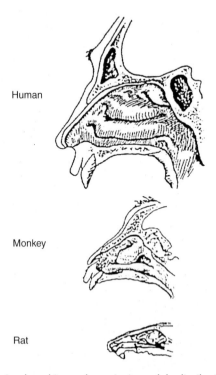

Human

Monkey

Rat

Fig. 24.1 Drawings comparing the architectural organization and the distribution of turbinates in the lateral walls of human, monkeys and laboratory rats. Humans have three primary turbinates on the lateral wall, monkeys have two such turbinates, and rats have two very large complex turbinates anteriorly and a highly complex ethmoturbinate posteriorly. Drawings courtesy of J.R. Harkema. (Harkema, 1992 reprinted with permission.)

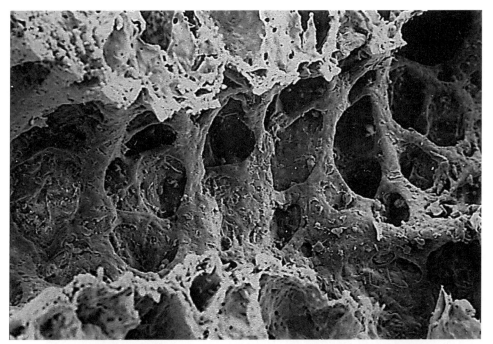

MMU A.D., 2 hrs exposure to 1 ppm Ozone

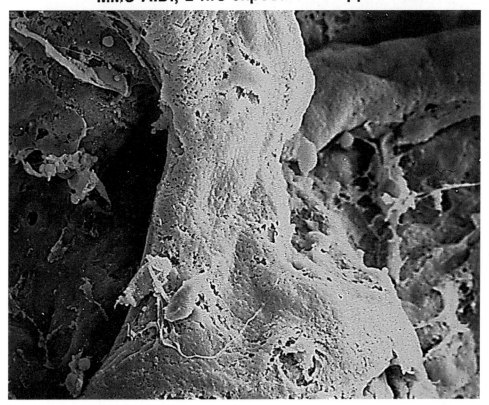

Fig. 24.2 Scanning electron micrographs of the proximate resiratory bronchiole of Rhesus monkeys following short-term exposure to ozone. After 2 h of exposure there is extensive infiltration of macrophages into alveolar areas and numerous exfoliating type I alveolar cells.

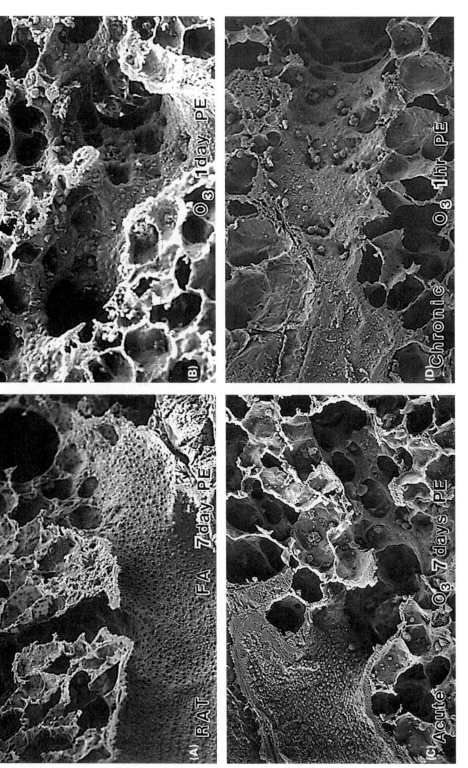

Fig. 24.3 Scanning electron micrograph comparing the cellular response of the central acinar region to ozone exposure (0.9 ppm) following a short-term (8-hour exposure) (B) and 7 days post exposure (C). Long-term exposure produces a chronic lesion characterized by alveolar accumulations of macrophages and extension of bronchiolar epithelium into alveolar ducts (D). In contrast, the bronchiolar epithelial boundary with alveolar tissue is abrupt in filtered air exposed animals (A).

In humans and other primates, the terminal bronchiole opens into a series of respiratory bronchioles that form the transition between the conducting airways and the alveolar gas exchange regions (Fig. 24.2). These transition zones are lined by bronchiolar epithelium with cuboidal non-ciliated bronchiolar cells contiguous with alveolar outpocketings.

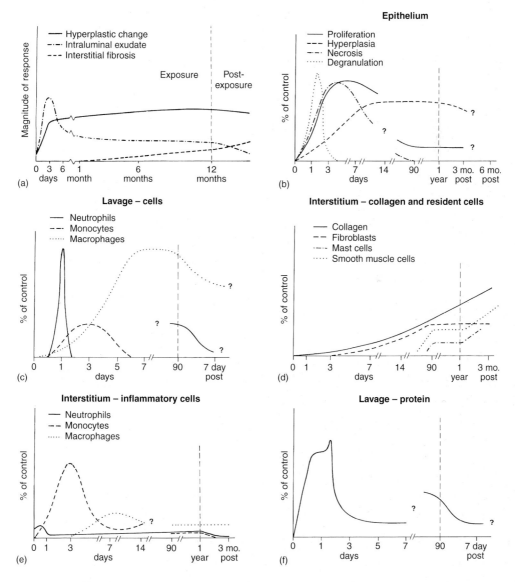

Fig. 24.4 Diagrammatic representation of the cellular and exudative changes occurring during the acute and chronic responses to short-term and long-term ozone exposure. The overall response is characterized by acute inflammatory events resulting in prolonged hyperplasia (a,b). The nature of the inflammatory response changes with time. Neutrophils predominate in the early response while macrophages persist throughout exposure (c). In the interstitium, collagen increases wih exposure and appears to remain elevated post-exposure (d). Inflammatory cells in the interstitium, mostly monocytes, peak early in exposure (e). Protein in lavage also peaks during the acute injury phase of exposure (f). Figure 4A reprinted with permission (Dungworth, 1989)

The normal rodent lung, however, contains few, if any, respiratory bronchioles. The terminal bronchiolar epithelium ends abruptly at the alveolar duct with no transition zone (Fig. 24.3). A recent study indicating that rats receive a lower than expected dose of ozone to the distal lung serves to underscore the importance of these anatomical and physiological differences. Interspecies differences can complicate and confound any attempt to compare these effects (Hatch *et al.*, 1994).

Response of the respiratory system to ozone exposure can be characterized in terms of (1) the initial response, (2) proliferation and repair of the epithelium, and (3) the response to continued exposure (Fig. 24.4b). Initial responses include injury and death of ciliated cells in the conducting airways and squamous epithelial cells in the parenchyma. In these first 8–12 h of exposure there are marked increases in intraluminal exudate initially containing primarily epithelial cells (Fig. 24.4c) and serum proteins (Fig. 24.4f), with minimal or no changes in the interstitium (Fig. 24.4a). The next stage is characterized by exfoliation of the epithelium and increased exudate. Inflammatory cells appear first in the interstitium, and then in the epithelium, before they are present in the exudate. The third stage is marked by proliferation of the epithelium, concurrent with downregulation of intraluminal exudates. Significant numbers of inflammatory cells still may be found migrating through the epithelium at this stage, but within 7 days the acute inflammatory response is typically almost completely resolved (Fig. 24.4e). At this time, the epithelial proliferation has ceased, epithelium is often hyperplastic, and proliferation of matrix components is in progress (Fig. 24.4d). After completion of this series of events, subsequent responses are dependent upon whether exposure continues. If exposure ceases, the assumption is that the affected compartments will revert to steady state within 7–10 days, which suggests that exposure during repair modifies this process. The effects of long-term exposure are characterized by continued hyperplasia and low-grade chronic inflammation with exudative cells, as well as synthesis of collagen in the matrix. The magnitude of the response is dependent upon the *inhaled* dose of ozone and varies with the time course of exposure. This is true also of the return to steady state. Return to normal appears less likely with prolonged exposure. Finally, it is important to consider that responses tend to be unique and species-specific. Responses also vary by site within the airway tree.

CENTRAL ACINAR REGION

Early Responses

Primate   In the initial 4–12 h of exposure to 0.8 ppm ozone, the respiratory bronchiolar epithelium (Fig. 24.5) in Rhesus monkeys exhibits extensive degeneration and necrosis of alveolar type I epithelial cells (Castleman *et al.*, 1980). The labeling index of the epithelium in these animals has increased by 18 h of exposure; the index reaches its maximum after 50 h of exposure. Cuboidal bronchiolar epithelial cells comprise most of the labeled and all of the mitotic cells. After 50 h of exposure, the respiratory bronchiolar epithelium is hyperplastic and macrophages are the predominant inflammatory cell present in the exudate. Complete resolution of the lesions was not observed during the period from 7 to 50 days post-exposure. The neutrophil influx following a 3-day ozone exposure resolves by 4 days post-exposure, but the number of macrophages and

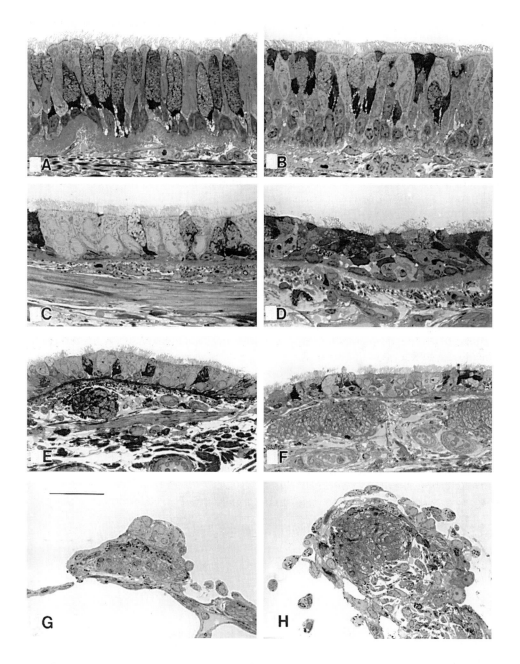

Fig. 24.5 High resolution histopathologic comparison of the epithelial composition lining the tracheobronchial airway tree in Rhesus monkeys and the cellular alterations occurring in these cell populations following short-term (4 h) exposure to 1 ppm ozone. Mucous cells, ciliated cells, and basal cells compose the majority of the epithelium of the trachea (A). In the trachea, ciliated cells are the primary cells injured (appear dense) following acute exposure (B). In proximal interpulmonary bronchi (C, D), necrotic cells are primarily ciliated cells, but there is a much higher level of infiltrated inflammatory cells in these airways than in the trachea early in the acute inflammatory response (D). In distal interpulmonary bronchi (generation 10–12) (E, F) fewer ciliated cells are injured and there are fewer infiltrating inflammatory cells than in more proximal bronchi. In the proximal respiratory bronchiole (G, H) cuboidal epithelium is intermixed with alveolar epithelium in adjacent alveoli. The short-term inflammatory response involves injury to epithelial cells of the alveolar areas, exfoliation, and infiltration of large numbers of inflammatory cells, primarily neutrophils (H). (Plopper et al. 1998 reprinted with permission.)

lymphocytes remains significantly elevated (Bassett *et al.*, 1988).

Laboratory Rodents   Early response of the terminal bronchiolar epithelium in the rat is characterized by necrosis of both ciliated cells in the bronchioles and type I cells in the centriacinar alveoli (Fig. 24.6). After 4 h of exposure to 1 ppm ozone, the terminal bronchiolar epithelium of rats exhibits necrosis, primarily of ciliated cells, prior to marked neutrophil migration (Pino *et al.*, 1992a). In these animals, total protein and number of neutrophils and epithelial cells in BALF increased with prolonged exposure (up to 24 h), while macrophages in BALF decreased with exposure duration. Depletion of neutrophils from rats with antibodies prior to ozone exposure nearly ablates the ozone-induced influx of neutrophils detectable in BALF (Pino *et al.*, 1992b). However, morphometry in normal and neutrophil-depleted exposed animals revealed no significant difference in volume of necrotic or degenerative epithelial cells per unit surface area of epithelial basal lamina in the central acinar region. The authors attribute the additional finding of significantly increased fibronectin in BALF to the absence of those neutrophil-derived enzymes responsible for normal degradation of fibronectin. In general, necrosis and exfoliation is most pronounced after 6–10 h of exposure and does not involve non-ciliated cells (Stephens *et al.*, 1974). The nonciliated cell response is degranulation, followed by dedifferentiation at the start of proliferation. Type I cells in the proximal alveoli exfoliate after as little as 2 h of exposure to 0.5 ppm and are replaced within 48 h of exposure by the proliferation of type II cells. Repair and adaptation follow the injury, reaching a maximum after about 24 h of exposure. At 48 h, lesions resulting from 8 h/day versus continuous exposure were indistinguishable. Recent data indicate that ozone exposure may also alter surfactant metabolism (Putman *et al.*, 1995). After rats were exposed for 2–12 h to 0.8 ppm ozone, investigators observed a significant increase in heavy subtype surfactant, rich in surfactant proteins A, B and C and in the ability to reduce surface tension, and a significant decrease in light subtype surfactant, virtually devoid of surfactant proteins A and B and less capable of reducing surface tension.

It appears that cells repopulating previously injured areas are as susceptible as those they replaced after the initial injury. Rats were exposed to 0.8 ppm for 3 days with a 6- or 27-day post-exposure period, followed by a 3-day re-exposure (Plopper *et al.*, 1979). After 6 days of recovery from the initial lesion, there were occasional intermediate cells lining the proximal alveoli and a few undifferentiated bronchiolar epithelial cells (Fig. 24.3b). Re-exposure 6 or 27 days following the initial exposure resulted in lesions that were indistinguishable from those observed in the initial exposure. Similarly, intermittent exposure of rats to 0.2, 0.5 or 0.8 ppm ozone for 8 h/day resulted in lesions with no substantial morphological differences from those observed in rats exposed for 24 h/day (Schwartz *et al.*, 1976). The lesions in the continuously exposed animals were generally more severe, as indicated by the number of inflammatory cells.

Acute inflammatory changes in guinea pigs are similar to those observed in rats. A 4-h exposure to 2 ppm ozone results in increased epithelial permeability as indicated by BALF albumin, which increases immediately following exposure and reaches a maximum level at 24 h post-exposure. By 7 days post-exposure there is a return to control levels (Schultheis and Bassett, 1994). Similarly, macrophages increase to levels twice that in control animals at 2 days post-exposure and return to control levels by 7 days post-exposure. In comparison to BALF, tissue analysis revealed that tissue inflammatory cells remained elevated 2–14 days post-exposure. While neutrophils in BALF increased immediately fol-

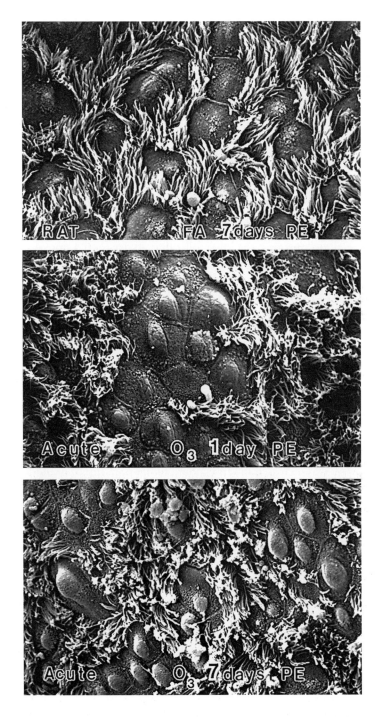

Fig. 24.6 Scanning electron micrograph comparing changes in epithelial surfaces of terminal bronchioles in the rat following short-term (8 h) exposure to ozone (1 ppm) and the recovery for 1 and 7 days following exposure.

lowing exposure and sustained these elevated levels for 3 days, tissue neutrophils increased immediately following exposure to levels nearly five times that of controls, but declined to control levels over the next 12 h. Although relying primarily on relatively indirect assessments such as BALF, acute injury studies in mice have identified substantial differences in the inflamatory response across mouse strains (Kleeberger, 1995). There appears to be strong genetic influence on the time course and nature of the early response; different loci influence different portions of the response (Kleeberger *et al.*, 1993a,b).

Late Responses

Primates   Based on reproducible sampling methods targeting first generation respiratory bronchioles, few significant concentration- or duration-dependent effects were identified morphometrically in monkeys following a 6-day exposure to 0.15 ppm, versus a 90-day exposure to 0.15 or 0.3 ppm ozone (Harkema *et al.*, 1993). While the mass of epithelium was increased in all dose groups, the value following 90 days at 0.3 ppm was significantly higher than that in other dose groups. These exposures resulted in mild to moderate thickening of the bronchiolar interstitium and marked (780–996%) increases in the number of cuboidal epithelial cells per unit surface area of basal lamina. The number of squamous epithelial cells per unit surface area increased 161–205% in 6- or 90-day exposures to 0.15 ppm ozone, but showed no change following a 90-day exposure to 0.3 ppm. Additionally, the respiratory bronchioles of monkeys chronically exposed to ozone exhibit a thicker wall and more narrow airway lumen. This thickening results from both a thicker epithelial component and a much thicker interstitial component (Hyde *et al.*, 1989). This study also reports that cyclical exposure to ozone (1 month exposure followed by 1 month in filtered air repeated for a total of 18 months) resulted in a significant increase in total lung collagen, while monkeys exposed daily for 18 months did not display this increase.

Laboratory Rodents   The arithmetic mean thickness of the air–blood barrier in centriacinar alveoli was increased in rats exposed to 0.8 ppm for 20 or 90 days (Boorman *et al.*, 1980). These rats also exhibited a significant increase in inflammatory cells at 0.5 and 0.8 ppm at all time points. While the epithelial changes and accumulation of macrophages at 90 days was similar but less severe than at 7 days, significant morphological alterations persisted throughout the 90 days at both 0.5 and 0.8 ppm ozone.

Following exposure to 0.12, 0.5 or 1 ppm ozone for 6 h per day, 5 days per week for 20 months, the key morphologic change in the central acinar region involves extension of bronchiolar epithelium into the acinus (Fig. 24.7) (Pinkerton *et al.*, 1995). This change was concentration-dependent and site-specific. Those with ventilatory units arising from shorter paths (e.g. cranial region) exhibited the greatest change. With this exposure protocol, morphometric analysis also revealed an increase in the volume of interstitium and epithelium along the alveolar ducts, with the increase in epithelial volume due to metaplastic changes resulting in a less squamous and more cuboidal phenotype (Chang *et al.*, 1995). Mild fibrotic changes also were noted at 1 ppm. These involved an increase in interstitial matrix components, including increases in collagen and elastin, as well as a slight increase in interstitial fibroblast volume. However, excess collagen was confirmed biochemically only in female rats (Last *et al.*, 1993). No mRNA signal for collagen, elastin, fibronectin or interstitial collagenase was observed in rats from the same protocol at 20 months (Parks and Roby, 1994). The authors conclude that matrix production must

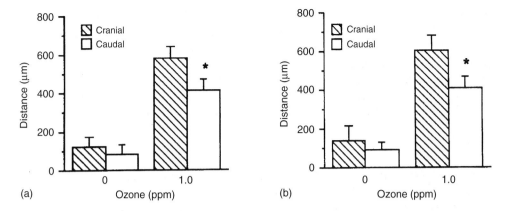

Fig. 24.7 Comparison of bronchiolar ciliated and nonciliated cells in alveolar ducts of rats exposed long-term (20 months) to ozone (1 ppm). There is a difference in the extent of bronchiolarization of alveolar ducts depending on whether the central acinar regions are determinations of short airway pass (cranial) or long airway pass (caudal). *Significantly different than cranial (p < 0.05). (Plopper et al. 1994a reprinted with permission.)

occur transiently early in the fibrotic response and must be turned off by 20 months of exposure. Quantitative increases in the cellular abundance of Clara cell secretory protein were present in terminal bronchioles and bronchiolarized alveolar ducts following 20 months of exposure (Dodge *et al.*, 1994).

Another study exposed rats to a simulated urban profile of ambient ozone for 78 weeks (Chang *et al.*, 1992). A loss of ciliated cells and differentiation of preciliated and Clara cells was observed initially. The acute responses observed after 1 week of exposure subsided after 3 weeks of exposure. These responses included epithelial inflammation, interstitial edema, interstitial cell hypertrophy and an influx of macrophages. Epithelial cell hyperplasia, fibroblast proliferation and accumulation of interstitial matrix components were the progressive responses. Type I cells also increased in number and became thicker. This change persisted throughout exposure and did not resolve after a recovery period of up to 17 weeks. Long-term exposure yielded structural changes suggestive of injury to both ciliated and Clara cells.

In view of reliance upon animal surrogates for human exposure, it is important to emphasize the differences in sensitivity that have been observed in primates and rodents. The biochemical and cellular response to ozone of the monkey and the rat have been directly compared under equivalent exposure conditions (Plopper *et al.*, 1991). While collagen metabolism was significantly increased in rat lungs following ozone exposure, a significantly larger increase was observed in monkeys exposed to the same concentration of ozone. Total epithelial thickness also increased by less than 10% in rats exposed to 0.25 ppm ozone for 8 h per day for 45 days, while exposure of monkeys to 0.15 ppm for 6 or 90 days resulted in a 150% increase in epithelial thickness compared with controls. These data suggest that the lungs of monkeys are more susceptible to damage by ozone exposure than are those of the rat.

Only limited information is available on the response of mice to ozone exposure. In general, cellular changes in mice are similar to those of rats. Shortened cilia were observed after 6 days of exposure to 0.8 ppm, loss of cilia after 10 days of exposure, and cilia regenesis occurred 10 days post-exposure (Ibrahim *et al.*, 1980).

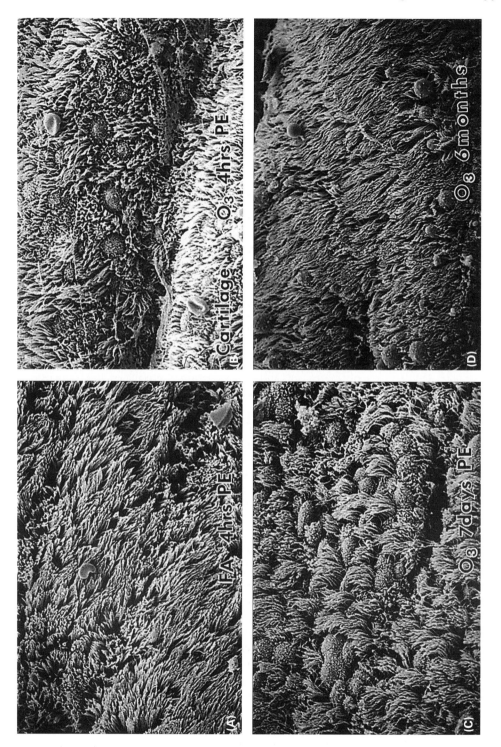

Fig. 24.8 Scanning electron micrograph comparing changes in surface epithelium of the trachea of Rhesus monkeys following short-term (8 h) exposure to ozone (0.9 ppm), the degree of recovery of steady state after 7 days post exposure (C), and the alterations following 6 months exposure.

Trachea and Bronchi

The acute responses of monkey tracheal epithelium to ozone are similar to those of rats and include a decrease in the number of ciliated cells, a loss of cilia (Fig. 24.8), focal epithelial stratification associated with an increase in the number of small mucous granule cells, and ultrastructural alterations in mucous goblet cell granules involving a decrease in density and irregularity in size (Wilson *et al.*, 1984). In those animals exposed to 0.64 ppm ozone, proliferation (based on labeling index) and number of necrotic ciliated cells was greatest after 3 days of exposure. Morphological changes in injured ciliated cells included irregular orientation of basal bodies in the apical cytoplasm, condensed electron dense staining mitochondria, and a pyknotic nucleus and swollen endoplasmic reticulum. An 8-h exposure to 0.96 ppm ozone resulted in significant epithelial necrosis in the trachea and bronchioles, as measured morphometrically 1 and 12 h post-exposure (Hyde *et al.*, 1992). Significant increases in necrotic epithelial cells also were observed in the bronchi, but only at 12 h post-exposure. In these animals, the labeling index was significantly elevated only in the bronchi and only at 12 h post-exposure. The timing, pattern and types of inflammatory cells migrating through the epithelium varies by airway level, and the time course for injury and repair appears to be different for the trachea and bronchioles when compared with the bronchi.

The tracheas of rats exhibited a significant dose-dependent decrease in the amount of stored mucin following exposure to ozone (0.12, 0.5 or 1 ppm) for 20 months (Plopper *et al.*, 1994a). However, the response in the bronchi differed by position within the airways.

Nasal Cavity

Laboratory Rodents

The time course of injury in the nasal cavity appears to differ significantly from that observed in the central acinar region. A transient influx of neutrophils and a significant loss of nasal transitional epithelial (NTE) cells occurs in rats 2–4 h following an 8-h exposure to 0.5 ppm ozone (Hotchkiss *et al.*, 1997). Neutrophilic rhinitis is restricted to the lateral meatus. By 12–20 h post-exposure, few neutrophils remain in the nasal mucosa. In those rats that were injected with BrdU 2 h prior to sacrifice, the labeling index was greatest at 20–24 h post-exposure and was diminished at 36 h post-exposure. The numeric density of NTE cells had returned to control levels by 20–24 h post-exposure and inflammatory cells were rarely observed. By 36 h post-exposure, only a few focal areas of mild basal cell hyperplasia remained. Comparison of nasal lavage fluid (NLF) and bronchoalveolar lavage fluid (BALF) from rats exposed to 0.12 ppm ozone for 6 h indicated a significant increase in neutrophils at 18 h post-exposure in NLF, with no concomitant changes in BALF (Hotchkiss *et al.*, 1989). The number of neutrophils in BALF increased with time, peaking at 42 h post-exposure and coinciding with a progressive decrease in the number of neutrophils in NLF. The number of neutrophils reached a minimum at 42 h post-exposure. A higher exposure concentration (1.5 ppm) yielded no significant increases in NLF neutrophils at any time post-exposure, but greatly increased the number of neutrophils in BALF at 3, 18 and 48 h post-exposure. These data suggest that higher concentrations of ozone result in a proportionately greater inflammatory response in the central acinar region, thus competing with and thereby attenuating the inflammatory response in the nose. Longer exposure (6 h per day for 3 days to 0.8 ppm ozone) resulted in significant hyperplasia and metaplasia of the nasal epithelium, which persisted after a 4-day post-exposure period (Hotchkiss *et al.*, 1991). However, this study found no significant differences in severity of lesion in rats exposed for 3 days with a 4-day post-exposure period compared with rats exposed for 7 days continuously. This suggests that once initiated, further ozone exposure is not required for this response pattern to progress. Exposure to 0.12 ppm ozone for 6 h per day for 7 days resulted in increased storage of mucosubstances in the nasal turbinate epithelium, with no change in the epithelium of the nasopharynx (Harkema *et al.*, 1989). This increase persisted in the turbinates and also was observed in the nasopharynx for 7 days following a 7-day exposure to 0.8 ppm ozone.

Chronic exposure to ozone alters the normal clearance function of the nasal epithelium. In regions of marked mucous cell metaplasia from rats exposed 20 months to 0.5 and 1 ppm ozone, decreased mucus flow rates were observed in the proximal third of the nasal airways, when compared with rats exposed to 0 or 0.12 ppm ozone (Harkema *et al.*, 1994). Even after 20 months exposure to 0.5 or 1 ppm ozone, the transitional epithelium was still hyperplastic and inflammatory cells were present in the proximal and middle nasal passages.

Primates

Studies conducted in primates have focused not only upon the kinetics of nasal epithelial cell loss, but also on the effects of ozone exposure on the composition of mucosubstances. While an 8 h per day, 6-day exposure to 0.15 ppm ozone in Bonnet monkeys resulted in a significant increase in both acidic and neutral glycoconjugates in the transitional and respiratory epithelium, a 90-day exposure resulted in significantly less mucosubstance than at 6 days (Harkema *et al.*, 1987). The increase at 6 days reflects an increase in the number of secretory cells containing this material, but the decrease at 90 days cannot be explained by a decrease in the number of secretory cells and therefore suggests changes in the production or storage of mucosubstances. Morphologic changes observed in the anterior nasal cavity at 6 days include secretory cell hyperplasia, deciliation and necrosis of ciliated cells as well as inflammatory cell infiltration.

PHYSIOLOGIC ALTERATIONS

In animal models, transient functional changes following ozone exposure have been attributed to inflammation in the distal lung, with persistent lesions rarely yielding a detectable functional change. The range of functions that have been tested include: ventilation (tidal volume and frequency), mechanics (dynamic compliance and resistance), airway reactivity and diffusing capacity.

VENTILATION AND MECHANICS

Exposure of monkeys to 0.5 or 0.8 ppm ozone 8 h per day for 90 days results in remodeling of the respiratory bronchiolar epithelium but does not produce any significant functional changes (Eustis *et al.*, 1981).

In the rat, increased pulmonary resistance was induced by a significantly lower dose of acetylcholine following a 2-h exposure to 4 ppm ozone, without a significant increase in neutrophils (Evans *et al.*, 1988). By 24 h post-exposure, the challenge dose required to elicit a similar change in resistance was not significantly different from controls. Increased frequency of breathing and decreased tidal volume was observed in rats on days 1 and 2 of a 5-day exposure to 0.35, 0.5 or 1 ppm ozone (Tepper *et al.*, 1989). These changes were diminished by the fifth day of exposure in rats exposed to 0.35 and 0.5 ppm, but persisted in those rats exposed to 1 ppm. Small but significant increases in functional residual capacity and residual volume and decreased carbon monoxide diffusing capacity were present in rats after a 52-week, 20 hour per day 0.5 ppm ozone exposure but returned to normal following a 3-month post-exposure period (Gross and White, 1987). However, exposure of rats for 78 weeks to an urban profile of ozone exposure resulted in significant small decreases in total lung capacity that did not completely resolve after a 17-week recovery in air (Costa *et al.*, 1995). Chronic exposure of monkeys (0.5 or 0.8 ppm ozone 8 h per day for 90 days) which results in remodeling of the respiratory bronchiolar epithelium, does not produce any significant functional changes (Eustis *et al.* 1981). Under the same exposure conditions (a 78-week exposure to an urban profile of ozone with a peak at 0.24 ppm), small but significant changes in breathing patterns were observed at 1, 3, 13, 52 and 78 weeks of exposure during a post-exposure challenge with carbon dioxide; the expected increase in ventilation did not occur (Tepper *et al.*, 1991). While no lesions were detected by light microscopy, an increase in respiratory resistance persisted throughout exposure and was most marked at 78 weeks. Pulmonary function of rats tested 1–3 days post-exposure to 0.12, 0.5 or 1 ppm ozone for 6 h per day, 5 days per week for 20 months, indicated small decreases in residual volume. A significant decrease was observed only in female but not male rats exposed to 0.5 ppm (Harkema *et al.*, 1996). The authors attribute this change to fibrosis and epithelial remodeling in the central acinar region, resulting in stiffened walls. Responsiveness of isolated airways taken from rats exposed under the same protocol was not significantly altered (Szarek, 1994).

Adaptation to ozone (an attenuation of the breathing response during the course of ozone exposure) observed immediately following an 18-month exposure to an urban profile of ozone, was not present after a 4-month post-exposure period (Weister *et al.*, 1995). This suggests that those alterations in response immediately following cessation of exposure to ozone which are termed 'adaptation' are really a reflection of the underlying proliferation and differentiation that is part of the repair process.

Airway Hyperresponsiveness

Guinea pig models for airway hyperresponsiveness following exposure to 3 ppm ozone for 1 h indicate a significant decrease in the dose of histamine required for bronchoconstriction during a post-exposure challenge (Vargas *et al.*, 1994). Most studies of airway hyperresponsiveness have utilized isolated perfused lungs; exposure of isolated perfused rat lungs to 1 or 2 ppm ozone for up to 3 h resulted in time- and concentration-dependent decrements in pulmonary function (increased pulmonary resistance, decreased dynamic compliance) (Pino *et al.*, 1992c). Morphologically, these lungs exhibited moderate necrosis and sloughing of epithelial cells, especially those in the larger bronchioles and bronchi. Neutrophils introduced experimentally in the isolated perfused rat lung exposed to 1 ppm ozone appear to have no effect on ozone-induced airway reactivity, but do appear to enhance epithelial injury and pulmonary function decrements. The latter effect may be due to microvascular leak (Joad *et al.*, 1993).

Primate hyperresponsiveness models have yielded seemingly inconclusive data. Anesthetized, intubated monkeys exposed to 0.5 ppm ozone for 5 min exhibited increased respiratory resistance (Fouke *et al.*, 1988, 1990). In a similar protocol, monkeys were anesthetized and intubated and exposed to 1 ppm ozone 2 h per week for 19 weeks (Johnson *et al.*, 1988). Hyperresponsiveness, including increased respiratory resistance and decreased dynamic compliance, was observed after 19 weeks of exposure. These changes persisted for more than 15 weeks post-exposure. However, given the means of ozone delivery, little can be concluded from the previous studies about the effect of ozone on airway reactivity. Exposure of conscious monkeys to 1 ppm ozone 6 h per day, 5 days per week for 12 weeks did not result in any significant changes from baseline pulmonary function measurements, nor did this exposure alter the methacholine challenge dose required to increase pulmonary resistance 200% (EC_{200} RL) (Biagini *et al.*, 1986).

A potential role for neurogenic modulation of epithelial injury is suggested by data indicating that depletion of C fibers results in severe acute interstitial inflammation and coagulative necrosis of airway epithelium, particularly in small peripheral airways and the parenchyma (Sterner-Kock *et al.*, 1996). C fibers depleted by capsaicin normally modulate the response to ozone exposure by effecting a change in breathing pattern, precipitating bronchoconstriction, and by releasing the neuropeptides that subsequently alter mucosal blood flow and permeability and enhance neutrophil chemotaxis. Thus, C fibers appear to modulate both the initial local exposure to ozone and the ensuing repair process.

METABOLIC CONSIDERATIONS

Whether exposure to ozone alters susceptibility to xenobiotics requiring metabolic activation is an issue that has not been thoroughly addressed. However, when evaluating data regarding the effects of ozone exposure on metabolic functions, there are several factors to consider:

(1) Does the study report increased protein/gene expression or increased pulmonary metabolic activity?
(2) What is the age of the animals upon commencement of exposure?
(3) What is the duration, concentration and periodicity of exposure?
(4) Is metabolism evaluated on a whole organ or site-specific basis?

Given the potential importance of these variables, it becomes clear why comparison of such studies is difficult and why many of the effects of ozone on pulmonary metabolism have yet to be elucidated.

Cytochrome P-450 Monooxygenases

Cytochrome P-450 monooxygenases are normally expressed in moderately low levels in the lung, but data suggest that expression and activity of this enzyme system may be altered by ozone exposure. Immunochemically detectable cytochrome P-450 2B1 (CYP 2B1) was significantly increased in whole lung microsomes from rats (8 weeks old) exposed continuously to 0.4 ppm ozone for 14 days (Takahashi *et al.*, 1994). CYP 2B1 positive cells were localized in the non-ciliated cells of distal bronchioles and the cuboidal cells of proximal alveolar epithelium. P-450 content and activity were assessed in lung microsomes of rats (22–24 weeks old) exposed to 0.4 and 0.8 ppm ozone and significant increases in both NADPH-cytochrome P-450 reductase activity (NADPH-reductase) and cytochrome P-450 content were observed following exposure to 0.8 ppm ozone (Takahashi *et al.*, 1985). P-450 activity, as measured by benzo[*a*]pyrene hydroxylase and 7-ethoxycoumarin-*O*-deethylase (diagnostic of isoforms 1A1 and 2B, respectively), also increased, with a significant increase observed following 7 days of exposure. Interestingly, 10-week-old rats exposed to a similar regimen (0.2 and 0.4 ppm continuously for 7 and 14 days) displayed significant increases in P-450 content and activity (measured by benzo[*a*]pyrene hydroxylase and ethoxycoumarin-*O*-deethylase) at both concentrations and exposure durations (Takahashi and Miura, 1985). These rats also exhibited significant increases in microsomal protein and NADPH-reductase activity at both concentrations following 14 days of exposure. Five-week-old rats were exposed continuously to 0.4 ppm ozone for 14 days and quantification of whole lung homogenates by immunoblotting demonstrated significant increases in expression of CYP2B1 at both 7 and 14 days of exposure (Suzuki *et al.*, 1992). Rats 19–22 weeks of age were exposed continuously to 0.2 or 0.4 ppm ozone for 4, 8 or 12 weeks, or to 0.1 for 4 weeks (Takahashi and Miura, 1987). This study also found significant increases in microsomal NADPH-reductase activity following exposure to 0.2 or 0.4 ppm for 4–12 weeks. Significant increases in microsomal P-450 content and activity, measured by benzo[*a*]pyrene hydroxylase, 7-ethoxycoumarin-O-deethylase and benzphetamine-*N*-demethylase, were observed at all concentrations and exposure durations. Significantly increased pulmonary microsomal CYP2F1 and 2F2 protein were detected in rats (19–22 weeks old) exposed to 0.2 ppm ozone for 14 days (Takahashi and Miura, 1990). Data on CYP2F activity following ozone exposure have yet to be reported. These studies rely upon microsomes prepared from whole lung homogenates and as such may not accurately reflect changes occurring at the primary site of injury, the central acinar region.

Microsomes prepared with whole lung homogenates or isolated Clara cells from 8-week-old rats exposed continuously to 0.8 ppm ozone for 7 days were compared for P-450 activity (Rietjens *et al.*, 1988). From the whole lung, significant decreases were observed in ethoxycoumarin-*O*-deethylase and ethoxyresorufin deethylase (EROD) activities, while pentoxyresorufin-*O*-dealkyation (PROD) activity was significantly increased in ozone exposed animals compared to controls. In contrast, isolated Clara cells from ozone exposed animals exhibited no change in PROD activity, but showed a significant decrease in EROD activity. This disparity in activity from whole lung versus isolated Clara cells underscores the need to assess activity on a more site-selective basis. Cytochrome P-450 2E1 activity in rats exposed to ozone was recently characterized on an airway-selective basis (Watt *et al.*, 1998). Rats (49–165 days old) were exposed to 1 ppm ozone for 8 h per day for 1–90 days. In 78- and 165-day-old rats exposed for 1-day and in 79-day-old rats

exposed for 10 days, a significant increase in 2E1 activity was observed in microsomes prepared from microdissected regions of the proximal intrapulmonary airways. These
increases returned to control levels within 2 days following cessation of exposure. In rats
(53 days old) exposed for 90 days, significant decreases in activity were observed in
microsomes prepared from microdissected segments of the proximal and distal intrapulmonary airways. No significant changes in activity were observed in the trachea with any
exposure.

Antioxidant Enzymes

Alterations also have been demonstrated in antioxidant enzyme systems following ozone
exposure. Exposure of weanling, preadult, or adult rats to 0.7 ppm ozone for 5 days
resulted in significant increases in whole lung homogenate activity of the following:
Cu,ZnSOD, MnSOD, catalase and glutathione peroxidase (Rahman et al., 1991).
Quantification of mRNA in adult rats indicated significant increases in message for
Cu,ZnSOD, catalase and glutathione peroxidase at 3 and 5 days of exposure. Significant
increases in glucose-6-phosphate dehydrogenase (G6P) and glutathione peroxidase activity in lung homogenates were observed after a 4-day recovery period following a 3-day
exposure to 0.75 ppm ozone (Bassett et al., 1988). Succinate oxidase and 6-
phosphogluconate (6PGD) dehydrogenase activity, in addition to G6P, were significantly
elevated in these rats immediately following the 3-day exposure period. Immediate
increases in activity are attributed to proliferation of the epithelium following acute
injury. Elevations observed 4 days post-cessation of exposure and still significant after correction for DNA may reflect an enrichment of these two antioxidant enzymes in the
epithelial cells. Exposure of guinea pigs to 0.4 ppm ozone for 2 weeks results in none of
the increased antioxidant enzyme activity found in rats (Ichinose and Sagai, 1989).
Rather, glutathione peroxidase is significantly decreased in guinea pigs. Glutathione
transferase is significantly decreased in both rats and guinea pigs under this exposure regimen. Rats in this study exhibited the previously observed increases in glutathione
peroxidase and G6P dehydrogenase activity, as well as increased levels of vitamin C and
non-protein sulfhydryls.

The previous studies present a uniform picture in the rat, indicating that superoxide
dismutase, glutathione peroxidase, catalase and G6P dehydrogenase are likely to be elevated by short-term (14 days or less) exposure to 0.4–0.7 ppm ozone. However, these
elevations characterize changes in whole lung homogenate. When specific regions of the
lung are evaluated by the use of microdissected airways, distal bronchioles from rats
exposed to 0.5 or 1 ppm ozone for 6 h per day, 5 days per week for 20 months (or to 0.12
or 1 ppm for 90 days) exhibited significantly increased glutathione transferase, glutathione peroxidase and superoxide dismutase activity (Plopper et al., 1994b). No changes
were observed in any other subcompartments of the lung, including proximal airways and
lung parenchyma.

When comparing responses of the lungs of rats and primates, the variation in antioxidant enzyme activities between species must be considered. In healthy, unexposed animals,
glutathione transferase activity varies by less than two-fold between lung compartments
(Duan et al., 1993). Determination of activity in different lung subcompartments (based
on samples selected by microdissection of airways) reveals that while glutathione peroxidase

is evenly distributed in the airways of the monkey, activity is slightly higher in proximal than in distal airways in rats. And while superoxide dismutase activity is similar for rats and monkeys in the subcompartments studied, catalase activity is relatively evenly distributed in the rat, while activity in the monkey is slightly higher in distal bronchioles and the lobar bronchus compared with the trachea. These differences in distribution of antioxidant activity do not explain the site-selective differences in susceptibility to ozone injury, but they underscore the need to consider interspecies differences when comparing responses to xenobiotics.

A more recent study examined the effects of ozone exposure on glutathione levels in various subcompartments of the rat and monkey lung following exposure to 0.4 or 1 ppm ozone for 2 h (Duan *et al.*, 1996). Glutathione (GSH) was decreased in the rat trachea after exposure to 0.4 ppm and increased in the lobar bronchus and distal bronchioles after exposure to 1 ppm. In the monkey, the distal intrapulmonary bronchi showed increases to 200% of control in GSH after exposure to 0.4 ppm, and the distal bronchioles showed decreases in glutathione to 55% of control after exposure to 1 ppm. These results do not suggest that steady-state glutathione status is the primary determinant in ozone-induced toxicity. However, this study also exposed rats and monkeys to 1 ppm ozone for 6 h per day, 5 days per week for 90 days and found that glutathione in the distal bronchioles of both rats and monkeys was increased to 164 and 165% of control, respectively, suggesting that alterations in cellular glutathione may play a role in the elevated resistance of this region to further injury with long-term exposure.

Increases in total superoxide dismutase and catalase activity by pretreatment with the phenyl urea compound EDU (*N*-[2-(2oxo-1-imidazolindinyl)ethyl]-*N*'-phenylurea) appear to attenuate neutrophil influx into the lungs of rats following a 2-day exposure to 0.85 ppm ozone (Bassett *et al.*, 1989). Neutrophil recovery in bronchoalveolar lavage from unexposed rats is 0.1 million cells/lung while rats exposed to ozone under the conditions of this study had a recovery of 5.54 million cells/lung. Those rats pretreated with EDU had a significantly lower recovery (2.12 million cells/lung). This study suggests that superoxide dismutase and catalase may be involved in preventing the amplification of ozone injury.

Effects of Dietary Deficiencies

Animal models also have provided information about the impact of dietary deficiencies on the predisposition of animals to ozone injury. In doing so, such studies also have provided insight into the potential involvement of antioxidant systems in the progression of ozone injury. A 7-day exposure (1.2 ppm ozone) of copper- or manganese-deficient weanling mice failed to elicit increases in Mn or Cu,ZnSOD activity (Zidenberg-Cherr *et al.*, 1991). Manganese (but not copper) deficiency prevented the elevation of glutathione peroxidase activity. Elevation of MnSOD activity as part of the cellular response to ozone is further suggested by studies using endotoxin pretreatment, selectively elevating MnSOD activity without concomitant increases in Cu,ZnSOD, catalase or glutathione peroxidase, to protect against ozone-induced lung edema (Rahman and Massaro, 1992).

While antioxidant enzyme levels were significantly increased in both normal and vitamin E (VE) deficient rats following a 7-day exposure to 0.1 ppm ozone, the number of rats exhibiting severe lesions appeared to decrease with increasing dietary VE (5/6 rats at 0 ppm

VE, 4/6 rats at 11 ppm VE, and 1/6 rats at 110 ppm VE) (Chow *et al.*, 1981). Rats pre-treated with 11 ppm VE also did not exhibit a significant increase in antioxidant enzymes following ozone exposure. Studies of normal and vitamin E-supplemented weanling (2-month-old) rats exposed to 0.1 or 0.2 ppm ozone for 7 days found little difference between groups in terms of prevalence of central acinar lesions (Plopper *et al.*, 1979).

Although ascorbate levels are elevated in BALF immediately following ozone exposure, no evidence suggestive of a protective role for ascorbate was observed in a comparison of normal and ascorbate-deficient guinea pigs exposed to ozone for 7 days (Kodavanti *et al.*,

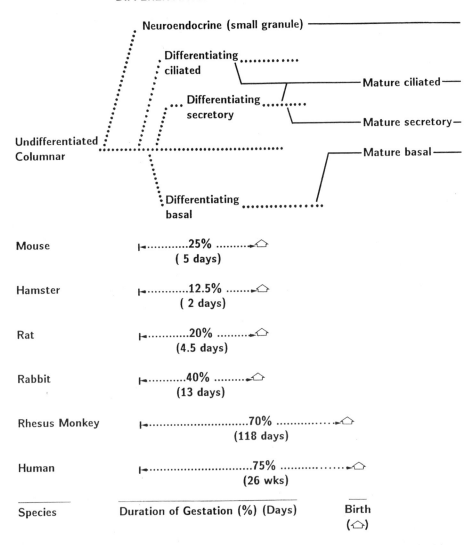

Fig. 24.9 Diagrammatic comparison of the pattern of epithelial differentiation in the trachea for each of the principal cell types. The approximate time during gestation in which these differentiation processes occur is compared for 6 species. (Plopper et al. 1992 reprinted with permission.)

1995). BALF parameters (including protein, albumin, total cells and neutrophils) were not significantly different for the two dietary groups.

IMPACT ON DEVELOPMENT

Given that lung development is a multievent process that continues postnatally in most species (including humans), the impact of ozone on developmental events critical for successful maturation of the respiratory tract may be considerable. Although the initial evagination of the tracheobronchial bud begins early in gestation, the majority of lung growth and development occurs postnatally. In humans, lung development and cellular differentiation continues through the first 8–10 years of postnatal life. Therefore, characterization of the response of the developing lung to ozone is an important part of any assessment of the health effects of ozone.

Normal Development

The critical difference between differentiation of the epithelial cell populations during development is the percentage of intrauterine life in which the differentiation occurs; the epithelium of the trachea is the earliest of all the epithelial populations to differentiate and is an instructive example (summarized in Fig. 24.9). In some species, this epithelium is relatively differentiated prior to birth, while in other species the majority of the differentiation occurs postnatally (Plopper et al., 1992). In most species, with the possible exception of the ferret, ciliated cells differentiate first. Non-ciliated cells with secretory granules appear next; basal and small mucous granule cells appear last. During the entire process of differentiation, a substantial proportion of cells of the epithelial population are undifferentiated and completely or nearly completely intermixed with mature epithelial cells. Cytodifferentiation occurs in a proximal to distal pattern, with the trachea and proximal bronchi being more fully differentiated at any time than the distal airways.

In the bronchioles, differentiation of the epithelial population involves the formation of two distinct cell types, the ciliated cell and the non-ciliated (or Clara) cell. In species such as the Rhesus monkey with extensive respiratory bronchioles, morphogenesis of the respiratory bronchiole also occurs during the period of cytodifferentiation (Tyler and Plopper, 1989; Tyler et al., 1988). Ciliogenesis occurs in the bronchioles in late gestation and continues well into the postnatal period in all the species evaluated to date. Clara cell cytodifferentiation is both a pre- and postnatal process, involving significant rearrangement and biogenesis of cellular organelles, differentiation of secretory activity, and a high level of epithelial cell turnover. Full differentiation of Clara cells in the bronchioles occurs over a 3- to 4-week period.

Cytodifferentiation of the epithelial populations in the gas-exchange area in the lung has been well characterized (Adamson, 1991). Two separate differentiation processes result in a gas-exchange area with two differentiated epithelial populations, the alveolar type I epithelial cells and the alveolar type II epithelial cells. The alveolar type I epithelial cells are the result of squamation of cuboidal epithelium, with the loss of the majority of cellular organelles. The type II cell differentiates from a glycogen-filled cuboidal cell with few organelles to become a cell filled with lamellar bodies. Overall growth of the lung represents a change in overall size that involves an increase in the volume of specific subcompartments associated with an increase in surface area. This requires continued proliferation, even of relatively well-differentiated populations, until overall body growth ceases.

Age-Related Susceptibility

In general, postnatal animals prior to weaning are less susceptible to pulmonary injury by ozone than are adults. Data supporting this are available for only a limited number of species but include the rat (Stephens *et al.*, 1978). The inflammatory response following ozone exposure may attenuate with age: the number of alveolar macrophages in neonatal (7 days old) rats was twice that observed in exposed juvenile (42 days old) rats (Barry *et al.*, 1985). Antioxidant enzyme levels remained essentially unchanged and lungs exhibited little morphological injury compared to controls in neonatal (0–20 days old) rats exposed for 72 continuous hours to 0.9 ppm ozone (Tyson *et al.*, 1982). Weanling (20–40 days old), and young adult (40–180 days old) rats exhibited increases in glucose-6-phosphate dehydrogenase (G6P), glutathione reductase (GR) and glutathione peroxidase (GPx) activity. The changes in young adults were statistically significant. Increased G6P and 6-phosphogluconate dehydrogenase activity also were reported by Elsayed *et al.* (1982). The question remains whether the increased activities observed here are merely reflective of the increased cell proliferation due to ozone injury.

Long-Term Consequences

There is a paucity of data regarding the long-term implications of ozone exposure during the postnatal developmental period. Following exposure of weanling (21–28 days old) rats to 0.2 ppm ozone continuously for 30 days, static pressure–volume measurements detected overdistension of the lung at high transpulmonary pressures (Bartlett *et al.*, 1974). This change is reflective of a decrease in lung elasticity, presumably due to interim collagen damage. The implications for long-term function are not known.

Development of the tracheal mucociliary apparatus in lambs following exposure to 1 ppm ozone 4 h per day for 5 days in the first week of life revealed significant differences in tracheal mucus velocity (TMV) at 2 weeks post-exposure (Mariassy *et al.*, 1990). The depression in TMV was attributed to increased tissue conductance (conductance normally decreases with age). While the depression in TMV was absent by 4 weeks, the mean value of this parameter remained lower in exposed lambs compared with unexposed lambs for 8–24 weeks post-exposure. (Tracheal mucus velocity is normally slow in newborns, reaching adult rates by 2 months of age.) No differences in normal age-related changes in tracheal dimensions were observed, but the composition of the airways was different. Normally, the mucous cell population decreases and the ciliated cell population increases in the first 2 weeks of life. In the ozone-exposed lambs, the percentages of mucous and ciliated cells remained at levels comparable to those of the newborn, thus suggesting that differentiation of the airway epithelium is altered after resolution of acute functional changes.

Persistent alterations in lung development were observed in the lungs of juvenile (6–7 months old) monkeys following exposure to 0.64 ppm for 8 h per day for 12 months

(Tyler *et al.*, 1991). Morphometric analysis revealed significant increases in volume fraction and volume of respiratory bronchioles in exposed animals compared to controls at 0 and 6 months post-exposure. While lesions were qualitatively similar in young (60 days old) and old (444 days old) rats exposed continuously to 0.35 or 0.80 ppm ozone for 72 h, the volume fraction of lesions in younger rats was significantly greater than that of older rats (Stiles and Tyler, 1988). The volume fraction of free inflammatory cells in young rats exposed to 0.35 ppm ozone compared with older rats at the same dose also was significantly increased. The remodeling associated with chronic ozone exposure may eventually alter airflow to the central acinar region, but whether this results in functional deficits is not yet known.

Critical questions regarding the long-term effects of ozone exposure early in postnatal life have yet to be addressed. While previous studies suggest that preweaning animals are less susceptible than adults to injury from exposure to equal concentrations of ozone, the manner in which these younger animals achieve repair has not been addressed. How might ozone injury in the postnatal developing lung affect the biology of the adult lung? Does exposure alter epithelial differentiation events that occur postnatally? Is the branching morphogenesis associated with postnatal alveolarization modified by ozone exposure? Additional questions remain regarding how developing lungs respond to concomitant exposures to ozone and other toxicants. Data that suggest induction of metabolizing enzymes by ozone exposure may vary with age further complicate the scenario, since children often are exposed to multiple airborne toxicants, of which ozone is only one (see Takahashi and Miura, 1985; Takahashi *et al.*, 1985).

SUMMARY

A large number of experimental studies have been directed at defining the risk, significant biological effects, and doses relevant to decrements in health following ozone exposure. The patterns which cellular compartments, metabolic processes, and physiological events follow after exposure are reasonably well understood. We know from animal studies that exposure history has a major impact on the nature of the biological response. The effect of fluctuating and episodic environmental patterns of air pollution also has been examined via animal studies but demands further study, particularly in higher primates.

Those cells that repopulate target areas following acute initial injury appear to be relatively resistant to further injury during the very earliest portions of the repair phase. However, continual exposure during repair alters the repair process, increasing the level of resistance of the cell population in target zones. The inflammatory response progresses through a series of different stages, extended and modified only to a limited extent by continual exposure. A limited number of studies indicate that the cells completing the repair phase within the first week following a single injury do not develop resistance to injury when re-exposed. They are apparently as susceptible as an initially exposed population. We do not know whether this is true in exposure situations where exposures over long periods are interrupted by intervals with little or no pollutants, as is the case with ambient ozone. It is clear that cell populations and the associated extracellular matrix vary by site within the respiratory system under steady-state conditions and that their responses to exposure are variable.

Modeling studies predict variations in local dose by anatomic location within the airway tree. However, most animal studies have not adequately addressed these unique microenvironment differences. Of the many biological properties of the respiratory system which appear to modify its response to ozone exposure, those predictive of human risk have not been identified. We also do not understand how these properties vary in certain subpopulations which may have potential for greater risk, such as aged animals and the very young. Since a substantial portion of lung development occurs postnatally in most mammals (including humans), subsequent studies should continue to address the impact of oxidant stress from ozone exposure on lung growth and development.

REFERENCES

Adamson IYR (1991) Development of lung structure. In: Crystal RG *et al.* (eds) *The Lung: Scientific Foundations.* New York: Raven Press, pp 663–670.

Barry B, Miller F and Crapo J (1985) Effects of inhalation of 0.12 and 0.25 ppm ozone on proximal alveolar region of juvenile and adult rats. *Lab Invest* **53**: 692–704.

Biagini RE, Moorman WJ, Lewis TR and Bernstein IL (1986) Ozone enhancement of platinum asthma in a primate model. *Am Rev Respir Dis* **134**: 719–725.

Bartlett D Jr, Faulkner CS II and Cook K (1974) Effect of chronic ozone exposure on lung elasticity in young rats. *J Appl Physiol* **37**(1): 92–96.

Bassett D, Bowen-Kelly E, Elbon C and Reichenbaugh S (1988) Rat lung recovery from 3 days of continuous exposure to 0.75 ppm ozone. *J Toxicol Environ Health* **25**: 329–347.

Bassett DJP, Elbon CL, Reichenbaugh SS *et al.* (1989) Pretreatment with EDU decreases rat lung cellular responses to ozone. *Toxicol Appl Pharmacol* **100**: 32–40.

Boorman GA, Schwartz LW and Dungworth DL (1980) Pulmonary effects of prolonged ozone insult in rats: morphometric evaluation of the central acinus. *Lab Invest* **43**(2): 108–115.

Castleman WL, Dungworth DL, Schwartz LW and Tyler WS (1980) Acute respiratory bronchiolitis: an ultrastructural and autoradiographic study of epithelial cell injury and renewal in Rhesus monkeys exposed to ozone. *Am J Pathol* **98**: 811–840.

Chang LY, Huang Y, Stockstill BL *et al.* (1992) Epithelial injury and interstitial fibrosis in the proximal alveolar regions of rats chronically exposed to a simulated pattern of urban ambient ozone. *Toxicol Appl Pharmacol* **115**: 241–252.

Chang LY, Stockstill BL, Menache MG *et al.* (1995) Part VIII. Morphometric analysis of structural alterations in alveolar regions. Consequences of prolonged inhalation of ozone on F344/N rats: collaborative studies. Health Effects Institute. Capital City Press, Montpelier, VT.

Chow CK, Plopper CG, Chiu M and Dungworth DL (1981) Dietary Vitamin E and pulmonary bio-chemical and morphological alterations of rats exposed to 0.1 ppm ozone. *Environ Res* **24**: 315–324.

Costa DL, Tepper JS, Stevens MA *et al.* (1995) Restrictive lung disease in rats exposed chronically to an urban profile of ozone. *Am Rev Respir Dis* **151**: 1512–1518.

Dodge DE, Rucker RB, Pinkerton KE *et al.* (1994) Dose-dependent tolerance to ozone: III. Elevation of intracellular Clara cell 10 kDa protein in central acini of rats exposed for 20 months. *Toxicol Appl Pharmacol* **127**: 109–123.

Duan X, Buckpitt AR and Plopper CG (1993) Variation in antioxidant enzyme activities in anatomic subcompartments within rat and rhesus monkey lung. *Toxicol Appl Pharmacol* **123**: 73–82.

Duan X, Buckpitt AR, Pinkerton KE *et al.* (1996) Ozone-induced alterations in glutathione in lung subcompartments of rats and monkeys. *Am J Respir Cell Mol Biol* **14**: 70–75.

Dungworth DL (1989). Noncarcinogenic responses of the respiratory tract to inhaled toxicants. In: McCellan RO, Henderson RF (eds) *Concepts in Inhalation Toxicology.* Hemisphere Publishing Corporation, New York, pp. 291.

Elsayed N, Mustafa M and Postlethwait E (1982) Age-dependent pulmonary response of rats to ozone exposure. *J Toxicol Environ Health* **9**: 835–848.

Eustis S, Schwartz L, Kosch P and Dungworth DL (1981) Chronic bronchiolitis in nonhuman primates after prolonged ozone exposure. *Am J Pathol* **105**: 121–137.

Evans TW, Brokaw JJ, Chung KF *et al.* (1988) Ozone-induced bronchial hyperresponsiveness in the rat is not accompanied by neutrophil influx or increased vascular permeability in the trachea. *Am Rev Respir Dis* **138**: 140–144.

Fouke JM, DeLemos RA and McFadden ER Jr (1988) Airway response to ultra short-term exposure to ozone. *Am Rev Respir Dis* **137**: 326–330.

Fouke JM, DeLemos RA, Dunn MJ and McFadden ER Jr (1990) Effects of ozone on cyclooxygenase metabolites in the baboon tracheobronchial tree. *J Appl Physiol* **69**: 245–250.

Gross KB and White HJ (1987) Functional and pathologic consequences of a 52-week exposure to 0.5 ppm ozone followed by a clean air recovery period. *Lung* **165**: 283–295.

Harkema JR (1992) Epithelial cells of the nasal passages. In: Parent RA (ed.) *Comparative Biology of the Normal Lung*. Boca Raton: CRC Press, pp. 27–36.

Harkema JR, Plopper CG, Hyde DM *et al.* (1987) Effects of an ambient level of ozone on primate nasal epithelial mucosubstances. *Am J Pathol* **127**: 90–96.

Harkema JR, Hotchkiss JA and Henderson RF (1989) Effects of 0.12 and 0.80 ppm ozone on rat nasal and nasopharyngeal epithelial mucosubstances: quantitative histochemistry. *Toxicol Pathol* **17**(3): 525–535.

Harkema JR, Plopper CG, Hyde DM *et al.* (1993) Response of macaque bronchiolar epithelium to ambient concentrations of ozone. *Am J Pathol* **143**(3): 857–866.

Harkema JR, Morgan KT, Gross EA *et al.* (1994) Part VII. Effects on the nasal mucociliary apparatus. Consequences of Prolonged Inhalation of Ozone on F344/N Rats: Collaborative Studies. Health Effects Institute. 65. Capital City Press, Montepelier, VT.

Harkema J, Mauderly J and Griffith W (1996) Pulmonary function alterations in F344 rats following chronic ozone inhalation. *Inhal Toxicol* **8**: 163–183.

Hatch GE, Slade R, Harris LP *et al.* (1994) Ozone dose and effect in humans and rats. *Am J Respir Crit Care Med* **150**: 676–683.

Hotchkiss JA, Harkema JR, Sun JD and Henderson RF (1989) Comparison of acute ozone-induced nasal and pulmonary inflammatory responses in rats. *Toxicol Appl Pharmacol* **98**: 289–302.

Hotchkiss JA, Harkema JR and Henderson RF (1991) Effect of cumulative ozone exposure on ozone-induced nasal epithelial hyperplasia and secretory metaplasia in rats. *Exp Lung Res* **15**: 589–600.

Hotchkiss JA, Harkema JR and Johnson NF (1997) Kinetics of nasal epithelial cell loss and proliferation in F344 rats following a single exposure to 0.5 ppm ozone. *Toxicol Appl Pharmacol* **143**: 75–82.

Hyde DM, Plopper CG, Harkema JR *et al.* (1989) Ozone-induced structural changes in monkey respiratory system. In: *Atmospheric Ozone Research and its Policy Implications*. Pittsburgh, PA: Elsevier Science Publishers, pp. 523–532.

Hyde DM, Hubbard WC, Wong V *et al.* (1992) Ozone-induced acute tracheobronchial epithelial injury: relationship to granulocyte emigration in the lung. *Am J Respir Cell Mol Biol* **6**: 481–497.

Ibrahim AL, Zee YC and Osebold JW (1980) The effects of ozone on the respiratory epithelium of mice. II. Ultrastructural alterations. *J Environ Pathol Toxicol* **3**: 251–258.

Ichinose T and Sagai M (1989) Biochemical effects of combined gases of nitrogen dioxide and ozone. III. Synergistic effects on lipid peroxidation and antioxidative protective systems in the lungs of rats and guinea pigs. *Toxicology* **59**: 259–270.

Joad JP, Bric JM, Pino MV *et al.* (1993) Effects of ozone and neutrophils on function and morphology of the isolated rat lung. *Am Rev Respir Dis* **147**: 1578–1584.

Johnson HG, Stout BK and Ruppel PL (1988) Inhibition of the 5-lipoxygenase pathway with piriprost (U-60, 257) protects normal primates from ozone-induced methacholine hyperresponsive small airways. *Prostaglandins* **35**(3): 459–466.

Kleeberger SR (1995) Genetic susceptibility to ozone exposure. *Toxicol Lett* **82–83**: 295–300.

Kleeberger SR, Levitt RC and Zhang LY (1993a) Susceptibility to ozone-induced inflammation. I. Genetic control of the response to subacute exposure. *Am J Physiol* **264**: L15–20.

Kleeberger SR, Levitt RC and Zhang LY (1993b) Susceptibility to ozone-induced inflammation. II. Separate loci control responses to acute and subacute exposures. *Am J Physiol* **264**: L21–26.

Kodavanti UP, Costa DL, Dreher KL *et al.* (1995) Ozone-induced tissue injury and changes in antioxidant homeostasis in normal and ascorbate-deficient guinea pigs. *Biochem Pharmacol* **50**(2): 243–251.

Last JA, Gelzleichter T, Harkema J *et al.* (1993) Effects of 20 months of ozone exposure on lung collagen in Fischer 344 rats. *Toxicology* **84**: 83–102.

Mariassy A, Abraham W, Phipps R *et al.* (1990) Effect of ozone on the postnatal development of lamb mucociliary apparatus. *J Appl Physiol* **68**: 2504–2510.

Mariassy AT (1992) Epithelial cells of the trachea and bronchi. In: Parent RA (ed.) *Comparative Biology of the Normal Lung.* Boca Raton: CRC Press, pp. 63–76.

Parks WC and Roby JD (1994) Part IV. Effects on expression of extracellular matrix genes. Consequences of Prolonged Inhalation of Ozone on F344/N Rats: Collaborative Studies. Health Effects Institute. 65. Capital City Press, Montpelier, VT.

Pinkerton KE, Menache MG and Plopper CG (1995) Part IX. Changes in the tracheobronchial epithelium, pulmonary acinus, and lung antioxidant enzyme activity. Consequences of Prolonged Inhalation of Ozone on F344/N Rats: Collaborative Studies. Health Effects Institute 65.

Pino MV, Levin JR, Stovall MY and Hyde DM (1992a) Pulmonary inflammation and acute epithelial injury in response to acute ozone exposure in the rat. *Toxicol Appl Pharmacol* **112**: 64–72.

Pino MV, Stovall MY, Levin JR *et al.* (1992b) Acute ozone-induced lung injury in neutrophil-depleted rats. *Toxicol Appl Pharmacol* **114**: 268–276.

Pino MV, McDonald RJ, Berry JD *et al.* (1992c) Functional and morphologic changes caused by acute ozone exposure in the isolated and perfused rat lung. *Am Rev Respir Dis* **145**: 882–889.

Plopper CG and Hyde DM (1992) Epithelial cells of the bronchioles. In: Parent RA (ed.) *Comparative Biology of the Normal Lung.* Boca Raton: CRC Press, pp. 85–93.

Plopper CG, Dungworth DL, Tyler WS and Chow CK (1979) Pulmonary alterations in rats exposed to 0.2 and 0.1 ppm ozone: a correlated morphological and biochemical study. *Arch Environ Health* **34**(6): 390–395.

Plopper CG, Harkema JR, Last JA *et al.* (1991) The respiratory system of nonhuman primates responds more to ambient concentrations of ozone than does that of rats. In: Berglund RL *et al.* (eds) *Trophospheric Ozone and the Environment.* Pittsburgh, PA: Air and Waste Management Association, pp 137–150.

Plopper C, St George J, Cardoso W *et al.* 1992) Development of airway epithelium: patterns of expression for markers of differentiation. *Chest* **101**(Suppl. 3): 2S–5S.

Plopper CG, Chu FP, Haselton CJ *et al.* (1994a) Dose-dependent tolerance to ozone. I. Tracheobronchial epithelial reorganization in rats after 20 months exposure. *Am J Pathol* **144**(2): 404–420.

Plopper CG, Duan X, Buckpitt AR and Pinkerton KE (1994b) Dose-dependent tolerance to ozone. IV. Site-specific elevation in antioxidant enzymes in the lungs of rats exposed for 90 days or 20 months. *Toxicol Appl Pharmacol* **127**: 124–131.

Putman E, Boere AJF, VanBree L *et al.* (1995) Pulmonary surfactant subtype metabolism is altered after short-term ozone exposure. *Toxicol Appl Pharmacol* **134**: 132–138.

Rahman I and Massaro D (1992) Endotoxin treatment protects rats against ozone-induced lung edema: with evidence for the role of manganese superoxide dismutase. *Toxicol Appl Pharmacol* **113**: 13–18.

Rahman IU, Clerch LB and Massaro D (1991) Rat antioxidant enzyme induction by ozone. *Am J Physiol (Lung Cell Mol Physiol)* **260**: L412–L418.

Rietjens IMCM, Dormans JAMA, Rambout PJA and van Bree L (1988) Qualitative and quantitative changes in cytochrome P450 dependent xenobiotic metabolism in pulmonary microsomes and isolated Clara cell populations derived from ozone exposed rats. *J Toxicol Environ Health* **24**(4): 515–531.

Schultheis AH and Bassett DJP (1994) Guinea pig lung inflammatory cell changes following acute ozone exposure. *Lung* **172**: 169–181.

Schwartz LW, Dungworth DL, Mustafa MG *et al.* (1976) Pulmonary responses of rats to ambient levels of ozone: effects of 7-day intermittent or continuous exposure. *Lab Invest* **34**(6): 565–578.

Stephens RJ, Sloan MF, Evans MJ and Freeman G (1974) Early response of lung to low levels of ozone. *Am J Pathol* **74**(1): 31–58.

Stephens RJ, Sloan MF, Groth DG *et al.* (1978) Cytologic response of postnatal rat lungs to O_3 or NO_2 exposure. *Am J Pathol* **93**: 183–200.

Sterner-Kock A, Vesely K, Stovall MY *et al.* (1996) Neonatal capsaicin treatment increases the severity of ozone-induced lung injury. *Am J Respir Crit Care Med* **153**: 436–443.

Stiles J and Tyler WS (1988) Age-related morphometric differences in responses of rat lungs to ozone. *Toxicol Appl Pharmacol* **92**: 274–285.

Suzuki E, Takahashi Y, Aida S *et al.* (1992) Alteration in surface structure of Clara cells and pulmonary cytochrome P450b level in rats exposed to ozone. *Toxicology* **71**: 223–232.

Szarek JL (1994) Part II: Mechanical properties, responses to bronchoactive stimuli, and eicosanoid release in isolated large and small airways. Consequences of prolonged inhalation of ozone on F344/N rats: Collaborative Studies. Health Effects Institute 65. Capital City Press, Montpelier, VT.

Takahashi Y and Miura T (1985) *In vivo* effects of nitrogen dioxide and ozone on xenobiotic metabolizing systems of rat lungs. *Toxicol Lett* **26**: 145–152.

Takahashi Y and Miura T (1987) A selective enhancement of xenobiotic metabolizing systems of rat lungs by prolonged exposure to ozone. *Environ Res* **42**: 425–434.

Takahashi Y and Miura T (1990) Responses of cytochrome P450 isozymes of rat lung to *in vivo* exposure to ozone. *Toxicol Lett* **54**: 327–335.

Takahashi Y, Miura T and Kubota K (1985) *In vivo* effect of ozone inhalation on xenobiotic metabolism of lung and liver of rats. *J Toxicol Environ Health* **15**: 855–864.

Takahashi Y, Aida S, Suzuki E *et al.* (1994) Cytochrome P450 2B1 immunoreactivity in bronchiolar and alveolar epithelial cells after exposure of rats to ozone. *Toxicol Appl Pharmacol* **128**: 207–215.

Tepper JS, Costa DL, Lehmann JR *et al.* (1989) Unattenuated structural and biochemical alterations in the rat lung during functional adaptation to ozone. *Am Rev Respir Dis* **140**: 493–501.

Tepper JS, Wiester MJ, Weber MF *et al.* (1991) Chronic exposure to a simulated urban profile of ozone alters ventilatory responses to carbon dioxide challenge in rats. *Fundam Appl Toxicol* **17**: 52–60.

Tyler NK and Plopper CG (1989) Cytodifferentiation of two epithelial populations of the respiratory bronchiole during fetal lung development in the rhesus monkey. *Anat Rec* **225**: 297–309.

Tyler NK, Hyde DM, Hendrickx AG and Plopper CG (1988) Morphogenesis of the respiratory bronchiole in rhesus monkey lungs. *Am J Anat* **182**: 215–223.

Tyler WS, Tyler NK, Magliano DJ *et al.* (1991) Effects of ozone inhalation on lungs of juvenile monkeys: morphometry after a 12-month exposure and following a 6-month post-exposure period. In: Berglund R *et al.* (eds) *Trophospheric Ozone and the Environment*. Pittsburg, PA: Air and Waste Management Association, pp 151–160.

Tyson CA, Lunan KD and Stephens RJ (1982) Age-related differences in GSH-shuttle enzymes in NO$_2$- or O$_3$-exposed rat lungs. *Arch Environ Health* **37**(3): 167–176.

US EPA (US Environmental Protection Agency) (1996) *Air Quality Criteria for Ozone and Other Photochemical Oxidants*. Research Triangle Park, NC: Office of Health and Environmental Assessment, Environmental Criteria and Assessment Office; Report No EPA/600/P-93/004cF. 3v.

Vargas MH, Segura P, Campos MG *et al.* (1994) Effect of ozone exposure on antigen-induced airway hyperresponsiveness in guinea pigs. *J Toxicol Environ Health* **42**: 435–442.

Watt KC, Plopper CG, Weir AJ (1998) Cytochrome P450 2E1 in rat tracheobronchial airways: response to ozone exposure. *Toxicol Appl Pharmacol* **149**: 195–202.

Weister MJ, Tepper JS, Doerfler DL and Costa DL (1995) Adaptation in rats after chronic exposure to a simulated urban profile of ozone. *Fundam Appl Toxicol* **24**: 42–51.

Wilson DW, Plopper CG and Dungworth DL (1984) The response of the macaque tracheobronchial epithelium to acute ozone injury. *Am J Pathol* **116**: 193–206.

Zidenberg-Cherr S, Han B, Dubick MA and Keen CL (1991) Influence of dietary-induced copper and manganese deficiency on ozone-induced changes in lung and liver antioxidant systems. *Toxicol Lett* **57**: 81–90.

OXIDES OF NITROGEN AND SULFUR

25

Epidemiological Effects of Oxides of Nitrogen, Especially NO$_2$

URSULA ACKERMANN-LIEBRICH and REGULA RAPP

Institute for Social and Preventive Medicine, University of Basel, Basel, Switzerland

INTRODUCTION

In the past two general types of outdoor air pollution problems have been distinguished. One relates to pollutants of long-standing public health concern such as sulfur dioxide and smoke or particulate pollution arising mainly from the combustion of coal or heavy oil for heating or power generation purposes. In developed countries emissions from these sources are typically highest in the winter months, especially when meteorologic conditions cause periods of air stagnation. Famous episodes like the London fog of 1952 occurred in the 1950s and 1960s with exceptionally high concentrations of these pollutants. These episodes are usually referred to as 'winter smog episodes'. Banning open coal fires, reducing stack emissions and diminishing the sulfur content of fuel reduced the high winter levels of sulfur dioxide and particles, and smog episodes have become rare in Western European countries.

In contrast, emissions from mobile sources have increased steadily. Road traffic produces a mixture of volatile hydrocarbons, fine particles, nitrogen dioxides, carbon monoxide and other substances throughout the year. The more recently emerging problem of photochemical pollution is of concern mainly on sunny days in summer and has led to the term 'summer smog'. Ozone is a principal component of it. Ozone and other photochemical species are secondary pollutants formed by reactions between oxides of nitrogen and volatile organic compounds in the presence of sunlight. In addition to such smog episodes, which result in undeniable increases in morbidity and/or mortality, findings of a broad range of epidemiological studies have shown that lower exposures, as occur on many days in urban areas, can also have adverse effects on public health.

AIR POLLUTION AND HEALTH
ISBN 0-12-352335-4

Emissions of nitrogen oxides mainly come from anthropogenic sources. They result from combustion processes at high temperatures, which form oxides of nitrogen through the oxidation of atmospheric nitrogen. The principal product is nitric oxide (NO), with a small proportion of NO_2 and other nitrogen oxides (NO_x). NO_2 is then formed by the gradual oxidation of NO. NO_2 is in the daily urban air pollution mixture and is a key component of the photochemical pollution.

Motor vehicles are not the only source of NO_x, but represent the most important source in urban areas. Except near industrial point sources, the highest concentrations of NO_x are generally found in busy streets. A typical diurnal pattern with peaks during the morning and evening rush hours is seen throughout the year, with only a slight difference between winter and summer. Average levels in European cities are 30–70 μg NO_2/m^3. NO_2 concentrations tend to be somewhat spatially homogeneous in the urban environment, but less so than the fine particle concentration. On days with stagnant weather conditions NO_2 can also spread widely across urban areas, as shown in a recent London episode in 1991 (Department of Health, 1993).

Substantial exposure to NO_2 takes place indoors, since outdoor NO_2 does penetrate indoors. When no indoor source is present, the outdoor-to-indoor ratio is typically about 0.5. This value is lower in winter than in summer, depending on the air exchange rate. However, gas cooking or heating, especially with unvented devices, add considerably to the indoor NO_2 burden. The use of a gas stove can result in peak hour concentrations of more than 1 mg/m^3 in the kitchen. As a consequence, in homes with a gas stove indoor levels are often higher than outdoor levels, particularly in winter. In addition, smoking, fireplaces, unvented water heaters and other open flame burners can increase the indoor NO_2 concentration.

As NO_2 is mainly a secondary air pollutant, deriving outdoors from motor vehicle traffic or indoors from gas cooking or heating, it almost never occurs on its own. It is always part of the respective mixture, be it emissions of the combustion process of petrol or of gas. NO_2 can easily be measured with continuous monitors or with portable passive samplers with a high degree of accuracy.

HEALTH EFFECTS OF HIGH NO$_2$ EXPOSURES

Early evidence for the potential dangers of nitrogen oxides occurred in accidents as the famous fire of x-ray films in a hospital in Cleveland 1929, where 26 of the 123 victims survived the first hours but died later owing to the effects of oxides of nitrogen from burning nitrocellulose. Other military and occupational events which generated substantial quantities of NO_2 (over 300 mg/m^3) resulted in acute pulmonary edema and death. The role of nitrogen oxides in the so-called silo-filler's disease with cough, dyspnea, hemoptysis and interstitial infiltrates in the roentgenogram after silage gas inhalation was recognized only in 1956. Some evidence of pulmonary dysfunction persisted after follow-up of such patients of months to years. Nowadays, high exposures to nitrogen dioxide in non-occupational settings can still occur in indoor ice-skating arenas which are poorly ventilated. The air of the rinks can become polluted with NO_2 from the ice resurfacing machines, which often have no emissions controls. Personal sampler measurements in school children in Sweden showed peak hour concentrations up to 8 mg NO_2/m^3

(Berglund *et al.*, 1994). There have been reported events with health consequences ranging from cough and hemoptysis to lung edema. While measurements have not been made at the time of the exposures, estimates suggest levels of at least 2 mg/m^3 (Karlson-Stiber *et al.*, 1996; Smith *et al.*, 1992; Soparkar *et al.*, 1992).

Controlled human exposures to NO$_2$ in the laboratory have found increases of inflammatory mediators and changes in cell counts and distributions with the technique of bronchoalveolar lavage after short exposures to levels of NO$_2$ down to 1 mg/m^3 (Jörres *et al.*, 1995; Rubinstein *et al.*, 1991). Consistent effects on health indicators, such as lung function parameters that can easily be measured in field studies of a broader population sample, have not been found.

HEALTH EFFECTS OF AMBIENT EXPOSURE LEVELS

Because of the availability of a well-developed measurement technique, NO$_2$ is a suitable indicator of traffic exhaust emissions as well as emissions from indoor combustion sources. However, the compositions of these respective mixtures are quite dissimilar. Epidemiological studies are not able to separate effects of components with the same time-course and geographic propagation. It may be difficult to separate the effects of NO$_2$ outdoors from the effects of particles or volatile organic compounds. From a public health perspective, identification of seemingly independent effects of these outdoor pollutants may not be as important as it seems: as the outdoor pollutants come mainly from the same sources, many source-directed control strategies will diminish the levels of all components. In studies of indoor pollution it may be difficult to separate effects of having or using a gas appliance from effects of NO$_2$ itself, if no measurements are made. Having gas as source of energy could be linked with higher humidity indoors, living in inner city areas, older homes, etc., confounding factors which are linked with poor health from other reasons. As a consequence caution is warranted in combining evidence on the health effects of NO$_2$ from studies of outdoor and indoor pollution.

Studies on Indoor Exposure to NO$_2$

Because indoor sources play a key role in determining personal exposure to nitrogen oxides, there have been many studies of the adverse health effects of indoor NO$_2$ exposure. After the publication of a study conducted in the UK during the 1970s, which observed increased symptoms of respiratory illness among children living in homes using gas stoves for cooking in comparison with those from homes with electric ranges, there was substantial concern with regard to the health effects of indoor NO$_2$ exposure. Most studies have compared respiratory symptoms or lung function in subjects from households with gas appliances with persons from households with electric ranges. The early studies used sources to classify exposure and had few, if any, indoor NO$_2$ measurements. In some studies measurements were made in a subgroup of participants over a short time, or even in comparable households of non-participants.

Studies on Children (Table 25.1)

Many studies on respiratory symptoms and illnesses have been done in children. The hypothesis that NO_2 may effectively impair the health by increasing the vulnerability to acute infectious disease is better studied in children, since acute respiratory illnesses are a major cause of childhood disease. These illnesses may in turn predispose to later chronic respiratory disease (Samet *et al.*, 1983).

In order to summarize the results of different studies on NO_2 exposure indoors and its effect on respiratory illness in children under the age of 12 years, a review and meta-analysis was conducted in 1992 by the US Environmental Protection Agency (Hasselblad *et al.*, 1992). There was relatively good agreement among the different studies and an estimated increase of the odds ratio of 1.2 indicating a 20% increase in respiratory illness in children associated with the use of gas cookers. This increase was put as an equivalent to an increase in exposure of 30 $\mu g/m^3$ NO_2 (Fig. 25.1). These exposure levels have been assessed usually as weekly averages. From these studies, it is not possible to separate the long-term effect of repeated peak exposures during cooking from those of an elevated average level which occurs in gas-heated indoor environments. This meta-analysis was based mainly on studies of school children; the two studies on infants included in the analyses showed no effect.

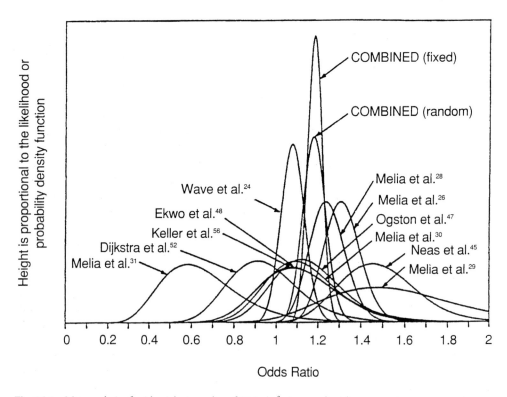

Fig. 25.1 Meta-analysis of epidemiologic studies of 30 $\mu g/m^3$ nitrogen dioxide exposure increase on respiratory illness in children ≤ 12 years old. Reproduced with permission from Hasselblad *et al.* (1992).

Since then, several studies have been published. A study in the Netherlands on about 1000 school children, which measured NO_2 levels in the kitchen, living room and bedroom, found a slight negative effect of exposure on lung function parameters, but overall not very consistently (Brunekreef *et al.*, 1990). The NO_2 measurements offered no additional information in the final regression model beyond the indicator variable source present or absent. Two prospective studies in the USA on 900 and 1200 infants, which were followed during their first 12 and 18 months, respectively, did not find more frequent respiratory symptoms or illnesses in children from homes with gas stoves (Aldous *et al.*, 1996; Samet *et al.*, 1993). An Australian study on 14 000 families with young children found an association of using natural gas stoves on asthma, wheezing, colds and hay fever, but no effect of other potential NO_2 sources as liquid petroleum gas stoves, flueless gas heaters, wood fires and wood heaters (Volkmer *et al.*, 1995). Another Australian study measured NO_2 concentrations in the classrooms of 388 school children during school hours in the heating period. In addition, the parents of those children from homes with gas use pinned a badge monitor on the clothing of their child during four evenings. Exposure to peak hourly levels of at least 150 µg/m^3 compared with exposure to background levels up to 34 µg/m^3 in schools was associated with an increase in sore throat, colds and absences from school (Pilotto *et al.*, 1997).

Studies on Adults (Table 25.2)

Target groups in studies with adults were mainly housewives, as they are most exposed to emissions from gas cooking stoves. Recent studies on chronic effects of gas stoves did find more frequent respiratory symptoms with gas use (Jarvis *et al.*, 1996; Ostro *et al.*, 1993; Viegi *et al.*, 1992). An English study on 1800 young adults found also a reduction of lung function in women, but not in men (Jarvis *et al.*, 1996).

Two panel studies, i.e. short-term cohort studies, have been published on groups considered to be susceptible to irritants. Daily spirometry results in heavy smokers did not correlate with daily levels of NO_2 measured with individual samplers over 14 days, and the study participants were not adversely affected by an intermittent chamber exposure to 580 µg NO_2 (Hackney *et al.*, 1992). On the other hand, a close follow-up of the consequences of daily use of gas stoves, wood stoves or fireplaces on 164 non-smoking asthmatics showed a close association between use of a gas stove and shortness of breath, cough or restrictions in activity (Ostro *et al.*, 1994).

There is a difference in the findings from studies on the effects of indoor NO_2 in infants on the one hand, and the results of studies in school-age children or in adults on the other hand. The long-term, prospective studies in infants could not find an effect, whereas studies in school children and adults with a cross-sectional design observed a slight, but not always consistent, increase in symptoms or diseases. This difference could be due to a long latency period for the effects of either gas use or NO_2. Another explanation may be the cross-sectional design of school children and adult studies. The positive findings in cross-sectional studies could be related to a confounder such as poorer housing in areas with gas use. In addition, it should be borne in mind that in cities, where gas appliances are more frequent than in rural areas, the contribution of NO_2 from outdoor pollution to indoor air is higher too. The observed effect in adults and school children in studies on gas stoves might therefore result from a combination of indoor and outdoor pollution.

Table 25.1 Studies on long-term effects of indoor NO_2 or of indoor gas use on respiratory health of children (only studies not in metaanalysis of Hasselblad et al., 1992)

Country	Sample	Exposure	Results	Comment	Author
England, USA, The Netherlands	Metaanalysis of 11 studies, mostly school children	Gas stove vs. electric stove, four studies with passive sampler measurements	Increase of 30 µg/m³ → about +20% respiratory illness	Effects based on studies with measurements stronger than based on comparison of gas vs. electric device	Hasselblad et al. (1992)
USA	1205 infants in Albuquerque, followed until 18 months of age	One-year indoor measurements: mean bedroom level in winter 49 µg/m³ and 13 µg/m³ in homes with gas and electric stoves, respectively	Lower respiratory symptoms OR 0.91 (0.81–1.04) in homes with gas stove		Samet et al. (1993)
USA	936 infants in Tucson, followed until 12 months of age	Gas heating, gas cooking	No significant correlation with lower respiratory illness in first year of life	No indoor measurements significant association with evaporative cooling	Aldous et al. (1996)
Australia	14 124 families with children 3 months to 5 years	Electric cooking and/or heating vs. natural gas, liquid petroleum gas, wood	Natural gas stove → OR 1.24 asthma / 1.16 wheezing / 1.14 excessive colds / 1.13 hay fever / Liquid petroleum gas stove → no effect / Flueless gas heater → OR 1.26 dry cough / Wood fire/wood heater → reduced cough and wheezing		Volkmer et al. (1995)

Table 25.1 *cont.*

Country	Sample	Exposure	Results	Comment	Author
The Netherlands	985 children, 6–9 years	Average of kitchen, living room and bedroom levels 21–40, 41–60 and over 60 μg/m^3, reference category 0–20 μg/m^3	Negative association between lung function parameters and NO$_2$, but few coefficients reached statistical significance. Reasonably consistent effect only on MMEF	NO$_2$ measurement offered no advantage over source presence as exposure variable	Brunekreef *et al.* (1990)
USA	925 parents of school children 5–9 years, Seattle	Parents reported use of a gas-, wood-, or kerosene-burning stove or wood-burning fireplace in the last 12 months	No association with physician-diagnosed asthma or parent-reported wheezing in the last 12 months	No indoor measurements significant association with household water damage and environmental smoke exposure	Maier *et al.*, (1997)
Australia	388 children, 6–11 years, Sydney	Average in schools with electric heating 15–43 μg/m^3, in schools with gas heating 33–244 μg/m^3	Exposure to peak hourly levels of >150 μg/m^3 compared with background levels of 34 μg/m^3 → increase in sore throat, colds, absences from school, statistically significant dose–response trends for mean rates for cough with phlegm, absences from school and sore throat		Piloto *et al.*, (1997)
England	486 children with asthma symptoms and 475 control children, 11–16 years, Sheffield	Methods of cooking	Use of gas for cooking not associated with wheezing in last 12 months	No measurements	Strachan and Carey (1995)

MMEF, maximum midexpiratory flow; OR, odds ratio.

Table 25.2   Studies on effects of indoor NO_2 on respiratory health of adults

Country	Sample	Exposure	Results	Comment	Author
USA	Case-control study, 213 women, non-smoking, 20–39 years with high or low FEV_1, Tecumseh	Gas stove vs. electric stove	No difference in lung function	No measurement	Jones et al. (1983)
Canada	Case-control study, 52 women, 25–44 years with cough and phlegm or healthy	24-h mean 49–207 µg NO_2/m^3, 2-h mean up to 3 mg/m^3	No effect on symptoms. Negative association between V25 and V 50 and log NO_2-indoor		Hosein and Corey (1986)
Italy	Cross-sectional study, 3866 adults in Pisa	Indicator variables for type of fuel, central heating or stove etc.	Cough more frequent in men without central heating or with gas cooker. Higher frequency of wheeze and shortness of breath in women with stove for heating	No measurement	Viegi et al. (1992)
England	Cross-sectional analysis of cohort study, 1864 adults 20–44 years	Reported use of gas for cooking, heating	In women, mainly using gas for cooking resulted in an OR of 2.07 (1.4–3.05) for wheezing, 2.32 for shortness of breath and 2.6 (1.2–5.6) for asthma attacks in past year, and a 3% reduction of FEV_1	No measurements. Stronger effect in women working in household or out of work; no effect in men	Jarvis et al. (1996)
USA	Cohort panel study 6 months, 321 adults in Los Angeles, non-smokers	Use of gas stove vs. electric stove	Presence of a gas stove in the house was associated with incidence of lower respiratory symptoms OR 1.23 (1.03–1.47) and with chronic respiratory condition OR 2.08 (1.49–2.9)	No NO_2 indoor measurements; NO_2 outdoor showed no association with symptoms	Ostro et al. (1993)

Table 25.2 *cont.*

Country	Sample	Exposure	Results	Comment	Author
USA	Combined laboratory and panel study during 2 weeks, 26 heavy smokers with COPD, 46–70 years, Los Angeles	Individual samplers 14-day mean 72–235 μg/m³, close correlation of daily levels with outdoor monitoring	No correlation of individual daily sampler concentration with daily spirometry results at home	Chamber exposure to 580 μg NO$_2$/m³ or to clean air during 4 h in the middle of each study week, no effect of chamber exposure	Hackney *et al.* (1992)
USA	Panel study during 3 months, 164 non-smoking asthma patients, 18–70 years, Denver	Daily use of gas stove, fireplace or woodstove	Reported use of gas stove → shortness of breath OR 1.6 (1.11–2.32), cough 1.71 (0.97–3.01), restrictions in activity 1.47 (1–2.16)	No indoor measurements	Ostro *et al.* (1994)

PEF, peak expiratory flow rate.

Studies on Outdoor Exposure to NO$_2$

Many early studies on health effects of air pollutants compared frequency of diseases in urban study sites with that in rural study sites. The effects of living in a city were attributed to the 'urban air pollution mix'; this was infrequently characterized in detail but was generally indexed by sulfur dioxide or particles. With the steady decrease of SO$_2$ in most cities in Western Europe the potential importance of NO$_2$ became of concern. Particularly in countries with lower levels of general pollution, less heavily polluting industries, and with small differences in living conditions between different population groups, NO$_2$ levels are associated with health effects.

Children (Table 25.3)

Assessment of lifelong exposure to outdoor NO$_2$ can be more readily accomplished in children than in adults, as children stay in or near their home or school and do not commute. An early study on outdoor NO$_2$ effects, the so-called Chattanooga School Children Study, was designed to investigate the effects of NO$_2$ emissions from a trinitrotoluene plant in Hamilton County, Tennessee, next to the city of Chattanooga. The ventilatory performance of second-grade school children in the high-NO$_2$ exposure area was significantly lower than the performance of children in the control areas, and the respiratory illness rates of the families in the exposed areas were consistently higher during the entire 24 weeks of study (Shy et al., 1970a).

Later studies in infants estimated the exposure to NO$_2$ not only with measurements at a fixed central station, but also by adding passive samplers or estimates from traffic density counts in the nearby road. They found longer duration of respiratory symptoms with increasing annual means of NO$_2$ (Braun-Fahrländer et al., 1992), more frequent hospitalizations for respiratory diseases with increasing 2-year mean levels (Walters et al., 1995) and more treatments in hospitals for lower respiratory illness with lifelong NO$_2$ exposure. In this last study the effect was seen only in girls (Pershagen et al., 1995).

Similar results were seen in Swiss school children from 10 communities, where the odds for chronic cough and respiratory infections as bronchitis or pneumonia in the last year increased with increasing annual means of NO$_2$ (Fig. 25.2). On the other hand allergic diseases, such as asthma and hay fever, showed no association with NO$_2$ (Braun-Fahrländer et al., 1997).

The effect of air pollution on lung function in 4000 US children and adolescents from 64 geographic sampling units was examined in the Second National Health and Nutrition Examination Survey (NHANES II: see Schwartz, 1989). After controlling for age, height, race, sex, body mass, smoking and respiratory symptoms, they observed a highly significant non-linear negative correlation of the lung function with increasing annual or biannual concentrations of NO$_2$, more pronounced over 75 μg/m^3. Demographic and geographic variables had little or no impact on the relationships, which also held when only persons still residing in their state of birth were considered.

Many studies on short-term effects in children have been based on symptom diaries or school absence records (Table 25.4). Diary studies record daily symptoms, medication use and sometimes lung function. Overall they could not find a consistent effect of the NO$_2$ level of the previous day on respiratory symptoms (Hoek and Brunekreef, 1994; Schwartz et al., 1994; Linn et al., 1996). Evaluation of alternative and longer time lags has been

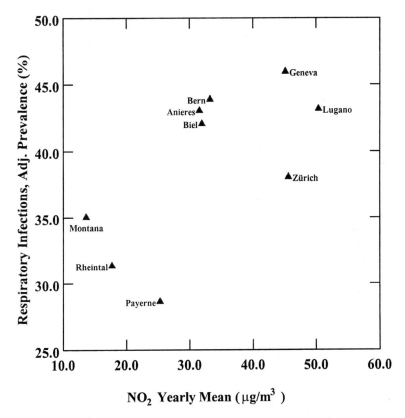

Fig. 25.2   Adjusted prevalence of respiratory infections in children of ten communities in Switzerland according to the mean NO₂ level. Reproduced from Braun-Fahrländer (1997).

proposed (Schwartz *et al.*, 1994). In contrast, the studies did show a decrease of lung function measures of around 1–2% of the forced vital capacity (FVC) per 100 µg NO_2/m^3 (Hoek and Brunekreef, 1994; Linn *et al.*, 1996), and perhaps a greater decrease in asthmatic children (Quackenboss *et al.*, 1991).

A time-series study using hospital admission data for infants and small children was conducted in five cities in Germany. There was an increase in the admission rate for croup cases of 28% for an increase of daily mean from 10 to 70 µg/m³; there was not a significant association with the admission for obstructive bronchitis (Schwartz *et al.*, 1991).

Adults (Table 25.5)

There are relatively few studies on the long-term effects of NO_2 in adults. The only prospective cohort study provides estimates of 10-year personal exposure for over 6000 Seventh Day Adventists in California (Abbey *et al.*, 1993). The estimates were based on indoor sources, derived from lifestyle and housing characteristics, and from the average of ambient concentrations of the three nearest outdoor monitors. Personal exposure to NO_2 was not significantly associated with risk for chronic obstructive bronchitis (relative risk

Table 25.3 Studies on associations of children's respiratory health with long-term NO_2 concentrations

Country	Sample	Exposure	Results	Comment	Author
USA	3922 children and adolescents 6–24 years from 64 sampling units throughout the US	Means of 365 days before spirometry: 10th and 90th percentile respectively 46 and 117 µg NO_2/m^3	Non-linear negative correlation of FVC, FEV_1, PEFR, more pronounced over 75 µg NO_2/m^3. Excluding children with respiratory symptoms or smokers did not change the results.	Identical relationships with NO_2 concentrations averaged over 2 years	Schwartz (1989)
Switzerland	625 preschool children 0–5 years in Basel and Zurich	Annual means 20–90 µg NO_2/m^3	Increase of 20 µg in annual means → duration of any respiratory episode OR 1.11 (1.07–1.16), duration of cough episode OR 1.09 (0.97–1.22), increase of 20 µg in 6-weeks home outdoor means → duration of any symptom OR 1.13 (1.01–1.27), no association of NO_2 with incidence of symptoms	Controlling for social class, day care, region season of participation	Braun-Fahrländer et al. (1992)
England	Hospitalization rate 1988–90 in children, younger and older (61+years) adults, in 39 electoral wards of West Midland (incl. Birmingham)	Not given	Increase of 30 µg/m³ in 2-years mean → standardized hospitalization rates of infants for respiratory diseases + 63% Association not significant in younger or older adults.	Possibly close correlation of social status (deprivation, Townsend score) and NO_2 levels. Social status associated with hospitalization but in adults, not in children	Walters et al. (1995)
Sweden	199 children with hospital-treated wheezing, bronchitis or asthma and 351 population controls, 4 months to 4 years	Estimated time-weighted 99th percentile for study subjects 20–205 µg/m³, mean 55 µg/m³	Lower respiratory illness associated with time-weighted mean outdoor NO_2 level in girls, not in boys on traffic intensity at home	Estimation of lifelong individual exposure to outdoor NO_2 with data address and day care centers	Pershagen et al. (1995)
Switzerland	4470 school children 6–15 years, in ten communities	Annual means 12–50 µg/m³	Increase between the most and least polluted community: OR for chronic cough 1.58 (1.14–2.18), nocturnal dry cough apart from colds 1.99 (1.51–2.61), bronchitis 1.35 (0.99–1.85), conjunctivitis 1.5 (1.15–1.96). No increase of wheeze, asthma ever, hay fever and diarrhoea with increasing levels of exposure	Controlling for fog. High correlation between averages of NO_2, PM_{10} and SO_2	Braun-Fahrländer et al. (1997)

(RR) 1.26, 95% confidence interval 0.58, 4.33, for an increase of 93 µg/m^3) but did reach statistical significance for exposure to outdoor NO$_2$, considering only the average of the city monitors (p=0.05, RR not given). This difference in association may be a result of the different function of NO$_2$ as an indicator of indoor pollution versus traffic-related pollution.

A cross-sectional study with 9650 Swiss adults in eight communities found increasing respiratory symptoms (but not asthma) and decreasing lung function in parallel with annual means of NO$_2$ (Ackermann-Liebrich *et al.*, 1997; Zemp *et al.*, 1999). The same was seen with particulate matter. In addition, lung function was also decreasing *within* the communities, which could not be explained by PM$_{10}$ (Schindler *et al.*, 1998). Annoyance, cough, throat and nose irritation were associated with the average 6-month mean of NO$_2$ in a Swedish study on 6100 adults in 55 communities (Forsberg *et al.*, 1997a).

The few panel studies on adults found no association of daily NO$_2$ levels with respiratory symptoms of healthy people (Ostro *et al.*, 1993) or asthmatics (Forsberg *et al.*, 1993). Other studies showed effects on bronchodilator use of asthmatics (Higgins *et al.*, 1995) and a heightened bronchial responsiveness on methacholine in asthmatics (Taggart *et al.*, 1996).

Studies on Registry Data (Tables 25.6 and 25.7)

Many studies have been conducted with registry or vital statistics data, looking at short-term effects on daily mortality counts or hospital admission rates.

In most studies, daily 24-h means or peak hour means of NO$_2$ measured at one fixed station were used as exposure measures for the population of a community. Because of common sources and similar effects of weather patterns, 24-hour means of NO$_2$ are often strongly correlated with daily PM$_{10}$ levels, and peak hour levels of NO$_2$ with peak hour levels of carbon monoxide. The geographic variation of NO$_2$, however, is greater than for PM$_{10}$. The strength of the association between NO$_2$ and a supposed health effect depends on how well the daily exposure variation of the population or study sample is characterized by the daily exposure variation at the measurement station. In general, the association of daily data from registries (as mortality data or hospital admissions) with NO$_2$ concentrations would be expected to be less strong than the association with fine particle concentration, independent of any basis in actual causal relations.

The European study on air pollution and health (APHEA), a coordinated study of the short-term effects of air pollution on mortality and hospital admissions using data from 15 cities, found significant positive associations between daily death and NO$_2$ (Touloumi *et al.*, 1997). An increase of 50 µg/m^3 (1-h maximum) was associated with a 1.3% increase in the daily number of deaths. The effect diminished when other pollutants such as black smoke were entered in the model, but the association with NO$_2$ remained significant. In other studies it was not possible to separate the effect of NO$_2$ from that of other pollutants. An increase in the daily mean of 100 µg NO$_2$/m^3 in cities was associated with an increase of around 2% in total mortality, while the same increase in daily peak hour levels was associated with an increase of 2–3% of total mortality (Anderson *et al.*, 1996; Kinney and Özkaynak *et al.*, 1991; Ostro *et al.*, 1996; Sunyer *et al.*, 1996; Touloumi *et al.*, 1997; Wietlisbach *et al.*, 1996).

Studies of the association of hospital admissions for all respiratory diseases with daily

Table 25.4 Panel cohort studies on outdoor NO_2 and children's respiratory health

Country	Sample	Exposure	Results	Comment	Author
USA	351 primary school children, in Chestnut Ridge, 8 months diary	Maximum daily peak hour 79 μg NO_2/m³, mean daily peak hour 40 μg/m³	No association of respiratory symptoms reported in diary or of peak flow rate, measured in school, with NO_2, nor with other pollutants	Very low NO_2 concentrations	Vedal et al. (1987)
USA	Time–activity diary and individually measured exposure during two weeks, 30 children with asthma, Tucson, AZ	Personal exposure 19.4 μg/m³, outdoor near home 25.1 μg/m³, central station: morning 37.2, evening 49 μg/m³	Increase of 20 μg/m³ in home outdoor week average level → decrement of morning PEFR of 40 l/min or 12%, reduced effect in evening PEFR. Central monitoring station levels also linked to morning and noon PEFR. No effect in non-asthmatic children	Time spent in various locations did not change the effects	Quackenboss et al. (1991)
Netherlands	Winter 1990/91, 112 children, 7–12 years in Wageningen	Maximum daily peak hour 127 μg NO_2/m³	No association of symptoms with pollutants. No association of spirometric lung function measurements with NO_2. Association of FVC and FEV_1 with SO_2 and particulate matter.	Very low NO_2 concentrations	Hoek and Brunekreef (1993)
Netherlands	Winter 1990/91, 73 children 6–12 years, with chronic respiratory symptoms, in Wageningen and Bennekom	Maximum daily peak hour 127 μg NO_2/m³	No association of symptoms or twice daily measured peak flow rate with NO_2. Small but consistent association of PEFR and prevalence of wheeze with PM_{10}, black smoke, and SO_2	Very low NO_2 concentrations	Roemer et al. (1993)
Netherlands	Winter 1987/88–1989/90, 1079 schoolchildren, 7–11 years, in Deurne, Enkhuizen, Zeist	Study mean 37 μg/m³, daily maximum 70 μg/m³	An increase of 50 μg NO_2/m³ was associated with a decrease of FVC and FEV_1 of 1%, and PEFR and MMEF of 2%. No association with respiratory symptoms		Hoek and Brunekreef (1994)

Table 25.4 *cont.*

Country	Sample	Exposure	Results	Comment	Author
USA	Diary study during one year 1844 school children in six cities	Study median 25 μg NO$_2$/m^3, maximum 82 μg NO$_2$/m^3	Increase of 18.5 μg/m^3 on the same day or previous day → cough incidence OR 1.1 (0.96–1.25), same increase previous 4 days OR 1.35 (1.11–1.63) up to 24 μg NO$_2$/m^3, afterwards no further increase. Increase of 18.5 μg/m^3 → incidence of lower respiratory symptoms OR 1.21 (0.98–1.47)	In restricted data set with two pollutants: 4 days NO$_2$, increase of 18.5 μg/m^3 OR 1.17 (0.94–1.46) and 3 days PM$_{10}$, increase of 30 μg/m^3 OR 1.2 (0.99–1.47).	Schwartz *et al.* (1994)
USA	269 school children, health testing during 6 weeks in different seasons, in three communities in South California	Central station mean 69 μg/m^3, range 22–156 μg/m^3	No association with symptoms. Loss of 22 ml of morning FVC per 1 μg increase in NO$_2$ level of previous day (95% CI, 0.03–0.39 ml). Predicted FVC loss over range of highest vs. lowest day level → 29 ml (1.2%). Morning to afternoon change of FVC decreased with increase of NO$_2$ measured over preceding 24 h		Linn *et al.* (1996)

Table 25.5 Cohort and cross-sectional studies on outdoor NO_2 and respiratory health of adults

Country	Sample	Exposure	Results	Comment	Author
USA	More than 5000 participants in two Californian communities	Annual means of daily maximum hour 60 and 211 μg NO_2/m^3 in the two areas	Prevalence of cough and phlegm higher in non-smokers of high pollution area, prevalence of wheezing in smokers higher in high pollution area	Comparison of two areas	Detels *et al.* (1981)
USA	6340 Seventh Day Adventists in California	10 years mean: 11th percentile of study population exposure 37 μg/m^3; 68th percentile, 74 μg/m^3; 88th percentile, 92 μg/m^3	Increase of 93 μg/m^3 in city monitor NO_2 data → increase in 10 years incidence in COPD (p=0.05, RR not given). Increase in adjusted (personal) NO_2 → RR for COPD 1.26 (0.58–4.33)	Personal exposure estimated using lifestyle and housing characteristics as indoor sources (and estimation of ambient concentrations from three nearest city monitors)	Abbey *et al.* (1993)
Switzerland	9651 adults 18–60 years in 8 communities	Annual mean 10–58 μg/m^3	Increase of 10 μg NO_2/m^3 → dyspnea stade I +6%; wheezing +6%; phlegm +9%; chronic cough +7%	Close correlation of NO_2, SO_2 and particulate matter	Zemp *et al.* (1999)
Switzerland	9651 adults 18–60 years in eight communities	Annual mean 10–58 μg/m^3	Increase of 10 μg NO_2/m^3 → lung function, FVC −1.18%, FEV_1 −0.7%	Close correlation of NO_2, SO_2 and particulate matter	Ackermann-Liebrich *et al.* (1997)
Switzerland	7656 adults 18–60 years in eight communities	Personal NO_2 measurements and estimation of average annual personal exposure 12.8–49.2 μg/m^3	Increase of 10 μg NO_2/m^3 → FVC −0.74%; FEV_1 −0.26%	Within-community coefficients smaller than the ones found between communities	Schindler *et al.* (1998)
Sweden	6109 adults 16–70 years, 55 communities	Average 6-month mean in winter 19 μg/m^3, (10–30 μg/m^3)	Annoyance, cough, throat and nose irritation correlated with NO_2. Prevalence increased in areas >22 μg/m^3		Forsberg *et al.* (1997a,b)

Table 25.6 Time-series studies on mortality data

Country	Sample	Exposure	Results	Comment	Author
USA	Daily mortality in Los Angeles County	Study mean 128 μg	Increase of 100 μg → increase of daily death counts of 2%		Kinney and Özkaynak (1991)
England	Daily mortality in London 1987–92	Average peak hour NO$_2$ 106 μg/m^3, daily mean 69 μg/m^3, max. daily mean 337 μg/m^3	Increase of daily mean 13–53 μg/m^3 in summer → increase of cardiovascular mortality of 2.54% (0.18–4.96%). Increase of max. hourly mean 35–89 μg/m^3 → increase of cardiovascular mortality 2.96% (0.8–5.17), increase of total mortality 1.73% (−0.2 to +3.3)	No effect in winter, correlation with ozone in summer could explain result	Anderson et al. (1996)
Chile	Daily mortality in Santiago 1989–91	Peak hour mean 103 μg/m^3, maximum 478 μg/m^3	Increase in peak hour of 103 μg/m^3 → +2% (1–4%) in total mortality, in summer +4% (0–7%)		Ostro et al. (1996)
Spain, APHEA	Daily mortality in Barcelona 1985–91	Peak hour mean in winter 88 μg/m^3, in summer 97 μg/m^3	Increase in peak hour of previous day of 100 μg/m^3 → total mortality + 3.4% (95% CI, 1.3–5.5%), cardiovascular mortality +3.8% (0.9–6.8%), no effect on respiratory mortality		Sunyer et al. (1996)
Switzerland	Daily mortality in Zurich, Basel, Geneva 1984-89		Increase of 109 μg/m^3 (3-day average) → total mortality in Zurich +0.48% SD 0.25%, in Basel 2.4% SD 0.48%, in Geneva +1.07% SD 0.5%, respiratory mortality in ZH +2.03% SD 0.6, BS +3.46% SD 1, GE +2.41% SD 0.8		Wietlisbach et al. (1996)
Europe, APHEA	Daily mortality in Athens, Barcelona, Köln, London, Lyon and Paris	Peak hour means 70–135 μg/m^3	Increase in peak hour of 50 μg/m^3 → increase in total mortality 0.5–2.7%, weighted mean 1.3% (95% CI, 0.9–1.8%)		Touloumi et al. (1997)

Table 25.7 Time-series series studies on hospital admission data

Country	Sample	Exposure	Results	Comment	Author
Canada	Hospital emergency visits for respiratory diseases and asthma of people 0–14,15–65 and over 65 years in Vancouver	Peak hour mean 74 μg/m³, maximum 240 μg/m³	Correlation of visits for respiratory diseases in people over 65 years with NO_2 in winter		Bates et al. (1990)
Greece	Emergency consultations and hospitalizations in greater Athens	Winter mean 94 SD 25 μg/m³, summer mean 111 SD 32 μg/m³	Increase of 76 μg/m³ (5th–95th percentile) in winter → increase of 11% in total, 15% in cardiac and 25% in respiratory admissions; no effect in summer	NO_2 only pollutant significantly associated with admissions	Pantazopoulou et al. (1995)
France, APHEA	Hospital admissions for respiratory diseases, COPD and asthma in Paris	Average daily mean 45 μg/m³	Increase of 100 μg/m³ in daily mean and peak hour level → admissions for asthma +17.5% (6–30%) and +8% (2–15%). No effect on COPD admissions or all respiratory admissions		Dab et al. (1996)
Canada	Emergency hospitalizations for cardiorespiratory disease in Toronto	Mean peak hour levels 71 μg/m³, maximum 150 μg/m³	One pollutant model: increase of 10.5 μg NO_2/m³ in 5 daytime (12-h) average → +4.8% respiratory admissions, same increase in 4 daytime average → +4.9% cardiac admissions three pollutants model (ozone, NO_2, SO_2): increase of 10.5 μg NO_2/m³ in 5 daytime average → +2.8% respiratory admissions	Particle mass and chemistry could not be identified as an independent risk factor for cardiorespiratory admissions beyond that attributable to climate and gaseous pollution	Burnett et al. (1997a)
USA	Hospital admissions of people over 65 years with pneumonia and COPD in Minneapolis	Daily mean (10th–90th percentile) 15–47 μg/m³	One pollutant model: increase of 18 μg/m³ on previous day → +2.2% admissions		Moolgavkar et al. (1997)
Europe, APHEA	Emergency hospitalizations for COPD in six European cities: Amsterdam, Rotterdam, Barcelona, London, Milano, Paris		Increase of daily mean of 50 μg/m³ on previous day → RR 1.02 (95% CI, 1.0–1.05)		Anderson et al. (1997)

Table 25.7 *cont.*

Country	Sample	Exposure	Results	Comment	Author
Spain	Emergency consultation of asthmatics (15–65 years, special register) in Barcelona	Average NO$_2$ level in summer 104 µg/m^3, in winter 101 µg/m^3	Increase in daily mean of 25 µg/m^3 $\rightarrow$ increase in consultations of 4.5% (summer) resp. 5.6% (winter)		Castellsague et al. (1995)
Canada	Emergency consultations for asthma in summer in St. John, New Brunswick	Mean peak hour level 47 µg/m^3	No effect of NO$_2$ as single pollutant and no significant effect in addition to ozone		Stieb et al. (1996)
USA	Emergency consultations for asthma in Santa Clara County, California	Mean peak hour level 113 µg/m^3, range 17–305 µg/m^3)	Same day NO$_2$ was associated with asthma consultations (β=0.013, p=0.024), became insignificant in regressions along with PM$_{10}$	NO$_2$ strongly correlated with PM$_{10}$	Lipsett et al. (1997)
Germany	6330 croup cases in children 0–6 years, 4755 cases of obstructive bronchitis in five cities	Study means 14–55 µg/m^3	Increase of daily mean 10–70 µg/m^3 $\rightarrow$ increase of croup cases 28%, no increase of admissions for bronchitis	Prospective sampling	Schwartz et al. (1991)
USA	Hospitalization of people 65+ years with cardiovascular diseases, in seven large cities	Mean peak hour levels 74–143 µg/m^3	Increase in peak hour of 185 µg/m^3 $\rightarrow$ OR for hospitalization in one pollutant model in LA 1.15 (1.1–1.19), in Chicago 1.17 (1.07–1.27), Philadelphia 1.03 (0.95–1.12), NY 1.07 (1.02–1.13), Detroit 1.04 (0.92–1.18), Houston 0.99, Milwaukee 1.05 (0.89–1.23)	Best indicator in all cities except NY was CO (in NY only NO$_2$ significant)	Morris et al. (1995)
Canada	Hospital admissions of people over 65 years for congestive heart failure in 10 cities	Average daily mean 42 µg/m^3, mean peak hour levels 72 µg/m^3	Log of same day average NO$_2$ associated with hospitalizations for heart failure in one pollutant model and in two polluted models together with CO	Best prediction model included CO peak hour levels, NO$_2$ daily mean levels, temperature and dew point temperature	Burnett et al. (1997b)

Table 25.7 *cont.*

Country	Sample	Exposure	Results	Comment	Author
USA	Hospital admissions of people over 65 years with cardiovascular diseases in Tucson, AZ	Average daily mean 36 $\mu g/m^3$	Increase of 10.5 $\mu g/m^3$ in daily mean → +0.69% ns (95% CI, −2.3 to +3.8)	Independent effects of PM_{10} daily means and CO peak hour means	Schwartz (1997)
Finland	Hospital admissions for cardiac and cerebrovascular diseases in Helsinki	Daily mean 39 μg NO_2/m^3, range 4–170 μg NO_2/m^3 and 91 μg NO/m^3, range 7–467 $\mu g/m^3$	Increase in previous days NO mean → increase in admissions for cardiac diseases, inconsistent results for NO_2		Pönkä and Virtanen (1996)

NO$_2$ levels have yielded inconsistent results (Bates *et al.*, 1990; Burnett *et al.*, 1997a, b; Dab *et al.*, 1996; Moolgavkar *et al.*, 1997; Pantazopoulou *et al.*, 1995). A meta-analysis with data on hospitalizations for chronic obstructive pulmonary disease (COPD) from six European cities (APHEA study) found an estimated relative risk of 1.02 for COPD admissions when the 24-h mean of NO$_2$ of the previous day had increased by 50 μg/m^3. The association was not statistically significant in the cities considered separately (Anderson *et al.*, 1997). For asthma, findings with a large effect of additional 2% admissions associated with an increase of 10 μg/m^3 of daily mean in Barcelona and Paris (Castellsague *et al.*, 1995; Dab *et al.*, 1996) contrast with findings from Saint John, New Brunswick (Stieb *et al.*, 1996). The same inconsistency is seen with emergency hospitalizations for cardiovascular diseases. NO$_2$ was the strongest or only pollutant significantly associated with cardiovascular admissions in Athens (Pantazopoulou *et al.*, 1995) and New York (Morris *et al.*, 1995). In studies in other cities, carbon monoxide or other pollutants were as closely associated as NO$_2$ (Los Angeles, Chicago: Morris *et al.*, 1995) or showed a better correlation than NO$_2$ (Tucson: Schwartz, 1997), or the best prediction model included NO$_2$ together with other gaseous pollutants (10 Canadian cities: Burnett *et al.*, 1997a,b).

Up until now there have not been enough studies on short-term or long-term effects of oudoor NO$_2$ to allow the calculation of a dose–effect relationship for distinct health outcomes. However, there does seem to be a qualitatively consistent association of longer term NO$_2$-emissions on respiratory symptoms and diseases, particularly in children. In both short-term and long-term studies NO$_2$ can be a good indicator of the pollutant mixture, depending on the main source of emissions.

CONCLUSION

There is still not enough evidence from epidemiological studies alone to establish whether the observed effects are causally related to NO$_2$. From the overall evidence, including findings from toxicological studies, it seems reasonable to assume that NO$_2$ is at least partially responsible for the observed health effects of urban pollution mixtures.

The World Health Organization Centre for Europe recommends in the revised Air Quality Guidelines for Europe as a long-term guideline for NO$_2$ an annual value of 40 μg/m^3 (21 ppb) and, based on clinical data, a 1-h guideline of 200 μg/m^3 (100 ppb). The directive of the Economic Commission for Europe proposes the same values as limit values to protect human health. The short-term value of 200 μg/m^3 averaged over 1 h must not be exceeded during more than 8 out of 8760 hours per calendar year; the long-term limit value will be 40 μg/m^3 averaged over a calendar year.

REFERENCES

Abbey DE, Colome SD, Mills PK *et al.* (1993) Chronic disease associated with long-term concentration of nitrogen dioxide. *J Exp Anal Environ Epidemiol* **3**: 181–202.

Ackermann-Liebrich U, Leuenberger P, Schwartz J and the SAPALDIA Team (1997) Lung function and long term exposure to air pollutants in Switzerland. *Am J Respir Crit Care Med* **155**: 122–129.

Aldous MB, Holberg CJ, Wright AL *et al.* (1996) Evaporative cooling and other home factors and lower respiratory tract illness during the first year of life. *Am J Epidemiol* **143**: 423–430.

Anderson HR, Ponce de Leon A, Bland JM *et al.* (1996) Air pollution and daily mortality in London, 1987–92. *Br Med J* **312**: 665–669.

Anderson HR, Spix C, Medina S *et al.* (1997) Air pollution and daily admissions for chronic obstructive pulmonary disease in 6 European cities: results from the APHEA project. *Eur Respir J* **10**: 1064–1071.

Bates DV, Baker-Anderson M and Sizto R (1990) Asthma attack periodicity: a study of hospital emergency visits in Vancouver. *Environ Res* **51**: 51–70.

Berglund M, Brabäck L, Bylin G *et al.* (1994) Personal NO$_2$ exposure monitoring shows high esposure among ice-skating schoolchildren. *Arch Environ Health* **49**: 17–24.

Braun-Fahrländer C, Ackermann-Liebrich U, Schwartz *et al.* (1992) Air pollution and respiratory symptoms in preschool children. *Am Rev Respir Dis* **145**: 42–47.

Braun-Fahrländer C, Vuille JC, Sennhauser FH and the SCARPOL Team (1997) Respiratory health and long-term exposure to air pollutants in Swiss schoolchildren. *Am J Respir Crit Care Med* **155**: 1042–1049.

Brunekreef B, Houthuijs D, Dijkstra L and Boleij SM (1990) Indoor nitrogen dioxide exposure and children's pulmonary function. *J Air Waste Manag Assoc* **40**: 1252–1256.

Burnett RT, Cakmak S, Brook JR and Krewski D (1997a) The role of particulate size and chemistry in the association between summertime ambient air pollution and hospitalization for cardiorespiratory disease. *Environ Health Perspect* **105**: 614–620.

Burnett RT, Dales RE, Brook JR *et al.* (1997b) Association between ambient carbon monoxide levels and hospitalizations for congestive heart failure in the elderly in 10 Canadian cities. *Epidemiology* **8**: 162–167.

Castellsague J, Sunyer J, Saez M and Anto JM (1995) Short-term association between air pollution and emergency room visits for asthma in Barcelona. *Thorax* **50**: 1051–1056.

Dab W, Medina S, Quenel P *et al.* (1996) Short term respiratory health effects of ambient air pollution, results of the APHEA project in Paris. *J Epidemiol Commun Health* **50**(suppl): S42–46.

Department of Health (1993) Advisory Group on the Medical Aspects of Air Pollution Episodes. Third Report: *Oxides of Nitrogen*. London: HMSO, pp 18–20.

Detels R, Sayre JW, Coulson AH *et al.* (1981) The UCLA population studies of chronic obstructive respiratory disease 4. Respiratory effect of long term exposure to photochemical oxidants, nitrogen dioxide and sulfates on current and never smokers. *Am Rev Respir Dis* **124**: 673–680.

Forsberg B, Stjernberg N, Falk M *et al.* (1993) Air pollution levels, meteorological conditions and asthma symptoms. *Eur Respir J* **6**: 1109–1115.

Forsberg B, Stjernberg N and Wall S (1997a) People can detect poor air quality well below guideline concentrations: a prevalence study of annoyance reactions and air pollution from traffic. *Occup Environ Med* **54**: 44–48.

Forsberg B, Stjernberg N and Wall S (1997b) Prevalence of respiratory and hyperreactivity symptoms in relation to levels of criteria air pollutants in Sweden. *Eur J Public Health* **7**: 291–296.

Hackney JD, Linn WS, Avol EL *et al.* (1992) Exposures of older adults with chronic respiratory illness to nitrogen dioxide – a combined laboratory and field study. *Am Rev Respir Dis* **146**: 1480–1486.

Hasselblad V, Kotchmar DJ and Eddy DM (1992) Synthesis of environmental evidence: Nitrogen dioxide epidemiology studies. *J Air Waste Manag Assoc* **42**: 662–671.

Higgins BG, Francis HC, Yates CJ *et al.* (1995) Effects of air pollution in symptoms and peak expiratory flow measurements in subjects with obstructive airways disease. *Thorax* **50**: 149–155.

Hoek G and Brunekreef B (1993) Acute effects of winter air pollution episodes on pulmonary function and respiratory symptoms of children. *Arch Environ Health* **48**: 328–335.

Hoek G and Brunekreef B (1994) Effects of low-level winter air pollution concentrations on respiratory health of Dutch children. *Environ Res* **64**: 136–150.

Hosein HR and Corey P (1986) Domestic air pollution and respiratory function in a group of housewives. *Can J Public Health* **77**: 44–50.

Jarvis D, Chinn S, Luczynska C and Burney P (1996) Association of respiratory symptoms and lung function in young adults with use of domestic gas appliances. *Lancet* **347**: 426–431.

Jones JR, Higgins ITT, Higgins MW and Keller JB (1983) Effects of cooking fuels on lung function of nonsmoking women. *Arch Environ Health* **38**: 219–222.

Jörres R, Nowak D, Grimminger F *et al.* (1995) The effects of 1 ppm nitrogen dioxide on bronchoalveolar lavage cells and inflammatory mediators in normal and asthmatic subjects. *Eur Respir J* **8**: 416–424.

Karlson-Stiber C, Höjer J, Sjöholm A *et al.* (1996) Nitrogen dioxide pneumonitis in ice hockey players. *J Intern Med* **239**: 451–456.

Kinney PL and Özkaynak H (1991) Associations of daily mortality and air pollution in Los Angeles County. *Environ Res* **54**: 99–120.

Leuenberger P, Künzli N, Ackermann-Liebrich U *et al.* (1998) Etude suisse sur la pollution de l'air et les maladies respiratories chez l'adulte (SAPALDIA). *Schweiz Med Wochenschr* **128**: 150–161.

Linn WS, Shamoo DA, Anderson KR *et al.* (1996) Short-term air pollution exposures and responses in Los Angeles area schoolchildren. *J Exp Anal Environ Epidemiol* **6**: 449–472.

Lipsett M, Hurley S and Ostro B (1997) Air pollution and emergency room visits for asthma in Santa Clara County, California. *Environ Health Perspect* **105**: 216–222.

Maier WC, Arrighi HM, Morray B *et al.* (1997) Indoor risk factors for asthma and wheezing among Seattle school children. *Environ Health Perspect* **105**: 208–214.

Moolgavkar SH, Luebeck EG and Anderson EL (1997) Air pollution and hospital admissions for respiratory causes in Minneapolis-St Paul and Birmingham. *Epidemiology* **8**: 364–370.

Morris RD, Naumova EN and Munasinghe RL (1995) Ambient air pollution and hospitalization for congestive heart failure among elderly people in seven large US cities. *Am J Publ Health* **85**: 1362–1365.

Ostro BD, Lipsett MJ, Mann JK *et al.* (1993) Air pollution and respiratory morbidity among adults in Southern California. *Am J Epidemiol* **137**: 691–700.

Ostro BD, Lipsett MJ, Mann JK *et al.* (1994) Indoor air pollution and asthma results from a panel study. *Am J Respir Crit Care Med* **149**: 1400–1406.

Ostro B, Sanchez JM, Aranda C and Eskeland GS (1996) Air pollution and mortality, results from a study of Santiago, Chile. *J Exp Anal Environ Epidemiol* **6**: 97–114.

Pantazopoulou A, Katsouyanni K, Kourea-Kremastinou J and Trichopoulos D (1995) Short-term effects of air pollution on hospital emergency outpatient visits and admissions in the greater Athens, Greece area. *Environ Res* **69**: 31–36.

Pershagen G, Rylander E, Norberg S *et al.* (1995) Air pollution involving nitrogen dioxide exposure and wheezing bronchitis in children. *Int J Epidemiol* **24**: 1147–1153.

Pilotto LS, Douglas RM, Attewell RG and Wilson SR (1997) Respiratory effects associated with indoor nitrogen dioxide exposure in children. *Int J Epidemiol* **26**: 788–796.

Pönkä A and Virtanen M (1996) Low-level air pollution and hospital admissions for cardiac and cerebrovascular diseases in Helsinki. *Am J Publ Health* **86**: 1273–1280.

Quackenboss JJ, Krzyzanowski M and Lebowitz MD (1991) Exposure assessment approaches to evaluate respiratory health effects of particulate matter and nitrogen dioxide. *J Exp Anal Environ Epidemiol* **1**: 83–107.

Roemer W, Hoek G and Brunekreef B (1993) Effect of ambient winter air pollution on respiratory health of children with chronic respiratory symptoms. *Am Rev Respir Dis* **147**: 118–124.

Rubinstein I, Reiss TF, Bigby BG *et al.* (1991) Effects of 0.6 ppm nitrogen dioxide on circulating and bronchoalveolar lavage lymphocyte phenotypes in healthy subjects. *Environ Res* **55**: 18–30.

Samet JM, Tager IB and Speizer FE (1983) The relationship between respiratory illness in childood and chronic air-flow obstruction in adulthood. *Am Rev Respir Dis* **127**: 508–552.

Samet JM, Lambert WE, Skipper BJ *et al.* (1993) Nitrogen dioxide and respiratory illness in infants. *Am Rev Respir Dis* **148**: 1258–1265.

Schindler C, Ackermann-Liebrich U, Leuenberger P and the SAPALDIA Team (1998) Associations between lung function and estimated average exposure to NO$_2$ in 8 areas of Switzerland (SAPALDIA). *Epidemiology* **9**: 405–411.

Schwartz J (1989) Lung function and chronic exposure to air pollution: a cross-sectional analysis of NHANES II. *Environ Res* **50**: 309–321.

Schwartz J (1997) Air pollution and hospital admissions for cardiovascular disease in Tucson. *Epidemiology* **8**: 371–377.

Schwartz J, Spix C, Wichmann HE and Malin E (1991) Air pollution and acute respiratory illness in five German communities. *Environ Res* **56**: 1–14.

Schwartz J, Dockery DW, Neas LM *et al.* (1994) Acute effects of summer air pollution on respiratory symptom reporting in children. *Am J Respir Crit Care Med* **150**: 1234–1242.

Shy CM, Creason JP, Pearlman ME *et al.* (1970a) The Chattanooga schoolchildren study: Effects of community exposure to nitrogen dioxide. 1. Methods, description of pollutants exposure and results of ventilatory function testing. *J Air Pollut Control Assoc* **20**: 539–545.

Shy CM, Creason JP, Pearlman ME *et al.* (1970b) The Chattanooga schoolchildren study: Effects of community exposure to nitrogen dioxide. 2. Incidence of acute respiratory illness. *J Air Pollut Control Assoc* **20**: 582–588.

Smith W, Anderson T, Anderson HA and Remington PL (1992) Nitrogen dioxide and carbon monoxide intoxication in an indoor ice arena – Wisconsin, 1992. *Mort Morb Weekly Rep* **41**: 383–384.

Soparkar G, Mayers I, Edouard L and Hoeppner VH (1993) Toxic effects from nitrogen dioxide in ice-skating arenas. *Can Med Assoc J* **148**: 1181–1182.

Stieb DM, Burnett RT, Beveridge RC and Brook JR (1996) Association between ozone and asthma emergency department visits in Saint John, New Brunswick, Canada. *Environ Health Perspect* **104**: 1354–1360.

Strachan DP and Carey IM (1995) Home environment and severe asthma in adolescence: a population based case–control study. *Br Med J* **311**: 1053–1056.

Sunyer J, Castellsague J, Saez M *et al.* (1996) Air pollution and mortality in Barcelona. *J Epidemiol Commun Health* **50**(Suppl. 1): S76–80.

Taggart SCO, Custovic A, Francis HC *et al.* (1996) Asthmatic bronchial hyperresponsiveness varies with ambient levels of summertime air pollution. *Eur Respir J* **9**: 1146–1154.

Touloumi G, Katsouyanni K, Zmirou D *et al.* (1997) Short-term effects of ambient oxidant exposure on mortality: a combined analysis within the APHEA project. *Am J Epidemiol* **146**: 177–185.

Vedal S, Schenker MB, Munoz A *et al.* (1987) Daily air pollution effects on children's respiratory symptoms and peak expiratory flow. *Am J Public Health* **77**: 694–698.

Viegi G, Carrozzi L, Paoletti P *et al.* (1992) Effects of the home environment on respiratory symptoms of a general population sample in Italy. *Arch Environ Health* **47**: 64–70.

Volkmer RE, Ruffin RE, Wigg NR and Davies N (1995) The prevalence of respiratory symptoms in South Australian preschool children I. Geographic location II. Factors associated with indoor air quality. *J Paediatr Child Health* **31**: 112–120.

Walters S, Phupinyokul M and Ayres J (1995) Hospital admissions rates for asthma and respiratory disease in the West Midlands: their relationship to air pollution levels. *Thorax* **50**: 948–954.

Wietlisbach V, Pope CA and Ackermann-Liebrich U (1996) Air pollution and daily mortality in three Swiss urban areas. *Soz Präventivmed* **41**: 107–115.

Zemp E, Elsasser S, Schindler C *et al.* (1999) Long term ambient air pollution and respiratory symptoms in adults (SAPALDIA). *Study J Resp J Crit Care Med* (in press).

Toxicology of Sulfur Oxides

RICHARD B. SCHLESINGER

Department of Environmental Medicine, New York University
School of Medicine, New York, NY, USA

INTRODUCTION

The group of air pollutants collectively termed sulfur oxides comprises both gaseous and particulate chemical species. There are four gas-phase compounds, namely sulfur monoxide, sulfur dioxide, sulfur trioxide and disulfur monoxide, but only sulfur dioxide (SO_2) occurs at sufficient concentrations in ambient atmospheres to be of public health concern. The particulate phase sulfur oxides consist of strongly to weakly acidic sulfates, namely sulfuric acid (H_2SO_4) and its products of neutralization with atmospheric ammonia: letovicite [$(NH_4)_3H(SO_4)_2$], ammonium bisulfate (NH_4HSO_4) and ammonium sulfate [$(NH_4)_2SO_4$]. Although most of the toxicological database for acidic sulfates involves sulfuric acid because it is the most acidic of the particulate sulfates, this species rarely occurs alone in ambient air, which generally contains some combination of the various sulfates. This chapter reviews the toxicology of inhaled sulfur dioxide and acidic sulfates based upon both laboratory animal and controlled human exposure studies.

DOSIMETRY AND BIOLOGICAL FATE OF SULFUR OXIDES

Due to its high solubility in water, sulfur dioxide can be very efficiently scrubbed from inspired air within the upper respiratory tract (Dalhamn and Strandberg, 1961; Frank *et al.*, 1969; Speizer and Frank, 1966). However, the extent of removal appears to depend upon the inhaled concentration. While the scrubbing efficiency is greater at high SO_2 concentrations, upper respiratory tract removal may be less efficient at the lower levels relevant to ambient air pollution (Amdur, 1966; Strandberg, 1964), resulting in increased penetration into the lungs. Penetration is also greater during oral rather than nasal breathing and with increased physical activity such as exercise. This change with exercise is due to the reduced residence time of the inspired air in the upper airways, increased minute ventilation and a shift from nasal to oronasal breathing.

Upon contacting airway lining fluid, SO_2 rapidly dissolves into the aqueous phase. As the gaseous anhydride of sulfurous acid, it readily dissociates into bisulfite and sulfite ions, which can then be translocated into the systemic circulation. These ions are eventually detoxified via sulfite oxidase and excreted in the urine as sulfate. While desorption of SO_2 may occur if the partial pressure of the gas on the mucosal surfaces exceeds that of inhaled air, only small amounts appear to be desorbed even with exposure to high concentrations (Speizer and Frank, 1966).

In the respiratory tract, inhaled sulfate particles are subjected to physical deposition mechanisms which are dependent upon the specific particle size distribution and ventilation characteristics. Like SO_2, greater penetration into the lungs generally occurs with oral rather than nasal breathing. However, sulfate particles have properties that can dynamically alter their size following inhalation, namely deliquescence (e.g. ammonium bisulfate) or hygroscopicity (e.g. sulfuric acid). This may result in substantial growth in diameter within the airways, often by factors of two to four times, while the particles are airborne in the humid atmosphere of the respiratory tract. Such particles will then deposit on airway surfaces according to their hydrated, rather than their initial dry, size (Martonen and Zhang, 1993; Morrow, 1986).

Inhaled sulfate particles may be further modified while still airborne by undergoing a neutralization reaction with endogenous respiratory tract ammonia (Larson *et al.*, 1977, 1982). Since ammonia concentrations are greater in the oral than in the nasal passages, the mode of inhalation influences the ratio of the specific sulfate species available for deposition in the lungs. Under most conditions, it is likely that only partial neutralization of inhaled sulfuric acid will occur prior to deposition. Deposited acid sulfate particles may be further modified by the action of buffers present in the fluid lining of the airways (Holma, 1985).

The extent of reaction between acidic sulfate particles and respiratory tract ammonia depends upon a number of factors (Larson *et al.*, 1993). One of these is inhaled particle size. For a given ammonia concentration, the extent of neutralization is inversely proportional to size, at least within the relevant ambient particle diameter range of 0.1–10 μm. Another factor is residence time within the airways, which is a function of ventilation rate. For example, a combination of high ammonia and low ventilation rate results in increased acid neutralization, while low ammonia and high ventilation rate results in decreased neutralization; these factors will modify the amount of acid available for deposition even if the acid concentration at the point of inhalation remains constant.

MARKERS OF EXPOSURE TO SULFUR OXIDES

It has been difficult to find a biomarker for exposure to either sulfur dioxide or acidic sulfates, since these chemical species are consumed via dissolution and/or chemical reaction at the point of contact with airway surfaces. While the bisulfite ion can react with various biomolecules, the only reaction that may provide a marker of inhalation exposure is that with disulfide (S—S) bonds (found in the protein components of nasal and airway fluid linings), forming S-sulfonates (Bechtold *et al.*, 1993; Gunnison *et al.*, 1987). This would be especially useful for assessing the SO_2 exposure of nasal passages, from which airway lining fluid can easily be obtained by lavage. The S-sulfonates that are formed do not appear to accumulate with repeated daily exposures, thus providing a reasonable recent exposure history (Bechtold *et al.*, 1993). There is currently no suitable biomarker of exposure for acidic sulfate particles.

TOXICOLOGY OF SULFUR DIOXIDE

Sulfur dioxide is a chemical irritant. Exposure to >26.2 mg/m³ (10 ppm) for more than a few minutes will result in irritation of the eyes, mucous membranes and throat. However, sulfur dioxide is not acutely lethal. Levels exceeding 157–262 mg/m³ (60–100 ppm) are generally needed for mortality, which results from respiratory insufficiency and concomitant effects on the central nervous system.

Pulmonary Mechanical Function

Normal healthy individuals show minimal, if any, significant alterations in pulmonary functional indices following exposure to SO_2 at ≤13.1 mg/m³ (5 ppm), for durations of <1–4 h. The only effects may be some upper respiratory symptoms (Sheppard, 1988; US EPA, 1982, 1986). On the other hand, when asthmatics and others with hyperreactive airways are exposed to SO_2 at 0.66–1.3 mg/m³ (0.25–0.5 ppm), they show a striking acute response, characterized by bronchoconstriction, associated increased airway resistance and decreased expiratory flow rates, as well as the clinical symptoms of wheezing and shortness of breath. While the magnitude of this response is quite variable among individual asthmatics, it is a very sensitive marker of SO_2 exposure.

With moderate exercise, bronchoconstriction will occur in asthmatics exposed to even lower SO_2 levels (Barnes, 1994; Linn *et al.*, 1987). This results from an increase in both ventilation rate and the relative contribution of oral to total ventilation (Bethel *et al.*, 1983). Oral ventilation not only allows more lower respiratory tract penetration of SO_2, but also results in airway drying, affecting gas absorption. SO_2-induced bronchoconstriction is also exacerbated by breathing cold or dry air (Bethel *et al.*, 1983; Sheppard *et al.*, 1984).

The time required for SO_2 to elicit bronchoconstriction in exercising asthmatics is short. Measurable changes may occur after only 2 min of exposure to 1–2.6 mg/m³ (0.4–1.0 ppm) and the maximal response may occur within 5–10 min after a short exposure (Balmes *et al.*, 1987; Gong *et al.*, 1995; Horstman *et al.*, 1988). Following a single SO_2 exposure during exercise, airway resistance in asthmatics appears to recover to normal levels within 1–2 h (Hackney *et al.*, 1984). There is some indication of attenuation of SO_2-induced bronchoconstriction, however, with repeated or continuous exposure (Kehrl *et al.*, 1987; Sheppard *et al.*, 1983).

Unlike many other pollutants, SO_2 does not appear to cause pronounced changes in overall non-specific airway reactivity (Hazucha *et al.*, 1984), i.e. an enhanced response to an exogenous bronchoconstrictor agent such as histamine. This suggests that any increased SO_2 responsiveness is not due simply to elevated non-specific airway reactivity but, rather, that asthmatics are actually more reactive than normals to SO_2 itself.

Pulmonary Defenses

Sulfur dioxide can affect various aspects of pulmonary defenses, such as mucociliary transport, alveolar clearance of deposited particles and pulmonary macrophage function. Exposure to 13.1 mg/m^3 (5 ppm) SO_2 for 2 h did not alter tracheobronchial mucociliary transport in resting humans (Wolff *et al.*, 1975), but exposure with exercise to the same level resulted in an increased tracheobronchial mucociliary clearance rate (Newhouse *et al.*, 1978). On the other hand, decreased nasal mucociliary clearance was noted with exposure of humans to 13.1 mg/m^3 (5 ppm) for 4–5 h (Andersen *et al.*, 1974). The difference between the nasal and bronchial response may be dose related; that in the nose may be greater than that in the airways. Higher levels of SO_2 are also needed to alter mucociliary clearance in animals with acute exposure (Riechelmann *et al.*, 1995; Spiegelman *et al.*, 1968).

Chronic exposure to sulfur dioxide also affects tracheobronchial mucociliary clearance. Dogs exposed for 1 year (1.5 h/day, 5 days/week) to 2.62 mg/m^3 (1 ppm) SO_2 showed a decrease in tracheobronchial mucociliary clearance rate (Hirsch *et al.*, 1974).

Clearance of particles from the alveolar region is affected by SO_2. It was accelerated in rats exposed to 0.26 mg/m^3 (0.1 ppm) for 10 to 25 days, but was depressed with exposure to 2.62 mg/m^3 (1 ppm) (Ferin and Leach, 1973). On the other hand, increased clearance was noted in rabbits exposed to 26.2 mg/m^3 (10 ppm) SO_2 for 16 h/day for 20 weeks (Holma, 1967).

The alveolar macrophage is important for clearance function, as well as for other aspects of lung defense. Exposure of rats to 2.62–52.4 mg/m^3 (1–20 ppm) SO_2 for 24 h resulted in increased *in vitro* phagocytosis by alveolar macrophages at $\geq$13.1 mg/m^3 (5 ppm) (Katz and Laskin, 1971). On the other hand, macrophages obtained from exercising hamsters exposed to 131 mg/m^3 (50 ppm) SO_2 for 4 h showed decreased motility and phagocytic function (Skornik and Brain, 1990). Examination of the lavage fluid of humans exposed to 21 mg/m^3 (8 ppm) SO_2 for 20 min during mild exercise showed an increase in lysozyme-positive macrophages, an indicator of cell activation (Sandström *et al.*, 1989).

In vitro exposure of human alveolar macrophages to 6.6–32.8 mg/m^3 (2.5–12.5 ppm) SO_2 for 10 min resulted in a dose-dependent increase in spontaneous production of reactive oxygen species; 0.5 h exposure to 32.8 mg/m^3 (12.5 ppm) had a cytotoxic effect and decreased the production of reactive oxygen species compared with 6.6 mg/m^3 (2.5 ppm) (Kienast *et al.*, 1993). Other studies that have examined the effects of SO_2 on alveolar macrophage functionality following *in vitro* exposure have indicated that SO_2 can alter cell migration via changes in chemotactic mechanisms, and also decrease production or release of tumor necrosis factor (TNF) and interleukin-1 (IL-1) (Knorst *et al.*, 1996a,b). The relevance of these results to *in vivo* exposures is unclear.

Most of the database involving effects of SO_2 on lung defense is concerned with non-specific, i.e. non-immunologic, mechanisms. The little available evidence on pulmonary humoral or cell-mediated immunity provides only equivocal indication of the ability of SO_2 to modulate the activity of cells involved in allergic responses (Riedel *et al.*, 1988).

Pulmonary Morphology and Biochemistry

Morphological changes in both the upper and lower respiratory tracts may follow exposure to SO_2, especially at high levels. For example, mice exposed to 26.2 mg/m³ (10 ppm) continuously for 72 h showed loss of nasal cilia and decreased mucosal thickness (Giddens and Fairchild, 1972). Guinea pig tracheas exposed to ≥ 15 mg/m³ (5.8 p.p.m) showed sloughing of the epithelium, intracellular edema, mitochondrial swelling, increased intercellular spaces and ciliary cytoplasmic extrusions (Riechelmann *et al.*, 1995).

Bronchial epithelial changes generally involve the mucus secretory cells. Goblet cell hypertrophy can occur with exposure of rats at high concentrations, leading to the suggestion that SO_2 produces a chronic bronchitis similar to that in humans (Lamb and Reid, 1968). But repeated exposures to ≥131 mg/m³ (50 ppm) are needed to elicit more long-term or permanent changes.

TOXICOLOGY OF ACIDIC SULFATES

Sulfuric acid is a corrosive irritant. Exposure to 0.5–2 mg/m³ will begin to elicit irritation of the eyes and upper respiratory tract in some individuals, while 3 mg/m³ will be detected by most exposed individuals and 6–10 mg/m³ will be perceived as very unpleasant, eliciting coughing. As with SO_2, H_2SO_4 is not acutely lethal, and the cause of death due to high level H_2SO_4 exposure is laryngeal or bronchial spasm. However, sulfuric acid is more acutely toxic than are its neutralization products.

Pulmonary Mechanical Function

Controlled clinical studies of healthy adults have shown no consistent effects on pulmonary function or respiratory symptoms with acute exposure to H_2SO_4 (0.1–1 μm diameter), even with exercise (Avol *et al.*, 1988a; Frampton *et al.*, 1992; US EPA, 1989). Acute exposures at higher concentrations using larger particles in the 10–20 μm range that are characteristic of acidic fogs have been associated with some increases in respiratory symptoms, e.g. cough, but no effect has been found in lung function parameters (Avol *et al.*, 1988a,b; Linn *et al.*, 1989). On the other hand, there is some evidence that asthmatics may be more sensitive than healthy individuals to the effects of acid sulfate aerosols on lung mechanical function. They also may experience modest bronchoconstriction following exposure to H_2SO_4 at <1 mg/m³ (Avol *et al.*, 1988a,b; Koenig *et al.*, 1993; Linn *et al.*, 1989; US EPA, 1989). But such effects are neither as consistent nor as dramatic as those associated with exposure to SO_2.

All asthmatics may not be equally sensitive to acid sulfates. While elderly asthmatics do not appear to be a susceptible population (Koenig *et al.*, 1993), adolescents with allergic asthma are. Small declines in FEV_1 were noted after only a 40 min exposure to H_2SO_4

(0.6 μm) at 0.068 mg/m^3 in one study of adolescents (Koenig *et al.*, 1983), although effects on lung function have not been consistently observed at exposure concentrations <0.1 mg/m^3 (Koenig *et al.*, 1992). It is likely that specific characteristics of those asthmatic populations studied may play a significant role in apparent sensitivity to acid exposure and to the resultant effective concentration.

The effects of acid sulfates on healthy humans are consistent with those observed in experimental toxicology assessments. Acute exposures of most species to H$_2$SO$_4$ at ≤1 mg/m^3 (<1 μm diameter) have not been shown to alter standard lung function tests (US EPA, 1989). The notable exception is the guinea pig, which appears to be especially sensitive and shows pulmonary functional changes at H$_2$SO$_4$ levels down to 0.1 mg/m^3.

While standard measures of lung function generally will not be affected in normals, an increase in airway reactivity was observed with bronchoconstrictor challenge following exposure to 1 mg/m^3 H$_2$SO$_4$ for 16 min, and in some adult asthmatics following exposure to 0.1 mg/m^3 (Utell *et al.*, 1983a). There also was some indication of a delayed response to exposure. Furthermore, it has been suggested that the degree of baseline nonspecific airway responsiveness in asthmatics may predict responsiveness to acid aerosols (Hanley *et al.*, 1992; Utell *et al.*, 1983b), i.e. the response to a constrictor agent was similar to that resulting from exposure to H$_2$SO$_4$, in contrast to the case for SO$_2$ (noted above). But there appears to be no consistent effect of acute exposure to H$_2$SO$_4$ on airway reactivity in either healthy or asthmatic individuals. Other studies have failed to show alterations in reactivity at concentrations up to 2 mg/m^3 with either acidic fog droplets or submicrometer acidic particles (Avol *et al.*, 1988a,b; Linn *et al.*, 1989).

The ability of H$_2$SO$_4$ to alter airway responsiveness has been assessed in a number of studies using various species. While acute exposure of healthy guinea pigs has yielded inconsistent results, with effective concentrations ranging from 0.2 mg/m^3 to >1 mg/m^3 (Kobayashi and Shinozaki, 1993; Silbaugh *et al.*, 1981), chronic exposure of healthy rabbits to 0.25 mg/m^3 (0.3 μm; 1 h/day, 5 days/week for 4 months) has induced nonspecific airway hyperresponsiveness (Gearhart and Schlesinger, 1986). Furthermore, a single exposure of rabbits to 0.075 mg/m^3 H$_2$SO$_4$ was shown to induce non-specific hyperresponsiveness using an *in vitro* airway preparation (El-Fawal and Schlesinger, 1994).

While controlled exposures to acid sulfates have produced transient changes in pulmonary function in asthmatics, including enhanced non-specific airway hyperresponsiveness in some cases, the contribution of chronic ambient sulfur oxide exposure to the development of airway hyperresponsiveness in normal individuals remains unclear. The development of hyperresponsive airways in healthy animals at exposure levels below that producing any change in standard lung function indices may have implications for the pathogenesis of airway disease.

Pulmonary Defenses

As in the case of SO$_2$, sulfuric acid can affect various aspects of the pulmonary defense system. Acute exposure to H$_2$SO$_4$ at levels as low as 0.1 mg/m^3 (0.3–0.6 μm diameter) will alter mucociliary transport in normal humans (Leikauf *et al.*, 1981; Spektor *et al.*, 1989). On the other hand, no special sensitivity of mucociliary transport in asthmatics has been clearly demonstrated (Laube *et al.*, 1993; Spektor *et al.*, 1985). Furthermore, such exposures result in qualitatively similar effects on mucociliary clearance in humans and various animal species; the nature of clearance change, i.e. slowing or speeding, was dependent on exposure concentration, with stimulation of clearance occurring at low concentrations and retardation at higher levels (Schlesinger, 1990). Furthermore, partially or totally neutralized sulfates are less effective than sulfuric acid in altering clearance function (Schlesinger, 1984).

Chronic exposure to acid sulfates also may affect tracheobronchial mucociliary clearance. Rabbits exposed to H_2SO_4 at 0.125–0.25 mg/m³ (0.3 μm diameter; 1–2 h/day, 5 days/week) showed acceleration or retardation of clearance rate, depending upon the concentration (Schlesinger, 1990).

Clearance of particles from the alveolar region is affected by exposure to acid sulfates. Exposure to H_2SO_4 results in either accelerated or retarded particle clearance from the alveolar region, depending upon the exposure regime and animal species examined; effective concentrations range down to 0.25 mg/m³ (Phalen *et al.*, 1980; Schlesinger, 1990). As with other biological end-points, the partially or totally neutralized sulfates are less effective in this regard (Schlesinger, 1989).

At concentrations ≤ 1 mg/m³, sulfuric acid has been shown to affect certain functions of pulmonary macrophages lavaged from various species following single or repeated inhalation exposures. These include phagocytic activity, surface adherence, random mobility, intracellular pH and the release or production of certain cytokines (e.g. TNF-α and IL-1α) and reactive oxygen species (Chen *et al.*, 1995; Schlesinger, 1990; Zelikoff and Schlesinger, 1992; Zelikoff *et al.*, 1994). Such effects may ultimately be reflected in alterations in the ability of these cells to adequately perform their role in host defenses, including particle clearance from the alveolar region and resistance to disease. However, the evidence that sulfuric acid reduces resistance to bacterial infection is conflicting and may depend upon the animal model used (US EPA, 1989; Zelikoff *et al.*, 1994).

As with SO_2, the database for the effects of acid sulfates on specific immunologic lung defenses is limited. There is some evidence to suggest that acid sulfates can enhance sensitization to antigens (Osebold *et al.*, 1980), or that they can modulate the activity of cells involved in allergic responses (Fujimaki *et al.*, 1992).

Pulmonary Morphology and Biochemistry

Acute or chronic exposures to H_2SO_4 at high levels ($\gg 1$ mg/m³) are associated with a number of characteristic responses in animals, e.g. alveolitis, bronchial and/or bronchiolar epithelial desquamation and edema (Brownstein, 1980; Schwartz *et al.*, 1977). However, the sensitivity of morphologic end-points is species-dependent. The rat is apparently the least sensitive and the guinea pig the most sensitive (Cavender *et al.*, 1977; Schwartz *et al.*, 1977; Wolff, 1986). In many cases, the nature of lesions in different species is similar, but the lesions differ in location; this is, perhaps, a reflection of interspecies differences in particle deposition patterns.

Acute exposure to H_2SO_4 at ≤ 1 mg/m³ does not produce evidence of inflammatory responses in humans or animals (Frampton *et al.*, 1992; US EPA, 1989). Although one study indicated a change in airway permeability with exposure of guinea pigs to 0.3 mg/m³ (3 h/day, 1 or 4 days) (Chen *et al.*, 1992), others using concentrations ≥ 1 mg/m³ reported no such effects (Warren and Last, 1987; Wolff *et al.*, 1986). Chronic exposure to ≤ 1 mg/m^3 can produce a response characterized by alterations in the epithelial secretory cells. For example, rabbits exposed to 0.125–0.5 mg/m^3 H$_2$SO$_4$ (0.3 μm diameter) for 1–2 h/day, 5 days/week showed increases in the relative number density of bronchial secretory cells extending to the bronchiolar level (Gearhart and Schlesinger, 1988; Schlesinger et al., 1992a). These changes began within 4 weeks of exposure, and persisted for up to 3 months following the end of all exposure.

MIXTURES CONTAINING SULFUR OXIDES

Ambient air pollution generally consists of complex mixtures rather than one chemical species. A number of controlled exposure studies have indicated that significant toxicological interactions may occur between sulfur oxides and other pollutants, the most notable of which is ozone (Frampton et al., 1995; Linn et al., 1994, 1995; Schlesinger et al., 1992b). However, the nature or extent of interaction appears to depend upon the exposure regimen, biological end-points and the animal species used. No unifying principles or consistent conclusions are evident (Schlesinger, 1995; Schlesinger et al., 1992b).

While the topic of pollutant interactions and complex mixtures is beyond the scope of this chapter, one specific mixture involving sulfuric acid bears mentioning. Within combustion-generated aerosols, sulfuric acid may exist as a coating on the surface of other particles, such as metals or carbon. Studies with guinea pigs suggest that, for comparable biological results, exposure levels of pure acid aerosols need to be an order of magnitude higher than when the acid is coated on a solid particle core (Amdur and Chen, 1989). Exposure to acid present as a coating produced biological effects at concentrations as low as 0.02 mg/m^3 (as H$_2$SO$_4$). While some confounders in these studies include differences in particle sizes between the coated and pure acid droplets and differences in the number concentration of particles in the different exposure atmospheres, it is likely that the physical nature of the inhaled acid sulfate particle is a key factor in determining the ultimate response to exposure.

MECHANISTIC BASES OF SULFUR OXIDE TOXICITY

The exact mechanisms underlying the toxicity of sulfur dioxide or acidic sulfates to the respiratory tract are not known with certainty, but the results of toxicological studies have generated a range of possibilities. Sulfur dioxide-induced apnea is probably due to activation of vagal C fibers and irritant receptors in the airways (Wang et al., 1996). SO$_2$-induced bronchoconstriction is due to contraction of smooth airway muscle via parasympathetic reflex pathways and/or following release of mediators such as histamine and tachykinins (Hajj et al., 1996; Nadel et al., 1965; Sheppard et al., 1980, 1981). This response can be reduced or inhibited by both β-agonists and mast cell stabilizers. It is likely, however, that bronchoconstriction results from more than one event. Direct damage to the airway may result in epithelial denudation and exposure of sensory nerves as well as exaggerated reflex bronchoconstriction upon stimulation. This may be coupled with a loss of barrier function and of epithelium-derived relaxing factor(s) or the involvement of other mediators (Jackson and Eady, 1988; McManus *et al.*, 1989; Norris and Jackson, 1989).

Whatever the mechanism, the reactive chemical species is very likely the bisulfite ion rather than the sulfite ion, and the former has been implicated as a more potent bronchoconstrictor (Fine *et al.*, 1987a). Bisulfite may act by altering acetylcholine receptors via sulfonation of disulfide bonds, thus modulating the contraction of smooth muscle (Atzori *et al.*, 1992).

Local alterations of airway lining fluid pH due to hydrogen ions (H$^+$) generated after SO$_2$ deposition do not seem to be involved in the bronchoconstrictive response to SO$_2$ (Fine *et al.*, 1987a); however, this may be a factor in SO$_2$-induced alterations in mucociliary clearance. *In vitro* exposure of guinea pig trachea to SO$_2$ resulted in an impairment of clearance and an associated change in ciliary beat frequency, suggesting depressed ciliary activity in an acidic environment (Kienast *et al.*, 1993). In most cases, however, ciliary beating is not directly altered by SO$_2$ (Knorst *et al.*, 1994). On the other hand, ciliary morphological changes were observed in humans following acute exposure to SO$_2$ (Carson *et al.*, 1987) and ciliary structural defects or alterations in ciliated epithelium following exposure (Leigh *et al.*, 1992; Man *et al.*, 1989) would probably result in compromised mucociliary function.

Other mechanisms underlying SO$_2$-induced changes in mucociliary transport might include alterations in mucus secretion – the formation of multiple points of adhesion of mucus strands between gland cells and the emergent mucus blanket (Majima *et al.*, 1986); or mucus hypersecretion due to hypertrophy of mucus glands (Scanlon *et al.*, 1987). It is known, for example, that the tracheobronchial epithelium of rats repeatedly exposed to a high concentration of SO$_2$ (1048 mg/m^3 (400 ppm)) shows an increased level of mucin mRNA, indicating a first step in the development of hypersecretion (Basbaum *et al.*, 1990). A change in the thickness of the mucus blanket due to hypersecretion, to hyperplasia of secretory cells or the differentiation of ciliated cells into mucus secreting cells could result in changes in mucociliary clearance function (Basbaum *et al.*, 1990; Fowlie *et al.*, 1990; Lamb and Reid, 1968).

The responses to acidic sulfate particles probably result from the deposition of H$^+$ on airway surfaces (Schlesinger, 1989; Schlesinger and Chen, 1994; US EPA, 1989). Examination of diverse biological end-points (mucociliary transport, alveolar clearance, pulmonary function and production of biological mediators) has shown that the relative potency of acidic sulfate aerosols is directly related to their degree of acidity, i.e. the H$^+$ content within the exposure environment (e.g. Koenig *et al.*, 1993; Schlesinger, 1984, 1989). What is not clear, however, is which metric for H$^+$ concentration better relates to these responses, i.e. whether it is the total available H$^+$ concentration, measured as titratable acidity in lung fluids following deposition, or free H$^+$ concentration, measured by pH (Fine *et al.*, 1987b).

Because the hydrogen ion is likely to be responsible for the toxicity of acid sulfates, the mode of inhalation will affect the potential for a biological response. For the same mass (ionic) concentration of acidic sulfates in an exposure atmosphere, oral inhalation results in more neutralization compared with nasal inhalation and, therefore, less H$^+$ available for deposition within the lower respiratory tract (Larson *et al.*, 1982). The extent of neutralization of inhaled acid sulfates has been shown to modulate toxic effects. Asthmatic subjects inhaling H$_2$SO$_4$ demonstrated greater responsiveness when exposure was conducted under conditions in which oral ammonia levels were low, compared to when they were high (Utell *et al.*, 1989).

If the responses to deposited acidic sulfate particles involve local changes in pH at those sites where these particles land (Hattis *et al.*, 1987; Last *et al.*, 1984), then it is likely that a critical number of such particles must be deposited at such sites in order to deliver sufficient H^+ to alter local pH. It has been demonstrated that the number of H^+-containing particles within an exposure atmosphere (rather than just the total mass concentration of H^+) is an important factor in determining whether any response follows inhalation of acidic sulfates (Chen *et al.*, 1995). Furthermore, there is a threshold for both number concentration and mass concentration.

Sulfuric acid is a bronchoactive agent that can produce constriction of smooth muscle. This results from irritant receptor stimulation following direct contact with deposited acid particles and/or from humoral mediators such as histamine, released as a result of exposure (Charles and Menzel, 1975). Furthermore, since hypo-osmolar aerosols can induce bronchoconstriction in some asthmatics, and acidity may potentiate bronchoconstriction caused by hypo-osmolar particles (Balmes *et al.*, 1988), the molarity of inhaled acid sulfate particles may be a factor in eliciting responses in susceptible individuals.

Acid sulfate-induced airway hyperresponsiveness may involve an increased sensitivity to mediators involved in control of airway smooth muscle tone. Guinea pigs exposed to H_2SO_4 showed a small degree of enhanced response to histamine, but a much more pronounced sensitivity to substance P, a neuropeptide having effects on bronchial smooth muscle (Stengel *et al.*, 1993). Inhalation of H_2SO_4 also has been shown to upset the balance of eicosanoid synthesis/metabolism, which is necessary for the maintenance of pulmonary homeostasis (Schlesinger *et al.*, 1990). Since some eicosanoids are involved in regulation of smooth muscle tone, this imbalance also may be involved in the development of airway hyperresponsivess. For example, incubation of rabbit tracheal explants in acidic media reduced the production of prostaglandin E_2 (Schlesinger *et al.*, 1990), an epithelial-derived mediator associated with bronchodilation and the inhibition of agonist-induced smooth muscle contraction.

Alterations in pulmonary pharmacological receptors may underlie some sulfur oxide-induced responses. In terms of airway hyperresponsiveness, this could involve interference with normal contractile/dilatory processes in the airways via modulation of the receptors involved in maintenance of muscle tone. A potentiation of the response of constrictor receptors and a concomitant diminution of the response of dilatory receptors was noted in the airways of rabbits following exposure to H_2SO_4 (El-Fawal and Schlesinger, 1994).

β-Adrenergic stimulation downregulates pulmonary macrophage function; in rabbits this has been shown to be influenced by short-term repeated exposure to H_2SO_4 (McGovern *et al.*, 1993). Any acid-induced enhanced downregulation – by affecting production of reactive oxygen species by macrophages – may create an environment conducive to secondary pulmonary insult (such as bacterial infections), especially in susceptible populations. In addition, alterations in mediator release from the macrophage due to receptor downregulation may in turn influence airway muscle tone.

The ability of sulfur oxides to affect pharmacological receptors in the airways also was demonstrated by the blunting of β-adrenergic function in human airway epithelial cells exposed *in vitro* to SO_2 (Fine *et al.*, 1991). This was accompanied by a decrease in cAMP in exposed cells. It is likely that the resulting imbalance between cAMP and cGMP (involved in the pharmacological control of airway muscle tone) may underlie any increased spasmogenic activity of bronchoconstrictor agents, such as acetylcholine or histamine, that is observed following sulfur oxide exposure.

As with SO_2, one of the mechanisms underlying acid sulfate-induced exacerbation of asthmatic symptoms may be increased airway epithelial permeability, with subsequent enhanced penetration of inhaled antigens to those submucosal cells involved in allergic reactions. Sulfate particles have been associated with permeability alterations (Kleinman, 1995).

Acidic particles which deposit upon airway surfaces can be buffered in airway fluids, as previously noted. However, the buffering capacity of mucus may be altered in individuals with compromised lungs. For example, sputum from asthmatics (which contains mucus and other fluids) was found to have a reduced buffering capacity compared with that from normals (Holma, 1985). This may, at least in part, underlie some of the former's greater sensitivity to acidic sulfates.

One of the most studied responses to acid sulfates is change in tracheobronchial mucociliary transport. High exposure concentrations of H_2SO_4 are needed to affect ciliary beating (Grose et al., 1980; Schiff et al., 1979), so any effects, especially at lower levels, are likely to be due to changes in the mucus lining itself, such as a change in its rheological properties. At alkaline pH, mucus is more fluid than at acidic pH, so a small increase in viscosity due to deposited acid could 'stiffen' the mucus blanket, altering the efficiency with which it is coupled to the beating cilia (Holma et al., 1977; Knorst et al., 1994).

Another mechanism by which acidic sulfates may affect mucociliary clearance is via alteration in the rate and/or amount of mucus secreted, a phenomenon noted above for SO_2. A small increase in mucus production could facilitate clearance, while more excessive production could result in a thickened mucus layer which would be ineffectively coupled to the cilia. Chronic exposure of animals to sulfuric acid has been shown to increase the number of airway secretory cells, especially in small airways where these cells are normally absent or few in number (Gearhart and Schlesinger, 1988). Furthermore, this increase has been associated with a change in the glycoprotein composition of the mucus, suggesting increased viscosity. The result of any increase in cell number could be an increase in secretory rate or volume. While mucus hypersecretion is a hallmark of chronic bronchitis, its importance in the development of chronic pulmonary disease from air pollution is not clear.

The airways actively transport ions. The interaction between transepithelial ion transport and consequent fluid movement is important in the maintenance of the mucus lining. A change in ion transport due to deposited acid particles may alter the depth and/or composition of the sol layer of the mucus blanket (Nathanson and Nadel, 1984), perhaps affecting clearance rate.

Changes in cell pH after sulfate exposure may be related to a number of physiological alterations. Intracellular pH is one of the major determinants of the rate of many cellular functions. It has been linked to control of vital cellular processes. Alveolar macrophages obtained from guinea pigs exposed to H_2SO_4 showed alterations in internal pH regulation, which was attributable to effects on the Na^+/H^+ exchanger located within the cell membrane (Qu et al., 1993). Deposited acid also may affect the internal pH of epithelial cells and other functions of these cells.

One of the more controversial biological responses to sulfuric acid exposure is the development of cancer. Occupational exposure data has suggested such a relationship

(Beaumont *et al.*, 1987; IARC, 1992; Siemiatycki, 1991; Soskolne *et al.*, 1989). Various potential mechanisms have been proposed as the underlying link to carcinogenesis, e.g. pH modulation of other xenobiotics or low pH-induced changes in cells in terms of mitotic and enzyme regulation (Cookfair *et al.*, 1985). However, the most likely one is irritation, the result of acid-induced chronic inflammation, resulting in increased cell proliferation. The ability of sulfuric acid to act as a tumor promoter was suggested by the assessment of occupational exposure to acid and its interaction with tobacco use (Ichinose and Sagai, 1992) as well as by a study of rats chronically exposed to sulfuric acid and nitrogen dioxide (Frampton *et al.*, 1995). The issue of sulfuric acid carcinogenicity is not resolved, especially since there have been no studies of the carcinogenicity of sulfuric acid in animals. Thus, any conclusions regarding the potential role of sulfuric acid in respiratory tract cancer must await experimental toxicological data.

SUMMARY

The toxicologically significant sulfur oxides are the gaseous sulfur dioxide (SO_2), produced primarily from combustion sources, and sulfuric acid and ammonium bisulfate, the secondarily derived particulate strongly acidic sulfates. Controlled clinical studies (Table 26.1) strongly suggest that asthmatics are more sensitive to SO_2, responding with bronchoconstriction at much lower exposure levels than do normal individuals. There also is some indication that asthmatics, particularly adolescents, are more responsive to acid sulfates as well. Toxicologically, exposure to high levels of SO_2 are needed before there are significant changes in pulmonary defenses, but exposure to lower levels of acidic sulfates will alter mucociliary transport, alveolar clearance of particles and airway reactivity, even in normal individuals. While the exact mechanisms underlying the toxicity of sulfur oxides are not known with certainty, the main response to SO_2, i.e. bronchoconstriction, is due to contraction of airway smooth muscle via the parasympathetic reflex and/or following release of humoral mediators. This is probably due most directly to the bisulfite ion produced upon dissolution of SO_2 in airway fluids. On the other hand, the toxicity of acid sulfates is most likely due to the deposition of H^+ on airway surfaces, and it appears that a threshold exists for both the number of deposited acid particles as well as the mass concentration needed to produce any biological response.

Table 26.1   Responses to sulfur oxides after controlled clinical acute exposures

Response	Sulfur oxide	Minimally effective concentration[a]
Pulmonary mechanical function (including non-specific airway reactivity)	SO_2	> 13.1 mg/m^3 (5 ppm) for normals < 1 mg/m^3 (0.4 ppm) for asthmatics
	H_2SO_4	> 1 mg/m^3 for normals < 0.1 mg/m^3 for asthmatics
Mucociliary transport	SO_2	13.1 mg/m^3 (5 ppm) (nasal passages)
	H_2SO_4	0.1 mg/m^3 (tracheobronchial tree)

[a] Exercise may result in a reduction in the effective level.

REFERENCES

Amdur MO (1966) Respiratory absorption data and SO_2 dose–response curves. *Arch Environ Health* **12**: 729.

Amdur MO and Chen LC (1989) Furnace-generated acid aerosols: speciation and pulmonary effects. *Environ Health Perspect* **79**: 147.

Andersen I, Lundquist GR, Jensen PL and Proctor DF (1974) Human response to controlled levels of sulfur dioxide. *Arch Environ Health* **28**: 31.

Atzori L, Bannenberg G, Corriga AM *et al.* (1992) Sulfur dioxide-induced bronchoconstriction in the isolated perfused and ventilated guinea-pig lung. *Respiration* **59**: 16.

Avol EL, Linn WS, Whynot JD *et al.* (1988a) Respiratory dose–response study of normal and asthmatic volunteers exposed to sulfuric acid aerosol in the sub-micrometer size range. *Toxicol Ind Health* **4**: 173.

Avol EL, Linn WS, Wightman LH *et al.* (1988b) Short-term respiratory effects of sulfuric aicd in fog: a laboratory study of healthy and asthmatic volunteers. *J Air Pollut Control Assoc* **38**: 258.

Balmes JR, Fine JM and Sheppard D (1987) Symptomatic bronchoconstriction after short-term inhalation of sulfur dioxide. *Am Rev Respir Dis* **136**: 1117.

Balmes JR, Fine JM, Christian D *et al.* (1988) Acidity potentiates bronchoconstriction induced by hypoosmolar aerosols. *Am Rev Respir Dis* **138**: 35.

Barnes PJ (1994) Air pollution and asthma. *Postgrad Med J* **70**: 319.

Basbaum C, Gallup M, Gum J *et al.* (1990) Modification of mucin gene expression in the airways of rats exposed to sulfur dioxide. *Biorheology* **27**: 485.

Beaumont JJ, Leveton J, Knox K *et al.* (1987) Lung cancer mortality in workers exposed to sulfuric acid mist and other acid mists. *J Nat Canc Inst* **79**: 911.

Bechtold WE, Waide JJ, Sandström T *et al.* (1993) Biological markers of exposure to SO_2: S-sulfonates in nasal lavage. *J Expos Anal Environ Epidemiol* **3**: 371.

Bethel RA, Erle DJ, Epstein J *et al.* (1983) Effect of exercise rate and route of inhalation on sulfur dioxide-induced bronchoconstriction in asthmatic subjects. *Am Rev Respir Dis* **128**: 592.

Brownstein DG (1980) Reflex-mediated desquamation of bronchiolar epithelium in guinea pigs exposed acutely to sulfuric acid aerosol. *Am J Pathol* **98**: 577.

Carson JL, Collier AM, Hu S-C *et al.* (1987) The appearance of compound cilia in the nasal mucosa of normal human subjects following acute, *in vivo* exposure to sulfur dioxide. *Environ Res* **42**: 155.

Cavender FL, Steinhagen WH, Ulrich CE *et al.* (1977) Effects in rats and guinea pigs of short-term exposure to sulfuric acid mist, ozone, and their combination. *J Toxicol Environ Health* **3**: 521–000.

Charles JM and Menzel DB (1975) Ammonium and sulfate ion release of histamine from lung fragments. *Arch Environ Health* **30**: 314.

Chen LC, Miller PD, Amdur MO and Gordon T (1992) Airway hyperresponsiveness in guinea pigs exposed to acid-coated ultrafine particles. *J Toxicol Environ Health* **35**: 165.

Chen LC, Wu CY, Qu QS and Schlesinger RB (1995) Number concentration and mass concentration as determinants of biological response to inhaled irritant particles. *Inhal Toxicol* **7**: 577–588.

Cookfair D, Wende K, Michalek A and Vena J (1985) A case–control study of laryngeal cancer among workers exposed to sulfuric acid. *Am J Epidemiol* **122**: 521.

Dalhamn T and Strandberg L (1961) Acute effect of sulfur dioxide on rate of ciliary beat in trachea of rabbit *in vivo* and *in vitro*, with studies on absorptional capacity of nasal cavity. *Int J Air Water Pollut* **4**: 154.

El-Fawal HAN and Schlesinger RB (1994) Nonspecific airway hyperresponsiveness induced by inhalation exposure to sulfuric acid aerosol: an *in vitro* assessment. *Toxicol Appl Pharmacol* **125**: 70.

Ferin J and Leach LJ (1973) The effect of SO_2 on lung clearance of TiO_2 particles in rats. *Am Ind Hyg Assoc J* **34**: 260.

Fine JM, Gordon T and Sheppard D (1987a) The roles of pH and ionic species in sulfur dioxide- and sulfite-induced bronchoconstriction. *Am Rev Respir Dis* **136**: 1122.

Fine JM, Gordon T, Thompson JE and Sheppard D (1987b) The role of titratable acidity in acid aerosol-induced bronchoconstriction. *Am Rev Respir Dis* **135**: 826.

Fine JM, Chen LC, Finkelstein M *et al.* (1991) Effects of SO_2 on cAMP levels in human airway epithelial cells. *Am Rev Respir Dis* **143**: A489.

Fowlie AJ, Grasso P and Benford DJ (1990) The short-term effects of carcinogens and sulphur dioxide on the nuclear size of rat nasal epithelial cells. *J Appl Toxicol* **10**: 29.

Frampton MW, Voter KZ, Morrow PE *et al.* (1992) Effects of H₂SO₄ aerosol exposure in humans assessed by bronchoalveolar lavage. *Am Rev Respir Dis* **146**: 626.

Frampton MW, Morrow PE, Cox C *et al.* (1995) Sulfuric acid aerosol followed by ozone exposure in healthy and asthmatic subjects. *Environ Res* **69**: 1–14.

Frank NR, Yoder RE, Brain JD and Yokoyama E (1969) SO₂ (^{35}S labeled) absorption by the nose and mouth under conditions of varying concentration and flow. *Arch Environ Health* **18**: 315–000.

Fujimaki H, Katayama N and Wakamori K (1992) Enhanced histamine release from lung mast cells of guinea pigs exposed to sulfuric acid aerosols. *Environ Res* **58**: 117.

Gearhart JM and Schlesinger RB (1986) Sulfuric-acid induced airway hyper-responsiveness. *Fundam Appl Toxicol* **7**: 681.

Gearhart JM and Schlesinger RB (1988) Response of the tracheobronchial mucociliary clearance system to repeated irritant exposure: effect of sulfuric acid mist on function and structure. *Exp Lung Res* **14**: 587.

Giddens WE and Fairchild GA (1972) Effects of sulfur dioxide on the nasal mucosa of mice. *Arch Environ Health* **25**: 166.

Gong Jr H, Lachenbruch PA, Harber P and Linn WS (1995) Comparative short-term health responses to sulfur dioxide exposure and other common stressors in a panel of asthmatics. *Toxicol Ind Health* **11**: 467–487.

Grose EC, Gardner DE and Miller FJ (1980) Response of ciliated epithelium to ozone and sulfuric acid. *Environ Res* **22**: 377.

Gunnison AF, Sellakumar A, Currie D and Snyder EA (1987) Distribution, metabolism and toxicity of inhaled sulfur dioxide and endogenously generated sulfite in the respiratory tract of normal and sulfite oxidase-deficient rats. *J Toxicol Environ Health* **21**: 141.

Hackney JD, Linn WS, Bailey RM *et al.* (1984) Time course of exercise-induced bronchoconstriction in asthmatics exposed to sulfur dioxide. *Environ Res* **34**: 321.

Hajj AM, Burki NK and Lee LY (1996) Role of tachykinins in sulfur dioxide-induced broncho-constriction in anesthetized guinea pigs. *J Appl Physiol* **80**: 2044.

Hanley QS, Koenig JQ, Larson TV *et al.* (1992) Response of young asthmatic patients to inhaled sulfuric acid. *Am Rev Respir Dis* **145**: 326.

Hattis D, Wasson JM, Page GS *et al.* (1987) Acid particles and the tracheobronchial region of the respiratory system: an 'Irritation-Signaling' model for possible health effects. *J Air Pollut Control Assoc* **37**: 1060.

Hazucha MJ, Kehrl HR, Roger LJ and Horstman DH (1984) Airway responsiveness to methacholine of asthmatics exposed to 0.25, 0.5, and 1.0 ppm SO₂. *Am Rev Respir Dis* **129**: A145.

Hirsch JA, Swenson EW and Wanner A (1974) Tracheal mucus transport in beagles after long-term exposure to 1 ppm sulfur dioxide. *Arch Environ Health* **30**: 249.

Holma B (1967) Lung clearance of mono- and di-dispersed aerosols determined by profile scanning and whole body counting. *Acta Med Scand* **473**(Suppl 1): 1.

Holma B (1985) Influence of buffer capacity and pH-dependent rheological properties of respiratory mucus on health effects due to acidic pollution. *Sci Total Environ* **41**: 101.

Holma B, Lindegren M and Andersen JM (1977) pH effects on ciliomotility and morphology of respiratory mucosa. *Arch Environ Health* **32**: 216.

Horstman DH, Seal Jr E, Folinsbee LJ *et al.* (1988) The relationship between exposure duration and sulfur dioxide-induced bronchoconstriction in asthmatic subjects. *Am Ind Hyg Assoc J* **49**: 38.

IARC (1992) Monograph for occupational exposures to mists and vapors from strong inorganic acids and other industrial chemicals, **54**: 89. International Agency for Research on Cancer, Lyon, France.

Ichinose T and Sagai M (1992) Combined exposure to nitrogen dioxide, ozone and sulfuric acid-aerosol and lung tumor formation in rats. *Toxicology* **74**: 173.

Jackson DM and Eady RP (1988) Acute transient SO₂-induced airway hyperreactivity: effects of nedocromil sodium. *J Appl Physiol* **65**: 1119.

Katz GV and Laskin S (1971) Effect of irritant atmospheres on macrophage behavior. In: Sanders C *et al.* (eds) *Pulmonary Macrophages and Epithelial Cells*, Springfield, VA: NTIS, p 358.

Kehrl HR, Roger LJ, Hazucha MJ and Horstman DH (1987) Differing response of asthmatics to SO_2 exposure with continuous and intermittent exercise. *Am Rev Respir Dis* **135**: 350.

Kienast K, Müller-Quernheim J, Knorst M *et al.* (1993) Reality-related *in-vitro* study on reactive oxygen-intermediates release by alveolar macrophages and peripheral blood mononuclear cells after short-term exposures with SO_2. *Pneumologie* **47**: 60.

Kleinman MT (1995) Cellular and immunologic injury with PM10 inhalation. *Inhal Toxicol* **7**: 589–602.

Knorst MM, Kienast K, Riechelman H *et al.* (1994) *In-vitro* evaluation of alterations in mucociliary clearance of guinea-pig tracheas induced by sulfur dioxide or nitrogen dioxide. *Pneumologie* **48**: 443.

Knorst MM, Kienast K, Gross S *et al.* (1996a) Chemotactic response of human alveolar macrophages and blood monocytes elicited by exposure to sulfur dioxide. *Res Exp Med* **196**: 127–135.

Knorst MM, Kienast K, Muller-Quernheim J and Ferlinz R (1996b) Effect of sulfur dioxide on cytokine production of human alveolar macrophages in vitro. *Arch Environ Health* **51**: 150–156.

Kobayashi T and Shinozaki Y (1993) Effects of exposure to sulfuric acid-aerosol on airway responsiveness in guinea pigs: concentration and time dependency. *J Toxicol Environ Health* **39**: 261.

Koenig JQ, Pierson WE and Horike M (1983) The effects of inhaled sulfuric acid on pulmonary function in adolescent asthmatics. *Am Rev Respir Dis* **128**: 221.

Koenig JQ, Covert DS, Larson TV and Pierson WE (1992) The effect of duration of exposure on sulfuric acid-induced pulmonary function changes in asthmatic adolescent subjects: a dose–response study. *Toxicol Ind Health* **8**: 285.

Koenig JQ, Dumler K, Rebolledo V *et al.* (1993) Respiratory effects of inhaled sulfuric acid on senior asthmatics and nonasthmatics. *Arch Environ Health* **48**: 171.

Lamb D and Reid L (1968) Mitotic rates, goblet cell increase, and histochemical changes in mucus in rat bronchial epithelium during exposure to sulphur dioxide. *J Pathol Bacteriol* **96**: 97.

Larson TV, Covert D, Frank R and Charlson R.J (1977) Ammonia in the human airways: neutralization of inspired acid sulfate aerosols. *Science* **197**: 161.

Larson TV, Frank R, Covert DS *et al.* (1982) Measurements of respiratory ammonia and the chemical neutralization of inhaled sulfuric acid aerosol in anesthetized dogs. *Amer Rev Resp Dis* **125**: 502.

Larson TV, Hanley QS, Koenig JQ and Bernstein O (1993) Calculation of acid aerosol dose. In: Mohr U (ed.) *Advances in Controlled Clinical Inhalation Studies.* Berlin: Springer-Verlag, pp 109–121.

Last JA, Hyde DM and Chang DPY (1984) A mechanism of synergistic lung damage by ozone and a respirable aerosol. *Exp Lung Res* **7**: 223.

Laube BL, Bowes III SM, Links JM *et al.* (1993) Acute exposure to acid fog: effects on mucociliary clearance. *Am Rev Respir Dis* **147**: 1105.

Leigh MW, Carson JL, Gambling TM and Boat TF (1992) Loss of cilia and altered phenotypic expression of ciliated cells after acute sulfur dioxide exposure. *Chest* **101**(Suppl 3): 16S.

Leikauf G, Yeates DB, Wales KA *et al.* (1981) Effects of sulfuric acid aerosol on respiratory mechanics and mucociliary particle clearance in health nonsmoking adults. *Am Ind Hyg Assoc* **42**: 273.

Linn WS, Avol EL, Peng R-C *et al.* (1987) Replicated dose–response study of sulfur dioxide effects in normal, atopic, and asthmatic volunteers. *Am Rev Respir Dis* **136**: 1127.

Linn WS, Avol EL, Anderson KR *et al.* (1989) Effect of droplet size on respiratory responses to inhaled sulfuric acid in normal and asthmatic volunteers. *Am Rev Respir Dis* **140**: 161.

Linn WS, Shamoo DA, Anderson KR *et al.* (1994) Effects of prolonged, repeated exposures to ozone, sulfuric acid, and their combination in healthy and asthmatic volunteers. *Am J Respir Crit Care Med* **150**: 431.

Linn WS, Anderson KR, Shamoo DA *et al.* (1995) Controlled exposures of young asthmatics to mixed oxidant gases and acid aerosol. *Am J Resp Crit Care Med* **152**: 885–891.

Majima Y, Okuyama H and Bang BG (1986) Effect of SO_2 on nasal secretory cells. *Acta Otolaryngol (Stockh)* **102**: 302.

Man SFP, Hulbert WC, Man G *et al.* (1989) Effects of SO_2 exposure on canine pulmonary epithelial functions. *Exp Lung Res* **15**: 181.

Martonen TB and Zhang Z (1993) Deposition of sulfate aerosols in the developing human lung. *Inhal Toxicol* **5**: 165.

McGovern TJ, Schlesinger RB and El-Fawal HAN (1993) Effect of repeated *in vivo* ozone and/or sulfuric acid exposures on β-adrenergic modulation of macrophage function. *Toxicologist* **13**: 49.

McManus MS, Altman LC, Koenig JQ *et al.* (1989) Human nasal epithelium: characterization and effects of *in vitro* exposure to sulfur dioxide. *Exp Lung Res* **15**: 849.

Morrow PE (1986) Factors determining hygroscopic aerosol deposition in airways. *Physiol Rev* **66**: 330.

Nadel JA, Salem H, Tamplin B and Tokiwa Y (1965) Mechanism of bronchoconstriction during inhalation of sulfur dioxide. *J Appl Physiol* **20**: 164.

Nathanson I and Nadel JA (1984) Movement of electrolytes and fluid across airways. *Lung* **162**: 125.

Newhouse MT, Dolovich M, Obminski G and Wolff RK (1978) Effect of TLV levels of SO$_2$ and H$_2$SO$_4$ on bronchial clearance in exercising man. *Arch Environ Health* **33**: 24.

Norris AA and Jackson DM (1989) Sulphur dioxide-induced airway hyperreactivity and pulmonary inflammation in dogs. *Agents Actions* **26**: 360.

Osebold JW, Gershwin LJ and Zee YC (1980) Studies on the enhancement of allergic lung sensitization by inhalation of ozone and sulfuric acid aerosol. *J Environ Pathol Toxicol* **3**: 221.

Phalen RF, Kenoyer JL, Crocker TT and McClure TR (1980) Effects of sulfate aerosols in combination with ozone on elimination of tracer particles by rats. *J Toxicol Environ Health* **6**: 797.

Qu QS, Chen LC, Gordon T *et al.* (1993) Alteration of pulmonary macrophage intracellular pH regulation by sulfuric acid aerosol exposures. *Toxicol Appl Pharmacol* **121**: 138.

Riechelmann H, Maurer J, Kienast K *et al.* (1995) Respiratory epithelium exposed to sulfur dioxide – functional and ultrastructural alterations. *Laryngoscope* **105**(3 Pt 1): 295–299.

Riedel F, Krämer M, Scheibenbogen C and Rieger CHL (1988) Effects of SO$_2$ exposure on allergic sensitization in the guinea pig. *J Allergy Clin Immunol* **82**: 527.

Sandström T, Stjernberg N, Andersson M-C *et al.* (1989) Cell response in bronchoalveolar lavage fluid after exposure to sulfur dioxide: a time-response study. *Am Rev Respir Dis* **140**: 1828.

Scanlon PD, Seltzer J, Ingram Jr RH *et al.* (1987) Chronic exposure to sulfur dioxide: physiologic and histologic evaluation of dogs exposed to 50 or 15 ppm. *Am Rev Respir Dis* **135**: 831.

Schiff LJ, Bryne MM, Fenters JD *et al.* (1979) Cytotoxic effects of sulfuric acid mist, carbon particulates and their mixtures on hamster tracheal epithelium. *Environ Res* **19**: 339.

Schlesinger RB (1984) Comparative irritant potency of inhaled sulfate aerosols – effects on bronchial mucociliary clearance. *Environ Res* **34**: 268.

Schlesinger RB (1989) Factors affecting the response of lung clearance systems to acid aerosols: role of exposure concentration, exposure time, and relative acidity. *Environ Health Perspect* **79**: 121.

Schlesinger RB (1990) The interaction of inhaled toxicants with respiratory tract clearance mechanisms. *Crit Rev Toxicol* **20**: 257.

Schlesinger RB (1995) The interaction of gaseous and particulate pollutants in the respiratory tract: mechanisms and modulators. *Toxicology* **105**: 315–325.

Schlesinger RB and Chen LC (1994) Comparative biological potency of acidic sulfate aerosols: implications for the interpretation of laboratory and field studies. *Environ Res* **65**: 69.

Schlesinger RB, Gunnison AF and Zelikoff JT (1990) Modulation of pulmonary eicosanoid metabolism following exposure to sulfuric acid. *Fundam Appl Toxicol* **15**: 151.

Schlesinger RB, Gorczynski JE, Dennison J *et al.* (1992a) Long-term intermittent exposure to sulfuric acid aerosol, ozone, and their combination: alterations in tracheobronchial mucociliary clearance and epithelial secretory cells. *Exp Lung Res* **18**: 505.

Schlesinger RB, Zelikoff JT, Chen LC and Kinney PL (1992b) Assessment of toxicologic interactions resulting from acute inhalation exposure to sulfuric acid and ozone mixtures. *Toxicol Appl Pharmacol* **115**: 183.

Schwartz LW, Moore PF, Chang DP *et al.* (1977) Short-term effects of sulfuric acid aerosols on the respiratory tracts. A morphological study in guinea pigs, mice, rats and monkeys. In: Lee SD (ed.) *Biochemical Effects of Environmental Pollutants*. Ann Arbor: Arbor Science, pp 257–271.

Sheppard D (1988) Sulfur dioxide and asthma – a double-edged sword? *J Allergy Clin Immunol* **82**: 961.

Sheppard D, Wong WS, Uehara CF *et al.* (1980) Lower threshold and greater bronchomotor responsiveness of asthmatic subjects to sulfur dioxide. *Am Rev Respir Dis* **122**: 873.

Sheppard D, Nadel JA and Boushey HA (1981) Inhibition of sulfur dioxide-induced bronchoconstriction by disodium cromoglycate in asthmatic subjects. *Am Rev Respir Dis* **124**: 257.

Sheppard D, Epstein J, Bethel RA *et al.* (1983) Tolerance to sulfur dioxide-induced bronchoconstriction in subjects with asthma. *Environ Res* **30**: 412.

Sheppard D, Eschenbacher WL, Boushey HA and Bethel RA (1984) Magnitude of the interaction between the bronchomotor effects of sulfur dioxide and those of dry (cold) air. *Am Rev Respir Dis* **130**: 52.

Siemiatycki J (ed.) (1991) *Risk Factors for Cancer in the Workplace.* Boca Raton, FL: CRC Press, p 156.

Silbaugh SA, Mauderly JL and Macken CA (1981) Effects of sulfuric acid and nitrogen dioxide on airway responsiveness of the guinea pig. *J Toxicol Environ Health* **8**: 31.

Skornik WA and Brain JD (1990) Effect of sulfur dioxide on pulmonary macrophage endocytosis at rest and during exercise. *Am Rev Respir Dis* **142**: 655.

Soskolne CL, Pagano G, Cipollaro M *et al.* (1989) Epidemiologic and toxicologic evidence for chronic health effects and the underlying biologic mechanisms involved in sub-lethal exposures to acidic pollutants. *Arch Environ Health* **44**: 180.

Speizer FE and Frank NR (1966) The uptake and release of SO_2 by the human nose. *Arch Environ Health* **12**: 725.

Spektor DM, Leikauf GD, Albert RE and Lippmann M (1985) Effects of submicrometer sulfuric acid aerosols on mucociliary transport and respiratory mechanics in asymptomatic asthmatics. *Environ Res* **37**: 174.

Spektor DM, Yen BM and Lippmann M (1989) Effect of concentration and cumulative exposure of inhaled sulfuric acid on tracheobronchial particle clearance in healthy human. *Environ Health Perspect* **79**: 167.

Spiegelman JR, Hanson GD, Lazarus JA *et al.* (1968) Effect of acute SO_2 exposure on bronchial clearance in the donkey. *Arch Environ Health* **17**: 321.

Stengel PW, Bendele AM, Cockerham SL and Silbaugh SA (1993) Sulfuric acid induces airway hyperresponsiveness to substance P in the guinea pig. *Agents Actions* **39**: C128.

Strandberg LG (1964) SO_2 absorption in the respiratory tract. *Arch Environ Health* **9**: 160.

US EPA (US Environmental Protection Agency) (1982) *Air Quality Criteria for Particulate Matter and Sulfur Oxides.* Office of Health and Environmental Assessment, Environmental Criteria and Assessment Office, Research Triangle Park, NC, EPA-600/8-82-029aF-cF. Springfield, VA: NTIS, PB84-156777.

US EPA (US Environmental Protection Agency) (1986) *Second Addendum to Air Quality Criteria for Particulate Matter and Sulfur Oxides (1982): Assessment of Newly Available Health Effects. Information.* Office of Health and Environmental Assessment, Environmental Criteria and Assessment Office, Research Triangle Park, NC, EPA-600/8-86-020F. Springfield, VA: NTIS, PB87-176574.

US EPA (US Environmental Protection Agency) (1989) *An Acid Aerosols Issue Paper: Health Effects And aerometrics.* Office of Research and Development, Research Triangle Park, NC, EPA-600/8-88-005f.

Utell MJ, Morrow PE and Hyde RW (1983a) Latent development of airway hyperreactivity in human subjects after sulfuric acid aerosol exposure. *J Aerosol Sci* **14**: 202.

Utell MJ, Morrow PE, Speers DM *et al.* (1983b) Airway responses to sulfate and sulfuric acid aerosols in asthmatics: an exposure–response relationship. *Am Rev Respir Dis* **128**: 444.

Utell MJ, Mariglio JA, Morrow PE *et al.* (1989) Effects of inhaled acid aerosols on respiratory function: the role of endogenous ammonia. *J Aerosol Med* **2**: 141.

Wang AL, Blackford TL and Lee LY (1996) Vagal bronchopulmonary C-fibers and acute ventilatory response to inhaled irritants. *Respir Physiol* **104**: 231–239.

Warren DL and Last JA (1987) Synergistic interaction of ozone and respirable aerosols on rat lungs. III. Ozone and sulfuric acid aerosol. *Toxicol Appl Pharmacol* **88**: 203.

Wolff. K (1986) Effects of airborne pollutants on mucociliary clearance. *Environ Health Perspect* **66**: 223.

Wolff RK, Dolovich M, Rossman CM and Newhouse MT (1975) Sulfur dioxide and tracheobronchial clearance in man. *Arch Environ Health* **30**: 521.

Wolff RK, Henderson RF, Gray RH *et al.* (1986) Effects of sulfuric acid mist inhalation on mucous clearance and on airway fluids of rats and guinea pigs. *J Toxicol Environ Health* **17**: 129.

Zelikoff JT and Schlesinger RB (1992) Modulation of pulmonary immune defense mechanisms by sulfuric acid: effects on macrophage-derived tumor necrosis factor and superoxide. *Toxicology* **76**: 271.

Zelikoff JT, Sisco MP, Yang Z *et al.* (1994) Immunotoxicity of sulfuric acid aerosol: effects on pulmonary macrophage effector and functional activities critical for maintaining host resistance against infectious diseases. *Toxicology* **92**: 269.

27

Acid Sulfate Aerosols and Health

FRANK E. SPEIZER

Brigham and Women's Hospital, Harvard Medical School and
Harvard School of Public Health, Boston, MA, USA

INTRODUCTION

In this chapter we review what is believed to be the health impact of acid aerosols produced as secondary pollutants resulting from the combustion of fossil fuels. After briefly reviewing the history of concern about acid aerosols, issues related to the measurement of acids in the air and evidence relating acid particulate exposure and health outcomes will be considered. Clearly there is a lack of sufficient data to establish firmly that acid aerosols produce permanent health damage; however, what is not so well understood is whether this lack of sufficient data relates to an inadequate number of investigations or to the fact that some of the data are conflicting. The chapter will end with recommendations for further research that may help to clarify the understanding of the potential impact of these common and seemingly increasingly widespread pollutants.

The concern about acid aerosols is not new and was certainly hinted at in the last quarter of the nineteenth century. Reports of fogs in London and in other parts of England exacerbated by the burning of fossil fuels appear in literature produced from the fourteenth century right up to modern times. Throughout the nineteenth century measures of the acidity of rainwater over large tracts of land were carried out. The concern related to understanding the potential impact of rainwater acidity on agricultural productivity. The first measures of the chemical composition of rainwater in metropolitan London came in the last quarter of the nineteenth century with the works of Smith (1872) and Russell (1884). By comparing data collected from monthly samples in sites in Central London with data from the less dense portions of the Greater London area, Russell documented approximately a three-fold gradient for chlorides and a two-fold gradient for sulfates in specimens obtained in a uniform manner.

No other significant number of measurements were made until approximately 1910

AIR POLLUTION AND HEALTH
ISBN 0-12-352335-4

when, under the auspices of the Coal Smoke Abatement Society, a Committee for the Investigation of Atmospheric Pollution was established in England. This Committee commissioned a network of deposition gauges that provided a continuous monthly record of undissolved solid particles and sulfates in London until the mid-1970s. However, it was soon realized that these techniques measured precipitation rather than actual air concentrations, and efforts to develop a real air concentration particle sampler began almost immediately. By the early 1920s, the smoke shade device that produced the values of 'British (Black) Smoke' were developed, and these measures became the standard of comparison until the mid-1960s.

It is not appropriate to explore the details of the development of instrumentation for measuring acid aerosols in this chapter. However, it is important to point out that, although these early measures estimated acidity as sulfate, the subsequent measures used throughout the world measured particle mass. It was not until the late 1980s that real mass/acidic measures were made (Koutrakis *et al.*, 1988; Pierson *et al.*, 1989). This delay related in part to the difficulties of collecting and preserving samples that would not become neutralized after collection. A further difficulty related to the fact that the studies of health impacts that had been proposed and carried out were not designed to deal appropriately with the complexities of the chemical composition of air pollutants, particularly as the primary pollutants were transformed into secondary acid aerosols by long-range transport, aging and chemical changes in the atmosphere. Thus, many of the early studies use measurements of levels of exposure from which investigators in later years were forced to infer the acid content of the putative air mass episodes. Few of the later studies use actual direct acid aerosol measures. Even in these studies, the authors generally admit that they are dealing with complex air mass mixtures of which acid aerosols are only one component.

In the USA, at the time of the Clean Air Act in 1970, sulfur dioxide (SO_2) was recognized as a primary pollutant resulting from direct emissions from the relatively short stacks associated with the production of power from the burning of fossil fuels. It was only with the response to limiting the production of SO_2 from local sources that the technique of using tall stacks to reduce local ambient air concentrations was introduced. This procedure resulted in emissions reaching higher altitudes and having a longer residence time in air. With longer transit times and higher altitudes, SO_2 is oxidized to sulfuric acids (Calvert *et al.*, 1985), with the resulting acid aerosol and acid rain production.

THE 'MODERN' EARLY STUDIES

Although deaths by week had been recorded since the mid-seventeenth century in the City of London and occasionally linked to what were variously called 'great stinking fogs' or the 'horrid smoke', it was not until the early 1930s that scientific concern about the health effects of acid air pollution health effects were formally presented. Firket (1931, 1936) reported on a five-day event that occurred in December 1930, in a narrow industrial valley of Belgium. Sixty people died after only a few hours of illness, a rate of mortality more than 10 times what would have been expected. During the episode many people developed respiratory symptoms of coughing and shortness of breath that also were accompanied by cardiovascular symptoms. Fifteen autopsies were performed that showed widespread congestion and irritation of the mucus membranes of the respiratory tract, along with fine particles of soot found deep in the lung. It is interesting to note that many cattle died or had to be slaughtered, while those that were saved had been driven up the hillside. Although no air pollution measures were made, because of the nature of the industrial emissions in the valley and the weather during the episode, Firket concluded that the fog must have been partially transformed into sulfuric acid aerosol. In his presentation to the Faraday Society in 1936, he further suggested that, if such a fog were to occur in the City of London, one could expect some 3200 extra deaths.

Firket's prediction proved prophetic when London experienced a severe stagnant air mass during the early weeks of December, 1952. Measures of sulfur dioxide and smoke in Kew Gardens and in other parts of London indicated a remarkable and persistent increase in pollution that correlated significantly with an excess in mortality that reached a peak on the fourth day of pollution (Wilkins, 1954). Increases in sudden death from respiratory and cardiac events were reported and there was widespread evidence of irritation of the respiratory tract. Hospital reports indicated frequent complaints of bronchial symptoms including cough and wheeze, as well as shortness of breath in many patients. Martin (1964) calculated the cumulative mortality during the episode. He confirmed Firket's estimates by suggesting that approximately 4000 extra deaths occurred during and following the episode. Again, no direct measures of acid aerosols were made; however, Waller (1963) later reported that sulfuric acid was one of the pollutants that may have been present. As in the Belgian incident, cattle were also affected. Many prize cattle entered in the Smithfield Fair in London were affected with respiratory symptoms that required veterinary treatment, and several had to be destroyed. Less prized animals in the Fair were not as severely affected, and it was later speculated that their pens were not cleaned as often and thus the ammonia from decomposing urine and feces neutralized the potential effect of the acid aerosols.

Other episodes throughout the industrialized world appear to have occurred from the late 1940s to the mid-1960s (Shy *et al.*, 1978). However, in each of these episodes no direct measures of aerosol acidity were available. In addition, what was clear in each episode was that the changes in air pollution were related to stagnation of the air mass in regions in which industrial or environmental sources were normally present.

MORE RECENT POPULATION-BASED ECOLOGIC AND COHORT STUDIES OF MORTALITY

Several analyses of routinely collected daily mortality data together with daily air pollution measures suggest a relationship between particle exposure and mortality. The US Environmental Protection Agency (US EPA) summarized a series of analyses that used mortality data from Standard Metropolitan Statistical Areas (SMSAs) for the period 1960–80 (see Table 27.1). These analyses supported previous findings of an association between mortality and total suspended particulates. However, because of what are considered relatively flawed measures of sulfates used at the time, as well as collinearity of the measures of particles and sulfates, few of these analyses showed a significant relationship between mortality and sulfates.

Table 27.1 Selected ecologic analyses of mortality data showing the relationship between
particulate/sulfate and mortality

Reference	Study description	Results and comments
Lave and Seskin (1977)	Total mortality, 1960, 117 US SMSAs	Combined effects of TSP and SO_4, dependent on how potential differences between communities considered
Chappie and Lave (1982)	Total mortality, 1974, 104 US SMSAs	No effect of combined TSP and SO_4
Lipfert (1984)	Total mortality, 1970, 69–111 US SMSAs (modeled as above)	TSP or SO_4, but not both in same model with borderline significance
Thurston *et al.* (1989)	Mortality in London, 1963–72 winters mean = 28 deaths/day	Excess mortality associated with PM, SO_2, H_2SO_4
Ito *et al.* (1993)	Further analysis of Thurston *et al.* data	Mean pollution effect of 2–7% but independent effect of SO_4 not detected
Spix *et al.* (1994)	Mortality in Erfurt, E. Germany, 1980–89, median = 6 deaths/day	SO_2 and particles (SP) associated with increased mortality; no acid or SO_4 measures
Özkaynak and Thurston (1987)	Total daily mortality in 1980, 98 US SMSAs	Significant correlations with both PM measures and SO_4; SO_4 measures questioned
Lipfert *et al.* (1988)	Total daily mortality in 1980, 122 US cities	No effect of TSP when SO_4 significant
Lipfert (1993)	Various cause mortality, 1980, 62–149 US SMSAs	Results depended on cause studied as well as how communities are modeled for potential differences

SMSA, Standard Metropolitan Statistical Areas
TSP, total suspended particulates
PM, particulate matter

Various estimates suggest that some 60% or more of the particle loads are sulfate based and could be considered as a surrogate for acid aerosols (Özkaynak *et al.*, 1996). Özkaynak and co-workers using pooled data from measurements made in several studies during the early to mid-1990s, showed that once population density was taken into account, relatively high correlations between levels of hydrogen ion acidity and sulfates were present throughout the USA, with the lowest correlations seen in the West (Fig. 27.1). Although crude, these correlations offer an opportunity to reassess much of the retrospective epidemiologic data and, as pointed out by Lippmann and Thurston (1996), can also be used, after adjustment for population density, as a measure of fine particles, defined as particles with a mass median diameter of 2.5 μm or less ($PM_{2.5}$). To date, such analyses have not been undertaken except in a very few instances.

An assessment that attempted to use such data was undertaken in the Harvard Six Cities Study by Dockery *et al.* (1993). The initial assessments of mortality by city using

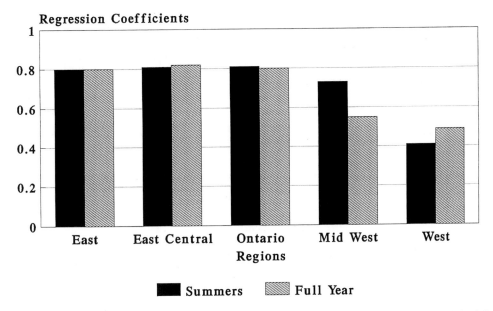

Fig. 27.1   Regression coefficients for H^+ versus SO_4 by regions combined over 30 sites for summers and for full year. Data from Özkaynak *et al.* (1996).

TSP or PM_{10} resulted in positive but variable dose–response relations. These measures were averaged over the years of the actual mortality assessment. If the TSP data available for the years before the follow-up period were used, then the associations were even stronger and less variable. Using the measures of $PM_{2.5}$ and sulfate available for a short portion of the period of observation gave an even stronger dose–response relation. However, in spite of these associations, no direct association was seen for hydrogen ion measures, although these were only measured for approximately 2 years during the course of the study.

The results of Six Cities Study led to a specific test of the hypothesis that sulfates were a surrogate of fossil fuel combustion and could be used to assess mortality over a wide range of regional differences. To test this hypothesis, Pope *et al.* (1995) carried out a similar analysis using the American Cancer Society database on approximately 500 000 volunteers living in 151 communities in which sulfate data could be derived from the national sulfate monitoring data network. The mortality effects noted were similar to those seen in the Harvard study, albeit over a wider range of sulfate levels. Again, no specific hydrogen ion data were available, and the results, although consistent with an acid aerosol effect, cannot be said to be more than suggestive.

These data on long-term and population-based annual assessments of mortality that compare different communities are only weakly suggestive of an acid aerosol effect. They are supplemented by an ever-increasing number of studies of daily mortality within communities, although these still are not specific for acid aerosol. In these studies daily variation in pollution levels as indicated by series of surrogates of air contaminated by products of local fossil fuel consumption (sources of NO_2, SO_2 and particles), as well as long-range transport (more likely sources for sulfates, acids and accumulation mode particles) find varying degrees of association with daily mortality. These studies have undergone intense scrutiny, and in general confirm that the major common theme of the associations found is in some way related to the smaller particles in the air. Except as indicated above, this does not mean that all these particles are in fact acid aerosols. Additional studies specifically designed to understand the potential direct effects of acids are warranted.

STUDIES OF MORBIDITY IN ASSOCIATION WITH ACID AEROSOLS

Although relatively few in number and in some ways inconsistent with some of the clinical studies, several investigations suggest that the chemistry of particulate load, and quite possibly the acidity, play an important role in the putative toxicity of ambient particles. Because measures of acid aerosols were not readily available before 1990, sulfate data have been used as surrogates for acid aerosols. Hospital admission data for large regions of southern Canada from exposures to long-range transport of sulfate air masses from the upper Midwest and northeastern parts of North America have suggested a gradient associated with sulfate levels (Bates and Sizto, 1983, 1987; Burnett et al., 1994). In these studies the authors suggested that the excess admissions for respiratory disease noted in summers in 79 hospitals in southern Ontario could not be attributed to either ozone or sulfate, but were most likely related to a pollutant that would travel in tandem with these pollutants – probably strong acid aerosols. This hypothesis was supported at the time by some anecdotal data suggesting that, while ozone and sulfates were present at normal levels over the summer, there were episodes of elevated hydrogen ion concentration that could account for the effects seen.

This Canadian proposal was supported by further work of Burnett and colleagues (1994), who extended these observations to 168 acute care hospitals in all of Ontario and used daily measures of ozone and sulfates recorded at nine monitoring stations throughout the region. In these analyses more sophisticated statistical models that adjusted for seasonal patterns, day-of-the-week effects, and individual hospital effects yielded positive associations between respiratory admissions and daily highs of both ozone and sulfates for day of admission and for up to 3 days before admission. No associations were seen for non-respiratory admissions. The authors suggested that because $PM_{2.5}$ and hydrogen ions are both highly correlated with sulfates, particularly during the summer months when the correlation exceeds 0.8, strong acid aerosols were likely to be implicated.

More recent studies with direct measures of acid aerosols and other putative pollutants have shown consistent relations between acid aerosols and hospital admissions for respiratory disease. Thurston et al. (1992, 1994) specifically set out to test the hypothesis that summertime haze measured as ozone and hydrogen ions was associated with excess hospital admissions for respiratory disease in regions with known acid aerosol episodes. Multiple statistical approaches were used to test the sensitivity of the analyses, and strong associations with both ozone and hydrogen ions were found to be robust in relation to both asthma and other respiratory disease admission rates. None of the other measures of pollution, including sulfur dioxide, nitrogen dioxide, fine particles, coarse particles, PM_{10} and TSP, remained significant after temperature was included in the models.

Although these studies provide ecologic and time-series information suggesting the association of acute morbidity with exposure to acid aerosols and, in particular, with the exacerbation of existing disease, the question of causality remains obscure. Part of the problem, as indicated earlier, was the lack of long-term acid measurement with which to characterize exposure before the onset of disease. In an attempt to deal with this criticism, several investigators have chosen to explore measurements made in children. This is because an estimate of chronic exposure may be more readily obtained in children, simply as a result of not having to extrapolate exposure for extended periods of time.

One such attempt was the effort to identify children who were lifelong residents of communities that had not undergone substantial changes in pollution sources and for whom monitoring of acid aerosols in the community over an extended period could be used to characterize chronic exposure. The Harvard 24 Cities Study was specifically designed to identify and monitor such communities (Speizer, 1989). The communities were selected from a combination of census data that would allow for relatively small and stable communities, as well as predicted levels of potential exposure to long-range transported fossil fuel burning sources with little in the way of local sources to confound the potential for acid aerosol exposures. Each community underwent a year-long sampling period during which representative 24-h particle sample measures were made every other day. These samples were subsequently analyzed for nitrous acid, nitric acid, sulfur dioxide and ammonia gases. In addition, the filters were analyzed for particle nitrate, sulfate and ammonia ions. Hydrogen ion concentration was determined from the pH of the extracts from the Teflon^R filters. A fine particle mass concentration was determined from a separate sampler which allowed separation of PM_{10} and $PM_{2.5}$. Daily ozone samples were also obtained.

There was approximately a three-fold variation in ozone levels across the sites and a tenfold range of sulfate levels, with an even higher range of acid particle concentration (Spengler *et al.*, 1996) (Table 27.2). The site selection proved appropriate in that none of the sites exceeded 40% of the sulfur dioxide ambient standard.

Table 27.2   Selected air pollution measurements in 24 US and Canadian communities, 1988–91 (means of 24-h values; ranges across communities)

Sites	Particle acidity	Total sulfate (SO_4)	Fine particles ($PM_{2.1}$)	Inhaled particles (PM_{10})	24-h ozone (O_3)
Sulfate belt[a]	40.7 (29–52)	67.3 (58–77)	17.2 (14–21)	28.1 (22–35)	28.8 (19–35)
Transport belt[b]	16.9 (6–26)	41.1 (34–48)	12.5 (10–14)	23.6 (18–31)	27.7 (25–30)
West coast[c]	12.6 (10–16)	20.2 (13–32)	14.1 (9–18)	20.2 (13–55)	28.2 (21–34)
Background[d]	4.1 (2–8)	12.0 (7–20)	7.3 (6–9)	20.1 (19–21)	23.6 (22–33)

[a] Sulfate belt sites: Hendersonville, TN; Oak Ridge, TN; Morehead, KY; Blacksburg, VA; Charlottesville, VA; Zanesville, OH; Athens, OH; Parsons, WV; Uniontown, PA; Penn Hill, PA; State College, PA; South Brunswick, NJ; Dunnville, ON.
[b] Transport belt sites: Newtown, CT; Leamington, ON; Pembroke, ON; Egbert, ON; Springdale, AR.
[c] West coast sites: Simi Valley, CA; Livermore, CA; Monterey, CA.
[d] Background sites: Aberdeen, SD; Yorkton, SK; Penticton, BC.

At the end of the year of ambient monitoring the parents of samples of children in the 4th and 5th grade in each community completed standardized respiratory symptom questionnaires about their children and the children underwent standard spirometric measures together with measures of height and weight (Dockery *et al.*, 1996; Raizenne *et al.*, 1996). Over a 3-year period more than 14 000 white children were studied. The prevalence of a history of asthma, asthma symptoms and episodes of bronchitis and bronchitic symptoms in the previous year were examined for associations with estimated levels of particle strong acidity from the aerometric data. After adjustment for sex, a history of allergies, parental asthma, parental education and current smoking in the household, particle strong acidity was associated with significantly higher reporting of bronchitis in the past year. None of the measures of exposure were associated significantly with higher reporting of asthma, attacks of wheeze, persistent wheeze, chronic cough or chronic phlegm production. Ozone was not associated significantly with any of the respiratory symptoms of interest.

Efforts to identify subgroups of children who might be more sensitive did not suggest that a history of any respiratory illness before age 2, a history of wheeze, lower levels of pulmonary function, parental history of asthma, maternal smoking during pregnancy, current smoking in the home, or any use of home humidifiers was associated with a different risk of bronchitis (Fig. 27.2).

In 22 of the 24 communities, the children had pulmonary function measured in a standardized manner, and among these children over 10 000 had at least two pulmonary function maneuvers that were recorded and considered acceptable by American Thoracic Society (ATS) criteria (Ferris, 1978). In general, the largest association found was that of percentage reduction of pulmonary function level with particle acidity (Table 27.3), although all of the particle measures were associated with a decline in mean differences across the communities. These declines were not significantly affected by inclusion of daily ozone measures in the analyses.

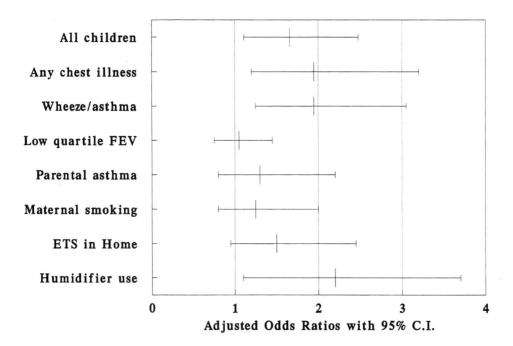

Fig. 27.2 Odds ratio and confidence interval for bronchitis among potentially susceptible subgroups. Modified from Dockery *et al.* (1996).

Table 27.3   Per cent decrement in city-specific spirometric pulmonary function measurements[a] associated with ranges of selected specific particulate pollutants (10 251 white children, aged 8–12, in 22 communities, USA and Canada, 1988–91; confidence interval of the percentage effect estimate in parentheses).

Pulmonary function measure	Particle acidity (52 nml/m^3)	Total sulfate (6.8 µg/m^3)	Fine particles (14.9 µg/m^3)	Inhaled particles (17.3 µg/m^3)
FVC	−3.45	−3.06	−3.21	−2.42
	(−4.9,−2.0)	(−4.5,−1.6)	(−5.0,−1.4)	(−4.3,−0.5)
FEV$_{1.0}$	−3.11	−2.63	−2.81	−2.09
	(−4.6,−1.6)	(−4.2,−1.0)	(−4.7,−0.9)	(−4.0,−0.1)
FEF$_{25-75}$	−3.47	−2.87	−2.73	−1.28
	(−6.5,−0.3)	(−5.9,+0.2)	(−6.3,+0.9)	(−4.8,+2.4)
PEFR	−3.71	−2.85	−3.28	−2.03
	(−7.1,−0.2)	(−6.2,+0.6)	(−7.1,+0.7)	(−5.8,+1.9)

[a] Adjusted for age, sex, weight, height and the interaction of sex and height.
Notes
FVC, forced vital capacity; FEV$_{1.0}$, forced expiratory volume in 1 s;
FEV$_{25-75}$, maximum mid expiratory flow rate;
PEFR, peak expiratory flow rate.
From Raizenne *et al.* (1996).

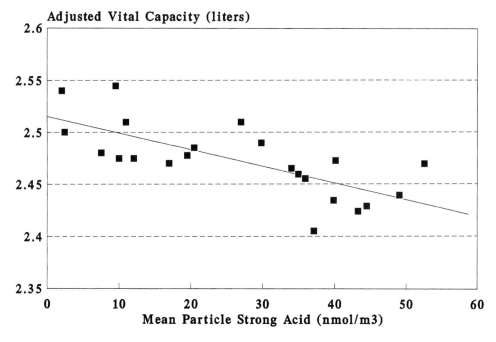

Fig. 27.3   Adjusted mean FVC for children in 22 cities by particle strong acidity. Modified from Raizenne *et al.* (1996).

An alternative way of assessing potential long-term impact on lung growth of these putative cumulative exposures was to assess the number of children in each community who had level of lung growth (as assessed by level of FVC) that was below 85% of predicted. Fig. 27.3 shows the relation between this indicator of lung size and level of community pollution expressed as particle acidity. There was a 2.5-fold increase (95% CI, 1.8–3.6) in the proportion of children with FVCs below 85% predicted across the range of particle acidity levels measured in the communities. This risk was not modified significantly by inclusion of daily ozone level in the analysis. Once again separate analyses of potentially susceptible subgroups of children did not identify groups of children at particularly higher risks.

Although these results suggested a chronic effect of particle acid exposure, the nature of the measures of exposure and the fact that this was a cross-sectional analysis of children about to enter early adolescence make it uncertain that these results would persist as normal growth continued, even if the indicators of chronic exposure were to remain constant. Thus, the data can only be considered as consistent with a chronic effect of acid aerosol exposure rather than proving that such an effect will persist. Certainly the respiratory morbidity suffered by these children cannot be considered trivial, but may reflect more acute or sub-acute phenomena that will be overcome by normal growth.

Other indicators of acute effects of acid aerosols have been found in studies of acute changes in pulmonary function among children in summer camps. Children studied repeatedly over several days to weeks with daily measures of peak expiratory flow rates and/or forced expiratory volumes when ozone and acid aerosols are measured show correlations with daily changes in pulmonary function consistent with changes in level of these pollutants. In some, but not all, of these studies, the effect of acid aerosols remains independently significant after adjustment for the effects of ozone (Table 27.4).

CONSIDERATIONS ABOUT MECHANISMS

There is substantial concern, given the neutralizing capability of the mouth and upper airways, that little of the strong acid formed in air actually penetrates into the deeper airways and lung. Studies in animals have shown substantial buffering capacity of the upper airways and effects on clearance require concentration of H^+ ions some 10 to 100 times greater than is normally found in the ambient environment. In contrast, clinical laboratory studies of young adult asthmatics exposed to concentrations of H^+ in the range of 100–450 $\mu g/m^3$ H_2SO_4 show changes in airways responsiveness (Koenig et al., 1983; Utell et al., 1983).

Impairment of mucociliary clearance in rabbits, mice, donkeys and humans follow exposure to strong acids (Schlesinger, 1990). The potential for inflammation to be caused from either acute or chronic exposure to particle acidity must be considered. With inflammation the potential exists for children chronically exposed to be unable (or unwilling because of pain or discomfort) to take a deep breath. This phenomenon would explain why those children living in regions with higher levels of chronic exposure to acid particles have lower levels of lung function.

Spengler et al. (1989) reported on the comparative cumulative concentrations of

Table 27.4 Selected summertime haze studies of children at summer camps

Author	No. of children	Time	Pulmonary function	Exposure measurement	Effect
Lippmann et al. (1983)	83	Two 5-day weeks summer	PEFR once/day	TSP, H$^+$, sulfate, O$_3$, temperature	Peak H$^+$ low; no H$^+$ effect; O$_3$ effect +
Lioy et al. (1987)	39	5 weeks summer	PEFR once/day	O$_3$, H$^+$, PM$_{15}$, PM$_{2.5}$, temperature	O$_3$ effect +; no H$^+$ effect
Spektor et al. (1988)	91	2–4 weeks summer	Spirometry plus PEFR once/day	O$_3$, SO$_4$, PM$_{15}$, PM$_{2.5}$, H$_2$SO$_4$, temperature	O$_3$ effect on PEFR and MMEF; no other effect significant
Raizenne et al. (1987)	52 (29 asthmatics)	1 week summer	Spirometry plus PEFR twice/day	O$_3$, SO$_2$, NO$_x$, SO$_4$, H$_2$SO$_4$, PM$_{10}$, PM$_{2.5}$, temperature	FVC, PEFR affected pH, by O$_3$, SO$_4$, PM$_{2.5}$, and temperature in non-asthma.
Raizenne et al. (1989)	112 girls	Three 13-day periods summer	Spirometry plus PEFR twice/day	O$_3$, SO$_2$, NO$_x$, SO$_4$, H$_2$SO$_4$, pH, PM$_{10}$, PM$_{2.5}$, temperature	FEV$_1$ dropped on high O$_3$ and H$_2$SO$_4$ days; no formal statistics.
Spektor and Lippmann (1991)	46	4 weeks summer	Spirometry plus PEFR twice/day	O$_3$, H$_2$SO$_4$, temperature	FEV$_1$ dropped with lagged O$_3$; no H$^+$ effect
Thurston et al. (1997)	52–58 asthmatic	Three 1-week periods over 3 years, summers	Spirometry plus PEFR and asthma medication use	O$_3$, H$^+$, temperature	Without exacerbations both O$_3$ and H$^+$ effects on PEFR. Increase in unscheduled medication use associated with O$_3$
Neas et al. (1995)	83	14–87 days summer	Spirometry plus PEFR twice/day	O$_3$, SO$_2$, PM$_{10}$ PM$_{2.5}$, SO$_4$, H$^+$, temperature	PEFR decline associated with H$^+$, SO$_4$; no effect O$_3$

exposure to acid aerosols in children in the camp studies compared with the concentrations received by subjects in the chamber studies mentioned earlier. They reported that the 1-h respiratory tract dose concentrations experienced by children exercising outdoors were approximately equivalent to those delivered to the asthmatics studied in the chambers. It therefore seems that for a given cumulative exposure, although in different settings, consistent results have been seen.

RESEARCH ISSUES

The following issues warrant further investigation.

(1) Research is needed to explore and identify the components of PM_{10} (fine particles, sulfates, acid aerosols or ultrafine particles) or combinations of these components that best explain the observation of excess mortality and morbidity.

Thurston suggests that this research would (a) address better the question of biologic plausibility, and (b) better protect the public's heath by more effectively identifying the potential adverse component, thus leading to a more targeted removal of this component from ambient pollution.

(2) Models of sulfate prediction of H^+ that take into account population density should be used with the retrospective data available to test for both morbidity and mortality.

(3) Further research is needed on the chemical composition and size distribution of particles in the air. Few studies have been able to take into account both size and chemistry.

(4) Follow-up of the studies of children as they complete their lung growth, or repeated studies of population groups in whom known chronic exposure to elevated levels of acid aerosols, are needed to determine whether chronic effects persist or are overcome by normal growth.

(5) Understanding of the effects of neutralization, or lack thereof, in the upper airways of selected subjects who appear to be responsive or non-responsive to acute exposure to acid particles may help explain why variations in responses are seen.

CONCLUSIONS

Although relatively scant compared with those available for other common ambient pollutants, the data appear to be sufficiently coherent and biologically plausible to warrant further investigation. In particular the potential chronic effects of regular exposure to acid aerosols may have deleterious effects not only on airway responsiveness and level of pulmonary function, but also on the potential for normal growth of lungs of children. The implication of those findings, if confirmed, could be substantial. Reduced lung growth as a result of chronic exposure could contribute to acute morbidity, since children with smaller lungs have smaller airways that would be more responsive to non-specific irritation (increased broncho-responsiveness). In addition, reduced lung function if it were to persist to adult life, is a recognized risk factor for excess morbidity and mortality. To determine the validity of such conclusions will require additional research, since direct measurement of acid aerosols is a relatively new science.

REFERENCES

Bates DV and Sizto R (1983) Relationship between pollutant levels and hospital admission in Southern Ontario. *Can J Public Health* **74**: 117–122.

Bates DV and Sizto R (1987) Air pollution and hospital admissions in southern Ontario: the acid summer haze effect. *Environ Res* **43**: 317–331.

Bates DV and Sizto R (1989) The Ontario air pollution study: identification of the causative agent. *Environ Health Perspect* **79**: 69–72.

Burnett RT, Dales R, Krewski D *et al.* (1994) Effects of low ambient levels of ozone and sulfates on frequency of respiratory admissions to Ontario hospitals. *Environ Res* **65**: 172–194.

Calvert JG, Lazrus A, Kok GL *et al.* (1985) Chemical mechanisms of acid generation in the troposphere. *Nature* **317**: 27–35.

Chappie M and Lave L (1982) The health effects of air pollution: a reanalysis. *J Urban Econ* **12**: 346–376.

Dockery DW, Cunningham J, Damokosh AI *et al.* (1996) Health effects of acid aerosols on North American children: respiratory symptoms. *Environ Health Perspect* **104**: 500–505.

Dockery DW, Pope CA, Xu X *et al.* (1993) An association between air pollution and mortality in six U.S. cities. *N Engl J Med* **329**: 1753–1759.

Ferris BG Jr (1978) Epidemiologic standardization project. III Recommended standard procedures for pulmonary function testing. *Am Rev Respir Dis* **118**: 55–88.

Firket M (1931) Sur les causes des accients survenus dans la vallée de la Meuse, lors des brouillards de decembre 1930. (The causes of accidents which occurred in the Meuse Valley during the fogs of December 1930). *Bull Acad R Med Belg* **11**: 683–741.

Firket M (1936) Fog along the Meuse Valley. *Trans Faraday Soc* **32**: 1192–1197.

Ito K, Thurston GD, Hayes C and Lippmann M (1993) Associations of London, England, daily mortality with particulate matter, sulfur dioxide, and acidic aerosol pollution. *Arch Environ Health* **48**: 213–220.

Koenig JQ, Pierson WE and Horike M (1983) The effects of inhaled sulfuric acid on pulmonary function in adolescent asthmatics. *Am Rev Respir Dis* **128**: 221–225.

Koutrakis P, Wolfson JM, Slater JL *et al.* (1988) Evaluation of an annual denuder/filter pack system to collect acidic aerosols and gases. *Environ Sci Technol* **22**: 1463–1468.

Lave LB and Seskin EP (1977) *Air Pollution and Human Health.* Baltimore, MD: The Johns Hopkins University Press.

Lioy PJ, Spektor D, Thurston G *et al.* (1987) The design considerations for ozone and acid aerosol exposure and health investigations: the Fairview Lake summer camp–photochemical smoke case study. *Environ Int* **13**: 271–283.

Lipfert FW (1984) Air pollution and mortality: specification searches using SMSA-based data. *J Environ Econ Manag* **11**: 208–243.

Lipfert FW (1988) Exposure to acid sulfates in the atmosphere: review and assessment. Palo Alto, CA: Electric Power Research Institute, report no EPRI EA-6150.

Lipfert FW (1993) Community air pollution and mortality: analysis of 1980 data from US metropolitan areas. I. Particulate air pollution. Upton, NY: Brookhaven National Laboratory, report no BNL-48446-R.

Lippmann M and Thurston GD (1996) Sulfate concentrations as an indicator of ambient particulate matter air pollution for health risk evaluation. *J Expo Environ Epidemiol* **6**: 123–146.

Lippmann M, Lioy P, Leikauf G *et al.* (1993) Effects of ozone on the pulmonary function of children. In: Lee SD, Mustafa MG and Mehlman MA (eds) *International Symposium on the Biomedical Effects of Ozone and Related Photochemical Oxidants,* March 1982, Pinehurst, NC. Princeton, NJ: Princeton Scientific Publishers, pp 423–446.

Martin AE (1964) Mortality and morbidity statistics and air pollution. *Proc R Soc Med* **57**: 969.

Neas LM, Dockery DW, Koutrakis P *et al.* (1995) The association of ambient air pollution with twice daily peak expiratory flow rate measurements in children. *Am J Epidemiol* **141**: 111–122.

Özkaynak H and Thurston GD (1987) Associations between 1980 U.S. mortality rates and alternative measures of airborne particle concentration. *Risk Anal* **7**: 449–461.

Özkaynak H, Xue J, Zhou H *et al.* (1996) Inter-community differences in acid aerosol (H⁺)/sulfate (SO_4^{2-}) ratios. *J Expo Anal Environ Epidemiol* **6**: 35–55.

Pierson WR, Brachaczek WW, Gorse RA Jr, *et al.* (1989) Atmospheric acidity measurements on Allegheny Mountain and the origin of ambient acidity in the northeastern United States. *Atmos Environ* **23**: 431–459.

Pope CA, Thun MJ, Namboodiri MM *et al.* (1995) Particulate air pollution as a predictor of mortality in a prospective study of US adults. *Am J Respir Crit Care Med* **151**: 669–674.

Raizenne M, Neas LM, Damokosh AI *et al.* (1996) Health effects of acid aerosols on North American children: pulmonary function. *Environ Health Perspect* **104**: 506–514.

Raizenne M, Stern B, Burnett R and Spengler J (1987) Acute respiratory function and transported air pollutants: observational studies. Presented at the 80th Annual Meeting of the Air Pollution Control Association, June, New York, NY. *Air Pollution Control Association* paper no 87-32.6.

Raizenne ME, Burnett RT, Stern B *et al.* (1989) Acute lung function responses to ambient acid aerosol exposures in children. *Environ Health Perspect* **79**: 179–185.

Russell WJ (1884) On London rain. Appendix I, *Monthly Weather Report,* April.

Schlesinger RB (1990) The interaction of inhaled toxicants and respiratory tract clearance mechanisms. *Crit Rev Toxicol* **20**: 257–286.

Shy CM, Goldsmith JR, Hackney JD *et al.* (1978) Health effects of air pollution. *ATS News* **6**: 1–63

Smith RA (1872) *Air and Rain.* London: Green & Co.

Speizer FE (1989) Studies of acid aerosols in six cities and in a new multi-city investigation: design issues. *Environ Health Perspect* **79**: 61–67.

Spektor DM and Lippmann M (1991) Health effects of ambient ozone on healthy children at a summer camp. In: Berglund RL, Lawson DR and McKee DJ (eds) *Tropospheric Ozone and the Environment*; papers from an international conference, March 1990, Los Angeles, CA. Pittsburgh, PA: Air & Waste Management Association, pp 83–89.

Spektor DM, Lippmann M, Lioy PJ *et al.* (1988) Effects of ambient ozone on respiratory function in active, normal children. *Am Rev Respir Dis* **137**: 313–320.

Spengler JD, Keeler GJ, Koutrakis P *et al.* (1989) Exposure to acid aerosols. In: Symposium on the Health Effects of Acid Aerosols, October 1997, Research Triangle park, NC. *Environ Health Perspect* **79**: 43–51.

Spengler JD, Koutrakis P, Dockery DW *et al.* (1996) Health effects of acid aerosols on North American children: air pollution exposures. *Environ Health Perspect* **104**: 492–499.

Spix C, Heinrich J, Dockery D *et al.* Summary of the analysis and reanalysis corresponding to the publication Air pollution and daily mortality in Erfurt, East Germany 1980–1989. Summary report for: Critical evaluation workshop on particulate matter – mortality epidemiology studies; November 1994, Raleigh, NC. Wuppertal, Germany; Bergische Universitat-Gesamthochschule Wuppertal.

Thurston GD, Ito K, Lippmann M and Hayes C (1989) Reexamination of London, England, mortality in relation to exposure to acidic aerosols during 1963–1972 winters. *Environ Health Perspect* **79**: 73–82.

Thurston GD, Ito K, Kinney PL and Lippmann M (1992) A multi-year study of air pollution and respiratory hospital admissions in three New York State metropolitan areas: results for 1988 and 1989 summers. *J Expo Anal Environ Epidemiol* **2**: 429–450.

Thurston GD, Ito K, Hayes GC *et al.* (1994) Respiratory hospital admissions and summertime haze air pollution in Toronto, Ontario: consideration of the role of acid aerosols. *Environ Res* **65**: 271–290.

Thurston GD, Lippmann M, Scott MB and Fine JM (1997) Summertime haze air pollution and children with asthma. *Am J Respir Crit Care Med* **155**: 654–660.

Utell MJ, Morrow PE, Speers DM *et al.* Airway responses to sulfate and sulfuric acid aerosols in asthmatics: an exposure–response relationship. *Am Rev Respir Dis* **128**: 444–450.

Waller RE (1963) Acid droplets in town air. *Int J Air Water Pollut* **7**: 773.

Wilkins ET (1954) Air pollution aspects of the London fog of December 1952. *Q J R Meteorol Soc* **80**: 267–271.

SUSPENDED PARTICULATES

28

Composition of Air Pollution Particles

FREDERICK D. POOLEY and MILAGROS MILLE

School of Engineering, Materials and Minerals Engineering Division, University of Wales Cardiff, Cardiff, UK

INTRODUCTION

Airborne particulates have been widely associated with health disorders by numerous studies. The terms used to describe and quantify airborne particulate pollution are 'suspended particulate matter' (SPM) and 'total suspended particles' (TSP). The former term refers to the total airborne particles, while the latter implies that a gravimetric procedure has been used to determine SPM. Terms such as 'respirable particles' and 'inhalable particles' are also employed to describe dusty atmospheres. These terms denote both a limited particle size and suggest possible health concerns. Inhalable particles are those particles that enter the respiratory system during breathing, while respirable particles are the ones that are capable of reaching the gas exchange regions of the lungs.

Recent attention has focused on the gravimetric measurement of the PM_{10} fraction of airborne particulate matter, because of the possible health impact that has been suggested by several recent studies (Pope, 1989; Royal Commission Report, 1994). The term PM_{10} stands for a fraction of airborne matter that contains particles with an aerodynamic diameter (which implies a specific gravity equal to one) of less than 10 μm. Technically speaking, the PM_{10} fraction of an airborne dust consists of particles that pass through a size-selective inlet with a 50% efficiency cut-off at 10 μm aerodynamic diameter (Warren Spring Laboratory, 1994). An interest in the fraction called $PM_{2.5}$ (which definition is analogous to PM_{10}) has also been given special consideration, since this fine fraction is suspected to be the major contributor to health effects because it can penetrate easily to the innermost regions of the lungs.

Other definitions concerning airborne particles include:

AIR POLLUTION AND HEALTH
ISBN 0-12-352335-4

(1) *dust:* particles formed by mechanical disintegration of solids;
(2) *aerosol:* airborne particles or droplets;
(3) *smoke:* particulate matter resulting from incomplete combustion processes (particulate size is less than 15 μm);
(4) *black smoke:* non-reflective particulate matter (the smoke stain measurement method is used with this matter); and
(5) *fume:* a condensation product of evaporated material (e.g. oxides of iron in welding) and smoke (particles from fuel combustion).

In industrialized countries, the daily deposition of PM_{10} particles in the lungs is roughly 250 μg/day (International Labour Office, 1991), which represents a small dose in terms of traditional toxicology studies. Studies of PM_{10} have considered this total material but have not asked how much its chemical or physical characteristics contribute to its total toxicity.

Airborne PM_{10} matter may not only be dangerous because of its inorganic chemistry but also because of the complex organic materials it contains. These include benzene, 1–3 butadiene, Polychlorinated Biphenyls and Polynuclear Aromatic Hydrocarbons (PAH). Many of these are known to be highly suspected potential carcinogens.

Concentrations of airborne aerosols are mainly localized over continental areas, specifically over urban and industrialized zones. Moreover, complexities in the physical and chemical characteristics of these materials make the effects they cause much more complicated than those caused by greenhouse gases. Mineral particles are often coated with sulfate and other soluble materials. This process is said to have originated from evaporating cloud drops which were once nucleated onto a sulfate cloud condensation nucleus (CCN). The presence of soluble material on mineral dust particles converts the latter into a CCN (Levin *et al.,* 1996). This theory is supported by the fact that droplets in the Northern Hemisphere are larger than those in the Southern Hemisphere, where aerosol concentrations are generally smaller.

In response to the obvious health risks airborne particulate matter causes, air quality standards for particulate matter have been set by the EPA (USA). These involve annual and 24-hour standards for both $PM_{2.5}$ and PM_{10}. For $PM_{2.5}$ the standards were set at 15 μg/m^3 the annual and at 65 μg/m^3 the 24-hour one. The annual and 24-hour PM_{10} standards have been set at 50 μg/m^3 and at 150 μg/m^3 respectively.

The standards described above represent PM_{10} and $PM_{2.5}$ mass limits. Unfortunately, no regulations for specific chemical components have been established, since few quantitative toxicological studies have been performed on PM_{10} and $PM_{2.5}$ chemical components to date. Nevertheless, an approximation of one fortieth the OEL (occupational exposure limits) for each substance carried by PM_{10} and $PM_{2.5}$ matter could be used in assessing the potential hazard of PM_{10} components.

In recognition of the health risks and the need for more severe limits, the Expert Panel on Air Quality Standards (EPAQS) recommended a PM_{10} limit of 50 ppb (μg/m^3) as a 24-h rolling average (November 1995) in the UK. The British Government has set this as a challenge to be met in an estimated 10-year period (Brow, 1995). During the last few years in the UK, there have been few occasions when the 150 μg/m^3 standard of the US EPA has been exceeded. However, the EPAQS recommendations are frequently exceeded in all cities that are part of the UK Air Pollution Network. Statistics show that the EPAQS recommendation of 50 μg/m^3 was exceeded for 27 days during 1993 in London

(Bloomsbury), in Belfast for 45, in Birmingham for 24, in Cardiff for 19, and in Edinburgh for 2 days.

SOURCES OF PM$_{10}$ MATERIAL

When examining the sources of PM$_{10}$ particulates, it is possible to make a distinction between what are referred to as primary particles and secondary particles. Primary particles are those emitted directly into the atmosphere from sources such as road traffic, coal burning, industry, windblown soil and sea spray. Secondary particles are those formed within the atmosphere by the chemical reaction of pollutant condensation of gases, as in the case of sulfate and nitrate salts which are the result of the oxidation products of sulfur dioxides and nitrogen oxides respectively.

As mentioned above, primary particles can have different sources. These can be natural (winds blowing up small particles of soil, dust released by explosions of volcanoes) or result from human activities.

Natural Sources of Primary Particles

There are different sources for SPM and PM$_{10}$ and they can vary greatly depending on the location (Burton *et al.*, 1996). Suspensions of dust originating from soil and also evaporation of sea water are considered as large sources of particles worldwide.

Suspension of soil dusts are caused by the action of the wind on dry loose soil. The amount of this dust in the Northern Hemisphere is estimated to be about 150 million tons per year; if the Sahara Desert were included in the calculation, this amount would double. The particles are generally coarse and may settle, but they can be resuspended in the air by such agents as the wind, dry weather or moving traffic (Quality of Urban Air Review Group, 1996). The size distribution and chemistry of the airborne dust of cities in the UK show that this source contributes significantly to SPM pollution.

For coastal areas, the contribution of salt particles to the SPM figure is very important. These salt particles result from the evaporation of minute droplets sent into the atmosphere by breaking waves. Other inorganic particles suspended in seawater also become airborne in this manner, while sea salt particles often have been measured in inland locations.

Primary Particles Derived from Combustion and Metallurgical Processes

Combustion and metallurgical processes, among other industrial activities, discharge particles into the atmosphere. The chemistry of these particles varies depending on the materials treated in these processes. During combustion, coal releases particles called fly ash (very fine particles contained in the coal). The combustion of diesel fuels contributes a large variety of particles, e.g. elemental carbon, organic carbon, sulfates, unburnt material, metals, etc. High temperatures are required during metallurgical and combustion processes, and together with refuse incineration these release particles rich in heavy metals formed by the condensation of metal fumes.

Particles Generated by the Transport Sector

The transport sector is a major source of particles in most urban locations. The particles contained in exhaust gases are the result of a combustion process and consist mainly of carbon and unburned or partially burned organic compounds from the fuel or lubricating oil. Light duty diesel vehicles emit 4–7 g of particulate matter per litre, heavy duty ones 7–14 g per litre, while conventional gasoline burning vehicles emit 0.65 g per litre. In the UK, most diesel is consumed by heavy duty vehicles which consume more fuel than automobiles (Hutchison, 1996).

It is thought that the transport sector is a major contributor of $PM_{2.5}$. Road traffic is on average responsible for 4.1% of the $PM_{2.5}$ mass. This is very consistent with the size distribution measurements of particles from vehicle exhausts which show that traffic related particles are very small in size being predominantly in the $PM_{2.5}$ fraction (Quality of Urban Air Review Group, 1996).

National Distribution of Primary Particles in the UK

UK emissions inventories for major sources of primary particle matter indicate the following contributions: road transport (25%); non-combustion processes such as construction, mining and quarrying and industrial processes (24%); industrial combustion plants and processes with combustion (17%); commercial, institutional and residential combustion plants (16%); and public power generation (15%). These percentages can vary considerably from one location to another, especially for urban locations where road traffic greatly increases the amount of airborne particles. In London, 86% of the particles come from this source (Quality of Urban Air Review Group, 1996).

Sources of Secondary Airborne Particulate Matter

Secondary particles are formed within the atmosphere, mostly by chemical oxidation of atmospheric gases. Ammonium sulfate and ammonium bisulfate are both common secondary particles in the air. Oxidation of sulfur dioxide to sulfuric acid and a subsequent total or partial neutralization with ammonium ions in the atmosphere leads to the formation of these substances.

Nitrates are also common secondary airborne particles, formed by the oxidation of nitrogen dioxide to nitric acid and a neutralization with ammonium cations to form ammonium nitrate. Ammonium chloride is also present in the air. It is formed by the reaction of hydrochloric acid gas (emitted during the combustion of coal and municipal incineration) and ammonium ions.

Formation of Aerosols

Aerosols are formed by single nucleation, condensation and by accumulation processes. Molecules of the same substance (present in a number that exceeds the saturation vapour pressure of the substance) meet together to form a very small particle, a nucleus. More molecules or particles (regardless of their chemical or physical characteristics) may meet the nucleus already formed and condense, causing growth in size. The result is very fine and light particles ranging from 1 to 10 nm in diameter. These tiny particles collide (during Brownian motion), adhere and agglomerate, forming larger particles. This gradually results in particles (accumulation mode) of roughly 0.05–2 µm diameter (Quality of Urban Air Review Group, 1996).

Some of the large particles contain mixtures of chemicals such as sulfate and sea salt. Other studies have shown that large particles tend to consist of many components rather than a single compound after the nucleus has been coated. The core composition is determined by the original primary aerosol composition, while the coating is identical to the chemical composition of the eutonic point of the airborne compounds (Ge *et al.*, 1996).

ANALYSIS OF AIRBORNE PARTICULATES

Determination of Physical Characteristics

The physical characteristics of airborne particulate matter that determine size, shape, surface and the structure arrangements of the particles can be assessed by electron microscopy, using either transmission or scanning microscopes.

Depending on the particle size range of interest to be analysed, size distribution analyses can be performed using inertial impactors (i.e. cascading impactors) and electrical/diffusion aerosol analysers. The inertial impactors can separate particles down to 0.4 µm in size and the most modern ones separate to below 0.1 µm. Meanwhile, to determine the size distributions of particles ranging below 0.5 µm, either electrical or diffusion aerosol analysers are used. The former measure the electrical mobility of particulate matter and relate that mobility to the particle size by calibration of the equipment with aerosols of known size; the latter separate particles according to their diffusion coefficients. No single method can give the complete size distribution of particles, hence several methods are required to determine the total size distribution of airborne particles (Hutchison, 1996).

Determination of Carbonaceous Material

Carbonaceous material consists of elemental carbon particles and carbon-based compounds (i.e. organic compounds). Several studies have characterized this material; many of these have focused on diesel particulates, which are known to be almost entirely composed of carbonaceous matter (Amann and Siegla, 1982).

The organic fraction of the particles can be characterized by using solvent extraction (e.g. dichloromethane, benzene–ethanol mixtures). The molecular weights for the extractable fraction are determined by techniques such as gel permeation chromatography. Thermogravimetric analyses are also performed. This involves recording the weight loss of a sample as its temperature is continuously increased. Samples are heated initially in a nitrogen atmosphere up to 500°C to drive off volatile components. From 500 to 700°C, heating is conducted in air, facilitating oxidation of the remaining combustible material (elemental carbon). For diesel samples, as much as 65% of the material has been shown to be volatile and 35% combustible (Amann and Siegla, 1982).

Other methods for determining the nature of the organic material include high-resolution capillary gel chromatography (GC). The organic material is thermally desorbed from the sample and then the vapour is transferred to a GC column for analysis.

Elemental Composition

The elemental composition of airborne particulate samples has been characterized using a variety of different methods, including X-ray fluorescence (XRF), energy-dispersive X-ray analysis (EDAX) and proton-induced X-ray emission (PIXE). A range of elements can be identified with these techniques, but no indication of the actual compounds of the sample is obtained. No previous sample treatment work is required, a fact that makes these techniques very convenient. For both XRF and EDAX, the preparation of standards is needed to obtain accurate results, otherwise only relative percentages are calculated.

Wet chemical techniques are also used to determine the general elemental composition of samples. Ashed particulate samples are extracted with mixtures of nitric and hydrochloric acids. The extract solutions can then be analysed for selected elements using atomic absorption (Thompson et al., 1970) or ICP techniques.

Mineralogical Analysis

The mineral or crystalline components of airborne particles can be determined by the use of X-ray powder diffraction (XRD). This technique identifies compounds based on their crystalline structures. The large amount of material needed for analysis is a major inconvenience. A minimum of 5 mg of material is usually required to produce reliable XRD results. Unfortunately, this amount of airborne material is not easily obtained during a sampling procedure. Single particle electron diffraction can also be used for identification when a transmission electron microscope is employed.

Water-soluble Material

Analysis of water-soluble material can be achieved by desorption of the particles via an ultrasonic agitation of the filter on which the particles have been collected. The suspension is filtered and the solution is analysed using several well-known techniques:

(1) *ion chromatography* (IC): permits the quantification of dissolved ions (e.g. sulfates, chlorides, sodium, calcium, etc.);
(2) *inductively coupled plasma spectroscopy* (ICP): facilitates the quantification of a large range of soluble metals at very low concentrations; and
(3) *atomic absorption spectroscopy.*

Particle Morphology of PM$_{10}$

The shape of airborne particles can vary greatly, from spherical particles to crystalline cubes (e.g. some particles of sodium chloride – sea salt). Particle shape depends on the chemical composition and the manner in which the particle has been formed. Some particles are totally flat and layered, some are globular, while others are fibrous (unlikely to be found in urban airborne dust samples). Particles can occur either on their own or aggregated to form agglomerates which at first glance look like compact single particles. Surfaces of particles also appear to have different textures. Some are smooth while others are very irregular and porous. Airborne particles from samples collected next to a road carrying heavy traffic are mainly aggregates rather than single particles. These aggregates have a random shape, but some of them can be characterized as small clumps while others are

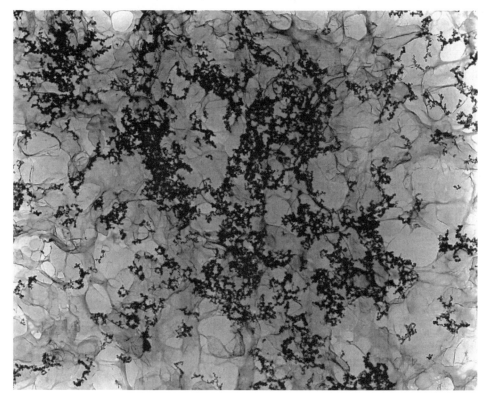

Fig. 28.1 Low magnification electron micrograph illustrating the appearance of an urban PM$_{10}$ sample.

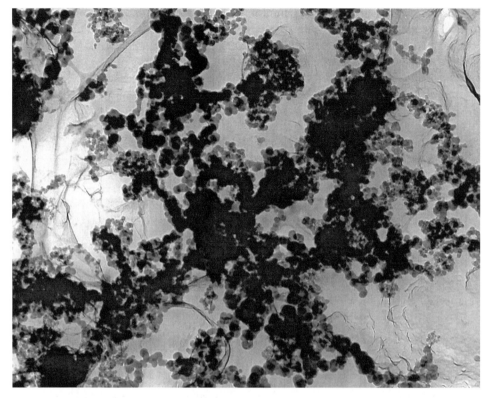

Fig. 28.2   High magnification electron micrograph of carbonaceous material observed in typical urban PM$_{10}$ samples.

chain-like (Hutchison, 1996). These particles are illustrated in Fig. 28.1, which consists of a low magnification image of this particulate material.

Particles of diesel exhaust have been described as spherical (called 'spherules') and appear as aggregates resembling either 'clusters of grapes' or 'chains of beads'. Each of them can contain as many as 4000 spherules, as shown by Fig. 28.2. The spherules vary in diameter between 10 and 80 nm in size, although most are in the 15–30 nm range. A particle size distribution analysis has shown two main groups of particles. The smaller size group (nuclei mode) are under 66 nm in size and are formed mainly by individual spherules. The larger sized group (accumulation mode) range from 100 and 440 nm in size and consist mainly of aggregates (Amann and Siegla, 1982).

Diesel particles are primarily carbonaceous soot. These particles are very porous and have a much larger surface area on which they can adsorb pollutant gases and other smaller particles. The chemistry of these particles therefore can be very diverse, containing different amounts of organic gases (unburnt or partially burnt hydrocarbons from oils and fuels), inorganic gases (as nitrogen or sulfur oxides) and other tiny particles with different chemical characteristics.

GENERAL CHEMICAL CHARACTERISTICS OF PM₁₀ MATERIAL

The chemistry of PM_{10} particulate matter is complex and variable. This complexity is a result of the amount and variety of substances that the particles are associated with. It has been suggested that there is a close relationship between particle size, shape and chemistry. For example, coarse particles will concentrate different chemical species than finer fractions and heavy metals from metallurgical industries will tend to have a rounded shape owing to condensation of metal vapours, as illustrated by Fig. 28.3. The chemistry is also variable because of the location of different sources and the time and rate of emission. There are few studies dealing with the total chemistry of PM_{10} particles; the majority have been concerned with only particular chemical characteristics of particular particles.

A range of different factors affect the chemistry of airborne particulate matter. These include type of source, chemical transformations in the atmosphere, long-range transport effects and removal processes. The airborne half-life of a specific material can be closely linked to its aerodynamic size distribution. For example, sulfates tend to stay in the air longer, showing a spatial uniformity in concentration. This contrasts with other substances such as quarry dust, which has a coarse aerodynamic size profile. Quarry dust has

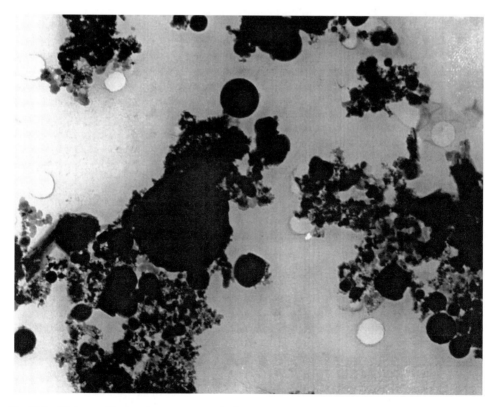

Fig. 28.3 High magnification electron micrograph of PM_{10} sample observed in sample collected adjacent to a steel works site.

a short residence time in the air and hence its concentration in the atmosphere does not tend to be uniform.

Several chemical species can be located more readily in the fine fraction of PM_{10} matter than in its coarse component. The most abundant species in the smaller fraction include nitrates, sulfates, heavy metals, organic compounds and elemental carbon. The coarser fractions are enriched with soil-derived elements at inland sites and with marine-derived species in coastal areas (Chow *et al.*, 1994).

The finer PM_{10} particles are mainly formed by:

(1) *secondary particles* such as sulfates, nitrates and chloride salts of ammonium, products of the reaction of SO_2, NO_2, hydrochloric acid and ammonium ions in atmospheric conditions;

(2) *heavy metals* originating from anthropogenic sources, such as smelting plants (iron, cadmium, zinc) and combustion processes (in urban locations, combustion of leaded fuel releases lead);

(3) *elemental carbon*, which is a product of incomplete combustion processes, including the combustion of fuels (for the transport sector), coal, biomaterial. The surface area of elemental carbon particles is considerably increased by their porous surface, hence their ability to receive more airborne chemical substances (such as organic compounds) is greatly enhanced;

(4) *organic compounds* that condense at the surface of tiny nuclei are adsorbed on carbon material, the small particles of which present a larger surface area for the condensation and/or adsorption of chemical species. Studies indicate that organic mutagenic compounds are mainly associated with the 0–2.5 µm particles (Watts *et al.*, 1988).

The coarse fraction of PM_{10} is formed mainly by minerals and natural salts coming from dust created by wind or marine aerosols. Some anthropogenic species also have been found in the coarse fraction, but in much smaller proportion when compared with the finer fraction. This is the case with carbon black, halogen (as bromine and iodine), sulfur, vanadium, nickel, zinc, indium and lead. These were recorded in average ratios (fine/coarse) of over 2 and up to 7 for carbon black and iodine (Maenhaut *et al.*, 1996).

ORGANIC CHEMISTRY OF AIRBORNE MATERIAL

The list of organic compounds in the air is extensive, since in addition to those compounds that are emitted, more are formed by chemical reactions in the atmosphere. The main sources of organic compounds are fuel evaporation or spillages, natural gas leakage, solvent usage, industrial and chemical processes, and the use of bitumen and road asphalt.

Organic compounds can be classed as volatile (a property associated with a low molecular weight), semivolatile and non-volatile. This characteristic, which is closely related to the molecular weight of the compound, will influence their behaviour in the atmosphere and their association with airborne particles. In general, the lower the molecular weight the more volatile the species. Low molecular weight compounds will remain in the gaseous phase in the atmosphere, until they are removed by natural oxidation processes in sunlight. As their molecular weight increases they will tend to condense on particles, preferring particles in the condensation mode rather than those in the nucleus mode owing to their larger surface area. Hence semivolatile compounds are likely to be present both in a gaseous form and adsorbed or absorbed onto particles in different proportions, depending mainly on their volatility. However, involatile compounds will remain in the particle form.

Organic compounds which usually exist only in gaseous form can be carried by PM_{10} particles. Coefficients of distribution between the phases (gaseous and particulate) have been calculated and show a decrease in the partition coefficient with increasing humidity, indicating a competition effect with water vapour for the adsorptive surface area (Whei-May and Lin-Yi, 1994). Organic compounds with high molecular weight are the ones that will be associated with particles, while those with lower molecular weight (and therefore more volatility) remain in the gaseous form.

Toxic organic micropollutants (TOMPs) have received special attention because of their health effects. These compounds include polynuclear aromatic hydrocarbons (PAH), polychlorinated biphenyls (PCB) and dioxins. They are frequently referred to in terms of toxic equivalent (TEQ). Their prevalence has been quantified for different urban locations in the UK (Quality of Urban Air Review Group, 1996).

The organic material in PM_{10} samples collected in the UK normally comprises 60–80% of its total carbon content (Quality of Urban Air Review Group, 1996). This organic pollution is associated mainly with road traffic, although other sources include fuel oil and coal combustion and incineration. In Birmingham, 88% of benzo[a]pyrene (PAH) came from road traffic (Harrison et al., 1996).

THE INORGANIC CHEMISTRY OF SUSPENDED PARTICULATES

The composition of particulate material in urban air samples can be divided into particulate elemental carbon or carbon black (which comes from road traffic and is associated with organic compounds in smoke particles) and water-soluble or insoluble inorganic compounds.

The elemental carbon particles result from the incomplete combustion of fuels and lubricating oils and are generally small, forming part of the finer fraction of PM_{10} matter. They have a porous surface that increases the area available for adsorption processes. Organic compounds, heavy metals and soluble inorganic material can be carried by elemental carbon particles. Health concerns were originally focused on this fact, but it is now recognized that elemental carbon can be dangerous by itself.

In the UK, diesel combustion is believed to be responsible for 80–95% of particulate elemental carbon (PEC). Measures made in Birmingham (Harrison et al., 1996) indicated a concentration of 3 $\mu g/m^3$ PEC which was 10% of the total PM_{10} mass.

Soluble inorganic compounds can be divided into several categories.

(1) *Aerosol acids.* Acidity can result from partial neutralization of the airborne acid species present in the atmosphere (nitric acid, sulfur acid, bisulfates) owing to an insufficient quantity of basic species, i.e. ammonium ions. These basic species are

present in sufficient quantity to neutralize the airborne acid species. Secondary particles such as ammonium nitrates and sulfates are formed through this process;

(2) *Sulfates.* Soluble sulfate salts ($CaSO_4$, Na_2SO_4, $(NH_4)_2SO_4$) are common in airborne particles. High percentages of sulfate (up to 85%) are likely to be found in PM_{10} samples, especially in the finer fraction $PM_{2.5}$. Most of the sulfate in urban areas originally comes from the oxidized SO_2, but sulfates of marine origin also contribute (mainly as Na_2SO_4). Sulfates have a long residence time in the atmosphere, and since they are widely distributed within an area good correlations between sampling sites can be achieved. Between 25 and 30% of airborne PM_{10} in the UK consists of soluble ionic material;

(3) *Nitrates.* The finest PM_{10} fraction has the majority of the nitrate content. This fraction is formed by an oxidation process like sulfates, but from NO_2. In the UK nitrates represent 25% of the total soluble ionic fraction. They are present in the form of NH_4NO_3 with some $NaNO_3$;

(4) *Chlorides.* Sodium chloride mainly enters the atmosphere from marine aerosols in coastal areas, and is mainly found in coarse particles. Ammonium chloride also is the result of a reaction between hydrochloric acid and the ammonium ions released by pollutant sources (industries). As a secondary particle, it is likely to be concentrated in the finest PM_{10} fraction. During winter months, sodium chloride can come from the salting of roads;

(5) *Ammonium compounds.* Ammonium is combined with sulfate, nitrate and chloride forming salts. These are secondary particles and hence are found in the $PM_{2.5}$ fraction;

(6) *Other cations.* Sodium, calcium and magnesium can be present. Sodium is widely associated with chloride, sea salt, typical marine aerosol. However, it also can occur in the form of nitrates and sulfates. Calcium occurs primarily as carbonates and sulfates. Magnesium is common near geological formations forming carbonates and chlorides close to the sea.

Insoluble inorganic material is typically associated with the coarse fraction of PM_{10} matter and is formed by soil-derived particles. It includes compounds such as α-quartz (SiO_2), calcite ($CaCO_3$), epsomite ($MgSO_4.7H_2O$), gypsum ($CaSO_4.2H_2O$), feldspar ($KalSi_3O_8$), chlorite, kaolinite and montmorillonite. In the UK this fraction can represent nearly 25% of the total PM_{10} mass.

Metals can be found in a variety of different compounds. They exist in both a soluble and insoluble elemental state. They may also be combined as organometallic compounds. Analysis of heavy metals in airborne samples tends to be difficult because of their low concentrations. Nevertheless, they can be detected. Their characterization and quantification are of special interest because of the health issues they raise.

The source of trace heavy metals is generally anthropogenic, e.g. the combustion of fossil fuels, waste incineration and high temperature processes (metal smelting industries). During the latter processes, heavy metals become vaporized and condense preferentially onto the surface of smaller ambient particles owing to their larger surface area/mass ratios. Metals are generally associated with particular processes. For example, airborne lead is related to the combustion of leaded gasoline, while cadmium is associated with the smelting of non-ferrous metals. The levels of these trace heavy metals have been measured at different sites in the UK. Zinc, lead, aluminium, iron, chromium, arsenic, cadmium, nickel, vanadium, mercury, silver, cerium, samarium and scandium have all been recorded (Quality of Urban Air Review Group, 1996).

COMBINED CHEMICAL COMPOSITION AND CHARACTERISTICS OF SUSPENDED PARTICULATES

General Chemistry of PM_{10}

The chemistry of PM_{10} varies with the location and source. Most global studies have focused on the gravimetric evaluation of the amount of PM_{10} matter and certain substances carried by it, but few studies have attempted to achieve a complete characterization of the physical and chemical characteristics of this material. Most surveys and sampling campaigns have covered the analysis of trace metals and organic pollutants associated with PM_{10} particles (Quality of Urban Air Review Group, 1996). These substances constitute only a small percentage of the total mass of PM_{10} matter. This also is the case in the UK, where studies have been performed at different locations and times using a range of techniques (Hutchison, 1996). PM_{10} matter does not have a constant composition, thus studies give us only a general idea of the composition of PM_{10} matter in the UK.

A typical breakdown of PM_{10} particulate matter in the UK has been provided by the Department of the Environment (Quality of Urban Air Review Group, 1996). This analysis reveals that some material can consist of ammonium 5%, sulfate, nitrate and chloride 30%, carbonaceous material 40%, metals 5% and insoluble material (minerals) 20%. This analysis was performed on the data from several studies since no survey covering all the components simultaneously has yet been undertaken in the UK (Quality of Urban Air Review Group, 1996). One chemical analysis of airborne PM_{10} at Cardiff yielded the following result: soluble inorganic matter 40.8%, insoluble inorganic matter 19.2%, organic matter 40% (Hutchison, 1996).

As described previously, the chemistry of ambient airborne particulate material is complex. The chemistry is also affected by physical characteristics of the particles, e.g. their size distribution. Complete characterization of a PM_{10} fraction is therefore a difficult process owing to both its material complexity and also to the fact that only very small samples of airborne particulates can be collected conveniently. Available analytical methods have not dealt with these inconveniences, since little attention has been given to the total identification of airborne particulate material.

In general, PM_{10} samples can be described most conveniently as a number of individual fractions, each of which is either soluble or insoluble in water or in organic solvent; the insoluble component is either carbon-based or inorganic material. This fractionation enables us to make some approximate division of the sample by source. For specific geographical locations, the contribution of any local industrial activity to the airborne particulate sample composition can be assessed and distinguished from the natural contributions of the sea and from secondary particulate formation due to atmospheric chemistry activity.

Figure 28.4 represents the fractionation and analysis of one such PM_{10} sample. This was collected at a central sampling station in a busy town centre (Cardiff, Wales), which is located close to the sea and surrounded by light industry.

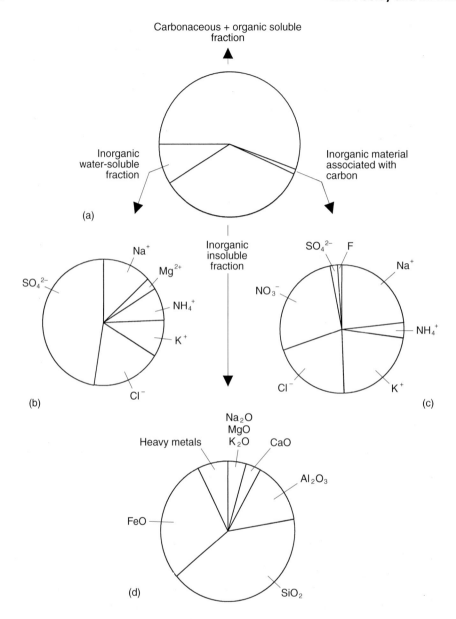

Fig. 28.4 Fractional semiquantitative chemical composition of an urban (Cardiff) PM_{10} sample. (a) Major fractional components of an urban PM_{10} sample. (b) Composition of water-soluble fraction. (c) Composition of inorganic material associated with carbon-based fraction. (d) Composition of inorganic fraction after removal of other components from the sample.

Table 28.1 Semiquantitative composition of an airborne particulate sample collected from the Cardiff area (expressed as parts per thousand on a mass basis)

	Aqueous soluble species	Aqueous soluble species after ashing	Inorganic insoluble species[a]	Fraction of total
Soluble organic and carbonaceous material	–	–	–	555
Na	12	2	2	16
NH_4^+	8	<1	–	8
Mg	3	–	6	8
K	9.	2	8	19
Ca	n.d.	–	12	12
F	n.d.	<1	–	<1
Cl	17	2	–	19
NO_3^-	<1	3	–	3
SO_4^{2-}	41	<1	–	41
Al	<1	<1	49	49
Si	<1	<1	144	144
Ti	–	–	8	8
Cr	<1	<1	4	4
Mn	–	<1	4	4
Fe	<1	<1	100	101
Co	<1	<1	–	<1
Ni	<1	<1	–	<1
Cu	–	<1	1	1
Zn	<1	<1	4	4
Pb	–	–	3	3
Totals	**90**	**9**	**345**	**1000**

n.d. = not detectable.
[a] The components of this fraction are expressed as elemental oxides. Crystalline and adsorbed water are neglected.

This semiquantitative analysis reveals that the sample consists of more than 50% carbon and organic soluble material and 10% water-soluble material. The remainder is insoluble inorganic material. The chemistry of the water-soluble fraction is consistent with the dissolution of secondary PM_{10} material and also with the contribution of sea salt particles. The inorganic insoluble fraction is predominantly silica rich and contains a mixture of naturally derived aluminium silicate particles. The high iron content of this fraction is consistent with particulate emissions from the steel industry in this region, while the insoluble lead comes from vehicular sources.

These proportions of carbon-based inorganic soluble and insoluble fractions can change dramatically from one sampling site to another. Very low levels of carbon-based material will be recorded in the close vicinity of heavy industries; their influence is diluted with distance and weather conditions.

REFERENCES

Amann CA and Siegla DC (1982) Diesel particulates – what they are and why. *Aerosol Sci and Technol* **1**: 73–101.

Brow P (1995) 25 deaths a day hastened by exhaust dust. *The Guardian* 9/11/95.

Burton RM, Suh HH and Koutrakis P (1996) Spatial variation in particulate concentrations within metropolitan Philadelphia. *Environ Sci Technol* **30**(2): 400–407.

Chow JC, Watson JG, Fujita EM *et al.* (1994) Temporal and spatial variations of $PM_{2.5}$ and PM_{10} aerosol in Southern California. *Atmos Environ* **28**(12): 2061–2081.

Ge ZZ, Wexler AS and Johnston MV (1996) Multicomponent aerosol crystallization. *J Colloid Interface Sci* **183**(1): 68–77.

Harrison RM, Smith DJT and Luhana L (1996) Source apportionment of atmospheric polycyclic aromatic hydrocarbons collected from an urban location in Birmingham. *UK Environ Sci Technol* **30**(3): 825–832.

Hutchison LM (1996) *Investigation of PM_{10}s Found in the Urban Environment and Human Lungs.* M. Phil Thesis, University of Wales, Cardiff.

Levin Z, Ganor E and Gladstein V (1996) The effects of desert particles coated with sulfate on rain formation in the Eastern Mediterranean. *J Appl Meteorol* **35**(8): 1151–1523.

Maenhaut W, Francois F, Cafmeyer J and Okunade O (1996) Size-fractionated aerosol composition at Gent, Belgium. *Nucl Instrum Meth Phys B* **109**: 476–481.

Pope III CA (1989) Respiratory disease associated with community air pollution and a steel mill – Utah Valley. *Am J Pub Health*, May, Vol. 79, No. 5, 623–628.

Quality of Urban Air Review Group (1996) *Airborne Particulate Matter in the United Kingdom.* ISBN0 9520771 3 2 Birmingham: University of Birmingham Quality of Urban Air Review Group.

Royal Commission on Environmental Pollution (1994) *Transport and the Environment*, 18th report. London: HMSO.

Saxton KE (1995) Wind erosion and its impact on off-site air quality in the Columbia plateau – an integrated research plan. *Trans Am Soc Agric Eng* **38**(4): 10310–10380.

Thompson RJ, Morgan GB and Purdue LJ (1970) Analysis of selected elements in atmospheric particulate matter by atomic absorption. *Atomic Absorption Newsletter* **9**(3): 53–56.

Warren Spring Laboratory (1994). *Air Pollution in the UK 1992–93.* Hertfordshire: Warren Spring Laboratory.

Watts R, Cupitt L and Zweidinger R (1988) Mutagenicity of organics associated with $PM_{2.5}$ and PM_{10} HiVol particles from a wood smoke impacted residential area. EPA/APCA Measurement of Toxic and Related Air Pollution. Int Symp, pp 879–886.

Whei-May, Lee G and Lin-Yi, Tsay (1994) The partitioning model of polycyclic aromatic hydrocarbon between gaseous and particulate (PM_{10}) phases in urban atmosphere with high humidity. *Sci Total Environ* **145**(1–2): 163–172.

Metals and Air Pollution Particles

ANDREW J. GHIO and JAMES M. SAMET

National Health and Environmental Effects Research Laboratory,
Environmental Protection Agency, Research Triangle Park, NC, USA

This report has been reviewed by the National Health and Environmental Effects Research Laboratory of the United States Environmental Protection Agency and approved for publication. Approval does not signify that the contents necessarily reflect the views and policies of the Agency nor does mention of trade names or commercial products constitute endorsement or recommendation for use.

INTRODUCTION

Metals are essential micronutrients utilized in almost every aspect of normal cell function and are particularly crucial for cellular metabolism. The transfer of electrons to O_2 is often catalyzed by metals because of their ability to rearrange orbital electrons into different spin states. It is the alterations in the spin states of the metal ion upon complex formation with O_2 that compensates for spin changes in O_2 during biological reactions and allows the activation energy of metabolic reactions to be met. As a result of this interaction with O_2 and because of their abundance in nature, metals have been selected in molecular evolution to carry out a wide range of biological functions and are required by all living organisms. Life has evolved with a dependency on metal availability and almost all living organisms require metals, including iron, copper, zinc, cobalt, manganese, chromium, molybdenum, vanadium, selenium, nickel and tin.

The same chemical properties that allow metals to function as catalysts in reactions involving molecular oxygen can make them a threat to life via the generation of O_2-based free radicals. Thus there exists a dilemma for aerobic organisms: in order to catalyze essential homeostatic and synthetic functions, the risk of damaging oxidative stress must be incurred.

The mechanism(s) of lung injury after exposure to ambient air pollution particles is not known. Injury has been postulated to be mediated by ultrafine particles, biological agents

AIR POLLUTION AND HEALTH
ISBN 0-12-352335-4

Metals

⇓

Oxidative stress

⇓

Signaling mechanisms

⇓

Mediator expression

⇓

Lung injury

Fig. 29.1 Metal-catalyzed oxidative stress and lung injury after exposure to air pollution particles.

(e.g. endotoxins), acid aerosols, and polyaromatic hydrocarbons present in the organic components of the particle. Another potential mechanism of injury to human tissues after exposures to air pollution particles is through metal-catalyzed oxidant generation. As discussed below, the oxidative stress catalyzed by the metals present in air pollution particles can result in phosphorylation-dependent cell signaling, and an activation of specific transcription factors such as nuclear factor (NF)-κB (Fig. 29.1). An increased expression of mediators can be effected. These mediators include those genes which have NF-κB binding sites in their promoter regions, such as interleukin (IL)-8, IL-6 and tumor necrosis factor (TNF). One possible consequence of the enhanced synthesis of mediators that results is inflammatory injury to the lung.

SPECIFIC METALS IN AIR POLLUTION PARTICLES

Measurement

The atmosphere constitutes a prime vehicle for the movement and redistribution of metals (Schroeder et al., 1987). Human activities have had a major impact on both the global and regional cycles of metals (Nriagu and Pacyna, 1988). As a result, there has been a significant contamination of air and water resources and an accumulation of metals in food chains.

Almost all metals in the atmosphere are associated with particles (Schroeder et al., 1987). Worldwide, the total particle production is $1–4 \times 10^9$ tons/year, with approximately 5–20% of that being anthropogenic in origin (Schroeder et al., 1987). Although each metal comprises only a small proportion of that figure, the total mass of each metal is still significant.

Since they are associated with particles, residence times of metals in the atmosphere are short, typically less than 40 days. Transport of metals nevertheless occurs over long distances, causing perturbations of ecosystems on a global scale (Nriagu and Pacyna, 1988).

To measure metals, air pollution particles must be collected. This is typically accomplished by employing either a filter or inertial separation. A number of specific, sensitive and accurate standardized techniques are available to quantify metal concentrations. However, there is no convention for choosing the procedures used to extract the metals for analysis. Atomic absorption spectrometry is the most widely employed technique for elemental analysis of airborne metals. Inductively coupled plasma emission spectroscopy also can be used for the simultaneous determination of up to 20 metals. Neutron activation analysis also can provide a nondestructive determination of up to 35 elements with high precision and accuracy. Finally, X-ray fluorescence spectrometry is a non-destructive technique that can simultaneously determine up to 30 elements. A major limitation of these analytical techniques is that they provide information on total metal content rather than specific compounds or chemical species.

Specific Metals

Air pollution particles from natural sources can be derived from terrestrial dust dispersed by wind and automobiles, from sea spray, biogenic emanations, volcanic dust, fires, pollen, plant debris and also from reactions between natural gaseous emissions. These particles usually have a mass median aerodynamic diameter (MMAD) greater than 2.5 μm in diameter and therefore can be isolated in the coarse fraction. In one rural area, coarse particles included clay minerals, quartz, gypsum and calcite (the four accounting for 50% of the mass of this fraction) (Mamane and Noll, 1985). In addition, there were spores, pollen and plant debris (25%). Finally, fly ash (9%) and sulfates (11%) also were found. This accumulation of sulfur in the coarse fraction of particles can reflect the heterogeneous nucleation of SO_2 on the surfaces of pollen and mineral oxides (Shaw and Paur, 1983).

The metals in air pollution particles originating from natural sources vary with the source of the particles (Schroeder et al., 1987). Those metals in highest concentrations in crustal dust (in order of abundance) are iron, manganese (550 μg/g), zinc, lead, vanadium, chromium, nickel, copper, cobalt, mercury and cadmium. While the ocean generally acts as a sink for toxic metals, ocean aerosol contains concentrations of metals which are very low, with the highest being iron (5 μg/g), manganese, lead, vanadium (0.0009 μg/g) and zinc. Volcanic ash can have great amounts of iron, manganese, vanadium, zinc, cobalt, arsenic and antimony (in decreasing order of importance). Finally, biological emanations contain low concentrations of zinc (from the leaves of many trees), mercury (from peas), and vanadium, chromium, manganese, iron, cobalt, nickel, copper, arsenic, lead, cadmium and antimony (from coniferous trees).

Metals are more frequently associated with air pollution particles of anthropogenic origin. These typically have values of MMAD less than 2.5 μm (the fine fraction) (Schroeder et al., 1987). Such particles originate from the incomplete combustion of carbon-containing materials at power plants, smelters, incinerators, cement kilns, home furnaces, fireplaces, and motor vehicles.

Acid sulfates also are a major constituent of fine particles. SO_2 (g), present in the atmosphere at concentrations approximating 140–840 nmol/m^3 (9–55 μg/m^3), is oxidized to SO_4^{2-} (Jacob *et al.*, 1989). Also prominent among the products of combustion in fine particles are soot and fly ash. The incomplete oxidation of carbonaceous materials can generate soot, which is a mixture of particulate carbon (C_8H; 90–98% carbon by weight) with organic and inorganic components. Fly ash also is generated. It is the inorganic residue that is left after burning such a fuel or waste substance. The composition of fly ash is not uniform, but it is usually characterized by a high content of Fe_2O_3, Fe_3O_4, Al_2O_3, SiO_2 and carbonaceous compounds that have undergone varying degrees of oxidation (Greenberg *et al.*, 1978).

Fine particles are removed by electrostatic precipitators, but some escape these devices and are released into the environment. The emission of metals from these combustion processes is determined by the level of metals in the fuel or waste being burned, the nature of the process, and the type of controls operating at the emitting facilities.

Iron is the metal present in greatest concentration in most emission source air pollution particles. An exception is oil fly ash, which can contain significant concentrations of vanadium and nickel compounds (Henry and Knapp, 1980). This fly ash is emitted by power plants and other industries that burn oil. It can also be unique in its pronounced quantities of other metals including zinc, lead, copper, arsenic, cobalt, chromium, manganese and antimony. However, it contains few or no organics or biological components. The characteristics of oil fly ash make it a useful model particle to use in studies which test the association between metals present in air pollution particles and various biological endpoints.

Fly ash from coal-fired power plants also has high concentrations of iron, zinc, lead, vanadium, manganese, chromium, copper, nickel, arsenic, cobalt, cadmium, antimony and mercury (Schroeder *et al.*, 1987). The incineration of municipal waste is becoming a more important atmospheric source of metals, including zinc, iron, mercury, lead, tin, arsenic, cadmium, cobalt, copper, manganese, nickel and antimony. Open-hearth furnaces at steel mills are a notable source of iron, zinc, chromium, copper, manganese, nickel and lead. Vehicle emissions also can contain relatively small concentrations of both zinc and iron. Prior to the banning of leaded gasoline in the USA, vehicle emissions represented a major source of atmospheric lead. The primary sources of this metal now are resuspended soil, oil burning and small-scale smelting. Among metals in emission air pollution particles, concentrations of each metal can correlate with the other metals and as one is increased in the specific particle, they all can be found in elevated quantities (Pritchard *et al.*, 1996).

As in the case of emission air pollution particles (both natural and anthropogenic), iron is the metal in highest concentration in ambient air pollution particles (Demuynck *et al.*, 1976; Dodd *et al.*, 1991; Mamane, 1990; Stevens and Dzubay, 1978; Vossler *et al.*, 1989). The concentration of iron in the atmosphere can regularly be found in concentrations several-fold greater than that of all other metals. This metal concentrates in the fogwater of urban settings, in cloud droplets and in aerosol particles and can approach concentrations of 1 mmol/l (Jacob *et al.*, 1989; Munger *et al.*, 1983; Waldman *et al.*, 1982).

Iron also is common in PM_{10}. Studies of the concentration of metals in air pollution particles sequestered on PM_{10} filters from sites around the USA reveal that iron is the metal present in greatest concentration. It has been found associated with the particles on every filter (Table 29.1). Concentrations of titanium, vanadium, chromium, manganese, nickel and copper are also significant in these sequestered particles, while levels of cobalt approach zero. The concentrations of individual metals in ambient air pollution particles correlated well with each other at a variety of geographic sites (Ghio *et al.*, 1996). This result is similar to that for emission source air pollution particles.

Table 29.1 Metal concentrations (µmol/l) associated with particulate air pollutants sequestered on filters after extraction of filters in acid

Filter	Metal							
	Ti	V	Cr	Mn	Fe	Co	Ni	Cu
Blank	0.0	1.6	2.4	0.5	13.6	0.0	4.7	0.5
Boston	7.0	7.0	0.5	7.7	122.4	0.0	4.3	18.9
Cottonwood Cyn.	0.5	2.2	2.5	0.9	23.1	0.0	4.0	23.4
Dallas	0.0	7.7	2.7	1.7	49.4	0.3	3.8	26.6
Denver	4.1	22.6	12.4	7.2	78.1	0.4	5.5	19.2
Lake Forest Park	0.0	4.4	2.6	1.3	39.1	0.0	4.1	7.7
Newark	2.0	27.3	21.6	10.4	93.9	0.6	5.7	15.1
Ogden	0.8	3. 7	2.6	1.7	30.2	0.0	4.0	6.2
Philadelphia	0.0	33.8	4.0	1.8	122.4	0.0	3.7	3.2
Pleasant Valley	1.0	3.9	3.2	1.1	40.1	0.0	4.0	4.2
Provo	13.1	147.2	38.2	100.3	1661.7	1.7	5.8	17.0
Salt Lake City	2.4	32.7	4.2	8.7	743.5	0.4	4.4	437.2
Trenton	0.0	74.3	5.3	11.2	111.1	0.4	3.9	1.0

Chemical Forms and Solubility

There is little information available regarding the chemical forms and physical/chemical transformations of metals in the atmosphere (Schroeder et al., 1987). It has been assumed that the metals of anthropogenic origin are present in the atmosphere as oxides. However, in numerous emission source air pollution particles and in several ambient air pollution particles, a majority of the metals occasionally exist in a soluble form rather than in highly insoluble oxide forms (e.g. oil fly ash, in which approximately 90% of compounds containing vanadium, nickel and iron-containing compounds such as vanadyl sulfate, nickel sulfate and ferric sulfate are water-soluble) (Faust and Hoffmann, 1986). Large quantities of the metal (12–56%) in fuel ash also can solubilize within 20 min in atmospheric water (Weschler et al., 1986). A significant mass of metal oxide appears to be photoreduced upon dissolution in an aqueous medium (Finden et al., 1984). This solubilization is dependent on the presence of a suitable ligand for the metal. Sulfate, a ligand available in the atmosphere in high concentrations, can complex many metals (e.g. estimates of stability constants between the ferrous and ferric ion with sulfate range between 10^2/mol and 10^5/mol) (Cavasino, 1968; Perrin, 1979). Consequently, the concentrations of sulfates and metals can be linearly related, suggesting that there is an association between them (Electric Power Research Institute, 1976; Moyers et al., 1977).

In addition to solubilization after photoreduction, arsenic, cadmium, manganese, nickel, lead, antimony, selenium, vanadium and zinc can volatilize at high temperatures during fossil fuel combustion. They also can condense uniformly on surfaces of entrained fly ash particles as the temperature falls below the temperature of the combustion zone. Significant coagulation and interaction occurs between ambient particles (both natural and anthropogenic in origin) and emitted species, as evidenced by the arsenic, copper and zinc that are found coated on clay or mineral agglomerates. All of these reactions affect the associations of metals in air pollution particles and influence their relative mobility and bioavailability in the atmosphere.

A majority of the metals associated with many ambient air pollution particles are insoluble in aqueous solution (i.e. the metal is not present as a water-soluble salt) (Pritchard *et al.*, 1996). It is therefore likely that components of atmospheric particles other than sulfates either (1) have a capacity to coordinate metals or (2) contain insoluble metal within a structural lattice (e.g. Fe_2O_3). Incompletely oxidized carbon has the capacity to complex metals at the surface where oxygen-containing functional groups (alcohol, carbonyl and carboxyl groups) exist. Quantification of total acidity and carboxylates in air pollution particles on a filter collected from several US cities identified one such carbonaceous product with a capacity to complex metals (Plate 2) (Ghio *et al.*, 1996). Furthermore, inorganic components of fly ash have this same capacity to coordinate metal cations incompletely. On the surfaces of mineral oxides included in fly ash there are numerous ligands which can coordinate metal cations (Schindler and Stumm, 1987). These functional groups can include —Ti—OH (e.g. rutile), —Si—OH (e.g. silica and silicates), —Mg—OH (e.g. silicates such as bentonite, cummingtonite, grunerite, talc, vermiculite), —Al—OH (e.g. Al_2O_3

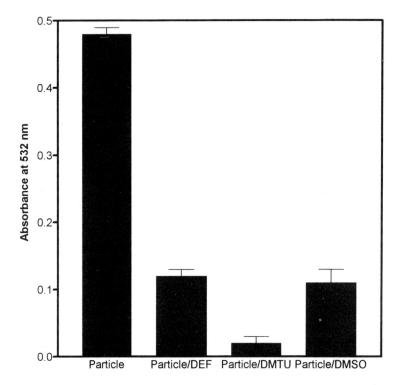

Fig. 29.2 Oxidant generation by residual oil fly ash. Radical production was measured using thiobarbituric acid (TBA) reactive products of deoxyribose. The reaction mixture containing 1.0 mmol/l deoxyribose, 1.0 mmol/l H_2O_2, 1.0 mmol/l ascorbate and 2.0 mg dust was incubated at 37°C for 30 min with agitation and then centrifuged at 1200 *g* for 10 min. One ml of both 1.0% (w/v) TBA and 2.8% (w/v) trichloroacetic acid were added to 1.0 ml of supernatant, heated at 100°C for 10 min, cooled in ice, and the chromophore concentration determined by its absorbance at 532 nm. Measurements were done in triplicate. The oxidation of deoxyribose associated with particle exposure was inhibited by both hydroxyl radical scavengers, dimethylsulfoxide (DMSO) and dimethylthiourea (DMTU) and the iron chelator deferoxamine (DEF), all at a final concentration of 1.0 mmol/l.

and silicates such as bentonite, feldspar, kaolinite and vermiculite), and —Fe—OH (e.g. iron oxides and silicates such as bentonite, cummingtonite, grunerite and vermiculite). The presentation of both soluble and incompletely complexed metal cations thus varies among air pollution particles. This property can be a key determinant of the ability of a particle to induce oxidant stress in cells.

OXIDATIVE STRESS AND METALS IN PARTICLES

In Vitro Oxidative Stress

Acellular

Metals which can exist in only one stable valence state (e.g. arsenic and cadmium) can present an oxidative stress either through a reaction with and depletion of sulfhydryls or through an interruption of normal iron metabolism (Stohs and Bagchi, 1995). In contrast, those metals which exist in more than one stable valence state can directly participate in electron transfer reactions. They consequently possess a potential to generate the oxidants that are key reactive intermediates in atmospheric chemistry (Behra and Sigg, 1990; Jacob *et al.*, 1989). A labile coordination site on the metal is required to allow its direct participation in electron transfer and lead to the generation of the highly reactive hydroxyl radical:

$$Fe^{2+} + H_2O_2 \Leftrightarrow Fe^{3+} + \cdot OH + {}^-OH$$

Metal in air pollution particles complexed to sulfate, nitrate, incompletely oxidized carbon fragments and functional groups at the surface of an inorganic oxide directly catalyzes oxidant generation. In addition to the metal, hydrogen peroxide and reductants may be required to drive free radical generation. Both hydrogen peroxide and reductants (including sulfites, aldehydes, and Cu^+) are available in the atmosphere (Jacob *et al.*, 1989). Certain atmospheric metal cations also can be photochemically reduced (Logan *et al.*, 1981). The resultant cycling through reduced and oxidized states to generate hydroxyl radicals has been shown to occur in the atmosphere (Behra and Sigg, 1990). Similarly, H_2O_2 and molecules with reducing potential are measurable in the lung environments of animals, including humans (Rom *et al.*, 1987; Slade *et al.*, 1985).

All the first row transition metals present in a number of emission source and ambient air pollution particles can catalyze oxidant generation when placed in a system containing hydrogen peroxide and ascorbate (Pritchard *et al.*, 1996). This free radical production in an acellular system containing air pollution particles correlated positively with concentrations of chromium, manganese, iron, cobalt, nickel and copper (Pritchard *et al.*, 1996). Deferoxamine inhibits oxidant production by these particles, consistent with metal participation in the catalysis of radicals (Fig. 29.2). Similarly, the hydroxyl radical scavengers dimethylthiourea (DMTU) and dimethylsulfoxide (DMSO) inhibited radical generation, implicating hydroxyl radical production (Fig. 29.2). Lastly, the direct hydroxylation of salicylate after *in vitro* exposure to air pollution particles strongly suggests metal-dependent ·OH production (Ghio *et al.*, 1996).

Cellular

Air pollution particles also have been associated with cellular generation of oxidants (Ghio *et al.*, 1997). *In vitro* exposure of rat alveolar macrophages to an emission source air pollution particle (residual oil fly ash) increases the cellular production of oxidants. This elevation in radical generation can be inhibited by the addition of either the metal chelator deferoxamine or the hydroxyl radical scavengers DMTU and DMSO. Addition of individual metals to alveolar macrophages as both soluble sulfates (and when complexed to the surface of a latex bead) similarly increased the oxidative burst in alveolar macrophages. This augmentation in oxidant production supports a potential role for metals in the increased production of oxygen-based radicals by phagocytes after exposure to an air pollution particle.

These findings are significant because phagocytes can be central in resident host defenses against particles. Metals associated with air pollution particles could potentially increase the oxidative burst by these cells, augment the production of destructive free radicals, and intensify injury to the respiratory tract.

In Vivo Oxidative Stress

Electron spin resonance (ESR) spectra support *in vivo* production of free radical(s) in the lungs of a Sprague Dawley rat exposed to an air pollution particle (Kadiiska *et al.*, 1997). The detection and identification of a chloroform-soluble radical adduct following instillation of a residual oil fly ash in the rat lung also has been demonstrated (Fig. 29.3). In addition, instillation in an animal of specific individual metal compounds present in the air pollution particle was associated with ESR spectra similar to those obtained after exposure of the animal to the air pollution particle (Fig. 29.4).

Preliminary data also support the *in vivo* generation of hydroxyl radical (Schapira *et al.*, 1997). This radical can initiate lipid peroxidation in the lower respiratory tract and causes the generation of alkyl radicals (Iwahashi *et al.*, 1991; Janzen *et al.*, 1987). After instillation of oil fly ash in an animal model, ESR analysis identified ethyl and pentyl radicals. These radicals result from the reductive decomposition of lipid peroxides. These data support the occurrence of lipid peroxidation in the lung, initiated by a metal-catalyzed hydroxyl radical after *in vivo* exposure to an air pollution particle.

CELLULAR EXPRESSION OF INFLAMMATORY MEDIATORS AND METALS IN AIR POLLUTION PARTICLES

Intracellular Signaling Mechanisms

Metals have been shown to disrupt differentially the signaling pathways that regulate cellular responses to stimuli, including toxic insults. For instance, cadmium, chromium and zinc ions have been shown to activate protein kinase C (PKC) (Bagchi *et al.*, 1997; Csermely and Somogyi, 1989). In contrast, exposure to mercury and lead inhibits PKC activity, apparently through a mechanism which involves the formation of dithiols

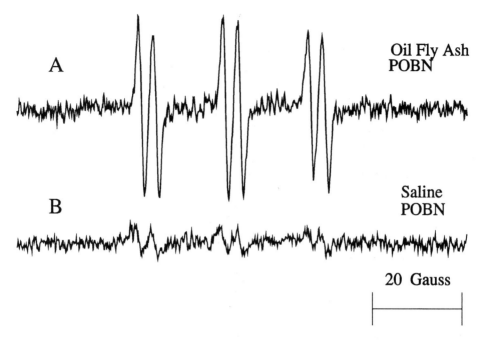

Fig. 29.3 (A) ESR spectrum of 4-POBN radical adducts detected in lipid extracts of lungs 24 h after intratracheal instillation of oil fly ash (500 μg per rat) and 1 h after intraperitoneal administration of 4-POBN (8 mmol/kg). (B) Same as in (A), but rats were instilled with saline. Instrumental conditions: microwave power, 20 mW; modulation amplitude, 1.33 G; time constant, 1 s; and scan rate, 5 G/min. Spectra reveal *in vivo* radical generation only after exposure to oil fly ash.

(Rajanna *et al.*, 1995). The heavy metals mercury, cadmium, copper and zinc also are reported to be inhibitors of PKA, although the mechanism may be different from that involved in PKC inhibition (Speizer *et al.*, 1989). Similarly, aluminum ions inhibit inositol phosphate accumulation and interfere with intracellular calcium homeostasis (Haug *et al.*, 1994). As discussed below, we also have shown that arsenic, vanadium and zinc potently activate the mitogen activated protein kinases (MAPKs) in human airway epithelial cells (Samet *et al.*, 1998a,b).

Intracellular signaling pathways share the phosphorylation of specific amino acid residues as a regulatory mechanism through which the transduction of intracellular signals is controlled. The balance between the activity of kinases that phosphorylate proteins and phosphatases that remove the phosphates determines the state of activation of the signaling pathway, and thus modulates the response of the cell to the stimulus (Hunter, 1995). Disruption of phosphorylation homeostasis appears to be a major mechanism of metal-induced signaling that leads to inflammatory mediator expression.

Vanadium is the best understood of the metals that are capable of disrupting protein phosphorylation. Tetravalent or pentavalent vanadium ions are potent inhibitors of protein tyrosine phosphatases and induce the massive accumulation of protein phosphotyrosines in a variety of cell types (Conde *et al.*, 1995; Hecht and Zick, 1992). Oil fly ash contains significant quantities of vanadium (Henry and Knapp, 1980). It is this element that is largely responsible for the toxicity of this air pollution particle. We have

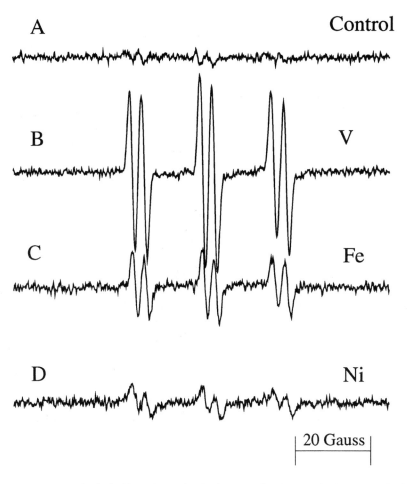

Fig. 29.4 ESR spectrum of radical adducts detected in lipid extract of rat lungs 24 h after intratracheal instillation of (A) 0.5 ml of saline, (B) 0.5 ml of 1.0 mmol/l vanadyl sulfate, (C) 0.5 ml of 1.0 mmol/l iron(III) sulfate, and (D) 0.5 ml of 1.0 mmol/l nickel sulfate and 1 h after intraperitoneal administration of 4-POBN (8 mmol/kg). Instrumental conditions: microwave power, 20 mW; modulation amplitude, 1.33 G; time constant, 1 s; and scan rate, 5 G/min. Spectra confirm *in vivo* radical generation after exposure to vanadium and iron.

shown that exposure to oil fly ash induces a rapid and marked increase in protein tyrosine phosphate accumulation in human airway epithelial cells, and that this effect is exclusively associated with its vanadium content (Samet *et al.*, 1997). Our studies also revealed a large decrease (>85%) in protein tyrosine phosphatase activity in cells exposed to oil fly ash, with no significant effect on tyrosine kinase activity.

Recent work conducted in our laboratory tested the effect of arsenic (III), chromium (III), chromium (VI), copper (II), iron (II), nickel (II), vanadium (IV) and zinc (II) ions on the phosphorylation and activation of the three major branches of the MAPK signaling pathways, ERK, JNK and P38, in the human airway epithelial cell line BEAS 2B (Samet *et al.*, 1998b). Of the metals studied, arsenic, vanadium and zinc induced a marked activation of the enzymes ERK, JNK and P38, while chromium and copper ions caused relatively small activations. Iron and nickel were without effect. Although the mechanisms responsible for the MAPK activations were not examined, it is interesting to note that, like vanadium, arsenic (Cavigelli *et al.*, 1996) and zinc (Wang and Pallen, 1992) have been reported to inhibit protein phosphatases.

Thus, phosphatase inhibition may represent an important initiating event in the activation of signal transduction pathways by metals. The mechanism involved in metal-induced inhibition of phosphatases is not entirely clear. Vanadate and arsenate are phosphate analogs that may function as steric inhibitors of phosphatases (Gordon, 1991; Goyer, 1996). Hydrogen peroxide also is a potent inhibitor of phosphatase activity which is generated by cells treated with transition metals such as vanadium, or by cells undergoing an oxidative burst in response to a stimulus. It has been proposed that the vanadium species that functions as a phosphatase inhibitor is actually a peroxovanadyl or vanadyl hydroperoxide ion formed by the reduction of vanadate by superoxide (Trudel *et al.*, 1991).

Signaling pathways such as the MAPK cascades culminate in the activation of transcription factors such as NF-kB, c-jun and ATF-2 (Israel, 1995; Treisman, 1996), which in turn regulate the transcriptional expression of a variety of genes, including those encoding inflammatory proteins (Davis, 1995; Munoz *et al.*, 1996; Tsai *et al.*, 1996). Our laboratory recently reported activation of NF-kB in human airway epithelial cells treated with oil fly ash (Quay *et al.*, 1997) and we recently observed the same effect in cells exposed to vanadium and arsenic (J.M. Samet, unpublished observations). We also have shown that the non-cytotoxic concentrations of arsenic, vanadium and zinc that cause MAPK activation also induce phosphorylation of c-jun and ATF-2, transcription factors that are substrates of JNK and P38, respectively (Samet *et al.*, 1998b). Therefore, metal-induced activation of signaling pathways leads to a functional activation of transcription factors capable of effecting the expression of inflammatory proteins.

Prostanoids

From a toxicological standpoint, the primary consequence of metal-induced activation of signaling pathways leading to transcription factor activation is the potential for enhanced expression of proinflammatory proteins, which may mediate the cardiopulmonary effects of exposure to particles. Inasmuch as oil fly ash may be considered to be a model of metallic constituents in air pollution particles, the findings discussed below suggest that increased expression of inflammatory mediators such as prostanoids and cytokines may mediate the pulmonary effects of ambient particles.

Exposure to oil fly ash results in marked increases in the cyclooxygenase (COX) metabolites of arachidonic acid PGE_2 and 15-HETE in BEAS cells and in primary cultures of human airway epithelial cells (Samet *et al.*, 1996). Striking increases in these COX products are evident within 24 h of continuous exposure of human airway epithelial cells to oil fly ash. Our investigation of the mechanism responsible for the increase in prostanoids showed significant increases in COX activity at 24 h of exposure, which was reflected by an increase in protein levels of COX2 (the inducible form of the enzyme). The increase in COX2 expression was confirmed by increased levels of COX2 mRNA, apparent after 2 h of exposure to oil fly ash, i.e. preceding the increase in COX2 protein, activity and the appearance of prostanoids in the culture media. In contrast, no alterations in the mechanisms that control arachidonic acid availability (phospholipase A_2 and ester-ification activities) or increases in free arachidonic acid levels were detected in cells exposed to the emission source air pollution particle (Samet *et al.*, 1996).

We have recently confirmed that COX2 metabolites play a role in the induction of lung injury after exposure to oil fly ash (Samet *et al.*, 1996). Intratracheal instillation of oil fly ash results in increased expression of COX2 protein detected immunohistochemically, and an accompanying elevation in the COX product PGE_2 recovered in bronchoalveolar lavage fluid. Paralleling these findings is an increase in protein in the lavage fluid (a marker of inflammatory lung injury) in animals exposed to oil fly ash. The increases in lavage protein and PGE_2 are mitigated by pretreatment of the animals with the specific COX2 inhibitor NS398, demonstrating a dependency of the lung injury on COX2 metabolites.

Cytokines

We have shown that metals also can effect the expression of cytokines in human airway epithelial cells (Carter *et al.*, 1997). Vanadium or oil fly ash induces significantly elevated expression of IL-8, IL-6 and TNF in primary human airway epithelial cell cultures, as determined by radioimmunoassay of the culture media and by reverse-transcriptase PCR analysis of cellular RNA preparations (Carter *et al.*, 1997). These effects were dependent on the presence of metals, as shown by the significant inhibition by the chelator deferox-amine. Moreover, this induction of cytokine expression was blocked by the radical scavenger dimethylthiourea, suggesting the involvement of reactive oxygen species. Interestingly, treatment with either nickel or iron ions did not result in increased cytokine expression in airway epithelial cells. We recently have demonstrated increased IL-8 expression in human airway epithelial cells treated with arsenite (J.M. Samet *et al.*, unpublished observations).

Dreher and colleagues have reported cytokine expression in rats instilled with oil fly ash (Dreher *et al.*, 1997). RT-PCR analyses showed that RNA transcripts of IL-1, IL-5, IL-6 and MIP-2 (but not TNF-α) were significantly elevated in the lungs of exposed animals relative to saline-exposed controls.

LUNG INJURY ASSOCIATED WITH METALS AFTER EXPOSURE TO AIR POLLUTION PARTICLES

Free radical generation catalyzed by metals present in the particle could contribute to lung injury after exposure to air pollution particles. Other models of oxidative stress, such as exposure to mineral oxide particles and oxidant gases, also can be associated with either the production or release of mediators of acute inflammation. In addition, lung exposure to metals and/or catalyzed oxidants can be associated with bronchoconstriction (O'Byrne *et al.*, 1984) and an elevated incidence of infections (Weinberg, 1974).

INFLAMMATORY LUNG INJURY

After instillation in an animal model, air pollution particles (emission source and ambient) provoked both an influx of neutrophils and an increase in the concentration of protein in lavage fluid (Pritchard *et al.*, 1996). Furthermore, both the cellular incursion and protein concentrations increased with the ionizable concentrations of metals in the particles (Fig. 29.5). Exposure of an animal to the soluble component of an oil fly ash, which contains almost all the metals present in unfractionated oil fly ash, reproduced the *in vivo* radical production and neutrophilic injury (Kadiiska *et al.*, 1997). Similarly, the instillation in a rat of both a mixture of metal sulfates, reflecting their concentrations in the oil fly ash, and the individual metal compounds included in the air pollution particle, was associated with oxidative stress and injury (Dreher *et al.*, 1997; Kadiiska *et al.*, 1997). The fact that metal salts can replicate the neutrophilic lung injury induced by instillation of an emission source air pollution particle strongly implicates a role for metals in tissue damage following exposure to particulate matter.

Airway Hyperreactivity

Elevated levels of air pollution can increase airway reactivity, especially in patients with pre-existing asthma. The National Health and Nutrition Examination surveys found an inverse relationship between ambient levels of total suspended particles and forced vital capacity and forced expiratory volume in 1 s (FEV$_1$) across a wide geographic and sociologic spectrum (Schwartz, 1989). Other studies also support the relationship between air pollution particles and decrements in pulmonary function (Dockery *et al.*, 1982). Instillation of air pollution particles in animals also results in changes in both lung

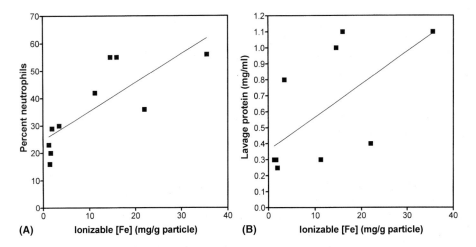

Fig. 29.5 The associations between ionizable iron concentrations in ten different emission source and ambient air pollution particles and lavage neutrophils (A) and protein (B) 4 days after intratracheal instillation (5000 µg) into rats. The percentage of both neutrophils and protein in the lavage increased with the concentration of iron. Similar associations between these end-points exist for numerous metals.

resistance and dynamic compliance (Pritchard *et al.*, 1996). These differences appear to be correlated with the concentrations of metals in the particles and the oxidant generation by the particle. Airway reactivity after oxidant exposures can be associated with an influx of inflammatory cells. This inflammation is believed to increase reactivity of the airways through several putative mechanisms, including wall edema, mediator release and epithelial damage (Holgate and Finnerty, 1988). The relationships of inflammation with both ionizable metal concentrations and oxidant generation are consistent with the increase in bronchial reactivity observed with these particles.

Respiratory Infections

Air pollution exposure has been associated with an increased susceptibility to respiratory infections (Florey *et al.*, 1979). Iron and other metals promote bacterial growth and virulence (Weinberg, 1974). The instillation of certain air pollution particles with high concentrations of metals enhances the virulence of a Streptococcal infection and leads to a higher mortality relative to saline-instilled controls (Pritchard *et al.*, 1996). These results suggest a potential participation of metals present in the air pollution particles in infectious complications following a variety of respiratory illnesses.

CONCLUSIONS

Metal cations are a constituent of both emission source and ambient air pollution particles. These metals have been shown to present an oxidative stress both *in vitro* and *in vivo*. Lung injury after exposure to air pollution particles corresponds to the concentrations of metals present in the particles and, in an animal model, can be reproduced by instillation of individual metals. The adverse health effects attributed to suspended sulfates also have been demonstrated to be significantly associated with atmospheric concentrations of transition metals capable of complexing with the oxidized sulfur groups. Similarly, the correlations between *in vivo* tissue damage and particle size may reflect the increased concentrations of catalytically active metals present in smaller particles.

REFERENCES

Bagchi D, Bagchi M, Tang L and Stohs SJ (1997) Comparative *in vitro* and *in vivo* protein kinase C activation by selected pesticides and transition metal salts. *Toxicol Lett* **91**: 31–37.

Behra P and Sigg L (1990) Evidence for redox cycling of iron in atmospheric water droplets. *Nature* **344**: 419–421.

Carter JD, Ghio AJ, Samet JM and Devlin RB (1997) Cytokine production by human airway epithelial cells after exposure to an air pollution particle is metal dependent. *Am J Respir Cell Mol Biol* **146**: 180–188.

Cavasino FP (1968) A temperature-jump study of the kinetics of the formation of the monosulfato complex of iron (III). *J Phys Chem* **72**: 1378–1384.

Cavigelli M, Li WW, Lin A *et al.* (1996) The tumor promoter arsenite stimulates AP-1 activity by inhibiting a JNK phosphatase. *EMBO J* **15**: 6269–6279.

Conde M, Chiara MD, Pintado E and Sobrino F (1995) Modulation of phorbol ester-induced respiratory burst by vanadate, genistein, and phenylarsine oxide in mouse macrophages. *Free Radic Biol Med* **18**: 343–348.

Csermely P and Somogyi J (1989) Zinc as a possible mediator of signal transduction in T lymphocytes. *Acta Physiol Hung* **74**: 195–199.

Davis RJ (1995) Transcriptional regulation by MAP kinases. *Mol Reprod Dev* **42**: 459–467.

Demuynck M, Rahn KA, Janssens M and Dams R (1976) Chemical analysis of airborne particulate matter during a period of unusually high pollution. *Atmos Environ* **10**: 21–26.

Dockery DW, Ware JH, Ferris BG Jr *et al.* (1982) Change in pulmonary function associated with air pollution episodes. *J Air Pollut Control Assoc* **32**: 937–942.

Dodd JA, Ondov JM, Tuncel G *et al.* (1991) Multimodal size spectra of submicrometer particles bearing various elements in rural air. *Environ Sci Technol* **25**: 890–903.

Dreher KL, Jaskot RH, Lehmann JR *et al.* (1997) Soluble transition metals mediate residual oil fly ash induced acute lung injury. *J Toxicol Environ Health* **50**: 285–305.

Electric Power Research Institute (1976) *Research Project 681*: Determination and possible public health impact of transition metal sulfite aerosol species.

Faust BC and Hoffmann MR (1986) Photoinduced reductive dissolution of alpha-Fe_2O_3 by bisulfite. *Environ Sci Technol* **20**: 943–948.

Finden DAS, Tipping E, Jaworski GHM and Reynolds CS (1984) Light-induced reduction of natural iron (III) oxide and its relevance to phytoplankton. *Nature* **309**: 783–784.

Florey DV, Melia RJW, Chinn S *et al.* (1979) The relationship between respiratory illness in primary schoolchildren and the use of gas for cooking. Nitrogen dioxide, respiratory illness and lung infection. *Int J Epidemiol* **8**: 347–353.

Ghio AJ, Stonehuerner J, Pritchard RJ *et al.* (1996) Humic-like substances in air pollution particulates correlate with concentrations of transition metals and oxidant generation. *Inhal Toxicol* **8**: 479–494.

Ghio AJ, Meng ZH, Hatch GE and Costa DL (1997) Luminol-enhanced chemiluminescence after *in vitro* exposure of rat alveolar macrophages to oil fly ash is metal dependent. *Inhal Toxicol* **9**: 255–271.

Gordon JA (1991) Use of vanadate as protein-phosphotyrosine phosphatase inhibitor. *Methods Enzymol* **201**: 477–482.

Goyer RA (1996) Toxic effects of metals. In Klaassen CD (ed.) *Toxicology. The Basic Science of Poisons.* New York: McGraw Hill, pp 691–736.

Greenberg RR, Zoller WH and Gordon GE (1978) Composition and size distributions of particles released in refuse incineration. *Environ Sci Technol* **12**: 566–573.

Haug A, Shi B and Vitorello V (1994) Aluminum interaction with phosphoinositide associated signal transduction. *Arch Toxicol* **68**: 1–7.

Hecht D and Zick Y (1992) Selective inhibition of protein tyrosine phosphatase activities by H_2O_2 and vanadate *in vitro*. *Biochem Biophy Res Commun* **188**: 773–977.

Henry WM and Knapp KT (1980) Compound forms of fossil fuel fly ash emissions. *Environ Sci Technol* **14**: 450–456.

Holgate ST and Finnerty JP (1988) Recent advances in understanding the pathogenesis of asthma and its clinical implications. *Q J Med* **66**: 5–19.

Hunter T (1995) Protein kinases and phosphatases: the yin and yang of protein phosphorylation and signaling. *Cell* **80**: 225–236.

Israel A (1995) A role for phosphorylation and degradation in the control of NF-kappa B activity. *Trends Genet* **11**: 203–205.

Iwahashi H, Albro PW, McGown SR *et al.* (1991) Isolation and identification of α-(4-pyridyl-1-oxide)-*N*-tert-butylnitrone radical adducts formed by the decomposition of the hydroperoxides of linoleic acid, linolenic acid, and arachidonic acid by soybean lipoxygenase. *Arch Biochem Biophys* **285**: 172–180.

Jacob DJ, Gottlieb EW and Prather MJ (1989) Chemistry of a polluted cloudy boundary layer. *J Geophys Res* **94**: 12975–13002.

Janzen EG, Towner RA and Haire DL (1987) Detection of free radicals generated from the *in vitro* metabolism of carbon tetrachloride using improved ESR spin trapping techniques. *Free Radic Res Commun* **3**: 357–364.

Kadiiska MB, Mason RP, Dreher KL *et al.* (1997) *In vivo* evidence of free radical formation after exposure to an air pollution particle. *Chem Res Toxicol* **10**: 1104–1108.

Logan JA, Prather MJ, Wofsy SC and McElroy MB (1981) Trophospheric chemistry: a global perspective. *J Geophys Res* **86**: 7210–7254.

Mamane Y (1990) Estimate of municipal refuse incinerator contribution to Philadelphia aerosol using single particle analysis. Ambient measurements. *Atmos Environ* **24B**: 127–135.

Mamane Y and Noll KE (1985) Characterization of large particles at a rural site in the Eastern United States: mass distribution and individual particle analysis. *Atmos Environ* **19**: 611–622.

Moyers JL, Ranweiler LA, Hopf SB and Korte NE (1977) Evaluation of particulate trace species in southwest desert atmosphere. *Environ Sci Technol* **11**: 789–797.

Munger JW, Jacob DJ, Waldman JM and Hoffmann MR (1983) Fogwater chemistry in an urban atmosphere. *J Geophys Res* **88**: 5109–5121.

Munoz C, Pascual-Salcedo D, Castellanos MC *et al.* (1996) Pyrrolidine dithiocarbamate inhibits the production of interleukin-6, interleukin-8, and granulocyte-macrophage colony-stimulating factor by human endothelial cells in response to inflammatory mediators: modulation of NF-kappa B and AP-1 transcription factors activity. *Blood* **88**: 3482–3490.

Nriagu JO and Pacyna JM (1988) Quantitative assessment of worldwide contamination of air, water and soils by trace metals. *Nature* **333**: 134–139.

O'Byrne PM, Walters EH, Gold BD *et al.* (1984) Neutrophil depletion inhibits airway hyperresponsiveness induced by ozone exposure. *Am Rev Respir Dis* **130**: 214–219.

Perrin DD (1979) *Stability Constants of Metal–Ion Complexes.* New York: Pergamon Press.

Pritchard RJ, Ghio AJ, Lehmann J *et al.* (1996) Oxidant generation and lung injury after exposure to particulate air pollutants are associated with concentrations of complexed iron. *Inhal Toxicol* **8**: 457–477.

Quay JL, Reed W, Samet J and Devlin RB (1997) Air pollution particles induce IL-6 gene expression via NF-κB activation in human airway epithelial cells. *Am J Respir Cell Mol Biol* **19**: 98–106.

Rajanna B, Chetty CS, Rajanna S *et al.* (1995) Modulation of protein kinase C by heavy metals. *Toxicol Lett* **81**: 197–203.

Rom WN, Bitterman PB, Rennard SI *et al.* (1987) Characterization of the lower respiratory tract inflammation of nonsmoking individuals with interstitial lung disease associated with chronic inhalation of inorganic dusts. *Am Rev Respir Dis* **136**: 1429–1434.

Samet JM, Reed W, Ghio AJ *et al.* (1996) Induction of prostaglandin H synthase 2 in human airway epithelial cells exposed to residual oil fly ash. *Toxicol Appl Pharmacol* **141**: 159–168.

Samet JM, Stonehuerner J, Reed W *et al.* (1997) Disruption of protein tyrosine phosphate homeostasis in bronchial epithelial cells exposed to oil fly ash. *Am J.Physiol (Lung Cell Mol Physiol)* **272**: L426–L432.

Samet JM, Ghio AJ and Madden MC (1998a) Increased expression of cyclooxygenase 2 mediates oil fly ash-induced lung injury. In preparation.

Samet JM, Graves LM, Quay J *et al.* (1998b) Activation of MAP kinases in human airway epithelial cells exposed to metals. *Am J Physiol (Lung Cell Mol Physiol)* **275**: L551–L558.

Schapira RM, Ghio AJ, Morrisey J *et al.* (1997) Hydroxyl radicals are formed in the lungs following instillation of air pollution particulates *in vivo. Am J Respir Crit Care Med* **155**: A244.

Schindler PW and Stumm W (1987) The surface chemistry of oxides, hydroxides, and oxide minerals. In Stumm W (ed.) *Aquatic Surface Chemistry. Chemical Processes at the Particle–Water Interface.* New York: John Wiley and Sons, pp 83–110.

Schroeder WH, Dobson M, Kane DM and Johnson ND (1987) Toxic trace elements associated with airborne particulate matter: a review. *J Air Pollut Control Assoc* **37**: 1267–1285.

Schwartz J (1989) Lung function and chronic exposure to air pollution: a cross sectional analysis of NHANES II. *Environ Res* **50**: 309–321.

Shaw RW Jr and Paur RJ (1983) Composition of aerosol particles collected at rural sites in the Ohio River Valley. *Atmos Environ* **17**: 2031–2044.

Slade R, Stead AG, Graham JA and Hatch GE (1985) Comparison of lung antioxidant levels in humans and laboratory animals. *Am Rev Respir Dis* **131**: 742–744.

Speizer LA, Watson MJ, Kanter JR and Brunton LL (1989) Inhibition of phorbol ester binding and protein kinase C activity by heavy metals. *J Biol Chem* **264**: 5581–5585.

Stevens RK and Dzubay TG (1978) Sampling and analysis of atmospheric sulfates and related species.

Atmos Environ **12**: 55–68.

Stohs SJ and Bagchi D (1995) Oxidative mechanisms in the toxicity of metal ions. *Free Rad Biol Med* **18**: 321–336.

Treisman R (1996) Regulation of transcription by MAP kinase cascades. *Curr Opin Cell Biol* **8**: 205–215.

Trudel S, Paquet MR and Grinstein S (1991) Mechanism of vanadate-induced activation of tyrosine phosphorylation and of the respiratory burst in HL60 cells. *Biochem J* **276**: 611–619.

Tsai EY, Jain J, Pesavento PA *et al.* (1996) Tumor necrosis factor alpha gene regulation in activated T cells involves ATF-2/Jun and NFATp. *Mol Cell Biol* **16**: 459–467.

Vossler TL, Lewis CW, Stevens RK *et al.* (1989) Composition and origin of summertime air pollutants at Deep Creek Lake, Maryland. *Atmos Environ* **23**: 1535–1547.

Waldman JM, Munger JW, Jacob DJ *et al.* (1982) Chemical composition of acid fog. *Science* **218**: 677–680.

Wang Y and Pallen CJ (1992) Expression and characterization of wild type, truncated, and mutant forms of the intracellular region of the receptor-like protein tyrosine phosphatase HPTP beta. *J Biol Chem* **267**: 16696–16702.

Weinberg ED (1974) Iron and susceptibility to infectious disease. *Science* **184**: 952–956.

Weschler CJ, Mandich ML and Graedel TE (1986) Speciation, photosensitivity, and reactions of transition metal ions in atmospheric droplets. *J Geophys Res* **91**: 5189–5204.

Particulate Air Pollution: Injurious and Protective Mechanisms in the Lungs

WILLIAM MacNEE

Unit of Respiratory Medicine, Royal Infirmary of Edinburgh, University of Edinburgh, Edinburgh, UK

KENNETH DONALDSON

Department of Biological Sciences, Napier University, Edinburgh, UK

ADVERSE HEALTH EFFECTS OF PM$_{10}$

There is increasing evidence for an association between adverse health effects and the levels of particulate air pollution. These effects are particularly associated with the levels of PM$_{10}$, i.e. the mass of particulate air pollution, collected by a convention that has a 50% efficiency for particles with an aerodynamic diameter of 10 μm. Epidemiological studies, which are not discussed here, but reviewed elsewhere (Pope *et al.*, 1995), have demonstrated a clear relationship between the levels of PM$_{10}$ and exacerbations of asthma and chronic obstructive pulmonary disease (COPD). Deaths, not only from respiratory causes but also from vascular causes, i.e. myocardial infarction and cerebrovascular accidents, are also related to levels of PM$_{10}$. The mechanisms of these effects are not well understood. The ability of the lung to protect itself against inhaled particles, and the susceptibility of individuals to the effects of particles, will determine the outcome in terms of the adverse effects of environmental particles. This chapter addresses these issues in relation to the adverse health effects of PM$_{10}$.

AIR POLLUTION AND HEALTH
ISBN 0-12-352335-4

TARGET TISSUES FOR THE ADVERSE HEALTH EFFECTS OF PM$_{10}$

The range of associations with mortality and morbidity described above indicate that a wide variety of tissues are affected by PM$_{10}$ in ways that lead to disease, and in addition that there are both local lung and systemic effects of PM$_{10}$. The target tissues for PM$_{10}$ are described in this section.

Airways

An important defence mechanism in the airways against inhaled particles is the muco-ciliary escalator. In the large proximal airways, goblet cells secrete mucus which traps deposited particles. The mucus with its trapped particles is then propelled upwards by cil-iated cells to be either expectorated or swallowed. Mucus secretion is controlled by at least eight mucin genes (Jeffery and Li, 1997). Different mucin genes are expressed by the same or distinct cells of the airway epithelium and glands (Audie *et al.*, 1993). Inflammatory mediators have been shown to upregulate mucin genes. An example of this is IL-6, which has been shown to increase the level of steady-state mRNA of the mucin 2 (MUC2) gene in normal human airway epithelial cells (Levine *et al.*, 1993). Prolonged exposure to the air pollutant sulfur dioxide (400 ppm 3 h/day, 5 days/week for 1–3 weeks) produces increased expression of the MUC2 gene in an animal model, associated with increased mucus secretion (Jany *et al.*, 1991). Particles may produce the same effect. Mucus has a major role in protecting the airways, particularly as it is a rich source of antioxidants (Cross *et al.*, 1994).

Mucus hypersecretion for most days over three consecutive months, for two consecu-tive years, is the defining feature of chronic bronchitis (Medical Research Council, 1965), but also occurs in some asthmatics. Chronic bronchitis forms part of the spectrum of chronic airways disease termed chronic obstructive pulmonary disease (COPD), which also includes emphysema, and some cases of chronic asthma (Celli *et al.*, 1995). In patients with COPD, the goblet cells, which are normally confined to the proximal air-ways, extend, although fewer in number, down into the bronchioles (Jeffery, 1998), where there is also evidence of a bronchiolitis, even in asymptomatic cigarette smokers (Niewohner, 1988). Thus although mucus hypersecretion may have a protective role under normal circumstances, both as part of the mucociliary escalator and as a result of its antioxidant properties, excessive hypermucus secretion, particularly when it occurs in smaller peripheral airways, results in mucus plugging, a feature which is commonly present in patients dying of COPD and asthma (Lamb, 1995).

In patients with COPD, and in cigarette smokers, there is damage to the cilia, which together with the excess amounts of mucus which are produced, overwhelms the mucocil-iary escalator and will reduce the ability of the lungs to deal adequately with inhaled particles.

The airway epithelial cells themselves are capable of responding to stimulation by the release of inflammatory mediators such as IL-8, and RANTES (Bayram *et al.*, 1998; Driscoll *et al.*, 1997) upon exposure to particles and other forms of air pollutants, such as nitrogen dioxide (Devalia *et al.*, 1997).

Macrophages are also present in the airway walls and on the surfaces of the airways and these cells can phagocytose particles. In doing so they may release inflammatory mediators. In airways diseases, particularly COPD, there are increased numbers of macrophages in the airway walls and in the airspaces (Jeffery and Li, 1997), which are already primed to release inflammatory mediators such as IL-8 and tumour necrosis factor (TNF). These inflammatory cytokines can be measured in increased amounts in the sputum of patients with COPD (Keatings *et al.*, 1996). Both in asthma and in patients with COPD, there are increased numbers of inflammatory cells in the airway walls, particularly T lymphocytes, although the subsets of T lymphocytes appear to differ in patients with these two conditions, being predominantly CD4 positive in asthmatics and CD8 positive in chronic bronchitis (Jeffery, 1998). The additional insult of an inhaled air pollutant could clearly exacerbate this background inflammation. Recent preliminary data suggest that bronchial epithelial cells derived from patients with asthma, and to some extent those from patients with COPD, release more proinflammatory cytokines than those from normal subjects (Davis *et al.*, 1998). Thus in patients with airways diseases such as asthma and COPD, lung epithelial cells may be more susceptible to the proinflammatory effects of inhaled particles than in normal subjects. The mechanism by which this occurs is not clear but may involve the upregulation of transcription factors for proinflammatory genes, such as NFκB and AP-1 (Rahman and MacNee, 1998). Recent preliminary data also suggest that $PM_{2.5}$ causes c-jun-dependent AP-1 activation (Timblin *et al.*, 1998). The signal transduction pathway for these events may be through oxidant-mediated activation of Ras/Mitogen-activated protein kinases (Janssen *et al.*, 1998).

Terminal Airways and Proximal Alveoli

Inhaled particles deposit in large numbers beyond the ciliated airways, in the terminal airways and proximal alveoli (Brody *et al.*, 1984), where the net flow of air is zero and where, for very small particles, deposition efficiency increases because of the high efficiency of deposition by diffusion (Anderson *et al.*, 1990). In this region, where cilia are sparse and the mucociliary escalator is consequently less formed, the macrophages play the most important role in removing particles. Macrophages phagocytose particles and eventually they migrate to the start of the mucociliary escalator and leave the lung with their cargo of particles, bound for the gut. Although some adverse effects associated with PM_{10} are clearly focused on the larger airways, inflammatory events in the alveolar region, beyond the ciliated airways, could be important for the systemic effects of PM_{10} inhalation. It is here that the mediators released as a result of the local inflammatory effects of particles in the airspaces are most readily transmitted to the blood. The resulting sequestration and migration of leukocytes into the lungs has the potential to cause lung injury and systemic cardiovascular effect.

The Pulmonary Interstitium and Lymph Nodes

If particles cross the epithelium and enter the lung interstitium they are no longer likely to be cleared by the normal processes and will either remain in the subepithelial regions close to key responsive cell populations, such as interstitial macrophages, fibroblasts and endothelial cells, or be taken to the draining lymph nodes. Interstitial inflammation is likely to be more potentially harmful than inflammation in the alveolar spaces. The effects of dust in the lymph node is not known but adjuvant effects might be anticipated.

The Liver

Cardiovascular deaths are an important aspect of the adverse health effects of PM_{10} (Pope *et al.*, 1995). Classically these are caused by the production of clots – in the coronary vessels in the case of myocardial infarction, and in the cerebral microvasculature in the case of stroke. Whilst intuitively an effect of inhaled particles on the lungs might be understandable, a link between the deposition of particles in the airways and effects that increase the likelihood of thrombosis is more problematic. We have hypothesized that the inflammation arising in the lungs of persons inhaling PM_{10} could impact on the coagulation system via the local production of procoagulant factors in the lungs, or as a result of the effects of mediators released from the lungs which act on the liver, to increase the levels of procoagulant factors (Seaton *et al.*, 1995). This hypothesis is shown diagrammatically in Fig. 30.1. Research is underway to test this hypothesis experimentally. There is epidemiological evidence to support this hypothesis, as shown by an increase in the number of individuals in the highest quintile of blood viscosity during a period of high air pollution (Peters *et al.*, 1997a).

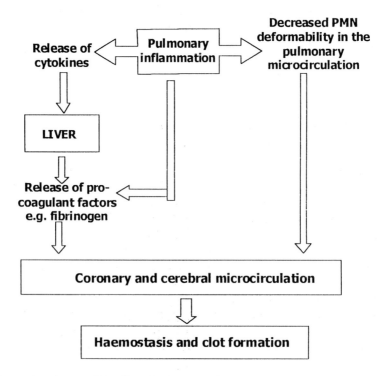

Fig. 30.1 Proposed mechanism of the effects of pulmonary inflammation following inhalation of PM_{10} and the systemic procoagulant effect.

Epidemiological studies have also shown a link between increased blood viscosity and increased procoagulant factors such as fibrinogen and factor VII and the risk of cardiovascular mortality (Heinrich *et al.*, 1994). A link has also been developed between respiratory infection or inflammation and cardiovascular disease owing to a systemic procoagulant effect (Seaton *et al.*, 1995). Thus alveolar inflammation may cause the release of interleukin-6 from macrophages and thus stimulate hepatocytes to secrete fibrinogen (Akira and Kishimoto, 1992).

PMN in the Pulmonary Microvasculature

Polymorphonuclear neutrophils (PMN) are important mediators of lung injury in acute and chronic inflammation because of their ability to release injurious substances such as proteases and reactive oxygen species. These toxic products of the PMN could be released whilst the PMN are in the vascular space as well as when they extravasate into tissue. Neutrophils are known to be held up (or sequester) in the pulmonary microcirculation under normal circumstances, their larger size necessitating deformation in order to negotiate the smaller pulmonary capillary segments (Selby and MacNee, 1993). PMN deformability is therefore a critical initiating factor, prior to increased PMN–endothelial adhesion of PMN sequestrated in the pulmonary microvasculature (Selby *et al.*, 1991a). Any factors which change cell deformability will change PMN sequestration in the pulmonary circulation (Selby *et al.*, 1991a). Airway inflammation, such as occurs in exacerbations of COPD, causes decreased PMN deformability, and hence increased PMN sequestration in the pulmonary microvasculature (Selby *et al.*, 1991b). This is associated with evidence of oxidant stress (Rahman *et al.*, 1996). Acute cigarette smoking also results in oxidative stress (MacNee *et al.*, 1989), which decreases neutrophil deformability (Drost *et al.*, 1992, 1993).

TOXIC POTENCY OF PM$_{10}$

The levels of PM$_{10}$ associated with adverse health effects are very low. Indeed, some studies have shown no threshold for the adverse health effects of PM$_{10}$ (e.g. Pope *et al.*, 1995). The relative toxicity of PM$_{10}$ is highlighted by considering the regulation of low toxicity or 'nuisance' particles in workplaces in the UK. The average PM$_{10}$ level in UK urban areas is below 50 μg/m^3 and levels are even less in country areas. By contrast, nuisance dusts are regulated in UK workplaces to 4 mg/m^3 of respirable dust. Since PM$_{10}$ appears to have adverse health effects without a threshold in some studies, even at very low levels as measured by mass, this suggests that PM$_{10}$ is a highly toxic material. However, the individual components of PM$_{10}$ are not particularly toxic at the levels present in ambient air, containing typically large proportions of carbon, salts and metals, as well as organic components (Clarke *et al.*, 1984). The potential role of transition metals on very small particles in mediating the effects of PM$_{10}$ are discussed below.

ULTRAFINE PARTICLES

Toxicity of Ultrafine Particles

Ultrafine particles are defined as those in the size range <100 nm in diameter. Research on ultrafine particles provides a possible explanation for the toxicity of PM_{10}. Ultrafine particles are highly toxic to the lungs, even when the particles are formed from materials that are not toxic as larger but still respirable particles, e.g. titanium dioxide (TiO_2) and carbon. For example the study of Ferin *et al.* (1992) reported the bronchoalveolar lavage inflammatory profile in rats exposed to the same airborne mass concentration of titanium dioxide as fine particles, approximately 250 nm in diameter (hereafter fine particles are defined as respirable particles that are larger than ultrafine particles) or as ultrafine particles, 20 nm in diameter. Although the rats were exposed to aerosols of 23 mg/m³ of both ultrafine TiO_2 and fine TiO_2, a marked inflammatory response was seen with the ultrafine TiO_2 and little effect with the fine TiO_2. Thus the same material showed a dramatic difference in pathogenicity when presented as ultrafine particles and as fine particles. We have reported a similar finding with fine (260 nm diameter) and ultrafine (14 nm diameter) carbon particles compared with the effects of PM_{10} (Fig. 30.2) (Li *et al.*, 1996). This suggests that ultrafine particles have toxicity that results from their small size, rather than from their chemical composition. The potential mechanisms of toxicity of ultrafine particles has recently been reviewed (Donaldson *et al.*, in press). The major mechanisms are:

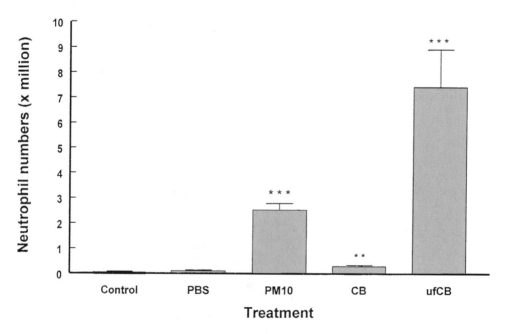

Fig. 30.2 The number of neutrophils in bronchoalveolar lavage (BAL) from rats 6 h after no instillation (control) or intratracheal instillation with PBS, PM_{10}, fine (CB) or ultrafine (ufCB) carbon black (n = 3–5, **p<0.01, ***p<0.001 compared with phosphate-buffered saline (PBS).

(1) Particle surface area
(2) Particle number
(3) Particle surface chemistry
(4) Oxidative stress
(5) Interstitialization of particles

Some of these mechanisms are discussed in more detail below.

Deposition of Ultrafine Particles

The deposition fraction is high for ultrafine particles, approaching 50% for 20 nm particles. Interestingly, the deposition efficiency is greater in a susceptible population, such as patients with COPD, than in normal subjects (Anderson *et al.*, 1990). This could be explained by the lower expiratory flows in patients with COPD that allow a longer residence time for the particles in the airspaces, which would favour deposition that depends largely on Brownian motion, as is the case for these very small particles (Anderson *et al.*, 1990).

In addition studies using radiolabelled particles indicate that particle deposition is uneven in patients with airflow limitation, resulting in concentration of particles in certain areas in the airways (Kim and Kang, 1997). Thus the dose of inhaled particles may be higher than expected in some parts of the airways. An additional factor which may lead to an underestimation of the dose of ambient inhaled particles results from the fact that particle levels in the air are measured as a 24-h running mean. However, preliminary studies, using personal particulate samplers, suggest that individual exposures may be much higher than those measured in ambient air (Watt *et al.*, 1995) (Fig. 30.3)

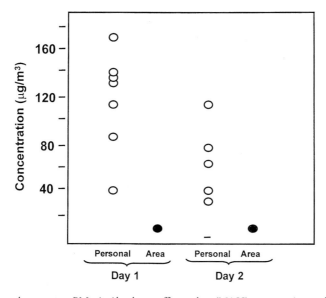

Fig. 30.3 Personal exposure to PM_{10} in Aberdeen traffic wardens (MASS) versus static sampler placed in Aberdeen City Centre (black smoke) on two consecutive days in November 1994. From Watt *et al.* (1995).

Evidence that PM₁₀ Contains Ultrafine Particles

There is considerable evidence that PM_{10} contains an ultrafine component (Oberdorster *et al.*, 1995), and indeed one report suggested that decrements in evening peak flow in a group of asthmatics was best associated with the ultrafine component of the airborne particles during pollution episodes (Peters *et al.*, 1997b). Diesel exhaust comprises singlet particles of around 50 nm diameters (Maynard and Waller, 1996). Several studies have described a substantial component, in number terms, of particles in the ultrafine range, although these represent a relatively small fraction of the total mass (Peters *et al.*, 1997b).

Classical Particle Overload

The rat is a species that has been shown to be a high responder to particles. Indeed, dusts considered to be 'non-toxic' in humans (e.g. carbon, diesel particles, TiO_2) can cause severe lung injury, culminating in fibrosis, epithelial hyperplasia and metaplasia and cancer in rats, if exposure is to high airborne concentrations such that a high lung dose is attained (Mauderly, 1996; Mauderly and McCunney, 1996). This phenomenon of *overload* lung injury appears to be confined to rats and is not seen in other rodents, even at similar lung doses (Mauderly and McCunney, 1996). The phenomenon of overload is associated with slowed clearance from the peripheral airspaces (macrophage-mediated clearance) and subsequent rapid accumulation of dose with ongoing exposure, culminating in the effects described above. The original hypothesis for the mechanism of overload focused on the volume of particles within the lung and specifically the load volume of particles inside macrophages (Morrow, 1988). This hypothesis suggested that when macrophages had phagocytosed a volume of particles equivalent to 6% of their internal volume, they began to show an impaired ability to move and carry their particle burden to the start of the mucociliary escalator for removal from the lungs. Morrow also calculated that by the time the average volume of particles inside macrophages reaches 60% of the total macrophage volume, their ability to move, and hence clearance, is completely inhibited (Morrow, 1988). This has been confirmed by Oberdorster and colleagues (1992, 1994). Recently, however, new data have allowed a revision of this hypothesis, by suggesting that that overload is best correlated to the metric of *surface area* of particles in the rat lung, not mass, volume or number of particles (Driscoll, 1996; Driscoll *et al.*, 1996; Oberdorster, 1996). A role for surface area appears intuitively likely for *toxic* particles since the interaction between particles and biological systems will occur with the surface, not the internal mass, of the particle. However, it is not immediately apparent why *non-toxic* particles might mediate their effects via their surface.

One hypothesis is that a large surface area allows absorption of substances on to the surface from the environment (Kasemo and Lausman, 1994), or from the lung epithelial lining fluid, which increases the reactivity of the particles. One such component may be iron, which can subsequently take part in Fenton chemistry to produce reactive oxygen species. In rats instilled intratracheally with fly ash the inflammatory response in the lungs was diminished by washing the particles in acid prior to instillation. This implies that iron complexed on the surface of the particles may account in part for their reactivity (Tepper *et al.*, 1994). The issue of whether particle number or surface area is the critical metric for the harmful effects of particulate air pollution is still a matter for debate (Ayres, 1998).

Ultrafine Particles and Overload

For a constant mass of monodispersed, (single diameter) particles, as the particle diameter reduces the surface area increases dramatically (Oberdorster *et al.*, 1995). Thus ultrafine particles may be more likely to cause overload at any given mass burden in the lungs because of their large surface area per unit mass. Amongst the most active of the 'low toxicity' dusts in causing lung overload tumours are the ultrafine particles (Driscoll, 1996, Driscoll *et al.*, 1996), presumably because they represent the biggest surface area per unit mass. Macrophages appear to be more adversely affected by loading of ultrafine TiO_2 than fine TiO_2. This is shown by the fact that the retention time of a radioactive marker particle in the lungs of rats exposed to ultrafine TiO_2 following inhalation exposure was increased about eight-fold compared with controls; however, lung burden data indicated that the macrophages were loaded with particles to a calculated volume of only 2.6% (Oberdorster *et al.*, 1994). By contrast, the calculated macrophage load volume was 9% for fine TiO_2 and this caused only a doubling of the retention time of the test particle, in keeping with Morrow's prediction (Morrow, 1988) on the volumetric index of dose. This supports the contention that volume is not the dose metric that best predicts impairment of macrophage function caused by ultrafine particles; particle number or particle surface area may be the most important index for ultrafine particles. Macrophage functions associated with clearance are substantially impaired when the cells contain small load volumes of ultrafine particles, although this could be a large surface area.

The term overload should not be applied to the effect of ultrafine TiO_2 in inhibiting clearance, since it occurs at a low lung burden. In relation to human risk assessment, it should be noted that although the phenomenon of overload occurs in rats, it is not clear whether it occurs in humans. In rats it occurs only at very high airborne mass concentrations, whereas PM_{10} toxicity occurs in humans at remarkably low airborne mass concentrations. Classical overload is not, therefore, the mechanism of lung injury caused by PM_{10}. Whereas the most important factors contributing to slowed clearance in classical overload is high macrophage burden of particles, this is clearly not the case with ultrafine particles and toxicity to the macrophages may be more important. Ultrafine particles can be highly toxic to the lungs, as shown by high levels of LDH in the lungs of exposed rats (Li *et al.*, 1996; Zhang *et al.*, in press), and toxic particles may cause slowed clearance by a mechanism involving frank toxicity to macrophages.

Particle Numbers

Macrophages attempting to phagocytose a large number of ultrafine particles could be stimulated, by the high particle load, to release inflammatory mediators such as TNF. The inability of macrophages to phagocytose the large numbers of ultrafine particles may also result in sustained stimulation of epithelial cells. This could cause release of chemokines, such as IL-8/MIP-1α, which would contribute to inflammation. Increased production of these chemokines has been demonstrated in rats inhaling ultrafine carbon black (Driscoll *et al.*, 1996) and also ultrafine particles of perfluoropolymer (Johnston *et al.*, 1996).

Transfer of Particles to the Interstitium

Interference with the normal process of phagocytosis and macrophage migration to the mucociliary escalator can lead to the adverse outcome of *interstitialization* (Morrow, 1988). Interstitialization is an adverse outcome because interstitial particles cannot then be cleared by the normal pathways and must either remain in the interstitium, where they can chronically stimulate interstitial cells, or transfer to the lymph nodes. Interstitialization of particles was a prominent correlate of the onset of inflammation for ultrafine TiO_2 in the study of Ferin and co-workers (1992), and interstitialization of particles was found to arise concomitantly with overload inflammation (Vincent and Donaldson, 1990). Interstitialization is likely to occur when there is failed clearance and failed clearance could result from: (a) particle-mediated macrophage toxicity or impairment of macrophage motility; or (b) overload. Both of these events would allow increased interaction between particles and epithelium that would favour interstitialization, and this could be further enhanced by increased epithelial permeability. In studies with rats it is clear that ultrafine particles and PM_{10} can cause increased epithelial permeability (Fig. 30.4) (Li *et al.*, 1996).

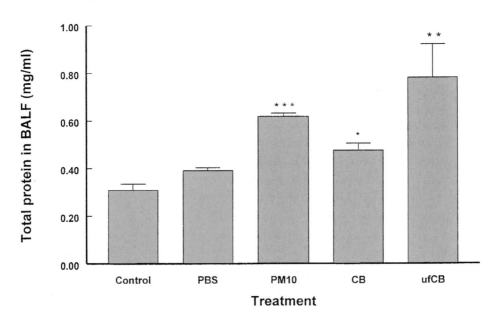

Fig. 30.4   The effect of intratracheal instillation of PM_{10}, fine (CB) and ultrafine (ufCB) carbon black on rat lung epithelial permeability *in vivo*, measured as total protein value in BALF 6 h after instillation ($n = 3.6$, $^*p<0.05$, $^{**}p<0.01$ and $^{***}p<0.001$ compared with PBS).

TRANSITION METALS

Transition Metals and Free Radicals in Particle Toxicity

The production of free radicals in the lungs has been seen as a general mechanism mediating the biological activity of a number of different pathogenic particles (Donaldson *et al.*, 1996; Kennedy *et al.*, 1989). These include quartz (Castranova *et al.*, 1997; Zay *et al.*, 1995), coalmine dust (Dalal *et al.*, 1989), residual oil fly ash (Dreher *et al.*, 1997), asbestos (Gilmour *et al.*, 1995) and synthetic mineral fibres (Brown *et al.*, 1998). The oxidative stress is considered to arise primarily from the particles themselves, owing to the

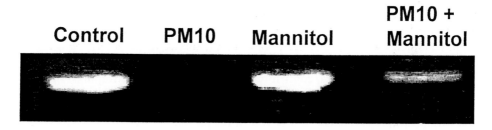

Fig. 30.5 (a) Typical result from a plasmid DNA injury assay. Note depletion of DNA by PM_{10} and partial protection with mannitol.

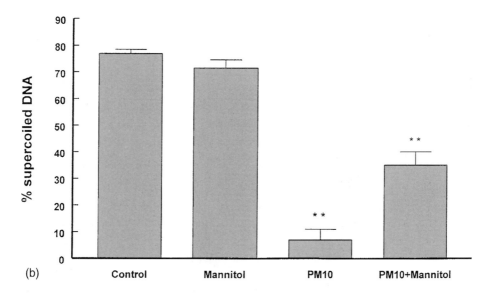

Fig. 30.5 (b) Quantified depletion of supercoiled DNA by blank filter (control), mannitol, PM_{10} and PM_{10} + mannitol. Data are shown as mean+SEM and there was a significant protection of the PM_{10} effect with mannitol ($p<0.05$).

localized release of high concentrations of transition metals. This action is supplemented by the release of reactive oxygen species by the inflammatory cell influx that results from the primary interaction between lung cells and particles. Oxidative stress is a general signalling mechanism within cells that produces the transcription of a number of proinflammatory genes for cytokines, antioxidant enzymes, receptors and adhesion molecules (Rahman and MacNee, 1998).

Free Radical Production by PM_{10}

To test this hypothesis which relates to oxidative stress produced by the release of free radicals from PM_{10}, we have collected PM_{10} in Edinburgh and London. PM_{10} was found to have the ability to generate hydroxyl radical activity, as shown in a supercoiled plasmid DNA scission assay (Fig. 30.5) (Gilmour et al., 1996). Further evidence for this was shown by its ability to form the hydroxylated derivative of salicylic acid – 2,3-dihydroxybenzoic acid (Donaldson et al., 1997). PM_{10} from both geographic sites contained a large proportion of iron and generated the hydroxyl radical, an effect that was blocked by iron chelators, confirming that Fenton chemistry is indeed the source of the hydroxyl radical (Gilmour et al., 1996). In addition the inflammatory response in the lungs was similar with both PM_{10} samples, supporting the hypothesis that the exact composition of the PM_{10} is not critical for its effects. The majority of the available iron was in the form of Fe^{3+}, but the presence in the lung of reductants such as superoxide anion and glutathione (GSH) would be able to initiate the reaction by reducing Fe^{3+} to Fe^{2+}.

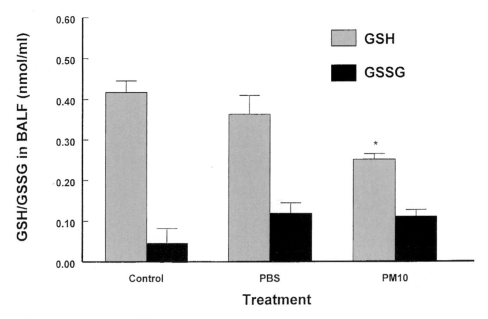

Fig. 30.6 The effect of intratracheal instillation of PM_{10} on GSH and GSSG concentrations in BALF 6 h after instillation in rat lungs (n = 3, *p<0.05 compared with PBS control).

Following the instillation of PM_{10} into the lungs of rats there was evidence of inflammatory neutrophil influx (Fig. 30.2) and oxidative stress as shown by depletion of GSH in lung lining fluid (Fig. 30.6) (MacNee *et al.*, 1997). Importantly, PM_{10} caused significantly more inflammation than a similar mass (125 μg) of carbon black that was 260 nm in diameter, i.e. not in the ultrafine size range; the inclusion of a mass bolus control is vital to the interpretation of this type of data.

Residual oil fly ash (ROFA) has been used as a surrogate for PM_{10}, although it is very different in many respects from PM_{10}. ROFA has been found to cause pulmonary inflammation after instillation, via a transition metal-mediated mechanism (Dreher *et al.*, 1997). Furthermore, in rats instilled with ROFA, an intraperitoneal injection of the free radical scavenger dimethyl-thiourea (DMTU) lowered the amount of PMN influx to the lung (Dye *et al.*, 1997). ROFA particles cause increased transcription of cytokine genes by human bronchial epithelial cells *in vitro* via a transition metal-mediated mechanism (Carter *et al.*, 1997), as shown by the fact that the effect could be blocked with the metal chelator deferoxamine. Interestingly, vanadium salts in solution could mimic the stimulation of cytokine production, which iron or nickel sulfate could not, suggesting a possible important role for vanadium.

Similarly, diesel oil particles have recently been shown in preliminary studies to enhance the release of cytokines from primary cultures of human bronchial epithelial cells (Mills *et al.*, 1998).

HYPOTHETICAL MECHANISM FOR THE CARDIOVASCULAR EFFECTS OF PM_{10}

The association of PM_{10} levels with the numbers of vascular deaths from strokes and myocardial infarction is one of the most puzzling of its adverse effects. We have suggested that the local pulmonary inflammation induced by PM_{10} could be translated into increased procoagulant status and conditions that could favour haemostasis in the cerebral and coronary microvasculature; these conditions would promote myocardial infarction and stroke, respectively (Seaton *et al.*, 1995). Two mechanisms could be operative:

(1) increased production of procoagulant factors in the liver or the lungs;
(2) decreased deformability of PMNs.

Lung and Liver as Sources of Procoagulant Factors

We have shown that, following deposition of ultrafine particles in the lungs, there is inflammation and oxidative stress (Li *et al.*, 1996, in press). In preliminary studies we have shown that inhalation of ultrafine particles in the rat increases the levels of factor VII in plasma (Li *et al.*, 1998b). Factor VII has been shown to be a risk factor for cardiovascular disease in population studies (Meade *et al.* 1986, 1993). This association is supported by two recent studies. The first showed that polymorphisms in the factor VII gene influence the plasma concentration of factor antigen and its activity, and in addition adversely influence the risk of myocardial infarction (Iacovelli *et al.*, 1998). The second, the use of warfarin to lower factor VII plasma concentrations, to those associated with low vascular risk in epidemiological studies was accompanied by significant protection from fatal vascular events (The MRC General Practice Research Framework, 1998). The principal site of synthesis of factor VII is the hepatocyte, but mRNA for both factor VII and tissue factor, which is also involved in coagulation, have been demonstrated in alveolar macrophages (McGee *et al.*, 1990). These cells are capable of synthesis of factor VII *in vitro* (Chapman *et al.*, 1985). Thus local inflammation in the lungs and the activation of alveolar macrophages could result in local and generalized release of pro-coagulant factors, which may enter the bloodstream and have a systemic effect.

Local inflammation in the lungs could also result in increased procoagulant activity from the liver. There are at least two possible mechanisms by which these effects can occur following inhalation of particulate air pollution:

(1) through the release of cytokines from inflammatory cells in the lungs;
(2) as a result of the development of systemic oxidant stress.

Both of these mechanisms could upregulate the genes for procoagulant factors in hepatocytes. We have shown clear evidence of systemic oxidant stress, measured as a decrease in the antioxidant capacity of the plasma, following instillation of PM_{10} particles; this was not apparent following instillation of the same mass of fine carbon particles (Li *et al.*, 1998a). Candidate cytokines, which may result in upregulation of procoagulant factors in the liver, are TNF-α and IL-6.

Alterations in Blood Rheology as a Cause of Increased Haemostasis

One unique feature of the pulmonary microcirculation is the close proximity of the distal airspace to the circulating blood, across the alveolar–capillary membrane, allowing easy access for inflammatory mediators in the airspaces to reach the blood. A recent study has shown increased plasma viscosity during an episode of air pollution confirming that a 'signal' from the lungs following exposure to particles can affect plasma indices (Peters *et al.*, 1997a). We have also shown that lung inflammation, which occurs during smoking or acute exacerbations of COPD, increases neutrophil sequestration in the lungs (MacNee *et al.*, 1989; Selby *et al.*, 1991b), very likely via an oxidant-induced decrease in cell deformability (Drost *et al.*, 1992). Thus oxidant-generating ultrafine environmental particles depositing in the distal airspaces may also produce increased neutrophil sequestration in the lungs, and so contribute to the initiation of lung inflammation. In addition the decrease in neutrophil deformability induced in cells in transit in the pulmonary microcirculation, or the release of less deformable cells from the bone marrow in response to particle-induced lung inflammation (Terashima *et al.*, 1997), may result in sequestration of these cells in the microcirculation of other organs, such as the heart and the brain, so contributing to local haemostasis and thrombotic events.

ACTIVATION OF NF-κB IN THE LUNGS AFTER INHALATION OF ULTRAFINE PARTICLES AS A CENTRAL INITIATING EVENT

The transcriptional activator NF-κB is a nuclear factor of the Rel family that is translocated to the nucleus to permit expression of a wide range of proinflammatory gene (Meyer *et al.*, 1994). The NF-κB heterodimer, comprising p65 and p50 proteins, is found in resting cells bound to its inhibitor IkB, which masks the nuclear translocation signal and so prevents its translocation to the nucleus (Fig. 30.7). Under oxidative stress or a range of other stimuli such as TNF, the IkB is phosphorylated and then degraded via the ubiquitin proteosome system, allowing the NF-κB to relocate to the nucleus (Brown *et al.*, 1995). Genes that have a κB binding site in their promoter include cytokines, growth factors, chemokines, and adhesion molecules and receptors (Rahman and MacNee, 1998). In addition the genes for tissue factor, which is important in the coagulation cascade, contain a κB binding site (Orthner *et al.*, 1995) and so may be susceptible to transcriptional activation during oxidative stress. We have demonstrated activation of NF-κB by oxidative stress in airspace epithelial cells (Mulier *et al.*, 1998). The deposition of particles that deliver an oxidative stress to the lungs may cause activation of NF-κB, and possibly of other oxidative stress-responsive transcription factors, which initiate a cascade of gene expression leading to a procoagulant and haemostatic state.

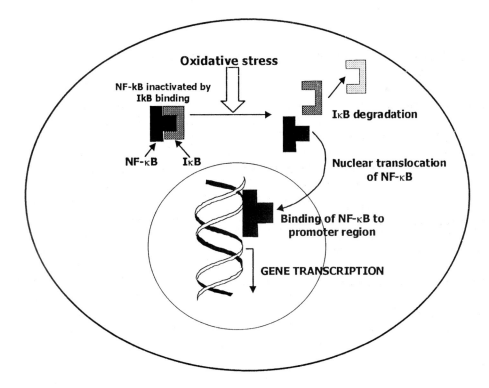

Fig. 30.7 The effect of particle-induced oxidative stress on gene transcription through activation of the transcription factor NF-κB

IMPLICATIONS OF AN OXIDATIVE STRESS-MEDIATED MECHANISM OF ACTION OF PM$_{10}$ FOR SUSCEPTIBLE PATIENTS WITH AIRWAYS DISEASE

The Central Role of the Epithelium

Since particles deposit on the epithelium prior to phagocytosis, it seems likely that the epithelium is a target for the PM$_{10}$ that may have a role in the observed increase in exacerbations of asthma and COPD (Fig. 30.8). Antigens for asthma are present in ambient air and to trigger an asthma attack antigen need only gain access to the subepithelial lymphoid tissue. There is evidence that various environmental particles such as ROFA (Dye *et al.*, 1997), PM$_{10}$ (Li *et al.*, 1996) and also ultrafine carbon black (Li *et al.*, in press) can compromise the epithelium by causing injury or oxidative stress. This results in the possibility that increased production of inflammatory mediators and increased permeability to antigen may be a mechanism for the induction of asthma attacks. In addition, the underlying inflammation in the airways of asthmatics means that they are in a 'primed' state for the further oxidative stress caused by depositing PM$_{10}$.

Existing Oxidative Stress in Susceptible Populations

The principal pulmonary effects of PM$_{10}$ are seen in susceptible populations, such as those with airways disease. If, as hypothesized here, PM$_{10}$ has its effect mainly by a mechanism that involves oxidative stress, then these susceptible populations might be susceptible because of pre-existing oxidative stress. We have utilized an assay (trolox equivalent antioxidant capacity, TEAC) that detects the global antioxidant defence in the plasma, and have demonstrated depleted antioxidant defences in patients with airways disease

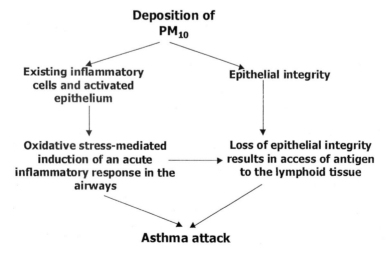

Fig. 30.8 Hypothesis for the central role of the epithelium following deposition of PM$_{10}$, which leads to attacks of asthma.

(Rahman *et al.*, 1996). Plasma from patients with asthma and COPD were found to have significantly lower TEAC values than normal subjects and these values were found to be further lowered during exacerbations of asthma and COPD. Clearly these patients would be susceptible to an oxidative insult such as that hypothesized here to emanate from PM_{10}.

REFERENCES

Akira S and Kishimoto T (1992) IL-6 and NF-IL6 in acute phase response and viral infection. *Immunol Rev* **127**: 25–50.

Anderson PJ, Wilson DJ and Hirsch A (1990) Respiratory tract deposition of ultrafine particles in subjects with obstructive or restrictive lung disease. *Chest* **97**: 1115–1120.

Audie JB, Janin A, Porchet N *et al.* (1993) Expression of human mucin genes in respiratory, digestive and reproductive tracts ascertained by *in situ* hybridization. *J Histochem Cytochem* **41**: 1479–1485.

Ayres J (1998) Particle mass or particle numbers. *Eur Respir Rev* **8**: 135–138.

Bayram H, Devalia JL, Sapsford RJ *et al.* (1998) The effect of diesel exhaust particles on cell function and release of inflammatory mediators from human bronchial epithelial cells *in vitro. Am J Respir Cell Mol Biol* **18**: 441–448.

Brody AR, Warheit DB, Chang LY *et al.* (1984) Initial deposition pattern of inhaled minerals and consequent pathogenic events at the alveolar level. *Ann NY Acad Sci* **428**: 108–120.

Brown K, Gerstberger S, Carlson L *et al.* (1995) Control of IkB-alpha proteolysis by site-specific, signal induced phosphorylation. *Science* **267**: 1485–1489.

Brown DM, Fisher C and Donaldson K (1998) Free radical activity of synthetic vitreous fibres: iron chelation inhibits hydroxyl radical generation by refractory ceramic fibre. *J Toxicol Environ Health A* **53**: 101–107.

Carter JD, Ghio AJ, Samet JM and Devlin RB (1997) Cytokine production by human airway epithelial cells after exposure to an air pollution particle is metal-dependent. *Toxicol Appl Pharmacol* **146**: 180–188.

Castranova V, Vallyathan V, Ramsey DM *et al.* (1997) Augmentation of pulmonary reactions to quartz inhalation by trace amounts of iron-containing particles. *Environ Health Perspect* **105**: 1319–1324.

Celli BR, Snider GL, Heffner J *et al.* (1995) Standards for the diagnosis and care of patients with chronic obstructive pulmonary-disease. *Am J Respir Crit Care Med* **152**: S–S.

Chapman HA, Allen CL and Stone OL (1985) Human alveolar macrophages synthesize factor VII *in vitro. J Clin Invest* **75**: 2030–2037.

Clarke AG, Willison MJ and Zeki EM (1984) A comparison of urban and rural aerosol composition using dichotomous samplers. *Atmos Environ* **18**: 1767–1775.

Cross CE, van der Vliet A, O'Neill CA *et al.* (1994) Oxidants, antioxidants and respiratory tract lining fluids. *Environ Health Perspect* **102**(Suppl 10): 185–191.

Dalal NS, Suryan MM, Vallyathan V *et al.* (1989) Detection of reactive free-radicals in fresh coal-mine dust and their implication for pulmonary injury. *Ann Occup Hyg* **33**: 79–84.

Davis RJ, Bayram H, Abdelaziz MM *et al.* (1998) Effect of diesel exhaust particles and release of inflammatory mediators from bronchial and epithelial cells of atopic asthmatic patients and non-atopic non-asthmatic subjects. *Am J Respir Crit Care Med* **157**: A743.

Devalia JL, Bayram H, Rusznak C *et al.* (1997) Mechanisms of pollution-induced airway disease: *in vitro* studies in the upper and lower airways. *Allergy* **52**: 45–51.

Donaldson K, Beswick PH and Gilmour PS (1996) Free-radical activity associated with the surface of particles – a unifying factor in determining biological activity. *Toxicol Lett* **88**: 293–298.

Donaldson K, Brown DM, Mitchell C *et al.* (1997) Free radical activity of PM_{10}: iron-mediated generation of hydroxyl radicals. *Environ Health Perspect* **105**(Suppl 5): 1285–1289.

Donaldson K, Li XY and MacNee W Ultrafine (nanometre) particle-mediated lung injury. *J Aerosol Sci*

Dreher KL, Jaskot RH, Lehmann JR *et al.* (1997) Soluble transition metals mediate residual oil fly ash induced acute lung injury. *J Toxicol Environ Health* **50**: 285–305.

Driscoll KE (1996) Role of inflammation in the development of rat lung tumours in response to chronic particle exposure. *Inhalation Toxicol* **8**(suppl): 139–153.

Driscoll KE, Carter JM, Howard BW *et al.* (1996) Pulmonary inflammatory, chemokine and mutagenic responses in rats after subchronic inhalation of carbon black. *Toxicol Appl Pharmacol* **136**: 372–380.

Driscoll KE, Carter JM, Hassenbein DG and Howard B (1997) Cytokines and particle-induced inflammatory cell recruitment. *Environ Health Perspect* **105**: 1159–1164.

Drost EM, Selby C, Lannan S *et al.* (1992) Changes in neutrophil deformability following *in vitro* smoke exposure: mechanism and protection. *Am J Respir Cell Mol Biol* **6**: 287–295.

Drost E, Selby C, Bridgeman MME and MacNee W (1993) Decreased leukocyte deformability following acute cigarette smoking in smokers. *Am Rev Respir Dis* **148**: 1277–1283.

Dye JA, Adler KB, Richards JH and Dreher K (1997) Epithelial injury induced by exposure to residual oil fly-ash particles: role of reactive oxygen species. *Am J Respir Cell Mol Biol* **12**: 625–633.

Ferin J, Oberdorster G and Penney DP (1992) Pulmonary retention of ultrafine and fine particles in rats. *Am J Respir Cell Mol Biol* **6**: 535–542.

Gilmour PS, Beswick PH, Brown DM and Donaldson K (1995) Detection of surface free-radical activity of respirable industrial fibers using supercoiled phi-x174 rf1 plasmid dna. *Carcinogenesis* **16**: 2973–2979.

Gilmour PS, Brown DM, Lindsay TG *et al.* (1996) Adverse health-effects of PM_{10} particles – involvement of iron in generation of hydroxyl radical. *Occup Environ Med* **53**: 817–822.

Heinrich J, Balleisen L, Schulte H *et al.* (1994) Fibrinogen and factor VII in prediction of coronary risk: results from the PROCAM study in healthy man. *Arterioscler Thromb Vasc Biol* **14**: 54–59.

Iacoviello L, Di Castelnuovo A, de Knijff P *et al.* (1998) Polymorphisms in the coagulation factor VII gene and the risk of myocardial infarction. *N Engl J Med* **338**: 79–85.

Janssen YMW, Macara I and Mossman BT (1998) Activation of NF-kappa B by reactive oxygen and nitrogen species in lung epithelial cells requires RAS/mitogen activated kinases. *Am J Respir Crit Care Med* **157**: A743.

Jany B, Gallup M, Tsuda T and Basbaum C (1991) Mucin gene expression in rat airways following infection and irritation. *Biochem Biophys Res Commun* **181**: 1–8.

Jeffery PK (1998) Structural and inflammatory changes in COPD: a comparison with asthma. *Thorax* **53**: 129–136.

Jeffery PK and Li D (1997) Airway mucosa: secretory cells, mucus and mucin genes. *Eur Respir J* **10**: 1655–1662.

Johnston CJ, Finkelstein JN, Gelein R *et al.* (1996) Characterisation of the early pulmonary inflammatory response associated with PTFE fume exposure. *Toxicol Appl Pharmacol* **140**: 154–163.

Kasemo B and Lausman (1994) Material–tissue interfaces: the role of surface properties and process. *Environ Health Prospect* **102**(Suppl 5): 41–45.

Keatings VM, Collins PD, Scott DM and Barnes PJ (1996) Differences in interleukin-8 and tumor-necrosis-factor-alpha in induced sputum from patients with chronic obstructive pulmonary disease or asthma. *Am J Respir Crit Care Med* **153**: 530–534.

Kennedy TP, Dodson R, Rao NV *et al.* (1989) Dusts causing pneumoconiosis generate OH and produce hemolysis by acting as Fenton catalysts. *Arch Biochem Biophys* **269**: 359–364.

Kim CS and Kang TC (1997) Comparative measurement of lung deposition of inhaled fine particles in normal subjects and patients with obstructive airway disease. *Am J Respir Crit Care Med* **155**: 899–905.

Lamb D (1995) Pathology. In: Calverley PMA and Pride NB (eds) *Chronic obstructive pulmonary disease*. London: Chapman & Hall, pp 9–34.

Levine SJ, Larivee P, Logun C *et al.* (1993) Corticosteroids differentially regulate secretion of IL-6, IL-8 and GM-CSF by a human bronchial epithelial cell line. *Am J Physiol* **265**(4)Part I: L360–L368.

Li XY, Gilmour PS, Donaldson K and MacNee W (1996) Free-radical activity and pro-inflammatory effects of particulate air-pollution (PM_{10}) *in-vivo* and *in-vitro*. *Thorax* **51**: 1216–1222.

Li XY, Donaldson K and MacNee W (1998a) Pro-inflammatory and oxidative activity of fine and ultrafine carbon black in rat lungs following instillation and inhalation. *Am J Respir Crit Care Med* **157**: A153.

Li XY, Ford I, Donaldson K *et al.* (1998b) Local and systemic effects of inhalation of fine and ultrafine carbon particles. *Thorax* **53**(Suppl 4): 181.

Li XY, Brown D, Smith S *et al.* (In press) Pro-inflammatory and oxidative activity of fine and ultrafine carbon black in rat lungs after intratracheal instillation. *Inhal Toxicol*

MacNee W, Wiggs B, Belzberg AS and Hogg JC (1989) The effect of cigarette smoking on neutrophil kinetics in human lungs. *N Engl J Med* **321**: 924–928.

MacNee W, Li XY, Gilmour PS and Donaldson K (1997) Pro-inflammatory effect of particulate air pollution (PM$_{10}$) *in vivo* and *in vitro*. *Ann Occup Hyg* **41**(Suppl I): 7–13.

Mauderly JL (1996) Lung overload: the dilemma and opportunities for resolution. *Inhal Toxicol* **8**(suppl): 1–28.

Mauderly JL and McCunney RJ (1996) Particle overload in the rat lung and lung cancer: implications for human risk assessment. Proceedings of a conference held at the Massachusetts Institute of Technology, March 1995. *Inhal Toxicol* **8**(Suppl): 298 pp.

Maynard RL and Waller RE (1996) Suspended particulate matter and health – new light on an old problem. *Thorax* **51**: 1174–1176.

McGee MP, Devlin R, Saluta G and Koren H (1990) Tissue factor and factor VII messenger RNAs in human alveolar macrophages: effects of breathing ozone. *Blood* **75**: 122–127.

Meade TW, Mellows S, Brozovic M *et al.* (1986) Haemostatic function and ischaemic heart disease: principal results of the Northwick Park Heart Study. *Lancet* **2**: 533–537.

Meade TW, Ruddock V, Chakrabarti R and Miller GJ (1993) Fibrinolytic activity, clotting factors, and long-term incidence of ischaemic heart disease in the Northwick Park Heart Study. *Lancet* **342**: 1076–1079.

Medical Research Council (1965) Definition and classification of chronic bronchitis for clinical and epidemiological purposes. *Lancet* **1**: 775–779.

Meyer M, Pahl HK and Bauerle PA (1994) Regulation of the transcription factors NF-kB and AP-1 by redox changes. *Chem Biol Interact* **91**: 91–100.

Mills PR, Sapsford RJ, Seemungal T *et al.* (1998) IL-8 release from cultured human bronchial epithelial cells (HBEC) of non-smokers with normal pulmonary function, and patients with COPD: the effect of exposure to diesel exhaust particles (DEP). *Am J Respir Crit Care Med* **157**(3): A743.

Morrow PE (1988) Possible mechanisms to explain dust overloading of the lungs. *Fundam Appl Toxicol* **10**: 369–384.

Mulier B, Watchorn T and MacNee W (1998) Intracellular GSH modulates H$_2$O$_2$-induced NFkB activation in alveolar epithelial cells. *Am J Respir Crit Care Med* **157**: A890.

Niewoehner DE (1988) Cigarette-smoking, lung inflammation, and the development of emphysema. *J Lab Clin Med* **111**: 15–27.

Oberdorster G (1996) Significance of particle parameters in the evaluation of exposure–dose relationships of inhaled particles. *Inhal Toxicol* **8**(Suppl): 73–90.

Oberdorster G, Ferin J and Morrow PE (1992) Volumetric loading of alveolar macrophages (AM): a possible basis for diminished AM-mediated particle clearance. *Exp Lung Res* **18**: 87–104.

Oberdorster G, Ferin J and Lehnert BE (1994) Correlation between particle size, *in vivo* particle persistence and lung injury. *Environ Health Perspect* **102**(Suppl 5): 173–179.

Oberdorster G, Gelein R, Ferin J and Weiss B (1995) Association of particulate air pollution and acute mortality: involvement of ultrafine particles. *Inhal Toxicol* **71**: 111–124.

Orthner CL, Rodgers GM and Fitzgerald LA (1995) Pyrrolidine dithiocarbamate abrogates tissue factor (Tf) expression by endothelial-cells – evidence implicating nuclear factor-kappa-b in Tf induction by diverse agonists. *Blood* **86**: 436–443.

Peters A, Doring A, Wichmann HE and Koenig W (1997a) Increased plasma viscosity during an air pollution episode: a link to mortality? *Lancet* **349**: 1582–1587.

Peters A, Wichmann HE, Tuch T *et al.* (1997b) Respiratory effects are associated with the number of ultrafine particles. *Am J Respir Crit Care Med* **155**: 1376–1383.

Pope CA, Bates DV and Raizenne ME (1995) Health-effects of particulate air-pollution – time for reassessment. *Environ Health Perspect* **103**: 472–480.

Rahman I and MacNee W (1998) Role of transcription factors in inflammatory lung diseases. *Thorax* **53**: 601–602.

Rahman I, Morrison D, Donaldson K and MacNee W (1996) Systemic oxidative stress in asthma, COPD and smokers. *Am J Respir Crit Care Med* **154**: 1055–1060

Seaton A, MacNee W, Donaldson K and Godden D (1995) Particulate air pollution and acute health effects. *Lancet* **345**: 176–178.

Selby C and MacNee W (1993) Factors affecting neutrophil transit during acute pulmonary inflammation – minireview. *Exp Lung Res* **19**: 407–428.

Selby C, Drost E, Wraith PK and MacNee W (1991a) *In vivo* neutrophil sequestration within the lungs of man is determined by *in vitro* 'filterability'. *J Appl Physiol* **71**: 1996–2003.

Selby C, Drost E, Lannan S *et al.* (1991b) Neutrophil retention in the lungs of patients with chronic obstructive pulmonary disease. *Am Rev Respir Dis* **143**: 1359–1364.

Tepper JS, Lehman JR, Winsett DW *et al.* (1994) The role of surface complexed iron in the development of lung inflammation and airway hyper-responsiveness. *Am J Respir Crit Care Med* **149**: A839.

Terashima T, Wiggs B, English D *et al.* (1997) Phagocytosis of small carbon particles (PM_{10}) by alveolar macrophages stimulates the release of polymorphonuclear leukocytes from bone marrow. *Am J Respir Crit Care Med* **155**: 1441–1447.

The Medical Research Council's General Practice Research Framework (1998) Thrombosis Prevention Trial: randomised trial of low-intensity oral anticoagulation with warfarin and low-dose aspirin in the primary prevention of ischaemic heart disease in men at increased risk. *Lancet* **351**: 233–241.

Timblin C, Berube KA and Mossman BT (1998) Particulater matter ($PM_{2.5}$) causes increases in c-jun, AP-1 dependent gene transcription and DNA synthesis in rat lung epithelial cells. *Am J Respir Crit Care Med* **157**: A154.

Vincent JH and Donaldson K (1990) A dosimetric approach for relating the biological response of the lung to the accumulation of inhaled mineral dust. *Br J Ind Med* **47**: 302–307.

Watt M, Godden D, Cherrie J and Seaton A (1995) Individual exposure to particulate air pollution and its relevance to thresholds for health effects: a study of traffic wardens. *Occup Environ Health* **52**: 790–792.

Zay K, Devine D and Churg A (1995).Quartz inactivates alpha(1)-antiproteinase – possible role in mineral dust-induced emphysema. *J Appl Physiol* **78**: 53–58.

Zhang Q, Kusake Y, Sato K *et al.* (1999) The pulmonary toxicity of standard nickel and ultrafine particles: surface reactivity as an additional factor in the toxicity of ultrafines. *Toxicol Appl Pharmacol* (in press).

Epidemiology of Particle Effects

C. ARDEN POPE III

Brigham Young University, Provo, UT, USA

DOUGLAS W. DOCKERY

Harvard School of Public Health and Harvard Medical School, Boston, MA, USA

INTRODUCTION

In the 1987 revision of the particulate air pollution standards, the US Environmental Protection Agency cited the results of three epidemiologic studies as providing the key supporting evidence for the proposed particle standard (US Environmental Protection Agency, 1987). There has been such a burst of epidemiologic studies of the health effects of particulate air pollution in the succeeding ten years that the Environmental Protection Agency, in promulgating a revised particulate air pollution standard in 1997, cited more than 80 epidemiologic studies (US Environmental Protection Agency, 1997). At the time of writing, there are at least 140 published epidemiologic studies of the health effects of particulate air pollution. More than 100 of these have appeared since 1992 (Fig. 31.1). In addition, there have been at least 17 reviews of these epidemiologic studies since 1993 (Brunekreef *et al.*, 1995; Dockery and Pope, 1994, 1996; Folinsbee, 1995; Jedrychowski, 1995; Katsouyanni, 1995; Lipfert, 1994; Lipfert and Wyzga, 1995; Moolgavkar and Luebeck, 1996; Ostro, 1993; Pope and Dockery, 1996; Pope *et al.*, 1995a,b; Schwartz, 1995; Thurston, 1996; US EPA, 1996; Vedal, 1997). Why has there been this recent proliferation of epidemiological studies of the health effects of particulate air pollution?

The link between cardiopulmonary disease and extremely high concentrations of particulate and sulfur oxide pollution had been well established by the 1970s. In the 1970s and 1980s some reviewers suggested that there might be important health effects at relatively low concentrations (Bates, 1980; Shy, 1979; Ware *et al.*, 1981). An extensive

AIR POLLUTION AND HEALTH
ISBN 0-12-352335-4

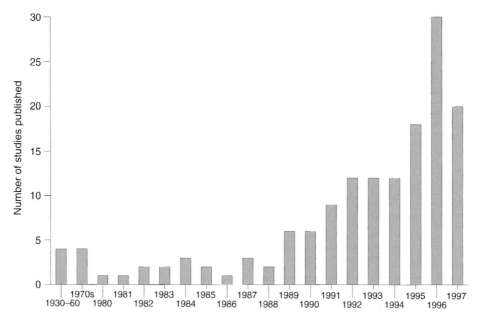

Fig. 31.1   Approximate number of published particle epidemiology studies.

research review by several prominent British scientists (Holland *et al.*, 1979), however, reflected the commonly held view that particulate air pollution at high levels posed a hazard to human health, but that the evidence was weak for health effects at the lower concentrations that existed in the USA and Britain by the 1970s.

In the 1970s through the mid-1980s a small number of original studies suggested adverse health effects at contemporary pollution levels (notably Dockery *et al.*, 1982; Lave and Seskin, 1970; Ostro, 1983; Özkaynak and Thurston, 1987; Samet *et al.*, 1981; Whittemore and Korn, 1980). Then, during the relatively short time period of 1989–92, results of three loosely connected epidemiological research efforts from the US were reported. These efforts included: (1) the Harvard Six Cities prospective cohort study that observed that long-term particulate exposure was associated with increased risk of respiratory illness in children (Dockery *et al.*, 1989) and with increased risk of cardio-pulmonary mortality (Dockery *et al.*, 1993); (2) a series of studies in Utah Valley that observed that particulate pollution was associated with a wide range of morbidity health end-points including respiratory hospitalizations (Pope, 1989, 1991), lung function and respiratory symptoms (Pope and Dockery, 1992; Pope *et al.*, 1991), school absences (Ransom and Pope, 1992) and mortality (Archer, 1990; Pope *et al.*, 1992); and (3) a series of studies that observed associations between daily changes in particulate air pollution and daily mortality (see reviews by Dockery and Pope, 1994; Ostro, 1993; Schwartz, 1994a; Thurston, 1996). Also during this time period, results of other studies from the USA (Ostro *et al.*, 1991; Thurston *et al.*, 1992), Germany (Wichmann *et al.*, 1989), Canada (Bates and Sizto, 1989), Finland (Pönkä, 1991) and the Czech Republic (Bobak and Leon, 1992) were published. Overall, these results suggested unexpectedly large health effects of relatively low concentrations of particulate air pollution. Although these studies

were controversial, the convergence of their reported results in such a short time period
resulted in a critical mass of evidence that prompted serious reconsideration of the contribution of particulate air pollution on human health. This scientific reconsideration
coupled with the US EPA's review of ambient standards for particulate air pollution
motivated additional research.

The recent growth in reported studies has been greatly facilitated by a number of
other factors. Study designs and analytic strategies that focus on individuals have been
applied. There have been advances in statistical analytic techniques and in the availability of statistical software implementing these techniques. Electronic access to pollution,
weather, mortality and medical data has been improved. The use of standardized analytic
techniques has allowed results to be compared across studies. Evaluation of the consistency
and coherence of results has provided understanding that was not possible with rote
reliance on measures of statistical significance.

In this review we will consider the developments in study designs, advances in analytic
methods, and availability of data and analytic tools which have led to this wave of new
studies. This discussion will not be framed as a formal meta-analysis as we and others have
done previously (Dockery and Pope, 1994; Ostro, 1993; Schwartz, 1994a). Rather it will
be framed in the context of the development of the literature, the basic inferences being
drawn from the literature, and the strengths and limitations associated with currently
available epidemiological evidence.

BASIC STUDY DESIGNS

Published epidemiologic studies of air pollution typically fall within two broad classifications: (1) acute exposure studies and (2) chronic exposure studies (see Fig. 31.2). The
acute exposure studies use short-term temporal changes in pollution as their source of
exposure variability. These studies evaluate short-term changes in health measures associated with short-term changes in air pollution. These studies range from simple
observations of changes in health over a single air pollution episode of one or more days,
to sophisticated statistical analyses of daily time-series over several decades. Because these
studies typically evaluate only short-term temporal relationships (usually 1–5 days), the
observed pollution effects are usually interpreted as health effects of acute exposure.

Available chronic exposure studies primarily use spatial differences in pollution as their
source of exposure variability. Chronic exposure studies, therefore, compare various health
outcomes across communities or neighborhoods with different levels of pollution. These
studies are principally cross-sectional in design and use longer-term pollution data (usually one year or more). These studies are often interpreted as evaluating chronic and/or
cumulative effects of exposure.

Nearly all currently published acute and chronic exposure studies can also be subdivided as population-based versus person-based studies (Fig. 31.2). Population-based
studies are often referred to as ecological studies where the units of comparison are entire
populations of communities or neighborhoods. Person-based studies are often referred to
as panel or cohort studies. They include information on individuals sampled from the
larger population. Although central-site community-based monitoring is typically used to
estimate pollution exposure in the panel or cohort-based studies, the units of comparison

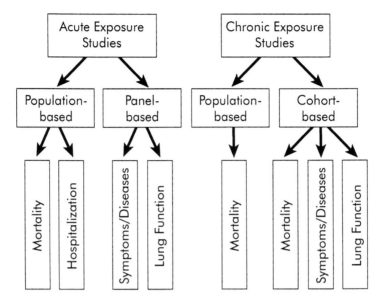

Fig. 31.2 Basic study designs of most of the published particle epidemiology studies.

for health outcomes and co-risk factors are individuals enrolled in a well-defined cohort, panel, or sample.

These studies can be further subdivided by the specific health outcomes evaluated. An appraisal of the strength of the overall epidemiological evidence of health effects of air pollution requires an evaluation of coherency (Bates, 1992). Effects of pollution should be evaluated across a range of related health outcomes. Cardiopulmonary health outcomes that have been evaluated include mortality, hospitalizations or health care visits for respiratory and/or cardiovascular disease, respiratory symptoms, measures of lung function, and restricted activity due to illness.

ACUTE EXPOSURE STUDIES

Population-based Mortality Studies

Table 31.1 lists and briefly summarizes more than 30 acute exposure, population-based, mortality studies. The earliest (beginning in the 1930s) and most methodologically simple of these studies focused on air pollution episodes. These studies evaluated changes in mortality before, during and after pollution episodes that lasted a few days or weeks. Early studies of severe episodes at the Meuse River valley in Belgium (Firket, 1931), Donora, Pennsylvania (Ciocco and Thompson, 1961), and London, England (Logan, 1953), demonstrated a clear linkage between cardiopulmonary mortality and morbidity with episodes of extremely elevated concentrations of particulate and/or sulfur oxide air pollution. Although the biological mechanisms were poorly understood, there remained

Table 31.1 Summary of selected acute exposure, population-based, mortality studies

Authors	Pub. year	Study area	Study period	PM measure	Comments
Episode studies					
Firket	1931	Meuse Valley, Belgium	1930	None	The first three studies reported the morbidity and mortality effects of episodes of extreme air pollution providing early quantitative evidence of severe adverse effects of air pollution. The three later studies observed smaller effects during episodes of less extreme pollution.
Ciocco and Thompson	1961	Donora, PA	1948	None	
Logan	1953	London, England	1952	Smoke	
Glasser and Greenburg	1967	New York, NY	1966	CoH	
Wichmann *et al.*	1989	North Rhine-Westfalia, Germany	1985	SP	
Anderson *et al.*	1995	London, England	1991	BS	
Early time-series mortality studies I					
Schimmel and Greenburg	1972	New York, NY	1963–68	CoH	Various researchers began to explore the use of daily time-series analysis to evaluate associations between air pollution and mortality. The pollution exposure data were often very limited; a variety of statistical methods were used; and the results were often not very robust. Associations between daily mortality counts and pollution, however, were generally observed.
Wyzga	1978	Philadelphia, PA	1957–66	CoH	
Mazumdar and Sussman	1983	Pittsburgh, PA	1972–77	CoH	
Ostro	1984	London, England	1958–72	BS	
Özkaynak and Spengler	1985	New York, NY	1963–76	CoH	
Shumway *et al.*	1988	Los Angeles, CA	1970–79	KM	
Schwartz and Marcus	1990	London, England	1958–72	BS	
Early time-series mortality studies II					
Fairley	1990	Santa Clara, CA	1980–86	CoH	These studies used increasingly formal time-series modeling including Poisson regression. Some researchers began to suggest the emergence of a consistent pattern. When mortality was analyzed by cause-of-death categories, particulate pollution had the largest effects on respiratory and cardiovascular mortality. Critics suggested that effects may be due to modeling techniques or due to confounding by long-term time trends, season, weather, or other pollutants.
Schwartz	1991	Detroit, MI	1973–82	TSP	
Kinney and Özkaynak	1991	Los Angeles	1970–79	KM	
Dockery *et al.*	1992	St Louis, MO / Kingston, TN	1985–86 / 1985–86	PM_{10}	
Schwartz and Dockery	1992a	Steubenville, OH	1978–84	TSP	
Schwartz and Dockery	1992b	Philadelphia, PA	1973–80	TSP	
Pope *et al.*	1992	Provo/Orem, UT	1985–89	PM_{10}	
Schwartz	1993a	Birmingham, AL	1985–88	PM_{10}	

Table 31.1 *cont.*

Authors	Pub. year	Study area	Study period	PM measure	Comments
Recent time-series mortality studies – replication and reanalysis of US and London studies					
Ito et al.	1993	London, England	1965–72	BS	Further analysis demonstrated that previous studies could be replicated. Similar associations were generally observed in other cities. Increasingly sophisticated models were employed including the use of non-parametric smoothing of time, weather, and pollution variables. The use of synoptic weather modeling was included in some of the studies. Air pollution effects generally persisted with various approaches to control for long-term time trends, seasonality, and weather. Exposure–response relationships with particulates were often near linear with little evidence of a threshold. The estimated effects of particulate pollution were sometimes sensitive to the inclusion of other pollutants in the models.
Schwartz	1994	Philadelphia, PA	1973–80	TSP	
Schwartz	1994b	Cincinnati, OH	1977–82	TSP	
Kinney et al.	1995	Los Angeles, CA	1985–90	PM_{10}	
Moolgavkar et al.	1995b	Steubenville, OH	1974–84	TSP	
Moolgavkar et al.	1995a	Philadelphia, PA	1973–88	TSP	
Styer et al.	1995	Cook Co., IL; Salt Lake City, UT	1985–92	PM_{10}	
Ostro	1995	Riverside, CA	1980–86	PM_{10}	
Samet et al.	1995	Several US cities	1970s–80s	TSP PM_{10}	
Schwartz et al.	1996	Six US cities	1970s–80s	$PM_{2.5}$ PM_{10}	
Pope and Kalkstein	1996	Provo/Orem, UT	1985–89	PM_{10}	
Ito and Thurston	1996	Cook Co., IL	1985–90	PM_{10}	
Anderson et al.	1996	London, England	1987–92	BS	
Samet et al.	1997	Philadelphia, PA	1973–88	TSP	

Table 31.1 *cont.*

Authors	Pub. year	Study area	Study period	PM measure	Comments
Recent time-series mortality studies – replication in other cities throughout the world					
Spix et al.	1993	Erfurt, Germany	1988–89	TSP	Although most of the early studies were conducted with data from US cities and London, recently there have been daily time-series studies conducted in more than 20 cities in other parts of the world. Remarkably, even with the large differences in socioeconomic, climate, environmental, and other conditions, positive associations with particulate air pollution and mortality were generally observed. The relative role of other pollutants, especially sulfur dioxide, remains unclear. The associations were generally stronger and more consistent in western European cities than in eastern European cities. Positive statistically significant associations were also observed in cities from China, Brazil, Chile, and Mexico.
Touloumi et al.	1994	Athens, Greece	1984–88	BS	
Xu et al.	1994	Beijing, China	1989	TSP	
Saldiva et al.	1995	São Paulo, Brazil	1990–91	PM_{10}	
Ostro et al.	1996	Santiago, Chile	1989–91	PM_{10}	
Wietlisbach et al.	1996	Three cities in Switzerland	1984–89	TSP	
Bachárová et al.	1996	Bratislava, Slovak Rep.	1987–91	TSP	
Zmirou et al.	1996	Lyons, France	1985–90	PM_{13}	
Wojtyniak and Piekarski	1996	Four cities in Poland	1977–90	BS	
Touloumi et al.	1996	Athens, Greece	1987–89	BS	
Spix and Wichmann	1996	Köln, Germany	1975–85	TSP	
Sunyer et al.	1996	Barcelona, Spain	1985–91	BS	
Dab et al.	1996	Paris, France	1987–92	PM_{13}	
Verhoef et al.	1996	Amsterdam, Netherlands	1986–92	PM_{10}	
Vigotti et al.	1996	Milan, Italy	1980–89	TSP	
Rahlenbeck and Kahl	1996	East Berlin, Germany	1981–89	SP	
Ballester et al.	1996	Valencia, Spain	1991–93	BS	
Borja-Aburto et al.	1997	Mexico City, Mexico	1990–92	TSP	
Katsouyanni et al.	1997	12 European cities	1975–92	PM_{10} BS	

TSP, total suspended particulate matter; SP, suspended particles; BS, British smoke; CoH, coefficient of haze; KM, measure thought to be similar to CoH; SO_4 sulfates; PM_{13}; PM_{10}; $PM_{2.5}$ airborne particles with an aerodynamic diameter equal to or less than 13, 10 or 2.5 micrometers respectively.

little disagreement that ambient air pollution at these very high levels could be an important risk factor for increased cardiopulmonary disease and early mortality (Bates, 1980; Holland *et al.*, 1979; Shy, 1979). Three additional studies of less severe episodes in New York (Glasser *et al.*, 1967), Germany (Wichmann *et al.*, 1989) and London (Anderson *et al.*, 1995) suggested smaller mortality effects associated with less extreme pollution.

In the 1970s and through the 1980s various investigators began to use daily time-series analysis to explore potential associations between daily mortality and air pollution at relatively low levels (see Early time-series mortality studies I, in Table 31.1). Time-series studies had the advantage of using mortality and pollution data over much longer time periods, rather than a few days before, during and after an episode. This approach potentially provided more statistical power to evaluate mortality effects of less extreme levels of pollution exposure. The earliest time-series mortality studies were limited by inadequate pollution data. A variety of time-series approaches were applied to these data which made the results difficult to compare. Nevertheless, associations between daily mortality counts and particulate air pollution were often observed.

In the early 1990s, several investigators reported time-series analysis based on Poisson regression modeling (see Early time-series mortality studies II, in Table 31.1). These studies observed changes in daily death counts associated with short-term changes in particulate air pollution. These studies took advantage of the readily available computerized records of air pollution, counts of daily mortality and weather. In addition, statistical software became available for Poisson regression analysis of count data. This permitted more appropriate statistical analysis of the daily counts of deaths, particularly in communities with small numbers of deaths per day. In addition, the use of consistent statistical methods had the effect of allowing direct, quantitative comparison of effect estimates between studies.

Some of these studies also provided a breakdown of mortality by broad cause-of-death categories. Particulate air pollution generally had the largest effect on respiratory disease mortality but effects on cardiovascular mortality were also observed. These studies did not observe a particulate pollution threshold. The relative risk of mortality increased monotonically with particulate concentrations – usually in a near-linear fashion. Also, these studies often observed a lead–lag relationship between air pollution and mortality. The results suggested that the increased mortality occurred concurrently or within 1–5 days following an increase in air pollution.

Results of the daily time-series mortality studies reported in the early 1990s generated substantial controversy. Some authors noted the lack of biological mechanisms for mortality effects at such low levels of particulate air pollution and suggested that the observed associations between daily mortality and particulate air pollution might be due to the selection of modeling techniques or due to confounding by long-term time trends, season, weather variables, other pollutants, or some other unknown factor (Lipfert and Wyzga, 1995; Moolgavkar and Luebeck, 1996). Numerous studies have since been conducted in an effort to reanalyze or replicate the results of earlier studies (see Table 31.1). These secondary reanalyses have focused on evaluating alternative explanations for the observed findings.

Ultimately, however, true replication depends on other investigators conducting independent assessments of these associations in other time periods and/or in other populations. A substantial number of studies have been conducted in other cities

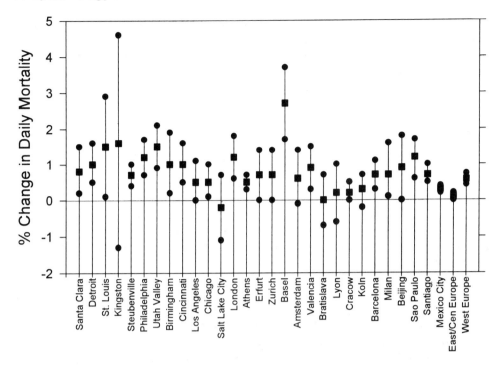

Fig. 31.3 Estimated per cent changes in daily mortality associated with a 10 µg/m³ increase in PM₁₀ (with 95% confidence intervals) for a number of cities.

throughout the world. Increasingly sophisticated models have been used to control for long-term time trends, seasonality, and weather. The early studies generally could be replicated and similar mortality effects were often, but not always, observed in other cities in the USA and around the world.

Because various measurements of particulate pollution were used, and because various modeling strategies were used, precise comparisons of effect estimates across all the studies are difficult. For illustrative purposes and to provide an idea of the heterogeneity of the studies, the estimated changes in daily mortality associated with a 10 µg/m³ increase in PM_{10} concentration are plotted for a number of cities in Fig. 31.3. Particulate measurements are converted to PM_{10} by assuming that $PM_{10} = 0.55 \times TSP$; $PM_{10} = CoH/0.55$; and $PM_{10} = BS$.

Population-based Hospitalization Studies

Daily counts of hospital admissions can be analyzed in the same way that daily counts of mortality have been assessed. Studies of the severe air pollution episodes of the 1930s to 1950s suggested increased morbidity, including hospitalization and related health care end-points. For example, in the 1952 London fog episode, requests for hospital beds increased dramatically during the episode. Hospital admissions were limited by available beds. This illustrates that there are constraints on hospital usage which are not found in the mortality data. In addition, records of hospital admissions are not as readily available as routinely collected public health records such as death records. Hospital admissions data have been more difficult to obtain and intrinsically have other constraints that make analysis harder than mortality time series. Thus most of the studies of air pollution that focused primarily on hospitalization and related health care end-points have been reported quite recently (see Table 31.2).

Earlier studies had looked at the effects of episodes of high air pollution on hospital admissions. In 1989, Pope reported on the effects of a unique natural experiment that occurred in the Utah Valley (Pope, 1989). During the winter of 1986–87, a labor dispute resulted in the closure of the local steel mill, the largest single source of particulate air pollution in the valley. During this winter, PM_{10} concentrations averaged 51 µg/m^3 with a high of 113 µg/m^3 compared with the previous winter's mean of 90 µg/m^3 with a high of 365 µg/m^3. During this winter, children's hospital admissions for respiratory disease were more than 50% lower than in adjacent years. Regression analyses (Pope, 1989, 1991, 1996) estimated a 4.2% decrease in asthma and bronchitis admissions of children associated with a 10 µg/m^3 decrease in 2-month mean PM_{10}.

In 1981 Samet *et al.* (1981) reported statistically significant but very small increases in respiratory emergency room visits in Steubenville, Ohio, associated with elevated levels of particulate pollution and sulfur dioxide. In 1983, based on a limited study of outpatient data from a clinic in Salt Lake City, Utah, Lutz (1983) reported that strong positive associations were observed between weekly particulate pollution levels and the percentage of patients with primarily respiratory illness. In 1987 Bates and Sizto (1987) first reported associations between respiratory hospital admissions with sulfates and ozone during summer months in Southern Ontario, Canada. This study was the first to use routinely collected billing information on hospital admissions to assess associations with short-term air pollution exposures.

Since these early studies in the 1980s, more than 30 time-series studies have reported associations between particulate air pollution and various respiratory hospitalizations or related health care end-points (Table 31.2). For example, using more rigorous statistical modeling, Burnett *et al.* (1994) largely replicated the associations between increased rates of respiratory hospital admissions in southern Ontario that Bates and Sizto originally reported to be associated with increased sulfate and ozone concentrations. Thurston and colleagues (1994) reported similar associations in Toronto, Ontario, and for several cities in New York State (Thurston *et al.*, 1992). The focus of these studies was the effects of acid aerosols, but estimates were reported for various measures of particle exposure. Schwartz (1994c,d,e) applied these methods to the assessment of billing data for hospital admissions available from the Health Care Financing Administration (HCFA) in the USA. The HCFA provides payments (MEDICARE) for all hospital admissions of elderly people in the USA. Merging these hospital usage data with air pollution and weather data provided the opportunity to assess the role of air pollution on a range of health indicators in the elderly population. Several studies also analyzed emergency department visits and found them to be associated with particulate air pollution. For example, particulate air pollution was associated with emergency department visits for asthma in Seattle, Washington (Schwartz *et al.*, 1993) and emergency department visits for chronic obstructive pulmonary disease in Barcelona (Sunyer *et al.*, 1993). Several recent studies from various European cities also reported associations between particulate air pollution and various respiratory hospitalizations or related health care end-points. As with the acute mortality studies, increasingly sophisticated time-series statistical regression models were used to control for long-term time trends, seasonality, weather, and other factors.

Table 31.2 Summary of selected acute exposure, population-based, hospitalization and related studies

Authors	Pub. date	Study area	Health end-points studied	PM measure
Hospital admissions				
Bates and Sizto	1987	Southern Ontario	Resp.	SO_4
Pope	1989	Provo/Orem, UT	Resp.	PM_{10}
Bates and Sizto	1989	Southern Ontario	Resp.	SO_4
Pönkä	1991	Helsinki, Finland	Asthma	TSP
Pope	1991	Provo/Orem UT	Resp.	PM_{10}
Lipfert and Hammerstrom	1992	Southern Ontario	Resp.	SO_4
Thurston *et al.*	1992	Three New York Cities	Resp	SO_4
Burnett *et al.*	1994	Southern Ontario	Resp.	SO_4
Schwartz	1994c	Birmingham, AL	Elderly resp.	PM_{10}
Schwartz	1994e	Minneapolis, MN	Elderly resp.	PM_{10}
Thurston *et al.*	1994	Toronto, Canada	Resp.	TSP, PM_{10}, $PM_{2.5}$, SO_4
Schwartz	1994d	Detroit, MI	Elderly resp.	PM_{10}
Burnett *et al.*	1995	Ontario	Resp., cardio	SO_4
Schwartz and Morris	1995	Detroit, MI	Cardio	PM_{10}
Schwartz	1995	Tacoma, WA New Haven, CT	Resp.	PM_{10}
Schwartz	1996	Spokane, WA	Elderly resp.	PM_{10}
Schwartz *et al.*	1996	Cleveland, OH	Resp.	PM_{10}
Dab *et al.*	1996	Paris, France	Resp.	PM_{13}, BS
Ponce de Leon *et al.*	1996	London, England	Elderly resp.	BS
Pönkä and Virtanen	1996	Helsinki, Finland	Elderly resp.	TSP
Schouten *et al.*	1996	Two Netherlands Cities	Elderly resp.	BS
Vigotti *et al.*	1996	Milan, Italy	Elderly resp.	TSP
Moolgavkar *et al.*	1997	Minneapolis, MN Birmingham, AL	Elderly resp.	PM_{10}
Burnett *et al.*	1997	16 Canadian cities	Resp.	CoH
Schwartz	1997	Tucson, AZ	Cardio.	PM_{10}
Anderson *et al.*	1997	Six European cities	COPD	BS, TSP
Poloniecki *et al.*	1997	London	Cardio.	BS
Emergency visits				
Samet *et al.*	1981	Steubenville, OH	Resp.	TSP
Bates *et al.*	1990	Vancouver, BC	Resp.	SO_4
Schwartz *et al.*	1993	Seattle, WA	Asthma	PM_{10}
Sunyer *et al.*	1993	Barcelona, Spain	COPD	BS
Hefflin *et al.*	1994	Southeast WA	Resp.	PM_{10}
Pantazopoulou *et al.*	1995	Athens, Greece	Resp., cardio	BS
Delfino *et al.*	1997	Montreal, Canada	Elderly resp.	PM_{10}, $PM_{2.5}$, SO_4
Lipsett *et al.*	1997	Santa Clara, CA	Asthma	CoH, PM_{10}
Clinic, outpatient, medical visits or care				
Lutz	1983	Salt Lake City, UT	Respiratory	TSP
Schwartz *et al.*	1991	Five areas in Germany	Croup, bronchitis	TSP
Xu *et al.*	1995	Beijing, China	Visits	TSP
Gordian *et al.*	1996	Anchorage, AL	Resp.	PM_{10}
Choudhury *et al.*	1997	Anchorage, AL	Asthma	PM_{10}
Other				
Ransom and Pope	1992	Utah Valley, UT	School absences	PM_{10}

Resp, respiratory; cardio, cardiovascular; COPD, chronic obstructive pulmonary disease.

However, the adequacy of these methods has been questioned by Moolgavkar *et al.* (1997) in a reanalysis of the Birmingham and Minneapolis hospitalization data.

Probably the most significant recent developments are findings of associations between particulate air pollution and hospitalizations for cardiovascular disease. Burnett *et al.* (1995) evaluated hospital admissions data for 168 hospitals in Ontario and observed significant positive associations between both respiratory and cardiac hospital admissions with sulfate particle concentrations. Associations between cardiovascular hospitalizations or emergency visits and particulate air pollution have also recently been observed in Detroit, Michigan (Schwartz and Morris, 1995), Athens, Greece (Pantazopoulou *et al.*, 1995), Tucson, Arizona (Schwartz, 1997), and London, UK (Poloniecki *et al.*, 1997).

Person-based Studies of Short-Term Exposures

Population-based studies generally have the weakness that they have little or no information on the individual characteristics of those being affected by air pollution. Personal characteristics change little, if at all, day to day, and therefore will not confound the associations with short-term air pollution exposures. As it is likely that the response to air pollution exposures varies among individuals depending on their personal characteristics, studies of individual responses offer the opportunity to identify personal risk factors associated with the deleterious health affects of exposure.

Panel-based Lung Function Studies

Lung function is an objective and potentially sensitive indicator of acute response to air pollution. Two studies in the 1970s postulated that episodes of air pollution would lead to reversible deficits in lung function. Stebbings and colleagues took repeated measurements of the lung function of children in Pittsburgh following a major air pollution episode (Stebbings and Fogleman, 1979). They hypothesized that the air pollution episode had produced lower lung functions that would return to normal in the weeks following exposure. These data were suggestive of such a recovery, but were not very convincing as they lacked lung function measurements before or during the episode.

In 1978, Dockery and colleagues undertook a prospective study of lung function in Steubenville, Ohio, an area with frequent fall air pollution episodes. Baseline lung function of children in four schools was measured. Upon the declaration of an air pollution alert by the local air pollution control agency, lung function measurements were repeated. Three weekly follow-up measurements assessed recovery. These panels of elementary school children were monitored during four episode periods in 1978 through 1980 (Dockery *et al.*, 1982). Declines in forced expired volume in 0.75 s ($FEV_{0.75}$) were observed following these episodes. While designed as a simple episode study, regression analyses suggested declines in lung function across the full range of exposures. This study first applied random effects models to assess the effects of air pollution exposures on individual children. This design and these analytic methods have been applied in many studies of repeated lung function measurements around the world. Since the early 1980s, at least 15 studies have used repeated measures of lung function to evaluate potential effects of particulate air pollution (see Table 31.3).

Table 31.3 Summary of selected acute exposure, cohort-based, lung function studies

Authors	Pub. date	Study area	Cohorts	PM measure	Lung function measure
Dockery *et al.*	1982	Steubenville, OH	Children	TSP	FVC, $FEV_{0.75}$
Johnson *et al.*	1982	Missoula, MT	Children	TSP	FVC, FEV_1, FEF_{25-75}
Lebowitz *et al.*	1985	Tucson, AZ	Children and adults	TSP	PEF
Dassen *et al.*	1986	Ijmond, Netherlands	Children	TSP	FVC, FEV_1, PEF
Johnson *et al.*	1990	Five Montana cities	Children	TSP, PM_{10}, $PM_{2.5}$	FVC, FEV_1, FEF_{25-75}, PEF
Pope *et al.*	1991	Utah Valley, UT	Asthmatics, children	PM_{10}	PEF
Pope and Dockery	1992	Utah Valley, UT	Children, symptomatic children	PM_{10}	PEF
Hoek and Brunekreef	1993	Wageningen, Netherlands	Children	PM_{10}	FVC, FEV_1, PEF
Koenig *et al.*	1993	Seattle, WA	Children	$PM_{2.5}$	FVC, FEV_1
Pope and Kanner	1993	Salt Lake City, UT	Adult smokers with COPD	PM_{10}	FVC, FEV_1
Roemer *et al.*	1993	Wageningen, Netherlands	Symptomatic children	PM_{10}, BS	PEF
Neas *et al.*	1995	Uniontown, PA	Children	PM_{10}, $PM_{2.5}$, SO_4	PEF
Neas *et al.*	1996	State College, PA	Children	PM_{10}, $PM_{2.1}$, SO_4	PEF
Romieu *et al.*	1996	Mexico City	Asthmatic children	PM_{10}, $PM_{2.5}$	PEF
Peters *et al.*	1997b	Sokolov, Czech Rep.	Asthmatic children	TSP, PM_{10}, $PM_{2.5}$	PEF
Peters *et al.*	1997c	Erfurt, Germany	Asthmatic	Size Frac. PM	PEF
Vedal *et al.*	1998	Port Alberni, BC	Children	PM_{10}	PEF

FVC, forced vital capacity; $FEV_{0.75}$, forced expiratory volume in 0.75 seconds; $FEV_{0.1}$, forced expiratory volume in 1 second; FEF_{25-75}, maximum midexpiratory flow; PEF, peak expiratory flow rate.

In a reanalysis of the Steubenville data, Brunekreef *et al.* (1991) found the strongest association with the mean particle concentrations over the previous 5 days. Similar decreases in forced expired volume in 1 s (FEV_1) were observed in school children following a particulate and sulfur oxide pollution episode in January 1985 in the Netherlands (Dassen *et al.*, 1986). Subsequent panel studies of school children in the Netherlands with weekly lung function measurements (Hoek and Brunekreef, 1993, 1994) have also shown decreased FEV_1 associated with daily PM_{10} concentrations. Effects of exposures lagged by up to 7 days were observed. Studies in Montana (Johnson *et al.*, 1982, 1990) observed similar declines in lung function associated with abnormal urban air pollution episodes, but comparable declines were not associated with a volcanic ash episode.

Pope and Kanner (1993) analyzed repeated FEV_1 measurements in a panel of adults with chronic pulmonary disease who were participating in the Lung Health Study. Measurements were taken 10–90 days apart. FEV_1 levels were found to be associated with PM_{10} concentrations. Koenig and colleagues (1993) studied the lung function (forced vital capacity (FVC) and FEV_1) of children in Seattle, Washington, with relatively low particulate air pollution levels. Lung function declines were associated with fine particulate air pollution for asthmatic children, but not for non-asthmatic children. Overall, these studies generally observed a decrease of less than 0.35% in FEV_1 associated with each 10 $\mu g/m^3$ increase in daily mean PM_{10}.

Several studies have used peak expiratory flow (PEF) measurements as an indicator of acute changes in lung function, including studies in the Netherlands, Utah Valley, Utah, Uniontown and State College, Pennsylvania, Mexico City, and Port Alberni, British Colombia (see Table 31.3). In these studies, a 10 $\mu g/m^3$ increase in PM_{10} was associated with very small, but often statistically significant, decreases in peak flow measurements. As with FEV_1, the strongest associations with peak flow included particulate pollution over the previous several days, allowing for a lag in effect.

Panel-based Symptom/Disease Studies

The use of daily diaries to record respiratory or other symptoms is a relatively inexpensive method of evaluating acute changes in respiratory health studies. In an early study, Whittemore and Korn (1980) analyzed daily asthma attacks as recorded on daily diaries of asthmatics in the Los Angeles, California, area. They observed increased asthma attacks on days with high oxidant and particulate air pollution. This study was important in introducing the concept of pooling the results of subject-specific regressions into a single effect estimate. This analysis also first addressed the issue of day-to-day correlations in response to air pollution or other stimuli.

Since this early study, there have been at least 15 additional studies that have similarly evaluated associations between daily respiratory symptoms and particulate air pollution (see Table 31.4). While some of these studies focused on asthmatics and exacerbation of asthma, many followed non-asthmatics and evaluated changes in acute respiratory health status more generally. Reported symptoms were often aggregated into upper respiratory symptoms (including such symptoms as runny or stuffy nose, sinusitis, sore throat, wet cough, head cold, hay fever, and burning or red eyes) and lower respiratory symptoms (including wheezing, dry cough, phlegm, shortness of breath, and chest discomfort or pain). In addition, a cough, the most frequently reported symptom, was often analyzed separately.

Small, often statistically insignificant, associations between particulate pollution and upper respiratory symptoms were observed. Associations with lower respiratory disease and cough, however, were typically larger and usually statistically significant. Exacerbation of asthma based on recorded asthma attacks or, in the case of three studies, increased bronchodilator use, were also associated with particulate air pollution.

Associations between particulate air pollution and more general measures of acute disease also have been observed. For example, Ostro (1983, 1987, 1990) and Ostro and Rothschild (1989) evaluated the timing of restricted activity days of US adult workers. Restricted activity due to respiratory morbidity was consistently associated with

Table 31.4 Summary of selected acute exposure, cohort-based, symptom studies

Authors	Pub. date	Study area	PM measure	Cohorts	Symptom
Whittemore and Korn	1980	Los Angeles, CA	TSP	Asthmatics	Asthma attacks
Vedal et al.	1987	Chestnut Ridge, PA	CoH	Children	URI, LRI
Krupnick et al.	1990	Los Angeles, CA	CoH	Children and adults	Various respiratory
Pope et al.	1991	Utah Valley, UT	PM_{10}	Asthmatic children	URI, LRI, asthma medication use
Ostro et al.	1991	Denver, CO	$PM_{2.5}$	Asthmatics	Asthma, cough dyspnea
Pope and Dockery	1992	Utah Valley, UT	PM_{10}	Children	URI, LRI, cough, asthma medication use
Braun-Fahrländer et al.	1992	Two Swiss Cities	TSP	Children	URI
Roemer et al.	1993	Wageningen, Netherlands	PM_{10}	Children	Cough, wheeze
Ostro et al.	1993	Southern, CA	CoH, SO_4	Adults	URI, LRI
Hoek et al.	1994	4 Netherlands cities	PM_{10}, SO_4	Children	URI, LRI, cough
Schwartz et al.	1994	Six US cities	PM_{10}	Children	URI, LRI, cough
Neas et al.	1995	Uniontown, PA	$PM_{10}, PM_{2.5}$	Children	Cough
Hoek and Brunekreef	1995	Deurne, Netherlands	PM_{10}	Children	URI, LRI, cough
Neas et al.	1996	State College, PA	$PM_{10}, PM_{2.5}$	Children	Cough
Romieu et al.	1996	Mexico City	$PM_{10}, PM_{2.5}$	Asthmatic children	LRI
Peters et al.	1997c	Eurfurt, Germany	Size frac. PM	Asthmatics	Respiratory illness, cough
Peters et al.	1997a	Sokolov, Czech Rep.	$TSP, PM_{10}, PM_{2.5}$	Asthmatic children	Various respiratory, asthma medication use
Vedal et al.	1998	Port Alberni, BC	PM_{10}	Children	URI, LRI, cough

particulate pollution. Morbidity was often more strongly associated with the fine, respirable, or sulfate component of particulate pollution. Ransom and Pope (1992) reported similar associations between PM_{10} and grade-school absences in children in Utah Valley, Utah. Lagged pollution effects of up to several weeks were observed for both restricted activity in adults and in school absences.

Summary of Acute Exposure Studies

The largest number of epidemiologic studies of the effects of particulate air pollution exposures address response to short-term exposures. The early studies of extreme air pollution episodes demonstrated clear short-term adverse health effects. The population-based mortality and morbidity time-series studies have developed as natural extensions of these episode studies. Initially they were based on routinely available population death records. The analytic methods have become fairly standardized. Because of the availability of statistical software packages which provide appropriate methods for regression analyses of count data, the ready availability of data, and the straightforward modeling, these daily mortality time-series studies have become common. These mortality studies were then expanded to use billing data on hospital admissions. We should expect that more of these types of data will become available through managed health care systems, and that the number of daily hospital admissions studies will expand.

Evidence from the large number of acute exposure studies briefly reviewed here suggests a coherence of effects across a range of related health outcomes and at least some consistency of effects across independent studies with different investigators and from different settings. Figure 31.4 presents a stylized summary of effect estimates of acute

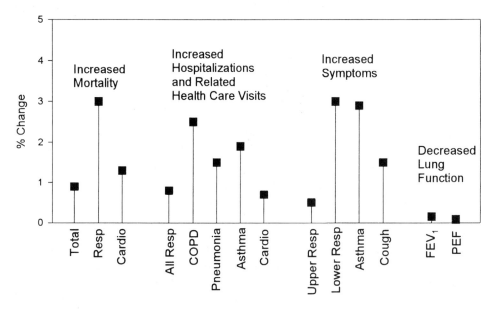

Fig. 31.4 Stylized summary of acute exposure studies, per cent change in health end-point per 10 μg/m³ change in PM_{10}.

exposure to particulate air pollution. Per cent changes in various health end-points associated with a 10 µg/m^3 increase in PM$_{10}$ are presented. The effect estimates provided in this figure clearly are not precise. Precise point estimates cannot be given because the recent rapid developments of the literature in this area make them a moving target and because of the difficulty of comparing across studies that used different measures of pollution, different models, and differently defined health end-points. Figure 31.4, therefore, should be considered as illustrative of the relative magnitude of the epidemiologic associations.

Acute exposure studies suggest that each 10 µg/m^3 increase in PM$_{10}$ is associated with approximately a 0.8% increase in daily mortality. Somewhat larger associations are observed for cardiovascular mortality and considerably larger associations for respiratory mortality. Particulate air pollution is also associated with increased hospitalization and related health care visits for respiratory disease and, to a somewhat lesser degree, cardiovascular disease. Particulate air pollution is associated with lower respiratory symptoms, exacerbation of asthma, and coughing. Weaker associations are observed for upper respiratory symptoms. Small, but usually statistically significant, declines in lung function have also been observed.

CHRONIC EXPOSURE STUDIES

Acute exposure studies provide little information about how much life is shortened, how pollution effects longer-term mortality rates, or pollution's potential role in the process of inducing chronic disease that may or may not be life threatening. Chronic exposure studies evaluate the effects of low or moderate exposure that persists for long periods of time as well as the cumulative effects of repeated exposure to elevated levels of pollution.

Population-based Mortality Studies

Many studies have suggested mortality effects of chronic exposure to air pollution. In 1964 Martin reported that overall annual respiratory mortality (as opposed to episodic mortality) in the Greater London region was significantly related to smoke (or particulate pollution) levels. In 1970 Lave and Siskin reported the results of one of the first serious attempts to measure long-term mortality effects of air pollution in the USA. They observed significant associations between annual mortality rates and particulate air pollution across US metropolitan areas. The work of Lave and Siskin was quite controversial, and other similar studies were conducted to replicate or refine their use of population-based cross-sectional study designs (Table 31.5). Most of these studies observed that mortality rates tended to be higher in cities with higher fine or sulfate particulate pollution levels. Regression techniques were used to evaluate cross-sectional differences in air pollution and mortality and to control for other ecological variables. In an attempt to control for other risk factors, population average values for demographic variables and other factors such as smoking rates, education levels, income levels, poverty rates, housing density, and others were included in the regression models. The basic conclusions from the population-based cross-sectional studies were: (1) mortality rates

Table 31.5   Selected chronic exposure studies

Authors	Pub. date	Study area	Study subjects	PM measure
Mortality (population-based)				
Lave and Seskin	1970	US SMSAs	Whole population	TSP, SO_4
Mendelsohn and Orcutt	1979	US county groups	Whole population	TSP, SO_4
Evans *et al.*	1984	US SMSAs	Whole population	TSP, SO_4
Lipfert	1984	US SMSAs	Whole population	TSP, SO_4
Ozkaynak and Thurston	1987	US SMSAs	Whole population	TSP, PM_{10}, $PM_{2.5}$, SO_4
Lipfert *et al.*	1988	US cities	Whole population	$PM_{2.5}$, SO_4
Bobak and Leon	1992	Czech Republic districts	Infants, postneonatal	PM_{10}
Mortality (cohort-based)				
Dockery *et al.*	1993	Six US cities	Adults	TSP, PM_{10}, $PM_{2.5}$, SO_4
Pope *et al.*	1995	50 and 151 US cities	Adults	$PM_{2.5}$, SO_4
Nishino *et al.*	1997	California	Non-smoking adults	Days >100 $\mu g/m^3$, PM_{10}
Woodruff *et al.*	1997	86 US cities	Postneonatal infants	PM_{10}
Symptoms/disease				
Dockery *et al.*	1989	Six US cities	Children	TSP, PM_{15}, $PM_{2.5}$, SO_4
Portney and Mullahy	1990	Cities in US annual Health Interview Survey (HIS)	Adults	TSP
Schwartz	1993b	53 US cities from NHANESI	Adults	TSP
Abbey *et al.*	1995	California	Non-smoking adults	TSP
Dockery *et al.*	1996	24 US and Canadian cities	Children	PM_{10}, $PM_{2.1}$, SO_4
Lung function				
Dockery *et al.*	1989	Six US cities	Children	TSP, PM_{15}, $PM_{2.5}$, SO_4
Raizenne *et al.*	1996	24 US and Canadian cities	Children	PM_{10}, $PM_{2.1}$, SO_4
Ackermann-Liebrich *et al.*	1997	Eight Swiss cities	Adults	PM_{10}

were associated with air pollution; (2) mortality rates were most strongly associated with fine or sulfate particulate matter; and (3) an average mortality effect of 3 to 9% of total mortality could be estimated.

These population-based studies dealt with mortality for all ages. However, one study focused on infant mortality (Bobak and Leon, 1992) in the Czech Republic. Infant mortality, especially respiratory postneonatal infant mortality, was strongly associated with particulate air pollution. After adjusting for various socioeconomic characteristics, relative risk of respiratory postneonatal mortality was approximately 3 for the most polluted areas versus least polluted areas.

Although population-based cross-sectional studies suggested that air pollution contributes to human mortality, these studies have been largely discounted because of the potential for uncontrolled confounding. Population-based cross-sectional studies cannot directly control for individual differences in cigarette smoking and other risk factors. Analytic studies of individuals in the population are required to address this issue.

Cohort-based Mortality Studies

Recently the results of several cohort studies which measure the survival of individuals prospectively have been reported (Table 31.5). These person-based studies use subject-specific information on risk factors to adjust for potential confounding of the air pollution associations. These studies provide some of the most compelling evidence about mortality effects of chronic exposure to air pollution.

Dockery and colleagues (1993) reported the results of a 14- to 16-year prospective follow-up of more than 8000 adults living in six US cities (Harvard Six Cities Study). Mortality, especially cardiopulmonary mortality, was significantly associated with sulfate and fine particulate air pollution – even after controlling for individual differences in age, sex, cigarette smoking, obesity, socioeconomic status, occupation and pre-existing chronic disease. Sulfate and fine particulates, combustion-source pollutants, were highly correlated in these cities. When either of these two pollutants were used directly in the analysis, the adjusted excess mortality risk increased linearly by 26% across the range of pollution in the study (see Table 31.6).

To test the hypotheses that mortality is associated with combustion-source particulate air pollution, a collaborative study was conducted linking individual risk factor data from the American Cancer Society (ACS) Cancer Prevention Study II (CPS-II) with national ambient air pollution data (Pope *et al.*, 1995c). Two indices of combustion-source particulate air pollution, mean sulfate and median fine particle concentrations, were chosen based on relevance and availability. Because sulfate and fine particle data were limited, an evaluation of the hypothesis that mortality is associated with combustion-source particulate air pollution would be strengthened by the use of two independently compiled indices of combustion source particles.

Prospective mortality data between 1982 and 1989 was available for more than 500 000 persons living in 151 different US metropolitan areas with sulfate measurements and 240 000 persons living in 50 metropolitan areas with fine particle measurements. When either of the two combustion-source particulate air pollution indices were used in

Table 31.6 Comparisons of mortality risk ratios (and 95% CI) for smoking and air pollution from the Six Cities and ACS prospective cohort studies

Cause of death	Current smoker[a]		Particulate air pollution (Most vs least polluted city)		
	Six Cities	ACS	Six Cities $(PM_{2.5})$	ACS $(PM_{2.5})$	ACS (SO_4)
All	2.00 (1.51–2.65)	2.07 (1.75–2.43)	1.26 (1.08–1.47)	1.17 (1.09–1.26)	1.15 (1.09–1.22)
Cardiopulmonary	2.30 (1.56–3.41)	2.28 (1.79–2.91)	1.37 (1.11–1.68)	1.31 (1.17–1.46)	1.26 (1.16–1.37)
Lung cancer	8.00 (2.97–21.6)	9.73 (5.96–15.9)	1.37 (0.81–2.31)	1.03 (0.80–1.33)	1.36 (1.11–1.66)
All others	1.46 (0.89–2.39)	1.54 (1.19–1.99)	1.01 (0.79–1.30)	1.07 (0.92–1.24)	1.01 (0.92–1.11)

[a] Risk ratios for current cigarette smokers with approximately 25 pack-years (about average at enrollment for both studies) compared with never smokers.

the analysis, adjusted excess mortality risk of 15 to 17% was observed across the range of pollution in the study. The estimated effects of smoking and the estimated effects of particulate air pollution on cardiopulmonary mortality were nearly the same for the two studies (Table 31.6). Inferences related to these similar results were further strengthened by their complementarity. Both the Harvard Six Cities and the ACS studies were based on prospective-cohort health and individual risk factor data and could control for individual differences in age, sex, race, cigarette smoking, and other risk factors. Nonetheless, these two studies had somewhat different strengths and limitations. The strengths of the Harvard Six Cities study were related to its balanced study design, the planned prospective collection of air pollution data, and the ability to present some of the basic analytic results in an easily understood graphical format. The primary limitation of the Harvard Six Cities study was the limited number of subjects in a limited study area, i.e. the mid- to north-eastern USA. In contrast, the ACS study sampled a large number of participants and cities across the USA. The ACS study did not include prospective collection of air pollution data, but used limited air pollution data that were compiled from other sources.

A third study (Nishino *et al.*, 1997) followed a cohort of 6338 non-smoking California Seventh-Day Adventists prospectively from 1977 to 1992. Concentrations of PM_{10} were estimated by interpolating measurements at fixed-site monitors. The principal measure of particle exposure was the estimated annual number of days when PM_{10} exceeded 100 $\mu g/m^3$. Risk factors controlled for in the analysis included education, past smoking, environmental tobacco smoke, occupational exposures, weekly physical exercise, dietary variables, body mass index, history of hypertension and parental history of cancer. Particulate pollution exposure was significantly associated with increased risk of mortality with contributing respiratory causes for individuals who were age 85 or less at time of enrollment into the cohort in 1977. No association was observed for the 136 persons over age 85 in 1977. The measure of exposure used in this study makes it difficult to compare results directly with other studies.

Woodruff *et al.* (1997) recently reported a study of post-neonatal mortality in the US. National Center for Health Statistics birth and death records for approximately four million infants born between 1989 and 1991 were linked with PM_{10} data from the EPAs Aerometric Database in 86 metropolitan areas. Unlike the cohort studies that followed adult subjects prospectively, this study obtained post-neonatal infant data retrospectively. In essence, all infants born in the study cities during the study period who did not die within a month of birth were followed-up (retrospectively) until they reached one year of age. The analysis compared post-neonatal mortality against average PM_{10} concentrations during the first two months of life. The analysis adjusted for individual differences in maternal race, maternal education, marital status, month of birth, maternal smoking during pregnancy and temperature. Particulate pollution exposure was associated with post-neonatal infant mortality for all causes, respiratory causes among normal birthweight children, and sudden infant death syndrome (SIDS). The estimated excess risk of mortality across the relevant range of pollution equaled 25%. This estimated excess risk is very similar to that estimated for adults in the Harvard Six Cities and ACS studies, even though the time frame of exposure for the infants was clearly far shorter than for the adults. This observation suggests that the relevant time frame of exposure is short (a few months versus years), or that infants are at greater risk to air pollution exposure.

Cohort-based Symptoms/Disease Studies

There have been several studies that have evaluated associations between particulate air pollution and chronic respiratory symptoms and disease. Early studies generally compared disease only between pairs of cities – polluted versus clean. More recent studies have evaluated five or more cities and are briefly summarized in Table 31.5. These include the Harvard Six Cities study's (Dockery et al., 1989) analysis of bronchitis in children, symptoms of never-smoking Seventh-Day Adventists in California (Abbey et al., 1995), analysis of data from the 1979 version of the US National Health Interview Survey (NHIS) (Portney and Mullahy, 1990), analysis using data from the first National Health and Nutrition Examination Survey (NHANES I) (Schwartz, 1993b), and analysis of the symptoms data from the 24-cities study of school children (Dockery et al., 1996). The effects of air pollution on respiratory disease or symptoms were estimated while adjusting for individual differences in various other risk factors. In all of these studies, statistically significant positive associations were observed between particulate air pollution and respiratory symptoms. Chronic cough, bronchitis and chest illness (but not asthma) were associated with various measures of particulate air pollution, suggesting that particulate air pollution was most consistently associated with bronchitic symptoms. The results suggest that a 10 $\mu g/m^3$ increase in PM_{10} was typically associated with a 5–25% increase in bronchitis or chronic cough.

Cohort-based, Lung Function Studies

In 1965 Holland and Reid reported a cross-sectional comparison of British male postal employees in London and in smaller country towns, where levels of SO_2 and particulate pollution were about half of those in the metropolis (Holland and Reid, 1965). Accounting for cigarette smoking, significant decrements of FEV_1 in London employees compared with those in the provinces were reported. There have been several more recent studies that have evaluated associations between measures of lung function (FVC, FEV_1, PEF) and particulate pollution levels in the USA. These studies include analysis of children's lung function data from the Harvard Six-City study (Dockery et al., 1989), analysis of data from both the first and second National Health and Nutrition Examination Surveys (NHANES I and NHANES II) (Chestnut et al., 1991; Schwartz, 1989), analysis of children's lung function from 24 US cities (Raizenne et al., 1996), and analysis from eight different areas in Switzerland (Ackermann-Liebrich et al., 1997). Each of these studies had information on individual persons in the samples. The effects of air pollution on lung function were estimated after adjusting for individual differences in age, race, sex, height and weight and controlling for smoking or restricting the analysis to never-smokers.

All of these studies observed small negative associations between lung function and particulate air pollution. In the Six Cities study, which had the least statistical power, the association was very weak and statistically insignificant. In each of the other studies, the association was small, but statistically significant. The results suggest that a 10 $\mu g/m^3$ increase in PM_{10} was associated with only small declines in lung function (typically about 1–3%). However, lung function measures have been shown to be important measures of health with remarkable predictive capacity for survival (Bates, 1989). Furthermore, as reported in the 24-Cities study, the risk of relatively large deficits in lung function (less than 85% predicted) was much higher in the more polluted cities, suggesting detrimental effects of respirable particulates or particulate acidity on normal lung growth and development.

Summary of chronic exposure studies

Evidence from the chronic exposure studies indicates that longer-term exposure to particulate air pollution is associated with a range of cardiopulmonary health outcomes. Figure 31.5 presents a summary of effect estimates of particulate air pollution. Percent changes in various health end-points associated with a 5 μg/m³ increase in $PM_{2.5}$ are presented. $PM_{2.5}$ is chosen as the index of particulate exposure because it was a primary measure of pollution used in more of the most rigorous and recent chronic exposure studies. Furthermore, especially in the mortality studies, $PM_{2.5}$ was more closely associated with the health outcomes than was PM_{10} or TSP. The effect estimates provided in Fig. 31.5 are not intended to be definitive point estimates, but estimates based on selected studies that are reasonably representative. The population-based mortality estimates are based on results presented by Evans *et al.* (1984), Lipfert (1984), Lipfert *et al.* (1988) and Özkaynak and Thurston (1987). The cohort-based estimates are based on Dockery *et al.* (1993), Pope *et al.* (1995c) and Woodruff *et al.* (1997). Bronchitis, children's lung function, and adults' lung function estimates are based on results reported by Dockery *et al.* (1996), Raizenne *et al.* (1996), and Ackermann-Liebrich *et al.* (1997), respectively.

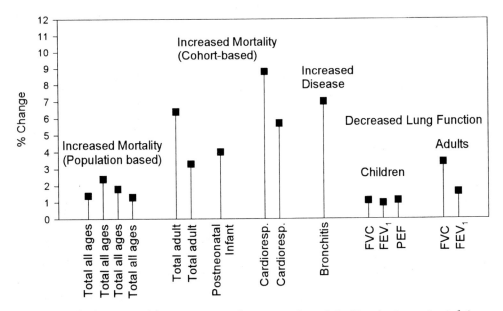

Fig. 31.5 Stylized summary of chronic exposure studies, per cent change in health end-point per 5 μg/m³ change in $PM_{2.5}$.

As illustrated in Fig. 31.5, total mortality is generally observed to be associated with chronic exposure to particulate air pollution by approximately 2–4% per 5 µg/m^3 increase in PM$_{2.5}$. Somewhat larger associations are generally observed for cardiopulmonary mortality. Remarkably, even after directly controlling for individual differences in age, sex, race, cigarette smoking and other risk factors, the prospective cohort studies estimated larger effect estimates than the population-based studies. Studies also observed that exposure to particulate air pollution was associated with increased respiratory disease. Small declines in lung function, including FVC, FEV$_1$ and PEF, were also observed.

LIMITATIONS OF THE EPIDEMIOLOGY

The fundamental strength of the currently available epidemiological evidence is its ability to evaluate health outcomes with real people, who are living in normal environments, and who are exposed to typical pollution. There are many studies using assorted study designs, conducted by various researchers, from numerous study areas. The reasonably consistent and coherent pattern of cardiopulmonary health effects associated with particulate air pollution that has been observed by the epidemiological studies currently is the strongest evidence of the potential health effects of this pollution.

A limitation or weakness of the epidemiology of particulate air pollution as presented here is that it is unbalanced in at least two important ways as illustrated in Table 31.7. First, the vast majority of the most relevant studies have been reported only recently, in the

Table 31.7 Number of primary studies on which this review is based separated by study design and time period published

Study design		1930s–1950s	1960s	1970s	1980s	1990s	Comments
Acute exposure							
Population	Mortality	3	1	2	5	24	Relatively inexpensive. Use retrospectively collected data. Usually single study area per study.
	Hospitalization	3	0	0	5	36	
Panel-based	Symptom	0	0	0	2	16	Data collected prospectively, but panels need not be extremely large and usually single study area per study.
	Lung function	0	0	0	4	13	
Chronic exposure							
Population	Mortality	0	1	3	5	2	Relatively inexpensive. Cross-sectional data collected retrospectively.
Cohort-based	Mortality	0	0	0	0	4	Expensive, time-consuming. Requires much data collection over long periods of time on relatively large cohorts or samples from multiple study areas.
	Symptom	0	0	0	1	4	
	Lung function	0	1	0	2	3	

last 10 years. Second, and more important, the large majority of the studies evaluate effects of acute exposures. The acute exposure studies are important and provide compelling evidence of cardiopulmonary health effects associated with acute exposure. Nevertheless, chronic exposure may be much more important in terms of overall public health. The large number of acute exposure studies relative to the number of chronic exposure studies has less to do with public health relevance than the relative ease and expense of conducting acute exposure studies.

In addition to the unbalanced development of this literature, there are at least four fundamental limitations that are inherent in the use of observational studies on real people, living in uncontrolled environments, exposed to complex particulate air pollution. First, epidemiology provides little information about biological mechanisms. The results of epidemiologic studies of particulate health effects provide a pattern that may be biologically germane. Biological plausibility is enhanced by the observation of a coherent cascade of cardiopulmonary health effects and by the fact that non-cardiopulmonary health endpoints are not typically associated with particulate pollution. The epidemiological evidence, however, is clearly limited on this subject. For example, both the acute and chronic exposure studies observed significant health effects, but linkages between acute and chronic effects in terms of biological mechanisms remain unclear.

A second basic limitation relates to the minimal information regarding coherence between ambient and personal exposures. Accurate measures of personal exposure to air pollutants for population-based studies or for studies of large cohorts is impractical and, for some applications, unnecessary. Exposures to air pollution has typically been estimated by using ambient air pollution data. Such an approach is useful because public policy and pollution abatement strategies typically (and often necessarily) focus on ambient concentrations of air pollutants.

A third basic limitation of the epidemiological studies involves the difficulty of disentangling independent effects or potential interactions between highly correlated risk factors. The difficulty of disentangling independent effects or interactions exists largely because alternative risk factors may be correlated with air pollution, resulting in potential confounding. Confounding may result when another risk factor that is correlated with both exposure and disease is not adequately controlled for in the analysis, resulting in spurious correlations. Although any single epidemiology study is highly limited in its ability to deal with all potential confounders, the broader body of epidemiological evidence provides important insights into potential confounding.

Cigarette smoking, for example, contributes to the baseline or underlying respiratory disease rates in a population, but it is not likely to serve as a common confounder across the epidemiologic studies of PM air pollution. Cigarette smoking would not be a confounder in the acute exposure studies for several reasons:

(1) Most of the lung function, respiratory symptoms, and school absences studies were conducted among non-smoking children.
(2) The largest association between respiratory hospitalizations and pollution was often with non-smoking children.
(3) Cigarette smoking does not change day-to-day, week-to-week, or month-to-month in correlation with air pollution. Furthermore, in recent cohort-based chronic exposure studies, the estimated pollution effects were observed after analytically controlling for cigarette smoking or after restricting the analysis to never smokers.

As with cigarette smoking, socioeconomic status in a population does not change day-to-day in correlation with air pollution. Therefore, socioeconomic variables are not likely confounders in the short-term time-series studies looking at lung function, respiratory symptoms, school absences, outpatient visits and mortality. Furthermore, recent cohort-based chronic exposure studies have controlled for various socioeconomic variables including sex, race and education levels.

In the acute exposure studies, confounding due to temporal correlations between pollution, weather and seasonal variables is a concern. However, independent pollution effects are typically observed even after using various approaches to control for weather variables in the regression model, and estimated pollution effects are reasonably consistent for areas with very different climates and weather conditions. Furthermore, daily, seasonal, or annual changes in weather are not potential confounders in the chronic exposure mortality and morbidity studies.

The potential for confounding by co-pollutants that are correlated with particulate pollution remains one of the most important limitations of the current epidemiology. Two basic approaches to evaluating the potential of confounding by co-pollutants have been used. One approach is to control for co-pollutants analytically in regression models, using statistical criteria such as significance levels or coefficient size and stability to evaluate the impact. Unfortunately, there are often strong correlations between the various pollutants, making analytic control techniques replete with statistical problems. Attempts to separate effects of a single pollutant are rarely conclusive for any single data set. Fortunately, across various study areas, there is substantial variability in the levels of co-pollutants and the degree of collinearity of these co-pollutants with PM. Therefore, a second and more compelling approach to evaluate for confounding by co-pollutants is to compare the estimated PM effects in areas with different potential for confounding by the co-pollutants. If the estimated PM effects are due to confounding by co-pollutants, then estimated PM effects would be larger in areas with higher potential for positive confounding by co-pollutants. Analyses of this type show little or no evidence of confounding by O_3 or SO_2 (Pope, 1995; Schwartz, 1994a). The potential for confounding by other measured or unmeasured pollutants remains unclear.

A fourth basic limitation of the epidemiological studies is the inability to fully explore the relative health impacts of various constituents of particulate pollution. Various measures or estimates of PM mass may be serving only as proxy variables for a primary toxic component or characteristic of PM, such as combustion-source particles, sulfates, ultra-fine particles or particulate acidity. Various physiologic and toxicologic considerations suggest that combustion-source particulate pollution may be a larger health concern than naturally occurring particles. Their size is such that they can be breathed most deeply in the lungs and they include sulfates, nitrates, acids, transition metals and carbon particles with various chemicals adsorbed onto their surfaces. Also, relative to coarse particles, indoor and personal exposure to combustion-source fine particles are much better represented by central site ambient monitors.

While the epidemiology suggests that combustion-source particulate pollution has a larger impact on cardiopulmonary health than comparable exposure to non-combustion related particles, it has substantial limitations with regards to characteristics or constituents of particulate pollution that are most likely responsible for the observed health effects. These limitations are largely due to PM measures being based mostly on mass, size cuts, sulfate concentrations or acidity. Real-world urban particulate pollution is a complex mixture. The currently available epidemiological data cannot reveal if the relative importance of combustion-source particles is due to the relative small size of these particles, their chemical composition, or both.

CONCLUSION

The overall epidemiologic evidence is consistent with the premise that particulate air pollution common to many urban and industrial environments is an important risk factor for cardiopulmonary disease and mortality. Although the biological linkages remain poorly understood, the results of the acute and chronic exposure studies are complementary. Much of the recent epidemiological effort has focused on the effects of acute exposure, primarily because of the relative availability of relevant time-series data sets. However, the effects of chronic exposure may be more important in terms of overall public health relevance. Long-term, repeated exposure increases the risk of chronic respiratory disease and the risk of cardiorespiratory mortality. Short-term exposures can exacerbate existing cardiovascular and pulmonary disease and increase the number of persons in a population who become symptomatic, require medical attention, or die. The pattern of cardiopulmonary health effects associated with particulate air pollution that has been observed by epidemiological studies is the strongest evidence of the health effects of this pollution. Nevertheless, the epidemiological studies have important limitations that stem largely from the use of people who are living in uncontrolled environments, and who are exposed to complex mixtures of air pollution. A more complete understanding of the health effects of particulate air pollution will require contributions from toxicology, exposure assessment, and other disciplines.

ACKNOWLEDGEMENTS

We would like to thank Frank McIntyre for his research assistance and Jeff Baliff for his preparation of the final manuscript.

REFERENCES

Abbey DE, Hwang BL, Burchette RL *et al.* (1995) Estimated long-term ambient concentrations of PM$_{10}$ and development of respiratory symptoms in a nonsmoking population. *Arch Environ Health* **50**: 139–152.

Ackermann-Liebrich U, Leuenberger P, Schwartz J *et al.* and SAPALDIA Team (1997) Lung function and long term exposure to air pollutants in Switzerland. *Am J Respir Crit Care Med* **155**: 122–129.

Anderson HR, Limb E, Bland JM *et al.* (1995) Health effects of an air pollution episode in London, December 1991. *Thorax* **50**: 1188–1193.

Anderson HR, Ponce de Leon A, Bland JM *et al.* (1996) Air pollution and daily mortality in London, 1987–1992. *Br Med J* **312**: 665–669.

Anderson HR, Spix C, Medina S *et al.* (1997) Air pollution and daily admissions for chronic obstructive pulmonary disease in 6 European cities, results from the APHEA project. *Eur Respir J* **10**: 1064–1071.

Archer VE (1990) Air pollution and fatal lung disease in three Utah counties. *Arch Environ Health* **45**: 325–334.

Bachárová L, Fandáková K, Bratinka J *et al.* (1996) The association between air pollution and the daily number of deaths: findings from the Slovak Republic contribution to the APHEA project. *J Epidemiol Community Health* **50**(Suppl 1): S19–S21.

Ballester F, Corella D, Pérez-Hoyos S and Hervás A (1996) Air pollution and mortality in Valencia, Spain: a study using the APHEA methodology. *J. Epidemiol Community Health* **50**: 527–533.

Bates DV (1980) The health effects of air pollution. *J Respir Dis* **1**: 29–37.

Bates DV (1989) *Respiratory Function in Disease*, 3rd edn. Philadelphia, PA: WB Saunders.

Bates DV (1992) Health indices of the adverse effects of air pollution: the question of coherence. *Environ Res* **59**: 336–349.

Bates DV and Sizto R (1987) Air pollution and hospital admissions in southern Ontario: the acid summer haze effect. *Environ Res* **43**: 317–331.

Bates DV and Sizto R (1989) The Ontario air pollution study: identification of the causative agent. *Environ Health Perspect* **79**: 69–72.

Bates DV, Baker-Anderson M and Sizto R (1990) Asthma attack periodicity: a study of hospital emergency visits in Vancouver. *Environ Res* **51**: 51–70.

Bobak M and Leon DA (1992) Air pollution and infant mortality in the Czech Republic, 1986–1988. *Lancet* **340**: 1010–1014.

Borja-Aburto VH, Loomis DP, Bangdiwala SI *et al.* (1997) Ozone, suspended particulates, and daily mortality in Mexico City. *Am J Epidemiol* **145**: 258–268.

Braun-Fahrländer, C., Ackermann-Liebrich U., Schwartz J. *et al.* (1992) Air pollution and respiratory symptoms in preschool children. *Am Rev Respir Dis* **145**, 42–47.

Brunekreef B, Kinney PL, Ware JH *et al.* (1991) Sensitive subgroups and normal variation in pulmonary function response to air pollution episodes. *Environ Health Perspect* **90**: 189–193.

Brunekreef B, Dockery DW and Krzyzanowski M (1995) Epidemiologic studies on short-term effects of low levels of major ambient air pollution components. *Environ Health Perspect* **103**(Suppl 2): 3–13.

Burnett RT, Dales RE, Raizenne ME *et al.* (1994) Effects of low ambient levels of ozone and sulfates on the frequency of respiratory admissions to Ontario hospitals. *Environ Res* **65**: 172–194.

Burnett RT, Dales RE, Krewski D *et al.* (1995) Associations between ambient particulate sulfate and admissions to Ontario hospitals for cardiac and respiratory diseases. *Am J Epidemiol* **142**: 15–22.

Burnett RT, Brook JR, Yung WT *et al.* (1997) Association between ozone and hospitalization for respiratory diseases in 16 Canadian cities. *Environ Res* **72**: 24–31.

Chestnut LG, Schwartz J, Savitz DA and Burchfiel CM (1991) Pulmonary function and ambient particulate matter: epidemiological evidence from NHANES I. *Arch Environ Health* **46**: 135–144.

Choudhury AH, Gordian ME and Morris SS (1997) Associations between respiratory illness and PM$_{10}$ air pollution. *Arch Environ Health* **52**: 113–117.

Ciocco A and Thompson DJ (1961) A follow-up on Donora ten years after: methodology and findings. *Am J Public Health* **51**: 155–164.

Dab W, Medina S, Quénel P *et al.* (1996) Short term respiratory health effects of ambient air pollution: results of the APHEA project in Paris. *J Epidemiol Community Health* **50**(Suppl 1): S42–S46.

Dassen W, Brunekreef B, Hoek G *et al.* (1986) Decline in children's pulmonary function during an air pollution episode. *J Air Pollut Control Assoc* **36**: 1223–1227.

Delfino RJ, Murphy-Moulton AM, Burnett RT *et al.* (1997) Effects of air pollution on emergency room visits for respiratory illnesses in Montreal, Quebec. *Am J Respir Crit Care Med* **155**: 568–576.

Dockery DW and Pope CA III (1994) Acute respiratory effects of particulate air pollution. *Annu Rev Public Health* **15**: 107–132.

Dockery DW and Pope CA III (1996) Epidemiology of acute health effects: summary of time-series studies. In: Wilson R and Spengler J (eds) *Particles in Our Air*. Cambridge, MA: Harvard University Press, pp 123–147.

Dockery DW, Ware JH, Ferris BG Jr *et al.* (1982) Change in pulmonary function in children associated with air pollution episodes. *J Air Pollut Control Assoc* **32**: 937–942.

Dockery DW, Speizer FE, Stram DO *et al.* (1989) Effects of inhalable particles on respiratory health of children. *Am Rev Respir Dis* **139**: 587–594.

Dockery DW, Schwartz J and Spengler JD (1992) Air pollution and daily mortality: associations with particulates and acid aerosols. *Environ Res* **59**: 362–373.

Dockery DW, Pope CA III, Xu X *et al.* (1993) An association between air pollution and mortality in six U.S. cities. *N Engl J Med* **329**: 1753–1759.

Dockery DW, Cunningham J, Damokosh AI *et al.* (1996). Health effects of acid aerosols on North American children: respiratory symptoms. *Environ Health Perspect* **104**: 500–505.

Evans J.S, Tosteson T and Kinney PL (1984) Cross-sectional mortality studies and air pollution risk assessment. *Environ Int* **10**: 55–83.

Fairley D (1990) The relationship of daily mortality to suspended particulates in Santa Clara County, 1980–1986. *Environ Health Perspect* **89**: 159–168.

Firket J (1931) The cause of symptoms found in the Meuse Valley during the fog of December, 1930. *Bull Acad R Med Belg* **11**: 683–741.

Folinsbee LJ (1995) Human health effects of air pollution. *Environ Health Perspect* **100**: 45–56.

Glasser M, Greenburg L and Field F (1967) Mortality and morbidity during a period of high levels of air pollution, New York, Nov 23 to 25, 1966. *Arch Environ Health* **15**: 684–694.

Gordian ME, Özkaynak H, Xue J *et al.* (1996) Particulate air pollution and respiratory disease in Anchorage, Alaska. *Environ Health Perspect* **104**: 290–297.

Hefflin BJ, Jalaludin B, McClure E *et al.* (1994) Surveillance for dust storms and respiratory diseases in Washington State, 1991. *Arch Environ Health* **49**: 170–174.

Hoek G and Brunekreef B (1993) Acute effects of a winter air pollution episode on pulmonary function and respiratory symptoms of children. *Arch Environ Health* **48**: 328–335.

Hoek G and Brunekreef B (1994) Effects of low-level winter air pollution concentrations on respiratory health of Dutch children. *Environ Res* **64**: 136–150.

Hoek G and Brunekreef B (1995) Effect of photochemical air pollution on acute respiratory symptoms in children. *Am Rev Respir Dis* **151**: 27–32.

Holland WW and Reid DD (1965) The urban factor in chronic bronchitis. *Lancet* **1**: 445–448.

Holland WW, Bennett AE, Cameron IR *et al.* (1979) Health effects of particulate pollution: reappraising the evidence. *Am J Epidemiol* **110**: 525–659.

Ito K and Thurston GD (1996) Daily PM_{10}/mortality associations: an investigation of at-risk subpopulations. *J Expos Anal Environ Epidemiol* **6**: 79–95.

Ito K, Thurston GD, Hayes C and Lippman M (1993) Associations of London, England, daily mortality with particulate matter, sulfate dioxide and acid aerosol pollution. *Arch Environ Health* **48**: 213–220.

Jedrychowski W (1995) Review of recent studies from central and eastern Europe associating respiratory health effects with high levels of exposure to traditional air pollutants. *Environ Health Perspect* **103**(Suppl 2): 15–21.

Johnson KG, Loftsgaarden DO and Gideon RA (1982) The effects of Mount St. Helens volcanic ash on the pulmonary function of 120 elementary school children. *Am Rev Respir Dis* **126**: 1066–1069.

Johnson KG, Gideon RA and Loftsgaarden DO (1990) Montana air pollution study: children's health effects. *J Off Stat* **5**: 391–408.

Katsouyanni K (1995) Health effects of air pollution in southern Europe: are there interacting factors? *Environ Health Perspect* **103**(Suppl 2): 23–27.

Katsouyanni K, Touloumi G, Spix C *et al.* (1997) Short term effects of ambient sulphur dioxide and particulate matter on mortality in 12 European cities: results from time series data from the APHEA project. *Br Med J* **314**: 1658–1663.

Kinney PL and Özkaynak H (1991) Associations of daily mortality and air pollution in Los Angeles County. *Environ Res* **54**: 99–120.

Kinney PL, Ito K and Thurston GD (1995) A sensitivity analysis of mortality/PM_{10} associations in Los Angeles. *Inhal Toxicol* **7**: 59–70.

Koenig JQ, Larson TV, Hanley QS *et al.* (1993) Pulmonary function changes in children associated with fine particulate matter. *Environ Res* **63**: 26–38.

Krupnick AJ, Harrington W and Ostro B (1990) Ambient ozone and acute health effects: evidence from daily data. *J Environ Econ Manag* **18**: 1–18.

Lave LB and Seskin EP (1970) Air pollution and human health. *Science* **169**: 723–733.

Lebowitz MD, Holberg CJ, Boyer B and Hayes C (1985) Respiratory symptoms and peak flow associated with indoor and outdoor air pollutants in the southwest. *J Air Pollut Control Assoc* **35**: 1154–1158.

Lipfert FW (1984) Air pollution and mortality: specification searches using SMSA-based data. *J Environ Econ Manag* **11**: 208–243.

Lipfert FW (1994) *Air Pollution and Community Health: A Critical Review and Data Sourcebook.* New York, NY: Van Nostrand Reinhold.

Lipfert FW and Hammerstrom T (1992) Temporal patterns in air pollution and hospital admissions. *Environ Res* **59**: 374–399.

Lipfert FW and Wyzga RE (1995) Air pollution and mortality: issues and uncertainties. *J Air Waste Manag Assoc* **45**: 949–966.

Lipfert FW, Malone RG, Daum ML *et al.* (1988) *A Statistical Study of the Macroepidemiology of Air Pollution and Total Mortality.* Washington, DC: US Department of Energy.

Lipsett M, Hurley S and Ostro B (1997) Air pollution and emergency room visits for asthma in Santa Clara County, California. *Environ Health Perspect* **105**: 216–222.

Logan WPD (1953) Mortality in London fog incident. *Lancet* **1**: 336–338.

Lutz LJ 1983 Health effects of air pollution measured by outpatient visits. *J Fam Pract* **16**: 301–313.

Martin AE (1964) Mortality and morbidity statistics and air pollution. *Proc R Soc Med* **57**: 969–975.

Mazumdar S and Sussman N (1983) Relationships of air pollution to health: results from the Pittsburgh study. *Arch Environ Health* **38**: 17–24.

Mendelsohn R and Orcutt G (1979) An empirical analysis of air pollution dose-response curves. *J Environ Econ Manag* **6**: 85–106.

Moolgavkar SH and Luebeck EG (1996) A critical review of the evidence on particulate air pollution and mortality. *Epidemiology* **7**: 420–428.

Moolgavkar SH, Leubeck EG, Hall TA and Anderson EL (1995a) Air pollution and daily mortality in Philadelphia. *Epidemiology* **6**: 476–484.

Moolgavkar SH, Leubeck EG, Hall TA and Anderson EL (1995b) Particulate air pollution, sulfur dioxide and daily mortality: a reanalysis of the Steubenville data. *Inhal Toxicol* **7**: 35–44.

Moolgavkar SH, Luebeck EG and Anderson EL (1997) Air pollution and hospital admissions for respiratory causes in Minneapolis-St. Paul and Birmingham. *Epidemiology* **8**: 364–370.

Neas LM, Dockery DW, Koutrakis P *et al.* (1995) The associations of ambient air pollution with twice daily peak expiratory flow rate measurements in children. *Am J Epidemiol* **141**: 111–122.

Neas LM, Dockery DW, Burge H *et al.* (1996) Fungus spores, air pollutants, and other determinants of peak expiratory flow rate in children. *Am J Epidemiol* **143**: 797–807.

Nishino N, Abbey DE and Burchette RJ (1997) Long-term ambient concentrations of PM_{10} and mortality in nonsmokers – The AHSMOG study. Presented at 1997 ISEE Conference (In press).

Ostro BD (1983) The effects of air pollution on work loss and morbidity. *J Environ Econ Manag* **10**: 371–382.

Ostro BD (1984) A search for a threshold in the relationship of air pollution to mortality: a reanalysis of data on London winter. *Environ Health Perspect* **58**: 397–399.

Ostro BD (1987) Air pollution and morbidity revisited: a specification test. *J Environ Econ Manag* **14**: 87–98.

Ostro BD (1990) Associations between morbidity and alternative measures of particulate matter. *Risk Anal* **10**: 421–427.

Ostro BD (1993) The association of air pollution and mortality: examining the case for inference. *Arch Environ Health* **48**: 336–342.

Ostro BD (1995) Fine particulate air pollution and mortality in two southern California counties. *Environ Res* **70**: 98–104.

Ostro BD and Rothschild S (1989) Air pollution and acute respiratory morbidity: an observational study of multiple pollutants. *Environ Res* **50**: 238–247.

Ostro BD, Lipsett MJ, Wiener MB and Selner JC (1991) Asthmatic responses to airborne acid aerosols. *Am J Public Health* **81**: 694–702.

Ostro BD, Lipsett MJ, Mann JK *et al.* (1993) Air pollution and respiratory morbidity among adults in southern California. *Am J Epidemiol* **137**: 691–700.

Ostro BD, Sanchez JM, Aranda C and Eskeland GS (1996) Air pollution and mortality: results from a study of Santiago, Chile. *J Expos Anal Environ Epidemiol* **6**: 97–114.

Özkaynak H and Spengler JD (1985) Analysis of health effects resulting from population exposures to acid precipitation precursors. *Environ Health Perspect* **63**: 45–55.

Özkaynak H and Thurston GD (1987) Associations between 1980 US mortality rates and alternative measures of airborne particle concentrations. *Risk Anal* **7**: 449–461.

Pantazopoulou A, Katsouyanni K, Kourea-Kremastinou J and Trichopoulous D (1995) Short-term effects of air pollution on hospital emergency and outpatient visits and admissions in the greater Athens, Greece area. *Environ Res* **69**: 31–36.

Peters A, Dockery DW, Heinrich J and Wichmann HE (1997a) Medication use modifies the health effects of particulate sulfate air pollution in children with asthma. *Environ Health Perspect* **105**: 430–435.

Peters A, Dockery DW, Heinrich J and Wichmann HE (1997b) Short-term effects of particulate air pollution on respiratory morbidity in asthmatic children. *Eur Respir J* **10**: 872–879.

Peters A, Wichmann HE, Tuch T *et al.* (1997c) Respiratory effects are associated with the number of ultrafine particles. *Am J Respir Crit Care Med* **155**: 1376–1383.

Poloniecki JD, Atkinson RW, de Leon AP and Anderson HR (1997) Daily time series for cardiovascular hospital admissions and previous day's air pollution in London, UK. *Occup Envrion Med* **54**: 535–540.

Ponce de Leon A, Anderson HR, Bland JM *et al.* (1996). Effects of air pollution on daily hospital admissions for respiratory disease in London between 1987–88 and 1991–92. *J Epidemiol Community Health* **50**(Suppl 1): S63–S70.

Pönkä A (1991) Asthma and low level air pollution in Helsinki. *Arch Environ Health* **46**: 262–270.

Pönkä A and Virtanen M (1996) Asthma and ambient air pollution in Helsinki. *J Epidemiol Community Health* **50**(Suppl 1): S59–S62.

Pope CA III (1989) Respiratory disease associated with community air pollution and a steel mill, Utah Valley. *Am J Public Health* **79**: 623–628.

Pope CA III (1991) Respiratory hospital admissions associated with PM_{10} pollution in Utah, Salt Lake, and Cache Valleys. *Arch Environ Health* **46**: 90–97.

Pope CA III (1995) Combustion-source particulate air pollution and human health: causal assocations or confounding? In: *Particulate Matter: Health and Regulatory Issues.* Pittsburg, PA: Air & Waste Management Association, pp 60–77.

Pope CA III (1996) Particulate pollution and health: a review of the Utah Valley experience. *J Expos Anal Environ Epidemiol* **6**: 23–34.

Pope CA III and Dockery DW (1992) Acute health effects of PM_{10} pollution on symptomatic and asymptomatic children. *Am Rev Respir Dis* **145**: 1123–1128.

Pope CA III and Dockery DW (1996) Epidemiology of chronic health effects: cross-sectional studies. In: R. Wilson and J. Spengler (ed), *Particles in Our Air.* Harvard University Press, pp 149–167.

Pope CA III and Kalkstein LS (1996) Synoptic weather modeling and estimates of the exposure-response relationship between daily mortality and particulate air pollution. *Environ Health Perspect* **104**: 414–420.

Pope CA III and Kanner RE (1993) Acute effects of PM_{10} pollution on pulmonary function of smokers with mild to moderate chronic obstructive pulmonary disease. *Am Rev Respir Dis* **147**: 1336–1340.

Pope CA III, Dockery DW, Spengler JD and Raizenne ME (1991) Respiratory health and PM_{10} pollution: a daily time series analysis. *Am Rev Respir. Dis* **144**: 668–674.

Pope CA III, Schwartz J and Ransom MR (1992) Daily mortality and PM_{10} pollution in Utah Valley. *Arch Environ Health* **47**: 211–217.

Pope CA III, Bates DV and Raizenne ME (1995a) Health effects of particulate air pollution: time for reassessment? *Environ Health Perspect* **103**: 472–480.

Pope CA III, Dockery DW and Schwartz J (1995b) Review of epidemiologic evidence of health effects of particulate air pollution. *Inhal Toxicol* **7**: 1–18.

Pope CA III, Thun MJ, Namboodiri MM *et al.* (1995c) Particulate air pollution as a predictor of mortality in a prospective study of US adults. *Am J Respir Crit Care Med* **151**: 669–674.

Portney PR and Mullahy J (1990) Urban air quality and chronic respiratory disease. *Regional Sci Urban Econ* **20**: 407–418.

Rahlenbeck SI and Kahl H (1996) Air pollution and mortality in East Berlin during the winters of 1981–1989. *Int J Epidemiol* **25**: 1220–1226.

Raizenne ME, Neas LM, Damokosh AI *et al.* (1996) Health effects of acid aerosols on North American children: pulmonary function. *Environ Health Perspect* **104**: 506–514.

Ransom MR and Pope CA III (1992) Elementary school absences and PM$_{10}$ pollution in Utah Valley. *Environ Res* **58**: 204–219.

Roemer W, Hoek G and Brunekreef B (1993) Effect of ambient winter air pollution on respiratory health of children with chronic respiratory symptoms. *Am Rev Respir Dis* **147**: 118–124.

Romieu I, Meneses F, Ruiz S *et al.* (1996) Effects of air pollution on the respiratory health of asthmatic children living in Mexico City. *Am J Respir Crit Care Med* **154**: 300–307.

Saldiva PHN, Pope CA III, Schwartz J *et al.* (1995) Air pollution and mortality in elderly people: a time-series study in São Paulo, Brazil. *Arch Environ Health* **50**: 159–163.

Samet JM, Speizer FE, Bishop Y *et al.* (1981) The relationship between air pollution and emergency room visits in an industrial community. *J Air Pollut Control Assoc* **31**: 236–240.

Samet JM, Zeger SL and Berhane K (1995) The association of mortality and particulate air pollution. In: *Particulate Air Pollution and Daily Mortality: Replication and Validation of Selected Studies.* The Phase I Report of the Particle Epidemiology Evaluation Project. Prepared for the Health Effects Institute, August 1995.

Samet JM, Zeger SL, Kelsall JE *et al.* (1997) *Air Pollution, Weather, and Mortality in Philadelphia 1973–1988.* Report to the Health Effects Institute on Phase IB (Particle Epidemiology Evaluation Project). Cambridge, MA: Health Effects Institute.

Schimmel H and Greenburg L (1972) A study of the relation of pollution to mortality New York City, 1963–1968. *J Air Pollut Control Assoc* **22**: 607–617.

Schouten JP, Vonk JM and De Graaf A (1996) Short term effects of air pollution on emergency hospital admissions for respiratory disease: results of the APHEA project in two major cities in The Netherlands, 1977–89. *J Epidemiol Community Health* **50**(Suppl 1): S22–S29.

Schwartz J (1989) Lung function and chronic exposure to air pollution: a cross-sectional analysis of NHANES II. *Environ Res* **50**: 309–321.

Schwartz J (1991) Particulate air pollution and daily mortality in Detroit. *Environ Res* **56**: 204–213.

Schwartz J (1993a) Air pollution and daily mortality in Birmingham, Alabama. *Am J Epidemiol* **137**: 1136–1147.

Schwartz J (1993b) Particulate air pollution and chronic respiratory disease. *Environ Res* **62**: 7–13.

Schwartz J (1994a) Air pollution and daily mortality: a review and meta analysis. *Environ Res* **64**: 36–52.

Schwartz J (1994b) Total suspended particulate matter and daily mortality in Cincinnati, Ohio. *Environ Health Perspect* **102**: 186–189.

Schwartz J (1994c) Air pollution and hospital admissions for the elderly in Birmingham, Alabama. *Am J Epidemiol* **139**: 589–598.

Schwartz J (1994d) Air pollution and hospital admissions for the elderly in Detroit, Michigan. *Am J Respir Crit Care Med* **150**: 648–655.

Schwartz J (1994e) PM10, ozone and hospital admissions for the elderly in Minneapolis-St. Paul, Minnesota. *Arch Environ Health* **49**: 366–373.

Schwartz J (1994f) What are people dying of on high pollution days? *Environ Res* **64**: 26–35.

Schwartz J (1995) Short term fluctuations in air pollution and hospital admissions of the elderly for respiratory diseases. *Thorax* **50**: 531–538.

Schwartz J (1996) Air pollution and hospital admissions for respiratory disease. *Epidemiology* **7**: 20–28.

Schwartz J (1997) Air pollution and hospital admissions for cardiovascular disease in Tucson. *Epidemiology* **8**: 371–377.

Schwartz J and Dockery DW (1992a) Particulate air pollution and daily mortality in Steubenville, Ohio. *Am J Epidemiol* **135**: 12–19.

Schwartz J and Dockery DW (1992b) Increased mortality in Philadelphia associated with daily air pollution concentrations. *Am Rev Respir Dis* **145**: 600–604.

Schwartz J and Marcus A (1990) Mortality and air pollution in London: a time series analysis. *Am J Epidemiol* **131**: 185–194.

Schwartz J and Morris R (1995) Air pollution and hospital admissions for cardiovascular disease in Detroit, Michigan. *Am J Epidemiol* **142**: 23–35.

Schwartz J, Spix C, Wichmann HE and Malin E (1991) Air pollution and acute respiratory illness in five German communities. *Environ Res* **56**: 1–14.

Schwartz J, Slater D, Larson TV *et al.* (1993) Particulate air pollution and hospital emergency room visits for asthma in Seattle. *Am Rev Respir Dis* **147**: 826–831.

Schwartz J, Dockery DW, Neas LM *et al.* 1994) Acute effects of summer air pollution on respiratory symptom reporting in children. *Am J Respir Crit Care Med* **150**: 1234–1242.

Schwartz J, Dockery DW and Neas LM (1996) Is daily mortality associated specifically with fine particles? *J Air Waste Manag Assoc* **46**: 927–939.

Schwartz J, Spix C, Touloumi G *et al.* (1996) Methodological issues in studies of air pollution and daily counts of deaths or hospital admissions. *J Epidemiol Community Health* **50**(Suppl 1): S3–S17.

Shumway RH, Azari AS and Pawitan Y (1988) Modeling mortality fluctuations in Los Angeles as functions of pollution and weather effects. *Environ Res* **45**: 224–241.

Shy CM (1979) Epidemiologic evidence and the United States air quality standards. *Am J Epidemiol* **110**: 661–671.

Spix C and Wichmann HE (1996) Daily mortality and air pollutants: findings from Köln, Germany. *J Epidemiol Community Health* **50**(Suppl 1): S52–S58.

Spix C, Heinrich J, Dockery D *et al.* (1993) Air pollution and daily mortality in Erfurt, East Germany, 1980–1989. *Environ Health Perspect* **101**: 518–526.

Stebbing JH Jr and Fogleman DG (1979) Identifying a susceptible subgroup: effects of the Pittsburgh Air Pollution Episode upon schoolchildren. *Am J Epidemiol* **110**: 27–40.

Styer P, McMillan N, Gao F *et al.* (1995) Effect of outdoor airborne particulate matter on daily death counts. *Environ Health Perspect* **103**: 490–497.

Sunyer J, Sáez M, Murillo C *et al.* (1993) Air pollution and emergency room admissions for chronic obstructive pulmonary disease: a 5-year study. *Am J Epidemiol* **137**: 701–705.

Sunyer J, Castellsagué J, Sáez M *et al.* (1996) Air pollution and mortality in Barcelona. *J Epidemiol Community Health* **50**(Suppl 1): S76–S80.

Thurston GD (1996) A critical review of PM_{10}–mortality time-series studies. *J Expos Anal Environ Epidemiol* **6**: 3–21.

Thurston GD, Ito K, Kinney PL and Lippmann M (1992) A multi-year study of air pollution and respiratory hospital admissions in three New York State metropolitan areas: results for 1988 and 1989 summers. *J Expos Anal Environ Epidemiol* **2**: 429–450.

Thurston GD, Ito K, Hayes CG *et al.* (1994) Respiratory hospital admissions and summertime haze air pollution in Toronto, Ontario: consideration of the role of acid aerosols. *Environ Res* **65**: 271–290.

Touloumi G, Pockock SJ, Katsouyanni K and Trichopooulos D (1994) Short-term effects of air pollution on daily mortality in Athens: a time-series analysis. *Int J Epidemiol* **23**: 957–961.

Touloumi G, Samoli E and Katsouyanni K (1996) Daily mortality and "winter type" air pollution in Athens Greece – a times series analysis within the APHEA project. *J Epidemiol Community Health* **50**(Suppl 1): S47–S51.

US Environmental Protection Agency (1987) Revisions to the national ambient air quality standards for particulate matter. *Federal Register* 1 July 1987. **52**: 248–254.

US Environmental Protection Agency (1996) *Air Quality Criteria for Particulate Matter*, vol II. EPA/600/P-95/0016F.

US Environmental Protection Agency (1997) National ambient air quality standards for particulate matter; final rule. *Federal Register* 18 July 1997. **62**(138): 38651–38701.

Vedal S (1997) Ambient particles and health: lines that divide. *J Air Waste Manag Assoc* **47**: 551–581.

Vedal S, Schenker MB, Munoz A *et al.* (1987) Daily air pollution effects on children's respiratory symptoms and peak expiratory flow. *Am J Public Health* **77**: 694–698.

Vedal S, Petkau J, White R and Blair J (1998) Acute effects of ambient inhalable particles in asthmatic and non-asthmatic children. *Am J Respir Crit Care Med* **157**: 1034–1043.

Verhoeff AP, Hoek G, Schwartz J and Van Wijnen JH (1996) Air pollution and daily mortality in Amsterdam. *Epidemiology* **7**: 225–230.

Vigotti AM, Rossi G, Bisanti L *et al.* (1996) Short term effects of urban air pollution on respiratory health in Milan, Italy, 1980–89. *J Epidemiol Community Health* **50**(Suppl 1): S71–S75.

Ware JH, Thidbodeau LA, Speizer FE *et al.* (1981) Assessment of the health effects of atmospheric sulfur oxides and particulate matter: evidence from observational studies. *Environ Health Perspect* **41**: 255–276.

Whittemore AS and Korn EL (1980) Asthma and air pollution in the Los Angeles Area. *Am J Public Health* **70**: 687–696.

Wichmann HE, Mueller W, Allhoff P *et al.* (1989) Health effects during a smog episode in West

Germany in 1985. *Environ Health Perspect* **79**: 89–99.

Wietlisbach V, Pope CA III and Ackermann-Liebrich U (1996) Air pollution and daily mortality in three Swiss urban areas. *Soz Präventivmed* **41**: 107–115.

Wojtyniak B and Piekarski T (1996) Short term effect of air pollution on mortality in Polish urban populations – what is different? *J Epidemiol Community Health* **50**(Suppl 1): S36–S41.

Woodruff TJ, Grillo J and Schoendorf KC (1997) The relationship between selected causes of post-neonatal infant mortality and particulate air pollution in the United States. *Environ Health Perspect* **105**: 608–612.

Wyzga RE (1978) The effect of air pollution upon mortality: a consideration of distributed lag models. *J Am Stat Assoc* **73**: 463–472.

Xu X, Gao J, Dockery DW and Chen Y (1994) Air pollution and daily mortality in residential areas of Beijing, China. *Arch Environ Health* **49**: 216–222.

Xu X, Dockery DW, Christiani DC *et al.* (1995) Association of air pollution with hospital outpatient visits in Beijing. *Arch Environ Health* **50**: 214–220.

Zmirou D, Barumandzadeh T, Balducci F *et al.* (1996) Short term effects of air pollution on mortality in the city of Lyon, France 1985–90. *J Epidemiol Community Health* **50**(Suppl 1): S30–S35.

The Health Effects of Diesel Exhaust: Laboratory and Epidemiologic Studies*

AARON J. COHEN

Health Effects Institute, Cambridge MA and Division of
Environmental Health, Boston University School of Public Health,
Boston, MA, USA

KRISTEN NIKULA

Biopersistent Particle Center, Lovelace Respiratory Research
Institute, Albuquerque, NM, USA

* The views expressed in this chapter are those of the authors and do not necessarily reflect those of the Health Effects Institute or its sponsors.

INTRODUCTION

The health effects of diesel engine exhaust have been a topic of research since the early 1950s, when diesel technology became the preferred power source in heavy-duty trucking and rail transport in the USA. Early toxicologic experiments by Kotin *et al.* (1955) provided the first evidence suggesting carcinogenicity of diesel exhaust, stimulating considerable additional research. Two decades later, Huisingh and colleagues (1978) demonstrated the direct mutagenicity of diesel exhaust particle extracts in the then newly developed Ames bacterial assay. Stimulated by this finding, researchers on the health effects of diesel exhaust received considerable impetus when the demands of oil-producing countries for higher prices led the US automobile industry to consider diesel technology as a fuel-efficient alternative for the private car fleet. Consequently, there is now an extensive toxicologic and epidemiologic literature, focused largely on the pulmonary carcinogenicity of diesel exhaust.

AIR POLLUTION AND HEALTH
ISBN 0-12-352335-4

In the 1990s, interest in the health effects of diesel exhaust is stronger than ever. Economic considerations, along with an increased awareness of the contribution of fossil fuel combustion to greenhouse gas emissions and the associated global warming (Intergovernmental Panel on Climate Change, 1996), have spurred a search for fuel-efficient alternatives to the current generation of gasoline engines. A consortium of US automakers and the US government have suggested that diesel technology might provide such an alternative (National Research Council, 1998). But over the last decade, the health effects of particulate air pollution, to which diesel exhaust emissions contribute in urban settings worldwide, have emerged as a major public health issue. A substantial body of epidemiologic research has demonstrated that particulate air pollution is associated with increased morbidity and mortality from cardiovascular and non-malignant respiratory disease (Pope and Dockery, 1996). These observations led the US EPA to promulgate new standards for ambient levels of particulate matter, with the focus on controlling fine particulate air pollution ($PM_{2.5}$) (US EPA, 1996), and the European Commission will soon be considering similar changes to their air quality directives. Motivated by these regulatory activities, as well as by the possibility of increasing use of diesel technology and continued scientific uncertainty, epidemiologists and toxicologists have begun to consider the impact of diesel emissions on health outcomes other than lung cancer, and to continue to study lung cancer risk.

This chapter reviews the toxicologic and epidemiologic data on the health effects of diesel emissions, and addresses the current evidence concerning diesel exhaust and non-malignant acute and chronic respiratory effects, as well as the better known, and often reviewed, data concerning lung cancer.

DIESEL EXHAUST: CHARACTERISTICS, SOURCES AND EXPOSURE

The popularity of the diesel engine in heavy-duty applications in the trucking, railroad, marine transport and construction industries is due to both its fuel efficiency and long service life relative to the gasoline engine. Compared with gasoline engine exhaust, diesel emissions are lower in CO, hydrocarbons and carbon dioxide, but higher in oxides of nitrogen (NO_x) and particulate matter.

The physical and chemical characteristics of diesel exhaust have been described recently by Sawyer and Johnson (1995) and Nauss and colleagues (Nauss *et al.*, 1995). Diesel exhaust is a complex mixture, composed of both a particulate and gaseous phase. Diesel exhaust particles have a mass median diameter of 0.05–1.0 μm, a size rendering them easily respirable and capable of depositing in the airways and alveoli (Snipes, 1989). The particles consist of a carbonaceous core with a large surface area to which various hydrocarbons are adsorbed, including carcinogenic polycyclic aromatic hydrocarbons (PAHs) and nitro-PAHs that have elicited the most concern with respect to human health. The gaseous phase contains various products of combustion and hydrocarbons, including some of the same PAHs present in the particle phase. Once emitted, components of diesel exhaust undergo atmospheric transformation in ways that may be relevant to human health. For example, nitro-PAHs created by the reaction of directly emitted PAHs with hydroxyl radicals in the atmosphere can be more potent mutagens and carcinogens, and more bioavailable, than their precursors (Winer and Busby, 1995). Diesel technology has changed over the past two

Table 32.1 Diesel exhaust: emission components, reaction products, and biological impacts

Emission component	Atmospheric reaction products	Biological impact[a]
Vapor-phase emissions		
Carbon dioxide	–	Major contributor to global warming
Carbon monoxide	–	Highly toxic to humans; blocks oxygen uptake
Oxides of nitrogen	Nitric acid, ozone	Nitrogen dioxide is a respiratory tract irritant and major ozone precursor. Nitric acid contributes to acid rain
Sulfur dioxide	Sulfuric acid	Respiratory tract irritation. Contributor to acid rain
Hydrocarbons		
Alkanes ($\leq C_{18}$)	Aldehydes, alkyl nitrates, ketones	Respiratory tract irritation. Reaction products are ozone precursors (in the presence of NO_x)
Alkenes ($\leq C_4$) (e.g. 1,3-butadiene)	Aldehydes, ketones	Respiratory tract irritation. Some alkenes are mutagenic and carcinogenic. Reaction products are ozone precursors (in the presence of NO_x)
Aldehydes		
Formaldehyde	Carbon monoxide, hydroperoxyl radicals	Formaldehyde is a probable human carcinogen and an ozone precursor (in the presence of NO_x)
Higher aldehydes (e.g. acrolein)	Peroxyacyl nitrates	Respiratory tract and eye irritation; causes plant damage
Monocyclic aromatic compounds (e.g. benzene, toluene)	Hydroxylated and hydroxylated-nitro derivatives[b]	Benzene is toxic and carcinogenic in humans. Some reaction products are mutagenic in bacteria (Ames assay)
PAHs (≤ 4 rings)[c] (e.g. phenanthrene, fluoranthene)	Nitro-PAHs (≤ 4 rings)[d]	Some of these PAHs and nitro-PAHs are known mutagens and carcinogens
Nitro-PAHs (2 and 3 rings) (e.g. nitronaphthalenes)	Quinones and hydroxylated-nitro derivatives	Some reaction products are mutagenic in bacteria (Ames assay)
Particle-phase emissions		
Elemental carbon	–	Nuclei adsorb organic compounds; size permits transport deep into the lungs (alveoli)
Inorganic sulfate	–	Respiratory tract irritation
Hydrocarbons (C_{14}–C_{35})	Little information; possibly aldehydes, ketones, and alkyl nitrates	Unknown
PAHs (≥ 4 rings) (e.g. pyrene, benzo[a]pyrene)	Nitro-PAHs (≥ 4 rings)[d] Nitro-PAH lactones	Larger PAHs are major contributors of carcinogens in combustion emissions. Many nitro-PAHs are potent mutagens and carcinogens
Nitro-PAHs (≥ 3 rings) (e.g. nitropyrenes)	Hydroxylated-nitro derivatives	Many nitro-PAHs are potent mutagens and carcinogens. Some reaction products are mutagenic in bacteria (Ames assay)

[a] Unless otherwise stated, the impact results from both the emissions components and the atmospheric reaction products.
[b] Some reaction products expected to partition into the particle phases.
[c] PAHs containing four rings are usually present in both the vapor and particle phases.
[d] Nitro-PAHs with more than two rings will partition into the particle phase.

Reproduced with permission from K.M. Nauss and the HEI Diesel Working Group (Nauss *et al.*, 1995)

decades, largely in response to regulation. Modern heavy-duty diesel engines emit substantially fewer particles, particle-associated organics and sulfates than their predecessors. However, because of the long-life of such engines, gains from technological advances accrue only as new engines slowly enter the fleet (Sawyer and Johnson, 1995).

Diesel exhaust is a ubiquitous air pollutant in urban settings worldwide. The proportional contribution of different sources to ambient concentrations, however, differs between Europe and the USA. In Europe, due largely to public policies intended to promote fuel efficiency, diesel-powered vehicles comprise a significant proportion of the passenger car fleet in addition to heavy-duty applications. In the USA, however, the widespread use of diesel technology in passenger vehicles predicted in the 1970s did not occur: diesel cars constituted 0.1% of US sales in 1991 in contrast to 25% of European sales in 1994 (Sawyer and Johnson, 1995). Consequently, in the USA, heavy-duty diesel use predominates: 99% of heavy-duty trucks and buses currently in use are diesel (US EPA, 1993).

Unfortunately, there are few detailed studies of the proportional contribution of diesel emissions to urban air pollution. Perhaps the most thorough study was conducted by Cass and Gray (1995), who measured diesel emissions and ambient concentration of fine particles (<2 μm in aerodynamic diameter) in the Los Angeles area in 1982. They estimated that diesel sources (on- and off-highway, marine and rail) were responsible for 67% of the 11.5 tons/day of fine elemental carbon emitted in the Los Angeles area, amounting to 3% of total particle emissions and 7% of all fine particle emissions. Cass and Gray calculated that the mean monthly ambient concentrations of diesel particles across the Los Angeles area ranged from 10 to 1.7 μg/m^3, depending on season and other factors. Using archived data from seven locations in Los Angeles from 1958 through the mid-1980s, they observed that ambient concentrations had decreased in some urban locations but had increased in areas that had undergone extensive urban development. The US EPA estimated annual average concentrations of 2 μg/m^3 for US urban locations in 1990 (US EPA, 1993), close to the value measured by Cass and Gray in less-polluted areas of Los Angeles.

Estimates of the concentrations of diesel exhaust to which human populations are exposed span three orders of magnitude: from 1 to 10 μg/m^3 in the general urban environment (as measured by Cass and Gray (1995) and US EPA (1993)) to greater than 1 mg/m^3 in some underground mining operations. The most extensive data on human exposure come from industrial hygiene studies, which have been reviewed by Watts (1995). Watts reported that mean exposure concentrations in the work environment, generally measured over a work shift and estimated using different measurement and analysis methods, ranged from 4 μg/m^3 in truck drivers to 1740 μg/m^3 in underground miners. The US National Institute for Occupational Safety and Health (NIOSH) estimates that 1.35 million US workers are exposed occupationally to diesel emissions (NIOSH, 1988). There are considerably fewer data on the magnitude of exposure to diesel exhaust in the general urban population. As noted above, long-term average ambient levels in urban settings are, in general, lower than occupational exposures, but in particular settings, such as near roadways or bus maintenance facilities, the ambient levels may be greater, as may short-term exposures on urban streets (Nauss et al., 1995). Clearly, many millions of people are exposed to diesel exhaust in the urban environment.

CARCINOGENICITY OF INHALED DIESEL EXHAUST

Lung Cancer

Epidemiology

Over 40 studies currently provide estimates of the risk of lung cancer associated with occupational exposure to diesel exhaust, and several reviews have recently been published (e.g. Bhatia *et al.*, 1998; Boffetta *et al.*, 1997; Cohen and Higgins, 1995; WHO, 1996)). The review by Bhatia and colleagues (1998) provides meta-analytic summary estimates of results of 29 studies grouped according to occupational group and key features of study design (e.g. control for smoking and study population). The studies of occupational exposure to diesel exhaust and lung cancer have consistently observed elevated lung cancer rates among exposed workers that cannot be readily attributed to known sources of bias or confounding. Unfortunately, however, no current study provides quantitative estimates of the past exposure of study subjects to any constituent of diesel exhaust;

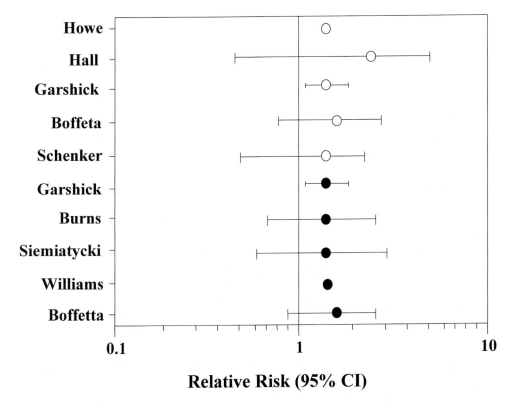

Fig. 32.1 Lung cancer and exposure to diesel exhaust in railroad workers. ●, RR adjusted for cigarette smoking; ○, RR not adjusted for cigarette smoking. Error bars show upper and lower 95% confidence intervals (CIs). For the two studies by Howe and Williams, CIs were not reported and could not be calculated. References cited are: Howe *et al* (1983), Hall and Wynder (1984), Garshick *et al* (1987), Boffetta *et al* (1988), Schenker *et al* (1984), Garshick *et al* (1988), Burns and Swanson (1991), Siemiatycki (personal communication), Williams *et al* (1997), Boffetta *et al* (1990).

therefore, the dose–response relation cannot be estimated with great accuracy from the available epidemiologic data.

The most frequently studied occupational groups have been railroad workers and truck drivers. The studies of truck drivers show a 20–50% excess incidence and/or mortality from lung cancer, which persists when cigarette smoking is accounted for in data analysis. Some studies included only small numbers of subjects and therefore offer imprecise estimates of the relative risk, as indicated by the width of the 95% confidence intervals. However, the upper bounds of the 95% confidence intervals indicate that few of the results are consistent with more than a tripling of risk and the larger studies (e.g. Steenland *et al.*, 1990) are consistent with less than a doubling. The summary estimate of the relative risk for occupation as a truck driver derived by Bhatia and colleagues from 10 studies was 1.49 (95% CI, 1.36, 1.64; Bhatia *et al.*, 1998).

Studies of railroad workers have also consistently observed excess relative risks on the order of 30–50%, after analytic control for cigarette smoking, comparable in magnitude to the estimates from the studies of truck drivers (Fig. 32.2). For railroad workers, the results of the largest study (Garshick *et al.*, 1988) is consistent with a doubling of risk at most. Bhatia and colleagues (1998) estimated a summary relative risk of 1.44 (95% CI, 1.30, 1.59).

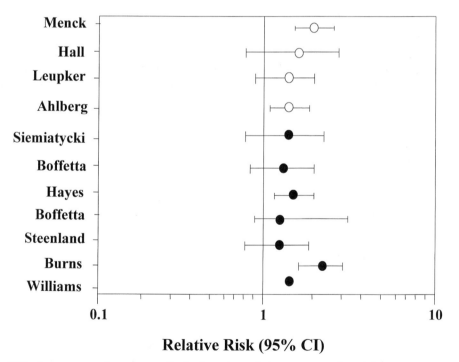

Fig. 32.2 Lung cancer and exposure to diesel exhaust in truck drivers. ●, RR adjusted for cigarette smoking; ○, RR not adjusted for cigarette smoking. Error bars show upper and lower 95% confidence intervals (CIs). For the two studies by Williams (1977), CIs were not reported and could not be calculated. For the study by Steenland (1990), the results were those based on union records for long-haul truckers. For the study by Boffetta (1988) the results were based on self-reports of diesel truck driving. For the Siemiatycki study the results were based on self-reports of heavy-duty truck driving (personal communication). Other references cited are: Menck and Henderson (1976), Hall and Wynder (1984), Leupker and Smith (1978), Ahlberg *et al* (1981), Hayes *et al* (1989), Boffetta *et al* (1990), Burns and Swanson (1991).

Lung cancer risk in bus garage workers has been studied in several European countries (Fig. 32.3). Studies of London Transport Authority workers in bus maintenance facilities during the 1950s and 1960s showed no consistent evidence of increased lung cancer risk (Raffle, 1957; Rushton *et al.*, 1983; Waller, 1981). However, these studies were plagued by several problems including inadequate time since first exposure, incomplete follow-up, and poor characterization of exposure. The recent Swedish study by Per Gustavsson and colleagues (1990) improved on the British studies in all of these design aspects and observed increases in both lung cancer mortality and incidence. Bhatia and colleagues (1998) estimated a summary relative risk of 1.24 (95% CI, 0.93, 1.64) for 'bus workers', including one study of bus drivers.

It is not clear to what extent heavy equipment operators working outdoors are exposed to diesel exhaust, and it is possible that exposures might vary widely.

A retrospective cohort study of heavy equipment operators by Wong and colleagues (1985) observed little evidence of increased lung cancer risk, although the American Cancer Society prospective cohort study reported by Boffetta and colleagues (1988) observed a 2.6-fold increase in mortality in heavy equipment operators after adjustment for age and cigarette smoking. Bhatia and colleagues (1998) estimated a summary relative risk of 1.11 (95% CI, 0.89, 1.38) for equipment operators.

Dock workers may be exposed to diesel exhaust in enclosed settings, such as the holds of ships or on loading docks, resulting, according to some reports, in higher levels of exposure to diesel exhaust than those estimated for truck drivers and railroad workers (Watts, 1995; Zaebst *et al.*, 1991).

Steenland and colleagues (1990) found no evidence of increased lung cancer mortality among US dock workers; however, diesel-powered cargo loading equipment has

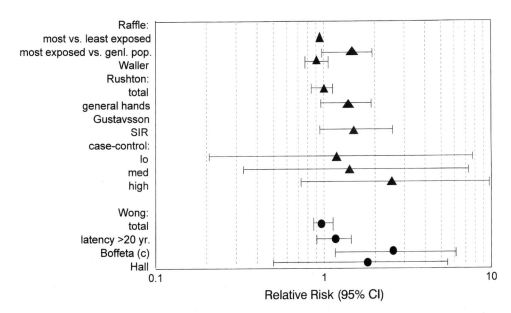

Fig. 32.3 Relative risk of lung cancer for bus garage workers (▲) and heavy equipment operators (●). Error bars show upper and lower 95% CI. References cited are Raffle (1957), Waller (1981), Rushton *et al.* (1983), Gustavsson *et al.* (1990), Wong *et al.* (1985) and Boffetta *et al.* (1988).

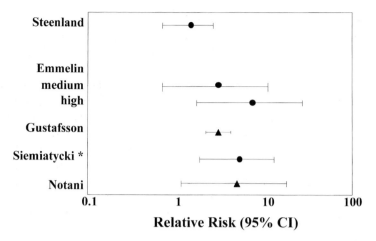

Fig. 32.4 Relative risk of lung cancer in dock workers ●, smoking adjusted; ▲, smoking not adjusted. Error bars show upper and lower 95% CI. References cited are Steenland *et al* (1990), Emmelin *et al* (1993), Gustafsson *et al* (1986), Siemiatycki (personal communication), Notari *et al* (1993).

only been in use in the USA since the early 1980s, and therefore this study, which only observed workers through 1983, provides little information about the possible effect of diesel exposure on US dock workers. Studies of dock workers in countries where diesel-powered equipment has been in use for decades, such as Canada (Siemiatycki, personal communication cited in Cohen and Higgins (1995)), Sweden (Emmelin and Wall, 1993; Gustafsson *et al.*, 1986), and India (Notani *et al.*, 1993) appear to indicate a two- to three-fold increase in lung cancer risk over that observed for truck drivers or railroad workers, but these estimates are imprecise (Cohen and Higgins, 1995).

Underground miners working with diesel-powered equipment are known to be exposed to levels of diesel exhaust that exceed those in most occupational settings; these exposures have been measured in milligrams of elemental carbon (Watts, 1995). While increased lung cancer occurrence has been observed in some studies, known carcinogens, such as radon, may account for those observations. Waxweiler's study of potash miners (Waxweiler *et al.*, 1973), who were not exposed to radon above background levels, observed no increased risk of lung cancer in two mines where diesel equipment had been used, but there may not have been sufficient time elapsed since the introduction of diesel equipment for exposure to have resulted in increased occurrence. Results of a recent study of miners exposed to diesel exhaust showed a weak association with diesel exposure in a cohort of 18 000 British coal miners (Johnson *et al.*, 1997). Controlling for cigarette smoking and allowing for a latency of 15 years, the lung cancer relative risk was 1.16 (95% CI, 0.90, 1.49) per gram hour per cubic meter of exposure to diesel exhaust. In a recently released but currently unpublished report, Säverin and colleagues (1998) observed 38 lung cancers in a cohort of 5000 German potash miners who were exposed to diesel exhaust in above-ground mining operations. Controlling for cigarette smoking, the investigators estimated a lung cancer relative risk of 2.2 (95% CI, 0.8, 6.0) comparing the least exposed and most exposed workers. A large new study of US underground miners in non-metal (e.g. potash) mines is currently underway, organized by NIOSH and National Cancer Institute (NCI). The study, which will require six years to complete, will ascertain the mortality of approximately 8000 miners through 1996, and will utilize historical data on workplace concentrations of diesel exhaust to estimate past exposures.

The possibility of confounding by cigarette smoking has been a major concern with respect to the interpretation of these studies. Although confounding by factors such as asbestos exposure, social class, diet and exposure to particulate matter from sources other than diesel exhaust may have confounded the results of individual studies (Cohen and Higgins, 1995; Nauss et al., 1995), confounding by cigarette smoking is the most plausible source of bias that might explain the consistent observation of small relative increases in lung cancer risk in several occupations and in studies of widely varying design. Two recent reviews have addressed this issue, and concluded that confounding by cigarette smoking could not explain the reported associations of lung cancer risk and occupational exposure. Cohen and Higgins (1995) presented graphical displays (Figs. 32.1 and 32.2) that indicated little difference between the relative risks in truckers and railroad workers estimated with and without control for smoking. They also used available data on the relative risk of smoking and lung cancer and the distribution of smoking prevalence in occupational groups to calculate the likely impact of confounding by cigarette smoking in two cohort studies (Garshick et al., 1988; Gustavsson et al., 1990). They found that uncontrolled confounding by cigarette smoking could not explain the elevated relative risks. Bhatia and colleagues (1998) estimated summary relative risks for 16 studies that adjusted for the effects of cigarette smoking and 13 that did not. The summary relative risks were 1.35 (95% CI, 1.20, 1.52) and 1.33 (95% CI, 1.20, 1.47), respectively.

Dose–response Analyses: the US Railroad Workers Cohort

Although the current epidemiologic literature has, by most recent accounts, identified occupational exposure to diesel exhaust as a risk factor for lung cancer, it has failed to quantitate the relationship between exposure and/or dose and lung cancer risk, because none of the current studies provide quantitative estimates of past exposure to diesel exhaust for study subjects (Cohen and Higgins, 1995; Silverman, 1998). In virtually all studies, exposure has been classified using job titles obtained from historical records or interviews. The more informative studies have used measurements of contemporary levels of exhaust constituents, such as respirable particulate (e.g. Garshick et al., 1987a, 1988) or elemental carbon (e.g. Steenland et al., 1990), combined with historical information on processes and working conditions (e.g. Gustavsson et al., 1990), to classify workers as to their past exposure. They have then, however, generally relied on measures such as duration of employment (Steenland et al., 1990) or other similar indices, such as diesel-years (Garshick et al., 1988), to quantify the relation between exposure and risk.

The case-control and cohort studies of US railroad workers conducted by Garshick and colleagues (1987a, 1988) provide some of the most detailed information that currently exists on occupational exposure to diesel exhaust and lung cancer risk. This is due, in large part, to the documentation of both contemporary and historical exposure that was provided by a series of industrial hygiene studies (Woskie et al., 1988a,b), and which enabled Garshick to classify the diesel exhaust exposure of study subjects more accurately than had been possible in any previous study. Despite this improvement in exposure characterization, attempts to estimate exposure response relations using data from the cohort study have been controversial.

Woskie and colleagues (1988a,b) conducted an industrial hygiene survey in four Northern US railroads in order to characterize the exposure of railroad workers to diesel exhaust. They identified 39 job titles that were thought to entail a broad range of exposure to diesel exhaust. As a surrogate for diesel exhaust they measured contemporary levels of respirable particles adjusted for cigarette smoke (adjusted respirable particles, or ARP) for each job title, and then collapsed the 39 job titles into five career exposure groups based on similarities in work environment and patterns of career advancement. Their goal was to provide an accurate relative ranking of diesel exhaust exposure for use in the epidemiologic study.

On the basis of Woskie's studies, Garshick and colleagues classified the jobs that cohort members held as exposed or unexposed. These workers generally held but one job throughout their career (Garshick *et al.*, 1988). This dichotomization may have introduced some misclassification of exposure because exposure varied even within the exposed and unexposed groups. However, if this misclassification did not differ in its magnitude between lung cancer decedents and controls, it would only serve to underestimate the relative risk (Cohen and Higgins *et al.*, 1995).

Garshick and colleagues ascertained the mortality through 1980 of 55 407 white male railroad workers who were aged 40 to 64 years in 1959, a date when most railroads had converted to diesel engines. Subjects had to have begun work between 10 and 20 years earlier and to have been employed in 1959 in one of 39 railroad jobs surveyed in the industrial hygiene study (Garshick *et al.*, 1987a, 1988). Of the 19 396 deaths in the cohort as of December 31, 1980, 1694 were due to lung cancer (International Classification of Diseases (8th Revision), 162 as either the underlying or contributing cause).

Lung cancer risk was inversely related to age in 1959; workers who were aged 40 to 44 in 1959 and who held exposed jobs in 1959 (and, therefore, had both the longest potential exposure to diesel exhaust and, potentially, the longest induction times between exposure and death) experienced a 46% increase (RR, 1.46; 95% CI, 1.12–1.90) in lung cancer mortality relative to members of their birth cohort who held unexposed jobs in 1959. Rate ratios appeared to increase with years of exposure (Table 32.2). Workers with more than 15 years employment in an exposed job experienced a 72% increase in lung cancer mortality (RR, 1.72; 95% CI, 1.27–2.33). However, the results of analyses of cumulative exposure appear to be sensitive to modeling assumptions. In subsequent analyses, more detailed adjustment for age made the apparent upward trend in the relative risk for cumulative exposure disappear and the relative risk associated with greater than 15 years exposure decreased from 1.7 to 1.4 (E. Garshick, personal communication).

As part of an assessment of diesel exhaust by the US EPA, Crump and colleagues (1991) re-analyzed the railroad workers cohort study to attempt to replicate the original results and to conduct new analyses using quantitative estimates of diesel exhaust exposure derived from the industrial hygiene measurements of Woskie and colleagues (1988a,b).

Crump successfully reproduced Garshick's major findings. In addition, he compared lung cancer mortality in relation to the job held in 1959 among the career exposure groups defined by Woskie and found increased lung cancer mortality among engineers

Table 32.2 Lung cancer relative risks and 95% confidence intervals (in parentheses) for the Railroad Retirement Board Cohort[a]

Relative Risk[b]	Age in 1959				
	40–44	45–49	50–54	55–59	60–64
Exposure in 1959	1.46	1.33	1.12	1.18	0.99
	(1.12, 1.90)	(1.03, 1.73)	(0.88, 1.42)	(0.93, 1.49)	(0.74, 1.33)
	Cumulative Years of Exposure				
	0	1–4	5–9	10–14	15+
Exposure	(1.0)	1.20	1.24	1.32	1.72
		(1.01, 1.44)	(1.06, 1.44)	(1.13, 1.56)	(1.27, 2.33)
Exposure with asbestos exposure excluded	(1.0)	1.34	1.33	1.33	1.82
		(1.08, 1.65)	(1.12, 1.58)	(1.10, 1.60)	(1.30, 2.55)

[a] Data are from Garshick *et al.* (1988), Fig. 1 and Table 6.
[b] Exposure classified dichotomously, that is, exposed or unexposed.
Reproduced with permission from K.M. Nauss and the HEI Diesel Working Group (Nauss *et al.*, 1995).

and firemen (RR, 1.7; 95% CI, 1.2, 2.4), and among brakemen, conductors, and hostlers (RR, 1.6; 95% CI, 1.2, 2.1) aged 40–44 in 1959, whose presumed exposures and potential induction times were greatest relative to unexposed workers. However, the inverse relation with age in 1959 observed by Garshick *et al.* (1988) was not evident for some exposed jobs, and the job-specific relative risks did not seem to correlate with the intensity of exposure as assessed by Woskie and colleagues. These latter observations were interpreted by Crump as evidence against an effect of diesel exhaust; however, Garshick (cited in Cohen and Higgins 1998, page 290, endnote 11) notes that exposure misclassification might well explain these apparent inconsistencies.

Crump also compared the lung cancer mortality of the cohort to that of the US population and found evidence that deaths that occurred between 1976 and 1980 had been underascertained by the Railroad Retirement Board, although analyses limited to the years 1959–76 still showed elevated relative risks. However, the missing data are from the later, and potentially most informative, years of follow-up, and could affect dose–response analyses.

When Crump attempted to use Woskie's (1988a,b) estimates of average respirable particulate levels for railroad worker job categories to calculate relative risks, the results were difficult to interpret. Crump constructed several different exposure metrics that combined measures of particulate levels with information on regional climate for the various areas of the USA and used these metrics and ages to fit 50 regression models to the lung cancer data. All but two models showed inverse relations between lung cancer mortality and the various exposure metrics: that is, those with the highest estimated exposures had the lowest risk of lung cancer mortality. He attributed these surprising findings to inadequate exposure data. As he noted in his report:

> . . . there are many limitations in using the data on markers of diesel exposure . . . to estimate exposure in the cohort of railroad workers studied by Garshick . . . These limitations are potentially of sufficient magnitude to obscure any relationship between exposure to [diesel exhaust] and lung cancer that might exist in the cohort'

(Crump *et al.*, 1991)

Animal Bioassays of Diesel Exhaust Carcinogenicity

The carcinogenicity of inhaled diesel exhaust has been studied extensively in rodent bioassays. These studies and their implications have been reviewed in detail (Busby and Newberne, 1995; Mauderly, 1992; McClellan, 1996). Bioassays have shown consistently that diesel exhaust is a pulmonary carcinogen in rats that repeatedly inhale high concentrations of whole diesel exhaust for 24 months or longer. Both the length of exposure and the exposure rate determine risk. The tumors occur late in life and an increased incidence of lung tumors occurs in a dose-related manner at weekly exposures above approximately 120 mg/h/m^3 (Mauderly, 1992). Levels of exposure associated with an increased incidence of lung tumors impair particle clearance causing a progressively increasing lung burden of soot (Vostal, 1986; Wolff et al., 1986, 1987). Non-neoplastic lesions including chronic, active inflammation, proliferation of alveolar epithelium, interstitial fibrosis and epithelial metaplasia consistently occur prior to and concurrent with the lung tumors (Henderson et al., 1988a,b; Nikula et al., 1995), reviewed in (Busby and Newberne, 1995; Mauderly, 1992, 1995; McClellan, 1996). Exposures to lower concentrations of diesel exhaust that do not result in markedly increased lung burdens of soot or the same constellation of non-neoplastic lesions do not increase lung tumor incidence (Mauderly et al., 1987a). Studies including groups of rats exposed to whole or filtered exhaust have shown that filtered diesel exhaust is not a pulmonary carcinogen in rats, thus substantiating that the carcinogenic response requires the presence of soot (Brightwell et al., 1989; Heinrich et al., 1986; Iwai et al., 1986).

Six inhalation bioassays of mice exposed to diesel exhaust have been conducted and reviews of these studies (Mauderly, 1992, 1995) have concluded that diesel exhaust is of questionable carcinogenicity in mice, with the strongest evidence in strains that are generally sensitive to chemical carcinogenesis. Increased incidence rates of lung tumors have been produced in some groups of Strain A and Sencar mice (Kaplan et al., 1983; Pepelko and Peirano, 1983), while decreased incidence rates or no change from controls have been found in other groups (Pepelko and Peirano, 1983). In strains of mice with high spontaneous rates of lung tumors, the results can be influenced by unusually low lung tumor rates in the control groups compared with historical controls (Heinrich et al., 1986, 1995; Pepelko and Peirano, 1983). One study using NMRI mice demonstrated an increased incidence of lung tumors in mice exposed to whole diesel exhaust and in mice exposed to filtered diesel exhaust (Heinrich et al., 1986), whereas another study in the same laboratory failed to reproduce this finding (Heinrich et al., 1985). Mauderly et al. (1996) found that diesel exhaust was not a pulmonary carcinogen in CD-1 mice exposed under conditions identical to those that were carcinogenic to F344 rats.

There have been five reported chronic inhalation bioassays of Syrian golden hamsters exposed to diesel exhaust (reviewed by Mauderly, 1992). None showed a diesel exhaust-related increase in lung tumors.

Animal Bioassays of the Carcinogenicity of Inhaled Particles

Data from several inhalation bioassays have led to the suggestion that the carcinogenic response of rats to inhaled diesel exhaust may be non-specific and related to the heavy lung burden of particulate material rather than to a unique, carcinogenic property of the soot-associated organic chemical fraction. Vostal (1986) first noted that the rat carcinogenic response to inhaled diesel exhaust was similar to that observed for a number of particulate materials, and that the exposure–response relationship for these materials seemed to exhibit a threshold. Further reviews on the carcinogenicity of inhaled solid particles in rats have been published (Hext, 1994; Mauderly, 1994a,b). As for diesel exhaust, exposures resulting in an increased incidence of lung tumors also impair macrophage-mediated clearance and the cancers tend to occur late, i.e. towards the end of exposures of two years duration. These exposures also cause pulmonary inflammation, fibrosis, alveolar epithelial proliferation and epithelial metaplasia. The exposure rates or specific lung burdens at which an increased incidence of lung tumors is observed in rats vary considerably among different types of particles, with particles of greater inherent cytotoxicity requiring lower exposures to induce tumors. For example, the lowest specific lung burdens associated with a significant increase in lung tumors can be compared for quartz and titanium dioxide. Rats exposed to quartz at a weekly exposure rate of 30 mg/h/m^3 had a specific lung burden of 0.4 mg/g wet lung at 24 months and a significant increase in lung tumors (Muhle *et al.*, 1991), whereas rats exposed to titanium dioxide at a weekly exposure rate of 1017 mg/h/m^3 had a specific lung burden of 7.4 mg/g wet lung at 24 months and a significant increase in lung tumors (Heinrich *et al.*, 1992).

Not all exposures that impair clearance cause lung tumors in rats. Exposure of rats to copying toner at 480 mg/h/m^3, which resulted in a specific lung burden of 9.2 mg/g at 24 months, impaired clearance (Bellmann *et al.*, 1991) but did not cause an increased incidence of lung tumors (Muhle *et al.*, 1992).

Studies of rats exposed to whole or filtered diesel exhaust demonstrated the importance of the particulate matter in causing lung tumors in rats, but they did not determine whether the particle-associated organic compounds or the particles themselves were responsible for the carcinogenesis. The finding that several non-fibrous solid particles caused lung tumors in chronically exposed rats raised the related issues of (1) the importance of the extractable organic chemical fraction in the carcinogenicity of diesel soot, and (2) if the organic fraction was not important to the rat response, the relevance of the rat data for predicting cancer risk in humans depositing much less soot per unit of lung tissue. To explore the importance of the diesel soot-associated organic compounds relative to the carbonaceous core of the particles, researchers chronically exposed rats to whole diesel exhaust or to aerosolized carbon black particles in studies at laboratories in the USA (Mauderly *et al.*, 1994, 1995) and in Germany (Heinrich *et al.*, 1995).

In the US study, male and female F344 rats were exposed for 24 months to diluted whole diesel exhaust or aerosolized carbon black for 16 h/day, 5 days/week, at particle concentrations of 2.5 mg/m^3 or 6.5 mg/m^3 or to filtered air. The carbon black served as a surrogate for the elemental carbon matrix of diesel soot. Considering both the mass fraction of solvent-extractable matter and its mutagenicity in the Ames *Salmonella* assay, the mutagenicity in revertants per unit particle mass of the carbon black was three orders of magnitude less than that of the diesel soot. Clearance was impaired in the lungs of exposed rats and both diesel soot and carbon black particles accumulated progressively, but the rate of accumulation was higher for diesel soot. Similar non-neoplastic lesions occurred in the rats exposed to diesel exhaust and carbon black. Both diesel exhaust and carbon black caused exposure-related increases in the occurrence of the same types of malignant and benign lung tumors (Mauderly *et al.*, 1994; Nikule *et*

Table 32.3   Non-fibrous solid particles causing lung tumors in rats[a] exposed by inhalation

Non-fibrous particles	References
Antimony ore	Groth *et al.* (1986)
Antimony trioxide	Groth *et al.* (1986)
Beryllium fluoride	IARC (1993), Schepers (1971)
Beryllium hydrogen phosphate	IARC (1993), Schepers (1961)
Beryllium metal	Finch *et al.* (1996), IARC (1993)
Beryllium sulfate	IARC (1993), Pelko (1984)
Beryl ore	Wagner *et al.* (1969)
Cadmium chloride	Glaser *et al.* (1990), IARC (1993)
Cadmium oxide	Glaser *et al.* (1990), IARC (1993)
Cadmium sulfate	Glaser *et al.* (1990), IARC (1993)
Cadmium sulfide	Glaser *et al.* (1990), IARC (1993)
Calcium chromate	IARC (1990)
Chromium dioxide	IARC (1990), Lee *et al.* (1988)
Carbon black	Heinrich *et al.* (1995), IARC (1996), Nikula *et al.* (1995)
Coal dust	Martin *et al.* (1977)
Coal tar aerosol	Heinrich *et al.* (1984), IARC (1984, 1987)
Diesel exhaust	Brightwell *et al.* (1986), Heinrich *et al.* (1986, 1995), IARC (1989a), Ishinishi *et al.* (1986), Mauderly *et al.* (1987), Nikula *et al.* (1995)
Nickel carbonyl	IARC (1990), Sunderman and Severi (1966)
Nickel metal	Heuper (1958), IARC (1990)
Nickel oxide	IARC (1990), National Toxicology Program (1994b)
Nickel subsulfide	IARC (1990), National Toxicology Program (1994a)
Oil shale dust	Holland *et al.* (1986), Mauderly *et al.* (1990)
Silica (crystaline)	IARC (1987b), Saffiotti *et al.* (1995)
Solvent-refined coal solids	Drozdowicz and Kelly (1989)
Talc (asbestos-free)	IARC (1987a), National Toxicology Program (1993)
Titanium dioxide	Heinrich *et al.* (1995), IARC (1989b), Lee *et al.* (1985)
Titanium tetrachloride	Lee *et al.* (1986)
Volcanic ash	Wehner *et al.* (1986)
Zinc manganese beryllium silicate	Schepers (1961)
Quartz	Dagle *et al.* (1986), Holland *et al.* (1986), Muhle *et al.* (1991)

[a] Significant increase in males or females or both.

al., 1995). The increases followed similar exposure–response relationships for the two types of particles.

In the German study, female Wistar rats were exposed for 24 months to diesel exhaust at 7.5 mg soot/m^3, or to carbon black or titanium dioxide at varying concentrations, or to filtered air. As in the US study, lung tumors of similar types were caused by all three exposures. The incidence of tumors was proportional to the cumulative exposure and, based on the incidence of tumors per unit of particle lung burden, the pulmonary carcinogenic potential was highest for titanium dioxide, intermediate for carbon black, and lowest for diesel soot. These studies suggest that the organic fraction of diesel exhaust does not play a significant role in the carcinogenicity of diesel exhaust in the rat bioassays. Certainly, the influence of the organic fraction would appear small in comparison to the three orders of magnitude difference between the mutagenic organic contents of the diesel soot and carbon black in the US study. As noted by McClellan (1996), if the organic compounds had an effect, it was at a low potency and below the limits of detection in the animal bioassay. Therefore, the dose of extractable, diesel soot-associated organic matter is unlikely to be an appropriate dose-equivalence term for extrapolating human unit risk estimates from rat lung tumor data.

Nauss *et al.* (1995) reviewed thresholds for particle-induced responses and found a threshold dose for impaired clearance, which depends on the exposure concentration, dose rate, length of exposure and the toxicity of the particle (Watson and Green, 1995). In their review of studies of rats chronically exposed to diesel exhaust, Watson and Green (1995) found that weekly exposures to diesel exhaust of approximately 70 mg/h/m^3 or greater were associated with chronic inflammation, epithelial cell proliferation, and depressed alveolar clearance. Resultant lung burdens were 0.5 mg/g of lung or greater and below these dose levels diesel exhaust-related effects were not detectable. Nauss and colleagues examined the data from eight rat inhalation bioassays of diesel exhaust and found that weekly exposures below approximately 200 mg/h/m^3 have not produced significant increases in lung tumors, whereas weekly exposures above 200 mg/h/m^3 generally have produced a statistically significant increase in tumor incidence. These data suggest that there may be an exposure threshold for impaired clearance, inflammation, epithelial proliferation, and subsequent carcinogenesis in rats exposed to diesel exhaust.

Mechanistic Studies of the Carcinogenicity of Diesel Exhaust and Particulate Materials

Both genotoxic and non-genotoxic mechanisms have been proposed for diesel exhaust-induced carcinogenesis. These classifications are convenient for discussion, but are overly rigid because pathways that are not directly genotoxic initially can eventually result in genetic damage, and also because the diesel exhaust may induce cancer through both genotoxic and non-genotoxic mechanisms.

The carcinogenicity of various chemical components of diesel exhaust has been reviewed (Scheepers and Bos, 1992) and the results of various short-term *in vitro* and *in vivo* test systems have been summarized (Shirnamé-Moré, 1995). The polycyclic aromatic hydrocarbons (PAHs) and the nitro-PAHs are among the most potent mutagens and carcinogens in diesel exhaust (IARC, 1989). Attention has focused on soot-associated organic chemicals because diesel soot particles are in the respirable range and contain mutagens

Table 32.4 Summary of genetic effects of diesel exhaust and its constituents in various short-term *in vitro* and *in vivo* assays[a]

Assay	Whole exhaust	Diesel particles	Particle extract	Gas phase	Benzo[a]-pyrene	1,6-Dinitropyrene
Mutagenicity (bacteria)	ND	+	+	+	+	+
Mutagenicity (mammalian cells)	ND	+	+	ND	+	+
Chromosomal damage	±	±	±	ND	+	+
Cell transformation	ND	ND	+	ND	+	+
DNA adducts	±	±	+	ND	+	+

ND = not determined, or only limited data available; + = positive in most studies; ± = equivocal data.
[a] Reproduced with permission from K.M. Nauss and the HEI Diesel Working Group (Nauss *et al.,* 1995).

Table adapted from Shirname-Moré (1995)

and carcinogens adsorbed onto their surfaces. Further, rat bioassays have demonstrated that the soot fraction is necessary for carcinogenesis. Typically, 10–40% of the diesel soot mass consists of organic chemicals, including PAHs and nitro-PAHs. An important consideration in estimating the potential dose of PAHs and nitro-PAHs delivered by inhaled diesel particles is the bioavailability of these compounds. The chemicals adsorbed to the particles may not be as mutagenic *in vivo* as they are in most *in vitro* assays because the material extracted using physiological fluids such as saline, serum, or lung lavage fluid is generally less genotoxic than that extracted using the non-physiological organic solvents employed in most *in vitro* assays (Brooks *et al.*, 1980; Siak *et al*, 1981). However, material extracted using an aqueous mixture that contained dipalmitoyl phosphatidylcholine, which was used to simulate pulmonary surfactant, was highly mutagenic (Keane *et al*, 1991).

Many other factors affect the bioavailability of carcinogens (Green and Watson, 1995). The degree of particle agglomeration (Gerde *et al.*, 1991) is one determinant of the release of organic chemicals. There may be a greater degree of agglomeration with intratracheal instillation and high exposure concentrations, used in animal bioassays, than of particles associated with typical occupational or environmental exposures. According to the simulation models of Gerde and co-workers (1991), PAHs would be released quickly from particles under low concentration exposure conditions, and tumors would be expected at the sites of initial deposition rather than at sites of prolonged retention. Schlesinger and Lippmann (1978) have drawn attention to the correlation between particle deposition in intrapulmonary bronchioles and primary bronchial tumors. Other factors affecting bioavailability are related to the physicochemical properties of the particles, the surrounding biological fluids, and cellular metabolism. The physicochemical properties of diesel exhaust particles have changed along with engine technologies, so previous data on bioavailability and genotoxicity of particle-associated organics may not apply to current engines. Organic chemicals generally occurred at higher concentrations and were less tightly bound to the particles from older engines, compared with those from current engines.

One approach to assessing the bioavailability and potential effects of genotoxic chemicals in diesel exhaust has been to assay for DNA adducts using ^{32}P post-labeling in lung tissues of rats exposed to diesel exhaust in short-term (Bond *et al.*, 1988, 1990a,b; Hsieh *et al.*, 1986) and long-term (Gallagher *et al.*, 1994; Randerath *et al.*, 1995; Wong *et al.*, 1986) inhalation studies. The first studies suggested a role for genotoxic chemicals in the rat lung tumor response: higher total DNA adduct levels were found in the lungs of rats exposed to high concentrations of diesel exhaust compared to controls (Bond *et al.*, 1988, 1990a,b; Wong *et al.*, 1986), and the highest levels were found in peripheral lung tissues, the site where tumors arise (Bond *et al.*, 1988). The finding that DNA adduct levels did not increase with exposure concentration (Bond *et al.*, 1990b) could not be readily explained. More recently, Gallagher and co-workers (1995) found no increase in total DNA adduct levels in the lungs of rats exposed to diesel exhaust, and Randerath and co-workers (1995) found small or variable differences in total DNA adducts between exposed and control rats. In addition, typical PAH or nitro-PAH DNA adducts were not found in the lungs of diesel exhaust-exposed rats. Several reasons for the lack of a consistent, dose–response relationship between diesel exhaust exposure and DNA adducts in rat lung tissue have been suggested (Nauss *et al.*, 1995). These include: (1) low bioavailability of PAHs and nitro-PAHs with resultant adduct levels below the assay sensitivity level; (2) finite DNA adduct lifetime so that adducts are markers of only recent exposures;

(3) dilution of adducts by particle-induced cell proliferation; and (4) altered carcinogen metabolism induced by the high concentration at which exposures were received so that activation of procarcinogens to active species was reduced.

One study has reported levels of DNA adducts in workers with occupational exposure to diesel exhaust. Hemminki and Pershagen (1994) found elevated aromatic-DNA adducts in lymphocytes from non-smoking bus maintenance and truck terminal workers. The main adducts were not benzo[a]pyrene-DNA or other PAH-DNA adducts, but they were not further identified.

The rat bioassays (Heinrich *et al.*, 1995; Mauderly *et al.*, 1994; Nikula *et al.*, 1995) demonstrating that inhaled carbon black, which contains negligible amounts of organic mutagens, is similarly carcinogenic to diesel exhaust, suggest that the mutagenic compounds in diesel exhaust do not play a significant role in the rat carcinogenic response to high concentrations of diesel exhaust. These results – coupled with the observations that diesel exhaust, as with several other inhaled particulate materials, only induces tumors in rats at the high exposure concentrations that impair macrophage-mediated clearance and induce inflammation, cell proliferation and fibrosis – have led to the suggestion that diesel exhaust induces tumors in rats through non-genotoxic or indirectly genotoxic mechanisms involving inflammation and cell proliferation (McClellan, 1996). Inflammation results in the release of leukotrienes, reactive oxygen species, proteolytic enzymes, cytokines and chemokines (TNF-α, IL-1, IL-4, IL-6, IL-8, MIP, MCP-1), and growth factors (fibronectin, TGF-β) that may interact in causing particle-induced carcinogenesis. Reactive oxygen species can have a genotoxic effect (Jackson *et al.*, 1989; Weitzman and Stossel, 1981, 1982). The other inflammatory mediators can amplify the release of reactive oxygen species through attraction and activation of neutrophils and monocytes, stimulate cell growth and mitogenic activity, or indirectly stimulate respiratory epithelial replication (compensatory regeneration and remodeling) through apoptosis and necrosis (Borm and Driscoll, 1996; Driscoll *et al.*, 1996). Enhanced epithelial cell proliferation increases the cell population at risk for mutations by endogenous and exogenous agents and increases the chance that oxidant-induced genetic damage becomes fixed in a dividing cell and is expanded clonally (Ames and Gold, 1990; Butterworth, 1990; Cohen and Ellwein, 1990; Preston-Martin *et al.*, 1990).

The work of Driscoll and colleagues (1995, 1996) provides recent experimental evidence supporting the hypothesis that inflammatory cell-derived oxidants and increased cell proliferation play roles in the pathogenesis of rat lung tumors induced by high levels of particles. First, these investigators showed that inflammatory cells lavaged from rats that had been instilled with carbon black *in vivo* induced mutations in the hypoxanthine-guanine phosphoribosyltransferase (*hprt*) gene of alveolar epithelial cells when the inflammatory cells were cocultured with a rat alveolar epithelial cell line *in vitro* (Driscoll *et al.*, 1995). Next, Driscoll and coworkers (Driscoll *et al.*, 1996) exposed rats to carbon black by inhalation and isolated epithelial cells after exposure. They found a significant exposure-dependent increase in *hprt* mutation frequency in alveolar epithelial cells after 13 weeks of exposure to 7.1 and 52.8 mg/m^3 carbon black, but not after exposure to 1.1 mg/m^3 carbon black. Clearance of carbon black was impaired after exposure to the two higher concentrations of carbon black, and lung tissue injury, inflammation, chemokine expression, epithelial hyperplasia and pulmonary fibrosis were observed after exposure to these higher concentrations but not after exposure to 1.1 mg/m^3 carbon black.

The exposure-dependent increase in *hprt* mutations in lung alveolar epithelial cells paralleled the exposure-dependent increase in neutrophils in bronchoalveolar lavage fluid (Driscoll *et al.*, 1996). These findings suggest that inherent particle genotoxicity is not a significant factor in the rat lung tumorigenic response to carbon black particles, and that rat lung tumors that develop in response to a number of poorly soluble particles may represent a generic response to lung overload and subsequent chronic inflammation and epithelial hyperplasia.

The rat lung tumor response to poorly soluble particles may be unique to this species. In addition to the studies of diesel exhaust listed earlier in this chapter, other particle exposures that induced tumors in rats but not mice include talc (National Toxicology Program, 1993), titanium dioxide (IARC, 1989), carbon black (Heinrich *et al.*, 1995), nickel subsulfide (National Toxicology Program, 1994a), cadmium sulfate (Glaser *et al.*, 1990; IARC, 1993), cadmium sulfide (Glaser *et al.*, 1990; IARC, 1993), and cadmium chloride (Glaser *et al.*, 1990; IARC, 1993). Further evidence that rats may be more prone than other species to inflammation and alveolar epithelial hyperplasia in response to particles comes from a number of studies in which rats, but not monkeys, developed active inflammation and alveolar epithelial hyperplasia in response to particulate material: petroleum coke (Klonne *et al.*, 1987), raw and processed shale dusts (MacFarland *et al.*, 1982), and beryl ore (Wagner *et al.*, 1969). Recently, Nikula and co-workers (1997) showed that retained coal dust and diesel soot in identically exposed rats and monkeys are distributed differently in the lung, and only the rats develop inflammation and alveolar epithelial hyperplasia in response to the particulate material.

The results of interspecies comparisons suggesting that the rat may respond differently to inhaled particles are supported to some extent by studies of occupational exposures to insoluble particles. Most notably, studies of lung cancer occurrence or mortality (IARC, 1997) among miners in the UK and the USA with long-term exposure to high concentrations of coal dust sufficient to cause high lung burdens of coal dust and increased rates of fibrotic lung disease (coal miner's pneumoconiosis or 'black lung') have failed to find evidence of increased lung cancer risk.

The IARC recently reviewed the epidemiologic evidence concerning occupational exposure to carbon black (IARC, 1996) and characterized it as inadequate. In particular, the evidence from two cohorts of carbon black production workers is conflicting. The most recent follow-up of the US multi-plant cohort (Robertson and Inman, 1996) reported a respiratory cancer relative risk of 0.8 (95% CI, 0.6, 1.1) based on 34 deaths. IARC reviewers commented that the mortality ascertainment in the US study may have been incomplete, potentially biasing the results downward. In contrast, a study of a British five-plant cohort (Hodgson and Jones, 1985) reported a lung cancer relative risk of 1.5 (95% CI, 1.0, 2.2). Neither study controlled for cigarette smoking, although a population-based case–control study of 857 incident lung cancers estimated a relative risk of 1.5 (95% CI, 0.6, 4.0) for exposure to carbon black after adjustment for smoking and other occupational exposures (Parent *et al.*, 1996).

Coal dust and carbon black differ in several ways that may be relevant to interpretation of the available epidemiologic data. Unlike coal dust, which is generated by a mechanical process, carbon black results from a combustion process, producing particles that are smaller and have organic compounds adsorbed on their surfaces, albeit tightly bound and in much smaller quantities than diesel particles. Two small studies of occupational exposure to non-asbestiform talc (Wergeland *et al.*, 1990) and titanium dioxide (Chen and Fayerweather, 1990) have not found excess lung cancer.

Summary

A large body of epidemiologic evidence indicates that workers exposed for prolonged periods to diesel exhaust in a variety of occupational settings may be at increased risk of lung cancer. Recent quantitative reviews of this literature estimate summary excess relative risks of 20–50% which cannot be explained by differences between exposed and non-exposed workers in cigarette smoking or other known causes of lung cancer. Unfortunately, none of the epidemiologic studies provides quantitative estimates of the levels of diesel exhaust particulate or other constituents to which workers were historically exposed. This has made it difficult to estimate accurately exposure–response relations.

Toxicologic experiments have also provided evidence of the carcinogenicity of diesel exhaust, but the evidence is difficult to interpret with respect to human lung cancer. Lifelong exposure to high levels of diesel particulate is consistently associated with increased incidence of lung tumors in rats, but not in other rodent species. Recent experiments by Mauderly and colleagues (1994a,b) suggest that the rat responds to the inhalation of high levels of particulate matter from any source in a qualitatively different manner from other species, and therefore that the results of the positive diesel exhaust bioassays in the rat are of questionable relevance to human beings exposed to diesel exhaust at much lower levels in most occupational settings and in the general ambient environment.

Risk Estimates for Human Lung Cancer

Over the past decade several agencies charged with the identification of human carcinogens and/or regulation of air quality have made determinations of the carcinogenicity of diesel exhaust. Some have also estimated the risk of lung cancer that diesel exposure poses to the general population. The insights derived from epidemiology and toxicology, and the remaining uncertainties, have both informed and complicated efforts to quantify lung cancer risk. In this section we review these efforts. Recent risk assessment efforts for diesel exhaust have been reviewed by Nauss and colleagues (Nauss *et al.*, 1995; Stayner *et al.*, 1998), and risk assessment of air pollutants is discussed more generally by Samet in this volume.

The earliest efforts to predict the risk of human cancer due to diesel exhaust date from the early 1980s (Nauss *et al.*, 1995). Because there were few epidemiologic or long-term toxicologic studies of diesel exhaust available, these risk assessments used the method of comparative potency to estimate lung cancer risk (Albert *et al.*, 1983; Cuddihy *et al.*, 1984). In this approach, the mutagenicity of diesel exhaust extracts is calculated relative to that of a known human combustion-source carcinogen, such as cigarette smoke. This estimate of relative mutagenicity (potency) is then applied to epidemiologic data for the known carcinogen to derive a risk estimate. Nauss *et al.* (1995) cite a range of risk estimates for lifetime exposure to 1 $\mu g/m^3$ of diesel exhaust from two excess cancers per million persons to four excess cancers per 100 000 persons that were derived using this approach. As the results of long-term bioassays and epidemiologic studies became available, the comparative potency method, which had a number of limitations, was abandoned in favor of more direct approaches (Nauss *et al.*, 1995).

IARC reviewed the evidence for the carcinogenicity of diesel exhaust in 1989 (1989) and designated diesel exhaust a 'probable human carcinogen'. IARC reviewers judged the evidence from animal data sufficient to conclude that diesel exhaust was a human car-cinogen, but considered the epidemiologic data limited in this regard. More recently, the World Health Organization (WHO) evaluated the currently available data and reached the same general conclusions (1996). However, they added that no human data were 'suit-able for estimating unit risk'.

Risk assessments have recently been conducted for truck drivers and underground miners. The truck drivers' risk assessment was conducted by Steenland and colleagues (Steenland et al., 1998) using data from a case–control study of US Teamsters (Steenland et al., 1990), along with information on historical trends in particulate emissions and char-acteristics of the US truck fleet. They estimated a lifetime risk of lung cancer of 1–2 per 100 persons exposed to 5 µg/m^3 of elemental carbon. Stayner et al. (1998) reviewed the results of a number of risk assessments that used either toxicologic experiments or epidemiologic studies to model lung cancer risk, and then derived estimated risks for underground miners based on their reported exposures. They concluded that all models predicted risks greater than 1 per 1000 persons for miners with long-term exposure to >1000 µg/m^3.

Two US regulatory agencies have attempted to estimate lung cancer risk due to diesel exposure in the general population. In its 1994 draft health assessment document the US EPA concluded that diesel exhaust was a 'probable human carcinogen' based on 'limited' epidemiologic evidence and 'sufficient' animal data (US EPA, 1994). Their attempts to use the data from Garshick's cohort study of US railroad workers to estimate a unit risk produced results that were difficult to interpret (US EPA, 1994), so they used data from rat experiments to calculate a unit risk of approximately three excess cancers per 100 000 people exposed to 1 µg/m^3 of diesel exhaust for their entire lives. More recently, the California EPA (Cal EPA) issued a draft assessment of diesel exhaust, which identified it as a toxic air contaminant, and concluded that a 'causal association' was a 'reasonable and likely explanation for the increased rates of lung cancer observed in the epidemiological studies' (Cal EPA, 1997). Using data from rat experiments, they estimated a lifetime risk of one excess cancer per 100 000 for a lifetime exposure to 1 µg/m^3. Unlike US EPA, however, Cal EPA considered the US railroad workers' data suitable for calculating a unit risk and estimated that risk to be two excess cancers per 1 000 people exposed to 1 µg/m^3. Both the US EPA and Cal EPA risk assessment documents have been subject to considerable critical discussion, and both are currently undergoing revision.

The questionable validity of using the results of high-dose rat studies to predict lung cancer risk in humans exposed to much lower concentrations continues to create obstacles to quantitative cancer risk assessment of diesel exhaust and complicate regulatory action based on such assessments, as does the inability to estimate with confidence the dose–response relation at those lower exposure levels from the current epidemiologic data.

Cancer of Other Organs

Epidemiology

Cohen and Higgins (1995) reviewed the evidence for increased risk of cancers other than lung and, for the most part, found that the epidemiologic studies which provided consistent evidence of increased risk of lung cancer failed to find such patterns of increased risk for cancer at any other site, although there were weak associations reported for various other cancers. The only possible exception is for bladder cancer, for which numerous case–control studies showed an increased risk associated with exposure to diesel exhaust, particularly among truck drivers. Several large cohort studies, however, did not find increased bladder cancer risk in other diesel-exposed occupations.

Retrospective cohort studies conducted by Howe and colleagues (1983) and Schenker and colleagues (1984) compared the mortality rates of Canadian and US railroad workers to their respective national populations. In both studies the highest relative risk estimate was for kidney cancer, which was the only cancer occurring in excess in both cohorts. The relative risks were 1.3 (95% CI, 1.0, 1.7) and 1.7 (95% CI, 0.6, 3.6) from the Canadian and US studies, respectively. No other cancer was in excess in both studies, although misclassification of exposure may have obscured other associations. This could be particularly the case in the Canadian study, in which only 25% of the person-time of follow-up were classified as probably exposed.

Two retrospective cohort studies examined the mortality of bus garage workers. Gustavsson and colleagues (1990) compared the cancer incidence of 695 Swedish bus garage workers with national cancer rates, and Rushton *et al.* (1983) compared the mortality of 8684 London bus garage workers with British national rates. Both studies reported small relative excesses of brain cancer, measured with considerable imprecision (relative risks 1.9; 95% CI, 0.7, 4.2 and 1.2; 95% CI, 0.5, 2.4, for the Swedish and British studies, respectively). No other cancer was elevated in both studies.

Two general population studies of occupational exposure to diesel exhaust reported increased risk for different cancers associated with exposure to diesel exhaust. Boffetta and colleagues (1988) reported the mortality experience of a cohort of US men characterized by their self-reported exposure to diesel exhaust in the occupations in which they had worked the longest. The study reported elevated risks of pancreatic cancer (RR, 1.4; 95% CI, 0.9, 2.0) and malignant melanoma (RR, 1.7; 95% CI, 0.8, 3.0). Siemiatycki and colleagues (1988) observed increased risks of colon cancer (RR, 1.3; 95% CI, 1.1, 1.6) and prostate cancer (RR, 1.2; 95% CI, 1.0, 1.5) in a case–control study of Montreal men, whose job histories were rated by chemists and industrial hygienists with respect to diesel exhaust exposure. Prolonged exposure to diesel exhaust (>10 years) was associated with higher relative risks for both cancers.

Thirteen case–control studies of bladder cancer provide estimates of the effect of occupational exposure to diesel exhaust (Brooks *et al.*, 1992; Coggon *et al.*, 1984; Hoar and Hoover *et al.*, 1985; Howe *et al.*, 1980; Iscovich *et al.*, 1987; Jensen *et al.*, 1987; Risch *et al.*, 1988; Silverman *et al.*, 1983, 1986; Steenland *et al.*, 1987; van Vliet *et al.*, 1997; Vineis and Magnani, 1985; Wynder *et al.*, 1985). These studies were conducted primarily in the USA and Europe using incident cases (Winer and Busby, 1995), and controls with other diseases or cancers (US EPA, 1996) or persons sampled at random from the base population (Pope and Dockery, 1996). Most relied on interviews to obtain occupational histories, and most (US EPA, 1993) controlled for cigarette smoking. Most of the case–control studies found increased relative risks of bladder cancer associated with employment in jobs that entailed exposure to diesel exhaust, mainly truck drivers and railroad workers (Cohen and Higgins, 1995). Longer duration of employment in exposed jobs (Hoar and Hoover, 1985; Jensen *et al.*, 1987; Silverman *et al.*, 1983, 1986; Steenland *et al.*, 1987); time since beginning such employment (Risch *et al.*, 1988); and regular employment in exposed jobs (Silverman *et al.*, 1986) were associated with higher relative risks. Cohort studies of railroad workers and heavy equipment operators (Howe *et al.*,

1980; Schenker *et al.*, 1984; Wong *et al.*, 1985) did not find increased risk of bladder cancer. These discrepant findings cannot be readily explained, and Cohen and Higgins (1995) suggest sources of biases and confounding that could either have exaggerated or understated the actual associations.

Animal Bioassays

Only one of the inhalation bioassays of rodents exposed to diesel exhaust showed an increased incidence of non-pulmonary tumors (Iwai *et al.*, 1986). Most studies focused on the lung, and only limited information on non-pulmonary tumors is available (reviewed in Mauderly, 1992). Lewis *et al.* (1989) provided the most complete information. They exposed rats to diesel exhaust at 2 mg soot/m^3, 7 h/day, 5 days/week for 2 years and examined 50 organs, with detailed data for tumors in 11 non-pulmonry organs, including the urinary bladder. They found no exposure-related difference in the incidence of tumors in any organ.

Summary

It is not unreasonable to hypothesize by analogy to another complex combustion source pollutant – cigarette smoke – that diesel exhaust might cause cancer at sites other than the lung. For no other cancer, however, is there the consistent pattern of association with diesel exhaust that has been observed for lung cancer. Bladder cancer is the sole exception to this general assessment, but here too the evidence is weak. There are few animal data; these are best characterized as showing absence of evidence, rather than evidence of absence. The epidemiologic data, though more numerous, are inconsistent.

NON-MALIGNANT RESPIRATORY DISEASE

Epidemiologic studies of occupational exposure

Two case reports suggest that acute exposure to high concentrations of diesel exhaust may trigger the onset of reactive airways disease. Kahn and coworkers (1988) reported that two of 13 railroad workers acutely exposed to high levels of diesel exhaust developed dyspnea and wheezing, but provided little additional information. Wade and Newman (1993) reported the onset of asthma in three railroad workers who were exposed acutely to exhaust from diesel locomotives. The workers were 40–50 years old, never- (two) or former- (one) smokers, with no prior history of lower respiratory tract disease. Two workers experienced the acute onset of dyspnea and wheezing shortly following self-reported exposure over several hours to high levels of diesel exhaust in an enclosed railway car located immediately behind a diesel locomotive. The third reported subacute onset of dyspnea and wheezing following the same type of exposure. No measurements of ambient levels of diesel exhaust in the railway cars were available. All three showed evidence of airway hyperreactivity and airflow obstruction responsive to bronchodilator therapy, and experienced recurrence of symptoms when re-exposed to diesel exhaust, and all remained symptomatic after 1 to 3 years. The authors noted the similarity in terms of both exposure conditions and clinical pictures to the Reactive Airways Dysfunction Syndrome described by Brooks and colleagues (1985). They reported the acute onset of 'an asthma-like illness' in 10 people exposed occupationally to single, high concentrations of different irritants under conditions of poor ventilation. The irritants included spray paints and sealants, but also combustion products.

The epidemiologic literature on diesel exhaust and acute and chronic non-malignant respiratory disease has recently been reviewed (Cohen and Higgins, 1995). Six studies measured the effects of acute exposure to diesel exhaust on changes in pulmonary function among miners (Ames *et al.*, 1982; Gamble *et al.*, 1978; Jorgensen and Svensson, 1970), bus garage workers (Gamble *et al.*, 1987b), and longshoremen (Ulfvarson and Alexandersson, 1990; Ulfvarson *et al.*, 1987). The results of these investigations were inconsistent. Several studies found evidence of small changes in pulmonary function related to diesel exhaust exposure. Gamble and colleagues (1978) studied a cohort of male US salt miners and observed small decreases in pulmonary function over an 8-h shift. The study observed small changes in forced expiratory volume in 1 s (FEV_1) in relation to nitrogen dioxide (NO_2) (–14 ml/ppm) and respirable particles (–18 ml/mg), but all estimates were imprecise. Ames and colleagues (1982) did not observe exposure-related decreases in pulmonary function overall among a cohort of US coal miners, but noted small but consistent decreases in lung volumes and flows among non- and ex-smokers. Gamble *et al.* (1987b) found no consistent declines in spirometric measurements adjusted for smoking in relation to either NO_2 or particles in a cohort of US bus garage workers.

Ulfvarson conducted both an observational study and a controlled experiment on the effects of acute exposure to diesel exhaust in cohorts of Swedish stevedores. Ulfvarson and colleagues (1987) studied changes in pulmonary function over an 8-h work shift in 47 stevedores exposed chiefly to diesel exhaust. The study found decreases in forced vital capacity (FVC) and FEV_1, but not in several other indices (FEV%, the percentage of forced expiratory volume, or MMEF, the maximal mid-expiratory flow rate). Because these associations were not related to levels of measured constituents of diesel exhaust such as NO_x or formaldehyde, the investigators speculated that some aspect of exposure to the particulate fraction of diesel exhaust might be responsible. To test this hypothesis, Ulfvarson and Alexandersson (1990) studied 24 stevedores who worked with standard diesel-powered vehicles over a 3-day period and 18 stevedores whose vehicles were equipped with particle filters. Workers exposed to unfiltered exhaust experienced a 5% reduction in FVC and a 2% increase in FEV%, while workers exposed to filtered exhaust experienced a 2% decrease in FVC and a 0.7% increase in FEV%. A group of 17 non-smoking control subjects unexposed to diesel exhaust experienced no changes in pulmonary function over the work shift.

Nine studies have examined the relationship between occupational exposure to diesel exhaust and the prevalence of chronic non-malignant respiratory disease. Of these, six reported the experience of underground miners (Ames *et al.*, 1984; Attfield *et al.*, 1982; Gamble *et al.*, 1983; Jacobsen *et al.*, 1988; Jorgensen and Svensson, 1970; Reger *et al.*, 1982), while the others examined railroad workers (Battigelli *et al.*, 1964), bus garage workers (Gamble *et al.*, 1987a), and longshoremen (Purdham *et al.*, 1987). In only one study (Ames *et al.*, 1984) were study subjects also followed over time to estimate the occurrence of chronic disease and loss of lung function. These studies do not provide consistent evidence for chronic, non-malignant respiratory effects associated with occupational exposure to diesel exhaust, but several studies are suggestive. Attfield and co-workers

(1982) observed no exposure-related differences in either respiratory symptoms or level of pulmonary function in analyses of cumulative NO_2 exposure among 630 miners at six potash mines that differed in the extent to which diesel equipment had been used. In contrast, Gamble and colleagues (1983) studied the pulmonary function of 259 salt miners in five salt mines that used diesel-powered equipment. They found small elevations in the prevalence of cough, phlegm and dyspnea when exposure was characterized either as cumulative exposure to NO_2 or respirable particulate (RP), or in which mine a subject had worked (ranked according to average levels of diesel exhaust surrogates, NO_2 and RP). Cumulative exposure to either NO_2 or RP, however, had no effect on either FVC or FEV_1.

Reger and colleagues (1982) measured the prevalence of respiratory symptoms and the level of pulmonary function among underground and 1646 surface miners at coal mines that differed with respect to the use of diesel-powered equipment. The prevalence of cough and phlegm, but not dyspnea, was elevated among underground miners in diesel mines relative to controls from non-diesel mines matched on age, height and years worked underground. Lung volumes and forced expiratory flow at 75% of FVC (FEF_{75}) were also reduced. The prevalence of obstructive disease, characterized by FEV_1/FVC less than 0.70, was the same in diesel and non-diesel miners (both underground and surface), but the prevalence of mild restrictive defects – per cent forced vital capacity (FVC %) predicted 0.66 to 0.80 – was greater among both underground and surface miners in diesel mines (4.1% versus 1.8% and 5.8 versus 3.1%, respectively).

Mortality from Nonmalignant Respiratory Disease

Cohen and Higgins (1995) reviewed 10 studies that provide information on mortality from nonmalignant respiratory disease among workers known or presumed to be exposed to diesel exhaust. For the most part, these are cohort studies of single occupational groups that examined multiple causes of death. Most studies did not find an increased risk of mortality from non-malignant respiratory disease, although the two studies which accounted for smoking (Boffetta *et al.*, 1988; Garshick *et al.*, 1987b) found small relative increases in COPD mortality of the same magnitude (about 20%), though both estimates were quite imprecise. The methodologic problems inherent in using death certificate date to study mortality from non-malignant respiratory disease are difficult and well described (Cohen and Higgins, 1995; Higgins and Thorn, 1989).

Epidemiologic Studies of General Populations

Until quite recently there were few studies that assessed the effects of diesel exhaust exposure, *per se*, in the general population, although the effects of exposure to undifferentiated (or poorly differentiated) vehicular traffic have been studied (Edwards *et al.*, 1994; Nakatsuka *et al.*, 1991; Weiland *et al.*, 1994; Wjst *et al.*, 1993). However, the high level of scientific interest in the health effects of fine particles has stimulated epidemiologic research on the non-cancer effects of diesel exhaust and encouraged the development and application of more specific indices of diesel exhaust exposure.

Investigators from University of Wageningen in the Netherlands have reported the results of a series of studies on the respiratory health of children attending schools near busy roadways (Brunekreef *et al.*, 1997; de Hartog *et al.*, 1997; van Vliet *et al.*, 1997). In these studies, the investigators have attempted to estimate the effects of exposure to diesel exhaust by relating health outcomes both to the volume of truck (vs. automobile) traffic and by measuring indoor levels of black smoke (BS); this is a reflectance measure of fine particulate black soot that is a surrogate for diesel exhaust under contemporary conditions in the Netherlands.

Brunekreef and colleagues (1997) studied the lung function of children living near heavily-trafficked roadways in six locations in the Netherlands. They measured lung function (FVC, FEV_1, PEF, and $FEF_{25-75\%}$) in 1200 children 7–12 years of age, and estimated the children's exposure to vehicular air pollution in terms of the volume of automobile and truck traffic on major roadways near their schools and indoor levels of particles – PM_{10} and black smoke (BS) and NO_2 – in their schools. Cross-sectional levels of FEV_1 were associated inversely with truck traffic density. All measured indices were inversely associated with the BS concentrations measured in the schools, and the strength of that relation itself varied inversely with proximity to the roadway. The associations were stronger among female students.

Van Vliet and colleagues (1997) conducted a cross-sectional study of respiratory symptoms in 1068 children in 13 schools located next to heavily-trafficked roads in the province of South Holland. The prevalence of chronic respiratory symptoms (cough, wheeze and physician-diagnosed asthma), ascertained by questionnaire, was inversely associated with distance from the roadway and both the density of truck traffic and in-school levels of BS were directly associated with reported symptom prevalence. As in Brunekreef's study, the reported associations were more pronounced in females. De Hartog and colleagues (1997) evaluated pulmonary function in relation to indices of traffic density and levels of pollution in this same population, and observed inverse associations between spirometric indices of lung function and distance of the school from the roadway, density of truck traffic and in-school levels of BS. They observed no association with automobile traffic density or NO_2.

Laboratory Studies Using Human Subjects

There are few laboratory studies of pulmonary function or non-allergic respiratory tract alterations in human subjects. Ruddell and colleagues conducted two controlled exposure studies. First, they exposed healthy volunteers in an exposure chamber to diesel exhaust and examined changes in lung function with dynamic spirometry. Finding no effect on FEV_1 and FVC (1994), they subsequently (1996) used more sensitive whole-body plethysmography to measure changes in lung function in healthy volunteers in an exposure chamber for one hour during light work. These volunteers were exposed to whole diesel exhaust or to diesel exhaust with a particle trap at the tail pipe. The particle trap reduced the number of particles by 46% without affecting other exhaust components. The main symptoms during exposure were eye and nose irritation and an unpleasant smell. Both airway resistance and specific airway resistance increased significantly during exposures. The particle trap did not reduce the symptoms or the amount of bronchoconstriction caused by diesel exhaust exposure. Thus, although exposure to diesel exhaust has been shown to cause temporary bronchoconstriction, the importance of the particles in this response and the long-term effects on pulmonary function have not been determined experimentally in humans.

Salvi and colleagues exposed 15 healthy volunteers (ages 21–28) to diluted diesel exhaust (PM_{10}=300 µg/m³, NO_2=1.6 ppm) and air for 1 h with moderate exercise. Lung function was measured immediately before and after the exposure, and bronchoscopy was performed and peripheral blood samples obtained 6 h following exposure. No exposure-related changes in pulmonary function were observed, but a pattern of inflammatory response, both pulmonary and systemic, was observed. The investigators reported increases in neutrophils, mast cells, CD4+ and CD8+ lymphocytes, up regulation of endothelial adhesion molecules, and increased LFA-1+ cells in bronchial tissue. They also reported increases in platelets and neutrophils in peripheral blood samples.

Laboratory Studies Using Animals

Watson and Green (1995) reviewed non-malignant effects of diesel exhaust reported in experimental animals in detail. Pulmonary function has been evaluated in a number of species. Both restrictive and obstructive patterns of dysfunction have been reported after chronic exposure. Decreases in lung volumes (Mauderly *et al.*, 1988; Moorman *et al.*, 1985; Vinegar *et al.*, 1981) and lung compliance (Heinrich *et al.*, 1986; Mauderly *et al.*, 1988) suggest restrictive disease and are compatible with parenchymal fibrosis. Decreases in flow rates (Lewis *et al.*, 1986) and increases in airway resistance (Heinrich *et al.*, 1986) suggest obstructive disease. However, exposure to diesel exhaust does not enhance bronchoconstrictive responses to acetylcholine (Heinrich *et al.*, 1986). Decreases in diffusing capacity (Mauderly *et al.*, 1988; Moorman *et al.*, 1985; Vinegar *et al.*, 1981) indicate inequalities in ventilation–perfusion relationships. In some – but not all – studies, the pulmonary function results correlate with the histopathology. Considering the focal nature of some of the histologic findings, the lack of correlation in some studies should not be surprising.

The effects of diesel exhaust exposure on non-specific host defenses have been studied in mice by assessing susceptibility to respiratory tract infections, with inconsistent results. Campbell and co-workers (1981) reported that acute and subacute exposures of mice to diesel exhaust caused an increase in mortality in response to β-hemolytic group C *Streptococcus pyogenes,* but not in response to A/PR8-34 influenza virus. Hahon and co-workers (1985) found no increase in mortality or other measures of influenza viral infection after one month of exposure to diesel exhaust; however, after longer exposure (3 and 6 months), pulmonary consolidation and virus growth were greater in diesel-exposed animals while interferon and hemagglutinin-antibody levels were depressed.

The most consistent finding relative to the potential of diesel exhaust exposure to depress nonspecific host defenses has been the decreased alveolar clearance of indicator particles (Chan *et al.*, 1984; Creutzenberg *et al.*, 1990; Griffis *et al.*, 1983; Heinrich *et al.*, 1986; Mauderly *et al.*, 1987b, 1989, 1994; Wolff *et al.*, 1987).

Watson and Green (1995) have reviewed the effect of chronic exposure to diesel exhaust on pulmonary inflammation, proliferation of epithelial cells, metaplasia and fibrosis; and this was mentioned earlier in this chapter relative to carcinogenesis in rat lungs. In general, inflammation, alveolar epithelial and bronchiolar hyperplasia, squamous metaplasia, bronchiolar-alveolar metaplasia, and fibrosis have been reported in multiple species chronically exposed to diesel exhaust. Rats tend to exhibit greater inflammation, epithelial hyperplasia and fibrosis than similarly exposed mice (Henderson *et al.*, 1988a,b), hamsters (Mauderly, 1994b), and monkeys (Nikula *et al.*, 1997).

Laboratory Studies of Diesel Exhaust and Aeroallergens

Diesel exhaust particles (DEP) can bind allergens such as the major grass pollen allergen under *in vitro* conditions (Knox *et al.*, 1997). Allergen molecules can attach to respirable particles and perhaps trigger respiratory allergies during episodes of air pollution.

Most laboratory studies of the potential role of diesel exhaust in allergic airway disease have focused on the ability of DEP and adsorbed polycyclic aromatic hydrocarbons to have an adjuvant activity. *In vitro* studies have shown that human peripheral monocytes and alveolar macrophages exposed to suspended DEP showed reduced phagocytosis and increased TNF-α release (Thomas *et al.*, 1995). Human airway epithelial cells exposed to DEP *in vitro* showed increased release of GM-CSF, a cytokine that influences eosinophil activity (Ohtoshi *et al.*, 1994). Organic extracts from DEP and phenanthrene, one of the major polycyclic aromatic hydrocarbons adsorbed to DEP, enhance *in vitro* production of IgE from IgE-secreting purified human B lymphocytes and from an Epstein–Barr virus-transformed human B lymphocyte line (Takenaka *et al.*, 1995; Tsien *et al.*, 1997). Pfützner and colleagues (1995) investigated the ability of DEP to influence the number and activity of IgE-secreting cells *in vitro*. They isolated peripheral blood mononuclear cells from patients with atopic eczema and from healthy, non-allergic individuals and exposed the cells to suspended DEP *in vitro*. The number of IgE-secreting cells was increased in some, but not all, of the DEP-exposed cell isolates from allergic eczema patients, and in none of the DEP cell isolates from controls. The IgE production paralleled the number of IgE-secreting cells. These results suggest that there may be individuals with enhanced susceptibility to DEP.

Human volunteers exposed to nasal challenges with DEP showed increased numbers of IgE-producing cells and IgE in nasal lavage (Diaz-Sanchez *et al.*, 1994). Diaz-Sanchez and colleagues (1994) have also shown enhanced nasal cytokine production in human volunteers exposed to nasal challenges with DEP. *In vitro* stimulation of cells recovered from nasal lavage fluid or stimulation of peripheral blood mononuclear cells with DEP did not increase cytokine mRNA production. These results show that DEP can act on non-B cells to enhance nasal IgE production, and they imply that the presence of these non-B cells in local tissues may be critical for the outcome of exposure to DEPs. The most recent work from this same laboratory (Diaz-Sanchez *et al.*, 1997) has shown that challenge with DEP combined with ragweed allergen enhances nasal ragweed-specific IgE and shifts nasal cytokine production to a T helper type 2 (TH$_2$) pattern in ragweed-sensitized patients.

Laboratory Studies Using Animals

The potential role of diesel exhaust in enhancing allergic rhinitis and asthma has been investigated experimentally. Kobayashi and Ito (1995) have shown that intranasal exposure of guinea pigs to diesel exhaust increases nasal airway resistance and augments nasal airway resistance and secretions in response to histamine. Most recent experimental work has focused on the possible adjuvant activity of diesel exhaust. Diesel exhaust particles increase the production of ovalbumin-specific IgE after repeated intranasal or intratracheal instillation in ovalbumin-sensitized and challenged mice (Takafuyi *et al.*, 1987; Takano *et al.*, 1992). Diesel exhaust particles also enhance antigen-specific IgE responses after repeated intraperitoneal injection of mice with DEP plus ovalbumin or Japanese cedar pollen (Muranaka *et al.*, 1986). Intranasal instillation of DEP and ovalbumin also caused increased bronchoalveolar lavage and lung tissue supernatant levels of IL-4, IL-5, GM-CSF and IL-2 (Takano *et al.*, 1997). Intranasal instillation of DEP and ovalbumin cause increased *in vitro* proliferation in response to ovalbumin and increased IL-4 production in cells from the cervical lymph nodes compared with mice instilled with ovalbumin alone (Fujimaki *et al.*, 1994, 1995). Inhalation of diesel exhaust has been shown to enhance the production of antigen-specific IgE antibody in mice through alteration of the cytokine network (Fujimaki *et al.*, 1997).

Attempts have been made to evaluate the relative importance of the particle core versus specific adsorbed chemicals to the adjuvant activity of diesel particles. Løvik and colleagues (1997) demonstrated that both DEP and carbon black injected in the foot pad in conjunction with ovalbumin can have an adjuvant activity on popliteal lymph node inflammation and systemic ovalbumin-specific IgE in mice. These results suggest that the particle core contributes to the adjuvant activity of DEP. However, Suzuki and colleagues (1993) showed that both pyrene and DEP have adjuvant activity on allergen-specific IgE antibody production when mice are immunized by intraperitoneal injection of ovalbumin or Japanese cedar pollen allergen with pyrene. These results suggest that chemical compounds contained in diesel soot can have adjuvant activity in mice. Maejima *et al.* (1987) exposed mice to intranasal particles (Kanto loam dust, fly ash, carbon black, DEP or aluminum hydroxide) followed by aerosolized Japanese cedar pollen allergens or dry pollen grains dropped onto the nose. They found that Japanese cedar pollen-specific IgE was produced earlier in mice treated with particles than in mice immunized with Japanese cedar pollen alone. There was no obvious relationship between particle composition, capacity to adsorb antigens, or particle size and the enhancement of IgE antibody production.

At the present time, the relative contributions of the particle core versus various adsorbed chemicals to the adjuvant activity of diesel particles are unresolved. Also unresolved is the relative TH_2 adjuvant potency or potential contribution of diesel particles compared with other inhaled particles in enhancing allergic respiratory disease. Lastly, and most importantly, it is not clear that inhaled diesel exhaust at environmental or occupational concentrations would have significant TH_2 adjuvant effects.

Sagai and colleagues (1996) investigated the potential for repeatedly instilled diesel exhaust particles to cause asthma-like symptoms. They found that diesel exhaust particles instilled intratracheally weekly for 16 weeks caused infiltration of inflammatory cells including eosinophils, proliferation of goblet cells, increased mucus secretion, and airway hyperresponsiveness to acetylcholine. These investigators did not determine if other particles could cause similar responses, but the results suggest that instilled diesel exhaust particles can cause asthma-like symptoms in mice.

Summary

Previous summaries of the toxicologic and epidemiologic evidence concerning non-cancer effects of diesel exhaust have identified effects of diesel exhaust exposure on a range of acute and chronic respiratory end-points, including inflammation, pulmonary function and respiratory symptoms (Cohen and Higgins, 1995; Watson and Green, 1995), although the evidence has, for some effects, been weak or inconsistent. We have also examined this evidence and, in addition, have reviewed new evidence on the possible interaction of diesel exhaust and aeroallergens and pulmonary function and respiratory symptoms in children. Overall, these new data provide additional evidence that acute and long-term exposure to diesel exhaust may be associated with non-malignant respiratory disease.

CURRENT ISSUES AND RESEARCH NEEDS

It is likely that current research on the health effects of diesel exhaust will continue to focus, at least to some extent, on lung cancer, due largely to the desire of US regulatory agencies to estimate a unit risk number for diesel exhaust as a basis for regulation. The questionable relevance of the rat data for most real-world exposure scenarios has highlighted the need for new epidemiologic studies whose results can be used to derive a unit risk estimate. New studies are needed that provide quantitative estimates of past exposures of study subjects in terms of some constituent of diesel exhaust. The NIOSH/NCI, collaborative study of underground miners will likely contribute such data for this highly exposed occupational group, but additional studies are needed to address exposures at lower levels, more akin to those experienced by the general population (Health Effects Institute, 1998).

Although lung cancer will remain a focus of research, it is likely that non-cancer health effects of diesel exhaust will assume a larger role in the future. The relative risks and benefits of diesel technology will need to be assessed in the context of diesel's contribution to particulate air pollution, where associations with acute and chronic non-malignant respiratory and, increasingly, cardiovascular disease have been shown. The work of Brunekreef and colleagues in the Netherlands on respiratory health in children and the emerging laboratory evidence on inflammatory responses and possible interaction with aeroallergens, both reviewed above, are recent examples of what will likely be a growing trend.

REFERENCES

Albert RE, Lewtas J, Nesnow S *et al.* (1983) Comparative potency method for cancer risk assessment: application to diesel particulate emissions. *Risk Anal* **3**: 101–117.

Ames BN and Gold LS (1990) Chemical carcinogenesis: too many rodent carcinogens. *Proc Natl Acad Sci USA* **87**(19): 7772–7776.

Ames RG, Attfield MD, Hankinson JL *et al.* (1982) Acute respiratory effects of exposure to diesel emissions in coal miners. *Am Rev Respir Dis* **125**(1): 39–42.

Ames RG, Reger RB and Hall DS (1984) Chronic respiratory effects of exposure to diesel emissions in coal mines. *J Occup Med* **39**: 389–394.

Attfield MD, Trabant GD and Wheeler RW (1982) Exposure to diesel fumes and dust at six potash mines. *Ann Occup Hyg* **26**: 817–831.

Battigelli MC, Mannella RJ and Hatch TF (1964) Environmental and clinical investigation of workmen exposed to diesel exhaust in railroad engine houses. *Ind Med* **3**: 121–124.

Bellmann B, Muhle H, Creutzenberg O *et al.* (1991) Lung clearance and retention of toner, utilizing a tracer technique, during chronic inhalation exposure in rats. *Fundam Appl Toxicol* **17**: 300–313.

Bhatia R, Lopipero P and Smith AH (1998) Diesel exhaust exposure and lung cancer. *Epidem* **9**: 84–91.

Boffetta P, Jourenkova N and Gustavsson P (1997) Cancer risk from occupational and environmental exposure to polycyclic aromatic hydrocarbons. *Cancer Cause Cont* **8**: 444–472.

Boffetta P, Stellman SD and Garfinkel L (1988) Diesel exhaust exposure and mortality among males in the American Cancer Society Prospective Study. *Am J Ind Med* **14**: 403–415.

Bond JA, Wolff RK, Harkema JR *et al.* (1988) Distribution of DNA adducts in the respiratory tract of rats exposed to diesel exhaust. *Toxicol Appl Pharmacol* **96**: 335–345.

Bond JA, Johnson NF, Snipes MB and Mauderly JL (1990a) DNA adduct formation in rat alveolar type II cells: cells potentially at risk for inhaled diesel exhaust. *Environ Mol Mutagen* **16**: 64–69.

Bond JA, Mauderly JL and Wolff RK (1990b) Concentration- and time-dependent formation of DNA adducts in lungs of rats exposed to diesel exhaust. *Toxicology* **60**: 127–135.

Borm PJA and Driscoll K (1996) Particles, inflammation and respiratory tract carcinogenesis. *Toxicol Lett* **88**: 109–113.

Brightwell J, Fouillet XLM, Cassano-Zoppi AL *et al.* (1986) Neoplastic and functional changes in rodents after chronic inhalation of engine exhaust emissions. In: Ishinishi N, Koizumi A, McClellan RO and Stober W (eds) *Carcinogenic and Mutagenic Effects of Diesel Engine Exhaust*. New York: Elsevier Science Publishing Co, pp 471–485.

Brightwell J, Fouillet X, Cassano-Zoppi AL *et al.* (1989) Tumors of the respiratory tract in rats and hamsters following chronic inhalation of engine exhaust emissions. *J Appl Toxicol* **9**: 23–31.

Brooks AL, Wolff RK, Royer RE *et al.* (1980) Biological availability of mutagenic chemicals associated with diesel exhaust particles. In: Pepelko WE, Danner RM and Clarke NA (eds) *Health Effects of Diesel Engine Emissions*. Cincinnati, OH: US Environmental Protection Agency. EPA/600/9-80/57a, pp 345–358.

Brooks DR, Geller AC, Chang J and Miller DR (1992) Occupation, smoking, and the risk of high-grade invasive bladder cancer in Missouri. *Am J Ind Med* **21**: 669–713.

Brooks SM, Weiss MA and Bernstein IL (1985) Reactive airways dysfunction syndrome (RADS). Persistent asthma syndrome after high level irritant exposures. *Chest* **88**(3): 376–384.

Brunekreef B, Janssen NA, de Hartog J *et al.* (1997) Air pollution from truck traffic and lung function in children living near motorways. *Epidemiology* **8**(3): 298–303.

Busby Jr WF and Newberne PM (1995) Diesel emissions and other substances associated with animal carcinogenicity. In: *Diesel Exhaust: A Critical Analysis of Emissions, Exposure, and Health Effects*. Cambridge, MA: Health Effects Institute.

Butterworth BE (1990) Consideration of both genotoxic and nongenotoxic mechanisms in predicting carcinogenic potential. *Mutat Res* **239**: 117–132.

California Environmental Protection Agency (Cal EPA) (1997) Public and Scientific Review Draft. *Health Risk Assessment for Diesel Exhaust*. Berkeley, CA: Office of Environmental Health Hazard Assessment, California Environmental Protection Agency.

Campbell KL, George EL and Washington Jr IS (1981) Enhanced susceptibility to infection in mice after exposure to dilute exhaust from light duty diesel engines. *Environ Int* **5**: 377–382.

Cass GR and Gray HA (1995) Regional emissions and atmospheric concentrations of diesel engine particulate matter: Los Angeles as a case study. In: *Health Effects of Diesel Engine Emissions: Characterization and Critical Analysis*. Cambridge, MA: The Health Effects Institute.

Chan TL, Lee PS and Hering WE (1984) Pulmonary retention of inhaled diesel particles after prolonged exposures to diesel exhaust. *Fundam Appl Toxicol* **4**: 624–631.

Chen JL and Fayerweather WE (1988) Epidemiologic study of workers exposed to titanium dioxide. *J Occup Med* **30**(12): 937–942.

Coggon D, Pannet B and Acheson ED (1984) Use of job-exposure matrix in an occupational analysis of lung and bladder cancer on the basis of death certificates. *J Natl Cancer Inst* **72**: 61–65.

Cohen AJ and Higgins MWP (1995) Health effects of diesel exhaust: epidemiology. In: *Diesel Exhaust: A Critical Analysis of Emissions, Exposure, and Health Effects.* A Special Report of the Institute's Working Group. Cambridge, MA: Health Effects Institute.

Cohen SM and Ellwein LB (1990) Cell proliferation in carcinogenesis. *Science* **249**: 1007–1011.

Creutzenberg O, Bellmann B, Heinrich U *et al.* (1990) Clearance and retention of inhaled diesel exhaust particles, carbon black, and titanium dioxide in rats at lung overload conditions. *J Aerosol Med* **21**(Suppl): S455–S458.

Crump KS, Lambert T and Chen C (1991) *Assessment of Risk from Exposure to Diesel Engine Emissions.* Report to the US Environmental Protection Agency. Washington, DC: Office of Health Assessment, U.S. Environmental Protection Agency. Contract 68-02-4601 (Work Assignment No 182, July).

Cuddihy RG, Griffith WC and McClellan RO (1984) Health risks from light-duty diesel vehicles. *Environ Sci Technol* **18**: 14a–21a.

Dagle GE, Wehner AP, Clarke ML and Buschbom RL (1986) Chronic inhalation exposure of rats to quartz. In: Goldsmith DF, Winn DM and Shy CM (eds) *Silica, Silicosis and Cancer.* New York: Praeger, pp 255–266.

de Hartog JJ, van Vliet PH, Brunekreef B *et al.* (1997) Relationship between air pollution due to traffic, decreased lung function, and airways symptoms in children (in Dutch). *Ned Tijdschr Geneeskd* **141**(38): 1814–1818.

Diaz-Sanchez D, Dotson AR, Takenaka H and Saxon A (1994) Diesel exhaust particles induce local IgE production *in vivo* and alter the pattern of IgE messenger RNA isoforms. *J Clin Invest* **94**: 1417–1425.

Diaz-Sanchez D, Tsien A, Casillas A *et al.* (1996) Enhanced cytokine production in human beings after *in vivo* challenge with diesel exhaust particles. *J Allergy Clin Immunol* **98**: 114–123.

Diaz-Sanchez D, Tsien A, Flemming A and Saxon A (1997) Combined diesel exhaust particulate and ragweed allergen challenge markedly enhances human *in vivo* ragweed-specific IgE and skews cytokine production to a T helper cell 2-type pattern. *J Immunol* **158**: 2406–2413.

Driscoll KE, Carter JM, Oberdörster G *et al.* (1995) Inflammation, cytokine expression, and mutagenesis in rat lung after carbon black inhalation. *Toxicology* **15**(Abstract): 46.

Driscoll KE, Howard BW, Carter JM *et al.* (1996) Alpha-quartz-induced chemokine expression by rat lung epithelial cells: effects of *in vivo* and *in vitro* particle exposure. *Am J Pathol* **149**(5): 1627–1637.

Drozdowicz BZ and Kelly CM (1989) Genetic and animal toxicity testing of solvent refined coal-I (SRC-I) products, intermediates, and waste materials. Final Report: DOE-OR-21446-T1, Appendix C, *Lifetime Inhalation Carcinogenicity Studies of First-Stage Solid Produce (SRC Solids).* Allentown, PA: Air Products and Chemicals, Inc.

Edwards J, Waters S and Griffiths RK (1994) Hospital admissions for asthma in pre-school children: relationship to major roads in Birmingham, United Kingdom. *Arch Environ Health* **49**: 223–227.

Emmelin A, Nyström and Wall S (1993) Diesel exhaust exposure and smoking: a case-referent study of lung cancer among Swedish dock workers. *Epidemiology* **4**: 237–244.

Finch GL, Hoover MD, Hahn FF *et al.* (1996) Animal models of beryllium-induced lung disease. *Environ Health Perspect* **104**(Suppl 9).

Fujimaki H, Nohara O, Ichinose T *et al.* (1994) IL-4 production in mediastinal lymph node cells in mice intratracheally instilled with diesel exhaust particulates and antigen. *Toxicology* **92**(1–3): 261–268.

Fujimaki H, Saneyoshi K, Nohara O *et al.* (1995) Intranasal instillation of diesel exhaust particulates and antigen in mice modulated cytokine productions in cervical lymph node cells. *Int Arch Allergy Immunol* **108**(3): 268–273.

Fujimaki H, Saneyoshi K, Shiraishi F *et al.* (1997) Inhalation of diesel exhaust enhances antigen-specific IgE antibody production in mice. *Toxicology* **116**(1–3): 227–233.

Gallagher J, Heinrich U, George M *et al.* (1994) Formation of DNA adducts in rat lung following chronic inhalation of diesel emissions, carbon black and titanium dioxide particles. *Carcinogenesis* **15**(7): 1291–1299.

Gamble J, Jones W, Hudak J and Merchant J (1978) Acute changes in pulmonary function in salt miners. *Industrial Hygiene for Mining and Tunneling.* Proceedings of an ACGIH Topical Symposium.

Gamble J, Jones W and Hudak J (1983) An epidemiological study of salt miners in diesel and nondiesel mines. *Am J Ind Med* **4**(3): 435–458.

Gamble J, Jones W and Minshall S (1987a) Epidemiological–environmental study of diesel bus garage workers: acute effects of NO_2 and respirable particulate on the respiratory system. *Environ Res* **42**: 201–214.

Gamble J, Jones W and Minshall S (1987b) Epidemiological–environmental study of diesel bus garage workers: chronic effects of diesel exhaust on the respiratory system. *Environ Res* **44**: 6–17.

Garshick E, Schenker MB, Muñoz A *et al.* (1987a) A case-control study of lung cancer and diesel exhaust exposure in railroad workers. *Am Rev Respir Dis* **135**: 1242–1248.

Garshick E, Schenker MB, Smith TJ and Speizer FE (1987b) A case-control study of respiratory disease mortality and diesel exhaust exposure in railroad workers (abstract). *Am Rev Respir Dis* **135**(4 Part 2): A339.

Garshick E, Schenker MB, Muñoz A *et al.* (1988) A retrospective cohort study of lung cancer and diesel exhaust exposure in railroad workers. *Am Rev Respir Dis* **137**: 820–825.

Gerde P, Medinsky MA and Bond JA (1991) Particle-associated polycyclic aromatic hydrocarbons – a reappraisal of their possible role in pulmonary carcinogenesis. *Toxicol Appl Pharmacol* **108**(1): 1–13.

Glaser U, Hochrainer D, Otto FJ and Oldiges H (1990) Carcinogencity and toxicity of four cadmium compounds inhaled by rats. *Toxicol Environ Chem* **27**: 153–162.

Green GM and Watson AY (1995) Relationship between exposures to diesel emissions and dose to the lung. In: *Diesel Exhaust: A Critical Analysis of Emissions, Exposure, and Health Effects.* Cambridge, MA: Health Effects Institute, pp 165–184.

Griffis LC, Wolff RK, Henderson RF *et al.* (1983) Clearance of diesel soot particles from rat lung after a subchronic diesel exhaust exposure. *Fundam Appl Toxicol* **3**(2): 99–103.

Groth DH, Stettler LE, Burg JR *et al.* (1986) Carcinogenic effects of antimony trioxide and antimony ore concentrate in rats. *J Toxicol Environ Health* **18**(4): 607–626.

Gustafsson L, Wall S, Larsson LG and Skog B (1986) Mortality and cancer incidence among Swedish dock workers – a retrospective cohort study. *Scand J Work Environ Health* **12**: 22–26.

Gustavsson P, Plato N, Lidstrom EB and Hogstedt C (1990) Lung cancer and exposure to diesel exhaust among bus garage workers. *Scand J Work Environ Health* **16**(5): 348–354.

Hahon N, Booth JA, Green F and Lewis TR (1985) Influenza virus infection in mice after exposure to coal dust and diesel engine emissions. *Environ Res* **37**(1): 44–60.

Health Effects Institute (1998) Request for Applications. Spring 1998 Research Agenda. RFA 98-3: Epidemiologic investigations of human populations exposed to diesel engine emissions: feasibility studies. Cambridge, MA: Health Effects Institute.

Heinrich U, Peters O, Creutzenberg O *et al.* (1994) Inhalation exposure of rats to tar/pitch condensation aerosol or carbon black alone or in combination with irritant gases. In: Mohr U, Dungworth DL, Mauderly JL and Oberdörster G (eds) *Toxic and Carcinogenic Effects of Solid Particles in the Respiratory Tract.* Washington, D.C.: ILSI Press, pp 433–442.

Heinrich U, Fuhst R, Dasenbrock C *et al.* (1992) Long term inhalation exposure of rats and mice to diesel exhaust, carbon black and titanium dioxide. Monterey, CA. In: *The Ninth Health Effects Institute Annual Conference Program.* Cambridge, MA: Health Effects Institute.

Heinrich U, Fuhst R, Rittinghausen S *et al.* (1995) Chronic inhalation exposure of Wistar rats, and two different strains of diesel engine exhaust, carbon black and titanium dioxide. *Inhal Toxicol* **7**: 533–556.

Heinrich U, Muhle H, Takenaka S *et al.* (1986) Chronic effects on the respiratory tract of hamsters, mice and rats after long-term inhalation of high concentrations of filtered and unfiltered diesel engine emissions. *J Appl Toxicol* **6**(6): 383–395.

Hemminki K and Pershagen G (1994) Cancer risk of air pollution: epidemiological evidence. *Environ Health Perspect* **102**(Suppl): 92.

Henderson RF, Leung HW, Harmsen AG and McClellan RO (1988a) Species differences in release of arachidonate metabolites in response to inhaled diluted diesel exhaust. *Toxicol Lett* **42**(3): 325–332.

Henderson RF, Pickrell JA, Jones RK *et al.* (1988b) Response of rodents to inhaled diluted diesel exhaust: biochemical and cytological changes in bronchoalveolar lavage fluid and in lung tissue. *Fundam Appl Toxicol* **11**(3): 546–567.

Heuper WC (1958) Experimental studies in metal cancerigenesis. IX. Pulmonary lesions in guinea pigs and rats exposed to prolonged inhalation of powdered metallic nickel. *AMA Arch Pathol* **65**: 600–607.

Hext PM (1994) Current perspectives on particulate induced pulmonary tumours. *Hum Exp Toxicol* **13**(10): 700–715.

Higgins ML and Thom T (1989) Incidence, prevalence, and mortality: intra- and intercountry differences. In: Hensley MJ and Saunders NA (eds) *Clinical Epidemiology of Chronic Obstructive Pulmonary Disease*. New York: Marcel Dekker.

Hoar SK and Hoover R (1985) Truck driving and bladder cancer mortality in rural New England. *J Natl Cancer Inst* **74**(4): 771–774.

Hodgson JT and Jones RD (1985) A mortality study of carbon black workers employed at five United Kingdom factories between 1947 and 1980. *Arch Environ Health* **40**(5): 261–268.

Holland LM, Wilson JS, Tillery MI and Smith DM (1986) Lung cancer in rats exposed to fibrogenic dusts. In: Goldsmith DF, Winn DM and Shy CM (eds) *Silica, Silicosis, and Cancer*. New York: Praeger, pp 267–270.

Howe GR, Burch JD, Miller AB *et al.* (1980) Tobacco use, occupation, coffee, various nutrients, and bladder cancer. *J Natl Cancer Inst* **64**(4): 701–713.

Howe GR, Fraser D, Lindsay J *et al.* (1983) Cancer mortality (1965–77) in relation to diesel fume and coal exposure in a cohort of retired railway workers. *J Natl Cancer Inst* 70: 1015–1019.

Hsieh LL, Wong D, Heisig V *et al.* (1986) Analysis of genotoxic components in diesel engine emissions. *Dev Toxicol Environ Sci* **13**: 223–232.

Huisingh JL, Bradow R, Jungers R *et al.* (1978) Application of bioassay to the characterization of diesel particle emissions. In: Waters MD, Nesnow S, Huisingh JL *et al.* (eds) *Application of Short-term Bioassays in the Fractionation and Analysis of Complex Environmental Mixtures*. New York: Plenum Press, pp 381–418.

Intergovernmental Panel on Climate Change (IPCC) (1996) Climate change 1995. The science of climate change. Contribution of Working Group I to the Second Annual Assessment Report of the IPCC. Cambridge, NY: Cambridge University Press.

International Agency for Research on Cancer (IARC) (1984) IARC Monographs on the Evaluation of Carcinogenic Risks to Humans, vol 34: *Polynuclear Aromatic Compounds*, Part 3. Industrial exposures in aluminum production, coal gasification, coke production, and iron and steel founding. Lyons, France: IARC.

International Agency for Research on Cancer (IARC) (1987a) IARC Monographs on the Evaluation of Carcinogenic Risks to Humans, Supple 7: *Overall Evaluations of Carcinogenecity, An Updating of IARC Monographs*. Lyons, France: International Agency for Research on Cancer, pp 1–42.

International Agency for Research on Cancer (IARC) (1987b) IARC Monographs on the Evaluation of Carcinogenic Risks to Humans, vol 42: *Silica and Some Silicates*. Lyons, France: IARC.

International Agency for Research on Cancer (IARC) (1989a) IARC Monographs on the Evaluation of Carcinogenic Risks to Humans, vol 46: *Diesel and gasoline engine exhausts and some nitroarenes*. Lyons, France: (IARC).

International Agency for Research on Cancer (IARC) (1989b) IARC Monographs on the Evaluation of Carcinogenic Risks to Humans, vol 47: *Some Organic Solvents, Resin Monomers and Related Compounds, Pigments, and Occupational Exposures in Paint Manufacture and Painting*. Lyon, France: World Health Organization, IARC.

International Agency for Research on Cancer (IARC) (1990) IARC Monographs on the Evaluation of Carcinogenic Risks to Humans, vol 49: *Chromium, Nickel, and Welding*. Lyons, France: IARC.

International Agency for Research on Cancer (IARC) (1993) IARC Monographs on the Evaluation of Carcinogenic Risks to Humans, vol 58: *Beryllium, Cadmium, Mercury, and Exposures in the Glass Manufacturing Industry*. Lyons, France: IARC.

International Agency for Research on Cancer (IARC) (1996) IARC Monographs on the Evaluation of Carcinogenic Risks to Humans, vol 65: *Printing Processes and Printing Inks, Carbon Black and Some Nitro Compounds*. Lyon, France: World Health Organization, IARC.

International Agency for Research on Cancer (IARC) (1997) IARC Monographs on the Evaluation of Carcinogenic Risks to Humans, vol 68: *Silica, Some Silicates, Coal Dust and Para-Aramid Fibrils*. Lyon, France: World Health Organization, IARC.

Iscovich J, Castelletto R, Esteve J *et al.* (1987) Tobacco smoking, occupational exposure and bladder cancer in Argentina. *Int J Cancer* **40**(6): 734–740.

Ishinishi N, Kuwabara N, Nagase S *et al.* (1986) Long-term inhalation studies on effects of exhaust from heavy and light duty diesel engines on F344 rats. *Dev Toxicol Environ Sci* **13**: 329–348.

Iwai K, Udagawa T, Yamagishi M and Yamada H (1986) Long-term inhalation studies of diesel exhaust on F344 SPF rats. Incidence of lung cancer and lymphoma. *Dev Toxicol Environ Sci* **13**: 349–360.

Jackson JH, Gajewski E, Schraufstatter IU *et al.* (1989) Damage to the bases in DNA induced by stimulated human neutrophils. *J Clin Invest* **84**(5): 1644–1649.

Jacobsen M, Smith TA, Hurley JF *et al.* (1988) *Respiratory Infections in Coal Miners Exposed to Nitrogen Oxides.* Cambridge, MA: Health Effects Institute. Research Report Number 18.

Jensen OM, Wahrendorf J, Blettner M *et al.* (1987) The Copenhagen case–control study of bladder cancer: role of smoking in invasive and non-invasive bladder tumours. *J Epidemiol Community Health* **41**(1): 30–36.

Johnston AM, Buchanon D, Robertson A and Miller BG (1997) Investigation of the possible association between exposure to diesel exhaust particulates in British coalmines and lung cancer. Edinburgh, UK: Institute of Occupational Medicine. TM/97/08. Technical Memorandum Series.

Jorgensen H and Svensson A (1970) Studies on pulmonary function and respiratory tract symptoms of workers in an iron ore mine where diesel trucks are used underground. *J Occup Med* **12**(9): 348–354.

Kahn G, Orris P and Weeks J (1988) Acute over-exposure to diesel exhaust: report of 13 cases. *Am J Ind Med* **13**: 405–406.

Kaplan HL, Springer KJ and MacKenzie WF (1983) Studies of potential health effects of long-term exposure to diesel exhaust emissions. San Antonio, TX: Southwest Research Institute. Final Report No 01-0750-103 (SWRI) and No 1239 (SFRE).

Keane MJ, Xing SG, Harrison JC *et al.* (1991) Genotoxicity of diesel-exhaust particles dispersed in simulated pulmonary surfactant. *Mutat Res* **260**(3): 233–238.

Klonne DR, Burns JM, Halder CA *et al.* (1987) Two-year inhalation toxicity study of petroleum coke in rats and monkeys. *Am J Ind Med* **11**(3): 375–389.

Knox RB, Suphioglu C, Taylor P *et al.* (1997) Major grass pollen allergen Lol p 1 binds to diesel exhaust particles: implications for asthma and air pollution. *Clin Exp Allergy* **27**: 246–251.

Kobayashi T and Ito T (1995) Diesel exhaust particulates induce nasal mucosal hyperresponsiveness to inhaled butamine aerosol. *Fundam Appl Toxicol* **27**(295): 202.

Kotin P, Falk HL and Thomas M (1955) Aromatic hydrocarbons: III. Presence in the particulate phase of diesel-engine exhausts and the carcinogenicity of exhaust extracts. *Ind Health* **11**:113–120.

Lee KP, Trochimowicz HJ and Reinhard CF (1985) Pulmonary response of rats exposed to titanium dioxide (TiO_2) by inhalation for 2 years. *Toxicol Appl Pharmacol* **79**: 179–192.

Lee KP, Kelly DP, Schneider PW and Trochimowicz HJ (1986) Inhalation toxicity study on rats exposed to titanium tetrachloride atmospheric hydrolysis products for 2 years. *Toxicol Appl Pharmacol* **83**: 30–45.

Lee KP, Ulrich CE, Geil RG and Trochimowicz HJ (1988) Effects of inhaled chromium dioxide dust on rats exposed for two years. *Fundam Appl Toxicol* **10**: 125–145.

Lewis TR, Green FHY, Moorman WJ *et al.* (1986) A chronic inhalation toxicity study of diesel engine emissions and coal dust, alone and combined. In: Ishinishi N, Koizumi A, McClellan RO and Stöber W (eds) *Carcinogenic and Mutagenic Effects of Diesel Engine Exhaust.* New York: Elsevier Science, pp 361–380.

Lewis TR, Green FHY, Moorman WJ *et al.* (1989) A chronic inhalation toxicity study of diesel engine emissions and coal dust, alone and combined. *J Am Coll Toxicol* **8**: 345–375.

Løvik M, Høgseth A-K, Gaarder PI *et al.* (1997) Diesel exhaust particles and carbon black have adjuvant activity on the local lymph node response and systemic IgE production to ovalbumin. *Toxicology* **121**: 165–178.

MacFarland HN, Coate WB, Disbennett DB and Ackerman LG (1982) Long-term inhalation studies with raw and processed shale dusts. *Ann Occup Hyg* **26**(1–4): 213–225.

Maejima K, Tamura K, Taniguchi Y *et al.* (1997) Comparison of the effects of various fine particles on IgE antibody production in mice inhaling Japanese cedar pollen allergens. *J Toxicol Environ Health* **52**: 231–248.

Martin JC, Daniel H and LeBouffant L (1977) Short- and long-term experimental study of the toxicity of coal-mine dust and of some of its constituents. In: Walton WH (ed) *Inhaled Particles* IV. Oxford: Pergamon Press, pp 361–370.

Mauderly JL (1992) Diesel exhaust. In: *Environmental Toxicants: Human Exposures and Their Health Effects*. New York: Van Nostrand Reinhold.

Mauderly JL (1994a) Contribution of inhalation bioassays to the assessment of human health risks from solid airborne particles. In: Mohr U, Dungworth DL, Mauderly JL and Oberdörster G (eds) *Toxic and Carcinogenic Effects of Solid Particles in the Respiratory Tract*. Washington, DC: International Life Sciences Institute Press, pp 355–365.

Mauderly JL (1994b) Noncancer pulmonary effects of chronic inhalation exposure of animals to solid particles. In: Mohr U, Dungworth DL, Mauderly JL and Oberdörster G (eds) *Toxic and Carcinogenic Effects of Solid Particles in the Respiratory Tract*. Washington, DC: International Life Sciences Institute, pp 43–55.

Mauderly JL (1995) Current assessment of the carcinogenic hazard of diesel exhaust. *Toxicol Environ Chem* **49**: 167–180.

Mauderly JL (1997) Relevance of particle-induced rat lung tumors for assessing lung carcinogenic hazard and human lung cancer risk. *Environ Health Perspect* **105**(Suppl 5): 1337–1346.

Mauderly JL, Jones RK, Griffith WC *et al.* (1987a) Diesel exhaust is a pulmonary carcinogen in rats exposed chronically by inhalation. *Fundam Appl Toxicol* **9**: 208–221.

Mauderly JL, Bice DE, Carpenter RL *et al.* (1987b) *Effects of Inhaled Nitrogen Dioxide and Diesel Exhaust on Developing Lungs*. Cambridge, MA: Health Effects Institute. Research Report No 8.

Mauderly JL, Jones RF, Henderson RF *et al.* (1988) Relationship of lung structural and functional changes to accumulation of diesel exhaust particles. In: Dodgson J, McCallum RI, Bailey MR and Fisher DR (eds) *Inhaled Particles* VI. Oxford: Pergamon Press, pp 659–669.

Mauderly JL, Bice DE, Cheng YS *et al.* (1989) *Influence of Experimental Pulmonary Emphysema on Toxicological Effect from Inhaled Nitrogen Dioxide and Diesel Exhaust*. Cambridge, MA: Health Effects Institute. Research Report No. 30.

Mauderly JL, Cheng YS and Snipes MB (1990) Particle overload in toxicological studies: friend or foe? *J Aerosol Med* **3**(1): S169–S187.

Mauderly JL, Snipes MB, Barr EB *et al.* (1994) *Pulmonary Toxicity of Inhaled Diesel Exhaust and Carbon Black in Chronically Exposed Rats*. Part I. Neoplastic and nonneoplastic lung lesions. Cambridge, MA: Health Effects Institute. Research Report No. 68.

Mauderly JL, Banas DA, Griffith WC *et al.* (1996) Diesel exhaust is not a pulmonary carcinogen in CD-1 mice exposed under conditions carcinogenic to F344 rats. *Fundam Appl Toxicol* **30**: 233–242.

McClellan RO (1996) Lung cancer in rats from prolonged exposure to high concentrations of carbonaceous hazard and human lung cancer risk. *Inhal Toxicol* **8**(Suppl): 193–226.

Moorman WJ, Clark JC, Pepelko WE and Mattox J (1985) Pulmonary function responses in cats following long-term exposure to diesel exhaust. *J Appl Toxicol* **5**: 301–305.

Muhle H, Bellmann B, Creutzenberg O *et al.* (1991) Pulmonary response to toner upon chronic inhalation exposure in rats. *Fundam Appl Toxicol* **17**: 280–299.

Muranaka M, Suzuki S, Koizumi K *et al.* (1986) Adjuvant activity of diesel-exhaust particulates for the production of IgE antibody in mice. *J Allergy Clin Immunol* **77**: 616–623.

Nakatsuka H, Watanabe T, Ikeda M *et al.* (1991) Comparison of the health effects between indoor and outdoor air pollution in Northeastern Japan. *Environ Int* **17**: 51–59.

National Research Council (NRC)(1998) Review of the research program of the partnership for a new generation of vehicles. Fourth Report. Washington, DC: National Academy Press.

National Toxicology Program (NTP) (1993) Toxicology and carcinogenesis studies of talc (CAS No 14807-96-6) (Non-asbestiform) in F344/N rats and B6C3F1 mice (Inhalation Studies). Research Triangle Park, NC: NTP Technical Report 421.

National Toxicology Program (NTP) (1994a) Toxicology and carcinogenesis studies of nickel subsulfide (CAS No 12035-72-2) in F344/N rats and B6C3F1 mice (Inhalation Studies). Research Triangle Park, NC: NTP Technical Report 453.

National Toxicology Program (NTP) (1994b) Toxicology and carcinogenesis studies of nickel oxide (CAS No. 1313-99-1) in F344/N rats and B6C3F1 mice (Inhalation Studies). Research Triangle Park, NC: NTP Technical Report 451.

Nauss K and The Diesel Working Group (1995) Critical issues in assessing the carcinogenicity of diesel exhaust: a synthesis of current knowledge. In: *Diesel Exhaust: A Critical Analysis of Emissions, Exposure, and Health Effects*. Cambridge, MA: The Health Effects Institute.

Nikula KJ, Snipes MB, Barr EB *et al.* (1995) Comparative pulmonary toxicities and carcinogenicities of chronically inhaled diesel exhaust and carbon black in F344 rats. *Fundam Appl Toxicol* **25**(1): 80–94.

Nikula KJ, Avila KJ, Griffith WC and Mauderly JL (1997) Sites of particle retention and lung tissue responses to chronically inhaled diesel exhaust and coal dust in rats and cynomolgus monkeys. *Environ Health Perspect* **105**(Suppl 5): 1231–1234.

NIOSH (1988) *Carcinogenic Effects of Exposure to Diesel Exhaust*. Atlanta, GA: Centers for Disease Control and Prevention. DHHS [NIOSH] Pub. No.88-116. Current Intelligence Bulletin 50.

Notani PN, Priyabala S, Kasturi J and Balakrishnan V (1993) Occupation and cancers of the lung and bladder: a case-control study in Bombay. *Int J Epidemiol* **22**(2): 185–191.

Ohtoshi T, Takizawa H, Sakamaki C *et al.* (1994) [Cytokine production by human airway epithelial cells and its modulation]. *Nippon Kyobu Shikkan Gakkai Zasshi – Jap J Thor Dis* **32**(Suppl 72)

Parent M-E, Siemiatycki J and Renaud G (1996) Case-control study exposure to carbon black in the occupational setting and lung cancer. *Am J Ind Med* **30**: 285–292.

Pepelko WE (1984) Experimental respiratory carcinogenesis in small laboratory animals. *Environ Res* **33**(1): 144–188.

Pepelko WE and Peirano WB (1983) Health effects of exposure to diesel engine emissions: a summary of animal studies conducted by the US Environmental Protection Agency's Health Effects Research Laboratories at Cincinnati. *J Am Coll Toxicol* **2**(4): 253–306.

Pfützner W, Thomas P and Przybilla B (1995) Influence of suspended diesel exhaust particles (DEP) on number and activity of IgE-secreting cells *in vivo*. *J Allergy Clin Immunol* **95**: 226.

Pope A and Dockery D (1996) Epidemiology of chronic health effects: cross-sectional studies. In: Wilson R and Spengler J (eds) *Particles in Our Air: Concentrations and Health Effects*. 7. Boston, MA: Harvard University Press, pp 149–167.

Preston-Martin S, Pike MC, Ross RK *et al.* (1990) Increased cell division as a cause of human cancer. *Cancer Res* **50**(23): 7415–7421.

Purdham JT, Holness DL and Pilger CW (1987) Environmental and medical assessment of stevedores employed in ferry operations. *Appl Ind Hygiene* **2**: 133–138.

Raffle P (1957) The health of the worker. *Br J Ind Med* **14**: 73–80.

Randerath K, Putman KL, Mauderly JL *et al.* (1995) *Pulmonary Toxicity of Inhaled Diesel Exhaust and Carbon Black in Chronically Exposed Rats*, Part II. DNA Damage. Cambridge, MA: Health Effects Institute, p 68.

Reger R, Hancock J, Hankinson J *et al.* (1982) Coal miners exposed to diesel exhaust emissions. *Ann Occup Hyg* **26**: 799–815.

Risch HA, Burch JD, Miller AB *et al.* (1988) Occupational factors and the incidence of cancer of the bladder in Canada. *Br J Ind Med* **45**: 361–367.

Robertson JM and Inman KJ (1996) Mortality in carbon black workers in the United States. *J Occup Environ Med* **38**(6): 569–570.

Rudell B, Sandstrom T, Hammarstrom U *et al.* (1994) Evaluation of an exposure setup for studying effects of diesel exhaust in humans. *Int Arch Occup Environ Health* **66**(2): 77–83.

Rudell B, Ledin MC, Hammarstrom U *et al.* (1996) Effects on symptoms and lung function in humans experimentally exposed to diesel exhaust. *Occup Environ Med* **53**(10): 658–662.

Rushton L, Alderson MR and Nagarajah CR (1983) Epidemiological survey of maintenance workers in London Transport Executive bus garages and Chiswick Works. *Br J Ind Med* **40**(3): 340–345.

Saffiotti U, Williams AO, Lambert ND *et al.* (1995) Carcinogenesis by crystalline silica: animal, cellular, and molecular studies. In: Castranova V, Vallyathan V and Wallace WE (eds) *Silica and Silica-Induced Lung Diseases*. Boca Raton, FL: CRC Press, pp 345–381.

Sagai M, Furuyama A and Ichinose T (1996) Biological effects of diesel exhaust particles (DEP). III. Pathogenesis of asthma like symptoms in mice. *Free Radical Biol Med* **21**(2): 199–209.

Salvi S, Blomberg A, Rudell B *et al.* Acute inflammatory responses in the airways and peripheral blood following short term exposure to diesel exhaust in healthy human volunteers. (Unpublished Manuscript).

Säverin R, Bräunlich A, Enderlein G and Heuchert G (1998) Cohort study on the effect of diesel motor emissions on lung cancer mortality in potash mining (Unpublished cohort study in German). NIH Library Translation. US National Institutes of Health. NIH-98-120.

Sawyer RF and Johnson JH (1995) Diesel emissions and control technology. *Diesel Exhaust: A Critical Analysis of Emissions, Exposure, and Health Effects.* Cambridge, MA: Health Effects Institute.

Scheepers PT and Bos RP (1992) Combustion of diesel fuel from a toxicological perspective. I. Origin of incomplete combustion products. *Int Arch Occup Environ Health* **64**(3): 149–161.

Schenker MB, Smith TF, Muñoz A *et al.* (1984) Diesel exposure and mortality among railway workers: result of a pilot study. *Br J Ind Med* **41**: 320–327.

Schepers GW (1961) Neoplasia experimentally induced by beryllium compounds. *Prog Exp Tumor Res* **2**: 203–244.

Schepers GW (1971) Lung tumors of primates and rodents. II. *Ind Med Surg* **40**(2): 23–31.

Schlesinger RB and Lippmann M (1978) Selective particle deposition and bronchogenic carcinoma. *Environ Res* **15**(3): 424–431.

Shirnamé-Moré L (1995) Genotoxicity of diesel emissions. Part 1. Mutagenicity and other genetic effects. In: *Diesel Exhaust: A Critical Analysis of Emissions, Exposure, and Health Effects.* Cambridge, MA: Health Effects Institute, pp 185–220.

Siak JS, Chan TL and Lees PS (1981) Diesel particulate extracts in bacterial test systems. *Environ Int* **5**: 243–248.

Siemiatycki J, Gerin M, Stewart P *et al.* (1988) Associations between several sites of cancer and ten types of exhaust and combustion products. *Scand J Work Environ Health* **14**: 79–90.

Silverman D (1998) Is diesel exhaust a human lung carcinogen? *Epidemiology* **9**: 4–6.

Silverman DT, Hoover RN, Albert S and Graff KM (1983) Occupation and cancer of the lower urinary tract in Detroit. *J Natl Cancer Inst* **70**: 237–245.

Silverman DT, Hoover RN, Mason TJ and Swanson GM (1986) Motor exhaust-related occupations and bladder cancer. *Cancer Res* **46**(4 Pt 2): 2113–2116.

Snipes MB (1989) Long-term retention and clearance of particles inhaled by mammalian species. *Crit Rev Toxicol* **20**(3): 175–211.

Stayner S, Dankovic D, Smith R and Steenland K (1998) Predicted lung cancer risk among miners exposed to diesel exhaust particles. *Am J Ind Med* **34**: 207–219.

Steenland K, Burnett C and Osorio AM (1987) A case–control study of bladder cancer using city directories as a source of occupational data. *Am J Epidemiol* **126**(2): 247–257.

Steenland NK, Silverman DT and Hornung RW (1990) Case–control study of lung cancer and truck driving in the Teamsters Union. *Am J Public Health* **80**: 670–674.

Steenland K, Deddens J and Stayner L (1998) Diesel exhaust and lung cancer in the trucking industry: exposure-response analyses and risk assessment. *Am J Ind Med* **34**: 220–228.

Sunderman FW and Severi L (1966) International Conference on Lung Tumors in Animals. Proceedings of the 3rd Quad Conference on Metastasizing Pulmonary Tumors in Rats Induced by the Inhalation of Nickel Carbonyl. Perugia, Italy: Perugia Division of Cancer Research.

Suzuki T, Kanoh T, Kanbayashi M *et al.* (1993) The adjuvant activity of pyrene in diesel exhaust on IgE antibody production in mice. *Arerugi – Jap J Allergology* **42**(8): 963–968.

Takafuji S, Suzuki S, Koizumi K *et al.* (1987) Diesel-exhaust particulates inoculated by the intranasal route have an adjuvant activity for IgE production in mice. *J Allergy Clin Immunol* **79**(4): 639–645.

Takano H, Yoshikawa T, Ichinose T *et al.* (1997) Diesel exhaust particles enhance antigen-induced airway inflammation and local cytokine expression in mice. *Am J Resp Crit Care Med* **156**(1): 36–42.

Takenaka H, Zhang K, Diaz-Sanchez D *et al.* (1995) Enhanced human IgE production results from exposure to the aromatic hydrocarbons from diesel exhaust: direct effects on B-cell IgE production. *J Allergy Clin Immunol* **95**(1:Pt 1): 103–115.

Thomas P, Maerker J, Riedel W and Przybilla B (1995) Altered human monocyte/macrophage function after exposure to diesel exhaust particles. *Environ Sci Pollut Res Int* **2**: 69–72.

Tsien A, Diaz-Sanchez D, Ma J and Saxon A (1997) The organic component of diesel exhaust particles and phenanthrene, a major polyaromatic hydrocarbon constituent, enhances IgE production by IgE-secreting EBV-transformed human B cells *in vitro*. *Toxicol Appl Pharmacol* **142**(2): 256–263.

Ulfvarson U and Alexandersson R (1990) Reduction in adverse effect on pulmonary function after exposure to filtered diesel exhaust. *Am J Ind Med* **17**: 341–347.

Ulfvarson U, Alexandersson R, Aringer L *et al.* (1987) Effects of exposure to vehicle exhaust on health. *Scand J Work Environ Health* **13**: 505–515.

US EPA (1993) Motor vehicle-related air toxics study. Ann Arbor, MI: Office of Mobile Sources. EPA 420-R-93-005.

US EPA (1994) Health risk assessment document for diesel emissions. EPA/600/8-90/057Ba and EPA/600/8-90/057Bb.

US EPA (1996) US Environmental Protection Agency (EPA), Office of Air Quality Planning and Standards. *Review of the National Ambient Air Quality Standards for Particulate Matter: Policy Assessment of Scientific and Technical Information.* OAQPS Staff Paper. Research Triangle Park, NC: U.S. Government Printing Office. EPA-452\R-96-013.

van Vliet P, Knape M, de Hartog J *et al.* (1997) Motor vehicle exhaust and respiratory symptoms in children living near freeways. *Environ Res* **74**(2): 122–132.

Vinegar A, Carson AI, Pepelko W and Orthoefer JG (1981) Effects of six months of exposure to two levels of diesel exhaust on pulmonary function of Chinese hamsters. *Fed Proc* **40**: 593.

Vineis P and Magnani C (1985) Occupation and bladder cancer in males: a case–control study. *Int J Cancer* **35**(5): 599–606.

Vostal JJ (1986) Factors limiting the evidence for chemical carcinogenicity of diesel emissions in long-term inhalation experiments. *Dev Toxicol Environ Sci* **13**: 381–396.

Wade JF and Newman LS (1993) Diesel asthma. Reactive airways disease following overexposure to locomotive exhaust. *J Occup Med* **35**(2): 149–154.

Wagner WD, Groth DH, Holtz JL *et al.* (1969) Comparative chronic inhalation toxicity of beryllium ores, bertrandite and beryl, with production of pulmonary tumors by beryl. *Toxicol Appl Pharmacol* **15**: 10–29.

Waller RE (1981) Trends in lung cancer in London in relation to exposure to diesel fumes. *Environ Int* **5**: 479–483.

Watson AY and Green GM (1995) Noncancer effects of diesel emissions: animal studies. In: *Diesel Exhaust: A Critical Analysis of Emissions, Exposure, and Health Effects.* Cambridge: Health Effects Institute, pp 139–184.

Watts CD (1995) *HEI Review.* Health Effects Institute.

Waxweiler RJ, Wagoner JK and Archer VE (1973) Mortality of potash workers. *J Occup Med* **15**(6): 486–489.

Wehner AP, Dagle GE, Clark ML and Buschbom RL (1986) Lung changes in rats following inhalation exposure to volcanic ash for two years. *Environ Res* **40**(2): 499–517.

Weiland SK, Mundt KA, Ruckmann A and Keil U (1994) Self-reported wheezing and allergic rhinitis in children and traffic density on street of residence. *Ann Epidemiol* **4**: 243–247.

Weitzman SA and Stossel TP (1981) Mutation caused by human phagocytes. *Science* **212**(4494): 546–547.

Weitzman SA and Stossel TP (1982) Effects of oxygen radical scavengers and antioxidants on phagocyte-induced mutagenesis. *J Immunol* **128**(6): 2770–2772.

Wergeland E, Andersen A and Baerheim A (1990) Morbidity and mortality in talc-exposed workers. *Am J Ind Med* **17**(4): 505–513.

Winer AM and Busby WF Jr (1995) Atmospheric transport and transformation of diesel emissions. In: *Health Effects of Diesel Engine Emissions: Characterization and Critical Analysis.* Cambridge, MA: The Health Effects Institute.

Wjst M, Reitmeir P, Dold S *et al.* (1993) Road traffic and adverse effects on respiratory health in children. *Br Med J* **307**(6904): 596–600.

Wolff RK, Henderson RF, Snipes MB *et al.* (1986) Lung retention of diesel soot and associated organic compounds. *Dev Toxicol Environ Sci* **13**: 199–211.

Wolff RK, Henderson RF, Snipes MB *et al.* (1987) Alterations in particle accumulation and clearance in lungs of rats chronically exposed to diesel exhaust. *Fundam Appl Toxicol* **9**(1): 154–166.

Wong D, Mitchell CE, Wolff RK *et al.* (1986) Identification of DNA damage as a result of exposure of rats to diesel engine exhaust. *Carcinogenesis* **7**(9): 1595–1597.

Wong O, Morgan RW, Kheifets L *et al.* (1985) Mortality among members of a heavy construction equipment operators union with potential exposure to diesel exhaust emissions. *Br J Ind Med* **42**: 435–448.

World Health Organization (1996) Diesel fuel and exhaust emissions. Environmental Health Criteria

171. International Programme on Chemical Safety. Geneva: WHO.

Woskie SR, Smith TJ, Hammond SK *et al.* (1988a) Estimation of the diesel exhaust exposure of railroad workers: II. National and historical exposures. *Am J Ind Med* **13**: 395–404.

Woskie SR, Smith TJ, Hammond SK *et al.* (1988b) Estimation of the diesel exhaust exposures of railroad workers: I. Current exposures. *Am J Ind Med* **13**: 381–394.

Wynder EL, Dieck GS, Hall NE and Lahti H (1985) A case–control study of diesel exhaust exposure and bladder cancer. *Environ Res* **37**(2): 475–489.

Zaebst DD, Clapp DE, Blade LM *et al.* (1991) Quantitative determination of trucking industry workers' exposures to diesel exhaust particles. *Am Ind Hyg Assoc J* **52**: 529–541.

CARBON MONOXIDE, LEAD AND AIR TOXICS

33

Carbon Monoxide

ROBERT L. MAYNARD and ROBERT WALLER

Department of Health, London, UK

The views expressed in this chapter are those of the authors and should not be taken as representing those of the UK Department of Health.

INTRODUCTION

Whilst there is still argument as to how many people die each year as a result of exposure to air pollutants such as ozone and particulate matter, there can be no doubt that carbon monoxide is responsible for many intentional and unintentional deaths. In the period 1979–88 some 11 547 unintentional deaths due to carbon monoxide poisoning occurred in the USA (Baker *et al.*, 1992). In England and Wales there are more than 60 deaths a year due to accidental exposure to carbon monoxide and some 500 admissions to hospital for treatment of such exposures. Suicides by carbon monoxide poisoning have declined in many countries as gas produced from coke (see below) has been replaced with natural gas which contains little carbon monoxide. Exhaust gases from petrol engines contain high concentrations of carbon monoxide and are still used as a means of committing suicide.

The literature on carbon monoxide poisoning is extensive and dates from the late nineteenth century. Earlier workers were aware of the poisonous nature of the fumes produced by the incomplete combustion of charcoal and coke. An excellent source of references to early studies of carbon monoxide and the literature until the 1960s has been provided by the US Department of Health, Education and Welfare: *Carbon Monoxide: A Bibliography with Abstracts* (Cooper, 1966). This invaluable compilation lists almost 1000 publications though many appear more than once. The authors point out that the poisonous nature of the fumes produced by burning charcoal were known in Rome in the time of Cicero when they were used as a means of capital punishment and of 'crossing the bridge of death', i.e., suicide.

AIR POLLUTION AND HEALTH
ISBN 0-12-352335-4

Modern studies of carbon monoxide began with Claude Bernard (1865), whose work is mentioned below in the discussion of the binding of carbon monoxide to haemoglobin. The English physiologist John Scott Haldane did much to explain the mode of action of carbon monoxide in the last years of the nineteenth and early twentieth century and the 'Haldane Equation' remains important in carbon monoxide toxicology (Douglas *et al.*, 1912; Haldane, 1895a,b). More recent work by Coburn and by Longo has improved our understanding of the binding of carbon monoxide to adult and fetal haemoglobin and will also be discussed later (Coburn *et al.*, 1965, 1973; Longo, 1970, 1977).

This chapter concentrates on the effects of exposure to ambient concentrations of carbon monoxide. Thus, only a brief description of the clinical signs and symptoms of severe poisoning and its treatment will be given. A detailed review by Lowe-Ponsford and Henry (1989) should be consulted for clinical details. Whilst it is unlikely that clinical poisoning by carbon monoxide (i.e. poisoning producing signs and symptoms and requiring treatment) occurs as a result of exposure to ambient outdoor concentrations of the gas, such exposures may play a role in precipitating deaths due to and hospital admissions for the treatment of myocardial infarction and heart failure (Morris *et al.*, 1995). Emphasis has thus been placed on these aspects of the problem.

Severe poisoning by carbon monoxide often results in lasting damage to the central nervous system (CNS). This may be delayed, progressive, irreversible and lethal. Though anoxia is undoubtedly important, autoimmune processes leading to a spongiform degeneration or leukoencephalopathy have been suggested (Norton, 1986). Leukoencephalopathy is not seen following acute exposure to low concentrations of carbon monoxide, but there is some dispute over whether more subtle effects on the CNS may occur as a result of repeated low level exposures. For this reason, some discussion of effects on the CNS is necessary.

It has recently been suggested that carbon monoxide may act as a transmitter substance (Barinaga, 1993; Dawson and Snyder, 1994; Ingi and Ronnett, 1995; Suematsu *et al.*, 1996; Verma *et al.*, 1993). This new insight into the possible functions of the endogenously produced gases nitric oxide and carbon monoxide will not be discussed.

It is not intended that this chapter should serve as a critical review of the large literature available on carbon monoxide in all its aspects. Instead, emphasis has been placed on studies of importance in decision-making about the effects of ambient concentrations of carbon monoxide on health and the setting of air quality standards.

PHYSICAL AND CHEMICAL PROPERTIES

Carbon monoxide is a colourless gas with virtually the same density as air. It is usually described as odourless though Haldane described a garlic like smell. Good accounts of its chemical properties can be found in textbooks of inorganic chemistry, the older books providing valuable information on the production of the various forms of gas used for heating and lighting (Partington, 1953).

The basic physical properties of carbon monoxide are shown in Table 33.1.

Table 33.1 Basic physical properties of carbon monoxide

Characteristic	Value
Molecular weight	28.01
Critical point	−145°C at 43.5 atmospheres
Boiling point	−191°C
Specific gravity relative to air	0.967
Solubility	
0°C	3.54 ml/100 ml
20°C	2.32 ml/100 ml
37°C	2.14 ml/100 ml

Conversion factors:

At 0 °C,	1 atm:	$1~mg/m^3$	=	0.8 ppm
		1 ppm	=	$1.250~mg/m^3$
At 25 °C,	1 atm:	$1~mg/m^3$	=	0.873 ppm
		1 ppm	=	$1.145~mg/m^3$

SOURCES OF CARBON MONOXIDE

Carbon monoxide is produced by the incomplete combustion of carbon-containing fuels:

$$2C + O_2 \rightarrow 2CO$$

In domestic coal fires it may be produced by the reduction of carbon dioxide produced at the base of the fire as it passes through the heated coal or coke above:

$$C + O_2 \rightarrow CO_2$$

$$CO_2 + C \rightarrow 2CO$$

Incomplete combustion is characteristic of petrol engines when the balance between air and fuel moves away from the stoichiometric ratio. This occurs during cold starting, idling and slow running. Diesel engines burn fuel in an excess of air and thus produce carbon dioxide rather than carbon monoxide.

In the UK, traffic now makes up the major outdoor source of carbon monoxide: 87% of the total inventory in 1990. The influence of traffic growth on the production of carbon monoxide in the UK is shown in Fig. 33.1.

Catalytic converters and improved engine design are playing an important part in reducing production of carbon monoxide in many countries. In the USA the reduction in production of carbon monoxide by passenger cars since 1970 has been dramatic and is shown in Fig. 33.2.

In the UK, production of carbon monoxide by motor vehicles is expected to fall sharply by the year 2005, the predicted fall in maximum running 8-h average concentrations at UK urban background and roadside sites is shown in Table 33.2.

In the UK and many other countries, the use of coal for domestic heating has declined rapidly during the past 50 years and no longer makes a substantial contribution to carbon monoxide production (QUARG, 1993). In 1970, 25% of the total production of carbon monoxide in the UK was due to the use of coal, while in 1990 the figure was 4%. In countries in transition and in some parts of developed countries, coal and wood are still

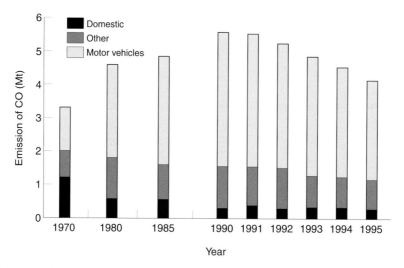

Fig. 33.1 Increasing contribution of road transport to emissions of carbon monoxide in the UK. From *The United Kingdom National Air Quality Strategy* (Department of the Environment, 1997). Reproduced with permission of the Department of the Environment and The Stationery Office.

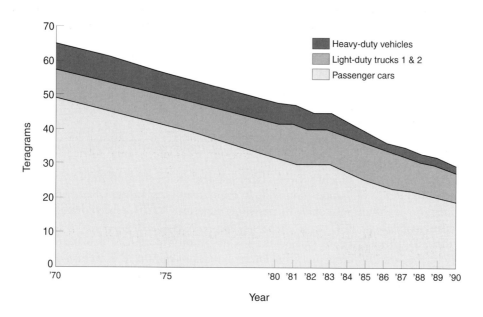

Fig. 33.2 Estimated emissions of carbon monoxide from gasoline-fueled highway vehicles, 1970 to 1990. Reproduced with the permission of the US EPA.

Table 33.2 Predicted maximum 8-h carbon monoxide running means at UK sites (ppm)

Site	2000	2005	2010
Stevenage	2.0–3.3	1.6–2.5	1.4–2.2
London, Victoria	3.8–7.6	2.9–5.8	2.6–5.1
London, Cromwell Road	4.3–10.5	3.3–8.1	2.9–7.1
London, Earls Court	2.8–10.7	2.1–8.2	1.9–7.3
London, Bloomsbury	2.4–4.4	1.8–3.4	1.6–3.0
Glasgow	3.2–8.5	2.4–6.5	2.2–5.7
Manchester	2.4–8.5	1.9–6.5	1.6–5.7
Sheffield	2.9–5.0	2.2–3.8	2.0–3.4
Belfast Centre	7.0–9.5	5.3–7.3	4.7–6.4
Birmingham Centre	2.6–7.3	2.0–5.6	1.7–5.0
Cardiff Centre	2.4–4.0	1.9–3.1	1.7–2.7
Edinburgh Centre	2.2–3.3	1.7–2.5	1.5–2.2

From The United Kingdom National Air Quality Strategy (Department of the Environment, 1997).

used on a large scale and are important sources of carbon monoxide (Khalil and Rasmussen, 1989).

Indoor concentrations of carbon monoxide may be particularly high where open solid fuel fired stoves are used for cooking in poorly ventilated rooms. This is currently a significant problem in China (Hong, 1991) and in Korea, where charcoal is used for heating and cooking.

Carbon monoxide is produced indoors in countries such as the UK by the use of gas cookers and gas fires (Cox and Whichelow, 1985; Spiller, 1987). If these are poorly maintained and inadequately ventilated, dangerous concentrations may occur. It should be noted that of those dying from non-occupational, accidental exposure to carbon monoxide, all receive their fatal exposure indoors.

Cigarette smoking is an important indoor source of carbon monoxide. Cigarette smoke contains about 5% carbon monoxide, while that of pipe and cigar smoke contains rather more, perhaps 10–15%, as a result of the lower temperature of combustion. Smoking causes a very significant increase in the percentage of total haemoglobin bound to carbon monoxide rather than to oxygen: the % carboxyhaemoglobin (COHb). Heavy cigarette smokers may have COHb levels of 10%, the non-smoker has a normal level of about 0.5–1.0%, though this may be raised by exposure to environmental tobacco smoke to 1.5% (WHO, 1979). Whether this moderate increase in concentration of COHb (to 1.5%) has long-term effects is unknown.

ENVIRONMENTAL CONCENTRATIONS OF CARBON MONOXIDE AND HOW THEY ARE MONITORED

Monitoring

The standard method for continuous automated monitoring of ambient concentrations of carbon monoxide is non-dispersive infrared photometry (NDIR). In 1993, a UK Expert Group commented on the method and pointed out the importance of appropriate calibration techniques. Interference by carbon dioxide was noted and it was recommended that this effect, which was observed not to be straightforward, should 'always be borne in mind when assessing NDIR data' (QUARG, 1993).

Gas chromatography can also be used to measure concentrations of carbon monoxide (Tesarik and Krejci, 1974). This is an accurate method and allows the determination of low concentrations. Continuous attendance by trained staff is, however, required.

Environmental Concentrations of Carbon Monoxide

Figures 33.3 and 33.4 show the hourly average concentrations of carbon monoxide recorded at two automated monitoring sites in London during 1994. The site in the Cromwell Road is an urban site where traffic density is high. The Bridge Place site, on the other hand, is in a quiet side street. Seasonal statistics for both sites are shown in Tables 33.3 and 33.4. Annual average concentrations in the UK, except close to busy roads, seldom exceed 1.0 ppm. Very much higher concentrations than those shown by these tables may be recorded during air pollution episodes: hourly average concentrations may exceed 15 ppm.

Urban concentrations of carbon monoxide vary with traffic density: the diurnal pattern related to the morning and evening rush hours is shown in Fig. 33.5.

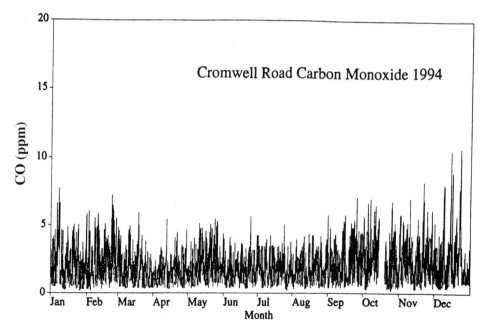

Fig. 33.3 Carbon monoxide concentrations alongside a busy road in London, 1994. Reproduced with the permission of the UK Department of the Environment.

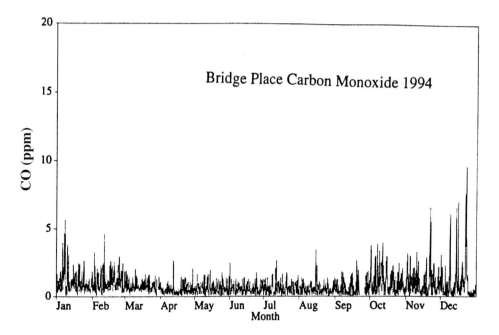

Fig. 33.4 Carbon monoxide concentrations in a quiet side street in London, 1994. Reproduced with the permission of the UK Department of the Environment.

Table 33.3 Seasonal statistics for carbon monoxide 1994, Cromwell Road (London)

Statistic	Units	Summer Apr–Sept 94	Winter Jan–Mar + Oct–Dec 94	Calendar year Jan–Dec 94
Geometric mean	ppm	1.5	1.7	1.6
Arithmetic mean	ppm	1.8	2.1	1.9
50th percentile	ppm	1.6	1.9	1.7
84th percentile	ppm	2.7	3.3	3.0
90th percentile	ppm	3.1	3.8	3.4
95th percentile	ppm	3.6	4.6	4.1
98th percentile	ppm	4.2	5.6	4.9
99th percentile	ppm	4.6	6.7	5.7
Maximum 15-min average	ppm	7.5	12.7	12.7
Maximum hourly average	ppm	7.1	10.7	10.7
Date of maximum hour	ppm	25/09/94	23/12/94	23/12/94
Maximum running 8-h	ppm	4.9	10.1	10.1
Maximum daily average	ppm	3.3	7.7	7.7
Data capture	%	100	97	98
Standard geometric deviation		1.7	1.9	1.8

Table 33.4 Seasonal statistics for carbon monoxide 1994, Bridge Place (London)

Statistic	Units	Summer Apr–Sept 94	Winter Jan–Mar + Oct–Dec 94	Calendar year Jan–Dec 94
Geometric mean	ppm	0.4	0.7	0.6
Arithmetic mean	ppm	0.5	1.0	0.7
50th percentile	ppm	0.5	0.7	0.6
84th percentile	ppm	0.8	1.4	1.1
90th percentile	ppm	0.9	1.7	1.4
95th percentile	ppm	1.2	2.2	1.8
98th percentile	ppm	1.4	3.3	2.5
99th percentile	ppm	1.7	4.8	3.3
Maximum 15-min average	ppm	3.9	9.7	9.7
Maximum hourly average	ppm	3.5	9.6	9.6
Date of maximum hour	ppm	15/08/94	23/12/94	23/12/94
Maximum running 8-h	ppm	2.3	8.9	8.9
Maximum daily average	ppm	1.2	6.7	6.7
Data capture	%	97	99	98
Standard geometric deviation		1.9	2.1	2.1

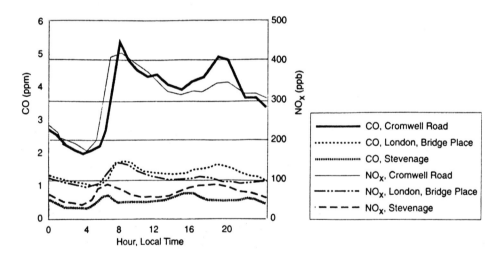

Fig. 33.5 Diurnal variations in concentrations of carbon monoxide at monitoring sites in the UK (1991). Only 'Cromwell Road' is a roadsite site. Reproduced with the permission of the UK Department of the Environment.

Concentrations of Carbon Monoxide Inside Vehicles

The interior of a motor car in slow moving or stationary traffic is a prime site for high exposure to carbon monoxide (Fernandez-Bremautz and Ashmore, 1995). Ventilation systems drawing air from around the vehicle effectively sample mid-carriageway concentrations of carbon monoxide (Colwill and Hickman, 1980). Such concentrations may be significantly higher than those recorded by roadside monitors. Chaney (1978) reported peak concentrations of 15 ppm in vehicles moving at less than 10 mph and concentrations of up to 45 ppm when traffic stopped. Concentrations as high as this present real dangers if exposure is prolonged. Long delays produced by 'grid-lock' are becoming more common and the importance of encouraging drivers to switch off their engines under such conditions is clear. In addition to carbon monoxide drawn in from around the car, the gas may leak directly into the vehicle from a damaged exhaust system. Pressure differences created when the back door of a van or estate-car is left open during driving can draw in exhaust fumes with a high concentrations of carbon monoxide. Children riding in the back of pick-up trucks may be exposed to, and poisoned by, carbon monoxide (Hampson and Norkool, 1992).

Indoor Concentrations of Carbon Monoxide

In many industrial countries people spend more time indoors than out. In the UK, on average people spend more than 80% of their lives indoors. Indoor exposure to pollutants may thus make up a significant proportion of total exposure. UK studies of concentrations of carbon monoxide in kitchens where gas cookers are in use show that they may exceed 9 ppm, though average concentrations are lower (Wade *et al.*, 1975).

Unventilated gas-fired water heaters (or geysers) may also generate high indoor concentrations of carbon monoxide. A study in The Netherlands showed that concentrations at breathing height could sometimes exceed 100 ppm when these devices were in use (Brunekreef *et al.*, 1982).

Exposure to Carbon Monoxide

It has been shown in the previous section that high concentrations of carbon monoxide may occur in busy roads, in motor vehicles and in homes. It would, however, be an error to assume that *exposure* at these sites is also high. It should be understood that carbon monoxide in the body and in the air are in contact across the lung. The gas diffuses freely, and thus, if the external concentrations remain constant, equilibrium will eventually be obtained. On moving to an environment with a lower concentration, the individual will slowly give off carbon monoxide as the system equilibrates to the lower concentration. Similarly, on moving to a location with a higher concentration the individual will take on carbon monoxide. Attainment of equilibrium is slow (see below), and thus blood concentrations lag behind environmental concentrations.

On a single day an individual is likely to pass through a number of microenvironments each characterized by its own concentration of carbon monoxide. The sequence: bedroom, bathroom, kitchen, car, street, office, street, car, living room, bathroom, bedroom would be a familiar one for many people. Depending on the time spent in each of these microenvironments, an individual may be sequentially taking up or giving out carbon monoxide. Predicting blood COHb levels during such a changing pattern of exposure is difficult, though the problem has been tackled by a number of workers (Johnson *et al.*, 1986; Ott and Mage, 1978; Ott *et al.*, 1982).

For some pollutants, e.g. ozone, essentially all that is inhaled is absorbed, and thus an estimate of total dose can be made by summing the products of concentration and time for each defined microenvironment encountered. Of course, attention should also be given to ventilation rate. Such an approach is not possible for carbon monoxide as it equilibrates across the lung.

In addition to the temporal pattern of uptake of carbon monoxide shown by individuals, there will be a considerable distribution of exposure patterns across the population. The delivery cyclist working all day in heavy traffic, the policeman on point-duty, and the traffic warden will all experience greater exposure to carbon monoxide than the office worker (Read and Green, 1990). Here again, errors can be made: it is always important to recall that carbon monoxide equilibrates between the blood and the air and thus the COHb concentration can only increase until equilibrium with the highest inspired concentration of carbon monoxide is reached.

Other occupational exposures may be high in garages, tunnels and some mines (Evans *et al.*, 1988; Rosenman, 1984). Studies reporting exposure under such conditions will not be discussed here, but it is interesting to recall that it was the problem of carbon monoxide in mines and tunnels that led Haldane to the study of this branch of toxicology before motor vehicles had been developed.

TOXICOKINETICS OF CARBON MONOXIDE

It is not possible to discuss the toxicokinetics of carbon monoxide without reviewing the role of haemoglobin in oxygen transport. Haemoglobin is, after DNA and RNA, perhaps the most interesting molecule found in the animal kingdom. Chlorophyll resembles it in that it, too, has a tetrapyrole ring, though the central metal atom is magnesium rather than iron. In mollusca haemocyanin, with a copper atom, replaces haemoglobin, and in some sea squirts vanadium is the equivalent of iron. In some polychaetes and annelids the iron-containing pigments chlorocruorin and haemerythrin take the place of haemoglobin.

The structure of haemoglobin is well known (Perutz, 1967–68). Each of the four amino acid chains (two α, two β) is folded and carries a haem group attached to a histidine residue at position 87 in the α chains and 92 in the β chains (Nunn, 1993). The haem group is a tetrapyrole ring with a central iron atom. The iron atom is in the reduced, ferrous (Fe^{2+}) state and is not oxidized during transport of oxygen. Strang (1977) has pointed out that 'iron atoms can accept electrons from up to six oppositely charged ions or neutral substances, known as ligands, to form complex ions'. In haemoglobin, each iron atom forms six bonds: four with the nitrogen atoms of the four pyrole rings, one with the histidine residue of either the α or β amino acid chain, and one with oxygen.

The iron atom is also loosely bound, or attracted, to other amino acid residues, e.g. histidine 58 in the α chain. This leads to folding of the whole molecule, thus creating crevices in which the haem groups lie partly hidden. The position of the haem groups in the crevices is most important and access to the haem groups is controlled by the quaternary structure of the whole molecule.

Quaternary structure is affected by the extent of formation of loose bonds between the haem groups and the amino acid chains. These loose bonds are themselves affected by the binding of oxygen to the iron atom of the haem group. Thus, as oxygen is taken up the quaternary structure of the haemoglobin changes. In the deoxygenated state haemoglobin is described as being in a 'tense' conformation. As molecules of oxygen are taken up, the

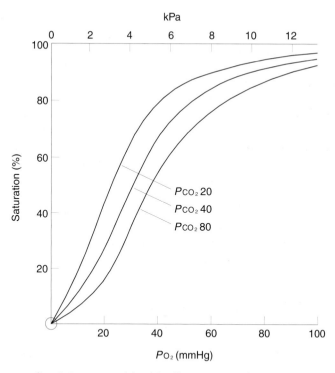

kPa

Fig. 33.6 The oxygen dissociation curve and the Bohr effect. From Saunders KB (1977) *Clinical Physiology of the Lung.* Oxford: Blackwell Scientific Publications. Reproduced with permission of the author and publisher.

quaternary structure changes and access of oxygen molecules to the remaining haem groups improves. The haemoglobin molecule is described as passing into a 'relaxed state'. In the tense state haemoglobin has a low affinity for oxygen; in the relaxed state the affinity is high.

It is this sequentially changing affinity of haemoglobin for oxygen which defines the shape of the oxyhaemoglobin (O_2Hb) dissociation curve. If only one amino acid chain with its haem group were considered, then the dissociation curve would describe a hyperbola. This is, indeed, the case with the single chain molecule myoglobin.

The standard O_2Hb dissociation curve is shown in Fig. 33.6. A number of equations have been devised to describe the curve: as might be expected, the more complex equations describe the curve well, but are difficult to use. The following equation by Severinghaus, Stafford and Thunstrom (1978) has been recommended by Nunn (1993):

$$So_2 = \frac{100\,(Po_2{}^3 + 2.667 \times Po_2)}{Po_2{}^3 + 2.667 \times Po_2 + 55.47}$$

where So_2 is oxygen saturation and Po_2 is partial pressure of oxygen in kilopascals (kPa).

Carbon monoxide molecules bind to the haem groups of haemoglobin in precisely the same way as do oxygen molecules but with about 245 times the affinity. If haemoglobin is exposed to carbon monoxide it takes up the CO in precisely the same way as it would take up oxygen and the dissociation curve for the resulting carboxyhaemoglobin is identical with that of the dissociation curve of O_2Hb. This is remarkable. Of course, the scales of the x axes of the two dissociation curves would be different: Hb is 100% saturated with oxygen at a pressure of 100 mmHg; 100% saturation with CO occurs at 100/245 = 0.4 mmHg. Note that the scales of the y axes, whether shown as percentage saturation or as volumes of O_2 or CO per 100 ml of blood, are identical: at 100% saturation 100 ml of blood with a normal haemoglobin concentration can carry either 20 ml of oxygen or 20 ml of carbon monoxide.

If haemoglobin is exposed to a mixture of CO and O_2, then the gases will compete for the haem groups. CO competes better because of its greater affinity for the haem groups. The relative concentrations of COHb and O_2Hb at equilibrium, are given by the following equation:

$$\frac{(COHb)}{(O_2Hb)} = M \times \frac{(Pco)}{(Po_2)}$$

where M is the relative affinity of CO and O_2 for Hb, i.e. 245 at pH 7.4 in humans. The equation is the Haldane Equation. M is described as the Haldane Coefficient or the Haldane Constant.

Comroe (1974) illustrated the relative affinity of haemoglobin for oxygen and carbon monoxide as follows:

(carbon monoxide) has an affinity for haemoglobin that is about 210 times that of O_2 for Hb. This means that O_2, to compete with CO on even terms for Hb, must be present in 210 times the concentration of CO. Since O_2 is almost 21% of air it is easy to remember that blood equilibrated with air + 0.1% CO (a 210:1 ratio of O_2:CO) will contain 50% HbO_2 and 50% HbCO; so will blood equilibrated with alveolar gas (14% O_2) containing 0.066% CO. [*245 is now accepted as a better estimate of M than 210.*]

It may be useful to consider here how rapidly equilibration with CO is likely to occur on exposure to a concentration of 0.1% (i.e. 1000 ppm) CO. The following analysis is taken from Comroe (1974):

Let blood volume be 6 litres.

At full saturation with O_2 each litre of blood would contain 200 ml O_2

Thus at 50% saturation with CO each litre of blood will contain 100 ml of oxygen and 100 ml CO.

Thus the total volume of CO carried by the blood at 50% saturation will be 600 ml.

Assume a minute volume of 6 litres and that all the inspired air becomes alveolar gas, then the maximum amount of CO reaching the alveoli will be 6 ml per minute (i.e. 0.1% of 6 litres).

Thus it would take 100 minutes to take aboard 600 ml CO (100 ml/l) and reach 50% saturation.

This calculation contains many assumptions and is only included for illustrative purposes. More accurate calculations are discussed below.

Effect of Carbon Monoxide on the O₂Hb Dissociation Curve

Consider again the O$_2$Hb dissociation curve shown in Fig. 33.6. Starting at 100% saturation, all the haem groups will be carrying a molecule of oxygen and the overall formula could be given as Hb$_4$(O$_2$)$_4$. As oxygen is given up the affinity of the haem groups for oxygen drops precipitously: over the 'shoulder' as it were of the dissociation curve. Thus a fall in Po$_2$ of about 60 mmHg is needed to release 5 ml of oxygen from 100 ml of blood. This loss of about 25% of the total O$_2$ carried could be represented as the loss of one molecule of oxygen from each molecule of Hb:

$$Hb_4(O_2)_4 \rightarrow Hb_4(O_2)_3 + O_2$$

Continuing down the dissociation curve a change of only 15 mmHg is needed to release the next molecule of oxygen:

$$Hb_4(O_2)_3 \rightarrow Hb_4(O_2)_2 + O_2$$

and a change of only 5 mmHg to achieve:

$$Hb_4(O_2)_2 \rightarrow Hb_4O_2 + O_2$$

The last oxygen molecule is not released under physiological conditions.

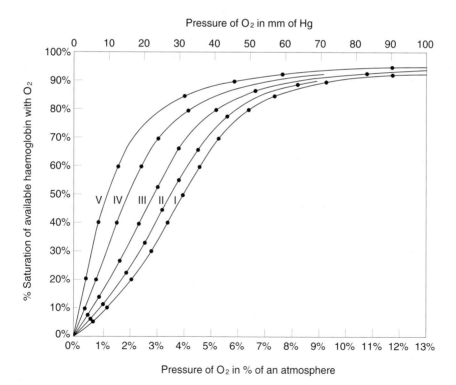

Fig. 33.7   Effect of carbon monoxide on the oxyhaemoglobin dissociation curve. Curve I, 0% saturation with CO; II, 10%; III, 25%; IV, 50%; V, 75%. Reproduced with the permission of The Physiological Society and publisher.

Now consider blood in which 50% of the Hb is saturated with CO. Of course, this does not mean that 50% of the Hb molecules are in the $Hb_4(CO)_4$ state and 50% in the $Hb_4(O_2)_4$ state. There will in fact be a spectrum and most molecules will be in the $Hb_4(O_2)_2(CO)_2$ state. The dissociation curves for blood at a range of saturations with carbon monoxide are shown in Fig. 33.7.

As the concentration of oxygen falls the CO molecules are not released, and thus the Hb molecule is maintained in a relaxed, high oxygen affinity state until lower than normal partial pressures of oxygen are reached. The dissociation curve is thus displaced to the left. This displacement is described as the Haldane Shift. Smith (1986) has explained this effect by pointing out that as there are, on average, only two molecules of oxygen bound to each molecule of haemoglobin, there can only be one opportunity for cooperative activity between oxygen molecules and thus only the first part of the O_2Hb dissociation curve is seen – albeit at lower Po_2 levels than normal because the CO molecules interfere with the release of oxygen.

It is useful to replot the dissociation curve in terms of volumes of oxygen per volume of blood rather than % saturation (see Fig. 33.8). Here the effects of 50% saturation and anaemia reducing the Hb concentration to 50% of normal are compared. It will be seen that the shape of the dissociation curve in anaemia is normal, the curve for 50% COHb is displaced to the left. The figure also illustrates the decrement of Po_2 needed to release 5 ml of oxygen from 100 ml of blood: the normal arteriovenous difference in oxygen content. Note that the venous points (V, V_1' and V_2') differ significantly, being at approximately 40, 27 and 14 mmHg, respectively. Thus binding of 50% of Hb to CO requires that to allow normal extraction of oxygen the Po_2 of venous blood must be significantly lower than that of normal blood and even than that of blood containing only 50% of the normal amount of haemoglobin. This explains the severe tissue hypoxia produced by carbon monoxide poisoning.

The left shift of the O_2Hb dissociation curve was described by Haldane and colleagues in 1912. It is interesting to note that the mathematical work was done by Haldane's son, JBS Haldane, whilst a schoolboy (Douglas et al., 1912; Haldane JBS, 1912). A letter from Haldane (senior) to the Head Master of Eton College asks for permission for his son to attend a meeting of the Physiological Society 'partly to fortify me against possible attacks from people who know the higher mathematics'!

Equilibration Between Blood and Ambient Carbon Monoxide

As has already been mentioned, prolonged exposure to CO leads to equilibration across the alveolar–capillary boundary. The equilibrium concentration is given by the Haldane Equation.

The rate of increase of COHb concentration diminishes as exposure to a constant concentration of CO continues and conforms to a standard 'wash in' exponential function. Such functions are described by the general linear differential equation:

$$\frac{dy}{dt} = k(y_\infty - y)$$

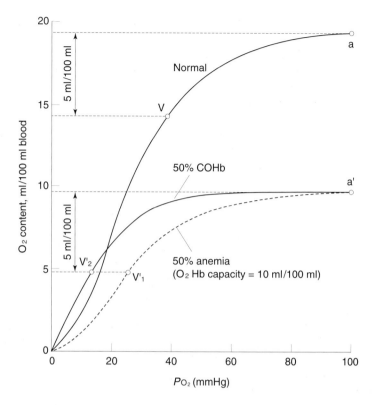

Fig. 33.8   Normal oxyhaemoglobin dissociation curve and curves for the case of a 50% anaemia and the case of a 50% carboxyhaemoglobinaemia. The delivery of 25% of the total oxygen content of fully oxygenated arterial blood (5 ml/100 ml blood) requires a drop in the P_{O_2} of about 60 mmHg (from point a to point V on the normal curve). Delivery of a comparable volume of oxygen in the case of a 50% anaemia requires a drop in the P_{O_2} of more than 75 mmHg (from point a' to point V'_1), but an even greater fall in the P_{O_2} is required to deliver the same volume of oxygen in the case of the curve distorted by the presence of carboxyhaemoglobin (from point a' to a point V'_2). From Smith RP (1986) Toxic responses of the blood. In: Klaasen CO, Amdur MO and Doull J (eds) Casarett and Doull's Toxicology, 3rd edn. New York: Macmillan Publishing Company.

As y_∞ is approached the term within the bracket approaches zero and the rate of increase of y slows towards zero. The value of y at time t is determined by integration:

$$y_t = y_\infty (1 - e^{-kt})$$

If the concentration of COHb is plotted against time on arithmetic axes then a curve will be generated. The curve becomes horizontal at infinite time, though long before this we may assume for all practical purposes that equilibrium has been reached.

A number of authors have produced equations based on measurements of COHb during exposure to CO which describe the uptake of the gas. Data from Forbes *et al.* (1945) were used to produce the series of curves shown in Fig. 33.9. These curves deserve careful study. Note that the dependence of the rate of uptake on minute ventilation has been illustrated by the series of scales available on the *x*, i.e. time, axis. Note also that the equilibrium concentration of COHb at a concentration of CO of 15 ppm is 2.4%.

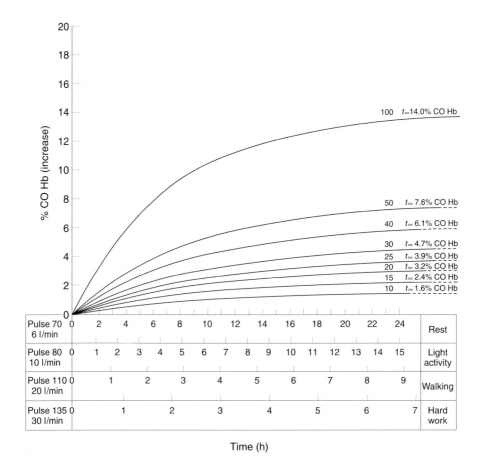

Fig. 33.9 Relationship between blood concentration of COHb and exposure concentration of CO (ppm) at different levels of exercise. Reproduced with the permission of the World Health Organization.

Mathematical models based on a detailed analysis of the physiological processes involved in the uptake of CO have been produced. The best known is that of Coburn, Forster and Kane (Coburn, 1979; Coburn *et al.*, 1965). The standard form of the Coburn Equation is:

$$\frac{V_B \, d(COHb)}{dt} = \dot{V}_{CO} - \frac{[COHb]\overline{P}_{cO_2}}{MB\,[O_2Hb]} + \frac{P_1CO}{B}$$

where

$$B = \frac{1}{DLCO} + \frac{P_L}{\dot{V}_A}$$

Here V_B is the blood volume in ml, (COHb) is measured as ml CO/ml blood, $\dot{V}_{CO}$ is the rate of endogenous production of CO in ml per minute (0.007 ml/min), $\bar{P}_{cO_2}$ is the mean partial pressure of O_2 in pulmonary capillaries expressed in mmHg, M is the Haldane coefficient, (O$_2$Hb) is the concentration of O$_2$Hb expressed in ml/100 ml, P_ICO is the partial pressure of CO in inspired air, $DLCO$ is the pulmonary diffusing capacity for CO expressed as ml/min/mmHg CO, P_L is the pressure of dry gases in the lung (713 mmHg) and $\dot{V}_A$ is the alveolar ventilation rate expressed as ml/min.

The form of the Coburn equation shown above is that of a linear differential equation. The derivation of the equation from first principles involves two important assumptions:

(1) that it is adequate to use average values for P_{O_2} and (O$_2$Hb) in pulmonary capillaries, though it is known that these vary along the capillaries in a complicated fashion;
(2) that (O$_2$Hb) is constant.

The latter assumption limits the application of the equation to low inspired concentrations of carbon monoxide. These assumptions have been addressed by Coburn and his colleagues and more complex solutions have been developed involving special computing techniques. Such methods are beyond the scope of this account, to say nothing of the mathematical ability of the present author [RLM]. For readers who, like the present author, are not overfamiliar with differential equations, one final point might be made. Though the differential equations discussed above are described as linear, the relationships they describe are exponential.

Rate of Uptake of Carbon Monoxide

The rate of uptake of CO is dependent on the difference in partial pressures of CO between the alveolar air and the blood. Removal of CO is dependent on the gradient between blood and the alveolar air. The gradient during removal of CO is generally less than that present during uptake and thus removal of CO is slower than uptake. The half-time for removal of CO from the body under resting conditions is 4–5 h. Model calculations by Hickman (1989) generated a series of graphs shown in Fig. 3.10a–c.

Note that during exposure to a fairly constant concentration of CO the concentration of COHb rose almost linearly. During exposure to a series of fluctuating concentrations the concentration of COHb rose in a stepwise manner and there was little evidence of removal of CO during simulated short duration exposure to low concentrations of CO.

Equilibration of Carbon Monoxide Across the Placenta and Effects on the Fetal O$_2$Hb Dissociation Curve

The affinity of fetal haemoglobin for oxygen is greater than that of adult haemoglobin and the dissociation curve is shifted to the left of the adult curve. Both these differences are illustrated in Fig. 33.11 (Longo, 1976).

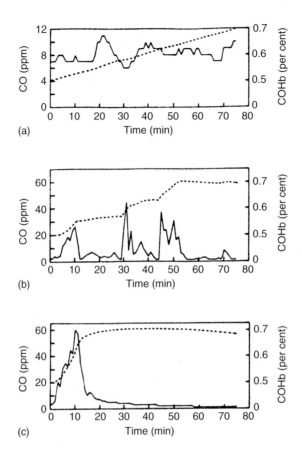

Fig. 33.10 (a), (b) and (c): Carboxyhaemoglobin produced from different types of exposure to carbon monoxide. ————, carbon monoxide; – – – – –, carboxyhaemoglobin. Reproduced with the permission of D.J. Hickman, Transport and Road Research Laboratory.

The increased affinity enhances O_2 uptake at the placenta: the highest fetal Po_2 level (found in umbilical vein blood) is only about 30 mmHg. Compensation for the low Po_2 is also provided by the greater concentration of fetal haemoglobin: 18 g/100 ml as compared with 13 g/100 ml in the mother, both at term. One of the reasons for the displaced position of the fetal O_2Hb is the reduced effect of the organic phosphate 2,3-DPG on fetal haemoglobin as compared with adult haemoglobin. In the adult, 2,3-DPG stabilizes haemoglobin in the low affinity (tense) state and thus effectively displaces the curve to the right.

Longo and colleagues have pointed out that fetal Hb also has an increased affinity for CO in comparison with oxygen and they have discussed in detail the effects of maternal exposure to carbon monoxide on the fetus (Garvey and Longo, 1978; Longo, 1970, 1976; Longo et al., 1967; Power and Longo, 1975). The relative affinity of fetal and adult haemoglobin for both carbon monoxide and oxygen is species dependent. In the sheep, dog and rabbit the ratio of the affinity of fetal Hb for oxygen to that of adult Hb for oxygen is greater than in humans. Though fetal haemoglobin has a higher affinity for carbon monoxide than for oxygen, the difference in affinities is less than seen in the adult. Thus Engel *et al.* (1969) demonstrated that the affinity ratio for fetal haemoglobin was about 170 as compared with 245 in the adult. The concentration of COHb in fetal blood as compared with that in maternal blood has been expressed by Longo as:

$$\frac{(COHb)_F}{(COHb)_M} = \frac{(O_2Hb)_F}{Po_{2_F}} \times \frac{Po_{2_M}}{(O_2Hb)_M} \times \frac{M_F}{M_M}$$

Where M_F and M_M are the Haldane coefficients for fetal and maternal blood, respectively. This equation may be simplified to:

$$\frac{(COHb)_F}{(COHb)_M} = \frac{A_FCO}{A_MCO}$$

where A_FCO and A_MCO are the affinities of fetal and maternal blood for CO, respectively (L.D. Longo, personal communication).

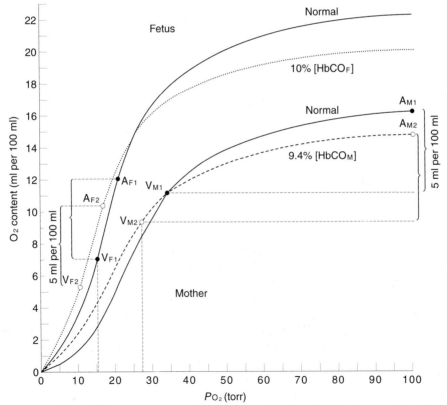

Fig. 33.11 Oxyhaemoglobin saturation curves of human maternal and fetal blood under control conditions and during steady-state conditions with 10% fetal and 9.4% maternal HbCO concentrations. The maternal and fetal haemoglobins contents were assumed to equal 12 and 16.3 g per 100 ml of blood, respectively. A normal O_2 consumption of 5 ml per 100 ml of blood was assumed for both the uterus and its contents and the fetus. Reproduced with the permission of the author and publisher.

Longo (1976) has pointed out that the left shift in the O_2Hb dissociation curve of fetal blood induced by CO may produce severe fetal hypoxia. This is shown in Fig. 33.11 from Longo's work.

It will be seen that 10% COHb will move the maternal venous point from 34 to 26 mmHg but that the fetal venous point moves from 16 to 11 mmHg. Thus the fetus, which already has a low oxygen tension and limited capacity to respond by increasing cardiac output, is placed at risk by exposure of the mother to concentrations of carbon monoxide which might do her comparatively little harm. The importance of this to understanding the effects of maternal cigarette smoking on the fetus is obvious.

ENDOGENOUS PRODUCTION OF CARBON MONOXIDE

Carbon monoxide is produced in the liver as a by-product of the breakdown of haemoglobin. During conversion of haemoglobin to bilirubin a carbon atom is released and oxidized to CO by hepatic microsomal haem-oxygenase (Berk, 1974). Other sources of endogenous CO include cytochromes, myoglobin and peroxidase enzymes. Endogenous production of CO leads to blood levels of COHb of about 0.5%. The rate of CO production is significantly increased during pregnancy and in the premenstrual period (Delivoria-Papadopoulos *et al.*, 1970; Lynch and Moede, 1972). Haemolytic disorders, fever and defects in erythropoiesis also lead to increased production of CO. Chronic haemolytic anaemia may lead to an endogenous COHb level of 3–4%.

CONCENTRATIONS OF COHb IN THE POPULATION

Healthy, non-smoking adults have COHb concentrations of the order of 0.5–1.0%. Cigarette smokers who are repeatedly exposed to high concentrations of carbon monoxide may show concentrations of up to 12% (Lawther and Commins, 1970). As might be expected, there is a statistical distribution of COHb concentrations in the population, with smokers and non-smokers forming two distinct sub-populations. Fig. 33.12 shows the cumulative frequency distributions for the US population (Radford and Drizd, 1982).

COHb concentrations in blood may be measured using spectrophotometric techniques or gas chromatography (GC). The spectrophotometric techniques (CO-Oximeter) are convenient and rapid but less accurate than the GC techniques at low concentrations. The important differences between the results obtained with these two methods at low concentrations of COHb are discussed below.

COHb concentrations can be estimated by measuring the CO concentration in alveolar air and applying the Coburn equation. This is a rapid technique of great use in clinics devoted to persuading people to give up smoking (Jarvis *et al.*, 1986).

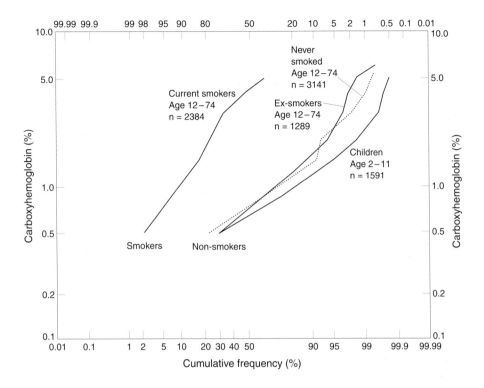

Fig. 33.12 Frequency distributions of carboxyhaemoglobin levels in the US population, by smoking habits. Reproduced with the permission of the US Department of Health and Human Services.

CARBON MONOXIDE POISONING

Symptoms and Signs of Poisoning

Carbon monoxide exerts its toxic effects by reducing the capacity of haemoglobin to transport oxygen and by interfering with the release of such oxygen as is carried. The hypoxia produced is particularly dangerous in that the partial pressure of oxygen is maintained despite a fall in the amount of oxygen per unit volume of blood. Thus, early stimulation of the respiratory centre is delayed. Haldane noted this in his comparisons of the effects of exposure to raised concentrations of carbon monoxide and to low concentrations of oxygen. He correctly noted that exposure to carbon monoxide was much more likely to lead to fainting on exertion than exposure to low concentrations of oxygen. Haldane's original description of the effects of breathing air with a high concentration of carbon monoxide are worth reading as they represent first-hand observations. The following remarkably disinterested and cool account is taken from his record of an experiment on himself (Haldane and Priestley, 1935):

Haldane found in these experiments that no particular effect was observed until the haemoglobin was about 20% saturated (with CO). At about this saturation any extra exertion, such as running up stairs, produced a very slight feeling of dizziness and some extra palpitation and hyperpnoea. At about 30% saturation very slight symptoms, such as slight increase in pulse rate, deeper breathing and slight palpitations, became observable during rest and running upstairs was followed in about half a minute by dizziness, dimness of vision and abnormally increased breathing and pulse rate. At 40% saturation these symptoms were more marked, and any exertion had to be made with caution for fear of fainting. At 50% saturation there was no real discomfort during rest, but the breathing and pulse rate were distinctly increased, vision and hearing impaired and intelligence probably greatly impaired. It was hardly possible to rise from a chair without assistance. Writing was very bad and spelling uncertain . . . In one experiment the saturation reached 56%. It was then hardly possible to stand and impossible to walk.

Haldane laconically recorded that deaths had occurred at saturations of 60%. Recovery was rapid on exposure to fresh air but accompanied by nausea, vomiting, headache and 'extreme depression'.

Fainting may be due to the peripheral vasodilatation which accompanies the hypoxia: a comparatively small increase in cardiac output is prevented by myocardial hypoxia. Convulsions may occur when COHb saturations reach about 60%.

Carboxyhaemoglobin is cherry red in colour and thus may produce a pink coloration of the nail beds and skin at high saturations, though the pink coloration is usually only apparent in the areas of post mortem lividity (pooling of blood on the lower surface of the body) which is seen in victims of acute carbon monoxide intoxication. It was this cherry red colour which led Claude Bernard to investigate the mechanisms of action of carbon monoxide. His account of his logical testing of hypotheses regarding this effect is still worth reading (Bernard, 1865).

Poisoning with carbon monoxide is often insidious with the victim falling asleep and not waking again if the exposure is prolonged. Table 33.5 lists the signs and symptoms of CO poisoning at a range of COHb saturations. This table provides only general guidance: the variation from patient-to-patient is large.

Haldane noted that small mammals, and especially small birds, were more rapidly affected by carbon monoxide than humans. This results from the higher specific metabolic rate of such creatures leading to more rapid equilibration between blood and air. Interestingly, the canary, used until recently to detect carbon monoxide in mines in the UK, is less sensitive than humans to low concentrations of carbon monoxide (Spencer, 1961). This is due to avian haemoglobin having a lower affinity for carbon monoxide than human haemoglobin. Thus at low concentrations of carbon monoxide the miner will be affected, possibly even fatally, before the bird.

Treatment of Carbon Monoxide Poisoning

Only a brief account will be provided, since detailed discussions are available in the literature: Crawford et al. (1990), Lowe-Ponsford and Henry (1989), Meredith and Vale (1988). Treatment depends on removing the casualty from the source of carbon monoxide and providing 100% oxygen via a well fitted face mask. Provision of oxygen increases the rate of decline of COHb concentrations: the $t_{1/2}$ of COHb falls from 4–5 h to 40 min in an atmosphere of 100% O_2. Hyperbaric oxygen will drive the $t_{1/2}$ yet lower: 20 min at 2 atmospheres of oxygen.

Table 33.5 Signs and symptoms at various concentrations of carboxyhaemoglobin
From 'Gossel T.A. and Brickler J.D. (1994) *Principles of Clinical Toxicology*. New York:
Reproduced with permission of authors and publisher.

% COHb	Signs and symptoms
0–10	No symptoms; asymptomatic
10–20	Tightness across the forehead, possibly slight headache, dilation of the cutaneous blood vessels, exertional dyspnea
20–30	Headache and throbbing in the temples, easily fatigued, possible dizziness
30–40	Severe headache, weakness, dizziness, confusion, dimness of vision, nausea, vomiting, and collapse
40–50	Same as above, a greater possibility of collapse, syncope, and increased pulse and respiratory rate
50–60	Syncope, increased respiratory and pulse rate, coma, intermittent convulsions, and Cheyne–Stokes respiration
60–70	Coma, intermittent convulsions, depressed heart action and respiratory rate, and possibly death
70–80	Weak pulse, slow respirations, respiratory failure, and death within a few hours
80–90	Death in less than an hour
90+	Death within a few minutes

From Gossel T.A. and Brickler J.D. (1994) *Principles of Clinical Toxicology*. New York: Raven Press.
Reproduced with permission of authors and publisher.

It has been suggested that the addition of 5–6% CO_2 to the oxygen used in treating CO poisoning is an aid in that it stimulates respiration and thus increases the rate at which CO is lost. Haldane advocated this but opinion today is against it. Leading UK experts on the management of CO poisoning have pointed out that though adding CO_2 did produce a reduction in the $t_{1/2}$ of CO, the risks of acidosis outweighed the benefits. They advised that CO_2 should not be given (Meredith *et al.*, 1987). It is interesting to note that all mention of the use of carbon dioxide has been deleted from the Third Edition of the *Oxford Textbook of Medicine* (Vale *et al.*, 1996).

Diazepam should be used to control convulsions and mannitol may have a place in reducing cerebral oedema.

Though not relevant to the consideration of CO as an air pollutant, it is useful to recall that poisoning with dichloromethane will lead to greatly increased endogenous production of CO since it is metabolized by hepatic microsomal enzymes to carbon monoxide and will raise COHb levels.

Sequelae of carbon monoxide poisoning

A range of CNS effects have been described in patients who have recovered from acute poisoning with CO. These are well described and accepted (Ginsberg, 1985; Min, 1986; Myers *et al.*, 1985; Shillito *et al.*, 1936). Whether CNS effects occur as a result of chronic exposure to low concentrations of CO (subclinical poisoning) is less certain. These effects are discussed separately.

CNS Effects Following Recovery from Acute Poisoning with Carbon Monoxide

A very wide range of neurological effects may develop following severe CO poisoning. Descriptions of these effects date back to 1888. Meyer (1960) noted that in a review published in 1906 some 200 references to the effects of carbon monoxide on the CNS had been listed. Such effects are rare in patients with relatively low COHb concentrations and seem to be prevented by the use of hyperbaric oxygen. The recorded incidence of long-term neuropsychological sequelae varies from study to study, from 0.3% to 30% (Spinnler *et al.*, 1980).

A delayed neurological syndrome has been widely described. The patient is typically discharged after management of the acute episode only to be re-admitted up to 40 days later with evidence of deteriorating neurological function (Werner *et al.*, 1985). This may be of rapid onset with headache, neck stiffness and vomiting and accurate diagnosis is dependent on linking these signs with the earlier episode of poisoning. The presenting signs are often protean and range from a loss of concentration, dullness or depression to a Parkinsonian syndrome. The list of symptoms and signs which have been described is now long, the following is an abbreviated summary:

> apathy, disorientation in time and space, amnesia, hypokinesia, mutism, irritability, apraxia, bizarre behaviour, manneristic behaviour, confabulation, insomnia, depression, emotional lability, delusions, echolalia, urinary and/or faecal incontinence, disturbances of gait, rigidity, tremor, chorea, dysarthria, and flaccid paralysis.

Accounts in the original literature should be consulted for details of the presentation of individual cases: Choi (1983); Davous *et al.* (1986); Hart *et al.* (1988); Lugaresi *et al.* (1990); Schwartz *et al.* (1985); Spinnler *et al.* (1980). The capacity of such a range of effects to mimic almost any disease of the CNS means that diagnosis rests largely on an adequate index of suspicion. Confusion with cases of Parkinson's disease and the Gilles de la Tourette syndrome may occur (Grinker, 1926; Pulst *et al.*, 1983). Cortical blindness, or the Kluver–Bucy syndrome, seems to be a well-reported effect. Sophisticated testing with neuropsychological test batteries may be necessary to reveal the full range of effects (Hönigl *et al.*, 1993, Vicente, 1980).

Such widespread effects as these suggest a widespread pathological process (Meyer, 1960). Demyelination has been described, as has axis cylinder (axon) damage. Lesions appear most commonly in the globus pallidus and it has been suggested that this is due to this area being at increased risk of hypoxia because it lies across a divide in arterial supply. Lesions of the frontal cortex are also common. CT and MRI scanning have recently proved valuable in detecting lesions, though these techniques seem to provide only limited assistance in estimating prognosis (Lee and Marsden, 1994; Zagami *et al.*, 1993).

A range of possible underlying mechanisms have been suggested. The label leukoencephalopathy has been applied and similar changes have been produced in animal models by repeated exposure to air containing low concentrations of oxygen. Proposed mechanisms include free radical activity and lipid peroxidation (Norton, 1986). Endothelial proliferation has also been suggested.

The reasons for the markedly delayed onset of neurological symptoms and signs in some cases seem to be unknown.

CNS Effects of Prolonged or Repeated Exposure to Low Concentrations of Carbon Monoxide: the So-called 'Chronic Occult Carbon Monoxide Poisoning Syndrome'

A chronic flu-like illness has been described and attributed to prolonged or repeated exposure to relatively low concentrations of CO (Baker *et al.*, 1988; Crispen, 1989). The common complaints of headache, irritability and malaise have been described in both adults and children. Kirkpatrick (1987) has pointed out that COHb concentrations in blood on admission to hospital did not correlate well with symptoms and attributed this to the decline in COHb levels once exposure had stopped. The high concentrations of COHb found in cigarette smokers may also obscure the diagnosis. Patients described by Kirkpatrick showed symptoms at measured COHb concentrations of only 5% and the author pointed out that these would not have been expected from experimental studies. As cigarette smokers often show COHb saturations in excess of 5%, it is often assumed that they develop a degree of tolerance to carbon monoxide.

It is not possible to quantify accurately the public health effects of exposure to low levels of carbon monoxide. In 1989 Halpern suggested that 3–5% of those who seek medical attention for headache or dizziness in an urban emergency department may be suffering from the effects of exposure to carbon monoxide (Halpern, 1989).

EFFECTS OF CARBON MONOXIDE ON THE CARDIOVASCULAR SYSTEM

The literature dealing with the effects of carbon monoxide on the cardiovascular system (CVS) is large and to some extent contradictory. Mechanistic studies in this area have been dominated and complicated by studies of the effect of cigarette smoking on the CVS. There is little doubt that cigarette smoking increases the risk of atherosclerosis and myocardial infarction. Whether the chronically raised levels of COHb found in smokers are a major causal factor remains uncertain: raised concentrations of COHb are a good marker of cigarette smoking but this does not prove a causal link. Nicotine has been suggested to play a role and has been shown to be associated with the development of Monckeberg's sclerosis of the arterial tunica media. Animal studies involving exposure to high concentrations of carbon monoxide or cigarette smoke have demonstrated damage to the arterial wall including subintimal oedema, increased endothelial permeability and proliferation of vascular smooth muscle. In many of these studies animals have been maintained on a high-fat diet. The changes which are characteristic of atherosclerosis have been reduced or not found in animals fed on normal diets (Astrup *et al.*, 1967; Bing *et al.*, 1980; Malinow *et al.*, 1976; Penn *et al.*, 1983; Rogers *et al.*, 1980; Stender *et al.*, 1977; Turner *et al.*, 1979).

The possible chronic and acute effects of exposure to CO on the CVS need to be carefully differentiated. It has been suggested that chronic exposure to CO may lead to the establishment of a disease state, atherosclerosis, whilst acute exposure could trigger events such as myocardial arrhythmias and infarction. The evidence that CO exposure, other than by smoking, can produce chronic effects is not strong. Studies in workers exposed to repeated high concentrations of CO, e.g. tunnel workers and firemen, have suggested effects: it will be understood that the relevance of such studies to predicting whether long-term exposure to ambient concentrations of CO produce similar effects is questionable (Ayres *et al.*, 1973a; Gordon and Rogers, 1969). In this chapter emphasis has been placed on the effects of exposure to ambient concentrations of CO. Exposure to such concentrations is unlikely to raise the COHb level above 5%; indeed, for the great majority of people an upper figure of 2% is more likely. Thus, attention has been paid to studies which have looked at the effects of COHb levels of this order on the CVS. If significant effects are demonstrated at such levels then we may expect the public health impact to be large: cardiovascular diseases are common and so is exposure to carbon monoxide.

Both clinical (volunteer) and epidemiological studies have concentrated on the effects of carbon monoxide on the myocardium. As blood passes through myocardial capillaries more oxygen is extracted than in most other tissues. The mixed coronary venous Po_2 may be of the order of 20 mmHg as compared with a more general mixed venous figure of 40 mmHg. The Haldane Shift of the O_2Hb dissociation curve produced by exposure to carbon monoxide might therefore be expected to reduce the myocardial Po_2 to unacceptably low levels. In tissue such as the endocardium, at perhaps already borderline levels of Po_2, such an effect might be damaging. Myoglobin has been shown to play a role in oxygen transport and storage in the heart. Binding of CO to myoglobin may add to its effects on oxygen transport.

It is not difficult, then, to imagine that a patient suffering from inadequate coronary arterial blood flow and experiencing anginal pain on exertion might be adversely affected by exposure to carbon monoxide. This is the case and volunteer studies have shown that exposure to concentrations of CO sufficient to raise the blood COHb level to 5% reduced the time to onset of exercise-induced angina in such patients. The critical question is: is there a demonstrable threshold of effect? As in much of environmental toxicology the answer is probably: yes – at an individual level – and no – at a population level. One might reasonably expect those with a severely impaired myocardial blood supply to be affected before those with only borderline impairment. Each patient might well have an individual threshold of COHb at which effects become apparent, and the true population threshold will be that of the most sensitive individual. Despite this, the 'search for a threshold' has been enthusiastically pursued via chamber studies of comparatively small numbers of volunteers.

Clinical Studies of Volunteers

These studies usually involve the exposure of patients who suffer from exercise-induced angina, or defined exercise-induced changes in their electrocardiogram (ECG) traces, to known concentrations of CO. Exposure concentrations tend to be higher than ambient, the object being to reach a predetermined COHb level rapidly. Once this level is reached, subjects are asked to exercise until either a pre-determined change in the ECG trace, e.g. a 1 mm depression in the ST segment, or anginal pain occurs. In general, COHb concentrations relate to the rate at which these end-points are reached. Many workers have addressed these problems and a large literature has accumulated. As a rather selective guide to original sources, the following classified list is offered:

- *Original reports of studies:* Adams *et al.*, 1988; Allred *et al.*, 1989a,b; Anderson *et al.*, 1973; Aronow, 1981; Aronow and Isbell, 1973; Dahms *et al.*, 1993; Mall *et al.*, 1985; Sheps *et al.*, 1987, 1990.
- *Reviews and commentaries:* Aronow, 1978; Ayres *et al.*, 1969, 1970, 1973b; Katzenstein, 1990; Kuller *et al.*, 1972; Penney, 1988; Rosenman, 1990; Ström *et al.*, 1995; Walden and Gottlieb, 1990.

A general criticism of the approach is that it relates all the effects of CO to a change in COHb concentration. Though this is likely to be the major cause of the observed effects, binding of CO to other molecules, for example myoglobin and mitochondrial cytochrome enzymes, should not be forgotten. Effects such as these may be more difficult to relate directly to concentrations of COHb and little is known of their significance. The need for further work in this area is clear.

Another generic problem with these studies relates to the methods used for monitoring COHb levels. Two techniques have been used: the CO-Oximeter and gas chromatography. These methods are compared in a most important multicentre study sponsored by the US Health Effects Institute and reported in 1989. It was shown that the CO-Oximeter produced higher values for COHb concentrations than the GC method. The implications of this finding were discussed by the HEI Review Committee (Allred *et al.*, 1989b). It was accepted that the CO-Oximeter gave 'results of limited accuracy at low COHb concentrations (under 5%)'. The desirability of using GC methods was pointed out, though the greater technical difficulties and lack of rapid readings with this method were recognized. It was recommended that further studies of how to deal with the difference in results produced by the two methods, defined as the 'offset', were needed.

Rather than discuss a series of studies, emphasis has been placed on the results of the HEI Multicenter Study. This large and carefully conducted study was set up specifically to look at possible effects at low concentrations of COHb. In 62 non-smoking volunteers, the mean pre-exposure concentration of COHb was 0.64% (SEM = 0.02), measured using gas chromatography. Exposure to CO was designed so that post-exercise COHb concentrations should be 2 or 4%. Allowance was made for elimination of carbon monoxide during exercise and pre-exercise concentrations were a little higher than noted above.

Tables 33.6a and 33.6b shows the results of the HEI Multicenter Study, and Tables 33.7 and 33.8 sum up the results of the key studies in this area. These tables are reproduced in full from the HEI report: the footnotes provide useful information on the methods used and the original report should be consulted for details.

It was concluded that exposure to CO sufficient to raise the concentration of COHb to 2% was sufficient to produce effects during exercise in patients with coronary artery disease. In addition, evidence of an exposure–response relationship was found: effects were increased at 4% COHb.

An interesting comparison of the results of the HEI Multicenter Study and other published studies is shown in Fig. 33.13.

Note that the COHb concentrations are all shown as measured with the CO-Oximeter technique. This was the standard method in all the reported studies except the HEI Multicenter Study where it was used in addition to gas chromatography. The results recorded by Aronow indicate effects at lower concentrations of COHb than those used in the HEI study. In 1983, an *ad hoc* US EPA committee looked critically at the results of the Aronow studies and concluded that they did not reach a reasonable standard of

Table 33.6 Key results of the Health Effects Institute Multicenter study.
(a) Effect of carbon monoxide on time to ST end-point (combined data)

Exposure day	Sample size		COHb levels at end of exercise pre- and post-exposure		Time to ST end-point pre- and post-exposure[a]		Change in time to ST endpoint post- vs. pre-exposure (s)		% Decrease between air and CO days			
			Mean %COHb[b]	SEM	Trimmed mean	SEM	Trimmed mean	SEM	Trimmed mean %[c]	p-Value[d]	90% Confidence interval	95% Confidence interval
Air	62	Pre	0.64	0.02	560.0	26.6	16.0	11.6				
		Post	0.62	0.02	575.9	26.6						
2%-COHb target	61	Pre	0.62	0.02	574.1	26.8	-16.3	13.0	5.1	0.01	1.46, 8.74	0.77, 9.43
		Post	2.00	0.05	557.8	25.4						
4%-COHb target	62	Pre	0.64	0.02	562.9	27.6	-52.7	12.7	12.1	≤0.0001	9.0, 15.3	8.4, 15.9
		Post	3.87	0.08	510.1	25.9						

[a] Median time to ST end-point: air, pre = 540, post = 570; 2%-COHb target, pre = 560, post = 524; 4%-COHb target, pre = 540, post = 500.
[b] CO measured by GC.
[c] Median percent decrease: air vs. 2%-COHb target = 6.5; air vs. 4%-COHb target = 13.2.
[d] One-sided p-values.

Table 33.6 Key results of the Health Effects Institute Multicenter study.
(b) Effect of carbon monoxide on time to angina (combined data)

Exposure day	Sample size	COHb levels at end of exercise pre- and post-exposure		Time to angina- pre- and post-exposure[a]		Change in time to angina post- vs. pre-exposure (s)		% Decrease between air and CO days			
		Mean %COHb[b]	SEM	Trimmed mean	SEM	Trimmed mean	SEM	Trimmed mean %[c]	p-Value[d]	90% Confidence interval	95% Confidence interval
Air	63	Pre 0.64	0.02	Pre 519.0	26.7	-17.4	10.9				
		Post 0.62	0.02	Post 501.6	24.6						
2%-COHb target	62	Pre 0.62	0.02	Pre 525.2	26.2	-42.8	10.6	4.2	0.027	0.66, 7.94	0.4, 8.74
		Post 2.00	0.05	Post 482.4	22.0						
4%-COHb target	63	Pre 0.64	0.02	Pre 515.0	26.5	-49.6	10.9	7.1	0.002	3.06, 10.94	5.18, 14.46
		Post 3.87	0.08	Post 465.4	24.1						

[a] Median time to angina: air, pre = 520, post = 489; 2%-COHb target, pre = 482, post = 460; 4%-COHb target, pre = 480, post = 440.
[b] CO measured by GC.
[c] Median percent decrease: air vs. 2%-COHb target = 4.2; air vs. 4%-COHb target = 9.0.
[d] One-sided p-values.

From Allred et al. (1989b). Reproduced with permission of the US Health Effects Institute.

Table 33.7 Comparison of HEI Multicenter study with other studies evaluating the effect of carbon monoxide on the time to onset of angina in subjects with coronary artery disease

Investigators	No. of subjects	Exposures	%COHb[a] (mean ± SD) Spectrophotometry	Gas chromatography	Effect of CO on time to angina[b] % decrease CO day vs. air day	SEM
Aronow and Isbell (1973)	10	Air/2 h[c]	0.8 ± 0.2[d]	—	—	
		50 ppm CO/2 h[c]	2.7 ± 0.2[d]	—	15	2
Aronow (1981)	15	Air/1 h[c]	1.0 ± 0.1[e]	—	—	
		50 ppm CO/1 h[c]	2.0 ± 0.2[e]	—	12	1
Anderson et al. (1973)	10	Air/4 h[c]	1.3 ± 0.4[f]	—	—	
		50 ppm CO/4 h[c]	2.9 ± 0.7[f]	—	21	6
		100 ppm CO/4 h[c]	4.5 ± 0.8[f]	—	21	6
Kleinman and Whittenberger (1985)	26	Air/1 h[c]	1.4 ± 0.5[d] (1.4 ± 0.5)[e]	—	—	
		100 ppm CO/1 h[c]	3.0 ± 0.5[e] (2.8 ± 0.5)[e]	—	7	4
Sheps et al. (1987)	23[b]	Air/about 1 h[i]	1.7 ± 0.1[e]	—	—	
		100 ppm CO/h[i] about 1 h	4.1 ± 0.1[e] (3.6 ± 0.1)[e]	—	1[g]	9

Table 33.7 *cont.*

Investigators	No. of subjects	Exposures	%COHb[a] (mean ± SD)		Effect of CO on time to angina[b]	
			Spectrophotometry	Gas chromatography	% decrease CO day vs. air day	SEM
HEI Multicenter Study	63	0–2 ppm CO[i] mean 0.69/ 50–70 min	1.4 ± 0.4[e] (1.2 ± 0.4)[c]	0.70 ± 0.19 (0.62 ± 0.19)	—	
		42–202 ppm CO[i] mean 117/ 50–70 min	3.2 ± 0.3[e] (2.6 ± 0.3)[c]	2.38 ± 0.40 (2.00 ± 0.41)	5	2
		143–357 ppm CO[i] mean 253/ 50–70 min	5.6 ± 0.5[e] (4.7 ± 0.4)[c]	4.66 ± 0.62 (3.87 ± 0.62)	7	3

[a] % COHb at end of exposure. We calculated SD for Kleinman and Whittenberger data. Numbers in parentheses are %COHb at the end of the exercise test after the exposure period.

[b] The design of the HEI study was the same as the two Aronow studies in that it had a pre- and post-exposure exercise test each day. For those studies, we calculated the pairwise percent change between the tests on each day and then, pairwise, subtracted the result on the CO day from the result on the air day. For the Kleinman and Sheps data, we calculated the pairwise percent change between the air and CO days (as did Anderson). We used the average of the two non-randomized air days for the two subjects for whom data from the randomized air day were not available. For comparison with the other studies, the percentages listed in this table for the HEI study are means (see Appendix C) rather than trimmed means (Table 14). We calculated SEMs for all studies.

[c] Mask exposure.

[d] IL 182 CO-Oximeter.

[e] IL 282 CO-Oximeter.

[f] Spectrophotometric methods (Buchwald, 1969).

[g] Not statistically significant by one-tailed T-test.

[h] Thirty subjects in study, but angina analysed in only 23 of them.

[i] Chamber exposure.

From Allred *et al.* (1989b). Reproduced with permission of the US Health Effects Institute.

Table 33.8 Comparison of subjects in studies of the effect of carbon monoxide exposure on occurrence of angina during exercise

Study	No. of subjects	Gender	Medication	Smoking history	Description of disease	Age (years)
					Subject characteristics	
Aronow and Isbell (1973)	10	male	Not described	No current smokers	Classic exertional angina, CAD with > 50% stenosis of 1 or more major vessels	40–55 (mean = 49)
Aronow (1981)	15	14 male 1 female	Not described	No current smokers	Stable angina pectoris with angiographically demonstrated CAD; 8 had prior MI	50.0 ± 7.2
Anderson *et al.* (1973)	10	male	1 subject took digitalis; drug therapy basis for exclusion	5 smokers (refrained for 12 h prior to exposure)	Stable angina pectoris, positive exercise test (ST changes); reproducible angina on treadmill	(mean = 49.9)
Kleinman and Whittenberger (1985)	26	male	14 on β-blockers, 19 on nitrates	No current smokers	Ischaemic heart disease, stable exertional angina pectoris	49–66 (mean = 59)
Sheps *et al.* (1987)	30 (23 with angina)	25 male 5 female	26 subjects on medication; 19 on β-blockers; 11 on Ca-channel-blockers; 1 on long-acting nitrates	No current smokers	Ischaemia during exercise (ST changes or abnormal ejection fraction response) and 1 or more of the following: (1) angiographically proven CAD; (2) prior MI; (3) typical angina	36–75 (mean = 58.2)
HEI Multicenter Study	63	male	38 on β-blockers; 36 on nitrates; 40 on calcium antagonists	No current smokers	Stable exertional angina and positive exercise test (ST changes) plus 1 or more of the following: (1) ≥ 70% lesion by angiography in 1 or more major vessels; (2) prior MI: (3) positive exercise thallium test	41–75 (mean = 62.1)

From Allred *et al.* (1989b). Reproduced with permission of the US Health Effects Institute.

scientific quality and thus should not be used as a basis for defining a lowest observed adverse effect level (LOAEL) for carbon monoxide. This criticism hinged on the heterogeneity of the subjects included in Aronow's studies as regards the extent of ST segment depression produced by exercise; the comparatively small number of subjects; and the lack of data on therapy being taken by the subjects.

What are the public health implications of the HEI study? In discussing the relationship between exposure concentrations of CO and COHb concentration earlier in this

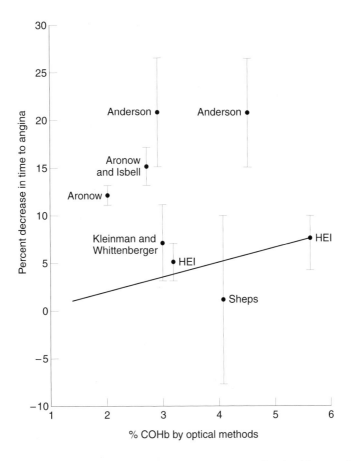

Fig. 33.13 Comparison of effect of CO exposure on time to angina in several studies. The regression line is based on data from the HEI study. Bars indicate standard errors of the mean that were calculated. Because of major protocol differences among these studies and the imprecision in optical measurements of COHb, these comparisons must be interpreted extremely cautiously. Note that in this figure, we refer to end-of-exposure %COHb levels rather than end-of-exercise levels. From Allred *et al.* (1989b). Reproduced with permission of the US Health Effects Institute.

chapter the Coburn Equation and the work of other authors was considered. In addition, the effects of exposure to changing concentrations of CO were considered. The US EPA in 1982 undertook an important sensitivity analysis which modelled a number of patterns of exposure (each conforming to the US National Ambient Air Quality Standard for CO) and a predicted normal range of physiological variables (Biller and Richmond, 1982). The results are shown in Table 33.9.

The key figure is the 10% of the population who might reach a COHb concentration of 2.1% on exposure to a pattern of CO concentrations which conformed to the US National Ambient Air Quality Standard: 9 ppm, 8-h average concentration. Given that coronary artery disease is common, it is clear that a substantial number of people might be placed at some increased risk even when CO concentrations conformed to the standard. The critical question is: how large an increased risk?

Table 33.9 Percentage of persons with carboxyhaemoglobin greater than or equal to specified peak value when exposed to air quality associated with 8-h daily maximum carbon monoxide standard[a,b,c]

Peak COHb %	9 ppm, 8-h; 1 expected exceedance		
	Low pattern	Midrange pattern	High pattern
3.7			
3.5			
3.3			
3.1			
2.9			<0.01
2.7			0.02
2.5		<0.01	0.2
2.3		0.02	2
2.1	<0.01	0.4	10
1.9	0.05	5	53
1.7	3	35	98
1.5	39	88	100
1.3	97	100	100
1.1	100	100	100

[a] COHb responses to fluctuating CO concentrations were dynamically evaluated using the Coburn model prediction of the COHb level resulting from one hour's exposure as the initial COHb level for the next hour. The series of 1-h CO concentrations used were from 20 sets of actual air quality data. Each pattern was proportionally rolled back or up so that its peak 8-h CO concentration equalled the level of the 8-h standard. Of the 20 selected patterns, results from three patterns are presented here. The low pattern tends to give the lowest peak COHb levels, the midrange pattern tends to give a midrange value, and the high pattern tends to give the highest value.

[b] Haldane constant = 218. Alveolar ventilation rate = 10 l/min. Altitude = 0.0 ft.

[c] The estimation of distributions for each of the physiological parameters used in the Coburn model and the Monte Carlo procedure used to generate these estimates are discussed in Appendix C of the Sensitivity Analysis.

From *Review of the NAAQS for Carbon Monoxide: Reassessment of Scientific and Technical Information.* US Environmental Protection Agency, Research Triangle Park, NC. This table has been modified and simplified.

Allred *et al.* (1989b) point out that the changes recorded in the HEI study at 2% COHb concentrations are similar to those 'considered clinically significant when evaluating the efficacy of anti-ischaemic therapeutic interventions'. The authors went on to point out that repeated episodes of myocardial ischaemia had been related to myocardial necrosis and might lower the fibrillation threshold, predispose to ventricular arrhythmias and may be related to sudden death. The possible public health effects should not then be taken lightly.

The authors also stressed possible effects on cigarette smokers. It has already been explained that a smoker with a COHb concentration of, say, 6% will, on exposure to 20 ppm CO, equilibrate downwards towards a COHb concentration of about 3%. This assumes that he or she stops smoking during the exposure to CO. If the person continues smoking, then the effects of smoking and exposure to CO will be additive and their COHb concentration will rise. Allred pointed out that the HEI study had shown an exposure–response relationship without an obvious threshold, and thus the additional COHb concentration contributed in smokers by exposure to low concentrations of CO could increase their risk of effects on the heart.

Dahms *et al.* (1993) looked at the effects of exposure to carbon monoxide on the frequency of ventricular ectopic beats in patients with a known history of such beats. Exposure to CO which raised COHb concentrations to 3% or 5% over 1 h failed to produce any increase in ectopic beats. This study casts some doubt on the possible association between exposure to carbon monoxide and sudden death due to ventricular fibrillation. This conclusion is supported by the work of Hinderleiter *et al.* (1989).

Epidemiological Studies

Two types of epidemiological study need to be considered:

(1) studies designed to investigate whether long-term exposure to carbon monoxide from sources other than smoking increases the likelihood of development of cardiovascular disease;

(2) studies designed to investigate whether acute exposure to raised ambient concentrations of carbon monoxide increases the likelihood of myocardial infarction or heart failure.

Studies Designed to Investigate Whether Long-term Exposure to Carbon Monoxide from Sources Other than Smoking Increases the Likelihood of Development of Cardiovascular Disease

The task of reviewing the literature in this area has been made lighter by a recent review by Ström, Alfredsson and Malmfors (1995). The authors undertook a systematic evaluation of both epidemiological and animal studies bearing on this topic: here only the epidemiological studies are considered. Studies in this area are dominated by those of the effects of occupational exposure to carbon monoxide. Of the 16 studies reviewed by Ström *et al.*, only two were studies of the effects of urban exposure to carbon monoxide. Ström *et al.* devised a weighting system for the studies reviewed. This allocated a mark: from 'X' to 'XXXXX', for increasing study quality depending on the quality of the study design and the suitability of the study for allowing 'profound and reliable' conclusions to be drawn. Both the studies of the effects of urban exposure scored only 'X'. Despite this, these studies require some examination.

Cohen examined case fatality reports for patients admitted to Los Angeles hospital for treatment of myocardial infarction (Cohen *et al.*, 1969). An increase in deaths was noted in heavily polluted areas during episodes of air pollution. The study did not control for smoking, diet, physical activity nor for social class. That an effect was seen only during episodes of pollution suggested that acute exposure to pollution could be playing a part. The study sheds little light on the effects of long term exposure to carbon monoxide.

Kuller reported an association between daily numbers of cases of myocardial infarctions and ambient concentrations of carbon monoxide in Baltimore (Kuller *et al.*, 1975). Again this seems to reflect, if anything, acute rather than long-term effects. The studies of repeated occupational exposures lead to a similar conclusion: one can make a case for effects of acute exposure but there are no satisfactory studies supporting an effect of chronic exposure. Designing a study which would provide persuasive evidence of such a causal relationship between chronic exposure to ambient concentrations of CO in either smokers or non-smokers will not be easy. Unfortunately, the well known Six Cities study in the USA (Dockery *et al.*, 1993) did not include measurements of carbon monoxide.

Recent studies in animals have cast doubt on the hypothesis that chronic exposure to CO increases the risk of development of atherosclerosis and so support the rather negative, though inadequate, epidemiological data (Bing *et al.*, 1980, Penn *et al.,* 1992).

Studies Designed to Investigate Whether Acute Exposure to Raised Ambient Concentrations of Carbon Monoxide Increases the Likelihood of Myocardial Infarction or Heart Failure

The clinical studies described above suggest that exposure to carbon monoxide might be a precipitating factor in diseases related to myocardial ischaemia. One study addressing this possibility has already been discussed briefly (Cohen *et al.*, 1969). In 1971, Hexter and Goldsmith (1971) reported an association between daily death rate (all causes) and daily average CO concentration in the Los Angeles County area. The statistical approach used was one of multiple regression with emphasis placed on the removal of secular trends. Day-to-day autocorrelations were also considered. The study showed that on a day when the average CO concentration was 20.2 ppm (the highest daily concentration recorded during 4 years), as compared with a day when the average concentration was 7.3 ppm (the lowest daily concentration recorded during 4 years), carbon monoxide contributed 11 to the total deaths occurring. The mean number of deaths occurring per day in the area studies was 159.43 (SD = 17.39). The association between oxidants and daily mortality was weaker than that described for carbon monoxide.

This study was reported more than 20 years ago and remarkably few studies have been reported since. Emphasis has been placed on other pollutants and carbon monoxide has been largely ignored.

One study which confirmed the findings of Hexter and Goldsmith was that of Kurt *et al.,* reported in 1976. The authors examined the daily frequency of attendances for treatment of cardiorespiratory complaints at emergency departments in Denver during a 3-month winter period. In all, some 8566 cases were analysed (average daily attendance = 93). A significant correlation between days of low or high CO concentration and attendances was demonstrated.

In 1995, Morris *et al.* reported an association between hospital admissions for congestive heart failure and ambient air pollution in a study of seven US cities. For each city admissions were positively associated with concentrations of carbon monoxide in both single and multiple pollutant models. The relative risk of admission associated with an increase of 10 ppm (maximum hourly concentration in day) ranged from 1.10 in New York to 1.37 in Los Angeles. The results of the authors' analysis are shown in Table 33.10.

The association with carbon monoxide is notably stronger than that found for other pollutants. The authors calculated that 5.7% of all admissions for management of congestive heart failure could be attributed to exposure to carbon monoxide. Based on standard costs of treatment, the authors further calculated that the annual cost of attributable admissions in the seven cities was $33 million. In this calculation only the direct cost of admissions for patients aged more than 65 years was included.

Studies in Los Angeles (Kinney *et al.*, 1995) and São Paulo (Saldiva *et al.*, 1995) have found positive associations between CO concentrations and daily death rates. In Detroit, a positive association between admissions for heart failure and daily CO concentrations has been demonstrated and found to be robust to the removal of the likely effects of a key co-pollutant, particulate matter, from the regression model (Pope *et al.*, 1995).

Table 33.10 Relative risks of congestive heart failure admission among Medicare patients as a function of ambient pollutant levels

Pollutant and city	Single-pollutant model		Multipollutant model	
	Relative risk	95% Confidence interval	Relative risk	95% Confidence interval
Carbon monoxide				
Los Angeles	1.36	1.28, 1.46	1.39	1.23, 1.56
Chicago	1.29	1.16, 1.44	1.23	1.07, 1.43
Philadelphia	1.17	1.05, 1.31	1.22	1.05, 1.41
New York	1.10	1.03, 1.18	1.05	0.97, 1.14
Detroit	1.24	1.11, 1.39	1.38	1.17, 1.63
Houston	1.11	0.97, 1.26	1.25	1.05, 1.49
Milwaukee	1.29	1.07, 1.57	1.26	0.89, 1.77
Nitrogen dioxide				
Los Angeles	1.15	1.10, 1.19	0.98	0.91, 1.05
Chicago	1.17	1.07, 1.27	1.06	0.92, 1.22
Philadelphia	1.03	0.95, 1.12	0.95	0.84, 1.08
New York	1.07	1.02, 1.13	1.08	1.01, 1.16
Detroit	1.04	0.92, 1.18	1.01	0.88, 1.15
Houston	0.99	0.88, 1.10	0.83	0.70, 1.00
Milwaukee	1.05	0.89, 1.23	0.87	0.68, 1.12
Sulfur dioxide				
Los Angeles	1.60	1.41, 1.82	1.00	0.81, 1.24
Chicago	1.05	1.00, 1.10	1.00	0.95, 1.06
Philadelphia	1.01	0.96, 1.06	0.99	0.94, 1.05
New York	1.04	1.01, 1.08	1.01	0.97, 1.06
Detroit	1.00	0.95, 1.06	0.98	0.91, 1.06
Houston	1.07	0.97, 1.17	1.11	0.99, 1.24
Milwaukee	1.07	0.99, 1.15	1.07	0.97, 1.19
Ozone				
Los Angeles	1.06	1.01, 1.11	1.03	0.97, 1.09
Chicago	1.03	1.93, 1.14	0.98	0.87, 1.10
Philadelphia	0.95	0.87, 1.05	0.98	0.89, 1.09
New York	0.89	0.81, 0.97	0.84	0.76, 0.92
Detroit	0.90	0.78, 1.05	0.92	0.76, 1.12
Houston	0.99	0.90, 1.08	0.98	0.87, 1.11
Milwaukee	1.00	1.90, 0.53	0.84	0.69, 1.02

Note: The multipollutant model included all four pollutants. All models included temperature, month, day of week, and year. Values refer to the relative risk associated with an increase of 10 ppm of carbon monoxide, 0.1 ppm of nitrogen dioxide, 0.05 ppm of sulfur dioxide, or 0.12 ppm of ozone.

From Morris *et al.* (1995). Reproduced with permission of authors and publisher.

A detailed discussion of the results of these recent studies has been provided by Schwartz (1995). He pointed out that the US National Ambient Air Quality Standard had been based on the results of chamber studies and argued that these studies might lead to an underestimate of the impact of carbon monoxide on public health. Schwartz raised the important question of whether the demonstrated association was causal and asked whether CO might be acting as a proxy for other pollutants: perhaps fine particles. This point does not yet appear to have been resolved.

Schwartz also pointed out that the correlation between mean population exposure to CO (unmeasured) and ambient concentrations as measured was likely to be stronger than the weak correlations reported between individual exposure and ambient concentration: time series studies using the day as the unit of analysis were thus recommended.

Given that there is a plausible mechanism of effect linking exposure to carbon monoxide with effects on the cardiovascular system, the results of the epidemiological studies described above deserve to be taken seriously. Further studies are clearly needed.

EFFECTS OF EXPOSURE TO AMBIENT CONCENTRATIONS OF CARBON MONOXIDE ON THE DEVELOPING CHILD

Given the already left-shifted dissociation curve of fetal O_2Hb, the relatively low fetal PO_2, the greater affinity of fetal haemoglobin than adult haemoglobin for CO and the limited capacity of the fetus to respond to low Po_2 levels it is hardly surprising that exposure of pregnant animals to high concentrations of CO has led to fetal damage. Studies of this problem have been driven largely by concerns about the effects of maternal smoking on the fetus (Butler *et al.*, 1972; Longo, 1977; Meyer and Comstock, 1972; Meyer *et al.*, 1976).

Animal Studies

A number of studies have shown that exposure of pregnant rats and rabbits to carbon monoxide or cigarette smoke leads to a reduction in birthweight of offspring, an increased incidence of still-births and a raised level of neonatal mortality. The concentrations of CO employed in such studies tend to be high and the results are not especially relevant to the possible effects of exposure to ambient concentrations of carbon monoxide (Astrup *et al.*, 1975).

Younoszi and colleagues, in 1969, reported a study in which pregnant rats were exposed to either cigarette smoke, smoke from burning lettuce leaves (no nicotine) or smoke from burning lettuce leaves with added nicotine for 4 min, on five occasions each day from the 3rd to the 22nd day of pregnancy. COHb concentrations in the adult rats were maintained at between 2 and 8%: no difference in the COHb concentrations were found between the groups. Effects on the birth weight of offspring were found in all groups, the largest effects being in the group exposed to cigarette smoke. It was noted that exposure reduced food intake and that this was a known cause of reduction in the birthweight of offspring. Reduction of maternal food intake failed, however, to reproduce the effects seen in the groups exposed to smoke. The results of this study are intriguing but do not seem to have been followed up.

Garvey and Longo (1978) found that exposure of pregnant rats to 90 ppm CO throughout pregnancy induced cerebral oedema in the fetuses. This was despite the fact that this level of exposure did not produce a reduction in the birthweight of the offspring. Such a decrement was produced by exposure of the mothers to 13% O_2 but no cerebral oedema was found. The authors argued that CO might have a differential effect on the development of some organs, e.g. the brain, and that this might not be explained by hypoxia alone.

A detailed study of different levels of exposure to pregnant mice to CO, continuously from day 7 to day 18 of gestation was reported in 1984 by Singh and Scott. Table 33.11 shows the key results. A clear gradient of effect with increasing exposure to CO is seen in each of the three measures of effect: % fetal mortality, number of dead or reabsorbed fetuses per litter and fetal birthweight. Fetal birthweight seemed the most sensitive

Table 33.11 Effects of prenatal carbon monoxide exposure (from gestation day 8 to 10) on implantation, % fetal mortality, number of dead or resorbed fetuses/litter, and fetal weight in mice

Treatment	No. of litters	No. of implantations	% Fetal mortality	No. of dead or resorbed fetuses/litter	Fetal body wt (g)
			$\bar{x}$ ± S.E.	$\bar{x}$ ± S.E.	$\bar{x}$ ± S.E.
0 ppm[a]	17	197	4.52 ± 1.89	0.53 ± 0.21	0.89 ± 0.02
65 ppm	17	210	5.89 ± 1.72	0.76 ± 0.21	0.86 ± 0.02
125 ppm	17	183	12.50 ± 3.37	1.23 ± 0.31	0.83 ± 0.02*
250 ppm	17	193	12.50 ± 3.76	2.05 ± 0.46	0.74 ± 0.02**
500 ppm	17	182	55.30 ± 7.93***	6.35 ± 1.02***	0.68 ± 0.01**

[a] Controls were exposed to the chamber environment and compressed airflows.
* Significantly different from control at $p < 0.05$.
** Significantly different from control at $p = < 0.01$.
*** Significantly different from control and rest of the CO treatment at $p = < 0.01$.

From Singh and Scott (1984). Reproduced with permission of authors and publisher.

indicator of effect with significant effects being recorded at 125 ppm CO. The authors argued that exposure to 65 ppm CO was equivalent to that occasionally encountered in an occupational setting and also by cigarette smokers. The data presented do not show a clear effect at 65 ppm though the recorded reduction in birthweight was regarded by the authors as 'suggestive of an effect'.

Human Studies

Human studies of the effects of carbon monoxide on the fetus can be divided into those dealing with the effects of poisoning with CO and those dealing with the effects of cigarette smoking: volunteer studies are, of course, impossible. The former deal with the effects of acute exposure to high concentrations of CO, the latter with the effects of chronically raised levels of maternal, and fetal, COHb. That these studies are not strictly comparable or perhaps relevant to the possible effects of exposure to ambient concentrations is obvious.

Studies of Poisoning

The lack of studies of low-grade poisoning was noted by Koren *et al.*, and discussed in an important paper in 1991. A study of 40 cases of CO poisoning during pregnancy was reported. The severity of the poisoning was graded 1–5, from alert and orientated to comatose. The findings are shown in Table 33.12.

The authors concluded that severe (grades 4 and 5) poisoning had clearly deleterious effects on the fetus. Grade 1 and 2 poisoning seemed to have no effects. Maternal concentrations of COHb in Grade 1 and 2 cases ranged up to 13.8% and thus covered the range seen in cigarette smokers. The authors speculated that acute exposure to low concentrations of CO were unlikely to affect the fetus adversely.

Though it sheds little light on the effects of exposure to ambient concentrations of CO on the fetus, a paper by Caravati *et al.* (1988) is of great clinical interest. Six cases of

Table 33.12 Relationship between measured or calculated COHb, clinical scoring and fetal outcome

COHb%	Time of exam (h) after exposure	Grade[a]	Rx	Length of Rx (h)	Outcome
40–50	2	5	HfO$_2$	2	Elective termination
39	2	4	HybO$_2$	2	Normal
26	1	4	HfO$_2$	3	Stillborn
25	2	4	HfO$_2$	2	Cerebral palsy
21	2	4	HybO$_2$	2	Normal
18	nk	1	HfO$_2$	12	Normal
14	nk	1	nil		Normal
13.8	1	2	HfO$_2$	7	
			+HybO$_2$	2	Normal
6.2	1.5	1	nil		Normal
2.4	nk	1	nil		Normal
0.8	2	1	nil		Normal
2	nk	1	nil		Normal
Indirect measures of exposure (calculated from ambient ppm)					
32[b]	2	1+	HfO$_2$	12	Normal development quotient
32		1+	nil		Fetal bradycardia
32		1	nil		Normal
14		1	nil		Normal
14		1	nil		36-week gestation
5		1	nil		Normal

[a] See original report for details of grading of level of poisoning.
[b] COHb measurement of affected son.
nk = not known.
HybO$_2$ = hyperbaric O$_2$.
HfO$_2$ = high-flow O$_2$.
From Koren *et al.* (1991). Reproduced with permission of authors and publisher.

poisoning were reported with three leading to loss of the fetus. The authors pointed out that maternal COHb concentrations on admission were not a good indicator of likely effects on the fetus. The clinical state of the mother at the time of discovery of poisoning, on the other hand, was a good indicator of likely effect. The authors pointed out that carbon monoxide is lost slowly from the fetus and, using a model developed by Hill *et al.* (1977), calculated that a woman would need treatment with 100% O$_2$ for five times as long as necessary to reduce her levels of COHb to normal in order to reduce fetal COHb levels to normal. This is a most important finding with respect to the treatment of CO poisoning during pregnancy.

Effects of Maternal Cigarette Smoking

It has been known for many years that maternal smoking leads to a decrease in birth-weight and increased prematurity (Garvey and Longo, 1978). In 1992, Anderson *et al.* reported a study which showed that though alcohol and caffeine had also been suspected of causing a reduction in birthweight, these effects were only seen in babies of mothers who smoked. The effects of smoking on the fetus have generally been attributed to carbon monoxide. Animal studies discussed above support this. However, in 1978 Krishna reported that still births, prematurity and low birthweight babies occurred more frequently than expected amongst mothers who chewed tobacco. The complex mixture of toxic substances in tobacco smoke makes it difficult to extrapolate from data relating to smoking to the effects of exposure to ambient concentrations of carbon monoxide.

Studies of the Effects of Exposure to Differing Ambient Concentrations of Carbon Monoxide

Alderman *et al.* (1987) reported a study of the effects of maternal exposure to different background concentrations of carbon monoxide on birthweight. Babies born in Colorado were grouped as either low or normal with regard to birthweight and their mothers' places of residence traced. If the mothers lived within 2 miles of a stationary carbon monoxide monitor the mother and child were included in the study. In all, 998 low birthweight babies and 1872 normal birthweight babies were followed up. After adjusting for confounding factors including race and maternal education, no association between low birthweight and maternal exposure to carbon monoxide could be demonstrated. Maternal smoking was not controlled for.

Hoppenbrouwers *et al.* (1981) demonstrated an association between the daily incidence of sudden infant death syndrome (SIDS) and daily mean concentrations of carbon monoxide in Los Angeles County. It was suggested that fetal anoxia caused by exposure to carbon monoxide could contribute to SIDS deaths. Positive associations were also demonstrated with daily concentrations of sulfur dioxide, nitrogen dioxide and hydrocarbons: all these pollutants tend to increase on high pollution days and little support for this hypothesis has appeared.

SETTING AN AIR QUALITY STANDARD FOR CARBON MONOXIDE

Ambient air quality standards are designed to protect the general population against the adverse effects of exposure to air pollutants. In setting standards it is appreciated that in any population there is likely to be a range of individual sensitivity to the pollutant in question and also a range of exposure to that pollutant. Standards are generally set so that even if sensitive individuals were to be exposed to the conditions specified by the standard, ill effects would be unlikely to occur.

In the case of carbon monoxide, two important but very different data sets need to be considered in setting the standard:

(1) data derived from clinical studies of the effects of carbon monoxide on those with impaired myocardial blood supply;
(2) data derived from epidemiological studies of the effects of day-to-day variations in levels of carbon monoxide on health.

It would be fair to say that more data were, at least until recently, available from the former studies. However, data from epidemiological studies are accumulating and give cause for concern.

Clinical Studies

The relevant studies have been discussed above. It seems clear that there is evidence that exposure to CO sufficient to raise the COHb level in non-smokers to about 2% is associated with adverse effects on the heart. The exact clinical significance of the reported effects may be debated, but it seems reasonable to conclude that because myocardial disease is common, and therefore even small adverse effects might have a large impact on public health, they should be avoided. No threshold of effect has been defined – effects have been reported at the lowest concentration of COHb studied: 2%. It may be reasonable to speculate that had subjects with more severe degrees of impaired coronary blood flow been studied, then effects would have been recorded at lower concentrations of COHb. Of course this is speculation, but not unreasonable speculation.

Epidemiological Studies

In comparison with other air pollutants such as ozone and particles, epidemiological studies showing effects of ambient concentrations of carbon monoxide are few. The available studies do not indicate a threshold of effect: indeed, it has been argued that the time-series approach (i.e. relating day-to-day concentrations of pollutants to day-to-day levels of effects) cannot demonstrate a threshold of effect. The lack of a demonstrable threshold of effect makes standard-setting difficult. This problem has been faced in the case of particles. No standards based on epidemiological studies have yet been set for carbon monoxide.

A number of countries have set outdoor air quality standards for carbon monoxide in terms of an 8-h average concentration. In the USA the value is 9 ppm, while in the UK a figure of 10 ppm was chosen (Department of the Environment, 1994). The World Health Organization has produced a series of guidelines for concentrations of carbon monoxide. These are:

100 mg/m^3	(90 ppm)	for	15 min
60 mg/m^3	(50 ppm)	for	30 min
30 mg/m^3	(25 ppm)	for	1 h
10 mg/m^3	(10 ppm)	for	8 h

Examination of the pattern of variation of concentrations of carbon monoxide in the UK suggested that if the 8-h standard of 10 ppm were met, then all the other guidelines recommended by WHO would also be met and thus only one standard was recommended. In the USA, a 1-h figure of 35 ppm is also used.

When setting an air quality standard at least six criteria should be specified:

(1) the measurement method
(2) the placing of the monitoring equipment

(3) the numerical value of the standard
(4) the relevant averaging time
(5) the allowed or acceptable rate of exceedance of the standard
(6) the date by which the standard specified by (3), (4) and (5) should be met.

In the UK these criteria have recently been defined for a range of air pollutants in the Government's Air Quality Strategy for the UK (Department of the Environment, 1997) and are considered further in Chapter 43, which deals with the development of air quality standards.

ACKNOWLEDGEMENTS

The authors are grateful to Mrs K Cameron and Professor J Henry for comments on the manuscript of this chapter. The authors also wish to thank Miss Julia Cumberlidge for help in preparing the manuscript.

REFERENCES

Adams KF, Koch GJ, Chatterjee B *et al.* (1988) Acute elevation of blood carboxyhemoglobin to 6% impairs exercise performance and aggravates symptoms in patients with ischemic heart disease. *J Am Coll Cardiol* **12**: 900–909.

Alderman BW, Baron AE and Savitz DA (1987) Maternal exposure to neighborhood carbon monoxide and risk of low infant birth weight. *Public Health Rep* **102**: 410–414.

Allred EN, Bleecker ER, Chaitman BR *et al.* (1989a) Short-term effects of carbon monoxide exposure on the exercise performance of subjects with coronary artery disease. *N Engl J Med* **321**: 1426–1432.

Allred EN, Bleecker ER, Chaitman BR *et al.* (1989b) Acute effects of carbon monoxide exposure on individuals with coronary artery disease. *Res Rep Health Effects Inst* **25**: 1–97.

Allred EN, Bleecker ER, Chaitman BR *et al.* (1991) Effects of carbon monoxide on myocardial ischemia. *Environ Health Perspect* **91**: 89–132.

Anderson EW, Andelman RJ, Strauch JM *et al.* (1973) Effect of low-level carbon monoxide exposure on onset and duration of angina pectoris. A study in ten patients with ischemic heart disease. *Ann Intern Med* **79**: 46–50.

Anderson HR, Bland MJ and Peacock JL (1992) The effects of smoking on fetal growth. In: Poswillo D and Alberman E (eds) *Effects of Smoking on the Fetus, Neonate and Child.* Oxford: Oxford University Press, pp. 89–107.

Aronow WS (1978) Effect of ambient level of carbon monoxide on cardiopulmonary disease. *Chest* **74**: 1–2.

Aronow WS (1981) Aggravation of angina pectoris by two percent carboxyhemoglobin. *Am Heart J* **101**: 154–157.

Aronow WS and Isbell MW (1973) Carbon monoxide effect on exercise-induced angina pectoris. *Ann Intern Med* **79**: 392–395.

Astrup P, Kjeldsen K and Wanstrup J (1967) Enhancing influences of carbon monoxide on the development of atheromatosis in cholesterol-fed rabbits. *J Atheroscler Res* **7**: 343–354.

Astrup P, Trolle D, Olson HM and Kjeldsen K (1975) Moderate hypoxia exposure and fetal development. *Arch Environ Health* **30**: 15–16.

Ayres SM, Mueller HS, Gregory JJ *et al.* (1969) Systemic and myocardial hemodynamic responses to relatively small concentrations of carboxyhemoglobin (COHb). *Arch Environ Health* **18**: 699–709.

Ayres SM, Giannelli S and Mueller H (1970) Myocardial and systemic responses to carboxyhemoglobin. *Ann NY Acad Sci* **174**: 268–293.

Ayres SM, Evans R, Licht D *et al.* (1973a) Health effects of exposure to high concentrations of automotive emissions. *Arch Environ Health* **27**: 168–178.

Ayres SM, Giannelli S, Mueller H and Criscitiello A (1973b) Myocardial and systemic vascular responses to low concentrations of carboxyhemoglobin. *Ann Clin Lab Sci* **3**: 440–447.

Baker MD, Henrettig FM and Ludwig S (1988) Carboxyhemoglobin levels in children with non-specific flu-like symptoms. *J Pediatr* **113**: 501–504.

Baker SP, O'Neil B, Ginsburg MJ and Li G (1992) *The Injury Fact Book*, 2nd edn. New York: Oxford University Press.

Barinaga M (1993) Carbon monoxide: killer to brain messenger in one step. *Science* **259**: 309.

Berk PD, Rodkey FL, Blaschke PL *et al.* (1974) Comparison of plasma bilirubin turnover and carbon monoxide production in man. *J Lab Clin Med* **83**: 29–37.

Bernard C (1865) *An Introduction to the Study of Experimental Medicine*. New York: Dover Publications Inc. (1957).

Biller WF and Richmond HM (1982) Sensitivity analysis on Coburn model predictions of COHb levels associated with alternative CO standards. Office of Air Quality Planning and Standards. Research Triangle Park, NC: US Environmental Protection Agency.

Bing RJ, Sarma JSM, Weishaar R *et al.* (1980) Biochemical and histological effects of intermittent carbon monoxide exposure in cynomolgus monkeys (*Macaca fascicularis*) in relation to atherosclerosis. *J Clin Pharmacol* **20**: 487–499.

Brunekreef B, Smit HA and Biersteker K (1982) Indoor carbon monoxide pollution in The Netherlands. *Environ Int* **8**: 193–196.

Butler NR, Goldstein H and Ross EM (1972) Cigarette smoking in pregnancy: its influence on birth-weight and perinatal mortality. *Br Med J* **2**: 127–130.

Caravati EM, Adams CJ, Joyce SM and Schafer NC (1988) Fetal toxicity associated with maternal carbon monoxide poisoning. *Ann Emerg Med* **17**: 714–717.

Chaney LW (1978) Carbon monoxide automobile emissions measured from the interior of a travelling automobile. *Science* **199**: 1203–1204.

Choi IS (1983) Delayed neurological sequelae in carbon monoxide intoxication. *Arch Neurol* **40**: 433–435.

Coburn RF (1979) Mechanisms of carbon monoxide toxicity. *Prev Med* **8**: 310–322.

Coburn RF, Forster RE and Kane PB (1965) Considerations of the physiological variables that determine the blood carboxyhemoglobin concentration in man. *J Clin Invest* **44**: 1899–1910.

Coburn RF, Ploegmakers F, Gondrie P and Abbond R (1973) Myocardial myoglobin oxygen tension. *Am J Physiol* **224**: 870–876.

Cohen SI, Deane M, Goldsmith JR and Berkeley C (1969) Carbon monoxide and survival for myocardial infarction. *Arch Environ Health* **19**: 510–517.

Colwill DM and Hickman AJ (1980) Exposure of drivers to carbon monoxide. *J Air Pollut Control Assoc* **30**: 1316–1319.

Comroe JH (1974) *Physiology of Respiration: An Introductory Text*, 2nd edn. Chicago: Year Book Medical Publishers Inc.

Cooper AG (1966) *Carbon Monoxide: A Bibliography with Abstracts*. Public Health Service Publication no. 1503. Washington, DC: US Department of Health, Education and Welfare.

Cox BD and Whichelow MJ (1985) Carbon monoxide levels in the breath of smokers and non-smokers: effect of domestic heating systems. *J Epidemiol Community Health* **39**: 75–78.

Crawford R, Campbell DGD and Ross J (1990) Carbon monoxide poisoning in the home: recognition and treatment. *Br Med J* **301**: 977–979.

Crispen C (1989) Carbon monoxide and flu-like symptom. *J Pediatr* **114**: 342.

Dahms TE, Younis LT, Wiens RD *et al.* (1993) Effects of carbon monoxide exposure in patients with documented cardiac arrhythmias. *J Am Coll Cardiol* **21**: 442–450.

Davous P, Rondot P, Marion MH and Gueguen B (1986) Severe chorea after acute carbon monoxide poisoning. *J Neurol Neurosurg Psychiatr* **49**: 206–208.

Dawson TM and Snyder SH (1994) Gases as biological messengers: nitric oxide and carbon monoxide in the brain. *J Neurosci* **14**: 5147–5159.

Deliveria-Papadopoulos M, Coburn RF and Forster RE (1970) Cyclical variations of heme destruction and carbon monoxide production (VCO) in normal women. *Physiologist* **13**: 178.

Department of the Environment (1994) Expert Panel On Air Quality Standards. *Carbon Monoxide*. London: HMSO.

Department of the Environment (1997) The Scottish Office. *The United Kingdom National Air Quality Strategy*. London: The Stationery Office.

Dockery DW, Pope CA, Xu X *et al.* (1993) An association between air pollution and mortality in six US cities. *N Engl J Med* **329**: 1753–1759.

Douglas CG, Haldane JS and Haldane JBS (1912) The laws of combination of haemoglobin with carbon monoxide and oxygen. *J Physiol* **XLIV**: 275–304.

Engel RF, Rodkey FL, O'Neal JD and Collison HA (1969) Relative affinity of human foetal haemoglobin for carbon monoxide and oxygen. *Blood* **33**: 37–45.

Evans RG, Webb K, Homan S and Ayres SM (1988) Cross-sectional and longitudinal changes in pulmonary function associated with automobile pollution among bridge and tunnel officers. *Am J Ind Med* **14**: 25–36.

Fernandez-Bremauntz AA and Ashmore MR (1995) Exposure of commuters to carbon monoxide in Mexico City – I. Measurement of in-vehicle concentrations. *Atmos Environ* **29**: 525–532.

Forbes WA, Sargent F and Roughton FJW (1945) The rate of carbon monoxide uptake by normal men. *Am J Physiol* **143**: 594–608.

Garvey DJ and Longo LD (1978) Chronic low level maternal carbon monoxide exposure and fetal growth and development. *Biol Reprod* **19**: 8–14.

Ginsberg MD (1985) Carbon monoxide intoxication: clinical features, neuropathology and mechanisms of injury. *Clin Toxicol* **23**: 281–288.

Gordon GS and Rogers RL (1969) *A Report of Medical Findings of Project Carbon Monoxide*. Washington, DC: Merkle Press.

Grinker RR (1926) Parkinsonism following carbon monoxide poisoning. *J Nervous Mental Dis* **64**: 18–28.

Haldane JBS (1912) The dissociation of oxyhaemoglobin in humans during partial CO poisoning. *J Physiol* 45 22P, pp. B434–436.

Haldane JS (1895a) The action of carbonic oxide on man. *J Physiol* **18**: 430–462.

Haldane JS (1895b) A method of detecting and estimating carbonic oxide in blood. *J Physiol* **18**: 463–469.

Haldane JS and Priestly JG (1935) *Respiration*, new edn. Oxford: Clarendon Press.

Halpern JS (1989) Chronic occult carbon monoxide poisoning. *J Emerg Nursing* **15**: 107–111.

Hampson NB and Norkool DM (1992) Carbon monoxide poisoning in children riding in the back of pick-up trucks. *J Am Med Assoc* **267**: 538–540.

Hart IK, Kennedy PGE, Adams JH and Cunningham NE (1988) Neurological manifestations of carbon monoxide poisoning. *Postgrad Med J* **64**: 213–216.

Hexter AC and Goldsmith JR (1971) Carbon monoxide: association of community air pollution with mortality. *Science* **172**: 265–266.

Hickman AJ (1989) *Personal Exposure to Carbon Monoxide and Oxides of Nitrogen*. Research Report 206. Crowthorne, Berkshire: Research Laboratory, Department of Transport.

Hill EP, Hill JR, Power GG and Longo LD (1977) Carbon monoxide exchanges between the human fetus and mother: a mathematical model. *Am J Physiol* **232**: H311–H323.

Hinderliter AL, Adams KF, Price CJ *et al.* (1989) Effects of low-level carbon monoxide exposure on resting and exercise induced ventricular arrhythmias in patients with coronary artery disease and no baseline ectopy. *Arch Environ Health* **44**: 89–93.

Hong CJ, Tao XG, Ma HB and Mao HQ (1991) Study on air pollution and its effects on health in Shanghai. *Proceedings of the China–Japan Symposium on Environmental Health*. Zhenzhou, China, pp. 172–174.

Hönigl D, Clebel H, Marlovits S and Kriechbaum N (1993) Subtle neuropsychiatric sequelae of carbon monoxide poisoning: a case report. *Eur J Psychiatr* **7**: 12–14.

Hoppenbrouwers T, Calub M, Arakawa K and Hodgman JE (1981) Seasonal relationship of sudden infant death syndrome and environmental pollutants. *Am J Epidemiol* **113**: 623–635.

Ingi T and Ronnett GV (1995) Direct demonstration of a physiological role for carbon monoxide in olfactory receptor neurons. *J Neurosci* **15**: 8214–8222.

Jarvis MJ, Belcher M, Vesey C and Hutchinson DC (1986) Low cost carbon monoxide monitors in smoking assessment. *Thorax* **41**: 886–887.

Johnson T, Capel J and Wijnberg L (1986) *Selected Data Analyses Relating to Studies of Personal Carbon Monoxide Exposure in Denver and Washington DC.* Environmental Monitoring Systems Laboratory. Contract no. 68-02-3496. Research Triangle Park, NC: US Environmental Protection Agency.

Katzenstein AW (1990) Carbon monoxide and myocardial ischemia. *N Engl J Med* **322**: 1086.

Khalil MAK and Rasmussen RA (1989) Urban carbon monoxide: contributions of automobiles and wood burning. *Chemosphere* **19**: 1383–1386.

Kinney PI, Ito K and Thurston GD (1995) A sensitivity analysis of mortality/PM_{10} associations in Los Angeles. *Inhalation Toxicol* **7**: 59–69.

Kirkpatrick JN (1987) Occult carbon monoxide poisoning. *West J Med* **146**: 52–56.

Krishna K (1978) Tobacco chewing in pregnancy. *Br J Obstet Gynecol* **85**: 726–728.

Koren G, Sharav T, Pastuszac A *et al.* (1991) A multi-centre prospective study of fetal outcome following accidental carbon monoxide poisoning in pregnancy. *Reprod Toxicol* **5**: 397–403.

Kuller L, Cooper M and Perper J (1972) Epidemiology of sudden death. *Arch Intern Med* **129**: 714–719.

Kuller LH, Radford EP, Swift D *et al.* (1975) Carbon monoxide and heart attacks. *Arch Environ Health* **30**: 477–482.

Kurt TL, Mogielnicki RP and Chandler JE (1976) Association of the frequency of acute cardiorespiratory complaints with ambient levels of carbon monoxide. *Chest* **74**: 10–14.

Lawther PJ and Commins BT (1970) Cigarette smoking and exposure to carbon monoxide. *Ann NY Acad Sci* **174**: 135–147.

Lee MS and Marsden CD (1994) Neurological sequelae following carbon monoxide poisoning. Clinical course and outcome according to the clinical types and brain computed tomography scan findings. *Movement Disorders* **9**: 550–558.

Longo LD (1970) Carbon monoxide in the pregnant mother and fetus and its exchange across the placenta. *Ann NY Acad Sci* **174**: 313–341.

Longo LD (1976) Carbon monoxide: effects on oxygenation of the fetus *in utero*. *Science* **194**: 523–525.

Longo LD (1977) The biological effects of carbon monoxide on the pregnant woman, fetus and new born infant. *Am J Obstet Gynecol* **129**: 69–103.

Longo LD, Power GG and Forster RE (1967) Respiratory function of the placenta as determined with carbon monoxide in sheep and dogs. *J Clin Invest* **46**: 812–828.

Lowe-Ponsford FL and Henry JA (1989) Clinical aspects of carbon monoxide poisoning. *Adverse Drug React Acute Poisoning Rev* **8**: 217–240.

Lugaresi A, Montagna P, Morreale A and Gallasi R (1990) 'Psychic akinesia' following carbon monoxide poisoning. *Eur Neurol* **30**: 167–169.

Lynch SR and Moede AL (1972) Variation in the rate of endogenous carbon monoxide production in normal human beings. *J Lab Clin Med* **79**: 85–95.

Malinow MR, McLaughlin P, Dhindsa DS *et al.* (1976) Failure of carbon monoxide to induce myocardial infarction in cholesterol-fed cynomolgus monkeys (*Macaca fascicularis*). *Cardiovasc Res* **10**: 101–108.

Mall T, Grossenbacher M, Perruchoud AP and Ritz R (1985) Influence of moderately elevated levels of carboxyhemoglobin on the course of acute ischemic heart disease. *Respiration* **48**: 237–244.

Mayer A (1960) Anoxias, intoxications and metabolic disorders. In: Greenfield JG, Blackwood W, McMenemy WH *et al.* (eds) *Neuropathology*. London: Edward Arnold Publishers.

Meyer MB, Jonas BS and Tonascia JA (1976) Perinatal events associated with maternal smoking during pregnancy. *Am J Epidemiol* **103**: 464–476.

Meyer MB and Comstock GW (1972) Maternal cigarette smoking and perinatal mortality. *Am J Epidemiol* **96**: 1–10.

Mennear JH (1993) Carbon monoxide and cardiovascular disease: an analysis of the weight of evidence. *Regulat Toxicol Pharmacol* **17**: 77–84.

Meredith T and Vale A (1988) Carbon monoxide poisoning. *Br Med J* **296**: 77–79.

Meredith TJ, Vale JA and Proudfoot AT (1987) Poisoning by inhalational agents: carbon monoxide. In: Weatherall DJ, Leddingham JGG and Warrell DA (eds) *Oxford Textbook of Medicine*, 2nd edn. Oxford: Oxford University Press, pp. 1093–1105.

Min SK (1986) A brain syndrome associated with delayed neuropsychiatric sequelae following acute carbon monoxide intoxication. *Acta Psychiatr Scand* **73**: 80–86.

Morris RD, Naumova EN and Munasinghe RL (1995) Ambient air pollution and hospitalization for congestive heart failure among elderly people in seven large US cities. *Am J Public Health* **85**: 1361–1365.

Myers RAM, Snyder SK and Emhoff TA (1985) Sub-acute sequelae of carbon monoxide poisoning. *Ann Emerg Med* **14**: 1163–1167.

Norton S (1986) Toxic responses of the central nervous system. In: Klaassen DD, Amdur MO and Doull J (eds) *Casarett and Doull's Toxicology*, 3rd edn. New York: McMillan Publishing Company, pp. 359–386.

Nunn JF (1993) *Nunn's Applied Respiratory Physiology*, 4th edn. London: Butterworth Heinemann.

Ott WR and Mage DT (1978) Interpreting urban carbon monoxide concentrations by means of a computerised blood COHb model. *J Air Pollut Control Assoc* **28**: 911–916.

Ott W, Thomas J, Mage D and Wallis L (1982) Validation of the simulation of human activity and pollutant exposure (shape) model, using paired days from Denver (Co). Carbon monoxide field study. *J Air Pollut Control Assoc* **32**: 826–833.

Partington JR (1953) *A Textbook of Inorganic Chemistry*, 6th edn. London: McMillan & Co Ltd.

Penn A, Currie J and Snyder C (1992) Inhalation of carbon monoxide does not accelerate arteriosclerosis in cockerels. *Eur J Pharmacol* **28**: 155–164.

Penn A, Butler J, Snyder C and Albert RE (1983) Cigarette smoke and carbon monoxide do not have equivalent effects upon development of arteriosclerotic lesions. *Artery* **12**: 117–131.

Penney DG (1988) A review: hemodynamic response to carbon monoxide. *Environ Health Perspect* **77**: 121–130.

Perutz MF (1967–68) The structure and function of haemoglobin. *Harvey Lectures*, Series 63, pp. 213–261.

Pope CA, Thun MJ, Namboodiri MM *et al.* (1995) Particulate air pollution as a predictor of mortality in a perspective study of US adults. *Am J Respir Crit Care Med* **151**: 669–674.

Power GG and Longo LD (1975) Fetal circulation times and their implications for tissue oxygenation. *Gynecol Invest* **6**: 342–355.

Pulst S-M, Walshe TM and Romero JA (1983) Carbon monoxide poisoning with features of Gilles de la Tourette's syndrome. *Arch Neurol* **40**: 443–444.

Quality of Urban Air Review Group (1993) *Urban Air Quality in the United Kingdom*. London: Department of the Environment.

Radford EP and Drizd TA (1982) *Blood Carbon Monoxide Levels in Persons 3–74 Years of Age: United States, 1976–80*. Hyattsville, MD: US Department of Health and Human Services, National Center for Health Statistics. DHSS Publication no. (PHS) 82-1250 (Advanced Data from Vital and Health Statistics: no. 76).

Read RC and Green M (1990) Carbon monoxide levels: individuals occupationally exposed to vehicle exhaust. *Environ Manag Health* **1**: 7–8.

Rogers WR, Bass RL, Johnson DE *et al.* (1980) Atherosclerosis-related responses to cigarette smoking in the baboon. *Circulation* **61**: 118–193.

Rosenman KD (1984) Cardiovascular disease and workplace exposure. *Arch Environ Health* **39**: 218–224.

Rosenman KD (1990) Environmentally related disorders of the cardiovascular system. *Med Clin N Am* **74**: 361–375.

Saldiva PHN, Pope CA, Schwartz J *et al.* (1995) Air pollution and mortality in elderly people: a time-series study in São Paulo, Brazil. *Arch Environ Health* **50**: 159–164.

Schwartz A, Hennerici M and Wegener OH (1985) Delayed choreoathetosis following acute carbon monoxide poisoning. *Neurology* **35**: 98–99.

Schwartz J (1995) Is carbon monoxide a risk factor for hospital admission for heart failure? *Am J Public Health* **85**: 1343–1344.

Severinghaus JW, Stafford M and Thunstrom AM (1978) Estimation of skin metabolism and blood flow with tcPO$_2$ and tcPCO$_2$ electrodes by cuff occlusion of the circulation. *Acta Anaesth Scand* **69**: 9–15.

Sheps DS, Adams JF, Bromberg PA *et al.* (1987) Lack of effect of low levels of carboxyhemoglobin on cardiovascular function in patients with ischemic heart disease. *Arch Environ Health* **42**: 108–116.

Sheps DS, Herbst MC, Hinderliter AL *et al.* (1990) Production of arrhythmias by elevated carboxyhemoglobin in patients with coronary artery disease. *Ann Intern Med* **113**: 343–351.

Shillito FH, Drinker CK and Shaughnessy TJ (1936) The problem of nervous and mental sequelae in carbon monoxide poisoning. *J Am Med Assoc* **106**: 669–674.

Singh J and Scott LH (1984) Threshold for carbon monoxide induced fetotoxicity. *Teratology* **30**: 253–257.

Smith RP (1986) Toxic responses of the blood. In: Klaassen DD, Amdur MO and Doull J (eds) *Casarett and Doull's Toxicology*, 3rd edn. New York: McMillan Publishing Company, pp. 223–244.

Spencer TD (1961) Effects of carbon monoxide on man and canaries. *Ann Occup Hyg* **5**: 231–240.

Spiller HA (1987) Carbon monoxide exposure in the home: source and epidemiology. *Vet Human Toxicol* **29**: 383–386.

Spinnler H, Sterzi R and Vallar G (1980) Amnesic syndrome after carbon monoxide poisoning, a case report. *Schweiz Arch Neurol Neurochir Psychiatri* **127**: 79–88.

Stender S, Astrup P and Kjeldsen K (1977) The effect of carbon monoxide on cholesterol in the aortic wall of rabbits. *Atherosclerosis* **28**: 357–367.

Strang LB (1977) *Neonatal Respiration: Physiological and Clinical Studies*. Oxford: Blackwell Scientific Publications.

Ström J, Alfredsson L and Malmfors T (1995) Carbon monoxide: causation and aggravation of cardiovascular diseases – a review of the epidemiologic and toxicologic literature. *Indoor Environ* **4**: 322–333.

Suematsu M, Wakabayashi Y and Ishimura Y (1996) Gaseous monoxides: a new class of microvascular regulator in the liver. *Cardiovasc Res* **32**: 679–686.

Tesarik K and Krejci M (1974) Chromatographic determination of carbon monoxide below 1 ppm. *J Chromatogr* **91**: 539–544.

Turner DM, Lee PN, Roe FJC and Goff KJ (1979) Atherogenesis in the White Carneau pigeon: further studies of the role of carbon monoxide and dietary cholesterol. *Atherosclerosis* **34**: 407–417.

Vale JA, Proudfoot AT and Meredith TJ (1996) Poisoning by inhalational agents: carbon monoxide. In: Weatherall DJ, Leddingham JGG and Warrell DA (eds) *Oxford Textbook of Medicine*, 3rd edn. Oxford: Oxford University Press.

Verma A, Hirsh DV, Glatt CE *et al.* (1993) Carbon monoxide: a putative neural messenger. *Science* **259**: 381–384.

Vicente PJ (1980) Neuropsychological assessment and management of a carbon monoxide intoxication patient with consequent sleep apnea: a longitudinal case report. *Clin Neuropsychol* **2**: 91–94.

Wade WA, Coat WA and Yocom JE (1975) A study of indoor air quality. *J Air Pollut Control Assoc* **25**: 933–939.

Walden SM and Gottlieb SO (1990) Urban angina, urban arrhythmias: carbon monoxide and the heart. *Ann Intern Med* **113**: 337–338.

Werner B, Bäck W, Akerblom H and Barr PO (1985) Two cases of acute carbon monoxide poisoning with delayed neurological sequelae after a 'free' interval. *Clin Toxicol* **23**: 249–265.

World Health Organization (1979) Environmental Health Criteria no. 13, *Carbon Monoxide*. Geneva: World Health Organization.

Younoszai MK, Peloso J and Haworth JC (1969) Fetal growth retardation in rats exposed to cigarette smoke during pregnancy. *Am J Obstet Gynecol* **104**: 1207–1213.

Zagami AS, Lethlean AK and Mellick R (1993) Delayed neurological deterioration following carbon monoxide poisoning: MRI findings. *J Neurol* **240**: 113–116.

34

Lead

ANDREW WADGE

Department of Health, London, UK

The views expressed in this chapter are those of the author and should not be taken to represent those of the Department of Health.

INTRODUCTION

The introduction of tetraethyl lead as an antiknock agent in gasoline in the 1920s, coupled with the massive growth in motor vehicle use since that time, has brought marked increases in lead emissions to air. Airborne lead from motor vehicle emissions has been dispersed globally throughout the atmosphere and is largely responsible for substantial increases in lead concentrations detected in polar ice caps (Murozumi et al., 1969). The impact of this technology has also been detected in humans. Although the effects of lead poisoning were well recognized at the time, the more insidious consequences of cumulative low-level exposure to lead such as subtle effects on neurobehavioural development in children were not appreciated by most regulators and industrialists. However, evidence accumulated over the last 20 years has shown that vehicular lead has made an important contribution to lead exposure in the general population which, together with other sources of lead exposure, has been associated with adverse effects upon health, most notably decrements in IQ in exposed children.

In many ways lead can be seen as an exemplar of how technological changes introduced by one industrial sector can have unforseen consequences for public health. Lead is now widely cited as a multimedia pollutant which is persistent in the environment and once absorbed can remain in the body for long periods. As a result of our experience with lead, regulators are now, rightly, cautious about the use of substances that are persistent and accumulate in the environment.

AIR POLLUTION AND HEALTH
ISBN 0-12-352335-4

LEAD IN THE ENVIRONMENT

Sources of Lead Emissions to Air

Motor vehicle emissions are the main source of lead in air, although the total emitted from this source is declining in many countries with the introduction of unleaded fuels. In the UK, estimated emissions of lead from motor vehicles was 1068 tonnes in 1995, which represents a substantial decline from an estimated total of 7500 tonnes in 1980, despite an increased consumption of gasoline during this period. Since 1988 the rapid increase in the use of unleaded fuels, representing 63% of the market share for gasoline in 1995, has brought steep falls in the total emissions from this source (Fig. 34.1). The increasing market share of diesel fuel will also have contributed to this decline since only trace amounts of lead are present in diesel.

Nevertheless, despite the substantial decline in lead emissions from motor vehicles, gasoline lead still accounted for about 73% of total lead emissions in the UK in 1995. Other sources include non-ferrous metal production, waste incineration and other high-temperature combustion processes (Fig. 34.2). Although not relevant for the majority of the population, industrial processes can make a substantial contribution to localized emissions of lead in air. This has been a particular problem with old, poorly controlled processes, but can be ameliorated to a large extent with the use of modern abatement equipment. Such equipment tends to be available only in the more affluent

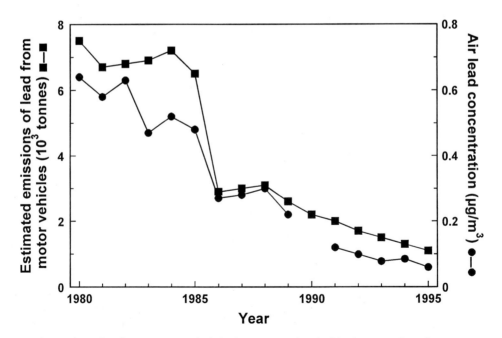

Fig. 34.1 Relationships between motor vehicle lead emissions and air lead levels in central London, UK. From Department of the Environment, Transport and the Regions (1997).

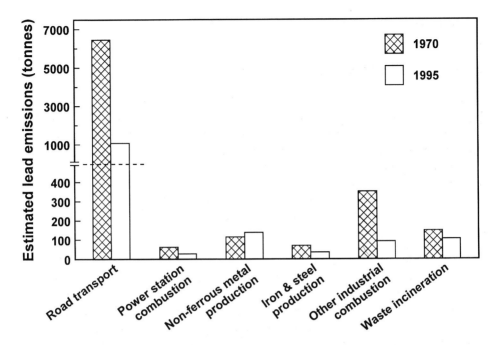

Fig. 34.2   Estimated UK emissions of lead in 1970 and 1995, tonnes. From Salway *et al.*, 1997.

parts of the world, while elsewhere localized pollution from industrial sources can remain a significant threat to the health of the population living in the vicinity of the plant.

Within the domestic environment, the legacy of old paint containing high concentrations of lead can be a serious hazard during renovation and redecoration (see below).

Concentrations of Lead in Air

Outdoor Air

Airborne lead is mostly present as submicrometre particles in the form of inorganic lead arising from combustion of tetraethyl and tetramethyl lead additives in gasoline. A fraction of this may be solubilized in acidic aerosols, while much larger particles of lead will be present from industrial emissions and re-suspension of crustal lead. A further fraction will be present as uncombusted organic lead.

Concentrations of lead in air range from about 5×10^{-5} $\mu g/m^3$ in remote parts of the world to about 10 $\mu g/m^3$ near to metal smelters which do not have modern pollution abatement equipment (WHO, 1995). In the UK, where most gasoline sold is unleaded, typical annual average concentrations of lead are between 0.1 and 0.2 $\mu g/m^3$ in urban areas and between 0.01 and 0.05 $\mu g/m^3$ in rural parts. The highest annual average concentrations in the UK are found near to industrial sites, where levels are typically between

0.2 and 1.0 µg/m^3 (Department of the Environment, Transport and the Regions, 1997). Levels of airborne lead have fallen considerably over the last 25 years as a result of progressive reductions in the maximum permitted concentration of lead in gasoline (in the UK this has been reduced from 0.8 g/l in 1971 to 0.4 g/l in 1981 and to 0.15 g/l in 1986) (Fig. 34.1). There has also been an increase in the use of unleaded gasoline. Similar airborne lead levels have been reported in other parts of Europe (WHO, 1995), while in the USA, concentrations of lead in urban areas had fallen below 0.07 µg/m^3 by 1990, reflecting the earlier introduction of unleaded gasoline in that country (US EPA, 1991). Lead concentrations in Australian cities are typically between 0.1 and 0.6 µg/m^3 (Department of the Environment, Sport and Territories, 1996), while concentrations in excess of 1.0 µg/m^3 are recorded in cities in developing countries where there is still a heavy reliance on leaded gasoline.

Indoor Air

Indoor air concentrations of lead were found to follow closely external concentrations in a study in Birmingham, UK, but they were usually lower. It was suggested that the external lead level, which is mostly associated with fine particulate matter, had a large influence on that indoors, but that the physical structure of the house was able to prevent admission of some of the lead associated with coarser, suspended particulate matter. Indoor lead levels averaged approximately 60% of those measured immediately outside the house (Davies *et al.*, 1990). Similar results have been reported from the USA (Yocom, 1982).

The presence of leaded paint in older housing can lead to elevation of indoor airborne lead concentrations during renovation and redecoration. In the Birmingham study, results for one house showed much higher lead concentrations indoors rather than those outside, in contrast to the general findings noted above. Further investigations revealed that a sanding machine had been used in this house to remove old paint, which had produced significant contamination of the house, and a dust lead concentration of 23 000 µg/g (Davies *et al.*, 1990).

LEAD EXPOSURE

Pathways of Lead Exposure

Lead in air can enter the body either directly through inhalation or indirectly via ingestion of lead deposited onto food crops or dust. Other sources of lead also make an important contribution to total exposure. For most people, the main sources are food and drinking water, while old leaded paint, various occupations and hobbies, and some traditional ethnic cosmetics and medicines can also give rise to lead exposure in some people. The relative importance of different sources will vary depending on the interplay of several different variables including age, sex and geographic location. For example, in a study of just under 100 2-year-old children in Birmingham, UK it was estimated that the mean total uptake of lead was 36 µg/day, of which 97% was from ingestion of dust, food and water and only 3% from inhalation (Davies *et al.*, 1990). The percentage of lead intake from different sources in 2-year-old infants and women of child-bearing age in the USA is illustrated in Fig. 34.3.

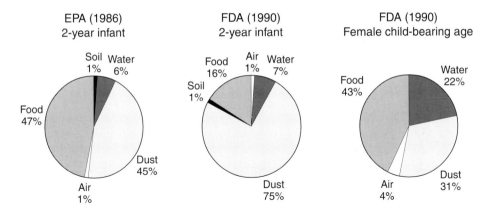

EPA (1986)
2-year infant

FDA (1990)
2-year infant

FDA (1990)
Female child-bearing age

Fig. 34.3   Percentage of lead intake from food and other sources in 2-year-old infants and women of child-bearing age in the USA. From WHO (1995).

Kinetics and Metabolism

The absorption of lead varies considerably with its physicochemical state, the route of entry into the body, and the age, physiological condition and nutritional status of the individual. Once absorbed, it is distributed rapidly into soft tissues (blood, kidney, liver) as well as bone and teeth. It is eliminated in urine and faeces and a variety of these tissues and excretion products have been used in attempts to assess exposure, although whole blood is most widely used for this purpose.

Absorption

As noted above, airborne lead is associated with submicrometre particular matter. A Task Group on Lung Dynamics predicted that 35–50% of inhaled lead with a mass median aerodynamic diameter of 0.5 µm is deposited in the respiratory tract (International Radiological Protection Commission, 1966). These are deposited primarily in the alveolar region of the respiratory tract. Actual deposition rates will vary with particle size and ventilation rates, but experimental observations support a deposition rate of airborne lead in the general adult population in the range of 30–50% (Chamberlain *et al.*, 1978; Kehoe, 1961; US EPA, 1986). Almost all the lead deposited in the lower respiratory tract will be absorbed into the body. Larger particles will be deposited in the upper airways with incomplete absorption. However, lead-bearing particles that are cleared from the lung may then be ingested and become available for absorption across the gastrointestinal tract. No experimental data are available on the absorption of inhaled lead in children, although James (1978) estimated a rate about twice that of adults, allowing for differences in body weight and the anatomy of the respiratory tract.

Lead is absorbed across the gastrointestinal tract at rates which vary with age, nutritional status, period of fasting and the chemical form of the lead. Experimental studies in adults indicate that about 10% of ingested lead is absorbed from the gastrointestinal tract (Kehoe, 1961; Rabinowitz *et al.*, 1976). This can increase up to about 50% in fasting subjects (Chamberlain *et al.*, 1978), while diets deficient in calcium, vitamin D and iron have been shown to increase lead absorption in laboratory animals (WHO, 1995). The fraction of lead absorbed in children has been reported to be 40–50%, although these estimates are based on a relatively small number of observations (Alexander *et al.*, 1973; Ziegler *et al.*, 1978).

Distribution

Lead absorbed into the body is distributed between blood, soft tissues (particularly liver and kidney) and mineralizing tissues (bone and teeth). The half-life of lead in these different compartments has been estimated as about 36 days, 40 days and 27 years for blood, soft tissues and bone, respectively (WHO, 1995). About 95% of the body lead burden in adults is located in bone, while in children about 70% of lead is in this compartment (Barry, 1975, 1981). Bone acts as a long-term storage site for lead where, because of its long half-life, it generally increases with age. However, it would appear that some of the lead in bone can be labile and act as an endogenous source, arising from metabolic activity in the bone. Mobilization of lead in pregnant rats has been reported (Buchet *et al.*, 1977), and it has been suggested that mobilization of lead from bone during pregnancy and lactation may present a risk to the fetus and the mother, particularly as lead freely crosses the placenta (Silbergeld, 1991). Using isotopic techniques, Gulson *et al.* (1995) has reported between 45% and 70% of lead in blood was derived from tissue stores among adult women immigrants from Eastern Europe into Australia. Since a fraction of blood lead is derived from non-contemporaneous sources, the rate of decline in blood lead levels following abatement of lead exposure is unlikely to be as rapid as the increase in blood lead following initial exposure.

Elimination

Lead is eliminated from the body in urine and faeces. Ingested lead, including airborne lead that has been swallowed and not absorbed, is excreted in faeces. Lead in the circulation may be excreted in urine, or in faeces via biliary excretion. It has been reported that between 40% and 70% of the absorbed dose is excreted (WHO, 1995).

Measurement of Body Burden

Lead measurements in hair, urine, teeth and bone have all been used as a marker of lead exposure. However, the most widely used marker of exposure is blood lead, despite its relatively short half-life (about 36 days). Most reports of lead exposure and its effects in environmental and occupational populations have used blood lead as a marker of exposure and there is evidence that a single blood lead measurement at 6 years of age gives a reasonable estimate of lifetime lead exposure for those children who have not had major changes in their environment (WHO, 1995).

   Tooth lead levels have been used in some epidemiological studies on the effects of lead in children. In these studies, measurements of tooth lead levels have been used to assess accumulation of lead from the period in the womb until loss of the teeth. However, concentrations of lead vary both within and between teeth, so it is not always possible to compare one study with another. Other tissues have not proved sufficiently reliable to be useful for large surveys of lead exposure.

Declining Blood Lead Levels

Substantial declines in blood lead levels have been recorded in many parts of the world. In the USA, the mean blood lead level fell by 78% from 12.8 to 2.8 µg/dl between 1976 and 1991 (Pirkle et al., 1994). The decline was observed across all sectors of the US population, although younger children, males, some ethnic groups and low income levels were all associated with higher blood lead levels. This decline was attributed to the virtual elimination of the use of lead in gasoline and the removal of lead from soldered cans. Similar decreases in blood lead levels since 1978, averaging about 5% and 10% per year, have been reported in Sweden (Strömberg et al., 1995) and Belgium (Ducroffe et al., 1990). In the UK, a mean blood lead level of 3.7 µg/dl was found in a survey of just under 7000 adults and children aged over 10 years conducted in 1995, while the mean blood lead level in a cohort of 2½-year-old children was 4.2 µg/dl (Golding et al., 1998; Primatesta et al., 1998). The prevalence of 2½-year-old children with blood lead levels of 10 µg/dl or above was 5.4%, almost the same as that for non-Hispanic white children aged 1–5 years in the US. Among adults in the UK, male sex, increasing age and consumption of cigarettes and alcohol was associated with increased blood lead levels. No difference was observed between urban and rural areas in England, although there were some regional differences, possibly related to lead in drinking water (Primatesta et al., 1998).

Contribution of Vehicular Lead to Total Body Burden

Since lead exposure can occur from many sources, establishing the contribution that gasoline lead makes to overall exposure has not been straightforward. As action has been taken to reduce exposure from a wide range of sources including gasoline, drinking water, air, food, industrial emissions, paint, cosmetics, ceramic glazes and toys, the contribution made to blood lead levels by any one source is difficult to estimate and will vary with location and age of the person.

Estimates of lead uptake from inhalation have been made in experimental conditions and in epidemiological studies. In some early studies, volunteers were exposed to airborne lead in experimental chambers (Griffin et al., 1975; Kehoe, 1961, 1966). Exposure was to lead oxide aerosols at concentrations between 3.2 and 40 µg/m^3 produced by burning tetraethyl lead in propane. The relationship between blood lead and airborne lead averaged for the volunteers ranged from 0.73 to 1.8 µg/dl per µg/m^3. The relevance of these results to motor vehicle exhaust has been questioned since the mode of generation of the lead aerosols was different. However, tracer experiments where volunteers were exposed to lead in vehicle exhaust found a similar rate of uptake of lead from the lung (Chamberlain et al., 1975). These investigators predicted an average blood lead contribution of 1 µg/dl following continuous exposure to airborne lead derived from vehicle exhaust at a concentration of 1 µg/m^3.

Experimental studies have the considerable advantage of standardization of techniques and control of other potentially confounding variables, but they do not allow for estimation of the total contribution of vehicular lead to body burden, only that from direct inhalation. Several studies have been published comparing blood lead levels in groups of individuals with the concentration of airborne lead in occupational and non-occupational environments. Whilst blood lead levels in such groups will reflect both direct and indirect (through ingestion of dust/soil) uptake of airborne lead, confounding factors such as occupation, geographical location, diet, housing conditions, exposure to leaded paint and/or lead plumbing can limit their usefulness.

The relationship between blood lead and airborne lead in occupational settings has been examined at concentrations between 9 and 450 $\mu g/m^3$ and for blood lead concentrations between 20 and 90 $\mu g/dl$ (Bishop and Hill, 1983; Gartside and Buncher, 1982; King and Conchie, 1979). This relationship has been described as curvilinear, with slopes between 0.02 and 0.08 $\mu g/dl$ per $\mu g/m^3$ (WHO, 1995). The curvilinear relationship is thought to be brought about by factors such as the increased renal clearance of lead at high blood lead concentrations (Chamberlain, 1983). Population studies of children and adults who are not occupationally exposed to lead show a somewhat higher impact of airborne lead upon blood lead, with relationships between 1.0 and 3.0 $\mu g/dl$ per $\mu g/m^3$ typically reported (Angle et al., 1984; Azar et al., 1975; Roels et al., 1980; Yankel et al., 1977).

Estimates of the relationship between airborne lead and blood lead which include the total contribution from air via indirect consumption of dust and soil as well as direct inhalation suggest a value between 3 and 5 $\mu g/dl$ per $\mu g/m^3$ (Brunekreef, 1984; US EPA, 1986).

Isotopic Studies

An isotopic lead experiment was conducted in Piedmont, Italy between 1977 and 1979 during which the isotopic ratio of ^{206}Pb:^{207}Pb used in gasoline was changed from 1.18 to 1.04. In Turin, there was close correlation between the isotopic ratio in air and that in gasoline, indicating that 90% of airborne lead was derived from gasoline, whereas in country areas outside Turin only 60% was gasoline derived. Resuspended lead and long-distance transport were thought to account for the remainder. In Turin, the average ratio ^{206}Pb:^{207}Pb in adult blood decreased from 1.1628 to 1.1325, which suggests that about 25% of the lead in blood was derived from gasoline, either directly or indirectly. In the surrounding rural areas, the fraction of adult blood lead originating from gasoline lead was less than 20% (Facchetti and Geiss, 1982).

Isotopic measurements have been performed on blood and teeth samples stored from population studies conducted in the 1980s in the UK. The mean contribution of gasoline lead to the blood lead of inner London children was between 30 and 40%, more than three times that for adults living in outer London. The high contribution (about 60%) made by water lead to blood lead in a plumbosolvent part of Scotland highlights the importance of not overlooking other variables such as housing, drinking water and localized industrial sources (Delves and Campbell, 1993).

Assuming typical lead intakes and absorption factors from food and drinking water, airborne lead has been estimated to account for between 17 and 67% of total intake among adults exposed to a mean air lead level of 0.3–3.0 $\mu g/m^3$. The corresponding figure for children was between 2 and 17% (WHO, 1987). Isotopic lead studies in the UK suggest that airborne lead (which is largely derived from motor vehicle emissions) contributed on average between 13 and 31% of dietary lead intake for adults in the early 1980s (Ministry of Agriculture, Fisheries and Food, 1982). These estimates imply that the contribution of airborne lead to total lead exposure is likely to be small except in areas of very high air lead levels. However, the substantial decline in blood lead levels, which has closely followed the fall in lead emissions from motor vehicles, suggests that these studies have underestimated the contribution from motor vehicle lead. In a regression analysis, Schwartz and Pitcher (1989) estimated that gasoline lead accounted for about 50% of blood lead in the late 1970s in USA, controlling for socioeconomic status, food intake, alcohol, smoking and occupation. It seems likely that the substantial decline in blood lead levels observed in many parts of the world since the elimination of lead from soldered cans has been largely brought about by the reduced consumption of lead in gasoline. This view is supported by analyses performed by Delves (1998), who has shown that the decline in median blood lead levels in both environmentally and occupationally exposed individuals in the UK over the period 1986 to 1995 is highly significantly correlated with emitted gasoline lead over this period. Other sources of environmental lead exposure such as drinking water (Watt *et al.*, 1996), contaminated house dust (Lanphear *et al.*, 1996), and industrial sources (Baghurst *et al.*, 1992; Junco-Muñoz *et al.*, 1996) will continue to make important contributions to blood lead levels in some areas, but controlling exposure from these sources will be harder to achieve.

EFFECTS OF LEAD UPON HEALTH

A considerable amount of information and knowledge on the effects of lead upon health has accumulated through occupational, environmental or accidental poisonings. There is also a substantial literature on the effects of lead on laboratory animals, although given the extent of human data the animal data is most useful as supporting evidence and in helping to identify mechanisms of toxicity.

Lead toxicity can be manifested in different ways, as shown in Table 34.1. Lead exerts its toxicity through it ability to form bonds with negatively charged groups, particularly sulfydryl groups, on a number of enzymes and proteins. Frank lead poisoning occurs at

Table 34.1   Lowest observed effect levels of lead in humans

Blood lead (μg/dl)	Effect
80	Severe encephalopathy, coma
70	CNS symptoms in chronic occupational exposure
60	Impaired renal function in adults
50	Anaemia in adults
45	Elevation of urinary aminolaevulinic acid in adult men
40	Impaired neurobehavioural test performance in workers; effects on sperm morphology and function; anaemia in children; increased excretion of coproporphyrin in urine; peripheral nerve dysfunction
35	Elevation of urinary aminolaevulinic acid in children
10	Inhibition of δ-aminolaevulinate dehydratase (ALA-D)
15	Elevation of protoporphyrin levels in blood
? 10	Effects of IQ in children
? 10	Increased blood pressure in adults

blood lead levels of at least 40 µg/dl and is associated with impaired renal function, anaemia, loss of CNS function and, at higher concentrations, coma and death. Such events do not occur as a result of typical environmental exposures to lead in air, food and drinking water, but are usually associated with acute exposures in poorly controlled workplaces for adults or ingestion of leaded paint, soil or dust by children.

Since the majority of the non-occupationally exposed population have blood lead concentrations below 10 µg/dl, this discussion is restricted to chronic effects which may occur at such exposure levels, namely effects on neurobehavioural development in children and effects on blood pressure in adults. These subtle effects on health have been recorded in a series of epidemiological studies in otherwise asymptomatic populations, often with conflicting findings. A comprehensive review of the effects of lead upon humans and animals is outside the scope of this chapter, but is available elsewhere (WHO, 1995).

Effects of Lead on the Haematopoietic System

Blood lead is mostly bound to haemoglobin, with less than 1% in the plasma. The process of haem biosynthesis and the way in which lead can disturb this has been well characterized. Lead interferes with the activity of three enzymes. It inhibits the activity of the cytoplasmic enzyme δ-aminolaevulinate dehydratase (ALA-D) and the mitochondrial enzymes coproporphyrinogen oxidase and ferrochelatase. Ferrochelatase is responsible for the last stage of haem synthesis, the incorporation of iron into the protoporphyrin, to produce haemoglobin. As a result haem synthesis is reduced, which in turn depresses the synthesis of the rate-limiting mitochondrial enzyme δ-aminolaevulinate synthase. The effect of this disturbance in enzyme activity can be seen in increased production and excretion of the precursors δ-aminolaevulinate and coproporphyrin and increased levels of protoporphyrin bound to zinc in the circulation (WHO, 1995).

The enzyme ALA-D is uniquely sensitive to lead toxicity. It has been measured in blood samples collected from the general population and these have shown a negative linear relationship between blood lead and ALA-D activity with no apparent threshold for inhibition of the enzyme (Roels and Lauwerys, 1987; Roels et al., 1976). It has been suggested that the blood lead level above which effects on ALA-D can be reliably demonstrated is 10 µg/dl (WHO, 1995). This would appear to be the most sensitive indicator of effects of lead on the haemopoietic system. Increases in levels of protoporhyrin in blood have been detected at blood lead levels as low as 15 µg/dl (Piomelli et al., 1982), although other studies suggest a somewhat higher threshold of about 25 µg/dl (Roels et al., 1976). The precise relationship between blood lead and protoporphyrin levels is influenced by iron status. In a study of 177 children, blood lead at concentrations between 12 and 120 µg/dl was associated with a decrease in the vitamin D metabolite, 1,25 dihydroxyvitamin D (Mahaffrey et al., 1982). A more recent study of children with lower blood lead levels and adequate nutritional status, found no effects of lead on 1,25 dihydroxyvitamin D at blood lead levels below 20 µg/dl (Koo et al., 1991).

Anaemia is a frequently recorded outcome of chronic occupational exposure to lead, as it is easily diagnosed and an accepted marker of lead toxicity. A threshold of effect of 50 µg/dl has been reported in adult lead workers (Tola et al., 1973), while in a group of 579 children living near to a lead smelter no effects of lead on anaemia were found at blood lead levels below 20 µg/dl (Schwartz et al., 1990).

Effects of Lead on Blood Pressure

The question of whether general environmental lead exposure is a causative agent in increasing blood pressure has been a subject of much debate and controversy. There have been several cross-sectional studies of lead exposure and blood pressure and a smaller number of prospective cohort studies and, as has been observed in studies on effects of lead on IQ (see below), these do not all show a consistent picture. The association between lead exposure and blood pressure has been the subject of a recent review (Hertz-Picciotto and Croft, 1993) and two meta-analyses have been carried out (Schwartz, 1995; Staessen et al., 1994). Overall, these studies point to a small, but statistically significant effect of lead on blood pressure which, if causal, would have important implications for cardiovascular morbidity and mortality.

In the first meta-analysis, Staessen et al. (1994) analysed 23 studies and found that a doubling of blood lead was associated with an increase of 1 mmHg in systolic pressure and an increase of 0.6 mmHg in diastolic pressure (1 mmHg ~133 Pa). Schwartz (1995) analysed 15 studies of blood lead on systolic pressure among men and reported a doubling of blood lead from 5 to 10 μg/dl was associated with an increase of 1.25 mmHg in systolic pressure. The results of four prospective studies are consistent in their overall sense of a small association between lead and blood pressure, but inconsistent in whether this is associated with diastolic or systolic pressure (Grandjean et al., 1989; Neri et al., 1988; Staessen et al., 1996; Weiss et al., 1986). Since the effect of lead, if any, is small, establishing the independence of this effect from other confounding factors (e.g. age, sex, socio-economic status, alcohol intake, nutritional status and pre-existing hypertension), especially since the measurements of lead exposure and blood pressure are both subject to variability, is difficult. This is particularly true for alcohol which may itself be contaminated with lead (Ministry of Agriculture, Fisheries and Food, 1989) as well as being associated itself with increased blood pressure. Nevertheless, the findings of meta-analyses and the evidence from studies in laboratory animals (Victery, 1988) are suggestive that lead exerts a small adverse effect on blood pressure. A recent cross-sectional survey of over 5000 adults in England was also supportive of a weak relationship between blood lead and increased blood pressure, although the association was only found to be consistent for men, not women, and for diastolic blood pressure not systolic blood pressure (Bost et al., 1998).

Effects of Lead on Children's IQ

Much of the debate over the last 20 years about the risks of low-level lead exposure and the calls for reductions in emissions of lead into the environment has been generated by a series of cross-sectional and cohort studies of the effects of lead on measures of intelligence in young children. Cross-sectional studies have been carried out in several parts of the world, including Europe, USA, Australia and New Zealand, and a smaller number of prospective cohort studies have been carried out subsequently in the USA and Australia in an attempt to improve assessments of exposure and control of potentially confounding variables (WHO, 1995).

On balance, these studies show an association between lead exposure and a reduction in measures of intelligence, although the effects are small compared with other influences on child intelligence (e.g. parental and social factors), and the studies have been hampered by difficulties in controlling for these and other confounding factors. A systematic review of 26 epidemiological studies of the effect of environmental lead on children's intelligence conducted since 1979 has been published (Pocock *et al.*, 1994). These include prospective cohort studies of birth cohorts, cross-sectional studies of blood lead, and cross-sectional studies of tooth lead. The cross-sectional studies showed an inverse association between lead exposure and IQ, with the use of tooth lead showing more consistency but a smaller magnitude of effect than blood lead. The prospective studies, with a total of over 1100 children, showed a small inverse association of blood lead at age 2 years with IQ, although no association of cord blood or antenatal maternal blood lead with subsequent IQ. Meta-analyses of the cross-sectional and prospective studies indicated that a doubling of blood lead concentration from 10 to 20 μg/dl is associated with a mean deficit of 1–2 IQ points (Figs. 34.4 and 34.5)

There has been considerable debate about whether the association between lead exposure and IQ is causal and, if so, the extent of the public health signficance of such a small effect on IQ. It has been suggested that children of lower IQ adopt behaviours which increase their exposure to lead through ingestion of non-food items containing lead (deSilva and Christophers, 1997), while others have argued against the so-called reverse causality hypothesis (McMichael, 1997). It is also possible that the published studies are unrepresentative, or there is inadequate allowance for confounders and/or selection biases in recruiting and following children (Pocock *et al.*, 1994). The epidemiological evidence

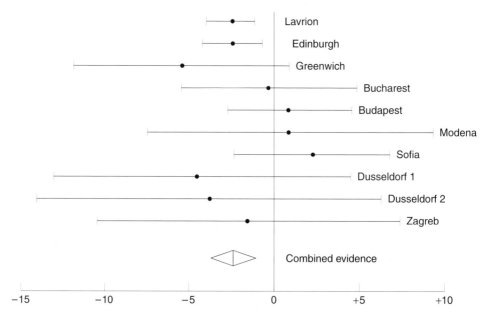

Fig. 34.4 Estimated mean changes in IQ for an increase in blood lead level from 10 to 20 μg/dl in cross-sectional studies. Bars show 95% confidence intervals. From WHO (1995).

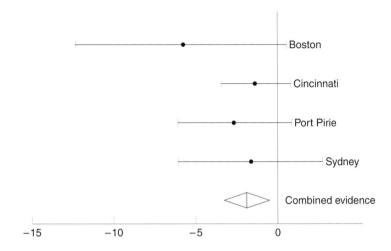

Fig. 34.5 Estimated mean changes in IQ for an increase in blood lead level from 10 to 20 μg/dl in prospective studies. Bars show 95% confidence intervals.s. From WHO (1995)

does not identify a threshold below which it is possible to be certain that effects, although small, are not occurring. The absence of a dose–response relationship from the epidemiological studies does not support the association being causal. Nevertheless, in light of the supporting animal evidence, most regulatory authorities have therefore adopted a precautionary approach and aim to reduce lead exposure whenever it is practicable.

CONCLUSIONS

Human activities have led to significant contamination of the environment with lead, arising from emissions to the atmosphere, especially from motor vehicles. The impact of these emissions can be detected in increased levels of lead in environmenal media throughout the world and elevations in blood lead levels during the middle part of the twentieth century. Interventions by governments to reduce lead emissions have brought about a decline in blood lead levels in recent years in many parts of the world. This has been achieved, in part, by controls on lead in food and drinking water, but much of the decline in lead exposure is a result of the elimination of lead from gasoline.

Whilst the ability of lead to cause severe morbidity and mortality in heavily exposed individuals has been well documented, its effects on the general population, who are exposed to much lower levels, has been the subject of considerable debate and controversy. Of most concern are a series of populations studies of lead on measures of childhood intelligence, but there is also evidence of a possible effect on blood pressure in adults. Uncertainties in both the measurement of lead exposure and the health end-points (IQ and blood pressure), as well as the profound difficulties in controlling adequately for confounding social and environmental factors in epidemiological investigations, restrict the interpretation of the causal nature of these associations. However, the overall consistency and coherence of findings with evidence in laboratory animals is suggestive of a small adverse effect on both IQ in children and blood pressure in adults. From a regulatory perspective, where it is desirable to obtain a margin of safety between exposure and effect levels, action to reduce emissions of lead to air has been a necessary measure to protect public health. The subsequent decline in measurements of lead exposure has demonstrated how reductions in risks to health can be brought about by effective controls on sources of air pollution.

REFERENCES

Alexander FW, Delves HT and Clayton BR (1973) The uptake and excretion by children of lead and other contaminants. In: Barth D, Berlin A, Engel R *et al.* (eds.) *Environmental Health Aspects of Lead.* Luxembourg: Commission of the European Communities.

Angle CR, Marcus A, Cheng IH and McIntire MS (1984) Omaha childhood blood lead and environmental lead: a linear total exposure model. *Environ Res* **35**: 160–170.

Azar A, Snee RD and Habibi K (1975) An epidemiological approach to community air lead exposure using personal air samplers. In: Griffin TB and Knelson JH (eds) *Lead.* New York: Academic Press, pp 254–290.

Baghurst PA, Tong SL, McMichael AJ *et al.* (1992) Determinants of blood lead concentrations to age 5 years in a birth cohort study of children living in the lead smelting city of Port Pirie and surrounding areas. *Arch Environ Health* **47**: 203–210.

Barry PSI (1975) A comparison of concentrations of lead in human tissues. *Br J Ind Med* **32**: 119–139

Barry PSI (1981) Concentrations of lead in the tissues of children. *Br J Ind Med* **38**: 61–71.

Bishop L and Hill WJ (1983) A study of the relationship between blood lead levels and occupational air lead levels. *Am Stat* **37**: 471–475.

Bost L, Dong W, Primatesta P and Poulter NR (1998) The relationship between blood lead and blood pressure in the English population. In: *IEH Report on Recent UK Blood Lead Surveys* (Report 9). Leicester: Institute for Environment and Health.

Brunekreef B (1984) The relationship between air lead and blood lead in children: a critical review. *Sci Total Environ* **38**: 79–123.

Buchet JP, Lauwerys R, Roels H and Hubermont G (1977) Mobilization of lead during pregnancy in rats. *Int Arch Occup Environ Health* **40**: 33–36.

Chamberlain AC (1983) Effect of airborne lead on blood lead. *Atmos Environ* **17**: 677–692.

Chamberlain AC, Clough WS, Heard MJ *et al.* (1975) Uptake of lead by inhalation of motor exhaust. *Proc R Soc Lond B* **192**: 77–110.

Chamberlain AC, Heard MJ, Little P *et al.* (1978) *Investigations into Lead from Motor Vehicles*, Report No AERE-R9198. Harwell: United Kingdom Atomic Energy Authority.

Davies DJA, Thornton I, Watt JM *et al.* (1990) Lead intake and blood lead in two-year-old UK urban children. *Sci Total Environ* **90**: 13–29.

Delves HT (1998) Overview of UK and international studies on trends in blood lead, and the use of lead isotope ratios to identify environmental sources. In: *IEH Report on Recent UK Blood Lead Surveys*, Report R9. Leicester: Institute for Environment and Health.

Delves HT and Campbell MJ (1993) Identification and apportionment of sources of lead in human tissue. *Environ Geochem Health* **15**: 75–84.

Department of the Environment, Sport and Territories (1996) *Australia State of the Environment 1996.* Canberra: State of the Environment Advisory Council.

Department of the Environment, Transport and the Regions (1997) *Digest of Environmental Protection and Water Statistics*, No 19. London: HMSO.

deSilva PE and Christophers AJ (1997) Lead exposure and children's intelligence: do low levels of lead in blood cause mental deficit? *J Paediatr Child Health* **33**: 12–17.

Ducroffe G, Claeys F and Bruaux P (1990) Lowering time trend of blood levels in Belgium since 1978.

Environ Res **51**: 25–34.

Facchetti S and Geiss F (1982) *Isotopic Lead Experiment: Status Report.* Luxembourg: Commission of the European Communities, Publication No. EUR 8352 EN.

Gartside PS and Buncher CR (1982) Relationship of air lead and blood lead for workers at an automobile battery factory. *Int Arch Occup Environ Health* **50**: 1–10.

Griffin TB, Coulston F, Wills H *et al.* (1975) Clinical studies on men continuously exposed to airborne particulate lead. In: Griffin TB and Knelson JH (eds) *Lead.* London Academic Press, pp 221–240.

Golding J, Smith M, Delves HT, Taylor H and the ALSPAC Study Team (1998) The ALSPAC study on lead in children. In: *IEH Report on Recent UK Blood Lead Surveys* (Report R9). Leicester: Institute for Environment and Health.

Grandjean P, Hollnagel H, Hedegaard L *et al.* (1989) Blood lead–blood pressure relations: alcohol intake and hemoglobin as confounders. *Am J Epidemiol* **129**: 732–739.

Gulson BL, Mahaffey KR, Mizon KJ *et al.* (1995) Contribution of tissue lead to blood lead in adult female subjects based on stable lead isotope methods. *J Lab Clin Med* **125**: 703–712.

Hertz-Picciotto I and Croft J (1993) Review of the relation between blood lead and blood pressure. *Epidemiol Rev* **15**: 352–373.

International Radiological Protection Commission (1966) Task Group on Lung Dynamics. Deposition and retention models for internal dosimetry of the human respiratory tract. *Health Physics* **12**: 173–207.

James AC (1978) *Lung Deposition of Submicron Aerosols Calculated as a Function of Age and Breathing Rate.* National Radiological Protection Board Annual Report. Harwell: National Radiological Protection Board.

Junco-Muñoz P, Ottman R, Lee JH *et al.* (1996) Blood lead concentrations and associated factors in residents of Monterrey, Mexico. *Arch Med Res* **27**: 547–551.

Kehoe RA (1961) The metabolism of lead under abnormal conditions. *J R Inst Publ Health* **24**: 101–143.

Kehoe RA (1966) Criteria for human safety from the contamination of the ambient atmosphere with lead. *Proc Int Congr Occup Health* Vienna, pp 83–98.

King E and Conchie D (1979) Industrial lead absorption. *Ann Occup Hyg* **22**: 213–239.

Koo WWK, Succop PA, Bornschein RL *et al.* (1991) Serum vitamin D metabolites and bone mineralization in young children with chronic to moderate lead exposure. *Pediatrics* **87**: 680–687.

Lanphear BP, Weitzman M, Winter NL *et al.* (1996) Lead-contaminated house dust and urban children's blood lead levels. *Am J Public Health* **86**: 1416–1421.

Mahaffrey KR, Rosen JF, Chesney RW *et al.* (1982) Association between age, blood lead concentration, and serum 1,25-dihydroxycholecalciferol levels in children. *Am J Clin Nutr* **35**: 1327–1331.

McMichael AJ (1997) Lead exposure and child intelligence: interpreting or misinterpreting, the direction of causality? *J Paediatr Child Health* **33**: 7–8.

Ministry of Agriculture, Fisheries and Food (1982) Survey of lead in food: second supplementary report. *Food Surveillance Paper* No. 10. London: HMSO.

Ministry of Agriculture, Fisheries and Food (1989) Lead in food: progress report. *Food Surveillance Paper* No 27. London: HMSO.

Murozumi M, Chow TJ and Patterson CC (1969) Chemical concentrations of pollutant lead aerosols, terrestrial dusts and sea salts in Greenland and Antarctic snow strata. *Geochim Cosmochim Acta* **33**: 1247–1294.

Neri LC, Hewitt D and Orser B (1988) Blood lead and blood pressure: analysis of cross-sectional and longitudinal data from Canada. *Environ Health Perspect* **78**: 123–126.

Piomelli S, Seaman C, Zullow D *et al.* (1982) Threshold for lead damage to heme synthesis in urban children. *Proc Natl Acad Sci* (USA) **79**: 3335–3339.

Pirkle JL, Brody DJ, Gunter EW *et al.* (1994) The decline in blood lead levels in the United States. The National Health and Nutrition Examination Surveys (NHANES). *J Am Med Assoc* **272**: 284–291.

Pocock SJ, Smith M and Baghurst P (1994) Environmental lead and children's intelligence: a systematic review of the epidemiological evidence. *Br Med J* **309**: 1189–1197.

Primatesta P, Dong W, Bost L *et al.* (1998) Survey of blood lead levels in the population in England, 1995. In: *IEH Report on Recent UK Blood Lead Surveys* Report 9. Leicester: Institute for Environment and Health.

Rabinowitz MB, Wetherill GW, Kopple JD (1976) Kinetics analysis of lead metablism in healthy humans. *J Clin Invest* **58**: 260–270.

Roels HA and Lauwerys R (1987) Evaluation of dose–effect and dose–response relationships for lead exposure in different Belgium population groups (foetus, child, adult men and women). *Trace Element Med* **4**: 80–87.

Roels HA, Buchet JP, Lauwerys R *et al.* (1976) Impact of air pollution by lead on the haem biosynthetic pathway in school-age children. *Arch Environ Health* **31**: 310–316.

Roels HA, Buchet JP, Lauwerys R *et al.* (1980) Exposure to lead by the oral and the pulmonary routes of children living in the vicinity of a primary lead smelter. *Environ Res* **22**: 81–94.

Salway AG, Eggleston HS, Goodwin JWL and Murrells TP (1997) *UK Emissions of Air Pollutants 1970–1995*. A Report of the National Atmospheric Emissions Inventory. Abingdon: AEA Technology.

Schwartz J (1995) Lead, blood pressure and cardiovascular disease in men. *Arch Environ Health* **50**: 31–37.

Schwartz J and Pitcher H (1989) The relationship between gasoline lead and blood lead in the United States. *J Off Stat* **5**: 421–431.

Schwartz J, Landrigan PL and Baker EL (1990) Lead-induced anemia: dose–response relationships and evidence for a threshold. *Am J Public Health* **80**: 165–168.

Silbergeld EK (1991) Lead in bone: implications for toxicology during pregnancy and lactation. *Environ Health Perspect* **91**: 63–70.

Staessen JA, Bulpitt CJ, Fagard R *et al.* (1994) Hypertension caused by low-level lead exposure: myth or fact? *J Cardiovasc Risk* **1**: 87–97.

Staessen JA, Roels H and Fagard R (1996) Lead exposure and conventional and ambulatory blood pressure: a prospective population study. *J Am Med Assoc* **275**: 1564–1570.

Strömberg U, Schütz A and Skerfving S (1995) Substantial decrease of blood lead in Swedish children, 1978–94, associated with petrol lead. *Occup Environ Med* **52**: 764–769.

Tola S, Hernberg S, Asp S and Nikkanen J (1973) Parameters indicative of absorption and biological effect in new lead exposure: a prospective study. *Br J Ind Med* **30**: 134–141.

US EPA (1986) *Air Quality Criteria for Lead*. Washington DC: Environmental Protection Agency, (EPA-600/8-83/028aF-dF).

US EPA (1991) *National Air Quality and Emission Trends Report*. Research Triangle Park, NC: US Environmental Protection Agency (EPA-450/4-91-023).

Victery W (1988) Evidence for effects of chronic lead exposure on blood pressure in experimental animals: an overview. *Environ Health Perspect* **78**: 71–76.

Watt GCM, Britton A, Harper Gilmour W *et al.* (1996) Is lead in tap water still a public health problem? An observational study in Glasgow. *Br Med J* **313**: 979–981.

Weiss ST, Muñoz A and Stein A (1986) The relationship of blood lead to blood pressure in a longitudinal study of working men. *Am J Epidemiol* **123**: 800–808.

WHO (1987) Lead. In: *Air Quality Guidelines for Europe*. Copenhagen: World Health Organisation.

WHO (1995) Inorganic lead. *Environmental Health Criteria* 165. Geneva: World Health Organisation.

Yankel AJ, von Lindern IH and Walter SD (1977) The Silver Valley lead study: the relationship of childhood lead poisoning and environmental exposure. *J Air Pollut Contr Assoc* **27**: 763–767.

Yocom JE (1982) Indoor–outdoor air quality relationships: a critical review. *J Air Pollut Control Assoc* **32**: 500–520.

Ziegler EE, Edwards BB, Jensen RL *et al.* (1978) Absorption and retention of lead by infants. *Pediatr Res* **12**: 29–34.

35

Selected Organic Chemicals

LESLEY RUSHTON

MRC Institute for Environment and Health, University of Leicester, Leicester, UK

KATHLEEN CAMERON

Department of the Environment, Transport and the Regions, London, UK

The views expressed in this chapter are those of the authors and should not be taken to represent those of the UK Department of the Environment, Transport and the Regions.

INTRODUCTION

The major class of organic compounds in the atmosphere is volatile organic compounds (VOCs). This class includes alkanes, alkenes, aromatics, aldehydes, ketones, alcohols, acids, ethers and halogenated species. Although methane is an important component of VOCs, it is normally excluded from consideration of general issues as its environmental impact derives principally from its contribution to global warming. VOCs are generated from a range of sources including motor vehicles, solvent use, and the chemical, oil and food industries. The major sources of VOCs, excluding methane, in the UK in 1995 are shown in Table 35.1.

Speciated VOCs emissions inventories (excluding methane) show that the largest contributions are from butanes and toluene. Butanes are emitted principally from petrol, solvents, petrol refining and distribution; toluene is mainly produced from petrol exhausts and solvents (DETR, 1997; QUARG, 1993).

VOCs in the atmosphere have two major health impacts: firstly, they are precursors to the photochemical production of ozone in the troposphere (see Chapter 26). Photochemical oxidation of VOCs is slow and thus the transport of VOCs away from

Table 35.1   Estimated emissions of volatile organic compounds (excluding methane) in the UK (1995) (DETR, 1997)

Source category	Estimated emissions (thousand tonnes)	Percentage of total
Power stations	5	–
Domestic	30	1.3
Commercial/public service	2	–
Refineries	2	–
Iron and steel	4	–
Other industrial combustion	9	0.4
Non-combustion sources	335	14.3
Extraction and distillation of fossil fuel[a]	334	14.3
Solvent use	700	29.9
Road transport[b]	690	29.5
Off-road sources	96	4.1
Military	1	–
Railways	8	0.3
Civil aircraft[c]	4	–
Shipping[d]	12	0.5
Waste treatment and disposal	26	1.1
Agriculture	–	–
Forests[e]	80	3.4
Total	2338	100

[a] Includes emissions for gas flaring, producers' own gas use and fugitive emissions from offshore installations. Also includes a small contribution from fuel oil use on offshore installations which is also included under shipping.
[b] Includes evaporative emissions from the petrol tank and carburettor of petrol-engined vehicles.
[c] Includes only those emissions associated with ground movement and take off and landing cycles up to 1 km from airport.
[d] Includes emissions from fishing, coastal shipping, oil exploration and production and includes fuel oil use on offshore installations. Includes only those emissions from marine bunker fuels from shipping within coastal waters (<12 miles).
[e] An order of magnitude estimate of natural emissions.

sources can result in the generation of ozone at considerable distances away. VOCs vary widely in their potential to promote ozone production and this has been used as a means of classifying them. Both butanes and toluene represent significant sources of ozone. Secondly, a number are directly toxic, for example, benzene and 1,3-butadiene, which are associated with causing cancer in occupationally exposed workers.

A number of other toxic compounds are present in the atmosphere including polycyclic aromatic hydrocarbons (PAHs), polychlorinated biphenyls (PCBs) and dioxins. These are generally considered separately from VOCs. Emissions of PAHs are primarily transport-related although there are some emissions from stationary combustion sources. PCBs are largely derived from the decommissioning of industrial plant while dioxins arise from the combustion of chlorine containing compounds.

This chapter addresses the human health effects of atmospheric exposure to some of the toxic pollutants listed above: benzene, 1,3-butadiene and dioxins. Toluene has also been included because of the large volumes emitted into the atmosphere.

BENZENE

Sources

The general population is exposed to benzene in the outdoor environment mainly through inhalation of polluted air from vehicles, and to a lesser extent, from industrial sources. In the indoor environment, tobacco smoke is the main source of benzene (Crump, 1997; EBS, 1996; Wallace, 1996). Emissions to ambient air have increased greatly since the 1960s due to the rapid increase in road traffic. During 1995, an estimated 35 kilotonnes of benzene were emitted to the UK environment (Salway *et al.*, 1997), with approximately 70% from gasoline use in road transport.

Exposure

Data on benzene concentrations in ambient air are available from both automatic and passive monitoring networks, and from *ad hoc* studies. Annual mean concentrations in the UK in 1995 ranged from 0.7 to 2.5 parts per billion (ppb) in urban sites, with 0.4 ppb for the one rural site monitored (AEA, 1995). The highest concentrations are usually found adjacent to roads with heavy traffic density (Department of the Environment, 1996). Concentrations are also higher in the winter than in the summer (AEA, 1995; Crump, 1997). EBS (1996) produce summary tables of many measurements made in ambient air throughout Europe. Annual averages range from 2 to 20 $\mu g/m^3$. A typical value in rural or suburban areas was 2 $\mu g/m^3$. Low annual averages (2–3.2 $\mu g/m^3$) were found for urban areas with low traffic density compared with high averages (8.8–15.5 $\mu g/m^3$) reported for urban areas with high traffic density. Higher winter values were generally noted. Measurements of nearly 130 $\mu g/m^3$ were found in ambient air near specific benzene industrial sources, such as coal coking and chemical manufacturing.

To put the contribution of ambient air to total benzene exposure into context, it is important to consider microenvironments in which the potential for exposure may be high. The three most important are refuelling of vehicles, driving and indoor air. In the UK, refuelling of vehicles is usually carried out by a vehicle occupant, rather than by a gasoline attendant. There have been several studies reporting benzene concentrations in service stations, both background levels and during refuelling (CONCAWE, 1994, 1996; Leung and Harrison, 1998b; Sperduto, 1993). Annual mean levels around a petrol station in the UK ranged from 0.5 to 2.2 ppb, with service station activities contributing 65% of the overall mean, 15% from adjacent road traffic and 20% from background levels (CONCAWE, 1996). Exposure through refuelling, which on average takes about one minute, varies by type of fuel and vehicle. A mean of 150 ppb was found for petrol station attendants in an Italian study from personal monitoring (Sperduto, 1993); a lower level (60 ppb annual running mean) was found in a UK study (Leung and Harrison, 1998b).

Levels of benzene inside vehicles can also be much higher than ambient levels experienced in urban areas. Mann *et al.* (1997), in a small study, found a mean in-vehicle concentration of 17 ppb compared with an outdoor background level of 1.6 ppb. In a larger study by Leung and Harrison (1998b), vehicle concentrations recorded whilst driving were similar to those measured on the outside of the vehicle, but, in urban traffic, were higher than roadside levels, perhaps due to pollution being drawn into the vehicle through the ventilation system. In-vehicle concentrations ranged from 5 to 38 times higher than suburban levels.

Although tobacco smoke only accounts for a tiny percentage of benzene emitted into the environment, it is an important source of human exposure for both smokers and, through passive smoking, non-smokers. Several studies have measured concentrations in the houses of smokers and non-smokers. Crump (1997) found a mean of 3 ppb in the living rooms of smokers compared with a mean of 2 ppb for non-smokers, and Leung and Harrison (1998a) reported daily exposures of 7.3 ppb for smokers relative to 3.6 ppb for non-smokers. Analyses of the breath and blood of smokers have confirmed that much higher levels of benzene are found in smokers than in non-smokers (Brugnone *et al.*, 1989; Hajimiragha *et al.*, 1989).

Effects

Benzene has been associated with a number of adverse health effects over a range of exposure levels, summarized in Table 35.2 (adapted from EBS (1996) and Courage and Duarte-Davidson (1998)). A brief overview of the effects resulting from high levels of exposure follows, with the focus then being on the effects at low levels.

The majority of absorbed benzene is eliminated unmetabolized through exhaled air, with a smaller proportion of metabolized benzene being excreted in the urine. The liver is the major site of primary benzene metabolism, but secondary oxidative metabolism of metabolites in the bone marrow is a prerequisite for the haematotoxic, genotoxic and

Table 35.2 Adverse health effects due to inhalation of benzene

Effect	Description	Length of exposure	Level (ppm)	Reference
Death		5–10 min	20 000	Thienes and Haley (1972)
Central nervous system	Vertigo, drowsiness headache, nausea	Hours	250–500	Clayton and Clayton (1994)
Haematological	Pancytopenia Aplastic anaemia Myelodysplastic syndrome	Years	100	Greenburg *et al.* (1939) Yin *et al.* (1987) Aksoy *et al.* (1972)
	Cytopenia	Years	35	Fishbeck *et al.* (1978) Kipen *et al.* (1989)
Mutagenic	Chromosome aberrations	Years	20-100	EBS (1996)
	Adduct formation	Years	13–63	Lin *et al.* (1996)
Carcinogenic	ANLL	Years	20–50	EBS (1996)

carcinogenic effects (Snyder and Kalf, 1994). Acute toxic effects on the central nervous system are experienced at high levels of exposure, usually as a result of accident or abuse. These range from vertigo, drowsiness, headache and nausea experienced at levels of 200–500 parts per million (ppm), through giddiness and staggered gait, and with continued exposure, unconsciousness, at concentrations of 1500 ppm, to death within 5–10 min at levels of 20 000 ppm (Clayton and Clayton, 1994; Thienes and Haley, 1972). Repeated exposure to benzene concentrations greater than 100 ppm over months or years has been reported from studies in industries such as shoe and tyre manufacturing to cause haematological conditions including pancytopenia, aplastic anaemia and myelodysplastic syndrome (Aksoy, 1985; Aksoy *et al.*, 1972, 1974; Paci *et al.*, 1989; Yin *et al.*, 1987). These are generally associated with a suppression of specific cell lines, such as white and red blood cells or platelets. Pancytopenia, for example, is characterized by a reduction in the ability of the bone marrow to produce all three blood constituents. Less severe haematological effects have been documented at lower benzene levels. These include cytopenias such as leukopenia (decrease in white blood cells counts), thrombocytopenia (decrease in platelet counts) and anaemia (decrease in red blood cell counts) (EPAQS, 1994a; Kipen *et al.*, 1989; Rothman *et al.*, 1996). A 'no observed affect of exposure level' (NOAEL) of 20 ppm for repeated 8-h time-weighted average (TWA) concentrations has been suggested for cytopenias (EBS, 1996).

Benzene has been shown to be a clastogen in both humans and animals (Sarto *et al.*, 1984; Snyder and Kocsis, 1975; Turkel and Egeli, 1994; Yardley-Jones *et al.*, 1990). Chromosomal aberrations may be predictors of carcinogenicity risk and exposure to benzene and subsequent leukaemia is thought to involve chromosome numbers 5 and 7, and perhaps chromosome numbers 8 and 12 (EBS, 1996; Irons and Stillman, 1996). In their review Paustenbach *et al.* (1993) concluded that benzene exposure at levels sufficient to cause haematoxicity may also produce clastogenic effects in human lymphocytes, i.e. prolonged exposure to concentrations above 20 ppm (EBS, 1996).

Benzene has long been known to be a human carcinogen, with the strongest evidence linking it with lymphohaematopoietic cancers, particularly acute non-lymphocytic leukaemia (ANLL). The vast majority of the evidence derives from industrial studies of workers exposed to benzene, often as a constituent of a complex mixture. These include the shoe-making, printing, petrochemical, chemical, coke production, and rubber manufacturing industries (Aksoy *et al.*, 1974; Christie *et al.*, 1991; Decoufle *et al.*, 1983; Hurley *et al.*, 1991; Paxton *et al.*, 1994a,b; Rushton, 1993; Tsai *et al.*, 1983; Vigliani and Forni, 1976; Yin *et al.*, 1989). Many of the populations in these studies were exposed to benzene levels that were extremely high compared to levels experienced in these industries today. Most had sparse information about the actual levels, and did not estimate exposures for the study individuals. A preponderance of acute myeloid leukaemia (AML) and related forms, such as acute monocytic leukaemia and erythroleukaemia, was revealed in these studies.

There have been a small number of cohort and nested case–control studies that have estimated exposure information for each worker. Based on these studies, attempts have been made to develop a dose–response relationship using cumulative exposure, i.e. assuming a symmetrical contribution of exposure concentration and exposure duration. The cohort that has been most often used in risk assessment is the 'Pliofilm' cohort, which included workers from three factories manufacturing rubber film in the USA. A wide range of exposures was encountered, with relatively few other chemicals involved.

There have been several publications from this cohort giving mortality results (Infante *et al.*, 1977; Paxton *et al.*, 1994a; Rinsky *et al.*, 1981, 1987; Wallace, 1996), and risk assessments using different methods of exposure estimation and different mathematical models (Austin *et al.*, 1988; Brett *et al.*, 1989; Byrd and Barfield, 1989; Crump, 1994; Crump and Allen, 1984; Paustenbach *et al.*, 1992; Paxton *et al.*, 1994a; Rinsky, 1989; Rinsky *et al.*, 1989; Schnatter *et al.*, 1996). The most recent follow-up (Paxton *et al.*, 1994b) reported 15 leukaemia deaths, and risks were estimated using three separate sets of exposure estimates (Crump and Allen, 1984; Paustenbach *et al.*, 1992; Rinsky *et al.*, 1981). The results are summarized in Table 35.3. Using the Paustenbach *et al.* (1992) exposure estimates, a significant risk was found for cumulative exposures above 500 ppm. In contrast, the other two estimates found a statistically significant risk at levels between 50 and 500 ppm-years or greater. Schnatter (1996) used these three estimates, together with their median exposure, and defined a long-term average concentration associated with the maximally exposed job for each worker in the Pliofilm cohort. Subsets of workers were identified who were only exposed to long-term-average (LTA) concentrations below a given concentration. Critical concentrations which influence risk were determined by examining subgroups of workers only exposed to successively larger concentrations. For acute monocytic and myeloid leukaemia (AMML), both the median and Crump and Allen (1984) exposure estimates indicated that concentrations above 50 ppm influenced risk. In contrast, the corresponding figures using the Rinsky *et al.* (1981) estimates were 20 ppm, and using the Paustenbach *et al.* (1992) estimates 140 ppm. Using the Rinsky *et al.* (1981) estimates, workers who were exposed to LTA concentrations above 20 ppm were exposed for an average of 6 years, while AMML cases were exposed for an average of 13 years, suggesting that these exposures need to be experienced for years (rather than days or weeks) to result in excess AMML risk.

Four other cohort studies have quantified benzene exposures for all workers (Bond *et al.*, 1986; Hayes *et al.*, 1997; Ireland *et al.*, 1995; Wong, 1987a). The results are summarized in Table 35.3 for total leukaemia. The Bond study is a follow-up of a study by Ott *et al.* (1978) and gives risk estimates among 956 Dow chemical workers exposed to benzene from ethyl cellulose and chlorobenzol, and alkyl benzene operations. There were three deaths from leukaemia compared with 1.9 expected. Wong (1987a,b) followed up 7676 chemical workers, 3536 of whom were continuously exposed to benzene. Six leukaemia deaths were reported in the exposed group. An update of a subgroup of this population has been reported by Ireland *et al.* (1997). LTA concentrations were only available for 666 production workers. There were three cases of leukaemia observed at exposures of 6 ppm or greater compared with 0.65 expected. Some attempt was made to address the issue of risks from intermittent exposure, and a statistically significant excess of all leukaemias was found for all employees experiencing 40 or more peaks of exposure, defined as a 15 min episode of exposure exceeding 100 ppm. Hayes *et al.* (1997) report mortality results from a very large cohort of Chinese workers exposed between 1972 and 1987 in a variety of occupations. Excess risks were found for all leukaemias at all levels of exposure and above 40 ppm for ANLL.

The results from two nested case–control studies of petroleum distribution workers are presented in Table 35.4 (Rushton and Romaniuk, 1997; Schnatter *et al.*, 1996b). The method of retrospective quantitative estimation of exposures used in these studies was essentially the same. Full work histories were obtained for each case and control, including

Table 35.3 The risk of leukaemia from cohort studies with quantified exposure estimates of benzene

Cohort studies	Exposure (ppm years)	Observed deaths	Expected deaths	Standardized mortality ratio	95% confidence interval
Pliofilm (Paxton *et al.*, 1994b)	0–5	3	1.5	1.97	0.41–5.76
(Rinsky *et al.* estimates, 1981)	5–50	3	1.3	2.29	0.47–6.69
	50–500	7	1.0	6.93	2.8–14.3
	>500	7	0.1	20.0	0.5–111.0
Pliofilm (Paxton *et al.*, 1994b)	0–5	1	1.1	0.88	0.02–4.89
(Crump and Allen	5–50	4	1.2	3.25	0.88–8.33
estimates, 1984)	50–500	6	1.2	4.87	1.79–10.6
	>500	3	0.3	10.30	2.13–30.2
Pliofilm (Paxton *et al.*, 1994b)	0–5	1	0.8	1.33	0.03–7.43
(Paustenbach *et al.*,	5–50	2	1.1	1.79	0.22–6.45
estimates, 1992)	50–500	4	1.4	2.80	0.80–7.16
	>500	7	0.6	11.9	4.8–24.44
Dow (Bond *et al.*, 1986)	0–42	2	1.2	1.67	0.14–4.9
	42–83	0	0.3	0.00	0–3.3
	83+	1	0.4	2.50	0–10.0
Chemical Manufacturers	0–14.9	2	2.1	0.97	0.12–3.49
Association (Wong, 1987a)	15–59	1	1.3	0.78	0.02–4.34
	60+	3	1.1	2.76	0.57–8.06
US Production Workers	0	5	4.35	1.1	0.4–2.6
(Ireland *et al.*, 1997)	<1	2	0.80	2.5	0.3–8.9
	1–6	0	0	0	0
	>6	3	0.65	4.6	0.9–13.4
Chinese Workers	<40	18	8.18	2.2	1.1–4.5
(Hayes *et al.*, 1997)	40–99	11	3.79	2.9	1.3–6.5
	>100	29	10.74	2.7	1.4–5.2

job titles, dates of starting and finishing each job, and the locations of work. For each job and distribution terminal, exposure estimates were derived from actual exposure measurements, which were adjusted, where appropriate, to take account of changes over time in factors which might influence exposure, such as technology. These estimates were then applied to each line of work history and summed to give a cumulative exposure for each study subject. No excess risk was found in any category of exposure in the study by Schnatter *et al.* (1996). Of the 90 cases of leukaemia identified in the Rushton and Romaniuk (1997) study, 31 were AMML. Risk from AMML did not increase with cumulative exposure analysed as a continuous variable, but a trend was observed when categorized into discrete ranges (test for trend significant at the 10% level). When the 5 and 10 years of most recent exposure were excluded the magnitude of the odds ratios increased and the trend became significant at the 5% level. Intermittent exposure was qualitatively estimated and, when included in the analyses, was shown to be related to risk and to reduce that associated with cumulative exposure.

Table 35.4 The risk of leukaemia determined from case–control studies with quantified exposure estimates of benzene

Case–control studies	Exposure (ppm)	Cases	Controls	Odds ratio	95% confidence interval
All leukaemias					
Canadian petroleum	<0.45	10		1.0	
distribution (Schnatter,	0.45–4.5	1		0.43	0.01–4.0
1996b)	4.5–45	1		0.16	0.0–1.3
	>45	2		1.47	0.2–13.1
UK petroleum distribution	<0.45	22	109	1.0	
(Rushton and Romaniuk,	0.45–4.5	47	172	1.42	0.8–2.6
1997)	4.5–45	20	69	1.48	0.7–3.0
	>45	1	4	1.35	0.1–12.8
AMML					
UK petroleum distribution	<0.45	7	46	1.0	
(Rushton and Romaniuk,	0.45–4.5	15	51	2.17	0.8–6.1
1997)	4.5–45	20	23	2.82	0.8–9.4
	>45	0	1	0	

A pooled analysis of the data from four studies (Ireland *et al.*, 1997; Paxton *et al.*, 1994a; Rushton and Romaniuk, 1997; Schnatter *et al.*, 1996) has been carried out (EBS, 1996), using several models. No dose–response pattern was revealed at levels of exposure below 1 ppm, and a cumulative exposure relationship with ANLL was suggested when concentrations were above approximately 20–50 ppm.

Assessment of the risk to the general population

The air quality guidelines recommended by the WHO are used as basis for setting EU standards (NSCA, 1997), and while these are not mandatory, they are generally accepted as being levels not to be exceeded if health air quality is to be maintained. Currently the WHO does not recommend a safe level for benzene owing to its carcinogenicity. The UK Expert Panel on Air Quality Standards (EPAQS) has recommended 5 parts per billion (ppb) running annual mean as an ambient air quality standard, basing their decision on the long-term risk of developing non-lymphocytic leukaemias from exposure to low levels of benzene (EPAQS, 1994a).

The epidemiological studies have revealed a critical lack of information about the risk at levels of benzene exposure below 20 ppm. However, the levels to which the general population are exposed are well below those reported in any of the above studies, and below the levels to which a risk has been definitely attributed. Thus the risk of any adverse health effects associated with current environmental levels of benzene is likely to be extremely small.

TOLUENE

Sources

Toluene is one of the most widespread hydrocarbons in the environment. Although it occurs naturally in crude oil, natural gas deposits and in emissions from volcanoes and forest fires, its presence is largely due to its high volume of production and extensive use. The main source of production is during the catalytic reforming process of petroleum refining. It is also produced as a by-product during styrene manufacture and coke-oven operations (Fishbein, 1988). Toluene is used in a mixture with benzene and xylene in gasoline blending. Isolated toluene is used in the production of benzene (IARC, 1989). It also has widespread use in solvent applications in the paint, rubber, printing, cosmetics and adhesive industries, and as a starting material for the synthesis of other chemicals, such as the production of toluene diisocyanate for use in polyurethane production (Hoff, 1983; Low *et al.*, 1988).

The main source of toluene in ambient air is through evaporation of gasoline and vehicle exhaust emissions. In 1991 an estimated 151 000 metric tonnes of toluene was emitted in the UK, 58% from petrol exhaust and a further 35% from solvents (QUARG, 1993).

Exposures

The UK hydrocarbon network makes automated hourly measurements on 26 individual hydrocarbons, including toluene, benzene and xylene (Harrison, 1996). Annual average toluene concentrations were, respectively, 0.501 ppb, 0.999 ppb and 1.495 ppb at Great Dun Fell (1989–91), West Beckham in East Anglia (1989–91) and Harwell, three rural areas, and in two urban areas were 1.431 ppb (at the National Physical Laboratory, Teddington, 1988–91) and 2.584 in Middlesborough (1992) (QUARG, 1993). However, maximum values can be much higher, with an hourly maximum reaching 52.9 ppb in London, compared to a mean of 3.47 (Harrison, 1996).

Levels in indoor environments may be much higher than those outdoors owing to the use of commercial products containing toluene, and the presence of tobacco smoke (DeBortoli *et al.*, 1984; Krause *et al.*, 1987). Gilli *et al.* (1994) found that active smokers were exposed to about four times the level of passive smokers. In a study by Hagimargh *et al.* (1984) smokers were found to have significantly higher blood concentrations of both toluene (median 2201 ng/l) and benzene (median 547 ng/l) than non-smokers (median 1141 ng/l and median 190 ng/l respectively). In a study which measured levels of benzene and toluene concurrently outdoors and indoors in children's bedrooms, the average indoor concentrations of both components was 2–3 times the outdoor concentrations. The ratios were 3:1 in the winter and 2:1 in midsummer (Porstmann *et al.*, 1994). The ambient concentrations were higher on streets with a higher traffic load and this impacted slightly on indoor levels. However, the indoor levels were most influenced by environmental tobacco smoke and redecoration. Levels in certain occupations, for example printing, painting and rubber manufacture, can also be high (greater than 200 ppm) (IARC, 1989).

Several biomarkers of toluene exposure have been developed using urine, end-expired air and venous blood, the last being the most reliable and sensitive, particularly at low levels of exposure (<1 ppm) (Mizunuma *et al.*, 1994). Initial elimination from blood after exposure terminates is rapid, but, because release of toluene from adipose tissue is slow (half-time about 80 h) detectable levels of toluene result even 2 weeks after exposure ceases (Brugnone *et al.*, 1995; Foo *et al.*, 1988; Nise and Orbaek, 1988).

Effects

Toluene exposure has been investigated in studies of spontaneous abortion and other adverse pregnancy outcomes, but differences between exposed and non-exposed groups were not great and toluene exposure was experienced in a mixture of other solvents. Studies of chromosomal aberrations also have been inconclusive. Case–control studies investigating carcinogenicity risks have not been able to evaluate risks related to toluene in isolation from other solvents (IARC, 1989).

Dysfunction of the central nervous system (CNS) is the adverse health effect of primary concern with exposure to toluene. Acute and subchronic inhalation of toluene in several animal species has been shown to produce CNS depression, disturbance of the circadian rhythm and behavioural effects (IARC, 1989). Most of the effects in humans are the result of exposure to toluene in association with other organic solvents, such as benzene, which may enhance or reduce any adverse effect.

Effects at high levels of exposure to toluene have been observed in volunteers and in cases of solvent abuse. In the latter group fatalities have been reported (Anderson *et al.*, 1985). Heavy accidental exposure can lead to coma (Bakinson and Jones, 1980). At about 3000 mg/m³ severe fatigue, pronounced nausea, mental confusion, staggering gait and strongly affected pupillary light reflexes have been observed (IARC, 1989). After three days of heavy exposure to a mixture of toluene, workers in a factory suffered memory disturbances that continued for months (Stollery and Flindt, 1988). Some of these symptoms, such as impaired coordination, enlarged pupils and fatigue, have been experienced in volunteers exposed to 750 mg/m³ for 8 h. Volunteers exposed to 100 ppm (377 mg/m³) for 6 h per day for 4 days suffered from headaches, dizziness and a sensation of intoxication (Andersen *et al.*, 1983).

Case studies and occupationally based cross-sectional studies have also reported similar effects. Examples include glue sniffers (Byrne and Zibin, 1991; Hormes *et al.*, 1986; Maas *et al.*, 1991), painters (Winchester and Madjar, 1986) and printers (Hanninen *et al.*, 1987). There have been several studies which have carried out measures of neurophysiological and neurobehavioural functioning, some of which attempt to derive concentration–response relationships.

Six healthy adults were exposed to 100 ppm toluene or air (control) for 6 h, with exposures separated by at least 14 days, and including 30 minutes exercise that quadrupled minute ventilation (Rahill *et al.*, 1996). Lung function was unchanged but composite 1 h complex performance test scores were lower during toluene exposure, with the score for the last hour reduced by 10%. Response times to neuropsychological tests were also reduced.

A study of neurobehavioural effects in male car painters and printers, exposed to toluene, xylene and methyl ethyl ketone, found a poor visual retention performance compared with that of a control group (Lee, 1993). However, a study of men exposed to paint solvents or working with toluene (Cherry *et al.*, 1985) and studies of printers exposed almost exclusively to toluene (Antti-Poika *et al.*, 1985; Struwe and Wennberg, 1983) found few effects on the CNS.

Another study addressed the issue of intermittent exposure (Baelum *et al.*, 1990), by comparing responses in 32 males and 39 females exposed for 7 h to either clean air, constant exposure to 100 ppm toluene, or a varying exposure with the same time-weighted average but with peaks of 300 ppm every 30 min. The subjects exercised in three 15-min periods with a load of 50 to 100 W. Exposure to toluene caused significantly greater complaints about poor air quality, altered temperature and noise perception, increased irritation in the nose and feeling of intoxication. There were only minimal effects on psychomotor or visual performance, but a tendency towards a lower score in a vigilance test. There were no differences in the effects found between the constant and intermittent exposures.

A similar study which exposed printers who had been working with solvents for up to 25 years, to 100 ppm of toluene or clean air for 6.5 h, found similar physical symptoms and decreased manual dexterity, colour discrimination and visual perception accuracy when exposed (Baelum *et al.*, 1985).

Foo *et al.* (1990, 1993) found that tests of psychomotor function were significantly lower in 30 females exposed to an 8-h time-weighted average (TWA) of 88 ppm compared with those from the same factory exposed to a TWA of 13 ppm.

Recent studies have suggested that some ethnic groups may be at more risk to the neurobehavioural effects of toluene owing to metabolic differences. Toluene is initially metabolized to benzyl alcohol by the microsomal mixed-function oxidase system. Subsequently oxidation to benzaldehyde and then to benzoic acid is carried out by the two enzymes alcohol and aldehyde hydrogenase, respectively (Greenberg, 1997). All populations have two forms of aldehyde dehydrogenase, one having a high K_m and one with a much lower K_m (K_m is a constant, defined for each form of a specific enzyme, which gives an indication of the speed of the enzyme reaction). Some Asian populations have a defective gene for the low K_m enzyme (Kawamoto *et al.*, 1994a). Toluene-exposed Japanese workers with the defective gene had lower levels of urinary hippuric acid than those without it, which suggests a lower rate of toluene metabolism (Kawamoto *et al.*, 1994b).

Assessment of the Risk to the General Population

The studies by Foo *et al.* (1990, 1993) were used by the US EPA to identify a 'lowest-observed adverse effect level' (LOAEL). This served as the basis for the derivation of the inhalation reference concentration (RfC) for chronic exposure of 0.1 ppm (0.41 mg/m^3) (IRIS, 1991) (RfC is the level of a continuous exposure that is likely to be without an appreciable risk of deleterious non-cancer health effects during a lifetime.)

There is no current ambient air quality standard for the UK or Europe. A European guideline level for CNS and reproductive effects of 0.26 mg/m^3 as a weekly average has been recommended, with an odour threshold of 1 mg/m^3 over the same time period (WHO, 1996).

Ambient levels of toluene in outdoor air are very much lower than the levels at which adverse health effects appear to be experienced. However, levels of toluene in indoor air can be much higher than those outdoors and account needs to be taken of the impact of potential genetic polymorphisms.

1,3-BUTADIENE

Sources in Ambient Air

1,3-Butadiene is an industrial chemical primarily used in the manufacture of synthetic rubber. It is also formed during combustion of fossil fuels, including the burning of petrol and diesel in motor vehicles. This is the major source of 1,3-butadiene in ambient air. Some commercial liquid petroleum gases contain up to 8% 1,3-butadiene by volume and it is also present in tobacco smoke.

Exposure and Atmospheric Levels

Human exposure occurs primarily through inhalation. Annual average levels of 1,3-buta-diene measured in ambient air in the UK range from 0.7 ppb in urban areas close to busy roads to 0.04 ppb in rural areas. During air pollution episodes peak hourly average levels of around 30 ppb have been recorded (EPAQS, 1994b).

Effects

Exposure to high levels of 1,3-butadiene causes irritation of the eyes, nose and throat and CNS effects (Himmelstein *et al.*, 1997). At ambient levels of exposure the major effect of concern is the risk of cancer.

1,3-Butadiene has been shown to be mutagenic following metabolic activation in a number of test systems (Fielder, 1996). Studies of genotoxicity in the bone marrow of mice have reported an increased frequency of sister chromatid exchanges at 6.25 ppm, micronuclei formation at 62.5 ppm and chromosome aberrations at 625 ppm after expo-sure for h per day, 5 days per week for 2 weeks. After exposure for 13 weeks, micronuclei were noted in peripheral blood at exposures of 6.25 ppm. In addition, 1,3-butadiene exposure has been reported to cause germ cell mutations in male mice (Adler *et al.*, 1995).

Long-term inhalation studies in rats and mice have shown that 1,3-butadiene causes cancer in a variety of organ sites. In male and female Sprague–Dawley rats exposed to 0, 1000 or 8000 v/v 1,3-butadiene for 6 h/day and 5 days per week for 2 years, increased tumour incidences and/or dose-response trends were reported at a number of sites, includ-ing pancreatic exocrine glands and Leydig cells in males. Similar studies in B6C3F1 mice exposed to concentrations ranging from 0 to 1250 ppm 6 h/day for 5 days per week for up to 2 years showed that mice were much more sensitive than rats, with increased tumour incidence being noted at much lower levels of exposure. For example, lung tumours occurred at concentrations of 6.25 ppm and mammary tumours at 20 ppm. At exposures of 625 ppm tumours were noted at several sites (lymphomas, haemangiosarcomas of the heart, lung, forestomach, Harderian gland and preputial gland) in mice after only 13 weeks of exposure (Melnick *et al.*, 1993).

1,3-Butadiene is metabolized by cytochrome P-450 mono-oxygenases to the epoxides 1,2-epoxy-3-butene and 1,2,3,4 -diepoxybutane, which are removed through conjugation with glutathione or hydrolysis by epoxide hydrolase. *In vitro* studies have shown that both these epoxides are mutagenic, with the diepoxide having 100-fold greater activity. The species differences in sensitivity to cancer have been attributed to differences in metabolism. Mice metabolize 1,3-butadiene to the mono-epoxide more quickly than rats and also have much greater capacity to further oxidize this to the diepoxide. Given the differences in species sensitivity, considerable research efforts have gone into understanding whether humans more closely resemble mice or rats. The data available at present suggest that metabolism in humans is more similar to that in rats. However, there are concerns that individuals with certain genetic polymorphisms may be more susceptible (EPAQS, 1994b; Himmelstein *et al.*, 1997).

A number of epidemiological studies of workers occupationally exposed to 1,3-butadiene have also been carried out. These have included several studies of a cohort of butadiene monomer workers (Divine, 1990, 1993; Divine and Hartman, 1996; Downs *et al.*, 1987), several cohorts of styrene-butadiene rubber (SBR) workers (Delzell *et al.*, 1996; Matanoski *et al.*, 1990; Meinhardt *et al.*, 1982) and a case–control study of lymphohaematopoietic cancer nested within one of the SBR cohorts (Santos-Burgoa *et al.*, 1992).

The cohort of 1,3-butadiene monomer production workers was first studied by Downs *et al.* (1987), who reported on 2586 workers employed for at least 6 months between 1943 and 1976 at the 1,3-butadiene monomer production facility supplying the SBR plants at Port Neches (see below). No exposure data were reported but workers were divided into four categories by job description and duration of employment. An excess of haematopoietic and lymphatopoietic cancers, particularly lymphosarcoma and reticulosarcoma, were reported. Leukaemia was also elevated. This cohort has been updated by Divine and colleagues (Divine, 1990, 1993; Divine and Hartman, 1996). In the most recent update to 1994, 2795 workers are included. Kidney cancer mortality was elevated (13 observed compared with 7.3 expected), a finding not seen in any of the other studies of 1,3-butadiene workers. Mortality from lymphosarcoma and 'other lymphatic cancers' was elevated compared with US rates. However, leukaemia rates were comparable to national rates. Analysis by year of employment showed that lymphosarcoma mortality was elevated for workers hired during World War II (7 observed compared with 2.9 expected) (Downs *et al.*, 1987).

Meinhardt *et al.* (1982) investigated cancer mortality in 2756 workers employed for at least 6 months between 1943 and 1976 at two SBR plants (A and B) in Port Neches, Texas. For plant A non-significant excess of lymphatic and haematopoietic cancers was observed (9 observed compared with 5.79 expected), but there was no excess at plant B. All of the cases at Plant A occurred in workers first employed before 1945 when the process changed from batch to continuous feed operation. Exposure estimates were not available for the study period. Data obtained after 1976 showed that mean TWA levels in Plant A were 1.24 ppm compared with 13.5 ppm at Plant B.

A second cohort of SBR workers investigated involved workers at seven plants in the USA and one in Canada (Matanoski *et al.*, 1990). The cohort was defined as all men who had been employed for at least one year between 1943, or whenever personnel records were first complete, and 1982. Employees were divided into four categories: production, maintenance, utility and other. No significant excess for cancer as a whole was demonstrated. Haematopoietic and lymphatic cancers were subdivided into a number of categories including leukaemia, for which there was an excess in black production workers (3 observed compared with 0.5 expected) but not in white production workers and 'other lymphatic', for which there was an excess in all groups. Limited exposure information was available.

Santos-Burgoa *et al.* (1992) further investigated this cohort in a nested case–control study. Exposure to 1,3-butadiene and styrene was categorized based on a scoring system for each job. Fifty nine cases were matched with 193 controls on plant, age, years of hire, duration of employment and survival until death of the case. Relative risks of 7–10 were reported for 'all leukaemia' for exposure to 1,3-butadiene compared with no exposure and 'high' versus 'low' exposure, taking account of exposure to styrene and other confounding factors. In contrast 'other lymphatic' cancer was associated with exposure to styrene and other factors besides exposure to 1,3-butadiene. A number of researchers have commented on the conflict between these results and the study of the base cohort (ECETOC, 1992; Himmelstein *et al.*, 1997).

In a study carried out for the International Institute of Synthetic Rubber Producers, Delzell *et al.* (1996) studied a cohort of SBR workers, including seven of the eight plants studied by Matanoski *et al.* (1990) and the two plants studied by Meinhardt *et al.* (1982). The study population included 15 649 men who had been employed for at least one year from 1943, or when personnel records were first complete, until 1991. Standardized mortality ratios (SMRs) for lymphopoietic cancers were elevated for leukaemia (48 observed compared with 37 expected), but not for lymphosarcoma or other lymphatic cancers. When the analysis was restricted to workers employed for at least 10 years and followed for at least 20 years, the SMR for leukaemia was further elevated (29 cancers compared with 14.5 expected), but, again, the same pattern was not seen for lymphosarcoma or other lymphatic cancer. A similar pattern with slightly higher SMRs for leukaemia was seen with hourly workers. Analysis by type of leukaemia showed excesses for chronic lymphocytic, chronic myelogenous and acute unspecified. Analysis by job type found excess leukaemia in workers in polymerization, maintenance labour and laboratories. Exposure was estimated for job categories and estimates made for 8-h TWAs and numbers of 15-min exposures over 100 ppm/year based on job history. The relative risk for leukaemia increased with increasing cumulative exposure to 1,3-butadiene and was increased in the intermediate category for peak exposures, but not in the highest category (Himmelstein *et al.*, 1997).

In addition to the epidemiological studies, a few groups have carried out cytogenetic analyses of peripheral blood lymphocytes. Sorsa *et al.* (1994) examined 40 individuals occupationally exposed to 1–3 ppm 1,3-butadiene. Although smoking-related effects were seen for sister chromatid exchanges, micronuclei formation, age-related increases in chromosome aberrations and micronuclei formation, none of the cytogenetic parameters correlated with exposure to 1,3-butadiene, which was measured by personal monitoring (Sorsa *et al.*, 1994). Au *et al.* (1995) reported similar findings. However, chromosomal alterations were increased in lymphocytes from exposed workers compared with controls following *in vitro* X-ray irradiation. Ward *et al.* (1994) have reported higher frequencies of lymphocytes with mutations at the HRPT locus in eight workers in high-exposure areas (average approximately 3.5 ppm) compared with five workers in low-exposure areas (0.03 ppm). This finding was correlated with levels of the butadiene-specific metabolite, 1,2-hydroxy-4-(N-acetyl cysteinyl-S-) butane in urine.

Assessment of Risk to the General Population

1,3-Butadiene is present in low concentrations in ambient air. It is clearly carcinogenic in experimental animals and is probably carcinogenic in humans. Species differences in sensitivity have been attributed in part to differences in metabolism. The epidemiological studies indicate increased mortality from haematopoietic and lymphopoietic cancers, although the patterns of cancers are not consistent throughout the studies. Generally, exposure information in these studies has been lacking and little account has been taken of exposure to other chemicals. IARC has classified 1,3-butadiene in category 2A – probably carcinogenic to humans (IARC, 1992). A number of quantitative risk assessments have been carried out for 1,3-butadiene. All of these have been based on the data from experimental animals because of the inadequacy of the epidemiological studies for quantitative risk assessment purposes. The additional risk of death from cancer varies from 0 to 2613 at 1 ppm 1,3-butadiene, depending on the tumour data (e.g. pooled, lung, heart), species and sex of animals and model used (Himmelstein *et al.*, 1997). In the UK an air quality standard of 1 ppb has been established for 1,3-butadiene (EPAQS, 1994b).

DIOXINS

Dioxins are stable compounds which are formed as unintended by-products of a number of chemical reactions. Strictly speaking the term dioxins refers to polychlorinated dibenzo-*p*-dioxins (PCDDs). However, the related compounds polychlorinated dibenzofurans (PCDFs) are usually included. More recently dioxin-like polychlorinated biphenyls (PCBs) have been considered in the same group. There are 210 dioxins and furans and 12 dioxin-like PCBs.

The toxicity of the various congeners varies considerably, with 2,3,7,8-tetrachlorodibenzo-*p*-dioxin (TCDD) being the most toxic. As dioxins and dioxin-like compounds are usually present in mixtures, the concept of 'toxic equivalency factors' (TEF) has been developed to express the toxicity of the congeners relative to 2,3,7,8-TCDD through a series of weighting factors. This allows reporting of mixtures and emissions in terms of toxic equivalents (TEQs). Most countries use the system proposed by NATO in 1988 (I-TEFs) (NATO/CCMS, 1988). However, WHO and IPCS are leading discussions on developing internationally agreed TEFs. TEFs have been assigned to the 17 dioxin and furan congeners which have the 2,3,7,8 substitution pattern and the dioxin-like PCBs. Those for dioxins and furans are shown in Table 35.5.

Table 35.5   I-TEFs for dioxins and furans

Dioxin congener	I-TEF	Furan congener	I-TEF
2,3,7,8-tetraCDD	1	2,3,7,8-tetraCDF	0.1
1,2,3,7,8-pentaCDD	0.5	2,3,4,7,8-pentaCDF	0.5
		1,2,3,7,8-pentaCDF	0.05
1,2,3,4,7,8-hexaCDD	0.1	1,2,3,4,7,8-hexaCDF	0.1
1,2,3,6,7,8-hexaCDD	0.1	1,2,3,6,7,8-hexaCDF	0.1
1,2,3,7,8,9-hexaCDD	0.1	1,2,3,7,8,9-hexaCDF	0.1
		2,3,4,6,7,8-hexaCDF	0.1
1,2,3,4,6,7,8-heptaCDD	0.01	1,2,3,4,6,7,8-heptaCDF	0.01
		1,2,3,4,7,8,9-heptaCDF	0.01
OctaCDD	0.001	OctaCDF	0.001
Other CDDs	0	Other CDFs	0

Sources

Dioxins are ubiquitous pollutants and can be found at low levels in all environmental compartments. They are persistent compounds formed as unintended by-products of a number of chemical reactions including combustion processes, thermal processing of metals, and manufacture of some chlorinated chemicals including some pesticides.

Evaluation of emissions in the UK in 1995 showed that a major source of emissions to air was municipal waste incinerators, accounting for 70% of emissions. Metal processing industries and coal and wood burning from industrial and domestic sources were the next major sources. Since 1995 controls on emissions from municipal waste incinerators have been introduced that have resulted in reductions of 90% from this source (Douben, 1997).

Releases to land and water in the UK are significantly higher than releases to air. The major sources are open use of chemicals, manufacture of pesticides and chlorophenols, accidental fires and disposal of sewage sludge and waste oil. The majority of discharges to land are placed in landfills, however wastes from some processes are disposed of by application to land (Dyke *et al.*, 1997).

Exposure and Atmospheric Levels

Humans can be exposed to dioxins through a variety of routes: via food, inhalation, absorption through the skin. It has been estimated that the food chain contributes the major part of the exposure. Dioxins are poorly water-soluble but can accumulate in fat, so are present in fatty foods such as milk and dairy products. Inhalation exposures are generally low.

A summary of typical concentrations in various media in developed countries is shown in Table 35.6 (summarized from IARC, 1997)

Dietary intake in the UK is comparable to that in other industrialized countries. Total daily intake in the UK has been estimated at 2.4–4.2 pg TEQ/kg bodyweight/day; this is comparable to intakes of 3–6 pg TEQ (including dioxin-like PCBs)/kg bodyweight per day in the USA (EPA, 1994; MAFF, 1997). Surveys of levels of dioxins in foods have shown that levels decreased 64% in the last 10–20 years (IARC, 1997; MAFF, 1997).

Table 35.6   Typical concentrations of dioxins in various media (developed countries)

Medium	I-TEQ	Comments
Air (pg/m³)		
Ambient	0.015–0.1	
Near emission sources	0.05– 0.2	Higher concentrations up to 2990 pg/m³ have been reported
Water (pg/l)		
River	1–5	Few measurements have been made
Soil (ng/kg)		
Rural	0.5– 2	
Industrial	10–15 000	Considerably higher values have been reported in a few instances
Food (ng/kg fat)		
Cows' milk	0.5–10 (mean 2.3)	
Meat	0.2–60 (mean 6.5)	Includes different meats and meat products
Poultry	1.6–5	
Eggs	0.7–8	
Fish	0.2–60 (mean 25)	

A number of analyses of body burdens of dioxins have been carried out. Background levels of TCDD in non-occupationally and non-accidentally exposed individuals are currently around 2–5 ng/kg fat. The sum of the penta- and hexa-chlorinated PCDF congeners commonly found in human tissues is around 10–100 ng/kg fat. Tissue concentrations of TCDD four orders of magnitude greater have been reported after accidental exposures (IARC, 1997).

Dioxins are excreted in human breast milk and a number of surveys have been carried out. Levels of 21–24 ng TEQ per kg milk fat were reported in 1993–94 in the UK (MAFF, 1997). This level is very similar to the mean value of 20 ng TEQ per kg milk fat reported by IARC in 1997. Calculation of intakes for nursing infants are in the range 20 pg TEQ/kg bodyweight to 130 pg TEQ/kg bodyweight. The amount ingested over a 6-month breast-feeding period has been calculated to be less than 5% of the quantity ingested over a lifetime, although this occurs during a period when susceptibility might be particularly high. There is clear evidence of a decrease in dioxin levels in human milk over time. However, estimated intake levels for babies consuming this milk far exceed recommended tolerable intakes (IARC, 1997; MAFF, 1997).

Effects

Much of the concern about possible effects of dioxins comes from detailed studies of experimental animals which have shown considerable variability in effects and species sensitivity. Information about the effects of human exposure to dioxins comes from studies of populations exposed after accidents, e.g. Seveso, BASF Germany, or from high-exposure situations, e.g. Vietnam veterans. Exposure to dioxins can cause chloracne, changes in liver enzymes and cancer. Effects have also been reported on the immune system and the reproductive system including developmental effects. Chloracne has only been seen after high levels of exposure. There is a large body of scientific literature on the effects of dioxins although most studies have concentrated on 2,3,7,8-TCDD. A brief outline is provided below.

Mechanisms

There is considerable evidence that the binding of dioxins and related compounds to the aryl hydrocarbon (Ah) receptor mediates dioxin-induced biochemical and toxic effects. However, the link between these is still unclear. As yet no natural ligand for the Ah receptor has been identified and its role is unknown, although there is some evidence suggesting that it may play a role in normal liver development (IARC, 1997; WHO, 1989).

Immune System

In animal studies suppression of the immune system by dioxins has been noted, characterized by thymic atrophy. The thymus appears to be the main target organ. Guinea pigs and some strains of mice are among the more sensitive animals to the immunotoxic effects of TCDD, while humans and other primates are less sensitive. Sensitivity to the immunotoxic effects of TCDD in experimental animals is greatest if exposure occurs during the perinatal period, when exposure to doses which are not overtly toxic or immunotoxic in adult animals can cause profound and prolonged immunosuppression. Studies of the functional integrity of the immune system have found that cell-mediated immunity is depressed and antibody production in response to T-dependent antigens is impaired (IARC, 1997; COT, 1995; Department of the Environment, 1989). Studies of host resistance indicate effects in rats and mice on susceptibility to pathogenic microorganisms (COT, 1995; IARC, 1997).

A number of studies have reported on testing of immune function in humans exposed to dioxins. With the exception of one small study, these have failed to demonstrate significant changes. However, the studies have a number of inadequacies including low numbers of subjects, lack of standardized testing procedures and absence of exposure information (COT, 1995; IARC, 1997; Tonn *et al.*, 1996; WHO, 1989).

Reproductive System and Developmental Effects

Many studies have been carried out in experimental animals to investigate the reproductive toxicity of dioxins. These have shown that TCDD is a developmental and reproductive toxicant. In adult animals reproductive toxicity occurs at levels that are overtly toxic. However, effects on the developing animal can occur at levels more than 100 times lower. Sensitive targets include the developing reproductive and immune systems.

Fetotoxic effects have been seen in Rhesus monkeys exposed to 25 ng/kg TCDD in the diet, with few offspring carried to term. Similar effects have been seen in other experimental species. Exposure of pregnant mice to 3–24 pg/kg 2,3,7,8-TCDD caused hydronephrosis and cleft palate in offspring. Hydronephrosis could also be induced in mice by exposure during lactation, although higher exposures were required. Mice seem to be particularly sensitive. Although renal effects could be induced in other experimental species, higher exposures causing maternal toxicity were required. In contrast, rats seem to be more sensitive to effects on immune function. In both species thymic atrophy and effects on the ratio of T lymphocyte subsets have been reported. In rats this has been associated with permanent suppression of delayed-type hypersensitivity (Gehrs *et al.,* 1998). Prenatal exposure of experimental animals has also resulted in decreased weights of accessory sex organs and decreased sperm counts in male offspring of rats exposed to 64 pg/kg. Delayed puberty and, in some studies, changes in sexual behaviour have also been noted. In female offspring delayed vaginal opening and vaginal clefting have been noted in both rats and hamsters (Gray *et al.,* 1995a,b; Mably *et al.,* 1992a,b,c).

In adult male animals exposure to TCDD causes changes in testes with loss of germ cells, spermatocytes, spermatozoa and reduction in numbers of tubules with mature sperm. It also causes a decrease in circulating testosterone, which has been suggested as the cause of the effects seen. In female animals, TCDD causes reduced fertility, reduced litter size, ovarian toxicity, suppression of the oestrous cycle and anovulation (IARC, 1997; WHO, 1989). Studies in Rhesus monkeys have shown ovarian dysfunction at high levels of exposure. Increased incidence and severity of endometriosis has also been reported at lower levels of exposure (25 ng/kg in the diet) (Rier *et al.,* 1993).

A number of studies have reported on reproductive effects in exposed workers, including changes in hormone levels, decreased sperm count and abnormal sperm, small increases in rates of miscarriage and numbers of malformed offspring (WHO, 1989; IARC, 1997). One study of the Seveso population reported an excess of female births in the 8 years following the accident compared with the following 10 years (Mocarelli *et al.,* 1996). No significant changes in incidence of developmental abnormalities has been reported after the Seveso accident.

Cancer

There have been several long-term studies to determine whether TCDD is a carcinogen in several species of experimental animals, including rats, mice and hamsters. All of these have been positive. 2,3,7,8-TCDD is not directly genotoxic. A number of mechanisms for its carcinogenic action have been proposed including Ah receptor-mediated alteration of gene expression, biochemical pathways and immune function.

A number of epidemiological studies have also been carried out on human populations exposed to TCDD. These have been extensively reviewed by IARC (IARC, 1997). Analyses have been carried out for all cancers and cancers at a number of sites including lung, gastrointestinal tract, non-Hodgkin lymphoma and soft tissue sarcoma. The IARC workgroup calculated overall SMRs for each of these, using data from studies of the most highly exposed populations (Becher *et al.,* 1996; Fingerhut *et al.,* 1991; Hooiveld *et al.,* 1996; Kogevinas *et al.,* 1997; Koopman Esseboom *et al.,* 1994). The calculated values are summarized in Table 35.7 (IARC, 1997).

The raised SMRs for all cancers and lung cancer were considered to be associated with the exposures to dioxins. Results for the other cancer sites were considered more equivocal. The SMR for non-Hodgkin lymphoma was also raised. However, there was no increased risk in the large cohort examined by Fingerhut *et al.* (1991) and the number of cases of soft tissue sarcomas observed was small.

Dose–response information for exposure to dioxins is available from three studies. Flesch-Janys *et al.* (1995, 1996) analysed risk for all cancers for approximately 1000 workers involved in the production of phenoxy herbicides and chlorophenols. Seven exposure categories were used based on TEQs in blood fat. A relative risk (RR) of 2.7 (95% CI, 1.7–4.4) was reported for the highest exposure category. RRs for all but the lowest exposure categories were raised, but not significantly. The overall test for trend

Table 35.7 summary of calculated standardized mortality ratios (SMR) and 95% confidence intervals (95% CI) from studies of high exposure to dioxins

Cancer type	No. of cases	SMR	95% CI	Significance value
All cancers	288	1.4	1.2–1.6	<0.001
Lung cancer	94	1.4	1.1–1.7	<0.01
Gastrointestinal cancer	61	1.2	0.9–1.5	0.23
Non-Hodgkin lymphoma	11	2.6	1.3-4.7	<0.01
Soft tissue sarcoma	3	4.7	–	–

yielded a significance value of less than 0.01, but largely as the result of a high RR for the highest exposure category. Ott and Zober (1996) analysed 243 workers accidentally exposed to 2,3,7,8-TCDD. Workers were divided into four categories of exposure based on measured and estimated 2,3,7,8-TCDD levels. Although none of the SMRs were significantly increased, there was a trend of increasing SMRs with exposure up to an SMR of 2.0 for the highest exposure category. The only study examining specific cancer sites is that of Kogevinas *et al.* (1997), which addressed two cancer sites, non-Hodgkin lymphoma and soft-tissue sarcoma, in the IARC cohort. Exposure to 2,3,7,8-TCDD was classified into four qualitative categories. Although none of the odds ratios were significant, tests for trend gave significance values of 0.1 for non-Hodgkin lymphoma and 0.04 for soft tissue sarcoma.

Generally the epidemiological studies show an increase in total cancers with exposure to dioxins. With regard to specific sites, the evidence is consistent for lung cancer and non-Hodgkin lymphoma. However, dose–response information for specific sites is limited. In 1997 IARC classified 2,3,7,8-TCDD as a human carcinogen (IARC, 1997).

Endocrine Effects

Dioxins are also listed among the chemicals in the environment that have endocrine-disrupting properties. Disruption of endocrine function can have a variety of subtle effects. In animals studies dioxins have been shown to act as antioestrogens, to interfere with thyroid function and to alter retinoic acid metabolism. In humans small changes in thyroid function have been noted in exposed adults. In nursing infants changes in circulating T3 and T4:thyroxine binding globulin ratio and plasma TSH have been noted related to intake of PCDD, TEQ and I-TEQ (Koopman Esseboom *et al.*, 1994; Pluim *et al.*, 1992, 1993). Recent studies have reported decreased testosterone levels in small numbers of workers occupationally exposed to TCDD with serum TCDD concentrations of 20 pg/g blood lipid (Egeland *et al.*, 1994).

Assessment of Risk to the General Population

A number of national and international organizations have established tolerable daily intakes (TDI) for dioxins. The TDI is the amount of a chemical contaminant that can be absorbed without appreciable risk. In 1990 WHO established a TDI of 10 pg/kg bodyweight/day for 2,3,7,8-TCDD. This has recently been revised to 1–4 pg/kg bodyweight/day for 2,3,7,8-TCDD (WHO, 1998). In general, exposure of adults is within this range. However, for some groups, particularly breast-fed infants, exposures can be considerably higher.

CONCLUSIONS

In this chapter we have reviewed some of the organic air pollutants to which humans are exposed. Of the pollutants chosen, two (benzene and 2,3,7,8-TCDD) are human carcinogens, while a third, 1,3-butadiene, is probably carcinogenic to humans. The reviews have shown that ambient air pollution is not always the most important source of exposure. For example, smokers are exposed to much higher levels of benzene in cigarette smoke than are present in ambient air, and the major route of exposure to dioxins is through the diet. We have also pointed out that exposure to air pollution can occur outdoors or indoors, at work or at home. Exposure can also occur in other microenvironments. This may be particularly important in terms of total exposure for some groups of people. For example for non-smokers exposure inside vehicles and at petrol stations may contribute a significant proportion of their total exposure to benzene.

REFERENCES

Adler ID, Cochrane J, Ostermann-Golkar S *et al.* (1995) 1,3-Butadiene working group report. *Mutat Res* **330**: 101–114

AEA (1995) *Air Pollution in the UK.* Abingdon: AEA Technology Plc.

Aksoy M (1985) Malignancies due to occupational exposure to benzene. *Am J Ind Med* **7**: 395–402.

Aksoy M, Dincol K, Erdem S *et al.* (1972) Details of blood changes in 32 patients with pancytopenia associated with long-term exposure to benzene. *Br J Ind Med* **29**: 56–64.

Aksoy M, Erdem S and Dincol G (1974) Leukemia in shoe-workers exposed chronically to benzene. *Blood* **44**: 837–841

Andersen I, Lundqvist GR, Molhave L *et al.* (1983) Human response to controlled levels of toluene in six-hour exposures. *Scand J Work Environ Health* **9**: 405–418.

Anderson HR, Macnair RS and Ramsey JD (1985) Deaths from abuse of volatile substances: a national epidemiological study. *Br J Ind Med* **290**: 304–307.

Antti-Poika M, Juntunen J, Matikainen E *et al.* (1985) Occupational exposure to toluene: neurotoxic effects with special emphasis on drinking habits. *Int Arch Occup Environ Health* **56**: 31–40.

Au WW, Bechtold WE, Whorton EB Jr, Legator MS (1995) Chromosome aberrations and response to gamma-ray challenge in lymphocytes of workers exposed to 1,3-butadiene. *Mutat Res* **334**: 125–130.

Austin H, Delzell E and Cole P (1988) Benzene and leukaemia – a review of the literature and a risk assessment. *Am J Epidemiol* **127**: 419–439.

Baelum J, Andersen I, Lundqvist GR *et al.* (1985) Response of solvent-exposed printers and unexposed controls to six-hour toluene exposure. *Scand J Work Environ* **11**: 271–280.

Baelum J, Lundqvist GR, Molhave L and Andersen NT (1990) Human response to varying concentrations of toluene. *Int Arch Occup Environ Health* **62**: 65–71.

Bakinson MA and Jones RD (1980) Gassing due to methylene chloride, xylene, toluene and styrene reported to Her Majesty's Factory Inspectorate. *Br J Ind Med* **42**: 184–190.

Becher H, Flesch-Janys D, Kauppinen T *et al.* (1996) Cancer mortality in German male workers exposed to phenoxy herbicides and dioxins. *Cancer Causes Control* **7**: 312–321.

Bond GG, McLaren EA, Baldwin CL and Cook RR (1986) An update of mortality among chemical workers exposed to benzene. *Br J Ind Med* **43**: 685–691.

Brett SM, Rodricks JV and Chinchilli VM (1989) Review and update of leukaemia risk potentially associated with occupational exposure to benzene. *Environ Health Perspect* **82**: 267–281.

Brugnone F, Perbellini L, Faccini GB *et al.* (1989) Benzene in the blood and breath of normal people and occupationally exposed workers. *Am J Ind Med* **16**: 385–399.

Brugnone F, Gobbi M, Ayyad K *et al.* (1995) Blood toluene as a biological index of environmental toluene exposure in the 'normal' population and in occupationally exposed workers immediately after exposure and 16 hours later. *Int Arch Occup Environ Health* **66**: 421–425.

Byrd DM and Barfield ET (1989) Uncertainty in the estimation of benzene risks: application of an uncertainty taxonomy to risk assessments based on an epidemiology study of rubber hydrochloride workers. *Environ Health Perspect* **82**: 283–287.

Byrne A and Zibin T (1991) Toluene-related psychosis. *Br J Psychol* **158**: 578.

Cherry N, Hutchins H, Pace T and Waldron HA (1985) Neurobehavioral effects of repeated occupational exposure to toluene and paint solvents. *Br J Ind Med* **42**: 291–300.

Christie D, Robinson K, Gordon I, Bisby J (1991) A prospective study in the Australian petroleum industry. II Incidence of Cancer. *Br J Ind Med* **48**: 511–514.

Clayton GD and Clayton FE (1994) *Patty's Industrial Hygiene and Toxicology.* New York: John Wiley and Son.

CONCAWE (1994) *Review of European Oil Industry Benzene Exposure Data 1986–1992.* Brussels: Oil Companies, European Organisation for Environment, Health and Safety.

CONCAWE (1996) *A Year Long Study of Ambient Concentrations of Benzene Around a Service Station.* Brussels: CONCAWE.

COT (1995) Statement on the US EPA Draft Health Assessment Document for 2,3,7,8-Tetrachlorodibenzo-*p*-dioxin and Related Compounds (Committee on the Toxicity of Chemicals in Food, Consumer Products and the Environment). London: Department of Health.

Courage C and Duarte-Davidson R (1998) *Benzene in the Environment: An Evaluation of Exposure of the UK General Population and Possible Adverse Health Effects.* Leicester: Institute for Environment and Health.

Crump DR (1997) *Indoor Air Pollution.* Cambridge: The Royal Society of Chemistry.

Crump KS (1994) Risk of benzene-induced leukemia derived from the Pliofilm cohort: effect of additional follow up and new exposure estimates. *J Toxicol Environ Health* **42**: 219–242.

Crump KS and Allen BC (1984) *Quantitative Estimates of Risk of Leukaemia from Occupational Exposure to Benzene.* Washington, DC: US Department of Labour.

DeBortoli M, Knoppel H, Pecchio E *et al.* (1984) Integrating 'real life' measurements of organic pollution in indoor and outdoor air of homes in Northern Italy. In *Proceedings of the 3rd International Conference of Indoor Air Quality and Climate*, Vol 4, pp 21–26, Stockholm.

Decoufle P, Blattner WA and Blair A (1983) Mortality among chemical workers exposed to benzene and other agents. *Environ Res* **30**: 16–25.

Delzell E, Sathiakumar M, Macaluso M *et al.* (1996) Follow up study of synthetic rubber workers. *Toxicology* **113**: 182.

Department of the Environment (1989) *Dioxins in the Environment* Pollution Paper No. 27. London: HMSO.

Department of the Environment (1996) *United Kingdom National Environment Action Plan.* London: HMSO.

DETR (Department of the Environment, Transport and the Regions) (1997) *Digest of Environmental Statistics* No 19. London: The Stationery Office.

Divine BJ (1990) An update on mortality among workers at a 1,3-butadiene facility – preliminary results. *Environ Health Perspect* **86**: 119–128.

Divine BJ (1993) Cancer mortality among workers at a butadiene production facility. *IARC Scientific Publications* **127**: 345–362.

Divine BJ and Hartman CM (1996) Mortality update of butadiene production workers. *Toxicology* **113**: 169–181.

Douben PET (1997) PCDD/F emissions to atmosphere in the UK and future trends. *Chemosphere* **34**: 1181–1189.

Downs TD, Crane MM and Kim KW (1987) Mortality among workers at a butadiene facility. *Am J Ind Med* **12**: 311–329.

Dyke PH, Foan C, Wenborn M and Coleman PJ (1997) A review of dioxin releases to land and water in the UK. *Sci Total Environ* **207**: 119–131.

EBS (1996) *Benzene Risk Characterisation.* East Mill Stone: Exxon Biomedical Sciences.

ECETOC (1992) *1,3-Butadiene*, Special Report No. 4. Brussels: European Chemical Industry Ecology and Toxicology Centre.

Egeland GM, Sweeney MH, Fingerhut M *et al.* (1994) Serum 2,3,7,8-TCDD effect on total serum testosterone and gonadotropins in occupationally exposed men. *Am J Epidemiol* **139**: 272–281.

EPA (1994) *Estimating Exposure to Dioxin-like Compounds* (EPA/600/6-88/005Cb). Washington, DC: Environmental Protection Agency.

EPAQS (Expert Panel on Air Quality Standards) (1994a) *Benzene*. London: HMSO.

EPAQS (Expert Panel on Air Quality Standards) (1994b) *1,3-Butadiene*. London: HMSO.

Fielder RJ (1996) Risk assessment of 1,3-butadiene in ambient air: The approach used in the UK. *Toxicology* **113**: 221–225.

Fingerhut M, Halperin WE, Marlow DA *et al.* (1991) Cancer mortality in workers exposed to 2,3,7,8-tetrachlorodibenzo-*p*-dioxin. *N Engl J Med* **324**: 212–218.

Fishbein L (1988) Toluene: uses, occurrence and exposure. In *Environmental Carcinogens Methods of Analysis and Exposure Measurement* (Fishbein L and O'Neill IK, eds) Vol 10, Publication 85. Benzene and alkylated benzenes. *IARC Sci Lyon*, IARC.

Flesch-Janys D, Berger J, Gurn P, Manz A, Nagal S, Waltsgott H, Dwyer JH (1995) Exposure to polychlorinated dioxins and furans (PCDD/F) and mortality in a cohort of workers from a herbicide producing plant in Hamburg, Federal Republic of Germany. *Am J Epidemiol* **142**: 1165–76.

Flesch-Janys D, Berger J, Gurn P *et al.* (1996) Erratum. *Am J Epidemiol* **144**: 716.

Foo SC, Phoon WH and Khoo NY (1988) Toluene in blood after exposure to toluene. *Am Ind Hyg Assoc J* **49**: 255–258.

Foo SC, Jeyaratnam J and Koh D (1990) Chronic neurobehavioral effects on toluene. *Br J Ind Med* **47**: 480–484.

Foo SC, Jeyaratnam J, Ong CN *et al.* (1993) Biological monitoring for occupational exposure to toluene. *Am Ind Hyg Assoc* **52**: 212–217.

Gehrs BD, Martin MM, Williams WC and Smialowicz RJ (1998) Effects of perinatal 2,3,7,8-TCDD exposure on the developing immune system of the F344 rat. *Toxicologist* **15**: 104.

Gilli G, Scursatone E and Bono R (1994) Benzene, toluene and xylenes in air, geographical distribution in the Piedmont region (Italy) and personal exposure. *Sci Total Environ* **148**: 49–56.

Gray LE, Kelce WR, Monosson E *et al.* (1995a) Exposure to TCDD during development permanently alters reproductive function in male LE rats and hamsters: reduced ejaculated and epididymal sperm numbers and sex accessory gland weights in offspring with normal androgenic status. *Toxicol Appl Pharmacol* **131**: 108–118.

Gray LE, Ostby JS, Wolf C *et al.* (1995b) Functional developmental toxicity of low doses of 2,3,7,8-tetrachlorodibenzo-*p*-dioxin and a dioxin-like PCB (169) in Long Evans rats and Syrian hamsters: reproductive, behavioural and thermoregulatory alterations. *Organohalogen Compounds* **25**: 33–38.

Greenberg M (1997) The central nervous system and exposure to toluene: a risk characterization. *Environ Res* **72**: 1–7.

Greenberg L, Mayers MR, Goldwater L *et al.* (1939) Benzene (Benzol) poisoning in the rotogravure printing industry in New York City. *J Ind Hyg Toxicol* **21**: 395–420.

Hajimiragha H, Ewers U, Brockhaus A and Boettger A (1989) Levels of benzene and other valatile aromatic compounds in the blood of non-smokers and smokers. *Int Arch Occup Environ Health* **61**: 513–518.

Hanninen H, Antti-Poika M and Savolainen P (1987) Psychological performance, toluene exposure and alcohol consumption in rotogravure printers. *Int Arch Environ Health* **59**: 475–483.

Harrison RM (ed.) (1996) Air pollution: sources, concentrations and measurement In: *Pollution: Causes, Effects and Control*. Cambridge: The Royal Society of Chemistry.

Hayes RB, Yin SN, Dosemeci M *et al.* (1997) Benzene and the dose-related incidence of hematologic neoplasms in China. *J Natl Cancer Inst* **89**: 1065–1071.

Himmelstein MW, Acquavella JF, Recio L *et al.* (1997) Toxicology and epidemiology of 1,3-butadiene. *Crit Rev Toxicol* **27**: 1–108.

Hoff MC (1983) Toluene. In: Kroschwitz JI, Howe-Grant M (eds) *Encyclopedia of Chemical Technology*, Vol 23, pp 246–273. Chichester: Wiley.

Hooiveld M, Heederik D and Bueno de Mesquita HB (1996) Preliminary results of the second follow-up of a Dutch cohort of workers occupationally exposed to phenoxy herbicides, chlorophenols and contaminants. *Organo Compds* **30**: 185–189.

Hormes JT, Filley CM and Rosenberg NLK (1986) Neurologic sequelea of chronic solvent vapour abuse. *Neurology* **36**: 698–670.

Hurley JF, Cherrie JW and Maclaren W (1991) Exposure to benzene and mortality from leukaemia: results from coke oven and other coal product workers (letter). *Br J Ind Med* **48**: 502–503.

IARC (1989) *Some Organic Solvents, Resin Monomer and Related Compounds, Pigments and Occupational Exposure in Paint Manufacture and Painting*. IARC Monographs on the Evaluation of Carcinogenic

Risks to Humans, Vol 47. Lyon: The International Agency for Cancer Research.

IARC (1992) *1,3-Butadiene in Occupational Exposure to Mists and Vapours from Strong Inorganic Acids; and Other Industrial Chemicals.* IARC Monographs on Evaluation of Carcinogenic Risks to Humans, Vol 54. Lyon: The International Agency for Cancer Research.

IARC (1997) *Polychlorinated Dibenzo-para-Dioxins and Polychlorinated Dibenzofurans* IARC Monographs on Evaluation of Carcinogenic Risks to Humans, Vol 89. Lyon: The International Agency for Cancer Research.

Infante PF, Rinsky RA, Wagoner JK and Young RJ (1977) Leukemia in benzene workers. *Lancet* 9 July: 76–78.

Ireland B, Collins JJ, Buckley CF and Riordan SG (1997) Cancer mortality among workers with benzene exposure. *Epidemiology* **8**: 813–320.

IRIS (1991) *Inhalation Reference Concentrations for Chronic Exposure: Toluene.* Washington DC: US Environmental Protection Agency.

Irons RD and Stillman WS (1996) The process of leukemogenesis. *Environ Health Perspect* **104**: 1239–1246.

Kawamoto T, Matsuno K, Kodama Y *et al.* (1994a) ALDH2 polymorphism and biological monitoring of toluene. *Arch Environ Health* **49**: 332–336.

Kawamoto T, Murata K, Koga M *et al.* (1994b). Distribution of urinary hippuric acid concentration by ALDH2 genotype. *Occup Environ Med* **51**: 817–821.

Kipen HM, Cody RP and Goldstein BD (1989) Use of longitudinal analysis of peripheral blood counts to validate historical reconstruction of benzene exposure. *Environ Health Perspect* **82**: 199–206.

Kogevinas M, Becher H, Benn T *et al.* (1997) Cancer mortality in workers exposed to phenoxy herbicides, chlorophenols and dioxins: an expanded and updated international cohort study. *Am J Epidemiol* **145**: 1061–1075.

Koopman Esseboom C, Morse DC, Weisglas-Kuperus N *et al.* (1994) Effects of dioxins and poly-chlorinated biphenyls on thyroid hormone status of pregnant women and their infants. *Pediatr Res* **36**: 468–473.

Krause C, Mailahn W, Nagel R *et al.* (1987) Occurrence of volatile organic compounds in air of 500 homes in the Federal Republic of Germany. In: *Proceedings of the 4th International Conference on Indoor Air Quality and Climate*, Berlin, 17–21 August, Berlin. Institute for Water, Soil and Air Hygiene, p 102

Lee SH (1993) A study on the neurobehavioral effects of occupational exposure to organic solvents in Korean workers. *Environ Res* **60**: 227–232.

Leung PL and Harrison RM (1998a) Evaluation of personal exposure to monoaromatic hydrocarbons. *Occ Env Med* **55**: 249–257.

Leung PL and Harrison RM (1998b) Roadside and in-vehicle concentrations of monoaromatic hydro-carbons (In press).

Low LK, Meeks JR and Mackerer CR (1988) Health effects of the alkyl benzenes: toluene. *Toxicol Ind Health* **4**: 49–75.

Maas EF, Ashe E, Spiegel P *et al.* (1991) Accquired pendular nystagmus in toluene addiction. *Neurology* 282–285.

Mably TA, Moore RW and Peterson RE (1992a) *In utero* and lactational exposure of male rats to 2,3,7,8-tetrachlorodibenzo-*p*-dioxin: 1 Effects on sexual androgenic status. *Toxicol Appl Pharmacol* **114**: 97–107.

Mably TA, Moore RW, Goy RW and Peterson RE (1992b) *In utero* and lactational exposure of male rats to 2,3,7,8-tetrachlorodibenzo-*p*-dioxin: 2 Effects on sexual behaviour and the regulation of luteiniz-ing hormone secretion in adulthood. *Toxicol Appl Pharmacol* **114**: 108–117.

Mably TA, Moore RW, Goy RW and Peterson RE (1992c) *In utero* and lactational exposure of male rats to 2,3,7,8-tetrachlorodibenzo-*p*-dioxin: 3 Effects on spermatogenesis and reproductive capability. *Toxicol Appl Pharmacol* **114**: 118–126.

MAFF (Ministry of Agriculture, Fisheries and Food) (1997) *Dioxins and Polychlorinated Biphenyls in Foods and Human Milk* (Food Surveillance Information Sheet 105). London: Ministry of Agriculture Fisheries and Food.

Mann HS, Crump DR and Brown VM (1997) The use of diffusive samplers to measure personal expo-sure and area concentrations of VOCs including formaldehyde. In *Healthy Buildings. Global issues and regional solutions.* Proceedings of IAQ 97, Washington DC, USA, pp 135–40.

Matanoski GM, Santos-Burgoa C and Schwartz L (1990) Mortality of a cohort of workers in the styrene-butadiene polymer manufacturing industry (1943–1982). *Environ Health Perspect* **86**: 107–117.

Meinhardt TJ, Lemen RA, Crandall MS and Young RJ (1982) Environmental epidemiologic investigation of the styrene-butadiene rubber industry. *Scand J Work Environ* **8**: 250–259.

Melnick RL, Shackelford CC and Huff J (1993) Carcinogenicity of 1,3-butadiene. *Environ Health Perspect* **100**: 227–236.

Mizunuma K, Horiguchi S, Kawai T *et al.* (1994) Toluene in blood as a marker of choice for low-level exposure to toluene. *Int Arch Occup Environ Health* **66**: 309–315.

Mocarelli P, Brambilla P, Gethoux PM *et al.* (1996) Change in sex ratio with exposure to dioxin. *Lancet* **348**: 409.

NATO/CCMS (1988) *International Toxicity Equivalency Factor (I-TEF) Method of Risk Assessment for Complex Mixtures of Dioxins and Related Compounds: Pilot Study on International Information Exchange on Dioxins and Related Compounds*. Brussels, NATO Report No. 16.

Nise G and Örbaek P (1988) Toluene in venous blood during and after work in rotogravure printing. *Int Arch Occup Environ Health* **60**: 31–35.

NSCA (1997) *1997 Pollution Handbook: The Essential Guide to UK and European Pollution Control Legislation*. Brighton: National Society for Clean Air and Environmental Protection.

Ott MG, Townsend JC, Fishbeck WA and Langner RA (1978) Mortality among individuals occupationally exposed to benzene. *Arch Environ Health* **33**: 3–10.

OHMG, Zober A (1996) Cause specific mortality and cancer incidence among employees exposed to 2,3,7,8-TCDD after a 1953 reactor accident. *Occup Env Med* **53**: 606–612.

Paci E, Buiatti E, Constantini AS *et al.* (1989) Aplastic anemia, leukemia and other cancer mortality in a cohort of shoe workers exposed to benzene. *Scand J Work Environ Health* **15**: 313–318.

Paustenbach DJ, Price PS, Ollison W *et al.* (1992) Reevaluation of benzene exposure for the Pliofilm (rubberworker) cohort (1936–1976). *J Toxicol Environ Health* **36**: 177–231.

Paustenbach DJ, Bass RD and Price P (1993) Benzene toxicity and risk assessment, 1972–1992: implications for future regulation. *Environ Health Perspect* **101** (Suppl 6): 177–200.

Paxton MB, Chinchilli VM, Brett SM and Rodericks JV (1994a) Leukaemia risk associated with benzene exposure in the Pliofilm cohort. I: Mortality update and exposure distribution. *Risk Anal* **14**: 147–154.

Paxton MB, Chinchilli VM, Brett SM and Rodricks JV (1994b) Leukaemia risk associated with benzene exposure in the Pliofilm cohort. II. Risk estimates. *Risk Anal* **14**: 155–161.

Pluim HJ, Koppe JG, Van der Slikke JW *et al.* (1992) Effects of dioxins on thyroid function in newborn babies. *Lancet* **339**: 1303.

Pluim HJ, de Vijlder JJM, Olie K *et al.* (1993) Effects of pre- and postnatal exposure to chlorinated dioxins and furans on human neonatal thyroid hormone concentrations. *Environ Health Perspect* **101**: 504–508.

Porstmann F, Boke J, Hartwig S *et al.* (1994) Benzene and toluene in children's bedrooms. *Staub Reinhaltung der Luft* **54**: 147–153.

QUARG (Quality of Urban Air Review Group) (1993) *Urban Air Quality in the United Kingdom*, First Report. London: Department of the Environment.

Rahill AA, Weiss B, Marrow PE *et al.* (1996) Human performance during exposure to toluene. *Aviat Space Environ Med* **67**: 640–647.

Rier SE, Martin DC, Bowman RE *et al.* (1993) Endometriosis in Rhesus monkeys (*Mucaca mulatta*) following chronic exposure to 2,3,7,8-tetrachlorodibenzo-*p*-dioxin. *Fundam Appl Toxicol* **211**: 433–441.

Rinsky RA (1989) Benzene and leukemia: an epidemiologic risk assessment. *Environ Health Perspect* **82**: 189–191.

Rinsky RA, Young RJ and Smith AB (1981) Leukaemia in benzene workers. *Am J Ind Med* **2**: 217–245.

Rinsky RA, Smith AB, Hornung R *et al.* (1987) Benzene and leukaemia – an epidemiologic risk assessment. *New Engl J Med* **316**: 1044–1050.

Rinsky RA, Hornung R and Landrigan PJ (1989) Re: Benzene and leukaemia: a review of the literature and a risk assessment. *Am J Epidemiol* **5**: 1084–1085.

Rothman N, Li GL, Dosemeci M *et al.* (1996) Hemotoxicity among Chinese workers heavily exposed to benzene. *Am J Ind Med* **29**: 236–246.

Rushton L (1993) Further follow-up of mortality in a UK oil distribution centre cohort. *Br J Ind Med* **50**: 561–569.

Rushton L and Romaniuk HM (1997) A case–control study to investigate the risk of leukaemia associated with exposure to benzene in petroleum marketing and distribution workers in the United Kingdom. *Occup Environ Med* **54**: 152–166.

Salway AG, Eggleston HS, Goodwin JWL and Murrells TP (1997) *UK Emissions of Air Pollutants 1970–1995*. Abingdon: AEA Technology Plc.

Santos-Burgoa C, Matanoski GM, Zeger S and Schwartz L (1992) Lymphohematopoeitic cancer in styrene-butadiene polymerization workers. *Am J Epidemiol* **136**: 843–854.

Sarto F, Cominato I, Pinton AM *et al.* (1984) A cytogenic study on workers exposed to low concentrations of benzene. *Carcinogenesis* **5**: 827–832.

Schnatter AR, Katz AM, Nicholich MJ and Theriault G (1993) A retrospective mortality study among Canadian petroleum marketing and distribution workers. *Environ Health Perspect* **101**: 85–99.

Schnatter AR, Nicholich MJ and Bird MG (1996a) Determination of leukemogenic benzene exposure concentrations: refined analyses of the pliofilm cohort. *Risk Anal* **16**: 833–840.

Schnatter AR, Armstrong TW, Nicolich MJ et al. (1996b) Lymphohaematopoietic malignancies and quantitative estimates of exposure to benzene in Canadian petroleum distribution workers. *Occup Env Med* **53**: 773–781.

Snyder R and Kalf GF (1994) A perspective on benzene leukemogenesis. *Crit Rev Toxicol* **24**: 117–209.

Snyder R and Kocsis JJ (1975) Current concepts of chronic benzene toxicity. *Crit Rev Toxicol* **3**: 265–288.

Sorsa M, Autio K, Demopoulos NA *et al.* (1994) Human cytogenetic biomonitoring of occupational exposure to 1,3-butadiene. *Mutat Res* **309**: 321–326.

Sperduto B (1993*) Evaluation of Benzene Exposure of Employees and Customers in Filling Stations of the AgipPetroli Sector*. Rome: AgipPetroli.

Stollery BT and Flindt MLH (1988) Memory sequelae of solvent intoxication. *Scand J Work Environ Health* **14**: 45–48.

Struwe G and Wennberg A (1983) Psychiatry and neurological symptoms in workers occupationally exposed to organic solvents – results of a differential epidemiological study. *Acta Psychiatr Scand* **67**: 68–80.

Thienes CH and Haley TJ (1972) *Clinical Toxicology*. Philadelphia: Lea and Febiger.

Tonn T, Esser C, Schneider EM *et al.* (1996) Persistence of decreased T-helper cell function in industrial workers 20 years after exposure to 2,3,7,8-tetrachlorodibenzo-*p*-dioxin. *Environ Health Perspect* **104**: 422–426.

Tsai SP, Wen CP, Weiss NS *et al.* (1983) Retrospective mortality and medical surveillance studies of workers in benzene area of refineries. *J Occup Med* **29**: 685–692.

Turkel B and Egeli U (1994) Analysis of chromosomal aberrations in shoe workers exposed long term to benzene. *Occup Environ Med* **51**: 50–53.

Vigliani EC and Forni A (1976) Benzene and leukemia. *Environ Res* **2**: 122–127.

Wallace L (1996) Enviornmental exposure to benzene: an update. *Environ Health Perspect* **104**: 1129–1136.

Ward JB Jr, Ammenhauser MM, Bechtold WE *et al.* (1994) hrpt mutant lymphocyte frequencies in workers at a 1,3-butadiene production plant. *Environ Health Perspect* **102**(suppl 9): 79–85.

WHO (1989) *Polychlorinated Dibenzo-para-Dioxins and Dibenzofurans* (International Programme on Chemical Safety, Environmental Health Criteria 88). Geneva: World Health Organization.

WHO (1996) *Update and Revision of the WHO Air Quality Guidelines for Europe*, vol 1, Organics. Geneva: World Health Organisation.

WHO (1998) Press Release June 1998. Copenhagen: World Health Organization Europe.

Winchester RV and Madjar VM (1986) Solvent effects on workers in the paint, adhesive and printing industries. *Ann Occup Hyg* **30**: 307–317.

Wong O (1987a) An industry wide mortality study of chemical woxrkers occupationally exposed to benzene: I. General results. *Br J Ind Med* **44**: 365–381.

Wong O (1987b) An industry wide mortality study of chemical workers occupationally exposed to benzene: II. Dose–response analyses. *Br J Ind Med* **44**: 382–395.

Yardley-Jones A, Anderson D, Lovell DP and Jenkinson PC (1990) Analysis of chromosomal aberrations in workers exposed to low level benzene. *Br J Ind Med* **47**: 48–51.

Yin SN, Li GL, Tain FD *et al.* (1989) A retrospective cohort study of leukemia and other cancers in benzene workers. *Environ Health Perspect* **82**: 207–213.

Yin SN, Li Q, Liu Y *et al.* (1987) Occupational exposure to benzene in China. *Br J Ind Med* **44**: 192–195.

ESTIMATING HEALTH AND COST IMPACTS

Air Pollution and Lung Cancer

JONATHAN M. SAMET

Department of Epidemiology, School of Hygiene and Public Health, The Johns Hopkins University, Baltimore, MD, USA

AARON J. COHEN

Health Effects Institute, Cambridge, MA and Department of Enviromental Health, Boston University School of Public Health, Cambridge, MA, USA

The contents of this chapter are the responsibility of the authors, and do not necessarily reflect the views and policies of the Health Effects Institute or its sponsors.

INTRODUCTION

Overview

In the early decades of the twentieth century, lung cancer was a rare disease. By mid-century, however, it was evident that an epidemic of lung cancer was occurring among males in the USA and a number of European countries. An epidemic of lung cancer soon followed among women. Initial hypotheses concerning the rising number of lung cancer cases and deaths focused on outdoor air pollution, although the earliest formal epidemiological studies quickly indicted cigarette smoking as the predominant cause of the disease (Doll and Hill, 1950; White, 1990). Nonetheless, there have been persistent concerns that air pollution may cause lung cancer and other cancers. These concerns have been prompted by the release of carcinogens into outdoor air from industrial sources, power plants and motor vehicles, and the recognition that indoor air is also contaminated by carcinogens.

During recent decades, a number of airborne carcinogens, viewed as potential threats to public health, have been particularly controversial. The energy crisis of the early 1970s led

AIR POLLUTION AND HEALTH
ISBN 0-12-352335-4

to increased manufacture of diesel-powered vehicles; recognition that diesel-soot particles were mutagenic raised concern that the increasing numbers of diesel-powered vehicles would increase lung cancer risk for the population. Three indoor carcinogens received widespread attention from the scientific community and the public during the 1980s and 1990s: tobacco smoke inhaled by non-smokers, radon, and asbestos fibers. Programs for reducing the risk of each of these carcinogens became extremely controversial. For these and other inhaled carcinogens, substantial research has been motivated by concern for the public's health and by the need to develop the scientific foundation for control strategies.

This chapter provides an overview of the evidence on outdoor air pollution and lung cancer. The evidence is now extensive and this review is selective, emphasizing the most recent findings, primarily from the epidemiologic literature. A 1990 monograph provides a more complete review (Tomatis, 1990) and a 1994 book comprehensively covers the epidemiologic literature on lung cancer, including indoor and outdoor air pollution (Samet, 1994).

This review focuses on outdoor, primarily urban, air pollution and lung cancer in developed countries. Mounting evidence indicates that populations in developing countries may have exposure to indoor and outdoor environments that rival or even dramatically exceed those found in developed Western countries (Samet, 1994). Indoor air pollution from coal combustion and cooking fumes has been linked to increased lung cancer risk in homes in China and Hong Kong (Smith and Liu, 1994). Rising pollution of outdoor air in the mega-cities of the developing world may also pose a risk for lung cancer. These topics are addressed elsewhere (Smith, 1988).

Perspective on Exposures to Inhaled Carcinogens

Exposures to inhaled carcinogens that may cause lung cancer take place in a variety of settings, including the home, the workplace and other public and commercial locations, and outdoors. Additionally, tobacco smoke contains more than 50 documented carcinogens and for most smokers the exposures received through active smoking are the dominant determinant of lung cancer risk. The heightened cancer risk of the active smoker may increase risk from other exposures through synergistic combined effects. Nevertheless, carcinogens in indoor and outdoor air are projected to contribute to the burden of lung cancer in both smokers and non-smokers.

The concept of total personal exposure provides a useful framework for conceptualizing exposures to inhaled carcinogens and evaluating the contribution of outdoor air pollution (National Research Council, 1991). Total personal exposure represents the integrated exposure to an agent, as that exposure is received in multiple microenvironments, i.e. environments having a relatively homogeneous concentration of the agent of interest during a specified time period. For an inhaled carcinogen such as benzo[a]pyrene, relevant microenvironments during the day might include outdoor air contaminated by vehicle exhaust and industrial emissions, workplace exposures from an industrial process, and air contaminated by tobacco smoke in a bar. For each microenvironment, the exposure received depends on the concentration of the carcinogen and the time spent in the microenvironment. Total personal exposure is thus the sum of the product of concentration with time spent in each microenvironment. The actual lung dose will further depend on ventilation rate, physical characteristics of the agent, and other factors determining the extent and site of lung deposition.

The relevant microenvironments will vary with the carcinogen of concern. Because most time is typically spent at home in Western societies, exposures to contaminants in indoor air of residences may be dominant for many pollutants. Based on time–activity patterns, the workplace is also a significant locus of exposure. Generally, adults spend little time outdoors but exposures to outdoor carcinogens could also take place indoors from the entry of outdoor pollutants.

The related concepts of total personal exposure and of microenvironments are also relevant to the development of control strategies. Some environments are shared and public and any initiative to control exposure requires action at a societal level; other environments are private and control lies with individuals. Thus, regulations may be needed to control air toxics released by industrial processes while education may be the cornerstone of initiatives to reduce exposures to indoor carcinogens, such as tobacco smoke, in homes.

Descriptive Epidemiology of Lung Cancer

In most countries of the world with available data, lung cancer mortality has risen across the twentieth century (Gilliland and Samet, 1994; IARC, 1986) (Table 36.1). Incidence

Table 36.1 Annual age-adjusted lung cancer mortality per 100 000, seven countries, 1953–1987

Country	Period of death						
	1953–57	1958–62	1963–67	1968–72	1973–77	1978–82	1983–87
Denmark							
Male	21.4	29.2	37.2	43.4	46.8	53.4	54.8
Female	4.4	5.3	6.9	8.8	11.0	14.9	20.7
Italy							
Male	15.1	21.5	29.6	38.2	44.9	52.9	57.9
Female	3.2	3.8	4.5	4.7	5.4	6.0	6.9
Japan							
Male	5.8	9.8	13.2	16.3	19.8	24.2	27.7
Female	2.1	3.7	4.6	5.3	6.0	7.0	7.7
Poland							
Male	–	18.1	28.1	36.7	45.6	54.6	64.7
Female	–	3.6	4.3	5.1	5.6	6.8	8.2
England and Wales							
Male	52.9	63.5	70.0	74.0	73.6	69.9	64.2
Female	6.7	8.1	10.2	12.5	14.9	17.3	19.3
USA							
White Male	25.8	31.9	38.5	46.3	51.2	54.7	55.3
White Female	4.1	4.7	6.2	9.4	12.9	17.3	21.5
Non-white Male	22.8	31.3	41.6	53.8	62.2	68.3	68.8
Non-white Female	4.2	5.0	6.6	9.7	12.7	16.3	19.3

From Gilliland and Samet (1994).

data, available only for the smaller number of countries and regions with cancer registration, generally show a comparable pattern (Gilliland and Samet, 1994). Because of the poor survival of persons with lung cancer, mortality and incidence rates are close and mortality rates provide a useful measure of the occurrence of new cases of lung cancer.

Within the USA, distinct patterns of occurrence are evident within population groups classified by age, sex and race (Gilliland and Samet, 1994; USDHHS, 1995). Incidence and mortality rates rise with age before declining in the oldest age groups and males have higher rates than females. African-American males have the highest rates, although the excess beyond the rates observed in whites is not readily explained by smoking patterns (Coultas *et al.*, 1994). Hispanics and Native Americans have lower incidence and mortality rates for lung cancer than whites, although there is substantial variation in rates among the different groups of Hispanics and the various tribes of Native Americans (Coultas *et al.*, 1994). Over recent decades, lung cancer incidence and mortality rates have increased in most locales throughout the world (Gilliland and Samet, 1994). However, rates have started to decline among younger males in the USA and the UK and age-adjusted lung cancer mortality is now declining among males in the UK.

Incidence and mortality rates by smoking status (never, current and former smokers) are available from several large cohort studies and estimates have been made for groups in the general population as well; trends of lung cancer in never-smokers are of particular interest with regard to air pollution as interpretation of the trends is not complicated by patterns of active smoking. Only limited information on lung cancer occurrence in never-smokers is available as calculation of either incidence or mortality rates requires an estimate of the population of never-smokers at risk for lung cancer during some time interval (the denominator of a rate) and a count of the numbers of cases of lung cancer in never-smokers during the time period of interest (the numerator of a rate). The two large, longitudinal or cohort studies conducted by the American Cancer Society have provided mortality rates from lung cancer and other diseases in never-smokers (US DHHS, 1989). Participants in the first study, now referred to as Cancer Prevention Study I (CPS I), were followed between 1960 and 1972; participants in the second, CPS II, were enrolled in 1980 and follow-up continues. The never-smoker rates from these studies increase with age in both sexes and rates for men tend to be higher than for women (US DHHS, 1997). In older males, rates are somewhat higher in CPS II, the later study; a similar pattern is evident among older women. These increases are likely to reflect more aggressive diagnosis of lung cancer in older persons; rates have not changed at younger ages.

As assessed by light microscopy with tissue stained by hematoxylin and eosin, primary cancer of the lung occurs as multiple histological types, with the most common being squamous cell carcinoma (epidermoid carcinoma), adenocarcinoma, large cell carcinoma, and small cell carcinoma (Churg and Samet, 1994). These four types of lung cancer account for approximately 30%; 25%; 10% and 20% of cases, respectively (Churg and Samet, 1994; Gilliland and Samet, 1994). A recent trend of an increasing proportion of adenocarcinomas has been documented in many regions throughout the world, leading to the hypothesis that risk factors for lung cancer may be changing, even though the etiologic and pathogenetic basis of the different types of lung cancer remain uncertain (Churg and Samet, 1994; Gilliland and Samet, 1994). Changes in classification and diagnostic approaches offer an alternative explanation.

Causes of Lung Cancer

The lung cancer risks associated with indoor and outdoor air pollutants need to be considered in the context of smoking and other causes of lung cancer. While cigarette smoking and other forms of tobacco smoking are the leading causes of lung cancer in the world, workplace exposures contribute substantially and other factors also determine risk for lung cancer, either independently or by modifying the risk of smoking. A number of occupational exposures have been causally linked to lung cancer and others are suspect causes. In their 1981 analysis of the causes of cancer for the USA, Doll and Peto estimated that 15% of lung cancer in males and 5% in females could be attributed to workplace exposures. The risk of lung cancer also appears to be increased by the presence of fibrotic lung diseases and of obstructive lung diseases (e.g. chronic bronchitic and chronic obstructive pulmonary diseases (Tockman and Samet, 1994)), although the mechanisms by which these diseases increase lung cancer risk are unknown. Lung cancer risk also varies with dietary factors; persons having lower intake of foods containing certain micronutrients, particularly β-carotene, are at increased risk, as are those having lower consumption of fruits and vegetables generally (Byers, 1994).

Epidemiologic evidence has indicated that persons with a family history of lung cancer are at increased risk (Economou *et al.*, 1994). The new techniques of molecular and cellular biology, brought to bear in the population setting using the research approach referred to as molecular epidemiology, have begun to provide insights concerning the mechanisms underlying genetically determined risk for lung cancer (Economou *et al.*, 1994; Vineis and Caporaso, 1995). Inherited patterns of carcinogen metabolism, DNA repair defects, and susceptibility genes may all contribute to the observed familial occurrence of lung cancer.

Interactions of Smoking with Other Agents

The combined effects of smoking with other agents merit particular consideration because of the strength of smoking as a cause of lung cancer. For exposure to a specific inhaled agent, the risk might be the same, greater, or lower in persons who smoke compared with never-smokers. An increased risk in smokers is referred to as 'synergism', while a reduced risk is termed 'antagonism'. Synergism implies that the smoker is more susceptible to the agent than the never-smoker, while antagonism implies the contrary. The combined effects of smoking and other agents are typically assessed using epidemiological data that include information on both smoking and the other exposure. A theoretical framework and related statistical methods have been developed to assess combined effects of two agents, such as cigarette smoking and an occupational exposure (Greenland, 1993).

Synergism with smoking has been documented for several occupational agents including asbestos and radon (Saracci and Boffetta, 1994). In the conceptual model used by epidemiologists, the finding of synergism implies that the lung cancer cases in the exposed population include those attributable to the independent actions of smoking and the occupational agent and also attributable to their joint action. This last group of cases is theoretically preventable by eliminating either smoking or exposure to the agent.

Combined effects of ambient air pollution and smoking have not been well characterized, although some reviews have noted a greater-than-additive relationship between air pollution and cigarette smoking which implies synergism (Speizer and Samet, 1994).

Estimates of the magnitude of the effect of joint exposure to ambient air pollution and cigarette smoking have been reported or can be derived from several studies (Table 36.2). Although these results appear to suggest a greater-than-additive relation between cigarette smoking and ambient air pollution, they are subject to error from inaccurate measurement of both air pollution exposure and cigarette smoking, and to substantial imprecision due to small numbers of non-smoking lung cancer cases.

Epidemiologic Approaches

Epidemiologists have used three approaches to examine the role of air pollution, whether indoor or outdoor, in the etiology of lung cancer: cohort, case–control and ecologic designs. Detailed discussions of these designs can be found in epidemiology texts such as Rothman and Greenland (1998) or the recent review by Morgenstern and Thomas (1993). The cohort studies of outdoor air pollution have involved the follow-up of populations, usually defined with respect to residence in an area or areas, over some time interval with assignment of exposure based on residence location. Cohort members are classified with respect to air pollution exposure and other factors related to lung cancer occurrence, such as age and cigarette smoking, and the rates of lung cancer incidence or mortality are then compared between exposed and unexposed subjects. The case–control approach provides investigators

Table 36.2 Proportion of lung cancer attributable to the joint effect of air pollution and smoking

Study	Air pollution/smoking measure	Air pollution	Air pollution/ smoking	Proportion attributable to joint exposure (EFI)[a]
		Rate ratios relative to non-smoking residents of low pollution areas		
Stocks and Campbell (1955)	Urban residents/1 pk per day	9.3	21.2	0.31
Haenszel et al. (1962)	Male residents of urban counties/>1 pk per day	1.1	5.7	0.30
Vena (1982)	Lifetime residents of high and medium pollution areas/≥40 pk-yr.	1.1	4.7	0.45
Barbone et al. (1995)	Residence in areas with high levels of particulate deposition (>0.298 g/m²/day)/≥40 cigarettes per day	3.7	59.6	0.21
Jedrychowski et al. (1990)	Residents in high pollution areas/ever smokers	1.1	6.7	0.27

[a] The EFI (etioloic fraction due to interaction) provides an estimate of the proportion of disease among those exposed to both high air pollution and smoking (either former or current) that is attributable to their joint effect (Walker, 1981).

pk, packet of 20 cigarettes. pk-yr, one pack a day for a year. Ever-smoker is anyone who has ever smoked and does not differentiate between current and former, or incorporate whether the person is a light, medium or heavy smoker.

with an efficient way to estimate the relative risk of lung cancer in relation to air pollution exposure without having to collect information on an entire population. In the case–control design, cases of lung cancer that have occurred in the population are ascertained and classified according to their exposure to outdoor or indoor air pollution; a sample of the study population – the controls – is selected and similarly classified according to their exposures. An estimate of the relative risk can then be calculated from these data.

In contrast to the cohort and case–control designs, ecologic, or aggregate-level, studies do not collect information on individual subjects. Instead they compare lung cancer incidence or mortality rates for geographic regions distinguished by different overall levels of air pollution, making use of routinely collected data on both lung cancer rates and region-wide measurements of air pollution. Lung cancer relative risks can be estimated from ecologic data, but the interpretation of these estimates is more complicated than for estimates derived from cohort and case–control studies. For example, in most cases, the data do not exist to allow the investigator to account adequately for intra-individual and between-region differences in other lung cancer risk factors. Ecologic studies are considered by most epidemiologists to be potentially subject to more biases and therefore to provide weaker evidence of air pollution effects; however, ecologic studies provided the initial basis for concern about outdoor air pollution and lung cancer.

All three designs suffer from a common problem when applied to the study of air pollution and lung cancer: the difficulty of characterizing accurately the subject's exposure to air pollution. As described above, an individual's exposure to carcinogens in outdoor and indoor environments may be complex and occur in multiple microenvironments, and therefore it may be difficult to characterize for the purpose of epidemiologic analysis. For example, early lung cancer studies often defined exposure to outdoor air pollution in terms of urban and rural location of residence. More recent investigations have used crude indicators of cumulative exposure to air pollution such as duration of residence in an area characterized by a particular level of pollution: the level is usually derived from routinely collected air monitoring data from one, or at most several, stationary monitoring sites. Although these approaches may serve to identify most truly exposed subjects, they will tend to classify some truly unexposed subjects as exposed.

The resulting misclassification of exposure may produce biased results in studies of air pollution and lung cancer and create the false impression that results from different studies are not in agreement (Lubin *et al.*, 1995; Rothman and Greenland, 1998). Such misclassification can spuriously elevate or diminish estimates of effects depending on how misclassification differs between subjects with and without lung cancer. When the extent of misclassification of exposure is the same for those with and without lung cancer, i.e. non-differential misclassification, estimates of effect are, in most cases, attenuated. When the risk of lung cancer increases directly and monotonically with exposure (sometimes referred to as a 'dose–response'), non-differential misclassification of exposure can obscure this pattern (Birkett, 1992; Dosemici *et al.*, 1990). Most studies attempt to collect data on other lung cancer risk factors, such as cigarette smoking, which could confound the air pollution relative risks. Errors in the measurement of potential confounders can introduce bias, even if air pollution exposures are estimated with relatively little error, and the result may be either over- or under-estimation of the air pollution effect (Greenland, 1980). Some recent studies have made use of newly developed methods from molecular biology to classify study subjects according to molecular markers of exposure to carcinogens in ambient air (Hemminki *et al.*, 1997). These methods may enable epidemiologists to reduce bias from misclassification of exposure, but they are in their infancy and will require careful study and more extensive application before their true utility is known.

Risk Assessment

The concerns about outdoor air pollution and lung cancer largely relate to the risk posed to populations, not to individuals. Risk assessment, a framework for organizing information about risks to populations, has been applied to the problem of outdoor air pollution and lung cancer or policy making purposes. Another chapter in this book provides an introduction; specific findings related to lung cancer are cited here.

CARCINOGENS IN OUTDOOR AIR

Exposures to Carcinogens in Outdoor air

Ambient air, particularly in densely populated urban environments, contains a variety of known human carcinogens, including organic compounds such as benzo[a]pyrene and benzene, inorganic compounds such as arsenic and chromium, and radionuclides (Table 36.3). These substances are present as components of complex mixtures, which may include carbon-based particles to which the organic compounds are adsorbed, oxidants such as ozone, and sulfuric acid in aerosol form. The combustion of fossil fuels for power generation or transportation is the source of most of the organic and inorganic compounds, oxidants and acids, and contributes heavily to particulate air pollution in most urban settings. The radionuclides result from fuel combustion as well as from mining operations.

Unfortunately, there are few long-term trend data for ambient levels of known

Table 36.3   Selected known carcinogens in urban and rural ambient air

Substance	Urban air	Rural air
Inorganic particulates (ng/m^3)		
Arsenic	2–130	<0.5–5
Asbestos	10–100	–
Chromium	5–120	<1–10
Nickel	10–1000	<10
Radionuclides (Ci/m^3)		
^{210}Pb	1×10^{-15}–30×10^{-15}	5.5×10^{-15}–10×10^{-15}
^{212}Pb	0.1×10^{-15}–4×10^{-15}	0.03×10^{-15}–0.06×10^{-15}
^{222}Rn	20×10^{-125}–1000×10^{-12}	0.1×10^{-12}–20×10^{-12}
Gaseous and particulate organic species (ng/m^3)		
Benzene	5–90	–
Benzo[a]pyrene	1–50	–
Benzene-soluble organics	1000–2000	200–300

From IARC (1987), Table 1.

carcinogenic products of fossil fuel combustion that could be used to estimate long-term exposures for epidemiologic purposes. Available data indicate that over the past 20–30 years there have been improvements in some indices of air quality. According to a 1980 report of the Council for Environmental Quality, levels of benzo[a]pyrene in urban air decreased 70% between 1970 and 1980. Daisey (1988) reported that levels of sulfates and of particulate-associated organic matter declined by 30–40% between 1964 and 1983 in two industrialized New Jersey, USA locations.

The US Environmental Protection Agency (US EPA) collects data on six pollutants for which the US Government has promulgated national air quality standards. Particulate matter with an aerodynamic diameter <10 μm, PM_{10}, is the criteria pollutant of greatest current interest with respect to lung cancer because particles of size 10 μm or less can be inhaled into the lung and carry carcinogenic substances on their surfaces. Comprehensive monitoring of PM_{10} has only been in place since 1988; prior to 1988 the US EPA monitored only total suspended particulates (TSP), which include particles too large to be inhaled into the lung. Decreases in TSP were observed over the 1970s and early 1980s, but little change was noted through 1990. From 1988 to 1992 the average annual mean concentration of PM_{10} fell by 17% (US EPA, 1993). The PM_{10} standard will soon be complemented by a $PM_{2.5}$ standard.

The data discussed above refer for the most part to ambient air pollution over relatively large geographic areas. However, the exposure of human populations to carcinogens in ambient air may be the result of proximity to more localized sources such as small businesses (e.g. automotive body or chrome plating shops), municipal facilities (e.g. waste incinerators), or areas with high vehicular traffic. For example, data collected by Cass and colleagues (1984) in Los Angeles indicate that elemental carbon levels, mostly derived from diesel exhaust, declined in five of seven areas in Los Angeles between 1958 and 1981, but increased in two areas undergoing rapid growth. Under the Clean Air Act Amendments of 1990, the US EPA is charged with evaluating the risks of 189 hazardous air pollutants, primarily carcinogens from paint sources.

Combustion Products

As noted above, the combustion of fossil fuels for transportation and power generation contributes to the presence of many known or suspected carcinogens in ambient air. Some of the potentially more significant pollutants in terms of exposure prevalence and/or lung carcinogenicity are described below.

Polycyclic Organic Matter

Polycyclic organic matter (POM), as defined by the US EPA in the Federal Clean Air Act, comprises a large and varied class of chemical compounds, including polycyclic aromatic hydrocarbons (PAH) and nitro-PAHs, which are known carcinogens and mutagens (IARC, 1987). The compounds that comprise POM have common chemical features (one or more benzene rings and a boiling point > 100°C), and are found in both the particulate and gas phases of ambient air, depending on their exact chemical structure (e.g. those with more than five benzene rings tend to be associated with the particle phase). In addition to those compounds released directly into the environment by combustion processes, others are created from primary combustion products, such as those emitted by diesel engines, via chemical and photochemical reactions in the ambient environment (Greenberg, 1988; Natusch, 1978; Winer and Busby, 1995). Neither the contribution of these latter, secondary, compounds, to total ambient levels of POM, nor their relative mutagenicity and carcinogenicity, are well understood (Greenberg, 1988; Winer and Busby, 1995).

Although the combustion of fossil fuels is a ubiquitous source of POM in the urban ambient environment, it is not the only source of human exposure to POM, and for some individuals it may not be the predominant source. Other human exposure to POM comes from inhaling wood and tobacco smoke, and from the diet (e.g. from the consumption of grilled meat). Unfortunately, for the conduct of epidemiologic research, there are currently no validated markers of exposure to POMs from specific sources, either in ambient air or biological material.

Urban air contains a mixture of polycyclic organic compounds, but certain specific constituents, such as Benzo[a]pyrene, have been extensively studied and are known to be carcinogenic. benzo[a]pyrene has been frequently used as a surrogate or marker for combustion source air pollution in epidemiologic studies and for risk assessment (see below). The literature on cancer risk in relation to occupational and environmental exposure to PAHs has recently been reviewed by Boffetta *et al.* (1997), who concluded that PAHs are associated with increased lung cancer risk in a variety of occupational settings, and with increased lung cancer risk in urban populations. Mixtures of polycyclic compounds encountered in occupational settings, such as by coke-oven workers in the steel industry and by coal gasification workers (Doll *et al.*, 1972; Redmond, 1983), also are known to cause increased occurrence of lung cancer in exposed workers (IARC, 1984). The levels of POM encountered in the ambient urban environment, however, are substantially less than those encountered in heavily exposed occupational settings.

Particles

Like POM, particulate air pollution is not a single entity, but rather a chemically and physically diverse group of pollutants derived from sources as diverse as crustal dust and sea spray and the combustion of diesel fuel (Koutrakis and Sioutas, 1996). As such, it is not possible to address the carcinogenicity of particles generically. Attention has focused on combustion-source particles in urban air as potential lung carcinogens, largely because the carbonaceous particles produced by the combustion of fossil fuels are in the respirable range (generally <1.0 μm in diameter) and have known human carcinogens, such as PAHs, adsorbed to their surfaces, However, recent evidence from animal experiments showed that rats exposed to high levels of relatively pure carbon particles developed lung tumors at the same rate as rats exposed to diesel exhaust particles, suggesting that particles *per se* might, under some conditions, be carcinogenic. The relevance of these findings for humans is controversial (Winer and Busby, 1995).

The combustion of fossil fuels for power generation and transportation produce gaseous pollutants, such as sulfur dioxide (SO_2) and oxides of nitrogen (NO_x), that are converted into fine particulate air pollution in the atmosphere. Epidemiologic studies provide no consistent evidence of increased lung cancer risk from occupational exposure to SO_2; however, the International Agency for Research on Cancer (IARC) has classified strong sulfuric acid aerosol as a known human carcinogen based on epidemiologic findings of increased lung and laryngeal cancer in heavily exposed occupational groups (IARC, 1992). A more extensive discussion of SO_2 and acid particles can be found in chapters elsewhere in this volume.

Diesel

Diesel exhaust is a ubiquitous component of urban ambient air pollution throughout the world, although few studies have estimated its proportional contribution. In one of the few studies that estimated the proportional contribution of diesel exhaust to ambient air pollution, Cass and Gray (1995) estimated that diesel exhaust contributed 7% of the fine particulate matter (<2 µm) in the Los Angeles air basin in 1982.

Based on evidence from animal experiments and epidemiologic studies of occupationally exposed groups, diesel exhaust is considered by IARC to be a probable human carcinogen (2A) (IARC, 1989), although, as noted above, the mechanism by which exposure to diesel exhaust might produce lung cancer in humans remains to be determined. The evidence for the carcinogenicity of diesel exhaust has recently been extensively reviewed (Cohen and Higgins, 1995), and is discussed more extensively by Cohen and Nikula in this volume (see Chapter 32).

Other

The combustion of fossil fuels contributes known or suspected individual carcinogenic chemicals to urban ambient air. In addition to such known human lung carcinogens as arsenic, chromium and nickel, several others are worthy of note; these are discussed briefly below. More extensive discussions of 1,3-butadiene and formaldehyde can be found in the chapter by Cameron and Rushton in this volume (see Chapter 35).

Radionuclides Alpha-emitting radionuclides can also be measured in outdoor air. These include isotopes of lead, radium, thorium and uranium (Natusch, 1978). They are naturally present in fossil fuels and are emitted as by-products of combustion. The contribution of this route of exposure to the radiation dose received by the general population is negligible (NRC, 1990).

1,3-Butadiene 1,3-Butadiene is a volatile organic compound that has been employed since the 1930s in the production of synthetic rubber, and industrial emissions contribute to its presence in some urban areas. However, 1,3-butadiene is also emitted in automotive exhaust, accounting for the preponderance of emissions in the USA, and for much of the human exposure as well. Levels of 1,3-butadiene in ambient air are generally in the range 1–10 ppb, about 1000-fold less than occupational exposure levels (Health Effects Institute, 1993; IARC, 1992).

1,3-Butadiene has been classified by IARC as a probable human carcinogen (2A) based largely on the results of animal experiments, which indicated increases in tumors at multiple sites, including the lung. Epidemiologic studies of occupationally exposed populations (rubber workers and butadiene monomer production workers) have consistently observed increases in hematopoietic cancers, but not cancers of the respiratory system (IARC, 1992).

Aldehydes Various aldehydes classified as hazardous air pollutants by the US EPA (e.g. formaldehyde and acetaldehyde) are present in urban ambient air largely due to the combustion of gasoline and diesel fuel (Health Effects Institute, 1993). Exposures to formaldehyde and acetaldehyde in outdoor air tend to be highly correlated. In the case of formaldehyde, outdoor concentrations have generally been observed to be in the range 1–20 $\mu g/m^3$, although under certain conditions (in heavy traffic or air pollution episodes) levels of 100 $\mu g/m^3$ have been observed (IARC, 1995). The combustion of alternative fuels such as methanol and of oxygenated fuels containing the additive MBTE results in greater aldehyde emissions. This contributes to increased ambient concentrations in locales where they are widely used (Health Effects Institute, 1993).

Formaldehyde has been classified by IARC as a probable human carcinogen (2A) (IARC, 1995), based on evidence from animal experiments and epidemiologic studies in occupational groups of exposure-related excess nasal and naso-pharyngeal cancer. There is no consistent evidence that occupational exposure to formaldehyde is associated with increased lung cancer risk (IARC, 1995).

Point Sources

Residential proximity to industrial point sources of air pollution is a potential source of exposure to known or suspected carcinogens. Fossil fuel-fired (i.e. coal, oil and natural gas) electrical power plants emit known or suspected carcinogens (Natusch, 1978), including metals such as chromium and nickel, radionuclides such as radon and uranium, and POM such as benzo[a]pyrene. Non-ferrous metal smelters emit surfur dioxide in addition to inorganic arsenic and other metals (Pershagen et al., 1977). Municipal solid waste incinerators emit heavy metals (e.g. lead and cadmium), PAHs, organic compounds (such as dioxins) and acidic gases (WHO, 1988). Unfortunately, these sources of air pollution are more frequently located in or near poor working class communities, whose residents may, for a variety of reasons, be more susceptible to the effects of these pollutants.

In 1990, Pershagen et al. reviewed a number of epidemiologic studies of lung cancer occurrence and residential proximity to industrial point sources of air pollution. Eleven studies estimated lung cancer risk associated with proximity to non-ferrous metal smelters. Of these, five ecologic studies observed relative risks in males between 1.2 and 2.0, but only one accounted for employment at the smelter itself, and data on smoking were not available. These studies did not consistently observe elevations in risk among women. Six case–control studies presented conflicting results: several showed no association with residential proximity, but did not account for either employment at the facility or smoking habits. Two studies that did account for these factors observed relative risks of 1.6 and 2.0 in males. Ecologic studies of residential proximity to diverse industrial sources (e.g. petrochemical plants and steel mills) have generally observed increased relative risks of lung cancer, but have been unable to control for level confounders at the individual level, such as cigarette smoking and employment at the industrial facility itself.

Elliott and colleagues (1996) recently reported the results of an ecologic study of cancer incidence among 14 million people living near 72 municipal solid waste incinerators in Great Britain. Cancer rates and residential proximity to the incinerators were measured at the level of postal code (which is roughly analogous to neighborhood); the relative risks (compared with national incidence rates) were adjusted for age, sex, geographic region and an index of socioeconomic status. For several cancers (stomach, colorectal, liver and lung) excess relative risk was inversely related to distance of the residence from the incinerator. Lung cancer relative risks (95% confidence intervals) were

1.08 (1.07, 1.09) and 1.06 (1.05, 1.07) for residence distance 0–3 km and 0–7.5 km, respectively. However, Elliott and colleagues also observed equal elevations in lung cancer risk in the areas proximal to the incinerators prior to the construction of the facilities, leading them to conclude that residual confounding by unmeasured characteristics of the postal codes accounted for the apparent associations with proximity to the incinerators.

Fibers

Asbestos fibers are present in ambient air in contemporary rural and urban environments. Apparently they have been present in the ambient environment for at least 10 000 years. The literature on levels of asbestos in ambient air was reviewed as part of a comprehensive report on the health effects of asbestos in public buildings (Health Effects Institute, 1991). The median levels of asbestos fibers in rural environments in which there were no known natural sources of asbestos were in the order of 0.01– 0.001 ng/m^3, and few individual measurements exceeded 1 ng/m^3. In urban environments, where there is both a greater prevalence of asbestos-containing materials and a higher frequency of release of fibers from sources such as building materials and vehicular brake linings, a considerably greater proportion of individual measurements exceeded 1 ng/m^3 and the median levels ranged from 0.02 to 10 ng/m^3 (Health Effects Institute, 1991).

Epidemiologic studies of occupational groups, such as asbestos miners and asbestos textile production workers who were exposed to asbestos at concentrations several orders of magnitude greater than those cited above, have consistently observed increased rates of lung cancer. Based on a large body of experimental and epidemiologic evidence, asbestos is considered to be a known human carcinogen (Health Effects Institute, 1991). There are, however, no studies of lung cancer occurrence in relation to exposure at levels observed in the ambient urban environment.

EPIDEMIOLOGIC EVIDENCE ON OUTDOOR AIR POLLUTION AND LUNG CANCER

Introduction

Nearly five decades have passed since investigators in the UK initiated studies to test whether outdoor air pollution increases lung cancer risk. Three lines of epidemiological evidence have been relevant: (1) studies of lung cancer risk in migrants to areas having differing lung cancer risk from the native country; (2) ecological studies comparing lung cancer rates in areas assumed to have differing air pollution exposures; and (3) case–control and cohort studies that have estimated lung cancer risk associated with exposures of individuals.

Studies of migrants, reviewed in detail elsewhere (Shy *et al.*, 1978; Speizer and Samet, 1994), have provided support for the general hypothesis that air pollution is associated with lung cancer risk. Migrants from countries with higher lung cancer rates to countries with lower lung cancer rates tend to develop lung cancer at rates higher than those of the new country of residence. Ecological studies (i.e. studies involving the population as the unit of analysis) have also supplied support evidence by comparing rates for urban and rural populations, and for populations estimated to have higher and lower exposures to urban air pollution.

Case–control and cohort studies have the advantage of offering information on potential confounding and modifying factors, such as cigarette smoking. However, a strategy is needed for estimating air pollution exposure using historical information. Approaches have been based around reported residence locations and use of various surrogates. Misclassification of exposure is an evident limitation of such studies and would tend to reduce risk estimates and statistical power.

Exposure biomarkers – that is, indicators of exposure or dose measured in biological material – offer a new approach to quantifying the lung cancer risk associated with air pollution. Potential biomarkers for lung cancer risk include actual levels of the putative carcinogen in biological materials, adducts of potential carcinogens or metabolites down to DNA, and antibodies against such adducts (NRC, 1989; Schulte and Perera, 1993). Biomarkers of exposure to respiratory carcinogens provide an intermediate outcome for investigation that may prove valid surrogates for risk. Levels of adducts, for example, may prove to be a predictor of risk and may serve to bridge from animal models and *in vitro* assays to human risk.

Biomarkers

Ambient air pollution contains a number of specific carcinogens released from combustion of fossil fuels. In investigating biomarkers of air pollution exposure, emphasis has been placed on one group of such carcinogens, the polycyclic aromatic hydrocarbons of which benzo[*a*]pyrene is the prototype. The concentration of benzo[*a*]pyrene can be measured in the air and a number of studies have examined the association between exposure to benzo[*a*]pyrene and other polycyclic aromatic hydrocarbons and marker levels. These studies have investigated occupationally exposed workers, including foundry workers and traffic police, and persons living in highly polluted urban environments. The markers investigated have included polycyclic aromatic hydrocarbon-DNA adducts, sister chromatid exchange, chromosome aberrations and oncogene proteins.

Perera and colleagues (Motykiewicz *et al.*, 1996; Perera *et al.*, 1992) have investigated biomarkers in residents of a highly industrialized region of Silesia in Poland. Levels of benzo[*a*]pyrene are markedly elevated due to industrial activity and coal combustion for residential heating. Airborne particulate matter collected in this region has been subjected to assays for mutagenicity and found to have genotoxic activity in a variety of short-term tests. In comparison with controls from a less polluted region, persons residing in Silesia have significant increases in carcinogen-DNA adducts, sister chromatid exchange and chromosomal aberrations. They also show a doubling of the frequency of *ras* oncogene overexpression. These results indicate the potential to estimate lung cancer risk by combining information on concentrations of carcinogens in outdoor air with levels of adducts in exposed persons; the levels of biomarkers could then be used to predict cancer risks, once the biomarkers are validated (Perera *et al.*, 1996). Workers in certain occupations, including foundries or jobs involving heavy vehicle exhaust exposure, have also been investigated for levels of biomarkers of exposure to polycyclic aromatic hydrocarbons. In a study of foundry workers in Finland, levels of DNA adducts with polycyclic aromatic hydrocarbons were associated with workplace exposure (Hemminki *et al.*, 1997; Perera *et al.*, 1994).

Studies have also been conducted of biomarkers in traffic police workers and bus drivers. In a study of traffic police workers in Italy, levels of micronuclei in peripheral blood lymphocytes were not increased in the police workers in comparison with controls (Merlo *et al.*, 1998). Urinary excretion of 1-hydroxypyrene, however, did appear to be a useful biomarker for exposure to air pollution by polycyclic aromatic hydrocarbons (Merlo *et al.*, 1998). In contrast, bus drivers working in central Copenhagen had higher levels of polycyclic aromatic hydrocarbon-DNA adducts in comparison with controls (Nielsen *et al.*, 1996).

Epidemiologic Studies of Ambient Air Pollution and Lung Cancer

The earliest studies of air pollution and lung cancer contrasted lung cancer rates between urban and rural environments (Table 36.4). Most studies found overall excesses on the order of 30–40% in the urban areas. The attribution of these results to differences in air quality was strengthened by evidence of urban/rural differences in ambient levels of carcinogens such as benzo[*a*]pyrene and by the frequent persistence of the urban/rural differences after adjustment for cigarette smoking. In addition, the studies of Buell *et al.* (1967), Haenszel and Taeuber (1964), Haenszel *et al.* (1962) and Samet *et al.* (1987) observed larger effects in relation to longer urban residence. Doll and Peto, in their widely cited 1981 monograph, *The Causes of Cancer*, cast doubt on the causal role of air

Table 36.4   Urban/rural differences in lung cancer

Studies	Population	Cases	Rate ratio
Cohort studies			
Hammond and Horn (1958)	US veterans (1952–55)	448	1.3
Buell *et al.* (1967)	California residents (1957–62)	304	1.3
Hammond (1972)	US residents (1959–65), unexposed/ exposed to dust and fumes	1510	1.1/1.3
Cederlof *et al.* (1975)	Swedish men (1963–72)	116	1.4
Doll and Peto (1981)	British physicians (1951–71)	401	1.0
Tenkanen *et al.* (1987)	Finnish men (1964–79), smokers/ non-smokers	233	1.1/1.9
Case–control studies			
Stocks and Campbell (1955)	British men (1952–54)	725	1.7
Haenszel *et al.* (1962)	US white men (1958)	2381	1.4
Haenszel and Taeuber (1964)	US white women (1958–59)	749	1.3
Dean (1966)	Irish and English men/women	3040	2.1/1.3
Dean et al. (1978)	non-smokers (1960–62)		
Hitosugi (1968)	Japanese men/women (1960–66)	259	1.8/1.2
Samet *et al.* (1987)	New Mexico residents >25 years in urban counties (1980–82)	422	1.2–1.4

pollution because early research had not accounted for the effects of urban dwellers having started smoking at younger ages as cigarette smoking became increasingly prevalent in the early twentieth century. However, Dean *et al.* (1978) controlled for age at beginning smoking and found that the urban/rural gradient persisted. Moreover, cancer incidence data assembled by the IARC over the past decade continue to show evidence of urban/rural differences (IARC, 1997). However, the so-called 'urban factor' may reflect other influences instead of, or in addition to, outdoor air pollution: these could include indoor air pollution, patterns of migration, or factors related to population density.

Ecologic studies have compared lung cancer rates across urban areas with differing levels of air pollution (Table 36.5). These studies showed relative excesses of lung cancer in the more polluted areas of similar or slightly higher magnitude than the urban/rural studies. Taking advantage of a 'natural experiment,' Archer (1990) analyzed respiratory cancer mortality in two Utah counties with very low smoking rates. The counties were similar in many respects, with low and nearly equal respiratory cancer mortality rates, until a steel mill constructed during World War II caused substantial

Table 36.5   Epidemiologic studies of outdoor air pollution and lung cancer

Study	Locale	Exposure classification	Rate ratio (95% CI)
Ecologic			
Henderson *et al.* (1975)	Los Angeles	Polycyclic aromatic hydrocarbon levels by geographic area	1.3
Buffler *et al.* (1988)	Houston	Total suspended particle levels by census tract	1.9
Archer (1990)	Utah	Mean levels of total suspended particulates by county	1.6
Case–control			
Pike *et al.* (1979)	Los Angeles	Residence in high pollution (benzo[*a*]pyrene) area	1.3 (NA)
Vena (1982)	Buffalo, NY	>50 years residence in elevated TSP areas	1.7 (1.0–2.9)
Jedrychowski *et al.* (1990)	Cracow	Residence in elevated TSP and SO$_2$ areas	1.5 (1.1–2.0)
Katsouyanni *et al.* (1990)	Athens	Lifelong residence in high pollution areas	1.1 (NA)
Barbone *et al.* (1995)	Trieste	Residence in areas with high levels of particulate deposition (>0.298 g/m^2/day)	1.4 (1.1–1.8)
Cohort			
Abbey *et al.* (1995)	California	PM$_{10}$ 42 day/year >100 mg/m^3 O$_3$ 500 h/year >100 ppb	1.5 (0.9–2.4) 2.3 (0.9–5.3)
Dockery *et al.* (1993)	6 US cities	Residence in high sulfate or fine particulate pollution areas	1.4 (0.8–2.3)
Pope *et al.* (1995)	151 US cities	Residence in high sulfate particulate pollution areas	1.4 (1.1–1.7)
		Residence in high fine particulate pollution areas	1.0 (0.8–1.3)

increases in air pollution in one of them. The subsequent differences in lung cancer were substantial within about 15 years after the increase in air pollution and have persisted. A third neighboring county, unaffected by pollution from the steel mill, but with higher smoking rates, had higher lung cancer rates than either of the other two counties, underscoring the profound effects of cigarette smoking on lung cancer risks. However, because incidence, exposure and covariate data were all on the aggregate, or ecologic, level, it was not possible to account adequately for intra-individual and between-area differences in other risk factors.

Several case–control and cohort studies used air pollution monitoring data to estimate the exposures of study subjects (Table 36.5). Most found relative increases of lung cancer risks after adjustment for age, smoking and occupational exposure similar to those observed in the urban/rural and ecologic studies. Dockery and colleagues (1993) recently reported the results of a cohort study of 8111 adults living in six US cities. Cohort members were followed for between 14 and 16 years and their mortality was ascertained through 1989. Lung cancer relative risks were estimated with respect to average levels in each city of various components of air pollution, including total and fine particulate mass, ozone and sulfate particles. After adjustment for differences in age, sex, cigarette smoking, obesity, education and occupational exposure among cohort members, researchers observed a 37% excess lung cancer risk for a difference in fine particulate mass equal to that of the most polluted versus the least polluted city.

Pope and colleagues (1995) linked ambient air pollution data from 151 US metropolitan areas with risk factor data for 552 138 adults who were enrolled in the American Cancer Society (ACS) Cancer Prevention Study II (CPS II), and then monitored for vital status from 1982 to 1989. Using multivariable regression, the investigators controlled for individual differences in age, sex, race, cigarette smoking, pipe and cigar smoking, exposure to passive cigarette smoke, occupational exposure, education, body mass index and alcohol use. Lung cancer mortality was associated with air pollution when sulfate particulate was used as the index of air pollution exposure, but not when fine particle mass was used, as in the Six Cities study. This discrepancy did not appear to be due to differences in air monitoring data across the 151 metropolitan areas and remains unexplained. When sulfate particulate pollution was used as the index of exposure, estimated pollution-related mortality risk was as high for never-smokers as it was for smokers and as high for women as it was for men. Therefore, although the increased risk associated with air pollution was small compared with that from cigarette smoking, results of this study suggest that the association between pollution and mortality was unlikely to be due to inadequate control of smoking because the associations with air pollution persisted when accounting for smoking habits and exposure to second-hand smoke. The associations were also as large or larger for never-smokers as they were for smokers, though they were statistically less precise owing to the small number of lung cancer deaths among non-smokers.

The cohort of California residents studied by Abbey and colleagues (1995) consists entirely of Seventh Day Adventists whose extremely low prevalence of smoking and uniform (and relatively healthy) dietary patterns reduce the potential for confounding by these factors. Excess lung cancer was observed in relation to cumulative exposure to both particle and ozone exposure.

Risk attribution

Reported estimates of the population attributable risk of lung cancer due to outdoor air pollution have been based on markedly different methods and their range spans an order of magnitude. Basing their estimate on past and then current estimates of benzo[*a*]pyrene in urban air and extrapolation from occupational studies of PAH-exposed workers, Doll and Peto (1981) estimated that less than 1% of future lung cancer cases would be due to air pollution from the burning of fossil fuels. They did note, however, that perhaps 10% of then current lung cancer in large cities might have been due to air pollution. In 1990, the US EPA (1990) estimated that 0.2% of all cancer, and probably less than 1% of lung cancer, could be attributed to air pollution. This estimate was obtained by applying the unit risks for over 20 known or suspected human carcinogens found in outdoor air to estimates of the ambient concentrations and numbers of persons potentially exposed. The unit risks were derived either from animal experiments or extrapolation from studies of workers exposed to higher concentrations. One group based their estimates on direct observation of populations exposed to ambient levels of air pollution. Karch and Schneiderman (1981), using data from the American Cancer Society (CPS I) study and US Census data, estimated that the 'urban factor' accounted for 12% of lung cancer in 1980. They predicted that 1980 levels of total suspended particulates would be associated with a lung cancer rate ratio of 1.3, slightly less than the 47% increase observed for total suspended particles in the recent report of findings in the Six Cities Study. Each of these estimates of attributable risk is subject to considerable error with respect to both the relative magnitude of effect and the proportion of the population assumed to be exposed, but there seems to be no compelling argument to prefer estimates based on extrapolation from animal experiments or occupational studies to direct observation of the populations at risk.

DEVELOPING COUNTRIES

Current knowledge about ambient air pollution and lung cancer is based largely on the experience of populations of Western industrialized nations. The populations of the developing countries, however, are exposed to levels of air pollution from combustion sources in both ambient and indoor environments that rival or exceed those commonly observed in the industrialized West. Within the developing countries, the highest exposures, particularly among women, have been to indoor air pollution from the combustion of coal and biomass fuels for cooking and heating (Smith, 1998). For example, typical concentrations of coal smoke in rural Chinese homes exceeded 500 $\mu g/m^3$ and frequently exceeded 1 $\mu g/m^3$ (Smith, 1998). Smith and Liu (1994) recently reviewed the epidemiologic literature on indoor air pollution and lung cancer in the developing countries and found consistent evidence of increased rates of lung cancer associated with indoor cooking and heating with coal in studies done largely in China. A much smaller group of studies revealed no consistent association of lung cancer with indoor use of biomass fuels.

There has been little research on ambient air pollution and lung cancer among urban residents of the burgeoning cities of the developing countries, although mounting levels of urban air pollution from local stationary and, increasingly, mobile sources (Prentice and Sheppard, 1995) is recognized as an important environmental problem by international public health and economic agencies. In the cities of the poorest developing countries, WHO's Global Environmental Monitoring System observed average ambient concentrations of total suspended particles of 300 μg/m^3 (Navidi *et al.*, 1993), although levels in locales where coal is used for fuel, such as poor communities in South Africa, may exceed 1g/m^3 (WHO, 1990). One might predict that the high levels of ambient air pollution found in cities in the developing world would be associated with greater excess lung cancer occurrence than has been observed in Western industrialized settings. Although there are currently few relevant investigations, a case–control study in Shenyang, China, observed a two-fold increase in lung cancer risk after adjustment for age, education and smoking, among residents in 'smoky' areas of the city and a 50% increase among those in 'somewhat or slightly smoky' areas (Prentice and Sheppard, 1995).

As a greater proportion of the world's population moves from rural communities to the rapidly expanding and highly polluted cities of Asia and the Southern Hemisphere, there is a clear need to address the large gap in epidemiologic research on outdoor air pollution and lung cancer in the developing world. These studies will present even greater challenges than those in the industrialized West. In addition to the generic problem of estimating long-term exposure to air pollution discussed above, the ambient air pollution mixture in urban centers in the developing countries is changing, due in part to the increase in vehicular traffic. Careful planning will be required to characterize these changes as they will have occurred over time, including choosing and measuring indicator pollutants for different pollution sources. In addition, the current dramatic increases in cigarette smoking in the developing world (Council on Scientific Affairs, 1990), and the thoroughly predictable consequences, will complicate the interpretation of studies of air pollution and lung cancer.

CONCLUSIONS AND RESEARCH NEEDS

There remains a persistent basis for concern that indoor air pollution may cause lung cancer in both smokers and non-smokers. Carcinogens can be measured in indoor and outdoor environments and toxicologic data indicate the potential for human carcinogenicity. Epidemiologic research shows evidence of effects of indoor and outdoor air pollution on lung cancer risk, albeit weak for some agents. On the other hand, levels of exposure to many agents would not be expected to greatly increase risk and the multifactorial etiology of lung cancer lowers the signal-to-noise ratio.

Research is still needed on air pollution and lung cancer to guide policies for protection of public health. Direct epidemiologic observation of exposed populations can provide the best information for evaluating the magnitude of outdoor air pollution-related excess lung cancer. In general, large-scale epidemiologic studies of air pollution and lung cancer are needed if we are to obtain sufficiently informative data. Large numbers of cases will be necessary to measure the effects of air pollution among women and ethnic minorities and to measure the joint effects of air pollution and other factors, such as occupation and smoking. Without improved epidemiologic methods, however, even large studies may fail to inform. Current development of biologic markers of exposure to, and molecular effects of, PAHs represents one approach to improving epidemiologic methods. Markers of genetic susceptibility are also needed. In addition, and of equal importance, methods for the retrospective estimation of lifetime exposure to air pollutants should be developed and tested, so that large case–control and retrospective cohort studies can be feasibly conducted. These methods could combine time–activity information with data from national aerometric databases, such as those maintained by the US Environmental Protection Agency. This effort should include development of methods to characterize, quantify and adjust for exposure measurement error. For lung cancer, urban and relatively unpolluted areas with established population-based tumor registries might be targeted.

The air pollution mixtures in various US population centers should be characterized both in terms of physical and chemical constituents and of sources of major constituents. If possible, retrospective characterization of levels of certain constituents could be accomplished. This information would aid greatly in the interpretation of between-city patterns of lung cancer occurrence.

New designs and statistical methods for air pollution studies may provide additional insights. Navidi and Thomas (1993) and Prentice and Sheppard (1995) have described hybrid studies which combine ecologic-level contrasts of air pollution effects between cities with individual-level data on covariates, combining the strengths of both ecologic and individual-level studies. Studies using these designs could contrast the effect on lung cancer of exposure to the pollutant mixtures of different cities while effectively controlling confounding by cigarette smoking, diet or other factors, and adjusting for exposure measurement error.

REFERENCES

Abbey DE, Lebowitz MD, Mills PK *et al.* (1995) Long-term concentrations of particulates and oxidants and development of chronic disease in a cohort of nonsmoking California residents. *Inhal Toxicol* 7(1): 19–34.

Archer VE (1990) Air pollution and fatal lung disease in three Utah counties. *Arch Environ Health* **45**: 325–334.

Barbone F, Bovenzi M, Cavalleri F and Stanta G (1995) Air pollution and lung cancer in Trieste, Italy. *Am J Epidemiol* **141**(12): 1161–1169.

Birkett NJ (1992) Effect of nondifferential misclassification on estimates of odds ratios with multiple levels of exposure. *Am J Epidemiol* **136**: 356–362.

Boffetta P, Jourenkova N and Gustavsson P (1997) Cancer risk from occupational and environmental exposure to polycyclic aromatic hydrocarbons. *Cancer Cause Cont* **8**: 444–472.

Buell P, Dunn JE and Breslow L (1967) Cancer of the lung and Los Angeles type air pollution. *Cancer* **20**: 2139–2147.

Buffler PA, Cooper SP, Stinnett S *et al.* (1988) Air pollution and lung cancer mortality in Harris County, Texas, 1979–1981. *Am J Epidemiol* **128**: 683–699.

Byers T (1994) Diet as a factor in the etiology and prevention of lung cancer In: Samet JM (ed.) *Epidemiology of Lung Cancer*, Chapter 13. New York: Marcel Dekker Inc, pp 335–352.

Cass GR and Gray HA (1995) Regional emissions and atmospheric concentrations of diesel engine particulate matter: Los Angeles as a case study. In: *Health Effects of Diesel Engine Emissions: Characterization and Critical Analysis.* Cambridge, MA: The Health Effects Institute.

Cass GR, Conklin MH, Shah JJ *et al.* (1984) Elemental carbon concentrations: estimation of an historical data base. *Atmos Environ* **18**: 153–162.

Cederlof R, Friberg L, Hrubec Z *et al.* (1975) *The Relationship of Smoking and Some Social Covariables to Mortality and Cancer Morbidity.* Stockholm, Sweden: Karolinska Institute, Department of

Environmental Hygiene.

Churg A (1994) Lung cancer cell type and occupational exposure. In: Samet JM (ed.) *Epidemiology of Lung Cancer*, Chapter 16. New York: Marcel Dekker Inc, pp 413–436.

Cohen AJ and Higgins MWP (1995) Health effects of diesel exhaust: epidemiology. *Diesel Exhaust: A Critical Analysis of Emissions, Exposure, and Health Effects*. A Special Report of the Institute's Working Group. Cambridge, MA: Health Effects Institute.

Coultas DB, Gong HJ, Grad R *et al.* (1994) Respiratory diseases in minorities of the United States. *Am J Respir Crit Care Med* **149**: S93–S131.

Council for Environmental Quality (CEQ) (1980) *Eleventh Annual Report of the Council on Environmental Quality*. Washington, DC: US Government Printing Office.

Council on Scientific Affairs (1990) The worldwide smoking epidemic. Tobacco trade, use, and control. *J Am Med Assoc* **90**(24): 3312–3318.

Daisey JM (1988) Chemical composition of inhalable particulate matter: seasonal and intersite comparisons. In: Lioy PJ and Daisy JM (eds) *Toxic Air Pollution – A Comprehensive Study of the Non-Criteria Air Pollutants*. Chelsea, Michigan: Lewis Publishers Inc.

Dean G (1966) Lung cancer and bronchitis in Northern Ireland, 1960–2. *Br Med J* **1**: 1506–1514.

Dean G, Lee PN, Todd GF *et al.* (1978) Report on a second retrospective mortality study in Northeast England. Parts 1 and 2. London: Tobacco Research Council.

Dockery DW, Pope CA III, Xu X *et al.* (1993) An association between air pollution and mortality in six US cities. *N Engl J Med* **329**(24): 1753–1759.

Doll R and Hill AB (1950) A study of the aetiology of carcinoma of the lung. *Br Med J* **2**: 740–748.

Doll R and Peto R (1981) *The Causes of Cancer*. Oxford: Oxford University Press.

Doll R, Vessey MP, Beasley RWR *et al.* (1972) Mortality of gasworkers. Final report of a prospective study. *Br J Ind Med* **29**: 394–406.

Dosemeci M, Washolder S and Lubin JH (1990) Does nondifferential misclassification of exposure always bias a true effect toward the null value? *Am J Epidemiol* **132**: 746–748.

Economou P, Lechner JF and Samet JM (1994) Familial and genetic factors in the pathogenesis of lung cancer In: Samet JM (ed.) *Epidemiology of Lung Cancer*, Chapter 14. New York: Marcel Dekker Inc, pp. 353–396.

Elliott P, Shaddick G, Kleinschmidt I *et al.* (1996) Cancer incidence near municipal solid waste incinerators in Great Britain. *Br J Cancer* **73**: 702–710.

Gilliland FD and Samet JM (1994) Incidence and mortality for lung cancer: geographic, histologic, and diagnostic trends. *Cancer Surv* **19**: 175–195.

Gilliland FD and Samet JM (1994) Trends in cancer incidence and mortality. *Lung Cancer*. London: Imperial Cancer Research Fund, 175 19.

Greenberg A (1988) Analyses of polycyclic aromatic hydrocarbons. In: Lioy PJ and Daisey JM (eds) *Toxic Air Pollution. A comprehensive study of the non-criteria air pollutants*. Chelsea, Michigan: Lewis Publishers Inc.

Greenland S (1980) The effect of misclassification in the presence of covariates. *Am J Epidemiol* **112**: 564–569.

Greenland S (1993) Basic problems in interaction assessment. *Environ Health Perspect* **101**(Suppl 4): 59–66.

Haenszel W and Taeuber KE (1964) Lung-cancer mortality as related to residence and smoking history: II. White females. *J Natl Cancer Inst* **32**: 803–838.

Haenszel W, Loveland DB and Sirken MG (1962) Lung-cancer mortality as related to residence and smoking history: I. White males. *J Natl Cancer Inst* **28**: 947–1001.

Hammond EC (1972) Smoking habits and air pollution in relation to lung cancer. In: Lee DHK (ed.) *Environmental Factors in Respiratory Disease*. New York: Academic Press, pp 177–198.

Hammond EC and Horn D (1958) Smoking and death rates report on forty-four months of follow-up of 187,783 men. Part I. Total mortality. Part II. Death rates by cause. *J Am Med Assoc* **166**: 1159–1172; 1294–1308.

Health Effects Institute (1991) Health Effects Institute, Asbestos Research Committee and Literature Review Panel. *Asbestos in Public and Commercial Buildings: A Literature Review and a Synthesis of Current Knowledge*. Cambridge, MA: Health Effects Institute.

Health Effects Institute (1993) *Research Priorities for Mobile Air Toxics*. Cambridge, MA: Health Effects Institute, HEI Communication Number 2.

Hemminki K, Dickey C, Karlsson S *et al.* (1997) Aromatic DNA adducts in foundry workers in relation to exposure, life style and CYP1A1 and glutathione transferase M1 genotype. *Carcinogenesis* **18**(2): 345–350.

Henderson BE, Gordon RJ, Menck H *et al.* (1975) Lung cancer and air pollution in Southcentral Los Angeles County. *Am J Epidemiol* **101**(6): 477–488.

Hitosugi M (1968) Epidemiological study of lung cancer with special reference to the effect of air pollution and smoking habit. *Bull Inst Public Health* **17**: 236–255.

IARC (International Agency for Research on Cancer) (1984) IARC Monographs on the Evaluation of Carcinogenic Risks to Humans, vol 34: *Polynuclear Aromatic Compounds*, Part 3. Industrial exposures in aluminum production, coal gasification, coke production, and iron and steel founding. Lyons, France: International Agency for Research on Cancer

IARC (International Agency for Research on Cancer) (1986) IARC Monographs on the Evaluation of the Carcinogenic Risk of Chemicals to Humans: *Tobacco Smoking*. Lyons, France: World Health Organization, IARC vol 38.

IARC (International Agency for Research on Cancer) (1987) IARC Monographs on the Evaluation of Carcinogenic Risks to Humans, Suppl 7: *Overall Evaluations of Carcinogenecity, an Updating of IARC Monographs*. Lyons, France: International Agency for Research on Cancer, pp 1–42.

IARC (International Agency for Research on Cancer) (1989) IARC Monographs on the Evaluation of Carcinogenic Risks to Humans, vol 46: *Diesel and Gasoline Engine Exhausts and Some Nitroarenes*. Lyons, France: International Agency for Research on Cancer.

IARC (International Agency for Research on Cancer) (1992) IARC Monographs on the Evaluation of Carcinogenic Risks to Humans, vol 54: *Occupational Exposures to Mists and Vapours from Strong Inorganic Acids; and Other Industrial Chemicals*. Lyons, France: International Agency for Research on Cancer

IARC (International Agency for Research on Cancer) (1995) IARC Monographs on the Evaluation of Carcinogenic Risks to Humans, vol 62: *Wood Dust and Formaldehyde*. Lyons, France: International Agency for Research on Cancer

IARC (International Agency for Research on Cancer) (1997) *Cancer incidence in five continents*. vol VII. Lyons: IARC Press, pp 143.

Jedrychowski W, Becher H, Wahhrendorf J and Basa-Cierpialek Z (1990) A case–control study of lung cancer with special reference to the effect of air pollution in Poland. *J Epidemiol Community Health* **44**: 114–120.

Karch NJ and Schneiderman MA (1981) *Explaining the Urban Factor in Lung Cancer Mortality*. A Report of the Natural Resources Defense Council. Washington, DC: Clement Associates Inc.

Katsouyanni K, Karakatsani A, Messari I *et al.* (1990) Air pollution and cause specific mortality in Athens. *J Epidemiol Community Health* **44**: 321–324.

Koutrakis P and Sioutas C (1996) Physico-chemical properties and measurement of ambient particles. In: Spengler J and Wilson R (eds) *Particles in Our Air, Concentrations in Health Effects*. Cambridge, MA: Harvard University Press, pp 15–39.

Lubin JH, Boice JD Jr and Samet JM (1995) Errors in exposure assessment, statistical power, and the interpretation of residential radon studies. *Radiat Res* **144**(3): 329–341.

Merlo F, Andreassen A, Weston A *et al.* (1998) Urinary excretion of 1-hydroxypyrene as a marker for exposure to urban air levels of polycyclic aromatic hydrocarbons. *Cancer Epidemiol, Biomarkers Prevent* **7**(2): 147–155.

Morgenstern H and Thomas D (1993) Principles of study design in environmental epidemiology. *Environ Health Perspect* **101**(S4): 23–38.

Motykiewicz G, Perera FP, Santella RM *et al.* (1996) Assessment of cancer hazard from environmental pollution in Silesia. *Toxicol Lett* **88**(1–3): 169–173.

National Research Council (1991) National Research Council (NRC) and Committee on Advances in Assessing Human Exposure to Airborne Pollutants. *Human Exposure Assessment for Airborne Pollutants: Advances and Opportunities*. Washington, DC: National Academy Press.

Natusch DF (1978) Potentially carcinogenic species emitted to the atmosphere by fossil-fueled power plants. *Environ Health Perspect* **22**: 79–90.

Navidi W, Thomas D, Stram D *et al.* (1993) *Design and Analysis of Multigroup Analytic Studies with Applications to the Study of Air Pollution*. Los Angeles, CA: U.S.C. School of Medicine, Department of Preventive Medicine, Division of Biostatistics, Technical Report #49.

Nielsen PS, de Pater N, Okkels H and Autrup H (1996) Environmental air pollution and DNA adducts in Copenhagen bus drivers – effect of GSTM1 and NAT2 genotypes on adduct levels. *Carcinogenesis* **17**(5): 1021–1027.

NRC (National Research Council) (1989) *Biologic Markers in Pulmonary Toxicology.* Washington, DC: National Academy Press.

NRC (1990) (National Research Council) and Committee on the Biological Effects of Ionizing Radiation. *Health Effects of Exposure to Low Levels of Ionizing Radiation: BEIR V.* Washington, DC: National Academy Press.

Perera FP, Hemminki K, Gryzbowska E *et al.* (1992) Molecular and genetic damage in humans from environmental pollution in Poland. *Nature* **360**(6401): 256–258.

Perera FP, Dickey C, Santella R *et al.* (1994) Carcinogen-DNA adducts and gene mutation in foundry workers with low-level exposure to polycyclic aromatic hydrocarbons. *Carcinogenesis* **15**(12): 2905–2910.

Perera FP, Mooney LA, Dickey CP *et al.* (1996) Molecular epidemiology in environmental carcinogenesis. *Environ Health Perspect* **104**(Supp 3): 441–443.

Pershagen G (1990) Air pollution and cancer. In: Vainio H, Sorsa M and McMichael AJ (eds) *Complex Mixtures and Cancer Risk.* Lyon, France: International Agency for Research on Cancer, pp 240–251.

Pershagen G, Elinder CG and Bolander AM (1977) Mortality in a region surrounding an arsenic smelting plant. *Environ Health Perspect* **19**: 133–137.

Pike MC, Jing JS, Rosario IP *et al.* (1979) Occupation: explanation of an apparent air pollution related localized excess of lung cancer in Los Angeles County. In: Breslow N and Whitemore A (eds) *Energy and Health.* Philadelphia, Pennsylvania: SIAM, pp 3–16.

Pope CA III, Thun MJ, Namboodiri MM *et al.* (1995) Particulate air pollution as a predictor of mortality in a prospective study of US adults. *Am J Respir Crit Care Med* **151**(3): 669–674.

Prentice RL and Sheppard L (1995) Aggregate data studies of disease risk factors. *Biometrika* **82**: 113–125.

Redmond CK (1983) Cancer mortality among coke oven workers. *Environ Health Perspect* **52**: 67–73.

Rothman KJ and Greenland S (1998) *Modern Epidemiology,* 2nd edn. Philadelphia: Lippincott-Raven.

Samet JM (1994) *Epidemiology of Lung Cancer.* New York: Marcel Dekker Inc.

Samet JM, Humble CG, Skipper BE and Pathak DR (1987) History of residence and lung cancer risk in New Mexico. *Am J Epidemiol* **125**: 800–811.

Saracci R and Boffetta P (1994) Interactions of tobacco smoking and other causes of lung cancer. In: Samet JM (ed.) *Epidemiology of Lung Cancer,* Chapter 18. New York: Marcel Dekker Inc, pp 465–493.

Schulte PA and Perera FP (1993) In: Schulte PA and Perera FP (eds) *Molecular Epidemiology: Principles and Practices.* New York: Academic Press.

Shy CM, Goldsmith JR, Hackney JD *et al.* (1978) Health effects of air pollution. *ATS News* **6**: 1–63.

Smith KR (1988) Air pollution, assessing total exposure in developing countries. *Environment* **30**: 16.

Smith KR and Liu Y (1994) Indoor air pollution in developing countries. In: Samet JM (ed.) *Epidemiology of Lung Cancer.* New York: Marcel Dekker Inc, pp 151–184.

Speizer FE and Samet JM (1994) Air pollution and lung cancer. In: Samet JM (ed.) *Epidemiology of Lung Cancer,* Chapter 6. New York: Marcel Dekker Inc, pp 131–150.

Stocks P and Campbell JM (1955) Lung cancer death rates among non-smokers and pipe and cigarette smokers. Evaluation in relation to air pollution by benzo[*a*]pyrene and other substances. *Br Med J* **2**: 923–929.

Tenkanen L, Hakulinen T and Teppo L (1987) The joint effect of smoking and respiratory symptoms on risk of lung cancer. *Int J Epidemiol* **87**(4): 509–515.

Tockman MS (1994) Other host factors and lung cancer susceptibility. In: Samet JM (ed.) *Epidemiology of Lung Cancer,* Chapter 15. New York: Marcel Dekker Inc, pp 397–412.

Tomatis L (1990) *Air Pollution and Human Cancer,* European School of Oncology Monographs. Berlin: Springer-Verlag.

US Environmental Protection Agency (1990) *Cancer Risk from Outdoor Exposure to Air Toxics,* vol 1, Final Report. Research Triangle Park, NC: US Government Printing Office, EPA-450/1-90-004a.

US Environmental Protection Agency, Office of Research and Development, and Office of Air and

Radiation (1993) Respiratory health effects of passive smoking: lung cancer and other disorders. Washington, DC: US Government Printing Office, Monograph 4.

USDHHS (US Department of Health and Human Services) (1989) A Report of the Surgeon General: *Reducing the Health Consequences of Smoking. 25 Years of Progress.* Washington, DC: US Government Printing Office.

USDHHS (US Department of Health and Human Services) (1995) *SEER Cancer Statistics Review 1973–1990.* Bethesda, MD: National Institutes of Health, 93-2789.

USDHHS (US Department of Health and Human Services) Public Health Service, and National Cancer Institute (NCI)(1997) Changes in cigarette-related disease risks and their implication for prevention and control. In: Burns DM, Garfinkel L and Samet JM (eds) *Smoking and Tobacco Control.* Bethesda, MD: U.S. Government Printing Office, NIH Publication No 97-4213.

Vena JE (1982) Air pollution as a risk factor in lung cancer. *Am J Epidemiol* **116**: 42–56.

Vineis P and Caporaso N (1995) Tobacco and cancer: Epidemiology and the laboratory. *Environ Health Perspect* **103**: 156–160.

Walker AM (1981) Proportion of disease attributable to the combined effect of two factors. *Int J Epidemiol* **10**: 81–85.

White C (1990) Research on smoking and lung cancer: a landmark in the history of chronic disease epidemiology. *Yale J Biol Med* **63**: 29–46.

Winer AM and Busby WF Jr (1995) Atmospheric transport and transformation of diesel emissions. In: *Health Effects of Diesel Engine Emissions: Characterization and Critical Analysis.* Cambridge, MA: The Health Effects Institute.

World Health Organization (1988) *Emissions of Heavy Metal and PAH Compounds from Municipal Solid Waste Incinerators: Control Technology and Health Effects.* Report on a WHO meeting, Florence. Copenhagen: WHO.

World Health Organization (1990) *Global Estimates for Health Situation Assessment and Projections.* Geneva: WHO, HST/90.2.

Controlled Exposures of Asthmatics to Air Pollutants

DAVID B. PEDEN

Center for Environmental Medicine and Lung Biology, University of North Carolina School of Medicine, Chapel Hill, NC, USA

INTRODUCTION

Examination of the effect of air pollutants in humans, together with the development of appropriate environmental interventions, is analogous to the steps involved in the development of new drugs. Epidemiologic techniques define the scope of the problem and identify potential candidate pollutants. *In vitro* cell culture techniques and *in vivo* animal experiments are important in determining the biological mechanisms by which pollutants exert their biological effects. However, to appreciate fully the mechanisms by which pollutants affect humans, controlled exposure experiments (analogous to clinical trials of new drugs) are needed. As human exposures are completed, new information is available for epidemiological studies, thus beginning a new cycle of investigation.

According to recent estimates, up to 5–10% of the US population suffer from asthma. Marked increases in asthma death and morbidity have been observed in the USA and Europe in recent years (Crain *et al.*, 1994; Weiss *et al.*, 1993). There is substantial epidemiologic evidence demonstrating that exposure to increased levels of ambient air pollutants is associated with these increases in asthma morbidity. Among the pollutants of interest are ozone, sulfur dioxide, nitrogen dioxide and respirable particulate matter (Anon, 1996; Diaz-Sanchez *et al.*, 1997; Jorres *et al.*, 1996; Koren and Bromberg, 1995; Peden 1997; Peters *et al.*, 1997; Peterson and Saxon, 1996; Rusnak *et al.*, 1996; Timonew and Pekkanen, 1997). Because asthma is characterized by increased airway reactivity, eosinophilic inflammation and association with allergy, information regarding the effect of pollutants on these symptoms is of great interest. This chapter reviews the controlled exposure studies of asthmatics to pollutants, examining the effect of these pollutants on lung function, airway responsiveness, airway inflammation and response to allergen challenge.

GENERAL METHODOLOGY

In general, controlled exposure studies of airborne pollutants are performed in a facility with an exposure chamber in which not only the level of the pollutant of interest is controlled, but also other factors, including temperature, relative humidity, light and sound levels. The actual dose of the pollutant depends on the concentration of the pollutant, the duration of exposure and the ventilation rate of the subject – which is usually manipulated by exercise (Anon, 1996; Koren and Bromberg, 1995; Peden, 1997).

Exercise also is important in determining whether nasal or oronasal breathing occurs. This may be important since the nose may alter delivery of the pollutant to the lower airway. It is hypothetically possible that nasal deposition of the study pollutant may be decreased with oronasal breathing, which is of concern if the nasal mucosa is the biological target of interest. Finally, when studying asthmatics it is necessary to account for the possibility of exercise-induced bronchospasm. Thus, most studies do not depend solely on comparison of pre- and post-exposure measurement of the study endpoints, but also include a clean air control exposure.

Additional methods of examining the effect of pollutants may include instilling suspensions of candidate particles into the nasal or bronchial airway. Such studies are especially appropriate when the components of particulate matter are of interest and they usually concentrate on inflammatory end-points. Finally, the effect of particulates can be assessed by placing them in suspension for nebulization, in a fashion analogous to inhaled allergen challenge.

INVESTIGATIONS OF SPECIFIC POLLUTANTS

Gaseous Pollutants

Sulfur Dioxide

Sulfur dioxide is a potent inducer of bronchospasm in asthmatics. While controlled exposure studies do not reveal an effect of SO_2 on pulmonary function of normal subjects at levels as high as 0.6 ppm, several investigators have demonstrated bronchoconstriction in asthmatics at concentrations as low as 0.25 ppm (Anon, 1996; Boushey, 1982; Folinsbee et al., 1985; Hackney et al., 1984; Horstman et al., 1986; Kehrl et al., 1987; Molfino, 1992; Witek and Schachter, 1985). This gas exerts a rapid effect on the lung function of asthmatics, with initial bronchoconstrictive responses beginning at 2 min and becoming maximal in 5–10 min. Spontaneous recovery occurs as early as 30 min after challenge, with a refractory period of up to 4 h. Acute tachyphylaxis to SO_2 can be induced by repeated exposure to low levels of the pollutant. In one study, repeated exercise during an

SO_2 exposure resulted in diminished airway responsiveness to SO_2. However, initial responsiveness returned after 6 h exposure (Gong, 1992; Horstman et al., 1988; Pierson and Koenig, 1992; Roger et al., 1985).

Sulfur dioxide induces decreases in forced expiratory volume in 1 s (FEV_1) and increases in airway resistance. FEV_1 can drop as much as 60% after exposure to SO_2. There is variability in the response of asthmatics to SO_2, with some experiencing bronchoconstriction at levels as low as 0.25 ppm and others showing no response at levels as high as 2.0 ppm. However, levels above 0.50 ppm will typically cause airway constriction in asthmatics. Exercise enhances SO_2-induced bronchoconstriction in asthmatics. The combined effect of exercise and SO_2 appears to be additive. Symptoms associated with SO_2 inhalation in asthmatics include wheezing, chest discomfort and dyspnea.

Interestingly, nasal breathing largely cancels the effect of SO_2 on pulmonary function. This may be due to absorption of the water-soluble gas by the nasal mucosa. Because most people shift from strictly nasal breathing to oronasal respiration with moderate exercise, increased levels of SO_2 may reach the bronchial airway in that fraction of inspired air which does not interact with the nasal mucosa during exercise (40% with heavy exercise). This may explain how exercise enhances the effect of SO_2. Asthmatics also are more likely to have nasal pathology as well (such as allergic rhinitis or sinusitis), which may further decrease nasal airflow during exercise and perhaps increase the percentage of inspired air that is not exposed to nasal tissue (Linn et al., 1983).

A number of pharmacologic agents attenuate the effect of SO_2 in asthmatics. These include β-agonists, anti-muscarinic agents, and cromolyn sodium and necdromil. β-agonists are the most effective at mitigating SO_2-induced bronchospasm. In contrast, chronic use of methyl xanthines and inhaled corticosteroids as well as acute use of antihistamines does not affect the sensitivity of asthmatics to SO_2. Many investigators currently believe that SO_2 acts as a non-specific bronchoconstricting agent, analogous to the action of inhaled histamine and methacholine in asthmatics (Bigby and Boushey, 1993; Heath et al., 1994).

When coupled with experiments demonstrating development of tolerance to SO_2, the rapid onset of bronchoconstriction after exposure, and increased sensitivity of asthmatics to SO_2 when compared to non-asthmatics, these pharmacologic studies suggest a cholinergically mediated neural mechanism of action for SO_2. However, recent observations by Sandstrom et al. (1989) demonstrate that very mild airway inflammation also may be induced by SO_2 exposure. Thus, the mechanisms by which SO_2 may influence asthma are not yet fully understood.

Sulfur dioxide exposure also augments responses to other environmental agents which exacerbate bronchospasm. Exposure to ozone prior to exposure to SO_2 increases bronchial sensitivity to SO_2 in asthmatics (Koenig et al., 1990). Exposure to cold dry air also exacerbates the bronchospastic effect of SO_2 in asthmatics (Pierson and Koenig, 1992). However, unlike ozone, SO_2 exposure has not been shown to excerbate allergen-induced bronchospasm. The question of whether there is a relationship between sensitivity to SO_2 and asthma severity has not been resolved, although some studies suggest that SO_2 causes similar effects in mild and moderate asthmatics.

Nitrogen Dioxide

NO_2 has not been consistently shown to alter airway function of either normal subjects or asthmatics. Conflicting reports of the effect of NO_2 on non-specific reactivity in asthmatics can be cited, with some showing that low levels of NO_2 enhance bronchial responses to cholinergic or cold air challenge while others find no such effect (Anto and Sunyer, 1995; Bauer *et al.*, 1986; Linn *et al.*, 1985; Pierson and Koenig, 1992). However, as with ozone, higher levels of NO_2 (4.0 ppm) appear to alter airway function in asthmatics.

NO_2 does appear to have an effect on airway inflammation. In normal subjects, NO_2 was found to induce airway PMN influx (Sandstrom *et al.*, 1991). NO_2 also has been shown to increase proinflammatory cytokine production in epithelial cells after *in vitro* pollutant exposure (Devalia *et al.*, 1993). Overall, these data suggest that NO_2 could influence airway function of asthmatics by increasing airway inflammation.

In a recent study of the effect of exposure to 0.4 ppm NO_2 for 6 h on allergen-induced nasal inflammation in asthmatics, NO_2 alone was not associated with mast cell degranulation, eosinophil activation, neutrophil activation or increases in interleukin-8 (IL-8). When NO_2 was followed by allergen challenge, ECP (but not mast cell tryptase) increased in the nasal lavage fluid, indicating that this gas may alter eosinophil function without having any effect on mast cell degranulation (Wang *et al.*, 1995).

NO_2 also has been shown to have an effect on allergen-induced bronchospasm. Tunnicliffe and colleagues observed that 1 h exposure to 0.4 ppm NO_2 did result in a small but significant decrease in FEV_1, observed with a fixed dose of allergen when compared with allergen challenge alone (Tunnicliffe *et al.*, 1994). It also has been shown that exposure to a combination of 0.2 ppm SO_2 and 0.4 ppm NO_2 for 6 h does enhance immediate sensitivity of mild asthmatics to inhaled allergen, as shown by a significant drop in allergen-related PD20. This was not observed with either gas alone (Rusnak *et al.*, 1996). Exposure to NO_2 also has been reported to enhance late phase responses of asthmatics to inhaled allergen (Strand *et al.*, 1997). Taken together, these studies demonstrate that NO_2 is able to alter both allergen and non-specific airway responses in asthmatics.

Ozone

Ozone is perhaps the most extensively examined pollutant in regard to its effect on airway function and inflammation in asthmatics. Epidemiologic studies demonstrate that increases in ambient ozone are associated with an exacerbation of asthma, with most studies showing that asthma morbidity is correlated with increased ambient ozone during the previous day (Balmes, 1993; Strand *et al.*, 1997). Insight into the possible mechanisms by which ozone alters airway function in asthma has come from comparison of controlled exposure studies of normal subjects and asthmatics.

In normal subjects, ozone has two principal effects. One is a reproducible decrease of both forced vital capacity (FVC) and FEV_1 (with increased non-specific bronchial responsiveness and substernal discomfort when taking a deep breath) (Anon, 1996; Bates and Hazucha, 1973; Bates and Sizto, 1983; Koren and Bromberg, 1995; McDonnell *et al.*, 1983; Peden, 1997). The second is the development of neutrophilic inflammation as quickly as 1 h after exposure (Anon, 1996; Koren *et al.*, 1991; Peden, 1997). Interestingly, these two effects are not correlated with each other, suggesting that separate mechanisms mediate these changes.

Ozone and Lung Mechanics   The effect of ozone on lung mechanics is dependent on the concentration of ozone, duration of exposure and the level of exercise (with corresponding increases in minute ventilation) (McDonnel *et al.*, 1983). Indeed, exposures to ozone *without* exercise generally reveal no effect on lung function at levels below 0.50 ppm (Anon, 1996; Bates and Hazucha, 1973; Folinsbee *et al.*, 1994; McDonnell *et al.*, 1983 1991). With exercise, ozone induces increases in respiratory frequency and the level of ozone required to cause decreases in FEV_1, FVC and airway resistance can be much lower (McDonnell *et al.*, 1991).

Examination of the action of a variety of pharmacologic agents on the effect of ozone on respiratory mechanics has given some insight into the mechanisms of this effect. Atropine inhibits ozone-related decreases in airway resistance (though not spirometry), indicating that vagal mechanisms are involved in the respiratory response to ozone (Beckett *et al.*, 1985). Ozone also could directly increase smooth muscle sensitivity to acetylcholine, consistent with the observation that sensitivity to inhaled methacholine is increased after ozone exposure (Beckett *et al.*, 1985; Holtzman *et al.*, 1983).

Ozone also effects eicosanoid responses in the airway, thus altering airway mechanics. Levels of PGE_2 in bronchoalveolar lavage (BAL) recovered from normal subjects after O_3 exposure correlate with observed lung function decrements (Koren *et al.*, 1991). Additionally, several laboratories have shown that cyclooxygenase inhibitors, such as ibuprofen and indomethacin can inhibit ozone induced decreases in spirometry, although they have little effect on inflammatory responses to ozone (Eschenbacher *et al.*, 1989; Hazucha *et al.*, 1996; Schelegle *et al.*, 1987; Ying *et al.*, 1990). Airway C fibers can be stimulated by PGE_2, suggesting a mechanism by which ozone-related increases in PGE_2 also may account for changes in lung function due to ozone (Coleridge *et al.*, 1976, 1993). Animal studies in anesthetized dogs also suggest that ozone may act via C fiber mediated mechanisms in the lung (Coleridge *et al.*, 1993). This finding is further supported by recent observations in normal subjects that ozone exposure causes depletion of substance P (but not CGRP) in airway neurons. Levels of substance P and PGE_2 in BAL fluid also are correlated with prior ozone exposure (Krishna *et al.*, 1997).

Pain responses also may play a role in the transient restrictive defect caused by ozone. When applied to the upper airway, lidocaine exerts a partial inhibition of ozone-induced decreases in spirometry, suggesting that it inhibits irritant receptors in the upper airway which blunts an ozone-induced reflex-mediated decrease in inspiratory effort (reviewed in Folinsbee, 1993). There also is more direct evidence that activation of opiate receptors may blunt ozone-induced respiratory symptoms, again indicating a pain response to this stimulant. This is supported by observations that sufentanyl (a short acting narcotic) rapidly normalizes lung function in subjects found to have decreases in FVC and FEV_1 after ozone exposure (Passannante *et al.*, 1995). Pain responses which are sensitive to opiates thus may partially account for the pulmonary response to ozone.

Interestingly, β-agonists have little influence on the immediate effect of ozone on lung function. While there is one report that β-agonists can block some ozone-induced decreases in lung function, most studies indicate that β-agonists have no effect on the deleterious effect of ozone on lung function (Folinsbee, 1993; Gong *et al.*, 1988). Thus, it is unlikely that β-adrenergic tone plays a substantial role in the response of humans to ozone exposure.

Early studies comparing lung function of normal subjects with that of mild asthmatics found that asthmatics were not more sensitive to ozone following low level exposure

(0.25 ppm or less with little or no exercise; Folinsbee, 1993). However, other investigators have reported increased sensitivity of asthmatics to ozone. Kriet and co-workers (1989) demonstrated that following exposure to 0.4 ppm ozone for 2 h with rigorous exercise, asthmatics had greater increases than normal subjects in both changes in airway resistance and in non-specific airway hyperresponsiveness. Silverman *et al.* (1979) also demonstrated that exposure to 0.25 ppm ozone for 2 h resulted in decreased pulmonary function in asthmatics. Other studies indicate that ozone not only causes the well-described decrease in FVC, but also causes true bronchospasm in asthmatics (Horstman *et al.*, 1984, 1995; Kreit *et al.*, 1989). Thus, ozone has been demonstrated to worsen airflow in asthma directly.

However, recent studies (Fernandes *et al.*, 1994; Weymer *et al.*, 1994) have revealed that persons with exercise-induced bronchospasm (EIB) show no effect on subsequent EIB when ozone exposure is followed by an exercise challenge (without ozone). Weymer and colleagues (1994) exposed mild asthmatics with EIB to 0.0, 0.1, 0.25 and 0.40 ppm ozone for 1 h with mild exercise (VE of 20–30 l/min), followed an hour later with an exercise challenge if their FEV_1 was back to 90% of pre-exposure baseline. While ozone exposure itself had a dose-dependent effect on both spirometry and subjective symptom scores (a similar response to that described for normal subjects), subsequent bronchoconstrictive responses were not effected by ozone. A separate study examining the effect of exposure to 0.12 ppm ozone at rest for 1 h in asthmatics with EIB also revealed no exacerbation of EIB attributable to ozone (Fernandes *et al.*, 1994).

Ozone and Airway Inflammation Ozone induces neutrophilic influx into the airway in normal subjects, as revealed in analysis of BAL fluid and in bronchial mucosal biopsies following exposures ranging from 0.10 to 0.4 ppm. This neutrophil influx appears to be maximal between 1 and 6 h after exposure and persists to 18 h (Folinsbee, 1993; Holtzman *et al.*, 1983; Koren *et al.*, 1991). Inflammatory mediators such as IL-6, IL-8, PGE_2, LTB_4, TXB_2, fibronectin, plasminogen activator and elastase also are increased by ozone (Anon, 1996; Koren *et al.*, 1991). Direct comparison of the time course of mediator response to ozone indicates that changes in IL-6 and PGE_2 peak 1 h after exposure, whereas fibronectin and plasminogen activator are higher 18 h after exposure (Koren *et al.*, 1991).

The epithelium is thought to be very important in airway response to ozone. Epithelial cell lines exposed to ozone generate the eicosanoids PGE_2, TXB_2, PGF_{2a}, and the proinflammatory cytokines IL-6 and IL-8, as well as the proinflammatory protein fibronectin (McKinnon *et al.*, 1992, 1993). Nasal epithelial cells placed into culture respond to high level (0.5 ppm) ozone by generating ICAM-1, the major adhesion molecule for neutrophils and eosinophils in the airway, as well as the cytokines IL-1, IL-6 and TNF-α (Beck *et al.*, 1994). Thus, the effect of ozone on epithelial cells probably plays a major role in the inflammatory response to this pollutant, whereas neural responses (C fiber, pain fibers and possibly cholinergic fibers) probably exert the major effect on lung function changes.

With the exception of PGE_2, none of the markers of inflammation noted above correlate with immediate changes in the lung function of normal subjects (Anon, 1996; Folinsbee, 1993; Peden, 1997). Nor has an association been made between ozone-induced inflammation and airway responsiveness in normal subjects. Thus, the clinical significance of ozone-induced inflammation in normal subjects remains unclear. It has been suggested that this proinflammatory effect of ozone could have a greater significance in asthma, a disorder characterized in part by an eosinophilic inflammation of the lower airways (Ball *et al.*, 1996; Molfino *et al.*, 1991).

In persons with allergic rhinitis or asthma, ozone exposure appears to induce both eosinophil and PMN influx. This was initially observed by Bascom and colleagues, who reported an increase in both neutrophils and eosinophils in nasal lavage fluid obtained from allergic rhinitics after a 4-h exposure to 0.5 ppm ozone (Bascom *et al.*, 1990). This was later shown to be true in mite-sensitive asthmatics as well, following exposure to 0.4 ppm ozone for 2 h. In the latter study, ozone also caused increases in eosinophil cationic protein, suggesting that eosinophils may have been activated (Peden *et al.*, 1995).

Studies of the effect of ozone on bronchial inflammation in asthmatics have yielded conflicting results. Two studies comparing the effect of similar ozone exposures on normal and asthmatic subjects report that asthmatics have greater PMN influx than normal subjects, with neither group manifesting an eosinophilic response. In the first of these studies, Basha *et al.* (1994) reported that five asthmatics had a greater PMN and IL-8 response to ozone when compared to the response of five normal subjects. Scannell *et al.* (1996) also performed a similar comparison with 19 asthmatic and 20 non-asthmatic volunteers employing an exposure regimen of 0.2 ppm ozone for 2 h with moderate exercise. Lavage fluid was collected 18 h later with using three different techniques: lavage from an isolated proximal airway; analysis of the first 10 ml of traditional BAL (labeled the bronchial fraction); and traditional BAL fluid. Asthmatics had increased levels of PMNs, LDH, total protein, MPO, fibronectin and IL-8 with each of the sampling techniques following exposure. In the BAL fluid, asthmatics also had a greater PMN response than non-asthmatics, with a trend for increased IL-8 levels as well. Thus, these two studies suggest that asthmatics may be more sensitive to neutrophilic inflammatory responses to ozone, with a notable absence of eosinophils.

In contrast, an examination of eight mite-sensitive asthmatics exposed to 0.16 ppm ozone for 8 h revealed a substantial increase in both PMNs and eosinophils in BAL fluid when compared with samples collected following exposure to clean filtered air. Bronchial samplings included traditional BAL fluid as well as separate examination of the first 20 ml recovered (the bronchial fraction), with the bronchial fraction yielding more striking differences between air and ozone exposure in asthmatics (Peden *et al.*, 1997). Consistent with this are preliminary data reported by Frampton and colleagues (1997), in which a two-fold increase in eosinophils was observed in asthmatics following ozone exposure. Thus, in these two groups of asthmatics, ozone caused both neutrophil and eosinophil influx to the airway. Taken together, these four studies indicate that asthmatics react differently to ozone than do normal subjects.

Interaction of Ozone and Allergen Challenge   The defining feature of allergic subjects is their propensity to respond to inhaled allergen. Any observed effect of a pollutant on airway responsiveness to allergen therefore represents a unique risk to this segment of the population. Exposure to ozone has been found to enhance both immediate and late-phase responses to inhaled allergen. The first indication that exposure to ozone altered reactivity to inhaled allergen came from Molfino *et al.* (1991), who observed decreased PD20 values to inhaled allergen from seven grass-sensitive subjects challenged immediately after exposure to 0.12 ppm ozone for one hour at rest. A similar study of 12 subjects employing this exposure protocol did not find a significant effect and also did not reveal enhancement of bronchial reactivity to allergen (Ball *et al.*, 1996).

However, exposure to 0.25 ppm ozone for 3 h with moderate exercise did enhance bronchial reactivity of both allergic asthmatics and allergic rhinitics to inhaled allergen when the allergen challenge occurred shortly after ozone exposure (Jorres *et al.*, 1996). Preliminary data from our group also suggest that exposure to 0.16 ppm ozone for 8 h with moderate exercise increased the sensitivity of mite-sensitive asthmatics when inhaled mite allergen was given 18 h after the exposure was complete.

Late-phase response to allergen also is altered by ozone. In nasal studies of allergic asthmatics, 0.4 ppm ozone was found to enhance allergen-induced eosinophil influx, ECP levels and IL-8 levels (Peden *et al.*, 1995). One study in which exposure to 0.2 ppm ozone for 3 h was followed by inhaled allergen challenge did not show any effect of allergen-associated levels of eosinophils, neutrophils, ECP or IL-8 recovered in nasal fluid. Taken together, these studies suggest that the effect of ozone on inhaled allergen challenge is dose related. It is important to note that all of these studies were performed with mild asthmatics. It is possible that persons with more severe disease may be more sensitive to the effect of ozone on both immediate and late-phase responses to allergen.

Particulate Pollutants

Particulate matter less than 10 μm in size also plays a significant role in asthma exacerbation (Pope, 1991; Pope *et al.*, 1995; Schwartz and Dockery, 1992a,b; Schwartz *et al.*, 1993). There is a vast array of particles which have been identified and shown to have biological effects in animal and *in vitro* studies. Active agents in particles include silica, metal ions (such as iron, vanadium, nickel and copper), organic residues such as poly-aromatic hydrocarbons found on diesel exhaust particles, acid aerosols and biological contaminants such as endotoxin (Anon, 1996). To date, few human exposure data are available for controlled exposures to metal-containing particulates. However, a number of human exposure studies have examined the effect of diesel exhaust particles, acid aerosols and endotoxins on airway function or inflammation. We turn now to that evidence.

Diesel Exhaust Particles

Numerous animal and *in vitro* studies have demonstrated that diesel exhaust particles enhance allergic inflammation and the airway hyperresponsiveness associated with allergen exposure (Muranaka *et al.*, 1986; Takafuji *et al.*, 1987, 1989a,b; Takano *et al.*, 1997). Specifically, this includes increasing IgE production and enhancing cytokines involved in eosinophilic or allergic inflammation, especially IL-4, IL-5 and GM-CSF. Perhaps most interesting is the effect of DEP on enhancing B lymphocyte immunoglobuln isotype switching to IgE (Diaz-Sanchez *et al.*, 1994). Further examination of DEP suggests that polyaromatic hydrocarbon residues on DEPs are largely responsible for this effect on allergic inflammation (Takano *et al.*, 1995) .

In humans, challenge studies examining the effect of DEP on allergic inflammation have typically employed a nasal challenge model. The initial report of the effect of DEP on IgE production in humans was offered by Diaz-Sanchez, Saxon and colleagues (1994), who reported that exposure of human volunteers (four atopic and seven non-atopic) to DEP yields remarkable increases in nasal IgE production 4 days after DEP challenge, without any effect on IgG, IgA or IgM production. There were shifts in the ratio of the five isoforms of IgE during this challenge as well. These findings were reported with doses of 0.3 mg DEP. No such effect was observed with doses of 0.1 or 1.0 mg.

The same investigators have reported that DEP challenge of the nasal mucosa also results in increased cytokine production by cells recovered in lavage fluid. In this study, subjects underwent lavage before and after challenge with 0.3 mg of DEP. The recovered cells were then processed and analyzed for mRNA analysis. Cells recovered before challenge were generally found to have detectable mRNA levels for γ-interferon, IL-2 and IL-13. In contrast, DEP challenge yielded cells which had detectable levels of IL-2, IL-4, IL-5, IL-6, IL-10, IL-13 and γ-interferon. IL-4 protein also was measured in post-challenge lavage (Diaz-Sanchez et al., 1997). While it is unclear which types of cell were present in lavage fluid before or after challenge, it was not thought to be due to increased lymphocyte number. Overall, regardless of whether the increased cytokine expression is due to activation of resident cells or recruitment of new cells to the nasal mucosa, DEPs do appear to enhance TH$_2$ type inflammation.

The effect of coupling DEP with ragweed allergen was compared with nasal challenge with ragweed alone on allergen-specific IgE and IgG in ragweed-sensitive subjects (Diaz-Sanchez et al., 1997). DEP plus allergen resulted in a significant increase in allergen-specific IgE and IgG$_4$ without an increase in total IgE and IgG. When compared with DEP alone, DEP plus allergen was associated with increased expression of IL-4, -5, IL-6, IL-10, IL-13 and decreased expression of γ-interferon and IL-2.

Compared with pollutants such as ozone or metal ions, DEP appears to be unique in its effect on IgE production. The *in vivo* effect of DEP on IgE isotype switch can be replicated *in vitro* in isolated B lymphocytes with extracts from DEP which contain the polyaromatic hydrocarbon (PAH) fraction from these particles, as well as the specfic PAH compounds phenanthrene and 2,3,7,8-tetrachlorodibenzo-*p*-dioxin (Diaz-Sanchez et al., 1994, 1997; Peterson and Saxon, 1996; Takenaka et al., 1995). Polyaromatic hydrocarbons, through their action on B lymphocytes, thus appear to play a central role in the effect of diesel exhaust on allergic inflammation.

Salvi and colleagues (Salvi et al., 1997) have recently presented preliminary evidence suggesting that chamber exposure to diluted diesel exhaust (generated by a standardized engine) induces airway responses in non-asthmatics. Perhaps the most interesting observation by this group was an increase in mast cell numbers associated with diesel exposure. Because diesel exhaust is a complicated mixture containing DEPs of varying sizes, NO$_2$, and CO (among other components) it is not possible to attribute the effect of this exhaust directly to DEP. However, these observations are consistent with the nasal experiments cited above.

Acid Aerosols

Acid aerosols also have been examined for their effect on lung function. Numerous studies in non-asthmatic adults show that acute exposure to H$_2$SO$_4$ up to 2000 μg/m^3 for 1 h does not alter lung function (Aris et al., 1991; Folinsbee, 1993; Koenig et al., 1991; Utell et al., 1983). In contrast, similar studies of asthmatics do suggest a bronchoconstrictive effect of such exposure in this population (Koenig et al., 1983, 1989). This may be particularly true for adolescents with asthma, in whom lung function decrements were observed at acid levels as low as 100 mg/m^3. Adolescents thus appear to be more sensitive than adult asthmatics to acid aerosol exposure (Koenig et al., 1989). However, substantial controversy persists regarding the significance of acid aerosol exposure in asthma. To summarize, it appears that the effect of inhaled aerosolized acid on lung function in asthmatics roughly correlates with baseline levels of airway hyperresponsiveness.

Combinations of acid aerosols and ozone may be more deleterious than the effect of acid alone. One report revealed mildly enhanced decreases in lung function and increased airway reactivity in asthmatics exposed to 0.12 ppm ozone and 100 mg/m^3 H$_2$SO$_4$, when compared with the effect of ozone alone. H$_2$SO$_4$ alone had no effect on the lung function of these subjects (Koenig et al., 1990). Exposures occurred over 6.5 h with exercise, so that ventilation was 29 l/min. However, other studies examining the effect of combined exposures of ozone, H$_2$SO$_4$, or HNO$_3$ have revealed no effect of acid aerosols on the lung function of asthmatics, either alone or in combination with ozone (Anon, 1996).

Endotoxin

Atopy alone does not appear to contribute to the risk of pulmonary disease associated with endotoxin. Recent studies suggest that airborne endotoxin may be an important contributor to asthma morbidity (Blaski et al., 1996). The strongest evidence to date that LPS contributes to asthma severity is contained in a recent report which demonstrates that in house dust mite-sensitive asthmatics, disease severity correlates with dual exposure to LPS and mite allergen as measured in recovered house dust, but not with mite allergen or LPS alone (Michel et al., 1996).

Michel and colleagues also have performed a series of challenge studies examining the role of endotoxin in asthma pathogenesis (Michel et al., 1992a, 1994, 1996). In mild asthmatics, a 7% decrease in FEV$_1$ was observed 30 min after inhalation of 22 222 ng of the lipid moiety of endotoxin. Five hours after the challenge, the FEV$_1$ was only partially recovered and there was increased non-specific bronchial reactivity (NSBR) to histamine. A subsequent study of 16 asthmatics (eight of whom were atopic) confirmed that endotoxin challenge resulted in bronchospasm and increased NSBR to histamine. These effects correlated with baseline histamine NSBR, but not with state of atopy or history of tobacco use (Crain et al., 1994). A third study, which entailed inhalation of 20 000 ng of LPS, resulted in a decrease in the FEV$_1$ 45 min after challenge (which persisted for 5 h). These investigators also observed increases in peripheral blood TNF-α after 60 min, histamine NSBR after 5 h, and peripheral blood neutrophils at 6 h post-LPS challenge. These results all suggest that respiratory responses to LPS involve an inflammatory process.

Inhaled LPS is clearly able to cause neutrophilic inflammation in both atopic and non-atopic subjects and is a key feature of many occupational lung diseases such as byssinosis (Koenig et al., 1989; Michel et al., 1992b). The apparent interaction between LPS and allergen exposure reported by Michel et al. (1996) indicates that endotoxin can modulate allergic-related airway inflammation. However, it remains unclear whether LPS has an effect on the eosinophilic inflammation characteristic of asthma.

Hunt and colleagues reported that endotoxin-contaminated allergen solutions caused a marked neutrophilic inflammatory response when instilled into a segment of the lower airway. This effect was observed on follow-up bronchoscopy 18 h later (Hunt et al., 1994). This contrasts with the more typical allergen-induced inflammation, characterized by influx of both neutrophils (PMNs) and eosinophils, that was observed when endotoxin-free allergen was used. These data suggest that LPS dominates rather than enhances allergen-induced inflammation in the airway. However, animal studies indicate that LPS does enhance eosinophilic inflammation.

Macari and colleagues (1996) have reported that while 30 ng of LPS injected into the skin of guinea pigs did not directly induce eosinophil influx, there was a three-fold increase in the eosinophil response to cutaneously applied allergen as well as leukotriene B4, PAF and C5a des arg, without an enhancement of PMN influx. Doses of LPS ranging from 50 to 1000 ng of LPS do cause a dose-dependent eosinophil influx in guinea pig skin 4 h after injection (Weg *et al.*, 1995). This activity may be mediated by *de novo* generation of TNF-α, since both actinomycin D (which inhibits protein synthesis) and infused soluble TNF-α fusion protein (which binds TNF-α) blunts LPS-induced eosinophil influx (Henriques *et al.*, 1996).

Similarly, Henriques *et al.* (1996), using a murine pleurisy model, have reported that LPS causes a significant influx of PMNs after 4 h and eosinophils after 24 h. LPS also has been reported to enhance eosinophil survival as well as cytokine release from these cells. The eosinophil survival effect was inhibited by anti GM-CSF antisera. Peripheral human blood monocytes stimulated with LPS also yield a supernatant that enhances eosinophil degranulation and survival *in vitro* (Kita *et al.*, 1995). Comparison of the LPS-induced cytokine responses of peripheral blood monocytes and alveolar macrophages recovered from asthmatics and normal subjects demonstrate increased cytokine responses in cells from the asthmatics (Hallsworth *et al.*, 1994). Likewise, depletion of lymphocytes with appropriate monoclonal antibody also blunts LPS-induced pleural eosinophilia in a rat model (Bozza *et al.*, 1994a,b). Thus, LPS appears to have a proinflammatory effect on the airway which may be greater in asthmatics than in non-asthmatics.

Finally, asthmatics acutely exposed to allergen may have enhanced responses to endotoxin because of increased levels of soluble CD-14 and LPS-binding protein in bronchoalveolar lavage fluid after allergen challenge (Dubin *et al.*, 1996). Taken together, these results demonstrate that LPS may be an important air contaminant to which asthmatics may be particularly sensitive.

SUMMARY

Controlled exposure studies of the effect of air pollutants in asthmatics have confirmed that candidate pollutants can exert an effect in asthma. A common feature of most of these pollutants is their ability to enhance or augment airway inflammation. Because airway inflammation, and particularly eosinophilic inflammation, is a key feature of airway responsiveness in asthma, such proinflammatory effects may be a central feature in the pollutant-related exacerbation of asthma.

An important criticism of most controlled exposure studies is that the level of the pollutant employed to show an effect is often much higher than that which is actually encountered with ambient pollution. However, it should be noted that, for ethical and practical reasons, most controlled exposure studies of asthmatics only include subjects with mild disease. Subjects with mild disease are likely to have less airway inflammation that those with more significant disease. If the effect of air pollutants is due to an ability to alter airway inflammation, then subjects with more significant disease (and those most likely to have medical consequences from pollutant exposure) are likely to respond to *lower* levels of pollutants. Likewise, the interaction of pollutants with allergen also is important and the level of pollutant needed to alter allergen-induced responses may be less than that required to have a more overt effect on the airway itself.

With the development of better airway sampling techniques (e.g. investigative bronchoscopy with biopsy, induced sputum) and a better understanding of the molecular basis of the effect of many pollutants, future pollutant challenge studies of asthmatics can be better designed to provide greater insight into the action of air pollutants in asthma.

REFERENCES

Anon (1996) Health effects of outdoor air pollution. Committee of the Environment and Occupational Health Assembly of the American Thoriacic Society. *Am J Respir Crit Care Med* **153**(1): 3–50.

Anto JM and Sunyer J (1995) Nitrogen dioxide and allergic asthma: starting to clarify an obscure association. *Lancet* **345**: 402–403.

Aris R, Christian D, Sheppard D *et al.* (1991) Lack of bronchoconstrictor response to sulfuric acid aerosols and fogs. *Am Rev Respir Dis* **143**: 744–750.

Ball BA, Folinsbee LJ, Peden DB and Kehrl HR (1996) Allergen bronchoprovocation of patients with mild allergic asthma after ozone exposure. *J Allergy Clin Immunol* **98**(3): 563–572.

Balmes JR (1993) The role of ozone exposure in the epidemiology of asthma. *Environ Health Perspect* **101**: 219–224.

Bascorn R, Nacierio RM, Fitzgerald TK *et al.* (1990) Effect of ozone inhalation on the response to nasal challenge with antigen of allergic subjects. *Am Rev Respir Dis* **142**: 594–601.

Basha MA, Gross KB, Gwizdala CJ *et al.* (1994) Bronchoalveolar lavage neutrophilia in asthmatic and healthy volunteers after controlled exposure to ozone and filtered purified air. *Chest* **106**: 1757–1765.

Bates DV and Hazucha MJ (1973) *The Short-Term Effects of Ozone on the Human Lung.* US Gov Print Off 93–15: 507–540.

Bates DV and Sizto R (1983) Relationship between air pollutant levels and hospital admissions in Southern Ontario. *Can J Public Health* **74**: 117–122.

Bauer MA, Utell MJ, Morrow PE *et al.* (1986) Inhalation of 0.30 ppm nitrogen dioxide potentiates exercise-induced bronchospasm in asthmatics. *Am Rev Respir Dis* **134**: 1203–1208.

Beck NB, Koenig JQ, Luchtel DL et al. (1994) Ozone can increase the expression of intercellular adhesion molecule-1 and the synthesis of cytokines by human nasal epithelial cells. Inhal Toxicol 6: 345–357.

Beckett WS, McDonnell WF, Horstman DH *et al.* (1985) Role of the parasympathetic nervous system in acute lung response to ozone. *J Appl Physiol Respir Environ Exercise Physiol* **59**: 1879–1885.

Bigby B and Boushey H (1993) Effects of nedocromil sodium on the bronchomotor response to sulfur dioxide in asthmatic patients. *J Allergy Clin Immunol* **92**: 195–197.

Blaski CA, Clapp WD, Thorne PS *et al.* (1996) The role of atopy in grain dust-induced airway disease. *Am J Respir Crit Care Med* **154**(2 Part 1): 334–340.

Boushey HA (1982) Bronchial hyperreactivity to sulfur dioxide: physiologic and political implications. *J Allergy Clin Immunol* **69**(4): 335–338.

Bozza PT, Castro-Faria-Neto HC, Penido C *et al.* (1994a) Requirement for lymphocytes and resident macrophages in LPS-induced pleural eosinophil acculumation. *J Leukocyte Biol* **56**: 151–158.

Bozza PT, Castro-Faria-Neto HC, Penido C *et al.* (1994b) IL-5 accounts for the mouse pleural eosinophil accumulation triggered by antigen but not by LPS. *Immunopharmacology* **27**(2): 131–136.

Coleridge HM, Coleridge JCG, Ginzel KH *et al.* (1976) Stimulation of 'irritant' receptors and afferent C-fibres in the lungs by prostaglandins. *Nature* **264**: 451–453.

Coleridge JCG, Coleridge HM, Schelegle ES *et al.* (1993) Acute inhalation of ozone stimulates bronchial C-fibres and rapidly adapting receptors in dogs. *J Appl Physiol* **74**: 2345–2352.

Crain EF, Weiss KB, Bijur PE *et al.* (1994) An estimate of the prevalence of asthma and wheezing among inner-city children. *Pediatrics* **94**(3): 356–362.

Devalia JL, Campbell AM, Sapsford RJ *et al* (1993) Effect of nitrogen dioxide on synthesis of inflammatory cytokines expressed by human bronchial epithelial cells *in vitro. Am J Respir Cell Mol Biol* **9**: 271–278.

Diaz-Sanchez D, Dotson AR, Takenaka H and Saxon A (1994) Diesel exhaust particles induce local IgE production *in vivo* and alter the pattern of IgE messenger RNA isoforms. *J Clin Invest* **94**: 1417–1425.

Diaz-Sanchez D, Tsien A, Fleming J and Saxon A (1997) Combined diesel exhaust particulate and ragweed allergen challenge markedly enhances human *in vivo* nasal ragweed-specific IgE and skews cytokine production to a T helper cell 2-type pattern. *J Immunol* **158**(5): 2406–2413.

Dubin W, Martin TR, Swoveland P *et al.* (1996) Asthma and endotoxin: lipopolysaccharide-binding protein and soluble CD14 in bronchoalveolar compartment. *Am J Physiol* **270**(5 Part 1): L736–L744.

Eschenbacher WL, Ying RL, Kreit JW *et al.* (1989) Ozone-induced lung function changes in normal and asthmatic subjects and the effect of indomethacin. *Atmos Ozone Res Policy Implic* 493–499.

Fernandes ALG, Molfino NA, McClean PA *et al.* (1994) The effect of pre-exposure to 0.12 ppm of ozone on exercise-induced asthma. *Chest* **106**: 1077–1082.

Folinsbee LJ (1993) Human health effects of air pollution. *Environ Health Perspect* **100**: 45–56.

Folinsbee LJ, Bedi JF and Horvath SM (1985) Pulmonary response to threshold levels of sulfur dioxide (1.0 ppm) and ozone (0.3 ppm). *J Appl Physiol Respir Environ Exercise Physiol* **58**: 1783–1787.

Folinsbee LJ, Horstman DH, Kehrl HR *et al.* (1994) Respiratory responses to repeated prolonged exposure to 0.12 ppm ozone. *Am J Respir Crit Care Med* **149**: 98–105.

Frampton MW, Balmes JR, Cox C *et al.* (1997) Mediators of inflammation in bronchaleolar lavage fluid from nonsmokers, smokers and asthmatic subjects exposed to ozone – a collaborative study. *Health Effects Inst Res Rep* **78**: 73–79.

Gong H Jr (1992) Health effects of air pollution. A review of clinical studies. *Clin Chest Med* **13**: 201–214.

Gong H Jr, Bedi JF and Horvath SM (1988) Inhaled albuterol does not protect against ozone toxicity in nonasthmatic athletes. *Arch Environ Health* **43**(1): 46–53.

Hackney JD, Lin WS, Bailey RM *et al.* (1984) Time course of exercise-induced bronchoconstriction in asthmatics exposed to sulfur dioxide. *Environ Res* **34**: 321–327.

Hallsworth MP, Soh CP, Lane SJ *et al.* (1994) Selective enhancement of GM-CSF TNF-alpha, IL-1 beta and IL-8 production by monocytes and macrophages of asthmatic subjects. *Eur Respir J* **7**(6): 1096–1102.

Hazucha MJ, Madden M, Pape G *et al.* (1996) Effects of cyclo-oxygenase inhibition on ozone-induced respiratory inflammation and lung function changes. *Eur J App Physiol Occup Physiol* **73**(1–2): 17–27.

Heath SK, Koenig JQ, Morgan MS *et al.* (1994) Effects of sulfur dioxide exposure on African-American and Caucasian asthmatics. *Environ Res* **66**: 1–11.

Henriques GM, Miotla JM, Cordeiro SB *et al.* (1996) Selectins mediate eosinophil recruitment *in vivo*: a comparison with their role in neutrophil influx. *Blood* **87**: 5297–5304.

Holtzman MJ, Fabbri LM, O'Byrne PM *et al.* (1983) Importance of airway inflammation for hyperresponsiveness induced by ozone. *Am Rev Respir Dis* **127**: 686–690.

Horstman D, Roger LJ, McDonnell W *et al.* (1984) Increased specific airway resistance (SRaw) in asthmatics exercising while exposed to 0.18 ppm ozone (O_3). *Physiologist* **27**: 212.

Horstman D, Roger LJ, Kehrl H *et al.* (1986) Airway sensitivity of asthmatics to sulfur dioxide. *Toxicol Ind Health* **2**(3): 289–298.

Horstman DH, Seal E Jr, Folinsbee LJ *et al.* (1988) The relationship between exposure duration and sulfur dioxide-induced bronchoconstriction in asthmatic subjects. *Am Ind Hyg Assoc J* **49**(1): 38–47.

Horstman DH, Ball BA, Brown J *et al.* (1995) Comparison of pulmonary responses of asthmatic and nonasthmatic subjects performing light exercise while exposed to a low level of ozone. *Toxicol Ind Health* **11**(4): 369–385.

Hunt LW, Gieich GJ, Ohnishi T *et al.* (1994) Endotoxin contamination causes neutrophilia following pulmonary allergen challenge. *Am J Respir Crit Care Med* **149**(6): 1471–1475.

Jorres R, Nowak D and Magnussen H (1996) The effect of ozone exposure on allergen responsiveness in subjects with asthma or rhinitis. *Am J Respir Crit Care Med* **153**(1): 56–64.

Kehrl HR, Roger LJ, Hazucha MJ *et al.* (1987) Differing response of asthmatics to sulfur dioxide exposure with continuous and intermittent exercise. *Am Rev Respir Dis* **135**: 350–355.

Kelly FJ, Blomberg A, Frew A *et al.* (1996) Antioxidant kinetics in lung lavage fluid following exposure of humans to nitrogen dioxide. *Am J Respir Crit Care Med* **154**(6 part 1): 1700–1705.

Kita H, Mayeno AN, Weyand CM *et al.* (1995) Eosinophil-active cytokine from mononuclear cells cultured with L-tryptophan products: an unexpected consequence of endotoxin contamination. *J Allergy Clin Immunol* **95**: 1261–1267.

Koenig JQ, Pierson WE and Korike M (1983) The effects of inhaled sulfuric acid on pulmonary function in adolescent asthmatics. *Am Rev Respir Dis* **128**: 221–225.

Koenig JQ, Covert DS and Pierson WE (1989) Effects of inhalation of acidic compounds on pulmonary function in allergic adolescent subjects. *Environ Health Perspect* **79**: 173–178.

Koenig JQ, Covert DS, Hanley QS *et al.* (1990) Prior exposure to ozone potentiates subsequent response to sulfur dioxide in adolescent asthmatic subjects. *Am Rev Respir Dis* **141**: 377–380.

Koenig JQ and Pierson WE (1991) Air pollutants and the respiratory system: toxicity and pharmacologic interventions. *Clin Toxicol* **29**(3): 401–411.

Koren HS and Bromberg PA (1995) Respiratory responses of asthmatics to ozone. *Int Arch Allergy Immunol* **107**(1–3): 236–238.

Koren HS, Devlin RB, Becker S *et al.* (1991) Time-dependent changes of markers associated with inflammation in the lungs of humans exposed to ambient levels of ozone. *Toxicol Pathol* **19**(4): 406–411.

Kreit JW, Gross KB, Moore TB *et al.* (1989) Ozone-induced changes in pulmonary function and bronchial responsiveness in asthmatics. *J Appl Physiol Respair Environ Exercise Physiol* **66**: 217–222.

Krishna MT, Springall D, Meng QH *et al.* (1997) Effects of ozone on epithelium and sensory nerves in the bronchial mucosa of healthy humans. *Am J Respir Crit Care Med* **156**(3 Part 1): 943–950.

Linn WS, Shamoo DA, Spier CE *et al.* (1983) Respiratory effects of 0.75 ppm sulfur dioxide in exercising asthmatics: influence of upper-respiratory defenses. *Environ Res* **30**: 340–348.

Linn WS, Shamoo DA, Spier CE *et al.* (1985) Controlled exposure of volunteers with chronic obstructive pulmonary disease to nitrogen dioxide. *Arch Environ Health* **40**(6): 313–317.

Macari DM, Teixeria MM and Hellewell PG (1996) Priming of eosinophil recruitment *in vivo* by LPS pretreatment. *J Immunol* **157**: 1684–1692.

McDonnell WF, Horstman DH, Hazucha MJ *et al.* (1983) Pulmonary effects of ozone exposure during exercise: dose–response characteristics. *J Appl Physiol Respir Environ Exercise Physiol* **54**(5): 1345–1352.

McDonnell WF, Kehrl HR, Abdul Salaam S *et al.* (1991) Respiratory response of humans exposed to low levels of ozone for 6.6 hours. *Arch Environ Health* **46**: 145–150.

McKinnon KP, Madden MC, Noah TC *et al.* (1993) *In vitro* ozone exposure increases release of arachidonic acid products from a human bronchial epithelial cell line. *Toxicol Appl Pharmacol* **118**: 215–223.

McKinnon KP, Noah T, Madden M *et al.* (1992) Cultured human bronchial epithelial cells released cytokines, fibronectin, and lipids in response to ozone exposure. *Chest* **101**: 22S.

Michel O, Ginanni R, Le Bon B *et al.* (1992a) Inflammatory response to acute inhalation of endotoxin in asthmatic patients. *Am Rev Respir Dis* **146**(2): 352–357.

Michel O, Ginanni R and Sergysels R (1992b) Relation between the bronchial obstructive response to inhaled lipopolysaccharide and bronchial responsiveness to histamine. *Thorax* **47**(4): 288–291.

Michel O, Duchateau J and Sergysels R (1994) Are endotoxins an etopathogenic factor in asthma? *Am J Ind Med* **25**:129–130.

Michel O, Kips J, Duchateau J *et al.* (1996) Severity of asthma is related to endotoxin in house dust. *Am J Respir Crit Care Med* **154**(6 Part 1): 1641–1646.

Molfino NA (1992) Effect of air pollution on atopic asthma. *Medicina* **52**: 363–367.

Molfino NA, Wright SC, Katz I *et al.* (1991) Effect of low concentrations of ozone on inhaled allergen responses in asthmatic subjects. *Lancet* **338**: 199–203.

Muranaka M, Suzuki S, Koizumi K *et al.* (1986) Adjuvant activity of diesel exhaust particulates for the production of IgE antibody in mice. *J Allergy Clin Immunol* **77**: 616–623.

Passannante A, Hazucha MJ, Seal E *et al.* (1995) Nociceptive mechanisms modulate ozone-induced human lung function decrements. *Anesth Analg* **80**: S371.

Peden DB (1997) Mechanisms of pollution-induced airway disease: *in vivo* studies. *Allergy* **52**(38) Suppl): 37–44.

Peden DB, Setzer RW Jr and Devlin RB (1995) Ozone exposure has both a priming effect on allergen-induced responses and an intrinsic inflammatory action in the airways of perennially allergic asthmatics. *Am J Resp Crit Care Med* **151**: 1336–1345.

Peden DB, Boehlecke B, Horstman D and Devlin RB (1997) Prolonged acute exposure to 0.16 ppm ozone induces eosinophilic airway inflammation in asthmatic subjects with allergies. *J Allergy Clin Immunol* **100**: 802–808.

Peters A, Dockery DW, Heinrich J and Wichmann HE (1997) Short-term effects of particulate air pollution on respiratory morbidity in asthmatic children. *Eur Respir J* **10**(4): 872–879.

Peterson B and Saxon A (1996) Global increases in allergic respiratory disease: the possible role of diesel exhaust particles. *Ann Allergy Asthma Immunol* **77**(4): 263–268.

Pierson WE and Koenig JQ (1992) Respiratory effects of air pollution on allergic disease. *J Allergy Clin Immunol* **90**: 557–566.

Pope CA (1991) Respiratory illness associated with community air pollution in Utah, Salt Lake and Cache Valleys. *Arch Environ Health* **46**: 90–97.

Pope CA, Dockery DW and Schwartz J (1995) Review of epidemiological evidence of health effects of particulate air pollution. *Inhal Toxicol* **7**: 1–18.

Roger LJ, Kehrl HR, Hazucha MJ *et al.* (1985) Bronchoconstriction in asthmatics exposed to sulfur dioxide during repeated exercise. *J Appl Physiol Respir Environ Exercise Physiol* **59**(3): 784–791.

Rusznak C, Devalia JL and Davies RJ (1996) Airway response of asthmatic subjects to inhaled allergen after exposure to pollutants. *Thorax* **51**(11): 1105–1108.

Salvi SS, Blomberg A, Rudell B *et al.* (1997) Acute changes in the airways of healthy human subjects following short term exposure to diesel exhaust. *Am J Respir Crit Care Med* **155**: A425.

Sandstrom T, Stjernberg N and Andersson M (1989) Cell response in bronchoalveolar fluid after exposure to sulfur dioxide: a time response study. *Am Rev Respir Dis* **140**: 1828–1831.

Sandstrom T, Stjernberg N, Eklund A *et al.* (1991) Inflammatory cell response in bronchoalveolar lavage fluid after nitrogen dioxide exposure of healthy subjects – a dose–response study. *Eur Respir J* **4**: 332–333.

Sandstrom T, Bjermer L and Rylander R (1994) Lipopolysaccharide (LPS) inhalation in healthy subjects causes bronchoalveolar neutrophilia, lymphocytosis, and fibronectin increase. *Am J Ind Med* **25**(1): 103–104.

Scannel CH, Chen LL, Aris R *et al.* (1996) Greater ozone-induced inflammatory responses in subjects with asthma. *Am J Respir Crit Care Med* **154**: 24–29.

Schelegle ES, Adams WC and Siefkin AD (1987) Indomethacin pretreatment reduces ozone-induced pulmonary function decrements in human subjects. *Am Rev Respir Dis* **136**: 1350–1354.

Schwartz J and Dockery DW (1992a) Particulate air pollution and daily mortality in Stubenville, Ohio. *Am J Epidemiol* **135**: 12–19.

Schwartz J and Dockery DW (1992b) Increased mortality in Philadelphia associated with daily air pollution concentrations. *Am Rev Respir Dis* **145**: 600–604.

Schwartz J, Slater D, Larson TV *et al.* (1993) Particulate air pollution and hospital emergency room visits for asthma in Seattle. *Am Rev Respir Dis* **147**: 826–831.

Silverman F (1979) Asthma and respiratory irritants (ozone). *Environ Health Perspect* **29**: 131–136.

Strand V, Rak S, Svartengren M and Bylin G (1997) Nitrogen dioxide exposure enhances asthmatic reaction to inhaled allergen in subjects with asthma. *Am J Respir Crit Care Med* **155**(3): 881–887.

Takafuji S, Suzuki S, Koizumi K *et al.* (1987) Diesel-exhaust particulates inoculated by the intranasal route have an adjuvant activity for IgE production in mice. *J Allergy Clin Immunol* **79**: 639–645.

Takafuji S, Suzuki S, Muranaka M and Miyamoto T (1989a) Influence of environmental factors on IgE production, IgE, mast cells and the allergic response. *Ciba Found Symp* **147**: 188–204.

Takafuji S, Suzuki S, Koizumi K *et al.* (1989b) Enhancing effect of suspended particulate matter on the IgE antibody production in mice. *Int Arch Allergy Appl Immunol* **90**: 1–7.

Takano H, Yoshikawa T, Ichinose T *et al.* (1997) Diesel exhaust particles enhance antigen-induced airway inflammation and local cytokine expression in mice. *Am J Respir Crit Care Med* **156**(1): 36–42.

Takenaka H, Zhang K, Diaz-Sanchez D *et al.* (1995) Enhanced human IgE production results from exposure to the aromatic hydrocarbons from diesel exhaust particles: direct effects on B-cell IgE production. *J Allergy Clin Immunol* **95**: 103–115.

Timonen KL and Pekkanen J (1997) Air pollution and respiratory health among children with asthmatic or cough symptoms. *Am J Respir Crit Care Med* **156**(2 Part 1): 546–553.

Tunnicliffe WS, Burge PS and Ayres JG (1994) Effect of domestic concentrations of nitrogen dioxide on airway responses to inhaled allergen in asthmatic patients. *Lancet* **344**(8939–8940): 1733–1736.

Utell MJ, Morrow PE, Speers DM *et al.* (1983) Airway responses to sulfate and sulfuric acid aerosols in asthmatics. *Am Rev Respir Dis* **128**: 444–450.

Wang JH, Devalia JL, Duddle JM *et al.* (1995) Effect of six-hour exposure to nitrogen dioxide on early-phase nasal response to allergen challenge in patients with a history of seasonal allergic rhinitis. *J Allergy Clin Immunol* **96**(5 Part 1): 669–676.

Weg VB, Walsh DT, Faccioli LH *et al.* (1995) LPS-induced 111IN-eosinophil accumulation in guinea pig skin: evidence for a role for TNF-alpha. *Immunology* **84**: 36–40.

Weiss KB, Gergen PJ and Wagener DK (1993) Breathing better or wheezing worse? The changing epidemiology of asthma morbidity and mortality. *Annu Rev Public Health* **14**: 491–513.

Witek TJ and Schachter EN (1985) Airway responses to sulfur dioxide and methacholine in asthmatics. *J Occup Med* **27**(4): 265–268.

Wyemer AR, Gong H, Lyness A *et al.* (1994) Pre-exposure to ozone does not enhance or produce exercise-induced asthma. *Am J Respir Crit Care Med* **149**:1414–1419.

Ying RL, Gross KB, Terzo TZ *et al.* (1990) Indomethacin does not inhibit the ozone-induced increase in bronchial responsiveness in human subjects. *Am Rev Respir Dis* **142**: 817–821.

Risk Assessment and Air Pollution

JONATHAN M. SAMET

Department of Epidemiology, School of Hygiene and Public Health, The Johns Hopkins University, Baltimore, MD, USA

Much of the information in this chapter appears in Applied Epidemiology, Theory to Practice, (1998) Brownson RC and Petitti DB (eds), Epidemiology and Risk Assessment by JM Samet and TA Burke. New York: Oxford University Press. Permission granted by the publisher and co-author.

INTRODUCTION

Overview

Risk assessment, an approach for systematically organizing evidence about risk and characterizing the hazard to the population of interest, has received only limited application to outdoor air pollution. It has been primarily applied to estimate the risk of lung cancer posed by air pollution in general or by specific agents, such as diesel exhaust. Risk assessment has played a greater role in regard to indoor air pollution, being applied to tobacco smoke, asbestos and radon. Risk assessment for air pollution has also been most prominent in the USA, where its use has been promoted in regulatory statutes and its methods advanced through academic and non-academic activities (National Research Council, 1983a, 1994, 1996; Rodricks, 1992). Risk assessment may be used to estimate the effect of a point source of pollution on the surrounding population or to assess the effect of air pollution or on an air pollutant in general. Risk assessment has also been used by advocacy organizations to motivate more stringent control measures on the basis of estimated public health burden.

In spite of relatively limited use to date in the context of outdoor air pollution, risk assessment can inform decision-makers about the management of the risks of air pollu-

AIR POLLUTION AND HEALTH
ISBN 0-12-352335-4

tion, providing an estimate of the attributable burden of disease and of the potential consequences of alternative control strategies. In the 1990 amendments to the US Clean Air Act, for example, risk assessments were mandated for specific toxic air pollutants (National Research Council, 1994). Risk assessment can be applied to air pollution coming from a single point source or a class of point sources, e.g. from coal-fired power plants, or to air pollution having multiple and diverse sources, e.g. particulate air pollution.

This chapter provides an introduction to risk assessment and illustrates its application to outdoor air pollution. The findings of risk assessments on lung cancer and diesel exhaust and air pollution generally are covered in the chapters on these topics. Key US documents on risk assessment include 1983, 1994 and 1996 reports of the US National Research Council. Several texts address risk assessment generally (Fan and Chang, 1996; Rodricks, 1992) and air pollution more specifically (Gratt, 1996).

Risk assessment is a term now widely used for a systematic approach to characterizing the risks posed to individuals and populations by environmental pollutants and other potentially adverse exposures. A landmark 1983 US National Research Council report, *Risk Assessment in the Federal Government: Managing the Process* (often called the 'Red Book' because of its cover), defined risk assessment as '. . . the use of the factual base to define the health effects of exposure of individuals or populations to hazardous materials and situations' (National Research Council, 1983b). This conceptualization of risk assessment is both qualitative and quantitative. The term 'risk,' as used in the context of risk assessment, conveys the same meaning as in common parlance: the probability of an event, e.g. disease occurrence, taking place.

The 1983 National Research Council report explicitly positioned risk assessment as a tool for translating the findings of research into science-based risk management strategies (Fig. 38.1). Risk assessment evaluates and incorporates the findings of all relevant lines of investigation, from the molecular to the population levels, through the application of a systematic process with four components: hazard identification, dose–response

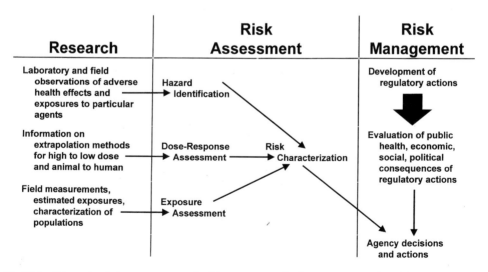

Fig. 38.1 Schematic relations among research, risk assessment and risk management. Adapted from NRC, 1983.

Table 38.1 The 'Red Book' paradigm: the four steps of risk assessment

Hazard identification	A review of the relevant biological and chemical information bearing on whether or not an agent may pose a carcinogenic hazard and whether toxic effects in one setting will occur in other settings
Dose–response	The process of quantifying a dosage and evaluating its relationship to the incidence of adverse health effects response
Exposure–assessment	The determination or estimation (qualitative or quantitative) of the magnitude, duration and route of exposure
Risk characterization	An integration and summary of hazard identification, dose–response assessment and exposure assessment presented with assumptions and uncertainties. This final step provides an estimate of the risk to public health and a framework to define the significance of the risk

assessment, exposure assessment, and risk characterization (Table 38.1). If there is no positive determination of the existence of a hazard at the step of hazard identification, then the subsequent elements are not warranted. Risk assessment also provides a comprehensive framework for bringing together all relevant information on the existence of a hazard to health and on the magnitude of the hazard. Thus, the hazard identification step could involve consideration of structure–activity relationships for a pollutant, laboratory findings from *in vitro* and *in vivo* experiments, and epidemiologic evidence. Dose–response assessment may also draw on multiple types of data, although evidence from epidemiologic studies and controlled human exposures has been most prominent for air pollution. While the schema separates research from risk assessment (Fig. 38.1), there is continued interplay between researchers and risk assessors as key gaps in evidence are identified and research is initiated to address them. For example, a 1998 report of the US National Research Council offers a long-term research agenda for airborne particulate matter that is targeted at uncertainties identified in a risk assessment framework (National Research Council, 1998).

Risk management follows and builds from risk assessment. Risk management involves the evaluation of alternative regulatory actions and the selection of the strategy to be applied. For air pollution, strategies relate primarily to source modification and to source use patterns. Risk assessment can be used to gauge the potential public health benefits of alternative control strategies. Risk communication is the transmission of the findings of risk assessments both to the many 'stakeholders' who need to know the results to participate in the policy-making process and to the general public. In this formalism, and in practice to some degree, those performing the risk assessment – risk assessors – and those managing the risks – risk managers – are separate groups of professionals and distinct from the researchers who develop the data used in risk assessments.

The Evolution of Risk Assessment

Risk assessment, as a formal method, has a brief and modern history (National Research Council, 1994; Rodricks, 1992). While many core concepts had been developed earlier, the origins of contemporary risk assessment can be traced to the 1970s when new environmental regulations in the USA called for information on risks in order to set policy. However, even earlier, the need to protect the general public and workers had led to the development of methods for setting exposure limits that inherently involved risk estimation. To protect workers, particularly against short-term toxicity, exposure limits were set that were below levels known or considered likely to have adverse effects. For example, threshold limit values (TLVs) were set by the American Conference of Government Industrial Hygienists (ACGIH). For foods, acceptable daily intakes (ADIs) of pesticides and food additives were set based on animal assays. The 'no observed effect level' (NOEL) in the assay, subsequently modified to the 'no observed adverse effect level' (NOAEL), was divided by a safety factor to yield the ADI for people.

In the 1960s and 1970s, mounting concern about environmental carcinogens increased the use of risk assessment by federal agencies in the USA, including the Food and Drug Administration, the Environmental Protection Agency, and the Occupational Safety and Health Administration. The widening use of risk assessment, as regulators attempted to manage increasing numbers of chemicals, motivated the Food and Drug Administration to support the National Research Council committee that wrote the Red Book. The committee's response to this charge continues to set the framework for risk assessment and risk management in the USA and to provide a model used by other countries. The committee recommended a clear conceptual distinction between risk assessment and risk management (Fig. 38.1) and formalized the risk assessment process into the four-component paradigm (Table 38.1). Scientific research provides the data substrate for risk assessment. The report acknowledged that uncertainties affect risk assessments and that gaps in knowledge need to be filled by making choices among plausible options. The committee also called for the development of uniform guidelines for selecting among such options, as the option selected could carry policy implications.

Subsequent to the Red Book, use of risk assessment at the federal and state levels increased in the USA. Guidelines for carcinogen assessment and other types of toxicity were published by the US Environmental Protection Agency (1986), which also developed guidelines for exposure assessment (US Environmental Protection Agency, 1992). Risk assessment was used as a priority-setting tool by the Department of Energy in implementing clean-up programs at its nuclear sites; the Agency for Toxic Substances and Disease Registry applied risk assessment approaches to contaminated sites throughout the USA. The Environmental Protection Agency took a risk-based approach in attempting to assign priorities to the many environmental hazards that it faced. Other countries, while not using risk assessment so extensively, are beginning to consider its role in managing environmental threats.

In the USA, the Clean Air Act Amendments of 1990 required a review by the National Academy of Sciences of methods used by the Environmental Protection Agency to estimate risk. The review, published in 1994 and entitled *Science and Judgment in Risk Assessment*, provides a summary of the state-of-the-art in risk assessment as of the early 1990s. It recommended the continued use of risk assessment but called for an iterative approach that better blended risk assessment with risk management. The Clean Air Act Amendments also mandated the establishment of a Presidential Commission on Risk Assessment and Risk Management that would 'make a full investigation of the policy implications and appropriate uses of risk assessment and risk management in regulatory programs under various Federal laws to prevent cancer and other chronic human health effects which may result from exposure to hazardous substances.' The final report of the commission comments that risk assessment has become more refined analytically, but notes that risk assessments done for regulation tend to give insufficient attention to risk reduction and improving health (Presidential/Congressional Commission on Risk Management, 1997a,b). It proposes a new framework for risk management that places collaboration with stakeholders at the center. Risk assessment remains key, but risks should be placed into the broad context of public health, and comparisons should be made with other risks to the population.

A similar broadening of the Red Book framework was proposed in the 1996 report of a committee of the National Research Council: *Understanding Risk. Informing Decisions in a Democratic Society.* This report extended the concept of risk characterization articulated in the Red Book. Like the report of the Commission on Risk Assessment and Risk Management, this report noted that a broad context needs to be set for risk characterization and recommended broad participation in risk characterization from all stakeholders. It called for an iterative process of analyses and deliberation and for determination of the concerns and perceived risks of stakeholders as the risk assessment is initiated. A risk characterization, to be informative, may need to be expressed along multiple dimensions, and not be limited to a simple numeric expression of harm, e.g. the number of excess cancers. It should be aimed at assisting in the decision process and solving problems.

As we approach the millennium, we seem poised for a broadening use of risk assessment in developing public policies, particularly those involving environmental regulation, including air pollution. Widening application to outdoor air pollution should be anticipated. For developed countries, risk assessment, combined with cost-benefit analysis, may become the foundation for control measures. For developing countries, risk assessment may prove useful for motivating governmental concern about the consequences of poorly controlled pollution. Regardless of the setting, risk assessment can be used as a priority-setting tool.

Uncertainty and Variability

Two concepts central to the interpretation and application of any risk assessment are uncertainty and variability. Assessment of risk involves the development of an underlying model with attendant assumptions that cover gaps in knowledge. Uncertainty refers to this lack of knowledge (National Research Council, 1994). Examples of sources of uncertainty include extrapolation of findings from animal experiments to humans; extrapolation from high-dose observable effects to the unobservable low-dose range; and use of models or assumptions to estimate population exposure indirectly, rather than with direct measurements. Analyses of uncertainty may be qualitative or quantitative. Qualitative analyses may involve expert judgments, whether accomplished informally or more formally using a systematic approach for achieving convergence among experts (National Council on Radiation Protection and Measurements, 1996). Quantitative assessments of uncertainty may use sensitivity analyses (varying model assumptions and assessing the consequences), or approaches based on a statistical model that characterizes the contributions of various sources of uncertainty to overall uncertainty.

Variability, although distinct from uncertainty, may also affect the interpretation of a risk assessment. There are many sources of variability that may affect a risk assessment (National Research Council, 1994). These include variability in exposures and susceptibility; together, these two sources of variability could lead to a wide range of actual risk in a population. Central estimates of risk, which do not address variation in risk across a population, may be misleading and may obscure the existence of a group at unacceptable risk that is hidden in the tail of the risk distribution.

THE FOUR COMPONENTS OF RISK ASSESSMENT

Hazard Identification

Hazard identification is the first step of a risk assessment, addressing the question of whether the agent or factor poses a risk to human health. This step is inherently integrative, as it may draw evidence from structure-activity relationships for chemical agents, *in vitro* evidence of toxicity, animal bioassays and epidemiologic data (National Research Council, 1983b). Epidemiologic data indicative of an adverse effect, when available, are strongly weighted in the evaluation of the weight of evidence to determine if an agent presents a hazard. Human data provide direct evidence of a hazard without the need to extrapolate from knowledge of toxicity in analogous agents or from another species. In fact, as we have gained a further understanding of the complexity of cross-species extrapolation from animal to humans, such extrapolations are viewed with less certainty, unless buttressed on an understanding of human and animal pathways of absorption and metabolism, and of mechanisms of action. Cross-species variation in responses, as in the instance of diesel exhaust and lung cancer, further complicates interpretation of animal studies. Furthermore, epidemiologic studies evaluate the impact of exposures received by the population, including complex pollution mixtures that may not be readily replicable in the laboratory. Epidemiologic research captures the consequences of interactions among agents, and investigations in populations may capture the full range of susceptibility. There is substantial epidemiologic evidence available for the principal outdoor air pollutants of concern with regard to human health (American Thoracic Society, 1996a,b). However, given the numbers of agents of concern, epidemiologic data have been available on only a small number of environmental contaminants and there is more often reliance on toxicologic evidence in identifying the hazard of specific air toxics.

In using epidemiologic and other data for the step of hazard identification, researchers' interpretation of the evidence is fully parallel to the assessment of the causality of an association between an exposure and an adverse health effect. There are no specific guidelines for interpretation of evidence in risk assessments that go beyond the conventionally applied criteria for causality. For cancer, guidelines for interpreting the strength of evidence have been published, for example, by the International Agency for Research on Cancer (WHO, IARC, 1972). However, these guidelines are not rigid criteria and, as with the widely applied criteria for causality, there may be disagreement on the proper classification of epidemiologic evidence for the purpose of hazard identification. For example, the interpretations of negative epidemiologic findings in the hazard identification process is a major source of disagreement.

Dose–Response Assessment

To characterize the risks posed by agents found to be hazardous, information is needed on the relationship between dose and response and also on the distribution of exposure. Dose–response assessment describes the quantitative relationship between dose and response. Dose, the quantity of material entering the exposed person, is not identical to exposure, which is defined as contact with a material at a potential portal of entry into the body: the skin, the respiratory tract and the gastrointestinal tract (National Research Council, 1991a). Typically, epidemiologic studies and controlled exposure studies of air pollution characterize the relationship between exposure, or a surrogate for exposure, and response. The dose–response relationship may be estimable if the relationship between exposure and dose can be established. For inhaled pollutants, the relationship between exposure and dose delivered to target sites in the respiratory tract may reflect physical characteristics of the pollutant, activity level and breathing pattern of the exposed persons, and aspects of pulmonary physiology. Dosimetric models of the respiratory tract can be used to model doses of gases and particles delivered to target sites in the lung (National Research Council, 1991b). For a risk assessment of air pollution, description of the exposure–response relationship may be sufficient as exposure can be linked to response, and concentrations along with time–activity patterns determine exposures. In combination with data on the distribution of exposure, the risk posed to a population by an agent can be estimated without moving to establish the dose–response relationship. Variation in dose at a given level of exposure, however, may contribute to variation in risk. Exercise, for example, may substantially increase risk at a particular level of exposure, and the relationship between exposure and dose may be affected by the presence of underlying lung disease.

For the purpose of risk assessment, characterization of the exposure–response relationship in the range of human exposures is needed. Findings from animal studies and controlled human exposures typically need to be extrapolated downwards. For some environmental agents, e.g. asbestos, epidemiologic studies of workers provide descriptions of exposure–response relationships, but downward extrapolation is needed for population application. Epidemiological data on outdoor air pollution reflect the consequences of exposures at levels of public health concern and consequently receive emphasis if available. However, epidemiologic studies on air pollution may not include comprehensive data on exposure or dose during the biologically relevant interval and exposures are often estimated from incomplete data. Consequently, misclassification of exposure may bias the description of the exposure–response relationship. Simple generalizations concerning the consequences of measurement error cannot be made (Armstrong et al., 1992). Random error or non-differential misclassification tends to blunt the exposure–response relationship, but other effects may also occur (Dosemeci et al., 1990). Non-random errors or differential misclassification may increase or decrease the gradient of the exposure–response relationship.

In using epidemiologic or other data to characterize the exposure–response relationship, there is a priori interest in determining if the exposure–response relationship is statistically significant – i.e. whether the null-hypothesis of a flat exposure–response relationship can be rejected – and in characterizing the shape of the relationship. The shape of the dose–response relationship has significant implications for risk assessment and risk management. Diverse dose–response relationships may be plausible (Fig. 38.2). Key

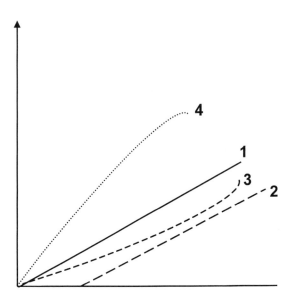

Fig. 38.2 Examples of dose–response models used for carcinogens: (1) linear non-threshold mode; (2) linear threshold model; (3) sublinear non-threshold model; (4) supralinear non-threshold model.

characteristics of the dose–response relationship include the presence/absence of thresh-old and the shape of the relationship in the range of interest, extending from the exposure at which data are available to those at which population exposures occur. Evidence for a threshold implies that some exposures do not increase risk, while evidence against a threshold suggests that any level of exposure should be considered as convey-ing some risk. The shape of the relationship provides the basis for extrapolating risks downward, from exposures at which data are available to those sustained by the popu-lation. Considerations as to the presence of a threshold and the shape of the dose–response relationship should be based on understanding of operative biologic mechanisms for the pollutants of interest. For example, the existence of effective repair processes would support a threshold, while irreparable damage would indicate a non-threshold relationship. Risk assessment offers a vehicle for integrating mechanistic evidence into population risk projections.

For air pollutants, carcinogens and non-carcinogens have been treated separately. Risk models for carcinogens, whether air pollutants or other types of agents, have generally incorporated linear non-threshold relationships between exposure or dose and response (Rodricks, 1992). The widespread application of the linear non-threshold model for car-cinogens originated with ionizing radiation and reflects a mechanism of action involving irreparable genetic damage by the carcinogen. There is less certainty concerning mecha-nisms for non-carcinogens, including some air pollutants.

For air pollutants, the exposure–response relationship might be estimated from epi-demiologic data in order to extrapolate from higher exposures where observations have been made, to lower levels where population exposures are occurring, or to describe the exposure–response relationship quantitatively at typical exposures. Various types of linear models are used for this purpose. Epidemiologic data may also be fitted with alternative models of the exposure–response relationship if there is uncertainty about the most appropriate shape. Model fit may be used to guide the selection of the 'best' model. However, epidemiologic data are rarely sufficiently abundant to provide powerful discrimination among alternative models, and sample size requirements for comparing fit of alternative models having different public health implications may be very high (Land, 1980; Lubin *et al.*, 1990).

Biomarkers of exposure, dose and response, and also of susceptibility, have been touted as possible solutions to the limitations of epidemiologic studies for characterizing the exposure–response relationship, particularly at low levels of exposure (Links *et al.*, 1994; Mendelsohn *et al.*, 1995). Use of biomarkers of exposure may reduce misclassification, while biomarkers of dose or response could potentially provide more proximate indicators of risk; biomarkers are also potential bridges for extending the results of animal studies to humans. A variety of biomarkers have been proposed for air pollution's adverse effects (National Research Council, 1989).

Exposure Assessment

To characterize risk comprehensively, information is needed on the full distribution of exposures in the population. Measures of central tendency may be appropriate for estimating overall risk to the population, but reliance on central measures alone may hide the existence of more highly exposed persons with unacceptable levels of risk. The US Environmental Protection Agency has recognized the need to characterize the upper end of the exposure distribution in its exposure assessment guidelines (1992). Driven largely by the needs of risk assessment, exposure assessment has matured: its underlying concepts have evolved and its methods have become more sophisticated. Key concepts of exposure assessment are presented in a 1991 National Research Council report (National Research Council, 1991b).

Modern exposure assessment is based on a conceptual framework that relates pollutant sources to effects, through the intermediaries of exposure and dose (Fig. 38.3) (National Research Council, 1991b, 1994). The concept of total personal exposure is central; that is, for health risk assessment, exposures received by individuals from all sources and media need to be considered. For a few air pollutants, e.g. lead, exposures may arise from multiple sources, media of exposure, and activities. Thus, lead may enter the body through ingestion of lead-paint-contaminated house dust, consumption of lead-contaminated foods and beverages, and inhalation of airborne lead.

For outdoor pollutants, exposures take place, of course, in outdoor and some transportation environments, but exposures may also take place in indoor environments for pollutants that penetrate indoors. For most outdoor air pollutants, however, inhalation is the sole route of exposure and air the sole medium. For air pollution, the tools of the exposure assessor include questionnaires to describe activities, monitoring devices for environmental and personal sampling, and biomarkers. Too often, data on exposures are limited, and model-based approaches may be substituted for actual population-based data. Modeling approaches, based on assuming an exposure model and statistical distributions for key model parameters, can yield exposure distributions. However, without validation, the results of such exercises are subject to substantial uncertainty.

For outdoor air pollution, population exposure distributions are generally derived

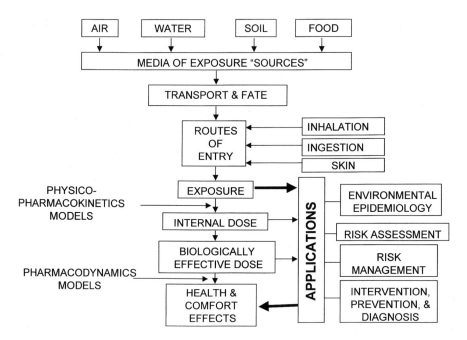

Fig. 38.3   Representation of the pathways from sources and media of exposure to health effects.

from concentrations measured by monitors sited for regulatory purposes. Typically these monitors are placed in an attempt to represent population exposures, although local sources may drive concentrations and validation studies are usually not performed to describe the relationship between measured concentrations and actual population exposures. For pollutants with broad regional sources, e.g. ozone and fine particles, one or more monitors in a regulatory network may provide a satisfactory assessment of exposure. For other pollutions having spatially more heterogeneous concentrations, e.g. carbon monoxide, regulatory monitors may not suffice for exposure assessment.

Air pollution models may also be used to estimate population exposure from point sources. Most of these models are based on Gaussian plume formulations (Gratt, 1996). As input, these models need the parameters of the source term, a site description and meteorological data. The models mathematically disperse pollution across the region in a pattern reflective of source operating conditions, meteorology and topography. The model output is typically pollutant concentrations over time at selected geographic points. Standard models are available for this purpose (see Gratt, 1996, for a review).

Risk Characterization

Risk characterization represents the final step in risk assessment. The data on exposure are combined with the exposure–response relationship to estimate the potential risk posed to exposed populations. The risk characterization becomes the basis for decision-making and for communication with stakeholders (National Research Council, 1996). Epidemiologic studies may provide a direct risk characterization if the findings can be linked to a specific population and the exposure assessment component of the study is sufficient. The population attributable risk, an indication of the burden of imposed morbidity or mortality, is an appropriate parameter for risk characterization, as would be years of life lost. This step puts the risk in perspective for risk managers and the public. The risk might be expressed as the increment in risk for population members across their lifetimes, the annual numbers of attributable events, or years of life lost. The Red Book approach emphasized the presentation of the probability of harm. However, a recent National Research Council report (1996) has called for risk characterization to be broader and to consider social, economic and political factors in describing risk and guiding risk management options.

APPLICATION OF RISK ASSESSMENT TO AIR POLLUTION

Overview

Risk assessment has been applied to both cancer and non-cancer outcomes associated with outdoor air pollution. For cancer, the emphasis has been on lung cancer, as urban air is known to contain carcinogens. Point sources emitting carcinogens, e.g. smelters, and mobile sources, particularly diesel exhaust, have also been of concern and their risks addressed with risk assessment. Diesel exhaust is addressed elsewhere in this volume; a recent report of the Health Effects Institute (1995) summarizes the various risk assessments.

This section provides examples of risk assessments of outdoor air pollutants; these risk assessments have been grouped as related to toxic air pollutants, power plants, and urban environments. For some pollutants emitted into the air, direct inhalation is not the principal route of exposure. For example, while dioxins are released into the air in incinerator emissions, exposure to humans comes principally through food chain contamination from deposited dioxins. This chapter focuses on ambient air pollutants, for which inhalation is the principal route of exposure.

Risk assessments of outdoor air pollution may be carried out with diverse rationales. In the USA, risk assessment of selected hazardous air pollutants is called for in the 1990 Amendments to the Clean Air Act (National Research Council, 1994). For the selected pollutants, the Environmental Protection Agency will base its regulatory decisions primarily on quantitative risk assessment. The agency will use technology-based approaches as a first stage and risk-based approaches in a second stage in regulating these pollutants. A quantitative risk assessment might also be directed at a specific point source to assess its impact on the surrounding population and to project the potential consequences of changing operations or of implementing controls. In the USA, risk assessment has also been widely used to influence policy-making by offering a quantitative measure, often a 'body count', that may influence policy-makers and the public.

Hazardous Air Pollutants

Hazardous or toxic air pollutants are general terms applied to a broad group of emissions that include carcinogens, mutagens, irritants and neurotoxins. In the USA, the hazardous air pollutants are explicitly distinguished from 'criteria pollutants' (lead, carbon monoxide, nitrogen dioxide, sulfur dioxide, particles, and ozone) based on the specific regulatory requirements of the Clean Air Act. The regulatory requirement for risk assessment has focused attention on approaches for quantifying the hazard posed by these agents. Risk assessment for the 189 hazardous air pollutants listed in the 1990 Clean Air Act Amendments was the focus for the 1994 National Research Council Report, *Science and Judgment in Risk Assessment.*

Both this report and the monograph by Gratt (1996) provide a perspective on risk assessment for these pollutants. Much of the evidence on toxicity comes from toxicological and not epidemiological evidence. In determining if a hazard is present, i.e. the hazard identification step, the risk assessor may be drawing inferences concerning toxicity to humans from animal exposures on short-term assays, such as the Ames test. For a few agents, e.g. benzene, human evidence or toxicity is available from studies of workers or other highly exposed populations. These same sources provide information on the dose–response relationship. Typically, mathematical models are fit to the data and then used to extend the dose–response relationship in the observed dose range to a lower range of doses relevant to the population. Model choice is critical to the magnitude of estimated risk; for example, non-threshold linear models, the typical default for carcinogens, project risk to the population at any exposure. Exposure to the hazardous air pollutants are generally estimated with models. Key elements include the emission factors for the source, meteorology, topography, and the population distribution in relation to the source. For some agents, routes of exposure other than inhalation may be relevant, adding additional elements to the exposure model. By combining the information on exposure with the exposure (dose)–response relationship, the population risk can then be calculated.

Selected Risk Assessments for Ambient Pollution

Natural Resources Defense Council

In a report entitled *Breath-Taking. Premature Mortality Due to Particulate Air Pollution in 239 American Cities* (Shprentz et al., 1996), the Natural Resources Defense Council (NRDC) estimated premature mortality due to particulate air pollution, finding a total of 64 000 premature deaths. This analysis was conducted as the US Environmental Protection Agency was reviewing the need for revising the particulate matter standard. To characterize exposure, data on particulate matter less than 10 μm in aerodynamic diameter (PM_{10}) were extracted from the Environmental Protection Agency's Aerometric Retrieval Information Service (AIRS) database for 1990–94 as available. The data were averaged to yield a single figure for use in the risk assessment. The PM_{10} data were found for 239 Metropolitan Statistical Areas (MSAs) and mortality data for selected cardiopulmonary causes in persons 25 years and above were collected from the National Center for Health Statistics.

To calculate the excess mortality from particulate air pollution, NRDC used the adjusted regression coefficient for cardiopulmonary mortality from the American Cancer Society's Cancer Prevention Study (CPS) II (Pope *et al.*, 1995). For approximately 500 000 participants in the study, mortality was examined in relation to current air pollution levels. This relative risk for mortality was 1.31, when the residents of the most polluted city (at 33.5 µg/m³ for fine particles) were compared with those living in the least polluted city (at 9.0 µg/m³ for fine particles). The increase in relative risk (RR) for each of the 239 MSAs was then calculated based on the difference between the average for the MSA and 9.0 µg/m³, the concentration in the least polluted location in CPS II. Attributable risk (AR) was then calculated as

$$AR = (\text{No. cardiopulmonary deaths}) \times (RR - 1)/RR$$

This calculation led to an estimate that about 64 000 people die prematurely from heart and lung disease each year. The actual extent of life-shortening cannot be determined from these calculations, although reliance on risk coefficients from long-term studies indicates that these effects should not be considered as representing a lifespan reduction of only a few days. There are abundant sources of uncertainty. These include the generalizability of the findings from CPS II, the appropriateness of a linear model, and the possibility of bias from uncontrolled confounding.

American Lung Association: Dollars and Cents. The Economic and Health Benefits of Potential Particulate Matter Reductions in the United States

In 1995, the American Lung Association published a report entitled *Dollars and Cents: The Economic and Health Benefits of Potential Particulate Matter Reductions in the United States* (Chestnut, 1995). The report estimates the potential health benefits of reducing particulate matter exposures from current levels to the California standards for PM_{10} (24 hr average of 50 µg/m³ and 30 µg/m³ annual average). The monitoring data were obtained from the AIRS database and gaps were completed using a random selection process. A selective literature review was conducted to identify studies providing exposure–response coefficients. Low, central and high estimates were listed and given probability weights using a somewhat *ad hoc* approach.

Calculations were then carried out with a simple multiplicative model that calculated the product of the risk coefficient, the exposed population, and the increment in pollution from the California standard to the current value. For mortality, it was assumed that 85% of the risk was sustained by persons 65 years and older. Morbidity measures considered included chronic bronchitis, respiratory hospital admissions, emergency room visits, asthma symptom days, restricted activity days, days with acute respiratory symptoms, and children with bronchitis.

For the respiratory hospital admissions, the data from the study in the Utah Valley by Pope (1991) were used. A coefficient was estimated for the daily risk of respiratory hospital admission per person per 10 µg/m³ PM_{10}. This model assumed that the relationship observed on a monthly basis could be extended to a daily basis. Coefficients were averaged for two sites studied by Pope: the Utah Valley and the Salt Lake Valley. High and low estimates were calculated using plus and minus one standard error from the central estimate.

The estimates for emergency room visits were based on the study of Samet *et al.* (1981) in Steubenville, Ohio. The risk coefficient was adjusted for the number of residents of Steubenville during the time of the study and high and low estimates were plus and minus one estimated standard error away from the central estimate. Aggravation of asthma was based on the model derived by Whittemore and Korn from analysis of a panel study in Los Angeles (1980).

The risk assessment provided estimates of the annual number of deaths and morbid events that could be prevented as well as associated costs (Table 38.2). As with the NRDC analysis, this risk assessment was intended to motivate policy. It encompassed multiple health outcome measures.

Environmental Protection Agency

In support of the revision of the NAAQS for Particulate Matter, researchers conducted a risk analysis for Philadelphia and Los Angeles (Deck *et al.*, 1996). The purpose was to estimate the potential benefit of reducing particulate levels to various targets that might be considered for a new standard. These locations were chosen because of the availability of pollution data and extensive prior research on air pollution and health carried out in both locations. An attributable risk methodology was used to estimate the consequences of the targeted reductions. The risk coefficients were derived by meta-analysis using a random effects model.

New York State Environmental Externalities Cost Study

This analysis of health impact was part of a larger project intended to develop a general methodology for estimating environmental externalities or costs (Rowe *et al.*, 1995). For estimating the impact on mortality, an attributable risk model is assumed, for persons above and below age 65 years as follows:

$$\text{Daily deaths for age group } a = R_a \times \Delta PM_{ij} \times POP_{aj}$$

Table 38.2 Estimated annual health benefits and associated savings: American Lung Association Risk Assessment for Particulate Matter

Health effect	Annual number of cases prevented	Annual monetary value
Premature mortality	1 964	$7071.9
Chronic bronchitis in adults (new cases)	12 767	$3064.0
Respiratory hospital admissions	1 964	$29.5
Emergency room visits (net)	36 729	$18.4
Asthma symptom days (diagnosed asthmatics)	447 650	$16.1
Restricted activity days (adults) (net)	6 627 657	$397.7
Days with acute respiratory symptoms (net)	20 387 479	$244.6
Acute bronchitis in children	139 717	$44.7
Total annual health benefits		$10 886.9

where R_a is the age-specific increase in daily death risk per 1 µg/m³ PM$_{10}$, ΔPM$_{ij}$ is the increase in PM$_{10}$ concentration on day i in location j, and POP$_{aj}$ is the population in age group a in location j. These daily increments were then summed to provide an annual total, without considering the possibility that some proportion of the daily increments represents short-term advancement of the time of death. A similar formulation was used for morbidity.

As in the *Dollars and Cents* report of the Lung Association, the risk coefficients were based on a selective literature review. For mortality, as an example, the coefficients were selected with adjustment from the analysis by Schwartz and Dockery (1992) of daily mortality in Philadelphia. Low, central and high estimates for the risk coefficients are provided in the report. The approach for morbidity closely follows that described above for the analysis of the American Lung Association.

American Lung Association: Breathless: Air Pollution and Hospital Admissions/Emergency Room Visits in 13 Cities

The American Lung Association commissioned this report to estimate the effect of O$_3$ exposure on both emergency hospital admissions and emergency room visits for respiratory illness in 13 selected urban areas in the USA. Hospital admissions data were obtained that provided the numbers of admissions and the characteristics of the admitted persons. A weighted coefficient was derived from published studies that estimated the increase in total respiratory hospitalizations for a 50 ppb increase in daily maximum 1-h O$_3$. This coefficient was then applied to the reported numbers of hospitalizations to derive the numbers of excess admissions. Upper and lower bounds were calculated using the 95% confidence limits of the pooled coefficient. The same coefficient was used for emergency room visits. This report attributed thousands of respiratory emergency room visits and hospitalizations to ozone.

Working Group on Public Health and Fossil Fuel Combustion

The 1997 report offers a global-scale risk assessment. Using scenarios of future energy consumption, as driven by climate change, the authors of the report estimate deaths attributable to air pollution, using particulate matter as the indicator pollutant. Combining projected emissions and attendant increments in particulate matter concentrations with exposure–response estimates from epidemiologic studies, they use an attributable risk approach to calculate pollution impact at the global level. They predict 700 000 avoidable deaths by the year 2020. This risk assessment is illustrative of the method's application for directing environmental policy.

CONCLUSION

A number of risk assessments on the health impact of outdoor air pollution have been published. They were conducted for various purposes, including informing policy-makers, demonstrating health risks and benefits of control, and projecting the consequences of scenarios of energy generation. In general, the methodologies are similar, combining literature-derived measures of risk with estimates of population exposures to pollutants. Of course, the findings of each risk assessment are reflective of underlying assumptions; each of the risk assessments, for example, has assumed a no-threshold relationship between exposure and disease risk. Under this assumption, any exposure conveys some risk to the population.

In managing environmental pollutants, including air pollution, risk assessment will play an increasing role. It can inform policy decisions and characterize the anticipated benefits of control measures. It can be used to rank environmental problems and assign priorities to various public health threats, including air pollution. Expanding use worldwide, building from the development of risk assessment in the USA and other countries, can be anticipated.

REFERENCES

American Thoracic Society, Committee of the Environmental and Occupational Health Assembly, Bascom R, Bromberg PA, Costa DA *et al.* (1996a) Health effects of outdoor air pollution. Part 1. *Am J Resp Crit Care Med* **153**: 3–50.

American Thoracic Society, Committee of the Environmental and Occupational Health Assembly, Bascom R, Bromberg PA, Costa DA *et al.* (1996b) Health effects of outdoor air pollution. Part 2. *Am J Resp Crit Care Med* **153**: 477–498.

Armstrong BK, Saracci R and White E (1992) *Principles of Exposure Measurement in Epidemiology*. New York: Oxford University Press.

Chestnut LG (1995) *Dollars and Cents: The Economic and Health Benefits of Potential Particulate Matter Reductions in the United States*. New York: American Lung Association.

Deck L, Post E, Weiner M *et al.* (1996) *A Particulate Matter Risk Assessment for Philadelphia and Los Angeles*. Research Triangle Park, NC: US Environmental Protection Agency.

Dosemeci M, Washolder S and Lubin JH (1990) Does nondifferential misclassification of exposure always bias a true effect toward the null value? *Am J Epidemiol* **132**: 746–748.

Fan AM and Chang LW (1996) *Toxicology and Risk Assessment. Principles, Methods, and Applications*. New York: Marcel Dekker Inc.

Gratt LB (1996) *Toxic Risk Assessment*. New York: Van Nostrand Reinhold.

Health Effects Institute (1995) *Diesel Exhaust: A Critical Analysis of Emissions, Exposure, and Health Effects*. A special report of the Institute's Diesel Working Group. Cambridge, MA: HEI.

Land CE (1980) Estimating cancer risks from low doses of ionizing radiation. *Science* **209**: 1197–1203.

Links JM, Kensler TW and Groopman JD (1994) Biomarkers and mechanistic approaches in environmental epidemiology. *Annu Rev Public Health* **16**: 83–103.

Lubin JH, Samet JM and Weinberg C (1990) Design issues in epidemiologic studies of indoor exposure to radon and risk of lung cancer. *Health Phys* **59**: 807–817.

Mendelsohn ML, Peeters JP and Normandy MJ (eds) (1995) *Biomarkers and Occupational Health. Progress and Perspectives*. Washington, DC: Joseph Henry Press.

National Council on Radiation Protection and Measurements (NCRP) (1996) *A Guide for Uncertainty Analysis in Dose and Risk Assessments Related to Environmental Contamination*. Bethesda, MA: NCRP.

National Research Council (NRC) and Committee on the Institutional Means for Assessment of Risks to Public Health (1983a) *Risk Assessment in the Federal Government: Managing the Means*. Washington, DC: National Academy Press.

National Research Council (NRC) and Committee on the Institutional Means for Assessment of Risks to Public Health (1983b) *Risk Assessment in the Federal Government: Managing the Process*. Washington, DC: National Academy Press.

National Research Council (NRC) (1989) *Biologic Markers in Pulmonary Toxicology*. Washington, DC: National Academy Press.

National Research Council (NRC) and Committee on Advances in Assessing Human Exposure to Airborne Pollutants (1991a) *Human Exposure Assessment for Airborne Pollutants: Advances and Opportunities.* Washington, DC: National Academy Press.

National Research Council (NRC) (1991b) *Frontiers in Assessing Human Exposures to Environmental Toxicants.* Washington, DC: National Academy Press.

National Research Council (NRC) and Committee on Risk Assessment of Hazardous Air Pollutants (1994) *Science and Judgment in Risk Assessment.* Washington, DC: National Academy Press.

National Research Council (NRC), Committee on Risk Characterization and Commission on Behavioral and Social Sciences and Education (1996) Stern PC and Fineberg HV (eds) *Understanding Risk. Informing Decisions in a Democratic Society.* Washington, DC: National Academy Press.

National Research Council (NRC) and Committee on Research Priorities for Airborne Particulate Matter (1998) *Research Priorities for Airborne Particulate Matter,* No. 1. Immediate priorities and a long-range research portfolio. Washington, DC: National Academy Press.

Pope CA III (1991) Respiratory hospital admissions associated with PM_{10} pollution in Utah, Salt Lake, and Cache Valleys. *Arch Environ Health* **46**: 90–97.

Pope CA III, Thun MJ, Namboodiri MM *et al.* (1995) Particulate air pollution as a predictor of mortality in a prospective study of U.S. adults. *Am J Respir Crit Care Med* **151**(3): 669–674.

Presidential/Congressional Commission on Risk Assessment and Risk Management (1997a) Final Report (vol 1): *Framework for Environmental Health Risk Management.* Washington, DC: Government Printing Office.

Presidential/Congressional Commission on Risk Management (1997b) Final Report (vol 2*): Risk Assessment and Risk Management in Regulatory Decision-Making.* Washington, DC: Government Printing Office.

Rodricks JV (1992) *Calculated Risks. Understanding the Toxicity and Human Health Risks of Chemicals in Our Environment.* Cambridge, UK: Cambridge University Press.

Rowe RD, Lang CM, Chestnut LG *et al.* (1995) *New York State Environmental Externalities Cost Study.* Dobbs Ferry, NY: Oceana Publications.

Samet JM, Speizer FE, Bishop Y *et al.* (1981) The relationship between air pollution and the emergency room visits in an industrial community. *J Air Pollut Control Assoc* **31**: 236–240.

Schwartz J and Dockery DW (1992) Increased mortality in Philadelphia associated with daily air pollution concentrations. *Am Rev Respir Dis* **145**(3): 600–604.

Shprentz DS, Bryner GC, Shprentz JS *et al.* (1996) *Breath-taking: Premature Mortality Due to Particulate Air Pollution in 239 American Cities.* New York: National Resources Defense Council.

US Environmental Protection Agency (EPA) (1986) Guidelines for carcinogen risk assessment. *Fed Reg* **51**: 3992–4003.

US Environmental Protection Agency (EPA) (1992) Guidelines for exposure assessment. *Fed Reg* **57**: 888–938.

Whittemore AS and Korn EL (1980) Asthma and air pollution in the Los Angeles area. *Am J Public Health* **70**(7): 687–696.

Working Group on Public Health and Fossil-Fuel Combustion (1997) Short-term improvements in public health from global climate policies on fossil-fuel combustion: an interim report. *Lancet* **350**: 1341–1349.

World Health Organisation (WHO) International Agency for Research on Cancer (IARC) (1972) *IARC Monographs on the Evaluation of Carcinogenic Risk of Chemicals to Man.* Geneva, Switzerland: International Agency for Research on Cancer, vol 1.

Estimating the Effects of Air Pollutants on the Population: Human Health Benefits of Sulfate Aerosol Reductions under Title IV of the 1990 Clean Air Act Amendments

BART D. OSTRO

Office of Environmental Health Hazard Assessment, California Environmental Protection Agency, USA

LAURAINE G. CHESTNUT and DAVID M. MILLS

Status Consulting, Inc., Boulder, CO, USA

ANN M. WATKINS

US Environmental Protection Agency, Washington DC,USA

INTRODUCTION

Title IV of the US Clean Air Act Amendments of 1990 calls for a 10 million ton reduction in annual emissions of sulfur dioxide (SO_2) in the USA by 2010. This represents an approximate 40% reduction in anthropogenic emissions from 1980 levels. Because electric utilities in the eastern USA use coal with a high sulfur content, about 85% of the emissions reductions called for under Title IV are expected to come from those utilities.

AIR POLLUTION AND HEALTH
ISBN 0-12-352335-4

Lower SO_2 emissions will result in lower gaseous SO_2 concentrations in areas close to major emissions sources, lower sulfate aerosol (SO_4) concentrations (including acid and non-acid aerosols), and lower acid precipitation throughout the affected region (which includes parts of southeastern Canada affected by the long-range transport of SO_2 from US emissions sources). The Clean Air Act Amendments require that benefits and costs of Title IV be quantified to the extent possible, given available scientific and economic information. The analysis reported here was undertaken as a contribution to the assessment of the human health benefits of Title IV. It focuses on estimating the health benefits by quantifying the number of specific health effects expected to be reduced and their associated monetary value as a result of the Title IV-related reductions in ambient sulfate aerosols. Ambient sulfate aerosol levels are the focus here because the potential human health benefits of reductions in these levels are expected to be substantial and because a quantitative assessment was feasible given available scientific and economic information.

An earlier version of this analysis was prepared for the US Environmental Protection Agency (US EPA), Acid Rain Division (Chestnut, 1995).* This chapter includes an update of the epidemiology literature review because of new studies that provide additional sulfate-specific results. This update benefited from the work done by the Health and Environment Impact Assessment Panel for Canada (Thurston *et al.*, 1997).

METHODS

Five quantification steps are required for this assessment. Steps 1 and 2 were conducted for the Acid Rain Division of the US EPA by other analysts. The results of these steps were used as inputs to steps 3, 4 and 5, which are the focus of this chapter, and address the overall question: What are the incremental health benefits achieved by Title IV's sulfate reductions relative to what would have happened without Title IV? The five steps are as follows:

(1) *Estimating changes in SO_2 emissions.* US EPA (1995b) reports estimates of what future SO_2 emissions from electric utilities in the eastern USA would have been both with and without Title IV (but with all other current emission requirements in place) through 2010.

(2) *Estimating changes in ambient sulfate concentrations.* SO_2 emissions estimates by county were partitioned into a grid with 80 square kilometer cells that is consistent with the US EPA's Regional Acid Deposition Model (RADM) (Dennis *et al.*, 1993), which covers the 31 easternmost states, the District of Columbia, and parts of southeastern Canada. The emissions estimates were used as inputs in RADM, which combines meteorological and atmospheric chemical modeling to predict the transport, transformation and ultimate deposition of SO_2 and the

---

* Other similar assessments have been conducted at US EPA, such as the Section 812 studies concerning the costs and benefits of the Clean Air Act as a whole (US EPA, 1997a), and the review of the National Ambient Air Quality Standards (NAAQS) for particulate matter (US EPA, 1997b). Although there are many similarities in the general approaches being taken in the health benefits components of these other assessments and in this assessment for Title IV, some of the details of the methods differ. Many of these differences stem from the fact that this assessment focused on SO_2 emissions and sulfate aerosols only, whereas the NAAQS assessment considers all sources of ambient particulate matter and the Section 812 studies consider all air pollutants regulated under the Clean Air Act.

secondary sulfate that forms as a result of SO_2 emissions. The RADM output used for this assessment was the ambient sulfate concentration (annual 50th percentile) within each of the RADM grid cells for 1997 and 2010 with and without Title IV (US EPA, 1995a).

(3) *Matching sulfate concentration changes to population.* Populations from the 1990 US Census and 1991 Canadian Census were matched to the 1330 RADM grid cells using a geographic information system to overlay census block group locations (USA) and enumeration area locations (Canada) with the RADM grid.

Table 39.1 Selected response coefficients for human health effects associated with sulfate aerosol concentration changes

Health effect category	Selected concentration–response: annual per capita incidence per $\mu g/m^3$ change in annual average SO_4 concentration (selected probability weights)		
Premature mortality			
Sources: Schwartz *et al.* (1996)	Low	1.8×10^{-5} (25%)	
Evans *et al.* (1984)	Low-central	2.4×10^{-5} (25%)	
Pope *et al.* (1995)	High-central	5.6×10^{-5} (25%)	
Dockery *et al.* (1993)	High	9.9×10^{-5} (25%)	
New case of chronic bronchitis	For population aged 25 and over:		
Source: Abbey *et al.* (1995)	Low	0.71×10^{-4} (25%)	
	Central	1.35×10^{-4} (50%)	
	High	2.00×10^{-4} (25%)	
Respiratory hospital admissions	Low	2.15×10^{-5} (25%)	
Source: Burnett *et al.* (1995)	Central	2.56×10^{-5} (50%)	
	High	2.96×10^{-5} (25%)	
Cardiac hospital admissions	Low	1.75×10^{-5} (25%)	
Source: Burnett *et al.* (1995)	Central	2.30×10^{-5} (50%)	
	High	2.85×10^{-5} (25%)	
Asthma symptom days	For population with asthma (4.7% of population):		
Source: Ostro *et al.* (1991)	Low	3.29×10^{-1} (25%)	
	Central	6.61×10^{-1} (50%)	
	High	9.89×10^{-1} (25%)	
Restricted activity days	For non-asthmatic population (95.3%) aged 18 and over:		
Source: Ostro (1990)	Low	1.53×10^{-2} (25%)	
	Central	2.66×10^{-2} (50%)	
	High	3.80×10^{-2} (25%)	
Days with acute (lower) respiratory symptoms	For non-asthmatic population (95.3%) aged 18 and over:		
	Low	4.75×10^{-2} (25%)	
Source: Ostro *et al.* (1993)	Central	14.2×10^{-2} (50%)	
	High	23.4×10^{-2} (25%)	
Child acute bronchitis	For population under age 18:		
Source: Dockery *et al.* (1996)	Low	2.70×10^{-3} (25%)	
	Central	4.40×10^{-3} (50%)	
	High	6.20×10^{-3} (25%)	

Because the assessment is for benefits in 1997 and 2010, expected population growth factors were applied (Bos *et al.*, 1994; US Bureau of the Census, 1997).

(4) *Estimating changes in numbers of cases of each type of sulfate-related health effect* (discussed below).

(5) *Estimating the monetary value of the changes in health effects* (discussed below).

Selection of Sulfate Concentration–Response Estimates

In step 4, low, central and high risk estimates for each health end-point associated with sulfate aerosols were selected from the available epidemiologic literature. A large body of epidemiologic literature examines the relationship between ambient sulfate aerosols and health effects. Table 39.1 shows the selected concentration–response relationships developed from this literature for this assessment.

An important underlying issue in interpreting available epidemiology results is whether the effect per unit of sulfate aerosol is different from that of other fine particulates. Although it is reasonable to expect that there may be differences, available evidence is limited in its ability to answer this question. Clinical and laboratory studies based on experiments where subjects are exposed to controlled amounts of sulfates alone provide evidence that at least some types of sulfate aerosols are harmful to the respiratory system. Thus, there is reason to believe that sulfates are contributing, at least in part, to the health effects observed in association with $PM_{2.5}$ and other particulate matter measures. The approach we take in this analysis is to select from the available literature those concentration–response functions that have been estimated using sulfate concentration data. In addition, the results from models that also included controls for ozone, which is often correlated with sulfate concentrations, are selected in preference to those that did not. This approach presumes that sulfate aerosols are at least a contributing causative constituent of $PM_{2.5}$, but does not assume that sulfate aerosols are the only causative constituent of $PM_{2.5}$.

To the extent that these epidemiology results are confounded by collinearity between sulfates and other types of particulate matter that may also be contributing to the observed health effects, the results may overstate the effects of changes in sulfate concentrations when only sulfates (and no other particulate matter constituents) are changed. To conduct a sensitivity test on this question, we multiply all the sulfate coefficients of the concentration–response functions by 0.4. This presents an alternative assumption that the sulfate coefficients reflect the effects of all other constituents of $PM_{2.5}$, as well as sulfates, and is based on the observation that in the eastern USA the sulfate concentrations average about 40% of $PM_{2.5}$ measured in the same locations. This is the maximum adjustment that would be needed. It probably substantially overadjusts, because sulfates and $PM_{2.5}$ are not, in fact, perfectly correlated.

Epidemiology Study Selection Criteria

Concentration–response functions for health effects identified and adapted from the available epidemiologic literature allow the estimation of the change in the number of cases of each health effect that would be expected as a result of changes in ambient sulfate concentrations. It was decided that epidemiologic studies would need to meet specified criteria before being used in this analysis.

First, a proper study design and a proper methodology were required. Studies were expected to include data based on continuous monitoring of the relevant pollutants, careful characterization and selection of exposure measures and minimal bias in study sample selection and reporting. In addition, the studies had to provide concentration–response relationships over a continuum of relevant exposures. The second criterion was that studies should recognize and attempt to minimize confounding and omitted variables. For example, studies that compared two cities or regions and characterized them as 'high' and 'low' pollution areas were not used because of potential confounding by other factors in the respective areas and a vague definition of exposure. Third, controls for the effects of seasonality and weather had to be included.

A fourth criterion for inclusion was that the study had to include a reasonably complete analysis of the data. Such analysis included a careful exploration of the primary hypothesis and preferably an examination of the robustness and sensitivity of the results to alternative functional forms, specifications and influential data points. Fifth, the study had to involve relevant levels of air pollution. Thus, studies that examined only high level pollution 'episodes' were not relied on for quantitative information. The final criterion was that studies should address clinical outcomes or changes in behavior to best lend themselves to economic valuation. Therefore, estimates for end-points such as changes in lung function that are difficult to link to clinically significant symptoms were not included. Also, preference was given to studies that focused on representative population groups to ensure the fullest possible coverage of the population. For example, hospital admissions studies that included visits made by all segments of the population were selected in preference to studies that examined only admissions of people with asthma.

Premature Mortality

Two types of long-term exposure studies have found statistically significant associations between mortality rates and sulfate levels in the USA. The first type is the ecologic cross-sectional study in which average annual mortality rates for various locations are compared to determine if there is a statistical correlation with average air pollutant concentrations in each location. Such studies have consistently found measurably higher mortality rates in cities with higher average sulfate concentrations. However, concern persists about whether these studies have adequately controlled for potential confounding factors. Özkaynak and Thurston (1987), Evans et al. (1984), and Chappie and Lave (1982) provide examples of ecologic cross-sectional studies. These studies each conducted a thorough examination of data for 100 or more US cities, including average sulfate concentrations for each city, with special emphasis on the effects of including or excluding potential confounding factors such as occupations or migration.

The second type of long-term exposure study is a prospective cohort study in which a sample of subjects with a range of pollution exposures is selected and followed over time. Dockery et al. (1993) report results of a 15-year prospective study of subjects in six US cities. Pope et al. (1995) report results of a 7-year prospective study of subjects in 151 US cities. These studies are similar in some respects to the ecologic cross-sectional studies because they rely on the same type of pollutant exposure measures, which are average pollutant concentrations measured at stationary outdoor monitors in given locations. However, the mortality data are for identified individuals, which permits better characterization of the study population and other health risks than when area-wide mortality data are used. Because they used individual-specific data, the authors of the prospective studies were able to control for premature mortality risks associated with differences in body mass, occupational exposures, smoking (present and past), alcohol use, age and gender.

The results from the ecologic and prospective cross-sectional studies with respect to sulfates fall between a 0.3% and 1.4% change in mortality per $\mu g/m^3$, with the exception of the sulfate result for the six-city prospective study, which is substantially higher. The results of the prospective studies are generally similar to or higher than the results of the ecologic studies, which supports the conclusion that the ecologic results are not just spurious statistical associations.

In recent years, a third type of epidemiologic study of air pollution and human mortality has reported effects of short-term fluctuations in air pollution exposures on daily mortality using time-series methods. The primary strength of time-series studies is that health and pollution variations in the same population (e.g. for a single city) are followed over time, so that the study population acts as its own 'control' and the effect of confounders is minimized. Time-series statistical models use respective day-to-day variations in ambient concentrations and mortality to determine whether mortality counts rise and fall with air pollution concentrations. This obviates the need to analyze separately comparison populations and to adjust statistically for differences in population characteristics (e.g. race, income, education). However, time-series studies are limited in that they generally fail to reflect potential effects of long-term exposures.

Many daily time-series studies for cities throughout the world have found statistically significant relationships between daily fluctuations in particulate matter concentrations and daily fluctuations in nonaccidental mortality rates. Schwartz *et al.* (1996) conducted a daily time-series analysis with pooled data from six US cities (the same six cities as in Dockery *et al.*, 1993) using sulfate measures; they also conducted analyses with $PM_{2.5}$ and PM_{10} data. This is the most comprehensive daily time-series mortality analysis conducted to date using sulfate data. The statistically significant relationship found between sulfates and daily mortality is somewhat higher than for previous time-series studies using PM_{10}, but lower than those found in the cross-sectional studies.

To calculate estimates of changes in incidence of premature mortality for this analysis, we selected a range of results from all three types of mortality studies. Premature mortality is a serious health end-point, and there is a large body of epidemiologic literature that has studied mortality as it relates to air pollutant exposure. However, there remain many uncertainties in specific quantitative interpretations of the results of the epidemiologic studies of the association between premature mortality and sulfate concentrations. We therefore selected a wider range of findings than those selected for the other health end-points quantified in this assessment. A series of estimates characterized as ranging from low to high were identified.

The Schwartz *et al.* (1996) time-series results were used to develop the low estimate of the sulfate mortality for this analysis because it is a recent analysis pooling results across six US cities and is based specifically on sulfate data. We select a low-central estimate based on Evans *et al.* (1984), which represents the low end of the ecological cross-sectional results for sulfates. The high-central estimate is based on Pope *et al.* (1995). This prospective study is confirmatory of many previous ecologic study findings using a study design that better controls for potential confounders (such as smoking) on an individual level. As a high estimate we selected the ecologic cross-section results reported by Chappie and Lave (1982). These results are consistent with $PM_{2.5}$-based results obtained in the six-city prospective study by Dockery *et al.* (1993). This is a conservative choice for the high estimate because the sulfate-based results from Dockery *et al.* (1993) indicate an even greater effect on mortality.

The study results are reported as percentage changes in mortality and must therefore be multiplied by average annual mortality to calculate the change in annual premature deaths per change in annual average sulfate concentration. For this we use the annual average US non-accidental mortality rate of about 8 per 1000.

Morbidity Health Effects

The available epidemiologic evidence shows a strong relationship between sulfate aerosols and a wide range of morbidity health effects. The types of illnesses range from severe acute and chronic illnesses to mild acute symptoms such as coughing and wheezing.

Abbey *et al.* (1995) report results of a 10-year prospective cohort study conducted at Loma Linda University in California with a large sample of non-smoking adults. New cases of chronic respiratory disease were analyzed in relation to pollution exposure for the matching 10-year period. The authors report a statistically significant relationship between new cases of airway obstructive disease (AOD) and average concentrations of sulfates. These results provide the basis for the adult chronic bronchitis coefficients used in this analysis. About 85% of AOD cases in this study included a diagnosis of chronic bronchitis.

Recent evidence indicates, after controlling for collinear ozone concentrations, an association between ambient sulfates and both respiratory hospital admissions (RHAs) and cardiac hospital admissions (CHAs). For this analysis, we derived specific quantitative estimates from Burnett *et al.* (1995), who studied the relationship between hospital admissions for respiratory and cardiac disease and both sulfates and ozone from 1983 through 1988 in Ontario, Canada. One-day lags of both ozone and sulfates were associated with respiratory admissions, and sulfates (but not ozone) were associated with cardiac admissions. The sulfate effects were observed in both the summer and winter quarters, in both males and females, and across all age groups. Results for RHAs and CHAs are used in this assessment from models that included both sulfates and ozone in the regression. For this analysis, we apply the estimated relative risks to the average per capita RHA and CHA (defined according to the International Classification of Disease (ICD) codes used by Burnett *et al.*, 1995) in the USA to calculate numbers of hospital admissions per unit change in SO_4.

Several studies have related air pollutant concentrations to exacerbation of asthma symptoms. Ostro *et al.* (1991) examined the association between several air pollutants, including sulfates, $PM_{2.5}$ and acidic aerosols, and increases in asthma symptom days among adults with diagnosed asthma during winter months in Denver, Colorado. A significant association was found between the probability of moderate or severe asthma symptom days (measured as shortness of breath) and sulfate aerosol levels, after controlling for temperature, day of week, previous-day illness, and use of a gas stove. Ozone was at background levels during the study period. The results from Ostro *et al.* (1991) are used for this assessment because they are not likely to be confounded by ozone and they reflect a wide range of ages of subjects. These results may overstate the year-round effect because they were obtained during winter months, when more frequent respiratory infections may aggravate asthma symptoms; the estimates are therefore halved for this analysis to approximate year-round conditions.

Ostro (1990) used national Health Interview Survey data to explore the association between adult restricted activity days (RADs) for respiratory conditions and several measures of particulate matter, including sulfates, over a three-year period. A statistically significant association was reported between respiratory-related RADs and sulfate concentrations for working adults. These results provide the basis of the RAD estimates for this analysis, and are applied to the adult population aged 18 and over.

Ostro *et al.* (1993) examined the association between air pollutants, including sulfate, and acute (lower and upper) respiratory symptoms reported in a daily diary study in Southern California. Using a logistic regression model and controlling for weather, gas stove use, day of study, gender, and the existence of chronic disease, a statistically significant association was found between sulfate concentrations and acute lower respiratory symptoms, defined as dry cough, cough with phlegm, shortness of breath, chest cold, croup, asthma, bronchitis, flu, or pneumonia.

Because daily symptom concentration–response functions for asthmatics are available based on studies focused specifically on those with diagnosed asthma, we exclude the asthmatic population from the calculations of RADs and acute (lower) respiratory symptom days (ARSs). Although asthmatics were not specifically excluded from the RAD and ARS studies, non-asthmatics are more representative of the response of the general population because only a small fraction of the general public has diagnosed asthma. We therefore apply the RAD and ARS concentration–response functions to the nonasthmatic portion (95.3%) of the population.

Days with acute respiratory symptoms may include restricted activity days. To avoid double-counting, we assume that all RADs are also ARSs and subtract a fraction of RADs from ARSs to obtain net ARSs. Ostro *et al.* (1993) report that 28% of the ARSs included a lower respiratory symptom. We therefore subtract 28% of RADs from ARSs to obtain net ARSs.

Dockery *et al.* (1996) report results of an analysis of data from 24 North American cities, using sulfate concentrations as a measure of air quality. Children aged 8 to 12 were assessed via questionnaire between 1988 and 1991 about incidents of acute bronchitis or other respiratory symptoms during the preceding 12 months. Among the cities, the annual prevalence rates for acute bronchitis ranged from 3 to 10%. The logistic regression analysis controlled for gender, history of allergies, parental asthma, parental education, and current smoking in the home. The study reported a statistically significant association between sulfate concentrations and incidence of acute bronchitis.

Selection of Monetary Values for Health Effects

Step 5 of this assessment uses the available economic literature to develop estimates of willingness to pay (WTP) for changes in incidence of specific health effects. Economic values for changes in human health should reflect the full costs to the affected individuals and to society, including financial losses such as medical expenses and lost income (referred to as the cost of illness), plus less tangible costs such as pain and discomfort, restrictions on non-work activities, and inconvenience to others. WTP is defined as the dollar amount that would cause the affected individual to be indifferent to experiencing an increase in the risk of the health effect or losing income equal to that dollar amount. WTP measures of monetary value for changes in health risks typically exceed health care and other out-of-pocket costs that are associated with illness or premature death, because WTP reflects these as well as other less-tangible effects of illness or premature death on a person's quality of life. Table 39.2 shows the WTP estimates selected for each health effect end-point for which concentration–response functions were developed.

The WTP estimates for mortality risks are relatively large, and so the mortality risk reduction benefits make up a large share of total estimated health benefits. There is considerable uncertainty about the accuracy of available WTP estimates for changes in mortality risks for this application. Available WTP estimates are based primarily on analyses of the labor market and on-the-job risks of fatal accidents. A few WTP studies have also considered mortality risks associated with transportation accidents, but very little empirical work has been done to estimate WTP to reduce mortality risks associated with exposures to environmental pollutants. Previous and on-going pollution control benefits assessments have all drawn on this same mortality risk valuation literature, but some fundamental questions remain about applying these estimates in an assessment of mortality risks associated with air pollution.

Table 39.2 Selected monetary values for mortality and morbidity effects

| Health effect | Estimate per incident (1995$) | | | Primary source | Type of estimate[a] |
	Low	Central	High		
Premature mortality	$2.1 million (33%)	$3.6 million (50%)	$7.3 million (17%)	Fisher et al. (1989) Cropper and Freeman (1991) Viscusi (1992) Jones-Lee et al. (1985)	WTP
Adult chronic bronchitis	$150,000	$220,000	$390,000	Viscusi et al. (1991) Krupnick and Cropper (1992)	WTP
Respiratory hospital admission	$7,000	$14,000	$21,000	Graves (1994)	Adjusted COI
Cardiac hospital admission	$7,500	$15,000	$22,500	Graves (1994)	Adjusted COI
Child acute bronchitis	$160	$320	$480	Krupnick and Cropper (1989)	Adjusted COI
Restricted activity day	$30	$60	$90	Loehman et al. (1979)	WTP and adjusted COI
Asthma symptom day	$13	$37	$60	Rowe and Chestnut (1986)	WTP
Acute respiratory symptom day	$6	$12	$17	Loehman et al. (1979) Tolley et al. (1986)	WTP
Selected probability weights for all morbidity effects	33%	34%	33%		

[a] WTP = willingness-to-pay estimate.
Adjusted COI = COI × 2 to approximate WTP.

Available studies provide estimates of WTP for small changes in mortality risk, which is, for example, the increment in income an individual is willing to forego (or must be paid) to obtain (or accept) a job with lower (or higher) risks of fatal accidents. This is the measure of WTP that is appropriate for a benefits assessment; however, there are some differences between the contexts in which most available WTP values for mortality risk changes were estimated and the context of mortality risk associated with exposures to environmental pollutants. Individuals at risk of on-the-job accident fatalities differ in two key ways from those at risk from air pollution exposure: the former are more likely to be working age adults and they may typically be in better health than those at risk from air pollution exposure. This means that the average loss in expected life-years may be less for those at risk because of air pollution than for those at risk of on-the-job accidents. To the extent that WTP is a function of expected remaining life-years (which is hard to say based on available information), available estimates may overstate the elderly's WTP to reduce pollution-related mortality risks. Based on available epidemiologic evidence of short-term exposure effects, we assume for this assessment that 85% of all deaths associated with sulfate aerosols are individuals over age 65. In addition, we assume that the WTP to reduce mortality risks for those over 65 is, on average, 75% of the WTP values for working-age adults. This adjustment is based on limited empirical data (Jones-Lee *et al.*, 1985) and subject to considerable uncertainty, but to the extent that expected life-years lost is an important factor in determining WTP to reduce mortality risk, using available WTP values without adjustment risks overstating the WTP value of reducing pollution-related mortality risks.

There are also differences in the nature of the risks of pollution-related mortality versus the risks of accidental deaths. There is not sufficient empirical evidence to make any quantitative adjustments in the WTP estimates, but there is evidence that suggests that deaths preceded by long and painful illness (e.g. cancer or chronic respiratory disease) are more greatly feared than quick deaths, which are common with accidents. Exposures to environmental contaminants outside the perceived control of the individual also appear to be considered worse than risks perceived to be somewhat under the individual's control, such as driving an automobile. Whether the expected upward effects on WTP of these differences in the nature of the risk would be enough to offset the expected downward effects on WTP of differences in the populations at risk is not possible to say without further empirical research.

WTP estimates are not available for some of the morbidity effects considered in this analysis. In these cases, cost-of-illness (COI) estimates are used and are adjusted upward by a factor of two to compensate for the expected ratio of WTP to COI. This adjustment is based on limited available evidence, but we believe the resulting adjusted health valuation estimates are less biased than those that would result if unadjusted COI estimates were used. The selected adjustment factor is based on the literature review provided by Rowe *et al.* (1995). In a practical sense, whether this adjustment is used or not makes little difference to the total benefits estimates because it applies to health end-points that make up only a small share of total monetary benefits.

Uncertainty Analysis

There are many uncertainties involved in estimating health benefits for improvements in air quality. Uncertainties arise from limitations in the scientific literature and from variations in the results of studies addressing air pollution health effects and associated economic values. Often, when we attempt to quantify the benefits of improvements in air quality, we are asking questions that the scientific and economic literature can answer only partially and only with considerable uncertainty. Uncertainty may be reduced as the underlying research progresses, but these are complex questions that we can never expect to answer with pinpoint accuracy. As a result, simply developing a 'best estimate' of benefits does not provide policy-makers with important information about the uncertainties and limitations of the estimate.

Assessments such as this have begun to incorporate various approaches for characterizing uncertainty in the assessment results. Early attempts to quantify uncertainty focused on bounding the estimates with highest and lowest possible values. Such ranges tended to overstate the uncertainty by failing to communicate the usually small likelihood that the extreme values might be correct relative to the central estimates. More recently, quantitative uncertainty analyses have begun to incorporate statistical information from the underlying literature to estimate both a range of possible benefit values and their expected probability of being correct. Quantitative uncertainty analyses are inevitably incomplete because they require a great deal of information; there remain many questions for which there is not sufficient information to quantitatively characterize uncertainty. A three-step approach is used to quantitatively specify uncertainty in this assessment:

(1) Select low, central, and high values for the parameters in the concentration–response and monetary valuation functions. Assign probabilities to those values to reflect the relative confidence in each estimate.
(2) Use the low, central, and high values and their assigned probability weights to calculate the uncertainty in benefits for each end-point such as premature mortality for each location.
(3) Calculate the cumulative uncertainty for the total benefits estimate across all end-points for each location and for the entire assessment area.

Selected central estimates generally reflect our 'best estimate.' Sometimes central estimates are from a single 'best' study for that parameter. In other cases, they may be an average or some other measure of central tendency across several studies. The high and low values are selected to be reasonably plausible alternatives to the central estimate based on literature and professional judgment; they do not represent the absolute highest and lowest values reported in the literature.

Each selected concentration–response and monetary value estimate is assigned a probability weight, with the weights summing to 100% for each quantified health effect and each monetary value estimate. These probabilities are used to cumulate uncertainty in the benefits estimates. The probability assignment represents that share of the probability distribution function that the low, central, and high values are assumed to represent (rather than the percentile point of the distribution). When the low, central, and high estimates are based on results from different studies all judged as equally reliable, an equal probability weight is given to each estimate. When only one study result is used, the range selected is usually plus and minus one statistical standard error of the selected central result. In this case, the probability weight given to the central estimate is 50%, with 25% each given to the high and low estimates. In a few cases, less (more) weight is given to a high or low estimate based on analyst judgment that the particular estimate is less likely (more likely) to be correct than the other available estimates. A probability distribution of total health benefits is calculated using the probability weights and the parameter values based on a Monte Carlo approach that orders 5000 random samples of concentration–response function values and monetary values, using the assigned probability weights to guide the frequency of selection combined with the change in estimated sulfate concentrations.

From the estimated health benefits distribution, we selected the mean value to represent the 'best' estimate of health benefits, and selected low and high values that represent the 20th percentile and the 80th percentile on the probability distribution of the total estimated health benefit. This means, for example, that there is a 60% probability that the 'true' value falls between these low and high results, given the magnitudes and the probabilities selected for each of the low, central, and high concentration–response and monetary value estimates. The 20th and 80th percentiles are selected for presentation rather than more extreme points on the probability distribution in the interest of conceptual consistency with the selection of the low, central, and high estimates of health risks and monetary values. An appropriate interpretation of this range is that the weight of evidence suggests that the 'true' value is likely to fall within this range, not that this range bounds all uncertainty in the estimates. The percentiles should also be interpreted cautiously with regard to any strict statistical meaning. These are not statistical confidence intervals in the same sense as those estimated in a single statistical model.

RESULTS

Table 39.3 summarizes the estimates of annual human health benefits, as numbers of cases and their monetary value, attributable to the sulfate aerosol reductions anticipated as a result of Title IV in 1997 and 2010 for the 31 easternmost states, the District of Columbia, and southern portions of the Canadian provinces of Ontario and Quebec. The mean total annual estimated health benefit for 1997 in the USA is $10.6 billion. The 20th percentile estimate for 1997 is $3.5 billion and the 80th percentile estimate for 1997 is $19.4 billion (all values in 1995 US dollars). The central 1997 benefits estimate for Ontario and Quebec is $0.9 billion. The mean total annual estimate of health benefits in the USA rises to $39.9 billion for 2010, when Title IV SO_2 requirements are expected to be fully implemented. The 20th percentile estimate for 2010 is $16.7 billion and the 80th percentile estimate is $69.7 billion. The mean value for Ontario and Quebec in 2010 is $1.0 billion. The estimates are dominated by premature mortality and chronic bronchitis. The numbers of cases in these health effects categories are relatively small, but the high monetary values per case result in large monetary benefits for these categories. Premature mortality reductions account for about 87% of the total health benefits. Chronic bronchitis reductions are an additional 10% of the total. Together they represent about 97% of the total health benefits.

This assessment indicates that as a result of the Title IV-related sulfate reductions, about 2700 premature deaths are avoided in 1997 in the USA and Canada. These annual estimates increase to about 10 500 premature deaths avoided by 2010. In addition, about 4800 new cases of chronic bronchitis are avoided in 1997, increasing to about 17 000 by 2010. The largest numbers of cases reduced are for asthma symptom days, restricted activity days, and days with acute lower respiratory symptoms.

Table 39.3 Estimates of annual human health benefits of Title IV sulfate aerosol reductions in the Eastern USA and Canada

	1997				2010			
	Annual number of cases prevented		Annual monetary value (millions 1995$)		Annual number of cases prevented		Annual monetary value (millions 1995$)	
Health effect	Eastern USA	Ontario & Quebec	Eastern USA	Ontario & Quebec	Eastern USA	Ontario & Quebec	Eastern USA	Ontario & Quebec
Premature mortality	2 455	212	$9 179	$792	9 244	236	$34 568	$881
Chronic bronchitis (new cases)	4 453	386	$1 136	$99	17 163	445	$4 377	$114
Respiratory hospital admission	1 273	110	$18	$2	4 796	122	$68	$2
Cardiac hospital admission	1 146	99	$17	$1	4 316	110	$65	$2
Asthma symptom days	1 545 414	133 343	$56	$5	5 819 866	148 276	$210	$5
Restricted activity days	952 419	79 567	$57	$5	3 664 846	91 120	$218	$5
Acute respiratory symptoms (net)	6 454 610	556 922	$75	$6	24 307 371	619 293	$283	$7
Child acute bronchitis	54 570	5 164	$18	$2	192 016	5 283	$62	$2
Total annual health benefits:								
Mean			$10 555	$911			$39 850	$1 017
20th percentile			$3 495	$364			$16 685	$427
80th percentile			$19 393	$1 612			$69 651	$1 775

There are many sources of uncertainty in the mean estimates of health benefits for Title IV reported here. Some specific sensitivity analyses were conducted to determine the potential effect on the results of changes in a few selected assumptions, reflecting two of the key uncertainties identified. The analyses reported in this section cover only the uncertainties in the concentration–response functions and in the monetary valuation of health effects. Additional uncertainties also exist in the estimates of changes in SO_2 emissions and ambient sulfate concentrations that were used as inputs to the benefits estimates.

The first sensitivity test concerns the question of whether there may be a health effects threshold for sulfate aerosols. There is considerable uncertainty about whether there is a 'safe' level of sulfate aerosol exposure – one that does not cause any harmful health effects. We selected three possible threshold levels to illustrate how this could affect the health benefit estimates. The existence of a threshold could only decrease, not increase, the results because further reductions in sulfate levels in areas already at or close to the threshold would not yield any health benefits. We selected alternative threshold assumptions of 5.0 µg/m^3, 3.6 µg/m^3, and 1.6 µg/m^3 annual median SO_4 concentrations. None has been identified as a true threshold, but each represents a mean or low-end value for the range of concentrations considered in one of the epidemiologic studies from which concentration–response functions were taken. The results indicate that when a threshold of 5.0 µg/m^3 SO_4 is applied in the calculations, while keeping all other factors the same, the mean annual health benefit estimate for 2010 ($16.5 billion) falls very close to the previous 2010 20th percentile estimate. A threshold of 3.6 µg/m^3 results in a health benefit estimate for 2010 ($29.2 billion) that falls about midway between the default mean and the 20th percentile default estimates. At a threshold of 1.6 µg/m^3 or lower, the health benefit estimate for 2010 is virtually unchanged. This sensitivity analysis illustrates the significance of the threshold question and shows that this continues to be an important research issue in evaluating the health benefits of pollution emission reductions.

There is a possibility that the sulfate-based concentration–response functions may be somewhat upwardly biased because of the typical collinearity between sulfates and other fine particulate constituents in the ambient air. For this sensitivity test we multiply the sulfate-based concentration–response functions by 0.4, which is the average ratio between measured sulfates and measured PM$_{2.5}$ in the eastern USA. This is the maximum adjustment that would be required if the sulfate coefficients represented the total effects of all PM$_{2.5}$, which would only occur if sulfate aerosol and PM$_{2.5}$ concentrations were perfectly correlated. This adjustment reduces the annual health benefit estimate to about $15.9 billion in 2010, which is slightly lower than the 20th percentile estimate with the default assumptions.

CONCLUSIONS

The results of this assessment show that the potential health benefits of reductions in exposures to sulfate aerosols in the eastern USA as a result of the SO_2 emissions reductions required by Title IV are substantial. Based on what we believe is a reasonable interpretation of the available epidemiologic and economic evidence on potential health effects of sulfate aerosols and their monetary value, we estimate that the annual health benefits of Title IV required reductions in SO_2 in 2010 in the eastern USA are likely to fall between

$17 billion and $70 billion, with an estimated mean value of $40 billion. There is reason to expect some possible upward bias at the higher end of this range, and the results of the sensitivity analyses suggest that the benefits in 2010 fall between $16 billion and $40 billion. The health benefits alone compare favorably with Title IV's estimated annualized cost of SO_2 emission controls: $1.3 billion for 1997 (US EPA, 1995b) and between $2.2 billion (US General Accounting Office, 1994) and $2.5 billion (US EPA, 1995b) for 2010.

We have been careful throughout this assessment to highlight key assumptions and uncertainties that exist in the quantification procedures, especially in the health effects quantification and valuation portions of the assessment, which are the focus of this chapter. Most of these uncertainties cannot be resolved without substantial new research on several topics. The most important empirical questions that remain in the health effects quantification are as follows: (1) What is the relative harmfulness of sulfate aerosols versus other fine particulate matter? (2) Is there a threshold for health effects from sulfate aerosols, and if so, what is it? (3) Is there sufficient evidence to presume that the observed association between sulfate concentrations and human health effects is causative? The most important questions in the monetary valuation of health effects are as follows: (1) Are WTP estimates for risks of accidental deaths in populations of average health status applicable to premature mortality risks associated with air pollutant exposures? (2) How do WTP values for premature mortality and other health risks vary for the elderly and for those whose health is already poor?

ACKNOWLEDGEMENTS

This research was supported by US EPA, in part, under Contract No. 68-D3-0005, sub-contract to ICF, Inc. The views and conclusions expressed are not necessarily those of US EPA. The authors thank Robin Dennis, Joe Kruger, Baxter Jones, Charlie Richman, Robert Rowe, Shannon Ragland, Sally Keefe and Angela Patterson for their contributions to various aspects of the assessment. Many reviewers at US EPA and elsewhere gave helpful comments. We especially thank the peer reviewers for the 1995 assessment: Morton Lippmann, A. Myrick Freeman, Bernard Weiss, David Bates, Gardner Brown and Lester Lave. Responsibility for remaining errors or omissions rests solely with the authors.

REFERENCES

Abbey DE, Ostro BD, Petersen F and Bruchette RJ (1995) Chronic respiratory symptoms associated with estimated long-term ambient concentration of fine particulates <2.5μm aerodynamic diameter ($PM_{2.5}$) and other air pollutants. *J Exp Anal Environ Epidemiol* **5**: 137–150.

Bos E, Vu MT, Massiah E and Bulatao RA (1994) *World Population Projections, Estimates and Projections with Related Demographic Statistics.* Baltimore, MD: Johns Hopkins University Press.

Burnett RT, Dales R, Krewski D *et al.* (1995) Associations between ambient particulate sulfate and admissions to Ontario hospitals for cardiac and respiratory diseases. *Am J Epidemiol* **142**: 1–8.

Chappie M and Lave L (1982) The health effects of air pollution: a reanalysis. *J Urban Econom* **12**: 346–376.

Chestnut LG (1995) Human Health Benefits from Sulfate Reductions under Title IV of the 1990 Clean Air Act Amendments. Report prepared for US Environmental Protection Agency, Acid Rain Division. Washington, DC: US Environmental Protection Agency.

Cropper ML and Freeman AM III (1991) Environmental health effects In Braden JB and Kolstad CD (eds) *Measuring the Demand for Environmental Quality.* New York: North-Holland.

Dennis RL, McHenry JN, Barchet WR *et al.* (1993) Correcting RADM's sulfate underprediction: discovery and correction of model errors and testing the corrections through comparisons against field data. *Atmos Environ* **37**A: 975–997.

Dockery DW, Pope CA III, Xu X *et al.* (1993) An association between air pollution and mortality in six US cities. *N Engl J Med* **329**: 1753–1759.

Dockery DW, Cunningham J, Damokosh AI *et al.* (1996) Health effects of acid aerosols on North American children: respiratory symptoms. *Environ Health Perspect* **104**: 500–505.

Evans JS, Tosteson T and Kinney PL (1984) Cross-sectional mortality studies and air pollution risk assessment. *Environ Internat* **10**: 55–83.

Fisher A, Chestnut LG and Violette DM (1989) The value of reducing risks of death: a note on new evidence. *J Policy Anal Manag* **8**: 88–100.

Graves EJ (1994) *Detailed Diagnoses and Procedures National Hospital Discharge Survey, 1992.* Hyattsville, MD: National Center for Health Statistics, Vital and Health Statistics.

Jones-Lee MW, Hammerton M and Philips PR (1985) The value of safety: results of a national sample survey. *Econom J* **95**: 49–72.

Krupnick AJ and Cropper ML (1989) *Valuing Chronic Morbidity Damages: Medical Costs, Labor Market Effects, and Individual Valuations.* Report to US Environmental Protection Agency. Washington, DC: Office of Policy Analysis.

Krupnick AJ and Cropper ML (1992) The effect of information on health risk valuations. *J Risk Uncertainty* **5**: 29–48.

Loehman ET, Berg SV, Arroyo AA *et al.* (1979) Distributional analysis of regional benefits and cost of air quality control. *J Environ. Econom Manag* **6**: 222–243.

Ostro BD (1990) Associations between morbidity and alternative measures of particulate matter. *Risk Anal* **10**: 421–427.

Ostro BD, Lipsett MJ, Wiener MB and Selner JC (1991) Asthmatic responses to airborne acid aerosols. *Am J Public Health* **81**: 694–702.

Ostro BD, Lipsett MJ, Mann JK *et al.* (1993) Air pollution and respiratory morbidity among adults in Southern California. *Am J Epidemiol* **137**: 691–700.

Özkaynak H and Thurston GD (1987) Associations between 1980 U.S. mortality rates and alternative measures of airborne particle concentration. *Risk Anal* **7**: 449–462.

Pope CA III, Thun MJ, Namboodiri MM *et al.* (1995) Particulate air pollution as a predictor of mortality in a prospective study of U.S. adults. *Am J Respir Crit Care Med* **151**: 669–674.

Rowe RD and Chestnut LG (1986) *Oxidants and Asthmatics in Los Angeles: A Benefits Analysis.* Report to US Environmental Protection Agency. Washington, DC: Office of Policy Analysis, EPA-230-09-86-018.

Rowe RD, Lang CM, Chestnut LG *et al.* (1995) *The New York Electricity Externality Study.* Dobbs Ferry, NY: Oceana Publications.

Schwartz J, Dockery DW and Neas LM (1996) Is daily mortality associated specifically with fine particles? *J Air Waste Manag Assoc* **46**: 927–939.

Thurston GD, Bates D, Burnett R *et al.* (1997) *Sulphur in Gasoline and Diesel Fuels.* Health and Environmental Impact Assessment Panel Report. Ottawa, Ontario: Health Canada.

Tolley GS, Babcock L, Berger M *et al.* (1986) *Valuation of Reductions in Human Health Symptoms and Risks.* Grant #CR-811053-01-0. Final report for US Environmental Protection Agency. Washington, DC: US Environmental Protection Agency.

US Bureau of the Census (1997) *Projections of the Population, by Age and Sex, of States: 1995 to 2025 and Projections of Households by Type: 1995 to 2010*, Series 1. <http://www.census.gov/gov/main/www/subjects.html> (9/20/97).

US Environmental Protection Agency (1995a). *Acid Deposition Standard Feasibility Study.* Report to Congress. Washington, DC: Office of Air and Radiation, Acid Rain Division.

US Environmental Protection Agency (1995b) *Economic Analysis of the Title IV Requirements of the 1990 Clean Air Act Amendments.* Washington, DC: Office of Air and Radiation, Acid Rain Division.

US Environmental Protection Agency (1997a) *The Benefits and Costs of the Clean Air Act, 1970 to 1990.* Report to Congress. Washington, DC: Office of Air and Radiation and Office of Policy Analysis.

US Environmental Protection Agency (1997b) *Regulatory Impact Analyses for the Particulate Matter and Ozone National Ambient Air Quality Standards and Proposed Regional Haze Rule.* Research Triangle Park, NC: Office of Air Quality Planning and Standards.

US General Accounting Office (1994) *Allowance Trading Offers an Opportunity to Reduce Emissions at Less Cost.* Washington, DC: US General Accounting Office.

Viscusi WK (1992) *Fatal Tradeoffs: Public and Private Responsibilities for Risk.* New York: Oxford University Press.

Viscusi WK, Magat WA and Huber J (1991) Pricing environmental health risks: survey assessments of risk-risk and risk-dollar trade-offs for chronic bronchitis. *J Environ Econom Manag* **21**: 32–51.

40

Costing the Health Effects of Poor Air Quality

DAVID MADDISON and DAVID PEARCE

Centre for Social and Economic Research on the Global Environment (CSERGE), University College London and University of East Anglia, UK

INTRODUCTION

If people's preferences are a valid basis upon which to make judgements concerning changes in human 'wellbeing', then it follows that changes in the risk of mortality should be valued according to what individuals are willing to pay (or, depending on the context, willing to accept as compensation) to forego the change in the risks that they face. The proper valuation of health impacts in monetary terms is necessary in order to allocate resources in an efficient manner between projects aimed at health and safety and other possible items of expenditure. This practice has long been followed in the planning of transportation systems in so far as the prevention of accidents are concerned.

Of course, the valuation of health impacts in monetary terms is not without controversial features. One major criticism of the monetary valuation, for example, is the view that health and safety ought in some way to be valued differently from other goods and services and be above monetary valuation. Nonetheless, people evidently do trade off risks to their own health and wellbeing against financial variables. In any event the monetary valuation is implicit in any decision, no matter how it is reached, since the financial costs of measures aimed at health and safety are readily observed. Another criticism is that 'since WTP [willingness to pay] is limited by ability to pay it is a fundamentally flawed measure of human preferences'. But this is nothing more than a manifestation of both the individual's and ultimately of society's overall resource constraint. Of course, one is not bound to adopt the view that individual preferences ought to be taken account of in the formation of public policy, but if not then one should have a good reason why not. One

AIR POLLUTION AND HEALTH
ISBN 0-12-352335-4

further objection is that the 'value of a statistical life' (VOSL) is infinite, since no individual would accept any finite sum in exchange for certain and immediate death. In fact VOSL is an unfortunate choice for a name since it incorrectly implies that what is being valued is life itself. In fact, what is being valued are small changes in risk aggregated over a small number of individuals. In many instances involving health and safety this appears to be the appropriate context.

Recent attention has turned to the health effects of air pollution with the publication of empirical studies identifying current ambient levels of air pollution as a significant cause of excess mortality (e.g. Pearce and Crowards, 1996). Those same estimates used to value accidental deaths have been used to value lives which, it is predicted, have been lost as a result of poor quality. The purpose of attaching values to these ill-effects is motivated by much the same set of concerns that prompt transport planners, namely a desire to allocate resources between air pollution abatement measures, investment in public transport systems, etc. and private consumption (particularly consumption of private transport). Putting a monetary value on the costs and benefits of air pollution control strategies provides a metric of human preferences with which the desirability of a range of options can be assessed. Merely enumerating the physical effects of air pollution control strategies is not on its own sufficient to assess their desirability. And yet for a number of reasons the valuation of the health effects of air pollution remains a contentious subject, even among those who accept the necessity of such valuation exercises in the case of designing transport systems.

In this chapter we do not question the appropriateness of the WTP methodology *per se* since those arguments have been aired extensively elsewhere. Nor do we discuss the epidemiological literature linking excess mortality to poor air quality. Instead, the purpose of this chapter is as follows:

(1) to review the evidence on the VOSL in the UK;
(2) to discuss the problems of transferring this transport/workplace-based literature into the context of deaths from air pollution;
(3) to illustrate the theoretical arguments for supposing that the VOSL is affected by the expected remaining life years;
(4) to speculate on a range of other factors that might affect the valuation of risk in the context of air pollution; and
(5) to illustrate a practical methodology for adjusting the available evidence on VOSL to render the figures more applicable to valuing deaths from air pollution.

The final section concludes with a call for yet more research into the valuation of morbidity effects and the WTP to avoid changes in the risk of mortality.

VALUING THE MORTALITY EFFECTS OF AIR POLLUTION

Everyday individual actions in which people trade money against a small reduction in personal safety can be used to estimate the value of a statistical life. Three alternative methodologies have been pursued in the context of determining the VOSL. These are labour market compensating differential (hedonic) analyses, contingent valuation (CVM) studies and market-based analyses. Labour market studies generally attempt to infer the compensation in exchange for increased risks associated with particular occupations, whilst standardizing for all other attributes of the job and the worker. The contingent valuation approach asks individuals hypothetical questions relating to their willingness to pay for reductions in the risk of their encountering particular hazards, whilst the market-based approach attempts to infer willingness to pay for reductions in risk by the purchase of goods whose only purpose is to reduce risks. A number of difficulties pervade the literature; the need to standardize for the many other characteristics of individuals or jobs which may affect the measured behaviour with regard to risk; the ability of an individual to comprehend risk scenarios with which he or she is confronted; and the assumption that a good's sole purpose is to reduce the risk of injury and also that the purchase price of a commodity is a true reflection of its cost to the individual.

Table 40.1 contains details of empirical studies into the VOSL conducted within the UK. These estimates are, in the case of the compensating wage differential studies, either averaged over the population of employed persons or, in the case of the single contingent valuation study by Jones-Lee (1992), across the whole population. The mean estimate for the VOSL based on the compensating differential studies is £3.7m, whereas the Jones-Lee CVM estimate is £2.6m. It is noteworthy that the methodology which deals with the VOSL of individuals of working age is higher than that dealing with the average VOSL of the entire population.

Table 40.1   The VOSL derived from published UK-based studies

Study	Year published	Type	Implied VOSL (£m) in 1993 prices
Melinek	1974	Comp. diff.[a]	0.5
Veljanovski	1978	Comp. diff.	5.5–7.6
Needleman	1980	Comp. diff.	0.2
Marin *et al.*	1982	Comp. diff.	2.4–2.7
Jones-Lee *et al.*	1985	CVM	2.6
Georgiou	1992	Comp. diff.	8.6

[a]   Compensating differential.
From OECD (1994) and Jones-Lee (1992).

In spite of the fact that there are so many studies into the VOSL, there are several areas of controversy which prevent the transfer of this body of literature into the context of premature deaths caused by air pollution. These stem mainly (but not exclusively) from the following propositions: first, that the risks from air pollution are not spread evenly across the population but are instead mainly focused on the over-65s; and second, that the remaining life-years of those over the age of 65 should have a lower value attached to them. The implication of this is clear enough:

> VOSL estimates should be adjusted to take account of the age differences which exist between those killed on the roads (average age 39 years) with those killed as a result of air quality (average age probably in excess of 65 years).

A later section of the chapter examines the evidence for the case that deaths are indeed greater in the over-65 age category, but the next section examines both the theoretical underpinnings of the view that the VOSL should be reduced for those with fewer remaining life years.

AGE EFFECTS, LATENCY AND THE VALUE OF A STATISTICAL LIFE

The most basic theoretical model assumes a utility function U with a single numeraire commodity X and a probability of survival equal to π. In the event of death the utility of the individual is assumed to be zero. Thus the expected utility of the individual $E[U]$, is given by:

$$E[U] = \pi\ U(X) \tag{1}$$

Totally differentiating this expression with respect to X and π and setting the resulting expression equal to zero gives:

$$MWTP = \frac{U(X)}{\pi \partial U(X)/\partial X} \tag{2}$$

Note that what has been computed here is the *Marginal* Willingness To Pay (MWTP) for a unit change in the probability of death (i.e. the VOSL itself). Note also that, because of the concavity of the utility function with respect to income, MWTP is an increasing function of the level of income.

And for exactly the same reasons, MWTP is a decreasing function of the probability of surviving the current time period. Without this individuals would be prepared to accept a finite sum in exchange for certain and immediate death – an intuitively implausible proposition.

So far it has been assumed that consumption possibilities are the same for all individuals (if they live). But the consumption possibilities available to a person are limited depending on the age of the person and his or her remaining life expectancy. Hence, there is the assumption that, other things being equal, older people have a lower WTP for the same change in the risk of death because they have fewer life-years of consumption ahead of them. This can be illustrated by generalizing the model presented above to cover multiple time periods:

$$E[U] = \sum_{t=1} \pi_t\, D^{t-1}\, U(X_t) \tag{3}$$

where:

$$D = \frac{1}{1+d} \tag{4}$$

and d is the subjective utility discount rate. Also:

$$\pi_t = \prod_{i=1}^{i=t} \pi_i \tag{5}$$

such that π_i is the probability of surviving through the current time period and π_t is the probability of surviving right through to the end of time period t. Totally differentiating the equation and setting the resulting expression equal to zero gives:

$$MWTP = \frac{U(X_1) + \sum_{t=2} \pi_t/\pi_1 D^{t-1} U(X_t)}{\pi_1 \partial U(X_1)/\partial X_1} \qquad (6)$$

This expression shows that the dependence of MWTP for a reduction in the risk of mortality depends upon the stream of future consumption possibilities.

Within the context of this model, it is also possible to derive the current MWTP for a small change in risk at a future point in time (denoted t').

$$MWTP = \frac{\sum_{t=t'} \pi_t/\pi_{t'} D^{t-1} U(X_t)}{\pi_1 \partial U(X_1)/\partial X_1} \qquad (7)$$

Thus, MWTP is always going to be smaller for future risks compared with contemporaneous risks and this difference will grow as the risk recedes further into the future. The intuitive reason is that a change in the current period survival probability affects the consumption possibilities from now onwards. An improvement in the probability of survival at some point in the future will, by contrast, improve only distant consumption possibilities which may be highly discounted.

This model can also be used to propose suitable adjustments to VOSL estimates derived from populations whose average age differs from the age group who are most at risk from air pollution. Suppose the VOSL is known for individuals of age j years. What is the relationship between VOSL for people of that age and the VOSL for individuals of age k years (where $k>j$)? Taking income as a constant per unit of time, the required adjustment is:

$$MWTP_k = MWTP_j \; \frac{\pi_j}{\pi_k} \; \frac{\sum_{t=k} \dfrac{\pi_t}{\pi_k} D^{k-1}}{\sum_{t=j} \dfrac{\pi_t}{\pi_j} D^{j-1}} \qquad (8)$$

Thus, other things being equal, the more terms inside the bracketed expression in the numerator the greater the MWTP is for a small reduction in the probability of surviving the current time period. So the VOSL of younger people should be higher.

It is interesting to note that the required adjustment is similar, although not identical to, the idea of adjusting the VOSL for age effects by multiplying by the ratio of 'discounted life-years'. The idea underlying the value of a life-year (or VOLY for short) is that each year of life has a fixed value associated with it, and that the discounted value of expected future life-years is the VOSL for example:

$$VOSL_k = \sum_{t=k} \pi_t D^{t-1} VOLY \qquad (9)$$

The attraction of this procedure is that if the MWTP is known for a person of average age, then, by taking the life expectancy at that age the value of a life-year can be computed. From this the MWTP for a person of any age (and associated life expectancy) can be computed as follows:

$$\text{MWTP}_k = \text{MWTP}_j \, \frac{\sum\limits_{t=k} \pi_t \, D^{k-1}}{\sum\limits_{t=j} \pi_t \, D^{j-1}} \tag{10}$$

But, as shown above, the VOLY concept cannot be derived from an intertemporal model of MWTP. More specifically, whilst the VOLY approach recognizes that consumption opportunities diminish with age, it fails to recognize that MWTP also depends positively upon the contemporaneous level of risk – which is also clearly dependent upon age. In this respect the VOLY approach appears to be unduly biased against the elderly.

SOME FURTHER CONSIDERATIONS

The model described above provides a way of adjusting a body of VOSL estimates which, such as those in Table 40.1, refer to individuals of prime age into values more appropriate for the more elderly in a manner consistent with a particular model of behaviour. It has also succeeded in showing that the VOLY approach is inconsistent with an intertemporal model of expected utility from a constant income stream. This model, however, is not very realistic, since the following are ignored when using the model to adjust VOSL estimates. In particular:

- The MWTP is affected by changes in earnings capacity and by borrowing and lending opportunities across the life cycle.
- Nowhere does longevity and the demand for life extension enter the utility function.
- Risks are not evenly spread over the entire population and marginal valuations may not be what is required.
- The existing health state of the victims is arguably poor and their life expectancy correspondingly low.
- Any deaths from chronic illnesses brought about by air pollution are likely to involve a period of extended suffering prior to death.
- MWTP estimates do not include the social costs or benefits of death.

To deal with the first point, note that MWTP may depend upon earnings capacity and whether one is able to borrow against future earnings when one is young or, alternatively, whether one is operating under a constraint whereby one is not allowed to be a net borrower. In fact, Shepard and Zeckhauser (1984) have shown that if one is unable to be a net borrower, then it is possible that WTP for a marginal change in risk could rise and then fall. The intuitive reason underlying the inverted U shape of the WTP function is that because at the beginning of the life cycle the high-income earning years are still far-distant, and towards the end of the life cycle expected lifetime consumption falls, so the amount people are willing to pay to reduce risk is relatively low during both these periods. But it is difficult to know what constraints individuals are operating under and what assumptions they make about their future earnings profile. It is also far from self-evident that the existence of capital market imperfections which prevent individuals from borrowing at the outset of their lives should be used to justify attributing to them a lower VOSL.

In the model developed above the assumption was made that only consumption motivates the will to live. Longevity itself has not appeared as an argument in the utility function of individuals and, given the same level of income, the level of utility should be the same in later years. This is a highly doubtful assumption. To understand what sorts of issues might be involved here, consider the case of an individual who is concerned for the welfare of his young family in the event of his death compared with the inclination of young single men to avoid risks (see Cropper and Sussman, 1988). Alternatively, even in the absence of arguments relating to a concern for the welfare of one's dependants, it is possible that the will to live anyway changes with age (this is best described as the 'old-age-can-be-very-fulfilling' argument). Such factors might well explain why the VOSL should peak in middle age instead of declining continuously. But whatever the motivation underlying changes in the demand for longevity, the magnitude of the effect that these have on MWTP is not something that theoretical reasoning or recourse to actuarial life tables can shed much light on. So it is hard to suggest suitable adjustments for these phenomenon.

Thus far the discussion has focused on the concept of *marginal* WTP. But it might be that the risks to some individuals are non-marginal in nature. Even if the changes in the risk of mortality for any one cohort of individuals is 'small' in comparison to the baseline risk that they face, it may be the case that the risks posed by air pollution are not evenly distributed within that group. Jones-Lee (1989) shows that a policy is worth more if it reduces a risk which is disproportionately borne by a small fraction of the population rather than equally between all individuals – even if the expected number of deaths is the same in both cases. The reason underlying this is that the 'compensation function' is a strictly convex increasing function of the probability of death.

An argument operating in the opposite direction is to suggest that the victims of air pollution are already suffering from poor health and that air pollution merely delivers the *coup de grâce*. If it could be demonstrated that this were indeed the case, then there might be an argument for significantly reducing the VOSL attributable to deaths from air pollution. The problem with this argument, however, is not one of valuation, but rather that the epidemiological evidence is insufficient to enable one to assess the health impairments suffered by the typical victim of poor air quality. Data on mortality counts typically record only the age of the individual and nothing more. Much more precise evidence on the extent of life lost is required before adjusting the value of lives lost through poor air quality on these grounds.

Another characteristic of air pollution is that it sometimes presents latent rather than contemporaneous risks. The preceding section showed how one might adjust for these effects (see also Cropper and Sussman, 1990). Again, the current state of epidemiology means that the latent effects of everyday exposure (as opposed to occupational exposure) to possible carcinogens cannot be ascertained. Yet another complexity is that it may be difficult to separate out issues relating to the quantity of life from those relating to the quality of life for latent risks. Considering the pain and suffering of a prolonged terminal illness, one might expect that the WTP to reduce these sorts of risks would be rather greater than to reduce risks of a death following an automobile accident. Jones-Lee (1989), for example, finds that the WTP for the avoidance in the risk of contracting a fatal cancer far outweighs WTP for the avoidance of a fatal road traffic accident.

The final consideration involves the existence of purely social costs. These costs are not included in measures of WTP since it is reasonable to assume that no-one is willing voluntarily to reduce the amount of risk which they expose themselves to in order to reduce, for example, their dependency upon the state and/or continue to make tax payments for the functioning of the state. Whilst it is assumed that these costs are likely to be small in relation to the private WTP to avoid the risk of death, in the context of individuals who are well beyond retirement age and thus most at risk from poor air quality this may no longer hold true. Obviously the social costs can be positive or negative, as they represent the difference between the present value of expected consumption and (gross) labour income for any one individual. What makes calculating this sum difficult is the fact that not all labour is marketed (i.e. the value of home-making services), making it necessary to use an imputed value; and the fact that some goods are provided free or at subsidized cost by the state (the most obvious example of which is medical care, the reliance upon which obviously increases with old age).

PRACTICAL METHODOLOGIES

Having rehearsed the theoretical arguments in favour of adjusting the VOSL used for appraising the benefits of preventing accidents before applying it to fatalities caused by air pollution, we now turn to the question of what feasible adjustments can be made to the VOSL to account for age effects.

If the VOLY proposal is ignored, there are two somewhat *ad hoc* possibilities for adjusting the VOSL estimates in Table 40.1 to reflect better the preferences of those at most risk from air pollution. The first is to use the results of the theoretical analysis coupled with tables of life expectancy. The problem here is clearly the realism of the assumptions underlying the model. The other possibility is to use the empirical results of Jones-Lee *et al.* (1985) in which the responses to WTP questions are categorized by different age groups. The problem with taking this purely empirical approach lies in the paucity of the available evidence. Furthermore, the question about how a 'typical' death from air pollution should be valued depends not only on the relative values one wishes to attach to deaths among the over-65s, but also on the prevalence of deaths in particular age ranges.

Evidence that particulate matter's effects are concentrated on the elderly is to be found in the work of Schwartz and Dockery (1992) for Philadelphia and Fairley (1990) for Santa Clara. In the study of Schwartz and Dockery the relative risk for under 65s is 1.049 per 100 $\mu g/m^3$ of PM_{10}, whereas for the over-65s it is 1.166. In the study of Fairley the relative risk for the under-70s is 1.033 whereas for those over 70 it is 1.076. There are also a number of other studies dealing exclusively with the over-65 age groups whose results are detailed in Table 40.2. Given that 84% of all deaths in the UK are accounted for by individuals over 65 years of age (Central Statistical Office, 1997), the relative risk ratios in the table imply that for every one person under the age of 65 who is killed from particulate emissions, approximately 12 are killed in the over-65 age category[*]. Alternatively put, 92% of the deaths attributable to particulate matter in the UK are of people in the over-65 age category.

---

[*] The calculation involved is $(0.097 \times 84\%)/(0.043 \times 16\%) = 11.8$.

Table 40.2 Relative risk ratios for particulate matter classified by demographic group

Demographic group	Relative risk ratio per 100 µg/m³ PM$_{10}$
Under 65 years	1.043 (0.008) two studies
Over 65 years	1.097 (0.014) five studies

From Maddison *et al.* (1997).

Suppose that the empirical results of Jones-Lee *et al.* (1985) are accepted, namely that the WTP of those in the over-65 age category is typically 75% of mean WTP of the entire population (see Table 40.3). Given that 16% of the population are in this age category (Central Statistical Office, 1997), this implies that the mean WTP of the under-65s is 105% of mean WTP. Combined with the result that 92% of those who die as a consequence of particulate concentrations are 65 years of age or older, this implies that the population-weighted average WTP to avoid the risks of death from poor air quality is £2.0m*. This can be compared with the mean WTP of the population as a whole of £2.6m based on the Jones Lee *et al.* (1985) estimate.

If, on the other hand, the adjustment is made on the basis of the theoretical model, and if we attribute a VOSL of £2.6m to an individual of average age (assumed to be 40 years), then we should attribute a value of £1.6m to the average individual over 65 years of age (assumed to be 75 years). This implies that, given the proportion of individuals in the different age categories, a VOSL of £2.8m should be attributed to the average individual under 65 years of age. Weighting these figures by the proportion of deaths gives a figure of £1.7m per life lost to pollution.

For comparison, the intertemporal WTP approach is compared with the VOLY approach in Table 40.4. It is evident that whereas the decline in MWTP relative to age 40 is somewhat attenuated in the case of the intertemporal WTP approach, in the case of the VOLY approach the ratio of discounted life-years falls more rapidly. This illustrates the extent to which that approach discriminates against the elderly.

Table 40.3 The ratio of VOSL by age to mean VOSL in the study of Jones-Lee *et al.* (1985)

Age group (years)	Ratio of VOSL by age to mean VOSL
<25	0.71
25–34	1.14
35–44	1.19
45–54	1.26
55–64	1.01
>64	0.75

From Jones-Lee *et al.* (1985). Note that these figures have been adjusted for income effects.

---

* The calculation is (0.92 × 0.75 × £2.6m) + (0.08 × 1.05 × £2.6m) = £2.0m.

Table 40.4   VOSL relative to VOSL for a male at age 40 based on discounted life-years and the WTP approach

Age	WTP/WTP at age 40	Discounted life-years/ discounted life-years at 40
20	1.26	1.27
30	1.14	1.15
40	1.00	1.00
50	0.84	0.81
60	0.68	0.59
70	0.59	0.38

From Central Statistical Office (1997). Note that the calculations are made using the 5-year survival probabilities based on the 1993–95 interim life tables. Note that the survival probabilities beyond the age of 85 are condensed into a single time period such that an 85-year-old male lives for exactly another 4.8 years and then dies.

Table 40.5 illustrates the willingness of a 20-year-old male to pay for future changes in the probability of survival. As one would expect, the WTP is markedly reduced over time. There are three reasons for this: first, there is the probability that the individual will in fact die before the future change in risk is scheduled to occur; second, future utility is heavily discounted; and third, WTP falls with age because the opportunities for consumption are in any case less (even though baseline risk is higher).

Table 40.5   WTP of a 20-year male to avoid latent risks

Age	WTP at age 20 to avoid a future risk/WTP at age 20 for a contemporaneous risk
20	1.00
30	0.68
40	0.44
50	0.27
60	0.15
70	0.09

CONCLUSIONS: WHERE NEXT FOR VALUING THE HEALTH IMPACTS OF AIR POLLUTION?

In order to value properly the health impacts of air pollution at least two key questions have to be answered. First, is it true that the majority of people who die as a consequence of poor air quality are over the age of 65? Second, even if they are well over 65 years of age, does this mean that they would express a greater preparedness to trade money for risk than that expressed by society as a whole? The answer to the first question appears to be that there is good evidence that the deaths from poor air quality are, in the UK, weighted in the ratio of 12 to 1 in 'favour' of the over-65s. The question about whether a lower value should be attached to their lives however is much more difficult.

The chapter demonstrates within the context of a simple model of expected utility that WTP ought to decline with age, but at the same time increase with the baseline probability of death (which also increases with death). Also, the chapter shows that the proposal to adjust VOSL estimates for individuals of a different age by the ratio of discounted future expected life-years is inconsistent with the simple intertemporal model of expected utility (Table 40.6).

Table 40.6 Conditional life expectancy and discounted life-years for a male

Age	Conditional life expectancy	Discounted life-years (@ 3% per annum)
20	52.8	28.0
30	43.2	25.4
40	33.7	22.0
50	24.5	17.8
60	16.2	12.9
70	9.6	8.3

From Central Statistical Office (1997). Note that the calculations are made using the 5-year survival probabilities based on the 1993–95 interim life tables. Note that the survival probabilities beyond the age of 85 are condensed into a single time period such that an 85-year-old male lives for exactly another 4.8 years and then dies.

The intertemporal model of WTP, however, is extremely simple in many respects, not least in that longevity does not appear as an argument in the utility function. This conflicts with much anecdotal evidence concerning the reasons why people's WTP for the avoidance of risk varies across their life cycle. This leads to the suggestion that adjustments to WTP should be based purely on empirical evidence. This empirical evidence is of two sorts: evidence from comparing the VOSL between studies examining different populations and the VOSL of different groups within the context of single studies. Evidence of either kind illustrates that during their working lives individuals have a higher WTP for reductions in risk than the mean WTP of the entire population. But there are also legitimate questions that can be asked about the empirical evidence. The reasons for this are that there are relatively few studies in which age effects are examined at all, and the empirical evidence may reflect individual constraints which governments may wish deliberately to override (such as the fact that young people express a lower WTP owing to their inability to borrow against future income). In either case, and despite the very different age profiles, the VOSL attached to a death from air pollution is shown to be not much lower than the value of deaths caused by transportation systems. Only one piece of evidence might overturn this view: a finding that the life expectancy of those who are killed by air pollution is measured in *months* rather than years. As yet there is very little evidence on the extent of life lost since all that mortality counts register is the age of the victim.

REFERENCES

Central Statistical Office (1997) *Annual Abstract of Statisfics.* London: HMSO.

Cropper M and Sussman F (1988) Families and the economics of risks to life. *Am Econom Rev* **78**: 255–260.

Cropper M and Sussman F (1990) Valuing future risks to life. *J Environ Econ Manag* **20**: 160–174.

Fairley D (1990) The relationship of daily mortality to suspended particulates in Santa Clara County 1980–86. *Environ Health Perspect* **89**: 159–168.

Jones-Lee M (1989) *The Economics of Safety and Physical Risk.* Oxford: Basil Blackwell.

Jones-Lee M (1992) The value of transport safety. *Oxford Rev Econom Policy* **6**: 39–59.

Jones-Lee M, Hammerton M and Philips PR (1985) The value of safety: results of a national sample survey. *Econ J* **95**: 49–72.

Maddison D, Johansson D and Pearce D (1997) *Damage Costs from Fuel Use in Major Urban Conurbations.* Washington, DC: Unpublished Report to the World Bank Environment Department.

OECD (Organizations for Economic Cooperation and Development) (1994) *Project and Policy Appraisal: Integrating Economics and the Environment.* Paris: OECD.

Pearce D and Crowards T (1996) Particulate matter and human health in the United Kingdom. *Energy Policy* **24**: 609–620.

Schwartz J and Dockery D (1992) Increased mortality in Philadelphia associated with daily air pollution concentrations. *Am Rev Respir Dis* **145**: 600–604.

Shepard D and Zeckhauser R (1984) Survival versus consumption. *Manag Sci* **30**: 423–439.

AIR QUALITY STANDARDS AND INFORMATION NETWORKS

41

Technology and Costing of Air Pollution Abatement

W. FRED DIMMICK and ALBERT H. WEHE

US Environmental Protection Agency, Research Triangle Park, NC, USA

The views expressed in this chapter are those of the authors and do not necessarily reflect the views or policies of the US Environmental Protection Agency.

INTRODUCTION

This chapter describes the basic approaches to abatement of air pollution. First, the basic science and engineering associated with air pollution abatement is presented. This includes an explanation of the background needed to understand and to make appropriate decisions about what kind of abatement is practical for particular situations. Second, pollution prevention or source reduction is described and examples are provided. In special circumstances, source reduction is superior to add-on technology abatement approaches. Source reduction can abate water pollution, solid waste generation and worker exposures along with air pollution. Next, this chapter outlines the basics and some effectiveness measures for several kinds of add-on technology or abatement systems. Performance assurance of these systems is discussed and approaches to ensure on-going high quality performance are outlined. In addition, this chapter explains the costing and economic aspects of many abatement systems. And finally, we provide suggestions for finding the latest information about abatement systems, their performance and their costs.

BASIC SCIENCE: THE PHYSICAL AND CHEMICAL ASPECTS OF AIR POLLUTION CONTROL

In selecting an abatement system, one must consider the air pollutant emissions, the gas stream containing the emissions and the required goal in emission reduction. In characterizing the emissions, its gas stream, and the requirements for removal, the first step is to develop enough background information to evaluate alternative abatement systems and to compute the cost and other impacts of these systems. With this background information, the engineer can evaluate the various abatement systems, select the appropriate system, and design the system (and ancillary equipment) in detail.

It is important first to develop a sufficient understanding of the emissions in order to determine what degree of control will be required. Public health considerations, regulations and laws typically determine the allowed emissions. Knowing the uncontrolled emissions and the allowed emissions allows the engineer to determine the required degree of control (emission reductions). The level of detail needed initially may be rudimentary compared to what is needed once the control system is selected and design of the system has begun. After identifying the degree of control, the next step is to characterize the emissions and gas stream completely and accurately based on the air pollutants that are to be destroyed or collected and the gas stream characteristics, both of which affect the effectiveness, size and operation of abatement systems.

The nature of the pollutants and the gas stream containing the pollutants is key to the design and effectiveness of the abatement system. Air pollutants can be generated and emitted in one of three physical states: gaseous, liquid or solid. Gaseous air pollutants can be reduced through chemical reactions such as combustion or through physical means such as condensation. In general, emissions of solid air pollutants are reduced through physical means such as impaction or filtration. Liquid air pollutants can be treated as solid air pollutants but also may be heated to change the liquid to a gas for abatement as a gas.

In addition to different physical states, air pollutants differ by their chemical complexity. They may be simple compounds such as oxides of nitrogen, or highly complex compounds such as pesticides. Some chemicals are easy to burn, while others are very difficult to burn. Other chemical and physical properties of the air pollutant are important when selecting an effective abatement system. Some chemicals are easily condensed, while others can only be condensed at very low temperature or at high pressures. In addition, most air pollution sources have multiple air pollutants of different natures, which in combination may require a different abatement approach than would be used for the individual pollutants.

Solid and liquid air pollution is generally called particulate matter (particulates). Particulates may be made of large particles (>10 μm), small particles, or fine particles (<1 μm). The particulates may be hard or sticky; they may be chemically active or inert. Gas streams containing particulates and gaseous pollutants often require dual abatement systems. The characteristics of the particulates affect the kind of abatement system or whether pre-treatment is needed to remove the particulates.

Gas streams carrying pollutant emissions vary in many ways that can in turn influence the selection and design of the abatement system. First, the gas stream may be small in volumetric flow with only a few 1000 cubic feet per minute (cfm) being moved.

Abatement systems for such flows are relatively small. In contrast, gas streams from large open operations, such as industrial ventilation systems, can reach 250 000 cfm or more. Abatement systems for such flows are very large, often taking several floors of plant space. Gas stream also can vary by others factors such as the temperature of the gas, the humidity, the presense of explosive concentrations of contaminants, and viscosity. These factors are considered in the final design of the abatement system.

POLLUTION PREVENTION OR SOURCE REDUCTION

Pollution prevention, perhaps more appropriately termed *source reduction*, is an abatement approach that addresses environmental pollution through targeting the generation of the pollution at its source. Source reduction may result in a complete elimination of the pollution or it may dramatically reduce the pollution. Source reduction typically involves all pollutant streams generated by a production process rather than only one pollution medium. If source reduction is not evaluated in a holistic manner, source reduction can result in the increase of certain kinds of pollutants in order to reduce other kinds of pollutants. In situations where source reduction does not provide more reductions than add-on control technology for air emissions, it is important to consider all the environmental reductions in selecting the appropriate approach.

Many source reduction examples involve significant changes to a production process, while others involve changes that are often considered typical 'technological solutions'. For example, the use of double seals on petroleum storage vessels (a technological solution) dramatically reduces the evaporation of the organic chemicals found in petroleum products. The reduction of evaporated organics reduces the emissions from this source. The use of leakless pump or valve technology essentially eliminates the loss of organics from such equipment.

Some approaches to source reduction involve the elimination of the pollutant of concern from the production process. Not using a pollutant to make the product is an obvious, although not necessarily easily achieved, way to reduce emissions. For example, removing lead from gasoline eliminates the pollution caused by lead emissions.

Some approaches to source reduction are based on modifying the production process. Such approaches change the process so that the generation of emissions is reduced. For example, the coating of surfaces with high volume, low pressure (HVLP) spray guns reduces the amount of organic solvents used to coat the surfaces of products such as wood furniture or many metal parts. HVLP spray guns are highly efficient at depositing a coating, but require that the operator learn how to use the spray gun properly. Some process modifications are more dramatic. For example, the primary production of aluminum has been designed based on several processes. The newest production processes are inherently less polluting and do not emit the quantities of polycyclic organic matter associated with the Sodeberg process that still is used by some companies.

ADD-ON CONTROL TECHNOLOGIES

Overview

Add-on control technologies are typically categorized by their mode of emissions removal into several groups: cyclones and other mechanical collectors, filters, electrostatic precipitators, scrubbers, adsorbers and combustors. Some abatement systems combine these technologies into integrated systems to achieve the required emission reductions, in particular where reductions in multiple pollutants are required. Cyclones are often used as precleaners to remove large particulates, before a gas stream is cleaned further of fine particulates with a fabric filter, or abated for gaseous pollutants. In addition, different technologies may be designed to function in series to achieve the required emission reductions. For example, gaseous emissions (e.g. Hg) may be captured by using an adsorbent, which is then captured with a filter.

Cyclones

Cyclones are abatement systems that separate particulates (either wet or dry) from a gas stream by applying a centrifugal force to the particulate by a vortex motion. The vortex motion is created by the design of the collector and separates the particulates by using the inertia of the particulates. The design of the equipment produces a spiraling gas (the vortex) where the particulates move outward under a centrifugal force until they impact the wall of the cyclone. Some cyclones collect the particulates in a hopper at the bottom while others are used to concentrate the particulates in a gas stream that flows to another separator.

Cyclones are relatively simple to design, construct and maintain. For large gas streams, multiple small cyclones are needed to handle the gas stream flow and to achieve optimal efficiency. They provide low-cost control for large particulates but offer only low efficiency for other particulates. They are often used as precleaners for other systems, where they protect and extend the useful life of these systems. As long as the interior of the cyclone remains clean, the energy needed to move the gas stream through the cyclone does not increase with time.

There are other simple mechanical abatement systems similar to cyclones. These devices collect particulates by gravity or centrifugal force, but unlike a cyclone do not depend upon a vortex. These devices include settling chambers, baffled chambers, louvered chambers and devices in which the carrier gas–particulate matter mixture passes through a fan in which separation occurs. In general, collectors of this class are of relatively low collection efficiency. They also are used as precleaners.

Fabric Filtration

Abatement systems that remove particulate matter from gas streams by capturing the particulates in or on a porous structure through which the gas stream flows are called filters. The porous structure is most commonly a woven fabric. A fabric filter unit consists of one or more separate compartments containing rows of fabric filter bags or tubes. The particulate containing gas passes along the surface of the bags and then radially through the fabric. Particulates are captured on the surface of the bags, while the cleaned gas stream is vented through the bags to the atmosphere. The filter is operated cyclically between relatively long periods of filtering and short periods of cleaning. During cleaning, dust that has accumulated on the bags is removed from the fabric surface and deposited in a hopper for subsequent disposal.

Fabric filters will collect particulate sizes ranging from submicrometer to several hundred micrometers in diameter at efficiencies generally in excess of 99 or 99.9%. Filters in general improve in removal efficiency as the pores of the fabric are filled by captured particulates and as the particulates themselves form a porous structure over the top of the fabric. This increase in filter thickness causes an increase in pressure drop across the filter. To address a decrease in gas stream flow through the filter (caused by an increased pressure drop), either the gas-moving equipment must be able to handle the increased pressure drop or the filter must be replaced or cleaned periodically.

Fabric filters also can be used to remove gaseous air pollutants by reaction with the particulates captured on the filter. In such a system, an absorbent or chemically active material is added to the gas stream before the stream reaches the fabric filter. The gaseous emissions either adsorb or chemically react with the material. With the gaseous material captured on the particulate, the fabric filter is used to remove both the material and the captured emissions.

Electrostatic Precipitators

Electrostatic precipitators (ESPs) use high-intensity electrical fields to cause particulates to acquire an electrical charge and then move to a collecting surface. The collecting surface of this abatement system can be either dry or wet. The electrostatic field influences the particulates (not the gas stream). After the particulates are collected on the plates, they must be removed from the plates without re-entraining them into the gas stream. This is usually accomplished by knocking them loose from the plates, thus allowing the collected layer of particles to slide down into a hopper, from which they are removed. Some precipitators remove the particles by intermittent or continuous washing with water.

Electrostatic precipitators operate with varying degrees of effectiveness. For a wide range of particulate loadings, ESPs provide a constant degree of removal efficiency. As a consequence, if the amount of emissions increases at the inlet to the ESP, the amount exiting the ESP also increases. Electrostatic precipitators remove particulates of different sizes with differing degrees of effectiveness. Overall, ESPs can be designed to remove 95% to 99% or more.

This abatement system has little resistance to the flow of the gas stream through the system and therefore the pressure drop across the precipitator is minimal. Pressure drop is very low and also does not tend to increase with time. In general, removal efficiency increases with the length of the passage through the precipitator; additional precipitator sections often are used in series to obtain higher collection efficiency.

Wet Scrubbing and Absorption Systems

Abatement systems that use contact with a liquid injected into the gas stream to capture emissions of the air pollutant are called absorbers or scrubbers. In general, absorbers are used to remove gaseous emissions from the gas stream and scrubbers are used to remove particulate emissions from a gas stream. Absorption is a process in which one or more soluble components of a gas mixture are dissolved in a liquid (i.e. a solvent). The liquid can be used to dissolve or react chemically with the emissions being removed. Physical absorption occurs when the absorbed compound dissolves in the solvent; chemical absorption occurs when the absorbed compound and the solvent react. Liquids commonly used as solvents include water, mineral oils, non-volatile hydrocarbon oils and aqueous solutions. For particulate emissions, the scrubbing liquid is used to intercept the particulates, potentially react with them, and then carry them away from the gas stream.

There are several ways to develop the contact between scrubbing liquid and gas stream. They include: spraying the liquid into open chambers with various forms of baffles or packing; bubbling the gas stream through tanks of a liquid; and producing liquid droplets and passing the gas stream around the droplets. The liquid can be recirculated to the absorber or scrubber or it may be necessary to treat all or part of the liquid as waste.

The relationship between collection efficiency and gas flow rate depends upon design, but removal efficiency usually increases as the pressure drop increases, assuming the liquid feed keeps pace with gas flow and carry-out of liquid with the gas stream is prevented. Removal efficiencies for gas absorbers vary for each pollutant–solvent system and with the type of absorber used. Most absorbers have removal efficiencies in excess of 90%, and packed tower absorbers may achieve efficiencies as high as 99.9% for some pollutant–solvent systems. For particulates, removal efficiencies vary widely depending on the particulate size and the solubility and reactivity of the particulate-scrubbing liquid.

Abatement of NO_x emissions from nitric and adipic acid manufacturing plants has occurred with extended absorption. Extended absorption or scrubbing reduces NO_x emissions by increasing absorption efficiency. This is achieved by either installing a single large tower, extending the height of an existing absorption tower, or by adding a second tower in series with the existing tower. Increasing the volume and the number of trays in the absorber results in more NO_x being recovered as nitric acid (1–1.5% more acid) and in reduced emissions. Extended absorption can be applied to new and existing plants; however, it is considered an add-on control only when applied to existing plants. Typically, retrofit applications involve adding a second tower in series with an existing tower. New plants are generally designed with a single large tower that is an integral component of the new plant design. New nitric acid plants have been constructed with absorption systems designed for greater than 99.7% NO_x recovery.

Adsorption Systems

Abatement systems in which gaseous air pollutants are captured on the surface of a bed of porous material through which the gas stream flows are called adsorbers. The material most often used is activated carbon. Other materials include silica gel, activated alumina, synthetic zeolites, fuller's earth and other clays. Because the physical structure of the bed is similar to that of a filter, the bed is typically protected from plugging by particulates by using a precleaning filter, so that the gas stream passing through the adsorbent is free of particulates. Pressure drop through an adsorber should not increase with time if the adsorber does not handle gas streams contaminated with particulate matter. The pressure drop does increase with an increase in gas stream flow rate.

There is typically little irreversible chemical reaction between the adsorbent and the gaseous pollutants in a well-designed system. The adsorbent is regenerated by using heat (hot gases or steam) or vacuum to de-absorb the captured gases. In some adsorbers, the adsorbent is regenerated in this manner for reuse, thus recovering the economic value of the captured gases. In some applications, the adsorbent is discarded and replaced with fresh adsorbent.

Four types of adsorption equipment have been used in collecting gases: (1) fixed regenerable beds; (2) disposable/rechargeable canisters; (3) traveling-bed adsorbers; and (4) fluid-bed adsorbers. Of these, the most commonly used in air pollution control are the fixed-bed and canister types. Fixed-bed units can be designed for controlling continuous, organic-containing streams over a wide range of flow rates, ranging from several hundred to several hundred thousand cfm. Canisters are best used where there is a predictable, modest quantity of organic emissions to be captured.

The organic concentration of streams that can be treated by adsorbers can be as low as several parts per billion by volume (ppbv) in the case of some toxic chemicals, or as high as 25% of the organic's lower explosive limit (LEL). For such organic-containing streams, carbon adsorbers can attain 80–95% removal, depending on the design. The design depends on the adsorbent, the depth of the adsorber bed, the velocity of the gas stream through the bed, the temperature of the system and other factors.

Fixed-bed adsorbers may be operated in either intermittent or continuous modes. In intermittent operation, the absorber removes organics for a specified time. After the absorber and the source are shut down, the unit begins the desorption cycle during which the captured organics are removed from the carbon. The desorption cycle includes: (1) regeneration of the carbon by heating, generally by blowing steam through the bed in the direction opposite to the gas flow; (2) drying the bed; and (3) cooling the bed to its operating temperature. At the end of the desorption cycle (which usually lasts 1–1½ h), the unit sits idle until the source starts up again.

In continuous operation, a regenerated carbon bed is always available for adsorption, so that the controlled source can operate continuously without shut down. For example, two carbon beds can be provided: while one is adsorbing, the second is desorbing/idled. Since each bed must be large enough to handle the entire gas flow while adsorbing, twice as much carbon must be provided than with an intermittent system handling the same flow. If the desorption cycle is significantly shorter than the adsorption cycle, it may be more economical to have three, four, or even more beds operating in the system. This can reduce the amount of extra carbon capacity needed, or provide some additional benefits relative to maintaining a low organic content in the gas stream.

Canister-type adsorbers differ from fixed-bed units in that they are normally limited to controlling low-volume (typically 100 cfm, maximum) intermittent gas streams, such as those emitted by storage tank vents. The carbon canisters are not intended for on site desorption. However, the carbon may be regenerated at a central facility. Once the carbon reaches a certain organic content, the unit is shut down, replaced with another, and disposed of or regenerated by the central facility.

Combustors

Control technology that uses heat to oxidize (combust) gaseous emissions are combustors. Combustors include flares, which are generally open to the atmosphere, and incinerators, which burn the gaseous emissions in a furnace with and without heat recovery. Flares and incinerators combust organic gases and produce carbon dioxide (CO_2) and water. Small amounts of nitrous oxides (NO_x), carbon monoxide (CO), other oxides of chemicals present in the emissions (e.g. sulfur oxides), and products of incomplete combustion also are produced.

Both flares and incinerators are more than 98% efficient if designed, constructed and operated properly. They provide destruction of virtually any gaseous organic emissions safely and cleanly, provided proper engineering design is used. One advantage of using combustion instead of recovery systems is that combustion devices are generally cheaper to install and operate. Combustion processes also are more efficient than recovery processes in reducing the organic emissions of the gas stream.

The primary drawback to using combustion processes is that they do not recover the organic emissions and they use fuel in the process. Where a heat recovery system is used in conjunction with an incinerator, the organic emissions and extra fuel can be used if the production plant needs the heat. In addition, the incinerating of chlorinated compounds may lead to the formation and emission of acid gases. These gases typically require additional control and incur added costs. Another drawback is that combustion devices are potential ignition sources. However, newly manufactured combustion devices have flame arresters to guard against flashback.

Flares

Flares combust organic emissions by igniting them as they pass through one or more burners. Flares may be either open or enclosed. An enclosed flare has fundamentally the same design as an open flare but adds a protective cylindrical shroud around the burners. Enclosed flares allow for control of operating temperature and residence time as well as emission testing. Flares are the least expensive combustion control system. Flares require little operator attention and will burn on their own as long as the incoming organic emissions contain enough hydrocarbon. Pilot burners are used to ensure that a flame is maintained in the event that the main flame goes out; the pilot burner is much smaller than the primary burners. Flares are usually more than 98% efficient as long as the combustion zone stays properly lighted. Flares, especially open ones, must be located away from people and equipment for safety reasons.

Incinerators

Incinerators operate by combusting organic emissions in a confined chamber under controlled conditions. Organic emissions enter the reaction chamber, combust, and then exit through an exhaust stack. Supplemental combustion air and fuel are added to the reaction mixture to maximize combustion efficiency. Combustion air is added to maintain an excess of oxygen, and supplemental fuel is added to maintain the desired operating temperature.

There are two basic designs of incinerators: recuperative and regenerative. Regenerative incinerators differ from recuperative incinerators in having heat-exchange media upstream and downstream of the reaction chamber. By periodically reversing the flow direction, regenerative incinerators can recover more energy than a recuperative incinerator equipped with a heat exchanger. In addition, incinerators can use a catalytic material to reduce the required temperature and yet achieve effective combustion.

When outfitted with a heat exchanger, recuperative incinerators can achieve up to 70% energy recovery. High destruction efficiencies (>99%) can be achieved by operating at higher temperatures or by increasing the residence time (via a larger combustion chamber), both of which result in increased fuel consumption. Recuperative incinerators are better suited to lower flows and richer gas streams that are more capable of supporting their own combustion.

Regenerative incinerators operate in much the same way as recuperative incinerators with heat exchangers. Regenerative incinerators, however, have direct-contact beds of silica gravel or ceramic burls that absorb heat from the exhaust gas. The direction of gas flow through the incinerator is reversed periodically. The bed that was being heated by the exhaust gas now pre-heats the incoming stream. By using this method of direct-contact heat exchange and reversing flow, energy recovery can be up to 95% efficient. Control efficiencies of up to 99% destruction have been achieved. The high-energy recovery makes regenerative incinerators well suited to lean, high-volume gas streams. In many cases, the pilot burner alone will supply the heat input necessary to maintain combustion.

Catalytic incinerators incorporate a bed of catalytic material that facilitates the overall combustion reaction. The catalyst has the effect of increasing the reaction rate, enabling combustion at lower reaction temperatures than in thermal incinerator units. The emissions and gas stream must be heated to a temperature sufficiently high (usually from 300 to 900°F) to initiate the oxidation reactions. The gas stream is preheated either directly in a pre-heated combustion chamber or indirectly by heat exchange with the incinerator's effluent or other process heat, or both. The pre-heated gas stream is then passed over the catalyst bed. The chemical reaction (combustion) between the oxygen in the gas stream and the gaseous pollutants takes place at the catalyst surface. Catalytic incineration in principle can be used to destroy essentially any oxidizable compound in the airstream. However, there are practical limits to the types of compounds that can be oxidized.

Particulate matter, including dissolved minerals in aerosols, can rapidly blind the pores of catalysts and deactivate them over time. Because essentially all the active surface of the catalyst is contained in relatively small pores, the particulate matter need not be large to blind the catalyst.

Work Practices for Equipment Leaks, Equipment Specifications for Storage Tanks, Design and Operational Requirements for Ventilation Systems and other Miscellaneous Emission Sources

Work practices are used to reduce emissions of leaks from equipment handling process fluids, typically organic liquids and gases. These fluids leak through the seals that cover drive shafts and other fittings that protrude through the equipment into the process fluids. Effective work practices are based on specifically searching for the presense of leaks with a portable organic vapor analyzer. If a leak is detected (as indicated by a measurement of 1000 ppm), then effective abatement occurs if the leak can be repaired within a specific time period, e.g. 5 days. For other equipment leaks, the specification of special seals (such as dual mechanical seals and closed sampling systems) is highly effective.

The use of advanced sealing systems represent effective abatement for storage vessels containing organic liquids, such as petroleum and chemicals. These sealing systems often involve the use of secondary seals that prevent evaporative losses above the normally installed primary seals on such vessels. Alternatively, the vessels can be closed and vented to an add-on control technology such as a carbon absorber or incinerator.

Most abatement systems are located some distance from the emission sources they control. The emissions and gas stream must be conveyed from the source to the abatement system device and from there to a stack before it can be released to the atmosphere. The kinds of equipment needed to carry the gas stream are the same for most kinds of control devices. These are: (1) hoods, (2) ductwork, (3) stacks, and (4) fans. Together, these items form the ventilation system. A hood is used to capture the emissions at the source; ductwork conveys them to the control device; a stack disperses them after they leave the device; and fans provide the energy for moving them through the control system. Properly designing and operating the ventilation system is critical to the effectiveness of abatement systems. In the case of many emission sources that are hooded to capture the emissions (especially for hot and windy conditions), if the hood is not adequate, then the emission will not be captured and significant abatement losses will occur.

Ensuring Continuous Performance

In selecting, designing, constructing and operating air pollution abatement systems, continuous performance of the system must be ensured. Systems that provide 90% removal efficiency 95% of the time are, in the end, more effective than systems that provide 99% removal only 50% of the time. As with all process equipment, proper operation and maintenance also is required to achieve the intended performance. The most certain way to ensure performance is to measure emissions from the abatement system and test the results continuously against the expected performance. Corrective measures must be implemented when performance begins to degrade. Where continuous emission monitors are not available, many surrogate measurement systems have been designed that can be correlated to the emissions performance of the abatement system. These parametric systems can be used to ensure proper operation and maintenance and compliance with the expected performance of the abatement system. To ensure overall management of the abatement system, it is also critical to provide records that allow for appropriate accountability of system performance. With a properly selected, designed and constructed system, management commitment ensures proper operation and maintenance and the achievement of the intended emission reductions.

COSTING OF AIR POLLUTANT ABATEMENT

Perspectives on Air Pollution Abatement Costing

Costing of the abatement approaches outlined above is very much like costing of other process equipment in a facility. Costing involves two principal phases: (1) a design phase in which the specifications are developed, the control technique is proposed, and the equipment is identified and sized to meet the specifications, and (2) a costing phase in which costs are developed either by obtaining vendor quotes or by using published cost functions or by a combination of these two. Cost functions are valid for a stated year and must be escalated to the desired year using the appropriate cost index (US EPA, 1995). Costing involves developing capital and operating costs, which may or may not be expressed as a total annualized cost by combining the annual operating costs with an annualized capital cost. The capital costs are annualized by using an acceptable equipment life and discount rate.

Designers of control equipment must consider both intensive and extensive parameters. Intensive parameters are independent of the size of the equipment and include such factors as temperature, pressure, concentration of air pollutant both before and after abatement, and composition of the stream. Extensive parameters are dependent on the size of the system and include such factors as mass or volumetric flow rate and pollutant flow rate.

The goal of the design and costing effort for air pollution abatement is normally to develop least-cost approaches to obtaining desired reduction in air pollution. These least-cost approaches, when plotted as a function of air pollutant emission reduction, will

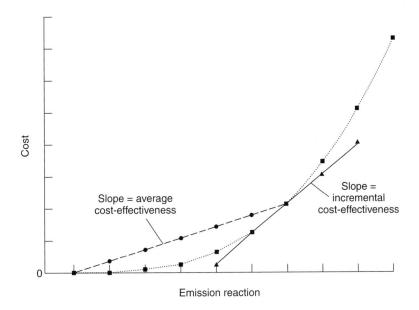

Fig. 41.1   Relationship between cost and emission reductions.

normally form a monotonically upward sloping curve, as shown conceptually in Figure 41.1. Points falling above this curve are considered inferior points, since the same emission reduction may be obtained at a lower cost by using an approach which falls on the curve. When considerations other than costs are important, it may be desirable to select an approach that is represented by an inferior point. The slope along the least-cost curve between a point and the immediately preceding point is termed the incremental cost-effectiveness, i.e. the cost per quantity of emissions reduced, for the approach represented by that point. Said another way, this slope represents the cost-effectiveness of using the approach represented by that point versus using the approach represented by the previous point. This incremental cost-effectiveness is normally more useful than an average cost-effectiveness, which is represented by the slope of the line drawn from the origin to a given point.

Comparisons of cost-effectiveness figures should be avoided between abatement measures that control different pollutants, between measures that apply nationwide and those that apply regionally, and between year-round and seasonal measures. Furthermore, because the cost and emission reduction data contain no information about human exposure or toxicity of the pollutants, cost-effectiveness figures provide little information about a control approach's health-related effectiveness.

Published Cost Functions for Add-on Abatement Equipment

Specific information with which to estimate 'study-type' estimates of add-on pollution control equipment are available in several published references. Study estimates are normally considered accurate to within ±30%. The US Environmental Protection Agency (EPA) 'OAQPS Control Cost Manual,' (Vatavuk, 1996) is a useful reference that outlines procedures, complete with cost functions, for the following types of add-on control equipment:

- thermal and catalytic incinerators
- carbon adsorbers
- fabric filters
- electrostatic precipitators
- flares
- refrigerated condensers
- gas absorbers
- hoods, ductwork and stacks.

In addition to individual chapters on these types of add-on control equipment, this reference contains two chapters that provide an overview of the manual and discuss cost estimating methodology. The cost manual also provides numerous references that pertain to control equipment costing.

Example Cost Information from the US Air Regulatory Program

The EPA's *Second Report to Congress on the Status of the Hazardous Air Pollutant Program under the Clean Air Act* (US EPA, 1997b) provides information on emission reductions

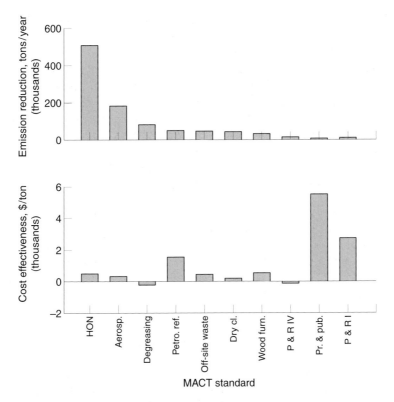

Fig. 41.2 Emission reductions and cost-effectiveness. Promulgated MACT standards, > 5000 tons/year reduction (2 and 4 year standards).

and total annualized cost of control for a number of regulations promulgated under Section 112 of the 1990 Clean Air Act. Section 112 concerns hazardous air pollutant regulation in the USA. This report shows that the costs under this legislation of controlling hazardous air pollution emissions from selected industries varies from about $230 million dollars per year to reduce about 500 million tons per year of air emissions from the organic chemical industry, to $14 million dollars per year to eliminate highly toxic chromium emissions from industrial cooling towers by using alternative cooling water treatment technologies. The report provides annualized cost of compliance for 12 hazardous air pollutant regulations that involve costs of >$10 million per year in the USA.

Figure 41.2 presents the emission reductions and average cost effectiveness for several regulations promulgated under Section 112. These average cost-effectiveness values range from cost savings as a result of de-greasing regulations to about $5400 per ton of emissions reductions for the regulation for printing and publishing. Most of the regulations have an average cost-effectiveness of less than $500 per ton of emissions reductions. An average cost-effectiveness for control from sources with small amounts of very toxic compounds such as hexavalent chromium or coke oven emissions is significantly higher than for those sources cited above.

INNOVATIVE APPROACHES TO AIR POLLUTANT ABATEMENT IN THE USA

No discussion of costing of air pollutant abatement is complete without a discussion of innovative approaches to reducing air pollutant emissions. Estimating the cost of these approaches to emission reduction goes beyond cost estimation using only engineering costing techniques. The result of employing these or similar innovative approaches may well be more sustainable, effective and efficient air pollutant abatement.

The EPA *Regulatory Impact Analysis for the Particulate Matter and Ozone National Ambient Air Quality Standards and Proposed Regional Haze Rule* (US EPA, 1997a) points out that in the USA:

> . . . some sweeping changes are occurring in the way environmental standards are being implemented. Several efforts are underway to create new regulatory processes that afford greater flexibility with the goal of lowering the costs of meeting environmental protection goals. These efforts include a variety of market-based incentive systems. Market-based systems to reduce pollutant emissions have been promoted for many years as an alternative to fixed regulatory standards. Such systems are expected to reduce the costs of compliance and induce more technological innovation in methods of reducing pollution.

> National and regional market-based programs such as emissions trading may achieve pollution control goals at dramatically less expense because they allow firms that face high costs to purchase 'extra' reductions from firms facing below-average control costs. This RIA models a SO_2 cap and trade program, but due to data limitations, does not attempt to model other potentially cost saving market-based programs. However, the lead and chlorofluorocarbon (CFC) phase-out plans and the Acid Rain program are all examples of the ability of national market-based programs to provide environmental protection at lower cost. With pollution control efforts pegged to the going price of allowances, rather than to the highest cost source, these market-based programs can promote both cheaper and faster compliance.

In addition,

> . . . market-based programs provide continuous and powerful incentives to develop new technologies while achieving emission reductions which otherwise would not be available under the typical regulatory approach.

INFORMATION ON ABATEMENT SYSTEMS

Several texts are available for general information on the application of abatement systems to specific processes and industries. With the assistance of the EPA the Air and Waste Management Association (AWMA) produced the *Air Pollution Engineering Manual* (Buonicore and Davis, 1992). This comprehensive manual covers the basics of abatement systems and the specific application of these systems to many industries.

The EPA maintains internet websites that provide a contact for further information. The EPA's program office with a specialty in air pollution abatement has a website located at www.epa.oaqps.gov, where several web pages are available. The EPA's technology transfer network (at http://www.epa.gov/ttn) and the Clean Air Technology Center (at http://www.epa.gov/oar/oaqps/ctc) can be used to download EPA's cost manual and many other air pollution abatement references. This information is available through the Internet without charge, except for the charges associated with Internet access.

The EPA produces a background information document and maintains a rule-making docket that contains detailed engineering analyses for each of the many rule-makings. These documents provide information on the current performance and costs associated with abatement systems in specific industries. The EPA also maintains a database of technological determinations that define, on a case-by-case basis, the best available control technology (BACT) and the lowest achievable emission rate (LAER).

REFERENCES

Buonicore AJ and Davis WT (1992) *Air Pollution Engineering Manual.* Air and Waste Management Association. New York: Van Nostrand Reinhold.

US EPA (1995) *Escalation Indexes for Air Pollution Control Costs*, EPA-453/R-95-006. This document is available on the World Wide Web, http://www.epa.gov/ttn/catc/products.html.#ccinfo

US EPA (1997a) *Regulatory Impact Analysis for the Particulate Matter and Ozone National Ambient Air Quality Standards and Proposed Regional Haze Rule.* Innovative Strategies and Economics Group, Office of Air Quality Planning and Standards, US Environmental Protection Agency, Research Triangle Park, NC. This document is available on the World Wide Web, http://www.epa.gov/ttn/oarpg/t1ria.html

US EPA (1997b). *Second Report to Congress on the Status of the Hazardous Air Pollutant Program under the Clean Air Act.* Reference: Clean Air Act of 1990, Section 112(s), EPA-453/R-96-015. This document is available on the World Wide Web, http://www.epa.gov/ttnuatw1/1125/1125.html

Vatavuk WM (1996) *OAQPS Control Cost Manual,* 5th edn, EPA-453/B-96-001. This document is available on the World Wide Web, http://www.epa.gov/ttn/catc/products.html.#ccinfo and from the National Technical Information Service, 5285 Port Royal Road, Springfield, VA, 22161 (phone: 1-800-553-6847).

United States and International Approaches to Establishing Air Standards and Guidelines

LESTER D. GRANT, CHON R. SHOAF and J. MICHAEL DAVIS

National Center for Environmental Assessment, RTP Division, US Environmental Protection Agency, Research Triangle Park, NC, USA

INTRODUCTION

During the past 30 years much progress has been made in the USA and elsewhere towards controlling air pollution and thereby reducing adverse impacts on human health and the environment. Contributing greatly to this progress has been (a) the codification, into national laws and/or international agreements, of approaches for establishing air quality guidelines or standards that define goals or targets to be met in reducing various ambient air pollutants and their precursors; (b) the development of specific guidelines or standards for particular substances; and (c) the implementation of control strategies that contribute toward meeting the established goals. The elaborate system established for developing and implementing US air pollution standards has proven to be successful in many regards and has often served as a model internationally, with numerous other nations having adopted or adapted various features of that system. The experiences and technical knowledge of many other industrialized and developing countries have also been shared and drawn upon extensively to facilitate worldwide efforts to ameliorate air pollution problems. Probably the most influential mechanisms by which such international sharing of information and expertise has been accomplished are environmental programs carried out under the auspices of one or more components of the United Nations, with certain elements of air pollution-related programs and activities conducted by the World Health Organization (WHO) being of special interest here.

Comprehensive discussion of the full range of elements comprising air pollution programs employed in the USA or internationally is beyond the purpose and scope of the present chapter. Rather, this chapter mainly focuses on providing a concise overview of key features of the US approach to establishing air pollution standards. It also provides, for comparison, a briefer overview of approaches employed by WHO for generating international air pollution guidelines. Only limited listings or discussions of guidelines or standards for specific substances are included, but the reader is directed in the chapter to other bibliographic sources or Internet websites for more detailed information on particular substances.

As for its organization, the chapter first orients the reader to the US Clean Air Act, which contains key legislative requirements mandated by the US Congress dealing with several major classes of air pollutants. These include certain widespread pollutants with potential to impact large segments of the general population and the environment of the USA (as is the case for these pollutants in many other countries). The discussion of these includes a brief description of the pertinent key Clean Air Act requirements. We also discuss the process used for periodic review of the standards, the current US standards, and the related air pollution alert system designed to protect against effects of acute exposure to 'criteria air pollutants', as they have come to be designated in the USA. The US approach for dealing with other hazardous air pollutants and associated potential health or environmental risks is next discussed. This is followed by an overview of certain US activities related to dealing with controversial issues concerning selected fuels and fuel additives of interest not only to the USA, but to the international community as well. Lastly, salient features of approaches employed by WHO in developing environmental health criteria and air quality guidelines for international use are also concisely summarized.

ORIENTATION TO THE US CLEAN AIR ACT

The first national legislation for abatement and prevention of air pollution in the USA (i.e. the US Clean Air Act; CAA) was enacted in 1963. Subsequent revisions to the Act increasingly expanded the role of the US Federal Government in the establishment and enforcement of air regulations. Of particular note, the Clean Air Act Amendments of 1970 (US Code, 1970) mandated that the Federal Government develop and promulgate National Ambient Air Quality Standards (NAAQS) specifying uniform, nationwide limits for certain major air pollutants. The Act, as amended in 1977 (US Code, 1977) and 1990 (US Code, 1991), contains a wide range of legislative requirements dealing with various other air pollution problems and issues. The Act is organized in terms of six main 'Titles' as shown in Table 42.1.

Of particular interest are those Clean Air Act sections that provide legislative requirements or guidance with regard to developing air quality standards for 'criteria air pollutants' (Title I, Part A, Sections 108 and 109); developing strategies for hazardous air pollutants (Title I, Part A, mainly Section 112); and dealing with selected mobile source pollutant issues of much current interest with regard to certain fuels and fuel additives (Title II, Part A, especially Section 211). Sections 108 and 109, as explained in more detail below, set forth basic requirements to be met in setting and revising nationwide air

Table 42.1   Organization of US Clean Air Act, as amended in 1990 (US Code, 1991)

Title I: Air Pollution Prevention and Control
   Part A   Air Quality and Emission Limitations
   Part B   Ozone Protection (replaced by Title VI)
   Part C   Prevention of Significant Deterioration of Air Quality
   Part D   Plan Requirements for Nonattainment Areas

Title II: Emission Standards for Moving Sources
   Part A   Motor Vehicle Emission and Fuel Standards
   Part B   Aircraft Emission Standards
   Part C   Clean Fuel Vehicles

Title III: General

Title IV: Acid Deposition Control

Title V: Permits

Title VI: Stratospheric Ozone Protection

More details on provisions included under each of the above Clean Air Act Titles are available via the Internet at http://www.epa.gov/oar/caa/contents.html.

quality standards for several of the most widespread air pollutants that have potential for posing risks to human health and the environment across many US areas. Section 112 addresses other 'hazardous air pollutants'. For these, attention is focused mainly on approaches to emissions control (rather than attainment of specific air quality standards for the subject pollutants), with resulting benefits in reducing potential health risks associated with the specific toxic air pollutants so controlled and in reducing contributions of some to formation of 'criteria air pollutants'. Analogously, Clean Air Act mobile sources provisions focus mainly on emissions limitations aimed both at minimizing risks associated directly with exposures to fuels or fuel additives or their combustion products in vehicle emissions and at minimizing exacerbation of 'criteria air pollutant' problems.

DEVELOPING US STANDARDS FOR 'CRITERIA AIR POLLUTANTS'

Section 108 of the 1970 Clean Air Act (and its 1977 and 1990 Amendments) directs the Administrator of the US Environmental Protection Agency (US EPA) to identify and issue air quality criteria for ubiquitous (widespread) air pollutants that may reasonably be anticipated to endanger public health or welfare. The 'air quality criteria' are to reflect the latest available scientific information on the nature and extent of all identifiable effects on public health or welfare that may be expected from the presence of the pollutant in ambient (i.e. outdoor) air. Section 109(a) of the CAA further directs the US EPA Administrator to propose and promulgate primary and secondary NAAQS for pollutants identified under Section 108. Primary standards for such 'criteria air pollutants' are to be set at levels (concentrations, averaging times, etc.) which, in the judgment of the EPA

Administrator, are necessary to protect the public health – i.e. to protect against adverse health effects in sensitive population groups with an adequate margin of safety. Section 109(b)(2) requires that the US EPA Administrator set secondary NAAQS to protect against welfare effects, e.g. damage to vegetation (agricultural crops, natural ecosystems, etc.), impacts on visibility and climate, damage to manufactured materials, etc. Also, periodic (at least every 5 years) review and revision of the criteria and, as appropriate, the primary and/or secondary NAAQS for a given pollutant are required by Section 109(d). Such periodic reviews of the criteria and NAAQS are to be carried out in accordance with previously noted Section 108 and 109 requirements. Section 109(d) also further mandates the establishment of a Clean Air Scientific Advisory Committee (CASAC), i.e. a panel of independent non-EPA experts, not only to advise the US EPA Administrator on the soundness of existing criteria and NAAQS, but also to identify research needed to improve the scientific bases for the NAAQS and to advise on other aspects related to monitoring and implementation of the NAAQS. The CASAC is administratively housed as part of US EPA's Science Advisory Board (SAB).

US EPA CRITERIA AND NAAQS REVIEW/REVISION PROCESS

US National Ambient Air Quality Standards were established in 1971 (Federal Register, 1971) for the following air pollutants (or pollutant classes): particulate matter (PM); ozone (O_3) and other photochemical oxidants; nitrogen dioxide (NO_2); sulfur dioxide (SO_2); carbon monoxide (CO); and hydrocarbons (later rescinded). Lead (Pb) was listed as a 'criteria air pollutant' in 1976, and the Pb criteria and NAAQS were finalized and promulgated in 1977. Since the passage of the 1977 Clean Air Act Amendments (US Code, 1977), the US EPA has undertaken periodic review and, as appropriate, revision of the criteria and NAAQS for these pollutants. The process by which the criteria are revised and US NAAQS decisions are proposed and promulgated (see Fig. 42.1) is somewhat complicated, but is necessary to meet provisions of both the US Clean Air Act and other applicable US statutes, e.g. the Federal Advisory Committee Act (FACA). The FACA requires, for example, that: (1) Federal advisory committees be chartered as standing committees; (2) committee meetings be announced in advance and open to the public; and (3) materials to be reviewed by Federal advisory committees be made available to the public and opportunities afforded to the public for comment on the materials.

   The research shown in Fig. 42.1 as being drawn upon by the US EPA to support development of criteria and standards for major (criteria) air pollutants is sponsored and/or conducted by the US EPA's Office of Research and Development (ORD) Laboratories; other US Federal, State and local government agencies; academic scientists; industry; foreign governments; international bodies (e.g. European Union, WHO, etc.); and other US and international stakeholders. Assessments of the latest available scientific information concerning the nature of effects due to given pollutants and other pertinent topics are presented in air quality criteria documents (AQCDs) prepared by ORD. Such documents are authored by ORD scientists and non-EPA experts. Initial drafts are typically peer-reviewed at EPA-sponsored expert workshops, then revised and released for multiple cycles of public comment and CASAC review before final publication. The AQCDs typically comprise an extensive assessment of the latest available scientific information on atmospheric processes; sources/emissions; air quality, human and ecological exposure; dosimetry; health effects (animal toxicology, controlled human exposure, epidemiology); welfare effects; and interpretive summaries. As the next step, often initiated in parallel with the final phases of AQCD preparation and CASAC reviews, a staff paper is prepared by the Office of Air Quality Planning and Standards (OAQPS) in the EPA's Office of Air and Radiation (OAR) with ORD scientific support. The staff paper includes interpretation of effects information based on the AQCD, combined with exposure analyses for risk characterization, and provides the rationale for policy options for consideration by the US EPA Administrator for proposing pertinent decisions (to retain, revise or add NAAQS for the given pollutant).

NAAQS proposals are then developed by OAQPS/OAR with inputs and assistance from ORD scientists. The proposals are published in the US Federal Register, followed by public hearings and written comments on the proposals. ORD provides scientific/technical support to OAR in evaluating and responding to public comments on the proposals. Final NAAQS decisions are made by the US EPA Administrator, taking into account public comments and CASAC advice. Promulgation of final NAAQS decisions includes publication in the US Federal Register of the rationale for selection of the level and form of the primary and/or secondary NAAQS for a given pollutant or pollutant class. It should be noted that, whereas the US EPA sets the NAAQS, individual states are responsible for implementing the standards. As part of this, State Implementation Plans (SIPs) are developed by each state that describe control strategies by which the state plans to bring non-attainment areas into compliance with the NAAQS for a given pollutant. The SIPs are submitted to and reviewed by the US EPA, and steps are taken by the

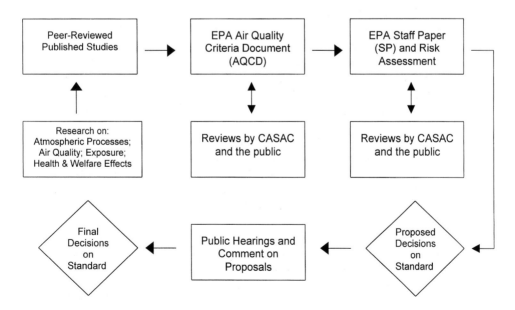

Fig. 42.1 US EPA criteria and NAAQS review/revision process.

US EPA to ensure that adequate SIPs are in place for each state to move non-compliance areas, not meeting given NAAQS, towards meeting the standard(s).

Table 42.2 lists the current US NAAQS for the seven air pollutants listed for regulation under US Clean Air Act, Sections 108 and 109. The most recent periodic review of criteria and NAAQS for given criteria air pollutants was completed by the US EPA in early 1996, providing key scientific inputs to ultimate promulgation in July 1997 of the current PM and O_3 NAAQS.

The last prior revision of the US PM NAAQS occurred in 1987, when the original 'Total Suspended Particulate' (TSP) NAAQS set in 1971 (Federal Register, 1971) were revised to protect against adverse health effects of inhalable airborne particles with an upper 50% cut-point of 10 μm aerodynamic diameter (PM_{10}), which can be deposited in the lower (thoracic) regions of the human respiratory tract (Federal Register, 1987). The US EPA PM AQCD (1996a) and PM Staff Paper (1996b) completed in 1996 characterize key elements of the scientific bases for the EPA's promulgation (Federal Register, 1996a) of decisions to retain, in modified form, the US PM_{10} NAAQS (150 μg/m³, 24 h; 50 μg/m³, annual average) and to add new $PM_{2.5}$ NAAQS (65 μg/m³, 24 h; 15 μg/m³,

Table 42.2　US National Ambient Air Quality Standards (NAAQS)[a]

Pollutant		NAAQS concentration	Standard type
Particulate matter < 10 μm (PM$_{10}$)			
24-h Average		150 μg/m³	Primary and secondary
Annual arithmetic mean		50 μg/m³	Primary and secondary
Particulate matter < 2.5 μm (PM$_{2.5}$)			
24-h average		65 μg/m³	Primary and secondary
Annual arithmetic mean		15 μg/m³	Primary and secondary
Ozone (O$_3$)			
24-h average	0.12 ppm	235 μg/m³	Primary and secondary
Annual arithmetic mean	0.08 ppm	157 μg/m³	Primary and secondary
Sulphur dioxide (SO$_2$)			
24-h average	0.14 ppm	365 μg/m³	Primary
Annual arithmetic mean	0.03 ppm	80 μg/m³	Primary
3-h average	0.50 ppm	1300 μg/m³	Secondary
Nitrogen dioxide (NO$_2$)			
Annual arithmetic mean	0.053 ppm	100 μg/m³	Primary and secondary
Carbon monoxide (CO)			
1-h average	35 ppm	40 μg/m³	Primary
8-h average	9 ppm	10 μg/m³	Primary
Lead (Pb)			
Quarterly average		1.5 μg/m³	Primary and secondary

[a]　For detailed information on scientific bases and policy considerations underlying decisions establishing the NAAQS listed here, see the AQCDs, staff papers, and NAAQS Promulgation notices cited in the text. Such information can also be obtained from several Internet websites (e.g. http://www.epa.gov/airs/criteria.html; http://www.epa.gov/oar/oaqps/publicat.html; and http://www.epa.gov/ncea/biblio.htm).

annual average) to protect against adverse health effects associated with exposures to fine particles. Some of the more salient aspects of these extensive recent assessments are summarized below.

The scientific bases for the 1997 PM NAAQS decisions include important distinctions between fine and coarse ambient air particles with regard to size, chemical composition, sources and transport. Also important are scientific issues concerning relative contributions of particles of ambient air origin (as measured at outdoor community monitoring stations) to total human exposures to PM, which also typically include indoor air exposures to particles generated from indoor sources as well as to particles of ambient origin that infiltrate into indoor microenvironments. Another key element was the assessment and interpretation of newly emerging findings on airborne particle health effects, especially as derived from newly published community epidemiologic studies. The assessment of extensive epidemiology information in the PM AQCD (US EPA, 1996a) and Staff Paper (US EPA, 1996b) highlighted over 80 key epidemiology studies, more than 60 of which found significant associations between premature mortality and/or morbidity and PM levels at or below the 1987 PM_{10} NAAQS. The main health effects of concern that the community epidemiology studies showed to be statistically associated with ambient PM exposures include evidence of increased mortality (especially among people over 65 years of age and those with pre-existing cardiopulmonary conditions) and morbidity (indexed by increased hospital admissions, respiratory symptom rates, and decrements in lung function).

Much interest and controversy have focused on the assessment of the methodological soundness and interpretation of the community epidemiology studies, as discussed extensively in the PM AQCD, Staff Paper, and published NAAQS Promulgation notice. As noted in the PM AQCD, several viewpoints have emerged on how best to interpret the epidemiology data. One sees PM exposure indicators as surrogate measures of complex ambient air pollution mixtures, with reported PM-related effects representing those of the overall mixture; another holds that reported PM-related effects are attributable to PM components (*per se*) of the air pollution mixture and reflect independent PM effects; or PM can be viewed both as a surrogate indicator as well as a specific cause of health effects.

Several aspects of the epidemiologic evidence (e.g. the consistency and coherence of the findings), as discussed in the PM AQCD and Staff Paper, support the conclusion that exposure to ambient PM, acting alone or in combination with other pollutants, is probably a key contributor to the increased mortality and morbidity risks observed in the epidemiology studies. Reliance was placed by the EPA on the relative risk (RR) levels for increased risks of mortality or morbidity reported in US and Canadian PM epidemiology studies, which provided the most directly pertinent quantitative risk estimates as inputs to US PM NAAQS decisions. These included relative risk estimates for mortality or morbidity (indexed by increased hospital admissions, respiratory symptoms, etc.) associated with 50 $\mu g/m^3$ increases in 24-h PM_{10} concentrations or with variable increases in fine particle indicators (e.g. 25 $\mu g/m^3$ increment in 24-h $PM_{2.5}$ concentrations) and analogous relative risk estimates for health effects related to specified increments in annual mean or median levels of fine particle indicators. The assessed study results were found to provide sufficient evidence for concluding that significant associations of increased mortality and morbidity risks were likely attributable to fine particles, as indexed by various fine particle indicators (e.g. $PM_{2.5}$, sulfates, etc.). However, it was also concluded that possible toxic effects of the coarse fraction of PM_{10} cannot be ruled out. PM_{10} coarse fraction particles do reach sensitive areas of the lower respiratory tract, and health effects of concern (e.g. aggravation of asthma and increased respiratory illness), particularly in children, are suggested by some epidemiology studies.

Both the PM AQCD (US EPA, 1996a) and Staff Paper (US EPA, 1996b) noted the very limited extent of available toxicologic findings by which (a) to identify key PM constituents of urban ambient air mixes that may be causally related to the types of mortality/morbidity effects observed in the community epidemiologic studies; or (b) to delineate plausible biological mechanisms by which such effects could be induced at the relatively low ambient PM concentrations evaluated in the epidemiologic studies. As discussed in the PM AQCD, several types of mechanisms have been shown to underlie toxic effects observed with acute or chronic exposures to various PM species or mixtures (e.g. acute lung inflammation; particle accumulation/overload; impaired respiratory function; impaired pulmonary defense mechanisms), but generally at much higher PM concentrations than currently encountered in the USA. As also discussed in the PM AQCD and Staff Paper, several fine particle constituents can be hypothesized as being likely important contributors to ambient PM effects. Those of much current interest include acid aerosols (indexed by sulfates; H+ ions, etc.); transition metals (e.g. Fe, Mn, etc.); and ultrafine particles (the latter mainly based on the possibility that the number or the larger total surface area of extremely fine particles, rather than particle mass alone, may be of much importance in inducing PM-related toxic effects).

The latest revisions to the O_3 NAAQS, reflected in Table 42.2, were carried out in parallel to the above-noted review of the PM NAAQS, with the retention of the earlier 1-h O_3 primary NAAQS. An additional new 8-h O_3 primary NAAQS was promulgated in July 1997 (Federal Register, 1997a). The secondary O_3 NAAQS, to protect against O_3 effects on sensitive crops and other vegetation, was also retained. The scientific bases and policy considerations taken into account in the most recent US EPA O_3 NAAQS decisions were assessed and described in the EPA O_3 AQCD (US EPA, 1996c) and Staff Paper (US EPA, 1996d) finalized in 1996. The original 1-h NAAQS promulgated in 1971 (Federal Register, 1971), i.e. 0.08 ppm (measured as total photochemical oxidants), was revised in 1978 to 0.12 ppm O_3 (Federal Register, 1979), based on then relatively limited data from controlled human exposure studies indicating increased bronchoconstriction and/or respiratory symptoms in exercising asthmatic subjects at O_3 concentrations as low as 0.15 ppm. Key health effects findings supporting the most recent O_3 primary NAAQS decisions include extensive new evidence from numerous controlled human exposure studies conducted since 1978 that substantiate the occurrence of bronchoconstrictive effects and respiratory symptoms in asthmatic subjects and some healthy individuals with 1-h O_3 exposures at > 0.12 ppm during moderate to heavy exercise. Promulgation of the new 8-h O_3 NAAQS was based primarily on evidence for similar effects and for indicators of lung inflammation being elevated with more prolonged 6–8 h exposures to O_3 at even lower (< 0.10 ppm) levels. Observation in field studies of analogous pulmonary function decrements among children exposed to somewhat lower O_3 concentrations in ambient air during outdoor exercise at summer camp provided further bases supporting the NAAQS decisions. An important contribution to the final NAAQS decisions came from results of animal toxicology studies showing altered collagen production and other effects associated with chronic exposures to O_3. Such results are seen mainly in laboratory animals at O_3 concentrations markedly higher than ambient and are difficult to extrapolate quantitatively to human equivalent exposures. Nevertheless, these results have raised concerns that potential alterations in lung structure (e.g. airway remodeling) associated with frequent repeated exposures of humans over many years to ambient O_3 concentrations could cause acute lung function decrements and/or inflammation. Children exposed to O_3 while exercising outdoors were highlighted as a susceptible population group at special risk, owing to the increased delivery of O_3 to lower respiratory tract tissues associated with increased breathing (ventilation) rates during normal play activities.

The primary SO_2 standards listed in Table 42.2 are the same as originally promulgated in 1971 (Federal Register, 1971), based primarily on epidemiologic studies associating increased mortality and/or morbidity with short-term (24-h) and long-term (annual average) ambient SO_2 exposures. US EPA assessments of new scientific information, contained in SO_2 AQCD materials (US EPA, 1986a, 1994a), have thus far not led the EPA to revise the SO_2 NAAQS. As discussed in the SO_2 NAAQS decision notice published in 1996 (Federal Register, 1996b), much attention has been focused in the 1990s on evaluation of the possible need for an even shorter-term (< 1 h) SO_2 NAAQS, to protect against exacerbation of respiratory effects (lung function decrements, respiratory symptoms) of the type observed with brief (5–10 min) experimental exposures of asthmatic subjects to SO_2 during moderate to heavy exercise. Coupling dose–response information for such effects with exposure analyses results, the US EPA obtained risk characterization projections that led the Agency to conclude that the likely low frequency of occurrence in the USA of ambient exposures to effective SO_2 concentrations of asthmatics while exercising outdoors at moderate or high exercise levels did not warrant the setting of any new shorter-term (< 1 h) SO_2 NAAQS. Rather than setting such nationally applicable NAAQS, other approaches are intended to be used for limiting localized, high concentrations of SO_2 of possible health concern.

The NO_2 primary NAAQS (0.053 ppm annual average) listed in Table 42.2 is also unchanged from 1971 (Federal Register, 1971). The epidemiologic analyses cited as the key bases for the 1971 standards were later judged by subsequent EPA reassessments as not yielding a credible basis for quantitative estimation of long-term ambient NO_2 concentrations associated with health effects in the general population. However, evaluations of newly available indoor air epidemiology studies (including by meta-analysis) presented in the most recent 1993 NO_x AQCD (US EPA, 1993a) and 1994 Staff Paper (US EPA, 1995a) suggest that an increased history of respiratory illness in children (6–12 years old) is likely associated with repeated indoor exposures to NO_2 from gas stoves. Other new controlled human exposure study data, which indicate increased bronchial reactivity due to short-term (< 3-h) NO_2 exposures at > 0.20 ppm, also contributed to the 1996 NO_2 NAAQS decision to retain the original primary standard (Federal Register, 1996c). US EPA analyses showed a very low probability of short-term (≤ 3 h) upward excursions of NO_2 to levels of possible health concern occurring at those community monitoring sites in the Los Angeles metropolitan area (the only US urban area out of NO_2 attainment) meeting the annual NO_2 NAAQS.

The key bases for the CO NAAQS promulgated in 1971 (Federal Register, 1971), i.e. certain neurobehavioral effects associated with relatively low carboxyhemoglobin (COHb) levels, were re-evaluated in subsequent US EPA assessments as not providing a credible basis for the NAAQS. Rather, the assessments in the latest EPA CO AQCD (US EPA, 1991a) and Staff Paper (US EPA, 1992a) have shifted attention to cardiovascular effects thought to be induced by hypoxic conditions due to CO binding to hemoglobin (elevated COHb levels in blood are used as a standard biomarker for CO exposure). Based mainly on several controlled human exposure studies demonstrating more rapid onset of angina in exercising heart patients as a result of CO exposures producing relatively low COHb levels, the US EPA promulgated in 1994 the primary NAAQS for CO shown in Table 42.2 (Federal Register, 1994a).

Lastly, the 1.5 $\mu g/m^3$ (90-day average) NAAQS for lead (Pb) shown in Table 42.2 remains unchanged from the level originally promulgated in 1977 (Federal Register, 1977). That NAAQS was based primarily on (1) evaluations in the 1977 Pb AQCD (US EPA, 1977) concluding that increased risks of 'subclinical' neuropsychological effects (e.g. IQ decrements) and sufficiently severe hemoglobin synthesis inhibition occurred in young children at blood lead (PbB) levels $\geqslant$ 30 $\mu g/dl$, and (2) other analyses parceling out average air lead and other, non-air (e.g. from water and diet) lead exposure contributions to then average US pediatric blood lead distributions. Subsequent US EPA periodic assessments (US EPA, 1986b, 1989a, 1990) of new, rapidly expanding Pb health effects information led to downward revision of Pb effect levels of concern from 30 $\mu g/dl$ to 10 $\mu g/dl$ PbB, a blood lead target level widely adopted by the US EPA and other Federal agencies for use in supporting multimedia lead regulations and enforcement actions. Actions taken to reduce or keep blood lead levels below 10 $\mu g/dl$ in young infants and children, as well as women of child-bearing age (due to risk to the fetus in case of pregnancy), are aimed at avoiding unacceptable risks of an array of lead-related pathophysiological effects (e.g. slowed neuropsychological development, impaired heme synthesis in blood cells and other tissues, etc.) of sufficient severity to be judged adverse as PbB levels reach or exceed 10–15 $\mu g/dl$. A 1990 US EPA Pb Staff Paper (US EPA, 1990) recommended consideration of revising the 1.5 $\mu g/m^3$ NAAQS to levels ranging down to as low as 0.5 $\mu g/m^3$. However, given marked decreases in ambient air lead levels in US cities mostly to below 0.25 $\mu g/m^3$ (mainly the result of phasing out lead in gasoline, which was the single most widespread former Pb exposure source for the general US population), the need for lowering the US Pb NAAQS was largely obviated. Instead, attention is being given to other, more focused approaches to dealing with scattered US air exposure situations of health concern.

US POLLUTANT STANDARD INDEX (PSI)

The US NAAQS also serve as the starting point in developing the bases for an important risk communication mechanism. That is, in addition to promulgating the NAAQS for 'criteria air pollutants', the US EPA has established an air pollution alert system. This serves to communicate to the general public and local authorities easily understandable information on potential health risks posed by acute exposures to elevated levels of one or another of the criteria air pollutants and to characterize actions that are advisable in order to lessen health risks for sensitive population groups. The alert system uses the US EPA-generated Pollutant Standard Index (PSI), which is summarized in Table 42.3. PSI values of 100 are assigned to short-term primary NAAQS concentration levels set to protect against acute exposure health effects. Pollutant concentrations below the NAAQS are assigned PSI values associated with appropriate descriptors (good, moderate), and those higher than the NAAQS are assigned increasingly higher PSI values associated with other appropriate descriptors (e.g. unhealthful, hazardous, etc.). Various PSI values and descriptors are also tied to (1) descriptions of general types of health effects anticipated to occur among susceptible population groups; and (2) cautionary statements recommending actions to lower the risk of such effects occurring. The US EPA is currently in the process of revising the PSI to include new entries appropriate for alert levels related to the new $PM_{2.5}$ and 8-h O_3 NAAQS promulgated in 1997. The PSI is also part of an expansion of Americans' 'right-to-know' about their environment, as embodied in the President's Environmental Monitoring for Public Access and Community Tracking (EMPACT) initiative. Information on the US PSI can be obtained via the Internet (http://www.epa.gov/ttnamti1/psi.html).

A number of other countries have adopted air pollution alert systems analogous to the US PSI. For example, in 1997 the Malaysian government effectively used an analogous Air Pollution Index (API) as a means to alert the general population in Kuala Lumpur and elsewhere in Malaysia to potential health risks associated with exposures to components of smoke haze from biomass fires in Indonesia exacerbated by El Niño meteorological conditions. Similarly, Mexican government authorities have employed analogous 'IMECA' values routinely in communicating information on the air pollution situation in Mexico City and elsewhere and, also, in calling alerts in spring 1998 due to smoke from biomass fires exacerbated by El Niño conditions.

US STRATEGIES FOR HAZARDOUS AIR POLLUTANTS

The 1990 Clean Air Act Amendments (CAAA; US Code, 1991) established a new US Federal Program for Hazardous Air Pollutants (HAP) which represented a dramatic change from earlier HAP provisions. The 1990 CAAA: (1) list certain toxic air pollutants for regulation as HAPs; (2) provide mechanisms for listing and delisting of HAPs; (3) provide authority for additional HAPs testing; and (4) include a major Federal regulatory program of emission standards to address HAPs from industrial sources, with strong incentives for industry innovation and pollution prevention. The 1990 CAAA also mandate determination of risk remaining after implementation of emission standards; a strategy to quantitate and ameliorate risks from urban air toxics; a role for the National Academy of Sciences in assessing the state of the science of risk assessment; and reports to Congress on environmental effects of HAPs on the Great Waters of the USA, on emissions from electric utility steam generating plants, and on hazards from mercury emissions. Activities in these areas are described below.

Listing/Delisting of HAPs

Section 112 of the 1990 CAAA mandates EPA to control hazardous air pollutants (HAPs). An initial list of 189 HAPs (now 188), included in Section 112(b)(1), identifies listed pollutants which present, or may present, through inhalation or other routes of exposure, a threat of adverse human health effects (including, but not limited to, substances known to be or reasonably anticipated to be carcinogenic, mutagenic, teratogenic,

Table 42.3 Comparison of PSI values with pollutant concentrations, descriptor words, generalized types of health effects and cautionary statements[a]

Index value	Air quality level	Pollutant levels					Health effect descriptor	General health effects	Cautionary statements
		PM_{10} (24-h) $\mu g/m^3$	SO_2 (24-h) $\mu g/m^3$	CO (8-h) ppm	O_3 (1-h) ppm	NO_2 (1-h) ppm			
500	Significant harm	600	2620	50	0.6	2.0	Hazardous	Premature death of ill and elderly. Healthy people will experience adverse symptoms that affect their normal activity	All persons should remain indoors, keeping windows and doors closed. All persons should minimize physical exertion
400	Emergency	500	2100	40	0.5	1.6	Hazardous	Premature onset of certain diseases in addition to significant aggravation of symptoms and decreased exercise tolerance in healthy persons	Elderly and persons with existing diseases should stay indoors and avoid physical exertion. General population should avoid outdoor activity
300	Warning	420	1600	30	0.4	12	Very unhealthful	Significant aggravation of symptoms and decreased exercise tolerance in persons with heart or lung disease, with widespread symptoms in the healthy population	Elderly and persons with existing heart or lung disease should stay indoors and reduce physical activity
200	Alert	350	800	15	0.2	0.6	Unhealthful	Mild aggravation of symptoms in susceptible persons, with irritation symptoms in the healthy population	Persons with existing heart or respiratory ailments should reduce physical exertion and outdoor activity
100	NAAQS	150	365	9	0.12	[b]	Moderate		
50	50% of NAAQS	50[c]	80[c]	4.5	0.06	[b]	Good		
0		0	0	0	0	0			

[a] More information on the US Pollutant Standard Index (PSI), including latest additions for new $PM_{2.5}$ NAAQS and new 8-h O_3 NAAQS, can be obtained from the Internet at http://www.epa.gov/ttnamtil/psi.html.

[b] No index values reported at concentration levels below those specified by 'Alert Level' criteria.

[c] Annual primary NAAQS.

neurotoxic, which cause reproductive dysfunction, or which are acutely or chronically toxic) or adverse environmental effects whether through ambient concentrations, bioaccumulation, deposition, or otherwise. An adverse environmental effect is defined in Section 112(a)(7) as any significant and widespread effect, which may reasonably be anticipated to affect wildlife, aquatic life, or other natural resources, including adverse impacts on populations of endangered or threatened species or significant degradation of environmental quality over broad areas. The HAPs list contains both specific chemical compounds and compound classes (e.g. arsenic compounds, glycol ethers, polycyclic organic matter, etc.) to be evaluated in order to identify source categories for which EPA will promulgate emissions standards (a complete list of the 188 HAPs is available via the Internet at http://www.epa.gov/ttn/uatw/188polls.txt). The EPA periodically reviews the HAPs list, and can be petitioned (Section 112(b)(3)(C)) to add or delete HAPs. To delete compounds, petitioners must: (1) demonstrate that no adverse human health or environmental effects are anticipated; (2) bear the burden of proof for potential adverse health and environmental effects of a given HAP; and (3) estimate potential exposures through inhalation or other routes resulting from emissions.

To date, only one HAP has gone through the entire delisting process. In response to a petition to delete caprolactam from the HAPs list, the EPA determined that the data provided were adequate for decision making. Exposure assessments were done based on known plant sites and their caprolactam emissions, and dose–response assessments were performed for acute and chronic exposures based on published studies. The resulting risk characterization indicated a very low level of risk around US caprolactam plants. Thus, a comprehensive review of the data submitted revealed that there were adequate data on health and environmental effects of caprolactam to find that emissions, ambient concentrations, bioaccumulation, or deposition of the substance would not reasonably be anticipated to cause any adverse human health or environmental effects. To help alleviate any public concern, detailed agreements were executed, under which it was agreed that, if caprolactam were delisted, emissions controls would be installed that are believed by the EPA to be equivalent to controls otherwise required had the EPA issued a standard to control these sources under Section 112. The controls, incorporated in federally enforceable operating permits for the affected facilities, will be in place years earlier than controls otherwise required.

HAPs Test Rule

Many listed HAPs chemicals have inadequate testing data to determine whether they may reasonably be anticipated to cause adverse health or environmental effects. Section 112(b)(4) of the CAAA permits the EPA Administrator to use any authority available to acquire information on health or environmental effects of HAPs. Authority under Section 4 of the Toxics Substances Control Act (TSCA) has been used to require by rule (US EPA, 1996e, 1997a) that manufacturers and processors of chemical substances conduct testing if it is found that: (1) the manufacture in commerce, processing, use, or disposal of a chemical substance or mixture, or that any combination of such activities, may present an unreasonable risk of injury to health or the environment; (2) a chemical substance or mixture is or will be produced in substantial quantities, and it enters or may reasonably be anticipated to enter the environment in substantial quantities or there is or may be significant or substantial human exposure to such a substance or mixture; (3) there are insufficient data and experience upon which the effects of the manufacture, distribution in commerce, processing, use, or disposal of such substance or mixture or of any combination of such activities on health or the environment can reasonably be determined or predicted; and (4) testing of such substance or mixture with respect to such effects is necessary to develop such data. Thus, once the EPA Administrator finds that a HAP may present an unreasonable risk of injury to health or the environment, or finds that a HAP is or will be produced in substantial quantities and either may enter the environment in substantial quantities or there may be significant substantial human exposure to the HAP, any type of health effects or environmental testing necessary to address unanswered questions about the effects of the HAP may be required. Further, if the EPA finds that data relevant to the determination of whether a HAP does or does not present an unreasonable risk of injury to health or the environment are insufficient, then testing may be required.

EPA is proposing a test rule under TSCA Section 4(a) which will require testing of 21 HAPs: biphenyl, carbonyl sulfide, chlorine, chlorobenzene, chloroprene, cresols (three isomers), diethanolamine, ethylbenzene, ethylene dichloride, ethylene glycol, hydrochloric acid, hydrogen fluoride, maleic anhydride, methyl isobutyl ketone, methyl methacrylate, naphthalene, phenol, phthalic anhydride, 1,2,4-trichlorobenzene, 1,1,2-trichloroethane, and vinylidene chloride. The EPA noted deficiencies in testing guidelines previously used and promulgated eleven harmonized guidelines (US EPA, 1997b) to be used in HAPs testing and future TSCA Section 4(a) test rules. The guidelines are for the following tests: acute inhalation toxicity with histopathology; subchronic inhalation toxicity; prenatal developmental toxicity; reproduction and fertility effects; carcinogenicity; bacterial reverse mutation test; *in vitro* mammalian cell gene mutation test; mammalian bone marrow chromosomal aberration test; mammalian erythrocyte micronucleus test; neurotoxicity screening battery; and immunotoxicity. Pharmacokinetic study data and other mechanistic data were invited to support route-to-route extrapolation and to inform the EPA of toxicity data from routes other than inhalation when scientifically defensible for use to estimate empirically inhalation risk. The pharmacokinetic proposals could form the basis for negotiation of enforceable consent agreements and replace some or all of the tests proposed under the HAPs test rule, as amended (US EPA, 1997a).

Emissions Standards

The CAAA requires the EPA to develop emission standards for all 188 listed HAPs, a notable increase over the eight HAPs previously identified by the EPA since 1970. Also, the EPA is required to identify all categories of major sources which emit these HAPs (i.e. sources which emit any one of the air toxics in amounts exceeding 10 tons per year or any combination of the toxics exceeding 25 tons per year). Standards based on best available technologies for reducing air toxic emissions are to be developed for all of these categories within 10 years of the 1990 CAAA enactment. Section 112(d) of the Act requires the standards to reflect the maximum degree of reduction in emissions of HAPs achievable, taking into consideration costs of achieving emission reductions, any non-air quality health and environmental impacts, and energy requirements. These control standards are commonly referred to as the maximum achievable control technology (MACT). The schedule for promulgation of MACT standards for the 174 source categories requires that 87 of them be developed by 15 November 1997. That schedule can be obtained from the Internet at www.epa.gov/ttn/uatw/eparules.html; www.epagov/ttn/uatw/2_4yrstds.html; and www.epa.gov/ttn/uatw/7_10yrstds.html. These MACT standards are to ensure that all major sources of air toxic emissions achieve the level of control already being achieved by better controlled and lower emitting sources in each category. The MACT for new sources must control emissions as well as the best controlled similar existing source in that category; and MACT for existing sources within a source category must be as good as the best performing 12% of existing sources for categories and subcategories with 30 or more sources, or the best-performing five sources for categories or subcategories with fewer than 30 sources. In setting these standards, the EPA will look into pollution control equipment and pollution prevention methods, such as substituting non-toxic materials for toxics currently used in the production process. It is estimated that this 10-year regulatory program will result in an annual reduction of over 1 million tons of emitted HAPs. State and local air pollution agencies will have primary responsibility to ensure that industrial plants meet the standards.

The 1990 CAAA favors setting of Federal standards which allow industry to determine how to meet the new standards, rather than specifically requiring certain equipment to be installed. For example, best technology standards for a factory currently emitting 100 tons of chloroform (one of the listed HAPs) annually, could require a 95% reduction in these emissions. The plant could meet this standard by eliminating use of chloroform in their processes, by modifying the production process to be more efficient and emit less chloroform, or by adding pollution control equipment. This flexibility allows industry to develop its own cost-effective ways of complying with the standards and still achieve significant reductions in HAPs. Section 112(g) of the CAAA allows industry to choose modifications of the process stream to achieve the mandated control. This can be done by offsetting emissions of more hazardous pollutants with less hazardous pollutants resulting from modifications of existing major sources. A hazard ranking was developed to determine if an increased pollutant emission may be offset by or traded for a decrease in a more hazardous emission (Shoaf *et al.*, 1994; US EPA, 1986c, 1994b). The hazard ranking methodology incorporates risk assessment and risk management decisions to divide the CAAA chemicals into 'non-threshold', 'threshold', 'high-concern', and 'unrankable' categories. Ranking within the non-threshold category was based upon weight of evidence for human carcinogenicity and potency. Ranking in the threshold and high-concern categories was based upon composite scores similar to the 'Reportable Quantities' methodology. No ranking was achieved within the unrankable category. Although this hazard ranking methodology was not promulgated as a rule, this guidance, along with documentation on determining *de minimis* emission rates (US EPA, 1994c), has become a useful tool for states in hazard identification and for the EPA in developing the National Urban Air Toxics Strategy.

Residual Risk

After MACT controls are in place, the EPA will review HAP emissions from each of the industrial categories and determine attendant 'residual risk'. Section 112(f) of the CAAA requires the EPA to investigate and report to Congress on: (1) methods of calculating the risk to public health remaining, or likely to remain, from sources subject to regulation under this section after the application of standards under subsection (d) (MACT standards); (2) the public health significance of such estimated remaining risk and the technologically and commercially available methods and costs of reducing such risk; (3) the actual health effects with respect to persons living in the vicinity of sources, any available epidemiological or other health studies, risks presented by background concentrations of hazardous air pollutants, any uncertainties in risk assessment methodology or other health assessment technique, and any negative health or environmental consequences to the community of efforts to reduce such risks; and (4) recommendations as to legislation regarding such remaining risk.

The four basic risk assessment steps defined in 1983 by the National Research Council will be used to calculate public health residual risk after MACT standards are implemented. Those steps are hazard identification, dose–response assessment, exposure assessment, and risk characterization. Hazard identification will determine whether the pollutants of concern can be causally linked to the health effects in question. Health effects to be considered are acute and chronic non-cancer effects, cancer effects with linear extrapolation (non-threshold), and cancer effects with non-linear extrapolation (threshold). Unlike the criteria air pollutants, which typically have sufficient data to determine complete dose–response relationships, air toxics often have insufficient data on health effects at low concentrations and therefore require extrapolation from higher concentrations with sufficient data to lower exposure concentrations where the dose–response relationship must be inferred. The proposed revision to US EPA cancer risk assessment guidelines (US EPA, 1996f) will be used in place of the original cancer guidelines (US EPA, 1996d).

Evaluating quantitative relationships between the concentration, exposure, or dose of a pollutant and associated health effects is dose–response assessment. The goal of the dose–response assessment for non-cancer effects is identification of a subthreshold dose or exposure level that humans could experience daily for a lifetime without appreciable probability of an adverse effect. The US EPA inhalation reference concentration (RfC) methodology (US EPA, 1994d) may be utilized for dose–response assessments for non-cancer effects from chronic exposure. This methodology may incorporate the single point 'no observed adverse effect level' (NOAEL) approach (US EPA, 1994d), the Benchmark approach (US EPA, 1995b), a Bayesian approach (Hasselblad and Jarabek, 1994), or other approaches such as the 'no statistical significance of trend' (Tukey *et al.*, 1985). For estimation of cancer dose–response curves, the revised Draft Cancer Guidelines (US EPA, 1996f) makes a fundamental distinction between linear and non-linear modes of action. A methodology similar to the non-cancer Benchmark dose is used to create a dose–response model, and the 95% lower confidence limit on dose associated with an estimated 10% increase in tumor response is designated as the point of departure for linear or non-linear modes of action. The linear mode of action dictates that a straight line be drawn from the point of departure to the origin, and risk at any concentration is determined from interpolation along the line. Gene mutation is a mode of action consistent with such linear extrapolations. For carcinogens with non-linear modes of action, a 'margin of exposure' (MOE) approach is used. In this approach, an MOE analysis is conducted which considers various other types of data to determine whether there is an adequate margin between the estimated exposures and the point of departure. Information such as the slope of the dose–response line, the nature of the response, the human variability in sensitivity, persistence of the agent in the body, and relative sensitivity of humans and animals should be considered. Default factors of 10-fold for human variability and species differences are proposed in the new cancer guidelines. For chronic non-cancer and cancer toxicity information to use in dose–response assessments, the preferred source of data is EPA's Integrated Risk Information System (IRIS). This data base provides toxicity information that has undergone internal peer review in the past and is now including external peer review in all of its assessments and can be accessed via the Internet at http://www.epa.gov/iris/.

For non-cancer effects from acute exposures, the EPA is developing a dose–response method that includes not only NOAEL or Benchmark approaches but a categorical regression approach (US EPA, 1994d) which accommodates the combination of data from different studies in order to evaluate the role of both exposure concentration and duration in producing the effect. The results of a categorical regression analysis are used the same as the NOAEL or Benchmark, as a point of departure for extrapolation to the human exposure of interest. Acute reference exposure (ARE) values are the preferred values to be used for residual risk assessments, when available. Other sources of acute toxicity data are the acute exposure guidance levels (AEGLs) developed by NRC guidelines, Emergency Response Planning Guidelines (ERPGs) developed by the American Industrial Hygiene Association, and levels of concern (LOCs) for extremely hazardous substances.

Exposure assessment should follow the Guidelines for Exposure Assessment (US EPA, 1992b) and perform a quantitative or qualitative evaluation of contact including such characteristics as intensity, frequency, and duration of contact. Major exposure assessment components are: (1) emissions characterization; (2) environmental fate and transport; (3) characterization of the study population; and (4) exposure calculation. The emissions of specific HAPs can be determined by using broad-scale emission inventories, specific data collection with particular industries, or information from State or local air toxics agencies. Air dispersion models are appropriate for determining air pollutant concentrations in ambient air. Meteorological and local topography information are also useful inputs to exposure assessments. Estimating population exposures requires information about the location of homes, workplaces, schools, and other receptor points. The Exposure Factors Handbook (US EPA, 1996g) contains statistical data on various factors used in assessing exposure.

Risk characterization is the final risk assessment step, bringing together and integrating information from the hazard identification, dose–response assessment, and exposure assessment. Risk characterization also includes discussion of uncertainty resulting from lack of knowledge and variability arising from true heterogeneity. Well-balanced risk characterizations both present risk conclusions and information about strengths and limitations of the assessment for other risk assessors, decision-makers, and the public. The EPA's Guidance for Risk Characterization (US EPA, 1995c) lists guiding principles for defining risk characterization in the context of risk assessment.

Section 112(f) also authorizes the EPA to consider adverse environmental effects in developing residual risk standards. An ample margin of safety must be provided to protect public health unless a more stringent standard is necessary to prevent (taking into consideration costs, energy, safety and other relevant factors) an adverse environmental effect. The EPA's Risk Assessment Forum developed a framework for ecological risk assessment (US EPA, 1992c) and has followed this with proposed guidelines (US EPA,

1996h). These Guidelines describe the ecological risk assessment framework as having three phases: problem formulation, analysis and risk characterization. The problem formulation step generates and evaluates preliminary hypotheses about why ecological effects have occurred. The analysis step involves characterization of exposure and ecological effects. The risk characterization step compares exposure levels and stressor–response profiles to estimate risk.

Assessment of the public health significance of residual risk after MACT standards are in place is mandated in Section 112(f)(1)(B). When residual risk assessments are completed for individual source categories, EPA will evaluate the significance as part of its decision-making process. No actual analyses have yet been completed, but the 'ample margin of safety' concept previously applied in the benzene standard (US EPA, 1989b) will be appropriate for the management of cancer risks. The EPA will use the 10^{-6} (one in a million) risk level as a guideline, not an absolute cut-off or 'bright line', for making risk decisions (US EPA, 1997c). The two-step process for developing national emission standards for HAPs (NESHAP) was to determine a safe or acceptable risk level above which exposure is clearly unacceptable (considering only public health factors), followed by setting an emission standard that provides an ample margin of safety considering relevant factors in addition to health (e.g. costs, economic impacts and feasibility). Acceptable risk levels are to be developed based on the assumption that an individual would be exposed to the maximum level of a pollutant for a lifetime, i.e. maximally exposed individual (MEI) risk. Generally an MEI of no higher than 10^{-4} (one in ten thousand) would be considered acceptable, but EPA would consider other health and risk factors (e.g. projected overall incidence of cancer or other serious health effects within the exposed population, numbers of people exposed within each individual lifetime risk range, estimation uncertainties underlying the risk estimates, etc.). The residual risk framework will also include criteria for non-cancer effects. Initial screening analyses are conducted for cancer and non-cancer risks followed by refined analyses. Screening analyses for non-cancer risks use a hazard index (upper end HAP exposure level divided by toxicity value) approach which does not distinguish between end-points associated with specific target organs, but target organ-specific hazard is used in the refined analysis. A hazard index less than 1 using target organ-specific hazard quotients is considered acceptable. Screening analyses for cancer assume linear mechanisms and that the additive individual cancer risk for all HAPs should be less than 10^{-6}. The intent is to try to ensure that the smallest possible number of people are exposed to the higher levels of risk and that the largest possible number are protected to a level of 10^{-6}. No individual risk greater than 10^{-4} should exist for any member of the general population, regardless of sensitivity. For non-linear cancer mechanisms, MOE analyses are to be done. To analyze cancer end-points for mixtures, ratios of individual HAP exposure levels to the corresponding departure point divided by the chemical-specific acceptable MOE is calculated. The sum of these ratios is indicative of potential hazard and is roughly analogous to treating the MOE as an uncertainty factor for determining a hazard quotient.

Section 112(f)(1)(C) requires EPA to assess and report on 'the actual health effects with respect to persons living in the vicinity of sources, any available epidemiological or other health studies . . .'. The EPA's approach is to evaluate scientific literature for published epidemiological studies related to the specific source categories, HAPs, and/or locations studied. Where epidemiological studies are not available, the EPA will consider examining raw health effects data for correlations between exposure and adverse effects. These may be obtained from State or national disease registries, hospital and other medical records, death certificates and questionnaires. Exposure to HAPs may be estimated based on city or county of residence; proximity to the emission source; data from stationary ambient monitors; mathematical modeling; or personal air monitors. Data from peer-reviewed published reports will be the best for decision-making. To comply further with Section 112 requirements for the EPA to investigate and report to Congress on '. . . any uncertainties in risk assessment methodology or other health assessment technique . . .', the Agency will follow policies set forth in guidance documents such as the Exposure Factors Handbook (US EPA, 1996g), which supports probabilistic approaches to the treatment of a number of commonly employed risk assessment input variables. Detailed recommendations for uncertainty and variability analysis will also be obtained from the Guidelines for Exposure Assessment (US EPA, 1992b), the Policy for Use of Probabilistic Analysis (US EPA, 1997d), and the Guiding Principles for Monte Carlo Analysis (US EPA, 1997e). The Policy for Use of Probabilistic Analysis provides technical guidance on uncertainty evaluation, and the Guiding Principles for Monte Carlo Analysis provides refined technical guidance as well as recommendations on presentation of uncertainty information to decision-makers.

National Academy of Sciences Report

Because of concerns about the state of the science of risk assessment, the 1990 CAAA required studies by the National Academy of Sciences (NAS) and EPA to review and improve the techniques for estimating the risks to public health associated with exposures to HAPs. In its January 1994 study, *Science and Judgement in Risk Assessment*, the NAS concluded that resources and data are not sufficient to perform a full-scale risk assessment on each of the 189 chemicals listed as HAPs in the CAAA, and often no such assessment is needed, since after MACT is applied, some of the chemicals will pose only *de minimis* risk (e.g. a cancer risk of adverse health effects of one in a million or less). For these reasons, the NAS believes that the EPA should undertake an iterative approach to risk assessment. An iterative approach would start with a relatively inexpensive screening technique, such as a simple, conservative transport model, and then move on to more resource-intensive levels of data-gathering, model construction, and model application for chemicals suspected of exceeding *de minimis* risk. To guard against serious underestimations of risk, screening techniques must err on the side of caution when there is uncertainty about the model assumptions or parameters.

Urban Air Toxics

The 1990 CAAA address a diverse array of problems associated with HAPs. The MACT standards address emissions from major sources (Section 112(d)), and mobile source emissions are addressed in Section 202(l). Non-major sources ('area' sources) found particularly in urban areas are addressed in Sections 112(c)(3) and 112(k). With the MACT program well underway, EPA is developing a program and strategy to address urban area source problems. According to Section 112(c)(3), the Administrator is responsible for listing categories of area sources to ensure that 90% of the area source emissions of the 30 HAPs that present the greatest threat to public health in the largest number of urban areas are subject to regulation. Congress has recognized that only a part of the air toxics problem results from large chemical and industrial facilities. A substantial portion of the air toxics problem may be caused by much smaller, diverse sources, such as wood stoves, apartment building boilers and other small combustion units, dry cleaners, service stations, publicly owned wastewater treatment works, drinking water supply systems, solid and hazardous waste landfills, pesticide applications, automobiles, trucks and buses. Congress recognizes in Section 112(k)(1) that emissions of HAPs from area sources may individually, or in the aggregate, present significant risks to public health in urban areas. Considering the large number of persons exposed and risks of carcinogenic and other adverse health effects from HAPs, ambient concentrations characteristic of large urban areas should be reduced to levels substantially below those currently experienced. The intent is to achieve a substantial reduction in emissions of area source HAPs and an equivalent reduction in public health risks of not less than 75% in cancer incidence attributable to emissions from such sources. Section 112(k)(3) therefore requires the EPA to submit to Congress a comprehensive strategy to control emissions of HAPs from area sources in urban areas. The strategy should identify not less than 30 HAPs which, as the result of emissions from area sources, present the greatest threat to public health in the largest number of urban areas.

The EPA will fulfill obligations under Section 202(l), 112(c), and 112(k) by developing an Integrated Urban Air Toxics Study and Strategy that addresses the urban air toxics risk from both stationary and mobile sources. Integration of the activities under these Sections of the CAAA will more realistically indicate the total outdoor exposure and will allow the EPA and the states to develop activities to address risks posed by toxic pollutants where the emissions are most significant and controls are most cost-effective. Specifically, the strategy will achieve the following: (1) develop a research strategy on air toxics, including research on toxicity of the urban HAPs, and develop monitoring and modeling improvements to better identify and address risks in urban areas; (2) identify at least 30 HAPs that occur in urban areas that present the 'greatest threat to public health'; (3) identify source categories or subcategories emitting the 30 HAPs and assure that 90% or more of the aggregate emissions are subject to standards; (4) develop a schedule for activities to reduce public health risk from area sources using all EPA and state/local authorities; (5) develop a requirement to achieve substantial reduction in risk (e.g. non-cancer and cumulative risk); (6) develop a schedule for activities to reduce the risk from cancer attributable to HAPs by 75%; (7) encourage state/local programs to reduce risk from particular HAP sources in urban areas; and (8) address the Section 202(l) requirement to determine if regulations containing reasonable and feasible requirement to control HAPs from motor vehicles or motor vehicle fuels are necessary.

The national efforts will be in the areas of research, planning, and standard setting. A research strategy to address deficiencies in the knowledge and technical information that is needed to identify and address fully the health risks posed by urban air toxics will be developed. Information on the cancer and non-cancer toxicity of the HAPs and development of risk assessment tools for State and local agencies will be identified. More studies and strategies necessary for determining whether additional fuel/vehicle emissions standards and the activities necessary to set those standards are required. The EPA is currently analyzing 11 cities for exposure to mobile source HAPs. Based on this initial exposure analysis, the EPA will determine if it is necessary or feasible to establish additional mobile sources standards. As part of the process to identify not less than 30 HAPs resulting from emissions from area sources and presenting the greatest threat to public health in the largest number of urban areas, the EPA has developed a draft list of HAPs. First, the HAPs were evaluated by using monitored ambient air concentrations along with Risk Based Concentrations which represent a measure of toxicity, and were then evaluated using nationwide emissions estimates combined with the Risk Based Concentrations. Next, the HAPs were evaluated based on the results of existing risk assessment studies, and the preliminary results from the EPA's Cumulative Exposure Project were utilized to prioritize the HAPs. Finally, the HAPs were evaluated based on potential for multi-pathway exposure concerns.

EPA reviewed and analyzed available data to revise the existing assessments of the sources in the USA that emit each of the identified priority HAPs. The quantities and emissions inventory for each HAP and source category were documented and used to provide emissions estimates for major, area and mobile sources. The base year for developing emissions data is 1990. The EPA is attempting to provide a centrally archived database for ambient monitoring data for air toxics by conducting a nationwide study to identify, catalogue and characterize all available ambient air quality data for toxics, and to make this data publicly accessible for analysis. For exposure assessments, the EPA will assess existing studies and data to provide some evaluation of potential risk. The EPA will analyze the emissions inventory to determine if 90% of the emissions of the identified HAPs are currently regulated and will regulate additional source categories to achieve this goal. To assess whether these source categories are already subject to regulations that address the emissions of the priority HAPs, EPA is soliciting additional information on any potentially affected HAPs that may be subject to State or Federal regulations. In summary, EPA intends to develop a strategy that has a strong national component by establishing national technology and fuel/vehicle standards and a schedule of activities to achieve risk reductions. Where these standards may not be adequate to achieve the risk goals, the EPA believes that it is critical to explore areas where State/local air pollution programs may be better able to address the risk in particular urban areas.

Technical Assistance

Section 112(l)(3) of the CAAA mandates that the EPA provide technical assistance in the area of air toxics. The Act requires the Administrator to maintain an air toxics clearinghouse and center with the purpose of providing technical information and assistance to State and local agencies and to others on control technology, health and ecological risk assessment, risk analysis, ambient monitoring and modeling, and emissions measurement and monitoring. The Air Risk Information Support Center (Air RISC) was created to provide three levels of technical assistance pertaining to health, exposure and risk assessments for air pollutants: (1) hotline assistance, (2) detailed technical assistance, and (3) general technical guidance. The Air RISC hotline provides an initial, quick response based upon health and exposure information available through the expertise of EPA staff, resources (documents, databases and other technical data), and contractors (US EPA, 1993b). The hotline operator puts the requestor in direct contact with experts in a variety of areas or identifies other appropriate information resources. Air RISC provides in-depth evaluation and/or retrieval of information when such detailed technical assistance is more appropriate than a phone response. When the requestor needs information retrieved immediately, Air RISC provides 'quick assistance'. Quick assistance technical guidance might be a review of and/or consultation on many health-based subjects, such as review of a site-specific risk assessment, a health and exposure summary, or a literature search and synthesis of retrieved materials. The response is typically delivered within several weeks of the request. General technical guidance is also provided by Air RISC on topics involving health, exposure and risk assessment issues of national interest. Guidance documents may take a year or longer to complete. Topics may be identified from information requested by State and local agency staff through the hotline or by suggestions from EPA staff or the Air RISC Steering Committee. Information on Air RISC services can be accessed via the Internet at http://www.epa.gov/epahome/hotline.htm.

Great Waters Program

Not enough is known to make judgements on how to address some HAP issues and problems. For example, some HAPs such as mercury persist in the environment and can be deposited in lakes and rivers. There is concern that this is already happening in the Great Lakes, Lake Champlain, Chesapeake Bay, and US coastal waters. During the past 30 years, scientists have collected convincing evidence showing that toxic chemicals released into air can travel long distances and be deposited on land or water. Polychlorinated biphenyls, for example, are present in the air above all five Great Lakes and the Chesapeake Bay. In Section 112(m) of the CAAA the US EPA Administrator is instructed to conduct a program to identify and assess the extent of atmospheric deposition of HAPs to the Great Lakes, the Chesapeake Bay, Lake Champlain and coastal waters. The program should monitor the Great Lakes, the Chesapeake Bay, Lake Champlain and coastal waters; investigate the sources and rates of atmospheric deposition; conduct research to develop and improve monitoring methods and determine the relative contribution of atmospheric pollutants to total pollution loadings; evaluate any adverse effects to public health or the environment caused by such deposition; sample for such pollutants in biota, fish and wildlife; and characterize the sources of such pollutants. The first Report to Congress on the deposition of air pollutants to the Great Lakes was submitted in May 1994. This report included a series of conclusions, recommendations and specific actions EPA would take to help protect these waters. It was noted in the report that the CAAA program to address the deposition of HAPs is only one part of a comprehensive program to reduce pollutant loadings to the Great Lakes. The EPA has worked with Congress to obtain discretionary authority, through a reauthorized Clean Water Act, to regulate or prohibit releases to any media that cause or contribute to a water quality impairment by highly toxic, persistent, bioaccumulative pollutants. The EPA is working internationally to explore alternatives to pesticides banned in the US, which are persistent in the environment and are transported over long distances. The EPA is also working with Canada to analyze samples from various Great Lakes monitoring stations.

Electric Utility Plants and Mercury Emissions

The EPA is also performing studies on human health and environmental effects of HAP emissions from electric utility power plants, anthropogenic sources of air emissions of mercury and the impact of these emissions on human health and the environment, and the scope and magnitude of the problem of HAPs in urban air. Section 112(n)(1)(A) of the CAAA mandates the US EPA Administrator to perform a study of the hazards to public health reasonably anticipated to occur as a result of emissions of listed HAPs by electric utility steam generating units. The study report, now in draft form, gives a description of the industry; an analysis of emissions data; an assessment of hazards and risks due to inhalation exposures of 67 HAPs; assessments of risks due to multipathway exposures to radionuclides, mercury, arsenic, and dioxins; and a discussion of alternative control strategies. The results of the study indicate that ambient levels of mercury are a potential health concern and that utilities are a significant source of mercury emissions, emitting 34% of anthropogenic mercury in the USA. Furthermore, the Administrator has concluded that there exists a plausible link between methylmercury concentrations in freshwater fish and mercury emissions from electric utility plants. Therefore, a program will be initiated to explore means of reducing mercury emissions from utilities. The utility study also indicated that emissions of dioxins and arsenic from utilities may be a concern to public health and that the inhalation cancer risk from nickel may be as high as 9×10^{-5} (9 in 100 000). The Administrator reported these results to Congress in 1998. The study report on HAP emissions from electric utilities, issued in draft form in December 1997, is planned for final issue to Congress in 1998.

Section 112(n)(1)(B) requires the EPA to study the impacts of mercury emissions from electric utility steam generating units, municipal waste combustion units, and other sources, including area sources. The Mercury Study Report to Congress (US EPA, 1997f) issued in 1997, addressed: (1) data on type, sources, and trends in emissions; (2) evaluation of the atmospheric transport of mercury to locations distant from emission sources; (3) assessment of potential impacts of mercury emissions close to the source; (4) identification of major pathways of exposure to humans and non-human biota; (5) identification of the types of human health consequences of mercury exposure and the amount of exposure likely to result in adverse effects; (6) evaluation of mercury exposure consequences for ecosystems and for non-human species; (7) identification of populations especially at risk from mercury exposure due to innate sensitivity or high exposure; and (8) estimates of control technology efficiencies and costs. The largest source of mercury emissions was identified as municipal waste combustors. The Report characterizes both the type and magnitude of health and ecological effects associated with airborne emissions of mercury from anthropogenic sources. It also validates methods and models, describes the basis for default options, articulates and prioritizes data needs, distinguishes between variability and uncertainty, and performs formal uncertainty analyses.

MOBILE SOURCE POLLUTANTS

Pollutants from vehicles and other non-stationary sources are primarily addressed under Clean Air Act Title II, 'Emission Standards for Moving Sources,' also known as the

'National Emission Standards Act' (42 U.S.C. 7401 nt). Under this title, the EPA Administrator is given authority and responsibility for prescribing 'standards applicable to the emission of any air pollutant from any class or classes of new motor vehicles or new motor vehicle engines, which in the Administrator's judgment cause, or contribute to, air pollution which may reasonably be anticipated to endanger public health or welfare.' This broad authority applies not only to passenger cars and light and heavy duty trucks, but to motorcycles, trains, aircraft, marine craft and off-road vehicles, as well as other portable small-engine devices such as lawn mowers, snow blowers, and leaf blowers.

Certain mobile source emissions are also addressed under other sections of the Clean Air Act. For example, motor vehicles have historically contributed substantially to emissions or formation of most of the 'criteria air pollutants' originally designated under Section 108, including carbon monoxide, hydrocarbons, lead, sulfur oxides, nitrogen oxides, ozone and particulate matter. Although mobile source contributions to some criteria pollutant emissions, such as lead and particulate matter, have declined greatly owing to various regulatory actions and control strategies adopted by EPA (e.g. the phase-down of lead in gasoline and improvements in emission control devices), motor vehicles are still significant contributors to ambient air levels of criteria pollutants such as carbon monoxide, nitrogen oxides and ozone. Other 'hazardous air pollutants' may also be emitted from mobile sources, including compounds such as 1,3-butadiene, formaldehyde, polycyclic aromatic hydrocarbons, and benzene; these are also addressed under Section 112 of the Clean Air Act.

Title II also pertains to the fuels used in motor vehicles. Clean Air Act Section 211 authorizes the EPA Administrator to require that designated fuels or fuel additives be registered by their manufacturers before their introduction into commerce. Section 211(b) also allows EPA to 'require the manufacturer of any fuel or fuel additive to conduct tests to determine potential public health effects of such fuel or additive (including, but not limited to, carcinogenic, teratogenic, or mutagenic effects) . . .' The authority to require health effects testing was discretionary when the 1970 Clean Air Act was enacted, but in 1977 Section 211(e) was added, which made implementation of testing requirements mandatory. In 1994, EPA implemented testing requirements by issuing the Fuels and Fuel Additives (F/FA) Rule (Federal Register, 1994b). Among the provisions of the F/FA Rule is that health effects testing requirements apply to both 'existing' and 'new' F/FAs. However, manufacturers of F/FAs that were already registered at the time the Rule was issued were allowed a 3-year period to comply with the requirements, whereas testing had to be completed before new F/FAs could be registered.

The testing requirements of the F/FA Rule are intended to determine potential adverse health effects of whole, complex mixtures of F/FA combustion and (separately) evaporative emissions rather than to examine the individual constituents of a combustion or evaporative emission mixture. The Rule is structured to allow information to be provided to the EPA in a graduated or tiered manner. Tier 1 of the F/FA Rule requires a literature search for health and ecological effects of the evaporative and combustion emissions. If adequate data are not already available, Tier 2 prescribes that 90-day inhalation toxicity studies be conducted for both evaporative and combustion emissions and that fertility/teratogenicity, neurotoxicity, and carcinogenicity/mutagenicity endpoints be evaluated with standard screening assays (see Table 42.4). The Rule also provides for more extensive testing through alternative Tier 2 requirements. Alternative Tier 2 may extend or add to the end-points specified in standard Tier 2. After either standard or alternative Tier 2 testing has been conducted, follow-up or more targeted testing can be required on a case-by-case basis under Tier 3 provisions of the Rule.

For F/FAs that are unlikely to be produced or used in substantial amounts, information obtained through Tiers 1 and 2 is expected to be sufficient to determine whether or not more extensive testing would be needed. If, for example, results from the standard screening assays in Tier 2 showed no indications of significant toxicity, and general population exposure to the product was quite limited, no further testing might be required. However, if information existed prior to Tier 2 testing to indicate, for example, that a F/FA could have significant toxicity or that exposures could be widespread, alternative Tier 2 testing could be required in lieu of standard Tier 2. This approach is intended to save time and resources in cases where standard Tier 2 testing is unlikely to provide sufficient information to address or resolve public health concerns. Alternative Tier 2 allows considerable flexibility to tailor testing requirements in relation to scientific considerations. To illustrate this point and other features of the F/FA Rule in application, two cases will be briefly

Table 42.4   Fuel/Fuel Additive (F/FA) Rule standard Tier 2 Tests

90-Day subchronic inhalation general toxicity Screening information on target organ toxicities and on concentrations useful for conducting chronic studies. Includes clinical signs and chemistry, opthalmological exam, gross and histopathology (especially respiratory tract). $N = 30$ animals per concentration per group; $N = 20$ for recovery group observed for reversible, persistent or delayed effects.

Fertility/Teratology Information on potential health hazards to fetus and on gonadal function, conception, and fertility. Observation for ≤ 13 weeks includes vaginal cytology, mating and fertility, gross necropsy (especially including reproductive organs), fetal anomalies, resorptions, histopathology of reproductive organs. $N = 25$ male/40 female animals per group; mating after 9 weeks of exposure, then exposure of females continues through gestational day 15. Limit test (if no effects at highest concentration, then skip lower concentrations).

In vivo *micronucleus* Detect damage to chromosomes or mitotic apparatus of cells (based on increase in frequency of micronucleated red blood cells); provides information on potential carcinogenic and/or mutagenic effects. $N = 5$ female/5 male animals per group. Positive control.

In vivo *sister chromatid exchange* Detect enhancement of exchange of DNA between two sister chromatids of a duplicating chromosome (using peripheral blood lymphocytes grown to confluence in cell culture); provides information on potential mutagenic and/or carcinogenic effects. $N = 5$ female/5 male animals per group. Positive control.

Neuropathology Provides data on morphologic changes in central and peripheral nervous system. Includes observations (e.g. body weight, movement disorders), brain size and weight, light (and possible EM) microscopy of sections, peripheral nerve teasing. $N = 10$ animals per group; $N = 20$ observed for reversible, persistent, or delayed effects. Positive control. Limit test.

Glial fibrillary acidic protein (GFAP) An indicator of neurotoxicity associated with astrocytic hypertrophy at site of damage. $N = 10$ animals per group.

Salmonella typhimurium *reverse mutation* In vitro microbial assay that measures histadine (*his*) reversions (*his⁻* to *his⁺*), which cause base changes or frameshift mutations in the genome; provides data on mutagenicity. Positive controls. Data presented as number of revertant colonies per plate, per kilogram of fuel, and per kilometer for each replicate and dose.

From Federal Register (1994b).

described: oxygenates and an organomanganese additive (MMT). Both of these F/FAs have been the focus of some debate about their potential impacts on public health (see 1998 reviews by Davis and Graham and by Davis), which is part of the justification for imposing more extensive alternative Tier 2 requirements for their evaluation.

Oxyfuels

Oxygenates are added to fuels to increase oxygen content and thereby reduce certain emissions; some are also used to increase the octane number of gasoline. The 1990 Clean Air Act Amendments (CAAA; US Code, 1990) require the use of oxygenates in two fuel programs administered by the EPA, one applying to areas exceeding the CO NAAQS and the other for areas exceeding the O_3 NAAQS. The CO-directed program requires the use of gasoline with an oxygen content of 2.7%-wt during cold-weather months. The ozone-directed program requires the use of gasoline with an oxygen content of 2.0%-wt year round. Although neither the US Congress nor the EPA has specified that a particular additive be used to achieve these oxygen requirements, the dominant chemical in the market has been methyl tertiary butyl ether (MTBE).

However, several other ethers and alcohols may also serve as oxygenates, including ethanol (EtOH), ethyl tertiary butyl ether (ETBE), tertiary amyl methyl ether (TAME), tertiary amyl ethyl ether (TAEE), diisopropyl ether (DIPE), dimethyl ether (DME), and tertiary butyl alcohol (TBA). Collectively, fuels containing oxygenate may be referred to as 'oxyfuels.'

Manufacturers of the same or similar F/FAs may form consortia to fund testing required by the F/FA Rule. In the case of oxyfuels, a consortium of about 150 producers has been created. Testing is to be conducted on baseline gasoline and a subset of oxyfuels containing MTBE, ETBE, TAME, DIPE, EtOH and TBA. These particular chemicals are produced by manufacturers that are willing to underwrite their testing. Although other oxygenates are also currently registered and thus may be legally used in US fuels (although apparently they are not used in significant amounts, if any), if testing is not conducted on these other products within the time frame specified by EPA, their registration may be terminated. The specific provisions of the alternative Tier 2 program for oxyfuels were established in 1998 by EPA (Federal Register, 1998), after the consortium and the public were afforded an opportunity to comment on the requirements before they were made final. Comments have been received on the proposed requirements, and the EPA is deliberating on what changes might be made before the final requirements are issued.

Although both evaporative and combustion emissions are covered by the F/FA Rule, the oxyfuels consortium has submitted a 'white paper' (Barter *et al.*, 1996) arguing in part that practical constraints on the use of engine exhaust for inhalation toxicity studies make such testing pointless. It was asserted that the dilution of the exhaust emissions necessary to avoid CO toxicity in the test animals would also lower the concentrations of other constituents of the exhaust to no-effect levels. If the exhaust concentrations were high enough to produce effects, the consortium argued that any observed effects would be dominated by the toxic effects of CO and nitrogen oxide, which have already been well characterized. Although this issue has not been finally resolved, the EPA has decided to defer testing on baseline gasoline and oxyfuel combustion emissions until more information is provided on human exposure to combustion emissions. Meanwhile, testing on evaporative emissions will be initiated.

Baseline gasoline is included in the oxyfuels testing program for multiple reasons. Given the extensive exposure to conventional gasoline, it is important to have an adequate database on the health effects of gasoline in its own right. Also, any attempt to evaluate the potential health risk associated with a particular oxyfuel should consider the risk in relation to that of baseline gasoline, for the ultimate question is not whether or not a given fuel poses some degree of health risk, but how that risk compares with the risk associated with another fuel that would be used in its place. The proposed alternative Tier 2 program specifies more extensive testing on baseline gasoline and MTBE-oxyfuel and less extensive testing on the other oxyfuels (blends of gasoline and ETBE, TAME, DIPE, EtOH or TBA). Specifically, the proposed requirements for baseline gasoline and MTBE-oxyfuel include a 2-year inhalation toxicity study in rats that would evaluate cancer and non-cancer end-points. This chronic bioassay is expected to provide valuable information for a quantitative cancer risk assessment of gasoline evaporative emissions. In addition, EPA has proposed that baseline gasoline and MTBE-oxyfuel be evaluated with a standard 90-day inhalation toxicity study; a two-generation reproductive toxicity study; a two-species developmental toxicity study; a neurotoxicity study including the GFAP assay (see Table 42.2), neuropathology, functional observational battery, and motor activity evaluations; and an immunotoxicity screening using the sheep red blood cell assay.

For the remaining oxyfuels, standard Tier 2 testing would be required, with the addition of the immunotoxicity assay and pharmacokinetic studies of the neat oxygenate vapors. Extensive pharmacokinetic studies on MTBE (including TBA as a metabolite of MTBE) have already been conducted and are ongoing, and therefore are not being proposed for MTBE-oxyfuel or for baseline gasoline. For the other oxyfuels, however, pharmacokinetic studies of the neat vapors would be an important adjunct to the toxicity assays on the mixtures and would be valuable in evaluating the relative toxicity of the various oxyfuels.

As noted above, toxicity testing of combustion emissions would be deferred while human exposure studies are conducted. Data on population exposures to emissions from conventional gasoline and MTBE-oxyfuel are limited. To address these data gaps, EPA has proposed that human exposures to emissions related to baseline gasoline and MTBE-oxyfuel be investigated in microenvironments that are likely to represent the upper end of the distribution of personal exposure levels. The studies would be conducted in cities with and without oxyfuels, and would examine several variables that could affect human exposures to evaporative and combustion emissions from fuels. This information will be an important component of health risk assessments of baseline gasoline and oxyfuels and in decision-making regarding the need for toxicity studies of combustion emissions.

As provided by Tier 3 of the F/FA Rule, follow-up studies may be required, depending in part on the results from alternative Tier 2 work. For example, if adverse effects were observed in the second generation, but not the first, of the two-generation reproductive toxicity study of MTBE-oxyfuel, this could warrant having the reproductive toxicity evaluations of the other oxyfuels extended from a one-generation study to a two-generation study. A positive effect in the fertility/teratogenicity screening tests for an oxyfuel could also justify requiring more extensive reproductive and developmental toxicity studies. Similarly, adverse effects in the second species tested in the two-species developmental toxicity studies of baseline gasoline and MTBE-oxyfuel could trigger a requirement for more extensive evaluation of developmental toxicity in other oxyfuels. The necessary and sufficient conditions for requiring contingent Tier 3 studies are difficult to specify *a priori*, but these hypothetical outcomes illustrate how the F/FA Rule may be applied in a graduated and flexible manner.

MMT

Another additive for which alternative Tier 2 testing requirements are currently being formulated by EPA is methylcyclopentadienyl manganese tricarbonyl (MMT), an organomanganese compound used to increase the octane number of gasoline. Since 1978, the manufacturer of MMT has petitioned EPA several times for a waiver that would make the use of MMT in unleaded gasoline legal (MMT was a legal additive to leaded gasoline from the 1970s until leaded gasoline itself was phased out). Section 211(f) of the Clean Air Act makes it unlawful for a manufacturer to introduce any fuel or fuel additive (F/FA) that is not substantially similar to F/FAs utilized in the certification of 1975 or later vehicle engines. This prohibition against new F/FAs may be waived if an applicant establishes that the F/FA will not cause or contribute to an exceedance of vehicle emission standards. Four waiver petitions submitted by the manufacturer were denied owing to concerns regarding increases in exhaust hydrocarbon emissions resulting from MMT use. During consideration of the third and fourth applications, the EPA also raised concerns regarding the possible adverse health effects of an increase in airborne manganese resulting from MMT use. In July 1994, the EPA Administrator denied the manufacturer's waiver petition specifically because of concerns about risks to public health (Federal Register, 1994c), but the manufacturer successfully challenged the denial of its petition in Federal court (*Ethyl Corporation* v. *US EPA*, 1995). The court ruled that Section 211(f)(4) of the Clean Air Act (CAA) provides no basis for EPA to deny the manufacturer's petition on any grounds other than whether MMT would cause or contribute to a failure of any emission control device or system. A remaining issue was whether MMT was already registered under the terms of the F/FA Rule, in which case the additive could be introduced before testing was completed. Although the EPA's position was that MMT had not been previously registered, the manufacturer successfully challenged that obstacle as well *(Ethyl Corporation v. Carol M. Browner of the US EPA and the US EPA, 1995)*. Consequently, since December 1995, the manufacturer has been marketing MMT to US oil refineries. However, as provided by the F/FA Rule, studies prescribed by EPA must still be conducted within a specified time frame, even though MMT is already in use.

Although EPA has not yet established alternative Tier 2 requirements for MMT, many aspects of a likely testing program have already been discussed and published. In 1991, EPA described a program of health and exposure studies that would provide the information needed to support an improved, quantitative assessment of the potential risks to public health associated with MMT (US EPA, 1991b). Because the maximum allowable concentration of MMT in gasoline (1/32 g Mn/gal) is so low and MMT itself is so readily photodegraded in air, the potential for human exposure to evaporative emissions was judged to be low. However, the potential for widespread inhalation exposure to the particulate emissions that resulted from the combustion of gasoline containing MMT was of concern, comparable to the situation that had existed when tetraethyl lead was widely used in gasoline and the average blood lead levels of the US population reflected exposure to the combustion emissions from leaded gasoline (Davis *et al.*, 1996). Limited information available had indicated that Mn_3O_4 was the predominant form of Mn emitted (Ter Haar *et al.*, 1972), but more recent (as-yet unpublished) evidence submitted to EPA by the manufacturer has raised questions about that conclusion. Notwithstanding a possible revision in the characterization of the combustion emissions related to MMT, the basic features of the health and exposure studies called for by the EPA in 1991 remain relevant for the purposes of the present discussion.

Proposed health effects studies would be intended to serve as a basis for deriving an inhalation reference concentration (RfC) specifically for the form or forms of manganese emitted from vehicles. The RfC is defined as an estimate (with uncertainty spanning about an order of magnitude) of a continuous inhalation exposure level for the human population (including sensitive subpopulations) that is likely to be without appreciable risk of deleterious non-cancer effects during a lifetime. The studies would be designed to determine LOAELs and NOAELs for inhaled manganese particles for neurotoxicological, reproductive, developmental and respiratory effects. They would be conducted with rodents and, particularly in the case of neurotoxicological effects, non-human primates. In addition, the studies would be intended to improve knowledge of sensitive subpopulations and provide information on Mn bioaccumulation to enable better extrapolation from subchronic to chronic exposure effects. The responsibility for developing detailed protocols to achieve the objectives of these studies is expected to be borne by the manufacturer, subject to EPA approval. Relatively standardized testing protocols are available for evaluating respiratory, reproductive, and developmental toxicity. However, given the apparently greater sensitivity of the central nervous system to manganese in the experimental and epidemiological literature, more extensive testing of neurotoxicological end-points will be needed to enhance confidence in the results if they should turn out to be negative.

Pharmacokinetic studies are also a likely aspect of the alternative Tier 2 program. In 1991, the EPA had recommended that the bioaccumulation of inhaled manganese be characterized, primarily to facilitate extrapolation from limited-duration, high-concentration exposures to lifetime, low-level exposures (US EPA, 1991b). Route-to-route pharmacokinetic studies were not advocated, because it was felt that such an extrapolation could introduce additional uncertainty in the derivation of an RfC. However, appropriate pharmacokinetic studies could provide useful dosimetric data for designing and interpreting toxicity studies, and the manufacturer has indicated a willingness and intention to conduct pharmacokinetic studies. It would be desirable for such studies to investigate manganese disposition in blood, excreta, and target tissues such as brain regions, lung, reproductive organs, liver, and kidney, as well as examine manganese disposition in relation to gender and to lifestage, from the prenatal period to senescence. Data on the pharmacokinetics of different oxidation states of Mn, particularly the divalent and trivalent forms, would also be useful because of the mixed forms present in different compounds of manganese.

For the exposure studies, in 1991 the EPA recommended measuring manganese exposure levels in areas where MMT is used in gasoline (US EPA, 1991b). At the time of this recommendation, Canada was the only country known to be using MMT in unleaded gasoline. However, it was recognized that differences between US and Canadian cities, including meteorology, traffic density and the concentration of MMT in gasoline, would need to be taken into consideration in the design and interpretation of the studies. A stratified probabilistic sampling design, with particular emphasis on representation of the upper tail of the exposure distribution (e.g. through oversampling of high-exposure subgroups), was recommended. The EPA also proposed that personal exposure monitoring and ambient monitoring (preferably located adjacent to each subject's residence) be conducted concurrently. Given seasonal differences in exposures, sampling during at least two different times or seasons of the year was advocated. As in the case of the health studies, it is expected that the manufacturer would take responsibility for developing detailed protocols for such exposure studies, subject to EPA approval. However, the manufacturer has already had one such exposure study conducted in Toronto, the unpublished results of which have been provided to the EPA.

DEVELOPING INTERNATIONAL GUIDELINES

As noted at the outset of this chapter, programs and activities carried out by components of the United Nations have been among the most widely influential mechanisms by which the international sharing of expertise and knowledge on control of air pollution has been accomplished. The World Health Organization (WHO) has been of particular importance in developing air quality guidelines for international use. Environmental Health Criteria (EHC) Documents containing assessments of pertinent scientific information and recommending air quality guidelines (AQG) for major air pollutants (PM, O_3, SO_2, CO, NO_2) were generated by WHO in the 1970s. Starting in the early 1980s, a newly established WHO-cosponsored International Program on Chemical Safety (IPCS), with its Central Unit housed at WHO Headquarters in Geneva, expanded production of EHC Documents and guidelines for numerous additional air pollutants.

The process typically employed for preparation of WHO/IPCS EHC documents is depicted in Figure 42.2. Such documents, like US EPA criteria documents, rely mainly on peer-reviewed, published literature reporting the results of pertinent research conducted worldwide. A first draft version of an EHC document on a given pollutant or pollutant class is prepared by an expert (at times, with inputs from other co-authors) selected by WHO/IPCS based on their internationally recognized expertise. The draft document is next circulated for comment to the collaborating institutions in numerous member countries that participate in the IPCS program. The IPCS collaborating institutions, typically national government units such as US EPA or the National Institute for Environmental Health Sciences (NIEHS) in the USA, review the draft document and/or call upon recognized experts from universities or other organizations in their country to assist in preparation of comments forwarded to WHO/IPCS. The comments and any newly identified pertinent research reports are then taken into account by the original author(s) in revising the draft document. Often included among new information provided by IPCS collaborating institutions are updated air monitoring results from various countries, useful in characterizing background information on air quality trends to help place the EHC health evaluation in an international context.

The revised, 'internationalized' draft EHC document next undergoes further peer-

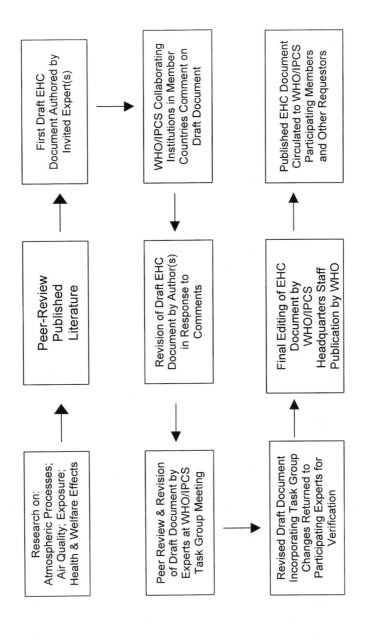

Fig. 42.2 Schematic representation of process for preparation of WHO/IPCS Environmental Health Criteria (EHC) documents.

review and revision at an IPCS-organized expert taskgroup meeting, which typically includes participation by both the principal author (or a co-author) and other IPCS-selected experts drawn from various countries. Revisions identified as needing to be made at the taskgroup meeting are either made by the taskgroup members during the meeting or, depending on how extensive, after the meeting. The chairman of the taskgroup, selected by the participating experts at the meeting, works with IPCS staff to ensure that the revisions agreed on at the taskgroup meeting are incorporated, and the resulting penultimate draft is typically circulated back to the participating taskgroup members to verify their agreement with the draft's content. Of particular importance, great care is taken not to alter the bottom line conclusions and air quality guidelines recommended by the taskgroup. Following final editing and other publication preparation steps by IPCS staff, the EHC document is published by WHO/IPCS, circulated to collaborating institutions in member countries, and made available to others.

The WHO/IPCS EHC documents that have been generated for particular substances are far too numerous to list here; WHO publication numbers for the series are now nearing 200. These include the addition of second edition update EHC documents on major air pollutants originally assessed in the 1970s as the new up-to-date assessments are completed (see, for example, the recently completed Second Edition EHC for Nitrogen Dioxide; WHO/IPCS, 1997). Listings of available WHO/IPCS documents can be ascertained via the Internet at http://www.who.ch/pll/dsa/.

It should be noted that the IPCS is in the process of implementing development of new, more concise international risk assessment documents, known as Concise International Chemical Assessment Documents (CICADs) (WHO/IPCS, 1996). The CICADs are to be derived from high-quality national review documents from participating IPCS member countries and to meet the same high standards as the well-established, more comprehensive WHO/IPCS EHC documents. For example, a Particulate Matter CICAD is in the process of being prepared, based on the recently completed 1996 US EPA PM AQCD (US EPA, 1996a) discussed earlier and updated with more recently published information.

The WHO Regional Office for Europe, in Copenhagen, had undertaken an earlier, highly successful effort at providing very concise assessments of a number of air pollutants considered to pose health and/or environmental threats for European countries. Those assessments, and air quality guidelines (AQGs) based on them, were published in 1987 in a WHO publication entitled Air Quality Guidelines for Europe, which has found wide use not only in Europe but elsewhere in helping to provide support for development of air standards in various nations (WHO, 1987). The process employed in producing the subject AQGs was analogous in many ways to that already described for developing the WHO/IPCS EHC documents. The process differed mainly in terms of taskgroups being convened, over the course of a few years, to take on review/revision of sets of individual chapters dealing with certain categories of pollutants (e.g. a taskgroup for major air pollutants, PM, O_3, NO_2, CO, SO_2; another for certain inorganic substances; another for certain organic compounds, etc.). During the past several years, the process has been repeated, resulting in a revised, updated version of the Air Quality Guidelines for Europe to be published in 1999, including assessment of an expanded list of air pollutants beyond those addressed in the 1987 edition (WHO, 1999). Information regarding the revised AQGs can be obtained via the Internet at http://www.who.ch/pll/dsa/.

It is important to note that the air quality guidelines, discussed above as being generated under WHO auspices (as contained in EHC documents, CICADs, or the last-described 1987 or 1999 AQGs), are not air standards, *per se*. Rather, the guidelines represent concentrations appropriate for certain averaging times (e.g. 1-h, 24-h, annual average) recommended by international experts for consideration by national or international authorities in promulgating air quality standards having the force of law for geopolitical entities falling under their jurisdiction.

REFERENCES

Barter RA, Twerdok LE, Sharp CC *et al.* (1996) *The Utility of Gasoline Engine Exhaust Emission Toxicology Testing.* Washington, DC: American Petroleum Institute.

Davis JM (1998) Methylcyclopentadienyl manganese tricarbonyl: health risk uncertainties and research directions. *Environ. Health Perspect* **106**(suppl 1): 191–201.

Davis JM and Graham JA (1998) Health effects of oxygenated fuel additives. In: Myers RA (ed) *Encyclopedia of Environmental Analysis and Remediation.* New York: John Wiley and Sons.

Davis JM, Elias RW and Grant LD (1996) Efforts to reduce lead exposure in the United States. In: Yasui M, Strong MJ, Ota K and Verity MA (eds) *Mineral and Metal Neurotoxicology.* Boca Raton, FL: CRC Press, pp 285–293.

Ethyl Corporation v. *Carol M Browner of the US Environmental Protection Agency and the US Environmental Protection Agency* (1995) No 94-1518, DC Circuit.

Ethyl Corporation v. *US Environmental Protection Agency* (1995) No 94-1505, DC Circuit.

Federal Register (1971) National primary and secondary ambient air quality standards. *Fed Reg* **36**: 8186–8201.

Federal Register (1977) Lead ambient air quality standard. *Fed Reg* **42**: 63076–63094.

Federal Register (1979) National primary and secondary ambient air quality standards: revisions to the National Ambient Air Quality Standards for photochemical oxidants. *Fed Reg* **44**: 8202–8237.

Federal Register (1987) Revisions to the national ambient air quality standards for particulate matter. *Fed Reg* **52**: 24634–24669.

Federal Register (1994a) National ambient air quality standards for carbon monoxide – final decision. *Fed Reg* **59**: 38906–38917.

Federal Register (1994b) Fuels and fuel additives registration regulations. *Fed Reg* **59**: 33042–33142.

Federal Register (1994c) Fuels and fuel additives; waiver decision/circuit court remand. *Fed Reg* **59**: 42227–42247.

Federal Register (1996a) National ambient air quality standards for particulate matter final rule. *Fed Reg* **62**: 38651–38752.

Federal Register (1996b) National ambient air quality standards for sulfur oxides (sulfur dioxide) – final decision. *Fed Reg* **61**: 100, 25566–25580.

Federal Register (1996c) National ambient air quality standards for nitrogen oxides; final rule. *Fed Reg* **61**: 196, 52852–52856.

Federal Register (1997) National ambient air quality standards for ozone final rule. Fed Reg **62**: 38856–38896.

Federal Register (1998) Health effects testing requirements for baseline gasoline and oxygenated non-baseline gasoline and approval of alternative emissions generator. *Fed Reg* **63**: 67877–67879.

Hasselblad V and Jarabek AM (1994) Dose–response analysis of toxic chemicals. In: Berry DA and Stangl DK (eds) *Bayesian Biostatistics.* New York: Marcel Dekker.

National Research Council (1983) *Risk Assessment in the Federal Government: Managing the Process.* Washington, DC: National Academy Press.

National Research Council (1994) *Science and Judgment in Risk Assessment.* Washington, DC: National Academy Press. 19 January 1994.

Shoaf C, Caldwell-Kendel JC and Siegel-Scott C (1994) A hazard ranking methodology for the Clean Air Act as a tool for comparative risk analysis. *Proceedings from the 4th U.S.–Dutch International*

Symposium on Comparative Risk Analysis and Priority Setting for Air Pollutants Issues. Jointly sponsored by the US EPA and the Dutch Ministry of Housing, Spatial Planning and Environment.

Ter Haar GL, Lenane DL, Hu JN and Brandt M (1972) Composition, size and control of automotive exhaust particulates. *J Air Pollut Control Assoc* **22**: 39–46.

Tukey JW, Ciminera JL and Heyse JF (1985) Testing the statistical certainty of a response to increasing doses of a drug. *Biometrics* **14**: 295–301.

US Code (1970) Clean Air Act. USC 42. PL 91–604.

US Code (1977) Clean Air Act. USC 42. PL 95–95.

US Code (1990) Clean Air Act, § 211, Regulation of fuels: (k) Reformulated gasoline for conventional vehicles; (m) Oxygenated fuels. USC **42**: § 7545.

US Code (1991) Clean Air Act, § 108, Air quality criteria and control techniques, § 109, National ambient air quality standards. USC 42 §§ 7408–7409.

US EPA (1977) *Air Quality Criteria for Lead.* Research Triangle Park, NC: Health Effects Research Laboratory, Criteria and Special Studies Office, report no. EPA-600/8-77-017. Available from: NTIS, Springfield, VA; PB-280411.

US EPA (1986a) Second Addendum to Air Quality Criteria for Particulate Matter and Sulfur Oxides (1982): *Assessment of Newly Available Health Effects Information.* Research Triangle Park, NC: Environmental Criteria and Assessment Office, report no. EPA/600/8-86-020-F. Available from: NTIS, Springfield, VA; PB-87-176574.

US EPA (1986b) *Air Quality Criteria for Lead.* Research Triangle Park, NC: Office of Health and Environmental Assessment, Environmental Criteria and Assessment Office, EPA report no EPA-600/8-83/028aF-dF. 4v. Available from: NTIS, Springfield, VA; PB87-142378.

US EPA (1986c) Technical background document to support rulemaking pursuant to CERCLA Section 102 vol. II August 1986. *Appendix: Methodology and Guidelines for Ranking Chemicals Based on Chronic Toxicity Data.* Cincinnati, OH: Office of Solid Waste and Emergency Response, Office of Research and Development, report no ECAO-Cin-R213.

US EPA (1986d) Guidelines for carcinogen risk assessment. *Fed Reg* **51**: 33992–34003.

US EPA (1989a) *Supplement to the 1986 EPA Air Quality Criteria for Lead,* vol I Addendum (pp A1–A67). Washington, DC: Office of Research and Development, Office of Health and Environmental Assessment, report no EPA/600/8-89/049A.

US EPA (1989b) National emission standards for hazardous air pollutants: benzene. *Fed Reg* **54**(177): 38044–38072, Rule and Proposed Rule. 14 September 1989.

US EPA (1990) *Review of the National Ambient Air Quality Standards for Lead: Assessment of Scientific and Technical Information.* Research Triangle Park, NC: Office of Air Quality Planning and Standards, report no EPA-450/2-89/022. Available from: NTIS, Springfield, VA; PB91-206185.

US EPA (1991a) *Air Quality Criteria for Carbon Monoxide.* Research Triangle Park, NC: Office of Health and Environmental Assessment, Environmental Criteria and Assessment Office, report no EPA/600/8-90/045F. Available from: NTIS, Springfield, VA; PB93-167492.

US Environmental Protection Agency (1991b) *Information Needed to Improve the Risk Characterization of Manganese Tetraoxide (Mn_3O_4) and Methylcyclopentadienyl Manganese Tricarbonyl (MMT).* Washington, DC: Office of Research and Development (December).

US EPA (1992a) *Review of the National Ambient Air Quality Standards for Carbon Monoxide; Assessment of Scientific and Technical Information* OAQPS Staff Paper. Research Triangle Park, NC: Office of Air Quality Planning and Standards, report no EPA-452/R-920-004. Available from: NTIS, Springfield, VA; PB93-157717.

US EPA (1992b) Guidelines for exposure assessment. *Fed Reg* **57**: 22888–22938.

US EPA (1992c). *Framework for Ecological Risk Assessment.* Washington, DC: Risk Assessment Forum, Office of Research and Development report no EPA/630/R-92/001 (February).

US EPA (1993a) *Air Quality Criteria for Oxides of Nitrogen.* Research Triangle Park, NC: Office of Health and Environmental Assessment, Environmental Criteria and Assessment Office; report nos EPA/600/8-91/049aF-cF. 3v. Available from: NTIS, Springfield, VA; PB95-124533, PB95-124525 and PB95-124517.

US EPA (1993b) *Air Risk Information Support Center, Status Report, Fiscal Year 1992.* Research Triangle Park, NC: Office of Air Quality Planning and Standards and the Environmental Criteria and Assessment Office (April).

US EPA (1994a) Supplement to the Second Addendum (1986) to the Air Quality Criteria for

Particulate Matter and Sulfur Oxides (1982): *Assessment of New Findings on Sulfur Dioxide Acute Exposure Health Effects in Asthmatic Individuals (1994)* Research Triangle Park, NC: Environmental Criteria and Assessment Office, report no EPA-600/FP-93/002.

US EPA (1994b) Technical background document to support rulemaking pursuant to the Clean Air Act Section 112(g): *Ranking of Pollutants with Respect to Hazard to Human Health*. Research Triangle Park, NC: Office of Air Quality Planning and Standards, report no. EPA-450/3-92-010.

US EPA (1994c) *Documentation of De Minimis Emission Rates* – Proposed 40 CFR Part 63, Subpart B, Background Document. Research Triangle Park, NC: Office of Air Quality Planning and Standards, report no EPA-453/R-93-035.

US EPA (1994d) *Methods for Derivation of Inhalation Reference Concentrations and Application of Inhalation Dosimetry*. Washington, DC: Office of Research and Development report no EPA/600/8-90/066F.

US EPA (1994e) *Methods for Exposure–Response Analysis and Health Assessment for Acute Inhalation Exposure to Chemicals: Development of the Acute Reference Exposure*. Draft Working Paper. Research Triangle Park, NC: Office of Health and Environmental Assessment.

US EPA (1995a) *Review of the National Ambient Air Quality Standards for Nitrogen Oxides: Assessment of Scientific and Technical Information*. OAQPS Staff Paper. Office of Air Quality Planning and Standards: EPA report no EPA-452/R-95-005 (September).

US EPA (1995b) *The Use of the Benchmark Dose Approach in Health Risk Assessment*. Washington, DC: Risk Assessment Forum, Office of Research and Development report no. EPA/630/R-94/007.

US EPA (1995c) *Guidance for Risk Characterization* ("Browner memorandum"). Washington, DC: Science Policy Council.

US EPA (1996a) *Air Quality Criteria for Particulate Matter*. Research Triangle Park, NC: National Center for Environmental Assessment – RTP report nos. EPA/600/P-95/001aF-cF. 3v. Available from: NTIS, Springfield, VA; PB96-168224.

US EPA (1996b) *Review of the National Ambient Air Quality Standards for Particulate Matter: Policy Assessment of Scientific and Technical Information*. OAQPS Staff Paper. Research Triangle Park, NC: Office of Air Quality Planning and Standards, report no. EPA/45/R-96-013. Available from: NTIS, Springfield, VA; PB97-115406REB.

US EPA (1996c) *Air Quality Criteria for Ozone and Related Photochemical Oxidants*. Research Triangle Park, NC: National Center for Environmental Assessment – RTP; report nos EPA/600/AP-93/004aF-cF. 3v. Available from: NTIS, Springfield, VA; PB96-185582, PB96-185590 and PB96-185608.

US EPA (1996d) *Review of National Ambient Air Quality Standards for Ozone Assessment of Scientific and Technical Information*. OAQPS Staff Paper. Research Triangle Park, NC: Office of Air Quality Planning and Standards; report no EPA/452/R-96/007. Available from: NTIS, Springfield, VA; PB96-203435/XAB.

US EPA (1996e) Proposed test rule for hazardous air pollutants. *Fed Reg* **61**: 33178–33200

US EPA (1996f) *Proposed Guidelines for Carcinogen Risk Assessment*. Washington, DC: Office of Research and Development report no EPA/600/P-92/003C.

US EPA (1996g) *Exposure Factors Handbook*. Washington, DC: Office of Research and Development report no EPA/600/P-95/002Bc.

US EPA (1996h) Proposed guidelines for ecological risk assessment. *Fed Reg* **61**: 47552. 9 September 1996. Washington, DC: Risk Assessment Forum, Office of Research and Development report no EPA/630/R-95/002B (August).

US EPA (1997a) Amended proposed test rule for hazardous air pollutants: extension of comment period. *Fed Reg* **62**: 67465–67485.

US EPA (1997b) Toxic Substances Control Act test guidelines – final rule. *Fed Reg* **62**: 43819–43864.

US EPA (1997c) Draft Residual Risk Report to Congress, Research Triangle Park, NC: Office of Air Quality Planning and Standards, 12 December 1997.

US EPA (1997d) *Policy for Use of Probabilistic Analysis in Risk Assessment*. 15 May 1997.

US EPA (1997e) Guiding Principles for Monte Carlo Analysis. EPA/630/R-97/001.

US EPA (1997f) *Mercury Study Report to Congress*. report no. EPA 452/R-97-003. (December).

WHO/IPCS (1996) *Report of the IPCS Second Steering Group Meeting on Concise International Chemical Assessment Documents (CICADs)*. Ottawa, Canada.

WHO/IPCS (1997) Environmental Health Criteria 188. *Nitrogen Oxides* (2nd edn) Geneva: WHO.

World Health Organization (1987) *Air Quality Guidelines for Europe.* Copenhagen, Denmark: Regional Office for Europe.

World Health Organization (1999) *Air Quality Guidelines for Europe*, 2nd edn. Copenhagen, Denmark: Regional Office for Europe.

43

Air Quality Guidelines and Standards

MORTON LIPPMANN

Nelson Institute of Environmental Medicine, New York University
School of Medicine, Tuxedo, NY, USA

ROBERT L. MAYNARD

Department of Health, London, UK

The views expressed in this chapter are those of the authors and should not be taken as representing those of the UK Department of Health.

CONCEPTS OF GUIDELINES AND STANDARDS

In broad terms, Air Quality Guidelines (AQGs) are non-binding recommendations prepared by knowledgeable professionals to assist other professionals and public health authorities in evaluating the nature and extent of health risks associated with exposures to airborne chemical agents. They are an essential part of the process that has become known in recent years as risk assessment. By contrast, air quality limits or emission limits having the force of law behind their enforcement are commonly known as Air Quality Standards (AQSs). Such standards are generally established and enforced by regulatory agencies in national governments. However, there are also consensus standards established by the International Standards Organization (ISO) and/or affiliated national standards organizations that only have the force of law behind them when they are also adopted by regulatory authorities. There are also so-called standards recommended by professional societies. For example, the American Society of Heating, Refrigeration, and Air Conditioning Engineers (ASHRAE) has published guidelines for indoor air quality that they have called standards. Some of these are legally binding in those parts of the USA that have included them in local codes.

AIR POLLUTION AND HEALTH
ISBN 0-12-352335-4

In terms of exposure limits for ambient air quality on the international level, the lead agency is the World Health Organization (WHO), which has, through its worldwide headquarters office in Geneva, established the limited number of AQGs listed in Table 43.1. A comprehensive list of AQGs was established in 1987 by the WHO Regional Office for Europe in Copenhagen, and these were updated and, in some cases, revised in 1997. The more comprehensive 1997 WHO-EURO Guidelines are currently under consideration by WHO-Geneva for adoption as worldwide guidelines.

The following is some background information on the WHO-EURO AQGs, and the role they are expected to play in the setting of AQSs, followed by a brief discussion of the roles that AQGs have played in the development of AQSs in the USA.

The purpose of the 1987 WHO-EURO Guidelines was clearly stated:

> . . . to provide a basis for protecting public health from adverse effects of air pollution and for eliminating, or reducing to a minimum, those contaminants of the air that are known or likely to be hazardous to human health and well being.

It was also stated that the Guidelines should provide background information for standard setting, though it was stressed that their use was not restricted to this. That the Guidelines were not intended, *per se*, as standards was made clear. It was pointed out that in moving from guidelines to standards, prevailing exposure levels and environmental, social, economic and cultural conditions should be taken into account. It was explicitly stated that:

> In certain circumstances there may be valid reasons to pursue policies which will result in pollutant concentrations above or below the guideline values.

This is often forgotten, especially by those who would criticise the governments of developing countries where levels of pollutants exceed the Guidelines.

Table 43.1 World Health Organization guideline values for carbon monoxide, lead, nitrogen dioxide, ozone and sulfur dioxide[a]

Compound	Guideline value ($\mu g/m^3$)	Averaging time
Carbon monoxide	100 000	15 min
	60 000	30 min
	30 000	1 h
	10 000	8 h
Lead	0.5	1 year
Nitrogen dioxide	200	1 h
	40	1 year
Ozone	120	8 h
Sulfur dioxide	1000	10 min
	250	24 h
	100	1 year

[a] No guideline value for particulate matter has been set. In the revision of the WHO Air Quality Guidelines for Europe it was agreed that a concentration–response curve should be defined: no threshold concentration is assumed. This will appear in the next edition of the WHO Air Quality Guidelines for Europe

It is generally accepted that an air quality standard is a description of a level of air quality or air pollution, which is adopted by a regulatory authority as enforceable. At its simplest, an air quality standard should be defined in terms of one or more concentrations and associated averaging times. In addition, information on the form of exposure and monitoring, which are relevant in assessing compliance with the standard, methods of data analysis and Quality Assurance and Quality Control requirements, should be added.

In some countries the standard is further qualified by defining an acceptable level of attainment or compliance. Levels of attainment may be defined in terms of the fundamental units of definition of the standard. Percentiles have been used: for example, if the unit defined by the standard is the day, then a requirement for 99% compliance allows 3 days exceedance of the standard in the year. The cost of meeting any standard is likely to depend critically on the degree of compliance required. Given this, it may be sensible not to debate for too long the basic level of the standard (the WHO Guideline might often be adopted), but to consider carefully the costs and benefits of different levels of compliance.

It is important to remember that the development of air quality standards is only part of an adequate air quality management strategy. Legislation, identification of authorities responsible for enforcement of emission standards and penalties for exceedances are all also necessary. Emission standards may play an important role in the management strategy, especially if exceedance of air quality standards is used as a trigger for abatement measures. These may be needed at both the national and the local level.

Air quality standards are also important in informing the public about air quality. Used in this way they are a double-edged weapon as the public may assume that once a standard is exceeded adverse effects on health will occur. This may not be the case. If standards are set at levels below the guidelines then it is even less likely that adverse effects will follow small exceedances.

The process of setting standards is simplified for many countries when the WHO offers a guideline value. In general, local review of the health effects data may be unnecessary. However, when published studies on associations between air pollutants and health effects in the local region are available, it is prudent for authorities responsible for setting national standards to give them due consideration in their evaluation of the adequacy of the WHO Air Quality Guidelines. If no single value is offered but rather a Unit Risk estimate or an exposure–response relationship is defined, then the following should be considered in setting standards:

(1) The nature of the effects indicated should be examined and decisions made as to whether they represent adverse health effects.
(2) Special populations at risk should be considered.

Sensitive populations or groups are defined here as those impaired by concurrent disease or other physiological limitations and those with specific characteristics making health consequences of exposure more significant (e.g. developmental phase in children). In addition, other groups may be judged to be at special risk because of their exposure patterns and due to increased effective dose for a given exposure (e.g. children). The sensitive populations may vary across countries owing to differences in the number of people with inadequate access to medical care, in the prevalence of certain endemic diseases, in the prevailing genetic factors, or in the prevalence of debilitating diseases or nutritional deficiencies. It is up to the regulator to decide which specific groups at risk should be protected by the standards.

In the USA, the concept of AQGs was officially adopted as part of the Clean Air Act (CAA) of 1963, and the procedures by which the various states would use them to establish AQSs was specified in the 1967 CAA. However, the 1970 CAA mandated National Ambient Air Quality Standards (NAAQS) for ubiquitous air pollutants, removing the authority of the states to set less stringent standards. To date, only the State of California has set AQSs more stringent than the NAAQS. There are only six pollutants having NAAQS, i.e. particulate matter (PM), sulfur dioxide (SO_2), nitrogen dioxide (NO_2), ozone (O_3), carbon monoxide (CO) and lead (Pb).

All other air pollutants are considered to originate primarily from a limited number of point sources, and air quality for these pollutants is controlled by the enforcement of National Emission Standards for Hazardous Air Pollutants (NESHAPs). Thus, there are no specific national AQGs or AQSs for these pollutants in the USA. However, individual states in the USA have adopted AQGs and/or AQSs for such air pollutants. When AQSs are issued, they are generally intended for use in the process of reviewing plans for industrial developments by local governments and planning boards.

The various states have followed various procedures in their individual adoption of AQGs and/or AQSs. Initially, most of them used the Threshold Limit Values (TLVs) recommended annually by the American Conference of Governmental Industrial Hygienists (ACGIH) for use by professional industrial and occupational hygienists as guides for the protection of the health of working men and women, and applied arbitrary safety factors (10 to 100) (ACGIH, 1996, 1997). The ACGIH TLVs, while intended to be used only as guidelines for the protection of the health of workers, are the largest body of recommended limits for inhaled materials, with over 750 substances listed in the latest annual list. They are based on the professional judgement of a committee of technical experts, and have a wide range of safety factors (for industrial workers) built into them. Those based on epidemiological data have relatively small safety margins (1.5 to 10), which are based on the strength and consistency of the evidence. Those based on animal toxicology have larger safety factors (10–100), to allow for high-to-low dose and interspecies differences. The additional 10 to 100 safety factors adopted when using TLVs as a basis for establishing AQGs is based on the extension of the exposure interval from 40 h/week to 168, and on the recognition that the general population includes the very young, the very old, and persons with serious acute or chronic disabilities, as well as healthy adults capable of full time work.

Other states have rejected this approach and followed an equally simplistic and misguided approach based on the reference concentrations (RfCs) adopted by the US Environmental Protection Agency (EPA). These are based primarily on carcinogen potency estimates from rodent cancer bioassay study data. They generally have several orders of magnitude of safety factor built into them, and are usually based on an acceptable lifetime risk of cancer in one in a million (10^{-6}).

In terms of emergency responses to unanticipated releases of chemicals into the atmosphere, such as the 3 December 1984 release of ~40 tons of methyl isocyanate at Bhopal, India, the American Industrial Hygiene Association (AIHA) has developed guidelines for emergency response planning. Such AQGs are viewed as:

(1) primarily useful for emergency planning and response;

(2) suitable for protection from health effects due to short-term exposures – but not for effects due to repeated exposure;

(3) not absolute levels demarcating safe from hazardous conditions;

(4) one element of the planning needed in developing a programme to protect a neighbouring community.

It was anticipated that these Emergency Response Planning Guidelines (ERPGs) would be most widely used for evaluating plausible release scenarios by air dispersion modelling for the estimation of impact concentrations, evacuation zones and other emergency responses.

In preparing their recommendations, the AIHA ERPG Committee defined three different concentration levels, as follows:

- *ERPG-1*: The maximum airborne concentration below which it is believed nearly all individuals could be exposed for up to 1 h without experiencing other than mild transient adverse health effects or perceiving a clearly defined objectionable odour.

- *ERPG-2*: The maximum airborne concentration below which it is believed nearly all individuals could be exposed for up to 1 h without experiencing or developing irreversible or other serious health effects or symptoms that could impair their abilities to take protective action.

- *ERPG-3*: The maximum airborne concentration below which it is believed nearly all individuals could be exposed for up to 1 h without experiencing or developing life-threatening health effects.

Recent ERPG values are summarized in Table 43.2.

The establishment of stringent AQSs cannot, by itself, ensure clean air in a community, region, or nation. It is merely one element of a strategy of reducing exposures to a population at risk of harm from such exposures. When it is necessary to reduce ambient concentrations below the established AQSs, there must be either reliance on emission controls on the sources of the pollutants, or separation of the source from the receptors, thereby allowing sufficient dilution. For controlling the ubiquitous pollutants in the USA, the original emphasis was on control of primary emission sources and/or on the use of high stacks for dilution of the concentration before impacting ground-based receptors. More recently, there has been recognition that discharges from high stacks, while limiting ground level maxima of SO_2 and NO_x, facilitated their atmospheric transformations to sulfuric acid and nitric acids and their widespread dispersion over wide geographic regions. Accordingly, the current emphasis is on emission controls at the source, e.g. through changes in combustion technology, evaporative vapours recycling, changes in the composition of fuels, stack gas cleaning, or catalytic oxidation of motor vehicle exhaust.

In dealing with ambient air concentrations exceeding the AQSs, public authorities usually begin with quantitative inventories of known sources. Control efforts are generally concentrated initially on those sources having the greatest impact and/or on those that can most readily be controlled without prohibitive costs or disruption of normal population activities. In the USA, such information is used in the development of a State Implementation Plan (SIP), outlining a plan and schedule that must be followed. The SIPs must be approved by the US EPA, and modified when they fail to meet their objectives.

Table 43.2　Current AIHA ERPGs (1998)

Chemical (CAS number)	ERPG-1		ERPG-2		ERPG-3	
Acetaldehyde (75-07-0)	10	ppm	200	ppm	1000	ppm
Acrolein (107-02-8)	0.1	ppm	0.5	ppm	7	ppm
Acrylic acid (79-10-7)	2	ppm	50	ppm	750	ppm
Acrylonitrile (107-13-1)	10	ppm	35	ppm	75	ppm
Allyl chloride (107-05-1)	3	ppm	40	ppm	300	ppm
Ammonia (7664-41-7)	25	ppm	200	ppm	1000	ppm
Benzene (71-43-2)	50	ppm	150	ppm	1000	ppm
Benzyl chloride (100-44-7)	1	ppm	10	ppm	25	ppm
Beryllium (7440-41-7)[a]		NA[c]	25	µg/m^3	100	µg/m^3
Bromine (7726-95-6)	0.2	ppm	1	ppm	5	ppm
1,3-Butadiene (106-99-0)	10	ppm	200	ppm	5000	ppm
n-Butyl acrylate (141-32-2)	0.05	ppm	25	ppm	250	ppm
n-Butyl isocyanate (111-36-4)	0.01	ppm	0.05	ppm	1	ppm
Carbon disulfide (75-15-0)	1	ppm	50	ppm	500	ppm
Carbon tetrachloride (56-23-5)	20	ppm	100	ppm	750	ppm
Chlorine (7782-50-5)	1	ppm	3	ppm	20	ppm
Chlorine trifluoride (7790-91-2)	0.1	ppm	1	ppm	10	ppm
Chloroacetyl chloride (79-04-9)	0.1	ppm	1	ppm	10	ppm
Chloropicrin (76-06-2)		NA[c]	0.2	ppm	3	ppm
Chlorosulfonic acid (7790-94-5)	2	mg/m^3	10	mg/m^3	30	mg/m^3
Chlorotrifluoroethylene (79-38-9)	20	ppm	100	ppm	300	ppm
Crotonaldehyde (4170-30-3)	2	ppm	10	ppm	50	ppm
Cyanogen chloride (506-77-4)[a]		NA[c]	0.4	ppm	4	ppm
Diborane (19287-45-7)		NA[c]	1	ppm	3	ppm
Diketene (674-82-8)	1	ppm	5	ppm	50	ppm
Dimethylamine (124-40-3)	1	ppm	100	ppm	500	ppm
Dimethyldichlorosilane (75-78-5)	0.8	ppm	5	ppm	25	ppm
Dimethyl disulfide (624-92-0)	0.01	ppm	50	ppm	250	ppm
Dimethylformamide (68-12-2)	2	ppm	100	ppm	200	ppm
Dimethyl sulfide (75-18-3)	0.5	ppm	500	ppm	2000	ppm
Diphenylmethane diisocyanate (MDI) (101-68-8)[a]	0.2	mg/m^3	2	mg/m^3	25	mg/m^3
Epichlorohydrin (106-89-8)	2	ppm	20	ppm	100	ppm
Ethylene oxide (75-21-8)		NA[c]	50	ppm	500	ppm
Fluorine (7782-41-4)[a]	0.5	ppm	5	ppm	20	ppm
Formaldehyde (50-00-0)	1	ppm	10	ppm	25	ppm
Furfural (98-01-0)[a]	2	ppm	10	ppm	100	ppm
Hexachlorobutadiene (87-68-3)	3	ppm	10	ppm	30	ppm
Hexafluoroacetone		NA[c]	1	ppm	50	ppm
Hexafluoropropylene (116-15-4)	10	ppm	50	ppm	500	ppm
Hydrogen chloride (7647-01-0)[a]	3	ppm	20	ppm	150	ppm[b]
Hydrogen cyanide (74-90-8)		NA[c]	10	ppm	25	ppm
Hydrogen fluoride (7664-39-3)	2	ppm	20	ppm	50	ppm
Hydrogen peroxide (7722-84-1)	10	ppm	50	ppm	100	ppm
Hydrogen sulfide (7783-06-4)	0.1	ppm	30	ppm	100	ppm
Iodine (7553-56-2)	0.1	ppm	0.5	ppm	5	ppm
Isobutyronitrile (78-82-0)	10	ppm	50	ppm	200	ppm
2-Isocyanatoethyl methacrylate (30674-80-7)		NA[c]	0.1	ppm	1	ppm
Lithium hydride (7580-67-8)	25	µg/m^3	100	µg/m^3	500	µg/m^3
Methanol (67-56-1)	200	ppm	1000	ppm	5000	ppm
Methyl bromide (74-83-9)		NA	50	ppm	200	ppm

Table **43.2** cont.

Chemical (CAS number)	ERPG-1		ERPG-2		ERPG-3	
Methyl chloride (74-87-3)		NA	400	ppm	1000	ppm
Methyl iodide (74-88-4)	25	ppm	50	ppm	125	ppm
Methyl isocyanate (624-83-9)	0.025	ppm	0.5	ppm	5	ppm
Methyl mercaptan (74-93-1)	0.005	ppm	25	ppm	100	ppm
Methylene chloride (75-09-02)	200	ppm	750	ppm	4000	ppm
Methyltrichlorosilane (75-79-6)	0.5	ppm	3	ppm	15	ppm
Monomethylamine (74-89-5)	10	ppm	100	ppm	500	ppm
Perchloroethylene (127-18-4)	100	ppm	200	ppm	1000	ppm
Perfluoroisobutylene (382-21-8)		NA[c]	0.1	ppm	0.3	ppm
Phenol (108-95-2)	10	ppm	50	ppm	200	ppm
Phosgene (75-44-5)		NA[c]	0.2	ppm	1	ppm
Phosphorus pentoxide (1314-56-3)	5	mg/m^3	25	mg/m^3	100	mg/m^3
Propylene oxide (75-56-9)	50	ppm	250	ppm	750	ppm
Styrene (100-42-5)	50	ppm	250	ppm	1000	ppm
Sulfur dioxide (7446-09-5)	0.3	ppm	3	ppm	15	ppm
Sulfuric acid (oleum [8014-95-7], sulfur trioxide [7446-11-9], and sulfuric acid [7664-93-9])	2	mg/m^3	10	mg/m^3	30	mg/m^3
Tetrafluoroethylene (116-14-3)	200	ppm	1000	ppm	10 000	ppm
Tetramethoxysilane (681-84-5)[a]		NA[c]	10	ppm	20	ppm
Titanium tetrachloride (7550-45-0)	5	mg/m^3	20	mg/m^3	100	mg/m^3
Toluene (108-88-3)	50	ppm	300	ppm	1000	ppm
1,1,1-Trichloroethane (71-55-6)	350	ppm	700	ppm	3500	ppm
Trichloroethylene (79-01-6)[a]	100	ppm	500	ppm	5000	ppm
Trichlorosilane (10025-78-2)[a]	1	ppm	3	ppm	25	ppm
Trimethoxysilane (2487-90-3)[a]	0.5	ppm	2	ppm	5	ppm
Trimethylamine (75-50-3)	0.1	ppm	100	ppm	500	ppm
Uranium hexafluoride (7783-81-5)	5	mg/m^3	15	mg/m^3	30	mg/m^3
Vinyl acetate (108-05-4)	5	ppm	75	ppm	500	ppm

[a] Included in 1998 update set.
[b] Value revised from earlier ERPG.
[c] NA, not appropriate.

AIR QUALITY STANDARD SETTING IN THE USA

Historical Development of Standards for Ambient Air Pollutants in the USA

Prior to 1971, there were no national standards for air pollutants in the USA. Control of air pollution was considered primarily a state's right and responsibility, and there was great diversity among the states and their local units in terms of which pollutants were regulated and the basis for the regulations.

The 1963 Clean Air Act (CAA) had directed the Secretary of the Department of Health, Education, and Welfare (HEW) to prepare Criteria Documents (CDs) summarizing scientific knowledge on air pollutants arising from widespread sources. The Air Quality Act of 1967 specified the process under which the CDs would serve in the development of standards by the states. The original CDs also provided the scientific basis for the promulgation, in 1971, of the initial National Ambient Air Quality Standards

(NAAQSs). Promulgation of these national standards was required by the CAA of 1970, which also authorized the EPA Administrator to revise the NAAQSs or to set additional NAAQSs as needed.

The CAA of 1970 also directed the EPA Administrator to regulate toxic air pollutants discharged by a limited number of point sources. These regulations, known as National Emission Standards for Hazardous Air Pollutants (NESHAPs), are based on applications of specified emission controls, rather than on strategies designed to limit ambient air concentrations. The health effects basis for each listing is presented in terms of a Health Assessment Document (HAD).

The lack of a programme for regular re-evaluation and revision of the original NAAQSs led Congress, in the CAA revisions of 1977, to mandate that the NAAQSs be revised or reauthorized by 31 December 1980, and at 5-year intervals thereafter.

The 1977 CAA Amendments also affected NESHAPs. They allowed EPA to promulgate a design, equipment, work practice, or operational standard, or combination thereof. This made it legal to have non-numerical controls as well as numerical emission rate limits (Tabler, 1984).

Two new NAAQS promulgations did take place before the end of 1980. A new NAAQS was promulgated for airborne Pb in 1978, and a revised ozone (O_3) NAAQS in 1979. However, both of these actions were stimulated, at least in part, by court litigation. Decisions were made in 1985 and 1996 to reaffirm the 1971 NAAQS for nitrogen dioxide (NO_2) in 1985 and 1994 to reaffirm the 1971 NAAQS for carbon monoxide (CO). The Congressional mandate to reissue or revise NAAQSs by 31 December 1980 and at 5-year intervals thereafter has not been met. Some of the reasons for the delays, and some CASAC recommendations for streamlining the process, were discussed by Lippmann (1987).

There are two, quite different processes followed by the EPA in establishing standards for ambient air pollutants. The process followed for the so-called criteria pollutants, i.e. those having ubiquitous sources and involving NAAQSs, is more extensive and formalized than that followed for the so-called toxic or hazardous air pollutants arising from a more limited number of definable point sources. It is characterized by more extensive participation by the scientific community. The major reasons for the more formal process for criteria pollutants having NAAQSs are the much more explicit and comprehensive statutory requirements, and the EPA responses to a series of court decisions regarding ambient air standards.

Ambient Air Quality Standards

Once a criteria pollutant is listed under the 1970 CAA, the Administrator has 12 months to issue a CD, which accurately reflects the latest scientific knowledge that may be useful in indicating the kind and extent of all identifiable effects on public health or welfare that may be expected from the presence of such pollutants in the ambient air in varying quantities. The CAA also defines primary and secondary standards as follows:

(1) National primary ambient air quality standards, prescribed under subsection (a), shall be ambient air quality standards the attainment and maintenance of which in the judgment of the Administrator, based on such criteria and allowing an adequate margin of safety, are requisite to protect the public health.

(2) Any national secondary ambient air quality standard, prescribed under subsection (a), shall specify a level of air quality the attainment and maintenance of which in the judgment of the Administrator, based on such criteria, is requisite to protect the public welfare from any known or anticipated adverse effects associated with the presence of such air pollutant in the ambient air.

Once a NAAQS is promulgated, each state and each territory has 9 months to prepare and submit a State Implementation Plan (SIP) for the way the standard will be achieved. The EPA has 4 months after the required date for submission to approve or disapprove the plan. After the plan is approved, the state has 3 years in which to apply the plan and meet the NAAQS. An extension of 2 years may be granted by the EPA if necessary. Should a state fail to submit a plan or revised a disapproved portion of one, the EPA must promulgate a plan for the state within 6 months.

The CAA Amendments of 1977 also established the Clean Air Scientific Advisory Committee (CASAC) to review all of the CDs prior to proposal and promulgation of NAAQS.

The 1977 CAA Amendments specifically required that CASAC be given the following functions and responsibilities.

- Review the criteria and the national primary and secondary ambient air quality standards and recommend to the Administrator any new NAAQS or revision of existing criteria and standards as may be appropriate.
- Advise the Administrator of areas where additional knowledge is required concerning the adequacy and basis of existing, new or revised NAAQS.
- Describe the research efforts necessary to provide the required information.
- Advise the Administrator of any adverse public health, welfare, social, economic or energy effects which may result from various strategies for attainment and maintenance of such national ambient air quality standards.

Although the EPA, by administrative order, decided to house the CASAC within its Science Advisory Board (SAB), the CASAC, as a committee established independently by statute, reports directly to the Administrator and to Congress.

The CASAC reviews Agency staff papers (SPs) prepared by the Office of Air Quality Planning and Standards (OAQPS) as well as CDs in public sessions. An SP discusses the key studies most relevant to setting standards in layman's terms and lays out the options for the Administrator's decision. Much of its content appears as a preamble in the *Federal Register* at the time a NAAQS is proposed, and also when standards are promulgated.

In response to a legal challenge of a NAAQS promulgation (*API v. Costle*), a Federal court had ruled that the EPA did not have to follow the advice of its science advisors, but simply receive it. The EPA began to take a safe position, however, in seeking scientific guidance from the CASAC before issuing a CD. The product of the CASAC review takes the form of a letter from the CASAC Chairman to the Administrator. While there is no legal requirement for this 'letter of closure', it demonstrates that the CASAC had conducted its review as specified by the law.

Because the CASAC reviews are public, the public, under provisions of the Federal Advisory Committee Act, also has access to copies of the materials being reviewed.

All comments received by the EPA as the result of the public review of the CDs and NAAQS proposals are catalogued and considered by the Agency. As required by the

1977 CAA Amendments, these comments from outside the Agency are placed into the public docket, and the file is open to the public for review. There is no legal requirement to extend a public review of criteria of the NAAQS proposals past a reasonable time (30–90 days). Because of external demand, however, the EPA began to make a practice of extending the time allowed for receipt of comments from the public.

From its first meeting in November 1978, the CASAC has played a significant role in the way the EPA operates its NAAQSs' decision-making process. The public reviews of the CD and SP are conducted as meetings of the CASAC, and comments received from both the public and the CASAC are utilized by the EPA in their document revisions and regulatory actions.

Standards for Hazardous Air Pollutants

The process of establishing the NESHAPs has also evolved. Once the Administrator lists a pollutant as requiring regulation, regulatory action is supposed to take place within 180 days. However, this statutory requirement proved to be unworkable. To allow more time for comment from interested or affected parties, the process was modified to begin with a notification of an intent to list. This was to permit and encourage advance preparation of public comment, which could then be submitted officially for consideration when the actual listing was made. With the concurrence of the SAB, the EPA staff screens Health Assessment Documents (HADs) and decides which ones to refer to the SAB for formal review. Following the SAB review and comment, a decision that the chemical does not require regulation might still be made. One other possible action by the EPA was taken in the case of acrylonitrile. In this case, the EPA decided that the HESHAPs were not warranted, but that the few individual states with significant point sources may want to regulate. The EPA referred regulatory actions to the states, offering assistance to them in formulating their own regulations.

The revised NESHAPs process, while accelerating the number of promulgations of emission standards, was judged by many in Congress to still be too slow and, in the 1990 CAA amendments, Congress mandated generic best available control technology requirements for 189 specific chemicals, with the later preparation of HADs for the residual risk following implementation of the generic technical controls.

Recent NAAQS Revisions

As noted earlier, the EPA Administrator is supposed to review the basis for its National Ambient Air Quality Standards (NAAQS) every 5 years and, if necessary for the protection of public health and/or welfare, issue new or revised standards. As shown in Table 43.3, the ozone (O_3) and PM NAAQS have been revised, albeit not at the specified 5-year intervals. The 1997 NAAQS revisions for PM and O_3 were unusual in a number of respects, including especially: (1) their simultaneous promulgations; (2) the tighter timetables involved in their preparations and public reviews; (3) the extraordinary controversy they engendered during the review process and since their promulgation; and (4) the substantial increases in the number of additional communities that will not be in compliance with the NAAQS for both PM and O_3.

Table 43.3 1997 revisions: US National Ambient Air Quality Standards (NAAQS)

I Ozone (revision of NAAQS set in 1979 and reaffirmed in 1993)

	1979 NAAQS	1997 NAAQS
Daily concentration limit (ppb)	120	80
Averaging time	maximum 1-h av.	maximum 8-h av.
Basis for excessive concentration	4th highest over 3 year period	3 year av. of 4th highest in each year
Equivalent stringency for 1 h max in new format (ppb)	~90	
Number of US counties exceeding NAAQS	106	280
Number of people in counties exceeding NAAQS	74×10^6	113×10^6

II Particulate matter (revision of NAAQS set in 1987)

	1987 NAAQS	1997 NAAQS	
	PM_{10}	PM_{10}	$PM_{2.5}$
Index pollutant	PM_{10}	PM_{10}	$PM_{2.5}$
Annual av. concentration limit ($\mu g/m^3$)	50	50	15
Daily concentration limit ($\mu g/m^3$)	150	150	65
Basis for excessive daily concentration	4th highest over 3 year period	> 99th percentile av. over 3 years	> 98th percentile av. over 3 years
Number of US counties expected to exceed NAAQS	41	14	~150
Number of people in counties exceeding NAAQS	29×10^6	$~9 \times 10^6$	$~6.6 \times 10^6$

The CAA requires the EPA Administrator to promulgate NAAQS that protects the health of 'sensitive segments' of the public with an adequate 'margin of safety'. The CAA is unique among the statutes that EPA enforces in that the primary NAAQS are supposed to be set without regard to the costs they impose on society. However, EPA does perform cost-benefit analyses for NAAQS to conform with a Presidential directive requiring such analyses for all regulations that significantly affect the national economy.

The exhaustive review process followed by EPA Staff and CASAC inevitably identifies important knowledge gaps that limit confidence in the NAAQS options as optimal choices. Recognizing these limitations, EPA Staff prepares lists of research recommendations for public review by CASAC, and CASAC summarizes and prioritizes research needs in a letter report to the EPA Administrator. It is hoped that EPA and other research sponsors will then arrange to support research that can close the critical knowledge gaps, so that the next cycle of NAAQS reviews can lead to more refined and well-targeted NAAQS.

Particulate Matter (PM)

In Europe, and elsewhere in the eastern hemisphere, particulate pollution has historically been measured as black smoke (BS) in terms of the optical density of stain caused by particles collected on a filter disc. However, it has been expressed in gravimetric terms ($\mu g/m^3$) based on standardized calibration factors. By contrast, US standards have specified direct gravimetric analyses of filter samples collected by a reference sampler built to match specific physical dimensions or performance criteria.

The initial PM NAAQS, established in 1971 used total suspended particulate matter (TSP) as the index pollutant and had an annual (geometric mean) limit of 75 $\mu g/m^3$ and a 24-h maximum of 260 $\mu g/m^3$, not to be exceeded more than once per year. The NAAQS was revised in 1987, replacing TSP as the index pollutant with particulate matter less than 10 μm in aerodynamic diameter (PM_{10}), and specifying an annual average concentration limit of 50 $\mu g/m^3$ and a 24-h maximum of 150 $\mu g/m^3$, based on the fourth highest value over a 3-year period (see Table 43.3). The 1997 PM NAAQS revision relaxed the basis for a 24-h NAAQS exceedance, and added new annual and 24-h NAAQS for particulate matter less than 2.5 μm in aerodynamic diameter ($PM_{2.5}$), as summarized in Table 43.3.

While justifications for the specific measurement techniques that have been used have generally been based on demonstrated significant quantitative associations between the measured quantity and human mortality, morbidity or lung function differences, it is fair to say that established biological mechanisms that could account for these associations are lacking, and that there is too little information on the relative toxicities of the myriad specific constituents of airborne PM. In addition to chemical composition, airborne PM also varies in particle size distribution, which affects the number of particles that reach target sites as well as the particle surface area. To date, there are no NAAQS for PM constituents (other than lead) or for the number or surface concentrations of the PM, although one recent study indicates that number concentration may correlate better with effects than does fine particle mass (Peters *et al.*, 1997).

A broad variety of processes produce PM in the ambient air, and there is an extensive body of epidemiological literature that demonstrates that there are statistically significant associations between the concentrations of airborne PM and the rates of mortality and morbidity in the human population. In those studies that reported on associations between health effects and more than one mass concentration, the strength of the association generally improves as one goes from total suspended particulate matter (TSP) to thoracic particulate matter, PM_{10}, to fine particulate matter, $PM_{2.5}$. The influence of a sampling system inlet on the sample mass collected is illustrated in Fig. 43.1.

The $PM_{2.5}$ distinction, while nominally based on particle size, is in reality a means of measuring the total gravimetric concentration of several specific chemically distinctive classes of particles that are emitted into or formed within the ambient air as very small particles. In the former category (emitted) are carbonaceous particles in wood smoke and diesel engine exhaust. In the latter category (formed) are carbonaceous particles formed during the photochemical reaction sequence that also leads to ozone formation, as well as the sulfur and nitrogen oxide particles resulting from the oxidation of sulfur dioxide and nitrogen oxide vapours released during fuel combustion and their reaction products.

The coarse particle fraction, i.e. those particles with aerodynamic diameters larger

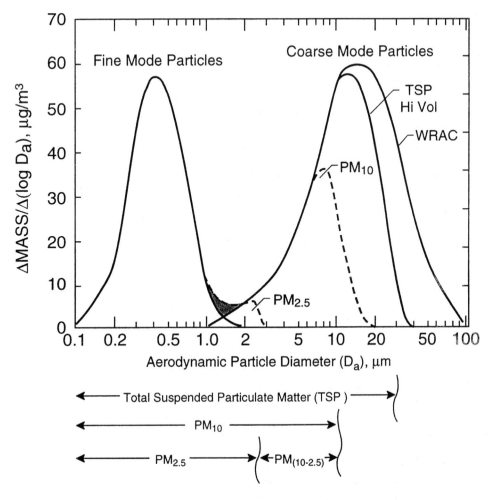

Fig. 43.1 Representative example of a mass distribution of ambient PM as function of aerodynamic particle diameter. A wide-ranging aerosol classifier (WRAC) provides an estimate of the full coarse mode distribution. Inlet restriction of the TSP high volume sampler, the PM_{10} sampler, and the $PM_{2.5}$ sampler reduce the integral mass reaching the sampling filter.

than ~2.5 μm, are largely composed of soil and mineral ash that are mechanically dispersed into the air. Both the fine and coarse fractions are complex mixtures in a chemical sense. To the extent that they are in equilibrium in the ambient air, it is a dynamic equilibrium in which they enter the air at about the same rate as they are removed. In dry weather, the concentrations of coarse particles are balanced between dispersion into the air, mixing with air masses and gravitational fallout, while the concentrations of fine particles are determined by rates of formation, rates of chemical transformation and meteorological factors. PM concentrations of both fine and coarse PM are effectively depleted by rainout and washout associated with rain. Further elaboration of these distinctions is provided in Table 43.4.

In the absence of any detailed understanding of the specific chemical components

Table 43.4   Comparisons of ambient fine and coarse mode particles

	Fine mode	Coarse mode
Formed from	Gases	Large solids/droplets
Formed by	Chemical reaction; nucleation; condensation; coagulation; evaporation of fog and cloud droplets in which gases have dissolved and reacted	Mechanical disruption (e.g. crushing, grinding, abrasion of surfaces); evaporation of sprays; suspension of dusts
Composed of	Sulfate, SO_4^{2-} nitrate, NO_3^-; ammonium, NH_4^+; hydrogen ion, H^+; elemental carbon; organic compounds (e.g. PAHs, PNAs); metals (e.g. Pb, Cd, V, Ni, Cu, Zn, Mn, Fe); particle bound water	Resuspended dusts (e.g. soil dust, street dust); coal and oil fly ash; metal oxides of crustal elements (Si, Al, Ti, Fe); $CaCO_3$, NaCl, sea salt; pollen, mould spores; plant/animal fragments; tyre wear debris
Solubility	Largely soluble, hygroscopic and deliquescent	Largely insoluble and non-hygroscopic
Sources	Combustion of coal, oil, gasoline, diesel, wood; atmospheric transformation products of NO_x, SO_2 and organic compounds including biogenic species (e.g. terpenes); high-temperature processes, smelters, steel mills, etc.	Resuspension of industrial dust and soil tracked onto roads; suspension from disturbed soil (e.g. farming, mining, unpaved roads); biological sources; construction and demolition; coal and oil combustion; ocean spray
Lifetimes	Days to weeks	Minutes to hours
Travel distance	100s to 1000s of kilometres	< 1 to 10s of kilometres

Source: EPA (1996c)

responsible for the health effects associated with exposure to ambient PM, and in the presence of a large and consistent body of epidemiological evidence associating ambient air PM with mortality and morbidity that cannot be explained by potential confounders such as other pollutants, aeroallergens, or ambient temperature or humidity, the EPA has established standards based on mass concentration within certain prescribed size fractions.

As indicated in Table 43.4, fine and coarse particles generally have distinct sources and formation mechanisms, although there may be some overlap. Primary fine particles are formed from condensation of high-temperature vapours during combustion. Secondary fine particles are usually formed from gases in three ways: (1) nucleation (i.e. gas molecules coming together to form a new particle); (2) condensation of gases onto existing particles; and (3) by reaction of absorbed gases in liquid droplets. Particles formed from nucleation also coagulate to form relatively larger aggregate particles or droplets with diameters between 0.1 and 1.0 μm, and such particles normally do not grow into the coarse mode. Particles form as a result of chemical reaction of gases in the atmosphere that lead to products that either have a low enough vapour pressure to form a particle, or react further to form a low vapour pressure substance. Some examples include: (1) the conversion of sulfur dioxide (SO_2) to sulfuric acid droplets (H_2SO_4); (2) reactions of H_2SO_4 with ammonia (NH_3) to form ammonium bisulfate (NH_4HSO_4) and ammonium sulfate

$((NH_4)_2SO_4)$; the conversion of nitrogen dioxide (NO_2) to nitric acid vapour (HNO_3), which reacts further with NH_3 to form particulate ammonium nitrate (NH_4NO_3). Although some directly emitted particles are found in the fine fraction, particles formed secondarily from gases dominate the fine fraction mass.

Figure 43.2 shows that the calculated relative acute mortality risks for PM_{10} are relatively insensitive to the concentrations of SO_2, NO_2, CO and O_3. The results are also coherent as described by Bates (1992). Figure 43.3 shows that the relative risks (RRs) for respiratory mortality are greater than for total mortality, and the RRs for the less serious symptoms are higher than those for mortality and hospital admissions.

While there is mounting evidence that excess daily mortality is associated with short-term peaks in PM_{10} pollution, the public health implications of this evidence are not yet fully clear. Key questions remain, including:

- Which specific components of the fine particle fraction $(PM_{2.5})$ and coarse particle fraction of PM_{10} are most influential in producing the responses?
- Do the effects of the PM_{10} depend on co-exposure to irritant vapours, such as ozone, sulfur dioxide, or nitrogen oxides?
- What influences do multiple day pollution episode exposures have on daily responses and response lags?
- Does long-term chronic exposure predispose sensitive individuals being 'harvested' on peak pollution days?
- How much of the excess daily mortality is associated with life-shortening measured in days or weeks as opposed to months, years or decades?

The last question above is a critical one in terms of the public health impact of excess daily mortality. If, in fact, the bulk of the excess daily mortality were due to 'harvesting' of terminally ill people who would have died within a few days, then the public health impact would be much less than if it led to prompt mortality among acutely ill persons who, if they did not die then, would have recovered and lived productive lives for years or decades longer.

While more research is needed on causal factors for the excess mortality and morbidity associated with PM in ambient air, and on the characterization of susceptibility factors, the EPA Administrator decided that her statutory responsibilities precluded waiting for the completion and peer review of this research. It is also clear that the evidence for adverse health effects attributable to PM challenges the conventional paradigm used for setting ambient air standards and guidelines, i.e. that a threshold for adversity can be identified, and a margin of safety can be applied. Excess mortality is clearly an adverse effect, and the epidemiological evidence is consistent with a linear non-threshold response for the population as a whole.

Thus, in risk management terms, the new PM NAAQS provides a modest degree of greater public health protection. Also, it represents an advance in scientific terms by its introduction of a more relevant index of exposure. The PM NAAQS may not be strict enough to fully protect public health, but there remain significant knowledge gaps on both exposures and the nature and extent of the effects that made the need for more restrictive NAAQS difficult to justify. It is essential that adequate research resources be applied to filling these gaps before the next round of NAAQS revisions during the first decade of the next century.

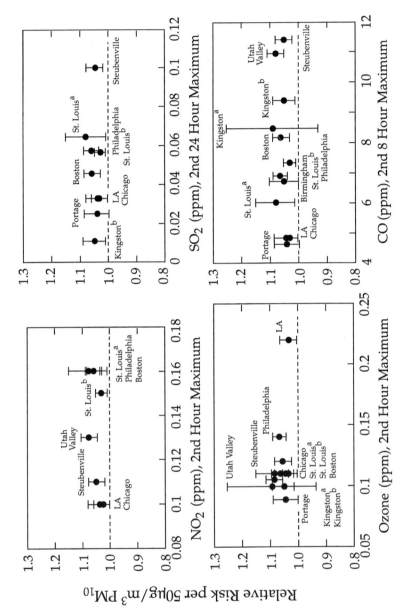

Fig. 43.2 Relationship between RR for excess daily mortality associated with PM₁₀ and peak daily levels of other criteria pollutants.

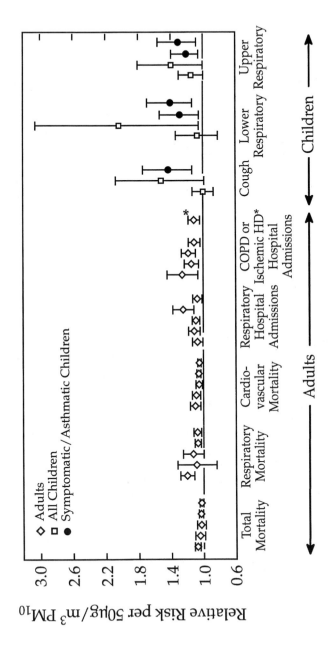

Fig. 43.3 Relationship between relative risks per 50 μg/m³ PM₁₀ and health effects. Source: US EPA (1996c).

Ozone

The US Occupational Safety and Health Administration's (OSHA) permissible exposure limit (PEL) for O_3 is 100 parts per billion (ppb), as a time-weighted average for 8 hours/day, along with a short-term exposure limit of 300 ppb for 15 minutes (US DOL, 1993). The American Conference of Governmental Industrial Hygienists (1977) threshold limit value (TLV) for occupational O_3 exposure is 100 ppb as an 8-h time-weighted average for light work; 80 ppb for moderate work; and 50 ppb for heavy work.

The initial primary (health-based) NAAQS established by the EPA in 1971 was 80 ppb of total oxidant as a 1-h maximum not to be exceeded more than once per year. The NAAQS was revised in 1979 to 120 ppb of O_3 as a 1-h maximum not to be exceeded more than 4 times in 3 years. The NAAQS is intended to prevent the health and welfare effects associated with short-term peaks in exposure and to provide protection against more cumulative damage that is suspected, but not clearly established.

The effects of concern with respect to acute response in the population at large are reductions in lung function and increases in respiratory symptoms, airway reactivity, airway permeability and airway inflammation. For persons with asthma, there are also increased rates of medication usage and restricted activities. Margin-of-safety considerations included: (1) the influence of repetitive elicitation of such responses in the progression of chronic damage to the lung of the kinds seen in chronic exposure studies in rats and monkeys; and (2) evidence from laboratory and field studies that ambient air co-pollutants potentiate the responses to O_3.

EPA initiated a review of the 1979 NAAQS in 1983, completed a Criteria Document for Ozone in 1986 and updated it in 1992 (US EPA, 1992). However, the Agency did not decide to either retain the 1979 standard or promulgate a new one until it was compelled to do so by an August 1992 court order. In response, the EPA decided, in March 1993, to maintain the existing standard and to proceed as rapidly as possible with the next round of review. This expedited review was completed with the publication of both a new criteria document (US EPA, 1996a) and staff paper in 1996 (US EPA, 1996b). In July of 1997 the EPA Administrator promulgated a revised primary O_3 NAAQS of 80 ppb as an 8-h time-weighted average daily maximum, with no more than four annual exceedances, and averaged over 3 years (see Table 43.3) (US EPA, 1997). The reason for the switch from one allowable annual exceedance to four was to minimize the designation of NAAQS non-attainment in a community that was triggered by rare meteorological conditions especially conducive to O_3 formation. The goal was to have a more stable NAAQS that allowed for extremes of annual variations of weather. The switch to an 8-h averaging time was in recognition that ambient O_3 in much of the USA has broad daily peaks, and that human responses are more closely related to total daily exposure than to brief peaks of O_3 exposure. Since the 120 ppb, 1-h average, one exceedance NAAQS would, on average, be equivalent to an 8-h NAAQS with four annual exceedances at a concentration a bit below 90 ppb, the 1997 NAAQS represents about a 10% reduction in permissible O_3 exposure. This new NAAQS will be difficult to achieve in much of the southern and eastern U.S.

Since O_3 is almost entirely a secondary air pollutant, formed in the atmosphere through a complex photochemical reaction sequence requiring reactive hydrocarbons, NO_2 and sunlight, it can only be controlled by reducing ambient air concentrations of hydrocarbons, NO_2 or both. Both NO and NO_2 are primary pollutants, known collectively as NO_x. In the atmosphere, NO is gradually converted to NO_2. Motor vehicles, one of the major categories of sources of hydrocarbons and NO_x, have been the target of control efforts, and major reductions (> 90%) have been achieved in the USA in hydrocarbon emissions per vehicle. However, there have been major increases in vehicle miles driven. Reductions in NO_x from motor vehicles and stationary-source combustion have been much smaller. The net reduction in exposure has been modest at best, with some reductions in areas with more stringent controls, such as California, and some increases in exposure in other parts of the USA. In 1988, there were record high levels of ambient O_3 with exceedance of the former 1-h maximum 120 ppb limit in 96 communities containing over 150 million people. Since then, there has been a gradual decrease of ambient O_3 in most of the USA.

We know a great deal about O_3 chemistry and have developed highly sophisticated O_3 air quality models (Seinfeld, 1998). Unfortunately, the models, and their applications in control strategies have clearly been inadequate in terms of community compliance with the NAAQS. We also know a great deal about some of the health effects of O_3. However, much of what we know relates to transient, apparently reversible effects that follow acute exposures lasting from 5 min to 6.6 h. These effects include changes in lung capacity, flow resistance, epithelial permeability and reactivity to bronchoactive challenges; such effects can be observed within the first few hours after the start of the exposure and may persist for many hours or days after the exposure ceases. Repetitive daily exposures over several days or weeks can exacerbate and prolong these transient effects. There has been a great deal of controversy about the health significant of such effects and whether such effects are sufficiently adverse to serve as a basis for the O_3 NAAQS (Lippmann, 1988, 1991, 1993).

Decrements in respiratory function such as forced vital capacity (FVC) and forced expiratory volume in the first second of a vital capacity manoeuvre (FEV_1) fall into the category where adversity begins at some specific level of pollutant-associated change. However, there are clear differences of opinion on what the threshold of adversity ought to be.

In terms of functional effects, we know that single O_3 exposures to healthy non-smoking young adults at concentrations in the range of 80–200 ppb produce a complex array of pulmonary responses including decreases in respiratory function and athletic performance, and increases in symptoms, airway reactivity, neutrophil content in lung lavage and rate of mucociliary particle clearance. The respiratory function responses to O_3 in purified air in chambers that occur at concentrations of 80, 100 and 120 ppb when the exposures involve moderate exercise over 6 h or more are illustrated in Fig. 43.4. Comparable responses require concentrations of 180 or 200 ppb when the duration of exposure is 2 h or less. On the other hand, Table 43.5 shows that mean FEV_1 decrements > 5% have been seen at 100 ppb of O_3 in ambient air for children at summer camps and for adults engaged in outdoor exercise for only 30 min. The apparently greater responses to O_3 in ambient air may be related to the presence of, or prior exposures to, acidic aerosol.

Further research will be needed to establish the interrelationships between small transient functional decrements, such as FEV_1 and mucociliary clearance rates, which may not in themselves be adverse effects, and changes in symptoms, performance, reactivity, permeability and neutrophil counts. The latter may be more closely associated with adversity in themselves or in the accumulation or progression of chronic lung damage.

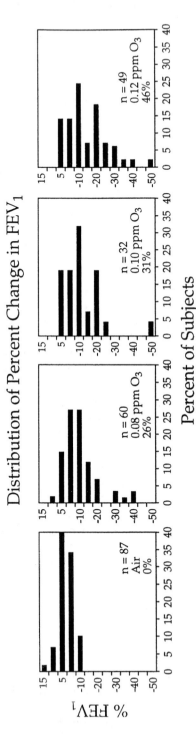

Fig. 43.4   Percentage change in FEV$_1$ in healthy non-smokers following 6.6-h exposures to clean air and O$_3$ at 80, 100 and 120 ppb during exercise lasting 50 min of each hour for studies performed at EPA's clinical research laboratory at Chapel Hill, NC. Each box shows the number of subjects studied and the percentage of subjects with reductions in FEV$_1$ that were greater than 10%.

Table 43.5 Population-based decrements in respiratory function associated with exposure to ozone in ambient air

| | Per cent decrement at 120 ppb O_3 | | | |
| | Camp children[a] | | Adult exercisers[b] | |
Functional index	Mean	90th percentile	Mean	90th percentile
Forced vital capacity	5	14	5	16
Forced expiratory volume in 1 s	8	19	4	12
FEF_{25-75}[c]	11	33	16	39
PEFR[d]	17	42	13	36

[a] 93 children at Fairview Lake, NJ, YMCA summer camp, 1984. Source: Spektor *et al.* (1988a).
[b] 30 non-smoking healthy adults at Tuxedo, NY, 1985. Source: Spektor *et al.* (1988b).
[c] Forced expiratory flow rate between 25% and 75% of vital capacity.
[d] Peak expiratory flow rate.

The plausibility of accelerated ageing of the human lung from chronic O_3 exposure is greatly enhanced by the results of chronic animal exposure studies at near ambient O_3 concentrations in rats and monkeys (Chang *et al.*, 1992; Huang *et al.*, 1988; Hyde *et al.*, 1989; Tepper *et al.*, 1991; Tyler *et al.*, 1988). There is little reason to expect humans to be less sensitive than rats or monkeys. On the contrary, humans have a greater dosage delivered to the respiratory acinus than do rats for the same exposures. Another factor is that the rat and monkey exposures were to confined animals with little opportunity for heavy exercise. Thus humans who are active outdoors during the warmer months may have greater effective O_3 exposures than the test animals (Spektor *et al.*, 1988a,b). Finally, humans are exposed to O_3 in ambient mixtures. The potentiation of the characteristic O_3 responses by other ambient air constituents seen in the short-term exposure studies in humans and animals may also contribute toward the accumulation of chronic lung damage from long-term exposures to ambient air containing O_3.

The lack of a more definitive database on the chronic effects of ambient O_3 exposures on humans is a serious failing that must be addressed with a long-term research programme. The potential impacts of such exposures on public health deserve serious scrutiny and, if they turn out to be substantial, strong corrective action. Further controls on ambient O_3 exposure will be extraordinarily expensive and will need to be very well justified.

STANDARD SETTING IN THE UK

Air quality standards form an integral and important part of the UK National Air Quality Strategy. The Strategy was published in early 1997 and the standards and objectives adopted by the Labour Government which came into power on 1 May 1997 (Department of the Environment, 1997). The Strategy sets out the way ahead for improving air quality in the UK and bases policies upon comparisons between monitored levels of air pollutants and levels prescribed as standards and objectives. Air quality standards are defined in the Strategy as ideals; objectives define how closely these ideals should be approached, in general, by 2005. The objectives have been derived by taking the standards and, on a cost-benefit basis, setting an acceptable level of achievement or attainment. In the case of pollutants for which standards are defined with averaging times of less than a year, a percentile approach has been adopted. Standards and objectives are shown in Table 43.6.

Details of the derivation of the standards may be found in the publications of the UK Expert Panel on Air Quality Standards (Department of the Environment, 1994a–d, 1995a,b, 1996; Department of the Environment, Transport and the Regions, 1998). These publications are in the form of booklets of about 20 pages which outline, at a level of detail designed to be appropriate for the intelligent layman, the basis for the standards. In addition, information on UK concentrations of the pollutants are provided with an outline of sources. It will be appreciated that such a brief presentation cannot include a detailed review of the evidence on the effects on health of the individual pollutants. This is provided in a series of more detailed accounts published by the UK Department of Health's Committee on the Medical Effects of Air Pollutants and the earlier Advisory Group on the Medical Aspects of Air Pollution Episodes (Department of Health, 1991, 1992, 1993, 1995). In recommending a standard, the Expert Panel has had access to these reports and there is a good deal of overlap between the memberships of the groups. In the case of carcinogens, such as benzene, the Expert Panel has had recourse to the opinions of the Department of Health's Committee on the Carcinogenicity of Chemicals in Food, Consumer Products and the Environment.

Table 43.6 UK National Air Quality Strategy summary of proposed standards and objectives

Pollutant	Standard		Objective – to be achieved by 2005
	Concentration	Measured as	
Benzene	5 ppm	Running annual mean	5 ppb
1,3-Butadiene	1 ppb	Running annual mean	1 ppb
Carbon monoxide	10 ppm	Running 8-h mean	10 ppm
Lead	0.5 $\mu g/m^3$	Annual mean	0.5 $\mu g/m^3$
Nitrogen dioxide	150 ppb 21 ppb	1-h mean Annual mean	150 ppb, hourly mean[a] 21 ppb, annual mean[a]
Ozone	50 ppb	Running 8-h mean	50 ppb, measured as the 97th percentile[a]
Fine particles (PM_{10})	50 $\mu g/m^3$	Running 24-h mean	50 $\mu g/m^3$ measured as the 99th percentile[a]
Sulfur dioxide	100 ppb	15-min mean	100 ppb measured as the 99.9th percentile[a]

ppm, parts per million; ppb, parts per billion; $\mu g/m^3$, microgram per cubic metre.
[a] These objectives are to be regarded as provisional.

It would be inappropriate here to provide details of all the pollutants for which the Expert Panel has recommended standards. Instead, some principles that have evolved during the life of the Expert Panel are outlined. It should be stated at the outset that no air quality standards had been set in the UK prior to the setting up of the Expert Panel in 1991, and thus the Panel has felt its way cautiously and recognized that different pollutants require different approaches.

When the Expert Panel was established it was made clear that a fairly rapid standard-setting process was required. It was anticipated that standards for the major 'classical' air pollutants should all be set within five or so years. This expectation precluded the exhaustive literature review undertaken by the US EPA in setting and reviewing the US National Ambient Air Quality Standards. Instead, more reliance was placed upon the judgement of the panel of experts, most of whom were familiar with the key literature and some of whom were undertaking research in the air pollution area. This approach has meant that a balance has been struck between thoroughness of review and speed. However, efforts have been made to ensure that the bases for recommendations are traceable to the evidence and that the standard-setting process is transparent, even to those who disagree with its conclusions.

It is useful to divide the pollutants for which standards have been set on the basis of their carcinogenicity, as opposed to standards set to protect against other health effects.

Effects Other than Cancer

Standards have been set for ozone, sulfur dioxide, nitrogen dioxide, carbon monoxide and particulate material measured as PM_{10} (particles generally less than 10 μm in aerodynamic diameter). For each pollutant the Panel attempted to identify a level of exposure at which adverse effects would not be expected to occur even amongst sensitive individuals. Sensitive individuals were defined as those falling into groups which had been recognized as being more sensitive than average to the pollutant in question. For example, asthmatic individuals have been demonstrated to be especially sensitive to sulfur dioxide, as have those suffering from impaired coronary blood flow to carbon monoxide. For ozone there is a marked spread of respiratory function responsiveness within the general population, though chamber studies have failed to demonstrate that patients with asthma or those with chronic obstructive pulmonary disease form an especially sensitive subgroup. This is surprising and not entirely consistent with evidence from epidemiological studies. These studies have shown that admissions to hospital for treatment of respiratory disorders are significantly related to ambient concentrations of ozone. It is likely that the majority of such patients suffer from long-standing respiratory disorders. This is discussed at length in Chapter 23.

Determining a 'lowest observed adverse effect level' (LOAEL) was successful for ozone, nitrogen dioxide, carbon monoxide and sulfur dioxide. Having determined such a level the Panel applied a safety factor to allow for the likelihood that especially sensitive individuals would have been unlikely to take part in chamber studies of the effects of these pollutants. It was also recognized that comparatively few studies had involved children though with their lower body weight, higher specific metabolic rate and likely considerable outdoor exposure these might form a group at increased risk. The determination of safety factors was done on the basis of expert judgement. Standard toxicological safety factors, usually in steps of 10, were considered inapplicable. For example, if it is accepted that the LOAEL for ozone is about 80 ppb for a 6.6-h exposure it would be ludicrous to apply a safety factor of 10 and derive a standard of 8 ppb. Such a standard would be simply impossible to reach, even if all anthropogenic production of ozone-forming primary pollutants ceased. In deciding on safety factors the Panel took into account the types and severity of effect seen at the LOAEL. In some cases these were regarded as only marginally 'adverse' and a small safety factor was assigned.

For ozone, carbon monoxide, nitrogen dioxide and sulfur dioxide the approach of the Panel led to standards not dissimilar to the relevant Air Quality Guidelines established by WHO Expert Groups in 1995–96. In the case of ozone and sulfur dioxide the UK standards are a little lower than the WHO figures; in the case of nitrogen dioxide the UK figure is a little higher than the WHO figure. These differences are seen as a result of different groups of experts looking at essentially the same data and coming to slightly different conclusions: this was not at all unexpected.

Particles presented a more difficult problem. It was accepted that recent epidemiological studies have demonstrated associations between daily average concentrations of particles and a wide range of health effects. The association has been accepted as likely to be causal. The published relationships give no indication of a 'no effect level' – thus effects are expected even at very low particle concentrations. Setting a standard at which no effects are expected is thus impossible and the Panel adopted an alternative approach.

A time-series study of the relationship between hospital admissions for respiratory diseases and daily PM_{10} levels was commissioned in Birmingham (UK). The study broadly confirmed the relationships reported from other studies and showed that on a day when the 24-h average concentration of particles rose from 20 $\mu g/m^3$ (just below the annual average concentration) to 50 $\mu g/m^3$ about one extra hospital admission was to be expected. The Panel noted that the population of Birmingham (UK) was about 1 million and considered that one extra hospital admission in such a population represented a small individual risk. It was appreciated that the risk should not be expressed as 1 in 1000 000 because the actual 'at risk' population was probably much less than 1000 000. On this basis a standard of 50 $\mu g/m^3$ was established. A caveat was added to the effect that efforts should be made to ensure that the annual average concentration of particles, measured as PM_{10}, should fall. Thus, in setting a standard for particles the Panel moved away from its approach of setting a safe level to setting a level that entailed a risk that the Panel regarded as acceptable.

Carcinogenicity

UK standards have been set for benzene and 1,3-butadiene. It was accepted that both compounds are genotoxic carcinogens and that no absolutely safe level of exposure could therefore be recommended. The Panel took the view that it should be possible to identify, from epidemiological studies of occupational exposure to these compounds, levels of exposure that would be unlikely to present a significantly increased risk. Taking this as a working 'no observed adverse effect level' (NOAEL), a series of standard safety factors were applied. These included factors of 10 to account for:

(1) differences between working-life exposure and whole-life exposure;

(2) distribution of sensitivity in the general population which is likely to be wider than that in a working population.

In addition, attention was paid to current levels of the pollutants with the Panel taking the precautionary view that increases in exposure to genotoxic carcinogens should be avoided as far as possible. In the case of 1,3-butadiene this led to a standard being set at 1 ppb: at about the current ambient concentration.

The approach adopted by the Panel could be criticized on the grounds that single studies were used to fix the working NOAEL. It is accepted that a meta-analysis of published studies might well have provided a different starting point for standard setting, though the lack of consistent design of studies makes formal meta-analysis difficult. The approach taken by the Panel is a pragmatic one in that a single study judged to be of acceptably high quality was chosen and used to derive a starting point for derivation of the standard.

From experience in dealing with both non-carcinogens and carcinogens it has become apparent that defining, in advance, an optimal strategy for setting a standard is difficult. A great deal depends on the characteristics of the individual pollutants and, perhaps more importantly, on the nature of the information available. A paper published in 1995 tried to draw together some of the thinking adopted in the UK for dealing with carcinogenic air pollutants (Maynard *et al.*, 1995). The UK experience in setting standards for carcinogenic air pollutants has been that though no absolutely safe level of exposure (and thus, ambient concentration) can be recommended it is possible to recommend a level at which the risk of effects, though unquantified, is likely to be very small. Though such an approach will always be susceptible to criticism, the recommendations provided by the Panel have been welcomed by a number of environmental groups in the UK and have attracted, as yet, only moderate criticism from industrial experts.

SETTING AIR QUALITY STANDARDS IN A DEVELOPING COUNTRY

The prominence given recently to the effects of air pollutants on health in developed countries has led developing countries to wish to set air pollution standards. Concern that concentrations of air pollutants in developing countries may exceed standards set in developed countries has fuelled this desire. In addition, the increases in levels of pollution occurring as a result of industry relocating from tightly regulated developed countries to less well-regulated developing countries have been identified and described as dumping of polluting industries on the 'Third World'. Whether this is actually occurring on a significant scale is debatable, but accidents such as that which occurred in Bhopal in 1984 have maintained interest and concern. A further cause for concern is the idea that industry may choose to sell low-grade polluting products in developing countries whilst unable to do so in countries with tight air pollution control. Until about 1993 motor cars manufactured in the UK for sale in the USA were notably less polluting than those offered for sale in the UK. This was because US emissions standards mandated the fitting of catalytic converters before these were required for the UK domestic market. Industrial competition tends to ensure that companies will not go further than their competitors in reducing production of pollutants and thus all tend to conform to, but not exceed, the requirements of national regulations.

It is often assumed that developing countries must have air pollution problems and are therefore in need of air quality standards. This is untrue: essentially agrarian economies may generate only low levels of air pollutants that pose a hazard to health. However, in expanding farming economies the rapid destruction and burning of native forest may produce acute air pollution problems. Climatic changes may add to the risks of extensive forest fires as were seen in South East Asia in the late 1997 and early 1998. In addition, long-range transport of air pollutants from neighbouring countries can cause problems in countries with only low domestic production of pollution.

In some developing countries indoor air pollution is a greater problem than outdoor air pollution. This may, of course, also be the case in developed countries as people tend to spend a much greater part of the time indoors than out. Despite this, emphasis in developed countries has been on outdoor air pollution on the grounds that the major sources of widespread air pollution occur outdoors and are, at least potentially, amenable to control. Setting outdoor air pollution standards and emission standards have been accepted as means of achieving this control.

Indoor sources in developing countries include the burning of biomass (wood, peat, dung) for warmth and cooking. In comparatively primitive and poorly ventilated dwellings very high concentrations of smoke and carbon monoxide may occur. This is the case, for example, in villages in India and Nepal. It is a moot point in such areas whether efforts should be focused on outdoor or indoor air pollution.

Steps in Setting Air Quality Standards

Before setting outdoor air quality standards the size of the problem posed by air pollution should be assessed. Monitoring of concentrations of air pollutants is required. Methods for monitoring the common air pollutants are discussed in Chapter 5. Special attention should be given to the siting of monitoring equipment. In all countries it is easy to record high concentrations of air pollutants simply by siting monitors adjacent to sources of pollution. For example, recording concentrations of nitrogen dioxide close to a busy road junction will be likely to yield high readings. The key point is that air pollutant monitoring should seek to inform about human exposure to pollutants. Additional monitoring close to sources may be required to check compliance with emission standards: such work should not be confused with that needed to discover whether levels of air pollution are likely to be a significant risk to health. Monitors placed in areas where people spend long periods out of doors, e.g. markets and thoroughfares, are most likely to provide useful data.

Before comparing the concentrations recorded with guidelines or standards, possible sources of confusion should be considered. For example, if PM_{10} (mass of particles generally less than 10 μm aerodynamic diameter) is recorded, then potential sources of particles and gas-phase precursors should be considered. In arid areas wind-blown dust may make a significant contribution to PM_{10}. Such dust tends to be comparatively large in diameter: between 2.5 and 10 μm. Recent studies have suggested that it may well be the sub-2.5 μm component of PM_{10} that is primarily responsible for the associations demonstrated between PM_{10} and indices of ill health (Schwartz *et al.*, 1996). If this is true, then the fraction in the size range 2.5–10 μm may be a comparatively inert diluent. The contribution of this fraction will be high in arid areas and thus high PM_{10} readings may be of less health significance than in other areas. Particles present a special problem as, unlike, for example, carbon monoxide or ozone, they do not represent one chemical compound. Expert analysis of particle composition is desirable before interpretation of measurements on a purely size/mass basis (e.g. PM_{10}) is undertaken.

Having measured air pollution concentrations, and taken into account other factors as discussed above, the results should be compared with the WHO Air Quality Guidelines (see Tables 43.7 and 43.8).

At the time of writing WHO is engaged in drawing up guidelines of world-wide applicability based on the WHO Air Quality Guidelines for Europe. It might be assumed, in principle, that guidelines (and recalling that the operative word is *guidelines* not *standards*) should be applicable the world over. Generally, this is likely to be true, though in some cases doubts may arise. Not all the WHO Air Quality Guidelines are expressed in terms of a concentration and an appropriate averaging time: some are expressed as risk estimates. The practice adopted has been to express the results of a risk analysis in terms of Unit

Table 43.7 World Health Organization Air Quality Guideline Values, 1987

Substance	Time-weighted average	Averaging time
Cadmium	1–5 ng/m^3	1 year (rural areas)
	10–20 ng/m^3	1 year (urban areas)
Carbon disulfide	10 μg/m^3	24 h
Carbon monoxide	100 mg/m^3	15 min
	60 mg/m^3	30 min
	30 mg/m^3	1 h
	10 mg/m^3	8 h
1,2-Dichloroethane	0.7 mg/m^3	24 h
Dichloromethane	3 mg/m^3	24 h
Formaldehyde	100 μg/m^3	30 min
Hydrogen sulfide	150 μg/m^3	24 h
Lead	0.5–1.0 μg/m^3	1 year
Manganese	1 μg/m^3	1 year
Mercury	1 μg/m^3 (indoor air)	1 year
Nitrogen dioxide	400 μg/m^3	1 h
	150 μg/m^3	24 h
Ozone	150–200 μg/m^3	1 h
	100–120 μg/m^3	8 h
Styrene	800 μg/m^3	24 h
Sulfur dioxide	500 μg/m^3	10 min
	350 μg/m^3	1 h
Sulfuric acid	–	–
Tetrachloroethylene	5 mg/m^3	24 h
Toluene	8 mg/m^3	24 h
Trichloroethylene	1 mg/m^3	24 h
Vanadium	1 μg/m^3	24 h

Table 43.8   Revised World Health Organization Air Quality Guidelines

Compound	Guideline	Averaging time	Unit risk estimate (unit concentration)
Arsenic			1.5×10^{-3} ($\mu g/m^3$)
Cadmium	5 ng/m^3	1 year	
Chromium (Cr^{IV})			4×10^{-2} ($\mu g/m^3$)
Fluoride	1 $\mu g/m^3$	1 year	
Lead	Not yet agreed		
Manganese	0.5 $\mu g/m^3$	1 year	
Mercury	No guideline proposed[a]		
Nickel			3.8×10^{-4} ($\mu g/m^3$)
Platinum	No guideline proposed[a]		
Benzene			6×10^{-6} ($\mu g/m^3$)
Butadiene	Not yet agreed		
Dichloromethane	3 mg/m^3	24 h	
	0.45 mg/m^3	1 week	
Formaldehyde	0.1 mg/m^3	30 min	
PAHs (as BaP)			8.7×10^{-5} ($\mu g/m^3$)
Styrene	70 $\mu g/m^3$	30 min	
Tetrachloroethylene	0.25 mg/m^3	Not yet specified	
Toluene	0.26 mg/m^3	1 week	
Trichloroethylene			4.3×10^{-7} ($\mu g/m^3$)
Carbon monoxide	100 mg/m^3	15 min	
	60 mg/m^3	30 min	
	30 mg/m^3	1 h	
	10 mg/m^3	8 h	
Nitrogen dioxide	200 $\mu g/m^3$	1 year	
	40 $\mu g/m^3$	1 h	
Ozone	120 mg/m^3	8 h	
Particulate matter	No single guideline recommended[a]		
Sulfur dioxide	500 $\mu g/m^3$	10 min	
	125 $\mu g/m^3$	24 h	
	50 $\mu g/m^3$	1 year	
PCBs	No guideline proposed[a]		
PCDDs and PCDFs	No guideline proposed[a]		

[a] Failure to recommend a guideline does *not* imply that the compound or compounds are of low toxicity. The 2nd edition of the WHO Air Quality Guidelines should be consulted for detailed explanations.

Risk, i.e. the likely added risk of, say, cancer occurring as a result of lifelong exposure to unit concentration of the individual pollutant. In some cases, e.g. benzene, the unit of concentration is 1 µg/m³; in others, e.g. PAH compounds, the unit of concentration is 1 ng/m³. PAH compounds present an example of where direct adoption of the Unit Risk estimate provided in the WHO Air Quality Guidelines for Europe might be unwise. In dealing with PAH compounds, which invariably occur as a mixture of many individual PAH compounds, WHO chose to use the concentration of benz[a]pyrene (BaP) as an index of the total carcinogenicity of the mixture and to base the risk assessment on data collected in studies of workers at coke ovens. Such an approach is predicated on the assumption that BaP makes the same relative contribution to the total carcinogenicity of the PAH mixture present in general ambient air as it does in the air at coke ovens. Analysis of the concentrations of individual PAH compounds in both environments suggests that this is likely to be at least partially true. In the 1987 WHO Air Quality Guidelines for Europe it was argued that as essentially the same fuel (coal or coal derivatives) was being burned both in coke ovens and domestic fires, the mix of PAH compounds produced was likely to be similar. That this is the case inside a dwelling where wood or dung is being burned is questionable and requires further study. In recommending an air quality standard for PAH compounds in the UK, an Expert Panel is currently pursuing this problem by examining the estimated contribution made by BaP to that of a mixture of seven key PAH compounds found both in the general environment and in industrial settings by examining both the relative concentrations and relative potencies of the individual PAH compounds. Such an approach requires detailed monitoring data on each PAH compound chosen. Comparison of measured concentrations with WHO Air Quality Guidelines is difficult in the case of particles. In the 1987 edition of the Air Quality Guidelines for Europe the following guidelines were produced:

Particles measured as black smoke (BS)	125 µg/m³ 24-h average
	50 µg/m³ annual average
Total suspended particulates	120 µg/m³ 24-h average
Thoracic particles (equivalent to PM_{10})	70 µg/m³ 24-h average.

These guidelines were derived from epidemiological studies of air pollution mixtures in which sulfur dioxide and particles invariably occurred together. During the revision of the Air Quality Guidelines for Europe in the mid-1990s the rapidly expanding database of time-series studies (see Chapter 20) played a large part in guiding expert groups and it was concluded that particulate matter should be regarded as a no-threshold pollutant. Instead of recommending a single figure guideline, a series of tables showing the observed ambient concentration–response relationships for a number of health end-points were provided: these are reproduced here as Tables 43.9 and 43.10. The straightforward interpretation of this approach is that there is no completely safe level of particulate air pollution: each level carries a penalty in terms of effects on health.

How, then, should a developing country assess its particle measurements against the WHO Air Quality Guidelines for Europe? The tables provided in the revised Guidelines allow, at least in principle, calculation of the number of health-related events expected at a given level of particulate air pollution. These calculations should be undertaken. In selecting end-points to examine, careful consideration should be given to the comparability of local conditions with those of the locations in which studies upon which the

Table 43.9 Summary of relative risk estimates for bronchodilator use, cough and LRS reporting, PEF changes and respiratory hospital admissions and daily mortality, associated with a 10 μg/m^3 increase in the concentration of PM$_{10}$ or PM$_{2.5}$.

End-point	Relative risk for PM$_{2.5}$ (95% CI)	Relative risk for PM$_{10}$ (95% CI)
Bronchodilator use	–	1.0337 (1.0205–1.0470)
Cough	–	1.0455 (1.0227–1.0687)
LRS lower respiratory tract symptoms	–	1.0345 (1.0184–1.0508)
Respiratory hospital admissions	–	1.0084 (1.0050–1.0117)
Mortality	1.0151 (1.0112–1.0190)	1.0070 (1.0059–1.0082)
PEF change (relative to mean)	Effects of PM$_{2.5}$ on PEF	Effects of PM$_{10}$ on PEF −0.13% (−0.17%, −0.09%)

Table 43.10 Estimated number of subjects, in a population of 1 million experiencing health effects over a period of 3 days characterized by a mean PM$_{10}$ concentration of 50 or 100 μg/m^3

Health effects indicator	No. of subjects affected by a 3-day episode of PM$_{10}$ at:	
	50 μg/m^3	100 μg/m^3
Mortality	3.5	7
Respiratory hospital admissions	3	6
Person-days of bronchodilator use	5 100	10 200
Person-days of symptom exacerbations	6 000	12 000

Guidelines are based were undertaken. For example, one would hardly expect a given level of particulate air pollution to have the same effect on hospital admissions in a country with few hospitals as in, say, New York or London. Daily deaths, on the other hand, may seem a more reliable end-point, though here too there are problems, especially in a population with a qualitatively and quantitatively different background pattern of disease than in the areas where the original studies were done. The relationship between daily concentrations of particles and daily deaths is acceptably linear at concentrations found in the USA and in Europe though in areas with higher average concentrations the line is significantly less steep. Subtle and unexplained differences occur today between the results of studies done in the USA and in Western Europe. On a mass for mass basis, particles in Western Europe seem to have less effect on health than they do in the USA (see Chapter 10 dealing with effects of particles). The reasons for this are obscure. Particle composition may vary as might the susceptibility of the population.

The problem is therefore difficult! In the authors' opinion, comparison of measured concentrations of particles with the WHO Guidelines for Europe as framed in 1987 should provide guidance as to whether a major problem exists. This is especially likely to be the case with regard to locations still experiencing coal-smoke air pollution.

Having made comparisons with WHO Guidelines, those pollutants that significantly exceed the Guidelines should be addressed and considered for standard setting. Scarce resources should not be wasted in setting standards for, and regulating, pollutants which do not pose significant public health problems.

Setting an Air Pollution Standard Based on a WHO Air Quality Guideline

It has been made clear that the WHO Air Quality Guidelines are not standards. Standards should be set with regard to local socioeconomic conditions. This implies that air quality standards might very sensibly vary from country to country. For the purposes of this account it will be assumed that the ambient level of pollution significantly exceeds the WHO Guidelines. The following scheme is proposed.

(1) *Adoption of a long-term objective or target.* It is perfectly reasonable to adopt the WHO Air Quality Guidelines as long-term objectives. The difficult decision is how rapidly these objectives should be approached.

(2) *Deciding on the rate of progress towards the objective.* When levels of air pollution are high, effects are obvious and obvious low-cost measures can be taken to reduce emissions: these should be undertaken without delay. This was the approach adopted in London in the 1950s. The Clean Air Act of 1956 was introduced and encouraged and aided the conversion from coal to gas, electricity and smokeless solid fuel for domestic heating in urban areas. Levels of coal-smoke pollution fell rapidly.

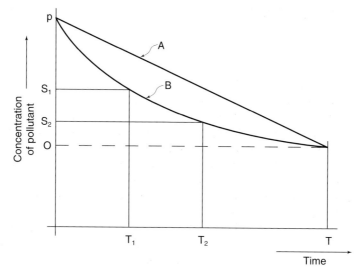

Fig. 43.5 Hypothetical decline in pollutant concentration with time and relevance to standard setting.

As levels of pollution fall, decisions about the need for further reductions become more difficult. It is here that standard setting can play an important part. Figure 43.5 shows, schematically, how reductions in air pollution might be scheduled. The present level of air pollution is indicated by point P on the Y axis. The long-term objective is shown as the dotted horizontal line. The time over which the objective should be achieved is shown as T on the x axis. If concentrations of pollutants declined linearly they would follow line A. However, the decline is more likely to follow line B in that early gains will be comparatively easy to achieve whilst later, as levels approach the objective, it is likely to become increasingly difficult to produce further reductions. The exact shape of line B will be dependent upon a number of factors:

- change in dominant sources of pollution with time;
- changing economic prosperity;
- increasing social awareness and concern about the effects of air pollution is likely to increase. This is discussed further below.

It is suggested that an air pollution standard S_1 should be established with the intention that this will be achieved by time T_1. A second standard S_2 to be achieved by time T_2 is also shown. These standards provide, not guarantees of the removal of the effects of air pollution on health, but rather a series of short-term targets that are within reach. Each standard should be based upon a rigorous cost-benefit analysis in which the costs of reducing levels of air pollution are weighed against the resultant benefits. As soon as this is attempted the difficult problem arises of quantifying the benefits that will emerge as a result of reducing levels of air pollutants. Economists, not surprisingly, find benefits such as reduced mortality or a reduction in hospital admissions difficult to handle unless they can be expressed in terms of money. To attempt such a valuation is repugnant to some people, though recent developments in the processes involved have removed many earlier crudities. Notable amongst these advances has been the development of the Willingness to Pay approach to determining the value placed by people upon a reduction in the risk of death or of episodes of illness (Pearce and Crowards, 1996). This methodology has refined estimates of the Value of a Statistical Life and takes into account factors such as life expectancy. The strict theoretical precepts of cost-benefit analysis should be supplemented by broader social and economic considerations. This process is sometimes described as 'Stakeholder Input'. The aim is to ensure as far as possible social equity and fairness to all involved parties. An adequate and early involvement of all concerned will increase the transparency of the process and is likely to increase the acceptability of the outcome. Expert political as well as scientific judgement must be applied.

It will be realized that the approach suggested above places a different meaning on the term standard than that often applied. Here the standard is seen not as an absolute value in the sense of assuring an acceptable level of safety but, rather, as a landmark on the route towards an objective. In many countries, e.g. the USA and countries of Western Europe, air quality standards are used in the sense of assuring safety.

Qualification of Standards

However an air quality standard is derived, it is unlikely that it will be met unfailingly. This has led many countries to quality their air quality standards with an acceptable level of exceedance. For example, the current European Community Directive, which specifies a short-term air quality standard (Limit Value) for nitrogen dioxide of 200 $\mu g/m^3$, is expressed as the 98th percentile of hourly average concentrations. This implies that 174 exceedances of the limit value of 200 $\mu g/m^3$ (hourly average concentration) are acceptable each year without breach of the Directive. Such a qualification can only be justified in purely health terms on the basis that multiple exceedances greatly increase the likelihood of an adverse response to a subsequent peak exposure. Clearly, the assertion that 174 exceedances would be harmless but that 175 exceedances would lead to damage to health would be ludicrous if interpreted literally. Such qualifications are based, at least in part, on cost-benefit analysis or on considerations of practicality of the standard.

Difficulties Likely to be Encountered in Developing Countries

It is probably true that no country currently regarded as developed, in the sense that it has become richer as a result of industrialization, has managed to achieve such a state without an initial increase in levels of air pollution. Rapidly developing industry and the use of fossil fuel as a source of energy inevitably leads to the likelihood of increased production of pollutants. To expect developing countries to pass through such a phase, whilst at the same time not experiencing at least a transient increase in air pollution, is to expect a great deal. Decisions in this area are difficult. For example, the use of leaded gasoline has come or is coming to an end in many developed countries. Despite this, expanding transport systems in developing countries may rely heavily on the use of older vehicles which require leaded petrol because of the nature of the valve seats in their engines. The alternatives are more expensive additives that will allow these vehicles to remain in use, albeit at a perhaps reduced level of performance, or an accelerated introduction of vehicles that can use unleaded gasoline. Either option may well be costly and should be weighed carefully against the extent of the expected effects of lead on health. Imposition of standards for lead in air that are not difficult to meet in developed countries is not the solution to this problem in developing countries.

CONCLUSION

Air Quality Guidelines and standards have an important role to play in the management of air quality. However they must be applied with care and should not be unthinkingly transferred from one country to another. Air quality can be improved without Air Quality Standards: the great improvement in air quality produced in the UK by the 1956 Clean Air Act were achieved without any air quality standards at all. However, in the later stages of air pollution controls, standards (goals) are needed to provide benchmarks of progress toward attaining a healthy environment.

ACKNOWLEDGEMENTS

Dr Lippmann's contribution to this chapter was undertaken as part of a Center program supported by Grant ES 00260 from the National Institute of Environmental Health Sciences. It has been based, in part, on material contained in the chapters on ambient particulate matter and ozone written by the author for the second edition of *Environmental Toxicants – Human Exposures and Their Health Effects*, being published by Wiley in 1999.

Noted Added in Proof

The UK National Air Quality Strategy was revised during 1998 and a revised version published for consultation on 13 January 1999: details can be found on the UK DETR web site. Updating of EC Air Quality Directions is underway at the time of writing.

REFERENCES

American Conference of Governmental Industrial Hygienists (1996) *Documentation of TLVs and BEIs*, 6th edn. Cincinnati: American Conference of Governmental Industrial Hygienists.

American Conference of Governmental Industrial Hygienists (1997) *Threshold Limit Values and Biological Exposure Indices for 1997*. Cincinnati: American Conference of Governmental Industrial Hygienists.

Bates DV (1992) Health indices of the adverse effects of air pollution: the question of coherence. *Environ Res* **59**: 336–349.

Chang L-Y, Huang Y, Stockstill BL *et al.* (1992) Epithelial injury and interstitial fibrosis in the proximal alveolar regions of rats chronically exposed to a simulated pattern or urban ambient ozone. *Toxicol Appl Pharmacol* **115**: 241–252.

Department of Health (1991) Advisory Group on the Medical Aspects of Air Pollution Episodes. First Report: *Ozone*. London: HMSO.

Department of Health (1992) Advisory Group on the Medical Aspects of Air Pollution Episodes. Second Report: *Sulphur Dioxide, Acid Aerosols and Particulates*. London: HMSO.

Department of Health (1993) Advisory Group on the Medical Aspects of Air Pollution Episodes. Third Report: *Oxides of Nitrogen*. London: HMSO.

Department of Health (1995) Committee on the Medical Effects of Air Pollutants. *Non-Biological Particles and Health*. London: HMSO.

Department of the Environment (1994a) Expert Panel on Air Quality Standards: *Carbon Monoxide*. London: HMSO.

Department of the Environment (1994b) Expert Panel on Air Quality Standards: *Ozone*. London: HMSO.

Department of the Environment (1994c) Expert Panel on Air Quality Standards: *1,3-Butadiene*. London: HMSO.

Department of the Environment (1994d) Expert Panel on Air Quality Standards: *Benzene*. London: HMSO.

Department of the Environment (1995a) Expert Panel on Air Quality Standards: *Sulphur Dioxide*. London: HMSO.

Department of the Environment (1995b) Expert Panel on Air Quality Standards: *Particles*. London: HMSO.

Department of the Environment (1996) Expert Panel on Air Quality Standards: *Nitrogen Dioxide*. London: The Stationery Office.

Department of the Environment (1997) *The United Kingdom National Air Quality Strategy.* London: The Stationery Office.

Department of the Environment, Transport and the Regions (1998) Expert Panel on Air Quality Standards: *Lead.* London: The Stationery Office.

Huang Y, Chang LY, Miller FJ *et al.* (1988) Lung injury caused by ambient levels of oxidant air pollutants. *Am J Aerosol Med* **1**: 180–183.

Hyde DM, Plopper CG, Harkema JR *et al.* (1989) Ozone-induced structural changes in monkey respiratory system. In: Schneider T, Lee SD, Wolters GJR and Grant LD (eds) *Atmospheric Ozone Research and its Policy Implications.* Amsterdam: Elsevier Science Publishers BV, pp. 523–532.

Lippmann M (1987) Role of science advisory groups in establishing standards for ambient air pollutants. *Aerosol Sci Technol* **6**: 93–114.

Lippmann M (1988) Health significance of pulmonary function responses to airborne irritants. *J Air Pollut Control Assoc* **38**: 881–887.

Lippmann M (1991) Air pollution. In: Hutzinger O (ed.) *The Handbook of Environmental Chemistry,* vol. 4, Part C. Heidelberg: Springer-Verlag, p. 31.

Lippmann M (1993) Health effects of tropospheric ozone: review of recent research findings and their implications to ambient air quality standards. *J Expo Anal Environ Epidemiol* **3**: 103–129.

Maynard RL, Cameron KM, Fielder R *et al.* (1995) Setting air quality standards for carcinogens: an alternative to mathematical quantitative risk assessment modelling – discussion paper. *Hum Toxicol* **14**: 175–186.

Pearce D and Crowards T (1996) Particulate matter and human health in the United Kingdom. *Energy Policy* **24**: 609–619.

Peters A, Wichmann E, Tuch T *et al.* (1997) Respiratory effects are associated with the number of ultrafine particles. *Am J Respir Crit Care Med* **155**: 1376–1383.

Schwartz J, Dockery DW and Neas L (1996) Is daily mortality associated specifically with fine particles? *J Air Waste Manag* **46**: 927–939.

Seinfeld JH (1998) Ozone air quality models. *J Air Pollut Control Assoc* **38**: 616–645.

Spektor DM, Lippmann M, Lioy PJ *et al.* (1988a) Effects of ambient ozone on respiratory function in active, normal children. *Am Rev Respir Dis* **137**: 313–320.

Spektor DM, Lippmann M, Thurston GD *et al.* (1988b) Effects of ambient ozone on respiratory function in healthy adults exercising outdoors. *Am Rev Respir Dis* **138**: 821–828.

Tabler SK (1984) EPA's program for establishing national emission standards for hazardous air pollutants. *J Air Pollut Control Assoc* **34**: 532–536.

Tepper JS, Wiester MJ, Weber MF *et al.* (1991) Chronic exposure to a simulated urban profile of ozone alters ventilatory responses to carbon dioxide challenge in rats. *Fundam Appl Toxicol* **17**: 52–60.

Tyler WS, Tyler NK, Last JA *et al.* (1988) Comparison of daily and seasonal exposures of young monkeys to ozone. *Toxicology* **50**: 131–144.

US DOL (1993) Occupational Safety and Health Administration (OSHA) Permissible Exposure Limits (PELs). Code of Federal Regulations (CFR), Title 29, Part 1910, Section 1000.

US EPA (1992) *Summary of Selected New Information on Effects of Ozone on Health and Vegetation:* Supplement to 1986 Air Quality Criteria for Ozone and Other Photochemical Oxidants. EPA/600/8-88/105F. PB92-235670. Springfield, VA: NTIS.

US EPA (1996a) *Air Quality Criteria for Ozone and Related Photochemical Oxidants.* EPA/600/P-93-004F. Research Triangle Park, NC: US EPA.

US EPA (1996b) *Review of the National Ambient Air Quality Standards for Ozone – Assessment of Scientific and Technical Information.* OAQPS Staff Paper. EPA/452/A-96-007. Research Triangle Park, NC: US EPA.

US EPA (1996c) *Review of the National Ambient Air Quality Standards for Particulate Matter: OAQPS Staff Paper.* EPA-452/R-96-013. Research Triangle Park, NC 27711: US EPA.

US EPA (1997) *National Ambient Air Quality Standards for Ozone; Final Rule.* Federal Register **62**: 38856–38896. Research Triangle Park, NC: US EPA.

Informing the Public about Air Pollution

R.L. MAYNARD

Department of Health, London, UK

S.M. COSTER

Department of the Environment, Transport and
the Regions, London, UK

The views expressed in this chapter are those of Drs Robert Maynard and Stephanie Coster and should not be taken as those of the UK Department of Health or of the Department of the Environment, Transport and the Regions.

INTRODUCTION

Control of air pollution is a multifaceted activity involving many groups of people. Of these, both as producers and as those affected by air pollution, the general public are perhaps the most important.

When considering provision of information on air pollution to the public, a number of questions need to be addressed. Since the type and nature of any information provided should take account of the requirements and experience of the recipients, the first, most obvious of these is – who are the public? The public in this context can include: individuals whose health, or family's health, may be affected by air pollution; people who wish to play their part when air pollution levels are high – perhaps by leaving their car at home; the interested lay person; and academic researchers. In addition, other individuals or groups such as health professionals, local authorities, environment and health groups may wish to receive, and in turn re-present the information in ways that better meet their, and their audience's, own needs.

The provision of information to the public can help to achieve a number of aims:

AIR POLLUTION AND HEALTH
ISBN 0-12-352335-4

- for members of the public to understand the impact of air pollution on their health or on the wider environment;
- to encourage the public to reduce emissions of pollutants within their control (for example by changing their driving habits);
- to allow those who may be affected by air pollution to take timely precautions to avoid such effects; and
- to enable the public to assess progress towards the achievement of national air quality objectives.

Providing such data is not easy and is not a matter of trivial cost: that it is important is beyond question.

Such information should be accurate, accessible and provided in a form that is easily understood. Air pollution information can be provided in a number of 'tiers' of increasing complexity and detail. This should both enable the lay public to take on board relevant air pollution information, and at the same time give air pollution specialists access to increasingly disaggregated data sets, including 'raw' data sets for those interested to use in their own data analysis. In addition, air pollution information can be provided in different formats and over different time frames – ranging from real-time data via the electronic media to publications in the form of periodic or annual reports, or occasional publicity material such as leaflets.

The UK has a comprehensive system of public information on air pollution, details of which are discussed later in this chapter. First, some of the key features of any air pollution information system are outlined below. The following are essential:

(1) an adequate system for monitoring concentrations of air pollutants to known quality standards;
(2) a centrally organized system for collecting the data and preparing it in a form that can be readily understood;
(3) a distribution system involving designated systems to allow dissemination in the desired time frame – such as permanent pages on televised information channels, the Internet, freephone services and access to the media: press, radio and television.

In addition, if people are to modify their behaviour on a day-to-day basis, they should be informed of the likelihood of pending air pollution problems. Therefore, when providing information in real time, the capacity to forecast levels of pollution – and particularly episodes of high pollution concentrations – is essential. There is much evidence to show that such air pollution episodes are associated with effects on health. Forecasting is therefore especially important for pollutants that have acute effects on health, e.g. ozone, sulfur dioxide, nitrogen dioxide and particles, but this is sometimes forgotten in the enthusiasm to provide hard data.

Episodes of elevated pollutant concentrations depend fundamentally on the balance between the emissions of pollutants and their dispersion in the air. The latter depends on the weather: in periods of still atmospheric conditions, often brought about by anticyclonic (high-pressure) weather systems, pollutants accumulate and concentrations rise. The former depend on the range of emitting sources some of which, such as motor vehicles, may be amenable to short-term control measures.

In 1997 the authorities in Paris introduced short-term restrictions on motor vehicle usage in response to an air pollution episode that was expected to persist. Whether or not such mandatory action is taken, forecasting of episodes is the key to achieving timely short-term action to reduce emissions. Choices regarding transport tend to be taken early in the day; this means that 'on the day' warnings can generally only influence the following day. Unless reliable fore-warning of air pollution episodes is provided, voluntary reduction in vehicle use is unlikely, and vehicle users may not be aware of mandatory restrictions in time.

Forecasting also allows those sensitive to air pollution to take preventive steps against the possible health effects of an episode, by modifying either their daily activity or, in selected cases, their medical therapy. For some individuals who know from experience that they are adversely affected by air pollution, it may be sensible to reduce the time spent out of doors on days of high pollution levels. For those suffering from asthma some alteration of their maintenance therapy may be useful. Again, these decisions depend upon adequate prior warning.

The next section of this chapter will focus primarily on the provision of real-time information to the public, with a particular focus on information for those with an interest in possible effect on health. Some essential requirements of such an air pollution information system are considered in more detail below.

KEY FEATURES OF AN AIR POLLUTION INFORMATION SYSTEM

An Adequate System for Monitoring Concentrations of Air Pollutants

Ideally, to best alert the public to the possible effects of air pollutants on their health, we should like to know the distribution of exposure of individuals. Of course, personal monitoring of individual exposure across the whole population is utterly impractical and the surrogate of monitoring at a number of fixed sites is generally adopted. Methods for monitoring air pollutants are considered in Chapter 5 and will not be considered in detail here. However some key features of an adequate monitoring system are worth listing:

(1) The system should provide a reliable flow of accurate on-line data. This implies that modern automated monitoring equipment is needed. Regular quality assurance/quality control (QA/QC) exercises are needed to validate and ratify the data collected. In some systems, including that in use in the UK, data provided on-line is regarded as provisional and, although subject to preliminary screening and validation, is not accepted as definitive until data records have been examined *post hoc* and ratified. In the UK this is undertaken by independent QA/QC units. The advantages in separating the responsibility for QA/QC from that for collecting the data are obvious. It will be understood that a public information system must provide the data as it is collected, i.e. pre-ratification. Because of this the highest standards of monitoring are required to ensure the dissemination of good quality real-time data takes place. This should minimize the risk of disseminating misleading or inaccurate data which may, inadvertently, trigger unnecessary alarm or disguise an air pollution problem.

(2) In many countries methods are available for providing on-line, and almost real-time, data on concentrations of ozone, oxides of nitrogen, sulfur dioxide, carbon

monoxide and particulate matter. These systems provide data on the concentrations of these pollutants averaged over predetermined and carefully specified periods. Where the monitoring system is to provide data for a health-based information service, these averaging periods must be specified with regard to the demonstrated effects of the individual air pollutants on health. Modern technology allows a great deal of data to be made rapidly available, and the risk of information overload is high in this area. 'Tiers' of information, of varying detail and complexity, can help here.

(3) It is clearly also valuable if data for the public can be provided in a form, in terms of units and averaging time, that allows ready comparison with air quality standards. For example, knowing the peak 15-minute concentration of benzene is of little value if the air quality standard for benzene is defined in terms of an annual average concentration. In some cases, e.g. particulate matter, the 24-h average concentration fits best with recorded effects on health and, irrespective of whatever other data are provided, this statistic should always be available.

(3) In addition to the adequacy of the monitoring equipment and the form in which the data are provided, coverage of the country is also important. National monitoring sites in the UK have been selected so as to provide an adequate indication of likely exposure of the population. Sites range from those representative of remote rural areas, suburban areas, urban background locations, through to 'hot spot' locations next to busy roads and in the vicinity of industrial sources. This allows the public to obtain information about the range of likely pollutant concentrations to which they may be exposed. More detail on the nature of these sites is given in Table 44.1.

Table 44.1 Air pollution monitoring site classifications used in the UK

Classification	Description
Kerbside	A site sampling within 1 m of the edge of a busy road.
Roadside	A site sampling between 1 m of the kerbside of a busy road and the back of the pavement. Typically this will be within 5 m of the road, with a sampling height of 2–3 m
Urban centre	A non-kerbside site, located in an area representative of typical population exposure in town or city centres (e.g. pedestrian precincts and shopping areas). This is likely to be strongly influenced by vehicle emissions, as well as other general urban sources of pollution. Sampling at or near breathing-zone heights will be applicable
Urban background	An urban location distant from sources and therefore broadly representative of city-wide background conditions, e.g. elevated locations, parks and urban residential areas
Urban industrial	An area where industrial sources make an important contribution to the total pollution burden
Suburban	A location type situated in a residential area on the outskirts of a town or city
Rural	An open country location, in an area of low population density, distanced as far as possible from roads, populated and industrial areas
Remote	A site in open country, located in an isolated rural area experiencing regional background pollution levels for much of the time
Special	A special source-orientated category covering monitoring studies undertaken in relation to specific emission sources such as power stations, garages, car parks or airports

A Centrally Organized System for Data Collection and Distribution

The importance of collecting on-line data from monitoring sites and preparing it in a form suitable for distribution has already been mentioned. That such a system should include regular examination of the data by trained and experienced staff is also desirable. Experience in the UK has shown that the first warnings of air pollution episodes frequently come from staff watching the patterns of data accumulate. This was particularly the case in 1991, when early warning of a very significant air pollution episode in London was provided by staff at the Warren Spring Laboratory – then the national centre for air Pollution Monitoring.

A Distribution System

Collecting and collating data on air pollution are technically complex tasks which, with sufficient investment, can be brought to a high standard. Bringing such data to the attention of the public, especially those most in need of such data, is much more difficult. Few countries would claim complete or even substantial success in this area.

The media clearly has a major role to play. However, the media tend to be most interested in periods when air pollution levels are high; persuading correspondents to carry routine information may, therefore, not be easy. This is a pity for to be effective a public information system on air pollution should be as familiar as the weather forecast. In the UK, information is available in real time via television text services, the Internet (http://www.environment.detr.gov.uk/airq/aqinfo.htm) and a freephone service. Many national newspapers also include a summary of the air pollution data on a daily basis. However, television weather forecasters tend to include forecasts of air pollution levels in their broadcasts only during periods when episodes of high air pollution are expected. Telling the public about likely episodes of air pollution is only one part of the whole service.

Televised information channels and the Internet are systems to which more and more people now have access. The Internet is a particularly effective means of bringing large quantities of data to the public. The entire UK air pollution database or archive, including some 20 million pieces of data has recently been made available via the Internet (see address above). This is, of course, a valuable resource for research workers in the air pollution field.

Other Forms of Dissemination of Information

Providing up-to-date information is an important part of an air pollution information service, but should, however, be supported by the publication of ratified data. These can usefully appear in annual updates and summaries of statistics. Publication in this form allows the inclusion of considered commentaries on the data and information on sources. Such information is important in permitting people to judge the effectiveness of air pollution control measures and strategies.

Description of Pollution Levels

Raw data on concentrations of air pollutants are of little use to the general public until they are either:

(a) placed in context – especially in the context of likely effects on health, or
(b) so familiar that, like temperature data, they can be instantly interpreted.

Air pollution concentrations tend to be reported either as mass per unit volume (e.g. $\mu g/m^3$) or as a volume mixing ratio: e.g. parts per billion. Neither system is much used by the public in everyday life and it is generally accepted that some system of description to indicate at the very least whether the information is 'good news' or 'bad news' is needed.

Whichever system is used, it is vital that the descriptors are easily understood and convey useful information about the likely consequences of exposure to current or predicted levels of air pollution. Many systems include a graded series of descriptors ranging from perhaps 'very good' to 'very poor' or from 'very low' to 'very high'. Describing the lower and upper ends of the scale is comparatively easy and most people can easily understand the various terms used. Satisfactory descriptors for intermediate levels of pollution have proved more difficult to find.

As discussed in Chapter 31, it is not possible to define, at a population level, a threshold of effect for ozone or particulate matter. Despite this, these are important air pollutants and most information systems deal with them. Details of the UK approach are given below. The problem is that at all levels of these pollutants some people may suffer adverse effects; the number of people affected rises with pollution levels. Terms such as 'satisfactory', 'good' or 'fairly good', all of which have been used or suggested for use, can, therefore, seem inappropriate. This is a criticism which has come to be applied in some countries, including the UK, to the concept of describing air quality in general as opposed to levels of air pollution. It is accepted that even low levels of air pollution can produce adverse effects and thus terms such as 'good air quality' are not ideal. A scale of air pollution levels with associated descriptors is perceived as less judgemental and, if provided with clear descriptions of likely effects, allows people to make up their own minds about likely effects on their health.

All systems based on a series of descriptors suffer from difficulties in dealing with the boundaries which separate the ranges or 'bands' of air pollutant concentrations associated with each descriptor. Air pollution toxicologists in all countries are familiar with the media attention that follows air pollution levels crossing from one band to another. It is important, but difficult, to explain that biological responses to air pollution do not include step changes as the boundaries between bands are passed. These boundaries are inevitably based on expert judgement, but in reality the system is more likely to be a continuum.

Of course, the problem is no different in principle from that associated with temperature, though the greater familiarity of the public and media with the temperature scale eases the difficulties. In some countries efforts have been made to provide some indication of the continuum of levels of pollution and associated effects by the use of a colour scale. Information is provided in this form to motorists in Finland by a roadside sign which shows the daily level of air pollution rising through a series of coloured bands.

PROBLEMS OF MULTIPLE POLLUTANTS

In many countries a number of common air pollutants are monitored each day and used as the basis of a public information system. Providing an adequate description of air pollution when only one or perhaps more than one pollutant occurs at unusually high concentrations is a problem. This is especially the case if a system based on describing 'air quality' is used; the problems seem to be less pressing if a system based on levels of 'air pollution' is used. For example, if air quality is described as 'good', 'satisfactory' or 'poor', how should the concentrations of different pollutants be combined in deciding on the appropriate descriptor for the day? Similarly, if air pollution levels are described as 'low', 'moderate' and 'high', should two pollutants reaching the 'moderate' level lead to a description of 'moderate' or one of 'high air pollution'? No completely satisfactory solution to this problem has been found. In several countries, including the UK (see below), the descriptor for the day is taken as the highest reached by any pollutant of the group that are monitored. Thus if only one pollutant reaches the 'moderate band' levels of air pollution are described as 'moderate'. If, say, four pollutants all reach the 'moderate band' air pollution is, again, described as 'moderate'. In the second case one might, however, expect a more significant effect on health than in the former.

In some countries indices of air pollution have been devised. These systems rely on relating measured or predicted concentrations of air pollutants to a scale running from perhaps zero to 100 or zero to 500. Verbal descriptors can be added to the unified scale in the same way as described above. This system has the advantage that the public is not asked to interpret a number of different concentrations – one for each air pollutant – and to recall that, for example, 200 ppb of ozone has very different effects than 200 ppb of carbon monoxide.

It can, however, be difficult for members of the public to obtain details of how to translate a unified pollution index back into the disaggregated 'real' pollutant levels. This means that unless international conventions are developed, it is impossible for a member of the public to understand how a level of pollution in their country compares to what they may be exposed to elsewhere in the world. In addition it can be difficult to use an index to compare pollutant levels with national or international standards or guidelines. The use of a standardized or unified scale does not solve the problems of how to report raised concentrations of a number of pollutants.

THE UK AIR POLLUTION BANDING SYSTEM

A system of banding of air quality was introduced in the UK in 1990. In 1997 this system was revised and replaced with one describing levels of air pollution. The original system was based, like the revised system, on an appreciation of the effects of air pollutants on health. A public consultation exercise was undertaken before the system was revised in 1997. This exercise revealed that the public felt it was easier to understand a system that dealt with the levels of air pollution than one which described air quality. During the revision the opportunity to relate the new banding system to UK Air Quality Standards that had appeared since the original system was devised was taken. Details of how the UK Air

Quality Standards were set may be found in the publications of the UK Expert Panel on Air Quality Standards (Department of the Environment, 1994a–d, 1995a,b, 1996), and in the National Air Quality Strategy (Department of the Environment, 1997).

Before considering the new system in detail, it may be useful to list some general points relating to it:

(1) The bands are based on effects on health. The boundary between the first two bands (low and moderate air pollution) for each pollutant was set at the relevant UK Air Quality Standard.

(2) The bands offer a broad guide to effects on health. It is stressed that the system does not indicate and should not be taken to indicate sudden changes in effects. Rather, a gradual increase in the risk of adverse effects on health is expected as concentrations of pollutants increase.

(3) The risk to healthy individuals is low at all levels of air pollution commonly experienced in the UK.

(4) It was appreciated that patients suffering from chronic obstructive pulmonary disease and/or cardiovascular diseases may be adversely affected by raised concentrations of air pollutants. However, it was stressed that, in general, such individuals should not modify their treatment without consulting their doctors.

It will be realized that these features of the system have a specifically UK-orientated slant. In some countries different levels of air pollution and a different distribution of individual sensitivity to air pollution might well require modification of some of the above. In world terms the UK does not have a very marked air pollution problem, even though significant effects on health occur. Point (3) above, for example, may well be inapplicable in a number of developing countries with more severe air pollution problems.

The bands used to describe air pollution levels in the UK are shown in Table 44.2. It will be noted that a four-band system has been devised. Each boundary or 'break-point' is also described. The first break-point is at the National Air Quality Standard as defined by the UK Expert Panel on Air Quality Standards, adopted by the Government in the National Air Quality Strategy. The remaining break-points provide an Information Threshold and an Alert Threshold, respectively. These latter terms were chosen adopting the terminology consistent with the European Community Directive on Ozone (92/72/EEC).

It will be noted that the averaging times for the different pollutants vary and are defined. Thus for sulfur dioxide a 15-min averaging time is specified; for carbon monoxide the averaging time is 8 h and for particles generally less than 10 μm in diameter (PM_{10}) it is 24 h. These averaging times are in accordance with the UK National Air Quality Standards and with current understanding of the effects of these pollutants on health.

It would be inappropriate to review in detail the evidence upon which the UK banding system is based. Detailed reviews for each pollutant will be found elsewhere in this volume. However, a short account may be of use to others devising banding systems. The following brief accounts are based upon recommendations and advice provided by the UK Committee on the Medical Effects of Air Pollutants and on those of its predecessor, the Advisory Group on the Medical Aspects of Air Pollution Episodes. Details may be found in the publications of these committees and in the UK Department of Health's Handbook on Air Pollution (Department of Health, 1991, 1992, 1993, 1995a–c, 1997, 1998).

Table 44.2   UK air pollution banding system

	Standard threshold	Information threshold	Alert threshold	
	Low	Moderate	High	Very High
Sulfur dioxide (parts per billion, 15-min average)	Less than 100	100–199	200–399	400 or more
Ozone (parts per billion)	less than 50 ppb (8-h running average)	50–89 (hourly average)	90–179 (hourly average)	180 or more (hourly average)
Carbon monoxide (parts per million, 8-h running	Less than 10	10–14	15–19	20 or more
Nitrogen dioxide (parts per billion, hourly average)	Less than 150	150–299	300–399	400 or more
Fine particles (PM_{10}) (24-h running average)	Less than 50	50–74	75–99	100 or more

Ozone

The UK Air Quality Standard is expressed as a running 8-h average concentration of 50 ppb. This was set on the basis of panel studies conducted largely in the USA and also on the results of chamber studies (Department of Health, 1991). The standard includes a margin of safety below the lowest level at which effects have been recorded in such studies: about 80 ppb for an approximately 7-h exposure. Effects seen at such concentrations are generally minor in terms of changes in indices of lung function, though evidence of an inflammatory response in the lung are less easily disregarded. In considering observed changes in indices of lung function the consistently observed person-to-person variation in response to a given exposure was noted. Epidemiological studies conducted both in the UK and elsewhere have suggested that no threshold of effect can be defined, though a study by Anderson et al suggested a threshold of effect at 40–50 ppb (maximum 8-h average concentration) (Ponce de Leon *et al.*, 1996). In reviewing ozone for the WHO (World Health Organization) Air Quality Guidelines for Europe, in 1996 an international expert panel recommended a guideline value of 60 ppb (8-h average concentration), but also provided a series of tables describing likely effects on a no-threshold basis (R.L. Maynard, personal communication). The difficulties experienced in devising a banding system on the basis of a no-threshold concentration–response relationship are discussed below in relation to particles and carcinogens.

With regard to the second and third break-points, it was accepted that at less than 90 ppb (1-h average) it is unlikely that anyone will notice any adverse effects, although they will be detectable by appropriate studies at a population level. It was further accepted that as concentrations rise towards 180 ppb (1-h average) some individuals, particularly those exercising out of doors, may experience eye irritation, coughing and discomfort on breathing deeply. At more than 180 ppb (1-h average) these effects may be expected to become more severe. Individuals suffering from asthma and other respiratory disorders associated with a reduction in respiratory reserve may experience earlier and more marked effects.

Sulfur Dioxide

Sulfur dioxide is a respiratory irritant (Department of Health, 1992). Inhalation of high concentrations rapidly causes bronchoconstriction and patients with asthma are notably more sensitive than other individuals. In recommending a standard for sulfur dioxide the Expert Panel on Air Quality Standards drew on the evidence provided by chamber studies of the effects of sulfur dioxide on patients suffering from asthma. A margin of safety was included in the standard as it was accepted that very sensitive individuals would be unlikely to participate in such studies. The standard recommended is 100 ppb (15-min average). This is therefore the first break-point in the banding system.

The second break-point, between 'moderate' and 'high' levels of air pollution, was set at 200 ppb (15-min average). At this level there is evidence from chamber studies that those suffering from asthma may experience some bronchoconstriction. The expected effects are small (modest changes in indices of lung function) and in many patients would be exceeded by responses to, for example, cold air or an upper respiratory infection.

The third break-point was set at 400 ppb (15-min average). At levels in excess of 400 ppb it was felt that patients suffering from asthma were at risk of significant bronchoconstriction and that they should ensure that they have ready access to their bronchodilator ('reliever') therapy, e.g. a salbutamol inhaler. Of course there is a considerable range of sensitivity to sulfur dioxide amongst asthma sufferers and some may require their bronchodilator therapy as concentrations of sulfur dioxide rise towards 400 ppb. The point that the break-points do not indicate sudden step changes in effects is strongly made.

Unlike ozone, sulfur dioxide is unlikely to have any effects on normal subjects at concentrations likely to be experienced in the UK.

Nitrogen Dioxide

Nitrogen dioxide is a common pollutant in urban areas. Both nitric oxide (NO) and nitrogen dioxide (NO_2) are emitted by both petrol- and diesel-engined motor vehicles. Nitric oxide is rapidly oxidized to the dioxide by ozone; it is oxidized much more slowly by oxygen. Despite intensive study, the database on the effects of nitrogen dioxide on health is still inconsistent and confusing. This is especially the case with regard to possible effects on exposure to concentrations of the order of 200–300 ppb for short periods. Epidemiological studies are also inconsistent, with nitrogen dioxide being much less strongly and convincingly linked to adverse health effects than, for example, particles, ozone or sulfur dioxide (Department of Health, 1993).

In recommending a UK standard for nitrogen dioxide emphasis was placed on the results of chamber studies. The study by Orehek *et al.* (1976) was examined closely. It was noted that this study, which reported effects on indices of lung function on exposure to

100 ppb NO_2, had not been replicated. It was therefore felt that it should be regarded as an outlier in the data set. A standard of 150 ppb (1-h average) was recommended. The first break-point in the banding system for nitrogen dioxide is thus at 150 ppb (1-h average).

The second break-point was set at 300 ppb (1-h average). Even at this level only small effects on indices of lung function are likely to occur. Remarkably, the exposure–response relationship derived from chamber studies becomes inconsistent at about 300–400 ppb with a number of studies failing to find clear responses at these concentrations. Epidemiological studies have, however, demonstrated effects at these concentrations. In setting the third break-point emphasis was placed on the findings of a study of an air pollution episode that occurred in London in 1991. During this episode concentrations of nitrogen dioxide rose to 423 ppb (1-h average). Anderson *et al.* (1995) reported a significant increase in all-cause deaths and general practitioner consultations during the episode. The third break-point was thus set at 400 ppb (1-h average).

Particulate Matter

Particulate matter has long been monitored in the UK by means of the black smoke (BS) method. During the last five or so years monitoring of PM_{10} (mass of particles generally less than 10 µm in diameter) has become widespread and continuous monitoring is now undertaken at more than 50 locations. A particularly rich database on the associations between daily levels of PM_{10} and daily numbers of deaths, hospital admissions, general practitioner consultations and symptoms amongst those suffering from respiratory disorders has accumulated, over the same period, in both the UK and elsewhere. This database is discussed in detail in Chapter 31 and in a recent report (Department of Health, 1995b). From the point of view of the air quality standard-setter the reported associations between daily levels of particles and effects on health are not easy to handle: no indication of a threshold of effect is provided.

In recommending a standard for PM_{10} the UK Expert Panel on Air Quality Standards commissioned a study of the relationship between daily concentrations of particles (PM_{10}) and daily admissions to hospital for respiratory diseases in Birmingham (UK) (Wordley *et al.*, 1997). It was found that an increase in PM_{10} from a level of about 20 µg/m^3 (close to, though a little below, the annual average concentration) to 50 µg/m^3 (both daily averages) was associated with about one extra admission to hospital for treatment of respiratory diseases per day. The Panel agreed that this was a small effect in a city of a million people and recommended 50 µg/m^3 (running 24-h average) as the standard. This concentration as a daily average was adopted as the first break-point of the air pollution banding system.

In devising the further break-points for the banding system for PM_{10} a linear extrapolation of the Birmingham data was undertaken and break-points at 75 and 100 µg/m^3 were recommended.

It was accepted that the choice of figures was in a sense arbitrary in that there was no evidence of a threshold of effect or of any step changes in response at higher levels. As concentrations of particles rise, effects increase. It was also noted that our understanding of the effects of particles on health is evolving rapidly and the break-points outlined above might need revision in the light of new evidence.

Carbon Monoxide

Of all the common air pollutants, the toxicology of carbon monoxide is perhaps the best understood (see Chapter 33). It is generally accepted that carbon monoxide exerts its effects by binding to haemoglobin and reducing the capacity of the blood both to transport oxygen and to release oxygen at the tissues. Effects are particularly severe at tissues where partial pressures of oxygen are already low, e.g. the endocardial aspect of the myocardium. Very recent evidence suggests that carbon monoxide may play a role as a transmitter substance in the body and may also interfere with production of endothelial nitric oxide (Bonn, 1997).

On binding to haemoglobin carbon monoxide produces carboxyhaemoglobin (COHb). The relationship between ambient carbon monoxide concentrations and COHb concentrations has been studied in volunteers and is well understood (World Health Organization, 1979). COHb is thus a biomarker both of exposure and of likely effects of carbon monoxide: such biomarkers are rare.

Chamber studies of patients suffering from angina have shown that exposure to CO leading to COHb concentrations of about 2% have adverse effects as indicated by a reduction in the 'time to pain' on exercise and on the time taken for a predetermined degree of depression of the ST segment of the electrocardiogram to occur (Allred et al., 1991). Eight-hour exposure to a CO concentration of 10 ppm leads to a COHb concentration of just under 2%. On this basis, and including a small margin of safety, the UK Expert Panel on Air Quality Standards recommended a standard for CO of 10 ppm (8-h average). It should be noted that recent epidemiological studies have shown associations between admissions to hospital for cardiovascular disorders and (unexpectedly low) ambient concentrations of carbon monoxide. Interpretation of these results remains difficult and may become clearer as studies accumulate (Poloniecki et al., 1997).

The first break-point in the banding system for carbon monoxide was therefore set at 10 ppm (8-h average). Exposure to 15 ppm for 8-h leads to a COHb concentration of 2.5%, a level at which there is evidence of adverse effects on patients suffering from angina. The third break-point was set at 20 ppm (8-h average), again on the basis of likely effects on patients with impaired coronary circulation. It was noted that such levels are very unlikely to be reached outdoors in the UK. High levels can occur indoors, however, when faulty or unflued gas appliances are in use.

Carcinogens

Genotoxic carcinogens such as benzene, 1,3-butadiene and polycyclic aromatic hydrocarbons (PAHs) present difficult problems to the regulatory toxicologist. It is generally accepted that no completely safe level of exposure to such compounds can be recommended. This has become a safe position to adopt. However, it is based on theoretical considerations and an absence of data on effects at low levels of exposure rather than on a clear demonstration of effects at such levels. Such a demonstration of effects is unlikely to be obtained as the effect of exposure to ambient concentrations of carcinogens is likely to be very small and probably impossible to demonstrate in any practicable epidemiological study. That effects occur at higher levels of exposure has been clearly demonstrated by epidemiological studies of workers occupationally exposed to benzene (Wong, 1987a,b).

Predicting effects at low levels of exposure inevitably involves the use of extrapolative techniques, sometimes referred to as the methods of mathematical quantitative risk assessment, often abbreviated as QRA. Such methods are neither universally accepted nor demonstrably reliable (Maynard *et al.*, 1995).

Standards for benzene and 1,3-butadiene have been recommended by the UK Expert Panel on Air Quality Standards using an approach based on the application of safety factors to data obtained from studies of the effects of occupational exposures. These standards (5 ppb for benzene and 1 ppb for 1,3-butadiene) are expressed in terms of running annual average concentrations. In providing the public with information about concentrations of these compounds no banding system has been produced: the running annual average is simply updated each hour and expressed as either above or below the air quality standard.

HEALTH ADVICE ON AIR POLLUTION TO THE PUBLIC

Advice to the UK public on likely health effects experienced in the four pollution 'bands' has been prepared by the Committee on Medical Effects of Air Pollutants. This is disseminated alongside the air pollution information and is provided in a deliberately user-friendly format. In terms of the bands described above the advice provided is as follows:

Low: Effects are unlikely to be noticed even by individuals who know they are sensitive to air pollutants.

Moderate: Mild effects unlikely to require action may be noticed amongst sensitive individuals.

High: Significant effects may be noticed by sensitive individuals and action to avoid or reduce these effects may be needed (e.g. reducing exposure by spending less time in polluted areas outdoors). Asthmatics will find that their 'reliever' inhaler is likely to reverse the effects on the lung.

Very high: The effects on sensitive individuals described for 'High' levels of pollution may worsen.

All systems that describe levels of air pollution in terms of a series of bands face the problem of the highest band having no defined upper limit. Exceptional levels of air pollution may occur and some special warning may be needed, even for individuals not usually sensitive to air pollution. In the UK a statement dealing with this has been prepared:

The risk to healthy individuals is very small at all levels of air pollution likely to be experienced in the UK. However, because the 'very high band' has no upper limit, it is possible that individuals not usually sensitive to air pollution may notice effects, including eye irritation, coughing and pain on breathing deeply when concentrations of air pollutants move well into the 'very high' band.

CONCLUSION

Providing the public with information about levels of air pollution should be an integral part of any Air Quality Strategy. Such information should be up-to-date and should include forecasts for the following day or days so that individuals can take steps: (a) to avoid the effects of pollution; and (b) to reduce activities that lead to raised levels of pollution. Many countries provide such advice. The approach outlined in this chapter is that taken in the UK. This approach is not offered as a perfect system, but rather as an example for those interested in setting up systems of their own.

REFERENCES

Allred EN, Bleecher ER, Chairman BR *et al.* (1991) Effects of carbon monoxide on myocardial ischemia. *Environ Health Perspect* **91**: 89–132.

Anderson HR, Limb ES, Bland JM *et al.* (1995) The health effects of an air pollution episode in London, December 1991. *Thorax* **50**: 1188–1193.

Bonn D (1997) New mechanism for CO poisoning uncovered. *Lancet* **350**: 1008.

Department of Health (1991) Advisory Group on the Medical Aspects of Air Pollution Episodes. First Report. *Ozone*. London: HMSO.

Department of Health (1992) Advisory Group on the Medical Aspects of Air Pollution Episodes. Second Report. *Sulphur Dioxide, Acid Aerosols and Particulates*. London: HMSO.

Department of Health (1993) Advisory Group on the Medical Aspects of Air Pollution Episodes. Third Report. *Oxides of Nitrogen*. London: HMSO.

Department of Health (1995a) Advisory Group on the Medical Aspects of Air Pollution Episodes. Fourth Report. *Health Effects of Exposures to Mixtures of Air Pollutants*. London: HMSO.

Department of Health (1995b) Committee on the Medical Effects of Air Pollutants. *Non-Biological Particles and Health*. London: HMSO.

Department of Health (1995c) Committee on the Medical Effects of Air Pollutants. *Asthma and Outdoor Air Pollution*. London: HMSO.

Department of Health (1997) Committee on the Medical Effects of Air Pollutants. *Handbook on Air Pollution and Health*. London: The Stationery Office.

Department of Health (1998) Committee on the Medical Effects of Air Pollutants. *Quantification of the Effects of Air Pollution on Health in the United Kingdom*. London: The Stationery Office.

Department of the Environment (1994a) Expert Panel on Air Quality Standards. *Ozone*. London: HMSO.

Department of the Environment (1994b) Expert Panel on Air Quality Standards. *1,3-Butadiene*. London: HMSO.

Department of the Environment (1994c) Expert Panel on Air Quality Standards. *Benzene*. London: HMSO.

Department of the Environment (1994d) Expert Panel on Air Quality Standards. *Carbon Monoxide*. London: HMSO.

Department of the Environment (1995a) Expert Panel on Air Quality Standards. *Sulphur Dioxide*. London: HMSO.

Department of the Environment (1995b) Expert Panel on Air Quality Standards. *Particles*. London: HMSO.

Department of the Environment (1996) Expert Panel on Air Quality Standards. *Nitrogen Dioxide*. London: The Stationery Office.

Department of the Environment (1997) *The United Kingdom Air Quality Strategy*. London: The Stationery Office.

Maynard RL, Cameron KM, Fielder R *et al.* (1995) Setting air quality standards for carcinogens: an alternative to mathematical quantitative risk assessment – discussion paper. *Human Exp Toxicol* **14**: 175–186.

Orehek J, Massari JP, Gayrard P *et al.* (1976) Effect of short-term, low-level nitrogen dioxide exposure on bronchial sensitivity of asthmatic patients. *J Clin Invest* **57**: 301–307.

Poloniecki JD, Atkinson RW, Ponce de Leon A and Anderson HR (1997) Daily time series for cardio-vascular hospital admissions and previous day's air pollution in London, UK. *Occup Environ Med* **54**: 535–540.

Ponce de Leon A, Anderson HR, Bland JM *et al.* (1996) Effects of air pollution on daily hospital admissions for respiratory disease in London between 1987–88 and 1991–92. *J Epidemiol Community Health* **33**(Suppl. 1): S63–S70.

Wong O (1987a) An industry wide mortality study of chemical workers occupationally exposed to benzene. I. General results. *Br J Ind Med* **44**: 365–381.

Wong O (1987b) An industry wide mortality study of chemical workers occupationally exposed to benzene. II. Dose response analysis. *Br J Ind Med* **44**: 382–395.

Wordley J, Walters S and Ayres JG (1997) Short term variations in hospital admissions and mortality and particulate air pollution. *Occup Environ Med* **54**: 108–116.

World Health Organization (1979) *Carbon Monoxide.* Environmental Health Criteria no. 13. Geneva: World Health Organization.

Index

Italic page numbers refer to figures and tables; main discussions are indicated by **bold** page numbers.